Textbook of Seismic Design

G. R. Reddy • Hari Prasad Muruva
Ajit Kumar Verma
Editors

Textbook of Seismic Design

Structures, Piping Systems, and Components

Springer

Editors
G. R. Reddy
Bhabha Atomic Research Centre
Mumbai, India

Hari Prasad Muruva
Bhabha Atomic Research Centre
Mumbai, India

Ajit Kumar Verma
Western Norway University of Applied Science
Haugesund, Norway

ISBN 978-981-13-3175-6 ISBN 978-981-13-3176-3 (eBook)
https://doi.org/10.1007/978-981-13-3176-3

Library of Congress Control Number: 2018960746

This Springer imprint is published by the registered company Springer Nature Singapore Pte Ltd.
The registered company address is: 152 Beach Road, #21-01/04 Gateway East, Singapore 189721, Singapore

Dedicated to Parents and Teachers

Preface

Earthquakes can happen anywhere in the world. In some places, the frequency of occurrence is very large maybe at the plate boundaries such as Himalayan region, and in some places, it is low like in the southern part of India. The knowledge has not been reached to predict the earthquakes precisely. Earthquake can cause large destruction in industrial structures, systems, and components (SSCs). For example, Bhuj earthquake 2001 caused large damage to the petrochemical industry, fertilizer industry, etc., and the loss due to the unavailability of the plants was also large. If the industry is handling large toxic and poisonous gases, the large damage will also affect the public around the plant. To minimize this, the loading generated due to the earthquake should be properly accounted for.

In view of the above, it is essential to get prepared for facing the earthquakes. Preparedness means to design the structures (e.g., residential buildings, industrial buildings, lifeline structures), systems (water supply lines, firefighting systems, industrial piping systems, electrical and control systems, communication systems, etc.), and components (industrial equipment, normal and fire water supply components, etc.) to withstand postulated design earthquakes. Failures or successes of the SSCs are not dependent on the location, and similar failures or successes of SSCs are seen anywhere in the world when they are subjected to earthquakes.

Safety and nonsafety-related SSCs are to be designed for earthquake loading as prescribed by the national or international standards as applicable. Typical industrial SSCs are shown in the figure below. This gives a fair idea about what to learn and the need of learning. This particular book is organized with the intention of explaining the basic mechanisms of earthquakes, generation of design basis ground motion, fundamentals of structural dynamics which are applicable to systems and components as well, geotechnical aspects which are related to earthquake design of SSCs, analysis of multi-degree-of-freedom systems, seismic design of RC structures and steel structures, component and system design which are located at the

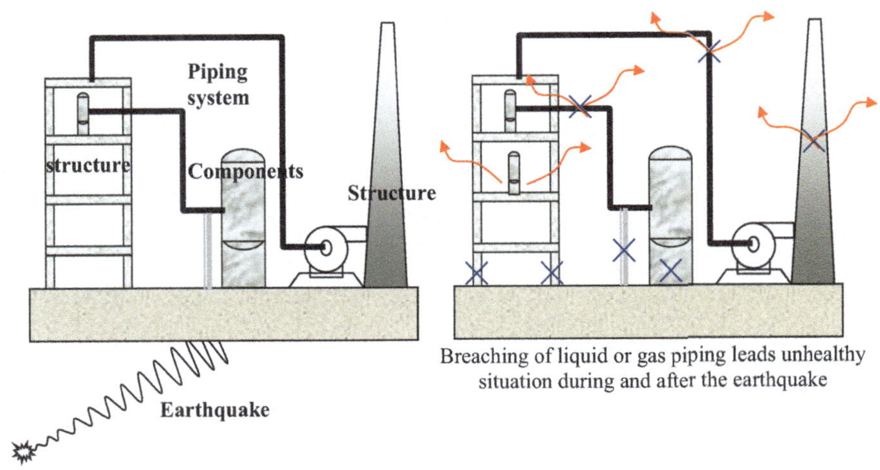

Schematic of industrial structures, systems, and components (SSCs)

ground level and at different floor levels of the structures, anchorage design of component and system supports, introduction to retrofitting, seismic response control, and lastly testing of SSCs.

In this book, sincere efforts are put to explain the subject of earthquake engineering with physical concepts. A good number of examples with the step-by-step approach are given on each topic to improve the understanding of the reader. This book is suitable for students, researchers, and designers including consultants.

Mumbai, India
July 2019

G. R. Reddy
Hari Prasad Muruva
Ajit Kumar Verma

Acknowledgement

This book is the outcome of the efforts, encouragement, and support received from various organizations and individuals. We are indebted to all of them.

We express our sincere thanks to Bhabha Atomic Research Centre, Department of Atomic Energy (DAE), for accepting the proposal of publishing a Textbook of Seismic Design. We like to thank Homi Bhabha National Institute (HBNI) for giving access to its resources. We also express our sincere thanks to Indian Nuclear Society (INS), Mumbai, for supporting idea of publishing this book which has basis of INS national workshop on Seismic Design of Industrial structures, equipment, and piping systems. The feedback received from various participants in the workshop for the past 10 years helped to improve the technical content of this book. We extend our thanks to INS Executive Committee, especially to Sh. H. S. Kushwaha and Sh. R. K. Singh for their encouragement.

We like to thank ASME for granting the permission to reproduce some of the excerpts from the codes as stated below: "Reprinted from ASME 2001 Section III, Division 1, Subsection NB, NC, ND, NF and Appendix N, by permission of the American Society of Mechanical Engineers. All rights reserved." We also like to thank Bureau of Indian Standards for according the approval for reproducing the extracts from Indian Standards as follows: "The extracts have been reproduced with the permission of the Bureau of Indian Standards."

Special thanks to all the authors/co-authors contributed towards each chapter for their sincere efforts made to bring a good shape and value to the book.

We acknowledge Sh. Y. S. Bhadauria, BARC, and Dr. P. Anbazhagan, IISc, Bangalore, for reviewing some of the chapters and providing valuable suggestions. We also like to thank Dr. D. Datta, BARC, HBNI, for providing student version of MATLAB and helping in coding.

Last but not least, we express our sincere thanks to all who have contributed directly or indirectly to the success of this book.

G. R. Reddy
Hari Prasad Muruva
Ajit Kumar Verma

Contents

Editors and Contributors

About the Editors

G. R. Reddy is a senior professor at Homi Bhabha National Institute, a teacher for Ph.D. in the University of Mumbai, and an outstanding scientist at Bhabha Atomic Research Centre, Mumbai. His research expertise is in the areas of structural dynamics and earthquake engineering. He has contributed to developing simple numerical modeling techniques for complex structures, structure–equipment interaction due to earthquakes, stochastic methods of analysis, dynamic substructuring techniques, etc. He was involved in developing seismic hazards at various sites of nuclear facilities. He has mastered seismic response control methods and developed passive response control devices including seismic base isolators which will eliminate snubbers in nuclear facilities, reduce the initial cost and maintenance cost, and improve the safety. He has developed simple seismic design procedures for equipment and piping supported on hysteretic supports. As a part of developing more realistic design procedures, he has performed a large number of experiments on beam–column joints, frames, and piping systems till collapse. He was involved in developing a methodology for seismic margin assessment in piping systems and structures. For the purpose of life extension of the existing facilities, he has developed methods for performing seismic retrofitting of structures using dampers, FRP, and steel jacketing. He has contributed significantly to analysis and design/requalification/retrofit of structures, piping systems, and components of old and new nuclear facilities. In addition, he has contributed to the design of a 30-m antenna for the Chandrayan project and of a large-scale gamma ray telescope. He has regularly given lectures on vibrations and earthquake engineering in training schools of Department of Atomic Energy and has involved in IS and departmental seismic code committees. He has guided a large number of M.Tech. and Ph.D. students and has authored more than 500 research publications.

Hari Prasad Muruva is a scientist at Bhabha Atomic Research Centre, Mumbai. His research primarily focuses on risk and reliability analysis, probabilistic safety assessment (PSA), seismic PSA, probabilistic seismic hazard analysis (PSHA), and risk-based inspection of power plants. He has collaborated on several national and international projects and has authored several research publications.

Ajit Kumar Verma is a professor (Technical Safety) at Western Norway University of Applied Sciences, Haugesund, Norway, has previously worked at IIT Bombay as a professor, was an adjunct at the University of Stavanger, and has been a guest professor at the Lulea University of Technology, Sweden. His publications include 3 edited volumes, 5 monographs, and more than 250 research articles in journals and conferences. He was EIC of OPSEARCH and is EIC of *Journal of Life Cycle Reliability and Safety Engineering* and Founder/Editor-in-Chief of *International Journal of Systems Assurance Engineering and Management*. He is also Springer series editor of three book series.

Contributors

M. K. Agrawal Bhabha Atomic Research Centre, Mumbai, India

Srijit Bandyopadhyay Bhabha Atomic Research Centre, Mumbai, India

Raj Banerjee Bhabha Atomic Research Centre, Mumbai, India

P. N. Dubey Bhabha Atomic Research Centre, Mumbai, India

M. Eswaran Bhabha Atomic Research Centre, Mumbai, India

B. K. Gangrade Bhabha Atomic Research Centre, Mumbai, India

D. K. Jha Bhabha Atomic Research Centre, Mumbai, India

A. R. Kiran Bhabha Atomic Research Centre, Mumbai, India

Binu Kumar Bhabha Atomic Research Centre, Mumbai, India

Praveen Kumar Bhabha Atomic Research Centre, Mumbai, India

Hari Prasad Muruva Bhabha Atomic Research Centre, Mumbai, India

T. Nagender Bhabha Atomic Research Centre, Mumbai, India

R. Parmar Bhabha Atomic Research Centre, Mumbai, India

Y. M. Parulekar Bhabha Atomic Research Centre, Mumbai, India

S. J. Patil Heavy Water Board, Mumbai, India

V. S. Phanikanth Bhabha Atomic Research Centre, Mumbai, India

M. K. Pradhan Bhabha Atomic Research Centre, Mumbai, India

G. R. Reddy Bhabha Atomic Research Centre, Mumbai, India

Ajit Kumar Verma Western Norway University of Applied Science, Haugesund, Norway

Gaurav Verma Bhabha Atomic Research Centre, Mumbai, India

R. K. Verma Bhabha Atomic Research Centre, Mumbai, India

Introduction to Earthquakes

1

Raj Banerjee, B. K. Gangrade, Srijit Bandyopadhyay, and G. R. Reddy

Failure of Structures due to earthquakes results in loss of life and economy

Symbols

M_L	Local Richter magnitude
M_w	Moment magnitude
m_b	Body wave magnitude
M_s	Surface wave magnitude
T	Wave period
Δ	Epicentral distance
E_0	Seismic energy
t_d	Decay time
t_m	Duration of strong motion
t_r	Rise time

1.1 Introduction

Seismology comes from the Greek term "Seismos" for earthquake and "Logos" for science which relates to the study of generation, propagation, and recording of waves in the earth and the sources that produce them. A complete description of these topics is found in numerous documents, such as Gutenberg and Richter [1], Richter [2], Bolt [3], and Gubbins [4]. Study of seismic waves generated due to earthquakes provides information about the seismic source as well as the interior of the Earth. There are two sources by which the seismic vibrations can be classified, namely natural sources and man-made sources. The different types of seismic events are given in Fig. 1.1.

R. Banerjee · B. K. Gangrade · S. Bandyopadhyay · G. R. Reddy (✉)
Bhabha Atomic Research Centre, Mumbai, India
e-mail: rssred@barc.gov.in

R. Banerjee
e-mail: rbanerjee@barc.gov.in

B. K. Gangrade
e-mail: gangrade@barc.gov.in

S. Bandyopadhyay
e-mail: srijit@barc.gov.in

G. R. Reddy et al., *Textbook of Seismic Design*,
https://doi.org/10.1007/978-981-13-3176-3_1

1.2 Earth's Interior, Plate Tectonics, and Faults

1.2.1 Earth's Interior

The greatest achievement of seismology was the determination of the internal structure of the earth. It is stated that the Earth was formed five billion years ago by massive conglomeration of space materials, and during the process, lot of heat energy was released. Dense materials sank to form core of the Earth, while lighter silicates, oxygen compounds, and water formed the outer shell of the earth surface. The earth is divided into four main layers: the inner core, outer core, mantle, and crust (Fig. 1.2a, b). The Earth is spherical in shape with an equatorial diameter of 12,740 km and polar diameter of 12,700 km, the former being high due to higher equatorial velocities of the earth. The outer core is liquid, while the inner core is solid. Even though the mantle is solid, it can deform gradually in a plastic manner.

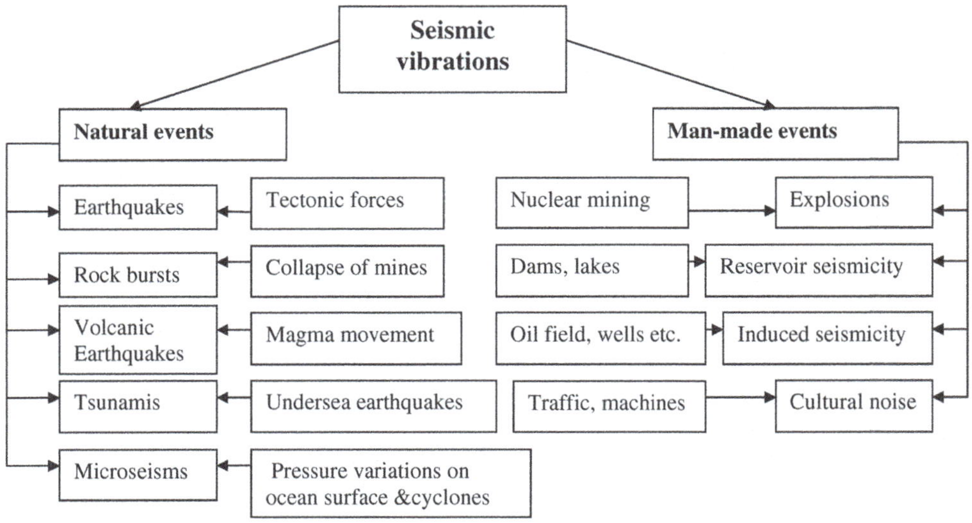

Fig. 1.1 Different types of seismic events

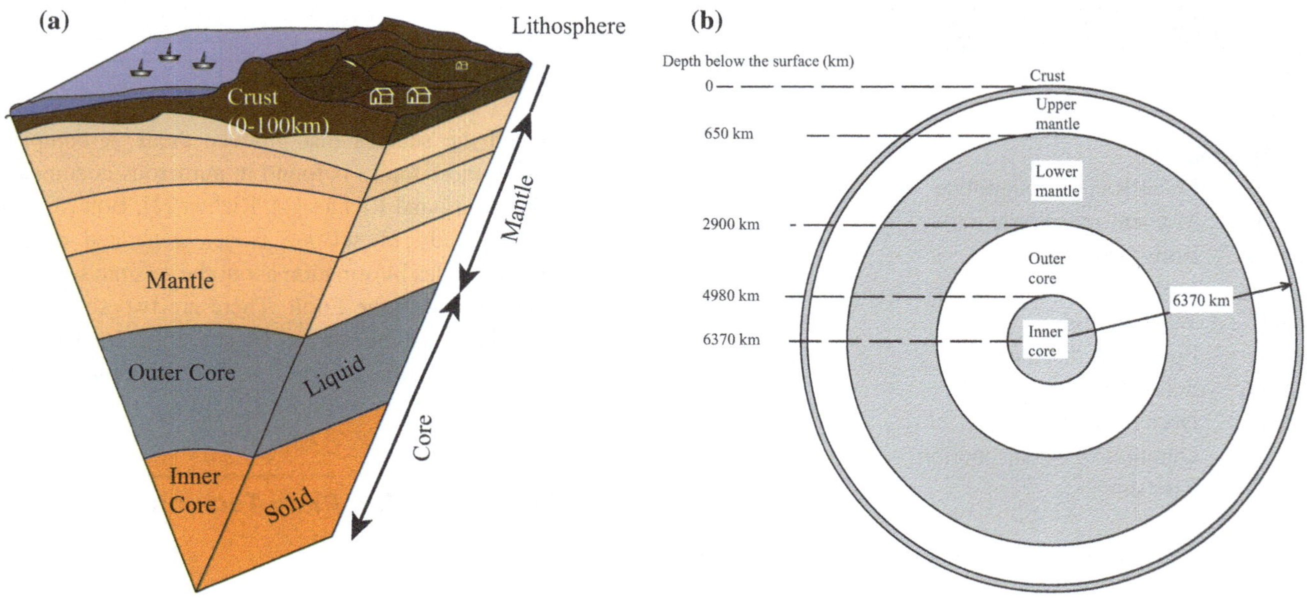

Fig. 1.2 **a** 3D view of the internal structure and **b** the cross section of the earth

The crust which is the topmost layer of the earth ranges about 100 km thick and forms the lithosphere as shown in Fig. 1.2a, on which the human beings live. Since the crust is exposed to the atmosphere, it is cooler than other layers of the earth. The mantle is 2850 km thick which is divided into upper and lower mantles. The upper mantle being close to the crust is relatively cooler as compared to the lower mantle although the average temperature being 4000 °F. The material of the mantle is normally in a viscous semi-molten state, but as soon as stress is applied in a rapid fashion, it behaves as a solid. Moreover, when subjected to long-term stresses, the material remains in a molten state. The outer core is 2260 km thick, made up of molten iron whose convective flow is the reason for the magnetic field of the earth, whereas the inner core is very dense made of nickel–iron material which is compressed under tremendous pressure, and the average uniform temperature is around 5000 °F.

Figure 1.3 shows the profile of the shear wave velocity across the interior of the earth. The outer core, which is in liquid state, causes no propagation of shear waves. The sudden drop of shear wave velocity near the boundary of the mantle is named as Gutenberg discontinuity.

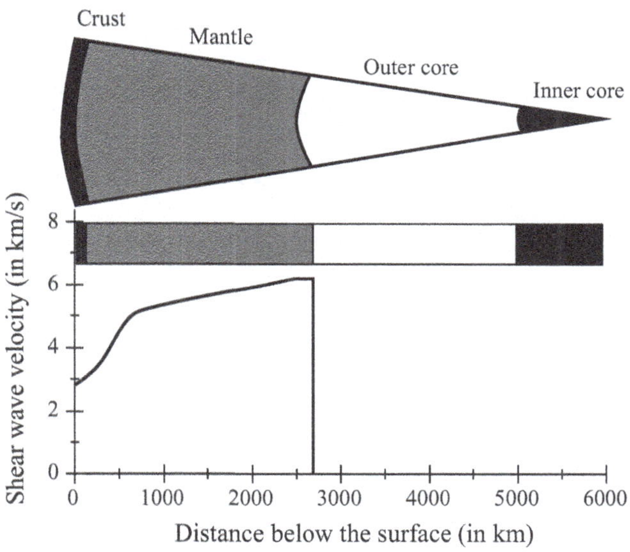

Fig. 1.3 Shear wave velocity profile with depth within the earth

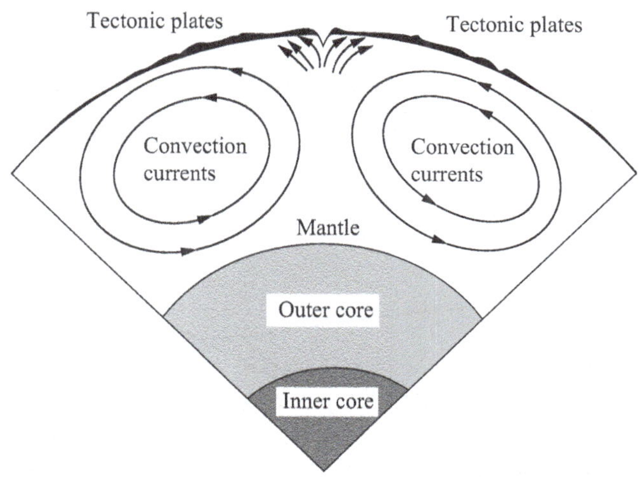

Fig. 1.5 Convection currents in mantle leading to the movement of the tectonic plate

1.2.2 Origin of Plate Tectonics

The theory of continental drift was not proposed until the twentieth century [5, 6]. It was the belief of Wegener that 225 million years ago the Earth was made of one large continent called Pangaea, and it slowly drifted into the present configuration of continents. The theory of continental drift was based on the images of the movement of massive continents through the ocean floor, but as the ocean floor is too strong to permit such a motion, the theory of plate tectonics evolved. As per this theory, the earth's crust is made up of a large number of intact blocks called plates which move with respect to each other with an average speed of a few centimeters per year due to convective forces in the mantle.

The crust of the Earth is made up of six continental-sized plates as shown in Fig. 1.4 and 14 subcontinental-sized plates such as Indian Plate, Arabian Plate, Philippine Plate. The upper portion of the mantle is connected to the relatively cooler crust with respect to the lower mantle which is nearer to the outer core. As a result, the temperature gradient exists in the mantle which results in the density gradient in the mantle due to which the cooler material flows to the bottom and the hot material moves to the top resulting in convection current. This phenomenon is known as thermomechanical equilibrium process and is illustrated in Fig. 1.5. This imposes shear stress on the bottom plates which drags them to the various directions of the earth. The differential motion between various plates results in the accumulation of energy, and the sudden release causes earthquakes.

Three types of plate boundaries or faults have been identified. The associated plate movements along with each of the boundaries will help in understanding the kinematics of plate tectonics.

Spreading ridge boundaries: This type of boundary is identified in certain areas in which the plates move apart from each other which results in the rising of the molten rock from mantle to the surface where it cools to become a part of the spreading plate. In this way, new crust is formed at the spreading ridge as shown in Fig. 1.6. It is estimated that the

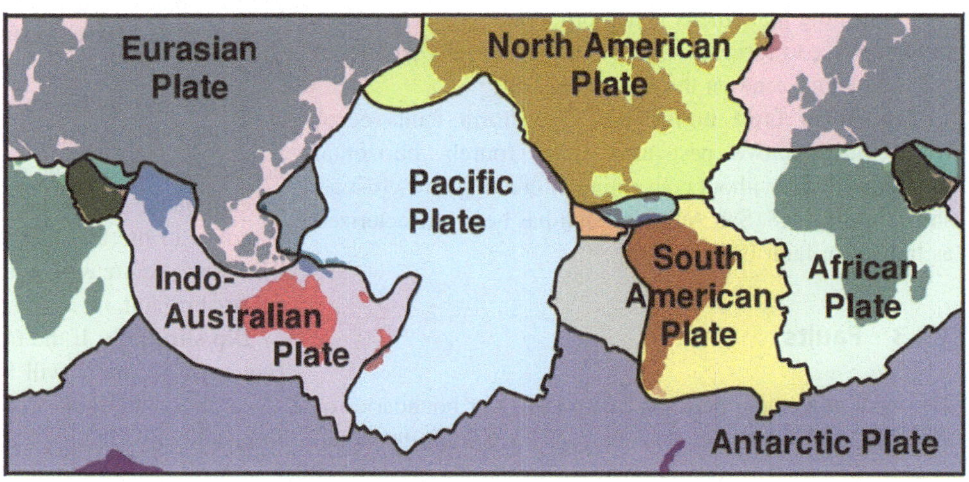

Fig. 1.4 Present-day tectonic plates [IITK-BMTPC]

Fig. 1.6 Spreading ridge boundary

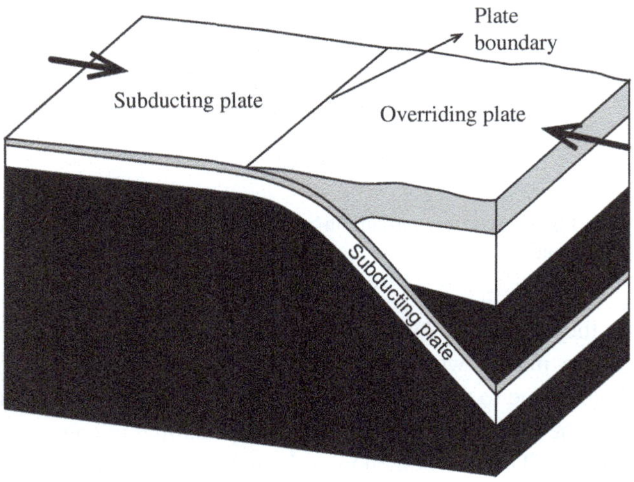

Fig. 1.7 Subduction zone boundary

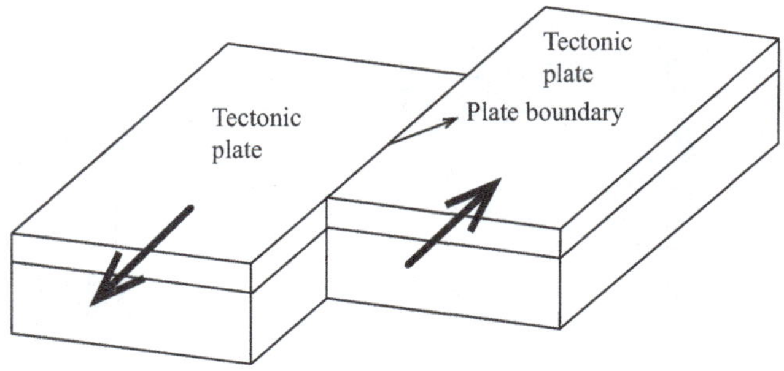

Fig. 1.8 Transform fault boundary

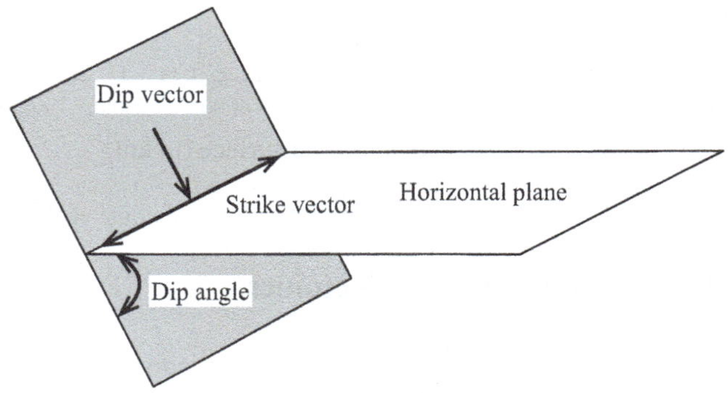

Fig. 1.9 Geometric notation for the orientation of fault plane

new oceanic crust is formed currently at the rate of 3.1 km^2/year worldwide [7]. Volcanic activity much of which occurs underwater is common near the vicinity of the spreading ridge boundaries.

Subduction zone boundaries: As the law of nature suggests that there must be a balance between the creation and its destruction, in the same line the creation of a new crust (or plates) at the location of the spreading ridges must be balanced by the consumption of the crust (or the plates) at the subduction zones in which one plate subducts beneath the other due to the relative movement of the plates toward each other, as shown in Fig. 1.7. The subducting plate gets warmer as it sinks and gradually becomes more and more ductile such that it is not capable of producing earthquakes. A portion of the subducting plate melts which produces magma that rises to the surface to form a line of volcanoes roughly parallel to the subducting zone on the overriding plate.

Transform fault boundaries: Transform faults occur where plates move past each other (purely horizontal shearing action) without consuming or creating new crust, as shown in Fig. 1.8. San Andreas Fault has been characterized as transform fault boundary [8].

1.2.3 Faults

The local movement between two portions or boundaries of the crust will occur on new or preexisting offsets in the geological structure known as faults. On a larger scale, the plate boundaries of the earth across which the movement of the plates takes place are all faults of different types.

1.2.3.1 Geometry of Fault

Standard geological notations used for describing the fault geometry in space are strike and dip. The strike of a fault is the horizontal line which is formed by the intersection of a horizontal plane with the fault plane, whereas the dip angle is the angle between the horizontal plane perpendicular to the strike and the fault plane as depicted in Fig. 1.9.

1.2.3.2 Types of Fault Movement

The faults are classified into two categories based on the way the rock blocks move on either side of the fault as depicted in Fig. 1.10. These include strike-slip fault and dip-slip faults.

Strike-slip fault: This involves shearing motion of the two sides of the fault parallel to the strike. If one looks across the fault and the other side moves to the right during an earthquake, it is called right-lateral strike-slip faulting. If it moves to the left, one has left-lateral strike-slip faulting. The San Andreas Fault in the USA is a right-lateral strike-slip fault.

Dip-slip fault: If the fault plane is dipping at some angle (not vertical), there will be some rock below the fault and some above the fault. The rock below the fault is called as footwall, while the rock above the fault is called as hanging

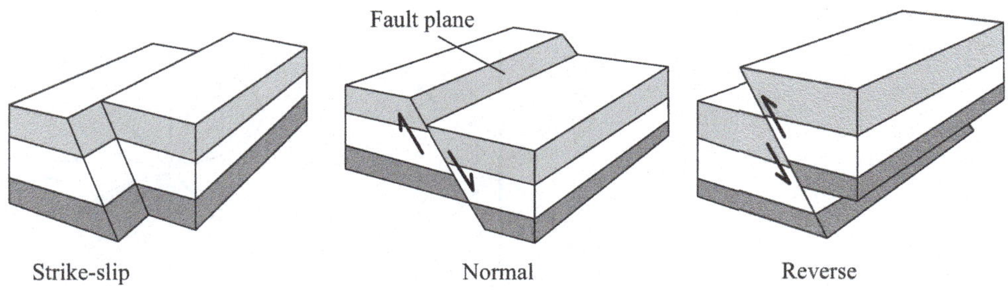

Fig. 1.10 Different types of faults

wall. The movement of the fault which occurs in the direction of the dip or perpendicular to the strike is known as dip-slip fault. Two types of dip-slip movement are identified, namely normal and reverse faults. If the faulting causes the hanging wall to move downward relative to the footwall, the fault is a *normal fault*. If the hanging wall moves upward relative to the footwall, it is called *reverse faulting*, whereas if the dip angle of the fault is less than 30°, then it is called *thrust faulting*. Reverse faulting and thrust faulting occur in regions of compression, where the surface is converging. This is common in subduction zones and in places where continents are colliding (Indian Plate and Eurasian Plate). In the subduction zones, the major earthquakes tend to be thrust faulting events.

The cause of an earthquake is a sudden release of elastic strain energy accumulated across a preexisting fault or fracture. The fracture, which originates at a point, spreads all around, terminates, and produces radiation of different modes of seismic waves such as P waves, S waves, and surface waves.

1.3 Seismic Waves

A fraction of the energy which is released from the total volume of the sliding rock is converted into seismic waves, and the rest of it dissipates as heat. The seismic waves which spread through the rock are of three kinds, viz. compressional (P), shear (S), and surface waves. P and S waves together are called body waves because they travel through the interior of the earth unlike the surface waves, which propagate along the surface.

1.3.1 P Waves

These are longitudinal waves which involve successive compression and rarefaction which is like the sound waves in which the particle motion is in the direction of the propagation of travel of wave (Fig. 1.11a, b). The direction of the oscillation of the particle is in the same direction in which the waves are propagating, resulting in both

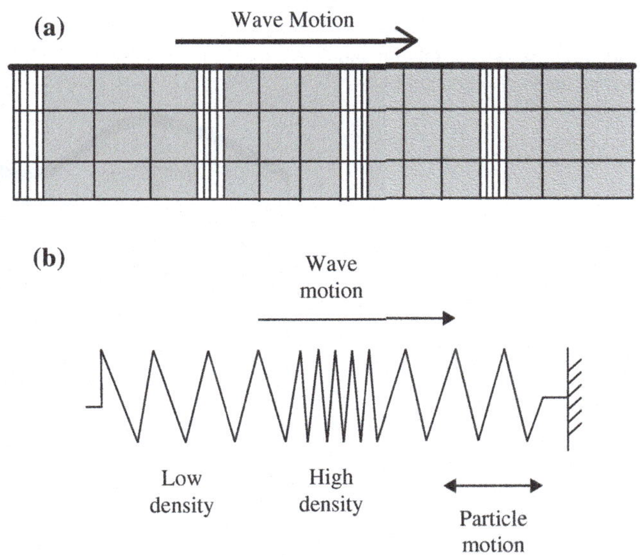

Fig. 1.11 **a** Particle motion and **b** spring analogy of P wave

compressive and tensile stresses. These waves can travel through both solids and fluids.

The variation of amplitude of elastic waves with distance from the source occurs due to both geometric spreading (radiation damping) and anelastic attenuation (material damping). The sketches in Fig. 1.12 show the radiation pattern of the P wave (where F is the fault plane). For P waves the radiation pattern shows that maximum amplitude for compression and dilatation occurs at 45° to the fault plane. The fault and auxiliary planes are nodal planes of zero amplitude. The radiation pattern of P wave leads to an earthquake source model represented by a pair of orthogonal couple perpendicular and parallel to the fault plane.

1.3.2 S Waves

In the S waves, oscillation of the particle is in the direction perpendicular to the wave propagation direction as shown in Fig. 1.13. The direction of the particle movement is used to distinguish between SV (vertical plane movement) and SH (horizontal plane movement) waves. As a well-known

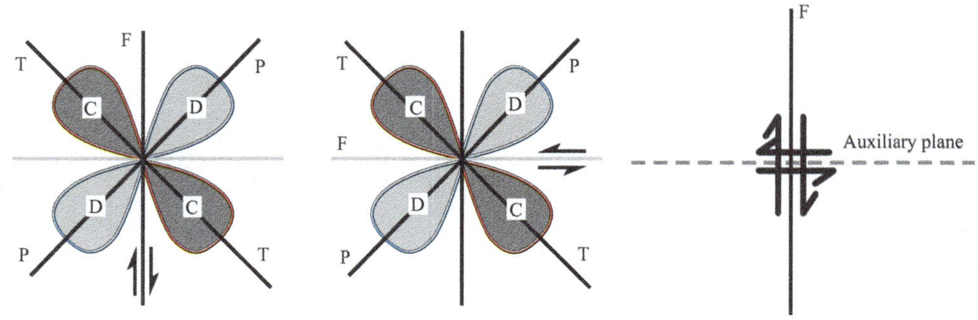

Fig. 1.12 Radiation pattern of P wave and orthogonal couple

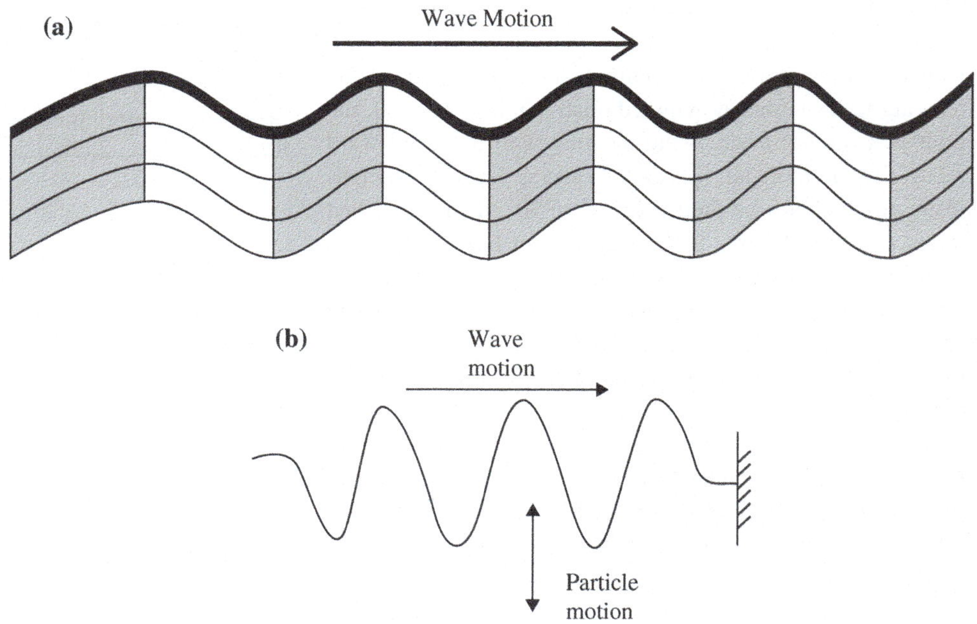

Fig. 1.13 **a** Particle motion and **b** rope analogy produced by S waves

phenomenon there is no effective restoring force due to shear in fluids causing no propagation of S waves in fluids. This explains the reason for the sudden drop in the shear wave velocity from mantle to the outer core. As the earthquake fault slides past each other, the seismic waves generated from the source are a combination of body and surface waves. As the material is stiffer in compression than shear, the velocity of P wave is faster than the velocity of S wave in a rock which results in a faster propagation of P wave front in comparison with S wave front.

As the Earth has different layers, as well as a free surface, the P and S waves can reflect inside the earth from the free surface which gives various phases of P and S waves as shown in Fig. 1.14. Thus, the ground motion recordings from earthquakes tend to be rather complex as shown in Fig. 1.14, with a sequence of arrivals of various distinct phases that are mainly controlled by the velocity profile of the interior structure of the earth. The source time function for an earthquake lasts only a few seconds, but the shaking of the ground along with the seismogram recording will be of a greater duration because the P and S waves travel with different velocities and there are many paths with different total travel times for the energy to get to the recording station.

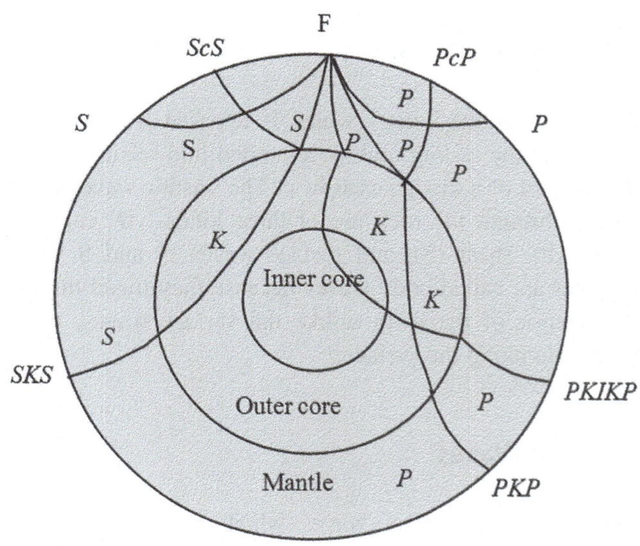

Fig. 1.14 Different ray paths inside the earth

1.3.3 Surface Waves

The interaction of P and S waves with each other along with the layering of the crust and mantle results in a pattern of vibration that is known as surface waves, which are mainly of two types Love and Rayleigh waves as shown in Figs. 1.15 and 1.16. Love waves are faster than Rayleigh waves, and they involve only horizontal motions of the earth perpendicular to the direction in which the wave is propagating.

Rayleigh waves are characterized by a retrograde, elliptical particle motion in a vertical plane which is a mixture of both P and S (SV) waves. As these waves propagate near the surface, the wave front of propagation is a cylinder rather than a spherical shell, and the amplitude of the waves decreases which is proportional to the distance from the source. Thus, the amplitude of the surface waves is larger than the body waves which is recorded in the seismogram. Thus, in the event of a nuclear or chemical explosion due to earthquake, surface waves play an important role.

Another kind of seismic wave that propagates in a crustal waveguide in the continental lithosphere is known as L_g or the surface shear wave whose energy departing downwards is wholly reflected back into the crust. The type of reflection occurring here is the total internal reflection. For a fixed source and receiver, there may be many reflection paths, all totally reflected and thus trapped within the crust. The radiation pattern of L_g waves is more isotropic than that of P and S waves. This feature adds to the usefulness of L_g waves as a magnitude estimator for small events due to the fact that full azimuthal coverage is not essential, and reliable magnitude estimation can be made from the data of only a few stations.

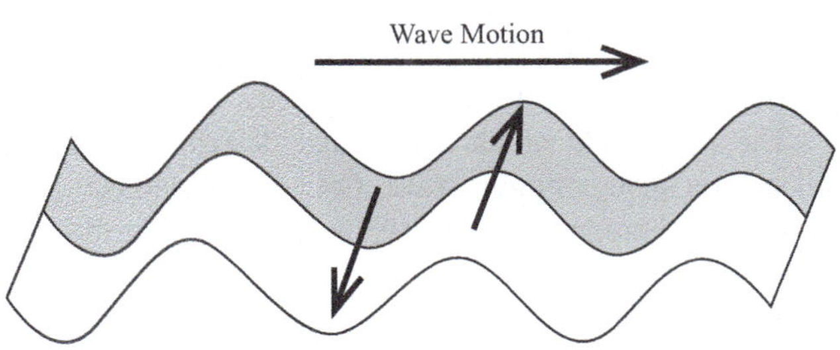

Fig. 1.15 Particle motion of love waves

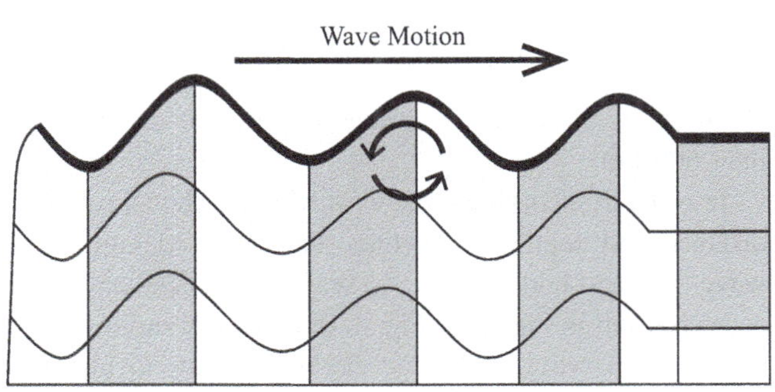

Fig. 1.16 Particle motion of Rayleigh waves

1.4 Seismometers, Seismographs, and Seismograms

The rupture of rock along a fault causes seismic waves to travel in all directions. The point at which the rupture originates is known as the focus of the earthquake (as shown in Fig. 1.17). The point on the ground surface which is directly above the focus is called the epicenter. The distance between the source of an earthquake (epicenter) and the receiver (or station) is known as the epicentral distance, and the distance between the receiver (or station) and the focus is called the focal distance or hypocentral distance.

Seismic waves attenuate in amplitude and energy while traveling over large distances due to geometrical spreading and material damping. However, sensitive detectors such as seismometers can sense these waves even if their amplitude ranges even a few nanometers. When these detectors are connected to a permanent recording system, they are known as seismographs. All the seismographs are based on a fundamental principle which records seismic waves based on the differential movement between a free mass and a fixed mass (which is fixed to the ground).

A single seismometer pendulum works in only one direction, and cannot give a complete picture of wave motions from other directions. To overcome this problem, modern seismograph stations have three separate instruments to record seismic waves, one to record motion in the vertical direction, and other two to record the motion in two mutually perpendicular directions in the horizontal plane, namely east–west and north–south components. Besides three-component instruments, clocks are an important part of a seismograph system. Modern seismographs use broadband

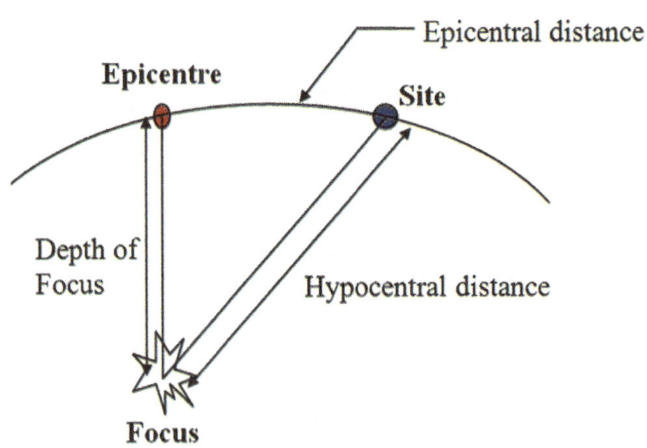

Fig. 1.17 Depth of focus and hypocentral distance

seismometers to acquire the data by computers and computer networks. A seismogram is the record of ground motion in terms of displacement or velocity or acceleration with respect to time.

1.5 Locating Earthquakes

When an earthquake occurs, it generates an expanding wave front from the earthquake focus at a speed of several kilometers per second. Various signals are recorded, with a network of seismographs on the earth surface. The times at which P waves arrive at each recording station are noted. One of the procedures to locate the earthquake source based on P-arrival times is as follows:

A guess is made about the epicenter, source depth, and origin time of the earthquake. Arrival time of P waves at each station is computed from the prior knowledge of seismic wave velocities inside the earth. These calculated arrival times are compared with the actual observed times at each station. The location is now changed a little so that difference between observed and calculated time reduces. This process is repeated till the time difference attains an absolute minimum. Mathematically, the problem is solved by setting up a system of linear equations, one for each station, and is solved by the method of least squares.

As the P waves travel faster than S waves, the time difference between the arrivals of the P wave and the S wave depends on the distance the waves traveled from the source to the recording station. Over the time, many such measurements have been made and travel-time curves (i.e., time vs. distance curves) for P, S, and many other waves have been developed for the earth. Therefore, the knowledge of S–P time difference on a seismogram will provide the information about the distance of the source from the station. For a given distance from the station one can draw a circle from the station with the radius as the distance from the station to the source, which shows that the source can be anywhere on the circle. If the information about the source distance is known from two stations, one can draw two circles with center as corresponding stations and the two intersecting points on the circles will be the possible locations of source. Similarly, if distances from three stations are known, the earthquake source can be unambiguously located. This is known as principle of triangulation. This is another method of locating earthquakes, which may be used to get a quick estimate of the earthquake location, as it requires data from only three stations. However, the former method is more accurate and adopted by various international data centers.

Recording of a three-component seismograph can also be used to obtain a crude estimate of the earthquake location. If the vertical motion of the P wave is upward, the radial component of the P wave is directed away from the epicenter, and if the vertical motion of the P wave is downward, the radial component of the P wave is directed back toward the epicenter. The amplitudes on the two horizontal components can then be used to obtain the vector projection of the P wave along the azimuth and to the seismic source. The distance of the seismic source is obtained from S–P time difference. Thus the knowledge of azimuth and distance from the recording station will help in identifying the event location. Earthquakes are measured in terms of intensity and magnitude, and the details are given below.

1.6 Intensity and Magnitude

The earthquake size which reflects the measure of amplitudes of ground shaking can be measured in both quantitative and qualitative manners. The first is a qualitative measure called intensity which is based on the evidences of the observed human reactions and damages occurred on its surroundings. This is actually a damage scale, in which the damage caused or level of shaking felt is categorized into twelve categories (Table 1.1). Higher intensity indicates higher damage which may be very near to the source. This does not use seismograms at all, but is very useful for historical events for which there are no seismic recordings available. However, magnitude is a quantitative measure of an earthquake. The amplitude of seismic waves recorded by the seismograms is used to determine the magnitude of an earthquake. Magnitude scales are represented in logarithmic scale based on powers of 10 which means that for each increase in the magnitude value by one unit, the ground motions are 10 times larger. The energy required to produce 10 times larger motions is about 30 times larger.

The magnitude of an earthquake describes its energy release on a logarithmic scale. One of the earliest scales which was introduced in the 1930s for measuring the magnitude of earthquakes is the Richter scale. This scale measures the maximum amplitude (A) of the signal recorded on a standard seismograph, which is then corrected for distance and instrument gain to obtain the magnitude. The formula for determining the Richter magnitude is given in Eq. (1.1).

$$M_L = \log(A) - \log(A_0) \tag{1.1}$$

where A_0 is the distance correction term for distances less than 600 km.

In order to incorporate the worldwide earthquakes for any distance and depth, the Richter scale evolved into surface wave (M_s) and body wave (m_b) scales. A typical anatomy of accelerogram is shown in Fig. 1.18 in which various phases of waves arriving at the accelerograms are identified as shown below.

Table 1.1 Relation between epicentral intensity and magnitude for an earthquake at 20 km depth [2]

Local magnitude	Intensity mm scale	Description of effect and damage
1.5	I	I. Not felt
3 3.9	II III	II. Felt by very few in the upper floors III. Felt noticeably in upper floors Vibrations like passing truck
4 4.9	IV V	IV. During day felt by many indoor. Windows, doors, dishes disturbed. Standing automobile rocked V. Some windows and doors broken. Many are awakened. Disturbance to tall objects like tree, pole
5 5.9	VI VII	VI. Heavy furniture moves. Damage to chimneys. People run outdoor VII. Considerable damage to poorly built buildings. Slight damage to good buildings
6 6.9	VIII IX	VIII. Partial collapse of ordinary buildings and damage in specially designed buildings. Changes in well water. Fall of tall structures like stacks, columns IX. Considerable damage/partial collapse of well-built structures. Ground cracked
>7	>X	X. Most of masonry and frame structures destroyed. Land slides. Rail bent XI. Few structures remain. Broad fissures in the ground. Earth slumps XII. Damage total. Waves seen on the ground. Objects thrown into air

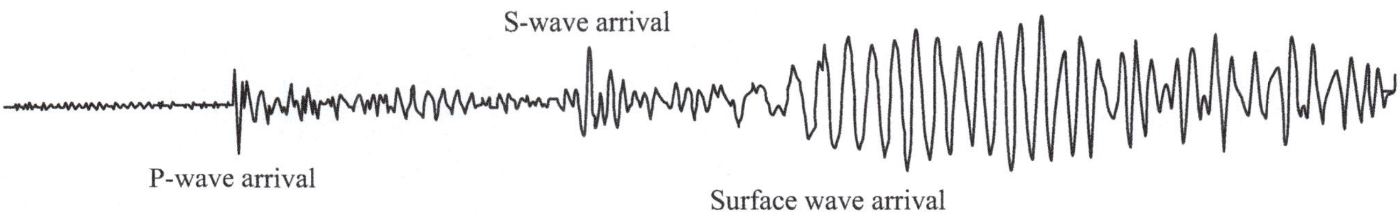

Fig. 1.18 A typical anatomy of an accelerometer reading (Stein and Wysession)

The body wave scale (m_b) is measured for shorter time period (or high frequency) and the surface wave scale (M_s) is measured for longer duration (or low frequency) which is illustrated from the vertical-component seismogram of Saguenay earthquake (25/11/1988) as shown in Fig. 1.19, in which the time history is divided into body and surface waves as per Fig. 1.19, and the different portions of the time history are converted into frequency domain as shown in Fig. 1.20.

By looking at the seismogram and the normalized spectra, it is observed that the surface waves contain longer-period energy than body waves; hence it gives us the justification of the period chosen for the above scales. For most of the shallow earthquakes, the amplitude of the earthquake is larger due to the propagation of surface waves; hence a different scale was required to measure the amplitude of these waves. After measuring the maximum surface wave amplitude, A (in microns), the surface wave magnitude, M_s (or the long-period magnitude), is obtained from the relationship given in Eq. (1.2) as given below:

$$M_s = \log\left(\frac{A}{T}\right) + 1.66 \log(\Delta) + 3.30$$

or,

$$M_s = \log(A) + 1.66 \log(\Delta) + 3.30 - \log(T), \qquad (1.2)$$

where A is the maximum displacement of the ground in micrometers, T is the measured wave period (generally equal to 20 s), and Δ is the epicentral distance in degrees,

Thus, putting $T = 20$ s in the above expression,

$$M_s = \log(A) + 1.66 \log(\Delta) + 3.30 - \log(20),$$

or,

$$\begin{aligned} M_s &= \log(A) + 1.66 \log(\Delta) + 3.30 - 1.30, \\ M_s &= \log(A) + 1.66 \log(\Delta) + 2.0 \end{aligned} \qquad (1.3)$$

To incorporate the earthquakes which occur deep inside the earth, a different scale was evolved which measures only the amplitude of body waves (the seismic waves that travel through the Earth's interior or body). To determine the magnitude of body waves m_b, a measurement of the maximum amplitude A from the first 10 s of P wave data of the seismogram records is required and the body wave

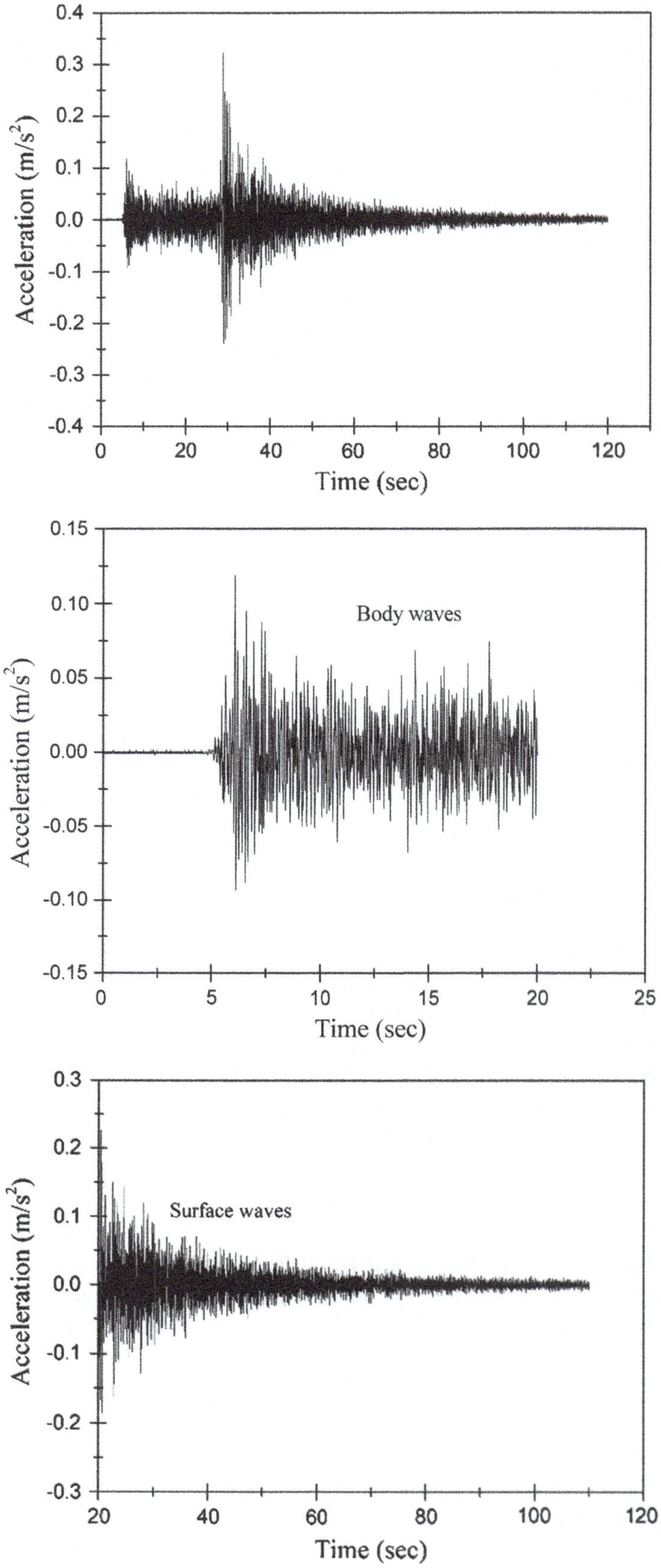

Fig. 1.19 Time history of Saguenay earthquake (1988) which is divided into body and surface waves [CESMD]

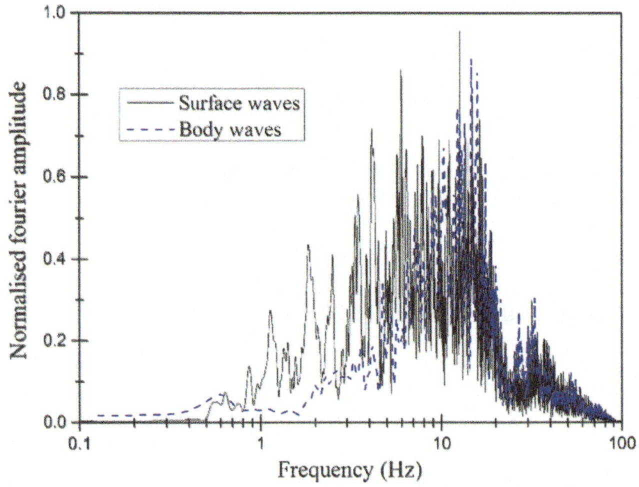

Fig. 1.20 Frequency domain representation of body and surface waves of Saguenay earthquake (1988)

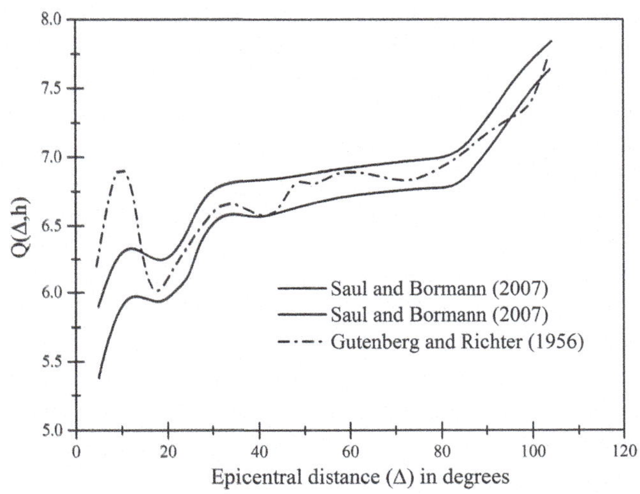

Fig. 1.21 The correction Q (Δ, h) that is applied to determine the body wave magnitude ([9], Saul and Bormann 2007)

magnitude m_b (or the short-period magnitude) can be obtained from Eq. (1.4) as given below:

$$m_b = \log\left(\frac{A}{T}\right) + Q(h, \Delta) \tag{1.4}$$

where T is the measured wave period and Q is an empirical function of focal depth 'h' and epicentral distance 'Δ'. The correction for distance and depth Q (h, Δ) is determined empirically. Figure 1.21 shows the values for Q (h, Δ). The correction is obtained from the contour for the appropriate depth and epicentral distance [9, 10].

Currently the m_b scale uses compressional body waves with a period of about 1 s, and the M_s scale uses Rayleigh surface waves with 18–22 s periods. In general, for any particular earthquake all these scales may give different magnitudes.

1.7 Magnitude–Energy Relationships

As described in the previous section, there exists a relationship between magnitude and energy. The relationship between the earthquake magnitude and energy can be represented as a linear one of the form as given in Eq. (1.5).

$$E_0 = A + BM \tag{1.5}$$

According to Gutenberg and Richter [1] the value of A is given as 11 and the value of B is given as 1.6. However, as suggested by Benioff [11] the linear form cannot be used for all the magnitude scales; hence a new equation has been proposed to replace Eq. (1.5) which fits all the observations accurately as given in Eq. (1.6).

$$\log(E_0) = 12.34 + 2q_0 + \log(t_0), \tag{1.6}$$

where E_0 is the seismic energy in ergs, and q_0 is equal to log (A/T) which is found out from different magnitudes of earthquake at various epicentral distances which is shown in Fig. 1.22.

It is observed that q_0 depends on the magnitude (M) and the relationship between q_0 and M is given as Eq. (1.7).

$$q_0 = \log\left(\frac{A}{T}\right) = -0.76 + 0.91M - 0.027M^2 \tag{1.7}$$

The data used for fitting the relation is taken from "United States Earthquakes" and from direct readings on seismograms of standard torsion instruments [9]. Similarly the term $\log(t_0)$ represents the duration of the wave group under consideration and is also a function of earthquake magnitude as shown in Fig. 1.23. The data can be represented by the Eq. (1.8).

$$\log(t_0) = -1.4 + 0.32M \tag{1.8}$$

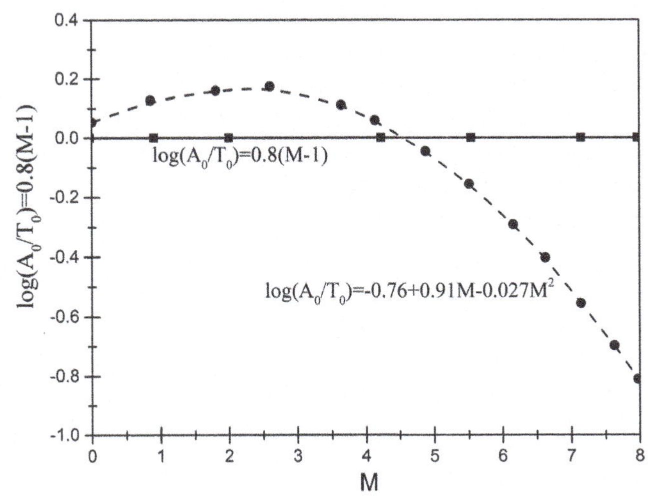

Fig. 1.22 Relationship between log(A/T) with magnitude (M) for different earthquakes

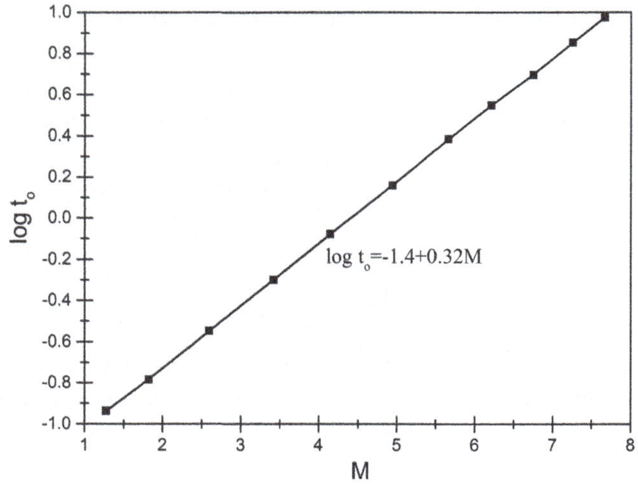

Fig. 1.23 Variation of duration of wave group with earthquake magnitude

Substituting the Eqs. (1.7) and Eq. (1.8) into Eq. (1.6), it reduces to Eq. (1.9), which is a quadratic variation of energy and magnitude in contrast to the linear relationship.

$$\log(E_0) = 9.4 + 2.14M - 0.054M^2 \tag{1.9}$$

For more details about the relationship between energy and magnitude the reader may refer to the Appendix 1 provided at the end of this chapter.

A scalar quantity which is useful for measurement of the source time function for an earthquake is the seismic moment. This quantity describes the amount of energy released during an earthquake, which depends on the geometry of the faulting, the ruptured area of the fault (which is taken equal to the total area of the fault), and the amount of slip that has taken place. The expression given in Eq. (1.10) is used to calculate the seismic moment of a fault:

$$M_0 = \mu A D \tag{1.10}$$

where D is the average displacement (or the slip of the fault) and A is the area of the fault. It is possible to relate the seismic moment with seismic energy. Kostrov [12] showed that the radiated seismic energy, E_0, is related to the stress drop (difference between the initial stress and the final stress) and can be represented as,

$$E_0 = \Delta\sigma D \frac{A}{2} \tag{1.11}$$

or

$$E_0 = \Delta\sigma \frac{M_0}{2\mu} \tag{1.12}$$

or

$$\log(E_0) = \log\left(\frac{\Delta\sigma}{2\mu}\right) + \log(M_0) \tag{1.13}$$

Assume that the stress drop is constant and equal to ~30 b, then Eq. (1.12) reduces to Eq. (1.14).

$$E = 5 \times 10^{-5} M_0 \tag{1.14}$$

Substituting E in Eq. 1.22 and replacing M_s by M_w one can get,

$$M_w = \frac{2}{3}\log_{10}(M_0) - 10.73 \tag{1.15}$$

This scale is tied to M_s but will not saturate because M_0 does not saturate. There are different ways to estimate M_0 such as from amplitudes of the long-period surface waves, amplitudes of seismic body waves, and spectra of seismic waves. Each scale of magnitude was designed for some specific type of seismic wave; for instance when the earthquake is shallow, it generates large surface waves for which the surface wave scale (M_s) is used for its measurement. This scale underestimates the size of the deep earthquakes. In contrast, body waves are well developed for both shallow and deep earthquakes, so 'm_b' can be used to compare them.

All the magnitude scales, except M_w, do not measure the size of large earthquakes correctly because amplitudes of the seismic waves tend to become constant with increasing magnitudes at the measuring frequencies (1.2 Hz for M_L, 1 Hz for m_b, and 0.05 Hz for M_s). This scale "saturation" occurs around magnitude 8 for the M_s scale and around magnitude 6.5 for the m_b scale. This means that a magnitude M_w = 9.0 event will have M_s estimate of ~8.0 and m_b estimate of ~6.5 only. Various magnitude definitions are summarized in Table 1.1 and 1.2.

The relationship between the various magnitude scales with moment magnitude (M_w) can be seen from Fig. 1.24.

As all other magnitude scales saturate at magnitude nearing 6.5, the most preferred scale of correlation with characteristics of ground motion is the moment magnitude (M_w), although it is the last one to have been developed.

1.8 Strong Ground Motion

An earthquake ground motion can be defined as the motion of sufficient strength to affect people and their environment. To name this ground motion as a strong motion, the threshold value of acceleration which is to be exceeded is not very clear. Usually an acceleration threshold of

Table 1.2 Magnitude definitions used by National Earthquake Information Centre, Denver, USA

Designator	Name	Magnitude formula
M_w	Moment magnitude	$M_w = (2/3) \log M_0 - 10.7$ where M_0 is the scalar moment for the best double couple in dyne cm
M_e	Energy magnitude	$M_e = (2/3) \log(Es) - 2.9$ where E_s is the radiated energy in Newton meters. M_e, computed from the high-frequency seismic data, is a measure of seismic potential for damage
M_s	Surface wave magnitude	$M_s = \log(A/T) + 1.66 \log D + 3.3$ where A is the maximum ground amplitude in micrometers of the vertical component of the surface wave within the period range $18 < T < 22$ s. T is period in seconds, and D is epicentral distance in degrees and $20° < D < 160°$ No depth correction for depths less than 50 km .
m_b	Compressional body wave magnitude	$m_b = \log(A/T) + Q\ (D, h)$ A is the ground amplitude in micrometers, and Q is a function of D in degrees and depth h in km
m_b, L_g	Body wave magnitude using L_g waves	$m_bL_g = 3.75 + 0.9 \log D + \log(A/T)$ for $0.5° < D < 4°$ $m_bL_g = 3.30 + 1.66 \log D + \log(A/T)$ for $4° < D < 30°$ where A is the ground amplitude in micrometers, T is the period in seconds from the vertical component of ~1 s L_g waves, and D is the distance in degrees
M_L	Local (Richter) magnitude	$M_L = \log(A) - \log(A_0)$ where A is the maximum amplitude in micrometers on short-period seismograph (usually S wave amplitude), and log (A_0) is the distance correction for distances less than or equal to 600 km.

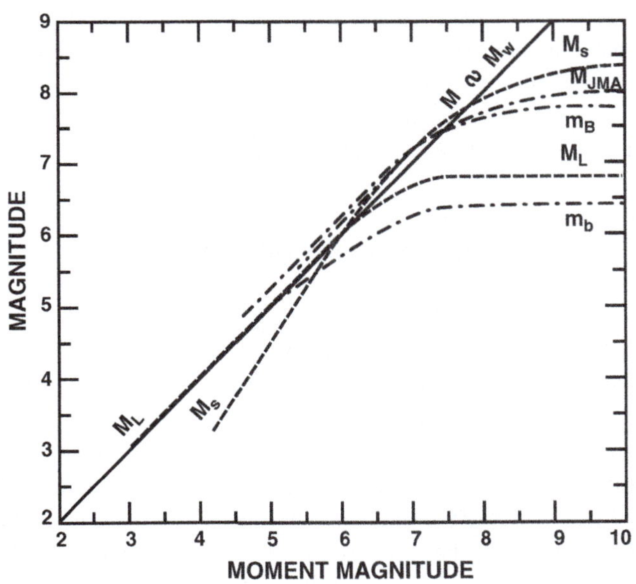

Fig. 1.24 Saturation of various magnitude scales of earthquake measurement

10–25 cm/s^2 is considered, since older strong motion recording instruments (accelerograph) are not able to resolve ground acceleration below 10 cm/s^2. The goal of the study of strong ground motion is to understand the physical processes that control the strong shaking and develop reliable estimates of ground motion for mitigation of earthquake effects through improved earthquake-resistant design and retrofit.

The strong ground motion records are of particular interest for engineering applications, since the ground acceleration is proportional to the force transmitted to the structures during earthquakes. The strong ground motion is recorded by an instrument which is known as strong motion accelerograph (SMA). Ground motion produced by an earthquake is very complicated in nature. One at a time, in one location ground motions are represented by three translations and three components of rotations. Generally rotational components are neglected. A typical strong motion record (such as the acceleration time history of January 17, 1994, Northridge earthquake) is shown in Fig. 1.25 [CESMD].

The ground motion represented in Fig. 1.26 has a time interval of 0.02 s. In general the ground motion is described with rise time, strong motion duration, decay time, and peak amplitudes. Details of measurements of strong motion are explained in the following sections.

1.8.1 Strong Motion Instruments

The first analog strong motion accelerograph was developed in the 1930s to record strong motion, and the first strong motion record came from Long Beach earthquake of

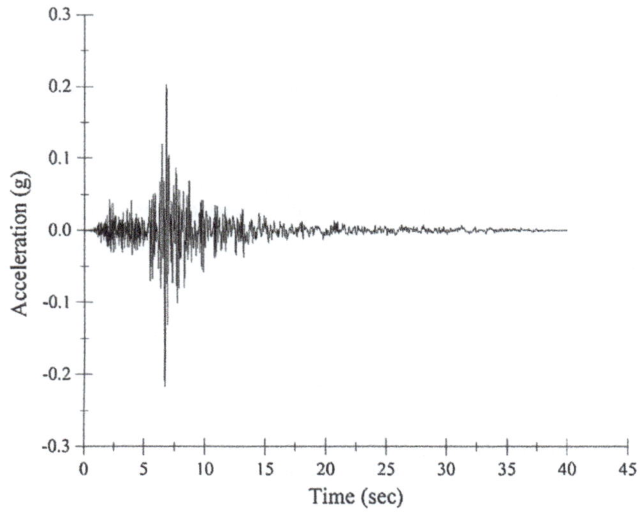

Fig. 1.25 Acceleration time history record of 1994 Northridge earthquake [CESMD]

California of March 10, 1933. Prior to 1990s, all strong motion accelerographs (SMA) were analog. With the advent of digital recording technology, the recording and processing of the strong motion have become easier. A digital SMA has several advantages over analog SMA and can replace the combined analog instruments, such as analog triaxial time history accelerograph (THA), passive triaxial response spectrum recorder (RSR), and triaxial peak acceleration recorder (PAR), normally used in critical facilities such as nuclear power plants. However, still about half of the world's SMAs are analog in nature.

An accelerogram is a box of size smaller than 1 feet on each side. The sensor, an accelerometer, is a spring–mass–damper system. The spring in the accelerometer is stiff, which provides a high natural frequency of the system (around 50 Hz), and the parameters are selected in such a way that the accelerations of 1–2 g will cause a full-scale deflection in the recording system. Stiffness is introduced electronically in a force-balance feedback loop that will keep the seismic mass motionless relative to the frame of the instrument. Such accelerographs are called force-balanced accelerographs (FBAs). The reason for keeping such a high natural frequency of the accelerometer system is that during the ground vibration, the frequency content is far lesser than that of its natural frequency; hence the deflection of the sensor becomes proportional to the acceleration. Every accelerograph contains three accelerometers in three mutually perpendicular directions to measure the three components of ground motion.

1.8.2 Measurements of Strong Motion

The accelerograms which are recorded on film formats are digitized. A great deal of efforts is required to process the records, because of the importance of accurate information for earthquake-resistant design. These accelerograms record acceleration, but the estimation of ground velocity and displacement is also important from the geophysical and engineering point of view; thus numerical integration of the acceleration record with proper baseline correction is required.

The processing of strong motion earthquake data involves the digitization of the record and correcting it for various errors. The digitization of records consists of converting a continuous trace into digital computer-compatible data on trace amplitude versus time. This can be accomplished manually, by semiautomatic and automatic machines. In manually or semiautomatic digitization, data is digitized for various peaks and troughs, i.e., at unequal time interval. Automatic machines allow digitization at equal time interval with about 50 samples/s. After digitization, the trace amplitudes are converted into acceleration and time values by multiplication of scale factor. The data for each peak and trough, which is at unequal time interval, is converted to equal time interval (typically, 0.02 s).

The true zero level of acceleration in the accelerogram is difficult to determine, both on the film record and on digital recordings. Currently, digital low-pass filtering is used with appropriate characteristics to make the baseline correction. From the film recordings, generally ground motion having frequency content greater than 0.2 Hz is recovered accurately, and the lower limit depends on the type of instrument. On analog SMAs, the ground motion with frequency content lower than 25 Hz is reliable, and the range of high frequency goes up to 50 Hz on modern digital SMAs.

1.8.3 Strong Motion Instrument Arrays Worldwide

The US Coast and Geodetic Survey installed the first network of 51 SMAs in 1935, and the total number remained low until the mid-1960s, in which the commercially produced SMAs were available. Most seismically active countries have at least a few SMAs in operation. Industrial nations with high seismic activity have extensive programs. Countries with the most extensive strong motion networks are Japan and Taiwan. USA has networks operated by the US Geological Survey (USGS) and several other organizations. In India, deployment of SMAs started in mid-1960s in Koyna Dam area, which recorded the world's largest reservoir-triggered earthquake of magnitude 6.5 and several of its aftershocks. In late 1960s, Department of Earthquake Engineering (DEQ), IIT Roorkee, had also started the indigenous development of strong motion instruments. In 1982, DEQ, IIT Roorkee, had installed the first array of 51 SMAs in Kangra region of Himachal Pradesh, and in 1985

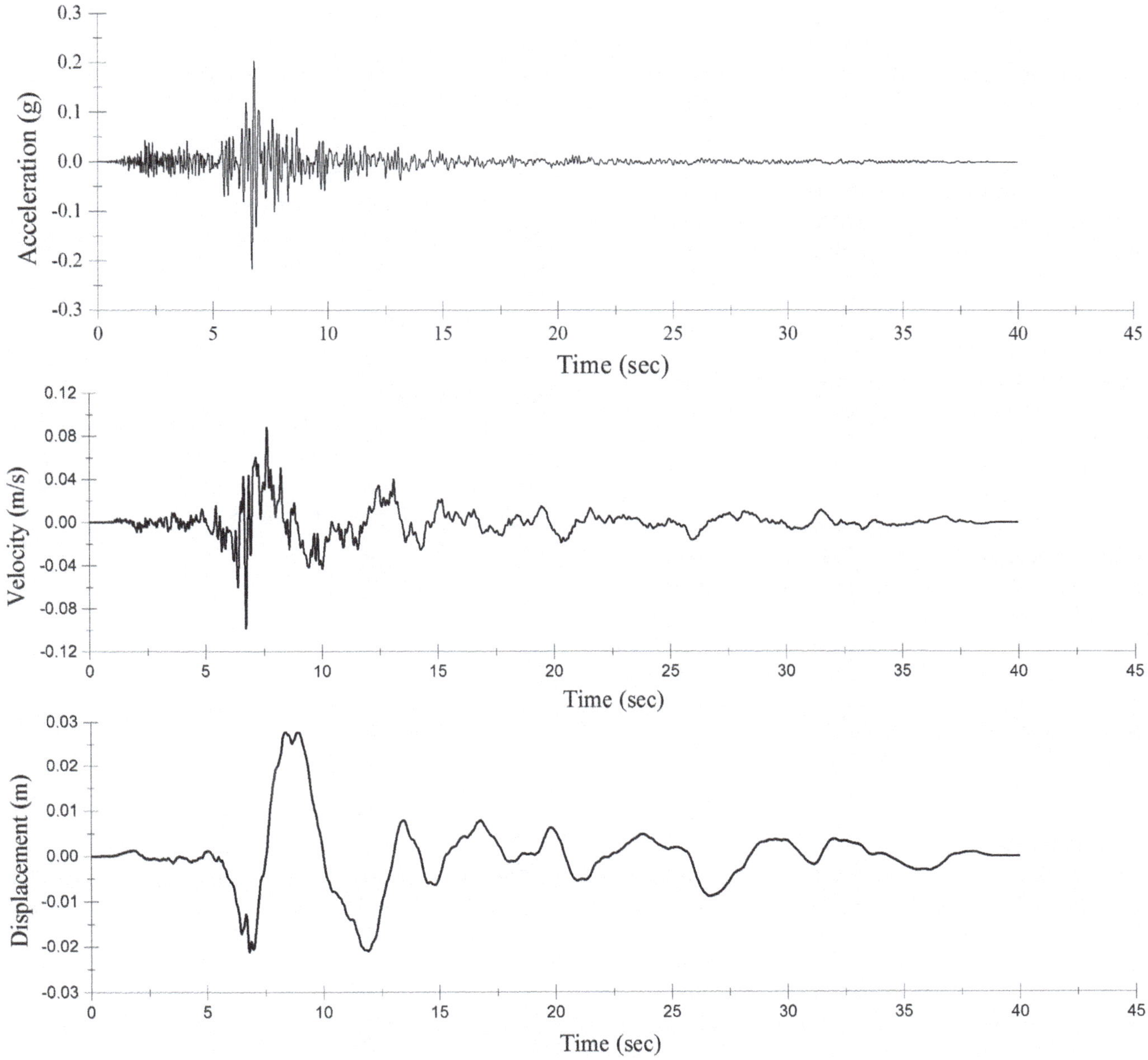

Fig. 1.26 Acceleration, velocity, and displacement time histories of 1994 Northridge earthquake (M_w = 6.7) [CESMD]

the second array of 45 SMAs was installed in Shillong region. At present, several organizations in India, such as Indian Meteorological Department (IMD), National Geophysical Research Institute (NGRI), Tehri Hydro Development Corporation (THDC), Sardar Sarovar Project (SSP), Nuclear Power Corporation of India Ltd (NPCIL), are operating strong motion instruments. However, still there is a need for more SMA networks to record strong ground motions due to medium-to-large earthquakes; for example there was no record for January 26, 2001, Bhuj earthquake due to no SMA stations around its epicenter.

Worldwide many different organizations are operating SMA networks. However, a single global system of access to strong motion data has not been achieved so far. The most thorough attempt was made by the World Data Center, which sought to compile and reproduce strong motion data on magnetic tapes in 1980s. Availability of strong motion data on Internet began in late 1990s. A large number of accelerograms are available through USGS Web site. In India, majority of strong motion data (1980s onwards) available are from Himalayan earthquakes.

1.8.4 Parameters to Describe Strong Ground Motion

A number of parameters such as peak amplitudes, strong motion duration, frequency spectrum, response spectrum of strong motion records are needed from engineering point of view. To explain these parameters, a well-known strong motion accelerogram of 1994 Northridge, California, earthquake of moment magnitude $M_w = 6.7$ is chosen (as shown in Fig. 1.26).

1.8.4.1 Parameters Related to Amplitude

The parameters related to amplitude such as peak acceleration, peak velocity, and peak displacement are in general used to describe the strong ground motion. These parameters are briefly described as follows:

Peak Acceleration

Most common measured parameter of a strong ground motion is peak ground acceleration (PGA) which is the absolute maximum value of ground acceleration which is recorded in the accelerogram. In the case of 1994 Northridge earthquake the PGA is 2.128 m/s^2 (or 0.217 g) as shown in Fig. 1.26. PGA is a very important parameter as it directly relates to inertial forces that will act on structures. Many researchers [13, 14] correlated earthquake magnitude with PGA. For a given earthquake there are two horizontal and one vertical components of strong motion. In general vertical component has less PGA compared to horizontal component. For this reason vertical acceleration received less attention than horizontal motions. Generally peak vertical ground acceleration is considered 2/3rd of peak horizontal acceleration (PHA).

Peak Velocity

The ground motion in terms of velocity can be obtained by integrating the acceleration time history. The velocity time history so obtained is also shown in Fig. 1.26. The peak horizontal velocity (PHV) is the maximum (absolute) value of the velocity time history, and it is 0.0983 m/s for 1994 Northridge earthquake. This is one of the important design parameters for structures having natural frequencies ranging from 4 to 10 Hz. This range of frequencies is referred as velocity control region of response spectrum details of which are described in the later part of this chapter.

Peak Displacement

Displacement time history can be obtained by double integrating the acceleration time history. The displacement time history so obtained is also shown in Fig. 1.26. Maximum (absolute) value of displacement time history is called peak displacement, and it is 27.7 mm for 1994 Northridge earthquake. Peak displacement is generally connected with low-frequency component of strong motion. This is also one of the important design parameters for structures having natural frequencies less than 0.5 Hz or time period more than 2 s. This range of frequencies or periods is referred as displacement control region of response spectrum details of which are described in the later part of this chapter.

1.8.4.2 Frequency Spectrum

Strong ground motion generated due to earthquake is a random signal having harmonics of various frequencies ranging from 0.1 to 50 Hz with different amplitudes. These characteristics are represented in terms of Fourier spectrum and its power spectrum details of which are explained below.

Fourier Spectra

If a function $\ddot{x}(t)$ is periodic with period T, then the strong ground motion can be represented as the sum of a series of harmonic terms which comprises of different amplitudes, frequencies, and phases as follows.

$$\ddot{x}(t) = a_0 + \sum_{m=1}^{\infty} a_m \cos(\omega_m t) + b_m \sin(\omega_m t) \quad (1.16)$$

where

ω_m $m\omega_0$
a_0 $\frac{1}{T}\int_0^T \ddot{x}(t)\,dt$
a_m $\frac{2}{T}\int_0^T \ddot{x}(t)\cos(\omega_m t)\,dt$
b_m $\frac{2}{T}\int_0^T \ddot{x}(t)\sin(\omega_m t)\,dt$

The infinite series which is shown in Eq. (1.16) is known as Fourier series. If a function $\ddot{x}(t)$ is non-periodic, then the Fourier transform of $\ddot{x}(t)$ is given as Eq. (1.17).

$$F(\omega) = \int_{-\infty}^{\infty} \ddot{x}(t)\,e^{-i\omega t}\,dt, \text{ or } \quad F(\omega) = \Delta t \sum_{k=1}^{N} \ddot{x}(t_d) e^{-i\omega t} \quad (1.17)$$

where

$e^{-i\omega t}$ $\cos(\omega t) - i\sin(\omega t)$,
t_d $k\Delta t$, where $k = 1, 2 \ldots N$ and
ω $n\Delta\omega 2\pi n/N\Delta t$,

$\ddot{x}(t) = 0$ when $t > t_d$ or $t < 0$ where t_d denotes the duration of the ground motion; hence, $F(\omega)$ can be written as Eq. (1.18).

$$F(\omega) = \int_0^{t_d} \ddot{x}(t)\cos(\omega t)\,dt - i\int_0^{t_d} \ddot{x}(t)\sin(\omega t)\,dt \quad (1.18)$$

Consequently, the amplitude and the phase of the Fourier spectrum of $\ddot{x}(t)$ are expressed by,

$$|F(\omega)| = \sqrt{\left[\int_0^{t_d} \ddot{x}(t)\cos(\omega t)\,dt\right]^2 + \left[\int_0^{t_d} \ddot{x}(t)\sin(\omega t)\,dt\right]^2} \quad (1.19)$$

$$\text{Phase angle, } |F(\omega)| = \arctan\left(-\frac{\left[\int_0^{t_d} \ddot{x}(t)\sin(\omega t)\,dt\right]}{\left[\int_0^{t_d} \ddot{x}(t)\cos(\omega t)\,dt\right]}\right) \quad (1.20)$$

As per ASCE 4-98, for the time histories which are not generated by the enveloping function, the duration of ground motion can be defined as the time required for the cumulative energy of the time history to rise from 5 to 75%. The cumulative energy of the time history can be defined as follows:

$$E(t) = \int_0^t \ddot{x}^2(t)\,dt \quad (1.21)$$

This formulation of duration is linked with Arias Intensity which is described in detail in Sect. 8.4.3 with a user subroutine. It is to be noted that the Fourier coefficients have units of the original variable multiplied with time. The plot of Fourier amplitude with frequency is called Fourier amplitude spectra. Using the above definition of the duration of ground motion, the time interval of 1994 Northridge earthquake in which 5–75% of the cumulative energy lies is between 4.4 and 8.0 s and the value of t_d is 3.6 s. Fourier amplitude spectra of strong motion provide clear information of frequency content of the earthquake and corresponding amplitude. Using the above information regarding the duration of ground motion, a broadband spectrum for 1994 Northridge (M_w = 6.7) ground motion is shown in Fig. 1.27.

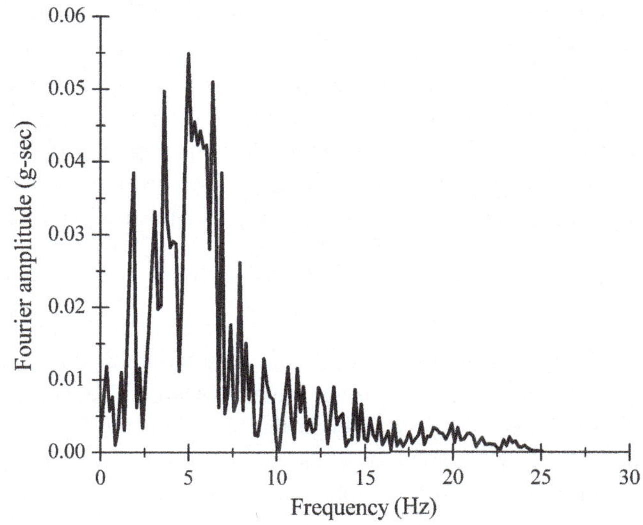

Fig. 1.27 Fourier spectrum of 1994 Northridge earthquake

Power Spectra

Power spectra is the variation of square of Fourier amplitude with frequencies. This describes the energy content of different harmonics of strong motion record and is given as:

$$P = \frac{1}{t_d}\int_0^{t_d} \ddot{x}^2(t)\,dt \quad (1.22)$$

where t_d is the duration of the ground motion as described in the previous section, and P is the average power (or the power spectral density) delivered by $\ddot{x}(t)$. Hence, the power spectral density of any non-periodic function $\ddot{x}(t)$ is defined in terms of the Fourier amplitudes of the time history by the relationship:

$$P = \frac{|F(\omega)|^2}{\pi t_d} \quad (1.23)$$

The similar relationship is also described in Chap. 12 with respect to the generation of artificial (or synthetic) generation of time history. This can also be utilized for characterizing the frequency content of ground motion. In the similar lines, the power spectral density for 1994 Northridge earthquake (M_w = 6.7) earthquake is shown in Fig. 1.28, and it is observed that the majority of the energy of the ground motion lies between 4 and 8 Hz. A user subroutine for computing the Fourier and power spectra of a given earthquake is provided in Appendix 2.

Response Spectra

Response spectra represent the maximum response of single-degree of freedom (SDOF) system (which has only

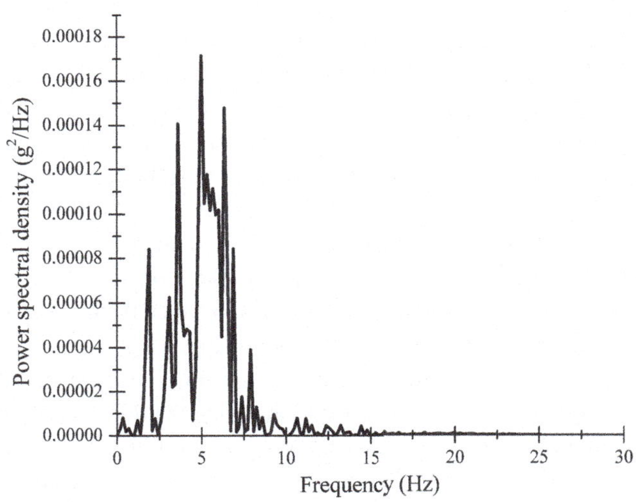

Fig. 1.28 Power spectral density of 1994 Northridge earthquake

mass and stiffness) for a given natural frequency (or time period) and damping ratio when it is subjected to strong ground motion. Response spectra are used as an input for the design of structure, systems, and components. The governing equation of motion for calculating the response of a SDOF subjected to strong ground motion for a given value of damping ratio is given in Eq. 1.24 (more details in Chap. 3).

$$m\ddot{x}(t) + c\dot{x}(t) + kx(t) = -m\ddot{x}_{\mathrm{g}}(t) \tag{1.24}$$

Figure 1.28 shows the details of construction of response spectrum for a given strong motion record. The strong motion record is passed through the SDOF of different frequencies and damping of 5% (typically), and response variation with time is obtained by solving Eq. (1.24). Response time histories for SDOF are shown in Fig. 1.29. As per the frequencies provided in Table 2.3-1 of ASCE 4-98, response histories can be obtained for various frequencies and one can pick up the maximum values from each response history. When these values are plotted in a graph with x-axis as frequency (or time period) and y-axis as maximum response, one can obtain the response spectrum.

In the similar lines the response spectrum of October 20, 1991, Uttarkashi earthquake for 2 and 5% damping is shown in Fig. 1.30 and 1.31, which has a PGA of 0.1 g.

1.8.4.3 Strong Motion Duration

Strong motion duration has a strong influence on the damage of structures. Many physical phenomena like development of pore water pressure in saturated soil and stiffness degradation of structure depends on number of cycle of stress reversal due to strong ground motion. As the magnitude of earthquake increases, the duration of strong motion also increases. Earthquake with long duration may produce adequate number of load reversal causing damages in the structure which can be observed from Table 1.3, where it is seen that as the earthquake duration increases (thereby increase in magnitude), which results in the increase in the number of equivalent cycles, in turn the number of load reversals increases. Duration of earthquake can be divided into three parts, namely rise time, duration of strong motion, and decay time, which are related to magnitude of earthquake (M_{w}) as per Table 2.3-1 of ASCE 4-98. Bolt [3] suggested bracketed duration, which is defined as the first and last exceedances of threshold acceleration (in this case the threshold acceleration is 0.01 g), and hence the bracketed duration is 19.60 s which is shown in Fig. 1.32.

Trifunac and Brady [14] calculated time duration based on the energy approach concept, in which the duration of an earthquake is the time interval between the points at which 5–95% of the energy has been released, which is based on Arias Intensity (in m/s) which is defined in Eq. (1.25).

$$I_x = \frac{\pi}{2g} \int_0^{t_{\mathrm{d}}} \ddot{x}^2(t)\,\mathrm{d}t \tag{1.25}$$

This parameter is based on the damage that is experienced by a structure during an earthquake event and is proportional to the energy dissipated by the structure during the total duration of an earthquake. A user subroutine for computing the Arias Intensity of a given earthquake is provided in Appendix 2. The Arias Intensity (in %) of 1994 Northridge earthquake is given in Fig. 1.33, from which the significant duration is calculated as 7.88 s.

Example 1.1 A vertical component of 25 November, Saguenay earthquake (1988) time history is taken from USGS (CESMD, Strong motion data, http://www.strongmotioncenter.org/), which is used for the determination of body wave magnitude. The details of the earthquake are shown in Table 1.4 as given below:

Determine the magnitude of the earthquake from the seismological recording, and compare with the published magnitude of this earthquake.

Solution The acceleration and velocity time history of the earthquake is plotted in Fig. 1.34:

For determination of body wave magnitude (m_{b}), the relationship developed by Gutenberg and Richter [9] for the body wave magnitude is given below:

$$m_{\mathrm{b}} = \log\left(\frac{A}{T}\right) + Q(h, \Delta)$$

The measurement of m_{b} is based on the measurement of displacement or velocity proportional seismograms. Hence, the term $(A/T)_{\max}$ is replaced by $(V_{\max}/2\pi)$ to determine mb from the peak velocity directly:

$$m_{\mathrm{b}} = \log\left(\frac{V_{\max}}{2\pi}\right) + Q(h, \Delta)$$

where $V_{\max}$ is the peak velocity (in µm/s) within the P wave train. Figure 1.18 gives an idea of different phases of wave arrivals at a given seismogram recording which helps to identify the P wave train in the velocity time history as given in Fig. 1.34b, and the initial portion of the time history (up to 10 s) is plotted both in time and frequency domains which is given below in Fig. 1.35:

From Fig. 1.35, the value of $V_{\max}$ is identified to be 0.0035 m/s (or 3500 µm/s) and epicentral distance of 134.80 km, or $\Delta = 1.21°$ (as shown in Fig. 1.38) the value of $Q(h, \Delta)$ for h = 29 km is found to be 4.2 (using the lower bound curve of Saul and Bormann (2007)) from Fig. 1.21,

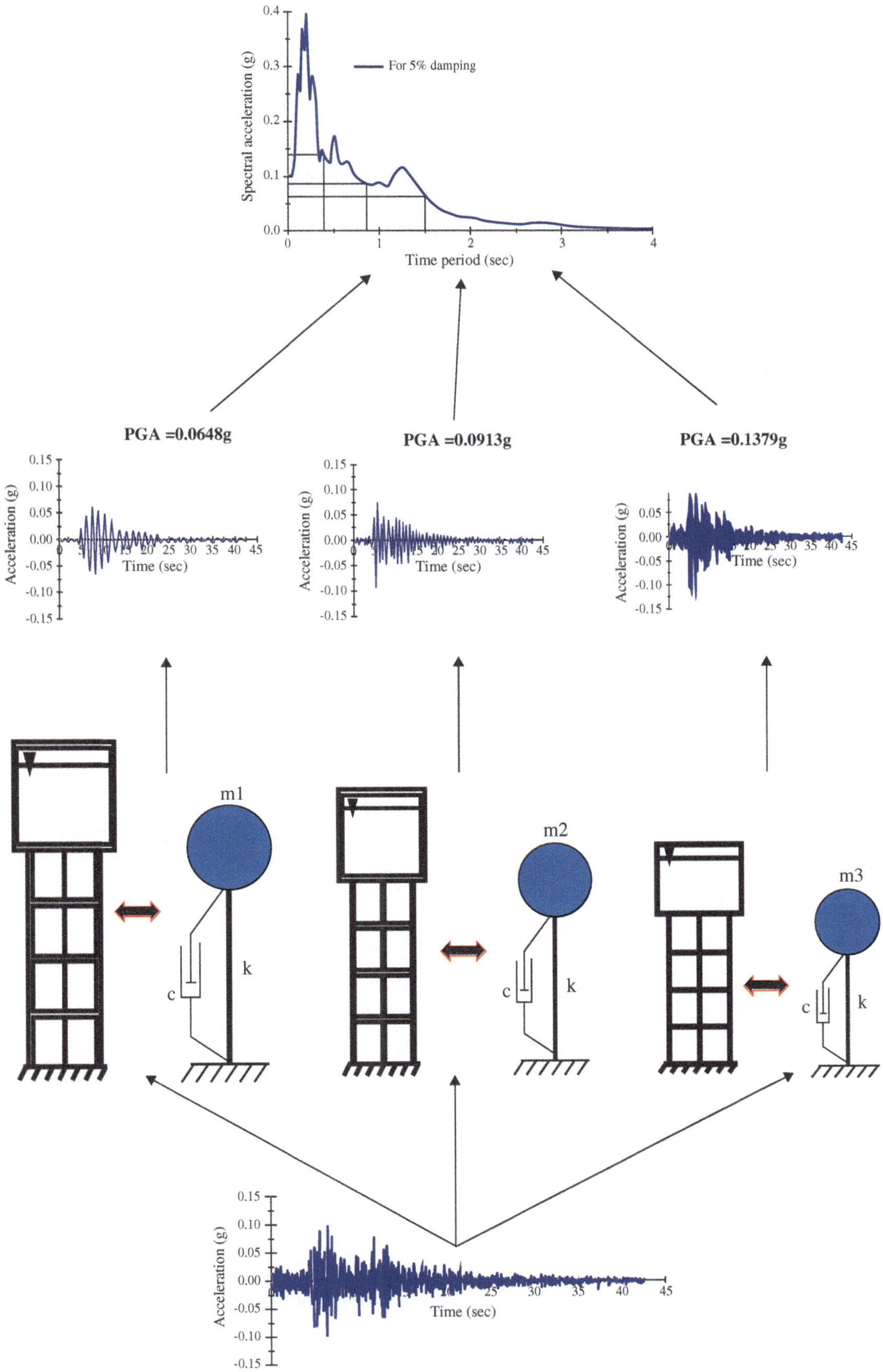

Fig. 1.29 Response spectrum for 1991 Uttarkashi earthquake

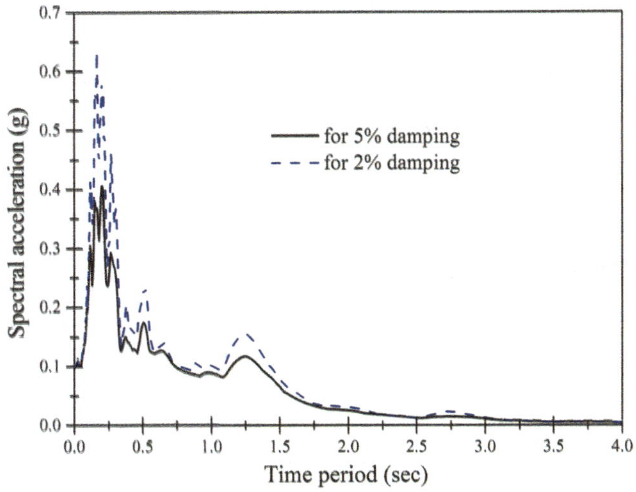

Fig. 1.30 Response spectrum for 1991 Uttarkashi earthquake in time domain

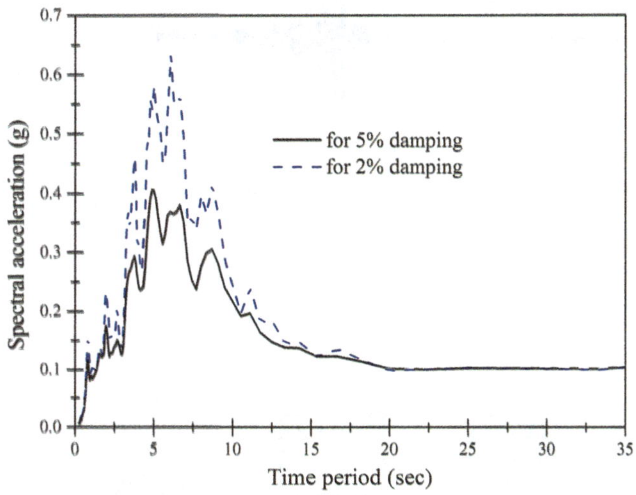

Fig. 1.31 Response spectrum for 1991 Uttarkashi earthquake in frequency domain

Table 1.3 Earthquake magnitude (M_w) with equivalent number of cycles [16]

Earthquake magnitude (M_w)	Number of significant stress cycles
5.25	2–3
6	5
6.75	10
7.5	15
8.5	26

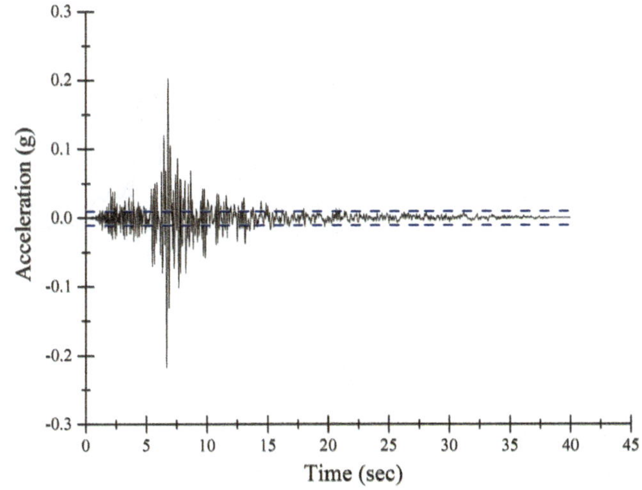

Fig. 1.32 Time history for 1994 Northridge earthquake showing bracketed duration [CESMD]

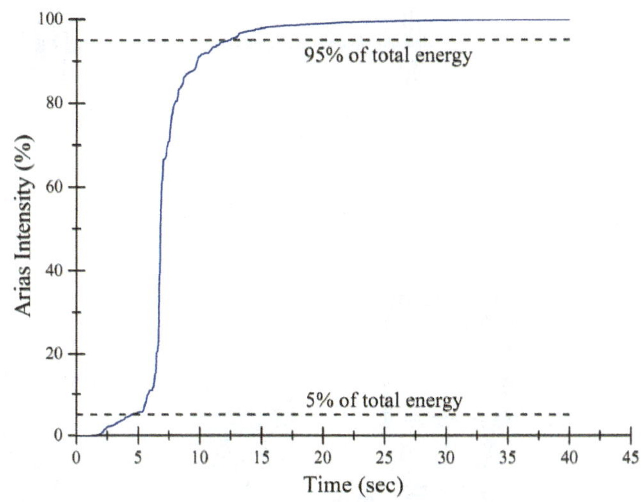

Fig. 1.33 Arias Intensity (in %) of 1994 Northridge earthquake

Table 1.4 Details of Saguenay earthquake

Name of the earthquake	Saguenay, 1988
Latitude and longitude of source (or event)	48.117°N, 71.183°W
Depth of focus	29 kms
Latitude and longitude of station	46.031°N, 68.206°W
Epicentral distance	134.68 kms (or $\Delta = 1.21°$)

which is found out by extrapolation. Hence, the value of m_b is found from the above expression which is 6.67 and the recorded body wave amplitude is 5.7, which gives us a rough estimation for finding out the body wave magnitude. The difference between the calculated and recorded amplitude is because as the epicentral distance lies between 5 and 20°, these calibration values (i.e., $Q(h, \Delta)$) are not reliable enough for global application (Saul and Bormann 2007). The corresponding moment magnitude is found out from Fig. 1.23 which is 8.57 and the recorded moment magnitude is 7.7 for this earthquake. For calculation of

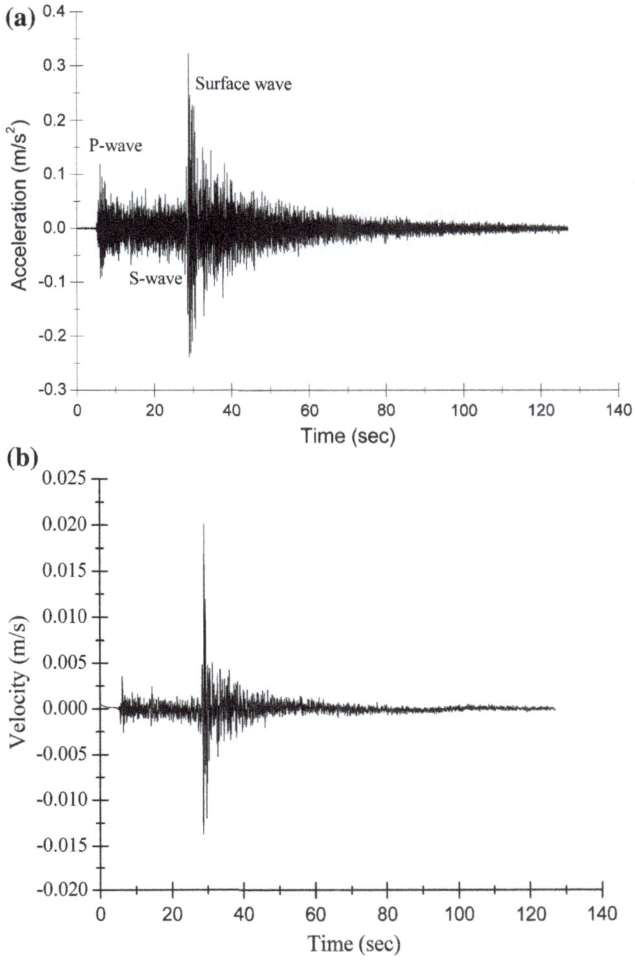

Fig. 1.34 **a** Acceleration time history of Saguenay earthquake (1988) [CESMD], **b** velocity time history of Saguenay earthquake (1988) [CESMD]

Table 1.5 Comparison of magnitudes from analytical solution and recorded data

Magnitude	Analytical	Recorded
Body wave magnitude	6.67	5.70
Moment magnitude	7.00	6.00
Seismic moment (dyne cm)	7.52×10^{17}	7.517×10^{16}
Seismic energy (in ergs)	5.4×10^{21}	1.98×10^{20}

seismic energy of the earthquake, Eqs. (1.6–1.8) are utilized, and putting the value of the moment magnitude (M_w) in these expressions gives the seismic energy (in ergs). From Eq. (1.4), the seismic moment for the given earthquake is calculated. The comparison of magnitudes, seismic moment, and energy which is found from analytical solutions and recorded data is shown in Table 1.5.

Exercise problems

1. In a general sense, what is an earthquake?
2. What are intraplate and interplate earthquakes?
3. What is the difference between the focus and epicenter of an earthquake?
4. What is the theory that explains the movement of tectonic plates?
5. What are the types of earthquake faults?
6. If during an earthquake a fault ruptures, would one gain or lose land if the fault type is (a) strike-slip (b) normal, or (c) thrust?
7. What is the difference between a foreshock, main shock, and aftershock in an earthquake?

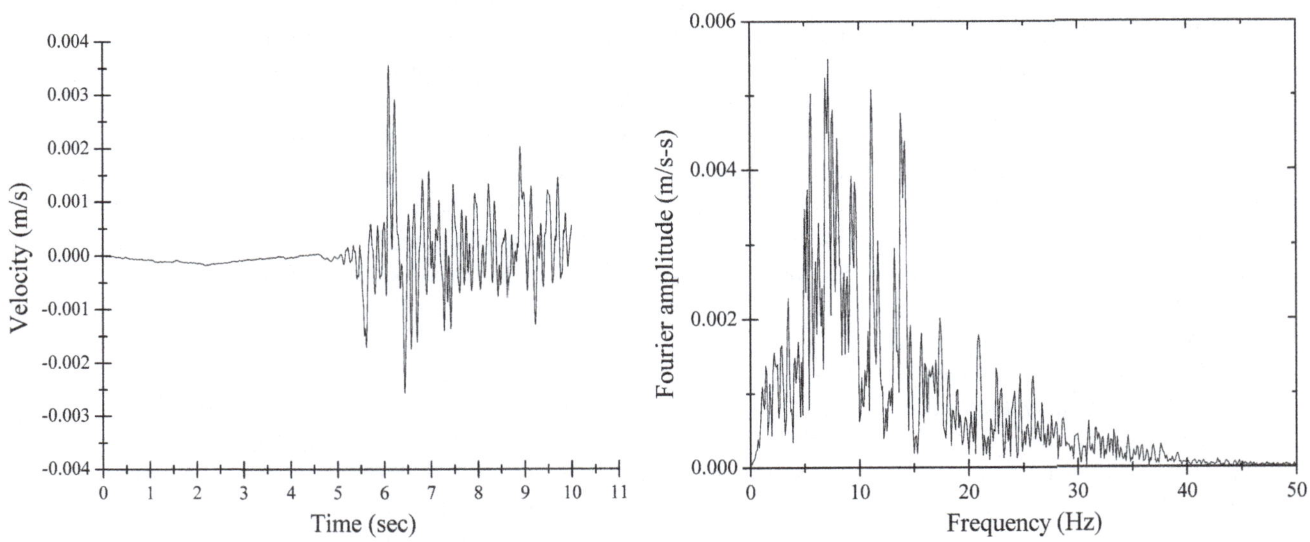

Fig. 1.35 Velocity history of Saguenay earthquake (2001) in both time and frequency domains, for P wave train [CESMD]

8. Estimate the moment magnitude of an earthquake generated at a fault that slips, on average, 6.2 m and ruptures an area 24 km wide and 82 km long.
9. An uncorrected acceleration record of May 17, 1976, Gazli earthquake is obtained from the reference given below as shown in Fig. 1.18. Integrate it to obtain the velocity and displacement time histories. Determine and comment on the value of the displacement obtained at the end of the displacement time history.
   (Reference: https://strongmotioncenter.org/vdc/scripts/default.plx)
10. Compute and plot the response spectrum (for 5% damping) and Fourier amplitude spectrum of the above acceleration record.
11. What is magnitude and intensity of an earthquake? With the help of diagram, define focus, epicenter, epicentral distance, and focal depth of an earthquake?
12. On what basis the seismic zoning map of India is prepared. Explain prominent zones and the significance of zoning map for public and industry.
13. Why intensity scales are used even today?
14. Why do earthquakes occur? Explain with relevance to India.
15. Explain major differences between "intraplate" and "interplate" earthquakes.
16. The recorded P wave amplitude of an earthquake (EQ1) is "100" micrometers, and its body wave magnitude (mb) is 5.0. What will be the magnitude (m_b) of another earthquake (EQ2) at the same place whose recorded amplitude is "200" micrometers.
17. Define with sketch diagrams different types of faults. Earthquakes on which type of fault may cause Tsunami and why?
18. State whether True or False. Give reasons.
    (a) Seismic Surface waves arrive before body waves.
    (b) Two earthquakes of same magnitude will cause similar damage in two different cities.
    (c) Surface waves can cause more damage to tall structures at large distances.
    (d) Moment magnitude is the most reliable indicator of earthquake energy.
19. The magnitude of an earthquake is reported as 7.9. This is most likely estimated on (give the most correct answer):
    (a) Surface wave magnitude scale
    (b) Moment magnitude scale
    (c) Body wave magnitude scale
    (d) Both (a) and (b).
20. An underground nuclear explosion may cause an earthquake?
21. A big earthquake in Australia can be the cause of an earthquake in Gujarat?
22. Why unsymmetrical geometry in both vertical and horizontal plans is not recommended to withstand earthquakes. Explain with illustration.
23. Explain the internal structure of the earth with sketch diagram showing approximate depths from surface and qualitative composition of different layers.
24. Explain why so many magnitude scales like mb, Ms, ML_g, Mw are used to measure the strength of earthquakes.
25. The geographical region where two plates of same type converge is called:
    (a) Trench
    (b) Ridge
    (c) Subduction zone
    (d) Transform fault.
26. The boundary between crust and mantle is called:
    (a) Mohorovicic discontinuity
    (b) Thrust boundary
    (c) Conrad discontinuity
    (d) Mantle–core boundary.
27. The hypocenter of an earthquake is fully defined if
    (a) Its latitude, longitude, and origin time are known
    (b) Its latitude, longitude, and depth are known
    (c) Its latitude, longitude, and distance are known
    (d) Its latitude, longitude, and magnitude are known.
28. Can underground nuclear explosions create earthquake?
29. A large earthquake can be prevented by creating many small earthquakes.
30. How vibrations are generated and explain briefly various types of vibrations.
31. What is the difference between P and S (SV and SH waves) during an earthquake event?
32. What are the types of earthquake faults?
33. What are intraplate and interplate earthquakes?
34. When it is said that an earthquake fault is inactive?
35. How can one determine the orientation of an earthquake-generating fault if one knows the direction of the first waves that arrive at a series of recording stations in the epicentral area?
36. Is volcanism associated with transform faults?
37. What is the difference between a foreshock, main shock, and aftershock in an earthquake?
38. Assuming that the rock has a shear (rupture) strength of 175 kPa, obtain the seismic moment and the radiated seismic energy of an earthquake.
39. Compute and plot the response spectrum (for 5% damping) and Fourier amplitude spectrum of the above acceleration record.

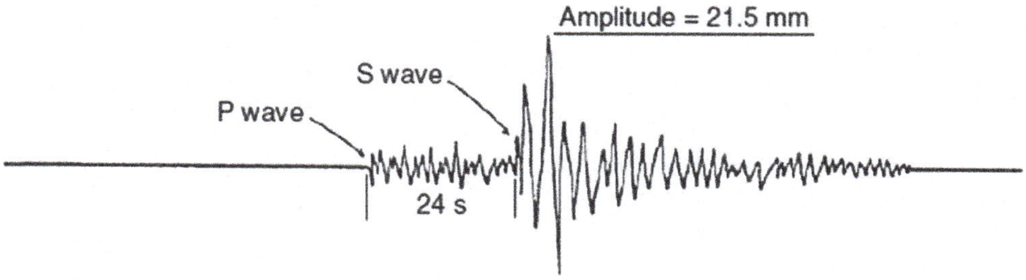

Fig. 1.36 Wood–Anderson seismograph

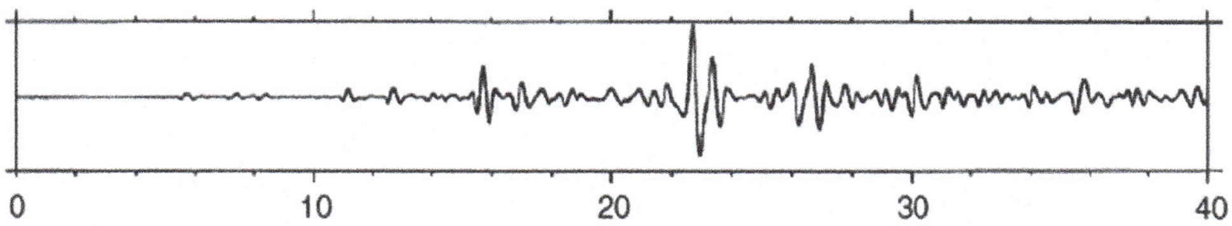

Fig. 1.37 Initial portion of a seismogram recorded in Australia from an earthquake in Oaxaca, Mexico

40. Compute the predominant period and Arias Intensity for the above acceleration record.
41. Determine and plot the variation of peak ground acceleration (PGA or zero period acceleration) with hypocentral distance for an earthquake of $M_w = 6.5$ for intraplate as well as interplate region using the attenuation relationship of NDMA (2007) and Campbell and Boroznia [16].
42. The average number of earthquakes exceeding a magnitude Mw in a year in a seismic region is given by $\log(N) = 4.87 - 0.54M_w$. What is the probability that an earthquake of magnitude 5.0 or greater occurs at least once in a 50-year interval in that region?
43. The seismogram shown in the figure was obtained from a Wood–Anderson seismograph located 100 km from the hypocenter of the recorded earthquake (Fig. 1.36),
44. Determine:
    (a) the average shear wave velocity for the region surrounding the recording station considering that the earthquake started at 07:19:32 and the P wave arrived at the recording site at 07:19:45 and
    (b) the average Poisson ratio for the region surrounding the recording station.
45. A figure below shows the initial portion of a seismogram recorded in Australia from an earthquake in Oaxaca, Mexico; identify in the figure the arrival times of the P, S, and surface waves, and explain the reasons for the selection (Fig. 1.37.

Appendix 1: Relationship Between Earthquake Magnitude and Energy from Seismogram Records

In this appendix the origin of the relationship between earthquake magnitude and energy from seismogram records is provided. The records of the seismographs for a given earthquake can be observed in twofolds: time reading and amplitude reading. The time reading gives an idea of the exact location, its focal depth, and the origin time of an earthquake, whereas the amplitude reading gives the information on the total energy released during an earthquake. There are two different ways in which the energy released in an earthquake can be quantified either from seismograms or from field observations in the epicentral area in combination with theoretical studies.

To bring more clarity on the different types of energy that is released during an earthquake, Yoshima (1963) has chosen a spherical source in a perfectly elastic solid as his model and has divided the energy released from the spherical source into three components, $W_R = E_0 + W_t + W_D$, where E_0 is the seismic wave energy, W_t and W_D represent the deformation energy. W_R is the outward flow of energy (or the strain energy) across a sphere of radius 'r' from a spherical source which can be obtained from Eq. (1.26) as given below:

$$W(r) = -4\pi r^2 \int_{t_1}^{t_2} \sigma_r \frac{\partial u}{\partial t} \mathrm{d}t \tag{1.26}$$

As σ_r is the radial stress per unit surface, the term $4\pi r^2$ (surface area of a sphere) comes into picture, and the distance corresponds to $\mathrm{d}u = \{\mathrm{d}u/\mathrm{d}t\}\ \mathrm{d}t$. The negative sign makes the energy positive when the direction for σ_r is chosen to be inward normal to the sphere. Similarly, the seismic wave energy, E_0, can be expressed by Eq. (1.27) as given below:

$$E_0 = 4\pi r^2 \rho V_p \int \left(\frac{\partial u}{\partial t}\right)^2 \mathrm{d}t \tag{1.27}$$

where V_P is the compression wave velocity. This is the expression of the seismic wave energy which can be used for the determination of wave energy from seismic records.

Consider a spherically symmetrical source of energy situated at a point F which emits P or S wave as shown in Fig. 1.38. The total seismic energy emitted from this source (only body waves) is $E_0 = E_P + E_s$ which is calculated from Eq. (1.28) as given below:

$$E_0 = 8\pi^3 \rho \left(h^2 + 4r_0(r_0 - h)\sin^2\left(\frac{\Delta}{2}\right)\right) e^{kR} \frac{(1+q)}{Q^2} \int_0^t V_p \left(\frac{A}{T}\right)^2 \mathrm{d}t \tag{1.28}$$

where

- ρ is the density of the medium,
- Q is the fraction of the recorded to the incident energy,
- h is the depth of the source,
- q E_s/E_p (fraction of S wave to P wave energy),
- E_0 $E_p(1 + q)$ or $E_0 = E_s(1 + 1/q)$,
- e^{kR} is the absorption factor (in which absorption of waves is taken into account),
- k is a value which varies with the type of wave, and
- t is the duration of the wave group ($=nT$) where n is the number of cycles (or number of wave trains)

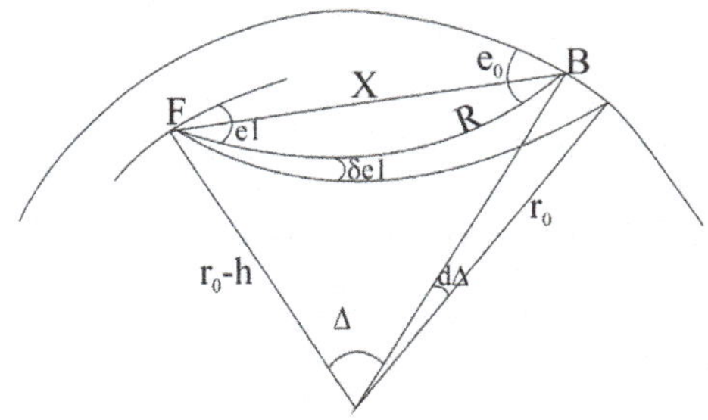

Fig. 1.38 A section through the earth which explains the calculation of energy of body waves

To make the calculations simple, the sum of all the wave trains are replaced by a single measurement of amplitude and time period, such that the energy computed remains the same. One of the most preferred choices is to measure the maximum amplitude (A_m) with its corresponding time period (T_m) within any given wave group [17]. The integrals over the wave trains (or the corresponding summations) can then be written as:

$$\int_0^t \left(\frac{A}{T}\right)^2 \mathrm{d}t = \left(\frac{A_m}{T_m}\right)^2 t_0 \tag{1.29}$$

It should be noted that (A_m) is the maximum amplitude in each wave group; i.e., for body waves, this means that (A_m) is not the amplitude of the first swing but can be measured up to about 10 s after the onset of the wave, just to correspond to the maximum in the group.

There may be an adequate risk to include some other phases of the body waves (as, e.g., there may be interference from pP for shocks shallower than about 40 km, or from PcP at distances beyond about 75–80° as given in Fig. 1.14) when the maximum amplitude measurement corresponds to 10-s window, but such complications are generally of no great consequence as the errors usually do not exceed those which anyway are inevitable in energy determinations. Another approach is that in practice, the period at which the magnitude of body waves is usually determined is 1 s [18]. Most of the calculations of the energy–magnitude relationships depend on the energy of a wave group emanating from a point source as shown below [9],

$$E_0 = 2\pi^3 h^2 V_p \rho \left(\frac{A_m}{T_m}\right)^2 t_0 \tag{1.30}$$

where t_0 is the duration of the wave group. This formulation applies at the epicenter ($\Delta = 0$) where h is the hypocentral distance. A factor of 3/2 is applied to the energy equation for calculating the total energy in terms of transverse waves. As at short distance, the amplitude of transverse wave dominates; hence it is more appropriate to represent Eq. (1.31) in terms of wave energy of S-group as shown below,

$$E_0 = 3\pi^3 h^2 V_s \rho \left(\frac{A_m}{T_m}\right)^2 t_0 \tag{1.31}$$

This form of equation is arrived in a similar fashion by putting ($\Delta = 0$, $q = 2$ and $Q = 2$) in Eq. (1.28), and the final energy expression in the form of transverse wave is,

$$E_0 = 3\pi^3 h^2 V_s \rho \left(\frac{A_m}{T_m}\right)^2 t_0 e^{kh} \tag{1.32}$$

where the integration of the wave group is carried out in terms of transverse waves by replacing $(1 + q)$ by $(1 + 1/q)$ in Eq. (1.28). This equation agrees with the equation given by Gutenberg and Richter [9] with the exception that it did not incorporate the absorption factor, e^{kh}.

As the energy varies over a large range, taking the logarithm on both sides of Eq. 1.28, the expression reduces to:

$$\log(E_0) = \log\left(8\pi^3 \rho V_p\right) + \log\left(h^2 + 4r_0(r_0 - h)\sin^2\left(\frac{\Delta}{2}\right)\right)e^{kR} + \log(1+q) + \log(t_0) + 2\log\left(\frac{A_{m1}}{T_{m1}}\right) \tag{1.33}$$

which is equivalent to the expression $\log(E_0) = a_1 + b_1 m_b$, and is taken as the definition of the body wave magnitude which is generalized as

$$m_b = \log\left(\frac{A_{m1}}{T_{m1}}\right) + F_1(\Delta, h) + C_{s1} + C_{r1} \tag{1.34}$$

where F_1 is a correction term for distance (Δ) and depth (h) of the source, C_{s1} is the station correction, taking the conditions at the respective stations into account, and C_{r1} is the regional correction, taking the focal mechanism (radiation pattern) and path properties into account. The function F_1 (Δ, h) has been determined by Gutenberg [19, 20] and Gutenberg and Richter [21] by a combination of theoretical and empirical findings, for the following waves: P vertical, P horizontal, PP vertical, PP horizontal, S horizontal. The constants a_1 and b_1 are then determined empirically from m and $\log E_0$. The combination of $F_1(\Delta, h) + C_{s1} + C_{r1}$ is $Q(\Delta, h)$ in which the density of the medium, depth of focus of earthquake, radius of the earth, and epicentral distances are all taken into consideration as shown in Fig. 1.36, and the plot of $Q(\Delta, h)$ is shown in Fig. 1.21.

In order to grasp the total energy of surface waves (Rayleigh and Love waves respectively), we have to carry out a threefold integration in space and time: along the circle through B around F (with circumference = $2\pi r_0 \sin(\Delta)$, along the duration of the wave train, and from the surface down to (theoretically) infinite depth. Hence, for surface waves the corresponding expression for seismic wave energy is given below in Eq. (1.35).

$$\log(E_0) = \log\left(4\pi^3 \rho V_L r_0 H\right) + \log\left(\sin(\Delta \Delta^n e^{k\Delta}\right)e^{kR}) + \log(1+q) + \log(t_0) + 2\log\left(\frac{A_{m2}}{T_{m2}}\right) \tag{1.35}$$

where $H = 1.1\lambda$ (λ is the wavelength of the wave), V_L is the velocity of the surface wave, Δ^n is the dispersion factor, and $q = E_R/E_L$ (fraction of Rayleigh wave to Love wave energy). This is equivalent to the expression $\log(E_0) = a_2 + b_2 M_s$ and is taken as the definition of the body wave magnitude which is generalized as:

$$M_s = \log\frac{A_2}{T_2} + F_2(\Delta, h) + C_{r2} + C_{s2}. \tag{1.36}$$

In Eq. (1.36), usually 'h' is omitted and surface waves are used for magnitude determinations only for shallow shocks. Another approach is that in practice, the period at which the magnitude of body waves is usually determined is around 20 s (Kramer 1996). The constants a_2 and b_2 are found out in a similar fashion as for body waves. The earthquake magnitude that is determined depends on seismic wave measured. There are different magnitude scales for P waves, for Rayleigh waves, and for different periods of motion.

Appendix 2: Subroutines

A.2.1 A User Subroutine for Computing the Fourier and Power Spectra of a Given Earthquake

```
%-----Input in excel in data(m/s2) format----------
s1=xlsread('C:\Users\Raj\Desktop\sinewave.xlsx',4,'F:F');
N=length(s1);
Td=5.6;  %equivalent duration based on ASCE 4-98
td=0.02;  %time step of the earthquake
fori=0:1:(N-1)
rsum=0;
csum=0;
for j=0:1:(N-1)
v=(6.28*i*j)/N;
rsum=rsum+(s1(j+1,1)*cos(v));
csum=csum+(s1(j+1,1)*-1*sin((2*3.14*i*j)/N));
end
wt(i+1,1)=rsum;
wt(i+1,2)=csum;
end
fori=1:1:N
ampl(i,1)=((sqrt((wt(i,1)*wt(i,1))+(wt(i,2)*wt(i,2))))*td); %Fourier amplitude
Power(i,1)=(ampl(i,1)*ampl(i,1))/(3.14*Td);                  %Power spectra amplitude
freq(i,1)=(i-1)/(N*td);                                      %frequency
end
figure (1);
plot(freq(1:(N/2)),ampl(1:(N/2)))  %plot of Fourier spectra with frequency
figure (2);
plot(freq(1:(N/2)),Power(1:(N/2))) %plot of Power spectra with frequency
```

A.2.2 A User Subroutine for Computing the Arias Intensity of a Given Earthquake

```
%-----Input in excel in time(sec)-data(m/s2) format----------
s1=xlsread('C:\Users\Raj\Desktop\sinewave.xlsx',4,'A:B');
%s1=xlsread('Earthquake.xlsx',1,'D:E');
N=length(s1);
td=s1(2,1)-s1(1,1);  %time step of the earthquake
fori=1:N             %Squaring of acceleartion time history
s11(i,1)=s1(i,2)*s1(i,2);
end
Z=(td*cumtrapz(s11)*3.14)/(2*9.81); %integration of the squared acceleration time history
fori=1:N
  Z(i,2)=((Z(i,1))/max(Z(:,1)))*100; %Arias Intensity (in %)
end
plot(s1(:,1),Z(:,2))  %plot of time vs Arias Intensity
```

A.2.3 A User Subroutine for Computing the Response Spectrum of a Given Earthquake

```
k=0.0;
Tmax=5;%max time period upto which spectral acceleration needed to be calculated
dT=0.005;%time step for period
s11=xlsread('C:\Users\Admin\Desktop\Acc_time.xlsx',1,'BQ:BQ');
Sa(1,1)=max(abs(s11));
%initial condition
u(1)=0.0;
v(1)=0.0;
nT=Tmax/dT;
n=length(s11);
t=linspace(0,63.74,n);  %total duration of the erathquake
p=-s11; %force=m*ag
T(1)=0.0;
g=.5; %gamma value used in newmarks beta algorithm
b=.25;%beta value used in newmarks beta algorithm
zi=0.05;
h=0.01;    %time step of the earthquake
df=1.0/(n*h);
f=0.0:df:(n*df);
count=0.0;
for j=2:nT  %loop for integration using Newmark beta
   count=count+1;
   T(j)=(j-1)*dT;
   F(j)=1/T(j);
   w=2*pi/T(j);
   a(1)=s11(1)-2*zi*w*v(1)-w*w*u(1);
    at(1)=a(1)+s11(1);
      K = (w*w + (g*2*w*zi)/(b*h) + 1/(b*h^2));
   A = 1/(b*h) + (g*2*zi*w)/b;
   B = 1/(2*b) + h*2*w*zi*((g/(2*b))-1);
   for i1=2:n
dp(i1-1)= p(i1)-p(i1-1);
     DP1(i1-1) = dp(i1-1)+A*v(i1-1)+B*a(i1-1);
     du(i1-1) = DP1(i1-1)/K;
     dv(i1-1) = (g*du(i1-1))/(b*h)-(g*v(i1-1))/b+h*a(i1-1)*(1-(g/(2*b)));
     da(i1-1) = du(i1-1)/(b*h^2)-v(i1-1)/(b*h)-a(i1-1)/(2*b);
     u(i1) = u(i1-1)+du(i1-1);
     v(i1) = v(i1-1)+dv(i1-1);
     a(i1) = a(i1-1)+da(i1-1);
     at(i1)=s11(i1)+a(i1);
   end
Sa(1,j)=max(abs(at));  %picking up the maximum absolute acceleration value
 end
for k=2:nT
   F(k-1)=1/T(k);  %converting time to frequency
end
plot(F,Sa)      %Plotting frequency vs spectral acceleration
```

References

1. Gutenberg B, Richter CF (1954) Seismicity of the earth and related phenomena. Princeton University Press, Princeton, p 310
2. Richter CF (1958) Elementary seismology. W. H. Freeman & Co., New York, p 768
3. Bolt BA (1993) Earthquakes. W.H. Freeman & Co, San Francisco, CA
4. Gubbins D (1990) Seismology and plate tectonics. University Press, Cambridge, p 339
5. Taylor FB (1910) Bearing of the tertiary mountain belt on the origin of the earth's plan. Bull Geol Soc Am 21:179–226
6. Wegener A (1915) Die entstehung der kontinente und ozeane. Vieweg, Branunschweig, Germany
7. Garfunkel Z (1975) Growth, shrinking and long term evolution of plates and their implications for flow patterns in the mantle. J Geophys Res 80:4425
8. Wilson JT (1965) A new class of faults and their bearing on continental drift. Nature 207:343–347
9. Gutenberg B, Richter CF (1956) Magnitude and energy of earthquakes. Bull Seismol Soc Am 1–14
10. Bormann P, Saul J (2008) The new IASPEI standard broadband magnitude mB. Seismol Res Lett. 79(5):699–706
11. Benioff H (1955) Seismic evidence of crustal structure and tectonic activity. Geol Soe Am Spec Pap 62(61):73
12. Kostrov BV (1974) Seismic moment and energy of earthquakes, and seismic flow of rock. Izv Acad Sci USSR, Phys Solid Earth (English Transl) 1:23–40
13. Krinitszky EL, Chang FK (1987) Parameters for specifying intensity-related earthquake ground motions, Miscellaneous paper S-73-1, Report 25, U.S. Army Corps of Engineers Waterways Experiment Station, Vicksburg, Mississippi, p 43
14. Trifunac MD, Brady AG (1975) On the correlation of seismic intensity with peaks of recorded strong ground motion. Bull Geol Soc Am 65:139–162
15. Seed HB, Idriss IM, Makdisi F, Banerjee N (1975) Representation of irregular stress time histories by equivalent uniform stress cycles in liquefaction analysis, EERC-75-29. Earthquake Engineering Research Center, University of California, Berkeley
16. Campbell KW, Bozornia Y (1994) Near source attenuation of peak horizontal acceleration from worldwide accelerograms recorded from 1957 to 1993. In: Proceedings 5th U.S. national conference on earthquake engineering, EERI, Berkeley, California, vol 1, pp 283–292
17. Ahrens LH, Press F, Runcorn SK, Urey HC (1966) Physics and chemistry of the earth. Pergamon Press, London
18. Kramer SL (1996) Geotechnical earthquake engineering, Prentice hall
19. Gutenberg B (1945b) Amplitudes of P, PP, and S and magnitude of shallow earthquakes. Bull Seism Soc Am. 35:57–69
20. Gutenberg B (1945c) Magnitude determination of deep-focus earthquakes. Bull Seism Soc Am. 35:117–130
21. Gutenberg B, Richter CF (1956) Earthquake magnitude, intensity, energy and acceleration. Bull Seismol Soc Am 105–145

Further Reading

22. Anderson JG (1991) Geotech News 9(1):35
23. Benjamin JR and Associates (1988) A criterion for determining exceedence of the operating basis earthquake. EPRI report NP-5930. Electric Power Research Institute, Palo Alto, California
24. Bullen KE, Bolt BA An introduction to the theory of Seismology. Cambridge University Press, Cambridge, p 499
25. Fowler CMR (1990) The solid earth. an introduction to global geophysics. Cambridge University Press, Cambridge, England, p 472
26. Kanamori H, Anderson D (1975) Theoretical basis of some empirical relations in seismology. Bull Seismol Soc Am 65:1073–1095
27. Kulhanek O Anatomy of seismograms. developments in solid earth geophysics 18, Elsevier, Amsterdam, p 78
28. Lay Thorne, Wallace TC (1995) Modern global seismology. Academic Press, London, p 521
29. Lee WHK, Kanamori H, Jennings PC, Kisslinger C. International handbook of earthquake and engineering seismology, Part A. Academic Press, London, p 933
30. Newmark NM (1973) A study of vertical and horizontal earthquake spectra. NM Newmark consulting engineering services, directorate of licensing. U.S. Atomic energy commision, Washington D.C
31. Kulhanek Ota (1997) Anatomy of seismograms. developments in solid earth physics. Elsevier, Amsterdam
32. Stein S, Wysession M (2003) An introduction to seismology, earthquakes and earth structure. Blackwell publishing, Oxford
33. Singh RP, Aman Ashutosh, Prasad YJJ (1996) Pure Appl Geophys 147(1):161–180
34. Veith KF, Clawson GE (1972) Magnitude from short-period Pwave data. Bull Seismol Soc Am 62:435–452
35. CESMD, Strong motion data. http://www.strongmotioncenter.org/
36. IITK-BMTPC, Earthquake tip. http://www.iitk.ac.in/nicee/EQTips/EQTip01.pdf

Design Basis Ground Motion

2

Hari Prasad Muruva, A. R. Kiran, Srijit Bandyopadhyay, G. R. Reddy, M. K. Agrawal, and Ajit Kumar Verma

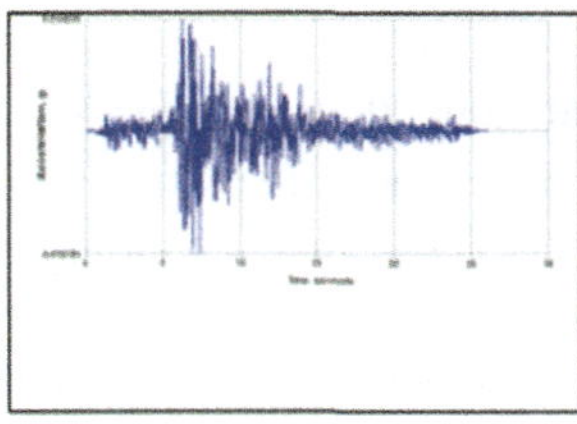

Design Basis Ground motion is represented as Acceleration with time or response of SDOF Vs Frequency

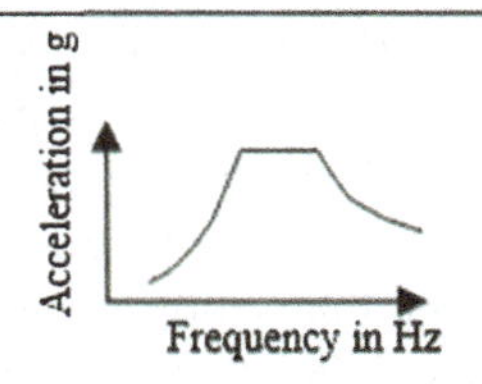

Symbols

N_m	Cumulative number of earthquakes of given magnitude or larger that are expected to occur during a specified period of time
a	The log number of earthquakes zero or greater expected to occur during same time
b	The slope of Gutenberg–Richter curve which characterizes a large portion of earthquakes
R	Hypocentral distance
M	Magnitude
$\upsilon_k(m_0)$	The annual frequency of occurrence of earthquakes on seismic source k whose magnitudes are greater than m_0 and below the maximum event size, m_u
$f_R(r)$	Probability density function of source-to-site distance 'R'
$f_M\ (m_i)$	Probability density function of magnitude 'M'
$P(A > a\|m_i, r_j)$	The probability that ground motion level 'a' will be exceeded, for a given earthquake of magnitude m_i at distance of r_j from the site
H	Depth of focus
t	Plant lifetime
M_o	Seismic moment
f_o	Corner frequency
β	Shear wave velocity
$\Delta\sigma$	Stress drop
$R_{\theta\psi}$	Radiation pattern
ρ	Density
K	Kappa
Q	Quality factor
$\frac{S_a}{g}$	The ratio of spectral acceleration at bedrock level to acceleration due to gravity

H. P. Muruva · A. R. Kiran · S. Bandyopadhyay · G. R. Reddy (✉) · M. K. Agrawal
Bhabha Atomic Research Centre, Mumbai, India
e-mail: rssred@barc.gov.in

H. P. Muruva
e-mail: hari_m78@yahoo.co.in

A. R. Kiran
e-mail: arkiran@barc.gov.in

S. Bandyopadhyay
e-mail: srijit@barc.gov.in

M. K. Agrawal
e-mail: mkagra@barc.gov.in

A. K. Verma
Western Norway University of Applied Science, Haugesund, Norway
e-mail: akvmanas@gmail.com

G. R. Reddy et al., *Textbook of Seismic Design*,
https://doi.org/10.1007/978-981-13-3176-3_2

2.1 Introduction

Earthquake can cause large destruction in industrial structures, equipment, and piping systems. To minimize this, the loading generated due to earthquake should be properly accounted for. Hence, it is essential to get prepared for facing the earthquakes. Preparedness means to design the structures (e.g., residential buildings, industrial buildings, lifeline structures, etc.), systems (water supply lines, fire-fighting systems, industrial piping systems, electrical and control systems, communication systems, etc.), and equipments (e.g., industrial equipment, normal and fire water supply components) to withstand postulated design earthquakes without damage. The ground shaking, for which the structures, systems, and equipment are designed, is referred to design basis ground motion. The design bases ground motion is described with various parameters such as peak ground acceleration, response spectrum, frequency content. Hence, while designing the industrial structures, systems, and equipments, it is very important to define the design basis ground motion parameters for which the structures are designed. If the design basis level increases, the cost of the plant increases due to the increase in sizes of the structural members, foundations, equipment sizes, piping sizes, etc.

Considering the case of nuclear facility, safety of the personnel and the surrounding environment of the facility should be ensured for various natural hazards including the earthquakes. By giving the high priority to the public safety, nuclear facilities are designed for earthquakes of low probable and high magnitude. The systems of nuclear facilities are broadly classified into structures, equipment, and piping with unique characteristics of their own. Brief explanation of various systems is given below.

The various structures of a NPP site are reactor building (e.g., containment structure), control building, reactor auxiliary building service building, waste management building, turbine building, spent fuel storage bay building, etc. These structures are directly founded on the ground and support the various equipment and piping systems. The NPP systems are classified into three categories. Category I systems (e.g., reactor containment structure, reactor core) are designed for S1 and S2 level earthquakes [1].

The S1 level earthquake can be reasonably expected at the site area once in the lifetime of the plant. This is also called the operating basis earthquake (OBE), whereas the S2 level earthquake has a very low probability of being exceeded at the site and can be expected once in 10,000 years. This is also called the safe shutdown earthquake (SSE).

In the case of conventional industries, there are two levels of design bases such as maximum considered earthquake (MCE) and design basis earthquake (DBE) (DBE is generally half of MCE). MCE has a return period of 2500 years, and DBE has a return period of 500 years. In this chapter, aspect of design bases ground motion (DBGM) generation is discussed in detail. DBGM can be evaluated either deterministically or probabilistically. This chapter gives details about both the analyses with example problems.

2.2 Deterministic Seismic Hazard Analysis

Seismic hazard analysis aims to determine design bases earthquake magnitude, peak ground acceleration, accelerogram (time history), and response spectrum. Design bases response spectrum or design bases time history is referred to DBGM. One can evaluate the above-mentioned parameters by deterministic seismic hazard analysis (DSHA). In DSHA, the seismic parameters are evaluated by postulating a specified size of earthquake at a specified location. These should be in consonance with the seismotectonic environment of the region around the site. Usually, an area with radius of 300 km is investigated [1] as shown in Fig. 2.1. DSHA involves the following steps:

1. Identification and analyzing the earthquake sources: Geotectonic setting of the region is identified by detailed geological investigations. These investigations will also help in identification of the maximum earthquake potential associated with each active tectonic feature. Lineaments and/or faults are identified and studied with respect to topography and geomorphology to find evidence of recent ground displacements and to ascertain their age and continuity. Local tectonics, fault structure, and correlation with historical earthquakes are studied. Figures 2.1 and 2.2 show the various sources around a site.

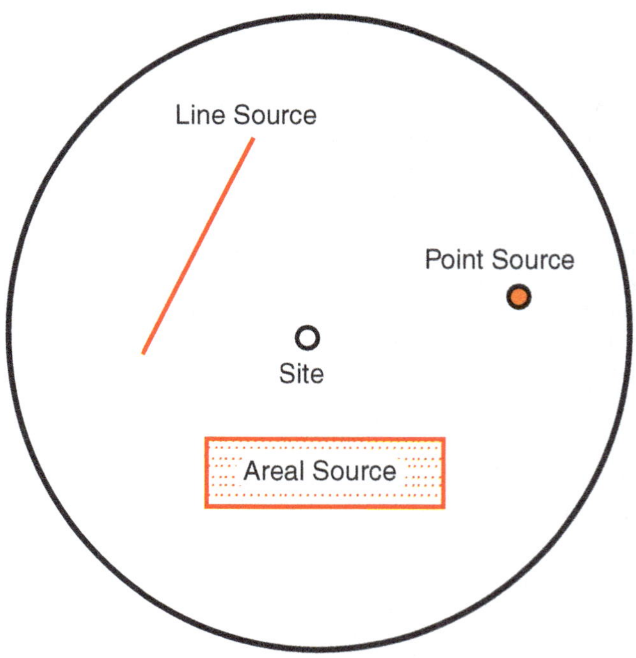

Fig. 2.1 Various sources of earthquake around the site

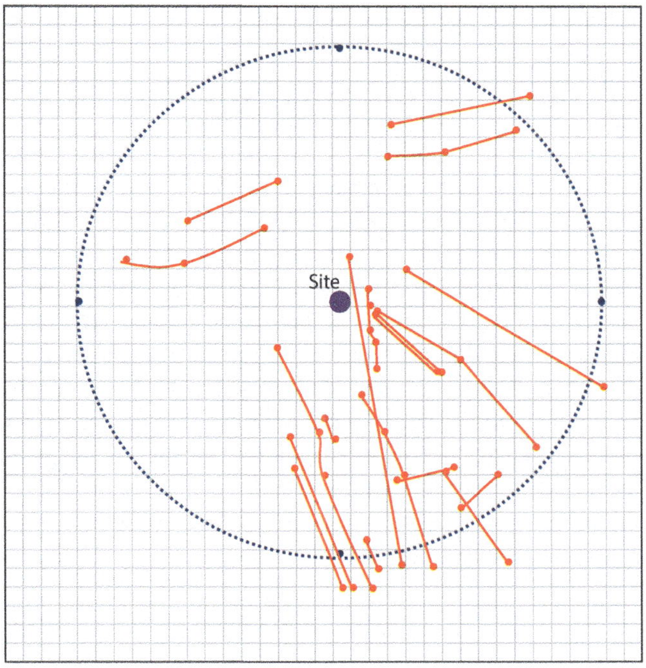

Fig. 2.2 Various faults around a site

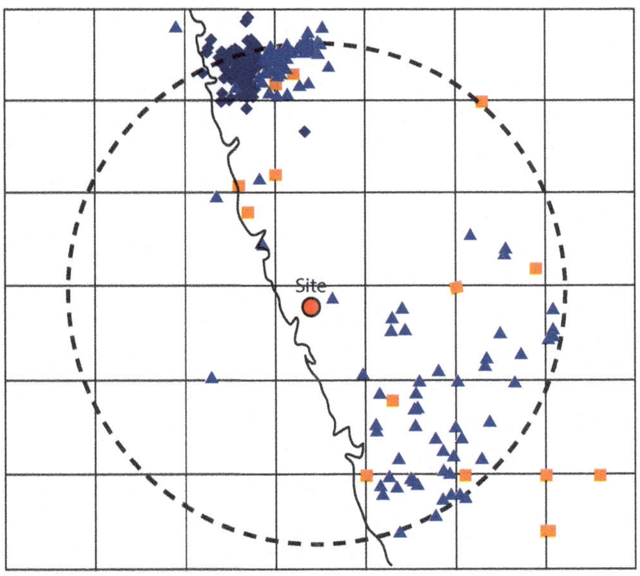

Fig. 2.3 Seismicity around the site

2. Studies on Past Earthquakes: Earthquakes of historical origin and instrumented data are collected (as shown in Fig. 2.3). These include primarily intensity or magnitude, depth of focus, epicenter, origin time, ground acceleration and velocity and the felt area. Based on this information, it is possible to assign a maximum possible earthquake to a particular fault or earthquake source.
3. In this analysis, it is assumed that a maximum possible earthquake on a particular source at a shortest distance from the site of interest would produce maximum shaking at the site. The shortest distance from site to source is calculated based on location of the site and the geometry of the source.
4. In the final step, the design bases response spectrum or design bases time history called hazard at the site is evaluated.

The above procedure is explained with the following example.

Example 2.1 Consider a site around which three earthquake source zones are located as shown in Fig. 2.4. The source characteristics are given in Table 2.1. Evaluate the design bases ground motion in terms of peak ground acceleration.

Table 2.1 provides the necessary information about the location, maximum magnitude, and source-to-site distance. Considering this information, the peak horizontal acceleration (PHA) at the site from various sources can be obtained by using an appropriate attenuation relationship (which is further explained in the upcoming sections). A sample attenuation relationship of Cornell et al. (1979) is given as follows:

$$\ln \text{PHA} = 6.74 + 0.859M - 1.80\ \ln(R + 25) \qquad (2.1)$$

The PHA values are shown in Table 2.1. From Table 2.1, one can select the controlling earthquake as occurrence of magnitude 7.0 at a distance of 25 km generated from source zone 2. One can evaluate the hazard at the site based on this

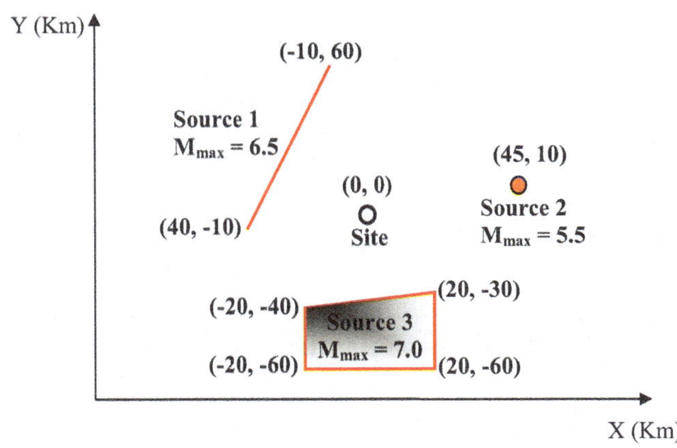

Fig. 2.4 Earthquake source zones around the site

Table 2.1 Source characteristics

Source zone	Maximum Magnitude (M_{max})	Distance, R (km)	PHA (g)
1	6.5	23.7	0.42
2	7.0	25.0	0.57
3	5.5	60.0	0.02

information. In the present case, the PHA that would be produced at the site is calculated as 0.57 g.

2.3 Generation of Design Bases Response Spectrum

In order to derive the design bases response spectrum, it is very important to know the procedure for evaluating response spectrum for given earthquake excitation or strong motion record (the reader may refer Chap. 1 for more details). The various steps involved in generating the design bases response spectrum are as follows:

1. Based on the attenuation relationships, the peak ground acceleration (PGA) shall be evaluated.
2. Within the radius of 300 km around the site, strong motion records are collected.
3. If the data is not sufficient, data of various sites of similar geology and seismology are collected. Typical strong motion records are shown in Figs. 2.5 and 2.6. Next, the strong motion history is normalized using the PGA values derived in step 1.
4. Evaluate the response spectra for the normalized time history. This will result in number of response values for a given Single Degree of Freedom (SDOF) system of certain frequency as shown in Fig. 2.7.
5. The mean and standard deviation of the response values are evaluated at each frequency.
6. Fix a design spectral value as $\mu + \sigma$ at different frequencies.

One of the design response spectra generated based on the above procedure is shown in Fig. 2.8. A typical design response spectrum for a site with various damping values developed is shown in Fig. 2.9. Figure 2.10 shows the average response spectral shapes (i.e., Sa/g) normalized to unit g for rock and medium soils and 5% damping as per IS 1893.

2.4 Probabilistic Seismic Hazard Analysis

Probabilistic seismic hazard analysis (PSHA) helps in the estimation of design bases ground motion parameter such as peak ground acceleration (PGA) that would be expected at a particular site based on probabilistic analysis. The final outcome of the analysis is the generation of site-specific hazard curves which is a representation of annual rate of exceedance of a particular level of PGA.

Like DSHA, PSHA also considers various earthquake sources around the site, past earthquake history, earthquake occurrence rate from various earthquake sources, maximum earthquake magnitude that can be expected on each source, ground motion attenuation from source to the given site, etc. However, these are dealt with probabilistic concepts in PSHA. The flow chart of various steps involved in PSHA is shown in Fig. 2.11 and is explained below.

As a first step in PSHA, the seismotectonic features such as lineaments, faults in an area of 300 km radius around the site are collected as shown in Fig. 2.1. Seismicity information in the aforesaid area is also obtained. This includes details of historic as well as instrumented earthquakes as shown in Fig. 2.3. In the second step, seismic recurrence characteristics for each source is obtained based on the available earthquake data. This in general represented by recurrence relationship, which indicates average occurrence of an earthquake of a given magnitude or larger on the given source over a period of time, usually one year. In general, Gutenberg and Richter's law is used as recurrence relationship [2, 3] which is given as follows:

$$\log_{10} N_m = a - bm \tag{2.1}$$

where N_m is the annual occurrence of total number of earthquakes whose magnitude is greater than or equal to 'm.' If one substitutes the value of 'm' as zero in Eq. (2.1), then one can obtain the value of 'a' as log of N_m which denotes log of annual occurrence of total number of earthquakes whose magnitude is greater than or equal to zero. If a graph is plot between $\log_{10} N_m$ and m (as shown in Fig. 2.12), it represents a straight line in which 'b' represents the slope of the curve. As the value of 'b' decreases, it may be possible that the region may experience higher magnitude of earthquakes.

In deterministic analysis, one controlling earthquake or one maximum earthquake is assigned for each source, whereas, in PSHA, earthquakes of all magnitudes are considered and are treated with probability distributions. However, in general, magnitude is chosen between a lower limit and an upper limit. It means each earthquake source is capable of producing all the possible magnitudes between a minimum and a maximum level of magnitude with a certain probability. The lower limit is chosen in such a way that sufficient data is not existed below that value or there is no interest from engineering point of view. Similarly, the maximum magnitude indicates that the probability of occurrence of an earthquake beyond this level is very low on a particular earthquake source.

In general, earthquakes of any magnitude can occur at any location on the earthquake source. Hence, one needs to consider occurrence of earthquake on the source at different distances from the site. However, in deterministic analysis, only the closest distance from source to the site is considered, whereas in the probabilistic analysis, a range of earthquake size, source-to-site distance pairs, and their associated probability of occurrence are taken into account.

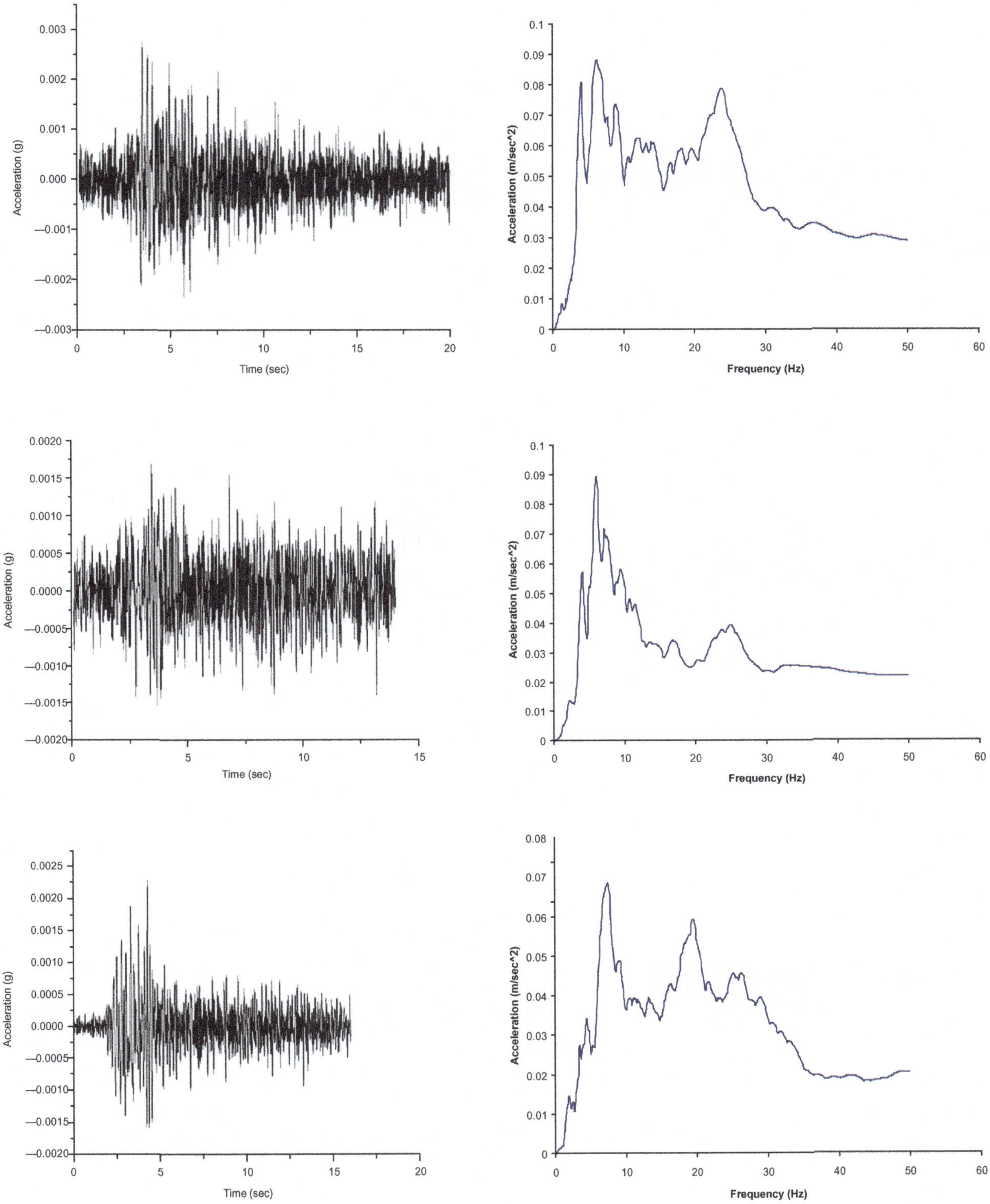

Fig. 2.5 Time histories and corresponding response spectrums of different earthquakes

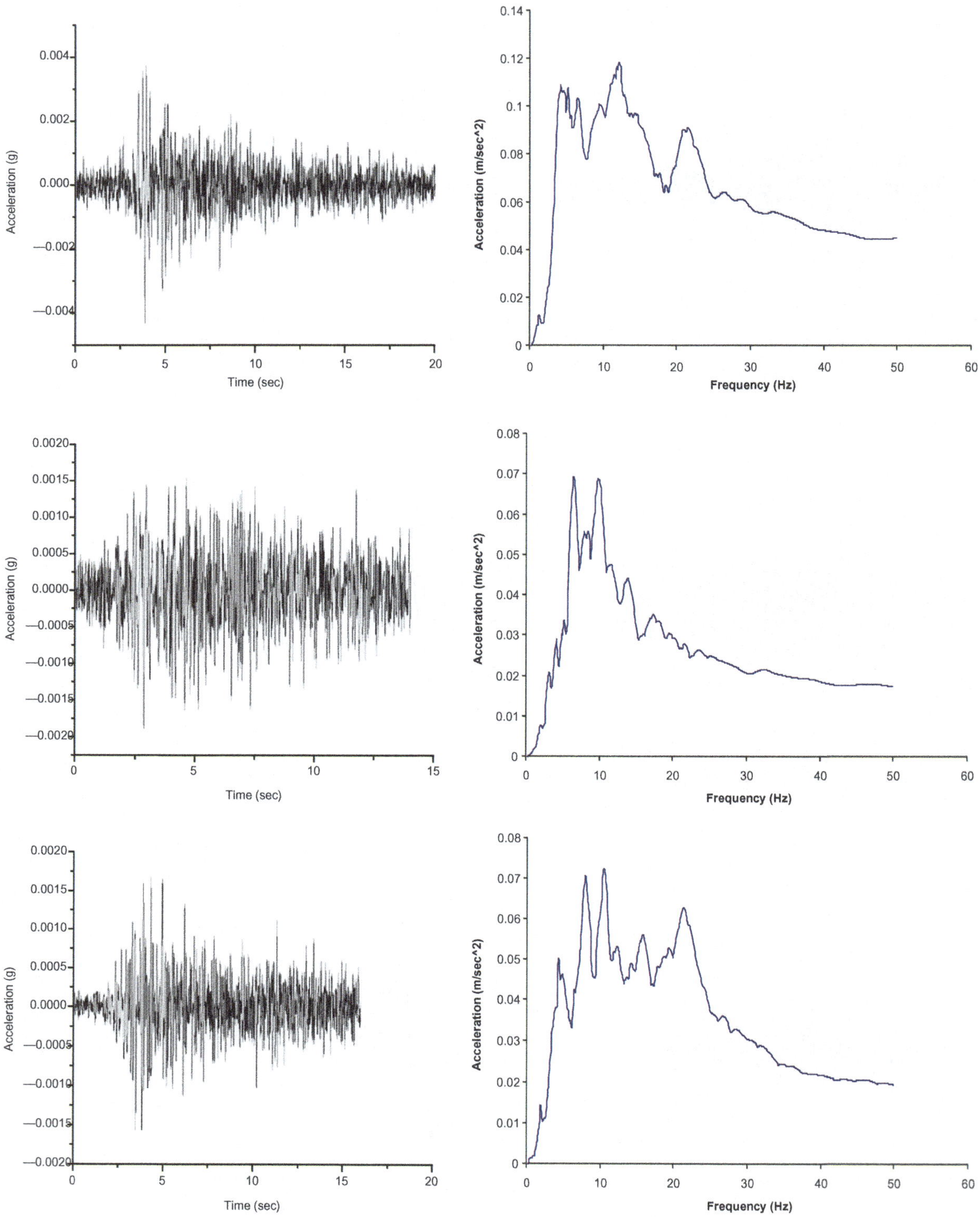

Fig. 2.6 Time histories and corresponding response spectrums of different earthquakes

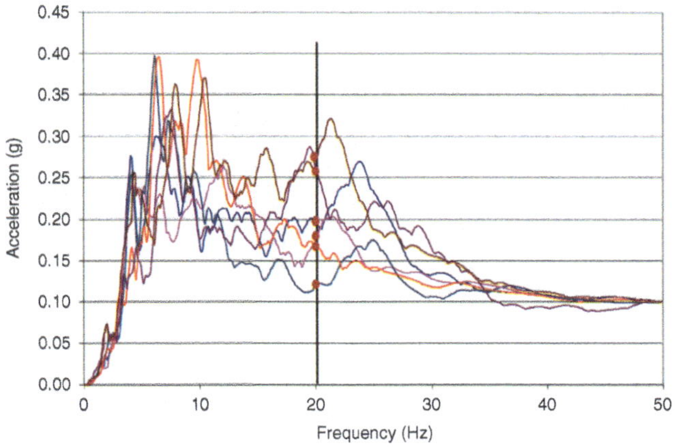

Fig. 2.7 Schematic of deriving design response spectrum

The details of this step are further explained under the subtitles probability density function of magnitude and distance in the following sections.

In the third step, effect of earthquake is estimated. In the deterministic analysis, only the effect of maximum possible earthquake occurring on the source at a shortest distance from the site is considered, whereas in the probabilistic analysis, occurrence of earthquakes of various magnitudes between a lower and upper limit depending on the source with different source-to-site distances is considered in estimating the ground motion parameter, such as peak ground acceleration. This can be estimated by using attenuation relationships which relates a ground motion parameter as a function of magnitude and distance. The generation of attenuation relationships is further explained in Sect. 2.7. One of the attenuation relationships which is available in the literature is given as follows:

Silva [8]

$$\ln(\text{PGA}) = c_1 + c_2 M + (c_6 + c_7 M) \times \ln(R + \mathrm{e}^{c_4}) + c_{10}(M-6)^2$$
$$c_1 = 3.54103 \quad c_2 = 0.18904 \quad c_4 = 2.7 \quad c_6 = -2.97418$$
$$c_7 = 0.19819 \quad c_{10} = -0.05814. \quad \sigma_{\ln y} = 0.84 \tag{2.2}$$

R Closest distance to the surface projection of ruptured surface

M Magnitude

In the final step, hazard at the site is estimated, which is different from the DSHA. In PSHA, the effect of all the earthquake sources along with different earthquake sizes depending on the capability of each earthquake occurring at various distances on the source from the site is considered. The hazard is represented in terms of hazard curves which is a graphical representation of annual rate of exceedance of a particular level of PGA. Hence, the annual rate of exceedance of ground motion level 'a' is given as

$$v(a) = \sum_{k=1}^{N} \upsilon_k(m_0) \int_{m_0}^{m_u} \int_{d}^{r_0} f_M(m) f_R(r) P[A > a|m, r] \mathrm{d}r\, \mathrm{d}m$$

or

$$v(a) = \sum_{k=1}^{N} \upsilon_k(m_0) \left[\sum_{i=1}^{N_M} \sum_{j=1}^{N_R} f_{Mk}(m_i) f_{Rk}(r_j) P(A > a|m_i, r_j) \Delta r \Delta m \right] \tag{2.3}$$

in which

$\upsilon_k(m_0)$ the annual occurrence rate of earthquakes of magnitude more than m_0 on the seismic source k

Fig. 2.8 Typical design response spectrum for 5% damping

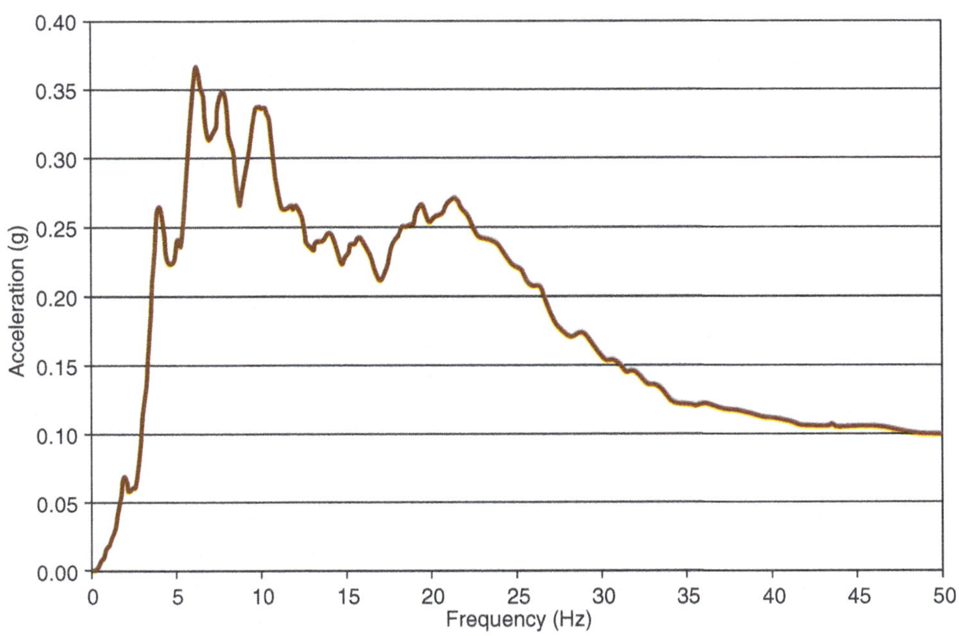

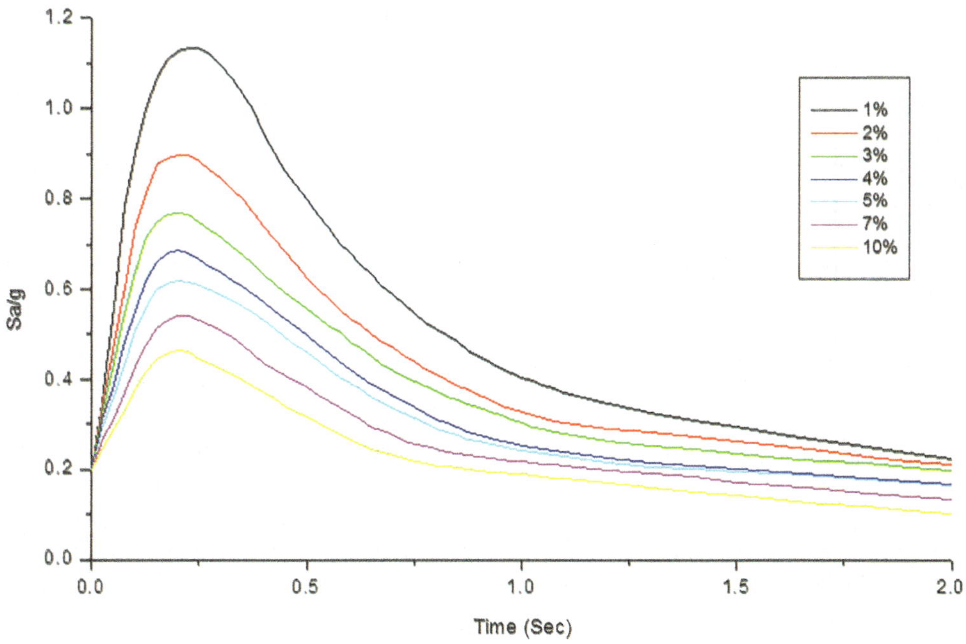

Fig. 2.9 Envelop spectra with 0.2 g ZPA

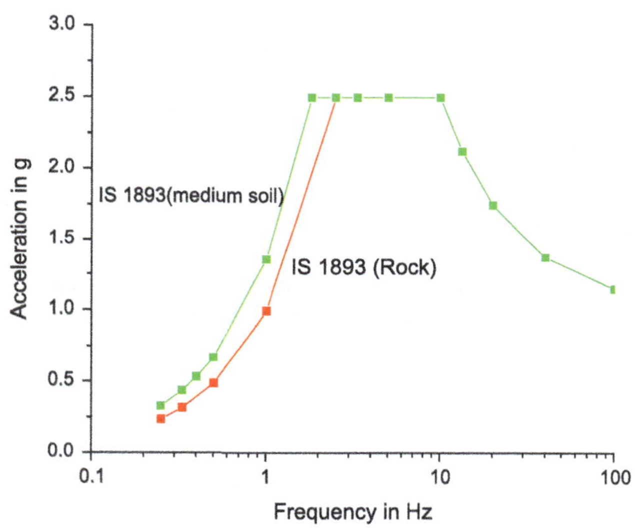

Fig. 2.10 Spectral shapes for rock and medium soils as per IS 1893

$f_R(r)$ probability density function of source-to-site distance 'R'

$f_M(m_i)$ probability density function of magnitude 'M'

$P(A > a | m_i, r_j)$ conditional probability of exceeding the ground motion level 'a' given the earthquake magnitude m_i and distance r_j

One can also calculate probability of exceedance for a given period of time (in general, it can be for different plant operating times 't') and also return period of certain level of PGA as follows [3]:

$$P[A > a] = 1 - e^{-\nu(a)t} \quad (2.4)$$

$$\text{Return Period} = \frac{-t}{\ln[1 - P(A > a)]} \quad (2.5)$$

From the above equation, it is clear that to estimate the annual frequency of exceedance of given ground motion level, one should have the information about the probability density function of magnitude, distance, and probability distribution function of ground motion level (can be derived from ground motion prediction model). These are further explained in detail in the following sections.

2.5 Probability Density Function of Magnitude

According to Gutenberg–Richter's recurrence relationship, the total number of earthquakes more than magnitude m_L in a year is given by

$$\begin{aligned} \log_{10} N_{m_L} &= a - bm_L \\ N_{m_L} &= 10^{a-bm_L} \end{aligned} \quad (2.6)$$

Similarly, the total number of earthquakes more than magnitude m_u in a year is given by

$$\begin{aligned} \log_{10} N_{m_u} &= a - bm_u \\ N_{m_u} &= 10^{a-bm_u} \end{aligned} \quad (2.7)$$

Hence, total number of earthquakes that can occur in a year between the limits m_L and m_u can be given as:

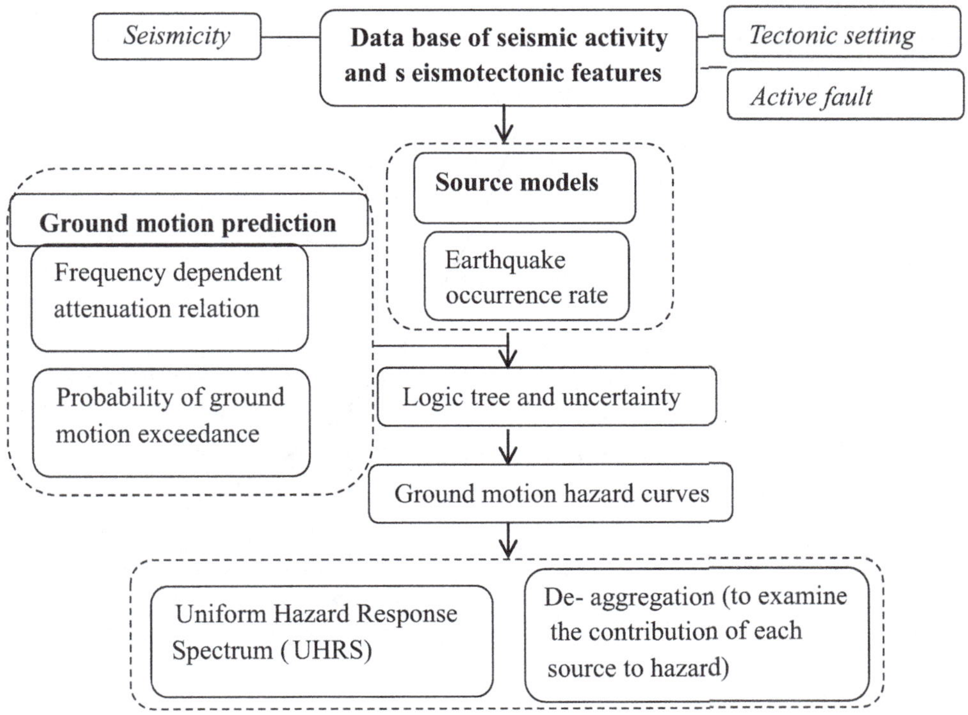

Fig. 2.11 Flow chart of PSHA

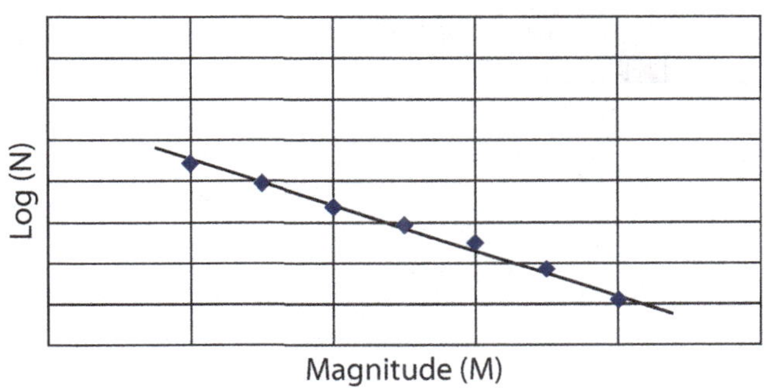

Fig. 2.12 Recurrence relationship

$$N_{m_u} - N_{m_L} = 10^{a-bm_u} - 10^{a-bm_L} \tag{2.8}$$

From the above information, one can easily obtain the cumulative distribution function (probability distribution function) of magnitude. The cumulative distribution function (cdf) of magnitude can be defined as the probability of occurrence of earthquakes whose magnitude is less than 'm'. In other words, it is nothing but the ratio between total number of earthquakes whose magnitude is less than magnitude 'm' and total number of earthquakes that can occur in a year. Accordingly, this can be given as:

$$P(M \leq m) = \frac{N_{m_L} - N_m}{N_{m_L} - N_{m_u}}$$
$$But\ N_m = 10^{a-bm}$$
$$N_{m_L} = 10^{a-bm_L}$$
$$N_{m_u} = 10^{a-bm_u}$$

$$P(M \leq m) = \frac{10^{a-bm_L} - 10^{a-bm}}{10^{a-bm_L} - 10^{a-bm_u}}$$
$$P(M \leq m) = \frac{10^{-bm_L} - 10^{-bm}}{10^{-bm_L} - 10^{-bm_u}} = F_M(m) \tag{2.9}$$

Now, probability density function (pdf) of magnitude can be obtained as follows:

$$f_M(m) = \frac{dF_M(m)}{dm}$$
$$f_M(m) = \frac{\beta 10^{-b(m-m_L)}}{1-10^{-b(m_u-m_L)}} \qquad m_L \leq M \leq m_u \tag{2.10}$$

where $\beta = b \log_e 10$

Example 2.2 Assume the parameters of Gutenberg–Richter's recurrence relationship as 'a' = 3.5 and 'b' = 0.86. Assume that the lower and upper limits of magnitude as m_L = 4.0 and m_u = 6.5. Plot probability density function of magnitude, and find the probability of occurrence of magnitude between 4.5 and 5.0.

Solution:

The pdf of magnitude 'm' can be given as follows from Eq. (2.10).

$$f_M(m) = \frac{\beta 10^{-b(m-m_L)}}{1-10^{-b(m_u-m_L)}} \qquad m_L \leq M \leq m_u$$

where $\beta = b \log_e 10$

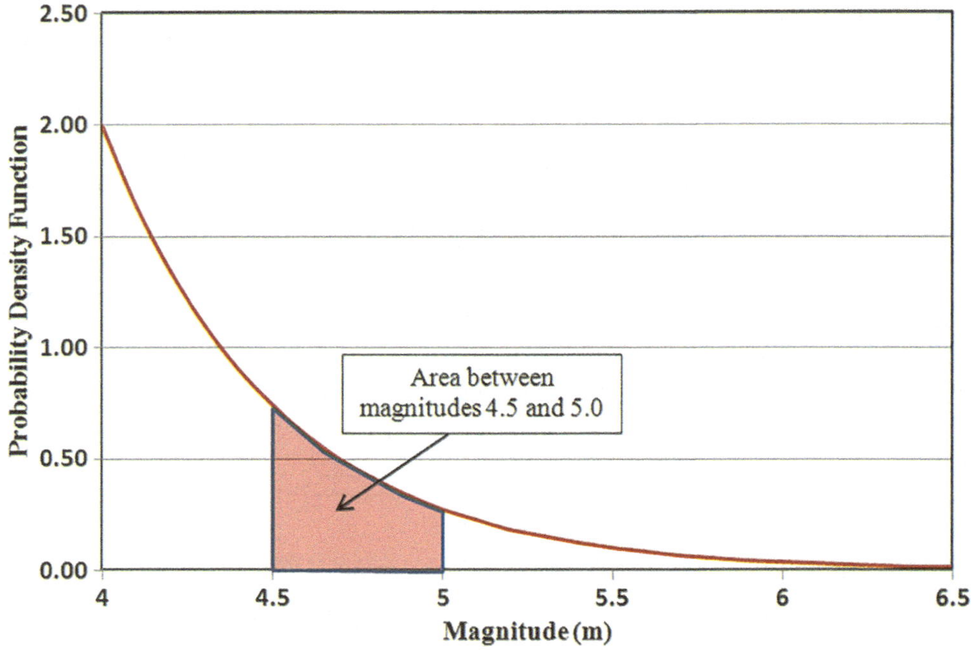

Fig. 2.13 Probability density function of magnitude 'M'

Substituting a = 3.5, b = 0.86, m_L = 4.0 and m_u = 6.5 in Eq. 2.7

$$\beta = b \log_e 10 = 0.86 \times 2.3026 = 1.9802$$

$$f_M(m) = \frac{1.9802 \times 10^{-0.86\,(m-4.0)}}{1 - 10^{-0.86(6.5-4.0)}} \qquad 4.0 \leq M \leq 6.5$$

$$f_M(m) = 1.9943 \times 10^{3.44-0.86m}$$

The probability density function is plotted in Fig. 2.13.

Now, one can calculate the probability of occurrence of magnitude between 4.5 and 5.0 as area under the pdf curve between 4.5 and 5.0. This is shown in Fig. 2.7. This can be obtained as follows:

$$f_M(m) = \frac{\beta 10^{a-bm}}{10^{a-bm_0} - 10^{a-bm_u}} m_0 \leq M \leq m_u$$

$$f_M(m) = \frac{dF}{dm} \Rightarrow \mathrm{d}F = f_M(m)\mathrm{d}m$$

$$\therefore F(m + \Delta m) - F(m) = f_M(m)\Delta m$$

$$\text{Here}\ m = 4.5,\quad \Delta m = 0.5 \Rightarrow m + \Delta m = 5.0$$

$$F(5.0) - F(4.5) = 0.5 f_M(4.5) = 0.5 \times 0.741 = 0.3705$$

2.6 Probability Density Function of Distance

This section discusses the mathematical derivation of pdf of distance from source to site. In estimation of hazard at the site, distance from source to site plays a major role. The general fault line model of an earthquake is shown in Fig. 2.14. Here, fault model is considered as a line source model which is located at a depth of 'd' from the surface. Length of the fault is assumed as 'L.' When the fault line is projected on to the surface, the site is located at a perpendicular distance of 'δ' from the fault line.

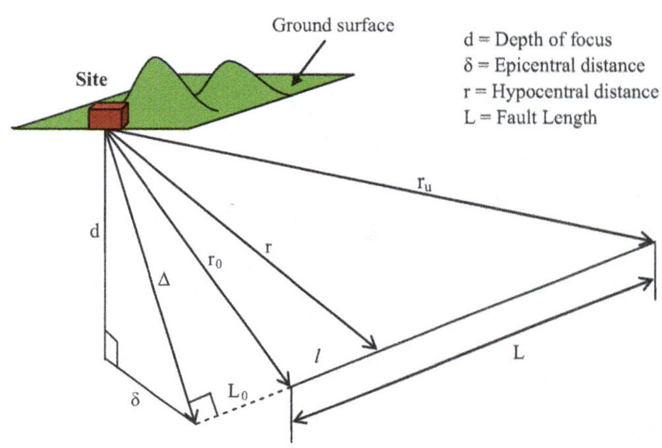

Fig. 2.14 Fault line model

In general, one can expect earthquake at any location on the fault. Hence, in the probabilistic analysis, earthquake occurrence at all the possible locations on the fault and the corresponding distances from source to the site are considered with a certain probability value. These probability values can be obtained from probability density functions. Consider R as a random variable which represents hypocentral distance. The minimum and maximum values of R are considered as r_0 and r_u as shown in Fig. 4.11. Now, earthquake can occur at any location on the fault with R varies from r_0 and r_u. Based on this information, one can find out the probability of occurrence of an earthquake on the fault at a distance less than certain hypocentral distance value, i.e., less than 'r'. This can be mathematically represented as follows:

$$P(R \le r/m) = \frac{l}{L} \quad r_0 < R < r_u$$
$$\text{where } l = \sqrt{r^2 - \Delta^2} - L_0$$
$$\Delta = \sqrt{d^2 + \delta^2}$$
$$r_0 = \sqrt{\Delta^2 + L_0^2} \tag{2.11}$$
$$r_u = \sqrt{\Delta^2 + (L + L_0)^2}$$
$$P(R \le r/m) = \frac{\sqrt{r^2 - \Delta^2} - L_0}{L} = F_R(r)$$

$$f_R(r) = \frac{dF_R(r)}{dr} = \lim_{\Delta r \to 0} \frac{\Delta F}{\Delta r} = \frac{F(r + \Delta r) - F(r)}{\Delta r} \tag{2.12}$$

Example 2.3 Consider the case of a fault line model as shown in Fig. 2.14. The various parameters of the model are given as: fault length = 200 km, depth of focus = 15 km and the epicentral distance = 30 km. Find the cumulative distribution the probability density functions of source-to-site distance for the given data.

Solution: The input information is given as:

$$\mathrm{L} = 200\,\mathrm{km} \quad \mathrm{d} = 15\,\mathrm{km} \quad \boldsymbol{\delta} = 30\,\mathrm{km} \quad L_0 = 10\,\mathrm{km}(\text{Assume})$$
$$F_R(r) = \frac{\sqrt{r^2 - \Delta^2} - L_0}{L}$$
$$\Delta = \sqrt{d^2 + \delta^2} = \sqrt{15^2 + 30^2} = 33.54\,\mathrm{km}$$
$$r_0 = \sqrt{\Delta^2 + L_0^2} = \sqrt{33.54^2 + 10^2} = 35\,\mathrm{km}$$
$$r_u = \sqrt{\Delta^2 + (L + L_0)^2} = \sqrt{33.54^2 + 210^2} = 212.66\,\mathrm{km}$$

As mentioned previously, R (hypocentral distance) is a random variable whose values vary from 35 to 212.66 km. This range can be divided into 30 equal subintervals with an interval size of 5.92 km. Based on Eqs. 2.8 and 2.9, the cdf and pdf can be computed as follows:

$$F_R(r) = \frac{\sqrt{r^2-\Delta^2}-L_0}{L}$$
$$\text{Let } r = 35\,\mathrm{km}$$
$$F_R(35) = \frac{\sqrt{35^2-33.54^2}-10}{200} = \frac{10-10}{200} = 0$$
$$\Delta r = 5.92\,\mathrm{km}$$
$$f_R(r) = \frac{F(r+\Delta r)-F(r)}{\Delta r}$$
$$f_R(35) = \frac{F(35+5.92)-F(35)}{5.92} = \frac{F(40.92)-F(35)}{5.92}$$
$$F_R(40.92) = \frac{\sqrt{40.92^2-33.54^2}-10}{200} = \frac{23.44-10}{200} = 0.067$$
$$F_R(35) = 0$$
$$f_R(35) = \frac{0.067-0}{5.92} = 0.01135$$

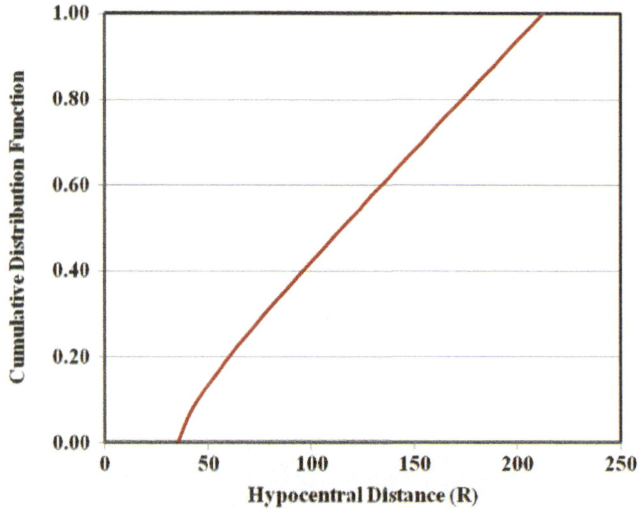

Fig. 2.15 Cumulative distribution function of hypocentral distance

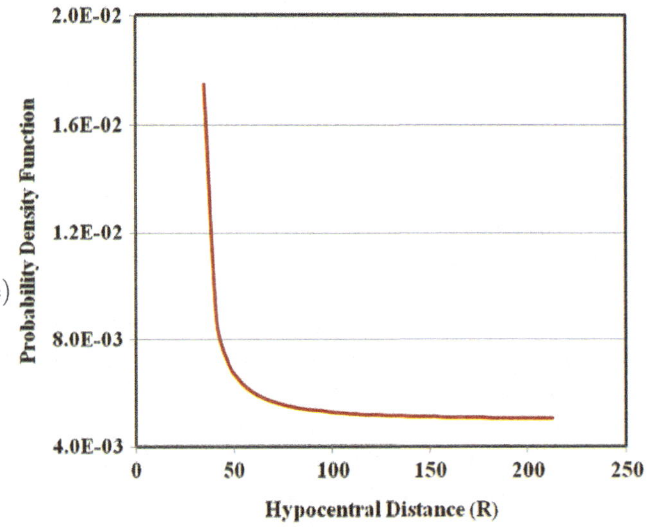

Fig. 2.16 Probability density function of hypocentral distance

Similar calculations can be repeated for all the intervals. The cdf and pdf curves are shown in Figs. 2.15 and 2.16, respectively.

2.7 Attenuation Relationships

One of the techniques to estimate the ground motion parameter (such as PGA)is by using the concept of attenuation relationship which estimates decay of ground motion from earthquake source to site. In general, attenuation relationships relate ground motion parameters with earthquake magnitude and the distance from source of occurrence of an earthquake to the site. Since in an earthquake the amount of energy released is related to its magnitude, the ground

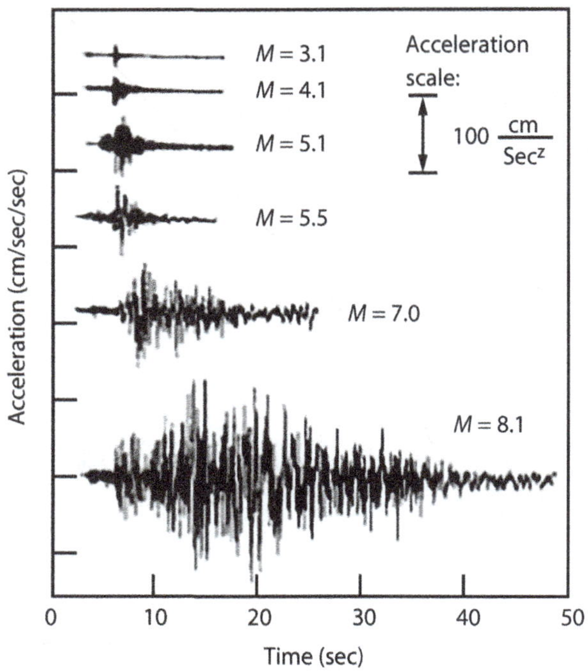

Fig. 2.17 Accelerograms of six earthquakes measured at the same distance from the source in the Pacific Coast of Mexico (Anderson 1991)

motion parameters are very much related to earthquake magnitude. Figure 2.17 illustrates the effects of earthquake magnitude on ground motion along the time domain, and each accelerogram was measured from the same distance from the source. The variations in amplitude and duration of motion with magnitude can be seen in Fig. 2.17.

Similarly, the specific energy (Energy/unit volume) decreases with increase of distance from hypocenter. Hence, the ground motion parameters are also related to distance from the source of earthquake. In general, the attenuation relationships are functional relationships between a dependent parameter and one or more independent parameters. Here, the dependent parameters can be spectral acceleration or peak ground acceleration, and the independent parameters can be magnitude, hypocentral distance or nearest source distance, type of faulting and soil type.

2.7.1 Functional Form of Attenuation Relationships

Attenuation relationships express ground motion parameters as a function of magnitude, distance, and other parameters such as source of earthquake and local site conditions. These can be developed either by regression analysis of earthquake records or by simulated acceleration time histories [2]. In general, attenuation relations are represented with a functional form which represents the physics of ground motion process and also which minimizes number of empirical constants. A typical functional form of attenuation relationships is given by:

$$\ln(\text{PGA}) = C_1 + C_2M + C_3M^{C_4} + C_5\ln(R + C_6\exp(C_7M)) + C_8R + f(\text{source/site}) \quad (2.13)$$

The reasons for choosing the above form are given below:

a. The ground motion parameters approximately follow lognormal distribution. Hence, the term ln (PGA) is chosen for regression.
b. It is usual practice to define magnitude as logarithm of some peak ground motion parameter. Hence, the term $C_1 + C_2M + C_3M^{C_4}$ is chosen.
c. As the seismic waves travel away from the earthquake source, the amplitude decreases according to $1/R^c$. Hence, the term $C_5 \ln R$ is included in the functional form.
d. Rupture area increases with magnitude. Accordingly, some waves arrive at distance R and some at a distance more than R. Hence, effective distance increases with magnitude.
e. Material damping causes reduction of ground motion amplitude. Hence, the term C_8R is considered.
f. Ground motion parameters are also influenced by site and source characteristics. Hence, the term $f(\text{source/site})$ is included in attenuation relations.

2.7.2 Procedure to Develop Attenuation Relationships

There are two approaches available for developing attenuation relationships. First one is regression from strong motion database, and the second one is from simulation. First method is suitable for interplate regions such as Western North America (WNA) where seismicity rate is high, and sufficient strong ground motion data is available for regression analysis. Second method is suitable for intraplate regions such as Peninsular India and Eastern North America (ENA) where seismicity rates are less. Finite-fault stochastic simulation is usually used for the generation of synthetic ground motion which in turn is used to develop attenuation relations for hard-rock site. The simulation of ground motion requires careful selection of seismological parameters such as quality factor, spectral decay parameter, and stress drop. The seismological parameters for simulation should be selected from the same region or regions having similar characteristics.

Stochastic simulation model is widely used to simulate artificial acceleration time histories and to develop attenuation relationships. In stochastic simulation method, first one needs to specify the Fourier Amplitude Spectrum (FAS) which in general is a function of distance and magnitude. The ω^2 shape spectrum is used to model FAS. The FAS at a point for an instantaneous shear dislocation can be given as follows [4]:

$$A(f) = \left\{CM_0(2\pi f)^2/\left[1+(f/f_0)^2\right]\right\} \times \{\exp(-\pi fk)\exp(-\pi fR/Q\beta)/R\} \quad (2.14)$$

where

M_0, R, f_0, and β are seismic moment, hypocentral distance, corner frequency, and shear wave velocity in km/s, respectively

Corner frequency, f_0 can be given by:

$$f_0 = 4.9\text{E}6\beta(\Delta\sigma/M_0)^{1/3} \quad (2.15)$$

where

$\Delta\sigma$ is stress drop in bars

M_0 is seismic moment (in dyne cm) which is related to Magnitude (M) by

$$\log M_0 = 1.5\,\text{M} + 16.1 \quad (2.16)$$

The constant C is given by:

$$C = R_{\theta\varphi}FV/4\pi\rho\beta^3 \quad (2.17)$$

where

$R_{\theta\varphi}$ is radiation pattern

F is free surface amplification

V is partition into two horizontal components

ρ is density.

To model upper crustal/near-surface attenuation and scattering processes, a high-cut filter such as $\exp(-\pi fk)$ is used, in which the parameter kappa (k) is known as spectral decay parameter, which represents the effect of intrinsic attenuation upon the wave field as it propagates from source to receiver through the crust. The quality factor, Q, is inversely related to an elastic attenuation. The implied $1/R$ geometric attenuation term is applicable for spreading of body wave in a whole space.

Usually, ground motion simulation is carried out using software based on stochastic finite-fault modeling in which fault plane is divided into $M \times N$ subfaults. The acceleration spectrum of ijth subfault, $A_{ij}(f)$ is described by:

$$A_{ij}(f) = \left\{CM_{0ij}(2\pi f)^2/\left[1+(f/f_{0ij})^2\right]\right\} \times \left\{\exp(-\pi fk)\exp(-\pi fR_{ij}/Q\beta)/R_{ij}\right\} \quad (2.18)$$

where M_{0ij}, f_{0ij}, and R_{ij} are the ijth subfault seismic moment, corner frequency, and distance from the site, respectively. The details of developing attenuation relationships are further explained in Appendix 1 with an example.

Several functional forms of ground motion attenuation have been proposed in the literature to reflect salient aspects of the spread of ground motion. Attenuation relationships currently available to estimate earthquake ground motions are broadly classified into interplate and intraplate types. This classification is based on the data set used to develop the correlation. For example, the correlations based on Western North America (WNA) data are interplate type, while the correlations from Eastern North America (ENA) data are intraplate type. These attenuation relationships provide estimates of earthquake ground motions at "rock sites."

2.7.3 Attenuation Relations for Interplate Regions

The earthquakes which occur at the boundary between two tectonic plates are called interplate earthquakes. Throughout the world, more than 90% of the seismic activity is related to this kind of earthquakes. At the interplate regions, the tectonic plates move past each other. During this process, the plates will be locked to each other. The stresses keep build up, and once the stresses reach sufficient level to break the lock, the plates will slip relative to each other. This slipping will create an earthquake, and the seismic waves will travel in all the directions through the Earth and along the Earth's surface. Tables 2.2 and 2.3 list some well-known

Table 2.2 Correlation parameters for various attenuation relations

Correlation parameter	1. Donovan (1974a)	2. Donovan (1973)	3. Esteeva (1970)	4. Esteeva and Rosenbleuth (1964)	5. Esteeva and Villaverde (1974)	6. McGuire (1974)
b_1	1.1	1.35	1.25	2.0387	5.71	0.472
b_2	0.5	0.58	0.8	0.8	0.8	0.28
b_3	1.32	1.52	2.0	2.0	2.0	1.3
D	25	25	25	0	40	25
σ_{lna}	0.722	0.699	1.39	1.09	0.793	0.77

Table 2.3 Correlation parameters for various attenuation relations

Correlation parameter	7. Orphab and Lahoud (1974)	8. Mickey (1971)	9. Donovan (1974b)	10. McGuire (1978)	11. Ghosh (1986)	12. Ghosh (1998)
b_1	0.066	0.000304	205.4	0.0306	1.04	4.63
b_2	0.4	0.74	0	0.89	0.483	0.528
b_3	1.39	1.4	1.83	1.17	1.2	1.6
D	0	0	25	0	40	40
$\sigma_{\ln a}$	0.815	1.75	1.25	0.861	0.731	0.678

correlations for interplate region to predict peak horizontal ground acceleration on rock sites. The correlations 1–5 and 10–12 are of the type:

$$a_p = b_1 \exp(b_2 M)(R+D)^{-b_3} \tag{2.19}$$

The correlations 6 to 8 are of the type:

$$a_p = b_1 10^{b_2 M}(R+D)^{-b_3} \tag{2.20}$$

where R is the hypocentral distance for correlation nos. 1 and 3–10 and R is the epicentral distance for correlation nos. 2 and 11–12.

2.7.4 Attenuation Relations for Intraplate Regions

An intraplate earthquake is an earthquake that occurs in the interior of a tectonic plate. Intraplate earthquakes do pose a significant hazard due to ground motions in stable continental regions (SCRs). These are fundamentally different from those from plate boundaries and zones of active deformation primarily in terms of lower rates of attenuation with distance due to the crust within an SCR being less fractured. The correlations for intraplate region are described below:

Atkinson and Boore [5]

Earthquake ground motion relations for hard-rock sites in Eastern North America (ENA) are given by Atkinson and Boore. This model utilizes the seismographic data obtained from ENA during 1811–2005. These correlations were developed using stochastic finite-fault methodology. The computer code extended finite-fault simulation (EXSIM) was used to perform the simulations. Attenuation relation given by Atkinson-Boore [5] is given in Eq. (2.21). These equations predict median horizontal ENA ground motion in terms of response spectra (for 5% damping) and peak ground acceleration (PGA) for hard-rock sites (near-surface shear wave velocity more than 2 km/s), as a function of moment magnitude and closest distance to the fault rupture.

$$\log(\mathrm{PSA}) = c_1 + c_2 M + c_3 M^2 + (c_4 + c_5 M) f_1 + (c_6 + c_7 M) f_2 + (c_8 + c_9 M) f_0 + c_{10} R_{cd} + S \tag{2.21}$$

where

$$f_0 = \max\left(\log\left(\frac{R_0}{R_{cd}}\right), 0\right) f_1 = \min(\log R_{cd}, \log R_1)$$

For hard-rock sites,

$$f_0 = \max\left(\log\left(\frac{R_{cd}}{R_2}\right), 0\right); R_0 = 10; R_1 = 70;$$
$$R_2 = 140;\ \text{and } S = 0$$

The stress drop is evaluated by:

$$\log \Delta\sigma = 3.45 - 0.2M$$

The reader may refer [5] for information about the coefficients of the equation.

NDMA [6]

An empirical attenuation relation for Peninsular India, proposed by NDMA (2010), is given in Eq. (2.22). In this equation, $\left(\frac{S_a}{g}\right)$ stands for the ratio of spectral acceleration at bedrock level to acceleration due to gravity. M and r refer to moment magnitude and hypocentral distance, respectively.

$$\ln\left(\frac{S_a}{g}\right) = C_1 + C_2 M + C_3 M^2 + C_4 r + C_5 \ln\left(r + C_6 e^{C_7 M}\right) + C_8 \log(r) f_0 + \ln(\epsilon) \tag{2.22}$$

where

$$f_0 = \max(\ln(r/100), 0)$$

The coefficients of the above equation are obtained from the simulated database of S_a by a two-step stratified regression. This correlation is valid for bedrock level, with shear wave velocity nearly equal to 3.6 km/s. One can refer [6] for the coefficients of Eq. (2.22).

Campbell [7]

The empirical ground motion relation given by Campbell [7] is suitable for estimating ground motion on ENA hard rock with a shear wave velocity of 2800 m/sec. This correlation was developed using hybrid empirical method in which ground motion relations for ENA were developed using western North America (WNA) empirical relations. Attenuation relation given by Campbell is given in Eq. (2.23) (refer [7] for constants). In this equation, $y_{br} = \left(\frac{S_a}{g}\right)$ stands for the ratio of spectral acceleration at bedrock level to acceleration due to gravity. M and r refer to moment magnitude and hypocentral distance, respectively.

$$\log(Y) = c_1 + f_1(M_w) + f_2(M_w, r_{rup}) + f_3(r_{rup}) \quad (2.23)$$

where

$$f_1(M_w) = c_2 M_w + c_3(8.5 - M_w)^2$$
$$f_2(M_w, r_{rup}) = c_4 \ln R + (c_5 + c_6 M_w) r_{rup}$$
$$R = \sqrt{r_{rup}^2 + [c_7 \exp(c_8 M_w)]^2}$$

and

$$f_3(r_{rup}) = \begin{bmatrix} 0 \quad \text{for } r_{rup} \le r_1 \\ c_7(\ln r_{rup} - r_1) \quad \text{for } r_{rup} \le r_1 \\ c_7(\ln r_{rup} - r_1) + c_8(\ln r_{rup} - r_2) \quad \text{for } r_{rup} > r_2 \end{bmatrix}$$

Silva [8]

The attenuation relation given by Silva (1997) is based on the regression analysis which provides best overall fit to simulations. The ground motion is simulated using stochastic method reflecting parametric variability for distances of 1, 5, 10, 20, 50, 75, 100, 200, and 400 km. At each distance, five magnitudes are used: M 4.5, 5.5, 6.5, 7.5, and 8.5. Attenuation relation given by Silva is given in Eq. (2.21).

$$\ln y = c_1 + c_2 M + (c_6 + c_7 M) * \ln(R + e^{c_4}) + c_{10}(M - 6)^2 \quad (2.24)$$

where R is taken as closest distance to the surface projection of the ruptured surface. For the coefficients of Eq. (2.24), one can refer [8].

Pezeshk [9]

Earthquake ground motion relations for hard-rock sites in Eastern North America (ENA) [9] are given by:

$$\begin{aligned}\log(y) = {} & c_1 + c_2 M_w + c_3 M_w^2 \\ & + (c_4 + c_5 M_w) \times \min\{\log(R), \log(70)\} \\ & + (c_6 + c_7 M_w) \times \max[\min\{\log(R/70), \log(140/70)\}, 0] \\ & + (c_8 + c_9 M_w) \times \max\{\log(R/140), 0\} + c_{10} R\end{aligned} \quad (2.25)$$

where

$$R = \sqrt{R_{rup}^2 + c_{11}^2}$$

and y is the median value of PGA or PSA (g).

The mean aleatory standard deviation of log(y) is given by:

$$\sigma_{\log(y)} = \begin{bmatrix} c_{12} M_w + c_{13} M \le 7 \\ -6.95\text{E} - 3 M_w + c_{14} M > 7 \end{bmatrix}$$

Refer [9] for the constants.

Toro (2013)

Attenuation relation given by Toro (2013) is given in Eq. (2.23). In this equation, Y is the spectral acceleration or peak ground acceleration (g). M and R_{JB} refer to moment magnitude and Joyner–Boore distance (closest horizontal distance to earthquake rupture), respectively. One can refer (Toro 2013) for the constants of Eq. (2.26). This correlation is derived from the predictions of stochastic ground motion model and is applicable for rock sites in Central and Eastern North America.

$$\begin{aligned}\ln(Y) = {} & c_1 + c_2(M - 6) + c_3(M - 6)^2 - c_4 \ln(R_M) \\ & - (c_5 - c_4)\max\left[\ln\frac{R_M}{100}, 0\right] - c_6 R_M + \\ & \epsilon_e + \epsilon_a\end{aligned} \quad (2.26)$$

where

$$R_M = \sqrt{R_{JB}^2 + c_7^2}$$

2.7.5 Comparison of Interplate and Intraplate Attenuation Relationships

More than 30 attenuation relationships have been proposed since 1965, some are for interplate, and some are for intraplate regions of different parts of the world. Similarly, different investigators use different definitions of source-to-site

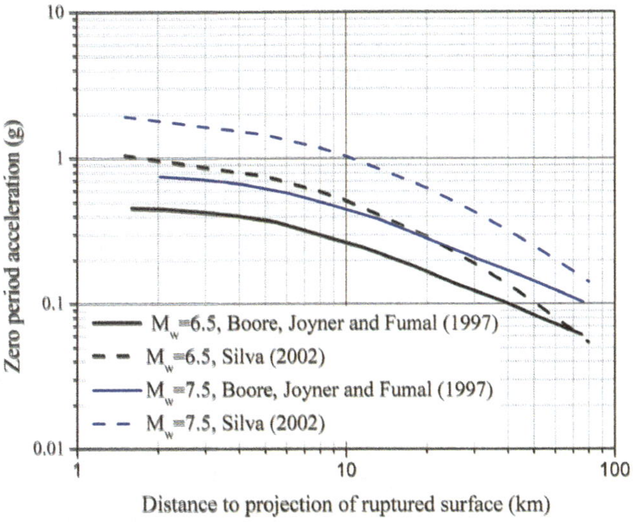

Fig. 2.18 Variation of peak ground acceleration with ruptured distance with moment magnitudes of 6.5 and 7.5 for interplate and intraplate regions

distance in these relationships, and it becomes difficult to compare the PGA attenuations with magnitude and distances for interplate and intraplate regions. Following the same vein, a comparison of intraplate and interplate PGA attenuation with magnitude and distance has been conducted using the relationships of Silva [8] and Boore et al. [11]. The comparison is made for moment magnitudes of 6.5 and 7.5 which is illustrated in Fig. 2.18. It is seen that at larger source-to-site distances the variation of PGA is not much for both interplate and intraplate regions, but as the distance reduces, the PGA observed at the intraplate region is higher in comparison to interplate regions.

2.8 Conditional Probability of Exceedance

It can be seen from the previous discussions that the ground motion parameter (such as PGA) for a given distance and magnitude represents a random variable. It means one can expect variation in the PGA values for a given distance and magnitude and not a fixed value. The variation in the PGA values can be represented with a distribution function such as lognormal distribution whose pdf can be represented as given below:

$$f_A(a) = \frac{1}{\sigma\sqrt{2\pi}\,a}\exp\left[-\frac{1}{2}\left(\frac{\ln a - \mu_{\ln a}}{\sigma_{\ln a}}\right)^2\right] \quad 0 \le A \le \infty \tag{2.27}$$

Here, 'A' represents a random variable of interest (peak ground acceleration). The pdf of PGA for a given distance and magnitude is shown in Fig. 2.19.

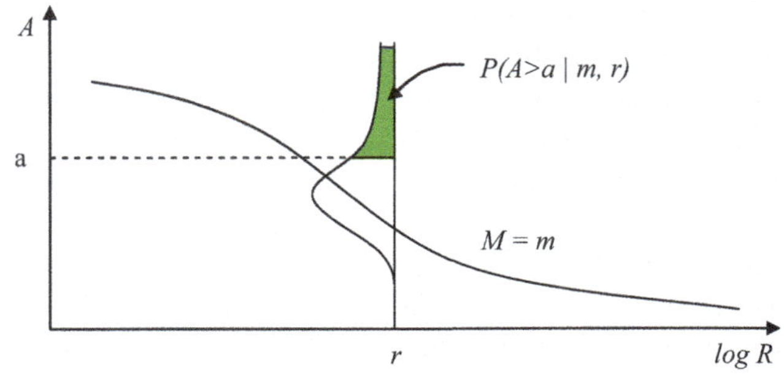

Fig. 2.19 Probability density function of ground motion parameter for a given magnitude and distance

The cdf (probability of non-exceedance of certain level of PGA) for the lognormal density function can be given by:

$$F = P(A \le a) = \Phi\left[\frac{\ln a - \mu_{\ln a}}{\sigma_{\ln a}}\right] \tag{2.28}$$

And the probability of exceeding a certain level of PGA is complementary to the cdf and can be given as follows:

$$P(A > a|m_i, r_j) = 1 - P(A \le a) = 1 - \Phi\left[\frac{\ln a - \mu_{\ln a}}{\sigma_{\ln a}}\right] \tag{2.29}$$

In general, the range for lognormal distribution is 0 to infinity (∞). Hence, the pdf of PGA is valid in the range of '0 to infinity (∞).' However, in reality, an earthquake of a given magnitude (smaller or bigger) at a given distance cannot produce infinity PGA value or in other words a smaller magnitude earthquake at a given distance can produce only a smaller PGA at the site. Hence, for a given distance and magnitude, it is realistic to have limits (lower and upper limits) on the PGA values. This can be represented with truncated pdf with modified limits for the random variable. This distribution can be obtained from the original function by applying transformations. In general, for normal or lognormal distribution, 3σ limits are applied to obtain the truncated distribution. According to this, the truncated lognormal distribution for PGA with 3σ limits can be represented as follows [3]:

$$P(A > a|m_i, r_j) = 1 - \Phi^1\left[\frac{\ln a - \mu_{\ln a}}{\sigma_{\ln a}}\right] = 1 - \Phi^1[u]$$

$$\text{where } \Phi^1(u) = \frac{\Phi(u) - \Phi(-3)}{1 - 2\Phi(-3)}$$

The lower and upper limits for the PGA can be obtained from its 3σ limits. The mean and standard deviation of PGA can be obtained from attenuation relations as discussed in the previous section.

Example 2.4 Consider an earthquake of magnitude 7.0 occurred at an interplate region. Estimate the PHA that would be produced at the site 100 km away from the fault source by using McGuire (1978) attenuation relationship. Also, estimate the probability of non-exceedance of PHA of 0.1 g at the site.

Solution:

i. McGurie (1978) attenuation relationship is given as:

$$a_p = b_1 \exp(b_2 M)(R + D)^{-b_3}$$
$$\text{where } b_1 = 1.04$$
$$b_2 = 0.483$$
$$b_3 = 1.2$$
$$D = 40$$
$$\sigma_{\ln a} = 0.731$$

Substituting $M = 7.0$, $R = 100$ km, one can get

$$a_p = 1.04 \exp(0.483 \times 7.0)(100 + 40)^{-1.2}$$
$$a_p = 0.0812\,\text{g}$$
$$\ln a_p = -2.5097$$
$$\therefore \mu_{\ln a_p} = -2.5097$$

From the above calculations, the PHA that would be produced at the site 100 km away from the fault source due to an earthquake of magnitude of 7.0 is obtained as 0.0812 g.

ii. The probability of non-exceedance is equivalent to cumulative distribution function. This can be estimated from Eq. (2.28) which is given as

$$F = P(A \le a) = \Phi\left[\frac{\ln a - \mu_{\ln a}}{\sigma_{\ln a}}\right] = \Phi[u]$$

The parameters needed for the calculation are

$$\mu_{\ln a} = -2.5097$$
$$\sigma_{\ln a} = 0.731$$
$$\text{u} = \frac{\ln a - \mu_{\ln a}}{\sigma_{\ln a}} = \frac{\ln(0.1) - (-2.5097)}{0.731} = 0.2834$$
$$\therefore P(A \le 0.1g) = \Phi[0.2834] = 0.6116$$

2.9 Estimation of Hazard at the Site

As per Eq. (2.2), the hazard at the site represents the annual rate of exceedance of particular level of PGA at the site of interest for various earthquake magnitudes that are postulated to occur at various sources and various distances from the site. In finding out the total hazard at the site, one needs to aggregate all the individual hazards from various earthquake sources around the site. Figure 2.20 shows contribution of individual hazards to the total hazard.

Example 2.5 Consider the two earthquake sources (source 1—line source, source 2—point source) which are located at a certain distance from the site under consideration as shown in Fig. 2.21. The characteristics of each fault are provided in Table 2.4. Develop the hazard curve for the given site.

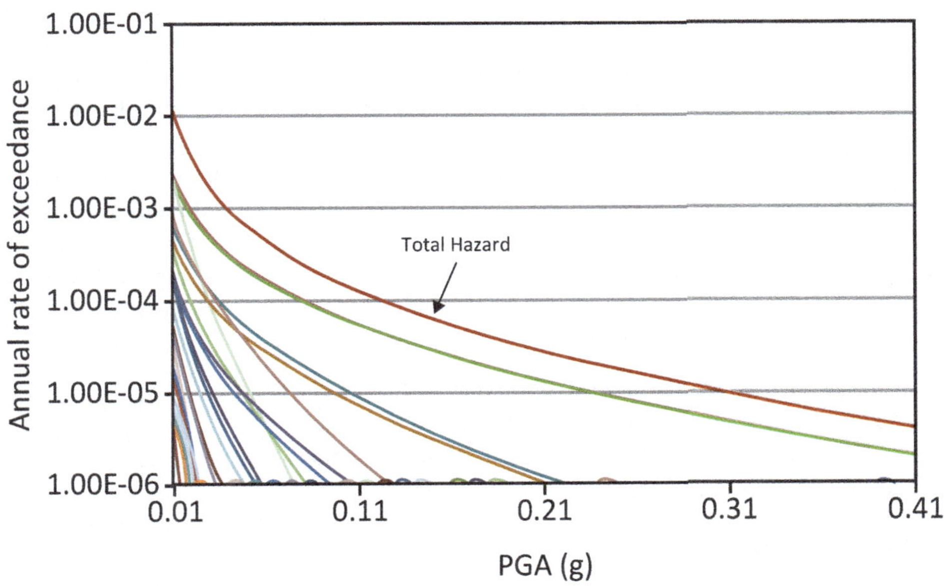

Fig. 2.20 Hazard curve of a typical site

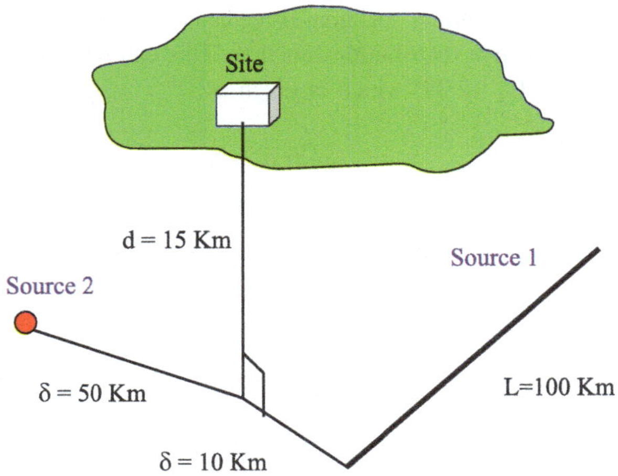

Fig. 2.21 Earthquake sources around a site

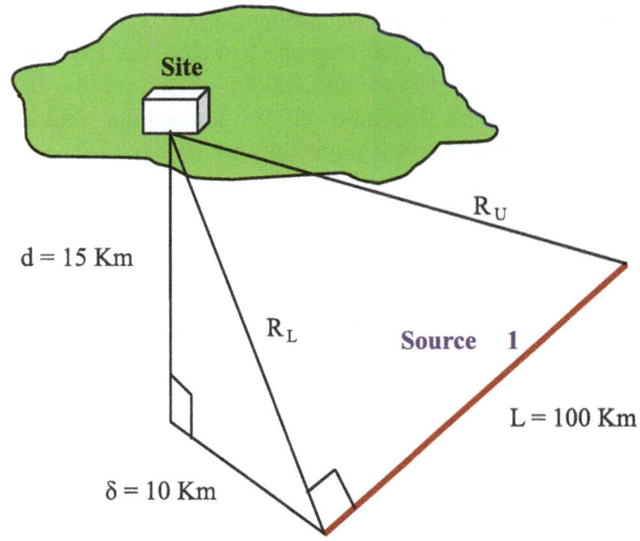

Fig. 2.22 Fault source 1: line source model

Table 2.4 Characteristics of the faults

	Source type	m_0	m_u	a	b	d (km)	δ (km)	L (km)
1	Line	3.0	5.0	3.21	1.13	15	10	100
2	Point	3.0	4.0	3.10	1.13	15	50	–

Solution:

As per Eq. 2.2, the total hazard at a site is represented as aggregation of hazard from individual sources and is given as follows:

$$v(a) = \sum_{k=1}^{N} \upsilon_k(m_0) \int_{m_0}^{m_u} \int_{d}^{r_0} f_M(m) f_R(r) P[A > a|m, r] \mathrm{d}r\, \mathrm{d}m$$

or

$$v(a) = \sum_{k=1}^{N} \upsilon_k(m_0) \left[\sum_{i=1}^{N_M} \sum_{j=1}^{N_R} f_{Mk}(m_i) f_{Rk}(r_j) P(A > a|m_i, r_j) \Delta r \Delta m \right]$$

The number of fault sources is 2(N = 2). To evaluate the above equation, the following variables should be determined:

- $v_k(m_0)$
- $f_{Mk}(m_i)$
- $f_{Rk}(r_j)$
- $P(A > a\,|m_i, r_j)$

In order to evaluate the hazard at the site first, we segregate the different sources and analyze the hazard from each source and at the end to get the total hazard one can combine the hazard from various sources.

Analyzing the Source 1

- Source type is line source.
- After segregating various sources, consider the source 1 separately as shown in Fig. 2.22.
- From the data, the depth of focus and epicentral distance for source 1 are given as
  d = 15 km
  δ = 10 km.
- The hypocentral distance from the site can be calculated as:

$$R_L = \sqrt{d^2 + \delta^2} = \sqrt{15^2 + 10^2} = 18.0278\,\text{km} = \Delta$$
$$R_U = \sqrt{R_L^2 + L^2} = \sqrt{18.028^2 + 100^2} = 101.612\,\text{km}$$

- Data given
  $m_0 = 3.0$ $m_u = 5.0$
  $a = 3.21$ $b = 1.13$
- Hence, the minimum and maximum earthquake magnitudes that can occur on source 1 are 3.0 and 5.0, respectively, at a hypocentral distance that can vary from a lower limit of 18.0278 km to an upper limit of 101.612 km.
- **Calculation of $v_1(m_0)$**
  - By using Gutenberg–Richter's recurrence relationship, one can find the annual frequency of exceedance of earthquake magnitude m_0 on fault 1 which is given as below:

$$\upsilon_1(m_0) = 10^{a - bm_0}$$
$$m_0 = 3.0$$
$$\upsilon_1(3) = 10^{3.21 - 1.13 \times 3.0} = 0.661/\text{year}$$

- **Calculation of $f_M(m)\Delta$m**
  - In this step, the aim is to find the probability of occurrence of a certain level of earthquake magnitude on the fault source In this case, magnitude is

considered as a random variable and is continuous in nature. As per probability theory, one cannot find probability value at a particular fixed value for continuous random variables. However, for a given interval, it is possible to find the probability value.

- Hence, in this example, the magnitude range is divided into some intervals and probability of occurrence for each interval is calculated.
- It is known that

$$f_M(m) = \frac{\beta 10^{a-bm}}{10^{a-bm_0} - 10^{a-bm_u}} \quad m_0 \le M \le m_u$$

where $\beta = b \log_e 10$

$$f_M(m) = \frac{dF}{dm} \Rightarrow dF = f_M(m)\,dm$$

$$\therefore f_M(m)\,\Delta m = F(m + \Delta m) - F(m) = P(m \le M \le m + \Delta m)$$

- The minimum and maximum values of magnitude that can occur on fault source 1 are given as:
  $m_0 = 3.0$
  $m_u = 5.0$
  Range = $m_u - m_0 = 5.0 - 3.0 = 2.0$
- Now, consider the number of intervals as 10. Then, interval size can be obtained as:

$$\Delta m = \frac{(m_u - m_0)}{10} = \frac{(5.0 - 3.0)}{10} = 0.2$$

- Now, one can find the value of $f_M(m)\Delta m$ either by directly obtaining from the pdf or from cdf. While finding out the value from density function, one has to choose a particular value of 'm'. In general one can choose mid-value of the interval. This is illustrated for first interval in the following steps.
- For the first interval, i.e., 3.0 − 3.2

$$m_1 = 3.0$$
$$m_2 = 3.2$$
$$\Delta m = 3.2 - 3.0 = 0.2$$
$$m = \frac{m_1 + m_2}{2} = \frac{3.0 + 3.2}{2} = 3.1$$

$$P(m_1 \le M \le m_2) = f_M(m)\Delta m$$
$$P(3.0 \le M \le 3.2) = f_M(3.1) \times 0.2$$

$$f_M(3.1) = \frac{\beta 10^{3.21-1.13\times 3.1}}{10^{3.21-1.13\times 3.0} - 10^{3.21-1.13\times 5.0}} \quad 3.0 \le M \le 5.0$$
$$\beta = b \log_e 10 = 1.13 \times 2.3026 = 2.602$$
$$f_M(3.1) = 2.017$$
$$f_M(m)\Delta m = f_M(3.1) \times 0.2 = 0.403$$

- The above calculations are repeated for various intervals, and the values are given in Table 2.5.

- **Calculation of $f_R(r)\Delta r$**
  - This step involves calculation of probability of occurrence of an earthquake of a given magnitude on the fault source 1 at a particular distance from the site.
  - This can be obtained from pdf of site to source distance. As the pdf is continuous in nature, it is only possible to calculate probability values for a given interval of distances. These calculations are same as explained for magnitude in the previous step.
  - From Fig. 2.24, the hypocentral distances from the site are calculated as:

$$R_L = 18.0278\,\text{km}$$
$$R_U = 101.612\,\text{km}$$

  - As per these values, the earthquake can happen on the source, at various hypocentral distances from the site which can vary from 18.023 km to 101.612 km.
  - The hypocentral distance range can be calculated as

$$R_L = 18.0278\,\text{km}$$
$$R_U = 101.612\,\text{km}$$
$$\text{Range} = R_L - R_U = 101.612 - 18.0278 = 83.5842\,\text{km}$$

  - Now, divide the range into ten equal intervals with each interval size as

$$\Delta r = \frac{83.5842}{10} = 8.3584\,\text{km}$$

Table 2.5 Probability of occurrence of magnitude in various intervals

S. No.	Interval	m	Δm	$f_M(m)$	$f_M(m)\ \Delta m$
1	3.0–3.2	3.1	0.2	2.0169E+00	4.0338E−01
2	3.2–3.4	3.3	0.2	1.1986E+00	2.3973E−01
3	3.4–3.6	3.5	0.2	7.1234E−01	1.4247E−01
4	3.6–3.8	3.7	0.2	4.2334E−01	8.4668E−02
5	3.8–4.0	3.9	0.2	2.5159E−01	5.0317E−02
6	4.0–4.2	4.1	0.2	1.4952E−01	2.9903E−02
7	4.2–4.4	4.3	0.2	8.8856E−02	1.7771E−02
8	4.4–4.6	4.5	0.2	5.2806E−02	1.0561E−02
9	4.6–4.8	4.7	0.2	3.1382E−02	6.2765E−03
10	4.8–5.0	4.9	0.2	1.86507E−02	3.7301E−03

- Now, one can find the value of $f_R(r)\Delta r$. This term represents the probability of occurrence of an earthquake in a particular distance interval 'r' and '$r + \Delta r$'. From Eqs. 2.8 and 2.9.

$$P(R \leq r/m) = \frac{\sqrt{r^2-\Delta^2}-L_0}{L} = F_R(r)$$
$$f_R(r) = \frac{dF_R(r)}{dr} = \lim_{\Delta r \to 0} \frac{\Delta F}{\Delta r} = \frac{F(r+\Delta r)-F(r)}{\Delta r}$$
$$\text{when } L_0 = 0$$
$$F_R(r) = \frac{\sqrt{r^2-\Delta^2}}{L}$$
$$\text{where}$$
$$\Delta = \sqrt{d^2+\delta^2} = 18.028\,\text{km}$$

- For the first interval,

$$r = 18.0278\,\text{km}$$
$$\Delta r = 8.3584\,\text{km}$$
$$r + \Delta r = 18.0278 + 8.3584 = 26.3862\,\text{km}$$

- From Eq. (2.9),

$$f_R(r) = \frac{F(r+\Delta r) - F(r)}{\Delta r}$$
$$f_R(r) = \frac{F(18.0278+8.3584) - F(18.0278)}{\Delta r}$$
$$f_R(r) = \frac{F(26.3862) - F(18.0278)}{\Delta r}$$
$$f_R(r)\Delta r = F_R(26.3862) - F_R(18.0278)$$

$$F_R(r) = \frac{\sqrt{r^2-\Delta^2}}{L}$$
$$F_R(26.3864) = \frac{\sqrt{26.3862^2-18.0278^2}}{100} = \frac{19.267}{100} = 0.19267$$
$$F_R(18.0278) = \frac{\sqrt{18.0278^2-18.0278^2}}{100} = 0$$
$$f_R(r)\Delta r = 0.19267 - 0 = 0.19267$$

- The above calculations are repeated for various intervals and are given in Table 2.6.
- It should be noted the PGA at the site is evaluated for a given value of magnitude and hypocentral/epicentral distance by using attenuation relationships. Hence, one should use point value of 'r,' not the interval. It is common practice to use mid-value of the interval. In the present case for the first interval PGA can be evaluated at 22.2070 km.

- **Attenuation relationship**
  - Attenuation relationships help ground acceleration at the site for a given earthquake magnitude and distance from the site.
  - As explained previously, several attenuation relationships are available depending on the region and site conditions. One should be careful in choosing these empirical relationships.
  - In the present example to find the peak ground acceleration (PGA), the attenuation relationship proposed by Silva [8] has been utilized and is given below

$$\ln y = c_1 + c_2 M + (c_6 + c_7 M) * \ln(R + e^{c_4}) + c_{10}(M-6)^2$$

where M is the magnitude, R is taken as closest distance to the surface projection of the ruptured surface

  - The coefficients for PGA are as follows: $c_1 = 3.54103$, $c_2 = 0.18904$, $c_4 = 2.7$, $c_6 = -2.97418$, $c_7 = 0.19819$, $c_{10} = -0.05814$ $\sigma_{\ln y} = 0.84$
  - For the purpose of illustration, consider a magnitude of 4.0 occurring at a hypocentral distance of 22.2070 km from the site.
  - In the above equation, R represents epicentral distance which can be calculated from hypocentral distance (r) and depth of focus (h) as follows:

$$R = \sqrt{r^2 - d^2}$$
$$R = \sqrt{22.2070^2 - 15^2} = 16.3753\,\text{km}$$

  - Now, the value of PGA can be found for a given magnitude ($M = 4.1$) and for a given epicentral

Table 2.6 Probability of occurrence of R in an interval

S. No.	Interval (km)		$F_R(r)$ I	$F_R(r + \Delta r)$ II	$f_R(r)\Delta r$ II–I
	r	$r + \Delta r$			
1	18.0278	26.3862	0.0000E+00	1.9267E−01	1.9267E−01
2	26.3862	34.7446	1.9267E−01	2.9702E−01	1.0434E−01
3	34.7446	43.1030	2.9702E−01	3.9152E−01	9.4503E−02
4	43.1030	51.4615	3.9152E−01	4.8200E−01	9.0485E−02
5	51.4615	59.8199	4.8200E−01	5.7039E−01	8.8383E−02
6	59.8199	68.1783	5.7039E−01	6.5752E−01	8.7129E−02
7	68.1783	76.5367	6.5752E−01	7.4383E−01	8.6316E−02
8	76.5367	84.8952	7.4383E−01	8.2959E−01	8.5757E−02
9	84.8952	93.2536	8.2959E−01	9.1494E−01	8.5355E−02
10	93.2536	101.6120	9.1494E−01	1.0000E+00	8.5056E−02

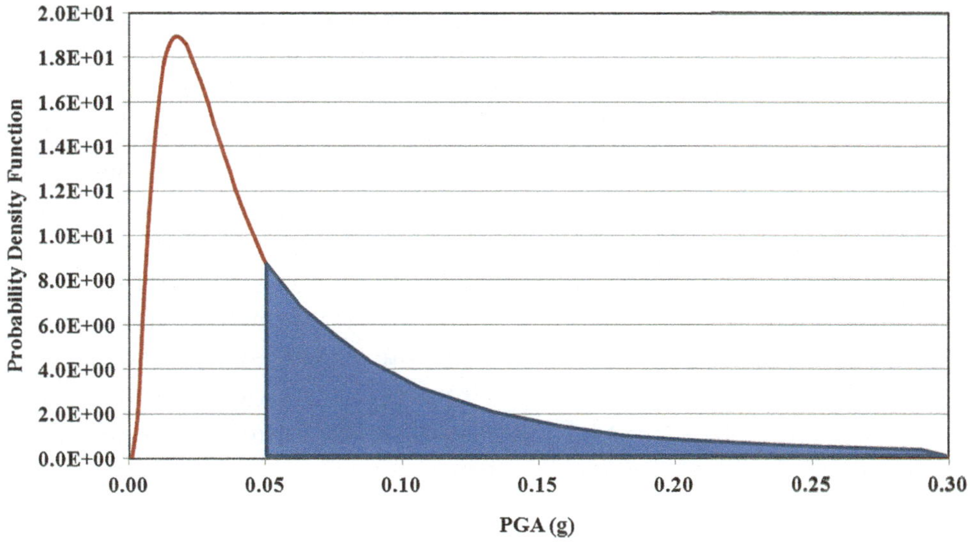

Fig. 2.23 Probability density function of PGA for $m = 4.1$ and $r = 22.2070$ km

distance ($R = 16.3753$ km) from Silva attenuation relationship as follows:

$$\begin{aligned}\ln y &= 3.54103 + 0.18904M \\ &\quad + (-2.97418 + 0.19819M) \times \ln(\mathrm{R} + \mathrm{e}^{2.7}) \\ &\quad - 0.05814 \times (M - 6)^2 \\ \ln y &= 3.54103 + 0.18904 \times 4.1 \\ &\quad + (-2.97418 + 0.19819 \times 4.1) \times \ln(16.3753 + \mathrm{e}^{2.7}) \\ &\quad - 0.05814 \times (4.1 - 6)^2 \\ \ln y &= -3.3344 \Rightarrow y = \mathrm{e}^{-3.3344} = 0.0356\,\mathrm{g}\end{aligned}$$

- The PGA value so calculated is a random variable whose density function follows lognormal distribution and has the following parameters:

$$\mu_{\ln y} = -3.3344$$
$$\sigma_{\ln y} = 0.84$$

- The probability density function of PGA is shown in Fig. 2.23

- **Computation of conditional probability of exceedance**
  - This step involves the calculation of probability of exceedance of PGA for a given value (assume 0.05 g) and can be calculated by using Eq. (2.27).

$$P(A > a|m_i, r_j) = 1 - \Phi^1\left[\frac{\ln a - \mu_{\ln a}}{\sigma_{\ln a}}\right] = 1 - \Phi^1[z]$$

$$\text{where } \Phi^1(z) = \frac{\Phi(z) - \Phi(-3)}{1 - 2\Phi(-3)}$$

$$z = \frac{\ln a - \mu_{\ln a}}{\sigma_{\ln a}} = \frac{\ln 0.05 + 3.3344}{0.84} = 0.4032$$
$$\Phi(z) = \Phi(0.4032) = P(A \le 0.4032) = 0.6566$$
$$\Phi(-3) = 0.00135$$
$$\Phi^1(z) = \frac{\Phi(z) - \Phi(-3)}{1 - 2\Phi(-3)} = \frac{0.6566 - 0.00135}{1 - 2 \times 0.00135} = 0.6570$$

$$\begin{aligned}P(A > 0.05|4.1, 22.2070) &= 1 - \Phi^1[z] = 1 - 0.6570 \\ &= 0.3430\end{aligned}$$

  - This value is shown in Fig. 2.10 as shaded area beyond the value 0.1 g.

- **Hazard at the site due to source 1**
  - The hazard at the site can be estimated by using Eq. (2.3). Here, hazard is represented as annual rate of exceedance of a particular level of PGA at the site due to occurrence of various magnitudes of earthquakes at various distances on various fault sources. This is given as follows:

$$v(a) = \sum_{k=1}^{N} \upsilon_k(m_0) \int_{m_0}^{m_u} \int_{d}^{r_0} f_M(m) f_R(r) P[A > a|m, r] \mathrm{d}r\, \mathrm{d}m$$

or

$$v(a) = \sum_{k=1}^{N} \upsilon_k(m_0) \left[\sum_{i=1}^{N_M} \sum_{j=1}^{N_R} f_{Mk}(m_i) f_{Rk}(r_j) P(A > a|m_i, r_j) \Delta r \Delta m\right]$$

  - Equation (2.3) can be expanded for various faults. In the present case, total number of faults is 2. Accordingly, Eq. (2.3) can be expanded for fault 1 as follows:

$$v_1(a) = \upsilon_1(3) \left[\sum_{M=3.0}^{M=5.0} \sum_{R=18.028}^{.101.612} f_M(m) f_R(r) P(Z > z/m_i, r_j) \Delta R \Delta m\right]$$

  - From the previous steps, the above parameters can be rewritten as

$$\upsilon_1(3) = 0.661/\text{year}$$
$$\text{for } m = 4.1 f_M(m)\Delta m = 2.9903 \times 10^{-2}$$
$$\text{for } r = 22.2070 f_R(r)\Delta r = 1.9267 \times 10^{-1}$$
$$\text{for } m = 4.1,\ r = 22.2070\,\text{km} \quad a = 0.0356\,\text{g}$$
$$P(Z > 0.05\,\text{g}/4.1, 22.2070) = 0.3430$$

Table 2.7 Annual rate of exceedance of 0.05 g at the site for different combinations of magnitudes and distances due to source 1

Magnitude	Hypocentral distance (km)					
	22.2070	30.5654	38.9238	47.2822	55.6407	63.9991
3.1	1.89E−03	1.04E−04	0.00E+00	0.00E+00	0.00E+00	0.00E+00
3.3	2.05E−03	1.69E−04	1.48E−05	0.00E+00	0.00E+00	0.00E+00
3.5	2.04E−03	2.22E−04	3.93E−05	1.35E−06	0.00E+00	0.00E+00
3.7	1.88E−03	2.58E−04	6.11E−05	1.29E−05	0.00E+00	0.00E+00
3.9	1.61E−03	2.71E−04	7.85E−05	2.33E−05	5.62E−06	0.00E+00
4.1	1.31E−03	2.63E−04	8.95E−05	3.21E−05	1.12E−05	3.19E−06
4.3	1.00E−03	2.37E−04	9.30E−05	3.84E−05	1.60E−05	6.43E−06
4.5	7.29E−04	2.00E−04	8.91E−05	4.15E−05	1.96E−05	9.23E−06
4.7	5.10E−04	1.59E−04	7.95E−05	4.12E−05	2.16E−05	1.13E−05
4.9	3.44E−04	1.20E−04	6.64E−05	3.79E−05	2.17E−05	1.25E−05

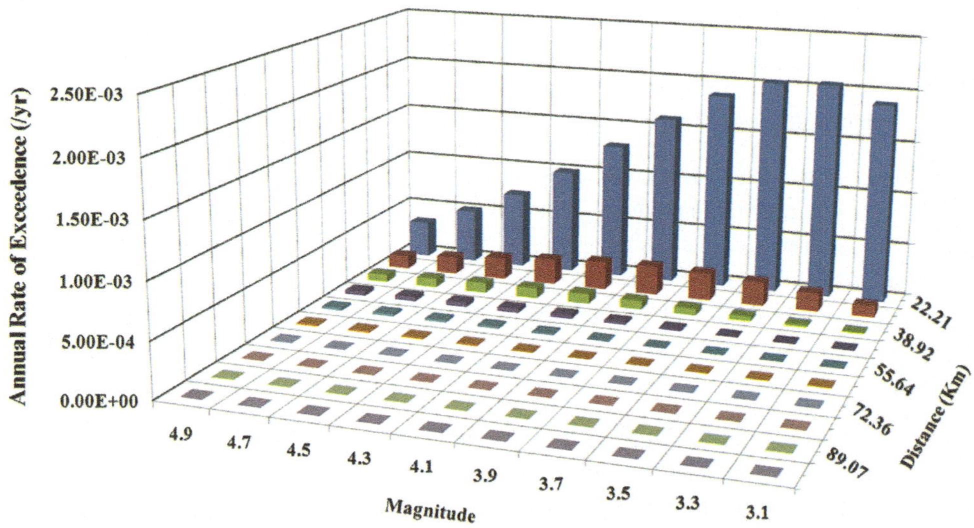

Fig. 2.24 Annual rate of exceedance at the site for PGA of 0.05 g for various magnitude and distance due to fault source 1

- Now, the frequency of exceedance of 0.05 g at the site due to fault source 1 can be calculated as:

$$\upsilon_1(0.05\text{ g}) = 0.661 \times 2.9903 \times 10^{-2} \times 1.9267 \times 10^{-1} \times 0.3430 = 1.3056 \times 10^{-3}/\text{year}$$

- These calculations can be repeated for various earthquake magnitudes occurring at various hypocentral distances on the source from the site which is given in Table 2.7 and also shown in Fig. 2.24 as magnitude-distance bins for PGA of 0.05 g.
- By summing up all the values, one can get annual rate of exceedance of 0.05 g at the site due to fault source 1. The value obtained is 1.6567×10^{-2}/year.
- The above procedure can be repeated for various values of PGA. The values of annual rate of exceedance for various values of PGA are given in Table 2.8 and a graph is shown in Fig. 2.25.

Analyzing the Source 2

- Source type is point source.
- After segregating various sources, consider the source 2 separately as shown in Fig. 2.26.
- From the data, the depth of focus and epicentral distance for source 2 are given as
  d = 15 km
  δ = 50 km
- The hypocentral distance from the site can be calculated as

$$R = \sqrt{d^2 + \delta^2} = \sqrt{15^2 + 50^2} = 52.20\,\text{km}$$

- Data given
  $m_0 = 3.0$ $m_u = 4.0$
  $a = 3.10$ $b = 1.13$
- Hence, the minimum and maximum earthquake magnitudes that can occur on source 2 are 3.0 and 4.0, respectively, at a hypocentral distance of 52.2 km from the site.

Table 2.8 Annual rate of exceedance with respect to PGA

PGA (g)	Annual rate of exceedance (/year)
0.005	2.5247E−01
0.010	1.4859E−01
0.015	9.9020E−02
0.020	7.0321E−02
0.025	5.2086E−02
0.030	3.9786E−02
0.035	3.1142E−02
0.040	2.4847E−02
0.045	2.0149E−02
0.050	1.6567E−02
0.055	1.3780E−02
0.060	1.1577E−02

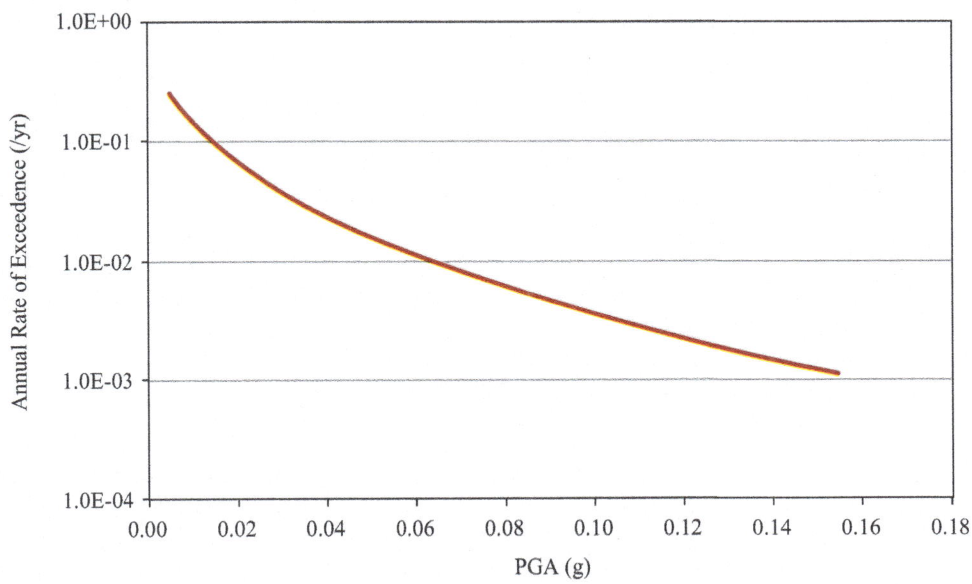

Fig. 2.25 PGA versus annual rate of exceedance due to fault 1

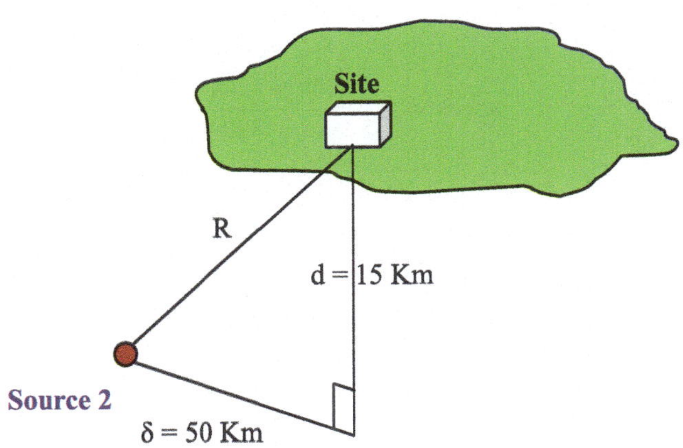

Fig. 2.26 Fault source 2—point source model

- **Calculation of $\nu_2(m_0)$**
  - Based on Gutenberg–Richter's relationship this value is calculated as below:

$$\upsilon_2(m_0) = 10^{a-bm_0}$$
$$m_0 = 3.0$$
$$\upsilon_2(3) = 10^{3.10-1.13\times3.0} = 0.513/\text{year}$$

- **Calculation of $f_M(m)\Delta m$**
  - This is similar to the calculations done in fault source 1. One can find the value of $f_M(m)\Delta m$ from the pdf of magnitude.

$$f_M(m) = \frac{\beta 10^{a-bm}}{10^{a-bm_0} - 10^{a-bm_u}} \quad m_0 \le M \le m_u$$
$$\text{where } \beta = b \log_e 10$$
$$f_M(m) = \frac{dF}{dm} \Rightarrow dF = f_M(m)\,dm$$
$$\therefore f_M(m)\,\Delta m = F(m+\Delta m) - F(m) = P(m \le M \le m+\Delta m)$$

Table 2.9 Probability of occurrence of M in an interval

S.No.	Interval	m	Δm	$f_M(m)$	$f_M(m)\ \Delta m$
1	3.0–3.2	3.1	0.2	2.1664E+00	4.3329E−01
2	3.2–3.4	3.3	0.2	1.2875E+00	2.5750E−01
3	3.4–3.6	3.5	0.2	7.6515E−01	1.5303E−01
4	3.6–3.8	3.7	0.2	4.5472E−01	9.0944E−02
5	3.8–4.0	3.9	0.2	2.7024E−01	5.4047E−02

- The minimum and maximum values of magnitude that can occur on fault source 2 are given as:
  $m_0 = 3.0$
  $m_u = 4.0$
  Range $= m_u - m_0 = 4.0 - 3.0 = 1.0$
- To have the same interval size as that of fault source 1, one needs to reduce the number of intervals. Hence, consider the number of intervals as 5. Then, interval size can be obtained as

$$\Delta m = \frac{(m_u - m_0)}{5} = \frac{(4.0 - 3.0)}{5} = 0.2$$

- The calculation of $f_M(m)\Delta m$ is similar to the calculations of fault source 1, and the results are given in Table 2.9.

- **Calculation of $f_R(r)\Delta r$**
  - This step involves calculation of probability of occurrence of an earthquake of a given magnitude on the fault source 2 at a particular distance from the site.
  - From Fig. 2.24, the hypocentral distance from the site is calculated as:

$$R = 52.2\,\text{km}$$

  - As this is point source, the probability of occurrence of an earthquake for a given magnitude anywhere on the point source is equal to 1 (because there is no variation in the hypocentral distance).
- **Attenuation relationship**
  - In the present example to find the peak ground acceleration (PGA), the attenuation relationship proposed by Silva [8] has been utilized and is given below

$$\ln y = c_1 + c_2 M + (c_6 + c_7 M) * \ln(R + e^{c_4}) + c_{10}(M - 6)^2$$

where M is the magnitude, R is taken as closest distance to the surface projection of the ruptured surface

  - The coefficients for PGA are as follows:
    $c_1 = 3.54103$, $c_2 = 0.18904$, $c_4 = 2.7$, $c_6 = -2.97418$, $c_7 = 0.19819$, $c_{10} = -0.05814$
    $\sigma_{\ln y} = 0.84$
  - In the above equation, R represents epicentral distance which is already given in the data as 50 km.
  - Now, the value of PGA can be found for a given magnitude ($M = 3.9$) and for a given epicentral distance ($R = 50$ km).

$$\begin{aligned}\ln y &= 3.54103 + 0.18904\text{M} \\ &\quad + (-2.97418 + 0.19819\,\text{M}) \times \ln(\text{R} + \text{e}^{2.7}) \\ &\quad - 0.05814 \times (\text{M} - 6)^2 \\ \ln y &= 3.54103 + 0.18904 \times 3.9 \\ &\quad + (-2.97418 + 0.19819 \times 3.9) \times \ln(50 + \text{e}^{2.7}) \\ &\quad - 0.05814 \times (3.9 - 6)^2 \\ \ln y &= -5.16 \Rightarrow y = \text{e}^{-5.16} = 0.0057\,\text{g}\end{aligned}$$

  - The PGA value so calculated is a random variable whose pdf follows lognormal distribution and has the following parameters:

$$\mu_{\ln y} = -5.16$$
$$\sigma_{\ln y} = 0.84$$

  - The pdf of PGA is shown in Fig. 2.27
- **Computation of conditional probability of exceedance**
  - This step involves the calculation of probability of exceedance of PGA for a given value and can be calculated by using Eq. 2.27.

$$P(A > a|m_i, r_j) = 1 - \Phi^1\left[\frac{\ln a - \mu_{\ln a}}{\sigma_{\ln a}}\right] = 1 - \Phi^1[z]$$

$$\text{where } \Phi^1(z) = \frac{\Phi(z) - \Phi(-3)}{1 - 2\Phi(-3)}$$

$$z = \tfrac{\ln a - \mu_{\ln a}}{\sigma_{\ln a}} = \tfrac{\ln 0.05 + 5.163}{0.84} = 2.5799$$
$$\Phi(z) = \Phi(2.5799) = P(A \le 2.5799) = 0.9951$$
$$\Phi(-3) = 0.00135$$
$$\Phi^1(z) = \tfrac{\Phi(z) - \Phi(-3)}{1 - 2\Phi(-3)} = \tfrac{0.9951 - 0.00135}{1 - 2 \times 0.00135} = 0.9964$$

$$\begin{aligned}P(A > 0.05|3.9, 52.2) &= 1 - \Phi^1[z] = 1 - 0.9964 \\ &= 3.6011 \times 10^{-3}\end{aligned}$$

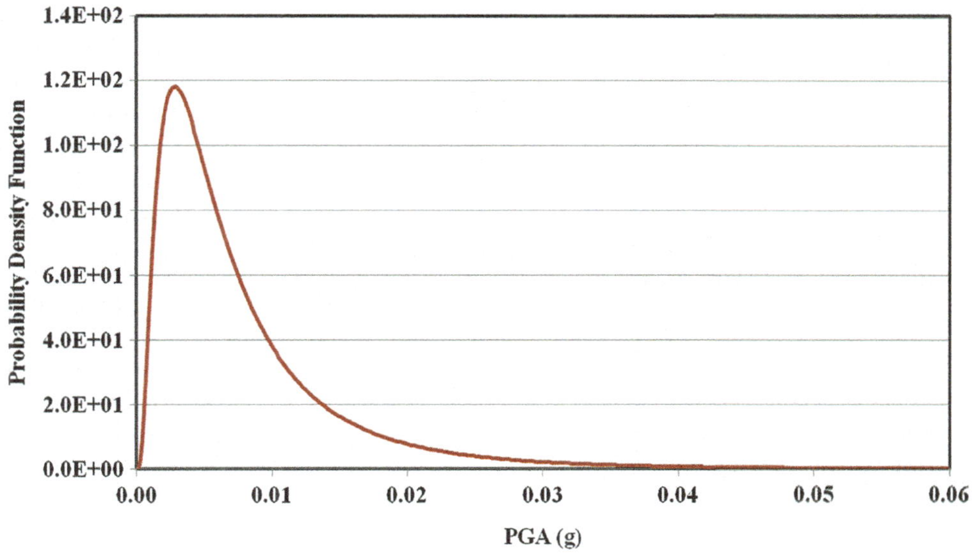

Fig. 2.27 Probability density function of PGA for m = 3.9 and r = 52.2 km

- **Hazard at the site due to source 2**
  - For fault source 2, Eq. (2.3) can be expanded as follows:

$$v_2(a) = v_2(3)\left[\sum_{M=3.0}^{M=4.0}\sum_{R=52.2} f_M(m) f_R(r)\, P(Z > z/m_i, r_j)\Delta r \Delta m\right]$$

  - As there is no variation in hypocentral distance $f_R(52.2)$ = 1 and the above equation becomes:

$$v_2(a) = v_2(3)\left[\sum_{M=3.0}^{M=4.0} f_M(m) P(Z > z/m_i, 52.2)\Delta m\right]$$
$$\because f_R(52.2) = 1$$

  - The values of parameters in the above equation can be taken from previous steps as follows:

$$\upsilon_2(3) = 0.513/\text{year}$$
$$\text{for}\, m = 3.9\ f_M(3.9)\Delta m = 5.4047 \times 10^{-2}$$
$$\text{for}\, m = 3.9\ \ r = 52.2\,\text{km}\ \ a = 0.0057\,\text{g}$$
$$P(A > 0.05\,\text{g}|3.9, 52.2) = 3.6011 \times 10^{-3}$$

  - The frequency of exceedance of 0.05 g at the site due to fault source 2 can be calculated as

$$\upsilon_2(0.05\,\text{g}) = 0.513 \times 5.4047 \times 10^{-2} \times 1 \times 3.6011 \times 10^{-3} = 9.9818 \times 10^{-5}/\text{year}$$

  - These calculations can be repeated for various earthquake magnitudes occurring at a hypocentral distance of 52.2 km on the fault source 2 from the site which is shown in Fig. 2.28.

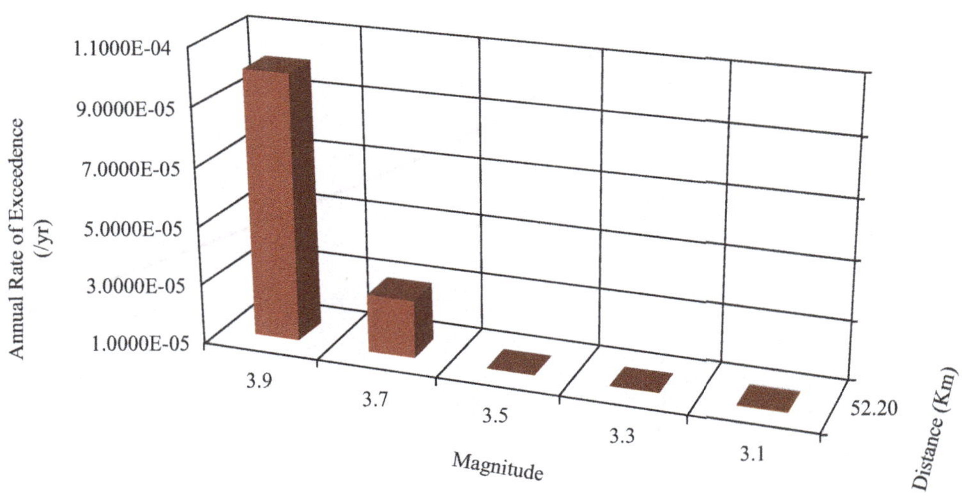

Fig. 2.28 Annual rate of exceedance at the site for PGA of 0.05 g for various magnitude and distance due to fault source 2

Table 2.10 Annual rate of exceedance with respect to PGA

PGA (g)	Annual rate of exceedance (/year)
0.005	1.2332E−01
0.010	3.6220E−02
0.015	1.3915E−02
0.020	6.1365E−03
0.025	2.8717E−03
0.030	1.4780E−03
0.035	7.8421E−04
0.040	4.2765E−04
0.045	2.3433E−04
0.050	1.3034E−04
0.055	6.2269E−05
0.060	3.4556E−05

- By summing up all the values, one can get annual rate of exceedance of 0.05 g at the site due to fault source 2. The value obtained is 1.3034×10^{-4}/yr.
- The above procedure can be repeated for various values of PGA. The values of annual rate of exceedance for various values of PGA are given in Table 2.10, and a graph is shown in Fig. 2.29.

- **Total hazard at the Site**
  - To obtain the total hazard due to all the fault sources around the site, one has to sum up all the hazards coming from each fault source. This is shown in Table 2.11.
  - De-aggregation of hazard with respect to each fault source and total hazard is shown in Fig. 2.30.
  - It can be seen from Fig. 2.30 the contribution for the total hazard is mostly coming from hazard due to fault source 1.
  - By using Eqs. (2.4) and (2.5), given the plant operating time (t), one can calculate probability of exceedance and also return period of a certain level of PGA as follows:

$$P[Z > z] = 1 - e^{-\nu(z)t}$$
$$\text{Return Period} = \frac{-t}{\ln[1 - P(Z > z)]}$$

  - The values are provided in Table 4.25 for various plant operating times and also return period against each PGA level (see Fig. 2.31 for the graph).

2.10 Treatment of Uncertainties

When dealing with the parameters whose characteristics are not well understood, it is always better to consider the uncertainty in those parameters to have an idea about the variation in the results. In such situations, the uncertainty at the parameter level should be propagated to get the uncertainty in the overall results. In general, uncertainty can be well treated with probabilistic concepts. In PSHA, the concept of "logic tree" is used to model the uncertainty in various parameters.

Logic tree is a graphical representation of various uncertain parameters with nodes and weighting factors in the form of a tree. In the logic tree method, one needs to identify all the uncertain parameters and variation in each

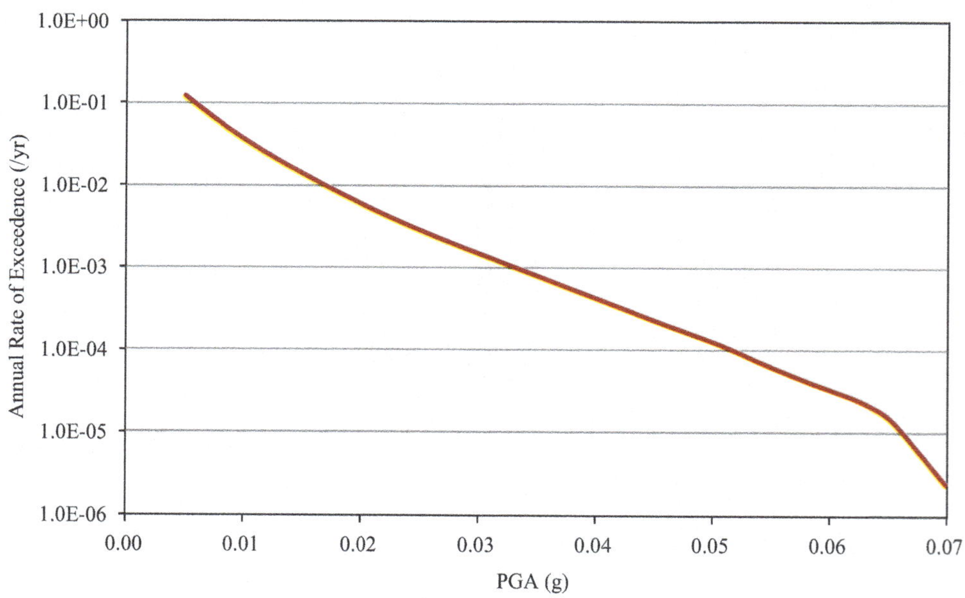

Fig. 2.29 PGA versus annual rate of exceedance due to fault 2

Table 2.11 Annual rate of exceedance with respect to PGA

PGA (g)	Annual rate of exceedance (/year)	Probability of exceedance			Return period (year) for t = 50 year
		t = 1 year	t = 50 year	t = 100 year	
0.005	3.7579E−01	3.1326E−01	1.0000E+00	1.0000E+00	2.6610E+00
0.010	1.8481E−01	1.6874E−01	9.9990E−01	1.0000E+00	5.4109E+00
0.015	1.1294E−01	1.0679E−01	9.9647E−01	9.9999E−01	8.8546E+00
0.020	7.6458E−02	7.3608E−02	9.7814E−01	9.9952E−01	1.3079E+01
0.025	5.4958E−02	5.3475E−02	9.3594E−01	9.9590E−01	1.8196E+01
0.030	4.1264E−02	4.0424E−02	8.7295E−01	9.8386E−01	2.4234E+01
0.035	3.1926E−02	3.1422E−02	7.9735E−01	9.5893E−01	3.1323E+01
0.040	2.5275E−02	2.4958E−02	7.1740E−01	9.2014E−01	3.9566E+01
0.045	2.0384E−02	2.0177E−02	6.3911E−01	8.6976E−01	4.9059E+01
0.050	1.6697E−02	1.6559E−02	5.6607E−01	8.1170E−01	5.9890E+01
0.055	1.3843E−02	1.3747E−02	4.9949E−01	7.4949E−01	7.2241E+01
0.060	1.1612E−02	1.1545E−02	4.4043E−01	6.8688E−01	8.6120E+01

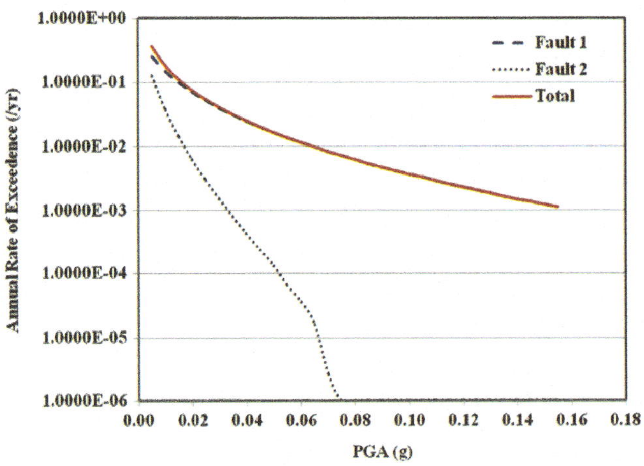

Fig. 2.30 Total hazard curve due to faults 1 and 2

parameter is specified with a certain weighting factor. As a general case, the uncertainty in the following parameters can be represented by using logic tree and is shown in Fig. 2.32.

- Source models,
- Regional recurrence parameters (a, b),
- Maximum magnitude (M_u),
- Apportionment of a and b,
- Depth (km),
- Attenuation relations.

In generating the hazard curve at the site by using logic tree method, one has to generate hazard curve for all the combinations of all the parameters that are presented in logic tree. For each combination of the logic tree, the

Fig. 2.31 PGA versus return period

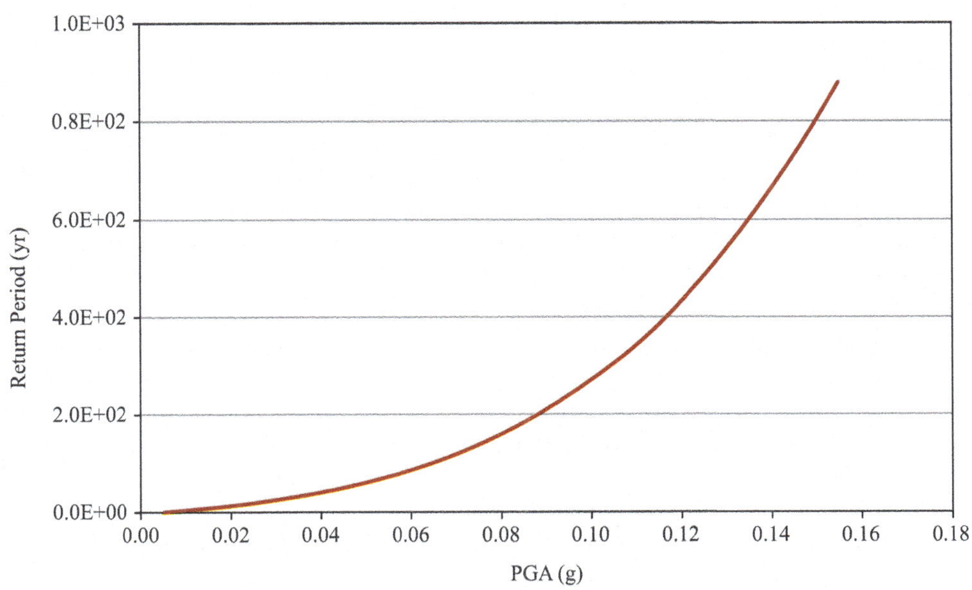

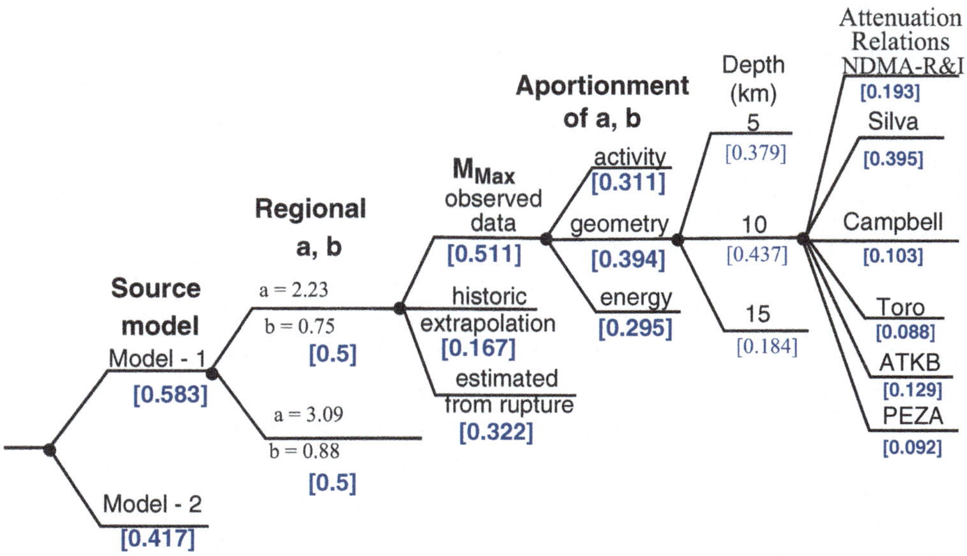

Fig. 2.32 Logic tree for a typical site

Table 2.12 Illustration of development of hazard curves based on logic tree method

Parameters	Combination (weightage)	
	1	2
Source model	Model 1(0.583)	Model 2(0.417)
Regional *a*, *b*	$a = 2.23$, $b = 0.75$(0.5)	$a = 2.23$, $b = 0.75$(0.5)
M_{max}	Observed data (0.511)	Observed data (0.511)
Apportionment of *a*, *b*	Energy (0.295)	Geometry (0.394)
Depth (km)	10 (0.437)	15 (0.184)
Attenuation relation	Campbell (0.103)	Silva (0.395)
Total weightage ($W = w_1 \times w_2 \times \ldots$)	1.9779E−3	3.0510E−3
Annual rate of exceedance for 0.1 g (v)	1.6785E−4	3.0235E−3
Annual rate of exceedance for 0.1 g considering total weightage ($v \times W$)	3.3199E−7 **I**	9.2246E−6 **II**
Annual rate of exceedance for 0.1 g considering combination 1 and 2 (**I** + **II**)		*9.5566E−6*

weightage factor of each parameter is multiplied to get the probability of getting that combination and this will be multiplied with the hazard value (annual rate of exceedance). The final hazard value for a given PGA will be sum of all the hazard values that are obtained with various combinations. This procedure has to be repeated for various PGA values to obtain the final hazard curve. This is explained in Table 2.12 for the logic tree shown in Fig. 2.5. Weightings are given in the brackets.

2.10.1 Uniform Hazard Response Spectrum

This section discusses the generation of response spectrum based on hazard curves that are developed for various frequencies. Uniform hazard response spectrum (UHRS) can be defined as the response spectrum that will have same hazard value at various frequencies. In general, in PSHA, the hazard curves are developed based on PGA or based on spectral acceleration. This can be developed with the help of attenuation relationships in which the values of coefficients change with frequency. So when a attenuation relationship is defined generally, all the coefficients will be defined for various frequencies and a separate case for PGA. It is known that response spectrum is a representation of spectral acceleration of a SDOF system with respect to its frequencies. Hence, one should have the information about the frequency and spectral acceleration. This can be obtained based on hazard curves. UHRS can be developed based on the following steps:

(a) In developing UHRS, first one need to develop hazard curves based on spectral acceleration for various frequencies by using proper coefficients for corresponding frequency in attenuation relationship.
(b) Now, one can plot all the hazard curves in one graph in which the *X*-axis is spectral acceleration and *Y*-axis is the annual rate of exceedance (hazard).
(c) If one is interested in obtaining the response spectrum for a particular hazard value (consider 1E−3/year), then draw a horizontal line parallel to *X*-axis and extract all the spectral accelerations corresponding to each frequency from the graph.
(d) Now, construct a graph between frequency (*X*-axis) and spectral acceleration (*Y*-axis). This will represent UHRS for a hazard value of 1E−3/year or for a return period of 1000 years.
(e) This way one can develop UHRS for various hazard values or for various return periods.

The above procedure is explained in Fig. 2.33. In the case of NPP, one can develop design basis response spectrum for OBE and SSE which are equivalent to UHRS for 100 years and 10,000 years return period, respectively, which is shown in Fig. 2.34 for a typical plant.

Exercise Problems

1. Consider a site around which two earthquake sources are located as shown in Fig. 2.35. Evaluate the design bases ground motion in terms of peak ground acceleration by using various attenuation relationships, and compare the results.
2. The parameters of Gutenberg–Richter's recurrence relationship are given as '*a*' = 4.2 and '*b*' = 0.75. Assume that the lower and upper limits of magnitude as $m_L = 3.0$ and $m_u = 7.0$.

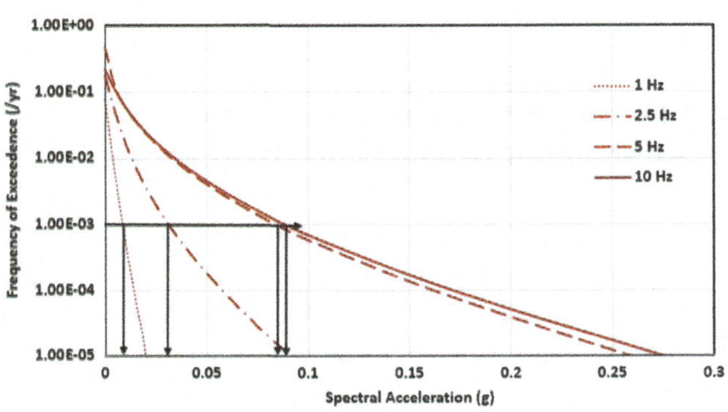

Fig. 2.33 Development of UHRS from hazard curves

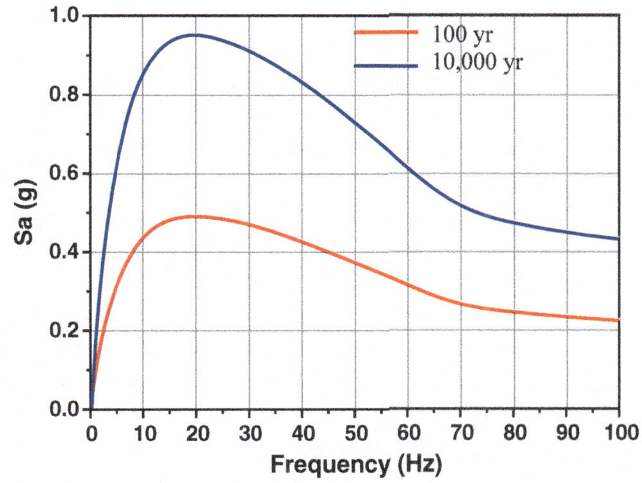

Fig. 2.34 UHRS for 100 and 10,000 years return period

a. Find the probability of occurrence of magnitude between 5.0 and 6.0.

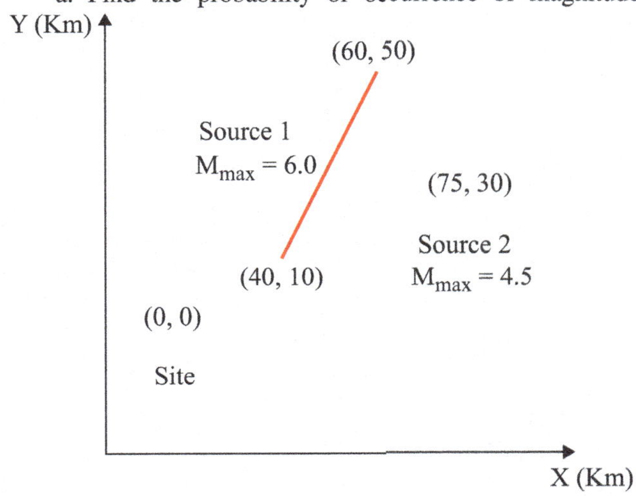

Fig. 2.35 Earthquake sources around the site

b. Find the annual occurrence of earthquakes whose magnitude is greater than 5.0.

3. In a seismically active region, earthquakes have been recorded over a 50 year period. All the recorded data is instrumental data. The distribution of earthquakes is as follows:

Moment magnitude	Number of earthquakes
3–4	1200
4–5	95
5–6	7
>6	2

(a) Estimate the Gutenberg–Richter parameters for the region.

(b) Find the probability of occurrence of magnitude between 4.0 and 5.0.

4. Consider an earthquake of magnitude 6.0 occurred in the intraplate region at a depth of 15 km. Estimate the PGA that would be produced at the site which is located at 50 km away from the fault source by using Silva [8] attenuation relationship. Also, estimate the probability of non exceedance of PGA of 0.05 g at the site.
5. Assume the parameters of Gutenberg–Richter's recurrence relationship as given in problem 2. Assume that the lower and upper limits of magnitude as $m_L = 3.5$ and $m_u = 6.5$. Plot probability density function of magnitude.
6. Derive the probability density function for source-to-site distance for the fault source as shown in Fig. 2.36.
7. Consider one earthquake source (line source,) which is located at a certain distance from the site under consideration as shown in Fig. 2.37. The characteristics of the fault are provided in the following Table 2.13. Develop the hazard curve for the given site.

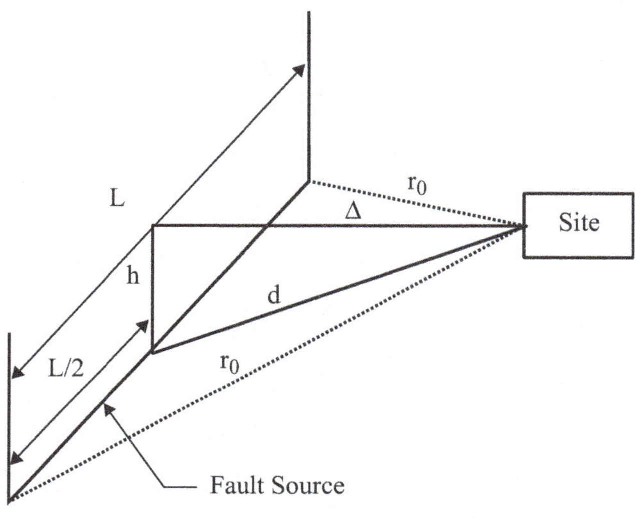

Fig. 2.36 Fault source and site

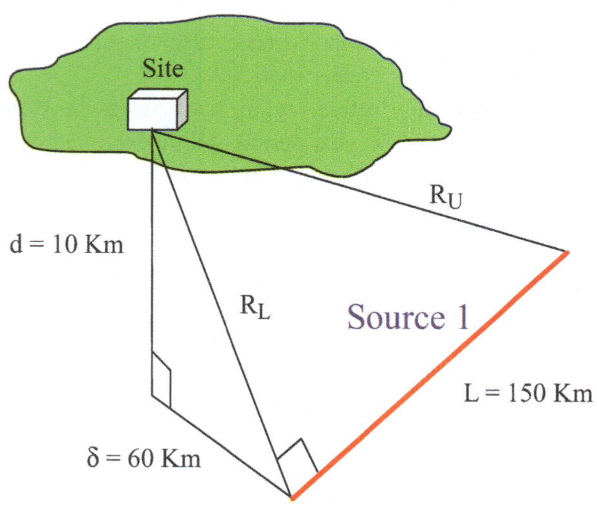

Fig. 2.37 Line source model

Table 2.13 Characteristics of the earthquake source 1

Source	Source type	m_0	m_u	a	b
1	Line	4.0	7.0	4.51	0.95

Appendix 1: Development of New Attenuation Relationships

A.1 Introduction

As explained in Sect. 2.7, there are two approaches available for developing attenuation relationships. First one is regression from strong motion database, and the second one is from simulation. The steps involved in development of new attenuation relationships are given below:

Step 1: Obtain acceleration time histories from either earthquake records or from simulation for various magnitudes and distances.
Step 2: Evaluate response spectra corresponding to required damping from the available acceleration time histories.
Step 3: For each frequency of the response spectrum, segregate the spectral acceleration data for all magnitudes and distances.
Step 4: Select a suitable form for attenuation relationship.
Step 5: Evaluate the coefficients of attenuation relationship for each frequency using the regression procedure explained in subsequent section.
Step 6: Evaluate the standard deviation for each frequency

A.2 Details of Evaluation of Coefficients of Attenuation Relationship Using Regression

After segregating the spectral acceleration for all magnitudes and distances corresponding to each frequency and selection of the form, the coefficients of attenuation relationship are evaluated using regression as described below:

Step 1: The first step in regression is to obtain a system of linear equations from the data and attenuation relationship form.
Step 2: Solution of set of linear equations to obtain the coefficients of attenuation relationship.
Step 3: Evaluate the standard deviation of the data.

Table 2.14 Sample data of PGA for various magnitudes and distances

S. No. (i)	Magnitude (M_i)	Distance (R_i) in kms	PGA_i(g)	log(PGA_i)
1	4.5	20	0.001326	-2.8774
2	4.5	80	0.00194	-2.7122
3	4.5	150	0.002218	-2.6540
4	5	30	0.001506	-2.8223
5	5	55	0.001774	-2.7510
6	5.5	70	0.001881	-2.7256
7	6	30	0.001506	-2.8223
8	6	200	0.002346	-2.6297
9	6.5	40	0.001633	-2.7870

This procedure is illustrated in the example given below:

Example A.1 From the data of PGA (g) for different magnitudes and distances given in Table 2.14, evaluate the coefficients of attenuation relationship given below. Also, estimate the standard deviation.

$$\log(\text{PGA}) = C_1 + C_2 M + C_3 M^2 + C_4 \log R$$

Step 1:
The first step in regression is to obtain a system of linear equations.

$$AX = B$$

In which A, B, *and* X *are* given as follows:

$$A = \begin{bmatrix} 1 & M_1 & M_1^2 & \log R_1 \\ 1 & M_2 & M_2^2 & \log R_2 \\ 1 & M_3 & M_3^2 & \log R_3 \\ 1 & M_4 & M_4^2 & \log R_4 \\ 1 & M_5 & M_5^2 & \log R_5 \\ 1 & M_6 & M_6^2 & \log R_6 \\ 1 & M_7 & M_7^2 & \log R_7 \\ 1 & M_8 & M_8^2 & \log R_8 \\ 1 & M_9 & M_9^2 & \log R_9 \end{bmatrix} = \begin{bmatrix} 1.0000 & 4.5000 & 20.2500 & 1.3010 \\ 1.0000 & 4.5000 & 20.2500 & 1.9031 \\ 1.0000 & 4.5000 & 20.2500 & 2.1761 \\ 1.0000 & 5.0000 & 25.0000 & 1.4771 \\ 1.0000 & 5.0000 & 25.0000 & 1.7404 \\ 1.0000 & 5.5000 & 30.2500 & 1.8451 \\ 1.0000 & 6.0000 & 36.0000 & 1.4771 \\ 1.0000 & 6.0000 & 36.0000 & 2.3010 \\ 1.0000 & 6.5000 & 42.2500 & 1.6021 \end{bmatrix}$$

$$B = \begin{bmatrix} \log(\text{PGA})_1 \\ \log(\text{PGA})_2 \\ \log(\text{PGA})_3 \\ \log(\text{PGA})_4 \\ \log(\text{PGA})_5 \end{bmatrix} = \begin{bmatrix} -2.8774 \\ -2.7122 \\ -2.6540 \\ -2.8223 \\ -2.7510 \\ -2.7256 \\ -2.8223 \\ -2.6297 \\ -2.7870 \end{bmatrix}$$

$$X = \begin{bmatrix} C_1 \\ C_2 \\ C_3 \\ C_4 \end{bmatrix}$$

Step 2: Obtain the least square solution for the set of linear equations, by solving
$X = A^{-1}B$ and X can be obtained as:

$$X = \begin{bmatrix} -3.2728 \\ 0.0325 \\ -0.0029 \\ 0.2444 \end{bmatrix}$$

Step 3: Evaluate the standard deviation of log(PGA), which is obtained as 0.0816.

References

1. AERB Safety Guide SG/S-11 (1990) Seismic studies and design basis ground motion for nuclear power plant site. AERB, Mumbai, India
2. Kramer SL (2003) Geotechnical earthquake engineering. University of Washington, Prentice-Hall International Series in Civil Engineering and Engineering Mechanics
3. Verma AK, Srividya A, Hari Prasad M (2015) Risk management of non-renewable energy systems, 1st edn. Springer series in Reliability Engineering, Springer International Publishing, Switzerland
4. Boore DM (2003) Simulation of ground motion using the stochastic method. Pure Appl Geophys 160:635–676
5. Atkinson GM, Boore DM (2011) Modifications to existing ground-motion prediction equations in light of new data. Bull Seismol Soc Am 101(3):1121–1135
6. Raghukanth STG and Iyengar RN (2010) Technical report of the working committee of experts (WCE) of NDMA on Development of probabilistic seismic hazard map of India
7. Campbell KW (2003) Prediction of strong ground motion using the hybrid empirical method and its use in the development of ground-motion (attenuation) relations in Eastern North America. Bull Seismol Soc Am 93(36):1012–1033
8. Silva W, Gregor N, Darragh R (2002) Development of regional hard rock attenuation relations for Central And Eastern North America. Unpublished report, Pacific Engineering and Analysis, El Cerrito, California
9. Pezeshk S, Zandieh A, Tavakoli B (2011) Hybrid empirical ground-motion prediction equations for Eastern North America using NGA models and updated seismological parameters. Bull Seismol Soc Am 101(4):1859–1870
10. Toro GR, Abrahamson NA, Schneider JF (1997) Model of strong ground motions from earthquakes in Central and Eastern North America: best estimates and uncertainties. Seismol Res Lett 68 (1):41–57
11. Boore DM, Joyner WB, Fumal TE (1997) Equations for estimating horizontal response spectra and peak acceleration from western North American earthquakes: A summary of recent work. Seismol Res Lett 68(1):128–153

Further Reading

12. Bommer JJ, Scott SG, Sarma SK (2000) Hazard-consistent earthquake scenarios. Soil Dyn Earthquake Eng 19:219–231
13. EPRI NP-4726 (1991) Seismic hazard methodology for the Central and Eastern United States. EPRI
14. IAEA-SSG-9 (2010) Seismic hazards in site evaluation for nuclear installations. International Atomic Energy Agency, Vienna
15. IAEA-TECDOC-724 (1993) Probabilistic safety assessment for seismic events. International Atomic Energy Agency, Vienna
16. Kennedy RP, Cornell CA, Campbell RD, Kaplan S, Perla HF (1980) Probabilistic seismic safety study of an existing nuclear power plant. Nucl Eng Des 59:315–338
17. Kijko A, Dessokey MM (1987) Application of the extreme magnitude distributions to incomplete earthquake files. Bull Seism Soc Am 77(4):1429–1436
18. Kijko A, Sellevoll MA (1989) Estimation of earthquake hazard parameters from incomplete data files Part I. Utilisation of extreme and complete catalogs with different threshold magnitudes. Bull Seism Soc Am 79(3):645–654
19. Kijko A, Sellevoll MA (1992) Estimation of earthquake hazard parameters from incomplete data files Part II. Incorporation of magnitude heterogeneity. Bull Seism Soc America 82(1):120–134
20. Reiter L (1991) Earthquake hazard analysis: issues and insights. Columbia University Press, New York
21. McGuire RK (1995) Probabilistic seismic hazard analysis and design earthquakes: closing the loop. Bull Seism Soc Am 85 (5):1275–1284
22. NUREG/CR-6372 (1997) Recommendations for PSHA. USNRC
23. Bazzurro P, Allin Cornell C (1999) Disaggregation of seismic hazard. Bull Seism Soc Am 89(2):501–520
24. RG-1.165 (1997) Identification and characterization of seismic sources and determination of safe shutdown earthquake ground motion. USNRC

Introduction to Structural Dynamics and Vibration of Single-Degree-of-Freedom Systems

3

M. Eswaran, Y. M. Parulekar, and G. R. Reddy

Simplest Mathematical way of representing system is SDOF

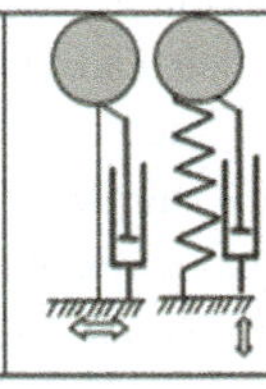

Symbols

M	Mass
K	Lateral stiffness
f_d	Damping Force
F_s	Spring Force
T	Total kinetic energy
V	Total potential energy
δ	Work done by non-conservative system
W_{nc}	Work done by non-conservative system
ω	Natural frequency of vibration
C_{cr}	Critical damping coefficient
A_X	Amplitude
ψ	Phase angle
x_{st}	Maximum value of static deformation
R_d	Dynamic response factor
E_d	Energy dissipated per cycle for linear viscous damping
ζ	Damping coefficient
γ and β	Newmark factor
τ	Extended time interval
$\ddot{x}$	Acceleration
$\Delta\tilde{F}$	Effective load
$\tilde{K}$	Effective stiffness matrix
σ_{max}	Maximum stress
$\dot{x}$	Velocity
Δt	Time step
a0, a1, a2, a3, a4, a5, a6, a7 and a8	Integration constants
AD	Amplitude decay
M_{max}	Bending moment
ω	Frequency
Δx	Incremental displacement
t	Time
t_d	Load duration
SDOF	Single degree of freedom
PE	Period elongation
$F_s(t)$	Elastic force
$F(t)$	Transient force
x	Displacement

M. Eswaran · Y. M. Parulekar · G. R. Reddy (✉)
Bhabha Atomic Research Centre, Mumbai, India
e-mail: rssred@barc.gov.in

M. Eswaran
e-mail: eswaran@barc.gov.in

Y. M. Parulekar
e-mail: yogitap@barc.gov.in

G. R. Reddy et al., *Textbook of Seismic Design*,
https://doi.org/10.1007/978-981-13-3176-3_3

3.1 Introduction

The structural dynamics topic is vital in the design and retrofit of structures to withstand severe dynamic loading due to earthquakes, strong winds, or to identify the occurrence and location of damage within an existing structure. The variation of response quantities such as displacement, velocity, acceleration, or forces of any system (e.g., structures, mechanical components, piping systems, rotating machines, machine tools subjected to time-varying loads) with respect to time is called vibration or dynamic behavior. The causes or types of vibrations in any structure are grouped into two categories, i.e., natural vibrations and man-made vibrations. Earthquakes, vibrations due to wind blow, tsunami, etc., come under natural vibrations, whereas vibrations of components, machines, musical instruments, automobiles, etc., come under man-made vibrations.

3.2 Different Kinds of Load on Structures

The frequency of the external force matches with one of the natural frequencies of the structures, and then resonance occurs which leads to large oscillations of structure. Failures of structures such as buildings, water pools, dams, bridges, rotors, and airplane wings have been associated with the occurrence of resonance in general and sever magnitudes of vibration in some cases. Different kinds of loads can be seen in day-to-day life. Here, few examples are discussed as follows:

a. In a rainy and windy day when we go out with our umbrella, the wind tries to blow the umbrella. When we resist this, the wind produces a lot of turbulence. With this, the umbrella vibrates as depicted in Fig. 3.1a.
b. The aircraft wing is designed to provide sufficient lift that enables aircraft to fly. But when it flies in turbulence, wind creates large vibrations as shown in Fig. 3.1b.
c. The tall chimneys vibrate due to the wind-induced oscillations, in the direction perpendicular to the flow of wind (Fig. 3.1c).
d. When a building structure is subjected to ground motion due to earthquake, the supported equipment, persons, and building itself start vibrating as represented in Fig. 3.1d.
e. In a pump as shown in Fig. 3.1e when there is some mass imbalance, it causes vibrations and needs to be cushioned by springs if the others near machines are to be saved.
f. In case of cross-flow heat exchangers, the fluid in the shell side flows perpendicular to the fluid in the tube side as shown in Fig. 3.1f. The cross-flow over the tubes causes vertices. These vertices cause the tubes to vibrate.

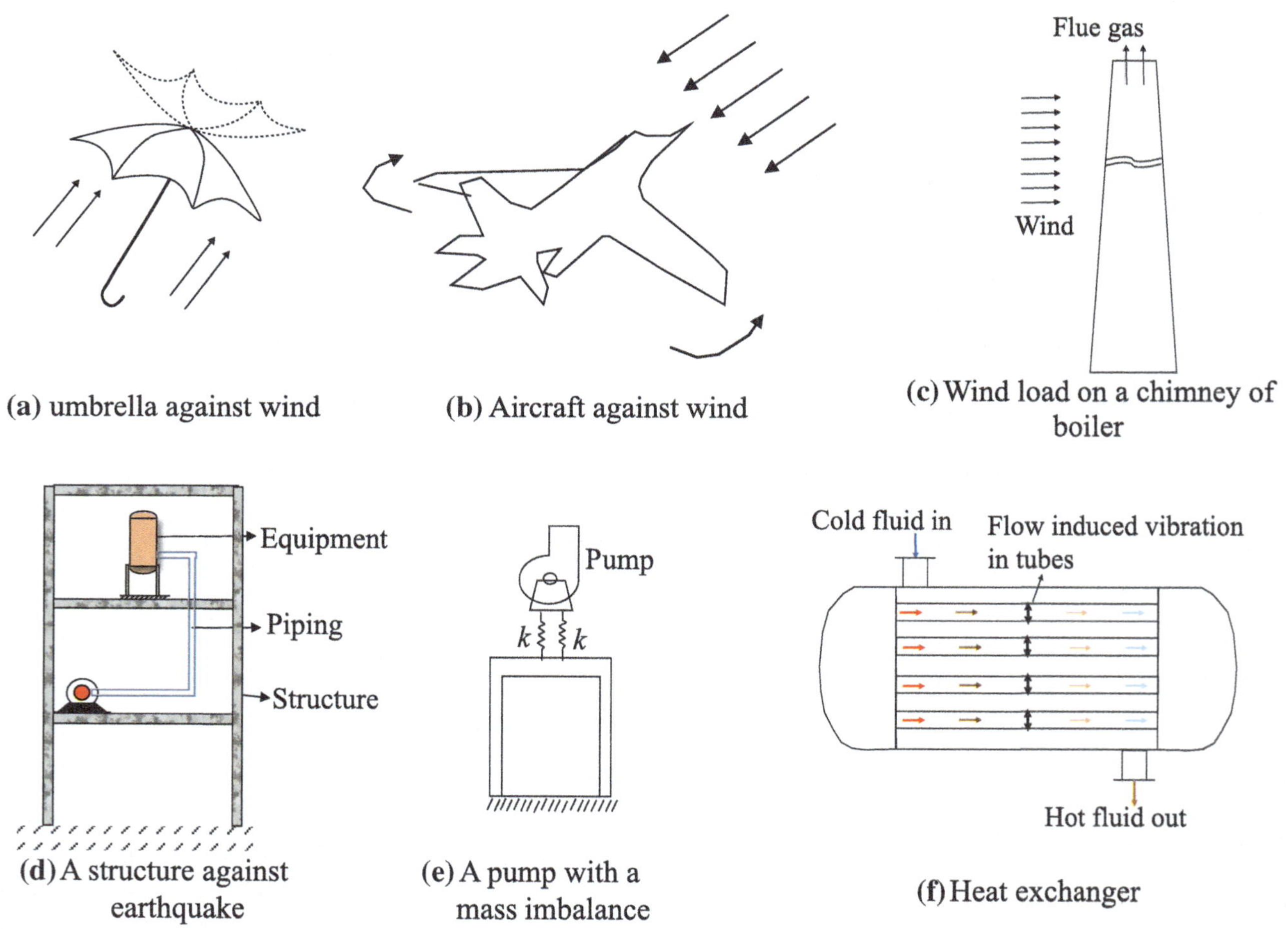

Fig. 3.1 Different kinds of loads

The frequency of vibration of these tubes is strongly related to the frequency of the vertices.

3.3 Idealization of Structures into SDOF Systems

To understand the behavior of structures subjected to dynamic loads, it has to be idealized into a mathematical model. The mathematical models are represented with differential equations which are continuous or discrete. The continuous differential equations are generally applicable for continuous structure which has uniform distributed mass, stiffness, and damping. However, general structures have complex mass, stiffness, and damping distribution. In this situation, mathematical models are represented as discrete differential equations, and mass, stiffness, and damping are lumped at various discrete points with different coordinates. Each coordinate will have time-dependent variables such as displacement, velocity, and acceleration. These further have three translations and three rotations which are defined as degree of freedom (DOF). Hence, if complex structure is represented with large coordinates, it will have large degrees of freedom which is called as a multi-degree of freedom system (MDOF).

However, simple structures shown in Fig. 3.2a and b can be idealized as a concentrated or lumped mass '*m*' supported by a massless structure with stiffness '*k*' in lateral direction. If these structures are subjected to horizontal motion, the mass will have one translation along the direction of motion ([1, 2]). These types of systems are called as a single-degree-of-freedom system (SDOF). A SDOF system requires only one coordinate to describe its position at any instant of time.

The SDOF model and corresponding equation of motion are shown in Fig. 3.3a–c. These are the cases when the damping of the SDOF systems is not considered. However, every structure has damping which depends on material, supports, etc. The cases with damping are dealt in the Subsect. 3.3.2.

3.3.1 Undamped Systems

If no energy is lost in the material and supports of the structure during vibration, the structure is known as undamped structure. If any energy is lost during vibration, it is called damped vibration. Consider a SDOF system as shown in Fig. 3.4 with no external force acting on it. This system can be idealized as a spring–mass system having lateral stiffness k and mass m as shown in Fig. 3.4. Differential equation governing lateral displacement $x(t)$ of this idealized system is given in Eq. (3.1) assuming that the lateral motion of the structure is small and within their elastic limit.

$$m\ddot{x}(t) + kx(t) = 0 \tag{3.1}$$

If such a system is given an initial displacement of $x(0)$ and then released to vibrate freely, the structure will ideally oscillate or vibrate back and forth about its initial equilibrium position as shown in Fig. 3.5a. However, in reality the structure would oscillate with ever decreasing amplitude and come to rest eventually. Thus, the motion of the structure will decay with time as shown in Fig. 3.5b.

3.3.2 Damped Systems

The process by which vibration steadily decreases in amplitude is called damped free vibration. The irreversible energy dissipated from the system results into these characteristics, and the corresponding energy is nothing but damping energy. In vibrating building structures or equipment, energy is dissipated in friction at steel connections, opening and closing of micro-cracks, friction between structure and non-structural elements like partition walls, bolted connections.

The dissipation of energy from the vibrating structure or part of vibrational energy is gradually transformed into heat or sound which is known as damping. Damping is classified as follows,

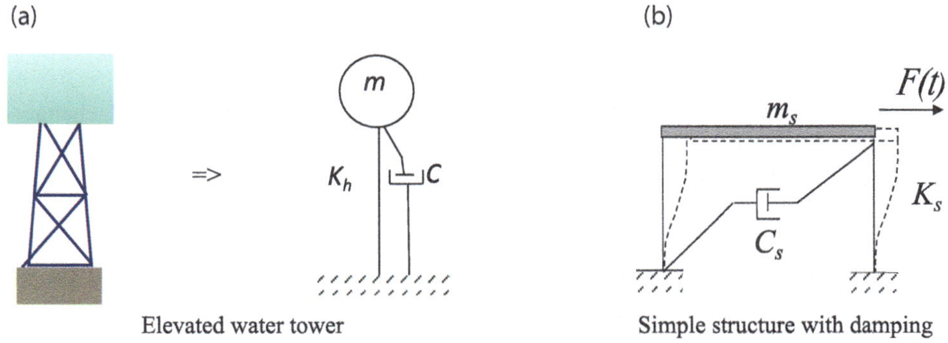

Fig. 3.2 Simple structures

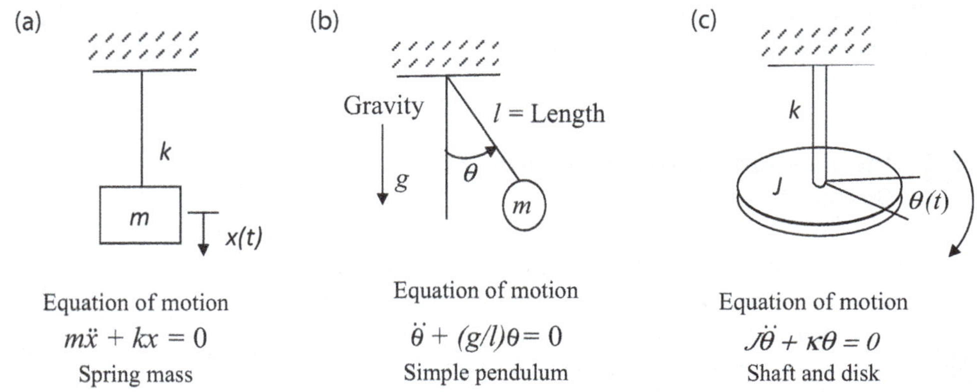

Fig. 3.3 SDOF models without damping

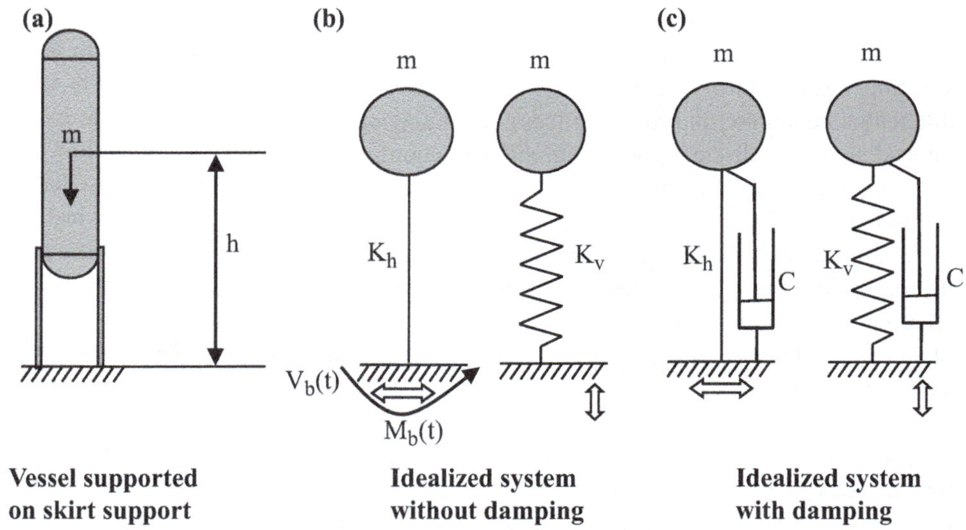

Fig. 3.4 Idealization of a vessel supported on skirt support as a SDOF

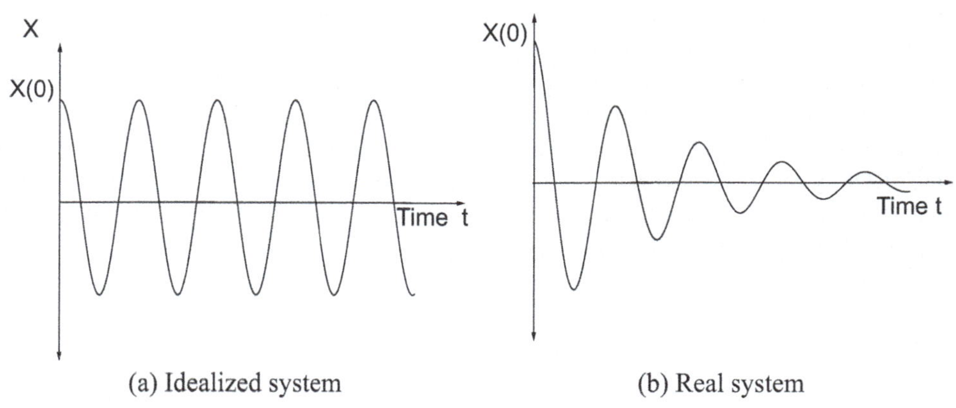

Fig. 3.5 Free vibration

(i) Viscous damping—Viscous fluid flows through a gap,
(ii) Hysteretic or material damping—The friction between two internal planes that slip or slide as the material deforms,
(iii) Dry friction of coulomb damper—When two bodies are in contact, the force required to produce sliding is proportional to the normal force acting in the plane of contact.

The hysteretic damping will change depending on the inertial and elastic condition. However, the work done will be equal to energy dissipated by damping. Figures 3.6 and 3.7 show the hysteretic damping at joints for steel and concrete.

It is not possible to identify or describe mathematically each of these energy-dissipating mechanisms. However, it is common practice to express the damping force in a structure

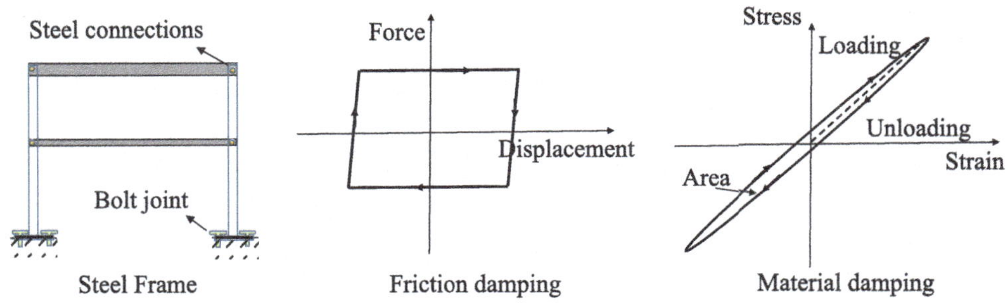

Fig. 3.6 Damping for steel structures

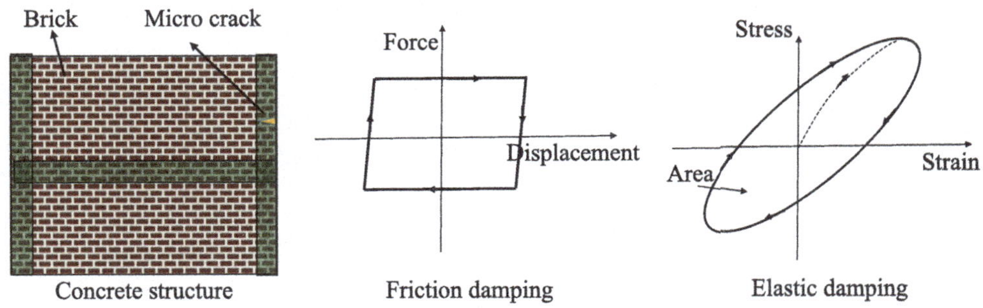

Fig. 3.7 Damping for concrete structures and non-structure elements

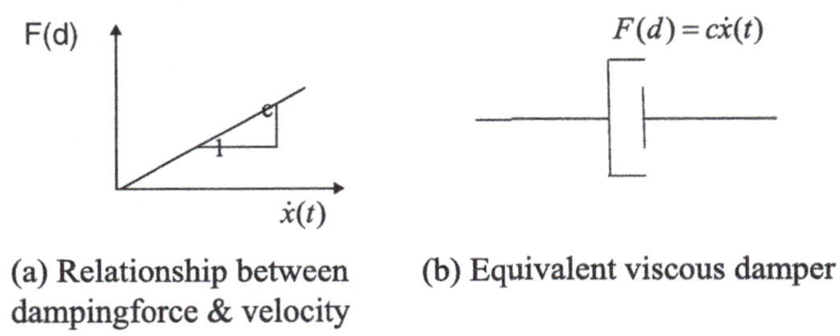

Fig. 3.8 Viscous damping force

proportional to the viscous damping force *F(d)* given as, $c\dot{x}(t)$, where *c* is the viscous damping coefficient of a structure. This force is proportional to the velocity of the structures as shown in Fig. 3.8. The procedure to evaluate the damping coefficient is explained in Sect. 3.6.

The equation of motion for free vibration of simple structure with mass m, lateral stiffness *k*, and damping coefficient c is given by Eq. (3.2).

$$m\ddot{x}(t) + c\dot{x}(t) + kx(t) = 0 \quad (3.2)$$

3.4 Structure Under Dynamic Loading

Almost all structures systems may be subjected to one form or another form of dynamic loading during its lifetime. Dynamic means time-varying load; therefore, the structural response to a dynamic load (i.e., resulting stresses and deflections) is also time-varying. Problems of dynamics can be categorized according to the type of loading. Dynamic loads can be periodic or non-periodic. Periodic loads are further classified as harmonic load or random loads. Examples for the periodic vibration are flow-induced vibration, wind loads with certain return period, machine vibration loads, etc. Figure 3.9a–c shows different types of loadings.

Periodic loading exhibits same time variation successively for large number of cycles, e.g., simple harmonic. Non-periodic loading may be short duration impulsive loadings (e.g., blast loading) or long duration general forms of loads (e.g., earthquake loading). Problems of blast loadings are wave propagation problems and can be solved by special simplified forms of analysis. Problems of earthquake loadings are structural dynamic problems and can be solved completely by general dynamic analysis procedures. Thus, a dynamic analysis becomes more complex and time-consuming in nature.

Fig. 3.9 Different types of loading

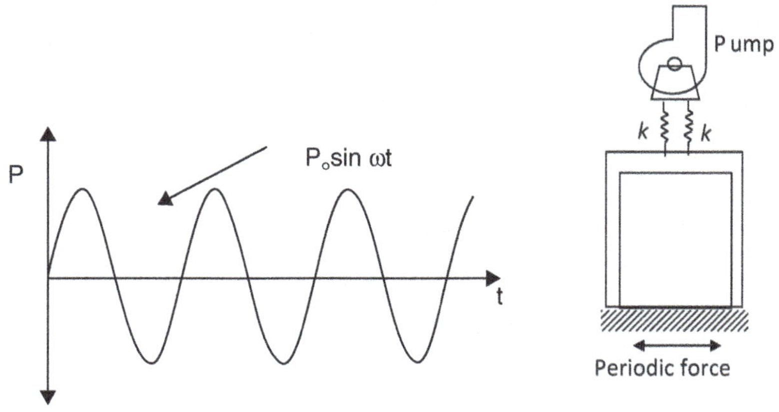

(a) Periodic (sinewave) loading

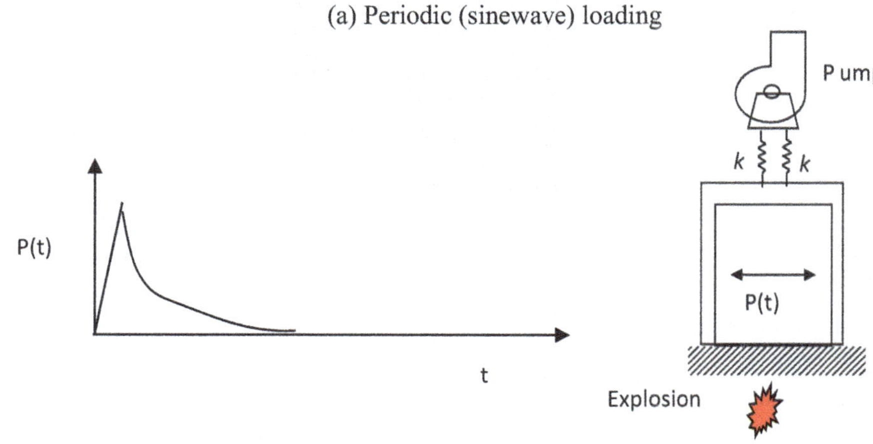

(b) Shock loading (explosion)

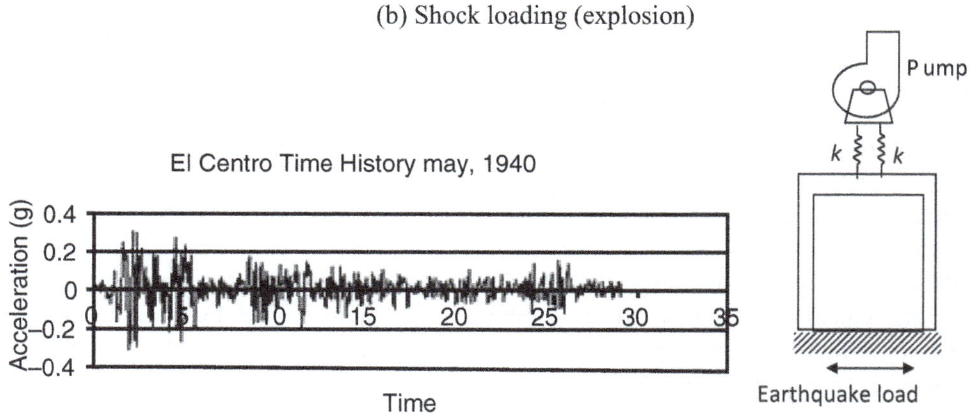

(a) Random loading (Earthquake)

3.5 Equation of Motion of SDOF Dynamic System

Consider a SDOF system subjected to externally applied dynamic force $F(t)$ in the direction of DOF (x) as shown in Fig. 3.10. The differential equation governing the displacements $x(t)$ can be derived by using Newton's second law of motion and D'Alembert's principle.

3.5.1 Formulation of the Equation of Motion

The mathematical expressions defining the dynamic displacements are called the equations of motion of the

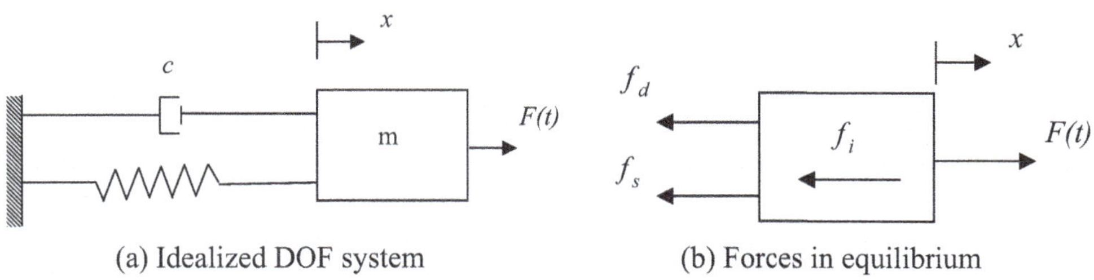

Fig. 3.10 SDOF system subjected to dynamic loading

structure. Formulation equations of motion can be derived from following three methods [3].

1. Direct equilibrium using D'Alembert's principle,
2. Principle of virtual displacement, and
3. Hamilton's principle.

3.5.2 Direct Equilibrium Using D'Alembert's Principle

The equation of motion of any dynamic system represents expression of Newton's second law of motion, which states that the rate of change of momentum of any mass m is equal to the force acting on it. This relationship can be stated mathematically as the differential equation,

$$F(t) = \frac{\mathrm{d}}{\mathrm{d}t}\left(m\frac{\mathrm{d}x}{\mathrm{d}t}\right) \tag{3.3}$$

where $F(t)$ is the applied force vector and $x(t)$ is the position vector of mass m. Generally, mass does not vary with time in structure mechanics. Equation (3.3) can be written as

$$F(t) = m\frac{\mathrm{d}^2x}{\mathrm{d}t^2} = m\ddot{x}(t), \tag{3.4}$$

when dot represents differentiation with respect to time. Again Eq. 3.4 can be written as,

$$F(t) - m\ddot{x}(t) = 0 \tag{3.5}$$

The second term $m\ddot{x}(t)$ in above Eq. (3.5) is called as inertia term, resisting the acceleration of the mass. A mass develops an inertia force proportional to its acceleration, and opposing force is known as D'Alembert's principle. The force $F(t)$ may be considered as many types of force acting on the mass such as elastic constrains, viscous force.

The forces acting in the direction of displacement DOF include the applied force $F(t)$ and three forces resulting from the motion, Inertia $F_i(t)$, damping $F_d(t)$, and elastic spring force $F_s(t)$. The equation of motion thus is merely an expression of the equilibrium of these forces, as follows,

$$F_i(t) + F_s(t) + F_d(t) = F(t) \tag{3.6}$$

As shown in Fig. 3.10b, forces are replaced as $F_s(t) = kx$, $F_i(t) = m\ddot{x}$, and $F_d(t) = c\dot{x}(t)$. In the case of earthquake analysis, the external force $F(t)$ can be changed to $-m\ddot{x}_g(t)$ expression, in which the negative sign indicates that the effective force opposes the sense of ground acceleration. Therefore, the equation of motion for a SDOF system shown in Fig. 3.10 subjected to earthquake load, where m is the mass, k is the spring stiffness, and c is the damping coefficient of the system can be represented as Eq. (3.7).

$$m\ddot{x}(t) + kx(t) + c\dot{x}(t) = -m\ddot{x}_g(t) \tag{3.7}$$

3.5.3 Principle of Virtual Displacement

The direct equilibrium of all forces acting on a structural system may be difficult, if the system is reasonably intricate and involved a number of interconnected mass points or bodies of finite size. Often, various forces involved may readily be expressed in terms of the displacement DOF, but their equilibrium relationships may be obscure. In this case, to formulate the equilibrium of motion as a substitute for the equilibrium relationships, the principle of virtual displacements could be used.

The total work done by force is zero, if the system which is equilibrium under the action of a set of forces subjected to a virtual displacement, i.e., any displacement compatible with system constrains. As per this principle, vanishing effect of the work done during a virtual displacement is equivalent to a statement of equilibrium.

The equations of motion are acquired by introducing the virtual displacements corresponding to each DOF and equating the work done to zero. The major benefit of this method is that the virtual work contributions are scalar quantities and can be added algebraically, when as force acting on the structure are vectorial and can be superposed vectorially.

If the mass is given a virtual displacement δx (the only displacement compatible with constrains), these forces will each do work. The total work done by the system can then be written as

$$\begin{gathered} -F_i(t)\delta x - F_s(t)\delta x - F_d(t)\delta x + F(t)\delta x = 0 \text{ or} \\ (-m\ddot{x}(t) - kx(t) - c\dot{x}(t) - m\ddot{x}_g(t))\delta x = 0 \end{gathered} \tag{3.8}$$

Since δx is nonzero, this can be easily put in the form of Eq. (3.7).

3.5.4 Hamilton's Principle

The most generally applicable variational concept is Hamilton's principle, which may be expressed as

$$\int_{t1}^{t2} \delta(T-V)\mathrm{d}t + \int_{t1}^{t2} \delta W_{nc}\mathrm{d}t = 0 \tag{3.9}$$

where T and V are total kinetic energy of the system and potential energy of the system. The W_{nc} and δ are work done by non-conservative forces acting on system, including damping and any arbitrary external loads and variation taken during indicated time interval, respectively. This uses scalar energy quantities in a variational form. Hamilton's principle states that the variation of the kinetic and potential energy plus the variation of the work done by the non-conservative forces considered any time interval t_1 to t_2 must equal zero. The application of this principle yields to the derivation of equation of motion for a system. This formulation has the advantage of dealing only with purely scalar energy quantities, whereas the forces and displacements used to represent corresponding effects in the virtual work analysis are all vectorial characters even though the work terms themselves are scalar.

The kinetic and potential energy of the system, by definition is given by $T = \frac{1}{2}mv^2$ and $V = \frac{1}{2}kx^2$, respectively. The non-conservative forces of the system are the damping force F_d and applied force F. The variation of the work done by these forces may be expressed as

$$\delta W_{nc} = F(t)\,\delta x - c\dot{x}\,\delta x \tag{3.10}$$

which is equivalent to the virtual work expression associated with these forces in equation. Now applying Hamilton's principle

$$\int_{t1}^{t2} (m\dot{x}\,\delta\dot{x} - c\dot{x}\,\delta x - kx\,\delta x + F(t)\delta x) = 0 \tag{3.11}$$

Now the first term can be integrated by parts as follows,

$$\int_{t1}^{t2} m\dot{x}\,\delta\dot{x}\,\mathrm{d}t = m\dot{x}\,\delta x|_{t1}^{t2} - \int_{t1}^{t2} m\ddot{x}\,\delta x\,\mathrm{d}t \tag{3.12}$$

Since it is assumed in Hamilton's principle that the variation δx vanishes at the limits of integration t_1 and t_2, Eq. (3.12) can be written as

$$\int_{t1}^{t2} (-m\ddot{x} - c\dot{x} - kx + F(t))\delta x\,\mathrm{d}t = 0 \tag{3.13}$$

Then the variation δx is arbitrary, and it is clear that Eq. (3.13) is satisfied only if the expression in brackets vanishes.

3.6 Undamped Free Vibration Using Classical Solution

The free vibration equation for undamped SDOF system shown in Fig. 3.11 is given in Eq. (3.14). Dividing this equation by m, one can get Eq. (3.15).

$$m\ddot{x}(t) + kx(t) = 0 \tag{3.14}$$

$$\ddot{x}(t) + \omega^2 x(t) = 0 \tag{3.15}$$

where $\omega = \sqrt{\frac{k}{m}}$ is the natural frequency of vibration.

Solution of Eq. (3.15) is

$$x(t) = A\cos(\omega t) + B\sin(\omega t) \tag{3.19}$$

Constants A and B are obtained by considering initial conditions. Free vibration is initiated by disturbing the system from its static equilibrium position by giving initial displacement $x(0)$ and initial velocity $\dot{x}(0)$. Substituting the value of $x(0)$ at $t = 0$ in Eq. (3.19), one can get the value of A. Differentiating Eq. (3.19) and then substituting the value of $\dot{x}(0)$ at $t = 0$, one can get the value of B. Therefore,

$$x(0) = A \quad \text{and} \quad \dot{x}(0) = B\omega \tag{3.20}$$

Substituting for A and B from Eq. (3.20) into Eq. (3.19), one can get the solution of Eq. (3.14) as

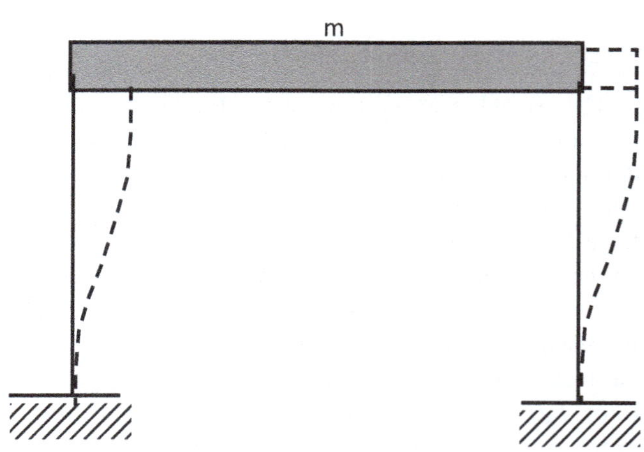

Fig. 3.11 Undamped SDOF system

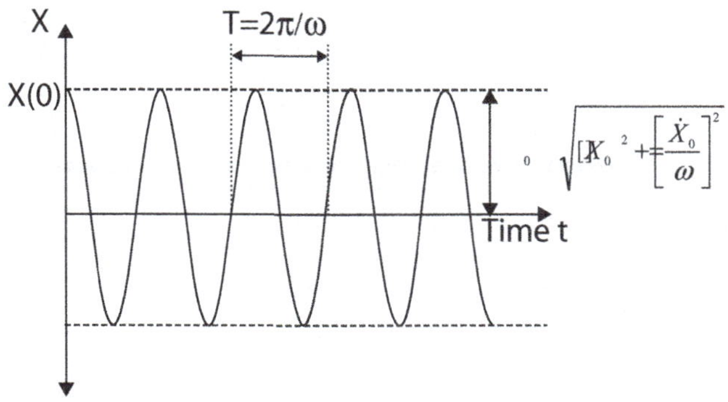

Fig. 3.12 Free vibration displacement response of undamped SDOF system

$$x(t) = x(0)\cos(\omega t) + \frac{\dot{x}(0)}{\omega}\sin(\omega t) \qquad (3.21)$$

Equation (3.21) and Fig. 3.12 show that the system undergoes vibratory motion about its equilibrium position (undeformed), and this motion repeats after $2\pi/\omega$ seconds.

3.7 Harmonic Vibration of Undamped Systems

Equation of motion for undamped SDOF system subjected to harmonic load shown in Fig. 3.13 is given in Eq. (3.22).

$$m\ddot{x}(t) + kx(t) = F_0\sin(\omega t) \qquad (3.22)$$

Solution to Eq. (3.22) is given as follows

$$x(t) = x_c(t) + x_p(t) \qquad (3.23)$$

Complementary solution to Eq. (3.23) is

$$x_c(t) = A\cos(\omega_n t) + B\sin(\omega_n t) \qquad (3.24)$$

Particular solution to Eq. (3.23) is

$$x_p(t) = \frac{F_0\sin(\omega_n t)}{K\left(1 - (\omega/\omega_n)^2\right)} \qquad (3.25)$$

Therefore, complete solution is given by Eq. (3.26)

$$x(t) = A\cos\omega_n t + B\sin\omega_n t + \frac{F_0\sin(\omega_n t)}{K\left(1 - (\omega/\omega_n)^2\right)} \qquad (3.26)$$

Substituting initial displacement $x(0)$ and velocity $\dot{x}(0)$ at $t = 0$ in Eq. (3.26), one can get

$$\begin{aligned}\text{Transient component} &= \left(x(0)\cos\omega_n t + \left[\frac{\dot{x}(0)}{\omega_n}\frac{F_0}{K}\frac{\omega/\omega_n}{(1-(\omega/\omega_n)^2)}\right]\sin(\omega_n t)\right)\\ &+\\ \text{Steady state component} &= \left(\frac{F_0\sin(\omega_n t)}{K(1-(\omega/\omega_n)^2)}\right)\end{aligned} \qquad (3.27)$$

The first part of Eq. (3.27) is called transient vibration which depends on the initial conditions, i.e., $x(0)$ and $\dot{x}(0)$. The latter part is called steady-state vibration as it is present because of the applied force, no matter what the initial conditions are. When there is no damping, transient component continues forever [4]. But in real systems, damping is inevitably present, and hence, the transient component vanishes with time and only steady-state component remains.

Example 3.1 A crane (rigid bar with mass m is hinged at one end and supported on two springs) of stiffness K_1 (N/m) at distance a (m) from support and K_2 (N/m) at other end as shown in Fig. 3.14. The total length of the bar is lm. Find the fundamental frequency of the system in Hz. Neglect the effect of left side cable.

Solution

Consider a small displacement x at the end of bar as shown in Fig. 3.14c.

Taking moment about A

$$m\ddot{x}\frac{l}{2} + K_1\frac{a}{l}xa + K_2xl = 0 \qquad (3.28)$$

$$m\ddot{x} + 2x\left(K_1\frac{a^2}{l^2} + K_2\right) = 0 \qquad (3.29)$$

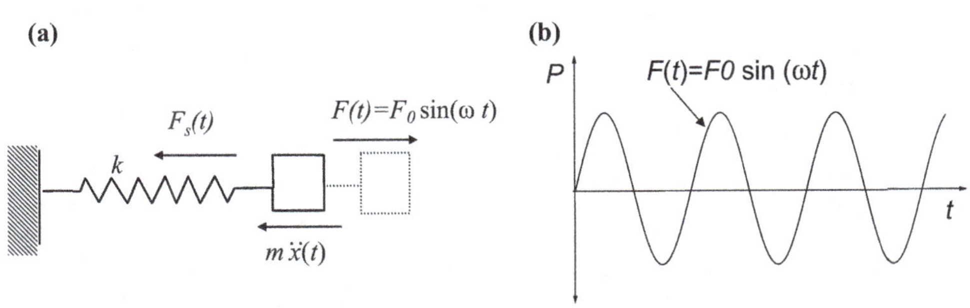

Fig. 3.13 Undamped SDOF system subjected to harmonic loads

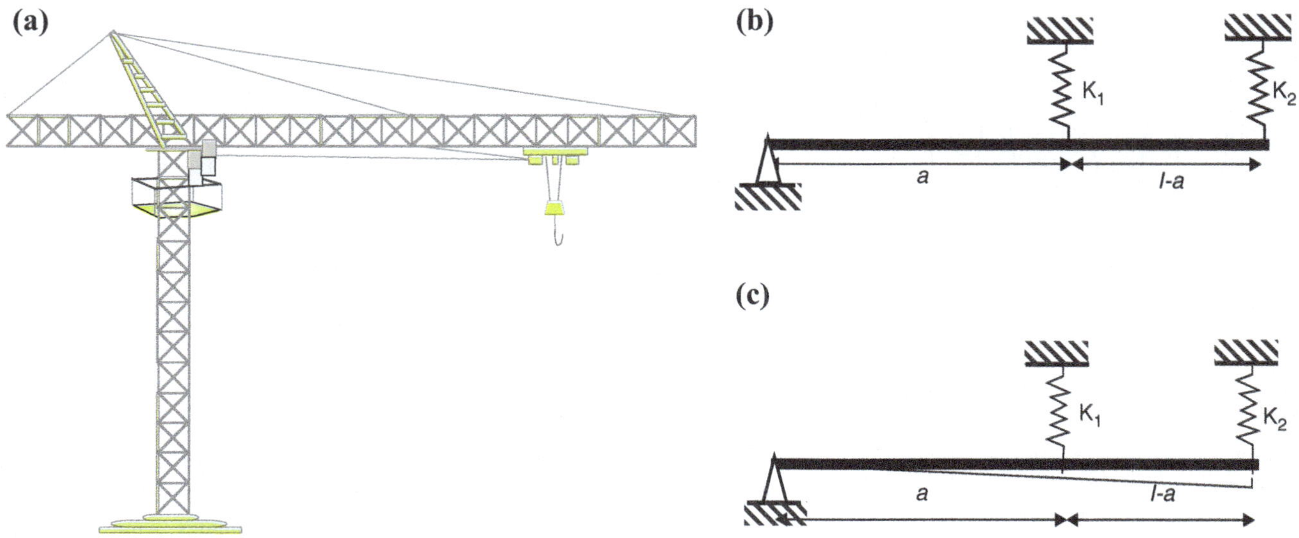

Fig. 3.14 **a** Crane **b** a rigid bar hinged at one end **c** a rigid bar hinged at one end given small displacement

Comparing with $m\ddot{x}(t) + kx(t) = 0$,

$$K = 2\left(K_1 \frac{a^2}{l^2} + K_2\right) \tag{3.30}$$

Therefore,

$$\omega = \sqrt{\frac{2K_1 \frac{a^2}{l^2} + K_2}{m}} \tag{3.31}$$

Frequency of the rigid bar in

$$\text{Hz} = \frac{1}{2\pi}\sqrt{\frac{2\left(K_1 \frac{a^2}{l^2} + K_2\right)}{m}} \tag{3.32}$$

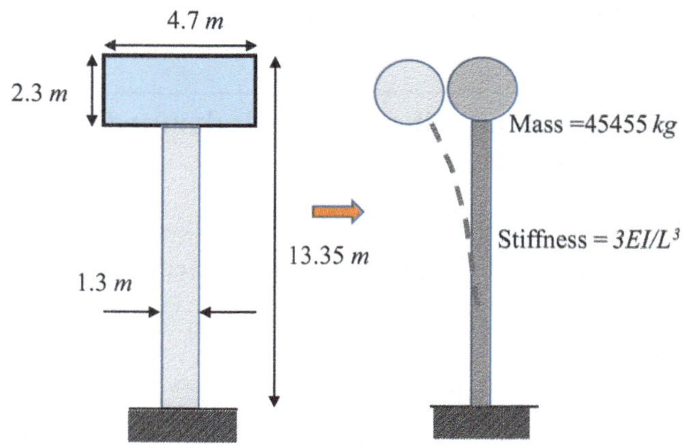

Fig. 3.15 A Water tank model

Example 3.2 A reinforced concrete water tank is situated on a 12.2-m-tall single concrete column as shown in Fig. 3.15. When the tank is completely filled with water, the weight of the tank with water is 45.45 tons. The concrete column is hollow with outer diameter 1.1 m and inner diameter 0.85 m. Calculate the fundamental frequency of the tank.

Solution

Young's modulus of concrete

$$E = 2.5 \times 10^4\ \text{N/mm}^2$$

The tank can be analyzed as a SDOF system with stiffness

$$K = 3EI/L^3$$

Moment of inertia of hollow concrete section

$$I = (\pi/64)\left(d_o^4 - d_i^4\right) = 4.6245 \times 10^{10}\ \text{mm}^4$$

Equivalent stiffness of the tank

$$\begin{aligned} K &= 3 \times 2.5 \times 10^4 \times 4.6245 \times 10^{10}/(13{,}350)^3 \\ &= 1457.74476\ \text{N/mm} = 1{,}457{,}744.76\ \text{N/m} \end{aligned}$$

$$\omega = \sqrt{\frac{k}{m}} = \sqrt{\frac{1{,}457{,}744.76}{45{,}455}} = 0.9\ \text{Hz}$$

The fundamental frequency of the tank is 0.9 Hz.

Example 3.3 A tank is mounted on the steel saddle supporting structure (as shown in Fig. 3.16) and which in turn supported on concrete walls. The mass of the tank is 138.125 tons. 2000 KN force is applied along all the nodes of the tank along X-direction and it is found that the

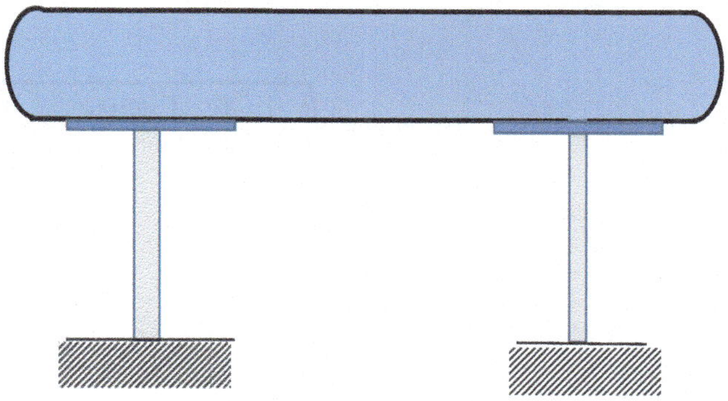

Fig. 3.16 A water tank model on the steel saddle supporting structure

deflection of the tank in that direction is 6.67 mm. Calculate the fundamental frequency of the tank in X-direction.

Solution

The frequency of the tank with its supporting structure is obtained by assuming the tank as a lumped mass and the supporting structure as a beam.

The stiffness of the tank in X-direction is

$$K_x = 2,000,000/0.00667 = 299.85 \times 10^6\ \text{N/m}.$$

Mass of the tank

$$m = 138,125\ \text{kg}.$$

Therefore, frequency of the tank in X-direction is

$$\omega = \sqrt{\frac{k}{m}} = \sqrt{\frac{299.85 \times 10^6}{138,125}} = 7.41\ \text{Hz}$$

The fundamental frequency of the tank in X-direction is 7.41 Hz.

3.8 Damped Free Vibration Using Classical Solution

The free vibration equation for damped SDOF system shown in Fig. 3.16 is given in Eq. (3.33). Dividing this equation by m, one can get Eq. (3.34).

$$m\ddot{x}(t) + c\dot{x}(t) + kx(t) = 0 \quad (3.33)$$

$$\ddot{x}(t) + 2\varsigma\omega\dot{x}(t) + \omega^2 x(t) = 0 \quad (3.34)$$

Solution to Eq. (3.34) is given by Eq. (3.35)

$$x(t) = e^{st} \quad (3.35)$$

Using Eq. (3.35) and substituting in Eq. (3.34), one can get

$$\left(s^2 + 2\zeta\omega_n s + \omega_n^2\right)e^{st} = 0, \quad s = \omega_n\left(-\varsigma \pm i\sqrt{1-\varsigma^2}\right) \quad (3.36)$$

Using the value of s and substituting in Eq. (3.35), $x(t)$ can be represented as Eq. (3.37)

$$x(t) = e^{-\zeta\omega_n t + i\omega_d t} \quad (3.37)$$

where

$$\omega_d = \omega_n\sqrt{1-\zeta^2} \quad (3.38)$$

$$x(t) = e^{-\zeta\omega_n t}(A\cos\omega_d t + B\sin\omega_d t) \quad (3.39)$$

Substituting initial displacement $x(0)$ and velocity $\dot{x}(0)$ at $t = 0$ in Eq. (3.39), one can get,

$$x(t) = e^{-\zeta\omega_n t}\left(x(0)\cos\omega_d t + \frac{x(0) + \zeta x(0)\omega_n}{\omega_d}\sin\omega_d t\right) \quad (3.40)$$

Displacement amplitude of a damped system decreases with every cycle of oscillation, and the amplitude decays exponentially with time as shown in Fig. 3.17.

If $\zeta = 1$, using Eq. (3.37) the solution of Eq. (3.40) will be reduced to

$$x(t) = e^{-\omega_n t} \quad (3.41)$$

Thus, the motion will become exponential and the system returns to equilibrium position without oscillating. The smallest value of damping coefficient which inhibits oscillation of the system is called as critical damping coefficient C_{cr}. Thus, when $C = C_{cr}$, $\zeta = 1$

$$C_{cr}/2m\omega_n = 1; \; C_{cr} = 2m\omega_n \quad (3.42)$$

If $\zeta > 1$ or $C > C_{cr}$, again the system does not oscillate and returns to equilibrium position. The systems for which $\zeta > 1$ are called as over critically damped systems (e.g., a door closer). If $\zeta < 1$ or $C < C_{cr}$, the system oscillates and the amplitude decreases with every cycle of oscillation. The systems for which $\zeta < 1$ are called under critically damped systems (e.g., all structures). It can be inferred from Eq. (3.38) that damping has an effect of lowering the natural frequency from ω_n to ω_d and lengthening the natural period of the structure. However, the range of damping ratio for most of the structures is below 0.2 for which these effects are negligible as shown in Fig. 3.18. Hence for most structures, damped frequency ω_d is approximately equal to undamped natural frequency ω_n.

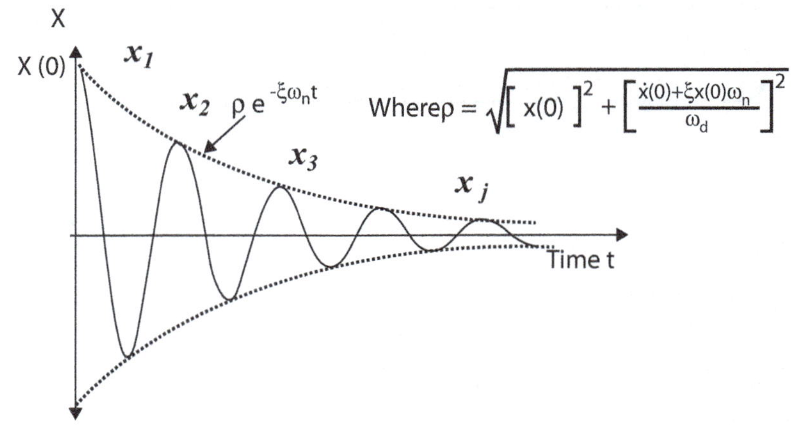

Fig. 3.17 Free vibration displacement response of damped SDOF system

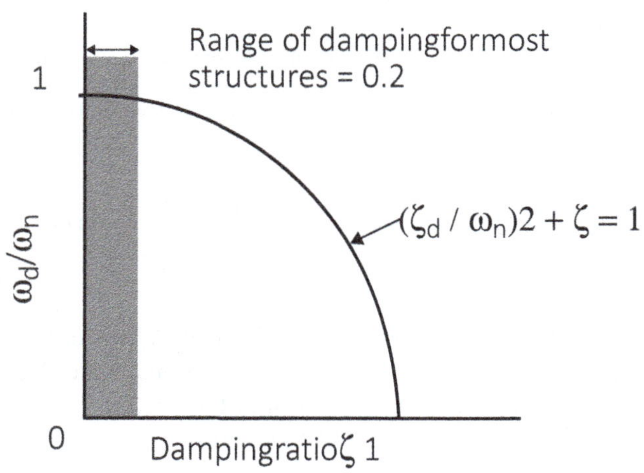

Fig. 3.18 Effects of damping on the natural frequency

3.8.1 Evaluation of Damping by Logarithmic Decrement Method

The displacement amplitude of a damped system follows logarithmic decay as shown in Fig. 3.17. Amplitude decay can be estimated by using Eqs. (3.43) and (3.44). Here x_1, $x_2 \ldots x_{j+1}$ represent the amplitude of every successive cycle.

$$\frac{x_1}{x_{j+1}} = e^{j\xi\omega_n T_D} as\, \omega_n \approx \omega_d \tag{3.43}$$

$$\ln\frac{x_1}{x_{j+1}} = 2\pi\xi j \tag{3.44}$$

3.8.2 Response of a Damped System Under Harmonic Load

Response of a damped system under a harmonic force and base harmonic force cases is discussed in this section. The equation of motion under forcing function can be given as

$$m\ddot{x}(t) + c\dot{x}(t) + kx(t) = F_0\cos(\omega t) \tag{3.45}$$

Assume that the particular solution of Eq. (3.45) as

$$x_p(t) = A_X\cos(\omega t - \varphi) \tag{3.46}$$

where A_X and φ are the amplitude and phase angle constant which need to be estimated. As shown in vector diagram in Fig. 3.19b,

$$A_X(k - m\omega^2)\cos(\omega t - \varphi) - A_X c\omega\sin(\omega t - \varphi) = F_0\cos(\omega t) \tag{3.47}$$

Solution of Eq. 3.47 gives

$$A_X = \frac{F_0}{\left[(k - m\omega^2) + c^2\omega^2\right]^{1/2}}$$

and

$$\varphi = \tan^{-1}\left(\frac{c\omega}{k - m\omega^2}\right) \tag{3.48}$$

Figure 3.19 shows the typical plots of the steady state and forcing function responses.

$$\omega_n = \sqrt{\frac{k}{m}};\ \zeta = \frac{c}{c_c} = \frac{c}{2m\omega_n};\ \delta_{st} = \frac{F_0}{k} \text{ and } r = \frac{\omega}{\omega_n} \tag{3.49}$$

Using the fundamental relations of vibration as given in Eq. (3.49), one can obtain

$$\frac{A_X}{\delta_{st}} = \frac{1}{\left\{\left[1 - \left(\frac{\omega}{\omega_n}\right)^2\right]^2 + \left[2\zeta\frac{\omega}{\omega_n}\right]^2\right\}^{1/2}}$$

and

$$\varphi = \tan^{-1}\left(\frac{2\zeta\frac{\omega}{\omega_n}}{1 - \left(\frac{\omega}{\omega_n}\right)^2}\right) \tag{3.50}$$

The relations mentioned in Eq. (3.50) can be converted into frequency ratio, and the effects of amplitude and phase angle over frequency ratio can be plotted as shown in Fig. 3.20.

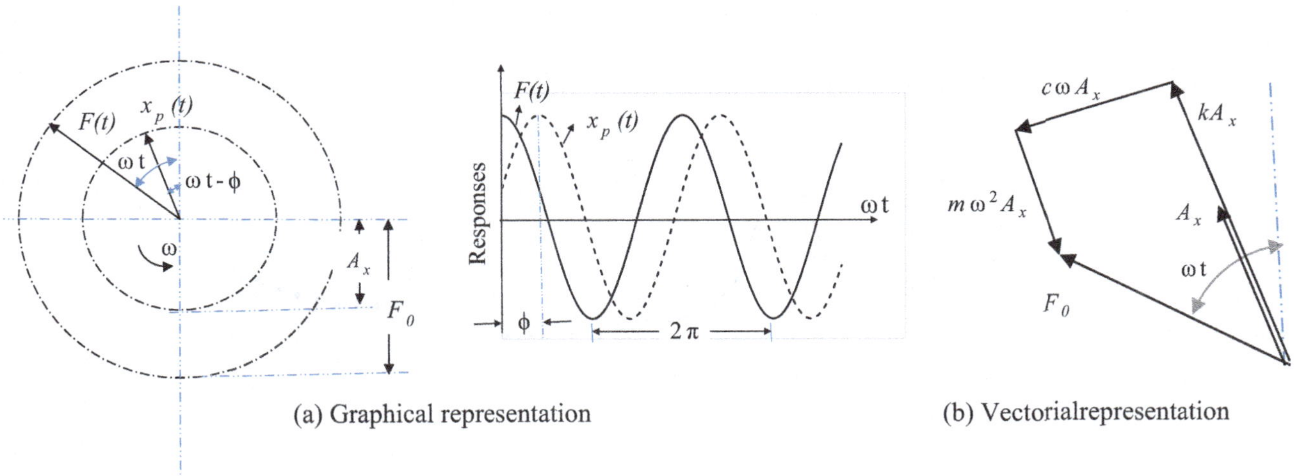

Fig. 3.19 Representation of forcing function and response

Fig. 3.20 Effect of amplitude and phase angle over frequency ratio under harmonic load

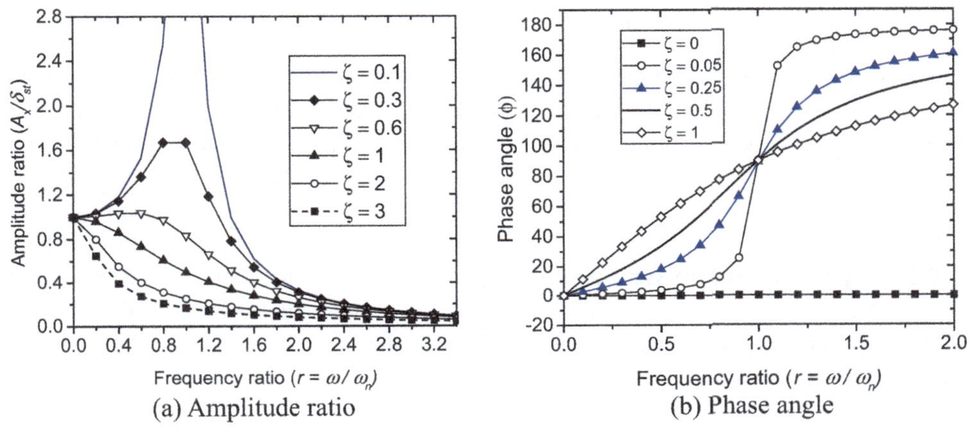

3.8.2.1 SDOF Undergoes Harmonic Excitation at Base

Consider a spring–mass system as shown in Fig. 3.21, which undergoes $y(t)$ displacement at base. Now the $x(t)$ is the displacement of mass. The free body diagram for this case is depicted in Fig. 3.21b. Now the equation of motion becomes,

$$m\ddot{x}(t) + c(\dot{x}(t) - \dot{y}(t)) + k(x(t) - y(t)) = 0 \quad (3.51)$$

Assume $y(t) = A_y\text{Sin}(\omega t)$, then

$$m\ddot{x}(t) + c\dot{x}(t) + kx(t) = ky(t) + c\dot{y}(t) \quad (3.52)$$

$$= A\sin(\omega t - \alpha)$$

where $A = A_y\sqrt{k^2 + (c\omega)^2}$ and $\alpha = \tan^{-1}\left[-\frac{c\omega}{k}\right]$. Solution of above Eq. (3.52) by assuming the particular solution method is,

$$x_p(t) = A_X \sin(\omega t - \phi) \quad (3.53)$$

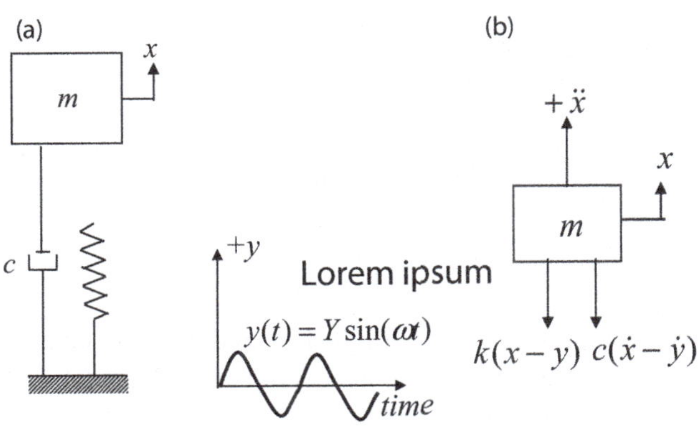

Fig. 3.21 SDOF system under base excitation

where A_X and ϕ are as follows

$$\frac{A_X}{A_y} = \left[\frac{k^2 + (c\omega^2)}{(k - m\omega^2) + (c\omega)^2}\right]^{1/2}$$

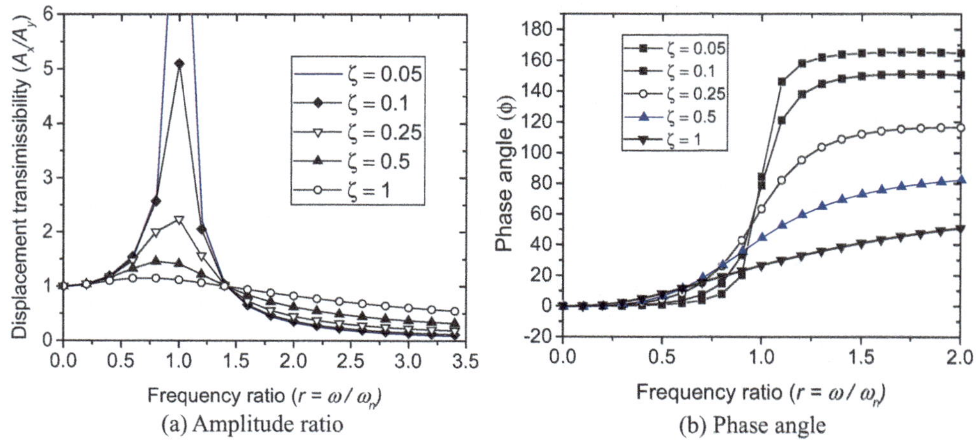

Fig. 3.22 Effect of amplitude and phase angle over frequency ratio under base harmonic load

and

$$\phi = \tan^{-1}\left(\frac{mc\omega^3}{k(k - m\omega^2) + (c\omega)^2}\right) \quad (3.54)$$

Above relations as mentioned in Eq. (3.54) can be converted into frequency ratio, and the effects of amplitude and phase angle over frequency ratio can be plotted as shown in Fig. 3.22.

3.8.3 Dynamic Response Factor

The steady-state dynamic response of sine load at forcing frequency is given by Eq. (3.55)

$$x(t) = \frac{F_0 \sin(\omega t)}{K(1 - (\omega/\omega_n)^2)} \quad (3.55)$$

Maximum value of static deformation is

$$(x_{st})_0 = F_0/k \quad (3.56)$$

Amplitude of vibratory deformation is

$$A = \frac{F_0}{K(1 - (\omega/\omega_n)^2)} \quad (3.57)$$

Dynamic response factor is defined as ratio of amplitude of vibratory deformation to the static deformation due to force *P*0.

Dynamic Response Factor,

$$R_d = \frac{1}{\left|1 - (\omega/\omega_n)^2\right|} \quad (3.58)$$

(Note: For $\omega < \omega_n$ displacement is in phase with applied force, and for $\omega > \omega_n$ displacement is out of phase with applied load). For damped system, Dynamic response factor can be calculated as given in Eq. (3.59).

$$R_d = \frac{1}{\sqrt{\left[1 - (\omega/\omega_n)^2\right]^2 + [2\zeta(\omega/\omega_n)]^2}} \quad (3.59)$$

A graph between dynamic response versus time for undamped system and damped system at resonance is shown in Figs. 3.23 and 3.24, respectively. It is observed from these figures that at resonance for undamped system, the response ratio increases infinitely as *t* increases, and for damped system, it increases and becomes equal to value (1/2ζ). Dynamic response factor is plotted against frequency ratio for damped and undamped systems as shown in Fig. 3.25. It is observed that at resonance (i.e., $\omega = \omega_n$) $R_d = \infty$ for undamped system, while for damped system $R_d = 1/2\zeta$.

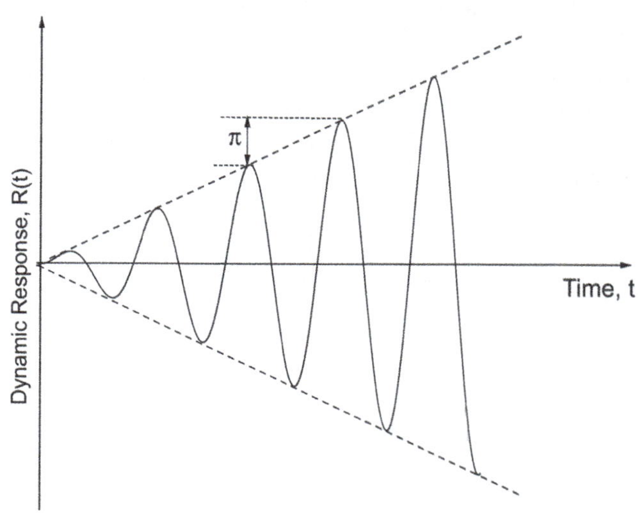

Fig. 3.23 Dynamic response versus time for undamped system at resonance

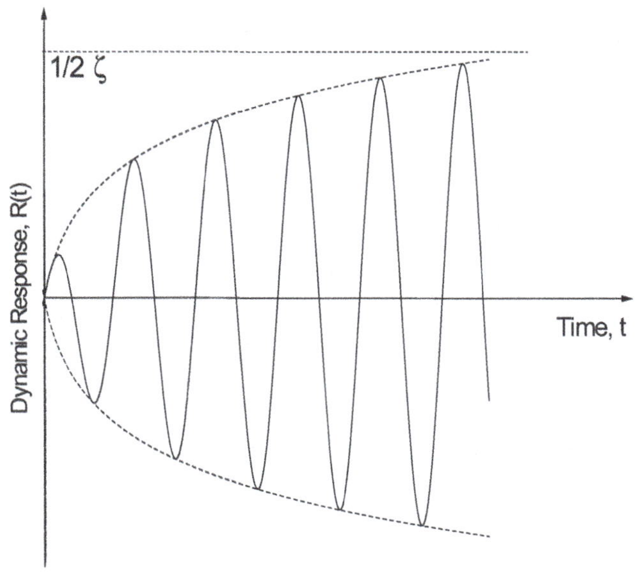

Fig. 3.24 Dynamic response versus time for damped system at resonance

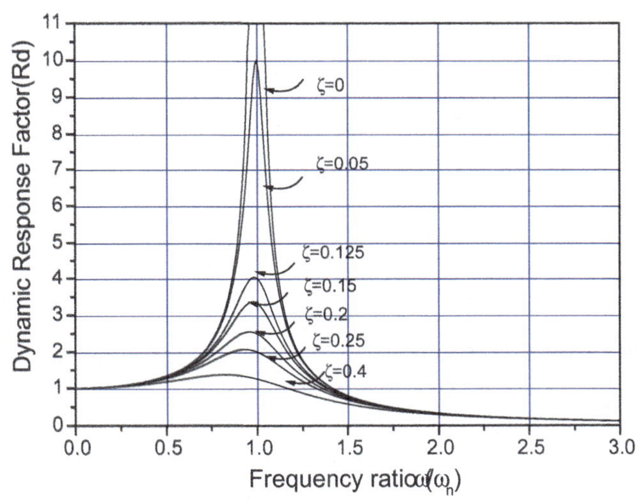

Fig. 3.25 Variation of dynamic response factor with frequency ratio

3.9 Energy Dissipated in Damping

At resonance, when the applied loading of harmonic force and the displacement response are 90° out of phase, the applied loading exactly balances the damping force. Thus, a plot of applied loading versus displacement will be interpreted as damping force versus displacement as shown in Fig. 3.26.

If the system possesses linear viscous damping, the force versus displacement diagram will be an ellipse. The maximum damping force can be calculated by using Eq. (3.60).

$$F_{\mathrm{dmax}} = c\dot{x}_{\mathrm{max}} \tag{3.60}$$

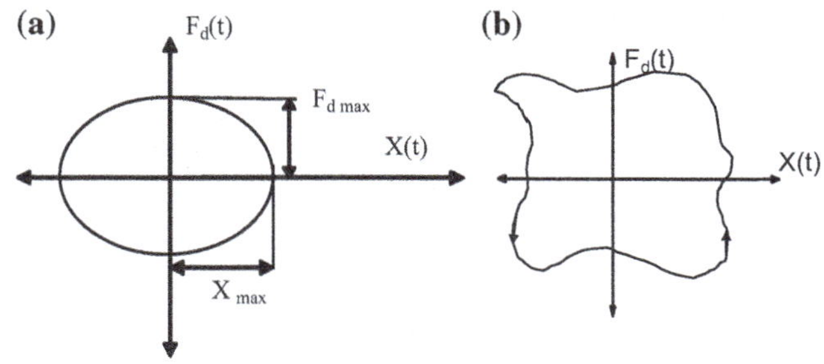

Fig. 3.26 a Applied loading versus displacement b generalized energy characteristics

However, if the damping is in nonlinear viscous form, then the shape of applied force versus displacement diagram will not be elliptical but will be of the shape of dotted line as shown in Fig. 3.26. For nonlinear viscous damping, the energy dissipated per cycle (E_d) is equal to the energy dissipated per cycle for linear viscous damping. Viscous damping is the damping associated with elliptical force–displacement which can be written as,

$$\begin{aligned}\text{Damping Force} &= F_{\mathrm{davg}} \times \dot{x}_{\mathrm{max}} \times \left(\frac{2\pi}{\omega}\right)\\ &= \frac{F_{\mathrm{dmax}}}{2} \times \dot{x}_{\mathrm{max}} \times \left(\frac{2\pi}{\omega}\right)\end{aligned} \tag{3.61}$$

The generalized energy characteristics are shown in Fig. 3.25b. The energy dissipation can be equated to area under the curve by integration $\left(\int F\mathrm{d}u\right)$. For the friction damper (as depicted in Fig. 3.7), equivalent viscous damping can be estimated. Equivalent viscous damping ratio is the damping ratio associated with elliptical force–displacement diagram having the same area E_d as the measured non-elliptical diagram.

$$E_{\mathrm{d}} = \varsigma_{\mathrm{eq}} m\omega^3 x_{\mathrm{max}}^2 \left(\frac{2\pi}{\omega}\right) \tag{3.62}$$

From Eq. (3.62), one can get

$$\varsigma_{\mathrm{eq}} = \frac{E_{\mathrm{d}}}{2\pi m\omega^2 x_{\mathrm{max}}^2} = \frac{E_{\mathrm{d}}}{4\pi\left(\frac{1}{2}Kx_{\mathrm{max}}^2\right)} = \frac{1}{4\pi}\left(\frac{E_{\mathrm{d}}}{E_{\mathrm{s}}}\right) \tag{3.63}$$

Thus, the equivalent viscous damping ratio is $(1/4\pi)$ times the ratio of damping energy to the total strain energy as given by Eq. (3.63).

Example 3.4 Weight of single-story building (as shown in Fig. 3.27) with rigid floors supported on four columns of negligible weight is 12 tons. The force required to displace the floor horizontally by 12 mm is 9 tons. The building is set into free vibration by giving an initial displacement of 12 mm. If the maximum displacement on return swing at the time of 0.75 s is 5 mm, determine the lateral stiffness, frequency, and damping of the structure.

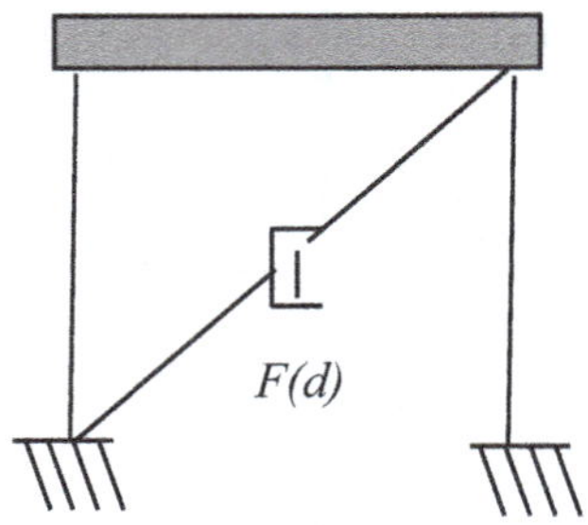

Fig. 3.27 Single-story building

Solution

$$\text{Lateral stiffness} = \frac{9000 \times 9.81}{0.012} = 7,500,000\,\text{N/m}$$

$$\text{Frequency} = \sqrt{\frac{k}{m}} = \sqrt{\frac{7,500,000}{12,000}} = 25\,\text{rad/s}$$

$$\omega_n = 25/2\pi = 3.98\,\text{Hz}$$
$$T_n = 1/\omega_n = 0.251\,\text{s}$$

For 0.75 s and 3 cycles,

$$\frac{x_1}{x_{j+1}} = e^{j\xi\omega_n T_D} \Rightarrow \ln\frac{x_1}{x_{j+1}} = 2\pi\xi j$$
$$\frac{1}{j}\ln\frac{12}{5} = 2\pi\xi \Rightarrow \xi = 0.046 = \frac{C}{C_{cr}}$$
$$C = \xi \times 2 \times m \times \omega = 27867.035\frac{\text{N}}{\text{m/s}}$$

Example 3.5 The steel frame shown in Fig. 3.28 supports a rotating machine which exerts a horizontal force at the girder level $F(t) = 400 \sin (5.3t)$ *N*. Assuming 5% damping, determine

(a) The steady-state amplitude of vibration,
(b) Maximum dynamic stress in the columns.

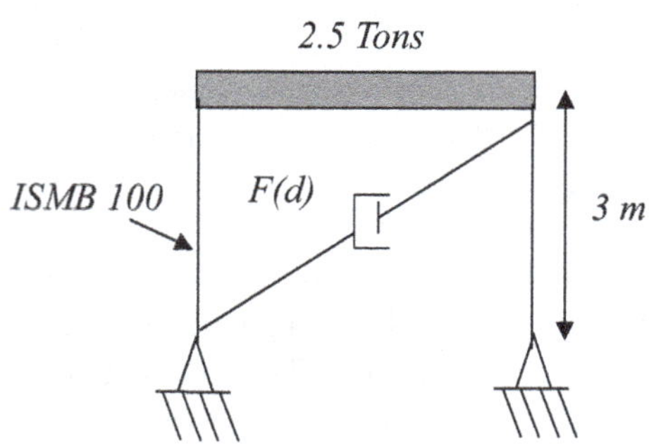

Fig. 3.28 Steel frame

Solution

Assume girder is rigid

$$k = \frac{3E(2I)}{l^3} = \frac{3 \times 2 \times 10^{11} \times 2 \times 257.5 \times 10^{-8}}{3^3}$$
$$= 114,444.44\,\text{N/m}$$

$$x_{st} = \frac{F_0}{k} = \frac{400}{114,444.44} = 0.003495\,\text{m}$$
$$\omega_n = \sqrt{\frac{k}{m}} = \sqrt{\frac{114,444.44}{2500}} = 6.76$$

Steady-state amplitude

$$x = \frac{x_{st}}{\sqrt{\left(1 - \left(\frac{\omega}{\omega_n}\right)^2\right)^2 + \left(2\left(\frac{\omega}{\omega_n}\right)\zeta\right)^2}} = 0.00631\,\text{m} \quad (3.66)$$

Maximum shear force in the columns

$$V_{max} = \frac{3EIx}{l^3} = 361.07\,\text{N} \quad (3.67)$$

Maximum bending moment in the columns

$$M_{max} = V_{max}L = 1083.31\,\text{N m} \quad (3.68)$$

Maximum stress in the columns

$$\sigma_{max} = \frac{M_{max}}{Z} = 21\,\text{MPa} \quad (3.69)$$

Example 3.6 If the frame in the above Example 3.5 is subjected to a sinusoidal ground motion $x_s(t) = 0.008 \sin 5.3t$. Determine

(a) The maximum shearing force in the supporting columns,
(b) The maximum stresses in the columns.

Solution

$$k = 114,444.44\,\text{N/m}$$
$$x_{st} = 0.00349\,\text{m}$$
$$\zeta = 0.05$$
$$X_0 = 0.008$$
$$\omega_n = 6.76\,\text{rad/s}$$
$$\omega = 5.3\,\text{rad/s},$$
$$\omega/\omega_n = 0.784$$

In terms of relative motion between mass and support, the equation of motion can be written as

$$m\ddot{x}(t) + c\dot{x}(t) + kx(t) = mx_0\omega^2 \sin \omega t \tag{3.70}$$

$$x(t) = \frac{mx_0\omega^2}{k} \frac{1}{\sqrt{\left[1 - \left(\frac{\omega}{\omega_n}\right)^2\right]^2 + \left[2\zeta\left(\frac{\omega}{\omega_n}\right)\right]^2}} \sin(\omega t - \phi) \tag{3.71}$$

Maximum relative displacement, X

$$X = x_0 \frac{\left(\frac{\omega}{\omega_n}\right)^2}{\sqrt{\left[1 - \left(\frac{\omega}{\omega_n}\right)^2\right]^2 + \left[2\zeta\left(\frac{\omega}{\omega_n}\right)\right]^2}} = 0.0125 \tag{3.72}$$

Maximum shear force in each column

$$V_{\max} = kX/2 = 715.2\,\text{N}$$

Maximum bending moment in the columns

$$M_{\max} = V_{\max}L = 2145.6\,\text{N–m} \tag{3.73}$$

Maximum stress in the columns

$$\sigma_{\max} = \frac{M_{\max}}{Z} = 41.67\,\text{MPa} \tag{3.74}$$

Example 3.7 Consider a SDOF system which is under sinusoidal excitation. At resonance frequency, the amplitude of displacement was measured to be 2.5 cm, while at one-tenth of the natural frequency, the amplitude of displacement is measured to be 2.5 mm. Estimate the damping ratio of the system.

Solution

Static displacement = 2.5 cm.

Dynamic amplification = x/x_0 = 2.5/0.25 = 10.

$$X = x_0 \frac{\left(\frac{\omega}{\omega_n}\right)^2}{\sqrt{\left[1 - \left(\frac{\omega}{\omega_n}\right)^2\right]^2 + \left[2\zeta\left(\frac{\omega}{\omega_n}\right)\right]^2}} \tag{3.75}$$

At resonance

$$\omega = \omega_n;$$
$$x = x_0/2\zeta$$

Therefore,

$$1/2\zeta = 10$$
$$2\zeta = 0.1$$

Damping ratio (ζ) = 0.05.

3.10 Duhamel Integral Method

Response to general dynamic loading can be obtained by Duhamel integral method considering the loading as succession of short duration impulses. Consider an arbitrary general loading $F(t)$ as represented in Fig. 3.29. Concentrate on impulsive load $F(\tau)$ acting in at time $t = \tau$. The loading represents a very short duration impulse $F(\tau)\mathrm{d}\tau$ on the structure.

Using impulse momentum relationship,

$$m\dot{x}(\tau) = F(\tau)\mathrm{d}\tau \tag{3.76}$$

After the termination of loading, the response is free vibration. Response at time t is given by Eq. (3.77)

$$x(t-\tau) = x(\tau)\cos\omega(t-\tau) + \frac{\dot{x}(\tau)}{\omega}\sin\omega(t-\tau) \tag{3.77}$$

Velocity $\dot{x}(\tau)$ is of the order of $\mathrm{d}\tau$; therefore, displacement $x(\tau)$ will be of the order of $(\mathrm{d}\tau)^2$, and hence, $x(\tau)$ will be very small and can be neglected. Substituting value of $\dot{x}(\tau)$ obtained from (3.76) and $x(\tau) = 0$ in Eq. (3.77)

$$x(t-\tau) = \frac{F(\tau)\mathrm{d}\tau}{m\omega}\sin\omega(t-\tau)\ \text{for}\ t > \tau \tag{3.78}$$

$x(t{-}\tau)$ represents time history response to differential impulse over entire time $t \geq \tau$. The entire loading can be considered as succession of such short impulses. Total response can then be obtained by summing up all the differential responses developed during the loading history.

By integrating the equation, one can get,

$$x(t) = \frac{1}{m\omega}\int_0^t F(\tau)\sin(\omega(t-\tau))\mathrm{d}\tau \tag{3.79}$$

Equation (3.79) is Duhamel Integral equation. It can be used to evaluate response of undamped SDOF system to any form of dynamic loading. For damped SDOF system, the response is evaluated by Eq. (3.80).

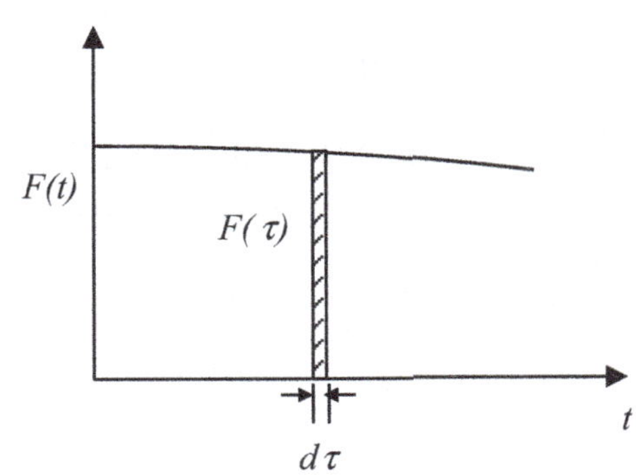

Fig. 3.29 Variation of loading $F(t)$ with time

$$x(t) = \frac{1}{m\omega}\int F(\tau)\sin\omega(t-\tau)e^{-\zeta\omega(t-\tau)}d\tau \quad (3.80)$$

Response to suddenly applied constant force for limited time duration, t_d, can be obtained from

$$x(t_d) = \frac{F_0}{k}(1-\cos\omega t_d); \quad \dot{x}(t_d) = \frac{F_0}{k}\omega\sin\omega t_d \quad (3.81)$$

For response after time t_d (i.e., $t > t_d$), apply equation for free vibration,

$$x(t) = \frac{F_0}{k}(1-\cos\omega t_d)\cos\omega(t-t_d) + \frac{F_0}{k}\sin\omega t_d\sin\omega(t-t_d) \quad (3.82)$$

$$x(t) = \frac{F_0}{k}\{\cos\omega(t-t_d) - \cos\omega t\} \quad (3.83)$$

Example 3.8 Estimate the response under the constant force as shown in Fig. 3.30 using Duhamel Integral approach

Solution

$$x(t) = \frac{1}{m\omega}\int_0^t F_0\sin[\omega(t-\tau)]d\tau = x_{st}\int_0^t \sin[\omega(t-\tau)]d\tau \quad (3.84)$$

$$x(t) = x_{st}\omega\left[\frac{\cos[\omega(t-\tau)]}{\omega}\right]_0^t = x_{st}(1-\cos\omega t) \text{ at } t = \pi/\omega \text{ or } t = T/2,\ x(t) = 2x_{st}. \quad (3.85)$$

Thus, a suddenly applied force produces twice the deformation it would have caused as a slowly applied force as shown in Fig. 3.31.

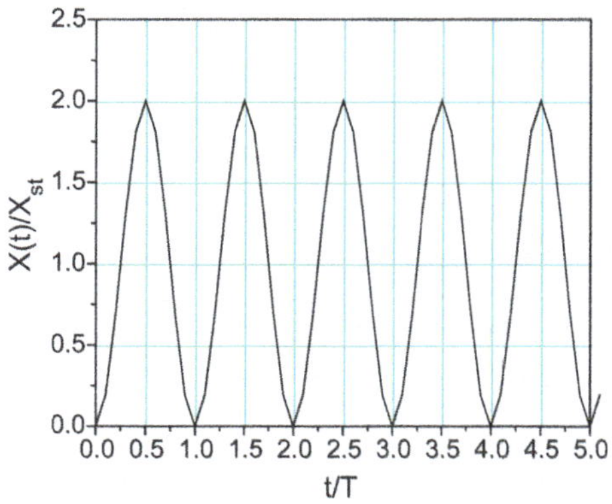

Fig. 3.31 Response for constant force

3.10.1 Dynamic Load Factor (DLF)

Dynamic load factor (DLF) is defined as the ratio between the displacement at time t to the static displacement. For the case of rectangular pulse load as shown in Fig. 3.32, the dynamic load factor can be calculated by using Eq. (3.86).

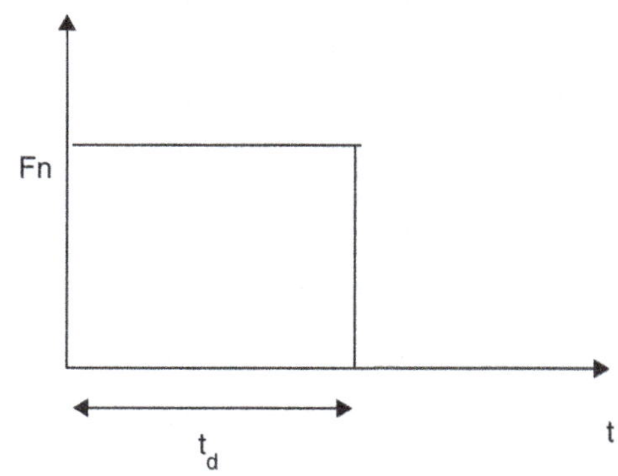

Fig. 3.32 Force time history (step load)

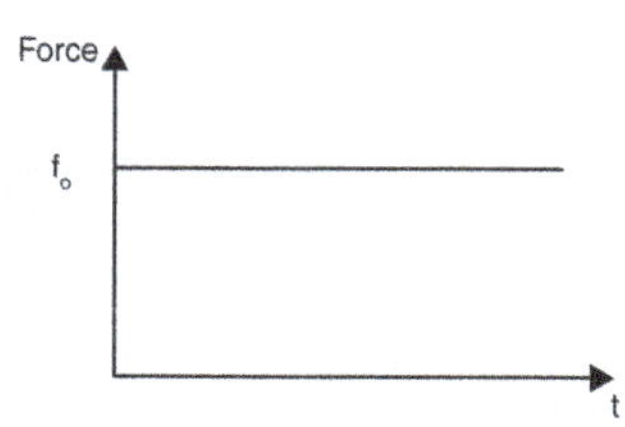

Fig. 3.30 Force time history

$$DLF = (1-\cos\omega t_d) = \left(1-\cos 2\pi\frac{t}{T}\right) \text{ for } t \le t_d \quad (3.86)$$

$$DLF = \cos\omega(t-t_d) - \cos\omega t = \cos 2\pi\left(\frac{t}{T}-\frac{t_d}{T}\right) - \cos 2\pi\frac{t}{T} \text{ for } t > t_d \quad (3.87)$$

The maximum DLF for loads of duration $t_d/T > 0.5$ is same as if the load duration had been infinite. For short duration loads, damping does not have significant effects on the response of the system. The maximum dynamic load factor usually corresponds to the first peak of the response. Figure 3.33 shows the maximum dynamic load factor for undamped SDOF system with rectangular pulse.

For the case of triangular pulse load, the dynamic load factor can be calculated by using Eq. (3.88).

$$DLF = \left(1-\cos 2\pi\frac{t}{T}\right) + \frac{\sin(2\pi t/T)}{2\pi t_d/T} - \frac{t}{t_d} \text{ for } t \le t_d \quad (3.88)$$

$$DLF = \frac{1}{(2\pi t_d/T)}\left\{\sin 2\pi\frac{t}{T} - \sin 2\pi\left(\frac{t}{T}-\frac{t_d}{T}\right)\right\} - \cos 2\pi\frac{t}{T} \text{ for } t > t_d \quad (3.89)$$

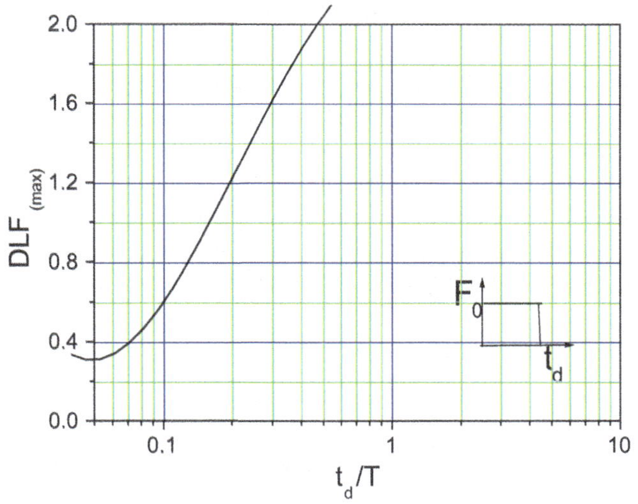

Fig. 3.33 Maximum dynamic load factor for undamped SDOF system with rectangular pulse

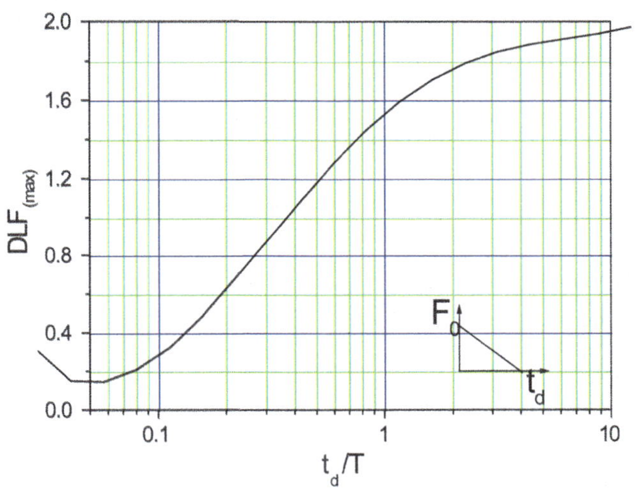

Fig. 3.34 Maximum dynamic load factor for undamped SDOF system with triangular pulse

The maximum value of DLF approaches 2 when duration t_d/T becomes large. Figure 3.34 displays the maximum dynamic load factor for undamped SDOF system with triangular pulse.

3.11 Time History Methods

It is a step-by-step procedure in which the loading and the response history are divided into sequence time intervals or steps. The response during each time step is then calculated from the initial conditions existing at the beginning of each time step and from the loading history during that step. These can be either explicit or implicit methods. Explicit methods are those in which new response quantities calculated in each step depend on quantities obtained in previous steps and analysis proceeds directly from one step to another. In implicit methods, the expressions giving the new values for a given step include one or more values pertaining to that same step, so that trial values of necessary quantities must be assumed and then refined by iterations [5].

This type of approach makes use of integration to step forward from the initial to the final conditions for each time step. The following equations represent the essential concept

$$\dot{x}_1 = \dot{x}_0 + \int_0^{\Delta t} \ddot{x}(\tau)\mathrm{d}\tau \tag{3.90}$$

$$x_1 = x_0 + \int_0^{\Delta t} \dot{x}(\tau)\mathrm{d}\tau \tag{3.91}$$

In order to carry out this type of analysis, first it is necessary to assume how the acceleration varies during the time step. Some of the time integration techniques discussed are central difference scheme, Runge–Kutta method, Newmark-beta technique, and Wilson theta technique.

3.11.1 Central Difference Scheme (CDS)

The advantages of CDS are that it is easy to understand and to implement, at least for simple relations, and that its convergence rate is faster than other differencing methods. CDS is based on the Taylor series expansion of X_{i+1} and X_{i-1} about the grid point i and divided into n equal parts of interval $h = \Delta t$ each as shown in Fig. 3.35.

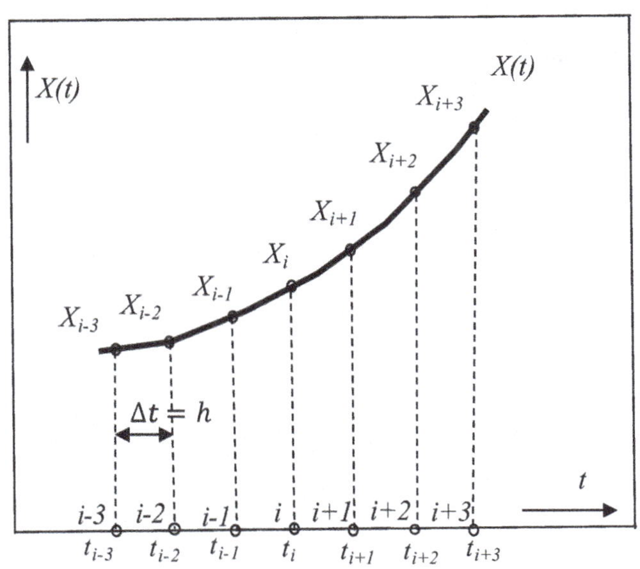

Fig. 3.35 Equally spaced grid points along the independent coordinate

$$X_{i-1} = X_i - h\dot{X}_i + \frac{h^2}{2}\ddot{X}_i - \frac{h^3}{6}\dddot{X}_i + \cdots. \tag{3.92}$$

$$X_{i+1} = X_i + h\dot{X}_i + \frac{h^2}{2}\ddot{X}_i + \frac{h^3}{6}\dddot{X}_i + \cdots. \tag{3.93}$$

Subtracting Eq. (3.92) from Eq. (3.93), one can obtain

$$\dot{X}_i = \frac{1}{2h}(X_{i+1} - X_{i-1}) \tag{3.94}$$

While summing Eqs. (3.92) and (3.93), one can obtain

$$\ddot{X}_i = \frac{1}{h^2}(X_{i-1} - 2X_i + X_{i+1}) \tag{3.95}$$

Substituting Eqs. (3.94) and (3.95) into equation of motion under forced vibration, one can get

$$M\left\{\frac{X_{i+1} - 2X_i + X_{i-1}}{(\Delta t)^2}\right\} + C\left\{\frac{X_{i+1} - X_{i-1}}{2\Delta t}\right\} + KX_i = F_i \tag{3.96}$$

The numerical procedure for CDS method is explained in Table 3.1.

3.11.2 Runge–Kutta Method

Runge–Kutta (RK) method is a self-starting method and has the advantage that no initial values are needed beyond the prescribed values. RK method can be applied directly to the first-order differential equations. In order to solve the second-order differential equation, it is first reduced to two first-order equations. Consider the differential equation for the SDOF system given in Eq. (3.97).

$$M\Delta\ddot{x} + C\Delta\dot{x} + k\Delta x = \Delta F \tag{3.97}$$

Equation 3.97 can be rewritten as

$$\ddot{X}_i = \frac{1}{M}\left[F(t) - C\dot{X} - KX\right] = f(X, \dot{X}, t) \tag{3.98}$$

By letting $U = X$ and $V = \dot{X}$, Eq. (3.98) is reduced to the following first-order equation (3.99).

$$\dot{U} = V;\ \dot{V} = f(X_1, X_2, t) \tag{3.99}$$

By defining,

$$\bar{X}(t) = \begin{Bmatrix} U(t) \\ V(t) \end{Bmatrix} \tag{3.100}$$

Table 3.1 Numerical procedure for CDS and Runge–Kutta methods

Central difference scheme (CDS)	Runge–Kutta (fourth order)
(a) **Initialization** (1) Initial conditions x_0, $\dot{x}_0$ and $\ddot{x}_0$ (2) Choice of Δt (3) $\ddot{x}_0 = \frac{F_0 - c\dot{x}_0 - kx_0}{m}$ (4) $x_{-1} = x_0 - \Delta t(\dot{x}_0) + \frac{\Delta t^2}{2}(\ddot{x}_0)$ (5) $\bar{k} = \frac{m}{\Delta t^2} + \frac{c}{2\Delta t}$ (6) $a = \frac{m}{\Delta t^2} - \frac{c}{2\Delta t}$ (7) $b = k - \frac{2m}{\Delta t^2}$ (b) **With each time step** (1) $\bar{F} = F_i - ax_{i-1} - bx_i$ (2) $x_{i+1} = \bar{F}_i/\bar{k}$ (3) Calculate $\dot{x}_i = \frac{x_{i+1} - x_{i-1}}{2\Delta t}$ and $\ddot{x}_i = \frac{x_{i+1} - 2x_i + x_{i-1}}{\Delta t^2}$	(a) **Initialization** (1) Initial conditions x_0, $\dot{x}_0$ and $\ddot{x}_0$ (2) Choice of Δt (3) Initial variables $v_0 = \dot{x}_0;\ \dot{v}_0 = \frac{1}{m}[F(0) - cv_0 - kx_0]$ (b) **With each time step** (1) $t = t_1;\ v_1 = v_1;$ $\dot{v}_1 = v_1 \frac{1}{m}[F(t) - cv_i - kx_i]$ (2) Fist midpoint of time interval $t = t_1 + \Delta t/2;\ x = x_{t+\Delta t/2} = x_i + v_i \frac{\Delta t}{2};$ $v_2 = v_{t+\Delta t/2} = v_i + \dot{v}_i \frac{\Delta t}{2}$ $\dot{v}_2 = v_{t+\Delta t/2} = \frac{1}{m}[F(t) - cv_2 - kx]$ (3) Second midpoint of time interval $t = t_1 + \Delta t/2;\ x = x_{t+\Delta t/2} = x_i + v_2 \frac{\Delta t}{2};$ $v_3 = v_{t+\Delta t/2} = v_i + \dot{v}_2 \frac{\Delta t}{2}$ $\dot{v}_3 = \frac{1}{m}[F(t) - cv_3 - kx]$ (4) Calculate displacements at the end $t = t_1 + \Delta t;\ x = x_{t+\Delta t} = x_i + v_3\Delta t;$ $v_4 = v_{t+\Delta t} = v_i + \dot{v}_3\Delta t$ $\dot{v}_4 = \frac{1}{m}[F(t) - cv_4 - kx]$ (5) Calculate displacement and velocity at the end of the time interval $x_{i+1} = x_i + \frac{\Delta t}{2}[v_1 + 2v_2 + 2v_3 + v_4]$ $v_{i+1} = v_i + \frac{\Delta t}{2}[\dot{v}_1 + 2\dot{v}_2 + 2\dot{v}_3 + \dot{v}_4]$

and

$$\bar{F}(t) = \begin{Bmatrix} V \\ f(x_1, x_2, t) \end{Bmatrix} \quad (3.101)$$

Equations (3.99) and (3.100) are used to find the solution at mesh or grid point t_i according to the fourth-order RK method as given in Eq. (3.102).

$$\bar{X}_{i+1} = \bar{X}_i + \frac{1}{6}[\bar{K}_1 + 2\bar{K}_2 + 2\bar{K}_3 + \bar{K}_4] \quad (3.102)$$

where

$$\begin{aligned} \bar{K}_1 &= h\bar{F}(\bar{X}_i, t_i); \\ \bar{K}_2 &= h\bar{F}\left(\bar{X}_i, \frac{1}{2}\bar{K}_1, t_i + \frac{1}{2}h\right); \\ \bar{K}_3 &= h\bar{F}\left(\bar{X}_i, \frac{1}{2}\bar{K}_2, t_i + \frac{1}{2}h\right); \\ \bar{K}_4 &= h\bar{F}\left(\bar{X}_i, \frac{1}{2}\bar{K}_3, t_{i+1}\right) \end{aligned} \quad (3.103)$$

Although the RK method does not essential the computation of derivatives beyond the first, its higher accuracy is attained by four evaluations of the first derivatives to get agreement with the Taylor series solution through terms of $O(h^4)$. This explicit method can be considered as an inherently stable method as time step can be simply altered between iterations. The numerical procedure for RK method is explained in Table 3.1.

The main disadvantage of the method is that each forward step needs numerous computations of the functions and accordingly it increases the computational cost [3]. The RK method is applicable and extendable to a system of differential equations. Since it uses only the information from the previous step, this method is called as single step method. The principle behind a multi-step method is to use past values, y and/or dy/dx, to construct a polynomial that approximate the derivative function. Multi-step methods are also available for better solution such as Adam Bashforth, Adam Moulton method, and predictor–corrector method.

3.11.3 Newmark-Beta Method

In Newmark's formulation, the basic integration equations for calculating final velocity and displacement are

$$\dot{x}(t_{i+1}) = \dot{x}(t_i) + (1 - \gamma)\ddot{x}(t_i)\Delta t + \gamma\ddot{x}(t_{i+1})\Delta t \quad (3.104)$$

$$x(t_{i+1}) = x(t_i) + \Delta t\dot{x}(t_i) + \left(\frac{1}{2} - \beta\right)\Delta t^2\ddot{x}(t_i) + \beta\Delta t^2\ddot{x}(t_{i+1}) \quad (3.105)$$

Factor γ provides linearly varying weighting between the influence of initial and final accelerations on change of velocity. And factor β provides linearly varying weighting between the influence of initial and final accelerations on change of displacement [6]. When $\gamma = 1/2$ and $\beta = 1/4$, Newmark's formulation reduces to constant acceleration method. When $\gamma = 1/2$ and $\beta = 1/6$, Newmark's formulation reduces to linear acceleration method.

Linear Acceleration Method

It is based on the assumption that the acceleration variation is linear during the time step. It is generally convenient for time integration methods to formulate the response in terms of the incremental equation of motion as given in Eq. (3.97). Considering linear variation of the acceleration as shown in Fig. 3.36, acceleration at any time, τ, can be given by Eq. (3.106)

$$\ddot{x}(\tau) = \ddot{x}(t_i) + \frac{\Delta\ddot{x}}{\Delta t}\tau \quad (3.106)$$

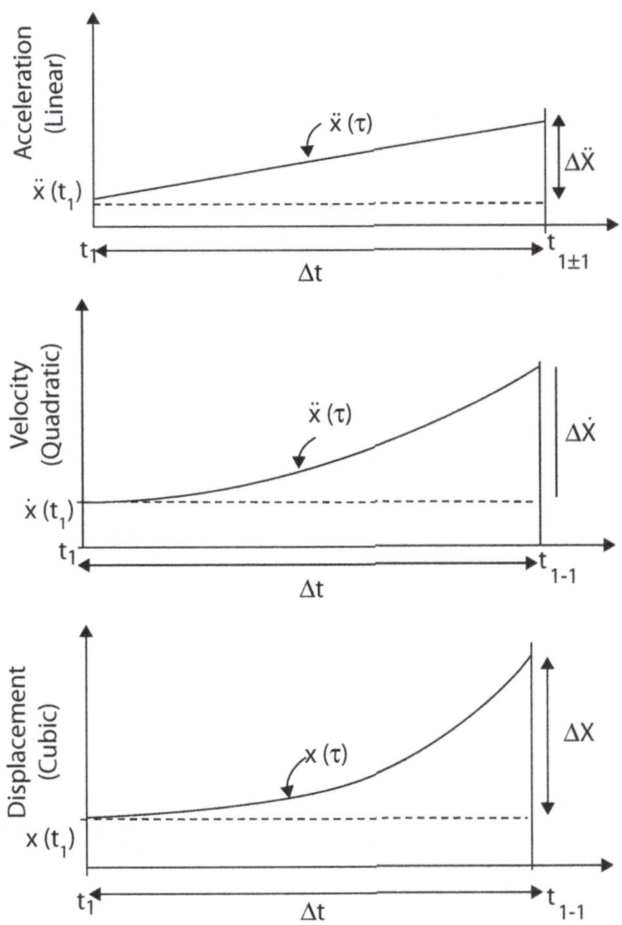

Fig. 3.36 Motion based on linearly varying acceleration

Similarly, the velocity at any time,τ, is obtained by integrating the acceleration Eq. (3.106) and is given by Eq. (3.107).

$$\dot{x}(\tau) = \dot{x}(t_i) + \ddot{x}(t_i)\tau + \frac{\Delta\ddot{x}}{\Delta t}\frac{\tau^2}{2} \quad (3.107)$$

Substituting the value of $\tau = \Delta t$ at time t_{i+1} in Eq. (3.107), one can get Eq. (3.108)

$$\Delta\dot{x} = \ddot{x}(t_i)\Delta t + \Delta\ddot{x}\frac{\Delta t}{2} \quad (3.108)$$

Similarly, displacement at any time,τ, is obtained by integrating the velocity Eq. (3.107) and is given by Eq. (3.109).

$$x(\tau) = x(t_i) + \dot{x}(t_i)\tau + \ddot{x}(t_i)\frac{\tau^2}{2} + \frac{\Delta\ddot{x}}{\Delta t}\frac{\tau^3}{6} \quad (3.109)$$

Substituting the value of $\tau = \Delta t$ at time t_{i+1} in Eq. (3.109), one can get Eq. (3.110)

$$\Delta x = \dot{x}(t_i)\Delta t + \ddot{x}(t_i)\frac{\Delta t^2}{2} + \Delta\ddot{x}\frac{\Delta t^2}{6} \quad (3.110)$$

Now using equation of motion and knowing initial velocity and displacement, initial acceleration can be calculated by using Eq. (3.111).

$$\ddot{x}(t_i) = \frac{F(t_i) - Kx(t_i) - C\dot{x}(t_i)}{M} \quad (3.111)$$

Using Eq. (3.110), the value of incremental acceleration is obtained in terms of incremental displacement as follows,

$$\Delta\ddot{x} = \frac{6}{\Delta t^2}\Delta x - \dot{x}(t_i)\frac{6}{\Delta t} - 3\ddot{x}(t_i) \quad (3.112)$$

Substituting Eq. (3.112) in Eq. (3.108), one can get the value of incremental velocity in terms of incremental displacement as given in Eq. (3.113),

$$\Delta\dot{x} = \frac{3}{\Delta t}\Delta x - 3\dot{x}(t_i) - \frac{\Delta t}{2}\ddot{x}(t_i) \quad (3.113)$$

Substituting (3.98) and (3.99) in (3.83),

$$\tilde{K}\Delta x = \Delta\tilde{F} \quad (3.114)$$

where

$$\tilde{K} = K + \frac{3C}{\Delta t} + \frac{6M}{\Delta t^2} \quad (3.115)$$

$$\Delta\tilde{F} = \Delta F + C\left[3x(t_i) + \frac{\Delta t}{2}\ddot{x}(t_i)\right] + M\left[\frac{6}{\Delta t}\dot{x}(t_i) + 3\ddot{x}(t_i)\right] \quad (3.116)$$

Once the incremental displacement is obtained, incremental velocity can be evaluated by using Eq. (3.113). Thus, velocity and displacement at the end of time increment Δt can be evaluated by using Eqs. (3.117) and (3.118)

$$\dot{x}(t_{i+1}) = \dot{x}(t_i) + \Delta\dot{x} \quad (3.117)$$

$$x(t_{i+1}) = x(t_i) + \Delta x \quad (3.118)$$

When this step has been completed, the calculation of response for this time increment is finished and the analysis is stepped forward to the next time interval. Numerical procedure for Newmark-beta method is given in Table 3.2.

Table 3.2 Numerical procedure for Newmark-beta and Wilson theta methods

Newmark-beta	Wilson theta
(a) **Initialization** (1) Initial conditions x_0, $\dot{x}_0$ and $\ddot{x}_0$ (2) Choice of Δt and γ, β (3) Assemble the stiffness matrices K and mass M (4) Form the effective stiffness matrix $\bar{k}$ (Eq. 3.115) (b) **With each time step** (1) Calculate the effective loading $\bar{F}$ (Eq. 3.116) (2) Compute incremental displacement, velocity (Eqs. 3.112–3.113) (3) Calculate displacement, velocity, acceleration at time $t + \Delta t$ (Eqs. 3.117–3.118)	(a) **Initialization** (1) Initial conditions x_0, $\dot{x}_0$ and $\ddot{x}_0$ (2) Choice of Δt and θ (3) Compute coefficients $a0 = 6/(\theta * \Delta t)$^2; $a1 = 3/(\theta * \Delta t)$; $a2 = 2 *$ a1; a3 = $\theta * \Delta t$ / 2; $a4$ = a0/ θ; $a5 = -a2/\theta$; $a6 = 1-3/\theta$; $a7 = \Delta t/2$; $a8 = \Delta t^2/6$ (4) Form the effective stiffness matrix $\bar{k} = k + (a0)m + (a1)c$ (b) **With each time step** (1) Calculate the effective loading $\bar{F}_{i+1} = F_i + m[(a0)x_i + (a2)\dot{x}_i + 2\ddot{x}_i] + c[(a1)x_i + 2\dot{x}_i + (a3)\ddot{x}_i]$ (2) $\bar{x}_{i+1} = \bar{k}_{i+1}/\bar{F}_{i+1}$ (3) Calculating displacements, velocities, and accelerations at time $t = t_1 + \Delta t$ $\ddot{x} = a4(x_{i+1} - x_i) + a5\dot{x}_i + a6\ddot{x}_i$ $\dot{x}_{i+1} = \dot{x}_i + a7(\ddot{x}_{i+1} + \ddot{x}_i)$ $x_{i+1} = x_i + \Delta t\dot{x}_{i+1} + a8(\ddot{x}_{i+1} + 2\ddot{x}_i)$

3.11.4 Wilson Theta Method

In light of excessive computation required for conditionally stable method, numerical procedure for multi-degree-of-freedom (MDOF) systems should be unconditionally stable. Constant acceleration method can be used, but Wilson theta method which is an unconditionally stable modification of linear acceleration method provides numerical damping which filters out response contribution of higher modes. Hence, this method is most commonly used to evaluate response of MDOF systems.

The assumption in this method is that acceleration varies linearly over the extended time interval $\tau = \theta\Delta t$, where $\theta = 1.4$. Thus, velocity and displacement at extended time step are given by Eqs. (3.119) and (3.120)

$$\hat{\Delta}\dot{x} = \ddot{x}(t_i)\tau + \hat{\Delta}\ddot{x}\frac{\tau}{2} \tag{3.119}$$

$$\hat{\Delta}x = \dot{x}(t_i)\tau + \ddot{x}(t_i)\frac{\tau^2}{2} + \hat{\Delta}\ddot{x}\frac{\tau^2}{6} \tag{3.120}$$

getting $\hat{\Delta}\ddot{x}, \hat{\Delta}\dot{x}$ in terms of $\hat{\Delta}x$ from equation

$$\bar{K}\hat{\Delta}x = \hat{\Delta}F \tag{3.121}$$

where $\bar{K}$ and $\Delta\bar{F}$ are evaluated for extended time interval τ. $\hat{\Delta}x$ obtained is substituted in (3.120) to obtain $\hat{\Delta}\ddot{x}$. The acceleration at normal time step is obtained by using interpolation

$$\Delta\ddot{x} = \frac{1}{\theta}\hat{\Delta}\ddot{x} \tag{3.122}$$

Numerical procedure for Wilson theta method is given in Table 3.2.

3.11.5 Stability and Accuracy

The numerical methods that lead to bounded solutions if the time step is shorter than some stability limit is known as conditionally stable and those which lead to bounded solutions irrespective of time step length are called unconditionally stable. Newmark's method is conditionally stable if

$$\frac{\Delta t}{T_n} \le \frac{1}{\pi\sqrt{2}}\frac{1}{\sqrt{\gamma - 2\beta}} \tag{3.123}$$

For constant acceleration method, $\gamma = 1/2$ and $\beta = 1/4$; $\Delta t/T_n < \infty$ hence this method is unconditionally stable. For linear acceleration method, $\gamma = 1/2$ and $\beta = 1/6$; $\Delta t/T_n <$ 0.551, hence this method is stable if $\Delta t < 0.551\ T_n$. However for adequate accuracy in numerical methods, $\Delta t/T_n$ should be less than 0.1.

3.11.6 Errors in Numerical Methods

The solution of equation of motion using numerical methods always have some errors. The nature of errors and useful way of minimizing and managing the errors must be identified. The two types of errors commonly observed in solution by numerical techniques are natural period elongation and the displacement amplitude decay with time. The two errors are explained in Fig. 3.37. The natural period is always elongated for Newmark-beta technique. If $\Delta t/T_n$ less than stability limit, period elongation for linear acceleration method is less than for constant acceleration method [1]. This property makes this method most suitable for SDOF systems. However, there is a rapid increase in period elongation in the linear acceleration method near $\Delta t/T_n = 0.551$. Wilson theta technique gives amplitude decay, i.e., numerical damping along with period elongation. Due to the inherent numerical damping given by Wilson theta technique for higher $\Delta t/T_n > 0.1$, the numerical damping is very high. Thus for modes which have higher frequencies, there is inherent numerical damping and this reduces the response of higher modes and hence this method is used for MDOF systems.

For adequate accuracy in numerical methods, we should choose Δt such that $\Delta t/T_n < 0.1$. The dynamic problem shall be solved with the time step that is thus reasonable, and then, the solution shall be repeated with smaller time steps [7]. The results of two solutions shall be compared and the process shall be continued until two successive solutions are close enough. This will always ensure the accuracy of the solution by numerical techniques.

Example 3.9 Consider a SDOF spring–mass system shown in Fig. 3.38a with $M = 0.2533$ kg, $K = 10$ N/m and damping ratio = 0.05. The force $P(t)$ is defined by the half-cycle sine pulse force as shown in Fig. 3.38b.

(a) Compute the displacement response of this system using Newmark-beta linear acceleration technique and compare with the theoretical response.

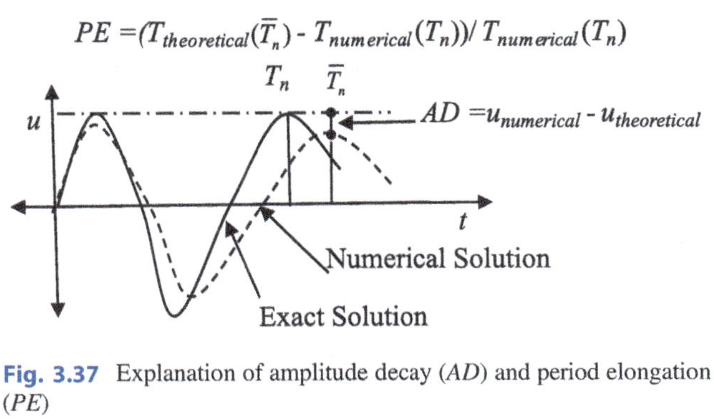

Fig. 3.37 Explanation of amplitude decay (*AD*) and period elongation (*PE*)

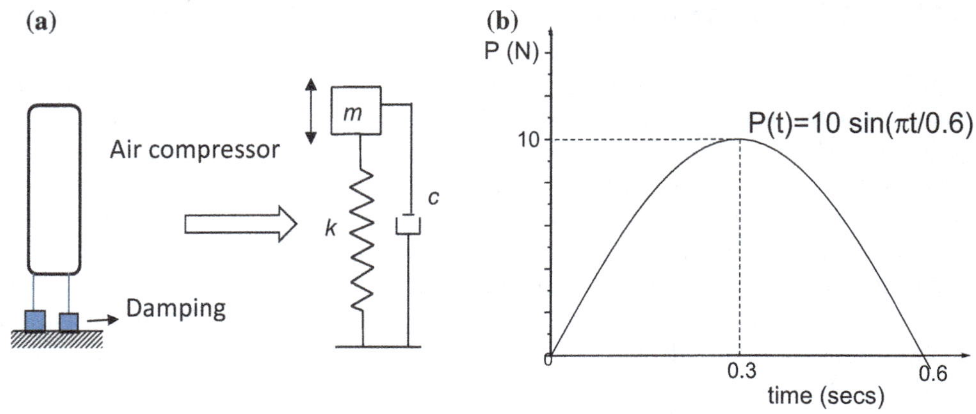

Fig. 3.38 **a** Damped SDOF spring–mass system **b** half-cycle sine pulse force

(b) Determine the response of this system using Wilson theta technique. Investigate the effect of $\Delta t/T_n$ parameter by varying from 0.001 to 0.3.
(c) Compare the responses from Newmark-beta, Wilson theta, Runge–Kutta (second order), Runge–Kutta (fourth order) and central difference schemes at $\Delta t/T_n = 0.1$.
(d) Evaluate the absolute error and relative error with theoretical response at $\Delta t/T_n = 0.1$.
(e) Estimate the period elongation and amplitude decay for various numerical schemes for SDOF system to half-cycle sine pulse force.
(f) Study the dynamic response to static response for the damped SDOF system under half-cycle sine pulse force.

Solution

Frequency $= (\mathrm{K/M})0.5 = 6.283\,\mathrm{rad/s}$
$\mathrm{T} = 1\,\mathrm{s}$, Therefore $\Delta t = 0.1\mathrm{T} = 0.1\,\mathrm{s}$
$C = 2M\omega\zeta = 0.1592\,\mathrm{N/m/s}$
$\mathrm{X}(0) = 0,$
Initial Acceleration $= 0$
$\Delta t = 0.1\,\mathrm{s}$

$$\tilde{K} = K + \frac{3C}{\Delta t} + \frac{6M}{\Delta t^2} = 166.8$$

$$\Delta\tilde{F} = \Delta F + C\left[3\dot{x}(t_i) + \frac{\Delta t}{2}\ddot{x}(t_i)\right] + M\left[\frac{6}{\Delta t}\dot{x}(t_i) + 3\ddot{x}(t_i)\right] = 5$$

$$\Delta x = \frac{\Delta\tilde{F}}{\tilde{K}} = 5/166.8 = 0.02995$$

$$\Delta\dot{x} = \frac{3}{\Delta t}\Delta x - 3\dot{x}(t_i) - \frac{\Delta t}{2}\ddot{x}(t_i) = (3/0.1)x\,0.03 = 0.8993$$

$$\dot{x}(t_{i+1}) = \dot{x}(t_i) + \Delta\dot{x} = 0.899\,3m/s; x(t_{i+1}) = x(t_i) + \Delta x = 0.03\,\mathrm{m}$$

The acceleration for this time step is obtained by substituting the values of velocity and displacement obtained as follows:

$$\ddot{x}(t_i) = \frac{F(t_i) - Kx(t_i) - C\dot{x}(t_i)}{M}$$

The analysis is further stepped forward for the next time step. The plot of displacement versus time for the SDOF system is shown in Fig. 3.39.

The accuracy of the numerical solution always depends on the size of the time step. If Δt is selected to be larger than stability limit, the method becomes unstable; that is, the truncation of higher order terms or rounding-off in the computer causes errors to grow and makes the dynamic response calculations meaningless. Effect of time step in solution of damped SDF system ($M = 0.2533$ kg, $K = 10$ N/m, $\zeta = 0.05$ and $P0 = 10$ N) to half-cycle sine pulse force by Wilson theta is shown in Fig. 3.40.

The damped SDF system to half-cycle sine pulse force is solved through CDS, RK second order, RK fourth order, and other discussed schemes. ($M = 0.2533$ kg, $K = 10$ N/m, $\zeta = 0.05$, $P0 = 10$ N and $\Delta t/T_n = 0.1$.) Results are shown in Figs. 3.41 through 3.42. Displacement of the SDOF under half-cycle sine pulse force is shown in Fig. 3.41b.

As discussed before, adequate accuracy of this problem is $\Delta t/T_n < 0.1$. However, at $\Delta t/T_n = 0.1$, Wilson theta and RK second-order solution accuracy are lower than other schemes

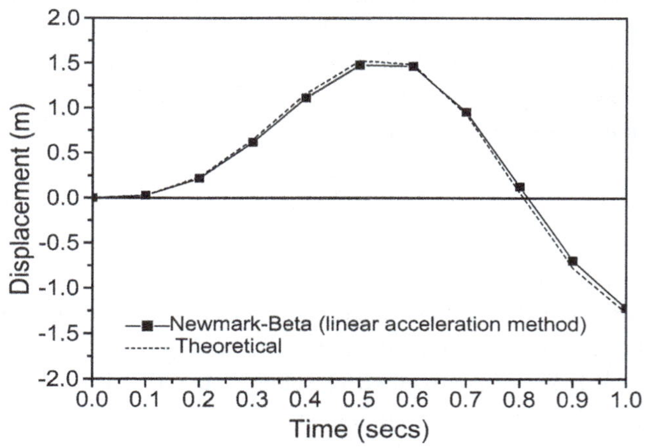

Fig. 3.39 Displacement curve through Newmark-beta

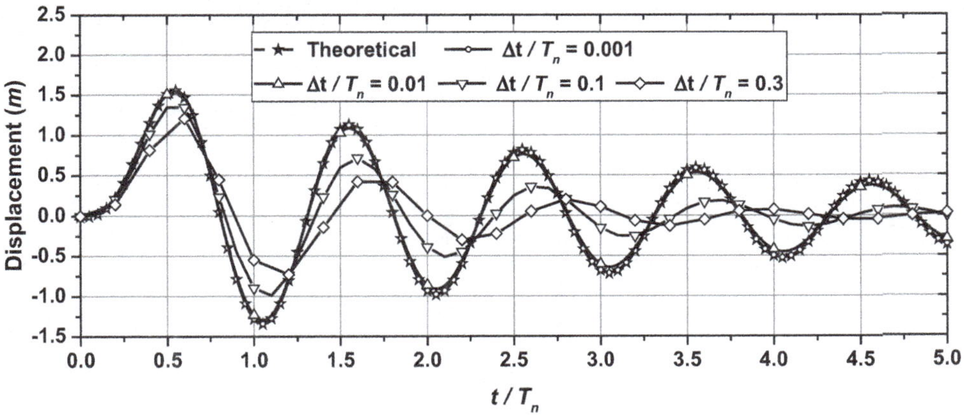

Fig. 3.40 Effect of time step in solution of damped SDF system to half-cycle sine pulse force by Wilson theta (M = 0.2533 kg, K = 10 N/m, ζ = 0.05, and $P0$ = 10 N)

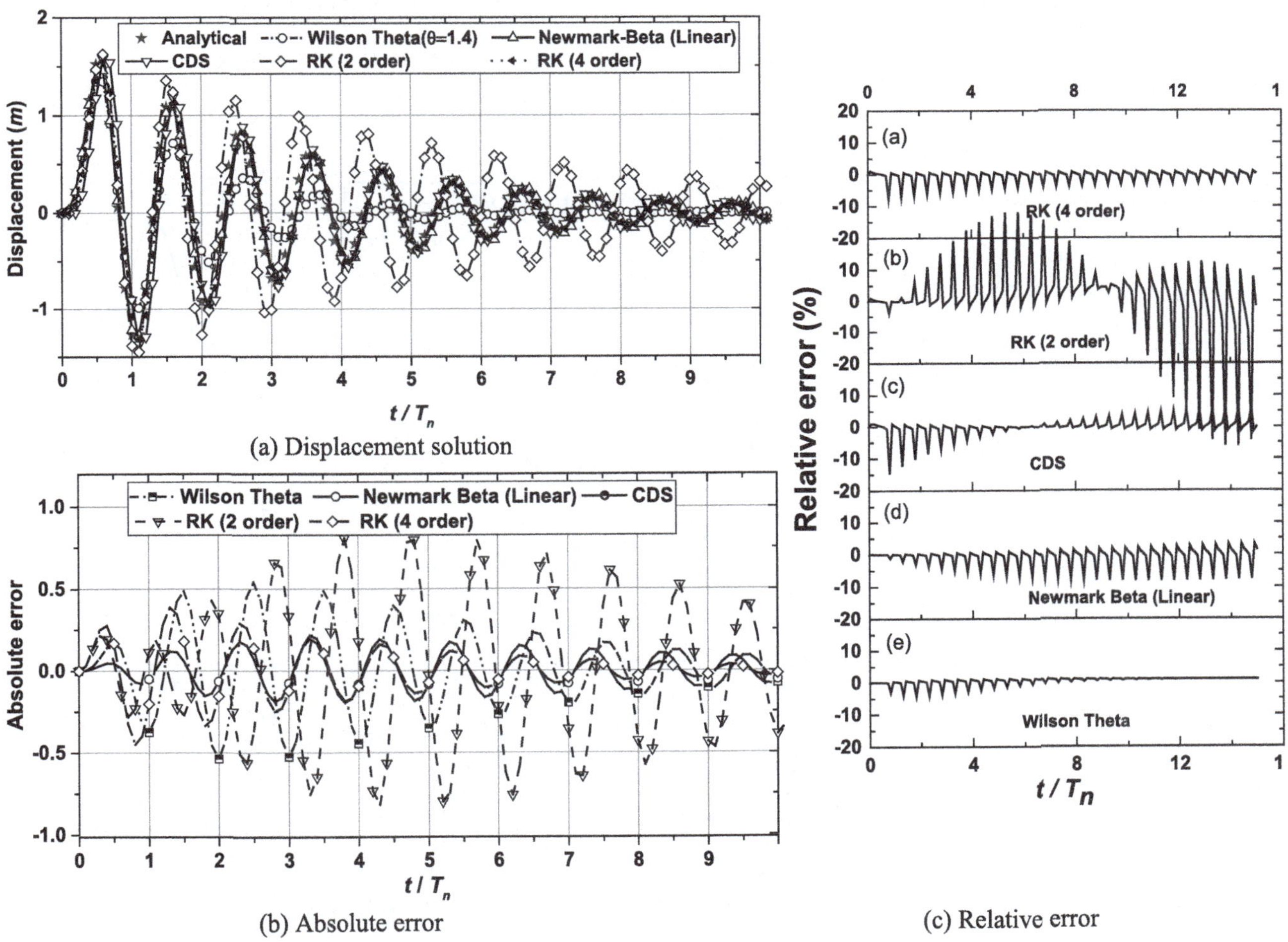

Fig. 3.41 Damped SDF system to half-cycle sine pulse force (M = 0.2533 kg, K = 10 N/m, ζ = 0.05, $P0$ = 10 N, and $\Delta t/T_n = 0.1$)

while compared to analytical solution. The absolute and relative errors are depicted in Fig. 3.41a–c. While looking into the error term, minimum relative error is found in Wilson theta and RK fourth order. A further detailed analysis is performed to understand the numerical schemes accuracy through period elongation (PE) and amplitude difference (AD). The PE and AD for numerical schemes are shown in Fig. 3.42a, b for damped SDF system to half-cycle sine pulse force (M = 0.2533 kg, K = 10 N/m, ζ = 0.05 and $P0$ = 10 N).

The dynamic response of damped SDF system to half-cycle sine pulse force (M = 0.2533 kg, K = 10 N/m, ζ = 0.05 and $P0$ = 10 N) is plotted in Fig. 3.43. The time variation of the normalized deformation, $u(t)/(u_{st})_o$ is

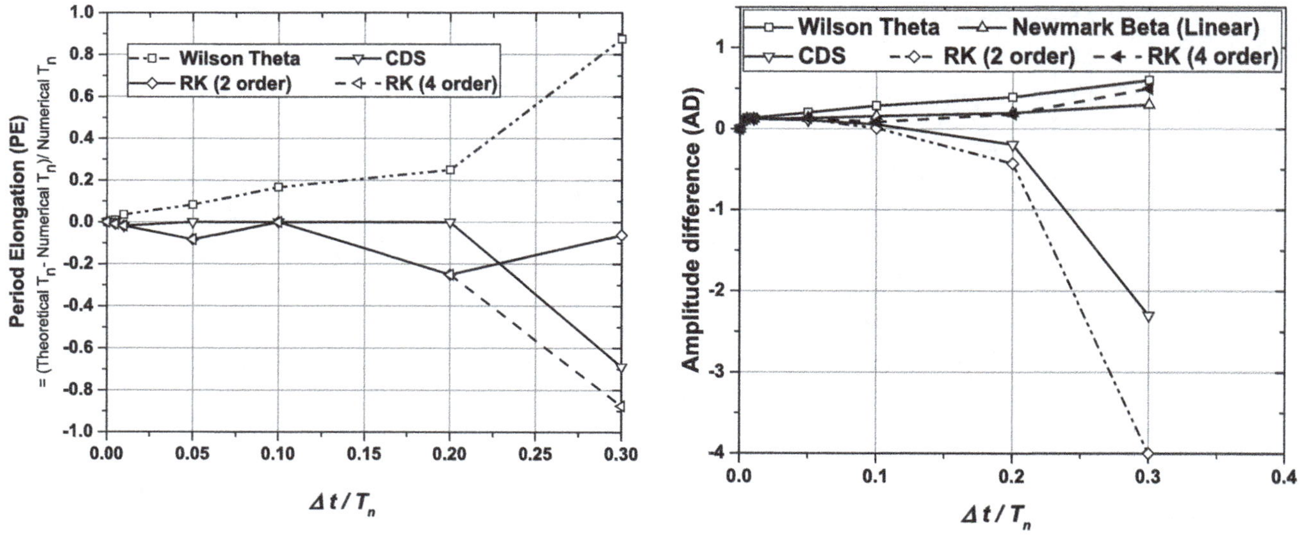

(a) Period elongation (b) Amplitude difference

Fig. 3.42 Credibility of numerical schemes in damped SDF system to half-cycle sine pulse force (M = 0.2533 kg, K = 10 N/m, ζ = 0.05 and $P0$ = 10 N)

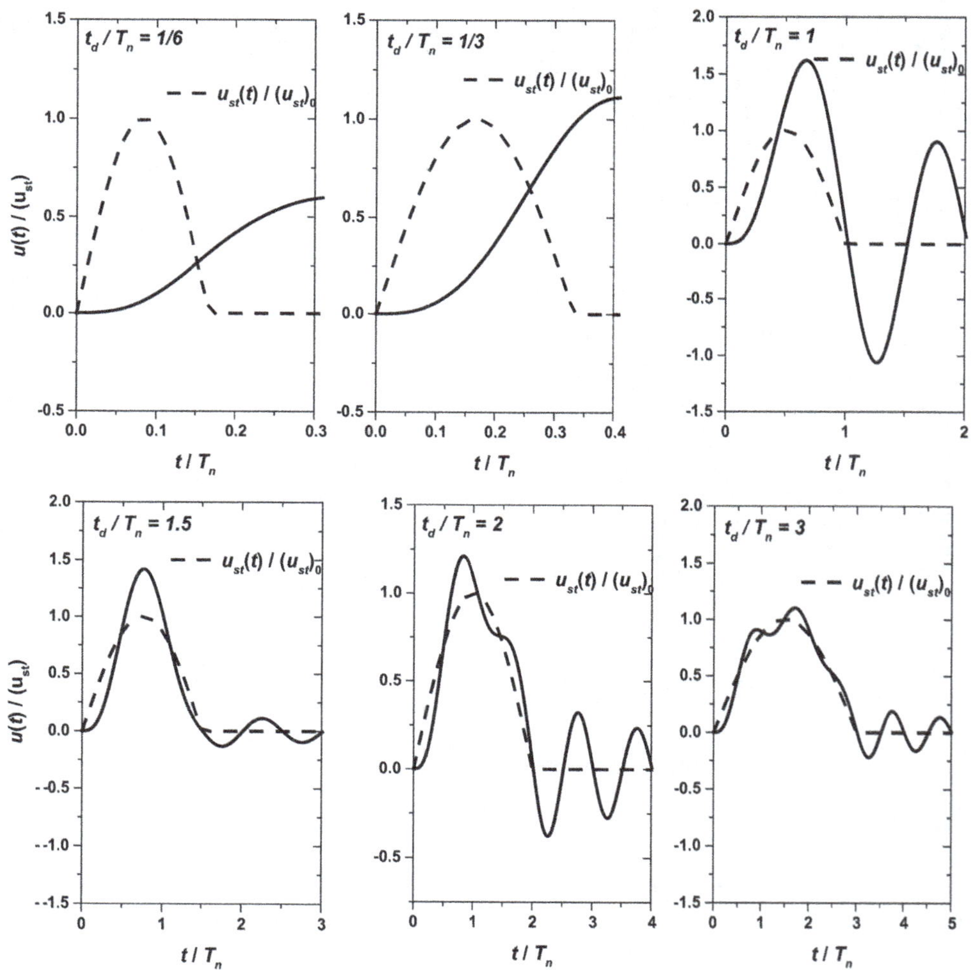

Fig. 3.43 Dynamic response of damped SDF system to half-cycle sine pulse force (M = 0.2533 kg, K = 10 N/m, ζ = 0.05, and $P0$ = 10 N)

illustrated for several values of t_d/T_n. The nature of the response is seen to vary greatly by changing just the duration t_d of the pulse. After the force pulse, the system oscillates freely about its undeformed configuration with constant amplitude for lack of damping. Static component of the analytical solution is also plotted. The difference between the two curves is an indication of the dynamic effects, which are seen to be small for higher t_d values. This implies that the force is varying slowly relative to the natural period T_n of the system.

Example 3.10 A SDOF system is shown in Fig. 3.37a with $M = 0.2533$ kg, $K = 10$ N/m. Use EL-centro earthquake as input time history. Find the displacement for damping ratio = 5%. And also plot an acceleration response spectrum.

Solution

Here, $M = 0.2533$ kg, $K = 10$ N/m and $C = 0.05$

The time response of a SDOF system under El-centro ground acceleration is governed by the differential equation:

$$m\ddot{x}(t) + c\dot{x}(t) + kx(t) = -m\ddot{x}_\mathrm{g} \tag{3.124}$$

It can be written as

$$\ddot{x}(t) + 2\varsigma\omega\dot{x}(t) + \omega^2 x(t) = -\ddot{x}_\mathrm{g} \tag{3.125}$$

To get the displacement response of SDOF system, Eq. (3.124) is integrated over time for 5% damping. For acceleration response, one should vary the natural period of vibration and represent the maximum value of each response for the fixed damping ratio. For the acceleration response, SDOF system displacement is multiplied with ω^2. For obtaining the solution, a MATLAB code is written (Appendix 1), and displacement response and acceleration response spectrum are displayed in Fig. 3.44.

Exercise Problems

1. Consider the system shown in Fig. 3.45. The bar is weightless and rigid. Obtain the natural frequency of the system.

   Ans: $\omega_n = \sqrt{\frac{k_1k_2l_1^2 + k_2k_3l_2^2}{m\left(k_1l_1^2 + k_2l_2^2 + k_3l_3^2\right)}}$

2. The column of the water tank is 90 m high and has reinforced concrete column with a tubular cross section of inner diameter 2.3 m and outer diameter 3 m. The tank mass when filled with water will be equal to 3 × 105 kg. By neglecting the mass of the column and assuming Young's modulus of reinforced concrete as 30 GPa, determine the following
   a. Natural frequency of transverse vibration of the tank.
   b. Response of water tank due to initial transverse displacement of 0.2 m and nil initial velocity.
   c. Maximum value of velocity and acceleration experienced by the tank.

   Ans: 1.035 rad/s, 0.2 sin $(1.035t + 0.5\pi)$, 0.207 m/s, 0.214 m/s^2.

3. A massless simply supported beam of flexural rigidity 450 N m^2 is attached to a spring of stiffness 50 N/m and 66.67 N/m as shown in Fig. 3.46. A mass of 10 kg is attached at one end. Obtain the angular frequency of the system. The system is excited by harmonic excitation F sin $10t$ applied in vertical direction. If this force is statically applied, the displacement of the system is 14 mm. Obtain maximum dynamic displacement of the system and the force F.

   Ans: (2 rad/s, 0.56 mm, 0.56 N)

4. Obtain the solution of free vibration of an oscillating spring–mass system with natural frequency ωn and over critical damping ratio ζ, where ωd = ωn$\sqrt{|1 - \zeta^2|}$ and A and B are constants.

   Ans: $x(t) = (A\sinh\omega_\mathrm{d}t + B\cosh\omega_\mathrm{d}t)\mathrm{e}^{-\varsigma\omega_n t}$

5. An empty elevated water tank has a fundamental period of vibration 0.5 s. If 2000 kg mass of water is filled in the tank, its fundamental period is lengthened to 1 s. What is the initial horizontal stiffness of the tank and the mass of the tank without water? If the damping ratio of tank is 0.05, find the damping coefficient.

   Ans. $k = 105278.8$ N/m, $m = 666.6$ kg, $c = 837.4$ N/m/s

6. Find the response of undamped SDOF system to symmetrical triangular pulse load as represented in Fig. 3.47 in terms of static displacement, xst using Duhamel integral method fort < td/2, td/2 < t < td.

   Ans. $x(t) = \frac{2x_\mathrm{st}}{t_\mathrm{d}}\left(t - \frac{\sin\omega t}{\omega}\right)$ for $(0 < t < t_\mathrm{d}/2)$
   $x(t) = \frac{2x_\mathrm{st}}{t_\mathrm{d}}\left[\left(t_\mathrm{d} - t + \frac{2\sin\omega(t - t_\mathrm{d}/2)}{\omega}\right) - \frac{\sin\omega t}{\omega}\right]$ for $t_\mathrm{d}/2 < t < t_\mathrm{d}$,

7. Write equation of motion for single-degree-of-freedom system without damping. Derive a solution of single-degree-of-freedom system without damping and without external force.

8. A 25 N weight is suspended from a spring which has a stiffness $k = 0.42$ N/mm. The weight is given an upwards velocity of 5 m/s, when it is 40 mm above its equilibrium position. Find its maximum upward displacement. Assume positive displacement is downward.

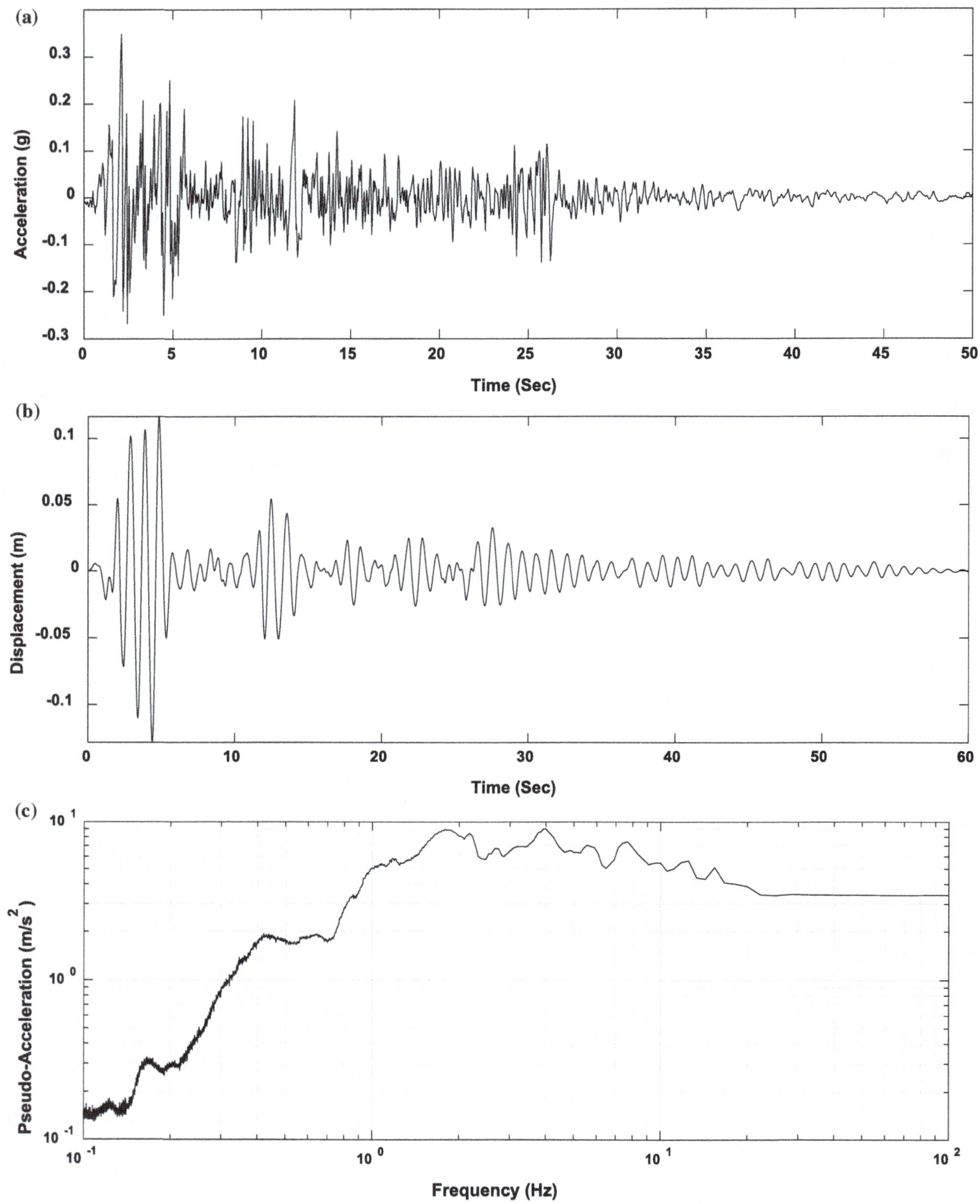

Fig. 3.44 Ground acceleration and response of SDOF. **a** Input time history (EL-centro earthquake), **b** displacement response of SDOF, **c** acceleration response spectrum for SDOF

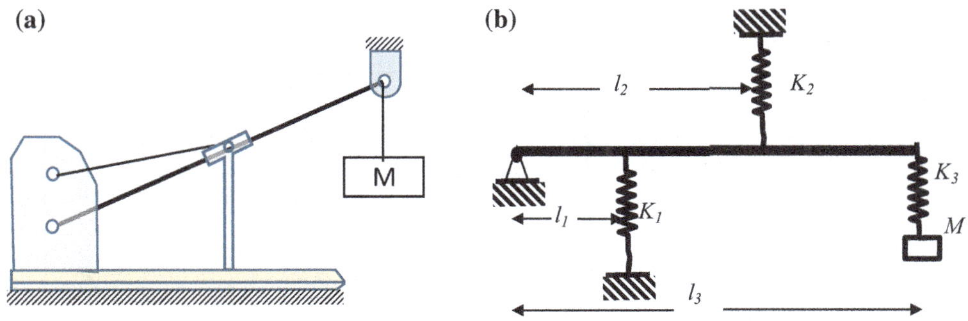

Fig. 3.45 **a** Crane **b** rigid bar system

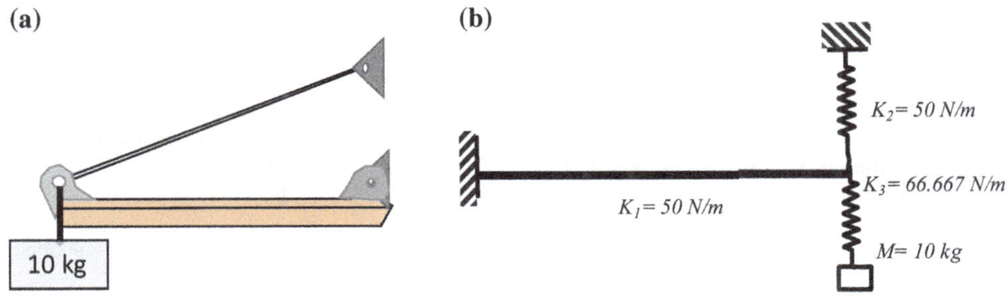

Fig. 3.46 **a** Crane **b** simply supported beam with spring

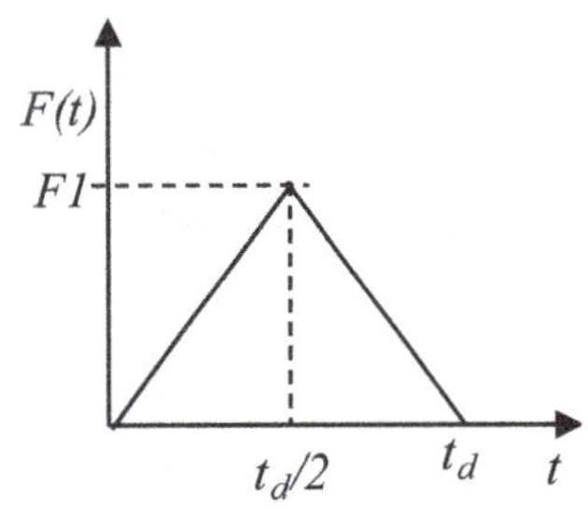

Fig. 3.47 Triangular pulse load

9. What is meant by underdamped, critically damped, and overdamped system? Write one example of each? The addition of damping to an initially undammed spring–mass system increases the period of vibration by 15%. Find the value of the damping ratio?
10. A mass of 2 kg is supported by a damped spring, where the damping constant (c) is 2.8 Ns/m. The mass is displaced 20 mm from its equilibrium position and released from rest, and the frequency is observed to be 2.5 Hz. Find
    (a) The displacement after five complete cycles
    (b) The time taken for the displacement to be reduced to 5% of the initial displacement.
11. What is dynamic response factor (DRF)? Plot the curve of dynamic response factor versus frequency ratio for a damped system subjected to harmonic excitation and give the value of this factor at resonance in terms of damping ratio (ζ).
12. Convert the following system into single-degree-of-freedom system and find the equivalent stiffness of the system (Fig. 3.48).

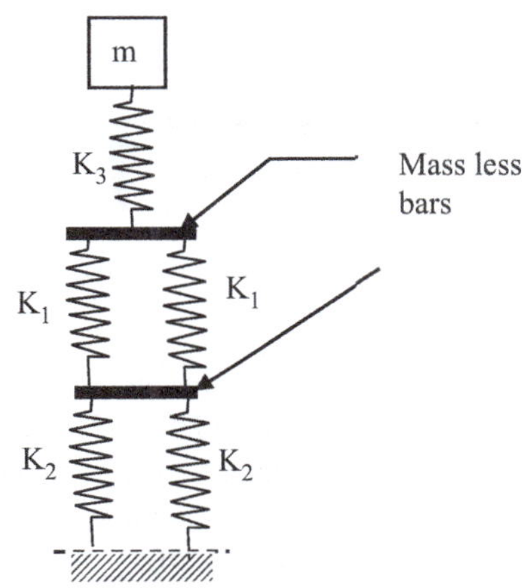

Fig. 3.48 Structure on two mass-less bars

13. Explain closely spaced modes? How the response is combined for closely spaced modes?
14. A massless cantilever beam with circular cross section is attached to a spring of stiffness 100 N/m as shown in the figure. A mass of 10 kg is attached at one end. Length of the beam = 2 m; Diameter of the beam = 0.05 m; Young's modulus of the material = 2 × 1011 N/m^2.
15. Assuming motion of free vibration as 'A sinωt', write the energy balance equation of the combined system. Using above formula, derive the natural frequency 'ω' of the system.
16. Carefully watch the following response of a single-degree-of-freedom system. Name the systems (Fig. 3.49).
17. What is resonance? At resonance, what is the ratio of dynamic to static amplitude of single-degree-of-freedom system with damping ratio ζ.

Fig. 3.49 Dynamic response curve for SDOF systems

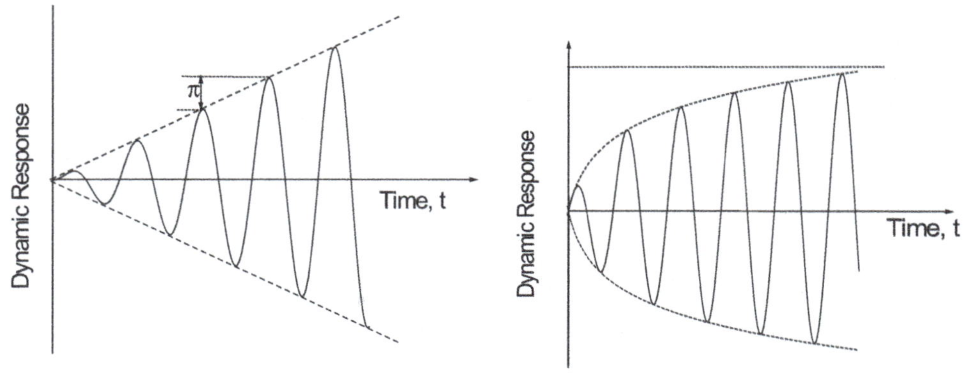

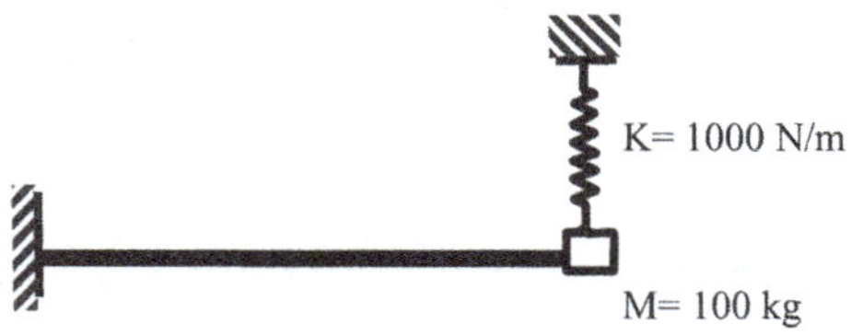

Fig. 3.50 Mass-less cantilever beam attached to a spring

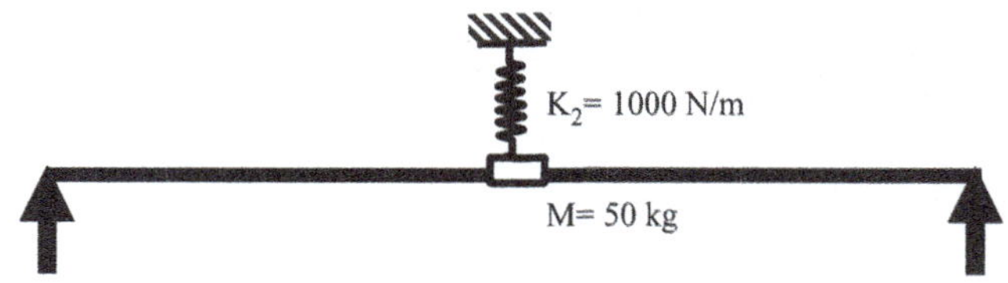

Fig. 3.52 A simply supported beam with a vertical beam

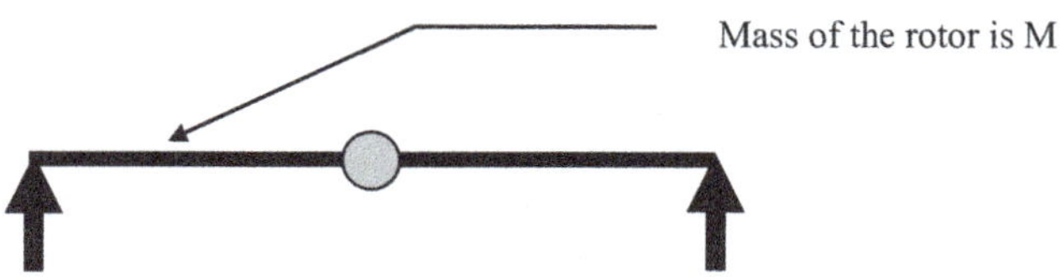

Fig. 3.51 A simply supported beam with a rotor

18. A massless cantilever beam attached to a spring of stiffness 100 N/m. A mass of 10 kg is attached at one end. Length of the beam = 2 m; geometry = circular; diameter of the beam = 0.05 m; Young's modulus of the material = 2E11 N/m^2 (Fig. 3.50).
19. Write equations of motion for single-degree-of-freedom system with damping. Convert the same in terms of frequency (ωn) and damping ratio ζ (damping/critical damping).
20. What are the complimentary function and particular integral? With the sketch explain their nature considering spring–mass–damper system.
21. Evaluate the natural frequency of the following system using energy principle and neglecting mass of the beam. Hint: Use simply supported beam stiffness. Length of the beam is l, moment of inertia is I, and Young's modulus is E (Fig. 3.51).
22. Considering mass per unit length of beam is m and assuming velocity is varying linearly, find the difference in natural frequency of the system with and without mass of beam.
23. A simply supported circular beam of length 2000 m and diameter of 50 mm (0.05 m) is attached to a spring of stiffness 50 N/m. Assuming the total mass of the beam of 50 kg is lumped at center, evaluate the frequency of the system. Hint: Use parallel spring approach and Young's modulus is 2E11 N/m^2 (Fig. 3.52).
24. Name the numerical methods for determining the eigenvalues and eigenvectors. Why implicit numerical integration methods are preferred over the explicit integration schemes?
25. Derive a solution of single-degree-of-freedom system without damping and without external force.
26. With graph, explain the differences between the motions of damped and undamped single-degree-of-freedom system with external force and different frequency ratios.
27. Evaluate the fundamental frequency for the below figure considerations with (a) 3 nodes, (b) 5 nodes, and (c) 9 nodes, and compare with its classical solution given by (Fig. 3.53).
28. What is Duhamel integration? Where it is applicable?

Fig. 3.53 A vertical box with holes

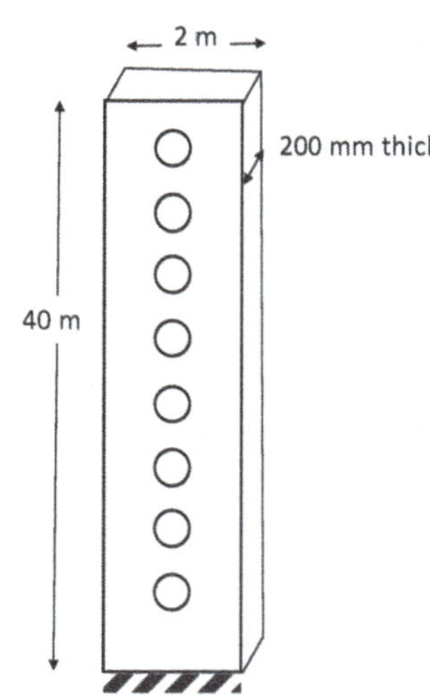

Appendix 1: Subroutines

A.1.1 MATLAB Code for Different Integration Schemes

MATLAB code for SDOF system for the following integration schemes is presented in this Appendix.

1. Newmark-beta—Linear acceleration,
2. Wilson Theta,
3. Runge–kutta fourth order,
4. Runge–Kutta second order,
5. Central difference scheme.

Function inputs: m—mass; K—stiffness; c—stiffness; p—force; x0—initial displacement; v0—final displacement; dt—time step; t—time

Function outputs: Displacement, velocity, and acceleration

1. **function [NBLx,NBLv,NBLaccx,t] = Newmark_linear (m,K,c,p,x0,v0,dt,t)**

```
gamma=1/2;betaL=1/6; % Linear acceleration

%Initial conditions
NBLx(1)=x0;
NBLv(1)=v0;
time=0;
[imin,imax]=size(t);
%newmark beta
NBLti=gamma/betaL;
NBLa=m/(betaL*dt)+NBLti*c;
NBLb=m/(2*betaL)+dt*c*((NBLti/2)-1);
NBLaccx(1)=(p(1)-K*x0-c*v0)/m;
NBLKbar=K+(3*c/dt)+(6*m/(dt*dt));

for i=1:imax-1
time=time+dt;
NBLdPbar=(p(i+1)-p(i))+NBLa*NBLv(i)+NBLb*NBLaccx
(i);
NBLdx=NBLdPbar/NBLKbar;
NBLdv=NBLti*NBLdx/dt-NBLti*NBLv(i)+dt*NBLaccx(i)*
(1-NBLti/2);
NBLv(i+1)=NBLv(i)+NBLdv;
NBLx(i+1)=NBLx(i)+NBLdx;
NBLdaccx=NBLdx/(betaL*dt*dt)-NBLv(i)/(betaL*dt)-
NBLaccx(i)/(2*betaL);
NBLaccx(i+1)=NBLaccx(i)+NBLdaccx;
End
```

2. **function [wx,wv,waccx,t] = Wilsontheta(m,K,c,p,x0, v0,dt,t)**

```
clc
%Wilson theta Method
theta=1.4;
%Initial conditions
x(1)=x0;
v(1)=v0;
time=0;
[imin,imax]=size(t);

ti=theta*dt;
a=m*6/ti+3*c;
b=3*m+(ti*c)/2;
accx(1)=(p(1)-K*x0-c*v0)/m;
kbar=K+3*c/ti+6*m/(ti^2);

for i=1:imax-1
time=time+dt;
dPbar= theta*(p(i+1)-p(i))+ a*v(i)+b*accx(i);
dxbar=dPbar/kbar;
accxbar=6*dxbar/(ti^2)-6*v(i)/(ti)-3*accx(i);
daccx=accxbar/theta;
dv=dt*accx(i)+dt*daccx/2;
dx=dt*v(i)+(accx(i)*dt^2)/2+(daccx*dt^2)/6;
v(i+1)=v(i)+dv;
x(i+1)=x(i)+dx;
accx(i+1)=accx(i)+daccx;
end
wx=x;
wv=v;
waccx=accx;
```

3. **function [RK4x,RK4v] = RK4(m,K,c,p,x0,v0,dt,t)**

```
%Initial conditions
RK4x(1)=x0;
RK4v(1)=v0;
time=0;
[imin,imax]=size(t);
dh=dt;

for i=1:imax-1
    time=time+dt;
    k0=dh*fRK(t(i),RK4x(i),RK4v(i));
    j0=dh*gRK(t(i),RK4x(i),RK4v(i),p(i),m,K,c);
    k1=dh*fRK(t(i)+dh/2,RK4x(i)+k0/2,RK4v(i)+j0/2);
    j1=dh*gRK(t(i)+dh/2,RK4x(i)+k0/2,RK4v(i)+j0/2,p(i),m,
    K,c);
```

```
    k2=dh*fRK(t(i)+dh/2,RK4x(i)+k1/2,RK4v(i)+j1/2);
    j2=dh*gRK(t(i)+dh/2,RK4x(i)+k1/2,RK4v(i)+j1/2,p(i),m,
    K,c);
    k3=dh*fRK(t(i)+dh,RK4x(i)+k2,RK4v(i)+j2);
    j3=dh*gRK(t(i)+dh,RK4x(i)+k2,RK4v(i)+j2,p(i),m,K,c);
    RK4x(i+1)=RK4x(i)+(k0+2*k1+2*k2+k3)/6;   %
    Displacement
    RK4v(i+1)=RK4v(i)+(j0+2*j1+2*j2+j3)/6; % Velocity
End
```

function [F] = fRK(t,x1,v1)

```
%UNTITLED5 Summary of this function goes here
% Detailed explanation goes here
F = v1;
End
```

function [G] = gRK(t,RK4x1,RK4v1,p1,m,K,c)

```
%UNTITLED7 Summary of this function goes here
% Detailed explanation goes here
G = (p1-c*RK4v1-K*RK4x1)/m;
End
```

4. **function [RK2x,RK2v] = RK2(m,K,c,p,x0,v0,dt,t)**

```
clc
%Initial conditions
RK2x(1)=x0;
RK2v(1)=v0;
time=0;
[imin,imax]=size(t);
dh=dt;

for i=1:imax-1
    time=time+dt;
    k0=dh*fRK(t(i),RK2x(i),RK2v(i));
    j0=dh*gRK(t(i),RK2x(i),RK2v(i),p(i),m,K,c);
    k1=dh*fRK(t(i)+dh,RK2x(i)+k0,RK2v(i)+j0);
    j1=dh*gRK(t(i)+dh,RK2x(i)+k0,RK2v(i)+j0,p(i),m,K,c);
    RK2x(i+1)=RK2x(i)+(k0+k1)/2; % Displacement
    RK2v(i+1)=RK2v(i)+(j0+j1)/2; % Velocity
End
```

function [F] = fRK(t,x2,v2)

```
%UNTITLED5 Summary of this function goes here
% Detailed explanation goes here
F = v2;
End
```

function [G] = gRK2(t,RK2x1,RK2v1,p2,m,K,c)

```
%UNTITLED7 Summary of this function goes here
% Detailed explanation goes here
G = (p2-c*RK2v1-K*RK2x1)/m;
end
```

5. **function [CDx,CDv,CDaccx,t] = CDS(m,K,c,p,x0,v0,dt,t)**

```
%Initial conditions
CDx(2)=x0;
CDv(1)=v0;
time=0;
[imin,imax]=size(t);

%central diffence
% CDti=gamma/beta;
CDa=m/(dt*dt)-c/(2*dt);
CDb=K-2*m/(dt*dt);
CDaccx(1)=(p(1)-K*x0-c*v0)/m;
CDx(1)=CDx(2)-dt*CDv(1)+dt*dt*CDaccx(1)/2;
CDKbar=c/(2*dt)+m/(dt*dt);
for i=1:imax-1
time=time+dt;

if(i<imax-1)
CDPbar=p(i)-CDa*CDx(i)-CDb*CDx(i+1);
CDx(i+2)=CDPbar/CDKbar;
CDv(i+1)=(CDx(i+2)-CDx(i))/(2*dt);
CDaccx(i+1)=(CDx(i+2)-2*CDx(i+1)+CDx(i))/(dt*dt);
end
end
```

A.1.2 MATLAB Code for Example 3.10

Integration schemes: ode45 solver
Function inputs: *time*=time, *acc*=acceleration, *fTn*=frequency time step, *k*=stiffness, *m*=mass, *C*=damping,*w*=frequency in *rad/sec.*
Function output: Displacement

```
fTn=sqrt(k/m)
%fTn = 1.;
[time,disp] = ode45(@find,t,y0,[],C,fTn,timein,accin);
plot(time,disp(:,1));
for fTn = 0:0.01:50
[time,disp] = ode45(@find,time,disp,[],C,fTn,timein,accin);
dispmax = max(abs(disp(:,1)));
dispmax(n) = dispmax;
n = n+1;
```

```
End

function dydt = find(time,acc,xi,fTn,timein,accin)
w= 2*pi/fTn;
acc = interp1(timein,accin,time);
dydt = [y(2); -acc -(2*w*xi)*y(2)-w^2*y(1)];
```

References

1. Chopra AK (2011) Dynamics of structures—theory and applications to earthquake engineering, 4th Ed. Pearson Education
2. Rao SS (2004) Mechanical vibrations, 4th Ed. Prentice-Hall, Upper Saddle River, NJ
3. Rajasekaran S (2009) Structural dynamics of earthquake engineering: theory and application using mathematica and Matlab, Woodhead Publishing in Materials, CRC Press, India
4. Jazar RN (2017) Vehicle dynamics theory and application, 3rd edn. Springer, New York
5. Bathe KJ (1996) Finite element procedures, 2nd edn. Prentice-Hall of India, New Delhi
6. Humar JL (1990) Dynamics of structures. Prentice Hall
7. Krishnamoorthy CS (1995) Finite element analysis: theory and programming, 2nd Ed. TATA—McGraw Hill

Analysis of Multi-degree-of-Freedom Systems

4

G. R. Reddy, R. K. Verma, Binu Kumar, and M. Eswaran

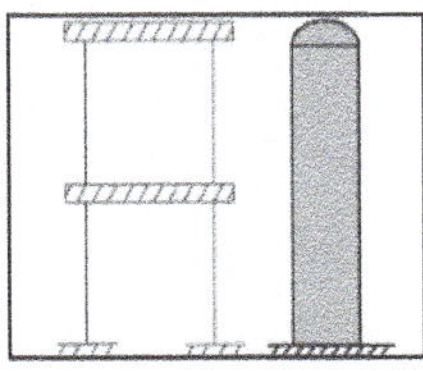

Practical Structures, Equipment and Piping Systems are multi-degree of Freedom System

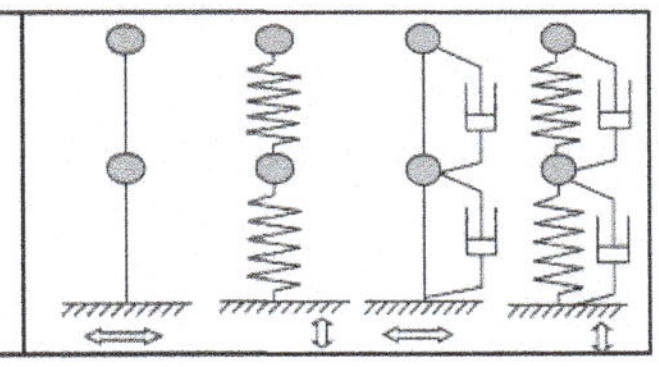

Symbols

$A(t)$	Pseudo-acceleration of mass
c	Damping coefficient
C	Damping coefficient matrix
E	Young's modulus
f_n	Frequency
f_s	Force
f	Spatial load vector
f_s	Static force
h	C.G. height
I	Moment of inertia
K	Stiffness
K_v and K_h	Stiffness of system along vertical and horizontal directions, respectively
m	Mass of the system
M_b	Overturning moment
S_a	Spectral value
t	time
T_n	Time period
V_b	Base shear
x	Relative displacement between ground and mass
$\dot{x}$	Velocity
$\ddot{x}_g$	Ground acceleration
ω_n	Circular frequency of system
ζ	Damping ratio of the system
ϕ	Mode shape
Γ_n	Mass participation factor

G. R. Reddy (✉) · R. K. Verma · B. Kumar · M. Eswaran
Bhabha Atomic Research Centre, Mumbai, India
e-mail: rssred@barc.gov.in

R. K. Verma
e-mail: rajeev@barc.gov.in

B. Kumar
e-mail: binu@barc.gov.in

M. Eswaran
e-mail: eswaran@barc.gov.in

G. R. Reddy et al., *Textbook of Seismic Design*,
https://doi.org/10.1007/978-981-13-3176-3_4

4.1 Introduction

The structures, equipment, and piping systems can be idealized as single-degree-of-freedom system (SDOF) or combination of multi-degree-of-freedom systems (MDOF). To evaluate the seismic response of these systems using idealized models, frequencies and mode shapes of the idealized systems need to be evaluated. In this chapter, the importance of frequencies, mode shapes, and procedures for evaluating these quantities is discussed in detail. Procedure to evaluate responses of the structures, equipment, and piping systems subjected to earthquake motion is also discussed.

4.2 Response of Single-Degree-of-Freedom System Due to Earthquake

Consider a single- storey building as shown in Fig. 4.1a or a vessel supported on skirt support as shown in Fig. 4.1b. The horizontal and vertical motions of these systems can be idealized as SDOF system as shown in Fig. 4.1c, d by lumping the mass at roof of the single-storey building or by

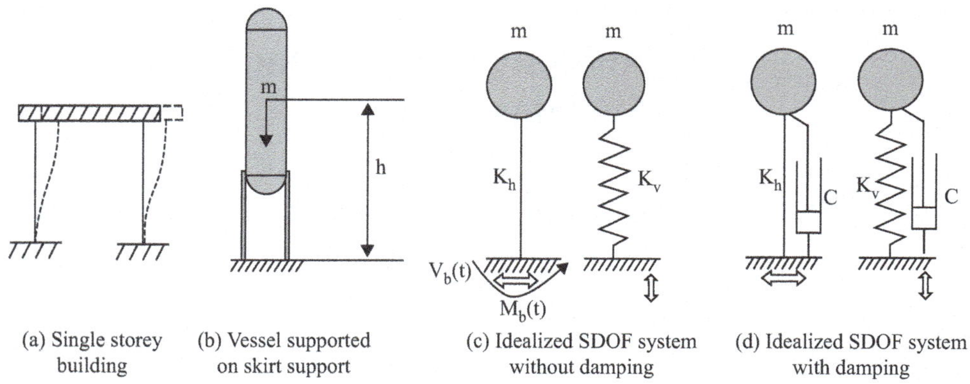

Fig. 4.1 Idealization of a single-storey building and a vessel

lumping the mass at the centre of the mass in the case of the vessel. For horizontal direction, both bending and shear deformations should be considered which depend on the aspect ratio of columns or skirt. If the height to the width ratio is more than 10, then the bending flexure deformations will dominate and if the aspect ratio is less than 10, then shear deformations are predominant. However, in the case of analysis by using standard finite element (FE) packages, both the effects can be considered all the time.

Considering as a linear SDOF system, the equation of motion for the base excitation explained in Chap. 3 is given as

$$m\ddot{x}(t) + c\dot{x}(t) + kx(t) = -m\ddot{x}_g(t) \tag{4.1}$$

where

m	mass of the system
x	relative displacement between ground and mass
c	damping coefficient
K_v & K_h	stiffness of the system along vertical and horizontal directions, respectively

The term "$-m\ddot{x}_g(t)$" represents the effective earthquake force. The force is equal to the mass times the ground acceleration acting on the system. The force applied depends on the inertia of the system and base acceleration to which the system is subjected. This process is called 'self-induced vibration'. Dividing Eq. (4.1) by mass m gives

$$\ddot{x}(t) + \frac{c}{m}\dot{x}(t) + \frac{k}{m}x(t) = -\ddot{x}_g(t) \tag{4.2}$$

The term $\frac{c}{m}$ can be represented by $2\zeta\omega_n$, where ω_n is the circular frequency of the system and ζ is the damping ratio of the system, which is represented as the percentage of the critical damping. $\frac{k}{m}$ can be replaced by ω_n^2. So, Eq. (4.2) can be represented as

$$\ddot{x}(t) + 2\zeta\omega_n\dot{x}(t) + \omega_n^2 x(t) = -\ddot{x}_g(t) \tag{4.3}$$

It is clear that for a given excitation $\ddot{x}_g(t)$, the deformation response $x(t)$ of the system depends only on the natural frequency ω_n or time period T_n of the system and its damping ratio ζ. Thus it can be written that

$$x = F_n(t, T_n, \zeta) \tag{4.4}$$

Thus any two SDOF systems having same value of T_n and ζ and subjected to same ground motion will have same deformation response $x(t)$, as response is dependent on time period or natural frequency and damping of the system. The maximum deformation of the system or the displacement of the mass relative to the moving ground is the quantity of the greatest interest in the structural engineering.

Once the deformation $x_{max}(t)$ has been evaluated by the dynamic analysis of the structure, the internal forces can be determined by static analysis of the structure at each time instant. The equivalent static force f_s is defined as

$$f_s(t) = kx_{max}(t) \tag{4.5}$$

where k is the stiffness of the system. Expressing k in terms of mass and natural frequency (ω_n),

$$f_s(t) = m\omega_n^2 x_{max}(t) = mA(t) \tag{4.6}$$

where $A(t) = \omega_n^2 x_{max}(t)$. Here $A(t)$ is termed as 'Pseudo-acceleration response' of the system and can be calculated from the deformation response $x_{max}(t)$. The base shear $V_b(t)$ and the base overturning moment $M_b(t)$ can be calculated as

$$V_b(t) = f_s(t)$$
$$M_b(t) = hf_s(t)$$

where h is the height of the mass above the base. Hence,

$$\begin{aligned} V_b(t) &= mA(t) \\ M_b(t) &= hmA(t) \end{aligned} \tag{4.7}$$

The displacement or pseudo-acceleration values can be obtained for various SDOF systems having different time periods or frequencies. As explained in Chap. 1, a plot of these quantities (maximum) as a function of natural vibration period (T_n) of the system or a related parameter such as circular frequency (ω_n) or natural frequency (f_n), is called the 'response spectrum' for that quantity. Each such plot for SDOF systems having a fixed damping ratio (ζ) and several such plots for different values of ζ are included to cover the range of damping values encountered in actual structures. In other words, the response spectra give the maximum responses of SDOF systems for a given damping.

4.3 Response of Multi-degree-of-Freedom System Due to Earthquake

Consider a two-storey structure as shown in Fig. 4.2a or a vessel as shown in Fig. 4.2b. The horizontal and vertical motions of these systems can be idealized as two-degree-of-freedom system as shown in Fig. 4.2c, d by lumping the mass at the roofs of the structure or by lumping the mass at top and middle in the case of the vessel.

At this stage, it is essential to discuss the procedure of discretization of the system. This is explained considering a cantilever and a simply supported beam as shown in Figs. 4.3 and 4.4.

For better accuracy, $(2m_n + 1)$ nodes are recommended to predict the eigenvalues and eigenvectors. Where 'm_n' is the number of required modes. Figure 4.3 shows the discretization of cantilever beam and Fig. 4.4 shows the discretization of simply supported beam. Considering as a MDOF system, the equation of motion for the base excitation is written as

$$[M]\{\ddot{x}\}+[C]\{\dot{x}\}+[K]\{x\}=-[M]\ddot{x}_g\{1\} \tag{4.8}$$

$$\text{Let } \{x\}=\{\phi\}X \tag{4.9}$$

where $\{\phi\}$ is the mode shape and X is the generalized displacement. Substituting Eq. (4.9) into Eq. (4.8), it modifies to

$$[M]\{\phi\}\ddot{X}+[C]\{\phi\}\dot{X}+[K]\{\phi\}X=-[M]\ddot{x}_g\{1\} \tag{4.10}$$

Multiplying Eq. (4.10) with $\{\phi\}^T$ on both sides, it modifies to

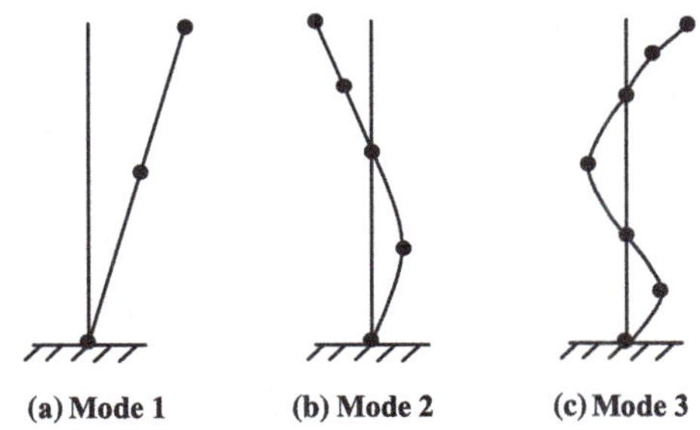

Fig. 4.3 Discretization of cantilever beam

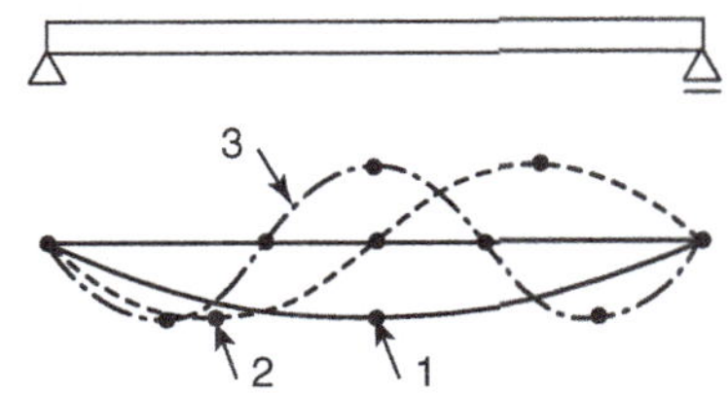

Fig. 4.4 Discretization of simply supported beam

$$\{\phi\}^T[M]\{\phi\}\ddot{X}+\{\phi\}^T[C]\{\phi\}\dot{X}+\{\phi\}^T[K]\{\phi\}X = -\{\phi\}^T[M]\ddot{x}_g\{1\} \tag{4.11}$$

Using orthogonal properties of mode shapes (i.e., $\{\phi\}^T[M]\{\phi\}=[1]$ and $\{\phi\}^T[K]\{\phi\}=[\omega_n^2]$), Eq. (4.11) can be written as

$$[1]\ddot{X}_n+[2\varsigma\omega_n]\dot{X}_n+\left[\omega_n^2\right]X_n=-\Gamma_n\ddot{x}_g \tag{4.12}$$

In the above Eq. (4.12), the matrices are diagonal and hence represents 'n' uncoupled equations of motion and each represents a SDOF system with a factor on right-hand side called participation factor. Equations (4.3) and (4.12) differ by this factor only. It means that if one knows the response of SDOF system and by multiplying it with participation factor, one can get the modal response of MDOF system as

$$\{\ddot{x}\}_n=\Gamma_n\{\phi\}_n\ddot{X}_n,\ \{\dot{x}\}_n=\Gamma_n\{\phi\}_n\frac{\ddot{X}_n}{\omega_n}\ \text{and}\ \{x\}_n=\Gamma_n\{\phi\}_n\frac{\ddot{X}_n}{\omega_n^2} \tag{4.13}$$

Fig. 4.2 Idealization of vessel and two-storey building

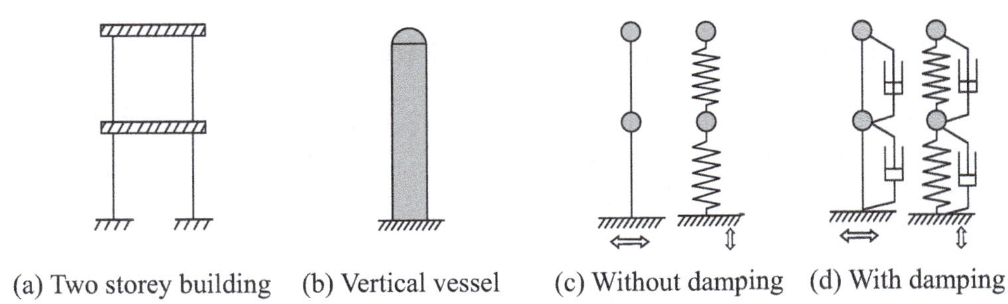

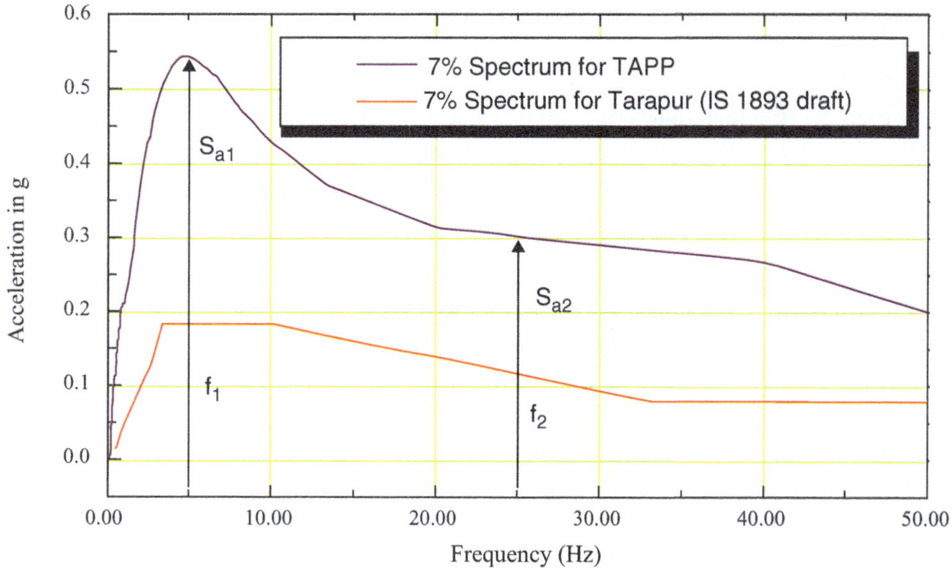

Fig. 4.5 Typical design response spectrum for 7% Damping

where $\ddot{X}_n$ are the spectral values such as S_{a1}, S_{a2}, etc., as shown in Fig. 4.5 for the corresponding frequencies f_1, f_2, etc.

Once the displacements are obtained at various nodes of the structural system or equipment, the forces can be obtained using Eq. (4.14).

$$[K]_e\{x\}_e = \{F\}_e \tag{4.14}$$

Hence, it can be concluded that once the frequencies are evaluated for MDOF system, the response accelerations can be obtained at those frequencies. Once the generalized coordinates (product of spectral accelerations and participation factors) are obtained, the responses at various nodes can be obtained by multiplying with mode shape.

Therefore, it is very important to evaluate the frequencies and mode shapes of a MDOF system more accurately. These can be obtained by solving the 'n' degree polynomial representing the MDOF system. Obtaining the roots of the polynomial is very difficult and hence, numerical procedures are used to evaluate the frequencies and mode shapes. These are also referred as eigenvalues and eigenvectors. Stiffness matrix for spring element and two-dimensional truss (only axial effect) are given in Annexure I. Natural frequencies and mode shapes of some common geometries are given in Annexure II.

4.3.1 Significance of Mass Participation Factors

As discussed in the previous section, from Eq. (4.12), participation factor can be represented as

$$\Gamma = \{\phi\}^T[M]\{1\}$$

and force vector can be represented as

$$\begin{aligned}\{F\} &= [M]\Gamma\{\phi\}^T S_a\{1\} \\ &= \Gamma\{\phi\}^T[M]\{1\}S_a \\ \{F\} &= \Gamma^2 S_a\end{aligned} \tag{4.15}$$

Square of the participation factor is the mass participated in the particular mode. Mass participation keeps reducing as the mode number increasing. This is explained with a cantilever beam as shown in Fig. 4.6.

The same concept may be expanded to complex structures in particular by subdividing into regular geometrical shapes as cantilever beam or plate, simply supported beam or plate, etc. These will have classical solution of eigenvalues and eigenvectors. Number of eigenvalues and eigenvectors may be decided based on excitation frequencies. In case of earthquake, it may be up to 50 Hz.

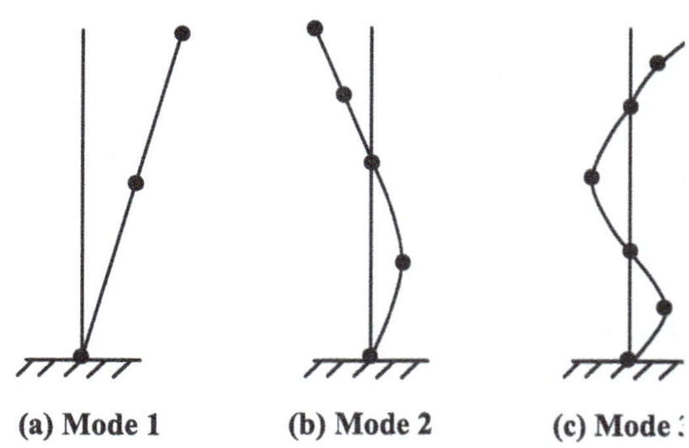

Fig. 4.6 Mode shape of a cantilever beam

4.4 Solution of Eigenvalues and Eigenvectors

As in the case of earthquake engineering, the eigenvalues and eigenvectors are called natural frequencies and mode shapes, which are free vibration characteristics of the structures or equipment. As explained in Chap. 3, natural frequencies do not depend on damping, and hence, Eq. (4.8) can be modified as

$$[M]\{\ddot{x}\} + [K]\{x\} = 0 \text{ or} \\ \{\ddot{x}\} + [M]^{-1}[K]\{x\} = 0 \tag{4.16}$$

Let

$$\{x\} = \{\phi\}\sin(\omega t) \\ \{\ddot{x}\} = -\omega^2\{\phi\}\sin(\omega t) \tag{4.17}$$

Substituting Eq. (4.17) in Eq. (4.16), then the free vibration characteristic of the system becomes

$$-\omega^2\{\phi\} + [M]^{-1}[K]\{\phi\} = 0 \tag{4.18}$$

$$[M]^{-1}[K]\{\phi\} = \omega^2\{\phi\} \tag{4.19}$$

Let

$$A = [M]^{-1}[K] \text{ and} \\ \lambda = \omega^2$$

Then Eq. (4.19) becomes

$$[A]\{\phi\} = \lambda\{\phi\} \tag{4.20}$$

Now the free vibration problem has been reduced to the general eigenvalue problem. Where ϕ is eigen vector and λ is eigen value which is square of ω_n. Generally, the solution of Eq. (4.20) is obtained using numerical techniques due to easiness and fast in computing by using computers. Commercially available finite element softwares use Subspace and Lanczos techniques due to its speed of solution. Hence, it is essential for the users to understand these techniques. Here a simplified technique for simple MDOF systems is discussed then followed by the detailed numerical methods.

4.5 Numerical Methods for Solving Eigenvalues and Eigenvectors

There are various numerical methods to obtain the eigenvalues and eigenvectors of Eq. (4.20) which has large number of degrees of freedom (i.e., large number of equations). Some of the methods used are:

(a) Simplified Technique,
(b) Inverse Iteration Technique,
(c) Simultaneous Iteration Technique,
(d) Subspace Iteration Technique,
(e) Lanczos Iteration Technique

Excluding simplified technique, the time required for computation of eigenvalues and eigenvectors reduces from technique b to e and it can be observed in the following explanation.

4.5.1 Simplified Technique

Consider a three-storey reinforced cement concrete (RCC) structure as shown in Fig. 4.7. The geometrical details and material properties of the structure are given in Table 4.1. The simplified model of the structure is shown in Fig. 4.8.

Neglecting the rotation, stiffness matrix of each element can be given as follows:

$$[K]_e = C\begin{bmatrix} 1 & -1 \\ -1 & 1 \end{bmatrix} \quad \text{where } C = \frac{12EI}{l^3}$$

Combined stiffness and mass matrices are given as follows:

$$[K] = 9 \times C\begin{bmatrix} 1 & -1 & 0 \\ -1 & 2 & -1 \\ 0 & -1 & 2 \end{bmatrix}; \ [M] = \begin{bmatrix} m_1 & & \\ & m_2 & \\ & & m_3 \end{bmatrix}$$

Mass of slab and beams considering columns are 1560, 1560, and 1560 kg at top, middle, and bottom of the structure, respectively. An additional mass of 500, 1000, and 500 kg also kept on top, middle, and bottom of the structure. Hence, the total mass is 2060, 2560, and 2060 kg at top, middle, and bottom of the structure. Initially, it can be assumed that the mode shape is of triangular shape and the values of the vector are shown in Fig. 4.9. Now, normalize the mode shape with mass as follows:

$$\phi^T M \phi = [\phi_1 \quad \phi_2 \quad \phi_3]\begin{bmatrix} m_1 & 0 & 0 \\ 0 & m_2 & 0 \\ 0 & 0 & m_3 \end{bmatrix}\begin{Bmatrix} \phi_1 \\ \phi_2 \\ \phi_3 \end{Bmatrix}$$

$$\phi^T M \phi = [1.00 \quad 0.67 \quad 0.33]\begin{bmatrix} 2060 & 0 & 0 \\ 0 & 2560 & 0 \\ 0 & 0 & 2060 \end{bmatrix}\begin{Bmatrix} 1.00 \\ 0.67 \\ 0.33 \end{Bmatrix}$$

$$\phi_n = \frac{\{\phi\}}{\sqrt{\phi^T M \phi}} = \begin{Bmatrix} 0.0171 \\ 0.0114 \\ 0.0056 \end{Bmatrix}$$

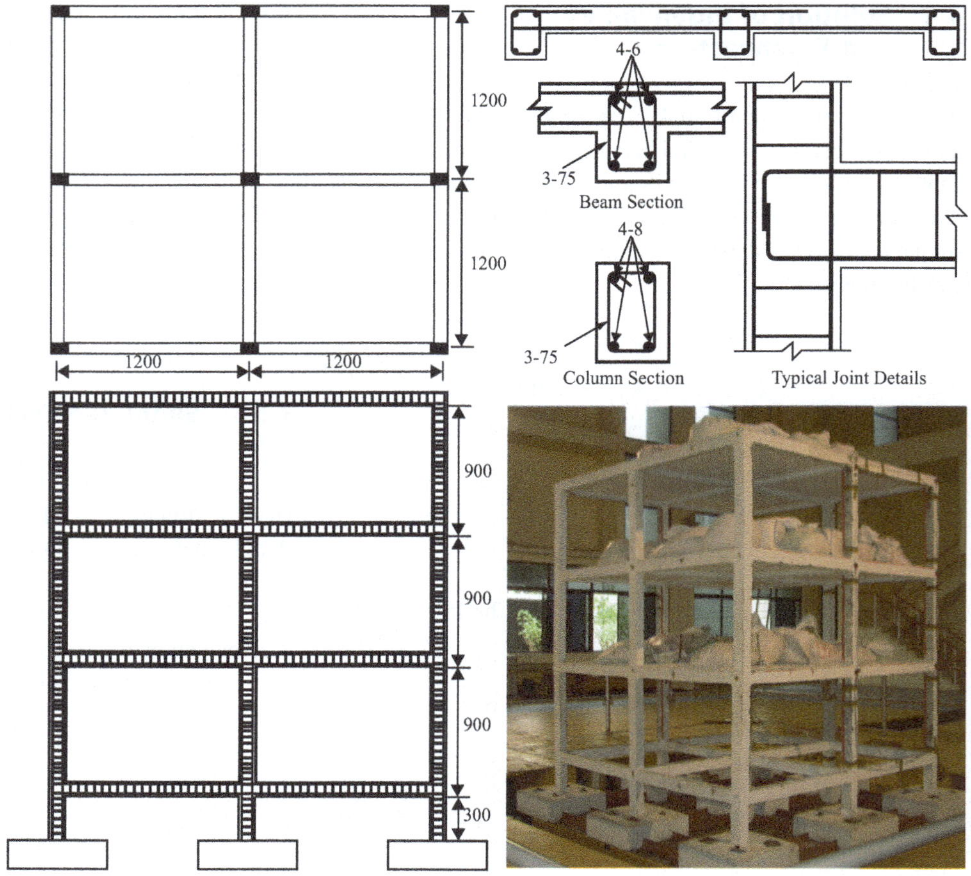

Fig. 4.7 Three-storey RCC structure

Table 4.1 Geometrical details and material properties of the RCC structure

Column sizes	100 × 75 mm
Young's modulus of concrete (E)	2.5×10^{10} N/m^2
Moment of inertia of the column section (I)	3.52×10^{-6} m^4
Height of the column	0.9 m

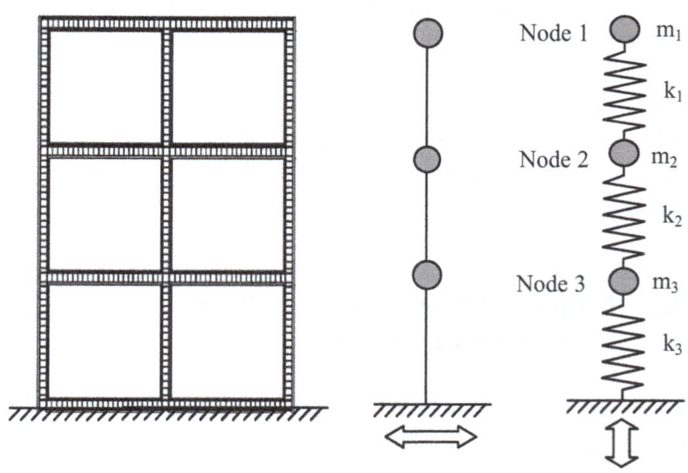

Fig. 4.8 Simplified model of the RCC structure

Fig. 4.9 Approximate eigenvector (mode shape) of the structure

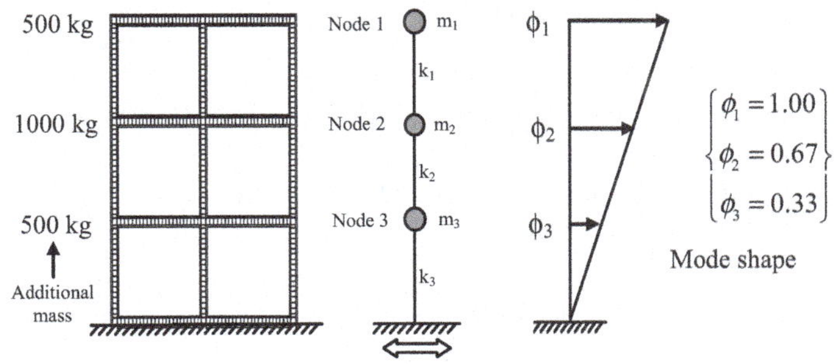

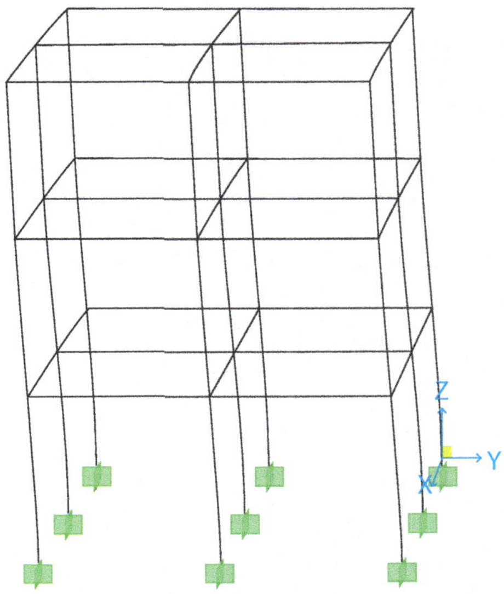

Fig. 4.10 Mode shape of the structure obtained with 3D FE model

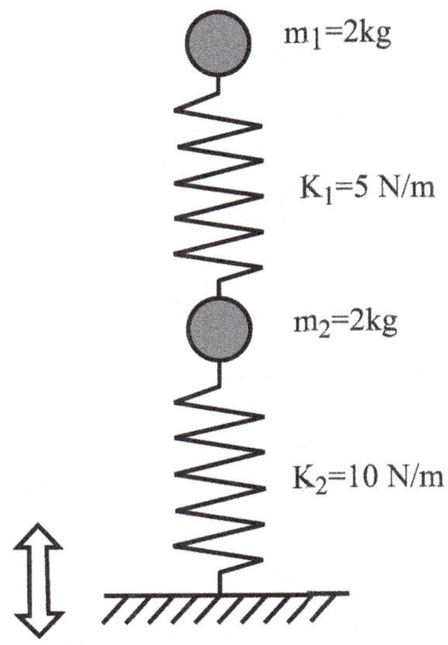

Fig. 4.11 2DOF system

Considering the orthogonal properties of mode shape as explained previously, one can obtain ω^2 as

$$\omega^2 = \{\phi\}_n{}^T[K]\{\phi\}_n = 1265.9$$

Hence, the frequency (ω) is 35.58 rad/s (5.96 Hz). The frequency of the structure with 3D finite element method considering the flexibility of floor is 4.25 Hz. The mode shape obtained is shown in Fig. 4.10 similar to triangular distribution what it was assumed in the simplified analysis. It says that for simple structures this kind of approximations will help to solve the problem quickly.

4.5.2 Inverse Iteration Technique

The technique of inverse iteration is very effectively used to calculate the eigenvector, and at the same time the corresponding eigenvalue can also be evaluated. This is explained with an example problem.

Example 4.1 Consider a 2DOF system as shown in Fig. 4.11. Evaluate the eigenvalues and eigenvectors using inverse iteration technique.

Solution

The free vibration equation of the system for vertical excitation can be written as follows:

$$\begin{bmatrix} 2 & 0 \\ 0 & 2 \end{bmatrix}\begin{bmatrix} \ddot{x}_1(t) \\ \ddot{x}_2(t) \end{bmatrix} + \begin{bmatrix} 5 & -5 \\ -5 & 15 \end{bmatrix}\begin{bmatrix} x_1(t) \\ x_2(t) \end{bmatrix} = 0$$

The characteristic equation can be written as

$$[K]\{X\} = \lambda[M]\{X\}$$

The above system has two eigenvalues and two eigenvectors and can be obtained using inverse iteration technique as follows:

i. **Evaluation of eigenvalues and eigenvectors of first mode**

Step 1: Assume $\{X\}$

$$\{X\} = \begin{Bmatrix} 1 \\ 1 \end{Bmatrix}$$

Step 2: Evaluate $\{R\} = [M]\{X\}$

$$\{R\} = \begin{bmatrix} 2 & 0 \\ 0 & 2 \end{bmatrix} \begin{Bmatrix} 1 \\ 1 \end{Bmatrix} = \begin{Bmatrix} 2 \\ 2 \end{Bmatrix}$$

Step 3: Evaluate new $\{X\}$ using $[K]\{X\} = \{R\}$

$$\begin{bmatrix} 5 & -5 \\ -5 & 15 \end{bmatrix} \begin{Bmatrix} X_1 \\ X_2 \end{Bmatrix} = \begin{Bmatrix} 2 \\ 2 \end{Bmatrix}$$

$$\begin{Bmatrix} X_1 \\ X_2 \end{Bmatrix} = \begin{bmatrix} 5 & -5 \\ -5 & 15 \end{bmatrix}^{-1} \begin{Bmatrix} 2 \\ 2 \end{Bmatrix} = \begin{Bmatrix} 0.8 \\ 0.4 \end{Bmatrix}$$

Step 4: Normalizing the mode shapes

$$\phi = \frac{\{X\}}{\sqrt{\{X\}^T[M]\{X\}}}$$

$$\phi = \begin{Bmatrix} 0.6325 \\ 0.3162 \end{Bmatrix}$$

Step 5: Evaluate eigenvalues

$$\{\phi\}^T[K]\{\phi\} = \lambda$$

$$\lambda = 1.5$$

$$\omega = 1.22$$

Consider $\{X\} = \{\phi\}$ and repeat the steps 1–4 till required convergence is reached. One suggested way of checking the convergence is

$$\varepsilon = \frac{\lambda_{i+1} - \lambda_i}{\lambda_{i+1}} \approx 0.0001$$

ii. **Evaluation of eigenvalues and eigenvectors of second mode**

Gram-Schmidt deflation technique is used to make mode shapes independent to avoid duplication of eigenvalues and eigenvectors which is explained as below.

Step 1: New Trial vector

$$\{X_1\} = \{X\} - \beta\{\phi\}$$

$$\{X_1\} = \begin{Bmatrix} 1 \\ 1 \end{Bmatrix} - \beta \begin{Bmatrix} 0.6325 \\ 0.3162 \end{Bmatrix}$$

$$\beta = \{\phi\}^T[M]\{X\}$$

$$\beta = [0.6325 \quad 0.3162] \begin{bmatrix} 2 & 0 \\ 0 & 2 \end{bmatrix} \begin{Bmatrix} 1 \\ 1 \end{Bmatrix} = 1.8974$$

$$\{X_1\} = \begin{Bmatrix} 1 \\ 1 \end{Bmatrix} - 1.8974 \begin{Bmatrix} 0.6325 \\ 0.3162 \end{Bmatrix} = \begin{Bmatrix} -0.2 \\ -0.4 \end{Bmatrix}$$

Now considering $\{X\} = \{X_1\}$ follow the steps 1–4 as explained above and evaluate eigenvalues and eigenvectors. After first iteration, one can obtain

$$\{\phi\} = \begin{Bmatrix} -0.2999 \\ 0.6000 \end{Bmatrix}$$

$$\lambda = 7.6498 \quad \omega = 2.766$$

Example 4.2 Consider a three-storey RCC structure as shown in Fig. 4.12. The geometrical details and material properties of the structure are given in Table 4.1. Mass of slab and beams considering columns are 1560, 1560, and 1560 kg at top, middle, and bottom of the structure. Evaluate the eigenvalues and eigenvectors using inverse iteration technique.

Solution

The characteristic equation can be written as

$$[K]\{X\} = \lambda[M]\{X\}$$

$$[K] = 9 \times C \begin{bmatrix} 1 & -1 & 0 \\ -1 & 2 & -1 \\ 0 & -1 & 2 \end{bmatrix} \quad C = \frac{12EI}{l^3} = 140.48559 \times 10^5$$

$$[M] = \begin{bmatrix} 1560 & 0 & 0 \\ 0 & 1560 & 0 \\ 0 & 0 & 1560 \end{bmatrix}$$

$$9 \times C \begin{bmatrix} 1 & -1 & 0 \\ -1 & 2 & -1 \\ 0 & -1 & 2 \end{bmatrix} \{X\} = \lambda \begin{bmatrix} 1560 & 0 & 0 \\ 0 & 1560 & 0 \\ 0 & 0 & 1560 \end{bmatrix} \{X\}$$

The system shown in Fig. 4.12 has three eigenvalues and three eigenvectors and can be obtained using inverse iteration technique as follows:

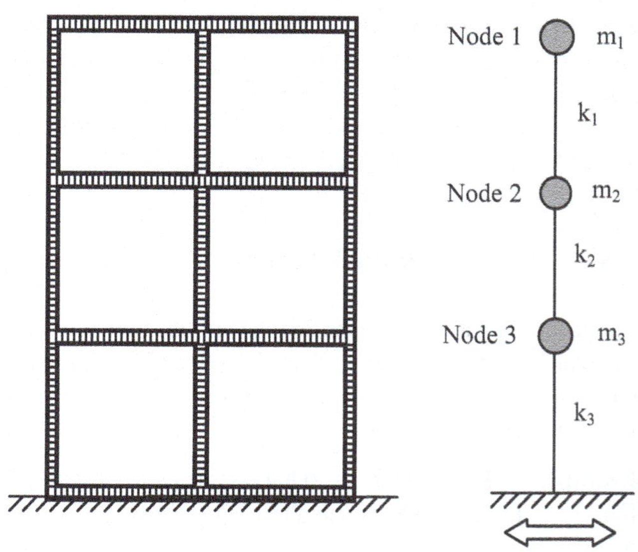

Fig. 4.12 Three-storey RCC structure

i. **Evaluation of Eigenvalues and Eigenvectors of first mode**

Iteration 1:

Step 1: Assume $\{X\}$

$$\{X\} = \begin{Bmatrix} 1.0 \\ 1.0 \\ 1.0 \end{Bmatrix}$$

Step 2: Evaluate $\{R\} = [M]\ \{X\}$

$$\begin{bmatrix} 1560 & 0 & 0 \\ 0 & 1560 & 0 \\ 0 & 0 & 1560 \end{bmatrix} \begin{Bmatrix} 1.0 \\ 1.0 \\ 1.0 \end{Bmatrix} = \begin{Bmatrix} 1560.0 \\ 1560.0 \\ 1560.0 \end{Bmatrix}$$

Step 3: Evaluate new $\{X\}$ using $[K]\ \{X\} = \{R\}$

$$9 \times C \begin{bmatrix} 1 & -1 & 0 \\ -1 & 2 & -1 \\ 0 & -1 & 2 \end{bmatrix} \begin{Bmatrix} X_1 \\ X_2 \\ X_3 \end{Bmatrix} = \begin{Bmatrix} 1560.0 \\ 1560.0 \\ 1560.0 \end{Bmatrix}$$

$$\{X\} = \begin{Bmatrix} X_1 \\ X_2 \\ X_3 \end{Bmatrix} = 1.0\text{e}{-3} \times \begin{Bmatrix} 0.7180 \\ 0.5983 \\ 0.3590 \end{Bmatrix}$$

Step 4: Normalizing the mode shapes

$$\phi = \frac{\{X\}}{\sqrt{[\{X\}^T[M]\{X\}]}}$$

$$\{\phi\} = \begin{Bmatrix} 0.0182 \\ 0.0151 \\ 0.0091 \end{Bmatrix}$$

Step 5: Evaluate eigenvalues

$$\{\phi\}^T[K]\{\phi\} = \lambda$$

$$\lambda = 1671.4$$

$$\omega = 40.8829$$

End of first iteration.

Now, consider $\{X\} = \{\phi\}$ and repeat the steps 1–4 till required convergence is reached. One suggested way of checking the convergence is:

$$\varepsilon = \frac{\lambda_{i+1} - \lambda_i}{\lambda_{i+1}} \approx 0.0001$$

Iteration 2: Repeating the steps 1–4, one can get

$$\{\phi\} = \begin{Bmatrix} 0.0186 \\ 0.0150 \\ 0.0084 \end{Bmatrix}$$

$$\lambda = 1655.5$$

$$\omega = 40.6875$$

Iteration 3: Repeating the steps 1–4, one can get

$$\{\phi\} = \begin{Bmatrix} 0.0187 \\ 0.0150 \\ 0.0083 \end{Bmatrix}$$

$$\lambda = 1655.2$$

$$\omega = 40.6844$$

Check convergence

$$\varepsilon = \frac{1655.2 - 1655.5}{1655.2} \approx 0.0001812$$

Repeat the steps 1–4 till required convergence is reached.

ii. **Evaluation of eigenvalues and eigenvectors of second mode**

Gram-Schmidt deflation technique is used to make mode shapes independent to avoid duplication of eigenvalues and eigenvectors which is explained as below.

Step 1: New Trial vector

$$\{X_1\} = \{X\} - \beta\{\phi\}$$

$$\{X_1\} = \begin{Bmatrix} 1.0 \\ 1.0 \\ 1.0 \end{Bmatrix} - \beta \begin{Bmatrix} 0.0187 \\ 0.0150 \\ 0.0083 \end{Bmatrix}$$

$$\beta = \{\phi\}^T[M]\{X\}$$

$$\beta = [0.0187 \quad 0.0150 \quad 0.0083] \times \begin{bmatrix} 1560 & & \\ & 1560 & \\ & & 1560 \end{bmatrix} \begin{Bmatrix} 1 \\ 1 \\ 1 \end{Bmatrix} = 65.4168$$

$$\{X_1\} = \begin{Bmatrix} -0.2201 \\ 0.0208 \\ 0.4560 \end{Bmatrix}$$

Now considering $\{X\} = \{X_1\}$ follow steps 1–4 of above and evaluate eigenvalues and eigenvectors

Iteration 1: After first iteration

$$\{\phi\} = \begin{Bmatrix} -0.0133 \\ 0.0047 \\ 0.0210 \end{Bmatrix} \quad \text{and}$$

$$\lambda = 13459$$

Iteration 2: After second iteration

$$\{\phi\} = \begin{Bmatrix} -0.0149 \\ 0.0061 \\ 0.0196 \end{Bmatrix} \quad \text{and}$$

$$\lambda = 13089$$

Iteration 3: After third iteration

$$\{\phi\} = \begin{Bmatrix} -0.0192 \\ 0.0030 \\ 0.0162 \end{Bmatrix} \quad \text{and}$$

$$\lambda = 12135$$

Check convergence

$$\varepsilon = \frac{\lambda_{i+1} - \lambda_i}{\lambda_{i+1}} \approx 0.0001$$

$$\varepsilon = \frac{12135 - 13089}{12135} \approx 0.078$$

Repeat the steps 1–4 till required convergence is reached.

4.5.3 Simultaneous Iteration Technique

The technique is explained with an example. Now, consider Example 4.2 and solve the problem with simultaneous iteration technique. The various steps in the technique are explained below.

Iteration-1

Step 1: Assume $\{X\}$

$$\{X\} = \begin{Bmatrix} 0 & 2 \\ 1 & 1 \\ 2 & 0 \end{Bmatrix}$$

Step 2: Evaluate {R} = [M] {X}

$$\begin{bmatrix} 1560 & 0 & 0 \\ 0 & 1560 & 0 \\ 0 & 0 & 1560 \end{bmatrix} \begin{Bmatrix} 0 & 2 \\ 1 & 1 \\ 2 & 0 \end{Bmatrix} = \begin{Bmatrix} 0 & 3120 \\ 1560 & 1560 \\ 3120 & 0 \end{Bmatrix}$$

Step 3: Evaluate new $\{X\}$ using $[K]\ \{X\} = \{R\}$

$$9 \times C \begin{bmatrix} 1 & -1 & 0 \\ -1 & 2 & -1 \\ 0 & -1 & 2 \end{bmatrix} \begin{Bmatrix} X_{11} & X_{12} \\ X_{21} & X_{22} \\ X_{31} & X_{32} \end{Bmatrix} = \begin{Bmatrix} 0 & 3120 \\ 1560 & 1560 \\ 3120 & 0 \end{Bmatrix}$$

$$\begin{Bmatrix} X_{11} & X_{12} \\ X_{21} & X_{22} \\ X_{31} & X_{32} \end{Bmatrix} = 1.0\text{e}{-3} \times \begin{Bmatrix} 0.4786 & 0.9573 \\ 0.4786 & 0.7180 \\ 0.3590 & 0.3590 \end{Bmatrix}$$

Step 4: Normalizing the mode shapes and making the second vector orthogonal to the first vector using Gram-Schmidt deflation technique

$$\phi = \frac{\{X\}}{\sqrt{[\{X\}^T[M]\{X\}]}}$$

$$X_1 = \begin{Bmatrix} X_{11} & X_{12} \\ X_{21} & X_{22} \\ X_{31} & X_{32} \end{Bmatrix} = \begin{Bmatrix} 0.0158 & 0.0172 \\ 0.0158 & -0.0035 \\ 0.0119 & -0.0182 \end{Bmatrix}$$

Step 5: Evaluate Eigenvalues

$$\{\phi\}^T[K]\{\phi\} = \lambda$$

$$\lambda_1 = 2038.5, \quad \lambda_2 = 12{,}758$$
$$\omega_1 = 45.149, \quad \omega_2 = 112.95$$

End of first iteration.

Iteration 2

$$\begin{Bmatrix} X_{11} & X_{12} \\ X_{21} & X_{22} \\ X_{31} & X_{32} \end{Bmatrix} = 1.0\text{e}{-3} \times \begin{Bmatrix} 0.0183 & 0.0150 \\ 0.0151 & -0.0070 \\ 0.0088 & -0.0192 \end{Bmatrix}$$

$$\lambda_1 = 1661.4, \quad \lambda_2 = 13024$$
$$\omega_1 = 40.76, \quad \omega_2 = 114.122$$

Iteration 3

$$\begin{Bmatrix} X_{11} & X_{12} \\ X_{21} & X_{22} \\ X_{31} & X_{32} \end{Bmatrix} = 1.0\text{e}{-3} \times \begin{Bmatrix} 0.0186 & 0.0179 \\ 0.0150 & -0.0051 \\ 0.0084 & -0.0172 \end{Bmatrix}$$

$$\lambda_1 = 1655.3, \quad \lambda_2 = 12641$$
$$\omega_1 = 40.685, \quad \omega_2 = 112.43$$

4.5.4 Subspace Iteration Technique

The technique is explained with an example. Now, consider Example 4.2 and solve the problem with subspace iteration technique. The various steps in the technique are explained below.

Step 1: Assume $\{X\}$

$$\{X\} = \begin{Bmatrix} 0 & 2 \\ 1 & 1 \\ 2 & 0 \end{Bmatrix}$$

Step 2: Evaluate $\{R\} = [M]\ \{X\}$

$$\begin{bmatrix} 1560 & 0 & 0 \\ 0 & 1560 & 0 \\ 0 & 0 & 1560 \end{bmatrix} \begin{Bmatrix} 0 & 2 \\ 1 & 1 \\ 2 & 0 \end{Bmatrix} = \begin{Bmatrix} 0 & 3120 \\ 1560 & 1560 \\ 3120 & 0 \end{Bmatrix}$$

Step 3: Evaluate new $\{X\}$ using $[K]\ \{X\} = \{R\}$

$$9 \times C \begin{bmatrix} 1 & -1 & 0 \\ -1 & 2 & -1 \\ 0 & -1 & 2 \end{bmatrix} \begin{Bmatrix} X_{11} & X_{12} \\ X_{21} & X_{22} \\ X_{31} & X_{32} \end{Bmatrix} = \begin{Bmatrix} 0 & 3120 \\ 1560 & 1560 \\ 3120 & 0 \end{Bmatrix}$$

$$\begin{Bmatrix} X_{11} & X_{12} \\ X_{21} & X_{22} \\ X_{31} & X_{32} \end{Bmatrix} = 1.0\text{e}{-3} \times \begin{Bmatrix} 0.4786 & 0.9573 \\ 0.4786 & 0.7180 \\ 0.3590 & 0.3590 \end{Bmatrix}$$

Step 4: Reduce the stiffness and mass matrix using above vectors

$$[M]_R = [X]^T[M][X] = \begin{bmatrix} 0.0009 & 0.0015 \\ 0.0015 & 0.0024 \end{bmatrix}$$

$$[K]_R = [X]^T[K\}[X] = \begin{bmatrix} 1.8666 & 2.2400 \\ 2.2400 & 4.1066 \end{bmatrix}$$

Step 5: Assume new trial vector

$$[X]_{k,j} = \begin{bmatrix} 1 & 0 \\ 0 & 1 \end{bmatrix}$$

Step 6: Considering load vectors $[R]_{k,j} = [M]_R\ \{X\}_{k,j}$

$$[R]_{k,j} = \begin{bmatrix} 0.0009 & 0.0015 \\ 0.0015 & 0.0024 \end{bmatrix} \begin{bmatrix} 1 & 0 \\ 0 & 1 \end{bmatrix}$$

$$[R]_{k,j} = \begin{bmatrix} 0.0009 & 0.0015 \\ 0.0015 & 0.0024 \end{bmatrix}$$

Step 7: Solving $[K]_R\ [X] = [R]_{k,j}$

$$[X] = 1 \times 10^{-3} \begin{bmatrix} 0.1269 & 0.2961 \\ 0.2961 & 0.4229 \end{bmatrix}$$

Normalizing first vector $[X]$ with $[M]_R$ modifies to

$$[X]_{k,1} = \begin{Bmatrix} 6.9096 \\ 16.1224 \end{Bmatrix}$$

First eigenvalue is given as $\{X\}_{k,1}^T[K]_R\ \{X\}_{k,1}$

$$\lambda_{11} = 1666.6$$

Step 8: To make the vectors independent, use Gram-Schmidt deflation technique as follows and obtain the second vector

$$\beta = [X]_{k,1}^T[M]_R[X]_{k,2}$$

$$= [6.9096 \quad 16.1226] \begin{bmatrix} 0.0009 & 0.0015 \\ 0.0015 & 0.0024 \end{bmatrix} \begin{bmatrix} 0.2961 \\ 0.4229 \end{bmatrix} \times 10^{-3}$$

$$\beta = 2.9946 \times 10^{-5}$$

$$\text{Modified} \quad \{X\}_{k,2} = \{X\}_{k,2} - \beta \begin{Bmatrix} 6.9096 \\ 16.1224 \end{Bmatrix}$$

$$\{X\}_{k,2} = \begin{Bmatrix} 0.9126 \\ -0.5507 \end{Bmatrix} \times 10^{-4}$$

Step 9: Normalizing second vector $[X]_{k,2}$ with $[M]_R$ modifies to

$$\{X\}_{k,2} = \begin{Bmatrix} 141.2504 \\ 85.2352 \end{Bmatrix}$$

Step 10: Second eigenvalue is given as $\{X\}_{k,2}^T[K]_R\ \{X\}_{k,2}$

$$\lambda_{22} = 13139.0$$

Considering these as new trial vectors and repeating steps 5-9, converged eigenvalues and vectors can be obtained satisfying the following convergence criteria.
Check for convergence

$$\varepsilon_j = \frac{\lambda_{j,i+1} - \lambda_{j,i}}{\lambda_{j,i+1}} \approx 0.0001$$

Step 11: Eigenvectors of the original problem can be written as follows:

$$[\phi] = [X][X]_R = 10^{-3} \times \begin{bmatrix} 0.4786 & 0.9572 \\ 0.4786 & 0.7179 \\ 0.3590 & 0.3590 \end{bmatrix} \times \begin{bmatrix} 6.9096 & 141.2501 \\ 16.1226 & -85.235 \end{bmatrix}$$

$$[\phi] = \begin{bmatrix} 0.0187 & -0.014 \\ 0.0149 & 0.0064 \\ 0.0083 & 0.0201 \end{bmatrix}$$

$$\phi^T K\phi = 10^4\begin{bmatrix} 0.1655 & -0.0039 \\ -0.0039 & 1.3142 \end{bmatrix}$$

$$\phi^T M\phi = \begin{bmatrix} 0.9999 & -0.0006 \\ -0.0006 & 1.0001 \end{bmatrix}$$

It also satisfied all the orthogonality conditions.

4.5.5 Lanczos Iteration Technique

The basic steps of the Lanczos iteration technique are to transform the generalized eigenproblem into a standard form with a tridiagonal coefficient matrix. The steps are illustrated below:

Step 1: Assume $\{X\}$
Step 2: Calculate $\{X_1\} = \{X\}/\gamma$
where $\gamma = \sqrt{\{X\}^T[M]\{X\}}$
Step 3: Let $\beta_0 = 0$;
Calculate $\bar{X}_i$, for i = 1 …, n by solving

$$[K]\{\bar{X}_i\} = [M]\{X_i\}$$

Step 4: Calculate α_i, for i = 1 …, n

$$\alpha_i = \sqrt{\{\bar{X}_{i1}\}^T[M]\{X_i\}}$$

Step 5: Calculate $\tilde{X}_i$

$$\tilde{X}_i = \overline{X_i} - \alpha_i X_i - \beta_{i-1}X_{i-1}, \quad \text{for } i = 1, \ldots, n-1$$

Step 6: Calculate β_i

$$\beta_i = \sqrt{\{\tilde{X}_i\}^T[M]\{\tilde{X}_i\}}$$

Step 7: Calculate X_{i+1}

$$X_{i+1} = \frac{\tilde{X}_i}{\beta_i}$$

Theoretically, the vectors X_i, $i = 1\ldots, n$, generated using above steps are M-orthogonal and the matrix $X_n = [X_i, \ldots X_n]$ satisfies the relationship

$$\{X_n\}^T\left([M][K]^{-1}[M]\right)\{X_n\} = T_n$$

where

$$T_n = \begin{bmatrix} \alpha_1 & \beta_1 & & \\ \beta_1 & \alpha_2 & \beta_2 & \\ & \beta_2 & \alpha_{n-1} & \beta_{n-1} \\ & & \beta_{n-1} & \alpha_n \end{bmatrix}$$

The eigenvalues of T_n are the reciprocals of the eigenvalues of $[K]\{X\} = \lambda[M]\{X\}$

Note: The eigenvalues of T_n can be obtained using any procedure explained above.

Example 4.3 Consider Example 4.2 and solve the problem with Lanczos iteration technique.

Solution

Step 1: Assume $\{X\}$

$$\{X\} = \begin{Bmatrix} 1.0 \\ 1.0 \\ 1.0 \end{Bmatrix}$$

Step 2: Calculate $\{X_1\} = \{X\}/\gamma$
where

$$\gamma = \sqrt{\{X\}^T[M]\{X\}}$$
$$\gamma = 68.41$$
$$\{X_1\} = \begin{Bmatrix} 0.0146 \\ 0.0146 \\ 0.0146 \end{Bmatrix}$$

Step 3: Let $\beta_0 = 0$;
Calculate $\bar{X}_i$, for $i = 1\ldots, n$ by solving

$$[K]\{\bar{X}_i\} = [M]\{X_i\}$$

for $i = 1$,

$$\{\bar{X}_1\} = 1.0\text{e}{-4} \times \begin{Bmatrix} 0.1049 \\ 0.0875 \\ 0.0525 \end{Bmatrix}$$

Step 4: Calculate α_i, for $i = 1\ldots, n$

$$\alpha_i = \sqrt{\{\bar{X}_i\}^T[M]\{X_i\}}$$

for $i = 1$,

$$\alpha_1 = 50.5841\text{e}{-004}$$

Step 5: Calculate $\tilde{X}_i$

$$\tilde{X}_i = \overline{X_i} - \alpha_i X_i - \beta_{i-1} X_{i-1}, \quad \text{for i } = 1, \ldots, \text{n} - 1$$

for $i = 1$,

$$\tilde{X}_1 = 10.0\text{e}{-}005 \times \begin{bmatrix} 00.2332 \\ 0.0583 \\ -0.2915 \end{bmatrix}$$

Step 6: Calculate β_i

$$\beta_i = \sqrt{\{\tilde{X}_i\}^T [M] \{\tilde{X}_i\}}$$

for $i = 1$,

$$\beta_1 = 10.4924\text{e}{-}004$$

Step 7: Calculate X_{i+1}

$$X_{i+1} = \frac{\tilde{X}_i}{\beta_i}$$

for $i = 1$,

$$\{X_2\} = \begin{Bmatrix} 0.0156 \\ 0.0039 \\ -0.0195 \end{Bmatrix}$$

Repeat steps 3–7 to get α_2, β_2
for $i = 2$,

Step 3:

$$\{\bar{X}_2\} = 1.0\text{e}{-}5 \times \begin{Bmatrix} 0.4207 \\ 0.2337 \\ 0.0 \end{Bmatrix}$$

Step 4:

$$\alpha_2 = 1.1681\text{e}{-}004$$

Step 5:

$$\tilde{X}_2 = 10.0\text{e}{-}006 \times \begin{bmatrix} 00.2003 \\ -0.3005 \\ 0.1002 \end{bmatrix}$$

Step 6:

$$\beta_2 = 10.480424\text{e}{-}005$$

Step 7:

$$\{X_3\} = \begin{Bmatrix} 0.0135 \\ -0.0203 \\ 0.0068 \end{Bmatrix}$$

Repeat steps 3–7 to get α_3, β_3

for $i = 3$,

Step 3:

$$\{\bar{X}_3\} = 1.0e - 6 \times \begin{Bmatrix} 0.8097 \\ -0.8097 \\ 0.0 \end{Bmatrix}$$

Step 4:

$$\alpha_3 = 40.2735\text{e}{-}005$$

Step 5:

$$\tilde{X}_3 = 10.0\text{e}{-}018 \times \begin{bmatrix} 00.1007 \\ 0.0817 \\ 0.0501 \end{bmatrix}$$

Step 6:

$$\beta_3 = 50.4912\text{e}{-}018$$

Step 7:

$$\{X_4\} = \begin{Bmatrix} 0.0183 \\ 0.0149 \\ 0.0091 \end{Bmatrix}$$

Now

$$T_n = \begin{bmatrix} 50.5841\text{e}{-}4 & 10.4924\text{e}{-}4 & \\ 10.4924\text{e}{-}4 & 10.1681\text{e}{-}4 & 10.4804\text{e}{-}5 \\ & 10.4804\text{e}{-}5 & 40.2735\text{e}{-}5 \end{bmatrix}$$

Note: The eigenvalues of T_n can be obtained by using any procedure explained above, and reciprocals of these will give the eigenvalues of original problem.

The eigenvalues of T_n are

6.041e−4
7.700e−5
3.690e−5.

Reciprocal of these are

1655.355,
12987.012
27100.27.

These are the eigenvalues of original problem.

Example 4.4 Consider a three-storey RCC structure as shown in Fig. 4.13a having column sizes of 100 × 75 mm. Moment of inertia of the column section is 3.52×10^{-6} m^4.

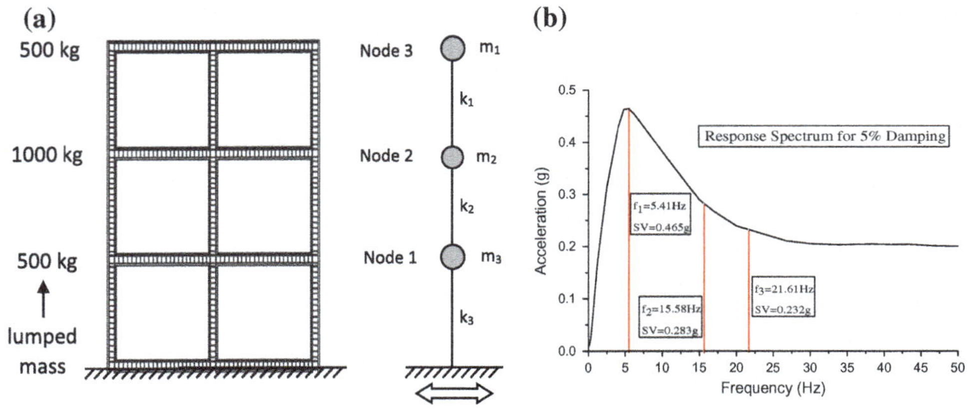

Fig. 4.13 Three-storey frame structure and site-specific response spectrum for 5% damping

Elastic modulus of concrete is 2.5×10^{10}. Determine the acceleration responses of all the floors due to site-specific response spectrum as shown in Fig. 4.13b by response spectrum method and time history method. Assume 5% damping for the RCC structure.

Solution

A. **Response Spectrum Method**

i. Estimate the natural frequencies and mode shapes using any one of the methods explained above.
Natural frequencies

$$f_1 = 5.408\ \text{Hz},\quad f_2 = 15.583\ \text{Hz}\quad \text{and}\quad f_3 = 21.607\ \text{Hz}$$

Mode shapes

$$\phi_1 = \begin{Bmatrix} 00.00695 \\ 00.01263 \\ 00.01545 \end{Bmatrix},\quad \phi_2 = \begin{Bmatrix} 00.01492 \\ 00.00724 \\ -00.01407 \end{Bmatrix}\quad \text{and}$$

$$\phi_3 = \begin{Bmatrix} 00.01465 \\ -00.01337 \\ 00.00699 \end{Bmatrix}$$

Participation factors

$$\Gamma_1 = 78.492,\quad \Gamma_2 = 20.286\quad \text{and}\quad \Gamma_3 = 10.363$$

ii. First mode, second mode, third mode acceleration response and total response are given in Tables 4.2, 4.3, 4.4 and 4.5, respectively.

B. **Time History Method**

Step 1: Generation of compatible time history
method to generate compatible time history is explained in Chap. 13 elaborately. An artificial time history compatible to site-specific response spectrum is shown in Fig. 4.14 (a). Its frequency content of input acceleration is depicted in 4.14(b).

Step 2: The equations of motion can be uncoupled, allowing them to be solved as SDOF equations using the orthogonality property of the natural mode shapes.

Step 3: Solution using Newmark-Beta

The Newmark-beta numerical method discussed in Chap. 3 has been used for time history analysis of the three-storey frame structure. Rayleigh damping has been used for time history analysis. Floor-wise displacement responses, frequency of displacement responses, acceleration responses, and elastic force responses of the structure are shown in Figs. 4.15, 4.16, 4.17 and 4.18, respectively.

Table 4.2 First mode acceleration response

Node	Mass participation factor	Mode shape	Spectral value	Modal response
1	78.492	0.00695	0.465	0.2537
2	78.492	0.01263	0.465	0.4610
3	78.492	0.01545	0.465	0.5639

Table 4.3 Second mode acceleration response

Node	Mass participation factor	Mode shape	Spectral value	Modal response
1	20.286	0.01492	0.283	0.0857
2	20.286	0.00724	0.283	0.0416
3	20.286	−0.01407	0.283	−0.0808

Table 4.4 Third mode acceleration response

Node	Mass participation factor	Mode shape	Spectral value	Modal response
1	10.363	0.01465	0.232	0.0352
2	10.363	−0.01337	0.232	−0.0321
3	10.363	0.00699	0.232	0.0168

Table 4.5 Total response using SRSS combination

Node	Total acceleration (g)
1	0.2701
2	0.4640
3	0.5699

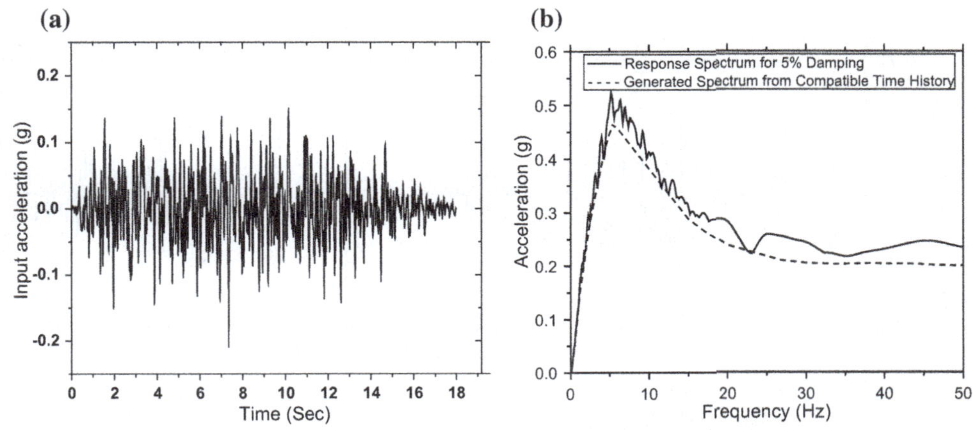

Fig. 4.14 **a** Compatible time history. **b** Comparison of original and generated response spectra

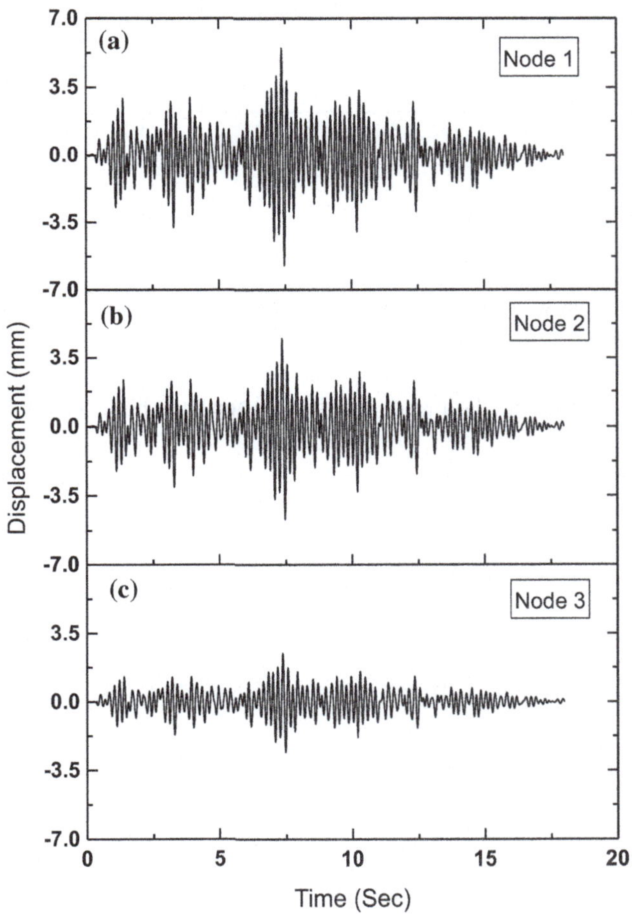

Fig. 4.15 Displacement responses at different floors of the structure **a** node 1 **b** node 2 **c** node 3

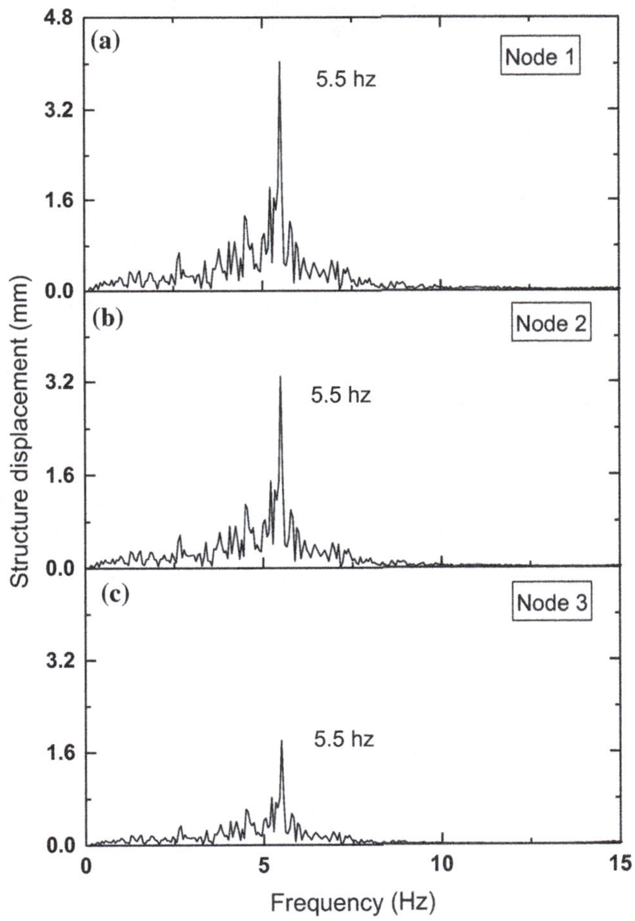

Fig. 4.16 Frequency of displacement responses at different floors of the structure **a** node 1 **b** node 2 **c** node 3

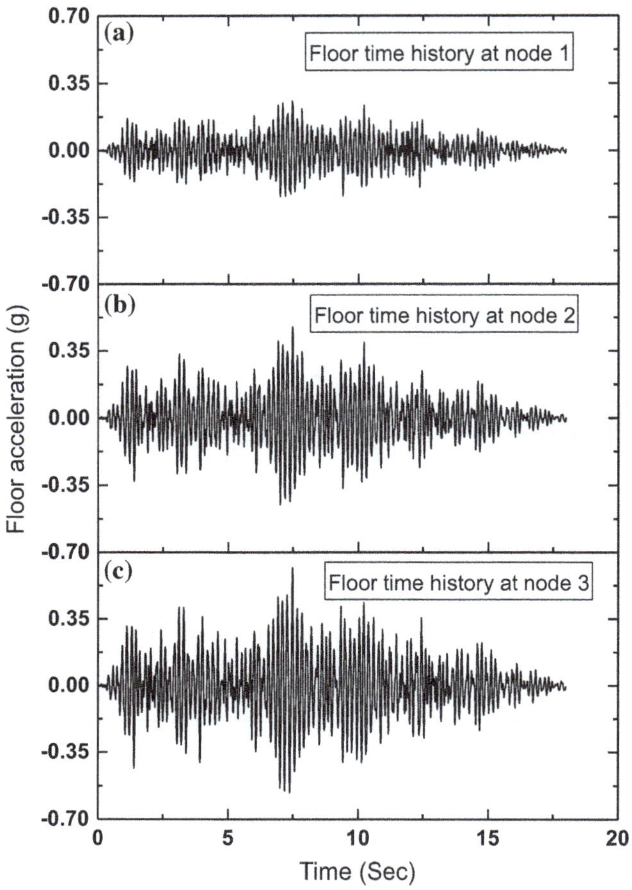

Fig. 4.17 Acceleration responses at different floors of the structure **a** node 1 **b** node 2 **c** node 3

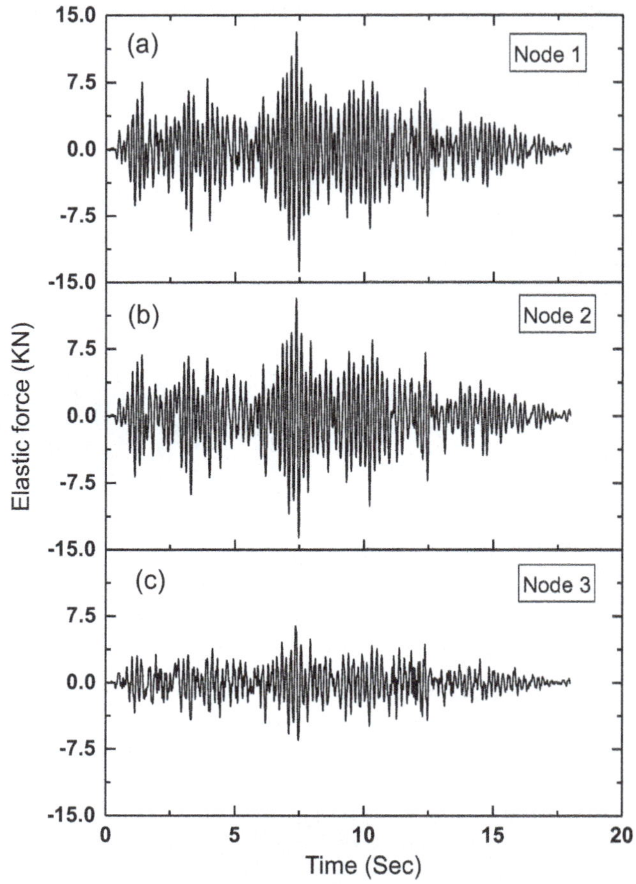

Fig. 4.18 Force responses at different floors of the structure **a** node 1 **b** node 2 **c** node 3

Step 4: Transform back to the normal coordinates.

Maximum responses obtained from time history analysis are compared with response spectrum method and details are shown in Table 4.6. Element forces are calculated based on the formulas given in [1, 2].

By comparing the response obtained from the response spectrum method and the time history method, it can be

Table 4.6 Comparison of response spectrum method and time history method

Node	Response spectrum	Time history	Difference (%)
Maximum acceleration (g)			
s1	0.2701	0.2572	-5.01555
2	0.464	0.4756	2.439024
3	0.5699	0.6187	7.887506
Maximum displacement (mm)			
1	2.158	2.4815	13.03647
2	3.919	4.496	12.83363
3	4.793	5.506	12.94951
Element force (KN)			
1	5.457	6.41	14.86739
2	11.655	13.19	11.6376
3	11.519	13.12	12.20274

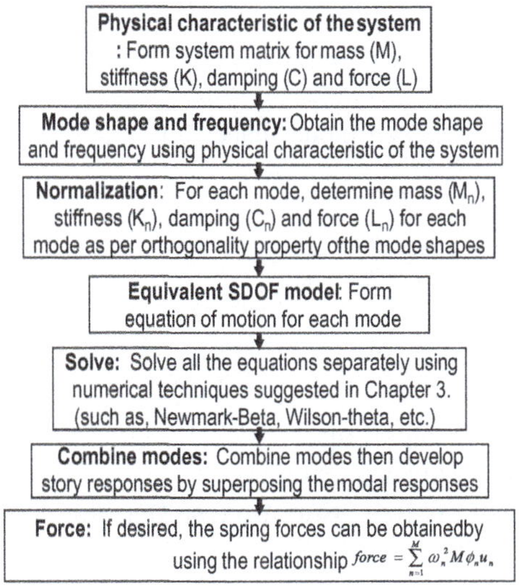

Fig. 4.19 Flow chart for performing time history analysis

concluded that the response spectrum method gives conservative results. The flowchart (Fig. 4.19) and MATLAB code are given below for time history analysis of Example 4.4.

MATLAB code

INPUT:
n—Number of DOFs;
nmode—Number of modes;
dt—Time Interval of force or earthquake;
M—Mass matrix of stories ($n \times n$);
K—Stiffness matrix of stories ($n \times n$);
zeta—Damping ratio (%);
w2—Eigen-values matrix ($nmode \times nmode$);
phi—Eigen-Vectors ($nmode \times nmode$);
F—Spatial load distribution vector.
OUTPUT: Story Response
dis—Displacement time history of stories (or DOFs);
Force—Elastic force time history of stories.

CODE:

```
function
[dis,acc,Force]=NewM(n,nmode,dt,F,ft,M,K,zeta,w2,phi)
Mn=phi'*M.*phi;
Kn=phi'*K*phi;
Ln=zeros(nmode);
for i=1:nmode
   Ln(i,i)=(phi(:,i))'*F/M.n(i,i);
end
u=zeros(length(ft),nmode);
v=zeros(length(ft),nmode);
ac=zeros(length(ft),nmode);
ft(end+1)=0;
for j=1:nmode
  m=M.n(j,j);
  k=Kn(j,j);
  c=2*m.*(w2(j,j))^.5*zeta/100;
  K=k+3*c/dt+6*m./(dt)^2;
  a=6*m./dt+3*c;
  b=3*m.+dt*c/2;
for i=1:length(u)-1
   u(1,j)=0;
   v(1,j)=0;
   ac(1,j)=0;
   df=Ln(j,j)*M.n(j,j)*(ft(i+1)-ft(i))+a*v(i,j)+b*ac(i,j);
   du=df/K;
   dv=3*du/dt-3*v(i,j)-dt*ac(i,j)/2;
   dac=6*(du-dt*v(i,j))/(dt)^2-3*ac(i,j);
   u(i+1,j)=u(i,j)+du;
   v(i+1,j)=v(i,j)+dv;
   ac(i+1,j)=ac(i,j)+dac;
end
end
ft(end)=[];
dis=zeros(size(u,1),n);
 acc=zeros(size(ac,1),n);
Force=zeros(size(u,1),n);
for st=1:n
for j=1:nmode
dis2=zeros(size(u,1),n);
accel=zeros(size(ac,1),n);
Force1=zeros(size(u,1),n);
dis2(:,st)=phi(st,j)*u(:,j);
accel(:,st)=phi(st,j)*ac(:,j);
Force1(:,st)=w2(j,j)*M(st,st)*phi(st,j)*u(:,j);dis=dis2+dis;
acc=accel+acc;
Force=Force1+Force;
end
end
```

4.6 MATLAB Code for Eigenvalues and Eigenvectors

In this section, MATLAB code is provided for finding the eigenvalues and eigenvectors by using various methods as explained in the previous sections with example problems.

Example 4.5 Consider a three-storey RCC structure is shown in Fig. 4.20 having column sizes of 100 × 75 mm. Moment of inertia of the column section is 3.52×10^{-6} m^4. Elastic modulus of concrete is 2.5×10^{10}. Find the eigenvalues and eigenvectors using simplified method, inverse iteration technique, simultaneous iteration technique, subspace iteration technique, and Lanczos iteration technique. (Neglect rotational stiffness).

A. **Modal analysis using Simplified Method**

```
clear all;
E=2.5*10^10;
I=3.52*10^-6;
L=0.9;
C=12*E*I/L^3;
% Global Stiffness and Mass matrices
K=9*C*([1,-1,0;-1,2,-1;0,-1,2]);
M=[2060,0,0;0,2560,0;0,0,2060];
disp('The stiffness matrix is')
K
disp('The mass matrix is')
M
% Assume first mode shape as triangular shape
n=size(K);
N=n(1);
for j=1:1:N
Q(j,1)=(N-(j-1))*(1/N);
end
q=Q'*M*Q;
qn=Q/sqrt(q);
Omega2=qn'*K*qn;
Freq=sqrt(Omega2)/(2*pi)
```

B. **Modal analysis using Inverse Iteration Technique**

```
clear all;
E=2.5*10^10;
I=3.52*10^-6;
L=0.9;
C=12*E*I/L^3;
% Global Stiffness and Mass matrices
K=9*C*([1,-1,0;-1,2,-1;0,-1,2]);
M=[1560,0,0;0,1560,0;0,0,1560];
disp('The stiffness matrix is')
K
disp('The mass matrix is')
```

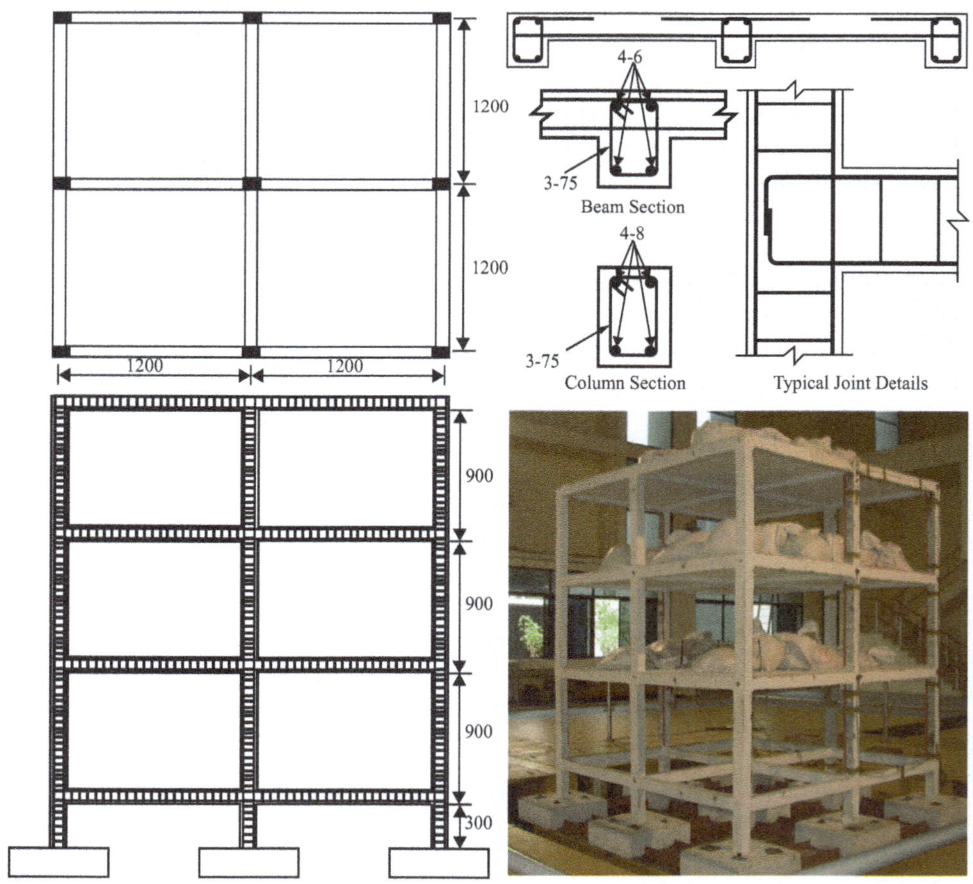

Fig. 4.20 Three-Storey RCC structure

```
M
A=size(K);
% Assume X for finding first mode
X=ones(A(1),1);
i=2;
Omega2(1)=0;
for i=2:1:5
R=M*X;
X_NEW=inv(K)*R;
Q=X_NEW'*M*X_NEW;
qn=X_NEW/sqrt(Q);
Omega2(i)=qn'*K*qn;
Freq(i)=sqrt(Omega2(i))/(2*pi);
E(i)=(Omega2(i)-Omega2(i-1))/Omega2(i);
% check convergence
if abs(E)>=0.0001
%n=n+1;
X=qn;
else
end
end
disp('The first modal frequency is')
Freq(5)
% Finding second eigen value and eigen vector
X1=ones(A(1),1);
qn2=qn;
BETA=qn2'*M*X1;
X2=X1-BETA*qn2;
for j=2:1:5
% Iteration
% checkconvergence
R2=M*X2;
X_NEW2=K\R2;
Q2=X_NEW2'*M*X_NEW2;
qn2=X_NEW2/sqrt(Q2);
Omega_2(j)=qn2'*K*qn2;
Freq2(j)=sqrt(Omega_2(j))/(2*pi);
E2(j)=(Omega_2(j)-Omega_2(j-1))/Omega_2(j);
% check convergence
if abs(E2(j))>=0.0001
X2=qn2;
else
end
end
disp('The second modal frequency is')
Freq2(j)
```

C. **Simulation Iteration Technique**

```
%Simultaneous iteration technique
clear all;
E=2.5*10^10;
I=3.52*10^-6;
L=0.9;
C=12*E*I/L^3;
% Global Stiffness and Mass matrices
K=9*C*([1,-1,0;-1,2,-1;0,-1,2]);
M=[1560,0,0;0,1560,0;0,0,1560];
disp('The stiffness matrix is')
K
disp('The mass matrix is')
M
% Assume X for finding first and second mode
n=size(K);
N=n(1);
for j=1:1:N
X(j,1)=j-1;
X(j,2)=N-j;
end
%
%X=ones(3,3)
n=5;
j=1;
Omega2_1= zeros(2,20);
for i=2:1:200
R=M*X;
X_NEW=K\R;
Q=X_NEW(:,1)'*M*X_NEW(:,1);
qn(:,1)=X_NEW(:,1)/sqrt(Q);
% A=size(K);
% X1=ones(A(1),1);
qn2=qn(:,1);
BETA=qn2'*M*X_NEW(:,2);
X_NEW2=X_NEW(:,2)-BETA*qn2;
% R1=M*X2;
% X_NEW2=K\R1;
Q2=X_NEW2(:,1)'*M*X_NEW2(:,1);
qn(:,2)=X_NEW2(:,1)/sqrt(Q2);
%qn(:,2)=X2(:,2)/sqrt(Q);
Omega2=qn'*K*qn;
Omega2_1(:,i)=diag(Omega2);
%for j=1:1:n
Freq=sqrt(Omega2)/(2*pi);
E(i)=(Omega2_1(1,i)-Omega2_1(1,(i-1)))/Omega2_1
(1,i);
j=j+1;
% check convergence
if abs(E)>=0.00001
%n=n+1;
aaaa(i)=abs(E(i))
X=qn
else
%end
end
end
disp('The first and second modal frequencies are')
diag(Freq)
disp('The eigen vectors are')
```

```
qn
```

D. **Subspace Iteration Technique**

```
clear all;
E=2.5*10^10;
I=3.52*10^-6;
L=0.9;
C=12*E*I/L^3;
% Global Stiffness and Mass matrices
K=9*C*([1,-1,0;-1,2,-1;0,-1,2]);
M=[1560,0,0;0,1560,0;0,0,1560];
disp('The stiffness matrix is')
K
disp('The mass matrix is')
M
% Assume X for finding first and second mode
n=size(K);
N=n(1);
for j=1:1:N
X(j,1)=j-1;
X(j,2)=N-j;
end
X_1=[1,0;0,1];
n=5;
j=1;
%for i=2:1:n
Omega2=zeros(2,10);
for j=2:1:n
R=M*X;
X_NEW=K\R
M_new=X_NEW'*M*X_NEW
K_new=X_NEW'*K*X_NEW
R_new= M_new*X_1
X_2=K_new\R_new
X_21=X_2(:,1)/sqrt((X_2(:,1)'*M_new*X_2(:,1)))
X_22=X_2(:,2)/sqrt((X_2(:,2)'*M_new*X_2(:,2)))
Omega(1,1)=X_21'*K_new*X_21;
Beta=X_21(:,1)'*M_new*X_2(:,2)
X_23=X_2(:,2)-Beta*X_21
X_24=X_23/sqrt((X_23'*M_new*X_23))
Omega(2,1)=X_24'*K_new*X_24;
Omega2(:,j)=Omega
%Omega3(i,j,k)=Omega2(i,j);
E2(j)=(Omega2(1,j)-Omega2(1,j-1))/Omega2(1,j)
% check convergence
if abs(E2(j))>=0.0001
X_1=X_2;
else
end
end
Freq=sqrt(Omega)/(2*pi);
disp('The First & Second modal frequenciesare')
Freq
```

E. **Lanczos Iteration Technique**

```
clear K;
clear M;
% Sample stiffness matrix K and mass matrix M
E=2.5*10^10;
I=3.52*10^-6;
L=0.9;
C=12*E*I/L^3;
% Global Stiffness and Mass matrices
K=9*C*([1,-1,0;-1,2,-1;0,-1,2]);
M=[1560,0,0;0,1560,0;0,0,1560];
A=size(K);
% Assume X for finding first mode
X=ones(A(1),1);
Y=sqrt(X'*M*X);
X11=X/Y
X1(:,2)=X11;
B(1)=0;
for i=2:1:4
X1BAR(:,i)=(inv(K)*M)*X1(:,i)
ALPHA(i-1)=(X1BAR(:,i)'*M*X1(:,i))
X_BAR(:,i)=X1BAR(:,i)-ALPHA(i-1)*X1(:,i)-B(i-1)
*X1(:,(i-1))
B(i)=sqrt(X_BAR(:,(i))'*M*X_BAR(:,i))
X2_BAR=X_BAR(:,i)/B(i)
X1(:,(i+1))=X2_BAR;
end
Tn=[ALPHA(1),B(2),0;B(2),ALPHA(2),B(3);0,B(3),
ALPHA(3)]
Omega_Tn=roots(poly(Tn))
for i=1:1:A(1)
Freq(i,1)= sqrt((1/Omega_Tn(i,1)))/(2*pi);
end
disp('The First, second & Third modal frequenciesare')
Freq
```

Example 4.6: Consider a four-storey RCC structure is shown in the Fig. 4.21 having column sizes of 100 × 75 mm. Moment of Inertia of the column section is 3.52×10^{-6} m^4. Elastic modulus of concrete is 2.5×10^{10}. Find the eigen values and eigen vectors using Inverse Iteration Technique. (Neglect rotational stiffness).

Solution

```
clear all;
E=2.5*10^10;
I=3.52*10^-6;
L=0.9;
C=12*E*I/L^3;
% Global Stiffness and Mass matrices
K=9*C*([1,-1,0,0;-1,2,-1,0;0,-1,2,-1;0,0,-1,2]);
M=[2060,0,0,0;0,2560,0,0;0,0,2560,0;0,0,0,2060];
```

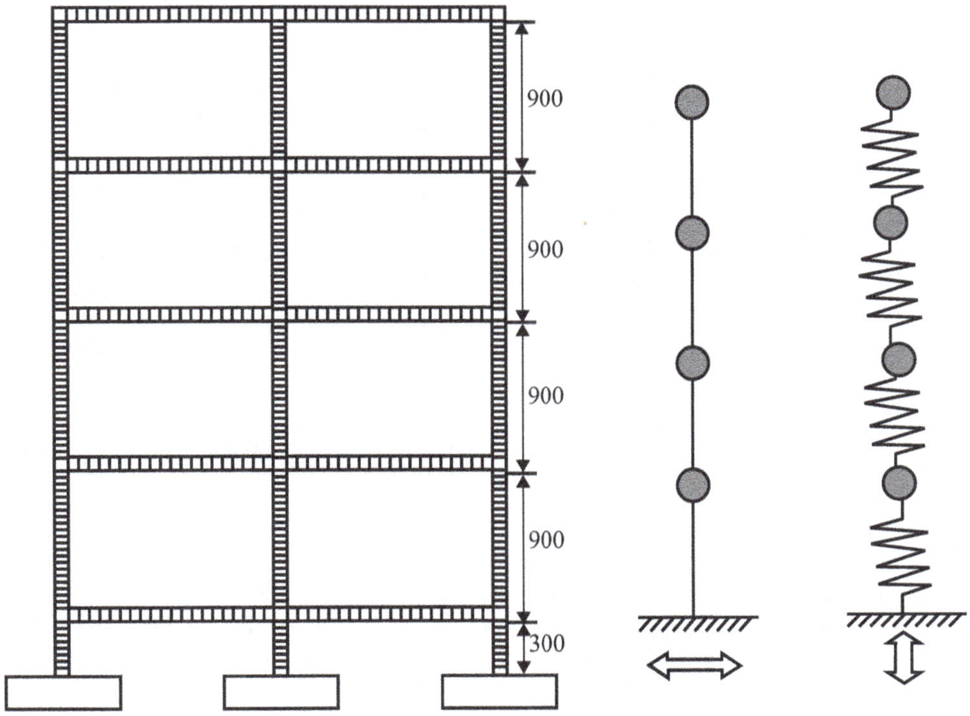

Fig. 4.21 Four-Storey RCC structure

```
disp('The stiffness matrix is')
K
disp('The mass matrix is')
M
A=size(K);

% Assume X for finding first mode
X=ones(A(1),1);
i=2;
Omega2(1)=0;
for i=2:1:5
R=M*X;
X_NEW=inv(K)*R;
Q=X_NEW'*M*X_NEW;
qn=X_NEW/sqrt(Q);
Omega2(i)=qn'*K*qn;
Freq(i)=sqrt(Omega2(i))/(2*pi);
E(i)=(Omega2(i)-Omega2(i-1))/Omega2(i);
% check convergence
if abs(E)>=0.0001
%n=n+1;
X=qn;
else
end
end
disp('The first modal frequency is')
Freq(5)
```

```
% Finding second eigen value and eigen vector

X1=ones(A(1),1);
qn2=qn;
BETA=qn2'*M*X1;
X2=X1-BETA*qn2;
for j=2:1:5

% Iteration
% check convergence
R2=M*X2;
X_NEW2=K\R2;
Q2=X_NEW2'*M*X_NEW2;
qn2=X_NEW2/sqrt(Q2);

Omega_2(j)=qn2'*K*qn2;
Freq2(j)=sqrt(Omega_2(j))/(2*pi);
E2(j)=(Omega_2(j)-Omega_2(j-1))/Omega_2(j);
% check convergence
if abs(E2(j))>=0.0001
X2=qn2;
else
end
end
disp('The second modal frequency is')
Freq2(j)
```

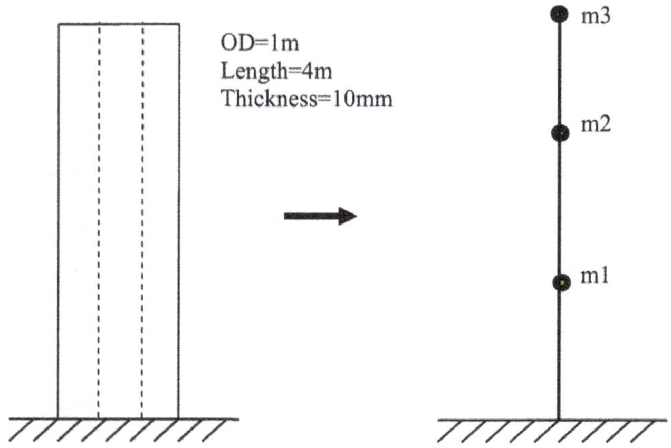

Fig. 4.22 Cylindrical vessel

Example 4.7 A cylindrical vessel shown in Fig. 4.22 has total mass of 400 kg, O.D. is 1m, length is 4m and thickness is 10 mm. Find eigen values, eigen vectors, modal participation factor, mass participation factor and significance of factor 1.5 in equivalent static method.

Solution

```
clear all;
E=2.11*10^12;
I=0.00381;
L=4/3;
C=12*E*I/L^3;

% Global Stiffness and Mass matrices
K=C*([1,-1,0;-1,2,-1;0,-1,2]); % check the matrix
%M=[133.2,0,0;0,133.2,0;0,0,66.6];
M=66.6*[1,0,0;0,2,0;0,0,2];
disp('The stiffness matrix is')
K
disp('The mass matrix is')
M
% Eigen Values
A=size(K);

% Lancozs Iteration Technique
% Assume X for finding first mode
X=ones(A(1),1);
Y=sqrt(X'*M*X);

X11=X/Y
X1(:,2)=X11;
X1(:,1)=X11;
B(1)=0;

for i=2:1:4
X1BAR=K\(M*X1(:,i))

ALPHA(i-1)=sqrt(X1BAR'*M*X1BAR)
X_BAR(:,i)=X1BAR-ALPHA(i-1)*X1(:,i)-B(i-1)*X1(:,
(i-1))
B(i)=sqrt(X_BAR(:,i)'*M*X_BAR(:,i))

X2_BAR=X_BAR(:,i)/B(i)
X1(:,(i+1))=X2_BAR;
end
Tn=[ALPHA(1),B(2),0;B(2),ALPHA(2),B(3);0,B(3),
ALPHA(3)]

Omega_Tn=roots(poly(Tn))

for i=1:1:A(1)
Freq(i,1)= sqrt((1/Omega_Tn(i,1)))/(2*pi);
end
disp('The First, Second & Third modal frequenciesare')
Freq

%Mass Participation
Q=[1.55E-2,-5.113E-2,-6.822E-2;5.17E-2,-
4.946E-2,4.882E-2;9.581E-2,6.997E-2,-3.067E-2];
% Modal Participation factor in first mode
R2_1=(Q(1,1)*133.3+Q(2,1)*133.3+Q(3,1)*66.6)^2
Percent_R2_1=R2_1*100/400;
R2_2=(Q(1,2)*133.3+Q(2,2)*133.3+Q(3,2)*66.6)^2
Percent_R2_2=R2_2*100/400;
R2_3=(Q(1,3)*133.3+Q(2,3)*133.3+Q(3,3)*66.6)^2
Percent_R2_3=R2_3*100/400;

disp(' Sr.no freq mass participation Cumulative mass
participation')
[1,0.2906863E+02,R2_1,Percent_R2_1;2,0.1640884E+03,
R2_2, Percent_R2_1+ Percent_R2_2;3,0.4086011E+03,
R2_3, Percent_R2_1+ Percent_R2_2+ Percent_R2_3]

% Calculation of frequency by Analytical method
E = 2.11e+12;
L = 4;
OD = 1;
ID= 0.98;
modes=zeros(3,1);
%mode shapes=zeros(3,40);
beta = [1.875 4.694 7.856];
I = 0.00381;
A = 0.0311;
mu = 400/A;
% loop over the three modes
for i=1:3,
%
```

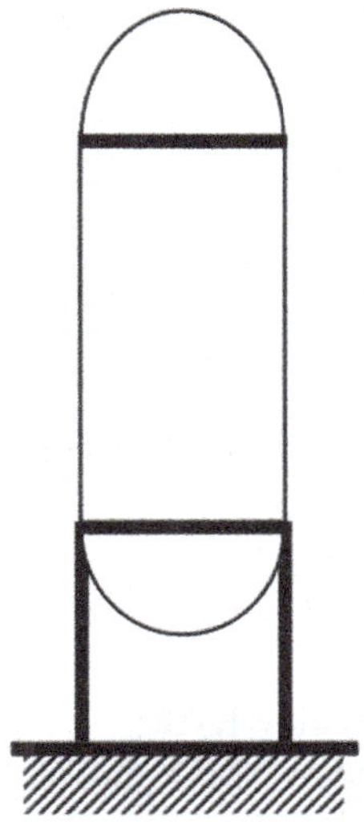

Fig. 4.23 Skirt-supported vertical tank

```
modes(i) = beta(i)^2 * sqrt(E*I/(mu*L^4));
modes(i) = modes(i)/(2*pi);
end
```

Example 4.8 A typical skirt supported vertical tank is shown in the Fig. 4.23. Find the fundamental frequency and stresses due to seismic and dead load. The support is of same thickness as that of tank.

Solution

```
% Calculation of Frequencies by Analytical method

rho =18631.83;
E = 2.1e+011;
L = 22;
OD = 4.334;
ID = 4.25;
modes=zeros(3,1);
modeshapes=zeros(3,3);
beta = [1.875 4.694 7.856];
I = 1.30415;
% finding first three modes
for i=1:3,
modes(i) = beta(i)^2 * sqrt(E*I/(rho*L^4));
modes(i) = modes(i)/(2*pi);
end
disp(' The first three mode frequency')
modes
% Stresses due to seismic
Ah=0.1;
W=4099000;
Vb=Ah*W;
Sigma_axial=((2/3)*Vb) /((pi/4)*(4.334^2-4.25^2));
Bending_stress=4508895.6*OD/(2*I);
% Stresses due to dead load
Axial_sress_dead=W/((pi/4)*(4.334^2-4.25^2));
Total_compressive_stress= Bending_stress+ Sigma_axial
+Axial_sress_dead;
```

Example 4.9: A cantilever beam with annular cross section having ID is 4.25 m, thickness is 40 mm and height is 53.59m is shown in Fig. 4.24. Find the frequencies, mode shapes and mass participation.

Solution

```
Di=4.25
Do=4.33
t=.04

% check R= radius of curvature and r= average radius
E=2.11e11
I=pi*(Do^4-Di^4)/64
A=pi*(Do^2-Di^2)/4
nu=0.3
L=53.59 % total height of the Tank
l=L/8
ro=7800

% Element Stiffness and Global Matrices
kb=(E*I/l^3)*[12 6*l -12 6*l;
6*l 4*l^2 -6*l 2*l^2;
-12 -6*l 12 -6*l;
6*l 2*l^2 -6*l 4*l^2]
mb=(ro*A*l/420)*[156 22*l 54 -13*l;
22*l 4*l^2 13*l -3*l^2;
54 13*l 156 -22*l;
-13*l -3*l^2 -22*l 4*l^2]

% Global Mass Matrix
M=[mb(1,1) mb(1,2) mb(1,3) mb(1,4) 0 0 0 0 0 0 0 0 0 0;
```

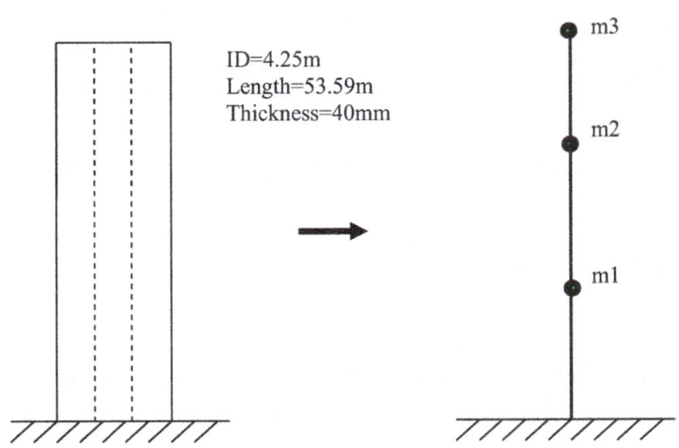

Fig. 4.24 Cantilever beam with annular cross section

```
mb(2,1) mb(2,2) mb(2,3) mb(2,4) 0 0 0 0 0 0 0 0 0 0;
mb(3,1) mb(3,2) mb(3,3)+mb(1,1) mb(3,4)+mb(1,2) mb
(1,3) mb(1,4) 0 0 0 0 0 0 0 0;
mb(4,1) mb(4,2) mb(4,3)+mb(2,1) mb(4,4)+mb(2,2) mb
(2,3) mb(2,4) 0 0 0 0 0 0 0 0;
0 0 mb(3,1) mb(3,2) mb(3,3)+mb(1,1) mb(3,4)+mb(1,2) mb
(1,3) mb(1,4) 0 0 0 0 0 0;
0 0 mb(4,1) mb(4,2) mb(4,3)+mb(2,1) mb(4,4)+mb(2,2) mb
(2,3) mb(2,4) 0 0 0 0 0 0;
0 0 0 0 mb(3,1) mb(3,2) mb(3,3)+mb(1,1) mb(3,4)+mb(1,2)
mb(1,3) mb(1,4) 0 0 0 0;
0 0 0 0 mb(4,1) mb(4,2) mb(4,3)+mb(2,1) mb(4,4)+mb(2,2)
mb(2,3) mb(2,4) 0 0 0 0;
0 0 0 0 0 0 mb(3,1) mb(3,2) mb(3,3)+mb(1,1) mb(3,4)+mb
(1,2) mb(1,3) mb(1,4) 0 0;
0 0 0 0 0 0 mb(4,1) mb(4,2) mb(4,3)+mb(2,1) mb(4,4)+mb
(2,2) mb(2,3) mb(2,4) 0 0;
0 0 0 0 0 0 0 0 mb(3,1) mb(3,2) mb(3,3)+mb(1,1) mb(3,4)
+mb(1,2) mb(1,3) mb(1,4);
0 0 0 0 0 0 0 0 mb(4,1) mb(4,2) mb(4,3)+mb(2,1) mb(4,4)
+mb(2,2) mb(2,3) mb(2,4);
0 0 0 0 0 0 0 0 0 0 mb(3,1) mb(3,2) mb(3,3)+mb(1,1) mb
(3,4)+mb(1,2);
0 0 0 0 0 0 0 0 0 0 mb(4,1) mb(4,2) mb(4,3)+mb(2,1) mb
(4,4)+mb(2,2)]

%Global Stiffness matrix
K=[kb(1,1) kb(1,2) kb(1,3) kb(1,4) 0 0 0 0 0 0 0 0 0 0;
kb(2,1) kb(2,2) kb(2,3) kb(2,4) 0 0 0 0 0 0 0 0 0 0;
kb(3,1) kb(3,2) kb(3,3)+kb(1,1) kb(3,4)+kb(1,2) kb(1,3) kb
(1,4) 0 0 0 0 0 0 0 0;
kb(4,1) kb(4,2) kb(4,3)+kb(2,1) kb(4,4)+kb(2,2) kb(2,3) kb
(2,4) 0 0 0 0 0 0 0 0;
0 0 kb(3,1) kb(3,2) kb(3,3)+kb(1,1) kb(3,4)+kb(1,2) kb(1,3)
kb(1,4) 0 0 0 0 0 0;
0 0 kb(4,1) kb(4,2) kb(4,3)+kb(2,1) kb(4,4)+kb(2,2) kb(2,3)
kb(2,4) 0 0 0 0 0 0;
0 0 0 0 kb(3,1) kb(3,2) kb(3,3)+kb(1,1) kb(3,4)+kb(1,2) kb
(1,3) kb(1,4) 0 0 0 0;
0 0 0 0 kb(4,1) kb(4,2) kb(4,3)+kb(2,1) kb(4,4)+kb(2,2) kb
(2,3) kb(2,4) 0 0 0 0;
0 0 0 0 0 0 kb(3,1) kb(3,2) kb(3,3)+kb(1,1) kb(3,4)+kb(1,2)
kb(1,3) kb(1,4) 0 0;
0 0 0 0 0 0 kb(4,1) kb(4,2) kb(4,3)+kb(2,1) kb(4,4)+kb(2,2)
kb(2,3) kb(2,4) 0 0;
0 0 0 0 0 0 0 0 kb(3,1) kb(3,2) kb(3,3)+kb(1,1) kb(3,4)+kb
(1,2) kb(1,3) kb(1,4);
0 0 0 0 0 0 0 0 kb(4,1) kb(4,2) kb(4,3)+kb(2,1) kb(4,4)+kb
(2,2) kb(2,3) kb(2,4);
0 0 0 0 0 0 0 0 0 0 kb(3,1) kb(3,2) kb(3,3)+kb(1,1) kb(3,4)
+kb(1,2);
0 0 0 0 0 0 0 0 0 0 kb(4,1) kb(4,2) kb(4,3)+kb(2,1) kb(4,4)
+kb(2,2)]

A=size(K);

% Assume X for finding first mode
X=ones(A(1),1);

% Modal analysis using Inverse Iteration Technique
i=2;
Omega2(1)=0;
for i=2:1:5
R=M*X;
X_NEW=inv(K)*R;
Q=X_NEW'*M*X_NEW;
qn=X_NEW/sqrt(Q);
Omega2(i)=qn'*K*qn;
Freq(i)=sqrt(Omega2(i))/(2*pi);
E(i)=(Omega2(i)-Omega2(i-1))/Omega2(i);

% check convergence
if abs(E)>=0.0001
%n=n+1;
X=qn;
else
end
end
disp('The first modal frequency is')
Freq(5)

% Finding second eigen value and eigen vector
X1=ones(A(1),1);
qn2=qn;
BETA=qn2'*M*X1;
X2=X1-BETA*qn2;
for j=2:1:5

% Iteration
% check convergence
R2=M*X2;
X_NEW2=K\R2;
Q2=X_NEW2'*M*X_NEW2;
qn2=X_NEW2/sqrt(Q2);

Omega_2(j)=qn2'*K*qn2;
Freq2(j)=sqrt(Omega_2(j))/(2*pi);
E2(j)=(Omega_2(j)-Omega_2(j-1))/Omega_2(j);
% check convergence
if abs(E2(j))>=0.0001
X2=qn2;
else
```

```
end
end

disp('The second modal frequency is')
Freq2(j)

disp('The first two mode shapes are')
Mode_Shape=[X,X2]
```

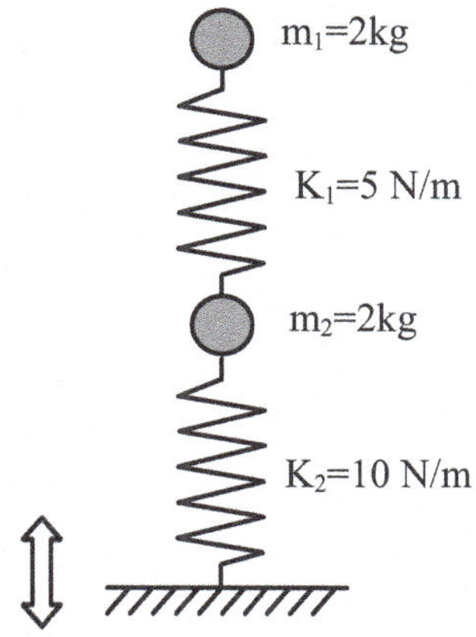

Fig. 4.26 Spring-mass system

Exercise Problems

1. Explain the significance of mass participation factors in dynamic analysis of multi-degree-of-freedom systems.
2. Find the eigenvalues and eigenvectors for the spring–mass system shown in Fig. 4.25.
3. Determine the maximum displacement of the masses shown in Fig. 4.26 due to site-specific response spectrum shown in Fig. 4.15 by response spectrum method.
4. A pipe of internal diameter of 100 mm and thickness of 4 mm with elbow of radius of 250 mm is shown in Fig. 4.27. The length of each segment is 1.25 m. Find the eigenvalue and vector of the system.
   Evaluate its eigenvalues and eigenvectors using the stiffness matrix explained in Annexure III for straight pipe and elbow, respectively. Masses may be lumped at the nodes along three directions.
5. Using the following data for the two-degree-of-freedom system, evaluate the following (Fig. 4.28).
   a. Evaluate the natural frequencies and mode shapes of the system.
   b. Evaluate the participation factors = $\frac{[\phi_n]^t[M][1]}{[\phi_n]^t[M][\phi_n]}$.
   c. Using the following response spectrum, evaluate the modal acceleration and modal displacement response and the final acceleration and displacement considering all the modes (Fig. 4.29).
6. Explain and compare Fourier transform and discrete Fourier transform of periodic random ground motion $\ddot{x}_g(t)$.
7. Using spectra of various dynamic loads explain various zones of response control.
8. Explain method of peak broadening of floor response spectra with reasons of broadening.
9. Frequency of the three-storey structures is 5 Hz. A pump with speed of 300 rpm needs to be placed on the top of building. Is it recommended or not? Explain.
10. Explain the procedure of calculating response of MDOF system using power spectral density function (PSDF) as design basis. Assume natural frequencies, mode shapes, participation factors, and damping values are known.
11. Write briefly on the following
    a. Power spectral density
    b. Frequency domain (Random vibration) analysis versus Time domain analysis
    c. D'Alembert's principle versus Principle of virtual displacements in formulating equation of motion.
12. Explain briefly the procedure of developing design basis response spectrum using deterministic approach
13. Explain clearly the procedure of generating compatible time history to the design basis ground response spectra.
14. Differentiate free and forced vibrations. What are the parameters which defines free vibration characteristics (Fig. 4.30)?
15. Write linear differential equations of motion for the following system. Also write the characteristic equation for evaluating eigenvalues and eigenvectors.
16. Convert the following system into single-degree-of-freedom system and give the equivalent stiffness of the system (Fig. 4.31).
17. Using the following data for the three-degree-of-freedom system, evaluate the following (Fig. 4.32).

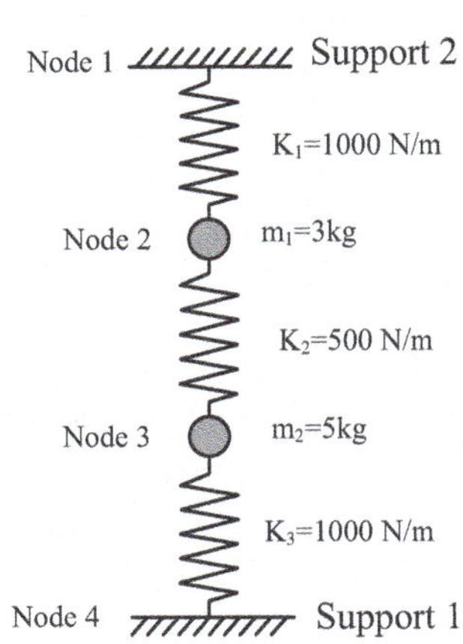

Fig. 4.25 Spring–mass system

Fig. 4.27 Piping system

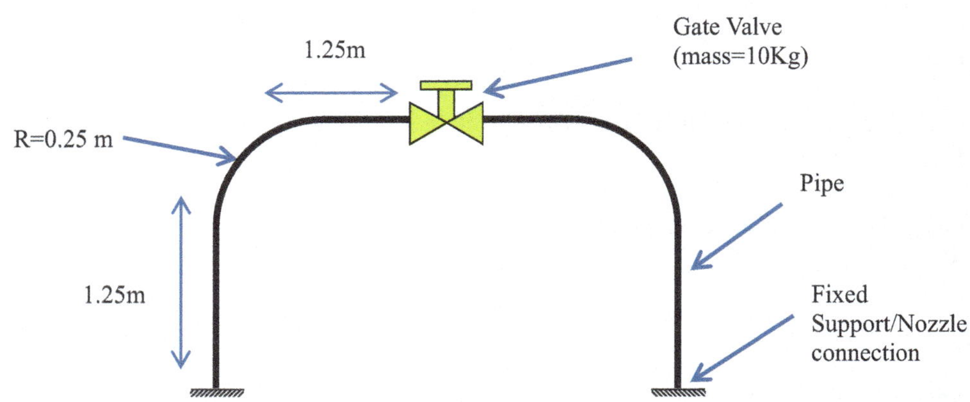

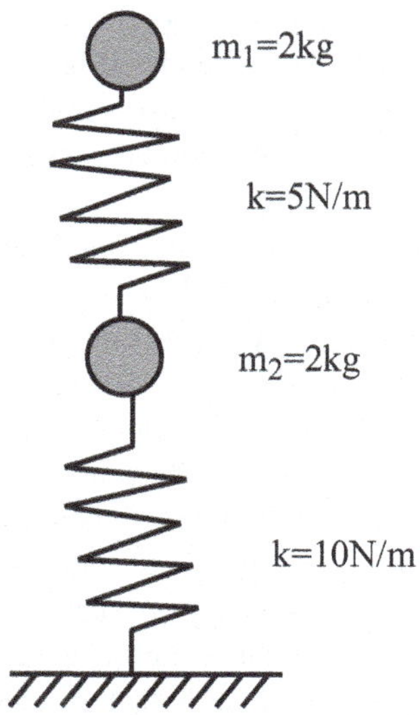

Fig. 4.28 Data for the two-degree-of-freedom system

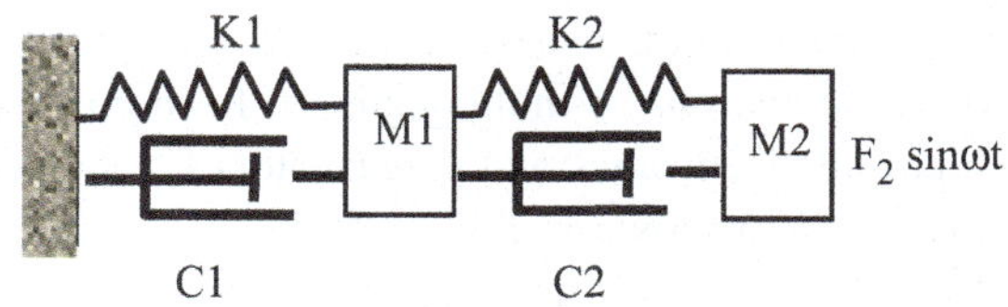

Fig. 4.30 Two degree of freedom system

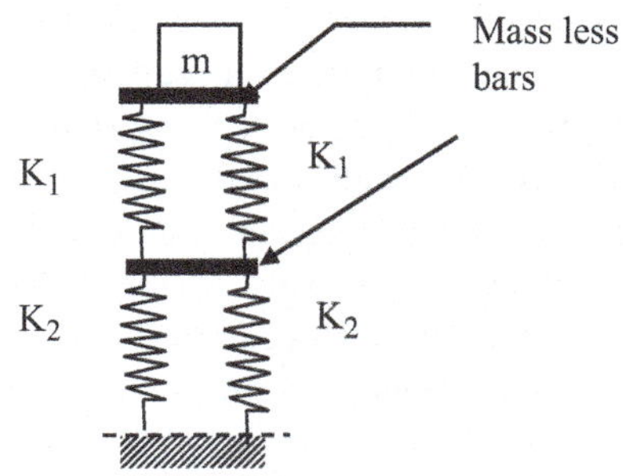

Fig. 4.31 Spring mass system

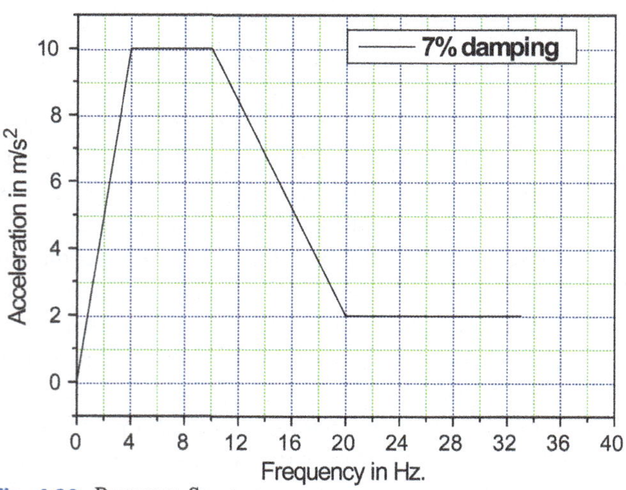

Fig. 4.29 Response Spectrum

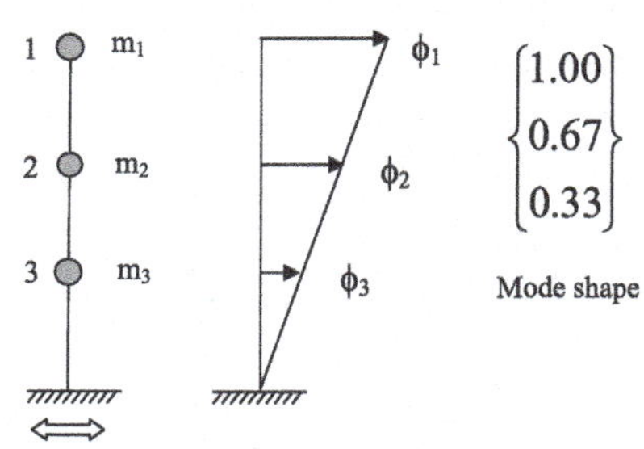

Fig. 4.32 Three degree of freedom system

$E = 2.5 \times 10^{10}$ N/m$_2$ (concrete) $C = \frac{12EI}{l^3}$
Moment of inertia of the column section I = 3.52 × 10^6 m^4
m^1 = m^2 = m^3 = 1560 kg; Length l = 1.5 m

(a) Normalize the mode shape with respect to mass
(b) Evaluate the natural frequency of the system
(c) Evaluate the participation factor = $\frac{[\phi_n]^t[M][1]}{[\phi_n]^t[M][\phi_n]}$
(d) Using the following response spectra, evaluate the modal acceleration and modal displacement

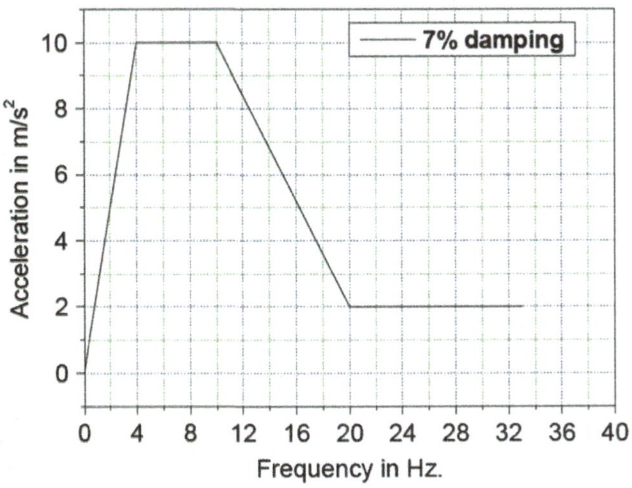

Fig. 4.33 Response Spectrum

response and the final acceleration and displacement considering all the two modes (Fig. 4.33).

18. Write equations of motion for the following multi-degree-of-freedom system for the free vibration condition (Fig. 4.34).
19. Write the equations of motion of the following system and derive the characteristic equation (Fig. 4.35).
20. Write the vector diagram of spring–mass–damper system when frequency of the system is equal to the excitation frequency and explain the significance.
21. Evaluate the natural frequency of the following system using energy principle.int: Fixed-Fixed Beam stiffness.
22. Length of the beam is l, Moment of inertia is I, and Young's modulus is E. Considering mass per unit length of beam is m and assuming velocity is varying linearly, find the difference in natural frequency of the system with and without mass of beam (Fig. 4.36).
23. A massless cantilever beam of stiffness 200 N/m is attached to a spring of stiffness 50 N/m as shown in the figure. A mass of 10 kg. is attached at one end. Evaluate the frequency of the system. Hint: use parallel spring approach (Fig. 4.37).
24. Using free vibration response of a system, explain how damping is evaluated using logarithmic decrement method.
25. What is square root of sum of squares (SRSS) combination method? Explain briefly.

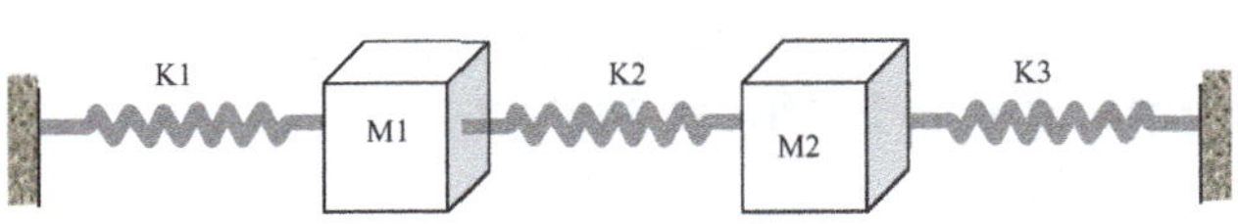

Fig. 4.34 Two degree of freedom system

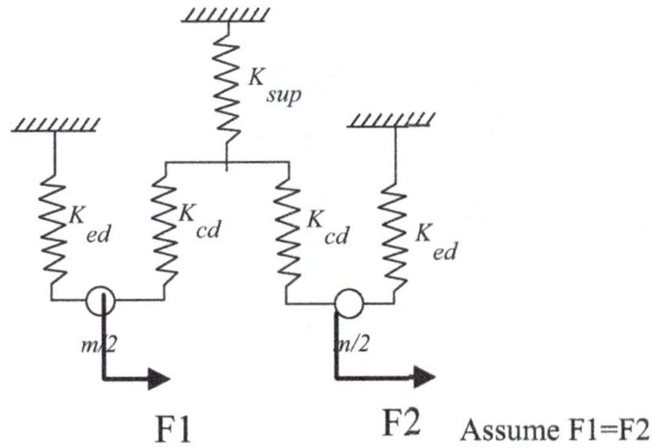

Fig. 4.35 Spring mass system

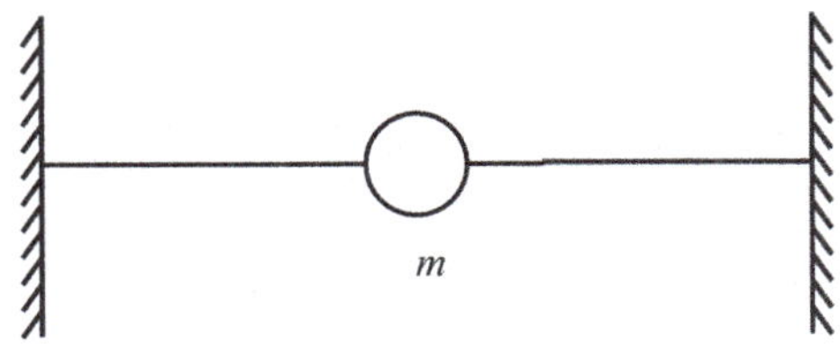

Fig. 4.36 Beam

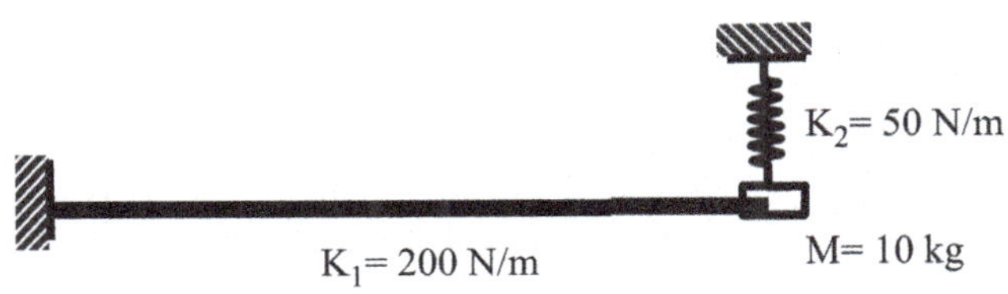

Fig. 4.37 Mass less cantilever beam supported on spring

26. Plot the frequency response curve for damped single-degree-of-freedom system subjected to harmonic excitation and explains peak value at resonance in terms of damping ratio ζ.
27. State the methods used for time history analysis. What time step size must be chosen to get accurate response using time history analysis? Which method shall be most suitable for analysis of multi-degree-of-freedom system and why?
28. Explain orthogonality and normalization of vectors. Find out the eigenvalues and corresponding vectors of $A = \begin{bmatrix} 5 & -5 \\ -5 & 10 \end{bmatrix}$.
29. For the two systems shown in Fig. 4.38, evaluate the frequency ratio.
30. Solve the following problem (Fig. 4.39).

$$K_1 = K_3 = 10\,\text{N/m}, \quad K_2 = 5\text{N/m}; \quad M_1 = 2\,\text{kg}, \quad M_2 = 4\,\text{kg}.$$

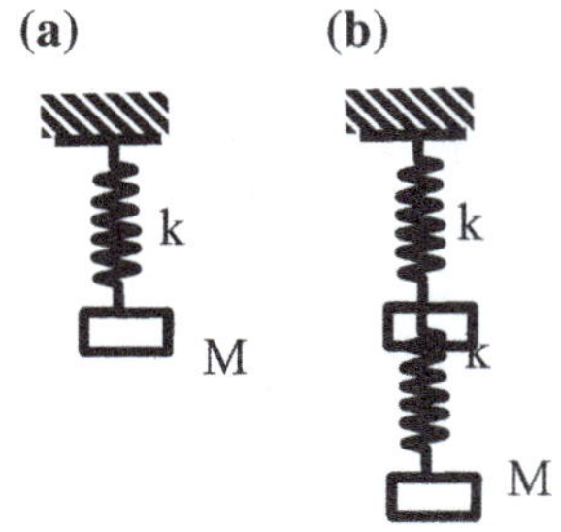

Fig. 4.38 Spring mass systems

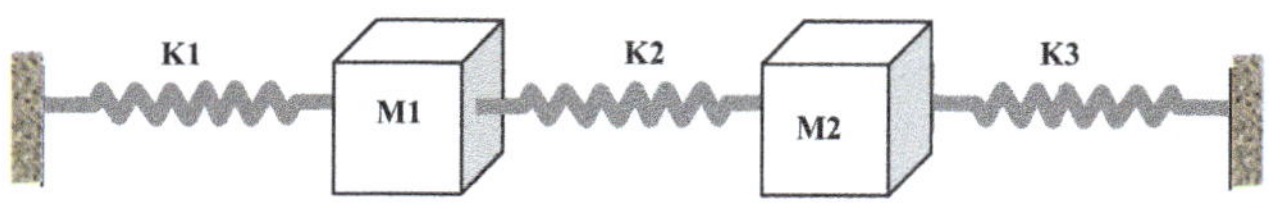

Fig. 4.39 Two degree of freedom system

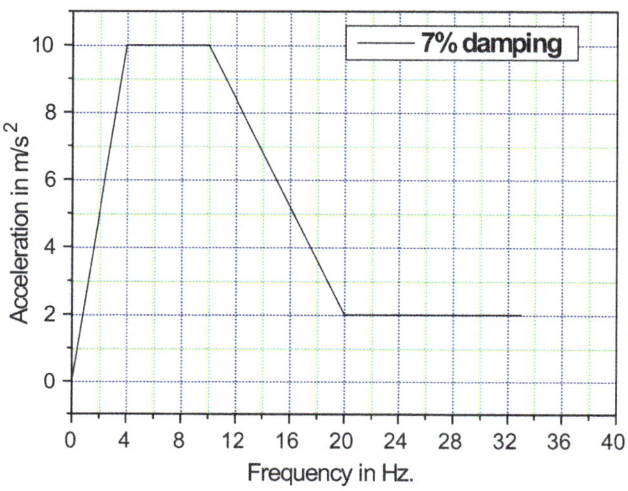

Fig. 4.40 Response Spectrum

The frequencies and the mode shapes of the structure are as follows:

$$\omega = \begin{bmatrix} 1.746 \\ 2.864 \end{bmatrix} \text{radians/sec} \quad \phi = \begin{Bmatrix} 0.5616 & 1 \\ 1 & -0.281 \end{Bmatrix}$$

Calculate the mass-normalized mode shapes.

31. Calculate the mass participation factors for the three modes using these normalized mode shapes using the formula $\Gamma_n = \frac{[\phi_n]^t[M][1]}{[\phi_n]^t[M][\phi_n]}$ Using the below response spectra, evaluate the modal acceleration and modal displacement response and the final acceleration and displacement considering all the modes (Fig. 4.40).
32. For a two-degree-of-freedom shear frame structure is shown in the figure below, determine the natural frequencies and mode shapes of the structure and draw the mode shapes. Calculate the mass-normalized mode shapes and prove the orthogonality property of the mode shapes (Fig. 4.41).

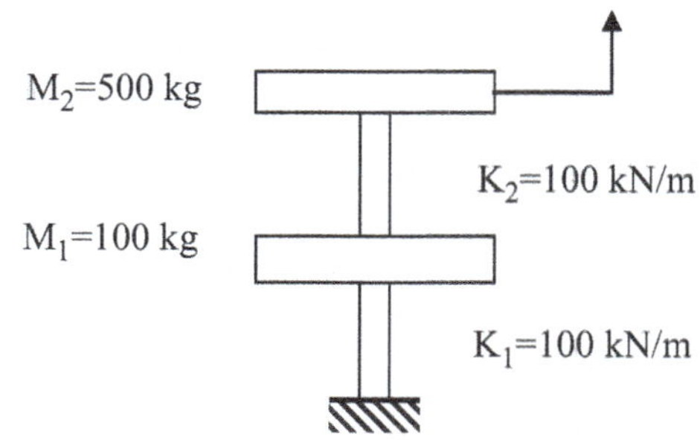

Fig. 4.41 Shear frame structure

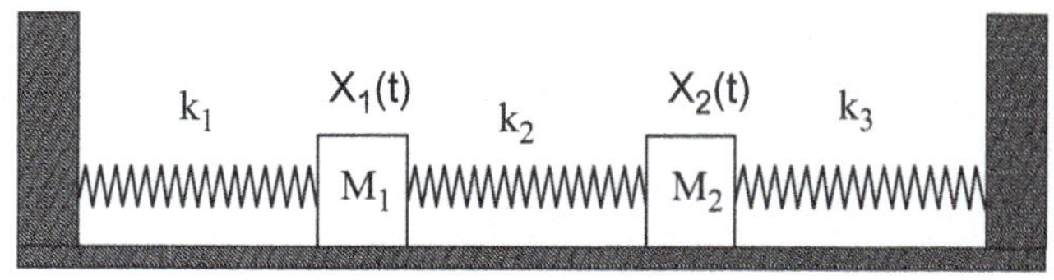

Fig. 4.42 Two degree of freedom system

33. What is meant by exploratory tests? Explain briefly estimating frequency and damping of SDOF structure or equipment with shock and sine sweep tests.
34. Explain the procedure of estimating the response of a structure or equipment (SDOF) with dampers and considering response spectrum method.
35. Explain the procedure for estimating frequency and damping of MDOF system using sine sweep tests.
36. Obtain equation of motion for two-degree-of-freedom system shown below, using Lagrangian approach (Fig. 4.42).
37. A three-storey shear building is shown in the figure below. Obtain the natural frequencies of the building using Lancotz iteration method (Fig. 4.43).
38. For the shear frame, shown below, find the peak values of base shear and top displacement between the first and second floors considering contributions from three modes only. Use SRSS rules of combinations and compare the results. Use a 5% response spectrum of the Bhuj/Kachchh earthquake as the basic input ground motion (Fig. 4.44).

Appendix 1: Stiffness Matrix

A.1.1 Stiffness Matrix for Spring Element

Stiffness matrix for spring element can be written as follows:

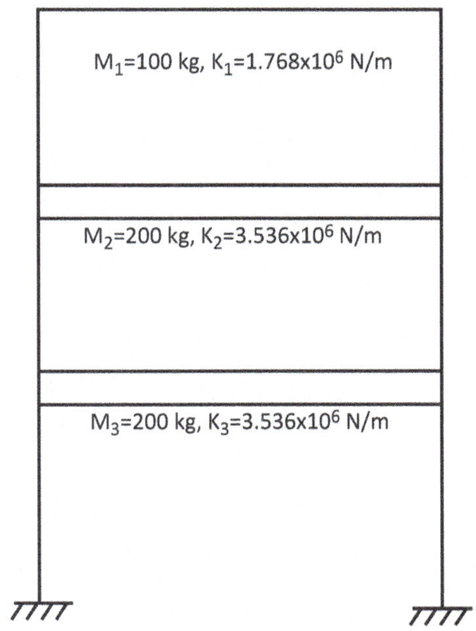

Fig. 4.43 Three storey shear building

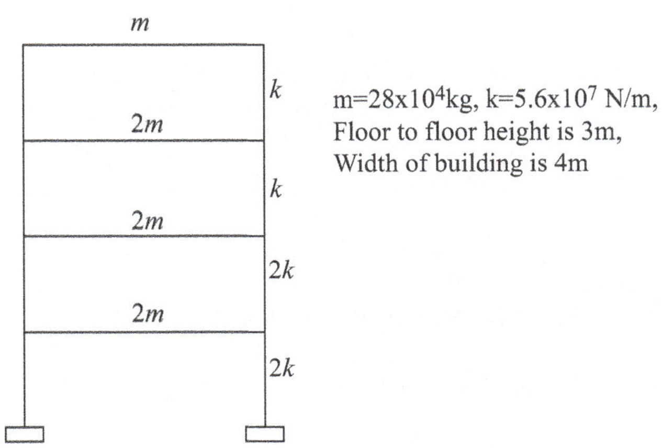

Fig. 4.44 Shear frame

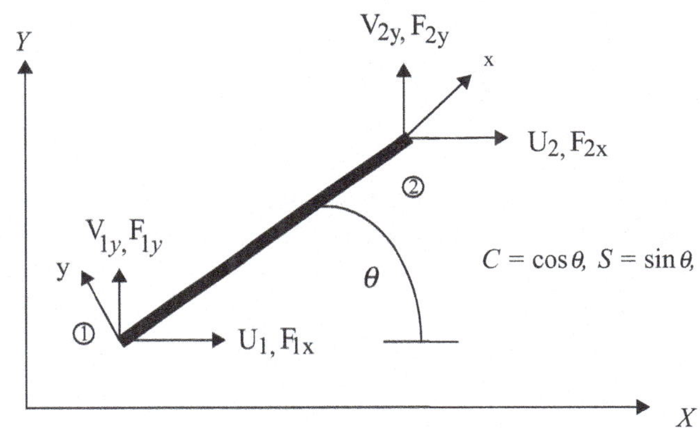

Fig. 4.45 2D truss element

$$[K]_{element} = \begin{bmatrix} k & -k \\ -k & k \end{bmatrix} \tag{4.21}$$

$$\text{where } k = \frac{AE}{L}$$

A.1.2 Stiffness Matrix for Two-Dimensional Truss (Only Axial Effect)

Generally, the structures, systems and components are in three-dimensional coordinate systems. The method of obtaining the stiffness and mass matrices is explained considering truss or spring elements (Fig. 4.45). The same method can be extended to other elements such as beam, plate and shell, and 3D brick.

Local coordinate system is specified along the element orthogonal direction as shown in Fig. 4.45. For example, along the axis of the element, it is local x-axis and perpendicular to it is the local y-axis and the global coordinate system refers to the total structure, systems, or components. x_1, y_1 and x_2, y_2 are local coordinates at nodes 1 and node 2, respectively, in local x-y system. u_{1x}, v_{1y} and u_{2x}, v_{2y} are the displacements in the local coordinate system at nodes 1 and 2, respectively.

X_1, Y_1 and X_2, Y_2 are global coordinates of the nodes 1 and 2, respectively in global X-Y system. U_{1x}, V_{1y} and U_{2x}, V_{2y} are the displacements in the global coordinate system at nodes 1 and 2, respectively.

For local coordinate system

$$\frac{AE}{L}\begin{bmatrix} 1 & -1 \\ -1 & 1 \end{bmatrix}\begin{Bmatrix} u_{1x} \\ u_{2x} \end{Bmatrix} = \begin{Bmatrix} f_{1x} \\ f_{2x} \end{Bmatrix} \tag{4.22}$$

It can also be written in general form as

$$\frac{AE}{L}\begin{bmatrix} 1 & 0 & -1 & 0 \\ 0 & 0 & 0 & 0 \\ -1 & 0 & 1 & 0 \\ 0 & 0 & 0 & 0 \end{bmatrix}\begin{Bmatrix} u_{1x} \\ v_{1y} \\ u_{2x} \\ v_{2y} \end{Bmatrix} = \begin{Bmatrix} f_{1x} \\ f_{1y} \\ f_{2x} \\ f_{2y} \end{Bmatrix} \tag{4.23}$$

From Fig. 4.45, the relation between local coordinates and global coordinates can be written as

$$\begin{aligned} x_1 &= X_1 \cos\theta + Y_1 \sin\theta \\ y_1 &= -X_1 \sin\theta + Y_1 \cos\theta \\ x_2 &= X_2 \cos\theta + Y_2 \sin\theta \\ y_2 &= -X_2 \sin\theta + Y_2 \cos\theta \end{aligned} \tag{4.24}$$

In the matrix form, it can be written as

$$\begin{Bmatrix} x_1 \\ y_1 \\ x_2 \\ y_2 \end{Bmatrix} = \begin{bmatrix} \cos\theta & \sin\theta & & \\ -\sin\theta & \cos\theta & & \\ & & \cos\theta & \sin\theta \\ & & -\sin\theta & \cos\theta \end{bmatrix}\begin{Bmatrix} X_1 \\ Y_1 \\ X_2 \\ Y_2 \end{Bmatrix} = [T]\begin{Bmatrix} X_1 \\ Y_1 \\ X_2 \\ Y_2 \end{Bmatrix} \tag{4.25}$$

where $[T]$ is transformation matrix. Similarly, the relation between the local and global displacement vector and forces vector can be related with transformation matrix $[T]$ as follows:

$$\begin{aligned} \{u\} &= \{T\}\{U\} \\ \{f\} &= [T]\{F\} \end{aligned} \tag{4.26}$$

The equilibrium equation in the local coordinate system would be

$$\begin{aligned} [K]_l\{u\} &= \{f\} \\ [K]_l[T]\{U\} &= [T]\{F\} \end{aligned} \tag{4.27}$$

Premultiplying by $[T]^{-1}$, the above equation reduces to

$$[T]^{-1}[K]_l[T]\{U\} = [T]^{-1}[T]\{F\} \tag{4.28}$$

But for transpose matrix, $[T]^T = [T]^{-1}$
Hence, the global stiffness matrix can be given by

$$[K] = [T]^T[K]_l[T] \tag{4.29}$$

$$[K] = \begin{bmatrix} C & -S & 0 & 0 \\ S & C & 0 & 0 \\ 0 & 0 & C & -S \\ 0 & 0 & S & C \end{bmatrix} * \frac{AE}{L}\begin{bmatrix} 1 & 0 & -1 & 0 \\ 0 & 0 & 0 & 0 \\ -1 & 0 & 1 & 0 \\ 0 & 0 & 0 & 0 \end{bmatrix} * \begin{bmatrix} C & S & 0 & 0 \\ -S & C & 0 & 0 \\ 0 & 0 & C & S \\ 0 & 0 & -S & C \end{bmatrix} \tag{4.30}$$

$$[K] = \frac{AE}{L}\begin{bmatrix} C^2 & CS & -C^2 & -CS \\ CS & S^2 & -CS & -S^2 \\ -C^2 & -CS & C^2 & CS \\ -CS & -S^2 & CS & S^2 \end{bmatrix} \tag{4.31}$$

The equilibrium equation in the global coordinate system can be given by

$$\begin{Bmatrix} f_{1x} \\ f_{1y} \\ f_{2x} \\ f_{2y} \end{Bmatrix} = \frac{AE}{L}\begin{bmatrix} C^2 & CS & -C^2 & -CS \\ CS & S^2 & -CS & -S^2 \\ -C^2 & -CS & C^2 & CS \\ -CS & -S^2 & CS & S^2 \end{bmatrix}\begin{Bmatrix} u_1 \\ v_1 \\ u_2 \\ v_2 \end{Bmatrix} \tag{4.32}$$

Similar approach can be followed for other elements such as beam, plate and shell, and brick. For more information, readers are requested to refer finite element text books such as Finite Element Analysis by C. S. Krishnamoorthy.

Appendix 2: Natural Frequencies and Mode Shapes of Some Common Geometries

A.2.1 Cantilever Beam

The first three undamped natural frequencies of a cantilever beam are given as

$$\begin{aligned} \omega_1 &= (10.875)^2 \times \sqrt{\frac{\text{EI}}{\text{mL}^4}}, \\ \omega_2 &= (40.694)^2 \times \sqrt{\frac{\text{EI}}{\text{mL}^4}}, \\ \omega_3 &= (70.855)^2 \times \sqrt{\frac{\text{EI}}{\text{mL}^4}} \end{aligned} \tag{4.33}$$

where E is the modulus of rigidity, I is the moment of inertia, ω is the circular natural frequency, L is the length (span), and m is the mass per unit length. First three mode shapes are shown in Fig. 4.46.

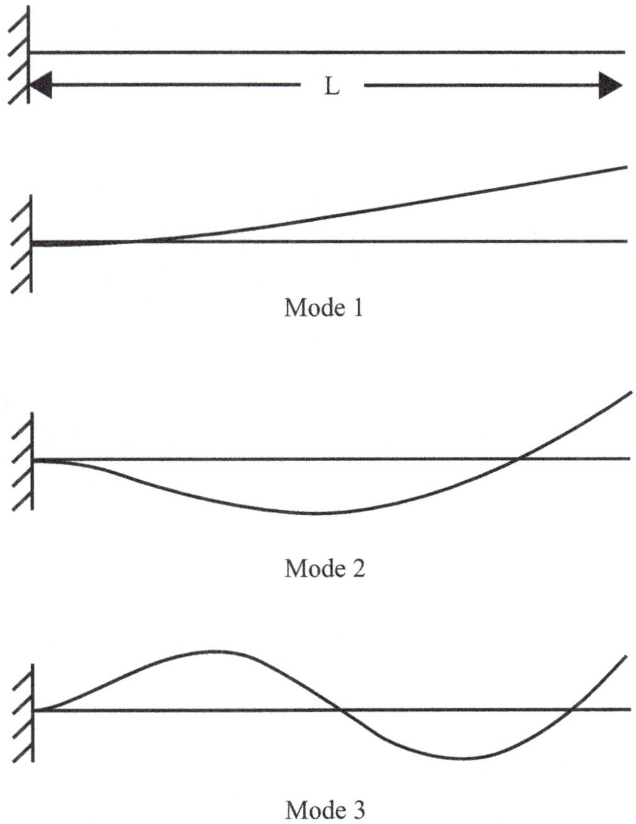

Fig. 4.46 Mode shapes of a cantilever beam

A.2.2 Simply Supported Beam

The undamped natural frequencies of a simply supported beam are given as

$$\omega_n = (n\pi)^2 \times \sqrt{\frac{EI}{mL^4}} \quad n = 1, 2, \ldots\ldots\ldots n \tag{4.34}$$

where E is the modulus of rigidity, I is the moment of inertia, ω is the circular natural frequency, L is the length (span), and m is the mass per unit length. First three mode shapes are shown in Fig. 4.47.

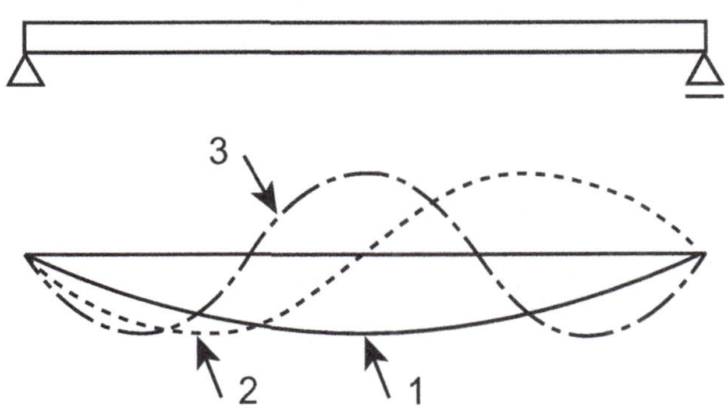

Fig. 4.47 Mode shapes of a simply supported beam

Appendix 3: Stiffness Matrices for 3D Elements

A.3.1 3D Pipe Elbow Stiffness Matrix

A systematic derivation of the 12 × 12 stiffness matrix for a three-dimensional pipe elbow element including the effects of transverse shear and pipe bend flexibility according to the ASME Boiler and Pressure Vessel Code is explained.

In Fig. 4.48, a pipe elbow cantilevered at one end (end B) and free at the other end (end A) is displayed. The free end, i.e., point A is subjected to forces and moments (three components) in radial, longitudinal, and normal directions. In Fig. 4.48, y-axis is tangential with the elbow placed in x-y plane. The radial and normal direction is represented with x- and z-axis, respectively. The forces and moments acting on the free-end A are represented with F_{x1}, F_{y1}, and F_{z1} and M_{x1}, M_{y1}, and M_{z1}, respectively. The reactions at the fixed end are denoted as F_{x2}, F_{y2}, and F_{z2}, and M_{x2}, M_{y2}, and M_{z2}, respectively. The angle of the elbow is denoted with angle θ with bend radius R. The displacements and rotations corresponding to the forces and the moments are denoted as d_x, d_y, and d_z, and θ_x, θ_y, and θ_z, respectively.

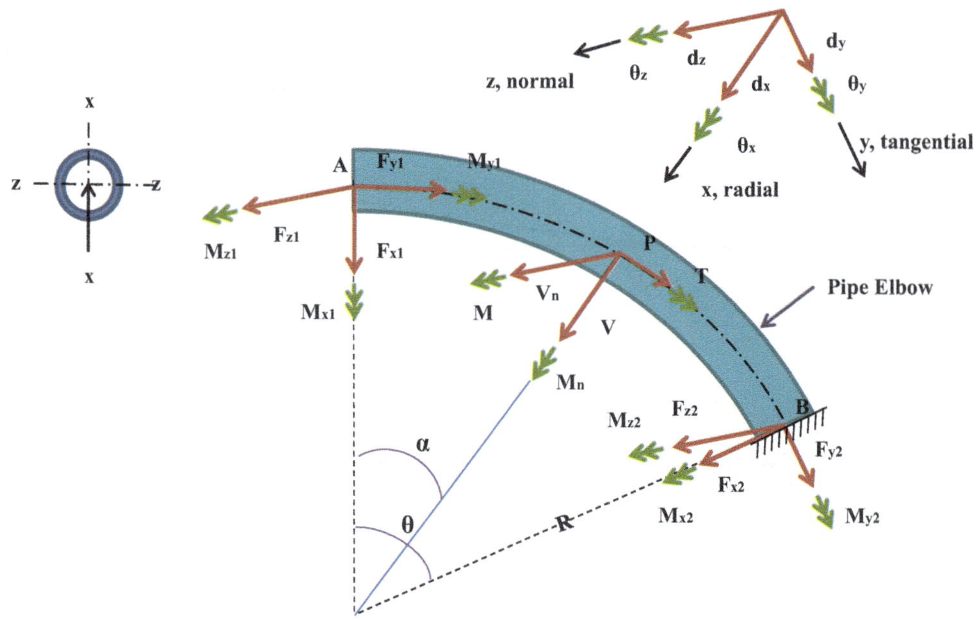

Fig. 4.48 Pipe elbow element

Assuming an arbitrary cross section at an angle α from free-end A where the internal forces and internal moments are denoted with V, P, and V_n and M_n, T, and M, respectively. Now, these internal forces and moments are written in the form of applied forces and moments at free-end A in Eqs. (4.35) and (4.36). It is worth noting that Eq. (4.35) expresses force equilibrium in-plane, whereas Eq. (4.36) expresses the equilibrium normal to the plane of the elbow.

$$\begin{bmatrix} V \\ P \\ M \end{bmatrix} = [U(\alpha)] \begin{bmatrix} F_{x1} \\ F_{y1} \\ M_{z1} \end{bmatrix} \tag{4.35}$$

$$\begin{bmatrix} V_n \\ M_n \\ T \end{bmatrix} = [V(\alpha)] \begin{bmatrix} F_{z1} \\ M_{x1} \\ M_{y1} \end{bmatrix} \tag{4.36}$$

where $[U(\alpha)]$ & $[V(\alpha)]$ is a 3 x 3 matrix defined as:

$$[U(\alpha)] = \begin{bmatrix} -\cos\alpha & \sin\alpha & 0 \\ -\sin\alpha & -\cos\alpha & 0 \\ -R\sin\alpha & R(1-\cos\alpha) & -1 \end{bmatrix} \tag{4.37}$$

$$[V(\alpha)] = \begin{bmatrix} -1 & 0 & 0 \\ R\sin\alpha & -\cos\alpha & \sin\alpha \\ R(1-\cos\alpha) & -\sin\alpha & -\cos\alpha \end{bmatrix} \tag{4.38}$$

Relations such as Eqs. (4.39) and (4.40) are achieved by suitable substitution of Equations (4.37) and (4.38) in the strain energy relation and by further integrating it.

$$\begin{bmatrix} d_{x1} \\ d_{y1} \\ \theta_{z1} \end{bmatrix} = [A] \begin{bmatrix} F_{x1} \\ F_{y1} \\ M_{z1} \end{bmatrix} \tag{4.39}$$

$$\begin{bmatrix} d_{z1} \\ \theta_{x1} \\ \theta_{y1} \end{bmatrix} = [AN] \begin{bmatrix} F_{z1} \\ M_{x1} \\ M_{y1} \end{bmatrix} \tag{4.40}$$

The 6 × 6 Elbow Flexibility Matrix using the standard assumption of small displacements and a linear elastic material is given in Eq. (4.41). The 6 × 6 matrix is inverted and the necessary force transfer matrices as shown in Eq. (4.41) are required to formulate the complete 12 × 12 stiffness matrix referenced to the radial, tangential, and normal elbow coordinate system.

The elbow diagonal partitioned flexibility matrix are given below

$$\mathrm{F} = \begin{bmatrix} [A] & [0] \\ [0] & [AN] \end{bmatrix} \tag{4.41}$$

where

$$[A] = \begin{bmatrix} A1 & A2 & A3 \\ A2 & A4 & A5 \\ A3 & A5 & A6 \end{bmatrix} \tag{4.42}$$

$$[AN] = \begin{bmatrix} AN1 & AN2 & AN3 \\ AN2 & AN4 & AN5 \\ AN3 & AN5 & AN6 \end{bmatrix} \tag{4.43}$$

$$A1 = \left(\frac{R^3}{EI} + \frac{R}{AE}\right)\left(\frac{\theta}{2} - \frac{\sin 2\theta}{4}\right) + \frac{R}{GA_S}\left(\frac{\theta}{2} + \frac{\sin 2\theta}{4}\right) \tag{4.44}$$

$$A2 = \left(\frac{R^3}{EI} + \frac{R}{AE} - \frac{R}{GA_S}\right)\left(\frac{\sin^2\theta}{4}\right) + \frac{R^3}{EI}(\cos\theta - 1) \tag{4.45}$$

$$A3 = \frac{R^2}{EI}(1 - \cos\theta) \quad (4.46)$$

$$A4 = \frac{R^3}{EI}(\theta - 2\sin\theta) + \left(\frac{R^3}{EI} + \frac{R}{AE}\right)\left(\frac{\theta}{2} - \frac{\sin 2\theta}{4}\right) + \frac{R}{GA_S}\left(\frac{\theta}{2} - \frac{\sin 2\theta}{4}\right) \quad (4.47)$$

$$A5 = \frac{R^2}{EI}(\sin\theta - \theta) \quad (4.48)$$

$$A6 = \frac{R}{EI}(\theta) \quad (4.49)$$

$$AN1 = \frac{R^3}{EI}\left(\frac{\theta}{2} - \frac{\sin 2\theta}{4}\right) + \frac{R^3}{GJ}\left(\frac{3\theta}{2} - 2\sin\theta + \frac{\sin 2\theta}{4}\right) + \frac{R\theta}{GA_S} \quad (4.50)$$

$$AN2 = \left(\frac{R^2}{GJ} + \frac{R^2}{EI}\right)\left(\frac{\sin^2\theta}{2}\right) + \frac{R^2}{GJ}(\cos\theta - 1) \quad (4.51)$$

$$AN3 = \frac{R^3}{EI}\left(\frac{\theta}{2} - \frac{\sin 2\theta}{4}\right) + \frac{R^2}{GJ}\left(\frac{\theta}{2} - \sin\theta + \frac{\sin 2\theta}{4}\right) \quad (4.52)$$

$$A4 = \frac{R}{EI}\left(\frac{\theta}{2} - \frac{\sin 2\theta}{4}\right) + \left(\frac{R^2}{GJ}\right)\left(\frac{\theta}{2} - \frac{\sin 2\theta}{4}\right) \quad (4.53)$$

$$AN5 = \left(\frac{R}{GJ} - \frac{R}{EI}\right)\left(\frac{\sin^2\theta}{2}\right) \quad (4.54)$$

$$AN6 = \frac{R}{EI}\left(\frac{\theta}{2} - \frac{\sin 2\theta}{4}\right) + \frac{R}{GJ}\left(\frac{\theta}{2} + \frac{\sin 2\theta}{4}\right) \quad (4.55)$$

where A_s is the area with shear factor, E is the Young's modulus, G is the modulus of rigidity, I is moment of inertia, and J is the torsional moment of inertia. In order to properly account for the cross-sectional ovalization in pipe elbows as per the ASME code, the flexibility factor is multiplied with the flexibility matrix. The in- and out-plane moment of inertia (I) is divided with the flexibility factor to incorporate elbow ovalization effect.

$$\text{Flexibility Factor} = \frac{1.65}{\frac{t*r}{R^2}} \quad (4.56)$$

Now, the construction of the 12 × × 12 elbow stiffness matrix is done with following steps

1. **Inversion of the 6 × 6 Elbow Flexibility Matrix**

$$F^{-1} = \begin{bmatrix} [B] & [0] \\ [0] & [C] \end{bmatrix} \quad (4.57)$$

where

$$[B] = [A]^{-1} \quad (4.58)$$

$$[C] = [AN]^{-1} \quad (4.59)$$

2. **Force Transfer Matrices**

In order to achieve the complete 12 × 12 stiffness matrix, the force transfer matrices which express equilibrium of the forces and moments at end A of the elbow in terms of the reactions at end B are given below

$$[U(\alpha)] = \begin{bmatrix} -\cos\alpha & \sin\alpha & 0 \\ -\sin\alpha & -\cos\alpha & 0 \\ -R\sin\alpha & R(1-\cos\alpha) & -1 \end{bmatrix} \quad (4.60)$$

$$[V(\alpha)] = \begin{bmatrix} -1 & 0 & 0 \\ R\sin\alpha & -\cos\alpha & \sin\alpha \\ R(1-\cos\alpha) & \sin\alpha & -\cos\alpha \end{bmatrix} \quad (4.61)$$

3. **The final 12 × 12 Elbow Stiffness Matrix *K* is given below**

$$K = \begin{bmatrix} [G1] & [0] & [G2] & [0] \\ [0] & [H1] & [0] & [H2] \\ [G2]^T & [0] & [G3] & [0] \\ [0] & [H2]^T & [0] & [H3] \end{bmatrix} \quad (4.62)$$

Where

$$[G1] = [B] \quad (4.63)$$

$$[G2] = [B][U]^T \quad (4.64)$$

$$[G3] = [U][B][U]^T \quad (4.65)$$

$$[H1] = [C] \quad (4.66)$$

$$[H2] = [B][V]^T \quad (4.67)$$

$$[H3] = [V][B][V]^T \quad (4.68)$$

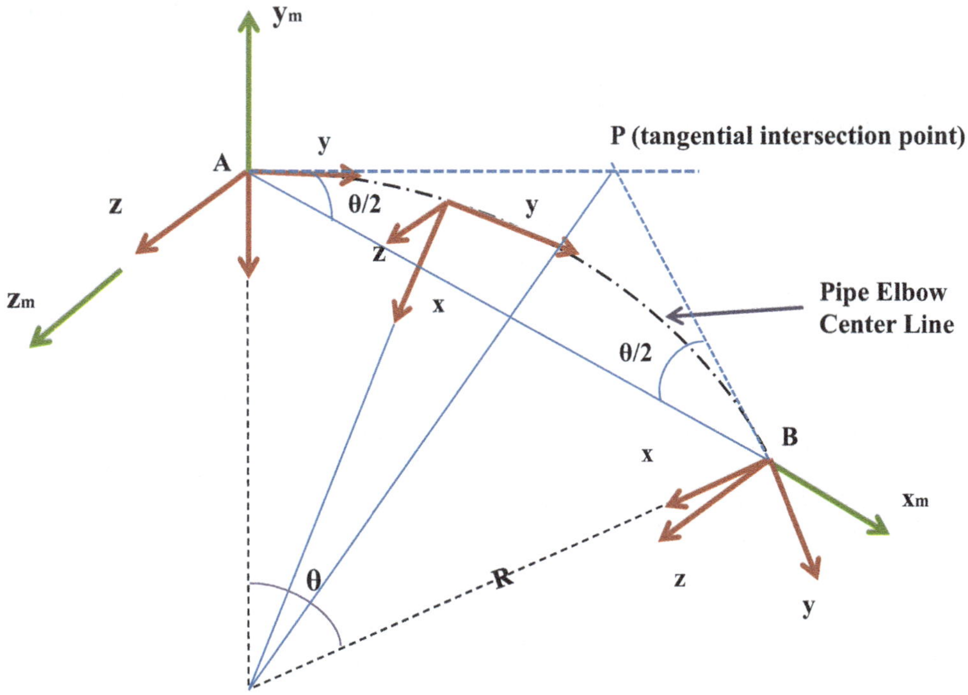

Fig. 4.49 Relationship between global and local axes in pipe elbow element

4. **Transformation of the 12 × 12 Elbow Stiffness Matrix**

Let the 12 × 12 stiffness matrix computed in Eq. (4.62) be denoted as [SM (x, y, z)] to call attention to the fact that it is written with respect to the x, y, and z axes at ends A and B as shown in Fig. 4.49. (Note that the directions x, y, and z at end A are not coincident with the directions x, y, and z at end B.) It is desired to formulate the elbow stiffness matrix in terms of the x_m, y_m, and z_m member axes, shown in Fig. 4.49, that are commonly used to describe the stiffness matrix for straight members directed from A to B.

If [SM (x_m, y_m, z_m)] denotes the 12 × 12 elbow stiffness matrix referred to the x_m, y_m, and z_m axes, then it can be shown that

$$[\mathrm{SM}(x_m, y_m, z_m)] = [R]^T [\mathrm{SM}(x, y, z)]\,[R] \qquad (4.69)$$

where

$$[R] = \begin{bmatrix} R(\theta) & 0 & 0 & 0 \\ 0 & R(\theta) & 0 & 0 \\ 0 & 0 & R(-\theta) & 0 \\ 0 & 0 & 0 & R(-\theta) \end{bmatrix} \qquad (4.70)$$

and

$$[R(\theta)] = \begin{bmatrix} \sin\frac{\theta}{2} & -\cos\frac{\theta}{2} & 0 \\ \cos\frac{\theta}{2} & \sin\frac{\theta}{2} & 0 \\ 0 & 0 & 1 \end{bmatrix} \qquad (4.71)$$

where θ being the elbow bend angle and 0 in Eq. (4.70) standing for the 3 × 3 null matrix.

The calculated stiffness matrix can be used to obtain eigenvalues, eigenvectors, and mass participation factors using general commercial packages.

A.3.2 3D Beam Element Stiffness Matrices

Figure 4.50 shows a beam segment with its 12 nodal coordinates numbered consecutively. The convention adopted is to label first the three translational displacements of the first joint followed by the three rotational displacements of the

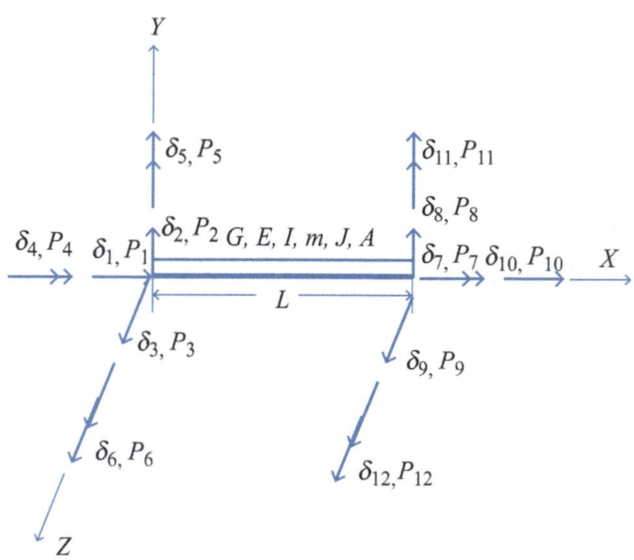

Fig. 4.50 Beam/Pipe element

same joint, then to continue with the three translational displacements of the second joint and finally, the three rotational displacements of this second joint. The double arrows used in Fig. 4.50 serve to indicate rotational nodal coordinates; thus, these are distinguished from translational nodal coordinates for which single arrows are used. The stiffness matrix for a three-dimensional uniform beam segment is readily written by the superposition of the axial stiffness matrix, the torsional stiffness matrix, and the flexural stiffness matrix. The flexural stiffness matrix is used twice in forming the stiffness matrix of a three-dimensional beam segment to account for the flexural effects in the two principal planes of the cross section. The final 12 × 12 stiffness matrix is given below, in which I_y and I_z are, respectively, the cross-sectional moments of inertia with respect to the principal axes labeled as y and z in Fig. 4.50, and L, A, and J are, respectively, the length, cross-sectional area, and torsional constant of the beam element. Based on the similar procedure explained in Annexure I, the local stiffness of beam or pipe element can be transferred to global axis or any desired axis using direction cosine matrix.

$$K = \begin{bmatrix}
\frac{EA}{L} & 0 & 0 & 0 & 0 & 0 & -\frac{EA}{L} & 0 & 0 & 0 & 0 & 0 \\
0 & \frac{12EI_Z}{L^3} & 0 & 0 & 0 & \frac{6EI_Z}{L^2} & 0 & \frac{-12EI_Z}{L^3} & 0 & 0 & 0 & \frac{6EI_Z}{L^2} \\
0 & 0 & \frac{12EI_Y}{L^3} & 0 & \frac{-6EI_Y}{L^2} & 0 & 0 & 0 & \frac{-12EI_Y}{L^3} & 0 & \frac{-6EI_Y}{L^2} & 0 \\
0 & 0 & 0 & \frac{GJ}{L} & 0 & 0 & 0 & 0 & 0 & \frac{GJ}{L} & 0 & 0 \\
0 & 0 & \frac{-6EI_Y}{L^2} & 0 & \frac{4EI_Y}{L} & 0 & 0 & 0 & \frac{6EI_Y}{L^2} & 0 & \frac{2EI_Y}{L} & 0 \\
0 & \frac{6EI_Z}{L^2} & 0 & 0 & 0 & \frac{4EI_Z}{L} & 0 & \frac{-6EI_Z}{L^2} & 0 & 0 & 0 & \frac{2EI_Z}{L} \\
\frac{-EA}{L} & 0 & 0 & 0 & 0 & 0 & \frac{EA}{L} & 0 & 0 & 0 & 0 & 0 \\
0 & \frac{12EI_Z}{L^3} & 0 & 0 & 0 & \frac{-6EI_Z}{L^2} & 0 & \frac{12EI_Z}{L^3} & 0 & 0 & 0 & \frac{-6EI_Z}{L^2} \\
0 & 0 & \frac{-12EI_Y}{L^3} & 0 & \frac{6EI_Y}{L^2} & 0 & 0 & 0 & \frac{12EI_Y}{L^3} & 0 & \frac{6EI_Y}{L^2} & 0 \\
0 & 0 & 0 & \frac{GJ}{L} & 0 & 0 & 0 & 0 & 0 & \frac{GJ}{L} & 0 & 0 \\
0 & 0 & \frac{-6EI_Y}{L^2} & 0 & \frac{2EI_Y}{L} & 0 & 0 & 0 & \frac{6EI_Y}{L^2} & 0 & \frac{4EI_Y}{L} & 0 \\
0 & \frac{6EI_Z}{L^2} & 0 & 0 & 0 & \frac{2EI_Z}{L} & 0 & \frac{-6EI_Z}{L^2} & 0 & 0 & 0 & \frac{4EI_Z}{L}
\end{bmatrix} \quad (4.72)$$

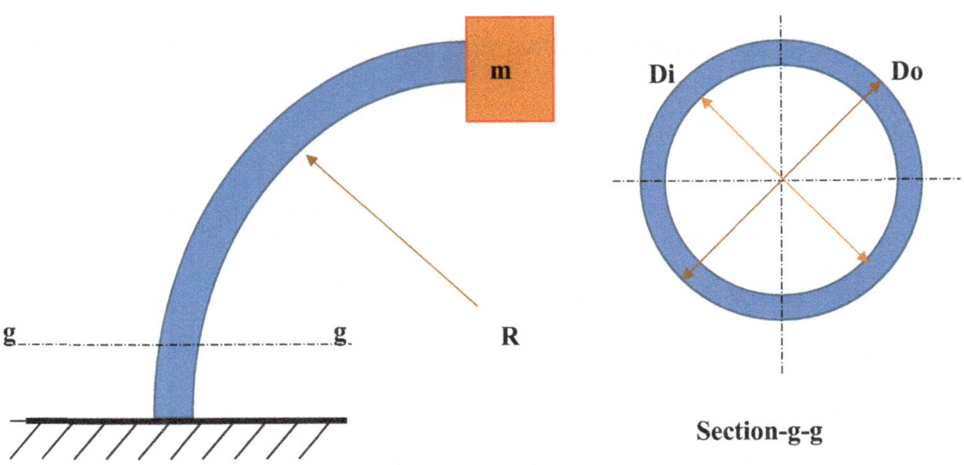

Fig. 4.51 90° massless elbow element with lumped mass at the free end with its cross section

Example 4.10 A 90° massless elbow element is fixed at one end with lumped mass at the free end. The elbow has a radius of 250 mm with internal diameter 100 mm and thickness 4 mm. A lumped mass of 12.014 kg is attached at the free end (Fig. 4.51). Determine the natural frequency and mode shapes of the system by formulating the stiffness matrix using Eq. (4.56) and the mass matrix of a two-noded elbow element discussed in the previous section.

MATLAB code:

```
% Natural frequency of the Pipe Elbow Element

clear all;
close all;
clc;

Di= 0.1;
Do= 0.108;
t= (Do-Di)/2;
R=0.25;
theta= pi/2;
E= 2.1e11;
G=7.7e10;
FF = 1.65*(((Do+Di)/2)^2)*(1/(4*t*R))          % Flexibility Factor
I= (pi/64)*(Do.^4 - Di.^4)*(1/FF);
I2= (pi/64)*(Do.^4 - Di.^4);
A= (pi/4)*(Do.^2 - Di.^2);                     % Area of Cross-Section
As = 0.504*A;                                  % Area with Shear Factor taken into
account
J=2*I2;
Rho=7800;
me= (A*2*pi*R*7850)/4

mt= me/2;

% For Creation of the Matrix

A1= ((((R.^3)/(E*I)) + (R/(A*E)))*((theta/2)-((sin(2*theta))/4))) + (R/(G*As))*((theta/2) +
((sin(2*theta))/4))

A2 = (((R.^3)/(E*I))+(R/(A*E))-(R/(G*As)))*((sin(theta)).^2/2) + ((R.^3)/(E*I))*(cos(theta)-
1)

A3 = ((R.^2)/(E*I))*(1-cos(theta))


A4=((R.^3)/(E*I))*((theta) - (2*sin(theta))) + ((((R.^3)/(E*I)) + (R/(A*E)))*((theta/2)
+((sin(2*theta))/4))) + (R/(G*As))*((theta/2)-((sin(2*theta))/4))

A5 = ((R.^2)/(E*I))*(sin(theta)-(theta))

A6 = ((R)/(E*I))*(theta)

% For creation of the AN Matrix

AN1=    (((R.^3)/(E*I))*((theta/2)-((sin(2*theta))/4)))    +    (R.^3/(G*J))*(((3*theta)/2)-
(2*sin(theta)) + ((sin(2*theta))/4))+((R*theta)/(G*As))

AN2 = ((((R.^2)/(G*J))-(R.^2/(I*E)))*(((sin(theta)).^2)/2) + (R.^2/(G*J))*(cos(theta)-1))
```

```
AN3= (((R.^2)/(E*I))*((theta/2) - ((sin(2*theta))/4))) + (R.^2/(G*J))*(((theta)/2) +
((sin(2*theta))/4) - ((sin(theta))))

AN4 = (((R)/(E*I))*((theta/2) + ((sin(2*theta))/4))) + (R/(G*J))*(((theta)/2) -
((sin(2*theta))/4))

AN5 = (((R)/(G*J))-(R/(E*I)))*((sin(theta)).^2/2)

AN6 = (((R)/(E*I))*((theta/2)-((sin(2*theta))/4))) + (R/(G*J))*(((theta)/2)+((sin(2*theta))/4))

A = [A1 A2 A3; A2 A4 A5; A3 A5 A6]

AN = [AN1 AN2 AN3; AN2 AN4 AN5; AN3 AN5 AN6]

B= inv(A)

C= inv(AN)

U= [-cos(theta) sin(theta) 0; -sin(theta) -cos(theta) 0; -(R*sin(theta)) (R*(1-cos(theta))) -1]

V= [-1 0 0; (R*sin(theta)) -cos(theta) sin(theta); (R*(1-cos(theta))) -sin(theta) -cos(theta)]

u= U'

v= V'

G1= B

G2= B*u

G3= U*B*u

H1= C

H2= C*v

H3= V*C*v

h1=H1'

h2=H2'

h3=H3'

H9= zeros(3,3)

% 12 x 12 Stiffness Matrix

K= [G1 H9 G2 H9; H9 H1 H9 H2; G2' H9 G3 H9; H9 H2' H9 H3]
```

```
% 12 x 12 Mass Matrix

M= zeros(12,12)

M(1,1)= 0

M(2,2)= 0

M(3,3)= 0

M(4,4)= 0

M(5,5)=0

M(6,6)= 0

M(7,7)= 12.014

M(8,8)= 12.014

M(9,9)= 0

M(10,10)= 12.014

M(11,11)= 0

M(12,12)= 0

% Since, the Elbow is fixed the 12 x 12 Stiffness and Mass
Matrices can be
% reduced effectively to 6 x 6 Matrices

% SM is the 6 x 6 Stiffness Matrix

SM= zeros(6,6);

SM(1,1)= K(7,7);

SM(1,2)= K(7,8);

SM(1,3)= K(7,9);

Sm(1,4)= K(7,10);

SM(1,5)= K(7,11);

SM(1,6)= K(7,12);

SM(2,1)= K(8,7);

SM(2,2)= K(8,8);

SM(2,3)= K(8,9);

SM(2,4)= K(8,10);

SM(2,5)= K(8,11);

SM(2,6)= K(8,12);

SM(3,1)= K(9,7);

SM(3,2)= K(9,8);

SM(3,3)= K(9,9);

SM(3,4)= K(9,10);

SM(3,5)= K(9,11);

SM(3,6)= K(9,12);

SM(4,1)= K(10,7);

SM(4,2)= K(10,8);

SM(4,3)= K(10,9);

SM(4,4)= K(10,10);

SM(4,5)= K(10,11);

SM(4,6)= K(10,12);

SM(5,1)= K(11,7);

SM(5,2)= K(11,8);

SM(5,3)= K(11,9);

SM(5,4)= K(11,10);

SM(5,5)= K(11,11);

SM(5,6)= K(11,12);

SM(6,1)= K(12,7);

SM(6,2)= K(12,8);

SM(6,3)= K(12,9);
```

```
SM(6,4)= K(12,10);

SM(6,5)= K(12,11);

SM(6,6)= K(12,12)

MA= zeros(6,6)                  % MA is the 6 x 6 Mass Matrix

MA(1,1)= 12.014;

MA(2,2)= 12.014;

MA(4,4)= 12.014

% To calculate Natural Frequency

[EV, E1]= eig (SM,MA)           % EV is the Eigenvector & E1 is the Eigenvalue

W=sqrt(E1)                      % W is the Angular Natural Frequency

freq= W/(2*pi)
```

Results

The eigenvalues (natural frequencies) and eigenvectors (mode shapes) are provided in the table below

	1st frequency	2nd frequency	3rd frequency
Eigenvalues (Hz)	98.89	109.6	521.8
Eigenvectors	−0.2083	0	−0.0732
	−0.1384	0	0.1101
	1.0000	0	1.0000
	0	0.3249	0
	0	0.9069	0
	0	1.0000	0

It is important to note that considering shear factor as 1 instead of 0.504 does not cause significant change in first and second eigen-values however it causes significant change in third eigen-value. The values then observed are as 99.59 , 110.83 , and 550.8 Hz, respectively.

References

1. Chopra AK (2007) Dynamics of structures, theory and applications to earthquake engineering, 2nd edn
2. Humar JL (1990) Dynamics of structures. Prentice Hall, Upper Saddle River
3. Bathe KJ (1996) Finite element procedures, 3rd Indian reprint
4. Krishnamoorthy CS (2007) Finite element analysis—theory and programming, 2nd edn
5. Chopra A (1995) Dynamic of structures. Prentice Hall, Upper Saddle River
6. MATLAB (2009) The MathWorks Inc.
7. Perry RF (1977) Pipe elbow stiffness coefficients including shear and bend flexibility factors for use in direct stiffness codes. In: IASMiRT-4, San Francisco, USA
8. Gan BS (2018) An iso-geometric approach to beam structures

Geotechnical Investigation and Its Applications in Seismic Design of Structures

5

Srijit Bandyopadhyay, M. K. Pradhan, Raj Banerjee, V. S. Phanikanth, and S. J. Patil

Geotechnical features of the site soil strata is an important element/interface between structures and earthquake waves

Symbols

$\Delta\sigma'$	Change in effective overburden pressure
σ'_0	Effective overburden pressure
R_r	Recovery Ratio
L_a	Actual length of sample in tube
L_t	Total length of sampling tube driven below the bottom of borehole
D_S	Inside diameter of sampling tube
D_C	Inside diameter of cutting edge
N	Observed at field SPT value
N_{60}	SPT (N) value corrected for 60% energy efficiency
η_H	Hammer energy transfer efficiency (%)
η_B	Bore hole diameter correction
η_S	Sampler correction
η_R	Rod length correction
A_c	Cross-sectional area of cone
Q_c	Tip load
A_f	Surface area of friction sleeve
c_u	Undrained shear strength
q_c	Cone penetration resistance
N_k	Cone factor
D_p	Diameter of probe at deflate condition
D_h	Diameter of the hole
R_c	Radius of the measuring cell at the deflated condition which is at starting of the test
r_0	Radius corresponding to point A
r_f	Radius corresponding to point B
B_z	Mass ratio
m	Mass of concrete block
ρ	Density of concrete block
ξ	Damping coefficient of soil
f_2, f_1	Two frequencies at which the amplitude is equal to $\mathbf{X_m}/\sqrt{2}$
X_m	Maximum amplitude
$\emptyset$	Angle of internal friction
γ	Unit weight of soil above water table
γ_{sub}	Submerged unit weight of soil
G	Specific gravity of soil particles
e	Void ratio
i_c	Critical hydraulic gradient

S. Bandyopadhyay (✉) · M. K. Pradhan · R. Banerjee · V. S. Phanikanth
Bhabha Atomic Research Centre, Mumbai, India
e-mail: srijit@barc.gov.in

M. K. Pradhan
e-mail: mpradhan@barc.gov.in

R. Banerjee
e-mail: rbanerjee@barc.gov.in

V. S. Phanikanth
e-mail: phani@barc.gov.in

S. J. Patil
Heavy Water Board, Mumbai, India
e-mail: sjp_patil@yahoo.co.in

G. R. Reddy et al., *Textbook of Seismic Design*,
https://doi.org/10.1007/978-981-13-3176-3_5

5.1 Introduction

Soil is a natural aggregate of mineral grains, with or without organic constituents which is formed by disintegrating of rock through gentle mechanical means such as wind, erosion due to precipitation. The process of weathering of the rock in

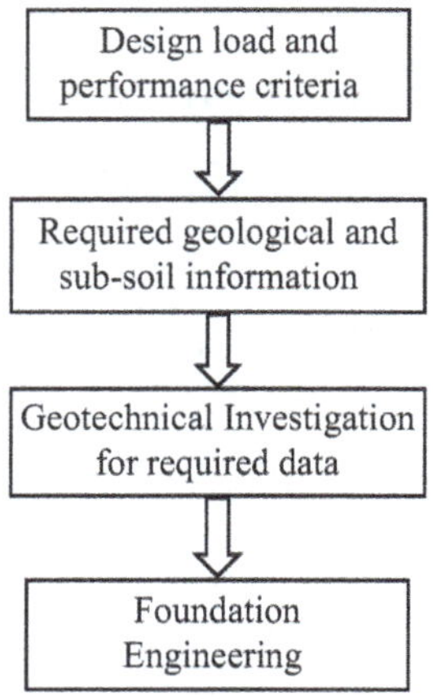

Fig. 5.1 Importance of geotechnical engineering

the presence of various external agencies as mentioned above decreases the cohesive forces binding the mineral grains and leads to the disintegration of bigger masses to smaller and smaller soil-like particle. Load-carrying capacity at different strain levels of soil depends upon type of soil such as cohesionless soil and cohesive soil. Hence, it is very important to identify the types of soil properties before starting the design of structures. In the design of the structure, apart from determination of grain size distribution and index properties of soil such as liquid limit, plastic limit soil, various chemical properties and engineering properties as explained below are also required. Determination of the location of groundwater table, chemical concentration of sulfate in soil, chloride content in water/soil, and various engineering properties of soil such as Young's modulus, shear wave velocity, shear modulus, strain of the soil are important inputs. Properties are evaluated through detailed geotechnical investigation which plays an important role in the design of structure as shown in the flow chart (Fig. 5.1).

5.2 Geotechnical Investigations

Geotechnical investigations are generally carried out before constructions of any structure with the following objectives:

i. To evaluate the general suitability of a given/considered site for construction of proposed structure such as buildings, chimney/stack, dam, road, bridge, etc.

ii. Acquiring required data of geotechnical parameters as explained above for safe and economical design structure,
iii. Evaluation of the ultimate and safe load-bearing capacity of the proposed foundation,
iv. Estimation of the probable settlement of a foundation considering the existing surface and subsurface soil of proposed site,
v. Assessment of liquefaction potential of the soil at a specific site,
vi. Establishment of the level of groundwater table at the site and its variations.

5.3 Phases/Stages of Geotechnical Investigation

The total process of geotechnical investigation from planning to the execution is mainly achieved in four phases/stages.

Stage 1: Reconnaissance of the area: Geological data collection and geophysical investigation are carried out by:

i. Visual inspection of the site,
ii. Study the available maps and various records,
iii. Watermark at nearby structure such as bridge, culvers, natural drains, etc.
iv. Groundwater levels from nearby water sources like well.

Stage 2: Collection of Preliminary Information

i. Obtaining information regarding the type of structure to be constructed and its intend use of the structure,
ii. General idea of topography of the area and type of soil,
iii. Few boring or test pits,
iv. Determine the depth, thickness, and composition of each stratum.

Stage 3: Detailed Investigations

i. Planning: number, depth, and spacing of boreholes,
ii. Sampling,
iii. Field tests,
iv. Laboratory tests,
v. Geophysical methods,
vi. Tests to determine dynamic properties of soils.

Stage 4: Preparation of report

i. A clear description about the soil exists at the proposed site,
ii. Method of exploration adopted during the soil investigation,
iii. Soil profile exists in site,
iv. Test methods and test results,
v. Location of the ground water.

5.4 Reconnaissance Survey

Site reconnaissance survey is the first step in a geotechnical investigation program which includes a visit to the proposed site and to study the maps and other relevant available records.

Reconnaissance survey helps in deciding future program of soil investigations, scope of work, method to be followed for further and detailed geotechnical investigation. It includes study of local topography, nearby excavations, quarries of Murom or minerals, evidence of erosion or landslides in rivers or water bodies, behaviors of existing structures at or near site, water level in streams and wells, flood marks, location of springs, nature of vegetation. Various relevant information are also gathered by inquiring the local people.

5.5 Detailed Soil Exploration Methods

The common methods adopted in detailed soil exploration process include:

i. Making open test pits
ii. Boring or drilling holes.

5.5.1 Open Test Pits

Excavated test pit is an open pit, which facilitates a detailed visual examination of the subsurface in situ condition. This test in general practice is conducted up to the depth of 4.00 to 5.00 ft from ground level.

5.5.2 Boreholes

Following methods are used for bore logs at a given site.

1. Auger boring,
   (a) Hand auger,
   (b) Power-driven auger.
2. Wash boring,
3. Percussion drilling,
4. Rotary boring.

The first three methods of the above are generally used in the soft type of soil, and last two methods are used in hard strata such as rock.

5.5.2.1 Auger Boring

In this method, auger is pushed and rotated until annular space of auger fills up then withdraws the auger and cleans it. There are two types of auger

1. Hand auger—for shallow depth (3.0–6.0 m): The hand augers are used for making 15–20 cm diameter borehole up to depth of 3–6 m. It is mostly used in soft soil.
2. Power-driven auger—for larger depth: Power-driven augers are used for making boreholes in hard strata.

5.5.2.2 Wash Boring

Wash boring is performed in soft to medium soil. Wash boring is done by two steps; initially, the test hole is bored up to a short depth by auger, and then, a casing pipe is inserted to prevent the side soil from collapse and caving in. The hole is then further advanced by the use of chopping bit fixed at the end of a hollow drill rod. The soil is loosened by the stream of water under pressure forced through the rod and the bit and the loosened soil with water flow up around the pipe. The water containing the loosened soil in suspension discharged into a tub. The change in stratifications of soil and tentative classification could be guessed from the rate of progress and color of wash water coming out from drilled hole.

5.5.2.3 Percussion Drilling

In this process, a heavy drilling bit is raised and dropped alternately in such a manner that underlying materials is powered by chipping and forms into slurry when mixed with water. The slurry is removed out of the hole periodically by means of bailer. Bentonite slurry or casing pipe is used to prevent the side soil from collapse and caving in. The change in stratification of soil and classification can be identified from the removed slurry and also from rate of chipping operation.

5.5.2.4 Rotary Drilling

Rotary drilling is another method of wash boring normally used in case of rock. In this method, a cutter bit which is attached to the end of an interconnected line of drill rods is rotated by a power rig. Power rotation of the coring bit varies as well the cutter bit also varies from metal bits to tungsten carbide or diamond bits depending upon the hardness of formation of soil. Diamond coring bits are the most versatile cutting bit among all above coring bits.

5.5.3 Number, Spacing, and Disposition of Boreholes/Open Test Pits

The lateral extent of exploration and the spacing of boreholes in geotechnical investigation depend mainly on the variation of the soil strata in the horizontal direction and the nature of structures to be built. The various recommendations of rule of thumb are:

1. As per IS: 1892-1979 (Reaffirmed 2002) recommendation, the number and spacing of boreholes depend on the extent of site and nature of structure coming on it.
2. Minimum one borehole is required for 0.4-hectare site area, and preferably one borehole at center and one at each corner of the building are adequate.

Thumb rules for approximate spacing of boreholes can be referred to given in Table 5.1.

Table 5.1 Spacing between boreholes for different types of structure

Types of project	Spacing (m)
Industrial building	25–30
Multi-story building	15–25
Highways	250–300
Earth dam	25–50

5.5.4 Depth of Boreholes

In geotechnical investigations, the depth of boreholes is governed by the depth of soil affected by the proposed structure/influence zone. The depth of influence zone depends upon the type of structure to be built and intensity of loading. The various codes such as IS 1892-1972 (Reaffirmed-2002), ASCE 1972, which provide the recommendations and the guidelines regarding fixing of the borehole depths are described below.

As per IS: 1892-1979, Reaffirmed 2002, depth of borehole is up to 1.5 times the least width of foundation from its base and in certain cases up to 2 times the depth of foundation.

As per ASCE 2000, variation of the net increase in effective stress ($\Delta\sigma'$) of soil under the foundation along the depth of soil is shown in Fig. 5.2. The depth D_1 is determined as the depth at which the value of net increase in effective vertical stress ($\Delta\sigma'$) is equal to 10% of the average load per unit area of the foundation (q).

$$\Delta\sigma' = \frac{10}{100} \times q \tag{5.1}$$

The depth D_2 is determined as the depth at which

$$\text{Net effective stress increase}(\Delta\sigma') = 0.05 \times (\sigma'_0) \tag{5.2}$$

where

σ'_0 effective vertical stress

The depth of boring is at least the minimum of D_1 or D_2 evaluated as above. Sowers and Sowers (1970) have provided the following empirical relationships for determination of the minimum depth of the boreholes for multi-storied buildings.

Approximate depth of boring(m) for narrow steel concrete building is

$$D = 3 \times (\text{Number of stories})^{0.7} \tag{5.3}$$

Approximate depth of boring(m) for heavy steel and narrow concrete building is

$$D = 6 \times (\text{Number of stories})^{0.7} \tag{5.4}$$

In case of end bearing pile foundation, the depth of soil exploration is 1.5 times the width of pile group (B) below the tip of the pile. But in case of frictional pile foundation, the soil exploration is 1.5 times the width of the pile group (B) measured from one-third length ($D/3$) of pile bottom as shown in Fig. 5.3.

Depth of boreholes during geotechnical investigation as recommended by IS 1892 [2], (Reaffirmed 2002), for different types of foundations is enumerated in Table 5.2.

5.5.5 Sampling Procedure

During the geotechnical investigation, two types of samples, i.e., disturbed and undisturbed, are collected from boreholes for laboratory testing.

5.5.5.1 Disturbed Samples

The natural structure of soil is disturbed at the time of drilling and exploration. These types of samples can be used

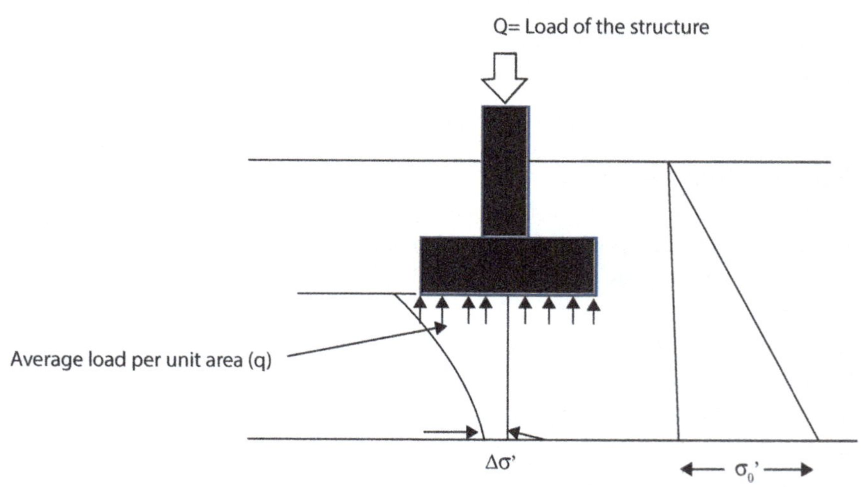

Fig. 5.2 Net increases in effective stress and variation of effective vertical stress with depth

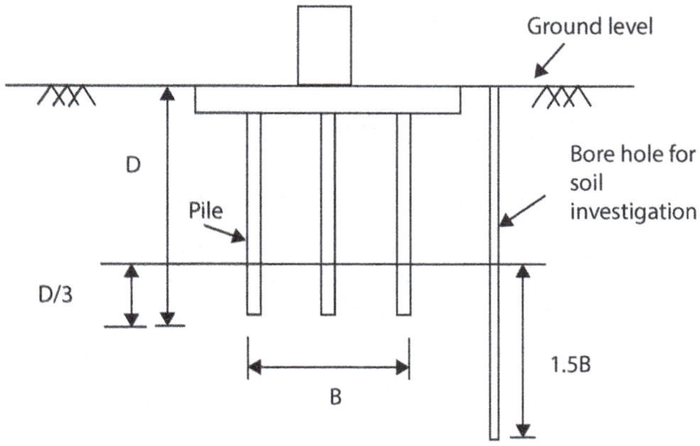

Fig. 5.3 Depth of soil investigation in case of pile foundation for friction pile

for grain size analysis, determination of liquid and plastic limit, determination of specific gravity, plasticity characteristics of soil, etc.

The degree of disturbance of a soil sample during drilling and extraction is expressed as:

$$R_r = \frac{L_a}{L_t} \tag{5.5}$$

where

R_r	recovery ratio
L_a	actual length of sample in tube
L_t	total length of sampling tube driven below the bottom of borehole
R_r	1 indicates no disturbance of the sample
R_r <1	indicates sample is compressed
R_r >1	indicates there is expansion of sample inside tube.

5.5.5.2 Undisturbed Samples

The soil samples retain in situ moisture, structure, and properties during drilling and explorations is called undisturbed sample. These types of samples can be used for compressibility, permeability, or shear strength test of the soil. The guidelines suggested by IS: 8763-1978 can be helpful in obtaining the undisturbed samples.

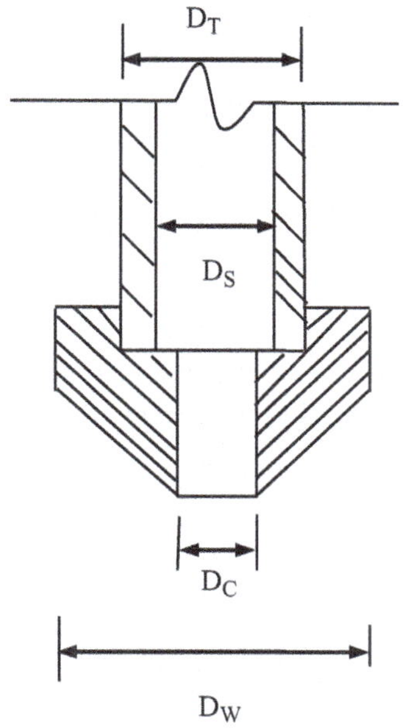

Fig. 5.4 Typical soil samplers

5.5.5.3 Features of Cutting Tools to Control Disturbance of Samples

The disturbance of the soil depends mainly upon following design criteria of cutting tools or soil sampler as shown in Fig. 5.4.

a. Inside clearance,
b. Outside clearance,
c. Area ratio.

These are briefly explained as below.

Inside clearance (C_I) is calculated as

$$C_I = \frac{D_S - D_C}{D_C} \tag{5.6}$$

where

D_S	inside diameter of sampling tube
D_C	inside diameter of cutting edge

Table 5.2 Depths of explorations [IS 1892 [2], Reaffirmed 2002]

Serial no	Type of foundation	Depth of explorations (*D*)
1.	Isolated spread footing or raft	1.5 times width of spread footing (B) or width of raft
2.	Adjacent footings with clear spacing less than twice the width	1.5 times length (L) of footing

C_I should be between 1 and 3% of internal diameter of sampler tube. C_I allows elastic expansion, reduces frictional drag, and helps retain core.

Outside clearance (C_O) is calculated as

$$C_O = \frac{D_W - D_T}{D_T} \tag{5.7}$$

where

D_W outside diameter of cutting edge
D_T outside diameter of sampling tube

Outside clearance (C_O) should not be much greater than inside clearance (C_I). Outside clearance (C_O) facilitates the withdrawals of sampler from ground. For reducing the driving force, outside clearance (C_O) should be as small as possible (0–3.00%).

Area ratio (A_r) is calculated as

$$A_r = \frac{D_W^2 - D_C^2}{D_C^2} \tag{5.8}$$

Where

D_W outside diameter of cutting shoe
D_C inside diameter of cutting shoe

For obtaining good quality undisturbed sample, A_r should be kept as low as possible.

i. For stiff formation, $A_r < 20\%$,
ii. For soft sensitive clay, $A_r < 10\%$.

5.5.5.4 Rock Core Drilling/Coring

In geotechnical investigation, rock coring is the process in which a sampler consisting of a tube (core barrel) with a cutting bit at its lower end cuts an annular hole in a rock mass, thereby creating a cylinder or core of rock which is recovered in the core barrel. Rock core is normally obtained by rotary drilling. Rock coring may be carried out by using diamond bits or tungsten carbide (T.C.) bits. The sizes of bit used in coring vary generally from 'AX' to 'NX' size though currently bigger size also used. The details of standard drill tools are given in Table 5.3.

Rock Core Recovery (RCR): Rock core recovery is defined as the ratio of the rock recovered in length (considering rock recovered of all lengths) to the length of runoff drilling or sampler.

Rock Quality Designation (RQD): Rock quality designation is defined as the percentage of rock cores that have length equal or greater than 100 mm over the total drill length as shown in Fig. 5.5. It is basically an assessment of degree of fracture of rock.

Example 5.1 During drilling the hole for geotechnical investigation, the cores recovered by the sampler of 1000 mm are marked as in Fig. 5.5. Details of length of rock core samples are given in Table 5.4. Calculate the RCR and RQD of the rock.

Table 5.3 Typical standard sizes of core barrels, drill rods, and compatible casing

Drill rod		Casing pipe		Core barrel		Core diameter	Hole diameter	
Designation	OD (mm)	Designation	OD (mm)	Designation	OD (mm)	mm	Designation	OD (mm)
A	41.27	AX	31.75	AWX	47.625	28.575	AX	47.625
				AWM	47.625	28.575	AX	47.625
B	47.625	BX	73.025	BWX	59.563	41.275	BX	60.325
				BWM	59.563	41.275	BX	60.325
N	60.325	NX	88.900	NWX	75.311	53.975	NX	76.200
				NWM	75.311	53.975	NX	76.200

Note symbol *X* indicates single barrel; *M* indicates double barrel

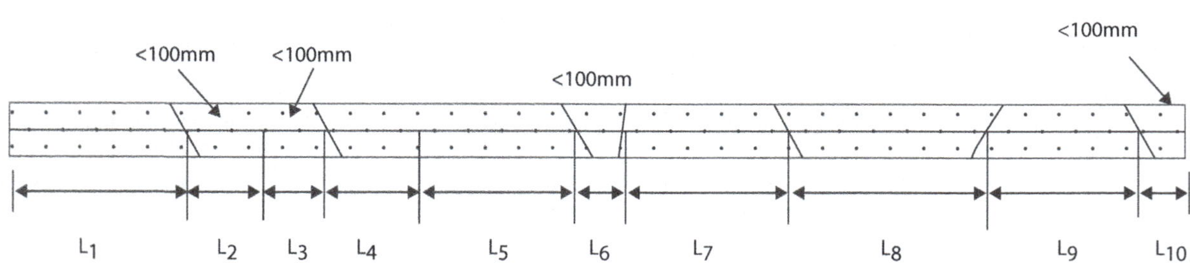

Fig. 5.5 Consideration of rock core in calculation of RQD and RCR

Table 5.4 Details of length of rock core samples

Piece number	Length (mm)	Piece considered for RCR	Piece considered for RQD	Remark
L_1	150	Yes	Yes	Length more than 100 mm
L_2	80	Yes	No	Length less than 100 mm
L_3	72	Yes	No	Length less than 100 mm
L_4	100	Yes	Yes	Length equal to 100 mm
L_5	120	Yes	Yes	Length more than 100 mm
L_6	55	Yes	No	Length less than 100 mm
L_7	110	Yes	Yes	Length more than 100 mm
L_8	153	Yes	Yes	Length more than 100 mm
L_9	105	Yes	Yes	Length more than 100 mm
L_{10}	45	Yes	No	Length less than 100 mm
Total	989			

Solution

$$\begin{aligned}\text{RQD} &= \frac{\sum L_i}{L} = \frac{L_1 + L_4 + L_5 + L_7 + L_8 + L_9}{L}\\ &= \frac{150 + 100 + 120 + 110 + 153 + 105}{1000} = \frac{738}{1000}\\ &= 73.80\%\end{aligned}$$

ΣL_i rock pieces of length >100 mm

L_2, L_3, L_6, and L_{10} are not considered as these are <100 mm

$$\begin{aligned}\text{RCR} &= \frac{\sum L_J}{L} = \frac{L_1 + L_2 + L_3 + L_4 + L_5 + L_6 + L_7 + L_8 + L_9 + L_{10}}{L}\\ &= \frac{150 + 80 + 72 + 100 + 120 + 55 + 110 + 153 + 105 + 45}{989}\\ &= \frac{989}{1000} = 98.9\%\end{aligned}$$

ΣL_J rock pieces of all lengths

The correlation between rock quality designations (RQD) and allowable contact pressure 'q_a' (Peck et al. 1974) is given in Table 5.5.

Preservation and retention of soil samples, cores, and records

During the geotechnical investigation, at the site of drilling, immediately on their withdrawals from the core barrels, the cores shall be placed in core tray and then transferred to core boxes specially made for this purposes. The boxes are generally made of compartments with standard width as shown in Table 5.6 for different size of core. In case of important projects, the core boxes are stored and maintained in the core library.

5.6 Testing of Soil

There are a variety of laboratory and field testing methods adopted in geotechnical investigations to obtain the various geotechnical parameters which are required for foundation

Table 5.5 Relation of RQD and in situ quality and allowable contact pressure q_a (Peck et al. 1974)

RQD (%)	Rock quality	q_a (tons/ft^2)
90–100	Excellent	200–300
75–90	Good	120–200
50–75	Fair	65–120
25–50	Poor	30–65
0–25	Very poor	10–25

Table 5.6 Width of compartments in core boxes for different sizes of cores [IS: 4078-1980 Reaffirmed 2004]

Sl. no	Core designation	Diameter of core (mm)	Width of compartment in core box (mm)
1	NX	54.7	61
2	BX	42.0	48
3	AX	30.0	36

analysis and design of the structure. During the field test of soil, the various properties are measured in their existing natural state and hence do not require sampling. Some field test can be performed from ground surface, i.e., SASW, MASW, ERT, and GPR, while others tests require the drilling or boreholes, i.e., SPT, cross-borehole test, etc. The brief of the various field tests and laboratory tests is briefed below and explained in more detail subsequently.

5.6.1 Field Testing of Soil

It is desired to evaluate various soil parameters based on the strain level the soil is experiencing. The soil in field may be subjected to the loads such as earthquakes, blast, or nuclear explosion which produces high level of strain in the range of 0.01–1.0%. The soil also may be subjected to the loads from the machines which produce low level of strain in the range of 0.0001–0.001%. The testing of soil in geotechnical investigation can be classified as high-strain field test and low-strain field tests based on the strain level of the soil during the test.

High-strain field tests (Stain 0.01–1.00%): The various field tests of soil conducted in high strains are:

1. Standard penetration test (SPT),
2. Cone penetration test (CPT),
3. Dynamic cone penetration test(DCPT),
4. Vane shear test (VST),
5. Pressure meter test(PMT).

Low-strain field tests (Stain 0.001–0.0001%): Low-strain tests are generally conducted at strain level below 0.001% of soil [8] which is not large enough to induce significant nonlinear stress–strain behavior in the soil. The various field tests of soil conducted in low-strain ranges are:

1. Seismic reflection test,
2. Seismic refraction test,
3. Seismic cross-hole test,
4. Seismic down-hole (up-hole) test.

5.6.2 Laboratory Testing

Laboratory tests of soil are generally performed on relatively small specimens of soil/rock that are assumed to be representative of a larger body. The following tests are commonly used for evaluating the various dynamic soil properties.

Low-Strain Laboratory Test

1. Resonant column test,
2. Ultrasonic pulse test,
3. Piezoelectric bender element test.

High-Strain Laboratory Test

1. Cyclic triaxial test,
2. Cyclic direct simple shear test,
3. Cyclic torsional shear test.

Laboratory tests of water sample collected from site during soil investigation are also conducted to determine various parameters such as chloride contents. In this chapter, only field testing methods of soil are discussed.

5.6.3 Standard Penetration Test (SPT)

Standard penetration test (SPT) is one type of in situ dynamic penetration test of soil which is very commonly used at site. Variations and classification in the various strata of soil are correlated with the number of blows required for unit penetration of standard sampler by a drive hammer of standard weight falling from specified height. The number of blows required to drive the sampler an additional 30 cm (after initial 15 cm penetration as seating penetration) is known as "standard penetration resistance" or "*N*-value." Under the following conditions, the penetration is referred to refusal and test is halted.

i. 50 blows are required for any 150 mm penetration.
ii. 100 blows are required for last 3000 mm penetration.
iii. 10 successive blows produce no advancement.

Purpose of SPT

The standard penetration test (SPT) is conducted at site in geotechnical investigation to determine geotechnical engineering properties of the soil such as relative density, angle of shearing resistance of cohesionless soil, unconfined compressive strength of soil, liquefaction potential of soil etc.

Test Components

During SPT testing, casing should be used in case of soft/non-cohesive collapsible nature of soil. Inner diameter of hole is generally kept between the range of 100–150 mm. Split-spoon sampler (IS 9640:1980) is used for sampling purpose. Drive weight assembly deployed for the SPT consists of a falling weight of 63.5 kg, which falls from a height of 75 cm.

Correction to Observed SPT Value

The 'N' value is a function of the soil type, soil density, and confining pressure but is also influenced by the various other factors such as test equipment and procedure IS 2131 [9] [Reaffirmed 2002]. Two types of corrections are normally applied to the observed SPT'N' values such as:

A. Correction due to overburden and correction for dilatancy,
B. Hammer energy, sampler, drill rod length, borehole diameter correction.

The correction for the overburden is applied first, and then, the correction for the dilatancy is applied to obtain the corrected SPT value from observed field SPT value.

Correction for overburden

The effects on penetration resistance to the SPT sampler are observed during the in situ testing due to the overburden pressure. For the same relative density, the soil sample with more overburden pressure will show more resistance to penetration which results in a greater SPT value as compared to less overburden pressure. Following methods and equations are applied for overburden correction.

As per Idriss. (2003 & 2008)

$$N' = \left(\frac{P_a}{\sigma'_v}\right)^m N \, if \left(\frac{P_a}{\sigma'_v}\right)^m < 1.7 \qquad (5.9)$$
$$m = 0.784 - 00768\sqrt{N}$$

where

σ'_V effective overburden pressure in kPa
N observed at field SPT value

Correction for dilatancy on saturated fine sand and silt

When dynamic loads are applied on silty and fine sandy soils in saturated state, the pore pressure in such soil will not be able to get dissipated due to low permeability. Hence, during dynamic loading (i.e., application of blows), the pore water will offer a temporary resistance to dynamic loads. This leads to higher value of N-value which is unsafe. Therefore, when SPT is performed in saturated silts and fine sands and if the observed N-value is more than 15, a correction has to be applied to reduce the observed values. This correction is applied on the N-value corrected for overburden pressure (N'). As per Terzaghi and Peck (1967), the corrected penetration number "N''"

$$N'' = 15 + \left(\frac{N' - 15}{2}\right) \text{if } N' > 15 \qquad (5.10)$$

$$N'' = 15 \quad if\ N' \le 15 \qquad (5.11)$$

As per Bazarra (1967),

$$N'' = 0.6 \times N' \qquad (5.12)$$

Hammer efficiency, drill rod, sampler, and borehole correction

The observed standard penetration value in SPT test is also influenced by the procedure adopted and the test equipment deployed for the testing. The correction to the observed site SPT number can be carried out as follows [14];

$$N_{60} = \frac{N\eta_H\eta_B\eta_S\eta_R}{60} \qquad (5.13)$$

where

N_{60} SPT (N) value corrected field procedure
η_H hammer energy transfer efficiency (%)
η_B borehole diameter correction
η_S sampler correction
η_R rod length correction

Hammer Efficiency Correction

Different types of hammers are used for driving the drill rod in standard penetration test. Types of hammer and corresponding hammer efficiency are [15]:

1. Donut with two turns of manila rope on the cathead $\eta_H = 45\%$,
2. Safety with two turns of manila rope on the cathead with hammer efficiency $\eta_H = 60\%$,
3. Rope pulley with cathead $\eta_H = 70–80\%$.

The correction factor applied in observed field standard penetration for borehole, sampler, and rod to obtain corrected SPT values proposed by Bowel 1996, is presented in Table 5.7.

Correlation between engineering properties of soil with SPT value

The value of Standard penetration test depends upon the unconfined compressive strength of cohesive soil and relative density (D_r) of cohesionless soil. The SPT value also

Table 5.7 Bore hole diameter, sampler, and rod length correction factor Idriss. (2003 and 2008)

Factor	Equipment variables	Value
Bore hole diameter factor (η_B)	65–115 mm	1.00
	150 mm	1.05
	200 mm	1.15
Sampling method factor (η_S)	Standard sampler	1.00
	Sampler without liner	1.20
Rod length factor (η_R)	3–4 m	0.75
	4–6 m	0.85
	6–10 m	0.95
	>10 m	1.00

Table 5.8 Correlation between SPT value, relative density, and angle of friction of granular soil [Ref: Terzaghi and Peck 1967]

N'' (corrected SPT)	Compactness	Relative density (D_r) %	Angle of friction ($\varnothing$)
0–4	Very loose	0–15	<28
4–10	Loose	15–35	28–30
10–30	Medium	35–65	30–36
30–50	Dense	65–85	36–41
>50	Very dense	>85	>41

Table 5.9 Corelations between SPT values and unconfined compressive strengths of clay soil [As per Peck et al. 1974]

Consistency	N (SPT value)	Unconfined compressive strength (q_u) kPa
Very soft	0–2	<25
Soft	2–4	25–50
Medium	4–8	50–100
Stiff	8–15	100–200
Very stiff	15–30	200–400
Hard	>30	>400

depends upon the angle of shearing resistance or angle of internal friction of cohesionless soil. The approximate values of various engineering properties of the soil can be predicted and correlated as follow:

SPT value related to relative density (D_r) of cohesion-less soil: The correlation between 'N' values (corrected) and relative density of granular soil and corresponding angle of internal friction suggested by Terzaghi and Peck (1967) is presented in Table 5.8.

SPT values related to consistency of clay soil: The correlation between 'N' values, consistency, and unconfined compressive strength of cohesive soil suggested by Peck et al. (1974) is given in Table 5.9.

Hera et al. [16] recommended the correlation between standard penetration number (N_{60}) and undrained shear strength of the clay (c_u) as:

$$\frac{c_u}{p_a} = 0.29 \times N_{60}^{0.72} \quad (5.14)$$

where

p_a is atmospheric pressure (=100 kN/m^2)

Correlation between modulus of elasticity and standard penetration number

In case of cohesionless soil or granular soil, the corelationship between modulus of elasticity (E_s) of sand and standard penetration number (N_{60}) recommended by Kulhawy and Mayne [17] is:

$$\frac{E_s}{P_a} = \alpha N_{60} \quad (5.15)$$

where

α = 5 for sand with fines

= 10 for clean normally consolidated sand

= 15 for clean overconsolidated sand

Table 5.10 Site classification and soil profile based on SPT values of soil shear wave and undrained shear strength

Site class	Soil profile	Average properties in top 100 ft		
		Standard penetration resistance	Soil shear wave velocity (V_s) in m/s	Soil undrained shear strength (S_u) in psf
A	Hard rock	N/A	$V_s > 1500$	N/A
B	Rock	N/A	$760 < V_s < 1500$	N/A
C	Very dense sand soft rock	$N > 50$	$360 < V_s < 760$	$S_u \geq 2000$
D	Stiff soil profile	$15 \leq N \leq 50$	$180 < V_s < 360$	$1000 \leq S_u \leq 2000$
E	Soft soil profile	$N < 15$	$V_s < 180$	$S_u < 1000$

Table 5.11 Types of soil mainly constituting the foundation based on SPT value of soil

Serial no.	Types of soil	Description of the soil	SPT range	Remark
1	I	Rock or hard soil	>30	Well-graded gravel and sand–gravel mixture with or without clay binder and clayey sand
2	II	Medium soil	10–30	All soils
			>15	Poorly graded sands or gravelly sands with little or no fines
3	III	Soft soil	<10	All soil

The value of E_s can be determined from the standard penetration number (N) using Eq. (5.16) (Schmertman 1970)

$$E_s = 766 \times \text{NkN/m}^2 \tag{5.16}$$

Soil classification based on SPT value of soil as per IBC and NEHRP

As per International Building Codes [19] and National Earthquake Hazard Reduction [20], the site is classified into various categories based on the observed SPT value of the soil as presented in Table 5.10.

There are also guidelines in IS 1893(Part 1): [3] for classifying the types of soil mainly constituting the foundation as described in Table 5.11.

5.6.4 Cone Penetration Test (CPT)

Static cone penetration test is a direct sounding test which is conducted at site to obtain a continuous record of soil characteristic with depth and to estimate various engineering properties of soil [21] (Reaffirmed 1997). In this test, a cone having an apex angle 60° and a base area of 10 cm^2 with a friction jacket/sleeve of an area of 150 cm^2 is pushed into the ground at a steady rate (pushing rate) of 10 mm/s. The static force required for unit penetration of the cone is correlated to various engineering properties of soil-like density, stratifications, etc. The test measures the cone resistance 'q_c' developed against the penetration of the cone and the frictional resistance 'f_c' developed between the sleeve and the surrounding soil. Friction ratio (%) evaluated as blow is the ratio of skin friction divided by the cone resistance.

$$\text{Tip Resistance}\,(q_c) = \frac{Q_C}{A_C} \tag{5.17}$$

where

Q_c tip load

Q_C load from pressure gauge reading + weight of cone + weight of connecting sounding rods

A_c cross-sectional area of cone = 10 cm^2

$$\text{Frictional Resistance}\,q_f = \frac{Q_T - Q_C}{A_f} \tag{5.18}$$

With friction sleeve $Q_T = Q_C + Q_F$

A_f surface area of friction sleeve

$$\text{Friction Ratio}\,f_r = \frac{q_f}{q_c} \tag{5.19}$$

The types of soil can be identified from the frictional ratio obtained from above calculation. The typical range of frictional ratio of soil and corresponding type of soil (as per Schmertmann et al. [6]) is presented in Table 5.12.

Relationship between q_c and c_u of clay

The cone penetration resistance q_c and undrained shear strength C_u may be related as [23]

Table 5.12 Typical range of soil for different types of soil (Schmertmann et al. [6])

Typical range of Frictional ratio (f_r)	Type of soil
0–5 (%)	Granular
5–10 (%)	Cohesive

Table 5.13 Value of cone factor N_k for different conditions of clay soil (Lunne and Kelven 1981)

Type of clays		Cone factor (N_k)
Normally consolidated		11–19
Overconsolidated	At shallow depth	15–20
	At deep depth	12–18

$$c_u = \frac{q_c - p_0}{N_K} \tag{5.20}$$

where

c_u undrained shear strength
q_c cone penetration resistance
N_k cone factor
$p_0 = \gamma z$ overburden pressure

Lunne and Kelven [23] suggested the value of cone factor N_k for both normally consolidated and overconsolidated clays as given in Table 5.13.

Correlation between SPT and CPT

Both standard penetration test and cone penetration test are conducted directly at site, and the brief idea about the types of soil can be obtained from both of the results. Meyerhof [24] presented correlation between standard penetration test (SPT) value and cone penetration test (CPT) value as given in Table 5.14.

Schmertman [25] proposed a relationship between cone resistance (q_c) and standard penetration value (N) as given in Table 5.15.

5.6.5 Vane Shear Test (VST)

Vane shear test is conducted at site by pushing the shear vane as shown in Fig. 5.6 into the ground at the bottom of

Table 5.15 Correlation between cone resistance 'q_c' and standard penetration value 'N' [Schmertman 1978]

Type of soil	Ratio of cone tip resistance and standard penetration value (q_c/N)
Sand and gravel mixture	6
Sandy silt	3
Clay–silt–sand mixture	2
Insensitive clay	1.5

the borehole to determine the shear strength of a cohesive soil in its natural condition. The soil is sheared along a cylindrical surface when a torque is applied through the handle at the top of the rod (Ref: IS4434, 1978). The torque required to shear the cylinder of the soil is measured. For a rectangular vane, the shear strength of soil 'S' is evaluated from Eq. (5.21).

$$S = \frac{T}{\pi\left(\frac{D^2H}{2} + \frac{D^3}{6}\right)} \times 10^6 \tag{5.21}$$

where

S shear strength in kN/m^2
T torque to shear the soil in N m
H height of the van in mm
D diameter of the soil cylinder sheared or overall diameter of vane in mm

For $H = 2D$,

$$S = \frac{3M}{11D^3} \times 10^6\,\text{kN/m}^2 \tag{5.22}$$

The vane shear test is very useful for evaluating the in situ shear strength of very soft and sensitive clay for which it is difficult to obtain undisturbed samples. However, vane shear test is not suitable for cohesionless soil.

5.6.6 Pressure Meter Test (PMT)

Menard developed a pressure meter for determining the stress–deformation characteristics of the soil in natural condition. The test procedure is also discussed in IS 1892-1972. The various parameters obtained in pressure

Table 5.14 Correlations between SPT and CPT [Meyerhof 1965]

State of sand	Relative density (D_r) (%)	SPT value N_{cor}	Static cone resistance Q_c (MPa)	Angle of internal friction Ø° (degree)
Very loose	<20	<4	<2.0	<30
Loose	20–40	4–10	2–5	30–35
Medium dense	40–60	10–30	5–10	35–40
Dense	60–80	30–50	10–20	40–45
Very dense	80–100	>50	>20	>45

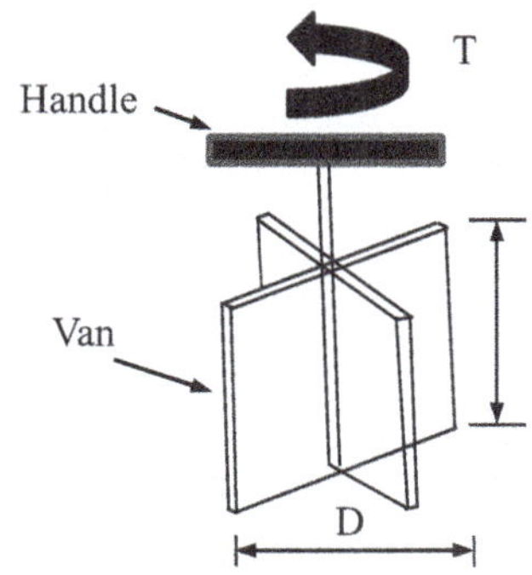

Fig. 5.6 Vane shear test

meter tests are Young modulus, shear modulus, undrained shear strength of soil, etc. The pressure meter test is carried out by drilling a hole, lowering the probe to the desired level, inflating the probe, and observing the expansion of soil or volume change of soil due to pressure. Menard suggested a guideline for the corelationship between diameter of hole and probe as:

$$1.03\,D_p < D_h < 1.2\,D_p \tag{5.23}$$

where

D_p diameter of probe at deflate condition
D_h diameter of the hole

The pressure meter consists of an inflatable cylindrical probe or flexible membrane which can be expanded by liquid or gas as shown in Fig. 5.7. The probe of the pressure meter consists of three cells in series out of which mid-cell is used as measuring cell and two end cells are used as exterior guard cells. The inflated guard cell seal prevents the measuring cell to expand toward the boreholes at both ends. In the pressure meter designed by Menard, the measuring cell is expanded by filling water, whereas the guard cells are expanded with filling gases such as air, carbon dioxide, or nitrogen from the control unit. Pressure gauges with the range 0–2500 kPa for soil and 0–10,000 kPa for rock are used to measure the water pressure flowing to the measuring cell.

In pressure meter test, a uniform radial stress/pressure is applied to the wall of boreholes after placing the probe at desired depth. The pressure to the surrounding soil is applied through the measuring cell by either equal pressure increment method or equal volume increment method. In the equal pressure increment method, ten equal increment of pressure is applied to reach the limiting pressure (p_l). Each equal increment pressure is held for approximately one minute and volume reading is observed after one minute. In the method of equal volume increment method, the volume of the probe is increased at a stage-wise at the rate 5% of the nominal volume (deflated volume) of the probe and held constant for 30 s. A graph is drawn with the volume of the water (v) as abscissa for each increment of pressure (p) as ordinate. The curve further corrected for various losses such as

i. Pressure losses p_c: occurs due to rigidity of probe membrane,
ii. Volume losses v_c: occurs due to expansion of tubing system and compressibility of any parts of testing equipment including probe,
iii. Hydrostatic correction p_w: occurs due to water column in the tube which exerts extra pressure.

$$p_w = \gamma_w H_w \tag{5.24}$$

where

γ_w unit weight of water
H_w difference in head between the measuring cell in borehole and pressure gauge in control unit

Corrected pressure and 'p' and volume 'v' of the curve are obtained as

$$p = p_r + p_w - p_c \tag{5.25}$$

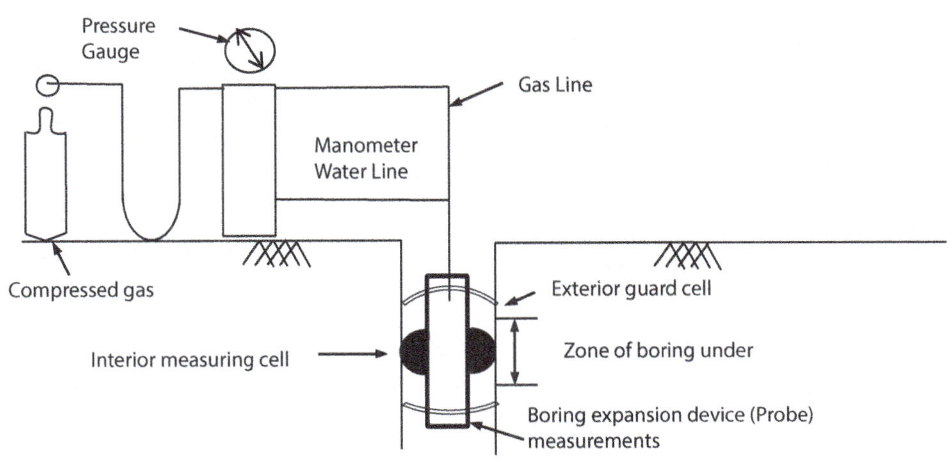

Fig. 5.7 Pressure meter test setup

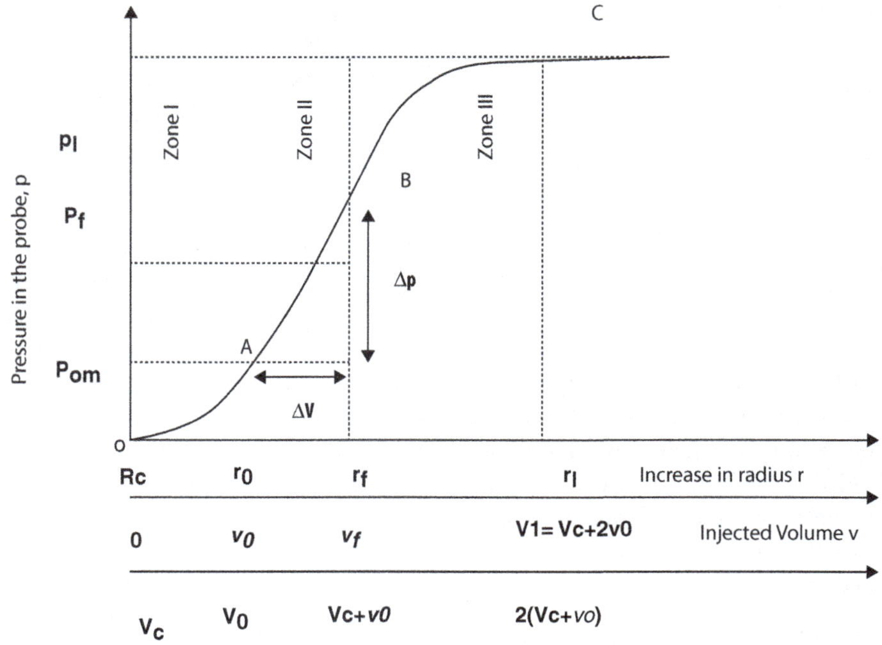

Fig. 5.8 Pressure meter versus cavity volume

where p_r is the actual pressure reading of gauge in control unit and

$$v = v_r - v_c \tag{5.26}$$

where v_r is the actual volume reading of the volume meter

A typical corrected curve is presented in Fig. 5.8, and there are three segments in the curve. The initial segment of the curve is OA, and the point 'A' represents P_0 and V_0.

$$V_0 = V_c + v_0 \tag{5.27}$$

The second segment (from A to B) is known as the pseudo-elastic phase of the test. The coordinate of the point 'B' is p_f and v_f where p_f is known as creep pressure. The segment BC is the final phase, and at point 'C' which represents the large deformation, the curve is asymptotic. The limit pressure 'p_l' is defined as the pressure required to double initial volume of the cavity, and limiting pressure is presented as:

$$v_l = V_c + 2v_0 \tag{5.28}$$

Calculation of Pressure Modulus

As the segment 'AB' of the curve is approximately straight line, it is assuming that the behavior of soil is linear. The equation for the radial expansion of a cylindrical cavity in infinite medium is expressed as

$$G = V \times \frac{\Delta p}{\Delta v} \tag{5.29}$$

where

G shear modulus
v volume of the cavity
p pressure in the cavity

The slope of AB is constant, and the volume (at point A) v_0 changes to v_f (at point B). For calculation of shear modulus 'G,' the midpoint volume V_m is considered.

$$V_m = V_c + \frac{v_0 + v_f}{2} \tag{5.30}$$

where V_c is the volume of the deflated measuring cell at zero reading in the voltmeter of control unit.

Now,

$$G = \frac{E_p}{2(1 + \mu_s)} \tag{5.31}$$

where μ_s is the Poisson's ratio of soil

The pressure meter modulus of the soil (E_p) is evaluated as:

$$E_p = 2(1 + \mu_s)(V_m)\left(\frac{\Delta p}{\Delta v}\right) \tag{5.32}$$

As suggested by Menard, by considering $\mu_s = 0.33$, the Menard modulus is

$$E_m = 2.66\, V_m\left(\frac{\Delta p}{\Delta v}\right) \tag{5.33}$$

where

$$V_m = V_c + \frac{v_0 + v_f}{2}$$
$$\Delta p = p_f - p_{0m}$$
$$\Delta v = v_f - v_0$$

μ_s Poison's ratio of soil

Substituting the values,

$$E_m = 2.66\left(V_c + \frac{v_f + v_0}{2}\right)\left(\frac{p_f - p_{0m}}{v_f - v_0}\right) \tag{5.34}$$

E_m in term of change in radius of the cavity

$$E_m = 2.66\left(R_c + \frac{r_f + r_0}{2}\right)\left(\frac{p_f - p_{0m}}{r_f - r_0}\right) \tag{5.35}$$

where

- R_c radius of the measuring cell at the deflated condition which is at starting of the test
- r_0 radius corresponding point A
- r_f radius corresponding to point B

Correlation between Menard modulus (E_m) and Young's modulus (E)
Menard [27] proposed the relationship between Menard Modulus (E_m) and Young's modulus (E) as:

$$E = \frac{E_m}{\alpha} \tag{5.36}$$

where

- E Young's modulus
- E_m Menard modulus
- α rheology factor (typical value of α for rock extremely fractured = 1/3, overconsolidated clay = 1, normally consolidated clay = 0.67

Menard [27] suggested the typical value of Menard modulus (E_m) and limiting pressure (p_l) for various types of soil is presented in Table 5.16.

The net limiting pressure p_l is presented as

$$\text{Limiting Pressure}\, p_l = p_l - p_{oh} \tag{5.37}$$

where p_{oh} is the at rest horizontal pressure at any depth and expressed as:

$$p_{oh} = (\gamma z - u)K_0 + u \tag{5.38}$$

where

- u pore pressure at depth z
- γ gross unit weight of soil
- K_0 coefficient of earth pressure at rest condition

Relationship between SPT value (N) and net limiting pressure (p_l) of granular soil
Baguelin et al. [28] suggested various typical range of value of standard penetration resistance (N) and limiting pressure as presented in Table 5.17.

Table 5.16 Typical value of Menard modulus (E_m) and limiting pressure (p_l)

Serial no.	Types of soil	E_m (MPa)	p_l (MPa)
1	Mud	0.2–1.5	0.02–0.15
2	Soft clay	0.5–3.0	0.05–0.3
3	Medium clay	3–8	0.3–0.8
4	Stiff clay	8–40	0.6–2.0
5	Loose silty sand	0.5–2.0	0.1–0.5
6	Silt	2–10	0.2–1.5
7	Sand and gravel	8–40	1.2–5

Table 5.17 Typical value of N and p_l of granular soil for different densities

Serial no.	Density of soil	N-Value	p_l (kPa)
1	Very loose	0–4	0–200
2	Loose	4–10	200–500
3	Medium dense	10–30	500–1500
4	Dense	30–50	1500–2500
5	Very dense	>50	>2500

5.6.7 Seismic Reflection Test

The stratification of soil or rock can be determined by geophysical methods such as seismic reflection and seismic refraction methods. Geophysical methods provide some additional information or missing information between widely spaced boreholes.

During seismic reflection testing at site, shock waves are generated either by blowing a mechanical heavy hammer at ground surface or by detonating of explosives. If the waves encounter a change in the physical properties of the soil material, in which they are traveling, undergo a change in direction. The wave will either penetrate deeper into the earth (refraction) or reflect to the surface (refraction). During testing, some energy is always transmitted, while some energy is reflected at a geological interface. Figure 5.9 shows the methodology for reflection mapping and layout of source and receivers in seismic reflection test. Geophones or detectors are installed in a straight line as shown in Fig. 5.9. The spacing between the geophones depends upon the depth of soil strata to be investigated and the level of details required.

Seismic reflection test produces an impulse at source usually rich in P waves, and measuring the time of arrival at receiver is useful for large-scale and/or very deep stratigraphy. Measuring the arrival times of the waves which follow direct path to receiver (on surface)

$$\text{Arrival time } t_{\text{d}} = \frac{\text{distance of travel}}{\text{wave velocity}} = \frac{x}{v_{p1}} \tag{5.39}$$

where

v_{p1} *P* wave velocity of upper layer

The primary wave velocity of the upper layer is calculated from horizontal distance '*x*' and the time of arrival 't_{d}'. Some part of energy moves downwards at an angle of inclination '*i*' and hits the interface horizontal layer and reflects back. The angle of incidence is:

$$i = \tan^{-1}\frac{x}{2H} \tag{5.40}$$

where '*H*' is the thickness of layer

Time taken to the wave to reflected back to reach the receiver on ground surface is [8]:

$$t_{\text{r}} = \frac{\text{distance of travel}}{\text{wave velocity}} = \frac{2\sqrt{\left(H^2 + (x/2)^2\right)}}{v_{p1}} = \frac{\sqrt{4H^2} + x^2}{v_{p1}} \tag{5.41}$$

Thickness of upper layer is:

$$H = \frac{1}{2}\sqrt{t_{\text{r}}^2\, v_{p1}^2} - x^2 \tag{5.42}$$

Advantages and disadvantages of reflection are presented in Table 5.18 (National Research Council, [29]).

Application: Seismic reflection test is used for determining the thickness of the layers of different types of soil.

Limitation: In seismic reflection test, interpretations of results for profile with low-velocity layers are difficult.

5.6.8 Seismic Refraction Test

Seismic refraction method of geophysical exploration which works on the principles of seismic waves has different velocities of travel in different types of soil or rock. The waves are refracted when it encounters the boundary of soil of different soil properties.

In seismic refraction method, shock waves are created into the soil, at ground level, or at a certain depth below it,

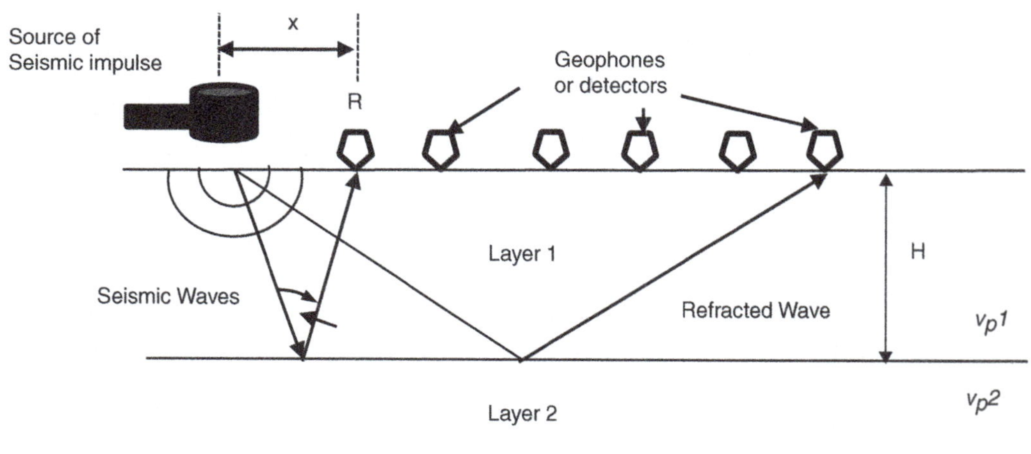

Fig. 5.9 Source and receiver layout and ray path in seismic reflection test

Table 5.18 Advantages and disadvantages of seismic reflection test [Ref: NRC 2000]

Serial no.	Advantages	Disadvantages
1	Observations are collected at small source–receiver offsets	Many source and receivers are used to produce meaningful images hence expensive to acquire
2	Method can work no matters how the propagation speed varies with depth	Processing is expensive as it is very computer intensive
3	Reflection observations are more readily interpreted in term of complex geology	Interpretation requires more sophistication and knowledge of the reflection process

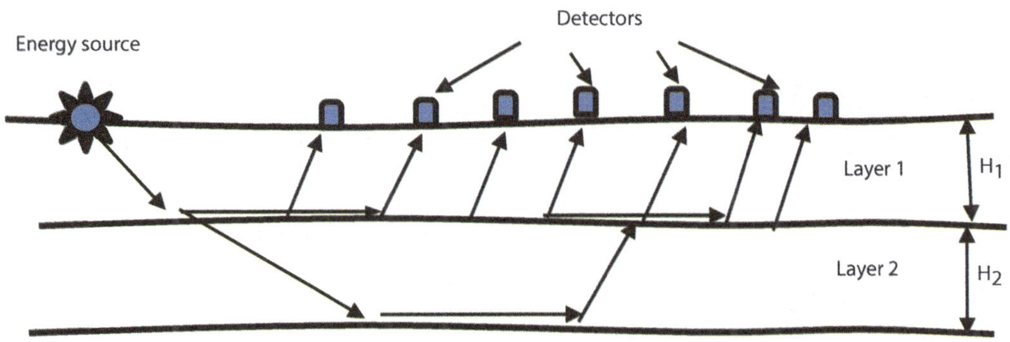

Fig. 5.10 Source and receiver layout and ray path in seismic refraction test

by striking a plate on the soil with a hammer or by exploding small charges in the soil. The radiating shock waves are picked up by the vibration detector (geophone), where the time of travel gets recorded. Either a number or geophones are arranged in a line as in Fig. 5.10 or shock producing device is moved away from the geophone to produce shock waves at given intervals. The distance between the geophones depends upon on the amount of details required and the depth of data to be investigated. In general, the distance between the first detector and last detectors is around 3–4 times the depth to be investigated. Some of the waves, known as direct or primary waves, travel directly from the shock points along the ground surface and are picked up first by geophone. If the soil consists of two or more distinct layers, some of primary waves travel downwards to the lower layer and get refracted at its surface. If the underlying layer is denser, the refracted waves travel much faster. As the distance between the shock point and geophone increases, the refracted waves reach the geophone earlier than the direct waves.

Application of test results: The primary applications of seismic refraction technique are to determine the depth of bed rock and bedrock structures as shown in Fig. 5.11.

If '*n*' numbers of receivers are placed in a line at certain distance from source '*S*,' the time taken by the direct wave to reach the *n*th receivers is:

$$t_{dn} = \frac{x_n}{v_1} \tag{5.43}$$

where

x_n is the distance of *n*th receiver from source
v_1 is the velocity of wave in first layer

In the two-layer soil system, some waves travel downwards toward the interface boundary. After that some portion of waves are reflected and some portions of waves are refracted

Fig. 5.11 Travel path and depth for first arrival calculation

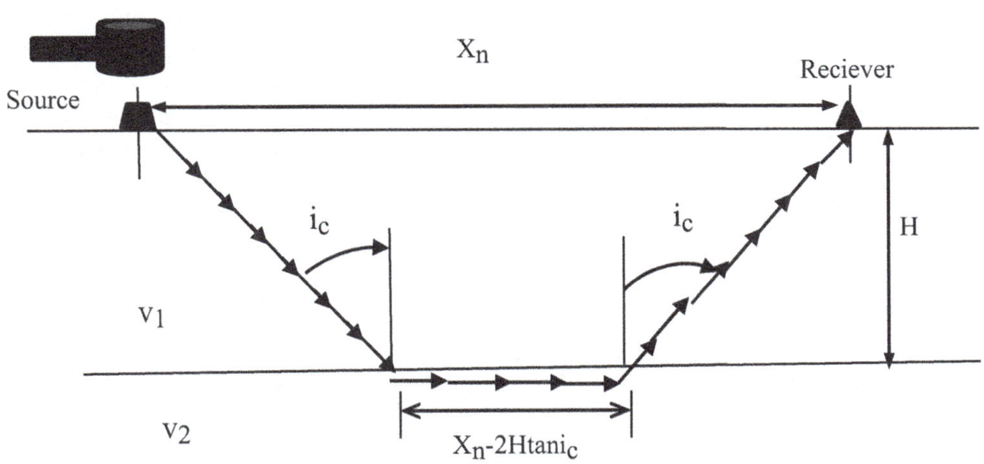

toward surface. At the critical angle of incidence (i_c), the refracted wave will move parallel to the boundary of interface. This refracted wave will produce a wave called as "head wave" which will move with a velocity v_1 in first layer in direction at 90-ic to the interface boundary. The direct wave reached faster in case of short distance between the source and receiver, but in case of the distance greater than critical distance x_c, the head waves reach earlier that the direct waves. The time taken by the head waves to arrive at nth receiver

$$t_{hn} = \frac{H}{v_1 \cos i_c} + \frac{x_n - 2H \tan i_c}{v_2} + \frac{H}{v_1 \cos i_c} \quad (5.44)$$

For critical incidence (as per Snell's law),

$$\sin_{i_c} = v_1/v_2 \text{ and } \cos^2_{i_c} = 1 - \sin^2_{i_c}$$

After rearranging,

$$t_{hn} = \frac{x_n}{v_2} + 2H\sqrt{\frac{1}{v_1^2} - \frac{1}{v_2^2}} \quad (5.45)$$

If the receiver is at critical distance 'x_c,' then

$$\frac{\boldsymbol{x_c}}{\boldsymbol{v_1}} = \frac{\boldsymbol{x_c}}{\boldsymbol{v_2}} + \boldsymbol{2H}\sqrt{\frac{1}{v_1^2} - \frac{1}{v_2^2}} \quad (5.46)$$

From which,

$$\boldsymbol{H} = \frac{x_c}{2}\sqrt{\frac{v_2 - v_1}{v_2 + v_1}} \quad (5.47)$$

For a typical three-layer soil system, the plot of test results of distance between the source–geophone and the arrival time is show in Fig. 5.12.

From Fig. 5.12 if the source–geophone spacing is more than the distance d_1 (distance between source and B), the direct wave reaches the geophone in advance the refracted wave and time–distance relationship is presented by straight line AB. If the source–geophone distance is greater than d_2, the refracted wave arrives in advance the direct waves and time–distance relationship is presented by straight line BC.

In general type of soil or rock, the depth of the top strata (H_1) can be evaluated from:

$$H_1 = \frac{d_1}{2}\sqrt{\frac{v_2 - v_1}{v_2 + v_1}} \quad (5.48)$$

The thickness of second layer (H_2) is estimated from:

$$H_2 = 0.85H_1 + \frac{d_2}{2}\sqrt{\frac{v_3 - v_2}{v_3 + v_2}} \quad (5.49)$$

As per Corp of Engineers, 1979, the thickness of kth layer, in case of multiple horizontal layers, as shown in Fig. 5.13, is given in Eq. 5.50.

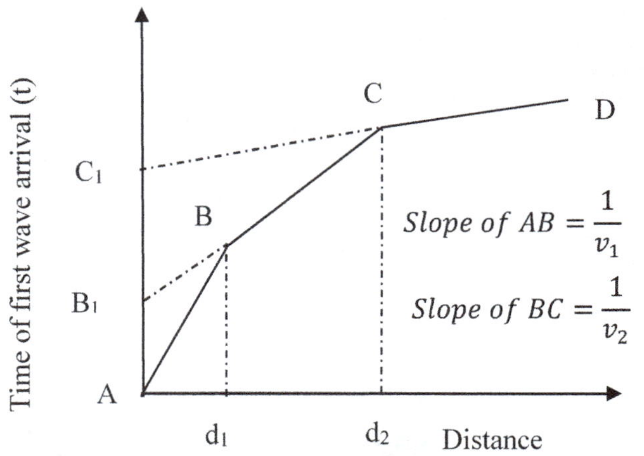

Fig. 5.12 Plot of test results of distance between the source–geophone and arrival time in seismic refraction test

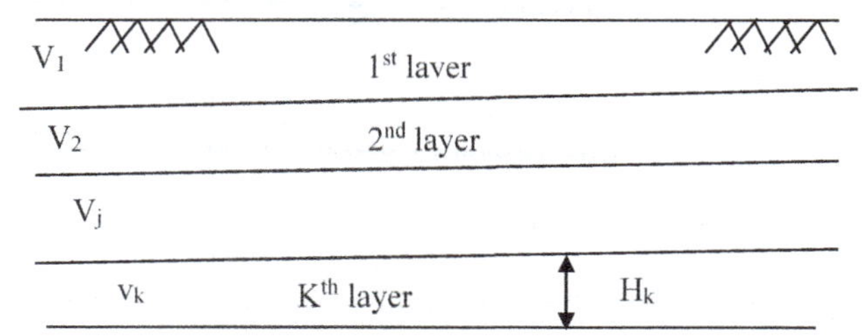

Fig. 5.13 Multi-layer soil system

$$H_k = \frac{x_{ck}}{2}\sqrt{\frac{v_{k+1} - v_k}{v_{k+1} + v_k}} + \sum_{j=1}^{k-1}\frac{H_j}{v_j} \times \frac{v_{k+1}\left(\sqrt{v_k^2 - v_j^2}\right) - v_K\left(\sqrt{v_{k+1}^2 - v_j^2}\right)}{\left(\sqrt{v_{k+1}^2 - v_j^2}\right)} \quad (5.50)$$

In all above expression, it is assumed that the travel velocity of wave of each layer is smaller than the velocity in the layer below. However, in many site conditions, the presumed soil condition may differ and hence the seismic refraction test mislead the test results. The presence of intermediate layer of insufficient layer thickness or insufficient velocity contrast, the head waves from the deeper layer can overtake the head wave of intermediate layer. The undetected thinner intermediate layer is called blind zone and will cause the depth of detected layer less than the actual depth. Advantages and disadvantages of refraction are presented in Table 5.19.

5.6.9 Cross-Hole Test

In the design and analysis of the foundation, subjected to dynamic loading, the various dynamic soil properties are required to be evaluated. The test cross-hole test is most

Table 5.19 Advantages and disadvantages of seismic Refraction test

Serial no.	Advantages	Disadvantages
1	Fewer source and receiver point required hence low cost	Observation required large source–receiver offset
2	Little processing is required	Works only when the speed of propagation increases with depth
3	Modeling and interpretation are straightforward	Observation generally interpreted in layer

commonly performed for obtaining the dynamic properties of the soil/rock, viz. shear wave velocity V_s, V_p, Young's modulus (E), shear modulus (G), strain(υ), etc.

The test is performed in the field using three numbers or five numbers of boreholes to measure wave propagation velocities along horizontal path. ASTM 4428/D4428M-07 provide the guidelines for conducting the cross-hole test. Boreholes are drilled simultaneously to the desired depth where test is being carried out. Velocity transducers (geophones) that have natural frequencies of 4–15 Hz are adequate for detecting (receiving) the shear waves as they arrive from the source. The steps involved in the test are:

1. A source capable of generating shear and compressional wave(seismic wave trains) at selected depth is lowered in one borehole.
2. A pair of matching three component geophone receiver is lowered to same depth in additional offset boreholes as shown in Fig. 5.14.
3. The time required for a shear wave to traverse the known distance between the boreholes is measured.

Application: The measured shear wave velocity and primary wave velocity are used in computing shear modulus, bulk modulus, Young's modulus, and Poisson's ratio of the soil between boreholes.

Calculation of Dynamic Properties of Soil: The various dynamic parameters such as Young's modulus, Poisson's ratio, bulk modulus, and shear modulus are evaluated based on the theories of elasticity, and these expressions are given below [Ref: ASTM 4428-07]

Poisson's ratio of the soil is:

$$\mu = \frac{m^2 - 2}{[2(m^2 - 1)]} \tag{5.51}$$

where

$$m = \frac{V_P}{V_s}$$

Young's modulus (E) of the soil is:

$$E = \rho V_P^2 \frac{(1+\mu)(1-2\mu)}{1-\mu} \tag{5.52}$$

where

ρ mass density of soil = γ/g
V_p P wave velocity
μ Poison's ratio

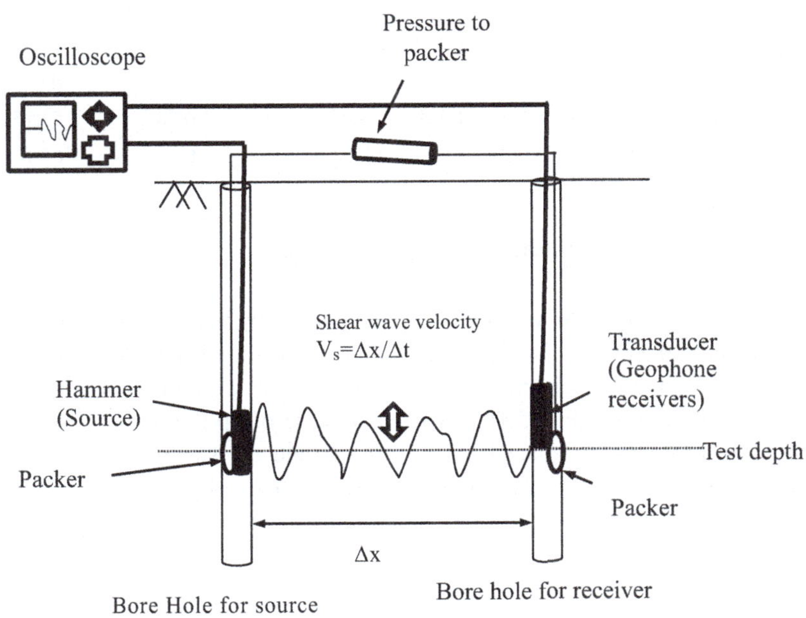

Fig. 5.14 Cross-hole test setup

Shear modulus (G) of soil is:

$$G = \frac{E}{2(1+\mu)} = \rho V_s^2 \tag{5.53}$$

Bulk modulus (K) of soil is:

$$K = \frac{E}{(3-6\mu)} \tag{5.54}$$

Example 5.3 During the cross-borehole test as per ASTM 4428-D4428M-07 of a power plant site, the seismic energy source was kept at depths at 19.5 and 40.00 m from ground level. Two three-dimensional transducers (receivers R_1 and R_3) are kept at horizontal distance of 3.00 m, and another two receivers (R_2 and R_4) are kept at 6.00 m horizontal distance from the source at a same depths of 19.5 and 40.00 m as source as shown in the plan view and sectional view Fig. 5.15. The bulk density of soil at depth 19.5 and 40.0 m is 2600 and 2900 kg/m^3, respectively.

The observations during the test are tabulated as in Table 5.20.

Calculate various parameters of the soil.

Solution

Shear wave velocity and primary wave velocity

	From R_1	From R_2	From R_3	From R_4	Average velocity
At depth of 19.50 m					
V_s (m/s)	2857.14	2803.7388	2727.27	2666.67	2760.00
V_p (m/s)	5263.157	5217.39	5263.15	5217.39	5218.00
At depth of 40.00 m					
V_s (m/s)	2912.62	2941.17	2912.62	2941.17	2931.00
V_p (m/s)	5882.35	5882.35	5660.37	5769.23	5823.00

Dept (h)	V_s (m/s)	V_p (m/s)	V_p/ V_s	Bulk density (kg/m^3)	Poisons ratio	Young's modulus (Gpa)	Shear modulus (Gpa)	Bulk modulus (Gpa)
19.5 m	2760.00	5218.00	1.76	2600.00	0.26	57.77	19.81	40.26
40.0 m	2931.00	5823.00	1.99	2900.00	0.33	66.10	24.91	65.22

5.6.10 Seismic Down-Hole/Up-Hole Test

Seismic down-hole and up-hole test are conducted by boring one hole. In the down-hole test, one or more sources are kept in the hole and the excitation source is placed on the ground surface as shown in Fig. 5.16b, whereas in case of up-hole test, the receivers are placed on the ground and the shear waves are generated at various depths in the borehole as shown in Fig. 5.16a.

5.6.11 Determination of Dynamic Properties of Soil

In the design and analysis of machine foundations such as turbogenerator, blowers for various power plants and structures which are subjected to dynamic loading, the various dynamic properties of the soil are to be evaluated by conducting various tests. The various dynamic parameters which are required to be evaluated are dynamic modulus (Young's modulus, shear modulus, bulk modulus), Poisson's ration, dynamic elastic constant (coefficient of elastic uniform compression), damping ratio, liquefaction parameters, etc. Ref: [12]. For evaluation of in situ dynamic and damping properties of soil, the following tests are carried out.

1. Block vibration test,
2. Cyclic plate load test.

IS 5249:1992(Reaffirmed in 1995) provides the guidelines for conducting the above tests and to evaluate dynamic parameters of the soil. In the following subsection, block vibration test is explained.

5.6.11.1 Block Vibration Test

In block vibration test, a pit of suitable size depending upon size of block is made. The size of pit may be 3 m × 6 m at bottom and a depth preferably equal to proposed depth of foundation. For the test block, a plain cement concrete (PCC) block is constructed as shown in Fig. 5.17. The size of block should be selected depending upon the subsoil condition. In ordinary soil, the size of the PCC block may be of size 1.00 m × 1.00 m × 1.50 m, and in dense soil, it may be of 0.75 m × 0.75 m × 1 m. Mass Ratio B_z:

$$B_z = \frac{(1-v)}{4} \times \frac{m}{\rho r_0^2} \text{ is always more than unity} \tag{5.55}$$

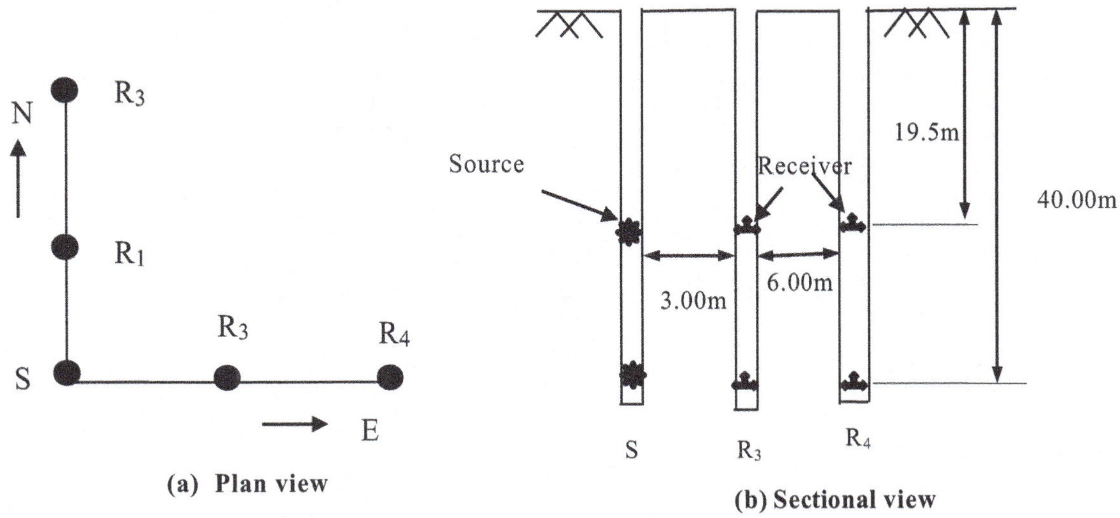

Fig. 5.15 Schematic diagrams of boreholes

Table 5.20 Observation of arrival times at all receivers at depth of 19.5 and 40.0 m

Serial no.	Depth (m)	Arrival time at R_1 (mS) (at distance 3.00 m)		Arrival time at R_2 (mS) (at distance 6.00 m)		Arrival time at R_3 (mS) (at distance 3.00 m)		Arrival time at R_4 (mS) (at distance 6.00 m)	
		V_s	V_p	V_s	V_p	V_s	V_p	V_s	V_p
1	19.5	1.05	0.57	2.15	1.15	1.10	0.57	2.25	1.15
2	40.0	1.03	0.51	2.04	1.02	1.03	0.53	2.04	1.04

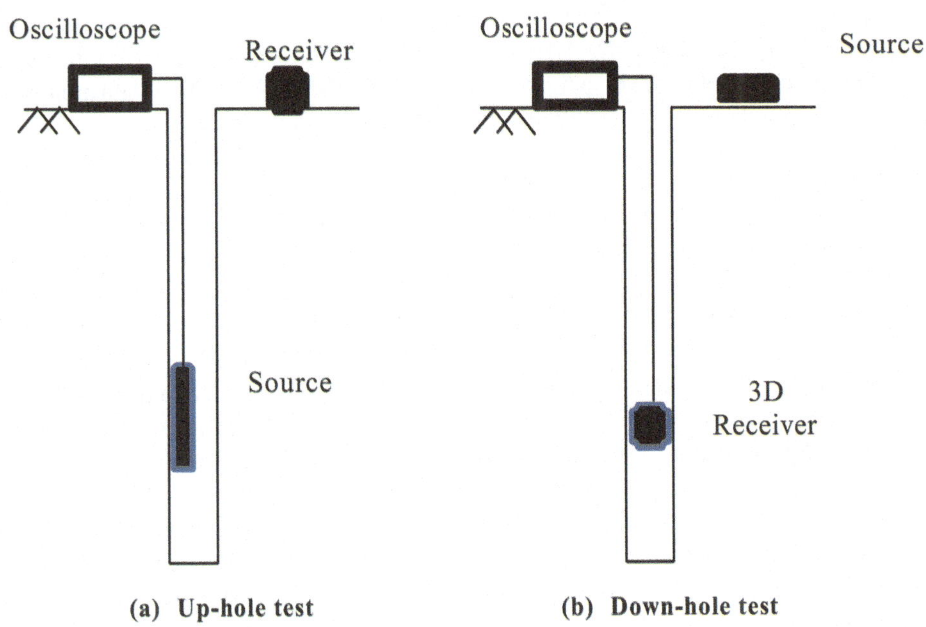

Fig. 5.16 Seismic up-hole and down-hole test

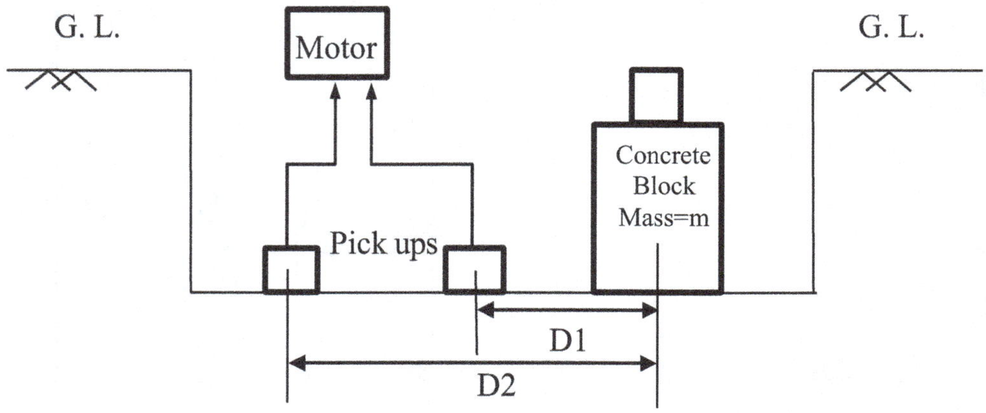

Fig. 5.17 Block vibration test setup

where

- B_z mass ratio
- m mass of concrete block
- ρ density of concrete block
- r_0^2 Effective radius

Foundation bolt in concrete block is provided on the top for fixing oscillator assembly. Vibrations pick-up are fixed at top of concrete block. In this test, the various apparatus (Ref: IS 5249 1992, Reaffirmed 1995) are used such as:

a. Mechanical oscillator: Oscillator to produce the vertical and horizontal excitation is mounted on the top of the concrete block and is driven by a D.C. motor. The mechanical oscillator should be capable of producing sinusoidal varying force and have a frequency range commensurate with the size of concrete block and types of soil.
b. D.C. motor: To run the above oscillator.
c. Acceleration pickup: Useful frequency 100 Hz or more and natural frequency should be 220 Hz undamped and 140 Hz damped.
d. Velocity pickup: Velocity pickup should be sensitive enough to record even feeble ground vibration. Natural frequency <10 Hz and damping less than 1% of the critical damping.
e. Displacement pickup: Displacement pickup should be of frequency response in the range of 0–100 Hz.
f. Geophones: Geophones should be sensitive enough to record even feeble ground vibration. Natural frequency <10 Hz and damping less than 1% of the critical damping.
g. Amplifier.
h. Oscillograph.
i. Hammer: A sledge hammer or drop hammer of weight 10 kg to impart blow to the concrete block to exiting under condition of free vibration or for generating waves in the ground.

Block Vibration Test-Types

Forced vibration test and free vibration tests can be performed with the above test set up and are detailed as below.

A. Force vibration test,
   (i) Vertical vibration test
      a. Determination of coefficient of elastic uniform compression of soil (C_u),
      b. Determination of damping coefficient of soil (ξ).
B. Free vibration test
   a. Determination of damping coefficient of soil (ξ),
   b. Evaluation of coefficient of attenuation of soil (α).

Forced vibration test in vertical vibration mode

The vibration pickup should be fixed at the top of the block such that it senses vertical motion of the block. The vibration exciter should be mounted on the block such that it generates purely vertical sinusoidal vibration and the line of action of the vibrating force passes through the center of gravity of block. The frequency of the exciter is increased in steps of small values (1–4 cycle/sec) up to the maximum frequency of the exciter. The amplitude of the vibration is observed by using vibration meter.

Determination of coefficient of elastic uniform compression of soil 'C_u'

The coefficient of elastic uniform compression 'C_u' of soil given by [IS 5249:1992 Reaffirmed in 1995] is:

$$C_u = \frac{4\pi^2 f_{nz}^2 m}{A} \tag{5.56}$$

where

f_{nz} = natural frequency

m = mass of block, exciter and motor, and other mountings

A = contact area of block with the soil or base is of concrete block

= 100 × 100 cm^2

The value of 'C_u' calculated from Eq. 5.58 corresponds to the base area of model block. From the value of $C_{u,}$ obtained from the test block of contact area A, the value of C_{u1} for actual foundation area A_1 up to base area of 10.00 m^2 can be calculated as:

$$\boldsymbol{C_{u1}} = \boldsymbol{C_u}.\sqrt{\frac{\boldsymbol{A}}{\boldsymbol{A_1}}} \tag{5.57}$$

For actual foundation of area larger than 10.00 m^2, the value of C_u remains constant (Barkan 1962) and C_{u1} obtained for foundation area of 10.00 m^2 can be used.

Example 5.4 Block vibration test is conducted for a blower foundation of a power plant. The concrete block size for the block vibration test was 0.75 × 0.75 × 1.00 m as the dense soil was available at proposed level of foundation. During the test, it was observed that the resonant frequency f_{nz} is 35 cycle per second. Determine the value of coefficient of elastic uniform compression (C_u) for a foundation of base area of 7.5 m × 4.5 m.

Solution

Coefficient of elastic uniform compression 'C_u' of soil

$$C_u = \frac{4\,\pi^2 f_{nz}^2 m}{A}$$

Resonant frequency = f_{nz} = 35.00 cycles/second

Base area A = 75 × 75 = 5625 cm^2 = 0.56 m^2

Mass of the block = 2400 × 0.75 × 0.75 × 1.00 = 1350 kg

$$C_u = \frac{4\pi^2 f_{nz}^2 m}{A} = \frac{4 \times \pi^2 \times 1225 \times 1350}{5632} = 11.58\,\text{kg/cm}^3$$

The base area of the proposed foundation = 7.5 × 4.5 = 33.75 m^2 > 10.00 m^2

For actual foundation of area larger than 10.00 m^2, the value of C_u remains constant

Hence, C_{u1} determined for 10.00 m^2 can be used for the proposed foundation (Eq. 5.59)

$$C_{u1} = C_u.\sqrt{\frac{A}{A_1}} = 11.58 \times \sqrt{\frac{0.56}{10.00}} = 11.58 \times \sqrt{0.056}$$
$$= 8.665\ \text{kg/cm}^3$$

Determination of damping coefficient of soil 'ξ'

From forced vibration test

In case of vertical vibration test, the value of damping coefficient ξ of soil is given by the following Eq. 5.60:

$$\xi = \frac{f_2 - f_1}{2f_{nz}} \tag{5.58}$$

where

ξ	damping coefficient of soil
f_2, f_1	two frequencies at which the amplitude is equal to $\frac{X_m}{\sqrt{2}}$ (Fig. 5.18)
X_m	maximum aptitude
f_{nz}	frequency at which the aptitude is maximum (resonant frequency)

From free vertical vibration test

The block is excited into free vertical vibrations by impact of sledge hammer or any suitable device as close to the center of the top of the block. The vibration is recorded to measure the frequency and amplitude of the vibration [Ref: IS 5249:1992]. The"damping coefficient" ξ may be obtained from free vibration test using Eq. (5.59).

$$\xi = \frac{1}{2\pi}\log_e \frac{\boldsymbol{X_m}}{\boldsymbol{X_{m+1}}} \tag{5.59}$$

X_m and X_{m+1} areas explained in Chap. 3.

The amplitude of vibration attenuates as the distance from the source of vibration increases. The coefficient which has the dimension of 1/distance used in the expression for determining the amplitude at any distance from the vibration.

The block vibration test setup is same as block resonance test discussed above. Pickup fitted on block removed and

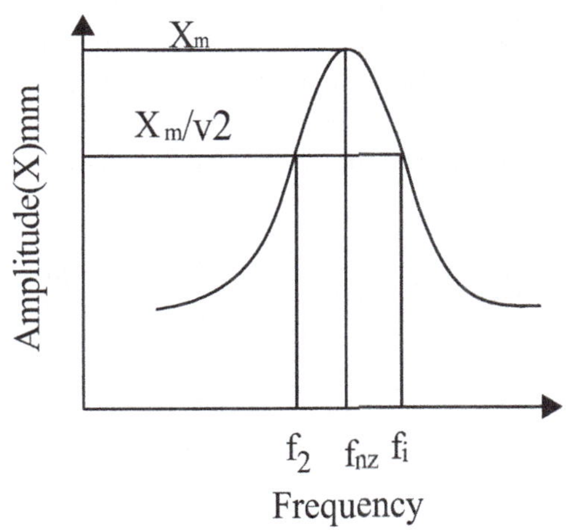

Fig. 5.18 Determination of damping from forced vibration test

installed at certain distance first at d_1 (approximately 30 cm) and subsequently at distance d_2 as shown in Fig. 5.18. Amplitude of vibration at two locations is measured for different frequencies. Coefficient of attenuation is calculated from the expression given in Eq. (5.60).

$$A_2 = A_1 \sqrt{\left[\frac{d_1}{d_2} e^{-\alpha(d_2 - d_1)}\right]} \quad (5.60)$$

where

A_2 amplitude at distance d_2
A_1 amplitude at distance d_1
α coefficient of attenuation

Values of coefficient of attenuation (α) of soil material recommended by Woods and Jedela (1985) and reported by Sreekala et al. (2008) for two values of vibration frequency (5 and 50 Hz) are presented in Table 5.21.

5.6.12 Chemical Test

Chemical tests of soils are carried out at laboratory, to provide design parameters for civil engineering materials in the ground. The various chemical tests are carried out to determine: pH value, sulfate, chloride and carbonate content, organic content in the soil. The concentration limit of sulfate and chlorides present in the soil is defined in IS 456-2000 (Reaffirmed 2008).

5.7 Observation of Water Table

The position of the water table in the ground of a particular location is having a great influence on the evaluation of settlement of foundation, bearing capacity of foundation, etc. The relative position of water table with respect to ground level varies significantly due to seasonal variation. To get a rough idea about the position of water table, the water level of nearby well can be observed. However, during the soil investigation process, the water table position is observed after keeping the groundwater inside the borehole undisturbed for duration of 24 h. For better idea borehole, water can be kept undisturbed for a week also.

5.8 Confirmatory Soil Investigation

During excavation of foundation, soil and rock samples are collected and various tests are carried out to confirm whether actual strata of soil (on site) during excavation matching/confirming or not with soil investigation carried out initially which considered in design of foundations. Confirmatory geotechnical investigations are carried out after excavation of strata to founding level which shows the presence of weaker zone, if any, which have not identified in original investigation.

5.9 Preparation of Soil Investigation Report

At the end of all subsoil exploration programs, appropriate laboratory testing, and the compilation of all of the required information, a soil investigation report is prepared for the use of the designer and for reference during future construction work at site. The information in the soil investigation report contains:

1. Agency deployed to carry out the soil investigation,
2. Location of proposed site with map and plan references,
3. Pit numbers,
4. Reduced levels (RL) of ground surfaces or other reference points,
5. Dates of starting and completion of soil investigation,
6. Scales of plans and elevations,
7. General descriptions of subsoil condition as determined from soil specimen, in situ test, and laboratory tests,
8. Recommendations for:
   a. Bearing capacity for shallow foundation,
   b. Settlement for shallow foundation.
9. Underground water level and borehole elevation.

Table 5.21 Values of coefficient of attenuation (α) of soil material (Woods and Jedela 1985)

Class	Material Damping coefficient (α) (m^{-1})		Description of material
	5 Hz	50 Hz	
I	0.01–0.03	0.1–0.3	Weak or soft soil ($N < 5$)
II	0.003–0.01	0.03–0.1	Competent soil ($5 < N < 15$)
III	0.0003–0.003	0.003–0.03	Hard soil ($15 < N < 50$)
IV	<0.0003	<0.003	Hard competent rock ($N > 50$)

5.10 Soil–Structure Interaction

The interaction between soil and structure plays an important role in engineering systems. Figure 5.19 shows the schematic of a building on soil foundations, which contains joints or interfaces.

In this figure, the response of the building will be different if it is supported on hard rock and/or soft soil. When a structure is founded on solid rock and it is subjected to strong ground motion, the rock constrains the motion to be very close to that of the free field motion due to its high stiffness. A ground motion refers as a *free field motion* when it is not affected by structural vibration or scattering of wave at and around the foundation. On the other hand, the same structure would respond in a different manner if the founding medium is a soft soil deposit. The inability of the foundation, which is placed over the soft soil deposit, to conform to the deformations of the free field motion would cause deviation from the free field response at the base of the structure; also, the dynamic response of the structure itself would induce deformation of the supporting soil. Thus in a nut shell, the process in which the response of the soil influences the motion of the structure and the response of the structure influences the motion of the soil is known as soil–structure interaction.

The nature of the soil media may influence the response of the structure in three ways:

i. Phenomenon of soil amplification/attenuation may occur in which the shear wave excitation at the base is amplified due to transmission of wave through the underlying soil/rock. The phenomena of soil amplification/attenuation occur due to impedance (which is defined as the product of wave velocity and density of the material, Pisal (2006)) contrast between the different layers of soil. Seismic waves propagate faster in hard rocks in compare to softer strata and sediments. As the waves travels from harder to softer strata, they become slow in velocity and gets bigger in amplitude for carrying the same amount of energy. Thus, shaking tends to be stronger at sites with softer surface layers, where seismic waves move more slowly.

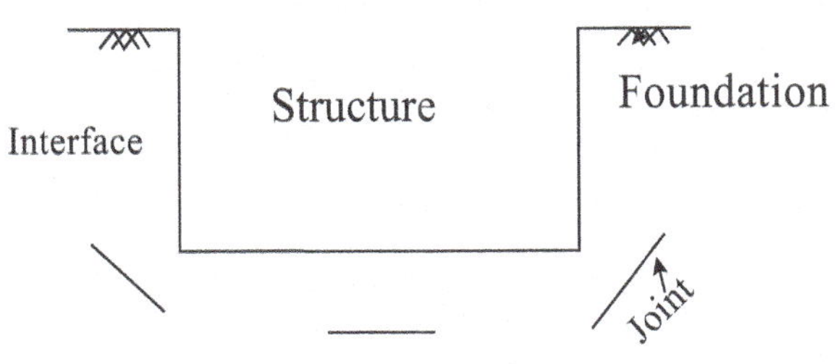

Fig. 5.19 Schematic diagram of soil–structure interaction

ii. The response of the structure may be significantly modified by the presence of the underlying rock/soil. This will include changes in mode shapes and periods of vibration resulting in the change in the forces acting on the structure.

iii. A significant part of the vibration energy of the flexibly supported structure may be dissipated by material damping and radiation damping in the supporting soil medium.

At this point in time, it has become clear that soil-interaction effects might not be important for regular, flexible buildings built on rock or stiff soil, but that they could be significant for stiff and massive structures such as nuclear power plants built on soft soils. Hence, whether the neglect of the SSI effects on the response of the structure is conservative or not depends entirely on the type of the problem at hand, and it has to be studied on a case-by-case basis.

5.10.1 Damping in Soil

In a homogeneous linear elastic medium, the waves travel in the medium without any decrease in its amplitude. This type of behavior does not occur in real materials, and the wave amplitude attenuates with the distance. This attenuation can be attributed to two sources: One is the geometric attenuation (or geometric damping), and the other is the anelastic attenuation (material damping).

Geometric damping is a measure of energy radiated away from the immediate region of the foundation. In this phenomena, as the energy propagates away from the location from where the disturbance originates, it means that the volume of the material affected simultaneously by a wave increases with distance travelled by a wave. As the disturbance releases a fixed amount of energy, it may be concluded that the energy absorbed per unit volume (specific energy) decreases with distance; hence, the amplitude of stresses and displacements decreases. This form of elastic energy dissipation in the form of waves traveling away from the foundation is called radiation (or geometric) Damping. This reduction is purely of geometric origin which results from the decrease in the specific energy of the wave. It is clearly distinguished from the material damping in which the elastic energy is actually dissipated by viscous, hysteretic, or other mechanisms.

Material damping is a measure of energy lost as a result of hysteresis effects. In case of real materials, a part of the energy transmitted by a wave is dissipated in the form of heat or permanent deformations. In other words, waves will attenuate with distance as a result of internal energy losses in the propagation medium. This additional energy dissipation mechanism may be observed when, for example, a soil sample is set into a state of free vibration since in such a case the vibration amplitude will decrease with each cycle and eventually die out. This type of energy dissipation due to friction is called hysteretic (or material) damping. The amount of material or internal damping in a soil deposit depends on the type of soil and varies thus over a wide range. It is primarily a function of strain amplitude and initial confining pressure. The average value of 0.05 may be thus added to the damping ratios due to radiation damping to obtain a rough estimate of the combined effects of radiation and internal damping. When a soil is loaded cyclically under medium to high strains, most of the soils exhibit a nonlinear hysteretic behavior. This behavior is presented in Fig. 5.20.

It may be observed, thus, that, in general, the shear modulus G of a given soil is not a constant, but it changes with strain level. At very low-strain shear modulus, value is G_{max} and it decreases with shear strain. An example of such an analytical model is that suggested by Ramberg and Osgood in 1943, which consists of the model with three parameters that defines the backbone curve (α, R, and τ_y), respectively, as well as a series of basic rules that govern the unloading and reloading behavior (Masing 1923). The stress and strain are related by the following Eq. (5.61):

$$\frac{\gamma}{\gamma_y} = \frac{\tau}{\tau_y}\left[1 + \alpha\left(\frac{\tau}{\tau_y}\right)^{R-1}\right] \tag{5.61}$$

where

α and R are the constants related to position and curvature
τ_y is the maximum shear strength of the soil ($c + \sigma \tan(\phi)$)
c and ϕ is the cohesion and internal friction of the soil
σ is the stress of the soil, and
γ_y $\tau_y/G_{max(s)}$, where, $G_{max(s)}$ is the secant modulus of the soil
$G_{max(s)}$ $\frac{\tau_c}{\gamma_c}$

Mathematically, the Masing criteria is expressed by Eq. (5.62) given below:

$$\begin{aligned} &\tau = f(\gamma), \text{for the backbone curve,} \\ &\frac{\tau - \tau_a}{2} = f\left(\frac{\gamma - \gamma_a}{2}\right), \text{for the unloading portion and,} \\ &\frac{\tau + \tau_a}{2} = f\left(\frac{\gamma + \gamma_a}{2}\right), \text{for the reloading portion.} \end{aligned} \tag{5.62}$$

The subscript 'a' stands for the stress/strain reversal for the unloading and reloading portion.

Rearranging Eq. (5.62) leads to the following expression:

$$\tau = \frac{\gamma}{\gamma_y}\frac{\tau_y}{\left[1 + \alpha\left(\frac{\tau}{\tau_y}\right)^{R-1}\right]} \tag{5.63}$$

$$\tau = \frac{\gamma G_{max(s)}}{\left[1 + \alpha\left(\frac{\tau}{\tau_y}\right)^{R-1}\right]} \tag{5.64}$$

$$\frac{\tau}{\gamma G_{max(s)}} = \frac{1}{\left[1 + \alpha\left(\frac{\tau}{\tau_y}\right)^{R-1}\right]} \tag{5.65}$$

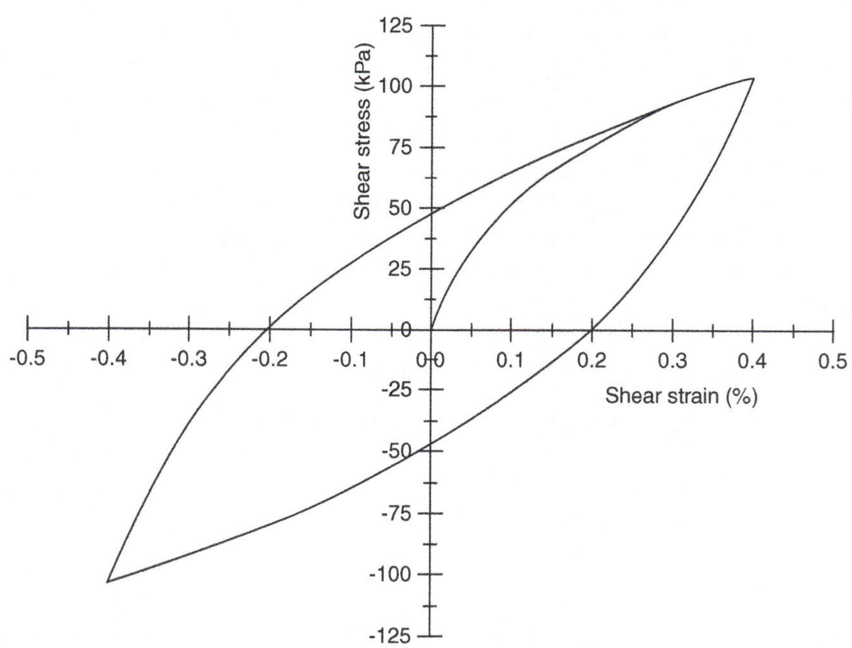

Fig. 5.20 Typical hysteretic behavior of soil

where $\frac{\tau}{\gamma}$ is equal to G (the current secant shear modulus) which substituted into the above equation gives

$$\frac{G}{G_{\max(s)}} = \frac{1}{\left[1 + \alpha\left(\frac{\tau}{\tau_y}\right)^{R-1}\right]} \quad (5.66)$$

It is known that the shakedown strength of the soil is dependent on the current strain level of the soil (strain controlled phenomenon). Hence, this equation has to be modified keeping this fact in mind. The present equation is modified in subsequent steps as given below:

$$\frac{G}{G_{\max(s)}} = \frac{1}{\left[1 + \alpha\left(\frac{\tau}{\gamma}\frac{\gamma}{\gamma_y}\frac{\gamma_y}{\tau_y}\right)^{R-1}\right]} \quad (5.67)$$

which is equivalent to

$$\frac{G}{G_{\max(s)}} = \frac{1}{\left[1 + \alpha\left(\frac{G}{G_{\max(s)}}\frac{\gamma}{\gamma_y}\right)^{R-1}\right]} \quad (5.68)$$

Equation (5.68) is a nonlinear equation which is strain dependent and must be solved in an iterative manner (method of bisection).

Absorption of energy (due to hysteresis behavior) occurs due to imperfect elastic properties of medium in which the particle of a medium does not react perfectly elastically with their neighbor and a part of the energy in the waves is lost instead of being transferred through medium, after each cycle. This anelastic absorption of energy can be quantified in terms of the area of the hysteresis loop, which is a measure of energy dissipation as shown in figure which can be described by the damping ratio (ξ) which is given by Eq. (5.69).

$$\xi = \frac{\Delta E}{E} = \frac{W_D}{4\pi W_S} = \frac{1}{2\pi}\frac{A_{\text{loop}}}{G\gamma_c^2} \quad (5.69)$$

where

W_D is the dissipated energy
W_S is the maximum strain energy, and
A_{loop} is the area of the loop

Hence, the damping ratio according to the Ramberg–Osgood model is given by Eq. 5.70.

$$\xi = \frac{2}{\pi}\frac{R-1}{R+1}\left(1 - \frac{G}{G_{\max(s)}}\right) \quad (5.70)$$

The user subroutine for calculating the shakedown strength of the soil ($G/G_{\max(s)}$) and damping ratio (ξ) with shear strain is given in Appendix 1. A typical shear modulus and damping degradation curve of soil (for $R = 2.5$ and $\gamma_y = 0.02$) are shown in Fig. 5.21.

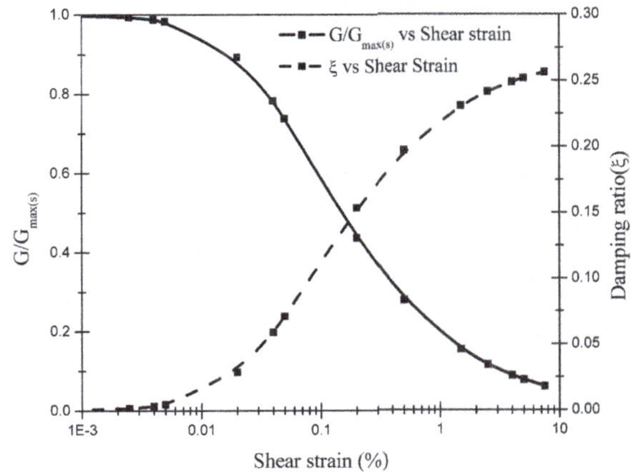

Fig. 5.21 Typical shear modulus and damping degradation curve of soil

5.10.2 Interaction Between the Structure and Subsoil During Dynamic Loading

There are two interaction effects that take place as a result of the presence of a structure at a soil site. These two effects are referred to as kinematic and inertial effects.

Kinematic Interaction: If a foundation (surface or embedded) impedes the free field deformation of the soil even if the foundation has no mass, then kinematic interaction is said to have taken place. This occurs due to the stiffness (or rigidity) of the foundation which influences the free field movement of the soil. The deformation due to kinematic interaction alone can be computed by if the foundation has only stiffness and is massless. This phenomenon is not always present in all situations. If a foundation rests on the surface and it is subjected to vertically propagating shear wave (or horizontal particle motion), the foundation would not impede the movement of the soil; hence, no kinematic interaction would exist.

Inertial Interaction: In this analysis, the structure and the foundation have masses which respond dynamically. If the flexibility and energy-dissipating capability of the supporting soil are accounted for, the foundation movements that would respond to a given ground motion in the case of the structure supported by the flexible medium would be substantially different from the system that would respond to the same ground motion when the same structure is considered fixed at its base. The deformation due to the inertial interaction is calculated by applying the inertial load only to the structure with the base of the soil considered stationary. The inertial loading is the combination of the base motion and the foundation input motion which reflects the influence of the kinematic interaction. In short, the kinematic interaction analysis produces the motion of the massless foundation and structure system. This motion is then combined with the base motion to represent the total kinematic motion of the foundation and structure system.

5.10.3 Fixed Base Analysis

A fixed base analysis may be assumed in modeling structures for seismic response analysis when the frequency obtained assuming a structure supported on soil springs representing the soil supporting medium is more than twice the dominant frequency obtained from fixed base analysis of the flexible structure representation.

5.10.4 Acceptable Methods for SSI

As per ASCE 4-98, there are two acceptable methods of SSI analysis: One is the direct method, and another one is impedance function approach.

5.10.4.1 Direct Method

In the direct method, the entire soil–structure system is modeled as one unit and analyzed in a single step under the ground motion prescribed under a free field. To encompass both the kinematic and inertial interaction, the direct method involves several steps. The first step involves the definition of ground motion at the bedrock level using the free field ground motion at a site. This is accomplished through the process of "inverse amplification" termed as deconvolution. Then, assuming that this bedrock motion is equal to the bedrock motion which is present directly below the structure, this motion is used as an input for the dynamic soil–structure interaction to evaluate the response of the structure and the soil domain. This method also involves the use of artificial (or absorbing) boundaries at the sides of the model to simulate a semi-infinite domain along the lateral direction as shown in Fig. 5.22.

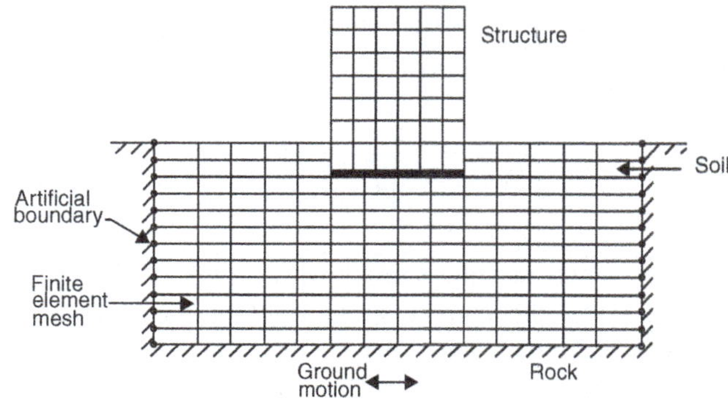

Fig. 5.22 Soil domain along with the semi-infinite boundaries

The solution for the equation of motion in direct method of analysis is done by two alternative methods. The first is a solution in the "frequency domain" and the second a solution in the "time domain." In the frequency domain method, the ground motion at bedrock is first transformed to the frequency domain by means of the Fourier transform. Then, the response of the soil–structure system in the frequency domain is obtained by multiplying the transfer function of the structure–soil system by the Fourier transform of the excitation. Thereafter, the response of the structure is obtained in the time domain by taking the inverse Fourier transform of such a product. The main advantage of a frequency domain solution is that frequency-dependent parameters can be easily incorporated into the analysis. Its main disadvantage is that it is limited to "linear systems" as it is based on the principle of superposition. For many problems, however, it is adequate to account for the nonlinear behavior of the system in an approximate way by performing an iterative analysis. In this iterative analysis, the soil's rigidity and damping are adjusted according to some characteristic measure of the strain obtained in a previous cycle as shown in figure. The designation "equivalent linear analysis" is used to describe this approach. The time domain analysis involves a step-by-step integration of the equations of motion. In this procedure, nonlinear stress train relationships can easily be incorporated into the solution.

5.10.4.2 Impedance Method

In impedance method, the vertical, horizontal, and rocking stiffness and damping of soil are represented as equivalent spring and dashpot. Description of spring constants in the elastic half-space model based on the calculation of impedance functions is described below. SSI analysis by the impedance function approach consists of the following steps:

(i) Determination of input motion to massless rigid foundation (kinematic interaction),
(ii) Determination of foundation impedance functions,
(iii) Analyze the soil structure system.

The controlled motion defined at the free field surface may be input to the massless rigid foundation.

Origin of the viscous dashpots and radiation damping in elastic "half-space"

As the soil is represented as "elastic half-space", which does not have any internal damping and therefore has no way to dissipate energy in the material itself. If the foundation had been on top of a closed elastic system with rigid boundaries surrounding the soil, the displacement of the foundation would have been infinite when vibrating at the natural frequency of the system. But because the system is open (half space) instead of closed, energy escapes in the form of waves propagating in the soil, with this energy never coming back, and this is why the displacement of the foundation is never infinite. This form of elastic energy dissipation in the form of waves traveling away from the foundation is called radiation (or geometric)damping, and it is the physical origin of the vertical and horizontal viscous dashpots, which are just approximate mathematical representations of the above phenomenon.

To complete the picture, it is important to add to the viscous dashpots calculated from Table 5.1, the contribution due to the energy dissipated internally in the soil, mainly due to friction (material damping), which is typically characterized by a soil damping ratio, ξ. The value of ξ depends on several factors, like the level of cyclic shear strain induced in the soil by the dynamic loading, the type of soil, and the plasticity index if the soil is a clay (Dobry 2014). The total dashpot including the effect of ξ is calculated from the expression given by Eq. 5.71

$$c_{\text{total}} = \left(\frac{k}{\pi f}\right)\beta + c_{\text{radiation}} \tag{5.71}$$

where k is the corresponding stiffness calculated at the same frequency for the elastic half space.

5.10.5 Analysis of Coupled Soil Structure Systems

The coupled soil–structure system shall include the structure or its modal representation, and the soil spring and dashpots anchored at the foundation level. The dynamic characteristics of the soil shall be defined by impedance functions computed from Tables 5.22, 5.23 and 5.24.

where

G	shear modulus of soil medium
v	Poisson's ratio of foundation medium
R	effective radius of contact area
η_ξ, η_z, and η_ψ	are horizontal, vertical, and rocking embedment coefficients, respectively
β_ξ, β_ζ, and β_ψ	are geometric coefficients
L	horizontal dimension of the footings, perpendicular to axis of rocking

Table 5.22 Equivalent spring constants for rigid circular and rectangular footings

Motion	Circular footings	Rectangular footings	
	Spring constant	Spring constant	Equivalent radius (R)
Vertical	$K_z = \frac{4GR}{1.-v}\eta_z$	$K_z = \frac{G\beta_z\sqrt{BL}}{1.-v}\eta_z$	$\sqrt{BL}/\pi$
Horizontal	$K_x = \frac{32.(1.-v)G_R}{7.-8.v}\eta_x$	$K_x = 2(1+v)G\beta_x\sqrt{BL}\eta_x$	$\sqrt{BL}/\pi$
Rocking	$K_\psi = \frac{8GR_o^3}{3.(1.-v)}\eta_\psi$	$K_\psi = \frac{G_s\beta_\psi BL^2\eta_\psi}{(1.-v_s)}$	$\sqrt{4}BL^3/3\pi$
Torsional	$K_\theta = \frac{16GR^3}{3.}$	$K_\theta = \frac{16GR_o^3}{3.}$	$\sqrt{4}BL(B^2+L^2)/6\pi$

Table 5.23 Embedment coefficients (η)

Mode of vibration	Embedment coefficient (η)
Vertical	$\eta_z = 1.0 + 0.99(H/R)$
Horizontal	$\eta_x = 1.0 + 3.36(H/R)$
Rocking	$\eta_\psi = 1.0 + 4.20(H/R) + 8.20(H/R)^3$
Torsional	$\eta_\theta = 1.0 + 0.77(H/R)$

Table 5.24 Equivalent damping constants

Motion	Circular base	Rectangular base
Vertical	$c_z = 0.85k_zR\sqrt{\rho/G}$	**Same as for circular base**
Horizontal	$c_x = 0.576k_xR\sqrt{\rho/G}$	
Rocking	$c_\psi = \frac{0.30}{1.+B_\psi}K_\psi R\sqrt{\rho/G}$	
Torsional	$c_\theta = \frac{\sqrt{K_\theta I_r}}{1.+2I_r/\rho R^5}$	

Table 5.25 Limiting values of damping ratio

Mode	Limiting value of total damping ratio
1. Horizontal translational springs	0.20
2. Vertical translational springs	0.30

B the remaining horizontal dimension
H embedment depth
ψ $3(1. - \nu_\sigma)I_o/8\rho^5$
ν_σ Poisson's ratio of soil
γ_σ unit weight of the soil
ρ mass density of the soil
R effective radius of contact area
I_o total mass moment of structure and base mat about the axis at the base, and
I_r polar mass moment of inertia of structure and base mat

The total damping ratio of the foundation medium will be computed adding the geometric damping ratio with the material damping ratio. Maximum composite modal Damping ratio computed from the equivalent strain Energy approach shall be restricted to 0.20 as per Cl.3.1.5.4 of ASCE 4-98. Hence, the total of material damping and the geometric damping values of the soil spring elements shall not exceed the threshold values given in Table 5.25.

Example 5.5 An elevated RCC cylindrical tank (Fig. 5.24) of 15 m internal diameter, 200 mm wall thickness, and 5 m height is supported on a cylindrical (hollow) shaft of 4 m internal diameter, 200 mm wall thickness, and 7 m high above foundation level. The structure is rested on stiff soil type founding strata with shear wave velocity of 200 m/s. The tank has a free board of 470 mm and the effective storage capacity of 800,000 L (800 m^3) of domestic/process water. The SDF representation of the tank is shown in Fig. 5.23 (Grade of c oncrete: M20 ($E = 2.24 \times 10^{10}$ N/m^2, unit weight = 25 kN/m^3)).

Fixed base analysis

Fundamental Natural Frequency (angular):

$$\omega = \sqrt{\frac{K}{M}} = 37.74\,\text{rad/s}$$

Hence, fundamental natural frequency (linear) $f = \frac{\omega}{2\pi} =$ 6.01 Hz and

fundamental time period = $T = 1/f = 0.166$ s

Design of raft foundation for the tank

The site is classified as "site class D" as per NEHRP classifications.

The value of $(N_1)_{60}$ is 20. The correlation of SPT N-value and foundation size of 4 m with its bearing capacity is obtained 170 kPa by Terzaghi and Peck (1996).

Considering the width of footing as 4 m, the allowable bearing pressure is 150 kPa, and

The footing radius required is calculated as: $\sqrt{\frac{(8005)}{(3.14X170)}} = 3.87\,\text{m}$. Hence, the footing radius of 4 m is provided.

Analysis considering SSI

The static soil stiffness in the horizontal direction is given by the formulation (Table 3.3-1, ASCE 4-98):

$$K_s = \frac{32(1-\vartheta)GR}{7-8\vartheta},$$

Solution

Weight of stored water in the tank = [Volume of tank] x [Unit weight of water]
= [{$\pi\,(7.5)^2$ x 4.53} m^3] x [10 kN/m^3] = 8005 kN

Moment of inertia,

$$I = \frac{\pi}{64}\left(4.4^4 - 4.0^4\right) = 5.83\,\text{m}^4 = 5.83 \times 10^{12}\,\text{mm}^4$$

Stiffness

$$K = 3\text{EI}/\text{L}^3 = \left(3 \times 2.24 \times 10^{10} \times 5.83\right)/7^3\,\text{N/m} = 7.65 \times 10^8\,\text{N/m}$$

where

R is the radius of the foundation (=4.0 m)
G is the shear modulus of the soil (=72,000 kPa), and
ν is the Poisson's ratio (=0.5)

The value of K_s is calculated for shear wave velocity of 200 m/s of soil:

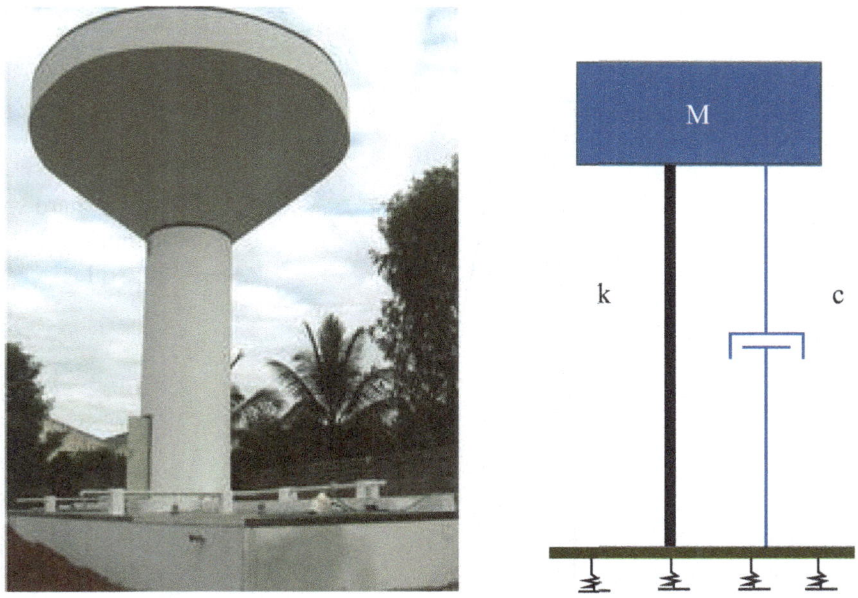

Fig. 5.23 Overhead tank and its equivalent SDOF representation

$$\text{For}\, V_s = 200\,\text{m/s},\ G = 72{,}000{,}000\,\text{N/m}^2,$$
$$K_s = (32(1-0.5)(4.0)(72{,}000{,}000))/(7-8(0.5))$$
$$= 15.36 \times 10^5\,\text{kN/m},$$

The equivalent stiffness of soil and structure is

$$K_{eq} = \left(\frac{1}{K_1} + \frac{1}{K_s}\right)^{-1}$$

and the value of K_{eq} is found to be 5.11 × 10^8 N/m.

Fundamental natural frequency (angular) with consideration of SSI is:

$$\omega = \sqrt{\frac{K}{M}} = 28.63\,\text{rad/s}$$

Hence, fundamental natural frequency (linear) $f = \frac{\omega}{2\pi} = 4.56\,\text{Hz}$ and

fundamental time period = $T = 1/f = 0.22$ s.

Fixed base frequency (Hz)	6.01 Hz	0.166 s
Frequency considering SSI (Hz)	4.56 Hz	0.220 s

Considering seismic input, response spectra is selected from the Fig. 5.24.

It is observed that the base shear does not change because of SSI as the spectral acceleration remains constant within the time period of 0.1–0.5 s (consider hard soil spectra), but there is an increase in the lateral displacement of the whole structure as the structure becomes more flexible. It is observed that the maximum lateral displacement at the top (i.e., for RCC cylindrical tank) with fixed base analysis is 6.67 mm and considering SSI, and the displacement increases to 16.2 mm.

Example 5.6 Seismic Analysis of G + 5 storey Structure is performed considering soil structure Interaction. In this example, earthquake analysis results considering soil–structure interaction of a G + 5 story building whose dimension approximately 29 m (X-direction) × 31 m (Z-direction) in plan and 23 m in height are discussed. Plan of the structure is shown in Fig. 5.25. Building is supported on 27 numbers of footings of dimension about 3 m × 3 m.

Solution

Seismic Categorization

The assigned design acceleration for firm bearing strata is considered as 0.2 g. Soil type is arrived 500 m/s as per average of shear wave velocity over 25.0 m below founding level. Five percentage damping is considered for evaluating the response spectrum.

Finite element modeling

A detailed finite element model has been developed by using FE packages. The 3D space frame structure of G + 5 story building has been discretized into finite element model using beam shell mass and boundary element as mentioned below.

(a) 3D general shell element for footing, basement walls floor slabs and roof slabs,
(b) 3D beam element for column and beam,
(c) 3D general translational spring element for modeling foundation stiffness (dynamic soil spring).

The finite element model of the structure is shown in Fig. 5.26. The number of beam elements and number of nodes of the model are 2369 and 3594, respectively. The damping of the structure is considered as 5%. The effects of modal damping are ignored in the present analysis.

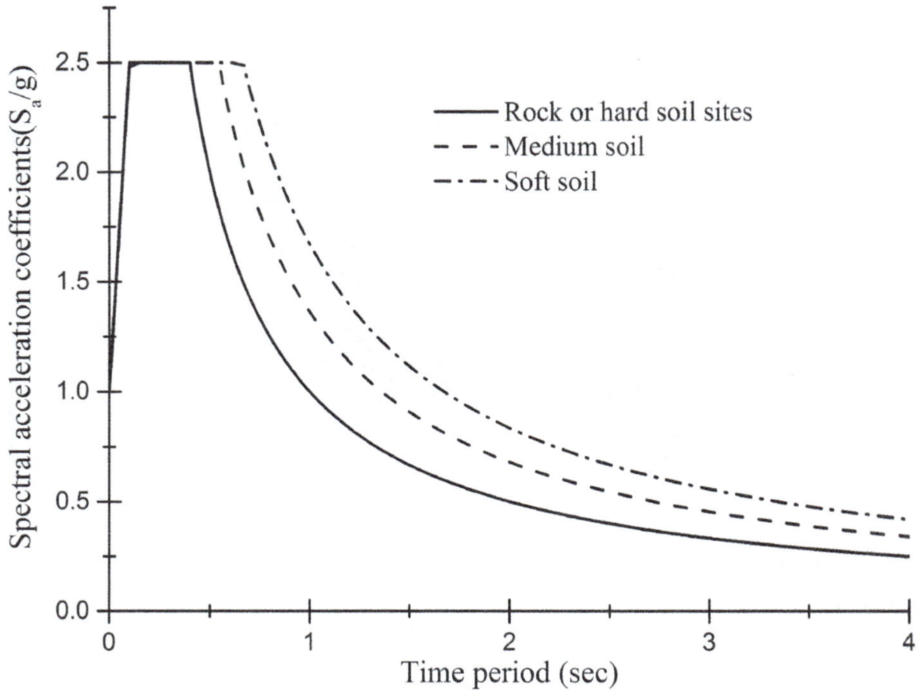

Fig. 5.24 Response spectrum as per 3

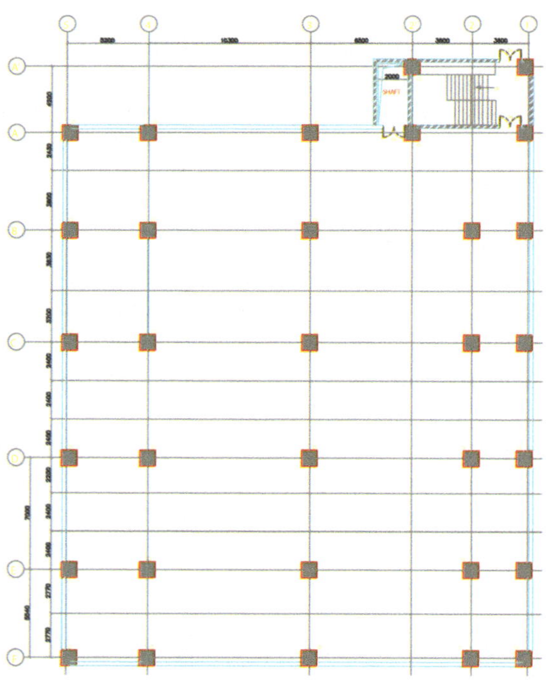

Fig. 5.25 Plan of the building

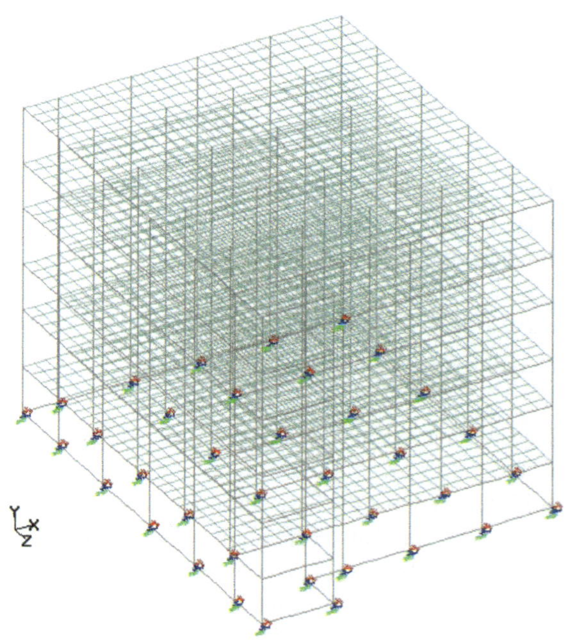

Fig. 5.26 FE model of the G + 5 story building

Material Properties

The structure is R.C.C. structure and has grade of concrete, M35. The Young's modulus is considered based on the expression $E_{concrete} = 5000\sqrt{f_{ck}}$, where f_{ck} is the characteristic compressive strength of concrete. Poisson's ratio is considered as 0.15.

Geotechnical Parameters

The borehole detail for the above structure is shown in Fig. 5.27. The soil is found to have a 5.6 m thick soft to stiff yellowish brown silt with sand and gravel, followed by a very dense residual sandy soil with weathered rock of 3.2 m thickness. The strata below this is found to consist of very weak greenish brown to gray, completely weathered rock, and is around 17.30 m thickness. The weighted average of shear wave velocity up to 25.0 m below founding level is 3300 m/s. The shear wave velocity at founding level is 500 m/s. for the analysis is $\gamma = 18.0$ kN/m^3. The groundwater table for design purposes is considered at the ground level, though the actual groundwater table is 3.0 m below ground level.

Modeling Base Soil Springs

Soil springs have been considered at the footing base in three translational directions using the following expressions based on ASCE: 4-98 as given in Eqs. (5.78) and (5.79).

$$K_z = \frac{G_s}{(1 - \nu_s)}\beta_z\sqrt{BL}\eta_z \quad (5.72)$$

$$K_x = 2(1 + \nu_s)G_s\beta_x\sqrt{BL}\eta_x 8 \quad (5.73)$$

where

K_z	spring constant in the vertical direction
K_x	spring constant in the horizontal direction
G_s	shear modulus of the soil medium
ν_s	Poisson's ratio of the soil medium
β_x, β_z	are geometric coefficients and are functions of *L/B* obtained as per ASCE: 4-98
B	width of base mat perpendicular to the direction of horizontal excitation
L	length of base mat in the direction of horizontal excitation

η_x, η_z are embedment coefficients defined as:

$$\eta_x = 1.0 + 3.36(H/R)$$
$$\eta_z = 1.0 + 0.99(H/R)$$

where

H	embedment depth and
R	radius of circular base mat

The above expressions are for rectangular bases and are applicable to rigid foundations. In the present analysis, spring constants are evaluated at each node based on contributory area. The spring stiffness is evaluated based on shear modulus at founding stratum. The base soil springs are evaluated using the above expressions and are given below:

Horizontal spring stiffness = $K_{FX} = K_{FZ}$ = 37,421,714.3 kN/m,
Vertical spring stiffness = K_{Fy} = 41,122,762.95 kN/m,
Soil–structure interaction also performed for other two shear wave velocities of V_s = 250 m/s and V_s = 1000 m/s,
For shear wave velocity of 250 m/s:
Horizontal spring stiffness = $K_{FX} = K_{FZ}$ = 9,355,428.3 kN/m,
Vertical spring stiffness = K_{Fy} = 10,280,690.74 kN/m.
For shear wave velocity of 1000 m/s:
Horizontal spring stiffness = $K_{FX} = K_{FZ}$ = 1.49E + 11 kN/m
Vertical spring Stiffness = K_{Fy} = 1.64E + 11 kN/m.

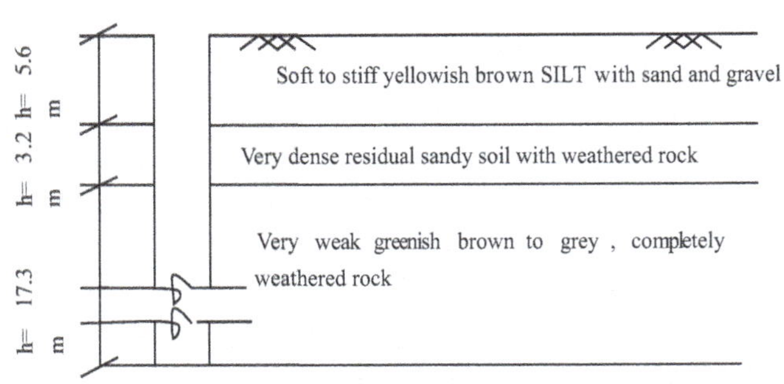

Fig. 5.27 Borelog data of the site where the structure is built

Dynamic Analysis of G + 5 Story building

The modal analysis for the above structure is carried out with the above FE model. The number of modes considered is such as to ensure 90% mass participation in each direction. In this case study, the same is achieved by considering thirty modes. The frequencies for first few modes have been presented in Figs. 5.28, 5.29 and 5.30, respectively. The significant frequencies where mass participation is more than 5% of mass of structure have been tabulated in Table 5.26. The modal participation factors for significant modes are given in Table 5.27.

Uncertainties in Soil–Structure Interaction

For realistic prediction of dynamic analysis results uncertainties in geotechnical parameters shall be accounted for, while performing the soil structure interaction. This is usually achieved by varying low-strain shear modulus. Thus, low-strain shear modulus shall be considered as the best estimate value times (1 + C_v) and the best estimate value divided by (1 + C_v). Hence, it is taken as 0.5, 1, and 2G. Change in frequency with soil–structure interaction is presented in Table 5.28 as well as shown in Fig. 5.31.

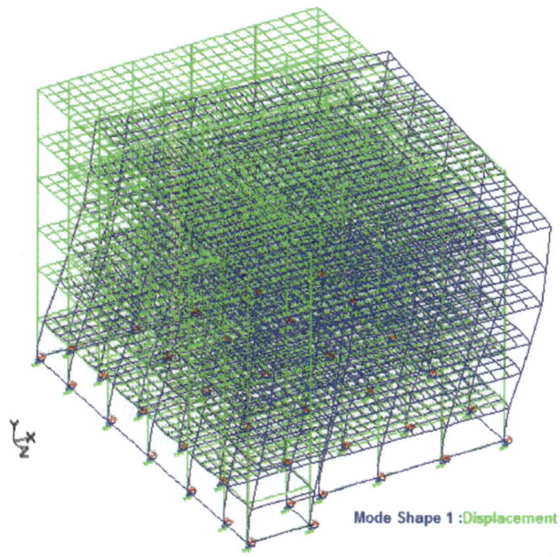

Fig. 5.28 First mode of the building (frequency = 1.29 Hz)

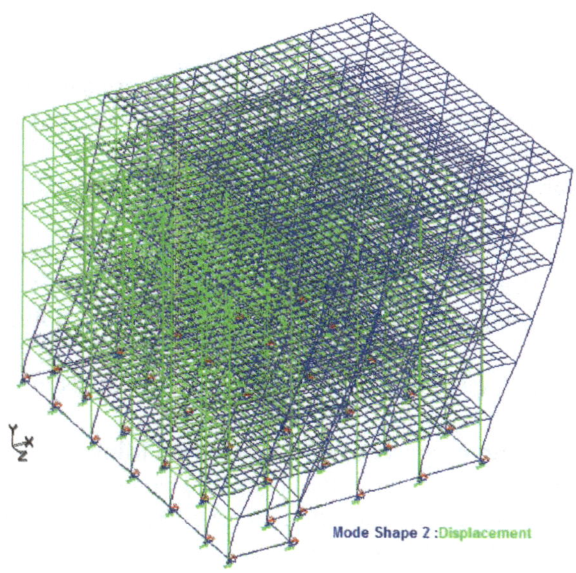

Fig. 5.29 Second mode of the building (frequency = 1.39 Hz)

For highly flexible structures, increase in shear modulus results in increase in frequencies. Consideration of uncertainties in soil–structure interaction is a must.

Discussion: The effect of soil–structure interaction has been considered in the case study of G + 5 story building in dynamic condition. The soil is replaced as equivalent translational springs at founding level, as per ASCE 4-98. The frequencies and base shears along each direction are evaluated using the soil–structure interaction. Uncertainties in soil–structure interaction have been accounted by varying shear modulus. The effect of shear modulus on frequencies in first three modes has been examined. It can be seen that by increasing shear modulus, the frequencies are increased.

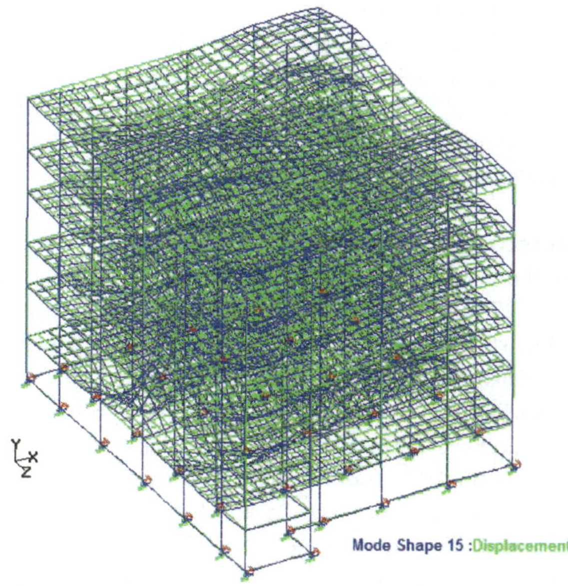

Fig. 5.30 Vertical mode of the building (frequency = 8.08 Hz)

5.11 Liquefaction of Soils

Liquefaction is one of the most interesting and controversial phenomena in geotechnical engineering. Its devastating effects came into existence when many of the structures, slopes, and foundation failed in the past. The cases of the foundation failures due to liquefaction are reported during the 1964 earthquake in Niigata, Japan, as shown in Fig. 5.32. Sand boils are generally observed during liquefaction, and a large number of such boils were observed during Dubri earthquake in Assam in 1930 and Bihar earthquake in 1934. In the last 30 years, significant studies have been performed by many researchers and different terminologies and methods of measurement of liquefaction potential of a site were proposed, but the road has not been smooth. This section will present conceptual understanding of liquefaction-related soil behavior and describe various methods by which liquefaction potential and hazards of a site can be evaluated.

5.11.1 Concepts of Liquefaction

The term liquefaction coined by Mogami and Kubo [4] has been extensively used in conjunction with soil deformation caused by monotonic, transient, or repeated disturbance of saturated loose cohesionless soil under undrained conditions. The primary cause of liquefaction is the tendency for densification of saturated cohesionless soil under undrained condition which in turn results in the generation of excess pore pressure and decrease in the effective stress. Liquefaction

Table 5.26 Significant frequencies obtained from modal analysis

Frequency	Frequency Shear wave velocity ratio = 1, V_s = 500 m/s	Frequency Shear wave velocity ratio = 0.5, V_s = 250 m/s	Frequency Shear wave velocity ratio = 2, V_s = 1000 m/s
Number	(Cycles/Sec)		
1	1.29	1.1311	1.3642
2	1.39	1.245	1.4628
4	4.6287	4.2578	4.6789
5	4.8985	4.3852	4.948

Table 5.27 Modal participation for significant modes

Mode	Participation factors (%)		
No.	X-direction	Y-direction	Z-direction
1	0	0	75
2	76.22	0	0.01
4	0	0	9.23
5	9.60	0	0
7	0	55.26	0
12	0	8.42	0

Table 5.28 Change in frequency with soil–structure interaction

Shear modulus ratio	Frequency (Hz)	% change from fixed base frequency
	Mode 1	
Fixed base	1.4257	–
2	1.36	4.6%
1	1.29	9.51%
0.5	1.13	20%

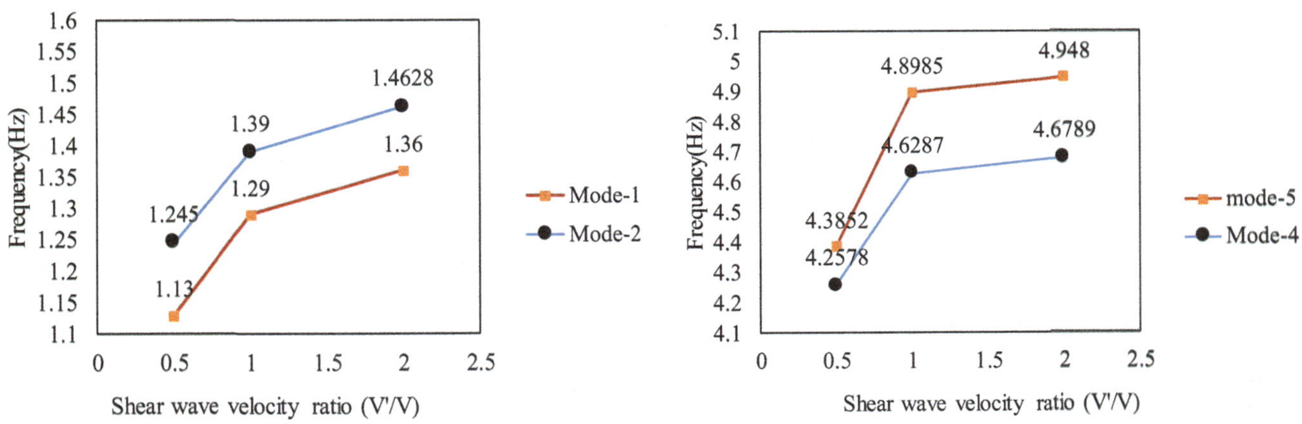

Fig. 5.31 Effect on the frequencies considering variation in shear modulus of the soil (V = 500 m/s)

phenomena can be divided into two parts: *flow liquefaction* and *cyclic mobility*. In the field, flow liquefaction occurs much less frequently than cyclic mobility, but its effects are much severe. But for assessing the liquefaction hazards and susceptibility, both the above phenomena should be carefully considered as they bear equal importance.

5.11.1.1 Flow Liquefaction

Flow liquefaction can occur when shear stress at static equilibrium of soil mass is greater than the shear strength of soil at the liquefied state (residual strength). This phenomenon produces the most dramatic effects and results in tremendous instabilities known as flow failures. It is

Fig. 5.32 Damage caused due to Niigata earthquake in Japan [13]

undrained flow of saturated, contractive soil when the static shear stress exceeds the residual strength of the soil. Failure may be triggered by cyclic or monotonic shear loading. The cyclic stresses may simply bring the soil to an unstable state at which strength drops sufficiently to allow the static stresses to produce the flow failure. The flow failures of Lower Fernando Dam caused in Feb 9, 1971 is an example of flow liquefaction in which the fluid nature of the liquefied soil was observed.

5.11.1.2 Cyclic Mobility

This phenomenon occurs when the static shear stress is greater than the shear strength of the liquefied state (residual strength). In comparison to flow liquefaction, cyclic mobility can occur in both cyclic and static shear stress application. The deformations produced by cyclic mobility failures develop incrementally during ground shaking. The mechanics of mobility may also be illustrated in Fig. 5.33. Consider behavior shown in the figure, when a fully saturated dilative sand starting, for example, point D is loaded monotonically (static) in the undrained condition. In that case, the point on the state diagram may move slightly to the left of point D. But then, it will move horizontally toward the steady-state line as load is applied. If one starts a new stress on point D but this time applies the cyclic loading, one can follow the behavior by plotting average void ratio and effective stress.

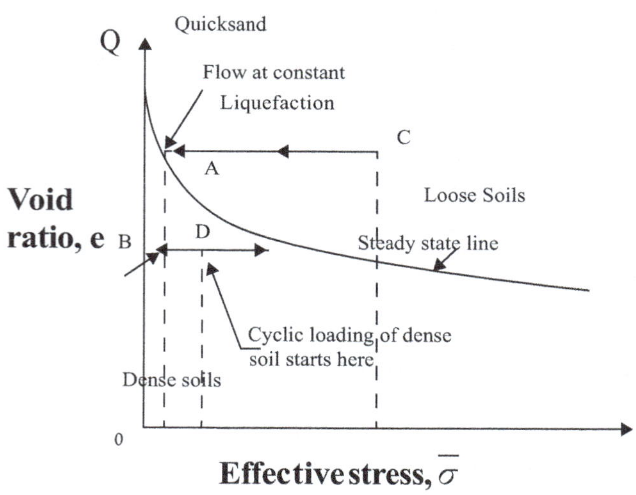

Fig. 5.33 Relationship between void ratio and effective stress in soil

5.11.2 Mechanism of Liquefaction

Liquefaction denotes a condition where a soil undergoes continued deformation at constant low residual stressor with no residual resistance due to the built up and maintenance of high pore water pressure which reduces the effective confining pressure to very low value; high pressure built up leading to liquefaction of this style may be due to static or cyclic stress application. Residual strength is the strength of the soil in its liquefied state. This can be observed in very loose state of sand in which the sample is subjected to undrained shearing in stress controlled environment. The sample exhibits a peak undrained strength at a very low level of strain, after which it collapses and undergoes steady state of deformation with a constant value of stress, and the effective confining pressure is a small fraction of the initial confining pressure.

In technical terms, liquefaction occurs when the pore water pressure (u) equals the total overburden stress (σ_v), i.e., [(σ_v − u)-u_{excess}) = (σ'_v-u_{excess}) = 0]. This implies that as the excess pore water pressure reaches the value of vertical effective stress at a particular depth, the soil loses all its shear strength.

5.11.3 Theory of Liquefaction

The strength of sand is primarily due to internal friction. In saturated state, it may be expressed as follows as shown in Fig. 5.34,

$$S = \bar{\sigma}_n \tan \emptyset \tag{5.74}$$

where

S shear strength of sand
$\bar{\sigma}_n$ effective normal pressure on any plane xx at depth $Z = \gamma h_w + \gamma_{sub}(Z - h_w)$
$\emptyset$ angle of internal friction

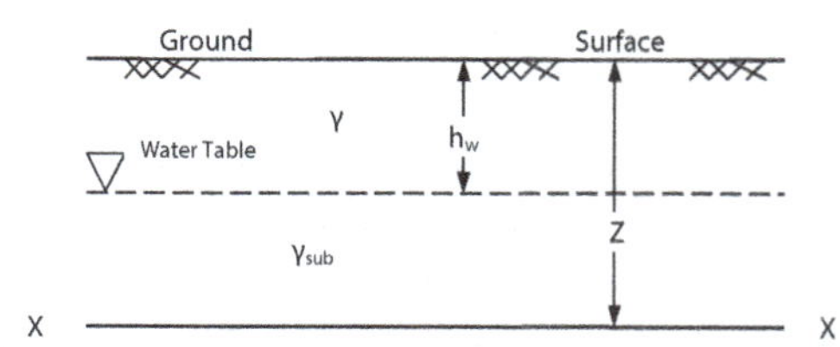

Fig. 5.34 Schematic diagram of sand with location of water table

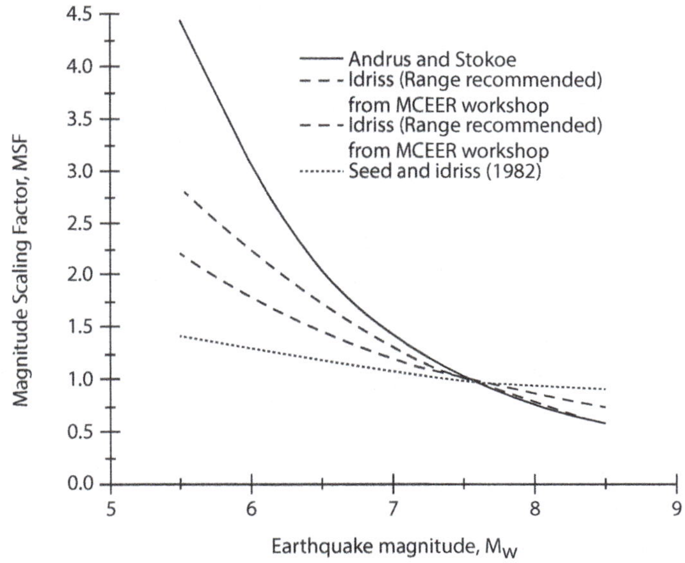

Fig. 5.35 Magnitude correction factor

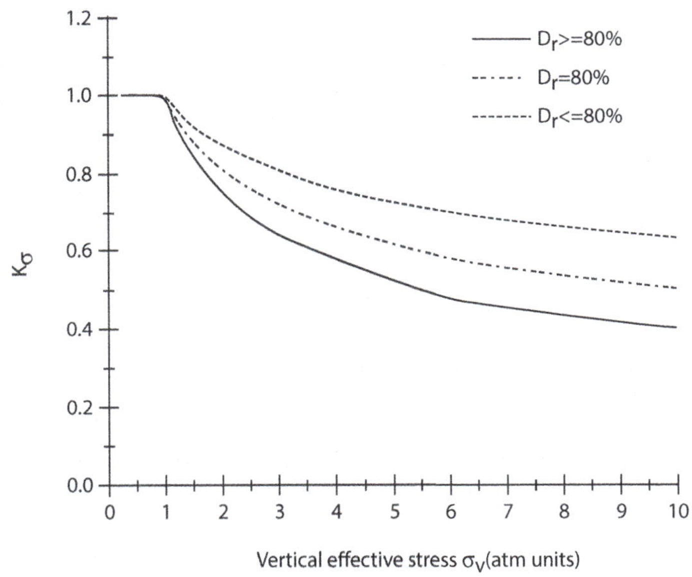

Fig. 5.36 Stress correction factor

γ unit weight of soil above water table

γ_{sub} submerged unit weight of soil

If a saturated sand is subjected to ground vibrations, it tends to compact and decrease in volume, if drainage is restrained the tendency to decrease in volume results in an increase in pore pressure. The strength may now be expressed as:

$$S_{dyn} = (\sigma - u_{dyn}) \tan \emptyset_{dyn} \quad (5.75)$$

where

S_{dyn} shear strength of soil

u_{dyn} excess pore water pressure due to ground vibrations

$\emptyset_{dyn}$ angle of internal friction under dynamic conditions

It is seen that with development of additional positive pore pressure, the strength of sand is reduced. In sands, φ_{dyn} is almost equal to φ, i.e., angle of internal friction in static conditions. For complete loss of shear strength, i.e., S_{dyn} is zero.

Thus,

$$(\sigma - u_{dyn}) = 0 \quad \text{or} \quad u_{dyn}/\sigma = 1 \quad (5.76)$$

Expressing u_{dyn} in terms of rise in water head, h_w and γ_{sub} as $\frac{G-1}{1+e}\gamma_w$,

the above equation can be written as:

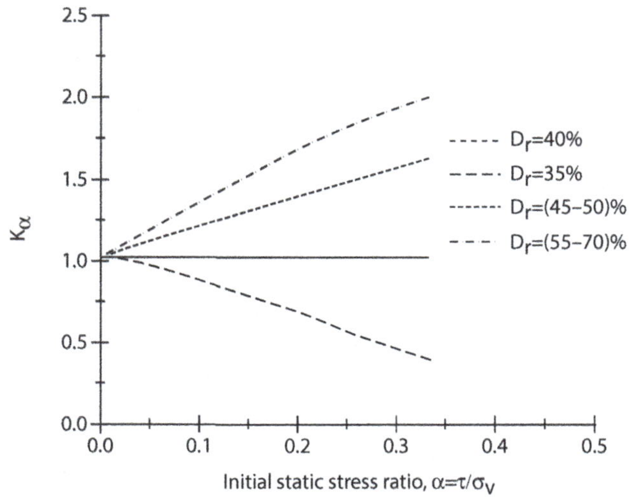

Fig. 5.37 Correction for initial static shear

$$\frac{\gamma_w h_w}{\frac{G-1}{1+e}\gamma_w Z} = 1 \tag{5.77}$$

or

$$\frac{h_w}{Z} = \frac{1+e}{G-1} = 1/i_c \tag{5.78}$$

where G specific gravity of soil particles
e void ratio
$1/i_c$ critical hydraulic gradient

In case of complete liquefaction, the effective stress is lost and the sand–water mixture behaves as a viscous material and surface settlement starts, resulting in closer packing of sand grains. Thus, the structures resting on such a material start sinking.

5.11.4 Procedure for Evaluation of Liquefaction Potential at a Site

Step 1: The subsurface data used to assess Liquefaction susceptibility should include the location of the water table, either SPT blow count (N) (or tip resistance of a standard CPT cone (q_c) or the shear wave velocity), mean grain size (D_{50}), unit weight, and fine content of the soil (percent by weight passing the IS Standard Sieve No. 75 μ).

Step 2: Evaluate the total vertical stress (σ_v) and effective vertical stress ($\sigma_{v'}$) for all potentially liquefiable layers within the deposit.

Step 3: The following equation can be used to evaluate the stress reduction factor r_d:

$$r_d = 1 - 0.00765\,z \text{ for } z <= 9.15\,\text{m and}$$
$$r_d = 1 - 0.0267\,z \text{ for } 9.15 < z <= 23\,\text{m} \tag{5.79}$$

where z is the depth below the ground surface in meters.

Step 4: Calculate the critical stress ratio induced by the design earthquake, CSR_{eq} as

$$CSR_{eq} = 0.65\frac{a_{max}}{g} r_d \frac{\sigma_v}{\sigma'_v} \tag{5.80}$$

where σ_v and σ'_v are the total and effective vertical stresses, respectively, at depth z, a_{max} is the peak ground acceleration, and g is the acceleration due to gravity.

Step 5: Correct the CSR_{eq} for earthquake magnitude (M_w) stress level and for initial static shear using correction factors: k_m, k_σ k_α, respectively, according to:

$$CSR_L = CSR_{eq} k_m k_\sigma k_\alpha \tag{5.81}$$

The correction factors are estimated using Figs. 5.35, 5.36 and 5.37, respectively.

Steps 6a, 6b, and 6c will be used for assessing Liquefaction susceptibility using the SPT, CPT, and the shear wave velocity, respectively.

Step 6a: Evaluate the standardized SPT blow count (N_{60}) which is the standard penetration test blow count for a hammer with an efficiency of 60%. Specifications of the "standardized" equipment corresponding to an efficiency of 60% are given in Table 5.29. If non-standard equipment is used, N_{60} is obtained from the equation:

$$N_{60} = N \cdot C_{60} \tag{5.82}$$

Table 5.29 Recommended "standardized" SPT equipment

Element	Standard specification
Sampler	Standard split-spoon sampler with: (a) outside diameter, O.D. = 51 mm, and inside diameter, I.D. = 35 mm (constant, i.e., no room for liners in the barrel)
Drill rods	A or AW-type for depths less than 15.2 m; N- or NW type for greater depths
Hammer	Standard (safety) hammer with (a) weight = 63.5 kg; (b) drop = 762 mm (delivers 60 of theoretical free-fall energy)
Rope	Two wraps of rope around the pulley
Borehole	100- to 130-mm-diameter rotary borehole with bentonite mud for borehole stability (hollow stem augers where SPT is taken through the stem)
Drill bit	Upward deflection of drilling mud (tricone or baffled drag bit)
Blow count rate	30–40 blows per minute
Penetration-resistant count	Measured over a range of 150–460 mm of penetration into ground

Table 5.30 Correction factors for non-standard SPT procedures and equipment

Correction for	Correction factor
Non-standard hammer type (DH = donut hammer; ER = energy ratio)	C_{HT} = 0.75 for DH with rope and pulley C_{HT} = 1.33 for DH with trip/auto and ER = 80
Non-standard hammer weight or height of fall (H = height of fall in mm; W = hammer weight in kg)	$C_{HW} = \frac{H.W}{63.5x762}$
Non-standard sampler setup (standard samples with room liners, but used without liners)	C_{SS} = 1.10 for loose sand C_{SS} = 1.20 for dense sand
Non-standard sampler setup (standard samples with room liners, but liners are used)	C_{SS} = 0.90 for loose sand C_{SS} = 0.80 for dense sand
Short Rod Length	C_{RL} = 0.75 for rod length 0–3 m
Non-standard borehole diameter	C_{BD} = 1.05 for 150 mm borehole diameter C_{BD} = 1.15 for 200 mm borehole diameter

Notes N = uncorrected SPT blow count
$C_{60} = C_{HT}\, C_{HW}\, C_{SS}\, C_{RL}\, C_{BD}$
$N_{60} = N\, C_{60}$
C_N = correction factor for overburden pressure
$(N_1)_{60} = C_N\, N_{60} = C_N\, C_{60}\, N$

where C_{60} is the product of various correction factors. Correction factors recommended by various investigators for some common non-standard SPT configurations are provided in Table 5.30. For SPT conducted as per IS: 2131-1981, the energy delivered to the drill rod is 60%, and hence, $C_{60} = 1$ is assumed.

Calculate the normalized standardized SPT blow count $(N_1)_{60}$. $(N_1)_{60}$ is the standardized blow count normalized to an effective overburden pressure of 96 kPa in order to eliminate the influence of confining pressure. This is obtained by the following equation:

$$(N_1)_{60} = C_N N_{60} \tag{5.83}$$

Stress normalization factor C_N is calculated from following expression:

$$C_N = \left(\frac{P_a}{\sigma'_v}\right)^{0.5} \tag{5.84}$$

subjected to $C_N <= 2$, where Pa is the atmospheric pressure. The critical resistance ratio (CRR) or the resistance of a soil layer against liquefaction is estimated from Fig. 5.38 depending on the $(N_1)_{60}$ value representative of the deposit.

Step 6b: Calculate normalized cone tip resistance $(q_{c1N})_{cs}$ using

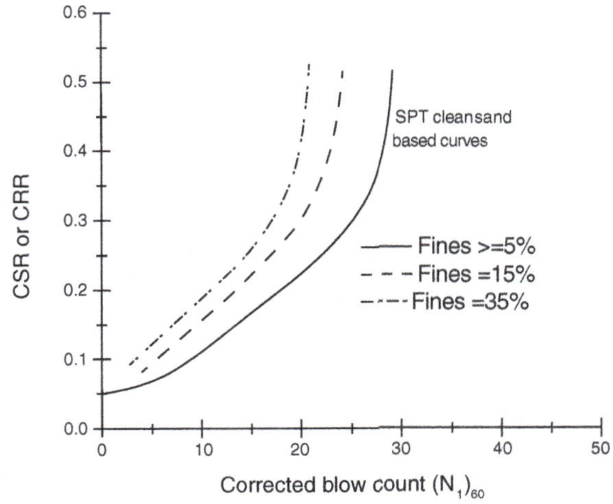

Fig. 5.38 Relationship between CRR and $(N_1)_{60}$ for sand for M_w, 7.5 earthquakes

$$(q_{c1N})_{cs} = K_c\left(P_a/\sigma'_v\right)^n (q_c/P_a) \tag{5.85}$$

where q_c is the measured cone tip resistance corrected for thin layers, exponent n has a value of 0.5 for sand and 1 for clay, and K_c is the correction factor for grain characteristics estimated as follows:

$$K_c = 1.0 \text{ for } I_c <= 1.64 \text{ and}$$
$$K_c = -0.403\, I_c^4 + 5.581\, I_c^3 - 21.63 f^2 + 33.75\, I_c - 17.88 \text{ for } I_c > 1.64 \tag{5.86}$$

The soil behavior type index I_c is given by

$$I_c = \sqrt{(3.47 - \log Q)^2 + (1.22 + \log F)^2} \tag{5.87}$$

where

$$Q = [(q_c - \sigma_v)/P_a]\left(P_a/\sigma'_v\right)^n, \quad F = f/(q_c - \sigma_v) \times 100, \tag{5.88}$$

f is the measured sleeve friction and n has the same values as described earlier.

Although soils with $I_c > 2.6$ are deemed non-liquefiable, such deposits may soften and deform during earthquakes. General guidance is not available to deal with such possibilities. Softening and deformability of deposits with $I_c >$ 2.6 should thus be treated on a material specific basis. The relationship between CRR and $(q_{n1c})_{cs}$ is shown in Fig. 5.39.

Step 6c: Calculate normalized shear wave velocity, V_{s1}, for clean sands using:

$$V_{s1} = V_s \times \left(P_a/\sigma'_v\right)^{0.25} \tag{5.89}$$

Assess liquefaction susceptibility of clean sands. Figure 5.40 shows the relationship between CRR and V_{s1}.

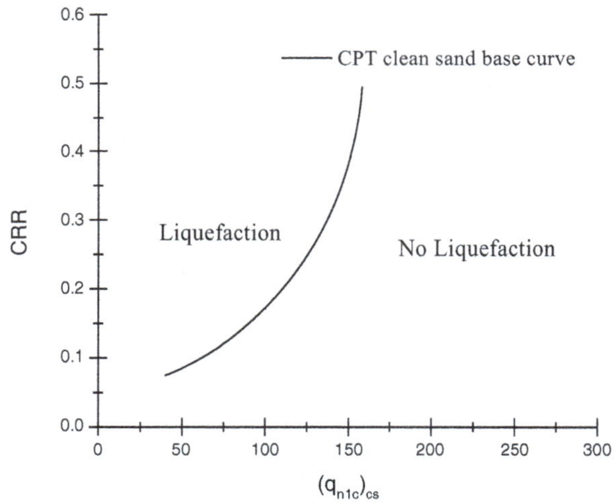

Fig. 5.39 Relationship between CRR and $(q_{n1c})_{cs}$ for M_w 7.5 earthquakes

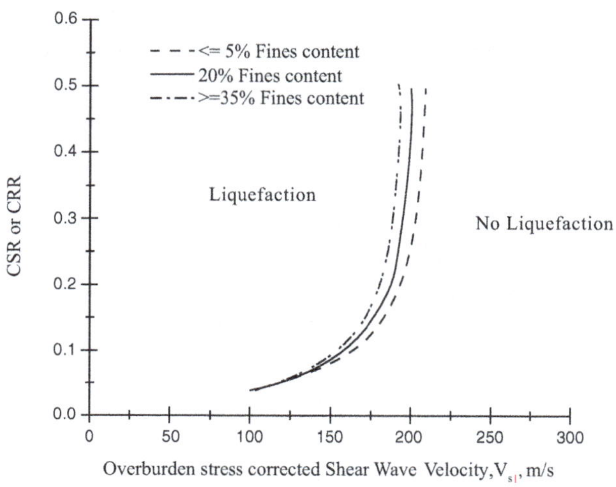

Fig. 5.40 Relationship between CRR and V_{s1} for M_w, 7.5 earthquakes

Step 7: Calculate the factor of safety against initial liquefaction FS_{liq} as

$$FS_{liq} = CRR/CSR_L \tag{5.90}$$

where CSR_L is as estimated in step 5 and CRR is from step 6a, 6b, or 6c.

Example 5.7 The measured SPT values in ash pond, results of sieve analysis for a site in Zone III are indicated in Table. 5.31. The water table is at 6 m below ground level. Estimate the liquefaction potential for 6.25 magnitude earthquake.

Solution

Site Characterization:

The SPT values range from 7 to 32. The site is located in Zone III. The peak horizontal ground acceleration value for the site will be taken corresponding to Zone III.

Liquefaction Potential of Underlying Soil

Step-by-step calculation for the depth of 6 m is given below. Detailed calculations for all the depths are given in Table 5.2.

$a_{max}/g = 0.211,\ M_w = 6.25,$
$\gamma_{sat} = 14\,\text{kN/m}^3,\ \gamma_w = 9.8\,\text{kN/m}^3$
Depth of water level below G.L. = 6.0 m,
Depth at which liquefaction potential is to be evaluated = 6.0 m.

Initial stresses:

$$\sigma_v = 6 \times 14 = 84\,\text{kPa}$$
$$u_o = 0$$
$$\sigma'_v = (\sigma_v - u_o) = 84\,\text{kPa}$$

Stress reduction factor:

$$r_d = 1 - 0.0765\,z = 1 - 0.00765 \times 6 = 0.954$$

Critical stress ratio induced by earthquake:

$$a\max = 0.211\,g,\ M\,w = 6.25$$
$$\text{CSR eq} = 0.65(a\,maz/g)r_d(\sigma_v/\sigma'_v)$$
$$= 0.65 \times (0.211) \times 0.954 \times (84/84)$$
$$= 0.13$$

Table 5.31 Result of the standard penetration test and sieve analysis

Depth (m)	N_{60}	Soil classification	Effective overburden pressure σ'_v	Total overburden pressure σ_v	Percentage fine
6	7	Sand	84	84	80
9	12	Sand	121	126	80
12	16	Sand	133	168	80
13.3	32	Sand	138	186	80

$CSR_L = 1.5 \times 1 \times 1 \times 0.13 = 0.195$

Calculation of CRR:

$(N_1)_{60} = C_N N_{60}$

$N_{60} = 7$

$C_N = \left(\frac{P_a}{\sigma'_v}\right)^{0.5}$

$P_a = 100\,\text{kPa}$

$\sigma'_v = 84\,\text{kPa}$

$C_N = 1.09$

$(N_1)_{60} = 7.63$

form CRR is 0.18, CRR < CSR_L so soil is **liquefiable.**

Example 5.8 The measured SPT values in sand, results of sieve analysis for a site in Zone IV are indicated in Table 5.32. The water table is at 6 m below ground level. Estimate the liquefaction potential for 7.5 magnitude earthquake.

Solution

Site Characterization:

This site consists of sand. The SPT values range from 7 to 32. The site is located in Zone III. The peak horizontal ground acceleration value for the site will be taken corresponding to Zone III.

Liquefaction Potential of Underlying Soil

Step-by-step calculation for the depth of 6 m is given below. Detailed calculations for all the depths are given in Table 5.2.

$a_{max}/g = 0.24, \quad M_w = 7.5,$

$\gamma_{sat} = 18\,\text{kN/m}^3,$

$\gamma_w = 9.8\,\text{kN/m}^3$

Depth of water level below G.L. = 6.0 m.

Depth at which liquefaction potential is to be evaluated = 6.0 m.

Initial stresses:

$\sigma_v = 6 \times 18 = 108\,\text{kPa}$

$u_o = 0$

$\sigma'_v = (P_o - u_o) = 108\,\text{kPa}$

Stress reduction factor:

$r_d = 1 - 0.00765z = 1 - 0.00765 \times 6 = 0.954$

Critical stress ratio induced by earthquake:

$a\ max = 0.24\ g,\ M\ w = 7.5$

$CSR_{eq} = 0.65(amaz/g)r_d(\sigma_v/\sigma'_v)$

$= 0.65 \times (0.24) \times 0.954 \times (108/108)$

$= 0.15$

Correction for SPT (*N*) value for overburden pressure:

$(N)_{60} = C_N \times N_{60}$

$C_N = (P_u/\sigma'_v)^{1/2}$

$= (96/108)^{1/2}$

$= 0.94$

$(N)_{60} = 0.94 \times 7 = 6.58$

Critical stress ratio resisting liquefaction:

For $(N)_{60} = 6.58$, fines content of 15%

CRR = 0.12 (Fig. 5.38)

Corrected Critical Stress Ratio Resisting Liquefaction:

$CSR_L = CSR_{7.5} k_m k_\sigma k_\alpha =$ k m Correction factor for earthquake magnitude other than 7.5 = 1.00 for $M\ w = 7.5$

Table 5.32 Result of the standard penetration test and sieve analysis

Depth (m)	N_{60}	Soil classification	Effective overburden pressure σ'_v	Total overburden pressure σ_v	Percentage fine
6	7	Sand	108	108	15
9	12	Sand	132.6	162	15
12	16	Sand	157.2	216	15
13.3	32	Sand	167.86	239.4	15

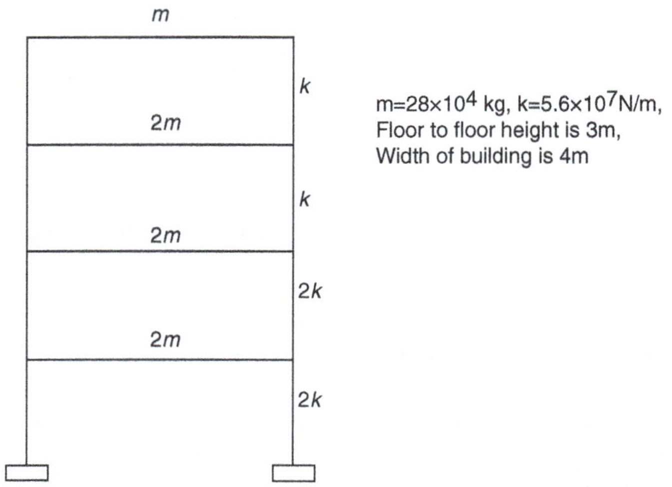

Fig. 5.41 Bhuj/Kachchh earthquake applied

$k\alpha$ = Correction factor for initial driving static shear
= 1.00, since no initial static shear

$k\sigma$ = Correction factor for stress level larger than 96 kPa
= 0.88

$CSR_L = 0.15 \times 1 \times 1 \times 0.88 = 0.106$

Factor of safety against liquefaction:

$FS_{liq} = CRR/CSR$

$= 0.12/0.106 > 1.0$

Hence, the strata are not liquefiable.

Exercise Problems

(1) What is geotechnical investigation and what is its purpose?
(2) What is standard penetration test? What are various corrections applied to the obtained SPT values from site? What's various applications of SPT in geotechnical investigation?
(3) Explain about seismic reflection, seismic refraction, and MASW methods of geotechnical investigation?
(4) What is cross-borehole test? What's are the various parameters of soil are determined from cross-borehole test?
(5) Explain about the test method and steps in dilatometer test in soil investigation?
(6) Describe about the block vibration test? What's various dynamic parameters are determined from block vibration test and how?
(7) What is confirmatory test in geotechnical investigation and why it is carried out?
(8) Discuss various guidelines about the number of holes, depth, and disposition of boreholes in geotechnical investigation?
(9) The frame shown in the figure below rests on footings of 1 m diameter. Soil properties are specified by VS = 250 m/s and ρ = 1800 kg/m^3. Using the method of direct analysis, obtain the time histories of base shear and total displacement of the top story for the Bhuj/Kachchh earthquake applied at the base of the frame (assume ξ = 5% for the frame) (Fig. 5.41).

Appendix 1: User Subroutines

The user subroutine for calculating the shakedown strength of the soil ($G/G_{max(s)}$) and damping ratio (ξ) with shear strain is given below.

```
  G=10000000;          %Initial shear modulus of soil
alfa=50;            %parameters of R-O model
exp_r=3.0;           %parameters of R-O model
g_ref=0.2;          %parameters of R-O model
count=1;
  st=[1e-4,5e-4,8e-4,1e-3,4e-3,8e-3,1e-2,4e-2,1e-1,3e-1,5e-1,8e-1,1,1.5,3,5,8,10];          %the
shear strains in which G/Gmax is calculated
for j=1:length(st(1,:))
g_oct=st(1,j);
erro=1.0;
    x=0.999999;
x_mensol=1.0;
while (erro>1e-6)    %method of bisection
    aux1= x_mensol-1/(1+alfa*(x_mensol*g_oct/g_ref)^(exp_r-1));
    aux2= x-1/(1+alfa*(x*g_oct/g_ref)^(exp_r-1));
x_pos=(aux2*x_mensol-aux1*x)/(aux2-aux1);
erro=abs(x-x_pos);
x_mensol=x;
    x=x_pos;
end
G_sec=1.0*x_pos*G;
G_curve(count)=G_sec/G;          %G/Gmax calculation
strain(count)=g_oct/g_ref;
damping(count)=(2*(exp_r-1)*(1-G_curve(count)))/(3.14*(exp_r+1));
        %D Ratio calculation
count=count+1;
end
figure (1);
plot(strain,G_curve);
figure (2);
plot(strain,damping);
```

References

1. ASCE 4-98 (2000) Seismic Analysis of Safety-Related Nuclear Structures and Commentary
2. IS 1892 (1979) Code of practice for subsurface investigation for foundations
3. IS 1893 (2002) Criteria for Earthquake Resistant design of structures
4. Mogami T, Kubo K (1953) The behavior of soil during vibration. In: Proceedings, 3rd international conference on soil mechanics and foundation engineering, Zurich, vol. 1, pp. 152–155
5. Casagrande (1976) Liquefaction and cyclic deformation of sands-a critical review. Harvard Soil Mechanics Series No. 88, Harvard University, Cambridge Press
6. Schmertmann JH, Hartman JP, Brown PR (1978) Improved strain influence factor diagrams. J Geotech Eng Div, ASCE 104 (8):1131–1135
7. Xenaki VC, Athanasopoulos GA (2003) Liquefaction resistance of sand-silt mixtures: an experimental investigation of the effect of fines. Soil Dyn Earthq Eng 23:183–194
8. Kramer SL (1996) Geotechnical earthquake engineering. Prentice Hall, Upper Saddle River, NJ
9. IS 2131 (1981): Method for standard penetration test for soils

10. Ishihara K (1993) Liquefaction of natural deposits during earthquakes. In: Proceedings 11th ICSMFE, San Francisco, 1, 321–376, vol. 2, pp 683–692
11. Sladen JA, D'Hollander RD, Krahn J (1985) The liquefaction of sands a collapse surface approach. Can Geotech J 26:103–113
12. Saran Swami (1999) Soil dynamics and machine foundations. Galgotia Publications, New Delhi
13. USGS, Niigata, Japan-Liquefaction Image. https://geomaps.wr.usgs.gov/sfgeo/liquefaction/image_pages/niigata_apts_bw.html
14. Skempton AW (1986) Standard penetration test procedures and the effects in sands of overburden pressure, relative density, particle size, ageing and over consolidation. Geotech 36(3)
15. Bowles JE (1996) Foundation analysis and design, 5th Edition, McGraw-Hill Book Co., New York, USA
16. Hera et al. (1974) Shear modulus and shear strength of cohesive soil, Soil and Found, 14(3):1–12
17. Kulhawy FH, Mayne PW (1990) Manual on estimating soil properties for foundation design, Electric power research institute, Palo, Alto, CA
18. Schmertmann JH (1970) Static cone to compute static settlement over sand. J Soil Mech and Found Div ASCE 96(SM3), p. 7302–1043
19. IBC (2000) International Building Code-2000, 5th Edition. International Code Council: Inc., Falls Church, VA
20. NEHRP (2000) 2000: Advantages and disadvantages of reflection
21. IS 4963 Part III:1976 (R2002) Indian Standard Method for subsurface sounding for soils, Static cone penetration test
22. Schmertmann JH (1975) In-Situ Measurement Of Shear Strength, State-of-the-Art Paper, Session III, Proceedings of the Conf. on In Situ Measurement of Soil Properties, Specialty Conf. of the Geotechnical Div., ASCE, North Carolina State University, Raleigh, Vol 1, pp. 57–138
23. Lunne T, Kleven A (1981) Role of CPT in North Sea Foundation Engineering, Symposium on cone penetration engineering division, ASCE, pp 49–75
24. Meyerhof (1965) Some recent research on bearing capacity of foundations. Canadian Geotech J, 1(1):16–26
25. Schmertmann JH, Hartman JP, Brown PR (1978) Improved strain influence factor diagrams. J Geotech Eng Div, ASCE, 104 (8):1131–1135
26. IS 4434 Indian standard Code of practice for in-situ vane shear test of soil
27. Menard L (1975) The Menard Pressure meter: Interpretation and Application of the Pressure Meter Test Results to Foundations Design, Sols-Soils, No. 26, Paris, France
28. Baguelin F, Jezequel JF, Shields DH (1978) The pressure meter and foundation engineering. TransTech Publications
29. NRC (2000). "Seeing into the Earth," Committee for Noninvasive Characterization of the Shallow Subsurface for Environmental and Engineering Applications, P. R. Romig, Chair, 129 pp
30. ASTM D4428/D4428 Standard test methods for cross hole seismic testing
31. IS:5249 1992 (R1995) Determination of dynamic properties of soil-Method of test
32. Wood and Jedela .. Sreekala et al (2008) (…) Values of coefficient of attenuation (α) of soil material
33. IS 456:2000 (Reaffirmed 2005) Indian Standard Plain and reinforced concrete-Code of practice
34. Jain SK, Murthy CVR (1893) "Proposed draft provisions and commentary on Indian Seismic code IS 1893 (Part I)", IIT (K) – GSDMA, India

Earthquake-Resistant Design of RC Structures

6

Praveen Kumar

	Earthquakes are natural Hazards which will not harm public in Open but the structures do.	

Symbols

DL	Dead load
LL	Live load
WL	Wind load
FRS	Floor response spectra
M_K	Modal mass
EPGA	Effective peak ground acceleration
ZPA	Zero period acceleration
e_{di}	Static eccentricity at floor
b_i	Floor plan dimension of floor i perpendicular to the direction of force
λ_k	Absolute value of quantity in mode k
R	No. of mode being considered

6.1 Introduction

Earthquake is recognized as one of the major causes of disaster on this planet. Problem associated with the design of structures to withstand earthquakes has engaged the attention of the professionals for a long time. There are a lot of difficulties arising out the catastrophic nature of the event. Further, in recent times, the development of nuclear power plants and their location in strategic areas which are prone to earthquakes has led to greater awareness of the problem of earthquake-resistant design and construction of the structures. This chapter covers the guideline provided by the code of practice, various aspects of earthquake-resistant design of structures, seismic analysis of building, design of various components of building and detailing as per the provisions of Indian standard code.

6.2 Types of Structures

A combination of members connected together in such a way to serve a useful purpose is called a structure. The structure can be classified based on structural arrangements and functions

(i) Classification based on structural arrangements,
   (a) Shell structure,
   (b) Reinforced concrete (RC) frame structure,
   (c) RC frame structure with shear wall,
   (d) Load bearing wall,
   (e) Steel frame structure,
   (f) Mass structure,
   (g) Cables and arches,

P. Kumar (✉)
Bhabha Atomic Research Centre, Mumbai, India
e-mail: praveen@barc.gov.in

G. R. Reddy et al., *Textbook of Seismic Design*,
https://doi.org/10.1007/978-981-13-3176-3_6

   (h) Grid,
   (i) Truss (pin-connected joints),
(ii) Classification based on functional category,
   (a) Commercial: office, store and shop, bank, public utility,
   (b) Residential: apartment, hotel, dormitory, hostel,
   (c) Industrial: warehouse, material, manufacturing process building,
   (d) Public assembly: theater, auditorium, restaurant, observation gallery,
   (e) Institutional: school, hospital, laboratory, library, museum, court of law, correctional institutions, religious edifices (temple, churches, mosques),
   (f) Special purpose: transport interface (air, rail, bus), garage, mausoleum.

6.3 Design Loads

Various types of loads to be considered in the analysis and design of the structure depending on the type and utility are given below:

(a) **Dead Load (DL)**:
These are permanent loads due to self-weight of RCC slabs, beams, walls, partitions, floor and roof finishes, waterproofing, and any other permanent load arising out of structure or its components. Equipment loads that are stationary and permanent are treated as dead load for the purpose of load combination. Unit weight of materials is considered as per IS: 875 (Part I) 1987.

(b) **Live Load (LL)**:
Live loads arise due to operating personnel, temporary placement of tools, tackles, movement of equipment, trolleys, any transient loads, etc. Live loads that act on the structure are computed as per IS: 875 (Part 2): 1987.

(c) **Seismic Load**:
Seismic loads that act on the structure are computed as per IS: 1893. The design seismic force depends on location, geotechnical strata at site, importance and type of structure, material used for structure, superimposed loads, etc.

(d) **Wind Load (WL)**:
Wind loads that act on the structure are computed as per IS: 875 (Part 3)-2015. The design wind speed and the design wind pressure depend upon type and location of the structure, risk coefficient, terrain, height, and structure size factor and topography factor and are calculated in accordance with the provision of IS: 875 (Part 3)-2015.

(e) **Soil/Earth Pressure**:
Underground structures or earth-retaining structure is subjected to lateral earth pressure due to backfill soil around the wall. The dynamic earth pressure under seismic condition is considered as per the provision of IS: 1893.

(f) **Static Liquid Pressure**:
Static liquid pressure is considered when the structure is retaining liquid inside or outside of the structure, e.g., water tank, dam, liquid storage vessel, etc.

(g) **Hydrodynamic Effect**:
The hydrodynamic effects occurs on the structure due to liquid inside or outside of the structure, and their effects are computed as per provision of IS: 1893.

(h) **Thermal Load**:
Thermal load on the structure will act when there is a difference in temperature across the wall thickness. The effects of temperature on structure and load combination can be computed as per provision of IS: 875 (Part 5)-1987.

(i) **Snow Load**:
Snow loads that act on the structure are computed as per the provision of IS: 875 (Part 4)-1987 in the zone, where snowfall occurs.

6.4 Generation of Seismic Force and IS: 1893 Code

6.4.1 Earthquake

Earthquakes are caused by sudden breaks in the earth's crust. They originate in the earth's crust (7–21 miles) and mantel (extending up to 1800 miles) and are classified as (a) deep focus (greater than 200 miles), (b) intermediate focus (40–200 miles), and (c) shallow focus (less than 40 miles). In certain parts of the world, it appears to be a continuous strain movement between one section of the earth crust and the adjacent one. When stress developed by such strains reaches the strength of the materials, a slip occurs between two portions of the crust and enormous amount of energy is released. The energy released during an earthquake travels as compressional or longitudinal waves called the P waves and shear or transverse waves called the S waves. These waves undergo reflection and refraction at interfaces and on the surface travels as Love and Raleigh waves. During the travel, the amplitude decreases with distance and short-period components decay with distance more than long-period components.

The resulting ground motion at any point on the surface may be very complicated because of the times of arrival of different types of waves and the direction from which they approach. The frequency of the vibrations may also vary over a wide range, and the resulting acceleration, velocity, and displacement of a given particle on the earth's surface may be highly variable. Strong motion recorders are used to measure ground motion during an earthquake. It provides three components of ground acceleration as a function of time. Two common measures of earthquakes are:

(a) **Magnitude (Richter)**: This is a measure of total energy released during the earthquake. This is defined as the common logarithm (base 10) of the trace amplitude (in microns) of a standard seismograph (Wood-Anderson-type torsion seismograph with damping 80% of critical, natural period 0.8 s, and magnification of 2800) located on firm ground at a distance of 100 km from the epicenter.
(b) **Intensity**: A subjective measure of the ground motion on the basis of the effect of earthquake on human beings and structures. This is measured on the 12-point Modified Mercalli scale (MM). IS: 1893 adopts this 12-point Modified Mercalli scale.

Richter magnitude (*M*) and Modified Mercalli (MM) intensity are related, and an approximate MM intensity for a Richter magnitude is given in Table 6.1. For more details about magnitude and intensity, the reader may refer to Chap. 1.

6.4.2 Major Earthquakes in India

Some of the major earthquakes recently occurred in India are given in Table 6.2.

Some of the failures reported during Bhuj earthquake, January 26, 2001, have been shown in Figs. 6.1, 6.2, 6.3 and 6.4.

6.4.3 Parts of IS: 1893 Code

Seismic codes are divided into five parts to cover the provision for different types of structures that can envisage in civil engineering field. These are

(a) Part 1: General provisions and buildings,
(b) Part 2: Liquid retaining tanks—elevated and ground supported,
(c) Part 3: Bridges and retaining walls,
(d) Part 4: Industrial structures including stack-like structures, and
(e) Part 5: Dams and embankments.

6.5 General Provisions for Buildings and Structural Analysis as per IS: 1893 (Part 1)-2016

It must be emphasized that the intention of the code is to ensure that structures possess minimum lateral strength to withstand earthquakes, but intention is not to prevent damage to structures due to the most severe shaking that a structure may be subjected to during its lifetime. Actual forces that appear on structures during earthquakes are much greater than the design lateral forces specified in the standard. However, ductility, arising from material behavior, detailing, over strength, additional reserve strength, etc., in structures over and above the design force are relied upon to account for this difference in actual and design lateral load.

6.5.1 The Following are the Major and Important Modifications Made in the Sixth Revision of IS: 1893 (Part 1)-2016

(a) Design spectra are added up to natural period of 6 s.
(b) Same design response spectra are specified for all buildings, irrespective of construction.
(c) Load combination is made consistent with other codes.

Table 6.1 Approximate relation between magnitude and MM intensity

Richter magnitude	5.0	6.0	6.5	7.0	7.5	8.0
MM Intensity (approximately max.)	VI–VII	VII–VIII	VIII–IX	IX–X	X–XI	XII

Table 6.2 Recent major earthquakes in India

Sl no.	Earthquake	Date	Magnitude	Death	Intensity
1	Uttarkashi	October 20, 1991	6.6	768	IX
2	Killari	September 30,1993	6.4	10,000	
3	Bhuj	January 26, 2001	7.7	18,600	X (MSK)
4	Sumatra (Indonesia, India, Sri Lanka & Maldives)	December 24, 2004	9–9.3	283106	IX
5	Muzafarabad (Pakistan & India)	October 08, 2005	7.6	>80,000	VIII
6	Andaman Island	August 11, 2009	7.5	26	
7	Nicobar Island	June 12, 2010	7.5	0	
8	Sikkim (India & Nepal Border)	September 18, 2011	6.9	118	
9	Nepal & India	April 25, 2015	7.8	9000	IX
10	North East India	January 3, 2015	6.7	11	VII

Fig. 6.1 Total collapse of a portion of a ten-story residential building in Ahmedabad [3]

(d) Temporary structures are added to this code.
(e) Importance factor provisions are modified based on the density of occupancy of building.
(f) A provision is introduced to ensure that all buildings are designed for at least minimum lateral force.
(g) Building with flat slab is added to this code.
(h) Effect of masonry infill walls has been included in the analysis and design of frame buildings.
(i) Arriving approximate natural period of buildings with basements, step-back buildings, and buildings on hill slope are introduced.
(j) Provisions on torsion have been simplified.
(k) Simplified method is introduced for liquefaction potential analysis.

6.5.2 Tectonic Map of India

The tectonic map of India has been developed based on the nature of geological formation of the bedrock in the earth crust revealing region characterized by structural features such as dislocation, distortion faults, folding, thrusts, volcanoes with their age of formation which are directly involved in the earth movement or earthquake resulting from the above causes. The tectonic map of India is shown in IS: 1893.

6.5.3 Seismic Zoning Map of India

An entirely scientific basis for zoning is not feasible in view of the scanty data available. Though magnitudes of different

Fig. 6.2 **a** Ground story collapse of a four-story building with open ground story and **b** collapse of column with open ground story at Bhuj [3]

Fig. 6.3 Collapse of open ground story RC frame residential building in Bhuj

Fig. 6.4 Collapse of intermediate story in a six-story RC frame commercial building at Bhuj [3]

earthquakes, which have occurred in the past, are known to a reasonable amount of accuracy. The intensities of the shocks caused by these earthquakes have so far been mostly estimated by damage surveys, and there is little instrumental evidence to correlate. The conclusions arrived at maximum intensity at different places can be fixed on a scale only on the basis of observations made and recorded after the earthquake, and thus a zoning map which is based on maximum intensities arrived at is likely to lead to incorrect conclusions in view of the following:

(a) Incorrectness in the assessment of intensities,
(b) Human error in judgment during the damage survey,
(c) Variations in quality and design of structures causing variation in type and extent of damage to the structures for the same intensity of shock.

Hence, this approach to zoning was abandoned, and in IS: 1893–2002, a rational approach to zoning map based on known magnitudes and the known epicenters assuming all other conditions being average and to modify such an average idealized isoseismal map in light of tectonics, lithology, and maximum intensities as recorded from damage surveys, etc., was adopted. The country in the zoning map is divided into four zones from Zone II to Zone V with increasing intensity of earthquake. The object of the map is to classify the area into number of zones in which one may reasonably expect earthquake shock of more or less the same intensity in the future. The seismic zoning map of India is shown in IS: 1893–2016.

The Modified Mercalli Intensity broadly associated with the various zones is given in Table 6.3.

6.5.4 Guidelines for Planning of Earthquake-Resistant Buildings

To perform well in a seismic event, a structure should possess four main attributes, namely simple and regular configuration, large lateral stiffness, adequate ductility, and adequate lateral strength. The structures having simple regular geometry and uniformly distributed mass and stiffness in plan as well as in elevation suffer much less damage than the structures with irregular configurations as shown in Fig. 6.5.

The general principles given below shall be observed in the planning of earthquake-resistant buildings:

(a) **Lightness**: The earthquake force is a function of mass, the building shall be kept as light as feasible consistent with structural safety and functional requirements. Roof and upper story of buildings generate higher lateral force; hence, in particular these floors should be designed as light as possible.
(b) **Continuity of construction**: At the extent possible, the parts of the building should be connected in such a manner that the building acts as a single unit. Parts of buildings between expansion joints should have floor slabs continuous throughout to the extent possible. Concrete slabs shall be rigidly connected or integrally cast with the support beams.
(c) **Projection and suspended parts**: Projection and suspended parts in the building should be avoided as far as possible. If they cannot be avoided, they shall be properly reinforced and firmly connected to the main structure. Suspended parts provided in the building should be as light as possible.

Table 6.3 Relation between zone and MMI

Zone	II	III	IV	V
MMI	≤ VI	VII	VIII	IX and above

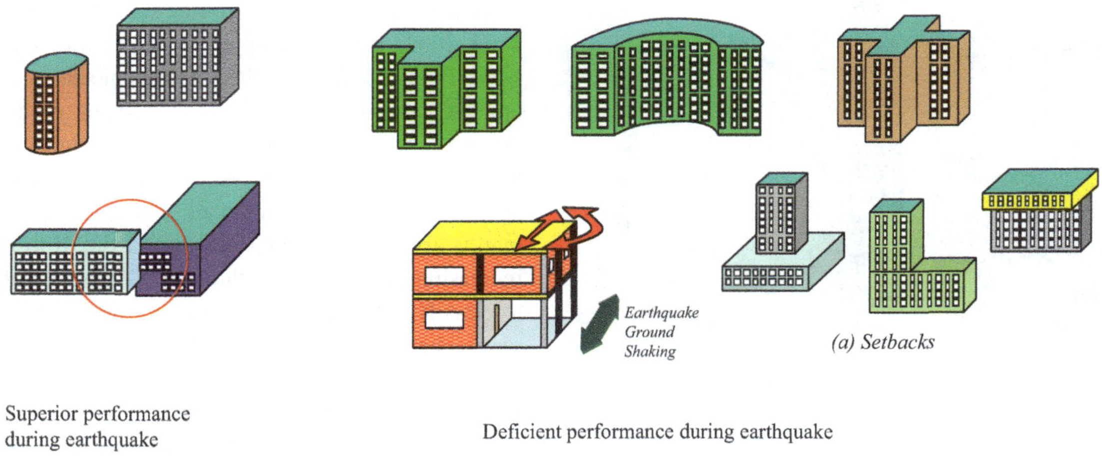

Fig. 6.5 Building with regular and irregular configuration [3]

(d) **Shape of the building**: To minimize torsion, the building should have a simple rectangular plan and shall be symmetrical with respect to both mass and rigidity so that the centers of mass and rigidity of the building coincide with each other. Buildings having plan with shapes like L, T, E, and Y shall preferably be separated into rectangular parts by providing separation gap at appropriate places to avoid the torsion in the building. Typical shapes of the buildings with separation gap are shown in Fig. 6.6.

6.5.5 Load Combination and Increase in Permissible Stresses

Load combinations are done to consider the set of loads which are likely to occur at the same time. In case of earthquake, permissible stresses in materials are increased.

6.5.5.1 Load Factors and Load Combination

(A) **Load factor for plastic design of steel structures**

(i) 1.7 (DL + IL)

(ii) 1.7 (DL ± EL)

(iii) 1.3 (DL + IL ± EL)

where

DL dead load

IL imposed load

EL earthquake load

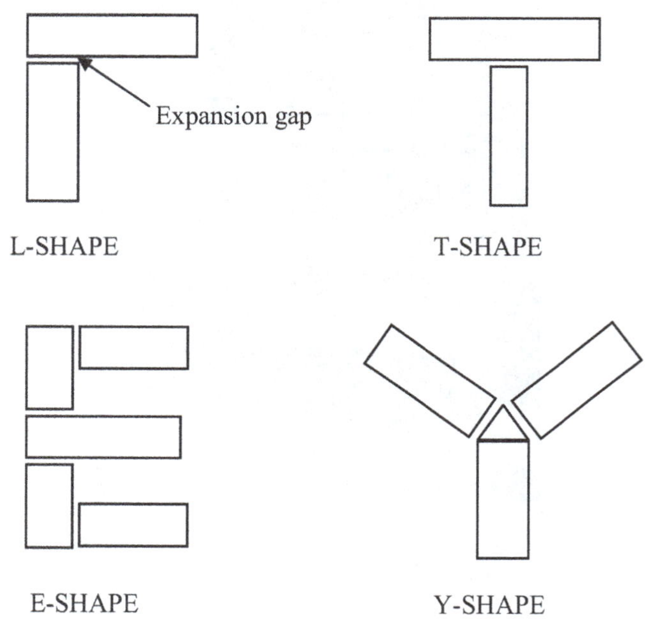

Fig. 6.6 Typical shapes of buildings with separation gap

(B) **Limit state design of RCC and prestressed concrete structure**

For limit state design of reinforced and prestressed concrete structures, the partial safety factor for the limit states of serviceability and collapse as given in IS: 456-2000 and IS: 1343-1980 are as follows:

(i) 1.5 (DL + IL)

(ii) 1.2 (DL + IL ± EL)
(iii) 1.5 (DL ± EL)
(iv) 0.9 (DL) ± 1.5 (EL)

6.5.5.2 Design Horizontal Earthquake Load

i. When lateral load resists, elements are oriented along the orthogonal direction: design for full earthquake load in one horizontal direction.
ii. When lateral load resists, elements are not oriented along the orthogonal direction: design for full earthquake load in one horizontal direction ±30% earthquake load in other direction.

6.5.5.3 Design Vertical Earthquake Load

i. Design vertical acceleration spectrum = 2/3 design horizontal acceleration spectrum.

6.5.5.4 Combination for Two or Three Component

(A) **100:30:30 combinations**

$$\begin{aligned} &\text{(i)} \quad \mathrm{EL} = \pm \mathrm{EL}_x \pm 0.3\mathrm{EL}_y \pm 0.3\mathrm{EL}_z \\ &\text{(ii)} \quad \mathrm{EL} = \pm \mathrm{EL}_y \pm 0.3\mathrm{EL}_x \pm 0.3\mathrm{EL}_z \\ &\text{(iii)} \quad \mathrm{EL} = \pm \mathrm{EL}_z \pm 0.3\mathrm{EL}_x \pm 0.3\mathrm{EL}_y \end{aligned} \tag{6.1}$$

(B) **Square root of the sum of the square (SRSS)**

$$\mathrm{EL} = \sqrt{(\mathrm{EL}_X)^2 + (\mathrm{EL}_Y)^2 + (\mathrm{EL}_Z)^2} \tag{6.2}$$

where x and y are two orthogonal directions and z is the vertical direction.

The combination procedure as defined above in (A) and (B) is applied to the same quantity (e.g., moment or shear) due to different components of ground motion. For two-component motions, delete the component which is not required in Eq. (6.1) or Eq. (6.2).

6.5.5.5 Increase in Permissible Stresses

Increase in permissible stresses in material

When earthquake forces are combined with normal design forces, the permissible stresses in materials may be increased by one-third for elastic method of design. However, steels having a definite yield stress, permissible stress may be limited to the yield stress. If steel is without a definite yield point, the permissible stress will be limited to 80% of the ultimate strength or 0.2% of proof stress, whichever is lesser. In prestressed concrete members, the tensile stress in the extreme fibers of the concrete may be permitted so as not to exceed two-third of the modulus of rupture of concrete.

Increase in permissible stresses in soils

To avoid liquefaction or excessive total and differential settlement, corrected standard penetration value (N) > 15 for $\leq$5-m depth below ground level and $N > 25$ for $\geq$10-m depth below ground level in Zone III, IV, V, and $N > 10$ for $\leq$5-m depth below ground level and $N > 20$ for $\geq$10-m depth below ground level in Zone II. Net bearing pressure and skin friction in soils shall be increased as per Table 6.4.

6.5.6 Information Required for Analysis of Structures

The following information is required for analysis of structures as per IS: 1893.

(a) Location of structure in the map of India (i.e., seismic zone and zone factor),
(b) Type of structure (steel or concrete, and if concrete, proposed grade of concrete) and corresponding importance factor,
(c) Fundamental period of structure,
(d) Structural configuration and loading,
(e) Type of soil constituting the foundation and type of foundation,
(f) Response reduction factor for structure,
(g) Modulus of elasticity,
    (i) For steel structures, the modulus of elasticity can be taken as 2.05×10^5 N/mm^2.
    (ii) For concrete structures, the modulus of elasticity shall be taken as the short-term static modulus of elasticity as outlined in IS: 456–2000, viz.

$$E_c = 5000\sqrt{f_{ck}} \tag{6.3}$$

where

E_c short-term static modulus of elasticity in N/mm^2 and
f_{ck} characteristic cube strength of concrete in N/mm^2.

6.5.6.1 Zone Factor

India is classified into four seismic zones to determining the seismic force. Based on the seismic zone classification, the zone factor (Z) refers to the zero-period acceleration value

Table 6.4 Increase in net bearing pressure and skin friction in soils (in %) [1]

Foundation system	Type A: rock or hard soils	Type B: medium or stiff soils	Type C: soft soils
Percentage increase allowable	50%	25%	0%

Table 6.5 Zone factor (Z) [1]

Seismic zone	II	III	IV	V
Seismic intensity	Low	Moderate	Severe	Very severe
Z	0.10	0.16	0.24	0.36

for the maximum considered earthquake (MCE) as given in Table 6.5. The design basis earthquake (DBE) is obtained by dividing the Z by a factor of 2.

6.5.6.2 Importance Factor

The importance factor depends upon the functional use of the structure characterized by hazardous consequences of its failure, post-earthquake functional needs, historical value, or economic importance. The minimum value of importance factor (I) for different building systems is given in Table 6.6. The designer may choose a value greater than the value mentioned in Table 6.6 depending on economy, strategy, and other considerations.

6.5.6.3 Fundamental Natural Period

The approximate fundamental translational natural period of vibration (T_a), in seconds, shall be estimated by the empirical expression as given below:

(a) Bare MRF building (without masonry infills)

$$T_a = \begin{cases} 0.075\, h^{0.75} & \text{For RC Frame Building} \\ 0.080\, h^{0.75} & \text{For RC-Steel CompositeMRF Building} \\ 0.085\, h^{0.75} & \text{For Steel Frame Building} \end{cases} \quad (6.4)$$

where 'h' is the height of the building, in meters. This excludes the basement stories, where basement story and walls are connected with the ground floor deck or fitted between the building columns, but includes the basement stories, when they are not so connected.

(b) Buildings with RC structural walls:

$$T_a = \frac{0.075h^{0.75}}{\sqrt{A_w}} \geq \frac{0.09\,h}{\sqrt{d}} \quad (6.5)$$

where $A_w = \sum_{i=1}^{N_w} \left[A_{wi}\left\{0.2 + \left(\frac{L_{wi}}{h}\right)^2\right\}\right]$

A_w total effective area (m^2) of walls in first story in building

A_{wi} effective c/s area of wall i in first story in building (m^2)

L_{wi} length of structural wall i in first story in the direction of lateral force (m)

d base dimension of the building at the plinth level, in m along the considered direction of the lateral force

N_w number of wall in the considered direction of the lateral force

(c) All other buildings:

$$T_a = \frac{0.09\,h}{\sqrt{d}} \quad (6.5a)$$

Table 6.6 Importance factor (I), [1]

Sl. no	Structure	I
1	Important service and community buildings or structure (critical governance buildings, school), signature buildings, monument buildings, lifeline and emergency buildings (buildings for hospital, bus station, metro rail, metro rail stations, telephone exchange, television/radio station), railway stations, airport, food storage buildings, fuel station buildings, power station buildings, fire station buildings, and large community buildings (cinema halls, shopping malls, assembly halls, and subway stations)	1.5
2	Residential and commercial building with occupancy more than 200 persons	1.2
3	All other buildings	1.0

6.5.6.4 Type of Soil and Corresponding Average Response Acceleration Coefficient (S_a/g)

Three types of soil mainly constituting the foundation have been classified based on N (standard penetration value). Design acceleration coefficient for three types of soil is arrived for equivalent static method and response spectrum method in IS 1893 (Part 1)-2016. Formulation for calculating average response acceleration coefficient for rock or soil sites based on appropriate natural period and damping of structure for response spectrum method are given in Table 6.7. A response spectrum for 5% damping given in IS code is shown in Fig. 6.7. The spectra can be obtained for different damping values by multiplying by factor mentioned in Table 6.8.

6.5.6.5 Seismic Weight

The seismic weight of the entire building is the sum of the seismic weights of all the floors. Any weight supported in between the stories shall be distributed to the floors above and below in inverse proportion to its distance from the floors. The seismic weight of each floor is its full dead load plus appropriate amount of imposed load. Imposed loads are considered, since during an earthquake they can enhance the load on the structure. The imposed load is specified in IS: 875 (Part-2) -1987. The percentage of imposed load to be considered in seismic weight calculation is given in Table 6.9. The imposed load on roof is not to be considered while calculating the design seismic forces.

6.5.6.6 Response Reduction Factor

The response reduction factor (R) depends on the perceived earthquake damage performance of the structure, characterized by ductile or brittle deformations. The ratio (R/I) shall be greater than 1.0. The response reduction factor represents the ductile capacity of structure. The response reduction values for different building systems are given in Table 6.10. In the table, ordinary moment-resisting frames (OMRFs) are those structures which are designed and detailed as per IS: 456-2000 or IS: 800-2007. Special moment-resisting frame (SMRF) is the special moment-resisting frame detailed to provide ductile behavior and comply with the requirements given in IS: 4326-1976 or IS: 13920-2016, or SP: 6(6)-1972. Buildings with dual systems consist of shear walls (or braced frames) and moment-resisting frames such that the two systems are designed to resist the total design force in proportion to their lateral stiffness considering the interaction of the dual system at all floor levels; the moment-resisting frames are designed to independently resist at least 25% of the design seismic base shear. In steel structures, members and their connections should be detailed such that high ductility is obtained, avoiding premature failure due to elastic or inelastic buckling of any type.

6.5.7 Method of Analysis of Structure

The IS code outlines two broad methods to be normally used for the computation of design seismic forces. These are:

(a) Seismic coefficient method/equivalent static method,
(b) Dynamic analysis (by response spectrum method/time history method).

6.5.7.1 Applicability of Seismic Coefficient and Dynamic Analysis Methods

Methods of analysis used for structure types are given in Table 6.11

6.5.7.2 Torsion

Provision shall be made in all buildings for increase in shear forces on the lateral force-resisting elements resulting from the horizontal torsional moment arising due to eccentricity between the center of mass and the center of rigidity. Negative torsional shear should be neglected. The torsion effect in buildings can be easily explained with the help of Fig. 6.8. It can be observed if one sits at one end of the cradle, it *twists* (i.e., moves more on the side he/she is sitting). Similarly, if the mass on the floor of a building is more on one side (for instance, one side of a building may have a store, equipment, or a library), then movement of the building on that side will be more.

Design eccentricity, e_{di}, at floor i can be given as in Eq. (6.6)

$$\begin{aligned} e_{di} &= 1.5e_{si} + 0.05b_i, \text{(or)} \\ e_{di} &= e_{si} - 0.05b_i \\ e_{di} &= \text{Maximum of above two values} \end{aligned} \tag{6.6}$$

Table 6.7 Formulation for calculating S_a/g value for 5% damping

Rocky soil sites	Medium stiff soil sites	Soft soil sites
$1 + 15T$, $0.00 \le T \le 0.10$ s	$1 + 15T$, $0.00 \le T \le 0.10$ s	$1 + 15T$, $0.00 \le T \le 0.10$ s
2.5, $0.10\text{ s} \le T \le 0.40$ s	2.5, $0.10\text{ s} \le T \le 0.55$ s	2.5, $0.10\text{ s} \le T \le 0.67$ s
$1.00/T$, $0.40\text{ s} \le T \le 4.00$ s	$1.36/T$, $0.55\text{ s} \le T \le 4.00$ s	$1.67/T$, $0.67\text{ s} \le T \le 4.00$ s
0.25, $T > 4.00$ s	0.34, $T > 4.00$ s	0.42, $T > 4.00$ s

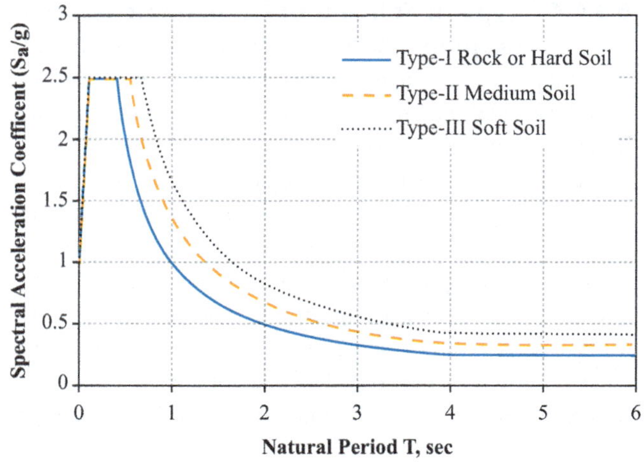

Fig. 6.7 S_a/g versus T for response spectrum method [1]

Table 6.8 Multiplying factor for 5% spectrum to obtain for other damping values [1]

Damping (% of critical)	0	2	5	7	10	15	20	25	30
Multiplying factor	3.2	1.4	1.0	0.9	0.8	0.7	0.6	0.55	0.50

Table 6.9 Percentage of imposed load to be considered in seismic weight [1]

Imposed uniformly distributed floor loads (kN/m^2)	% of imposed load
Up to and including 3.0	25
Above 3.0	50

where

e_{di} static eccentricity at floor i (distance between center of mass and center of rigidity)

b_i Floor plan dimension of floor i perpendicular to the direction of force

For highly irregular building, additional static eccentricity = $\pm 0.05 b_i$.

6.5.7.3 Deformations

Story drift limitation: The story drift in any story due to minimum specified design lateral force, with partial load factor of 1.0, shall not exceed 0.004 times the story height (Fig. 6.9).

Separation between adjacent units: Two adjacent buildings or units of the same structure with separation joint in between shall be separated by a distance of R times the sum of calculated story displacements (Fig. 6.10).

6.5.7.4 Cantilever Projections

Vertical: Towers, water tanks (Fig. 6.11), parapets, stacks (chimneys), and other cantilever projections attached to structures and projecting above the roof shall be designed for five times the design horizontal acceleration spectrum.

Horizontal: All horizontal projections like cornices and balconies (Fig. 6.12) shall be designed for five times the design vertical acceleration spectrum value.

6.5.7.5 Outline of Seismic Coefficient Methods

This method involves the following steps:

i. Compute the weight at each floor level.
ii. Compute the fundamental natural period of structure as per Sect. 6.5.6.3.
iii. Calculate the design horizontal/vertical seismic coefficient (A_h) by using Eq. (6.6).

$$A_h = \left(\frac{Z}{2}\right)\left(\frac{I}{R}\right)\left(\frac{S_a}{g}\right) \tag{6.6}$$

Design horizontal seismic coefficient, for any structure with $T_a \leq 0.1$ s, can be calculated as

$$A_h = Z/2 \tag{6.7}$$

iv. Compute the design base shear (V_B).
The total base shear in any principle direction shall be determined by the following expression:

$$V_B = A_h W \tag{6.8}$$

Table 6.10 Response reduction factor (*R*) for building systems [1]

Sl. No.	Lateral load resisting system	*R*
Moment frame systems		
1.	RC building with ordinary moment-resisting frame (OMRF)	3.0
2.	RC building with special moment-resisting frame (SMRF)	5.0
3.	Steel building with ordinary moment-resisting frame (OMRF)	3.0
4.	Steel moment-resisting frame designed as per SP:6(6)	5.0
5	Building with ordinary braces frame (OBF) having concentric bracing	4.0
6	Building with special braces frame (SBF) having concentric bracing	4.5
7	Building with special braces frame (SBF) having eccentric bracing	5.0
Structural wall systems		
8.	Load bearing masonry buildings	
	(a) Unreinforced masonry without horizontal RC seismic bands	1.5
	(b) Unreinforced masonry with horizontal RC seismic bands	2.0
	(c) Unreinforced masonry with horizontal RC seismic bands and vertical reinforcing bars at corners of rooms and jambs opening.	2.5
	(d) Reinforced masonry	3.0
	(e) Confined masonry	3.0
9.	Building with ordinary RC structural walls	3.0
10.	Building with ductile RC structural walls	4.0
Building with dual systems		
11.	Building with ordinary RC structural walls and RC OMRFs	3.0
12.	Building with ordinary RC structural walls and RC SMRFs	4.0
13.	Building with ductile RC structural walls and RC OMRFs	4.0
14	Building with ductile RC structural walls and RC SMRFs	5.0
Flat slab structural wall systems		
15	RC building with condition as per IS: 1893 Part 1 2016	3.0

Table 6.11 Criteria for choice of seismic analysis

Sr. no.	Structure types	Analysis to be carried out
1.	**Regular buildings**: buildings less than 15 m in heights in Zone II	Seismic coefficient method
2.	**Regular buildings and irregular buildings**: all buildings except in Sl. no. 1	Linear dynamic analysis

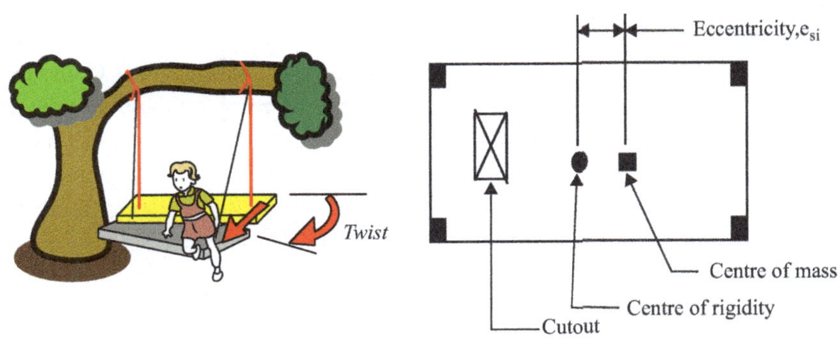

Fig. 6.8 Even if vertical members are placed uniformly in plan of building, more mass on one side causes the floors to twist

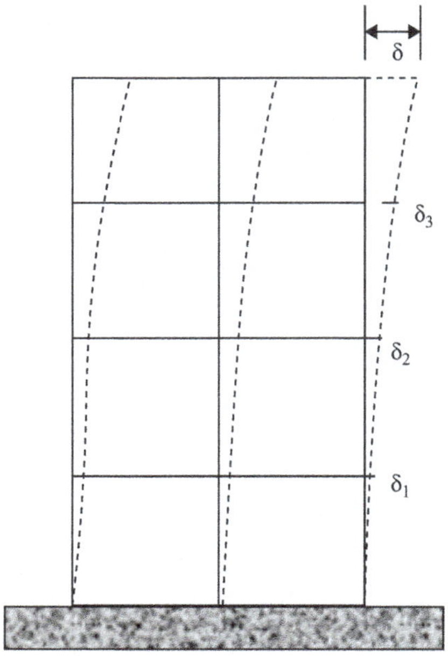

Fig. 6.9 Story drift of a four-story building

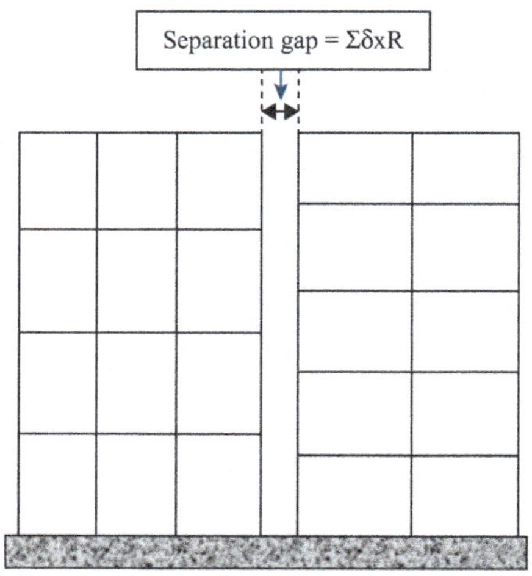

Fig. 6.10 Separation gap between two buildings

Fig. 6.11 Water tank above the building

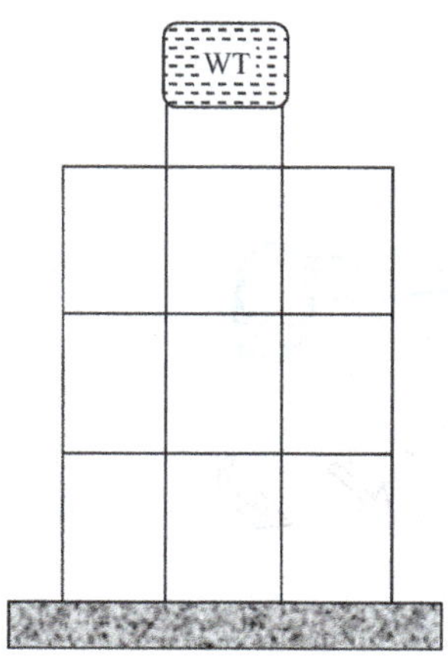

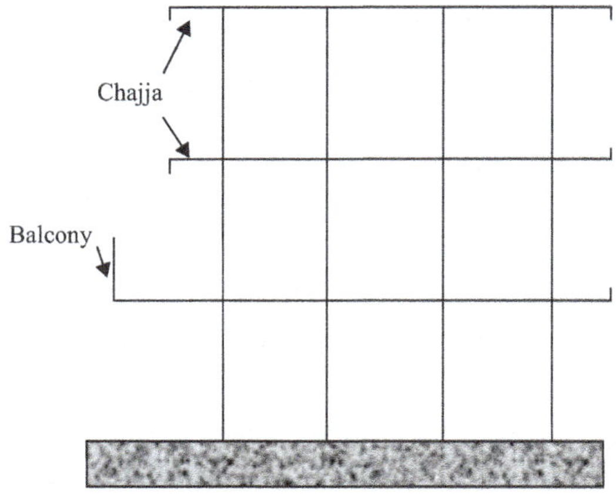

Fig. 6.12 Balcony and chajja

where

A_h design horizontal seismic coefficient using the fundamental natural period T_a in the considered direction of vibration

W seismic weight of the building

v. Distribute the design base shear to obtain forces in floor levels.

Vertical distribution of base shear to different floor levels

$$Q_i = V_B \frac{W_i h_i^2}{\sum_{j=1}^{n} W_j h_j^2} \quad (6.9)$$

where

Q_i design lateral force at floor i

W_i seismic weight of floor i

h_i height of floor i measured from base

n number of stories in the building is the number of levels at which the masses are located.

6.5.7.6 Outline of Dynamic Analysis Method

This method involves the following steps:

i. Compute the masses at each floor level.
ii. Compute the natural frequencies and mode shapes.
iii. Compute the modal participation factors.
iv. For the time periods of vibration and assumed damping factors, calculate the spectral accelerations for each mode.
v. Modify the spectral accelerations to take into account of soil type, importance factor of the structure, seismicity of the zone, and obtain the forces at story levels for each mode of vibration.

vi. Calculate the shears for each mode and combine modal shears.
vii. Calculate the forces from the design shears obtained.

A. **Dynamic Analysis**

Dynamic analysis can be performed either by time history method or response spectrum method. However, in either method, the base shear (V_B) shall be compared with base shear $(\overline{V_B})$ calculated using fundamental time period T_a. When $\overline{V_B}$ is less than V_B, all the response quantities obtained from the analysis shall be multiplied by $\overline{V_B}/V_B$. Design lateral force at each floor in each mode can be calculated by using Eq. (6.10).

$$Q_{ik} = A_k \phi_{ik} P_k W_i \tag{6.10}$$

where

A_k design horizontal acceleration for period T_k
P_k modal participation factor

Modal participation factor can be calculated by using Eq. (6.11).

$$P_k = \frac{\sum_{i=1}^{n} W_i \phi_{ik}}{\sum_{i=1}^{n} W_i \phi_{ik}^2} \tag{6.11}$$

where

Φ_{ik} mode shape coefficient at floor in mode k
W_i seismic weight of floor i

Story shear forces in each mode can be calculated by using Eq. (6.12).

$$V_{ik} = \sum_{j=i+1}^{n} Q_{ik} \tag{6.12}$$

Story shear force due to all modes considered: The peak story shear force (V_i) in story i due to all modes considered is obtained by combining those due to each mode in accordance with Eq. (6.13).

B. **Modal Combination**

(a) **Complete Quadratic Combination (CQC)**

$$V_i = \sqrt{\sum_{j=1}^{r} \sum_{k=1}^{r} V_{ij} \rho_{jk} V_{ik}} \tag{6.13}$$

where

r number of mode being considered
ρ_{jk} cross-modal coefficient

$$\rho_{jk} = \frac{8\varsigma^2(1+\beta)\beta^{1.5}}{(1+\beta^2)^2 + 4\varsigma^2\beta(1+\beta)^2} \tag{6.14}$$

where

ζ % critical damping
β frequency ratio

$$\beta = \frac{\omega_j}{\omega_i} \tag{6.15}$$

where

ω_j circular frequency in jth mode
ω_i circular frequency in ith mode

(b) **Square Root of the Sum of the Square (SRSS)**

$$\lambda = \sqrt{\sum_{K=1}^{r} (\lambda_k)^2}$$

where

λ_k absolute value of quantity in mode k
r number of mode being considered

Lateral forces at each story due to all modes considered

$$F_i = V_i - V_{i+1} \tag{6.16}$$

$$F_{\text{roof}} = V_{\text{roof}} \tag{6.17}$$

6.6 Provision of IS: 1893 (Part 4)-2015: Industrial Structures Including Stack-like Structures

IS: 1893 (Part 4)-2015 deals with earthquake-resistant design of the industrial structures (plant and auxiliary structures) including stack-like structures. Here, it will be disused only with respect to industrial building. Additional sub-clauses specified with respect to industrial building in IS: 1893 (Part 4)-2015 are illustrated here. All subclause of IS: 1893 (Part 1)-2016 is also applicable in industrial building. The design approach adopted in this standard is to ensure that the structures

- Possess minimum strength to withstand minor earthquakes, (<DBE) which occur frequently, without damage,
- Resist moderate earthquakes (DBE) without significant structural damage though some non-structural damage may occur, and
- Withstand a major earthquake (MCE) without collapse.

6.6.1 Categorization of Structures

To perform well in an earthquake, the industrial structure should possess adequate strength, stiffness, and ductility. Structures are classified into the following four categories:

Category 1: Structures whose failure can cause conditions that can lead directly or indirectly to extensive loss of life/property to population at large in the areas adjacent to the plant complex.

For example, control building (blast resistant), cryogenic storage tank (doubled walled), process column on elevated structures.

Category 2: Structures whose failure can cause conditions that can lead directly or indirectly to serious fire hazard/extensive damage within the plant complex. Structures, which are required to handle emergencies immediately after an earthquake, are also included.

For example, control buildings, fire station, fire water pump house, Diesel Generator hall, Demineralised water treatment plant etc.

Category 3: Structures whose failure although expensive does not lead to serious hazard within the plant complex.

For example, overhead water tank, intake structure, effluent treatment plants.

Category 4: All other structures.

For example, administration building, laboratory building, other non-plant building, and utility structure.

The term 'failure' used in this context means loss of function and not complete collapse.

6.6.2 Design Basis Earthquake (DBE)

Design basis earthquake (DBE) for a specific site is to be determined based on either site-specific studies or in accordance with provisions of IS: 1893 (Part 1)-2015. For all-important projects, and all industries dealing with highly hazardous chemicals, site-specific spectra for earthquake with probability of exceedance of 2% in 50 years (MCE) and 10% in 50 years (DBE) shall be evaluated. All Category 1 industrial structures shall be analyzed using site-specific spectra. However, if site-specific studies are not carried out, the code-specified spectra may be used. Structures in Category 1 shall be designed for MCE (which is twice of DBE). Structures in Category 2, 3, and 4 shall be designed for DBE.

For site-specific spectra, horizontal seismic coefficient is

$$A_{\mathrm{h}} = \left(\frac{I}{R}\right)\left(\frac{S_{\mathrm{a}}}{g}\right) \tag{6.18}$$

For code-specified spectra, horizontal seismic coefficient is

$$A_{\mathrm{h}} = \left(\frac{Z}{2}\right)\left(\frac{I}{R}\right)\left(\frac{S_{\mathrm{a}}}{g}\right) \tag{6.19}$$

Importance factor (I) given in Table 6.12 is relative importance assigned to the structure to account the consequences of its damage of the structure.

6.6.3 Damping

Damping factor given in Table 6.13 depends upon the material and type of construction of the structure and the strain level during earthquake.

6.6.4 Analysis Techniques Based on Category of Structure and Seismic Zone

Methods of analysis used for structure types are given in Table 6.14.

Detailed analysis: In addition to detailed dynamic analysis, secondary effects (influence of P-Δ effect) and torsion

Table 6.12 Importance factors (I) for various industrial structures [2]

Categories of structures	Importance factor
Structures in category 1	2.0
Structures in category 2	1.5
Structures in category 3	1.25
Structures in category 4	1.0

Table 6.13 Damping ratio coefficient for various materials for DBE and MCE conditions [2]

Material	DBE	MCE
Aluminum	0.02	0.04
Steel	0.02	0.04
Reinforced concrete	0.05	0.07
Prestressed concrete	0.03	0.05
Water and other liquids (contained in tank)	0.005	0.005
Granular materials contained in tanks	0.10	0.10

Note For combined structure, a composite damping ratio coefficient shall be evaluated. If a composite damping ratio coefficient is not evaluated, material having lower damping shall be considered

Table 6.14 Analysis technique [2]

Category of structure	Seismic zone	Analysis technique
Category 1	All seismic zones	Detailed analysis
Category 2	III, IV, and V	Detailed analysis
Category 3	IV and V	Detailed analysis
Category 2	II	Simplified analysis
Category 3	II and III	Simplified analysis
Category 4 (for building)	All seismic zone	Simplified analysis

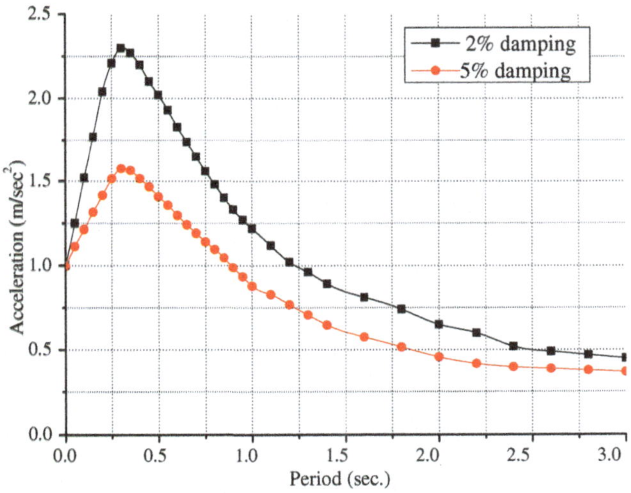

Fig. 6.13 Acceleration response spectra

effect are considered. Also, seismic analysis shall be performed for the three orthogonal directions (two horizontal and one vertical) component of earthquake motion. The responses from the earthquake motion in each direction shall be combined as specified in Sect. 6.5.4.4.

Simplified analysis: Simplified analysis shall be carried out by applying equivalent static lateral loads along each of the three principal directions.

Static load at each node = mass of influence area at the node × design spectral acceleration value

In the above method of analysis effect of soil–structure interaction and interaction effects between structure and equipment shall be considered.

Example 6.1 Read the ordinates of the acceleration spectra of Fig. 6.13 for natural period 0.3 s, and for

(A) Damping is 5% of critical.
(B) Damping is 2% of critical.

What is the ratio of the maximum response for the above two values of damping?

Solution

S_a (5%) = 1.6 m/s^2

S_a (2%) = 2.3 m/s^2

Ratio of maximum response = 1.4375

Example 6.2 A cycle stand of mild steel can be modeled as a SDOF system is shown in Fig. 6.14a having weight of roof is 60 kN. Obtain the maximum base shear and base moment. Assume natural period of cycle stand is 0.4 s (refer Fig. 6.13 in Example 6.1).

Solution

Ht = 2.5 m (Fig. 6.14a)

S_a (2%) = 2.2 m/s^2

Mass = weight/g = 6.12

Base shear = inertia force = $F = M \times S_a$ = 13.46 kN

Base moment = $F \times$ Ht. = 33.65 kN-m

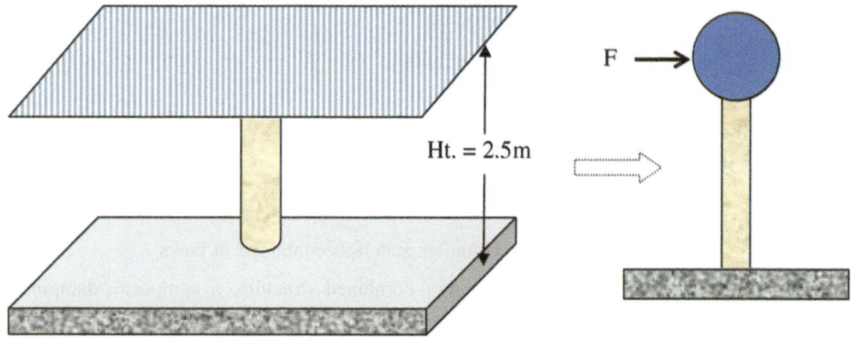

(a) Cycle Stand   Equivalent SDOF System

Fig. 6.14 **a** Cycle stand and **b** equivalent SDOF system

Example 6.3 Loads on the floor slab are shown in Fig. 6.15. Locate the center of mass of floor slab.

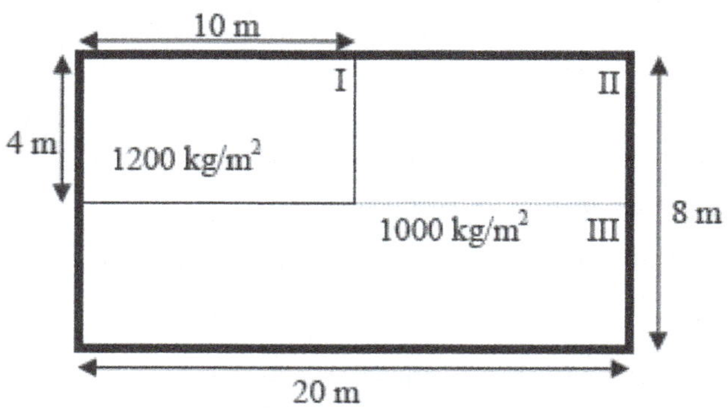

Solution

i. Mass of Part I is 1200 kg/m^2.
ii. Mass of parts II and III is 1000 kg/m^2.
iii. Let origin be at point A, from the left bottom corner, and the coordinates of the center of mass be at (X, Y).

$$X = \frac{(10 \times 4 \times 1200) \times 5 + (10 \times 4 \times 1000) \times 15 + (20 \times 4 \times 1000) \times 10}{(10 \times 4 \times 1200) + (10 \times 4 \times 1000) + 20 \times 4 \times 1000)}$$
$$= 9.76\,\text{m}$$
$$Y = \frac{(10 \times 4 \times 1200) \times 6 + (10 \times 4 \times 1000) \times 6 + (20 \times 4 \times 1000) \times 2}{(10 \times 4 \times 1200) + (10 \times 4 \times 1000) + 20 \times 4 \times 1000)}$$
$$= 4.1\,\text{m}$$

iv. Hence, coordinates of center of mass are (9.76, 4.1).

Example 6.4 Find the location of center of stiffness of frame structure as shown in Fig. 6.16. Size of all column and beam is the same.

Solution

i. Let the lateral stiffness of each transverse frame be k.
ii. Y-coordinate of center of stiffness is located symmetrically, i.e., at 4.0 m from the left bottom corner.
iii. X-coordinate of center of stiffness is located unsymmetrically

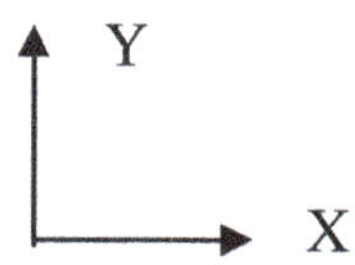

$$X = \frac{k \times 0 + k \times 4 + k \times 8 + k \times 16}{k + k + k + k}$$
$$= 7.00\,\text{m}$$

iv. Hence, coordinates of center of stiffness are (7.00, 4.0).

Example 6.5: A single-story building (Fig. 6.17) has two shear walls in each direction. All the shear walls are 250 mm thick and 5 m long. Assume effect on columns is negligible. Story height is 4.0 m. Floor consists of cast in situ reinforced concrete. Design shear force on the building is 200 kN in both X- and Y-directions. Determine design shear forces on shear walls using the torsion provision of IS: 1893(Part 1)-2016.

Solution

i. Center of mass (CM) shall be the geometric center of the floor slab, i.e., (10.0, 5.0).
ii. Let the lateral rigidity of each transverse wall be k.

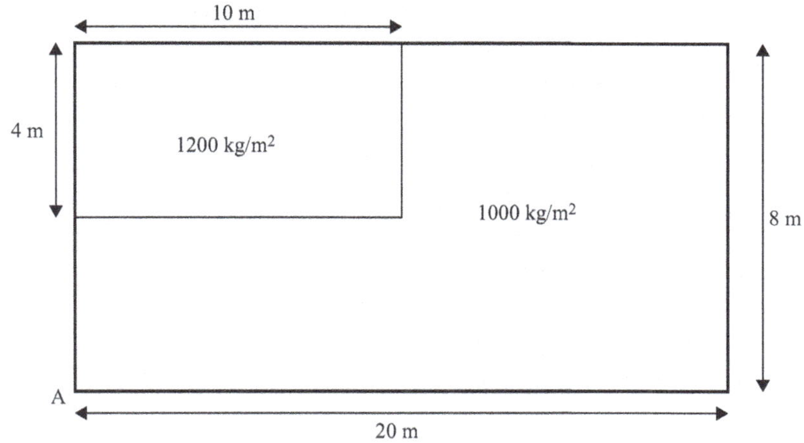

Fig. 6.15 Plan of floor slab

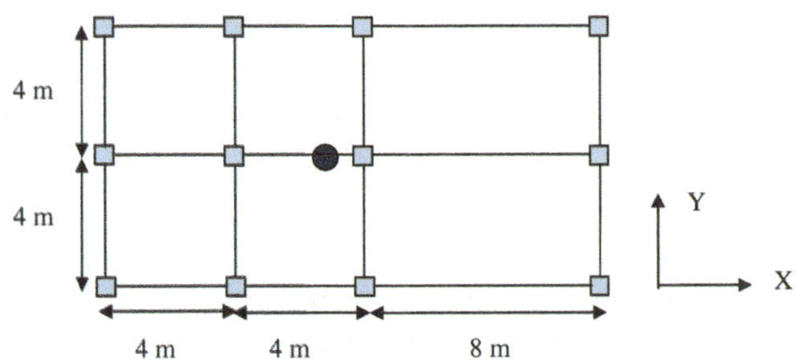

Fig. 6.16 Plan of frame structure

iii. Y-coordinate of center of stiffness is located symmetrically, i.e., at 5.0 m from the left bottom corner.

iv. X-coordinate of center of stiffness is located unsymmetrically

$$X = \frac{k \times 0 + k \times 5 + k \times 5 + k \times 15}{k + k + k + k}$$

$$X = 6.25 \text{ m}$$

v. Hence, coordinates of center of rigidity (CR) will be at (6.25, 5.0).

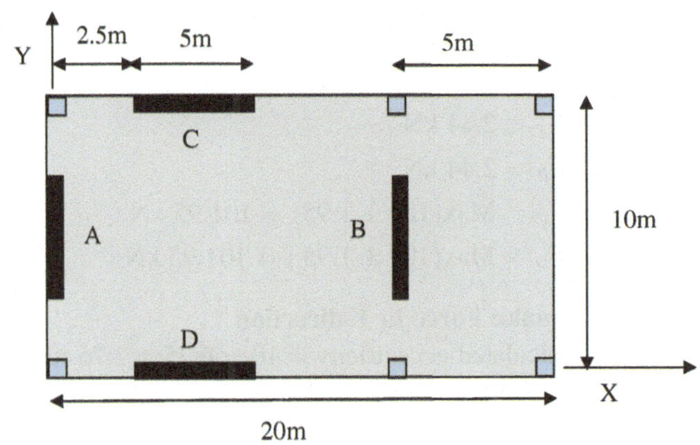

Fig. 6.17 Plan of frame structure

A. **Earthquake Force in *X*-direction**:

i. Because of symmetry in this direction, calculated eccentricity = 0.0 m,
ii. Design eccentricity:
$e_d = 1.5 \times 0.0 + 0.05 \times 10 = 0.5$ or
$e_d = 0.0 - 0.05 \times 10 = -0.5$ (Clause 7.9.2 of IS: 1893-2016)
iii. Lateral forces in the walls due to translation:

$$F_{CT} = \frac{K_C}{K_C + K_D} F = 100.0 \text{ kN}$$

$$F_{DT} = \frac{K_D}{K_C + K_D} F = 100.0 \text{ kN}$$

iv. Lateral forces in the walls due to torsional moment:

$$F_{iR} = \frac{K_i r_i}{\sum_{i=A,B,C,D} K_i r_i^2} (F e_d)$$

where r_i is the distance of the shear wall from center of rigidity.
All the walls have same stiffness as $KA = KB = KC = KD = k$, and
$r_A = -6.25$ m; $r_B = -6.25$ m; $r_C = 5.0$ m; $r_D = -5.0$ m; and $e_d = \pm 0.5$ m

$$F_{AR} = \frac{r_A k}{(r_A^2 + r_B^2 + r_C^2 + r_D^2)k} (F e_d) = \pm 2.44 \text{ kN}$$

Similarly, $F_{BR} = \pm 2.44$ kN; $F_{CR} = \pm 1.95$ kN; $F_{DR} = \pm 1.95$ kN

v. Total lateral forces in the walls due to seismic load in *X*-direction:

$$F_A = 2.44 \text{ kN}$$
$$F_B = 2.44 \text{ kN}$$
$$F_C = \text{Max}(100 \pm 1.95) = 101.95 \text{ kN}$$
$$F_D = \text{Max}(100 \pm 1.95) = 101.95 \text{ kN}$$

B. **Earthquake Force in *Y*-direction**:
i. Calculated eccentricity = 10 − 6.25 = 3.75 m,
ii. Design eccentricity:

$$e_d = 1.5 \times 3.75 + 0.05 \times 20 = 6.625 \text{ m or}$$
$$e_d = 3.75 - 0.05 \times 20 = 2.75 \text{ m}$$

iii. Lateral forces in the walls due to translation:

$$F_{AT} = \frac{K_A}{K_A + K_B} F = 100.0 \text{ kN}$$

$$F_{BT} = \frac{K_B}{K_A + K_B} F = 100.0 \text{ kN}$$

iv. Lateral force in the walls due to torsional moment: when $e_d = 6.625$ m

$$F_{AR} = \frac{r_A k}{(r_A^2 + r_B^2 + r_C^2 + r_D^2)k} (F e_d) = 32.31 \text{ kN}$$

Similarly,
$F_{BR} = 32.31$ kN; $F_{CR} = -25.85$ kN; $F_{DR} = 25.85$ kN
v. Total lateral forces in the walls:

$$F_A = 100 - 32.31 = 67.69 \text{ kN}$$
$$F_B = 100 + 32.31 = 132.31 \text{ kN}$$
$$F_C = -25.85 \text{ kN}$$
$$F_D = 25.85 \text{ kN}$$

vi. Similarly, when $e_d = 2.75$ m, then the total lateral forces in the walls will be

$$F_{AR} = \frac{r_A k}{(r_A^2 + r_B^2 + r_C^2 + r_D^2)\, k} (F e_d) = 13.41 \text{ kN}$$

Similarly,
$$F_{BR} = 13.41 \text{ kN}; F_{CR} = -10.73 \text{ kN}; F_{DR} = 10.73 \text{ kN}$$
$$F_A = 100 - 13.41 = 86.59 \text{ kN}$$
$$F_B = 100 + 13.41 = 113.41 \text{ kN}$$
$$F_C = -10.73 \text{ kN}$$
$$F_D = 10.73 \text{ kN}$$

vii. Maximum forces in walls due to seismic load in *Y*-direction:

$$F_A = \text{Max}(67.69,\ 86.59) = 86.59 \text{ kN};$$
$$F_B = \text{Max}(113.41,\ 132.31) = 132.31 \text{ kN};$$
$$F_C = \text{Max}(25.85,\ 10.73) = 25.85 \text{ kN};$$
$$F_D = \text{Max}(25.85,\ 10.73) = 25.85 \text{ kN};$$

C. **Combining the Forces Obtained from Seismic Loading in *X*- and *Y*-directions**:

$$F_A = 86.59 \text{ kN}$$
$$F_B = 132.31 \text{ kN}$$
$$F_C = 101.95 \text{ kN}$$
$$F_D = 101.95 \text{ kN}.$$

However, as per Clause 7.9.1 of IS: Part 1:2016 also states that "negative torsional shear shall be neglected." Hence, wall 'A' should be designed for not less than design shear force of 100 kN.

6.7 Ductility Requirement in Frame Structure

After performing the detailed analysis and design of the structure, detailing in the structure elements plays a very important role in the event of earthquake. Basic dimension required for beam and column is given in Table 6.15; however, details shall be referred in IS: 13920-2016 and ACI 318 M-11 for the frame structure.

Example 6.6: A six-storied reinforced concrete-framed structure has plan dimensions 32.0 m × 14.0 m and is shown in Fig. 6.18. The sizes of exterior columns (9 each on grid A and C) are 300 mm × 500 mm and the interior columns (9 on grids B) are 300 mm × 600 mm for the bottom three floors and, respectively, 300 mm × 400 mm and 300 mm × 500 mm for the top three floors. The height between floors is 3.5 m. The intensity of normal live load is 3.0 kN/m^2. Building is located in Mumbai, India. Foundation of building is resting on hard soil, and its net safe bearing capacity of soil at founding level is 250.0 kN/m^2. Concrete is of Grade M30 and steel Fe415. Determine the seismic load on the structure by seismic coefficient method (static analysis method). Draw the distribution of force and shear along the height of structure (Figs. 6.19, 6.20, 6.21 and 6.22). Figures 6.19, 6.20 shows the cross elevation and longitudinal elevation of the structure respectively.

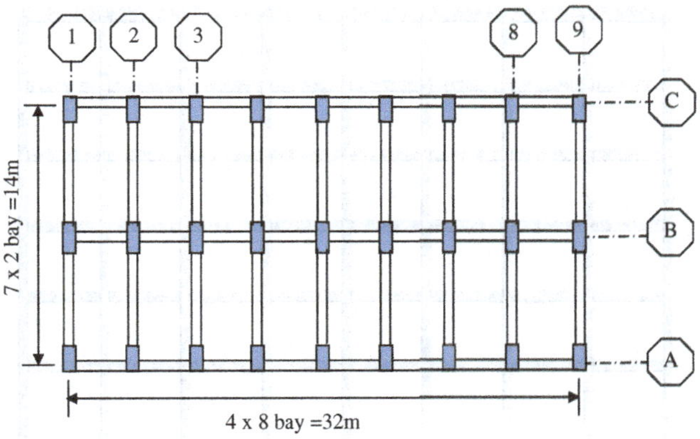

Fig. 6.18 Plan of structure

Fig. 6.19 Cross-elevation

Design Parameter

i. Mumbai falls in Seismic Zone III, and the zone factor Z is 0.16 (Table 6.5). Being this as office building, the importance factor, I, is 1.0 (Table 6.5). Building is provided with details as per IS: 13920-2016; hence, the response reduction factor, R, is 5 (Table 6.10).

Load Calculation on Structure

i. Load on the structure consists of self-weight of structural element, superimposed load on the structural

Table 6.15 Ductility provision as per IS: 13920-1993 and ACI 318 M-11

	IS: 13920-2016 "Ductile detailing of reinforced concrete structures subjected to seismic forces—code of practice" [10]	ACI 318 M-11 "Building code requirements for structural concrete and commentary" [11]
Applicability	Seismic Zones III, IV, V	Moderate and high seismic risk and structure assigned for intermediate and high performance
Flexural members	• Factored axial stress on member under EQ $\leq 0.1 f_{ck}$ where f_{ck} = compressive strength of cube • Width-to-depth ratio >0.3 • Min. width = 200 mm	• Factored axial compressive force on the member, P_u, shall not exceed $A_g f_c'/10$ where A_g = gross cross section and f_c' = compressive strength of cylinder • Clear span for member shall not be less than four times its effective depth • Width-to-depth ratio >0.3 • Min. width = 250 mm
Column and frame members subjected to bending and axial load	• Factored axial stress on member under EQ $\geq 0.1 f_{ck}$ • Width-to-depth ratio >0.4 • Min. dimension = 200 mm subjected to • Min dimension = 300 mm (beam of span >5 m) • Min dimension = 300 mm (column of unsupported length >4 m)	• Factored axial compressive force Pu exceeding $A_g f_c'/10$ • Width-to-depth ratio > 0.4 • Min. dimension = 300 mm • Resisting EQ-induced forces

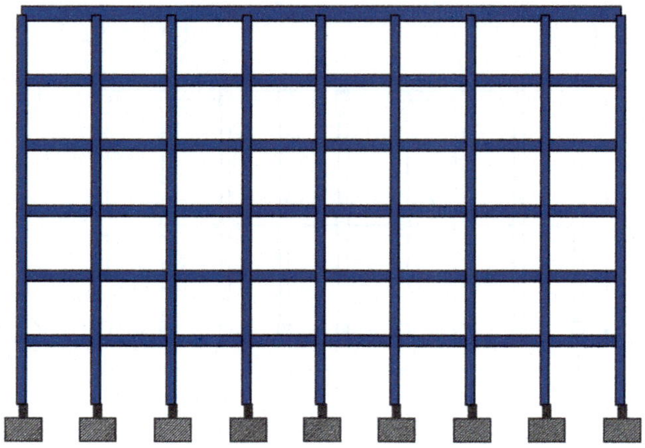

Fig. 6.20 Longitudinal elevation

member, live load, wind load, earthquake load, and snow load. In the present example, dead load consists of self-weight of floor slab/roof slab, floor finish at floor level, waterproofing course at the roof level, column, beam, and brick partition wall.

ii. Following structural parameters are considered for load calculation:
Thickness of floor and roof slab = 150 mm.
Thickness of floor finish = 50 mm.
Average thickness of waterproof course = 120 mm.
Height of parapet brick wall (230 thk.) = 1200 mm.
iii. Floor beam and roof beam 230 × 650 for 7-m beam span and 230 × 400 for 4-m beam span,

$$\text{Total DL on each floor} = (\text{Self wt. of slab} + \text{Floor finish}) + (\text{Column} + \text{Beam} + \text{Brick partition wall})$$

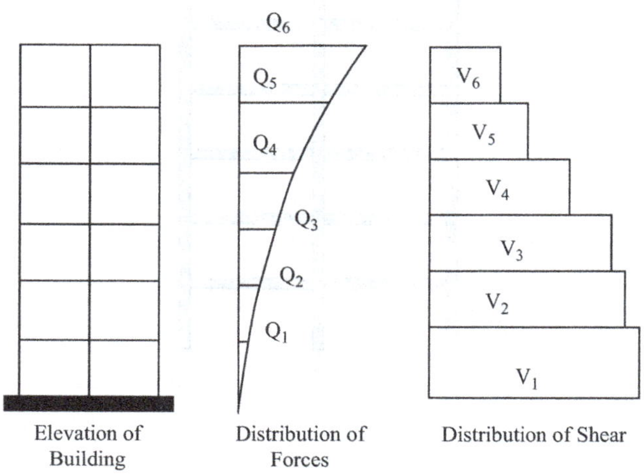

Fig. 6.21 Distribution forces and shear with the building height

Structural components	Size × density × length/area	Load (kN)
Self-wt. of floor slab	0.15 × 25 × 32 × 14	1680.00
Floor finish	0.05 × 25 × 32 × 14	537.60
Column (300 × 500)	0.3 × 0.5 × 25 × 3.5 × 18	236.25
Column (300 × 600)	0.3 × 0.6 × 25 × 3.5 × 9	141.75
Beam (230 × 400) in *X*-dir.	0.23 × 0.4 × 25 × 87.9	202.17
Beam (230 × 650) in *Y*-dir.	0.23 × 0.65 × 25 × 111.6	417.10
Brick wall (230 thk.) in *X*-dir.	0.23 × 3.1 × 18 × 87.9	1128.10
Beam (230 × 650) in *Y*-dir.	0.23 × 2.85 × 18 × 111.6	1316.80
Total DL on each floor		5660.00

Total design live load for EQ on each floor = 25% of IL = (0.25 × 3) × (32 × 14) = 336 kN

Fig. 6.22 Mode shapes for first, second, and third mode

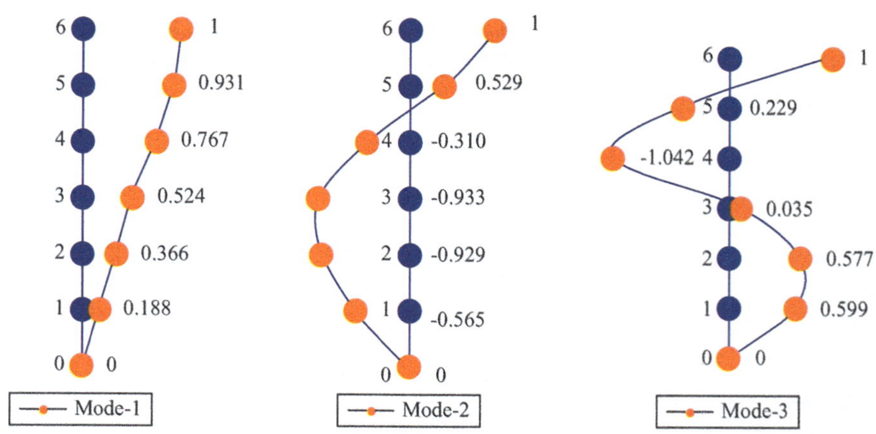

Total Weight on each floor = 5660 + 336 = **5996 kN**

Total DL on Roof = (Self Wt.of Slab + Water proof course) + (Beam + Parapet Wall)

Structural components	Size × density × length/area	Load (kN)
Self-wt. of roof slab	0.15 × 25 × 32 × 14	1680.00
Waterproof course	0.12 × 20 × 32 × 14	1075.20
Beam (230 × 400) in *X*-dir.	0.23 × 0.4 × 25 × 87.9	202.17
Beam (230 × 650) in *Y*-dir.	0.23 × 0.65 × 25 × 111.6	417.10
Parapet brick wall (230 thk.) in *X*-dir.	0.23 × 1.2 × 18 × 64	317.95
Parapet brick wall (230 thk.) in *Y*-dir.	0.23 × 1.2 × 18 × 28	139.10
Total DL at roof		3832.00

Total design live load for EQ on roof = 0% of IL on roof = 0.0

Total weight on roof = 3832 + 0=**3832 kN**

Total weight '*W*' = 5 × 5660 + 3832 = **32132 kN**

iv. Fundamental period = 0.075 $h^{0.75}$ = 0.075*$(21)^{0.75}$ = 0.735 s.

$S_a/g = 1.3605, Z = 0.16(\text{Zone} - \text{III}), I = 1.0,\ R = 5(\text{For, SMRF})$

$$A_h = \frac{ZIS_a}{2Rg}$$

$A_h = 0.021768$

v. Design seismic base shear, $V_B = A_h W$, = 0.021768 × 32,132 = 699.44 kN,

vi. Lateral load and shear force at various floor levels are given in Table 6.16.

$$Q_i = V_B \frac{W_i h_i^2}{\sum_{j=1}^{n} W_j h_j^2}$$

Vertical distribution of base shear at different floor levels, Q_i

Example 6.7: For the structure in Example 6.6, calculate the dynamic properties (natural period and mode shape) for vibration in *X*-direction and *Y*-direction. Obtain the design seismic force in *X*-direction and *Y*-direction by dynamic analysis method as per IS: 1893, and distribute it along the building height. Compare shear force by the seismic coefficient method and dynamic analysis method.

(a) **Equivalent Spring**

i. Equivalent k_c of single column (300 × 500),

$$k_c = 12E_cI/L3$$

where $E_c = 27.38 \times 10^6 \text{kN/m}^2$

$B = 0.3\text{m}$

$D = 0.5\text{m}$

$L = 3.5\text{m}$

$I = BD3/12$

$k_c = 23947.5\ \text{kN/m}$

ii. Equivalent k_c of single column (300 × 600), k_c = 42,875 kN/m,

iii. Equivalent k_c of single column (300 × 400), k_c = 12,261 kN/m,

iv. Total *k* per floor (for bottom three floors) = 18 × 23,947.5 + 9 × 42,875 = 816,930 kN/m,

v. Total *k* per floor (for top three floors) = 18 12,261 + 9 × 23,947.5 = 436,226 kN/m.

Equivalent spring–mass system for vibration along the shorter direction is given in Table 6.17.

(b) **Free Vibration Characteristics**

i. Free vibration characteristics, i.e., natural period and mode shapes, could be determined by the numerical methods (Holzer's method and Stodola's method).

Table 6.16 Lateral load and shear force

Mass no.	W_i (kN)	h_i (m)	$W_i\ h_i^2$	$\frac{W_i h_i^2}{\sum W_j h_j^2}$	Q_i (kN)	V_i (kN)
6 (roof)	3832	21.0	1,689,912	0.307	214.78	214.78
5	5660	17.5	1,733,375	0.315	220.30	435.08
4	5660	14.0	1,109,360	0.202	140.99	576.07
3	5660	10.5	624,015	0.113	79.31	655.38
2	5660	7.0	277,340	0.050	35.25	690.63
1 (bottom floor)	5660	3.5	69,335	0.013	8.81	699.44
$\sum W_j\ h_j^2$			5,503,337			

Table 6.17 Value of mass and stiffness

Mass no.	Mass, $m = W_i/g$ kN	k = 12EI/L^3 kN/m
6 (roof)	390.6	436,226
5	576.9	436,226
4	576.9	436,226
3	576.9	816,930
2	576.9	816,930
1 (bottom floor)	576.9	816,930

Table 6.18 Value of mode shapes

Parameter	First mode	Second mode	Third mode
Period (s)	0.7148	0.2739	0.1695
Mode participation factor $P_k = \sum m\phi / \sum m\phi^2$	1.34	−0.493	0.246
Φ_6	1.000	1.000	1.000
Φ_5	0.931	0.529	−0.229
Φ_4	0.767	−0.310	−1.042
Φ_3	0.524	−0.933	0.035
Φ_2	0.366	−0.919	0.577
Φ_1	0.188	−0.565	0.599

ii. Various quantities obtained for the first three modes of vibration are given in Table 6.18.

(c) **Earthquake Response**

i. Spectral acceleration for 5% damping of above periods

$$S_a/g(1) = 1.399,\ A_k = 0.02238$$
$$S_a/g(2) = 2.5, A_k = 0.04$$
$$S_a/g(3) = 2.5, A_k = 0.04$$

ii. Design lateral load and shear force at each floor in each mode are obtained by formula

$$Q_i^1 = A_k \Phi_1 P_k W_1 = 0.02238 \times 1 \times 1.34 \times 3832 = 114.92$$

(d) **Total Response**

i. Total response is obtained as a combination of the response in three modes,

ii. Peak story shear force,

$$V_i = \sum (\lambda_k)2,$$

where

λ_k = absolute value of quantity in mode k.

Peak story shear force for the structure is calculated and given in Table 6.19. Comparison of shear force by seismic coefficient method and dynamic method is given in Table 6.20.

Table 6.19 Peak story shear force

Mass no.	Mode I		Mode II		Mode III		Peak story shear force (V_i) (kN)
	Q	V	Q	V	Q	V	
6 (roof)	114.92	114.92	−75.57	−75.57	37.71	37.71	142.61
5	158.03	272.95	−59.04	−134.61	−12.75	24.95	305.36
4	130.19	403.14	34.60	−100.01	−58.03	−33.08	416.67
3	88.94	492.08	104.14	4.13	1.95	−31.13	493.08
2	62.12	554.20	102.57	106.70	32.14	1.00	564.38
1 (bottom floor)	31.91	586.11	63.06	169.76	33.36	34.37	611.17

Table 6.20 Comparison of shear force by different methods

Mass no.	Shear force (kN)	
	Seismic coefficient method	Dynamic method
6 (roof)	214.78	142.61
5	435.08	305.36
4	576.07	416.67
3	655.38	493.08
2	690.63	564.38
1 (bottom floor)	699.44	611.17

Example 6.8: For the structure in Example 6.6, model the structure using finite element method and obtain the dynamic properties (frequency, natural period and mode participation factor, mode shape) for vibration in *X*-direction, *Y*-direction, and *Z*-direction. Obtain the design seismic forces in *X*-direction, *Y*-direction, and *Z*-direction by dynamic analysis method as per IS: 1893 with the help of commercial software in structural elements. Calculate the load combination of structural members as per the IS: 1893 and IS: 456-2000. Design the structural elements like footing, column, beam, and slab. Draw the reinforcement detailing of structural elements as per the provision of IS: 13920-2016.

Solution

The building consists of superstructure and substructure. When the structure is not supported on a rock or rocklike soil foundation material, soil–structure interaction (SSI) shall be considered. In the present structure, the foundation is resting on hard strata; hence, fixed-based analysis may be assumed. The building shall be modeled in two parts as superstructure and substructure. Three-dimensional FE model of structure is shown in Fig. 6.23.

A. **Superstructure**

i. The complete structure is idealized as three-dimensional space frame using commercial software. Beam and column are idealized using beam element.

ii. Static analysis has been carried out for various loads like self-weight and other permanent loads using 3D finite element model. As the structure is founded on rock, boundary condition at base of

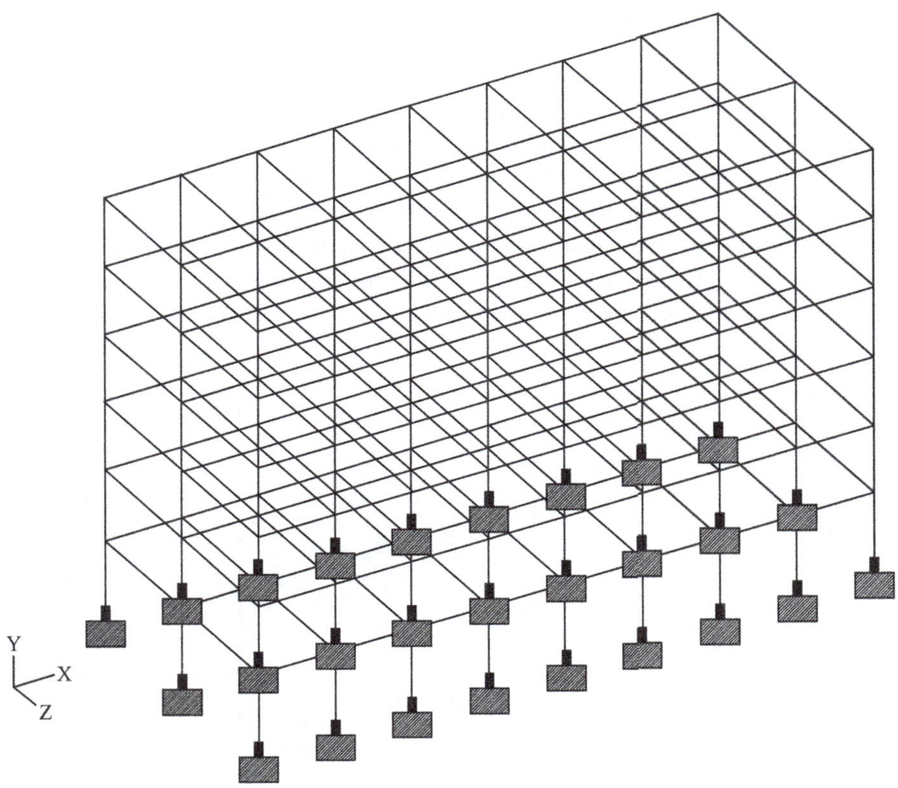

Fig. 6.23 Three-dimensional FE model of structure

Table 6.21 Frequency, time period, and mass participation factor

Mode	Frequency (Cycle/s)	Period (s)	Mass participation factor in %		
			Sum-*X*	Sum-*Y*	Sum-*Z*
1	0.706	1.41665	84.404	0.000	0.000
2	0.843	1.18635	84.404	0.000	80.091
10	2.266	0.44127	94.170	0.000	80.507
64	8.978	0.11138	100.000	51.716	99.046
71	9.944	0.10056	100.000	76.106	99.767
100	13.058	0.07658	100.000	85.174	100.000
125	24.717	0.04046	100.000	89.044	100.000
150	27.956	0.03577	100.000	93.140	100.000
180	29.528	0.03387	100.000	95.87	100.000

columns has been assumed to be fixed in all directions.

iii. Dynamic analysis will be carried out as per input data given in IS: 1893–2016 using response spectrum method. Dynamic analysis shall be carried out to assess the structural response due to earthquake motion in three orthogonal directions (two horizontal and one vertical) simultaneously. Seismic analysis shall be performed by response spectrum method using the acceleration response spectrum as per IS: 1893-2016 for Soil Type I. Response due to ground motions in three orthogonal directions shall be combined using SRSS method. Each direction modal response will be combined by SRSS method. In case of closely spaced modes, the response will be combined by CQC method.

iv. The number of modes considered shall be such that the highest mode corresponds to frequency of 33 Hz, or it is ensured that 90% of the total mass gets excited in each coordinate direction or inclusion of higher modes up to frequency given above does not increase response by more than 10%. Missing mass correction will be done for unanticipated mass.

v. These member/element forces due to earthquake are combined with the member/element forces due to other loads, using load factor given in Sect. 6.5.6.1.

B. **Substructure**

i. Substructure consists of isolated footing. Isolated footings are provided below the columns to transmit load to approved strata below.

ii. Table 6.21 shows the frequency, time period, and mass participation factor of the building for the dominant mode of participation in *X*-direction, *Y*-direction, and *Z*-direction. In the *X*-direction, 84.404% mass has been participated in first mode, in the *Y*-direction 80.091% mass has been participated in second mode, and in the *Z*-direction 51.716% mass has been participated in 64th mode. Normally, the mass participated in the vertical direction is at higher mode due to the fact that lateral stiffness is quite large in the vertical direction. Mode shape of the structure's first three modes is shown in Fig. 6.24b–d.

iii. Displacement in *X*-direction and *Z*-direction is shown in Fig. 6.25a, b, respectively. Maximum displacements observed at the top floor are 9.091 and 7.952 mm in *X*-direction and *Z*-direction, respectively. From Table 6.22, horizontal

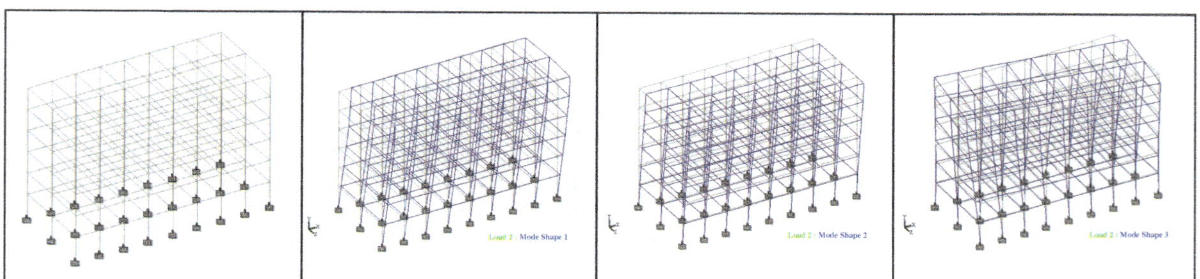

(a) FE Modal | (b) First Mode Shape | (c) Second Mode Shape | (d)Third Mode Shape

Fig. 6.24 FE model and mode shape of structure

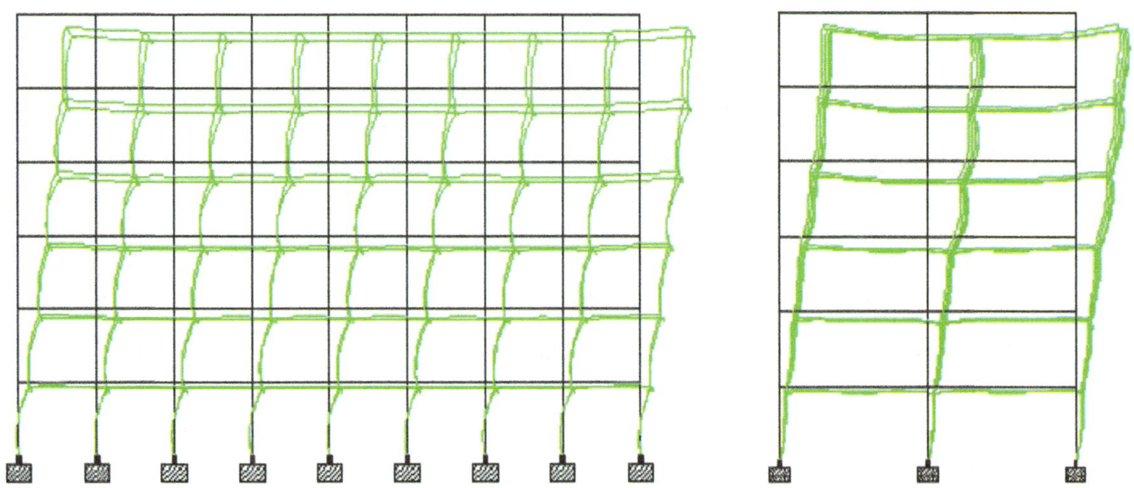

Fig. 6.25 **a** *X*-displacement (load case 14) and **b** Z-displacement (load case 7)

displacement (story drift) is within the allowable horizontal displacement as per the IS: 1893.

C. **Design of Structural Members**

i. Design of structural members is performed based on the result obtained from the analysis concept of design of structural member adopted as the provision of IS: 456-2000 and is given below.

D. **Design Philosophy**

i. The design of all structural members shall be carried out in accordance with IS: 456-2000. The method of design shall be based on limit state concept. The structure shall be designed to withstand safely all loads liable to act on it. It shall also satisfy serviceability requirements such as limitations on deflection and cracking.

ii. All relevant limit states are considered in design to ensure adequate degree of safety and serviceability. In general, the structure shall be designed on the basis of most critical limit state and shall be checked for other limit states. Following limit states are considered:

(a) **Limit State of Collapse**: The resistance to bending, shear, torsion, and axial loads at every section shall not be less than the appropriate value at that section produced by the probable most unfavorable combination of loads on the structure using the appropriate partial safety factors.

(b) **Limit State of Serviceability**: Limit of serviceability of structural elements is checked for limit state of deflection and cracking. Requirement of limit state of deflection and cracking as the code is given below.

Table 6.22 Story drift between two stories

Between two stories	Horizontal displacement as per analysis		Allowable horizontal displacement as per IS: 1893 ($0.004 \times H$)	
	X-dir.	*Z*-dir.	*X*-dir.	*Z*-dir.
	mm	mm	mm	mm
6–5	0.566	0.621	14	14
5–4	1.142	1.139	14	14
4–3	1.624	1.592	14	14
3–2	1.815	1.673	14	14
2–1	2.124	1.790	14	14
0–1	1.820	1.137	14	14

Table 6.23 Max. deflection allowed [9]

Load combination	Measured from	Max. deflection allowed
All load including the effects of temperature, creep, and shrinkage	As-caste level of floors, roofs, and all horizontal members	Span/250
All load including the effects of temperature, creep, and shrinkage after erection of partitions and the application of finishes	As-caste level of floors, roofs, and all horizontal members	Span/350 or 20 mm whichever is less

Table 6.24 Number of load case [1]

Sl. No	Load combination	Maximum no. of load cases
1	1.5 (DL + IL)	01
2	1.2 (DL + IL ± EL)	24
3	1.5 (DL ± EL)	24
4	0.9 (DL) ± 1.5(EL)	24 (stability check)
Total no. of load case for design		49

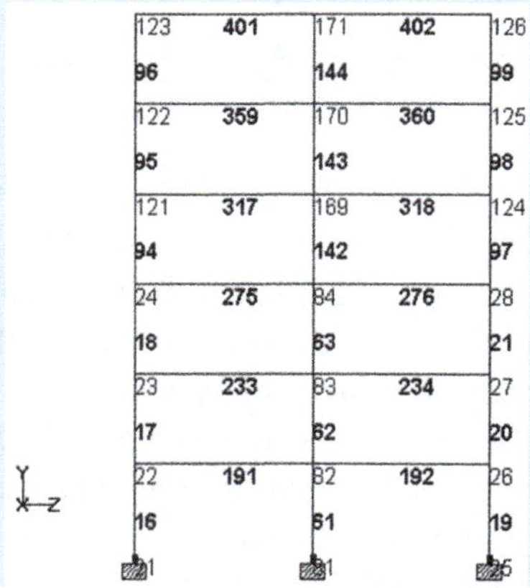

Limit State of Deflection

The vertical deflection as per Clause 23.2 of IS: 456-2000 shall be limited as given in Table 6.23.

The horizontal deflection (story drift) shall be limited to 0.004 * H (H = height between two stories).

Limit State of Cracking

- The crack width is to be calculated by relevant equations IS: 456:200 or ACI-224 R-90, wherever required.

E. **Design Load Case**

   i. Load case to be considered for analysis and design of structure and frame along grid 7 is given in Table 6.24.

F. **Design of Footing**

   i. There are different types of foundation system supporting the structure at the base. Types are isolated footing, combined footing, raft, strip footing, pile foundation. Type of foundation to be adopted to support the structure mainly depends on the types of soil strata below foundation, load coming at the base of structure.

   ii. In this problem, isolated footing has been design for node 81 for 49 numbers of load cases and maximum size works out from the above load case is considered as the final size of footing. Net safe bearing capacity of soil at 3.00 m below ground level is 250.0 kN/m^2.

   iii. For footing with moments or eccentricity about both axes, the soil pressure at is given by the equation:

$$p_{\max/\min} = \frac{P}{A} + \frac{M_x}{Z_y} + \frac{M_y}{Z_x}$$

where

P	total load from column
M_x	moment about X-direction
M_y	moment about Y-direction
Z_x & Z_y	section modulus

In case of earthquake condition, bearing pressure can be increased by 50%. Maximum and minimum bearing pressure due to different possible load combination with load case are tabulated in Table 6.25.

$$\text{Net bearing pressure} = \text{Bearing pressure due to total load} - \text{pressure due to footing \& soil above it}$$
$$= 256.11(\text{from load case10}) - (2.5*3.5*0.5*25 + 2.5*3.5*2.25*18)/2.5*3.5$$
$$= 256.11 - 52.99$$
$$= 203.12\ \text{kN/m}^2$$

iv. **Calculation along *Y*-direction**

$$\text{Max. BM at face of column} = pl^2/2 = 203.12 \times (1.45)^2/2 = 213.53\ \text{kN-m/m width of footing}$$

$$M_u/bd^2 = (1.2 \times 213.53 \times 10^6)/(1000 \times 630^2) = 0.645$$
$$P_t = 0.185\%\ (\text{From SP} - 16,\ \text{For M30 \& Fe 415})$$
$$A_{st} = 0.185 \times 1000 \times 630/100 = 1166\ \text{mm}^2/\text{m width of footing}$$

v. **Calculation along *X*-direction**

$$\text{Max. BM at face of column} = pl^2/2 = 203.12 \times (1.1)^2/2 = 122.9\ \text{kN - m/m width of footing}$$

$$M_u/bd^2 = (1.2 \times 122.9 \times 10^6)/(1000 \times 630^2) = 0.372$$
$$P_t = 0.112\%\ (\text{From SP} - 16,\ \text{For M30 \& Fe 415})$$

Hence, provide min reinforcement = 12% of gross area (Fig. 6.26)

$$A_{st} = 0.12 \times 1000 \times 700/100 = 840\ \text{mm}^2/\text{m width of footing}$$

vi. **Check for shear**

(a) **One-way Shear**: For one-way shear (beam action), the critical section is taken at a distance d from the face of column.
Nominal shear (τ_v) should be less than shear stress of concrete (τ_c).

(b) **Punching Shear**: For two-way action (punching shear), the critical section located all round is at a distance *d*/2 from the face of column.
Nominal shear (τ_v) should be less than allowable shear stress of concrete in punching = k_s (τ_c), where k_s = (0.5 + β_c)
$\tau_c = 0.16\sqrt{f_{ck}}$ (working state design)
$\tau_c = 0.25\sqrt{f_{ck}}$ (limit state design).

One-way shear and punching shear are found to be within permissible limits.

Analysis of footing is given in Table 6.26. Details of footing are given in Table 6.25, and typical reinforcement detail is shown in Fig. 6.27.

G. **Design of Column**

i. Column is designed for axial load and biaxial bending for 49 numbers of load cases. The

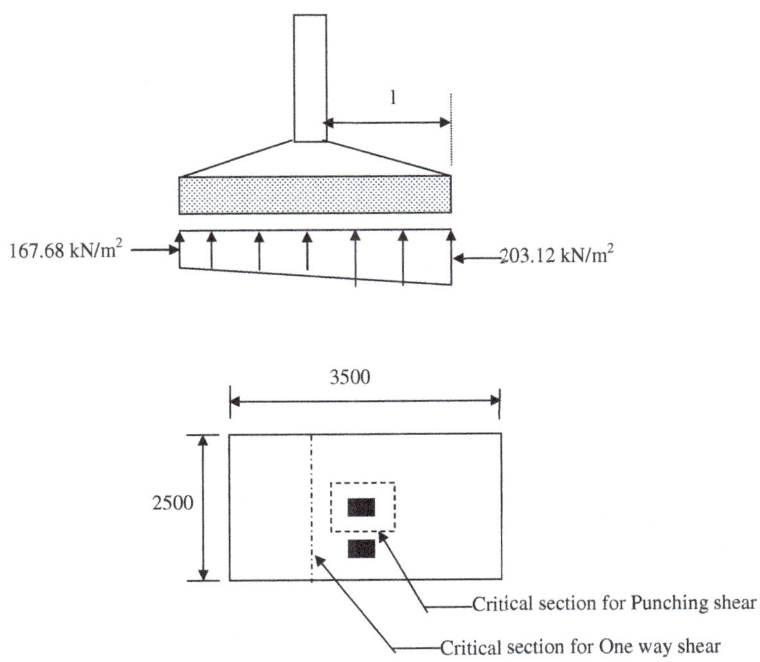

Fig. 6.26 Pressure diagram

Table 6.25 Details of footing

Size of footing	p_{max}	p_{min}	Reinforcement in y-direction	Reinforcement in x-direction
2.5m × 3.5m × 0.07/0.2	256.11 kN/m^2	220.67 kN/m^2	A_{st} = 2915 mm^2 16 Tor @160 c/c	A_{st} = 2940 mm^2 12 Tor @110 c/c

simplified method adopted by the code is based on Bresler's formulation for the 'load contour,' whereby an approximate relationship between $M_{ur,\,x}$ and $M_{ur,\,y}$ (for a specified $P_u = P_{ur}$) is established. This relationship is conveniently expressed in a non-dimensional form as follows:

$$[M_{ux}/M_{ux1}]^{\alpha n} + [M_{uy}/M_{uy1}]^{\alpha n} < 1.0$$

where

M_{ux}, M_{uy} moments about x- and y-axes due to design loads

M_{ux1}, M_{uy1} maximum uniaxial moment capacities for an axial load of P, bending about x- and y-axes, respectively, and

αn an exponent whose value depends on P_u/P_{uz} (see table)

where

$$\mathrm{P_u = 0.45 f_{ck} A_c + 0.75\, f_y A_{sc}}$$

P_u/P_{uz}	αn
≤ 0.2	1.0
≥ 0.8	2.0

ii. For intermediate values of $P_u/P_{uz,}$ linear interpolation may be done. Some of the critical load cases computed on the member No. 61 are given in Table 6.27.

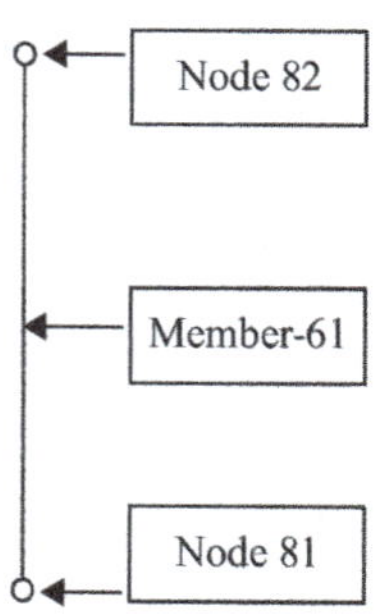

iii. Pitch and diameter of lateral tie will be least of the following:

(a) Least lateral dimension of compression member = 300 mm,

(b) 16 times the smallest dia of longitudinal reinforcement bar = 16 × 16 = 256 mm,

(c) 300 mm.

Hence, provide 8 tor @ 250 c/c spacing of tie bar.

iv. Details of reinforcement from critical load case of column as per the above formulation are calculated below:

Size of column	Longitudinal reinforcement required	Longitudinal reinforcement provided	No. of longitudinal bar in column	Tie bar
300 mm × 600 mm	A_{st} = 1800 mm^2	A_{st} = 1884 mm^2	6–20 Tor	8 tor @ 250 c/c spacing

Figure 6.28 shows the typical reinforcement detail of column.

H. **Design of Beam**

i. Envelope of bending moment diagram and shear force diagram is shown in Figs. 6.29 and 6.30, respectively. Design of beam is done as per the provisions of IS: 456-2000 [9] and is given in Table 6.28. Typical reinforcement details are shown in Fig. 6.31.

I. **Design of Slab**

i. A slab panel is bounded by main beams/secondary beams and is designed for combination of vertical static loads and the loads caused due to the vertical excitation of earthquake. A typical panel of slab is shown in Fig. 6.32.

ii. The vertical response acceleration at 33 Hz for 5% damping for floor under consideration is used for calculating the vertical seismic load for all slabs at that level. Design of slab is done as per the provisions of IS: 456-2000 [9] and is given in Table 6.29.

iii. Calculation of load due to earthquake

$$A_v = 2/3\, A_h = 0.0444$$

Considering peak value of S_a/g for structural time period in the absence of FRS value (Fig. 6.32)

Table 6.26 Analysis results of footing

Sl. No	Load case	Axial force, $F_y =$	Moment about *x*, $M_x =$	Moment about *y*, $M_y =$	Length of footing along *x*, $b_x =$	Length of footing along *y*, $b_y =$	Depth of footing, $D =$	Depth of soil above footing, $h =$	SBC of soil	Upward pr	Area of footing required	$p_{max} =$	$p_{min} =$
		kN	kN-m	kN-m	m	m	m	m	kN/m^2	kN/m^2	m^2		
1	DL + LL	1591.73	0	0.192	2.5	3.5	0.75	2	250	236.6624	8.28318	236.7	236.62
2	DL + LL + EQX + 0.3EQY + 0.3EQZ	1607.58	17.2075	34.0992	2.5	3.5	0.75	2	250	238.4738	8.34658	249.87	227.07
3	DL + LL + EQX + 0.3EQY − 0.3EQZ	1607.58	17.2067	34.0992	2.5	3.5	0.75	2	250	238.4738	8.34658	249.87	227.07
4	DL + LL + EQX − 0.3EQY + 0.3EQZ	1577.83	17.2075	34.0467	2.5	3.5	0.75	2	250	235.0738	8.22758	246.46	223.68
5	DL + LL + EQX − 0.3EQY − 0.3EQZ	1577.83	17.2075	34.0467	2.5	3.5	0.75	2	250	235.0738	8.22758	246.46	223.68
6	DL + LL + 0.3EQX + EQY + 0.3EQZ	1641.5	17.2075	10.22	2.5	3.5	0.75	2	250	242.35	8.48225	249.07	235.63
7	DL + LL + 0.3EQX + EQY − 0.3EQZ	1641.5	17.2075	10.22	2.5	3.5	0.75	2	250	242.35	8.48225	249.07	235.63
8	DL + LL − 0.3EQX + EQY + 0.3EQZ	1640.92	17.2067	10.3008	2.5	3.5	0.75	2	250	242.2833	8.47992	249.02	235.55
9	DL + LL − 0.3EQX + EQY − 0.3EQZ	1640.92	17.2075	10.3008	2.5	3.5	0.75	2	250	242.2833	8.47992	249.02	235.55
10	DL + LL + 0.3EQX + 0.3EQY + EQZ	1606.83	57.3575	10.1583	2.5	3.5	0.75	2	250	238.3881	8.34358	256.11	220.67
11	DL + LL + 0.3EQX − 0.3EQY + EQZ	1577.17	57.3567	10.1058	2.5	3.5	0.75	2	250	234.9976	8.22492	252.71	217.29
12	DL + LL − 0.3EQX + 0.3EQY + EQZ	1606.25	57.3567	10.3625	2.5	3.5	0.75	2	250	238.3214	8.34125	256.08	220.56
13	DL + LL − 0.3EQX − 0.3EQY + EQZ	1576.58	57.3567	10.415	2.5	3.5	0.75	2	250	234.931	8.22258	252.7	217.16
14	DL + LL − EQX − 0.3EQY − 0.3EQZ	1575.83	17.2075	34.3558	2.5	3.5	0.75	2	250	234.8452	8.21958	246.3	223.39
15	DL + LL − EQX − 0.3EQY + 0.3EQZ	1575.83	17.2067	34.3558	2.5	3.5	0.75	2	250	234.8452	8.21958	246.3	223.39
16	DL + LL − EQX + 0.3EQY − 0.3EQZ	1605.58	17.2075	34.3033	2.5	3.5	0.75	2	250	238.2452	8.33858	249.69	226.8
17	DL + LL − EQX + 0.3EQY + 0.3EQZ	1605.58	17.2067	34.3033	2.5	3.5	0.75	2	250	238.2452	8.33858	249.69	226.81
18	DL + LL − 0.3EQX − EQY − 0.3EQZ	1541.83	17.2075	10.4758	2.5	3.5	0.75	2	250	230.9595	8.08358	237.73	224.19
19	DL + LL − 0.3EQX − EQY + 0.3EQZ	1541.83	17.2067	10.4758	2.5	3.5	0.75	2	250	230.9595	8.08358	237.73	224.19
20	DL + LL + 0.3EQX − EQY − 0.3EQZ	1542.5	17.2075	10.045	2.5	3.5	0.75	2	250	231.0357	8.08625	237.72	224.35
21	DL + LL + 0.3EQX − EQY + 0.3EQZ	1542.5	17.2067	10.045	2.5	3.5	0.75	2	250	231.0357	8.08625	237.72	224.35
22	DL + LL − 0.3EQX − 0.3EQY − EQZ	1576.58	57.3575	10.415	2.5	3.5	0.75	2	250	234.931	8.22258	252.7	217.16
23	DL + LL − 0.3EQX + 0.3EQY − EQZ	1606.25	57.3575	10.3625	2.5	3.5	0.75	2	250	238.3214	8.34125	256.08	220.56
24	DL + LL + 0.3EQX − 0.3EQY − EQZ	1577.17	57.3575	10.1058	2.5	3.5	0.75	2	250	234.9976	8.22492	252.71	217.29
25	DL + LL + 0.3EQX + 0.3EQY − EQZ	1606.83	57.3575	10.1583	2.5	3.5	0.75	2	250	238.3881	8.34358	256.11	220.67

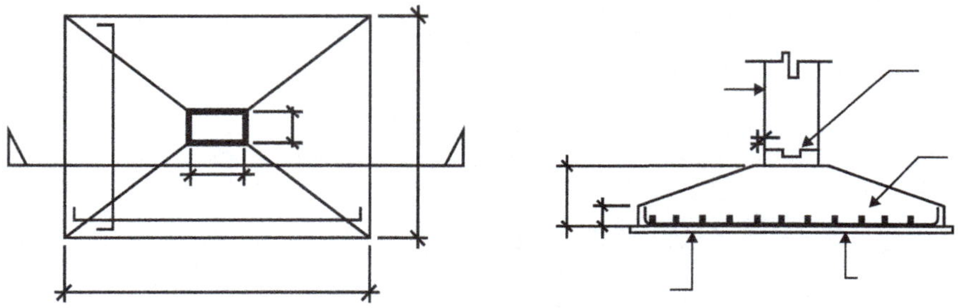

Fig. 6.27 Typical reinforcement detail of footing

Table 6.27 Load on joint 81 and 82 various load cases (kN and m)

Member	Load case		JT	Axial	Shear-Y	Shear-Z	Torsion	Mom-Y	Mom-Z
61	1	Dead load and live load	81	1591.7	0	0.08	0	−0.13	0
			82	−1592	0	−0.08	0	−0.15	0
	2	Seismic load in x-direction	81	0.99	0	18.64	0	34.2	0
			82	0.99	0	18.64	0	31.02	0
	3	Seismic load in y-direction	81	49.54	0	0.07	0	0.09	0
			82	49.54	0	0.07	0	0.16	0
	4	Seismic load in z-direction	81	0	25.6	0	0.37	0	57.36
			82	0	25.6	0	0.37	0	32.32
	5	1.5DLLL	81	2387.6	0	0.12	0	−0.19	0
			82	−2388	0	−0.12	0	−0.22	0
	6	1.2DLLL + 1.2EQX + 0.36EQY + 0.36EQZ	81	1929.1	9.22	22.48	0.13	40.92	20.65
			82	−1891	9.22	22.29	0.13	37.11	11.64
	7	1.2DLLL + 1.2EQX + 0.36EQY − 0.36EQZ	81	1929.1	−9.22	22.48	−0.13	40.92	−20.65
			82	−1891	−9.22	22.29	−0.13	37.11	−11.64
	8	1.2DLLL + 1.2EQX − 0.36EQY + 0.36EQZ	81	1893.4	9.22	22.43	0.13	40.86	20.65
			82	−1927	9.22	22.24	0.13	37	11.63
	9	1.2DLLL + 1.2EQX − 0.36EQY − 0.36EQZ	81	1893.4	−9.22	22.43	−0.13	40.86	−20.65
			82	−1927	−9.22	22.24	−0.13	37	−11.64
	10	1.2DLLL + 0.36EQX + 1.20EQY + 0.36EQZ	81	1969.8	9.22	6.89	0.13	12.26	20.65
			82	−1850	9.22	6.7	0.13	11.18	11.63
	11	1.2DLLL + 0.36EQX + 1.20EQY − 0.36EQZ	81	1969.8	−9.22	6.89	−0.13	12.26	−20.65
			82	−1850	−9.22	6.7	−0.13	11.18	−11.64
	12	1.2DLLL − 0.36EQX + 1.2EQY + 0.36EQZ	81	1969.1	9.22	−6.53	0.13	−12.36	20.65
			82	−1851	9.22	−6.72	0.13	−11.15	11.63
	13	1.2DLLL − 0.36EQX + 1.20EQY − 0.36EQZ	81	1969.1	−9.22	−6.53	−0.13	−12.36	−20.65
			82	−1851	−9.22	−6.72	−0.13	−11.15	−11.64
	14	1.2DLLL + 0.36EQX + 0.36EQY + 1.2EQZ	81	1928.2	30.72	6.83	0.44	12.19	68.83
			82	−1892	30.72	6.64	0.44	11.05	38.78
	15	1.2DLLL + 0.36EQX − 0.36EQY + 1.2EQZ	81	1892.6	30.72	6.78	0.44	12.13	68.83
			82	−1928	30.72	6.59	0.44	10.94	38.78
	16	1.2DLLL − 0.36EQX + 0.36EQY + 1.2EQZ	81	1927.5	30.72	−6.59	0.44	−12.43	68.83
			82	−1893	30.72	−6.78	0.44	−11.29	38.78
	17	1.2DLLL − 0.36EQX − 0.36EQY + 1.2EQZ	81	1891.9	30.72	−6.64	0.44	−12.5	68.83
			82	−1928	30.72	−6.83	0.44	−11.4	38.78

(continued)

Table 6.27 (continued)

Member	Load case		JT	Axial	Shear-Y	Shear-Z	Torsion	Mom-Y	Mom-Z
	18	1.2DLLL − 1.2EQX − 0.36EQY − 0.36EQZ		1891	−9.22	−22.29	−0.13	−41.23	−20.65
			82	−1929	−9.22	−22.48	−0.13	−37.46	−11.64
	19	1.2DLLL − 1.2EQX − 0.36EQY + 0.36EQZ	81	1891	9.22	−22.29	0.13	−41.23	20.65
			82	−1929	9.22	−22.48	0.13	−37.46	11.63
	20	1.2DLLL − 1.2EQX + 0.36EQY − 0.36EQZ	81	1926.7	−9.22	−22.24	−0.13	−41.16	−20.65
			82	−1893	−9.22	−22.43	−0.13	−37.34	−11.64
	21	1.2DLLL − 1.2EQX + 0.36EQY + 0.36EQZ	81	1926.7	9.22	−22.24	0.13	−41.16	20.65
			82	−1893	9.22	−22.43	0.13	−37.34	11.63
	22	1.2DLLL − 0.36EQX − 1.20EQY − 0.36EQZ	81	1850.2	−9.22	−6.7	−0.13	−12.57	−20.65
			82	−1970	−9.22	−6.89	−0.13	−11.53	−11.64
	23	1.2DLLL − 0.36EQX − 1.20EQY + 0.36EQZ	81	1850.2	9.22	−6.7	0.13	−12.57	20.65
			82	−1970	9.22	−6.89	0.13	−11.53	11.63
	24	1.2DLLL + 0.36EQX − 1.2EQY − 0.36EQZ	81	1851	−9.22	6.72	−0.13	12.05	−20.65
			82	−1969	−9.22	6.53	−0.13	10.8	−11.64
	25	1.2DLLL + 0.36EQX − 1.20EQY + 0.36EQZ	81	1851	9.22	6.72	0.13	12.05	20.65
			82	−1969	9.22	6.53	0.13	10.8	11.63
	26	1.2DLLL − 0.36EQX − 0.36EQY − 1.2EQZ	81	1891.9	−30.72	−6.64	−0.44	−12.5	−68.83
			82	−1928	−30.72	−6.83	−0.44	−11.4	−38.78
	27	1.2DLLL − 0.36EQX + 0.36EQY − 1.2EQZ	81	1927.5	−30.72	−6.59	−0.44	−12.43	−68.83
			82	−1893	−30.72	−6.78	−0.44	−11.29	−38.78
	28	1.2DLLL + 0.36EQX − 0.36EQY − 1.2EQZ	81	1892.6	−30.72	6.78	−0.44	12.13	−68.83
			82	−1928	−30.72	6.59	−0.44	10.94	−38.78
	29	1.2DLLL + 0.36EQX + 0.36EQY − 1.2EQZ	81	1928.2	−30.72	6.83	−0.44	12.19	−68.83
			82	−1892	−30.72	6.64	−0.44	11.05	−38.78

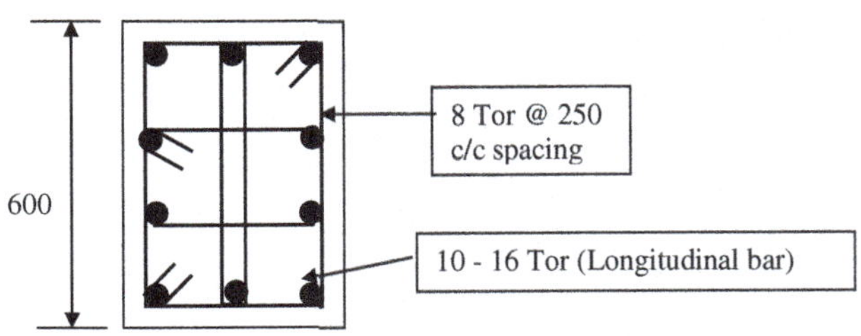

Fig. 6.28 Reinforcement detail of column

Additional force due to seismic $= A_v \times W \times g$

$$= 0.0444 \times ((0.15 \times 1 \times 1 \times 25 + 0.75(\text{Finishes}) + 25\% \text{ of } 3(\text{IL})) \times g$$

$$= 0.0444 \times 5.25 \times 9.81$$

$$= 0.2.28\ \text{kN/m}^2$$

iv. Design load combination:

(a) 1.5(DL + LL) = 11.25 kN/m^2

(b) 1.2(DL + LL + EQ) = 11.74 kN/m^2

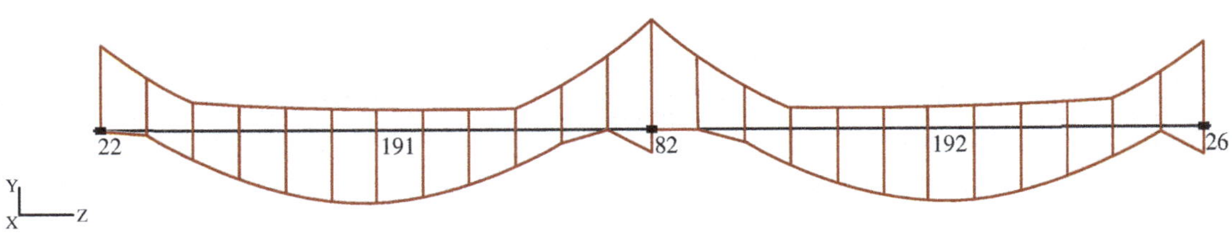

Fig. 6.29 Envelope of bending moment diagram

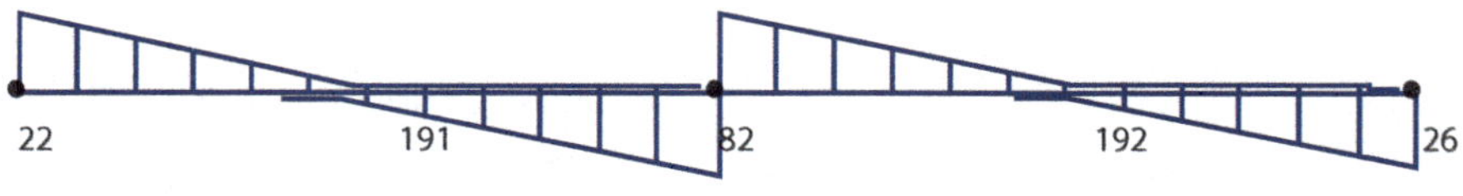

Fig. 6.30 Envelope of shear force diagram

Table 6.28 Calculation of reinforcement in beam

Member no.	Max. moment (kN/mm^2)		Span moment (kN/mm^2)	Max shear force kN		Longitudinal reinforcement			Shear reinforcement
	Left support	Right support		Left support	Right. support	Left	Right	Span	
191	137.6	+175.2 −36.7	−112.3 +41.2	121.5	135.4	665 mm^2	875 mm^2	540 mm^2	8 tor @125 C/C up to 2d on both support Remaining span 8 tor @250 c/c
192	137.6	+175.2 −36.7	−112.3 +41.2	135.4	121.5	875 mm^2	665 mm^2	540 mm^2	8 tor @125 C/C up to 2d on both support Remaining span 8 tor @250 c/c

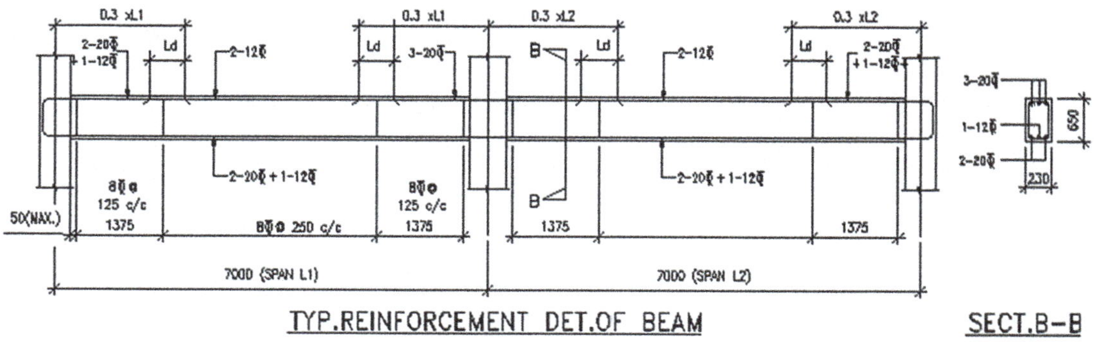

Fig. 6.31 Typ. reinforcement det. of beam

Table 6.29 Calculation of reinforcement in slab

Mark no. Slab thk. Case no.	Aspect ratio= ly/lx	Total load (kN/m^2)	α-x	α-y	M_{ux} (kN-m)	M_{uy} (kN-m)		A_{st}. req (mm^2)
S-1, 150 mm	1.75	11.74	−0.06	−0.032	−11.27	−6.01	0.75; pt = 0.214	257
Panel size = 4 m × 7 m			+0.045	+0.024	+8.45	+4.51	0.67; pt = 0.20	240
(interior panel)								

A_{st}. prov. (mm^2)	M_{uy}/bd^2	A_{st}. req. (mm^2)	A_{st}. prov. (mm^2)	Check for deflection
8 tor @ 180c/c (A_{st} = 279)	0.4; pt = 0.113	180	8 tor @ 200c/c (A_{st} = 251)	d = span/1.3 × 28
8 tor @ 180c/c (A_{st} = 279)	0.3; pt = 0.12(Min)	180	8 tor @ 200c/c (A_{st} = 251)	d = 110 < 120, OK

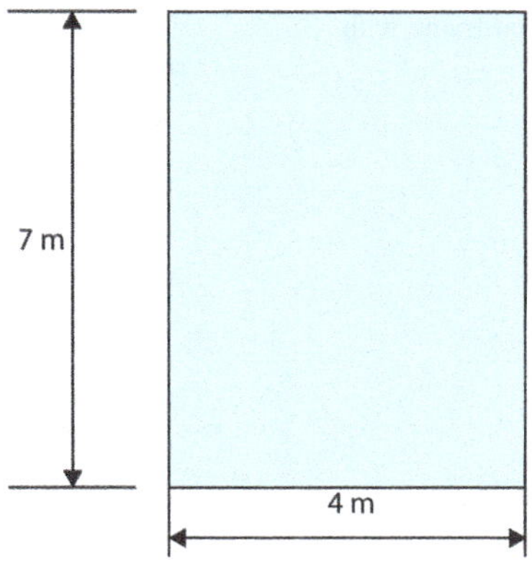

Fig. 6.32 Plan of one panel of slab

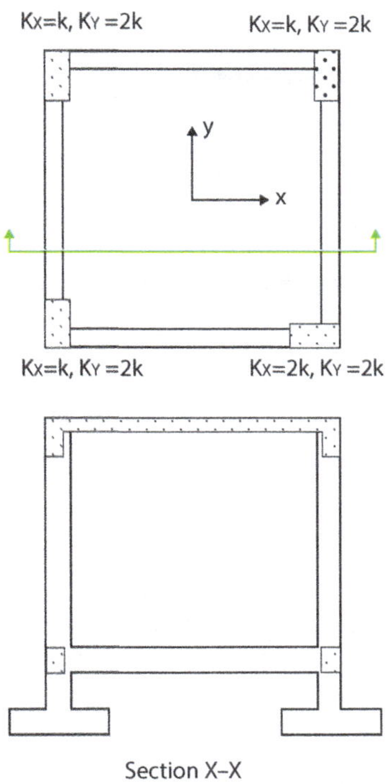

Fig. 6.33 RC frame

Exercise Problems

1. A three-storied reinforced concrete-framed structure has plan dimensions 20.0 m × 15.0 m. The sizes of exterior columns are 300 mm × 400 mm and the interior columns are 300 mm × 500 mm for all the floors. The height between floors is 3.0 m. The floors are to design to cater for live load of 4.0 kN/m^2 on floor and 1.5 kN/m^2 on the roof. Building is located in Delhi, India. Foundation of building is resting on medium soil, and its net safe bearing capacity of soil at founding level is 200.0 kN/m^2. Concrete is of Grade M30 and steel Fe415. Determine the seismic load on the structure using simplified analysis as per IS: 1893.
2. For the structure in Exercise 6.1, calculate the dynamic properties (natural period and mode shape) for vibration in *X*-direction and *Y*-direction. Obtain the design seismic force in *X*-direction and *Y*-direction by dynamic analysis method as per IS: 1893, and distribute it along the building height.
3. Mark true/false against the following statements:
   (a) A RC structure is designed for life safety (LS) performance objective. If the performance objective on the capacity curve lies outside demand curve, then the structure is safe.
   (b) If ductility factor of Structure A and Structure B is 1.0 and 4.0, respectively, then Structure A will perform better than Structure B during strong earthquakes.
4. Find the irregularities in reinforced concrete (RC) frame as shown in Fig. 6.33 and RC frame with shear wall in Fig. 6.34, and write how to modify structures to rectify them.
5. Write ductile detailing provisions (with suitable diagrams) for transverse reinforcement of RC column as per IS: 13920-2016.
6. For the RC beam section shown in Fig. 6.35, calculate the moment and curvature at initiation of cracking. Use the following data:
7. Draw the pushover curve for the portal frame assuming beam as rigid (Fig. 6.36). Moment-rotation properties of RC columns are shown in Fig. 6.37.
8. Explain any five types of structural irregularities in reinforced concrete buildings and their effect on the performance of building during earthquakes.
9. Define ductility and ductility factor. Explain how ductility affects the seismic performance of structures?
10. Draw and explain ductile detailing of reinforced concrete beam as per IS: 13920-2016.

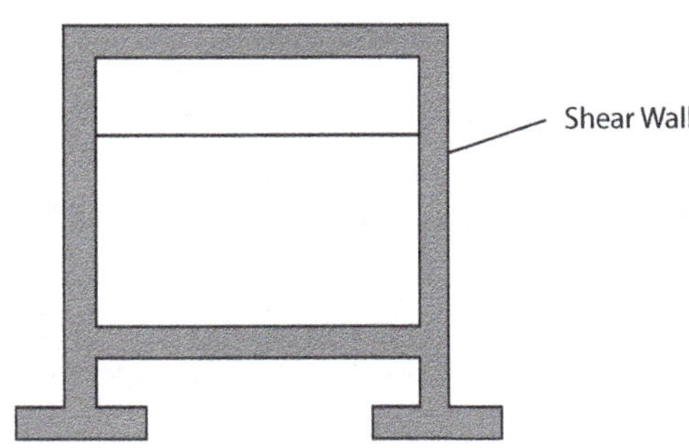

Fig. 6.34 RC frame with shear wall

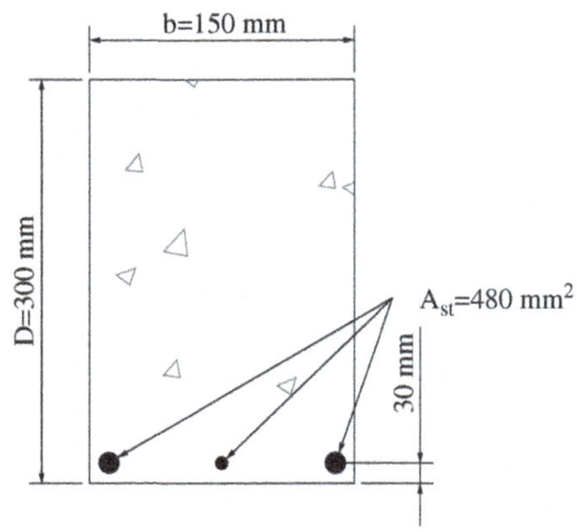

Fig. 6.35 RC beam section

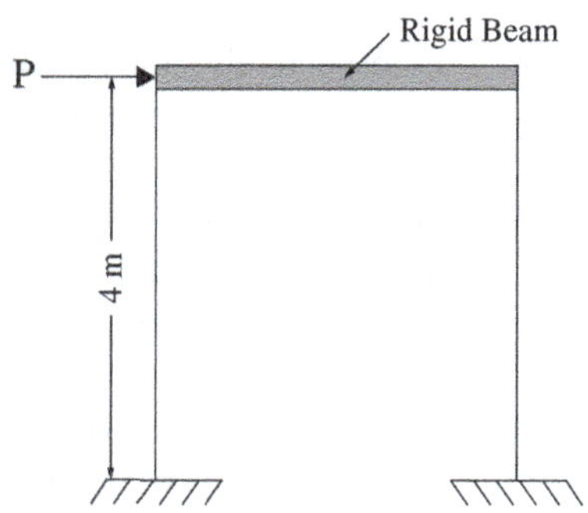

Fig. 6.36 Portal frame

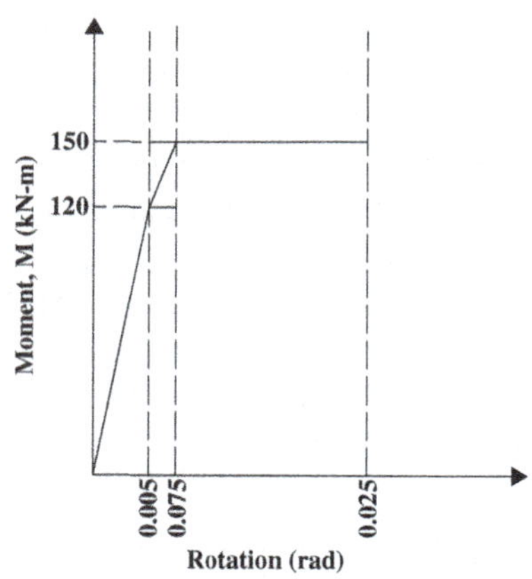

Fig. 6.37 Moment-rotation curve for RC column section

11. Explain the performance levels of reinforced concrete (RC) structures, and show these levels on load–deflection graph.
12. Explain the method to convert the capacity curve to capacity spectrum.
13. List various steps involved in calculation of performance point with figures.

Fig. 6.38 Portal frame with lateral load

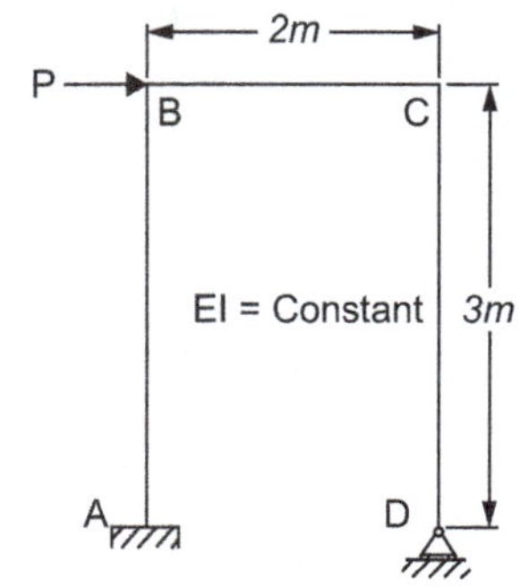

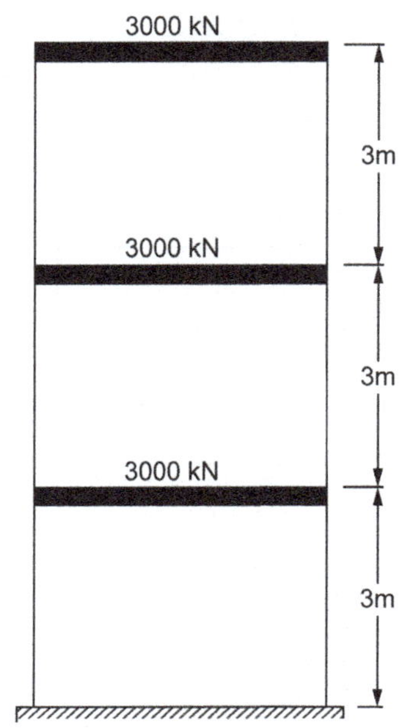

Fig. 6.39 Three-story reinforced concrete frame building

14. What are the assumptions made while evaluating moment–curvature relationship of a RC beam section?
15. Why weak beam–strong column design is more preferred than strong beam–weak column?
16. A lateral load applied on a portal frame as shown in Fig. 6.38 makes the structure to reach collapse prevention level. Mark the number of hinges and their locations?
17. Explain method to determine the moment-rotation from moment–curvature characteristics.
18. A three-story reinforced concrete frame building as shown in Fig. 6.39 is situated in Mumbai. The height between the floors is 3 m, and total height of building is 9 m. The total load of 3000 kN is lumped at each floor. The soil below the foundation is assumed to be hard rock. Determine the total base shear and shear force distribution along the height of building as per IS 1893-2002 (zone factor for Mumbai = 0.16, important factor for building = 1, response reduction factor = 3.0)

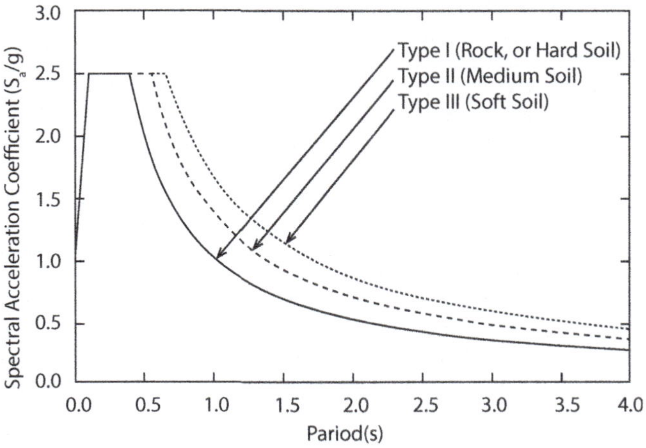

Fig. 6.40 Response spectra

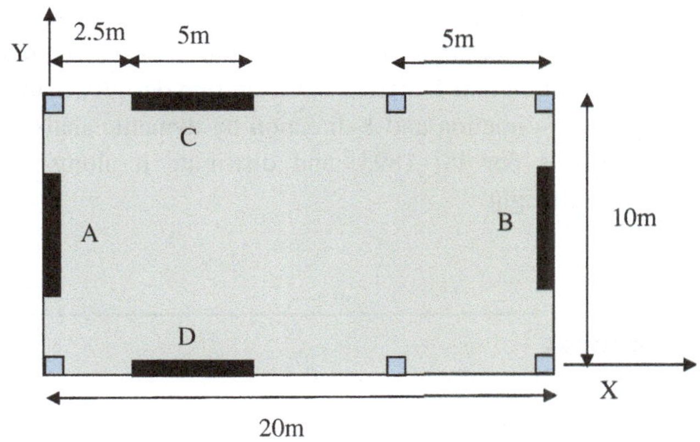

Fig. 6.42 Plan of frame structure

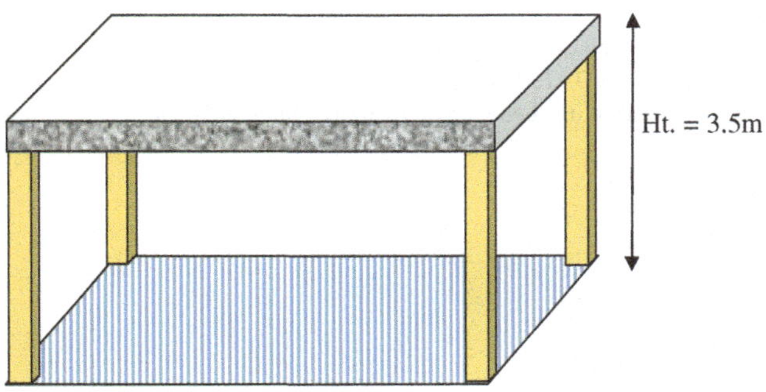

Fig. 6.41 Single-story structure

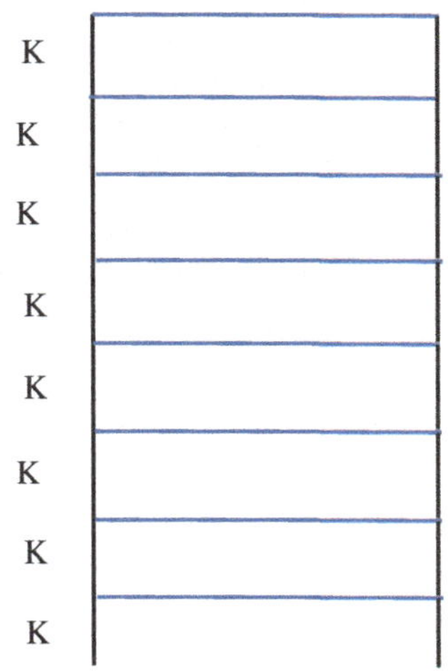

Fig. 6.43 Plan of frame structure

(hint: evaluate fundamental natural period as per IS: 1893 (part-1)-2016, use equivalent static lateral force method, and use response spectra as shown in Fig. 6.40).

19. A single-story RCC structure can be modeled as a SDOF system is shown in Fig. 6.41 having weight of roof (including dead load and live load) is 100kN. Obtain the maximum base shear and base moment. Assume natural period of cycle stand is 0.5 s, hard soil, Zone V as per IS 1893.
20. A single-story RCC building (Fig. 6.42) has two shear walls in each direction. All the shear walls are 250 mm thick and 5 m long. Size of column is 400 mm × 400 mm. Story height is 4.5 m. Floor consists of cast-in situ reinforced concrete. Design shear force on the building is 250 kN in both *X*- and *Y*-directions. Determine design shear forces on shear walls using the torsion provision of IS: 1893(Part 1)-2016.
21. An eight-story RC residential building (Fig. 6.43) is to be constructed in an area of Seismic Zone V having hard soil. The plan dimension of the building is 20 m × 30 m with story height of 3.5 m. Dead load and live load on floors are 10 and 5 kN/m^2, respectively. Dead load and live load on roof are 8 and 1.5 kN/m^2, respectively. Take the interstory lateral stiffness of floors, i.e., $k1 = k2 = k3 = k4 = 700 \times 10^6$ N/m and $k5 = k6 = k7 = k8 = 350 \times 10^6$ N/m. Determine the base shear as per the IS: 1893(part-1)-2016 code. Use both seismic coefficient and response spectrum approaches.
22. A three-storied reinforced concrete-framed structure has plan dimensions 20.0 m × 15.0 m. The sizes of exterior columns are 300 mm × 400 mm and the interior columns are 300 mm × 500 mm for all the floors. The height between floors is 3.0 m. The floors are to design to cater for live load of 4.0 kN/m^2 on floor and 1.5 kN/m^2 on the roof. Building is located in Delhi, India. Foundation of building is resting on medium soil, and its net safe bearing capacity of soil at founding level is 200.0 kN/m^2. Concrete is of Grade M30 and steel Fe415. Determine the seismic load on the structure using simplified analysis as per IS: 1893.

23. For the structure in Exercise 22, calculate the dynamic properties (natural period and mode shape) for vibration in *X*-direction and *Y*-direction. Obtain the design seismic force in *X*-direction and *Y*-direction by dynamic analysis method as per IS: 1893, and distribute it along the building height.

References

1. IS: 1893 (part-1)-2016, Indian standard for criteria for earthquake resistant design of structures (General provisions and buildings). Bureau of Indian Standards, New Delhi, India
2. IS: 1893 (part-4)-2015, Indian standard for criteria for earthquake resistant design of structures (industrial structures including stack like structures). Bureau of Indian Standards, New Delhi, India
3. IITK-BMTPC, Earthquake tip (from www.nicee.org)
4. IS: 875(Part-1) 1987, (Reaffirmed 2003), Indian standard for code of practice for design loads (other than earthquake) for building and structures- dead loads-unit weights of building materials and stored materials. Bureau of Indian Standards, New Delhi, India
5. IS: 875(Part-2) 1987, (Reaffirmed 2003), Indian standard for code of practice for design loads (other than earthquake) for building and structures-imposed loads. Bureau of Indian Standards, New Delhi, India
6. IS: 875(Part-3) 2015, Indian standard for code of practice for design loads (other than earthquake) for building and structures. Bureau of Indian Standards-wind loads, New Delhi, India
7. IS: 875(Part-4) 1987, (Reaffirmed 2003), Indian standard for code of practice for design loads (other than earthquake) for building and structures-snow loads. Bureau of Indian Standards, New Delhi, India
8. IS: 875(Part-5) 1987, (Reaffirmed 2003), Indian standard for code of practice for design loads (other than earthquake) for building and structures-Special loads and combinations. Bureau of Indian Standards, New Delhi, India
9. IS: 456-2000, Plain and reinforced concrete-code of practice. Bureau of Indian Standards, New Delhi, India
10. IS:13920–2016 (Reaffirmed 2003), Ductile detailing of reinforced concrete structures subjected to seismic forces-Code of practice. Bureau of Indian Standards, New Delhi, India
11. ACI 318 M-11, Building code requirements for structural concrete and commentary. American Society of Civil Engineers, Washington, USA
12. IS: 4326-1993, (Reaffirmed 2003), Indian standard for code of practice for earthquake resistant design and construction of buildings. Bureau of Indian Standards, New Delhi, India
13. SP6 (6)-1972, ISI handbook for structural engineers. Bureau of Indian Standards, New Delhi, India
14. IS: 800-2007, Indian standard for code of practice for general construction in steel. Bureau of Indian Standards, New Delhi, India

Design of Reinforced Concrete Chimneys 7

D. K. Jha and V. S. Phanikanth

	Earthquakes generate self-induced oscillations in the structures, equipment and piping systems Safety of the plant personnel and the nearby population hinges on the exhaust structures integrity due to earthquakes.	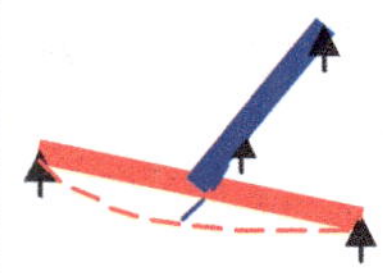

Symbols

RC	Reinforced concrete
IS	Indian standard
DL	Dead load
IL	Imposed load
TL	Temperature load
HMW	Hourly mean wind
$\sum$	Summation
$\geq$	Greater than or equal to
∞	Infinity
$\int$	Integration
C_t	Coefficient depending on slenderness ratio, k
h	Height of structure above base
E_C	Modulus of elasticity of concrete
g	Acceleration due to gravity
W_t	Total weight of structure including the lining and contents above base
r_e	Radius of gyration of stack shell at base section
k	Slenderness ratio
A_h	Design horizontal seismic coefficient
Z	Zone factor
R	Response reduction factor
S_a/g	Spectral acceleration coefficient for rock and soil sites
C_V	Coefficient depends on slenderness ratio k
D_V and D_M	Distribution factors for shear and moment, respectively, at a distance x from top
R	Response reduction factor
ζ	Damping ratio
I	Importance factor
A	Area of cross-section at base of stack
Z	Sesmic zone and zone factor
ψ_{ra}	Assumed deflection profile of mass M_r
N	Number of modes
P_i	Participation factor
C_d	Drag coefficient
R_e	Reynolds number
C_L	Lift coefficient
S_n	Strouhal number
p_z	Design wind pressure
z	Height of any section of chimney in m above base
η_{oi}	Peak tip due to vortex shedding in the ith mode of vibration in m
K_{si}	Mass damping factor for ith mode of vibration
φ_{zi}	Mode shape function normalized with respect to dynamic amplitude at top of chimney in ith mode of vibration
m_{ei}	Equivalent mass per unit length in Kg/m in the ith mode of vibration

D. K. Jha (✉)
Bhabha Atomic Research Centre, Mumbai, India
e-mail: dkjhabarc@gmail.com

V. S. Phanikanth
Bhabha Atomic Research Centre, Mumbai, India
e-mail: phani@barc.gov.in

G. R. Reddy et al., *Textbook of Seismic Design*,
https://doi.org/10.1007/978-981-13-3176-3_7

σ	Mass density of air to be taken as 1.2 kg/m^3
d	Effective diameter taken as average diameter over top 1/3rd of chimney
V_{cri}	Critical wind speed
F_{zoi}	Sectional shear force at any height z_o, for *i*th mode of vibration
M_{zoi}	Sectional bending moment at any height z_o, for *i*th mode of vibration
F_{zm}	Wind load in N/m due to MHW
F_{zf}	Wind load in N/m due to fluctuating component of wind at height z
G	Gust factor
g_f	Peak factor
B	Background factor indicating the slowly varying component of wind fluctuation
S	Size reduction factor
E	A measure of available in the wind at the natural frequency of chimney
$\vec{V}_{10}$	Hourly wind speed in m/sec
f_1	Natural frequency of chimney in Hertz
B	Structural damping as a fraction of critical damping to be taken as 0.016
K_a	Aerodynamic damping coefficient to be taken as 0.5
L	Correlation length (diameter) which may be taken as 1.0 in the absence of field data
C_L	RMS lift coefficient to be taken as 0.12
p_z	Design wind pressure at height z in N/m^2
M_{oe} or M_{oi}	External and internal ring moments
r_m	Mean radius of the shell at section under considerations in m
Y_{max}	Lateral deflection
d_{cw}	Design crack width
a_{cr}	Distance from the point considered to the surface to the surface of the nearest longitudinal bar
C_{min}	Minimum cover to longitudinal bar
ε_m	Average steel strain at the level considered

7.1 Introduction

Chimneys are tall and slender structures primarily made for discharging gases at higher elevations so that after dilution due to atmospheric turbulence, their concentration and that of their entrained solid particles are achieved within the acceptable limits. While handling a prescribed quantity of gases, the major factors which influence dimension of chimneys are:

(i) Draft requirements,
(ii) Environmental regulations,
(iii) Structural considerations.

The present chapter covers the design aspects of RC chimneys as per structural considerations. Seismic design aspects of RC Chimneys are described here as per Indian standard IS1893 (Part 4): 2015. Wind load assessment and design parameters are discussed as per Indian standard IS4998: 2015 provisions.

The base dimension of RC Chimney is mainly governed by the structural considerations. These dimensions of chimney are so chosen that the stresses due to defined load combinations due to wind, temperature, earthquake, dead loads, live loads, and other loads are safely withstood. The height of the stack should be such that the primary purpose of keeping the efflux in the higher atmosphere as long as possible to allow wide diffusion and reduction in the concentration of the pollutant is achieved. The physical height is first determined such that the draft is adequate to overcome the friction and other head losses as the gas flow up and out of stack. Top diameter of the stack is fixed such that a given volume of gases can be discharged at the design exit velocity. In practice, these dimensions are to be adjusted to cater for a diminished velocity along the walls.

7.2 Typical Loads on Chimney

Following loads shall be taken into account for design of chimney structures:

(i) Dead loads,
(ii) Imposed loads,
(iii) Wind loads (lateral component and circumferential component),
(iv) Earthquake/Seismic loads, and
(v) Temperature effects (vertically and circumferentially).

Imposed loads on chimney structures need not be considered for design of its shell and foundation. However, for design of individual structural elements such as platforms and for local strengthening of the shell, appropriate imposed loads shall be considered as per relevant code of practices. Wind and earthquake loads shall not be considered as acting simultaneously. The foundation design should also be checked with minimum weight of chimney under shell alone condition combined with coexisting lateral loads. For a lined chimney, factor of safety against overturning of foundation shall not be less than 1.5 for chimney shell alone case and shall not be less than 2.0 for the complete chimney. The various types of loads to be considered and relevant Indian standard (IS) code of practices are given in Table 7.1.

Table 7.1 Types of loading and relevant Indian codes of practices for assessment of loads

Type of loads	Relevant IS code	Remarks
Dead load (DL)	IS 875 (Part 1): 1987	Self-weight of chimney shell, liner (if any), liner supports and other accessories may be treated as DL
Imposed load (LL)	IS 875 (Part 2): 1987	
Wind load (WL)	IS 875 (Part 3): 2015 and IS 4998: 2015	
Earthquake load (EL)	IS 1893 (Part 1): 2016 and IS 1893 (Part 4): 2015	
Temperature load (TL)	IS 4998: 2015	

7.3 Seismic Analysis

Chimneys are, in particular, more vulnerable to earthquakes because they are tall slender structures and shall be designed for earthquakes by means of response spectrum method. A detailed dynamic analysis using time history shall be required for site-specific ground motion. In the present chapter, seismic design of chimney in accordance with IS 1893 (Part 4): 2015 and simplified modal analysis technique are discussed. The mathematical model of stack-like structures should be able to represent appropriately the variation in its stiffness (variation in cross section and thickness of chimney shell structure), lining mass, and its foundation. The discretization shall be such to capture the variation of stiffness and mass of the system appropriately. In case of chimneys with liner, stiffness contribution due to liner is ignored, however, the mass of lining above any corbel is assumed to be lumped at the corbel level.

The fundamental time period T_1 for stack-like structures (the structures in which mass and stiffness are more or less uniformly distributed along height) when fixed at base is given by:

$$T_1 = C_T\sqrt{\frac{W_t h}{E_c A g}} \tag{7.1}$$

in which

C_T coefficient based on slenderness ratio of the structure (the values of C_T against different slenderness ratio's are presented in Table 7.2)
W_t total weight of the structure including the lining and contents above the base
h height of the structure above base
E_c modulus of elasticity of concrete (material of construction for chimney shell)
A cross-sectional area at the base of stack
g acceleration due to gravity = 9.81 m/sec^2

in which

r_e radius of gyration of shell of stack at its base section
k slenderness ratio

7.3.1 Design Horizontal Seismic Coefficient (A_h)

Using Eq. (7.1) of fundamental time period (T_1), the design horizontal seismic coefficient A_h shall be obtained by the following expression as per IS 1893 (Part 4): 2015:

$$A_h = \frac{\left(\frac{Z}{2}\right)\left(\frac{S_a}{g}\right)}{\left(\frac{R}{I}\right)} \tag{7.2}$$

where

Z seismic zone factor
R response reduction factor
I importance factor
S_a/g spectral acceleration coefficient for rock and soil sites

The seismic zone factor Z defined for various seismic zones in India as per IS1893 (Part 4): 2015 is given in Table 7.3. Response reduction factor R, and importance factor I, applicable to stack-like structures as per IS 1893 (Part 4): 2015 are presented in Tables 7.4–7.5, respectively.

Appropriate damping values as given in IS 1893 (Part 4): 2015 shall be applicable for arriving at A_h values. The spectral acceleration for different types of soil is obtained in accordance with IS1893 (Part 1): 2016. Figure 7.1 shows the 5% spectra for rocky and soil sites.

Table 7.2 Values of C_T and C_V [IS 1893 (Part 4): 2015]

$k = h/r_e$	C_T	C_V
5	14.4	1.02
10	21.2	1.12
15	29.6	1.19
20	38.4	1.25
25	47.2	1.30
30	56.0	1.35
35	65.0	1.39
40	73.80	1.43
45	82.80	1.47
≥ 50	1.8 k	1.50

Table 7.3 Seismic zone factor, Z [IS 1893 (Part 4): 2015]

Seismic Zone of India	II	III	IV	V
Seismic intensity	Low	Moderate	Severe	Very severe
Z	0.10	0.16	0.24	0.36

Table 7.4 Response reduction factor R, applicable to stack-like structures [IS 1893 (Part 4): 2015]

Type of structure	R
Steel chimney	2.0
Reinforced concrete TV towers	1.5
Reinforced concrete ventilation stack	1.5
Reinforced concrete chimney	1.5
Reinforced brick masonry chimney	1.25
Unreinforced brick masonry chimney	1.00

Table 7.5 Importance factor I, applicable to stack-like structures [IS 1893 (Part 4): 2015]

Type of structure	I
Reinforced concrete ventilation stacks	1.50
Reinforced concrete chimneys	1.50
Reinforced brick masonry chimney for industry	1.50
Unreinforced brick masonry chimney for industry	1.00
Steel stack	1.50

7.3.2 Evaluation of Shear Force & Bending Moment Using Simplified Method

The design shear force V, and design bending moment M, for stack-like structure at a distance x from the top of stack, shall be calculated as per IS 1893 (Part 4): 2015 as follows:

$$V = C_V A_h W_t D_v$$
$$M = A_h W_t h D_m \tag{7.3}$$

where

C_V coefficient depends upon slenderness ratio k (the values of C_v against different slenderness ratio's are presented in Table 7.2)

A_h design horizontal seismic coefficient

W_t total weight of the chimney including the lining and contents above

h center of gravity (CG) of the chimney above its base

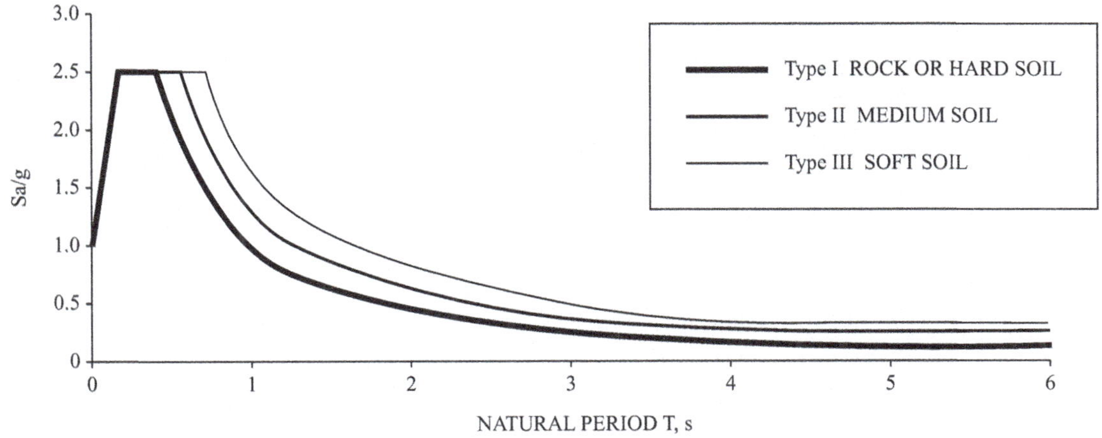

Fig. 7.1 Design acceleration coefficient (S_a/g) corresponding to 5% damping [Spectra for Response spectrum method as per IS 1893(Part 1): 2016]

Table 7.6 Distribution factors for shear force and bending moment[As per IS 1893 (Part 4): 2015]

Soil foundation	D_m	D_v
Fixed base or raft on hard soil (based on N[a] values)	$0.4\left(\frac{x}{h}\right)^{0.5}+0.6\left(\frac{x}{h}\right)^{4}$	$1.1\left(\frac{x}{h}\right)^{0.5}+0.75\left(\frac{x}{h}\right)+0.9\left(\frac{x}{h}\right)^{4}$ but ≤ 1.0
Raft on hard soil (based on N[a] values)	$0.6\left(\frac{x}{h}\right)^{0.5}+0.4\left(\frac{x}{h}\right)^{4}$	$1.1\left(\frac{x}{h}\right)^{0.5}-0.75\left(\frac{x}{h}\right)+0.65\left(\frac{x}{h}\right)^{4}$
Pile foundation	$0.5\left(\frac{x}{h}\right)^{0.5}+0.5\left(\frac{x}{h}\right)^{4}$	$0.66\left(\frac{x}{h}\right)^{0.5}-0.20\left(\frac{x}{h}\right)+0.54\left(\frac{x}{h}\right)^{4}$

[a]Here, N is the standard penetration resistance (required number of blows per 30 cm penetration of soil layer) obtained by conducting standard penetration test (SPT) as per [18]

D_v and D_m distribution factors for shear and moment, respectively, at a distance 'x' from the top and are given in Table 7.6.

7.3.3 Seismic Analysis by Simplified Method: An Illustration

Adopting the simplified method and using the modal analysis technique the seismic response of a RC chimney with the input data listed in Table 7.7 are evaluated in this illustration.

i. Weight of the chimney (W_t)

$$\begin{aligned} W_t &= \text{Cross - sectional Area}(A) \times \text{Height}(h) \\ &\quad \times \text{Weight density of concrete}(\gamma) \\ &= 8.16 \times 130.0 \times 25.0 \\ &= 26.52 \times 10^3 \text{ kN} \end{aligned}$$

ii. Radius of gyration (r_e)

$$\begin{aligned} r_e &= \sqrt{\frac{I}{A}} \\ &= \sqrt{\frac{87.9}{8.16}} \\ &= 3.28\,\text{m} \end{aligned}$$

iii. Slenderness ratio (k)

$$k = h/r_e = 39.63$$

iv. Coefficients C_T and C_V

$C_T = 73.1$ and

$C_V = 1.427$(interpolated from Table 7.2 with $k = 39.63$)

v. Fundamental time period (T_1)
Using Eq. (7.1)

Table 7.7 Input data for the seismic analysis of RC chimney

Parameter	Value
Height of chimney (h)	130.0 m
Area of cross section (A)	8.16 m^2
Moment of inertia (I)	87.9 m^4
Young's modulus of M30 concrete (E_c)	2.74×10^7 kN/m^2
Weight density of concrete (γ)	25.0 kN/m^3
Seismic zone and zone factor (Z)	0.16 (Zone III)
Importance factor (I)	1.75
Response reduction factor (R)	3.0
Damping factor (ζ)	5%
Founding stratum	Rock

$$T_1 = C_T\sqrt{\frac{W_t h}{E_c A g}}$$

Upon substituting the parameters,

$$T_1 = 2.89\,\text{s}$$

vi. Spectral acceleration coefficient (S_a/g)
For T_1 = 2.89 s, using Fig. 7.1
S_a/g is found to be = 0.345 (for rocky site)

vii. Design horizontal seismic coefficient

$$\begin{aligned} A_h &= (Z/2) \times (I/R) \times S_a/g \\ &= (0.16/2) \times (1.75/3) \times 0.345 \\ &= 0.0161 \end{aligned}$$

viii. Shear force and Bending moment

$$\begin{aligned} V &= C_V A_h W_t D_v \\ &= 1.427 \times 0.0161 \times 26.52 \times 10^3 \\ &= 609.28\,\text{kN} \end{aligned}$$

$$\begin{aligned} M &= A_h W_t \bar{h} D_m \\ &= 0.0161 \times 26.52 \times 10^3 \times (130/2) \times 1.0 \\ &= 27.753 \times 10^3\,\text{kN m} \end{aligned}$$

7.3.3.1 Modal Analysis

Energy method based on Rayleigh's principle may be used for modal analysis in a simplified way, and the procedure is outlined below [2].

Inertial Forces

The deflection profile φ_i for ith mode is usually assumed, and the lateral deflection of mass M_r would be $A\ \varphi_{ri}$ for mode i, where A is an arbitrary constant. The inertial force F_{ri} then will be:

$$F_{ri} = AM_r\varphi_{ri}\omega_i^2 = (\text{constant})(M_r\varphi_{ri}) \tag{7.4}$$

It is not possible to determine absolute value of F_{ri} and also it is not of significance. Therefore,

$$F_{ri} = M_r\varphi_{ri} \tag{7.5}$$

Frequency Determination

Under the effect of forces F_{ri}, the characteristic deflection profile of the structure is say φ_i'. A structure is assumed to undergo harmonic motion about its original position ($\varphi_i = 0$) with a maximum amplitude φ_i' in the ith mode. Using Rayleigh's principle and equating the kinetic energy of the system (masses M_r) at $\varphi_i = 0$ to its potential energy at amplitude $A'\varphi_i$, where A' is any arbitrary constant, one can get

$$\sum_{r=1}^{j}\frac{1}{2}M_r(A'\varphi_{ri}'\omega_i)^2 = \sum_{r=1}^{j}\frac{1}{2}F_{ri}(A'\varphi_{ri}')$$

$$\omega_i^2 = \frac{\sum_{r=1}^{j} F_{ri}(\varphi_{ri}')}{A'\sum_{r=1}^{j} M_r(\varphi_{ri}')^2} \tag{7.6}$$

Mode Shape Coefficient

Any assumed deflection profile can be built up by a combination of modal shapes and is expressed mathematically as below:

$$\varphi_{ra} = \sum_{i=1}^{N}\psi_i\varphi_{ri} \tag{7.7}$$

where

φ_{ra} assumed deflection profile of mass M_r and
N number of modes

Multiplying both sides of Eq. (7.7) by $M_r\varphi_{\text{rq}}$ (where φ_{rq} is the characteristic deflection profile of the qth mode) and on summation for all the masses,

$$\sum_{r=1}^{j} M_r\varphi_{\text{ra}}\varphi_{\text{rq}} = \sum_{r=1}^{j}\sum_{i=1}^{N} M_r\varphi_q\psi_i\varphi_{\text{ri}} \qquad (7.8\text{a})$$

Using the orthogonally condition, one can get

$$\psi_q = \frac{\sum_{r=1}^{j} M_r\varphi_{\text{ra}}\varphi_{\text{rq}}}{\sum_{r=1}^{j} M_r\varphi_{\text{rq}}^2} \qquad (7.8\text{b})$$

Higher Modes

To obtain the next higher mode, the assumed deflection profile for the higher mode is purified of earlier modes. For example, if the deflection profiles for two modes a and b have been determined and φ_c is the deflection profile for next higher mode, then the purified profile will be

$$\varphi_{\text{rap}} = \varphi_{c} - \psi_a\varphi_{a} - \psi_b\varphi_b \qquad (7.9)$$

Participation Factor

The participation factor P_i for the ith mode for a structure with j masses is given below:

$$P_i = \frac{\sum_{r=1}^{j} M_r(\Phi_{\text{ri}})}{\sum_{r=1}^{j} M_r(\Phi_{\text{ri}})^2} \qquad (7.10)$$

7.3.3.2 Simplified Procedure

The simplified procedure for modal analysis is described below:

i. Step 1: The mass of chimney is lumped at different nodes by dividing the chimney into appropriate segments.
ii. Step 2: A deflection profile is assumed and for this assumed profile inertial forces are evaluated using Eq. (7.5) for F_{ri}. The deflections are then computed using these inertial loads. These deflections are then compared to the initially assumed deflections. If different, the inertial forces are worked out again based on the actual deflection profile, and for these inertial loads, deflections are again computed. The above procedure is repeated until the assumed and computed deflection profile match with desired accuracy.
iii. Step 3: For the mode shape obtained upon convergence, the frequency is evaluated by using Eq. (7.6) for ω_i.
iv. Step 4: To obtain the next higher mode, a deflection profile is again assumed and the mode coefficient is obtained by using Eq. (7.8) for ψ_q. Using this, the assumed profile of the earlier modes is purged.
v. With this purified/purged profile, the iterative computation is being repeated till the mode shape converges, as described above in Steps 2–4.
vi. The participation factor P_i for each mode is then computed by using Eq. (7.10).
vii. The modal shear forces and moments at each mass level are determined, and then these modal parameters are combined to obtain the design values of shear force and moments.

Example 7.1

Calculate the total base shear and total base moments of a RC chimney for the above-mentioned illustration using the input from Table 7.7 adopting simplified modal analysis procedure.

Solution

i. For a cantilever structure loads as shown in Fig. 7.2b, the deflection at any point is given by:

$$EI_y = [(F/6)[(a-x)^3 + 3a^2x - a^3] \quad \text{if } x<a$$
$$= F\left[(a^2x/2) - a^3/6\right] \quad \text{if } x > a$$

ii. Deflection coefficients due to a unit load placed at load points 1–4 are given in Table 7.8. Iterations of first and second modes are terminated after two and three cycles, respectively, since the purpose is only to explain the procedure.
iii. **First mode:** Initially, the deflection profile φ_{ra} is assumed and the corresponding inertial loads F_r are obtained. With these loads acting on the beam, deflections at all node points are determined using the coefficients in Table 7.8. For example:
   (a) Deflection at node 2

$$= (H^3/E_cI) \times [270 \times 0.2347 + 426.3 \times 0.1707 + 223.2 \times 0.0933$$
$$+ 74.4 \times 0.0224]$$
$$= 158.630 \times H^3/E_cI$$

The resulting deflections (φ_r') are shown in Table 7.9. This profile is used as the assumed profile for the next cycle, and the procedure is repeated to arrive at the improved profile φ_r' (Table 7.10).

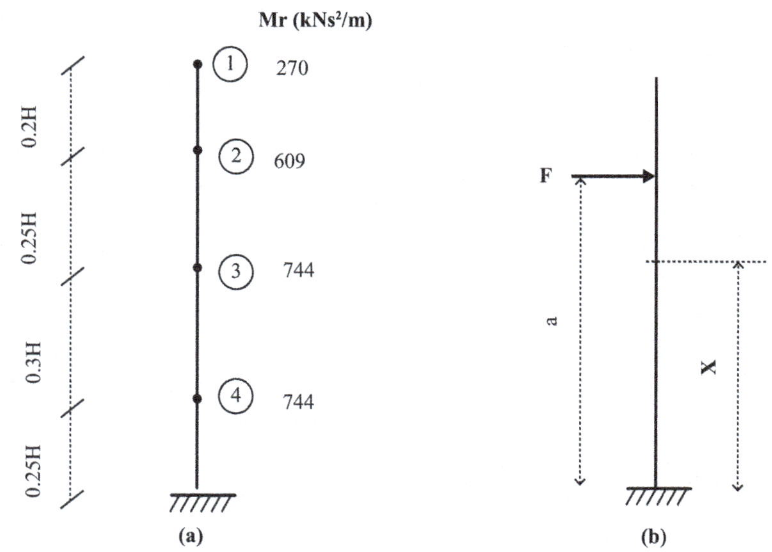

Fig. 7.2 Lumped mass representation of stack-like structure

Table 7.8 Deflection coefficients

	Deflection coeff at node 1	Deflection coeff at node 2	Deflection coeff at node 3	Deflection coeff at node 4
Unit load at node 1	0.3333	0.2347	0.1235	0.0286
Unit load at node 2	0.2347	0.1707	0.0933	0.0224
Unit load at node 3	0.1235	0.0933	0.0555	0.0146
Unit load at node 4	0.0286	0.0224	0.0146	0.0052

Deflection = Deflection coefficient × H^3/E_cI

Table 7.9 Modal analysis: First mode

	Load at node	M_r (kNs^2/m)	Φ_{ra}	F_r (= $M_r\Phi_{ra}$)	Deflection × E_cI/H^3	Φ'_r			
First cycle	1	270	1.00	270.0	219.74	1.00	ω_1= 2.17 rad/s		
	2	609	0.70	426.3	158.63	0.722	P_1= 1.512		
	3	744	0.30	223.2	86.60	0.394	T_1= 2.89 s		
	4	744	0.10	74.4	20.92	0.095			
	Load at node	M_r (kNs^2/m)	Φ_{ra}	$F_r = M_r\Phi_{ra}$	Deflection × E_cI/H^3	Φ'_r	$M_r\Phi'_r$	$M_r(\Phi'_r)^2$	$F_r\Phi'_r$
Second cycle	1	270	1.000	270.0	231.41	1.000	270.00	270.00	270.0
	2	609	0.722	439.70	167.36	0.723	440.31	318.34	317.9
	3	744	0.394	293.14	91.67	0.396	294.62	116.67	116.08
	4	744	0.095	70.68	22.22	0.096	71.42	6.86	6.79
						$\sum$=	1076.35	711.87	710.77

Table 7.10 Modal analysis: Second mode

	Load at	M_r kNs²/m	Φ_{r1}	Φ_{ra}	$M_r\Phi_{r1}\Phi_{ra}$	$\Psi_1\Phi_{r1}$	$\Phi_{rap}=\Phi_{ra}-\Psi_1\Phi_{r1}$	$F_r = M_r\Phi_{rap}$	Deflection × E_cI/H^3	Φ'_r
First cycle	1	270	1.00	1.00	270.00	0.2041	0.7959	214.89	4.651	1.00
	2	609	0.723	0.100	44.03	0.1476	−0.0476	−28.99	−0.157	−0.0338
	3	744	0.396	−0.500	−147.31	0.0808	−0.5808	−432.12	−3.621	−0.7785
	4	744	0.096	−0.300	−21.43	0.0196	−0.3196	−237.78	−2.049	−0.4406
				$\sum=$	145.29	Ψ_1 = 145.29/711.87 = 0.2041				
Second cycle	1	270	1.00	1.00	270.00	−0.008	1.008	272.16	6.113	1.00
	2	609	0.723	−0.0338	−14.88	−0.0058	−0.028	−17.052	−0.1819	−0.0298
	3	744	0.396	−0.7785	−229.36	−0.0032	−0.7753	−576.823	−4.7702	−0.7803
	4	744	0.096	−0.4406	−31.47	−0.0008	−0.4398	−327.211	−2.7213	−0.4452
				$\sum=$	−5.71	Ψ_1 = −5.71/711.87 = −8.02e−03				
	Load at	M_r	F_{r1}	Deflection× E_cI/H^3	Φ'_r	$F_r\ \Phi'_r$	$M_r\Phi'_r$	$M_r(\Phi'_r)^2$		
Third cycle	1	270	271.825	6.1288	1.00	271.825	270.00	270.00		
	2	609	−15.17	−0.1795	−0.0293	0.445	−17.844	0.523		
	3	744	−578.549	−4.7834	−0.7805	451.558	−580.692	453.23		
	4	744	−330.745	−0.4458	−0.4458	147.446	−331.675	147.86		
					$\sum=$	871.274	−660.211	871.614		

(b) Time period

$$E_cI = 2.407 \times 10^9\ \text{kN m}^2$$

$$\omega_1 = \sqrt{\left(\frac{\sum_{r=1}^{j} F_{ri}(\varphi'_{ri})}{A' \sum_{r=1}^{j} M_r(\varphi'_{ri})^2}\right)} = \sqrt{\left(\frac{710.77 \times 2.407\text{e}09}{231.41 \times 711.87 \times 130^3}\right)} = 2.17\ \text{rad/s}$$

$$T_1 = 2\pi/\omega_1 = 2.89\ \text{s}$$

The spectral acceleration for this time period for rocky site using Fig. 7.1 is $S_a/g = 0.345$

(c) Participation factor for ith mode is given as,

$$P_i = \frac{\sum_{r=1}^{j} M_r(\Phi_{ri})}{\sum_{r=1}^{j} M_r(\Phi_{ri})^2}$$

Participation factor for the first mode using the above expression is obtained as 1.512, and the modified acceleration $S_{a(mod)}$ for the time period of 2.89 s corresponding to first mode is

$$\begin{aligned} S_{a(mod)1} &= P_1 \times (S_a/g)_1 \times (z/2) \times (I/R) \times g \\ &= 0.345 \times (0.16/2) \times (1.75/3) \times 9.81 \\ &= 0.2387\ \text{m/s}^2 \\ &= 0.24\ (\text{Approximate value}) \end{aligned}$$

(d) Using this relationship, inertia forces at all node points are calculated. For instance, at node 2:

$$\begin{aligned} F_{21} &= 609 \times 0.723 \times 0.24 \\ &= 105.67\ \text{kN} \end{aligned}$$

(e) After evaluating the inertial forces, base shear and moment for the first mode are evaluated and shown in Table 7.11.

(f) For first mode:

$$\begin{aligned} \text{Base shear force} &= 64.80 + 105.674 + 70.71 + 17.14 \\ &= 258.352\ \text{kN} \\ \text{Base moment} &= 130 \times (64.80 \times 1.0 + 105.674 \times 0.8 \\ &\quad + 70.71 \times 0.55 + 17.14 \times 0.25) \\ &= 2.5027 \times 10^4\ \text{kN m} \end{aligned}$$

Table 7.11 Modal shear forces

Node	M_r (kNs^2/m)	Shear forces			
		S_{a1} = 0.24 (first mode)		S_{a2} = −0.737 (second mode)	
		Φ_1	F_1	Φ_2	F_2
1	270	1.00	64.800	1.000	−198.99
2	609	0.723	105.674	−0.0293	13.1508
3	744	0.396	70.7098	−0.7805	427.97
4	744	0.096	17.1418	−0.4458	244.445
		$\sum$=	258.325	$\sum$=	486.575

iv **Second mode:**

(a) Calculations for this mode are summarized in Table 7.10. The first mode shape coefficient is shown as Φr_1 and the initially assumed second mode shape as Φ_{ra}. Mode coefficient ψ_1 is worked out, and assumed mode shape Φ_{ra} is purged of the first mode to give purified mode shape $\Phi_{rap} = \Phi_{ra} - \psi_1\, \Phi_{r1}$; with this as mode shape, inertial loads and actual deflection profile Φ_{ra} are obtained as explained for the first mode. This profile is taken as the assumed profile for the next cycle.

(b) Time period

$$\omega_2 = \sqrt{\left(\frac{\sum_{r=1}^{j} F_{ri}(\varphi'_{ri})}{A' \sum_{r=1}^{j} M_r(\varphi'_{ri})^2}\right)}$$

$$= \sqrt{\left(\frac{871.274 \times 2.407e09}{6.1288 \times 871.614 \times 130^3}\right)} = 13.37\,\text{rad/s};$$

$T_2 = 2\pi/\omega_2 = 0.47\text{s}$ and

$P_2 = -660.211/871.614 = -0.757$

(c) Using equation for participation factor P_i as given above, participation factor for the second mode is mode is obtained as $P_2 = -0.757$ and time period is evaluated as $T_2 = 0.47$ s modified acceleration $S_{a(\text{mod})}$ for the second mode is obtained as:

$$S_{a(\text{mod})2} = P_2 \times (S_a/g)_2 \times (z/2) \times (I/R) \times g$$
$$= -0.757 \times 2.127 \times (0.16/2)$$
$$\times (1.75/3) \times 9.81 = -0.737\,\text{m/s}^2.$$

(d) For second mode:

$$\text{Base Shear} = 571.74\ \text{kN}$$
$$\text{Base moment} = 130 \times (-198.99 \times 1.0 + 13.15 \times 0.8$$
$$+ 427.97 \times 0.55 + 244.45 \times 0.25)$$
$$= 1.4044 \times 10^4\text{kN m}$$

(e) The rest of the calculations are similar to that for the first mode.

(f) The base shear and moment for the first two modes are then combined (as per SRSS) to give total values.

$$\text{Total shear force at base} = \sqrt{(258.525^2 + 486.575^2)}$$
$$= 550.99\,\text{kN}$$

$$\text{Total base moment} = \left[\sqrt{(2.5027^2 + 1.4044^2)}\right] \times 10^4$$
$$= 2.8698 \times 10^4\,\text{kN m}$$

7.4 Wind Load Analysis

The chimney shall also be analyzed for lateral loads induced due to wind forces. Tall concrete chimney structures especially with circular cross sections are wind sensitive, and they shall be designed to resist both the along-wind and across-wind effects wherever necessary. In the present section, wind load analysis of tall chimneys by conventional simplified approach is discussed in brief along with methods suggested in IS 4998:2015.

The wind load per unit height on a chimney in wind direction (along-wind load) at any height may be estimated conventionally by multiplying the "design wind pressure" at the considered height by the drag coefficient (C_d) and its sectional diameter (d_z). C_d depends on the shape of the chimney and also on Reynolds number of fluid (air). Also, lift coefficient $\sqrt{C_L^2}$ and Strouhal number S_n are dependent on Reynolds number R_e and their variations for a cylinder is shown in Fig. 7.3. The value of C_d for different cross sections is shown in Fig. 7.4.

7.4.1 Wind Load Analysis of RC Chimney by Conventional Method

The wind load analysis by conventional method by calculating the design wind pressure as per IS875 (Part 3): 2015 is discussed below.

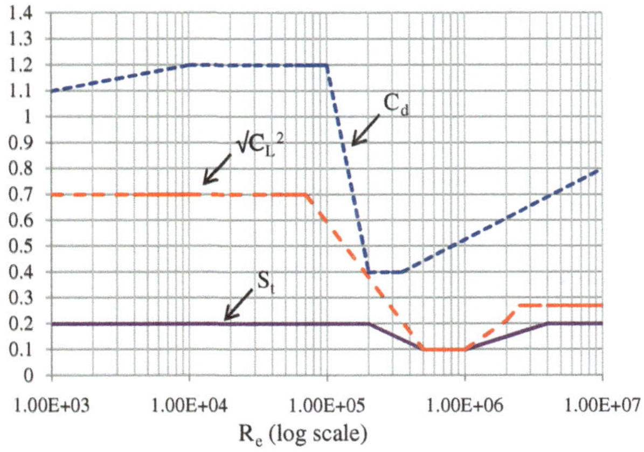

Fig. 7.3 Drag coefficient (C_d), lift coefficient ($\sqrt{C_L^2}$) and Strouhal number (S_t) for a cylinder [2]

7.4.1.1 Along-Wind Loads

The along-wind load or drag force per unit height of the chimney at any level are obtained by using the following expression.

$$F_z = p_z C_d d_z \tag{7.11}$$

where

p_z design wind pressure obtained in accordance with IS 875 (Part 3): 2015
z height of any section of the chimney in 'm' above base
d_z diameter of chimney at height 'z' in 'm'

7.4.1.2 Across-Wind Loads

Across-wind loads are generated due to the transverse oscillations caused by alternate shedding of vortices from the air flow at opposite sides of the stack. The vortices are of equal strength but of opposite rotation, and it is of this alternate shedding of each of the vortices when released from the side of the stack that causes the generation of a periodic transverse force known as the lift force, which causes oscillations. At the critical wind velocity, the frequency of vortex shedding coincides with the natural frequency of the stack. At this velocity, the stack oscillates at right angles to the direction of wind.

The amplitude of vortex exited oscillation perpendicular to direction of wind for any mode of oscillation conventionally shall be calculated by the following expression:

$$\eta_{oi} = \left\{ \frac{\int_0^H d_z \varphi_{zi} \mathrm{d}z}{\int_0^H \phi_{zi}^2 \mathrm{d}z} \right\} \times \frac{C_L}{4\pi S_n^2 K_{si}} \tag{7.12a}$$

where, $K_{si} = \frac{2 m_{ei} \delta_s}{\sigma d^2}$

$$m_{ei} = \frac{\int_0^H m_z \varphi_{zi}^2 \mathrm{d}z}{\int_0^H \phi_{zi}^2 \mathrm{d}z}$$

η_{oi} peak tip deflection due to vortex shedding in the ith mode of vibration in m
C_L peak oscillatory lift coefficient to be taken as 0.16
H height of chimney above ground level, in m
K_{si} mass damping parameter for the ith mode of vibration
S_n Strouhal number to be taken as 0.2
ϕ_{zi} mode shape function normalized with respect to dynamic amplitude at top of the chimney in the ith mode of vibration
m_{ei} equivalent mass per unit length in kg/m in the ith mode of vibration
σ mass density of air to be taken as 1.2 kg/m^3
D effective diameter taken as average diameter over top one-third height of chimney, in m

For a chimney with little or no tapering, i.e., if the average taper over one-third height is less than or equal to 1 in 50, where taper is defined as $2(d_{av}–d_{top})/H$.

The modal response, at a critical wind speed $V_{cri} = f_i\, d/S_n$, is obtained as

$$\eta_{oi} = \left\{ \frac{\frac{1.25 \bar{C}_L \mathrm{d} \phi_{zi} h}{\pi^2 S_n^2} \times \frac{\sigma d^2 \sqrt{\sqrt{\{\pi L/2(\cap + 2)\}}}}{m_{ei}}}{\left[1/H \int_0^H \phi_{zi}^2 \mathrm{d}z\right]^{1/2} \left(\beta - k_a \sigma d^2/m_{ei}\right)^{1/2}} \right\} \tag{7.12b}$$

in which

$\bar{C}_L$ RMS lift coefficient to be taken as 0.12
$\cap$ equivalent aspect ratio = H/d
L correlation length (diameter), which is taken as 1.0 in the absence of field data
K_a aerodynamic damping coefficient to be taken as 0.5.

7.4.1.3 Stability Analysis of Reinforced Concrete Chimney Against Wind Load

The parameters influencing the dynamic instability of a chimney are the aspect ratio (H/D_{co}) and non-dimensional damping parameter (K_s). The extensive experimental investigations carried out suggest that for each aspect ratio there is a critical wind velocity which if exceeded, can result in dynamic instability (Fig. 7.5). The graph shown in Fig. 7.5 is helpful to the designer on preliminary assessment of dynamic instability. If the design wind velocity at the site is lower than the critical wind velocity, the structure is likely to be dynamically stable.

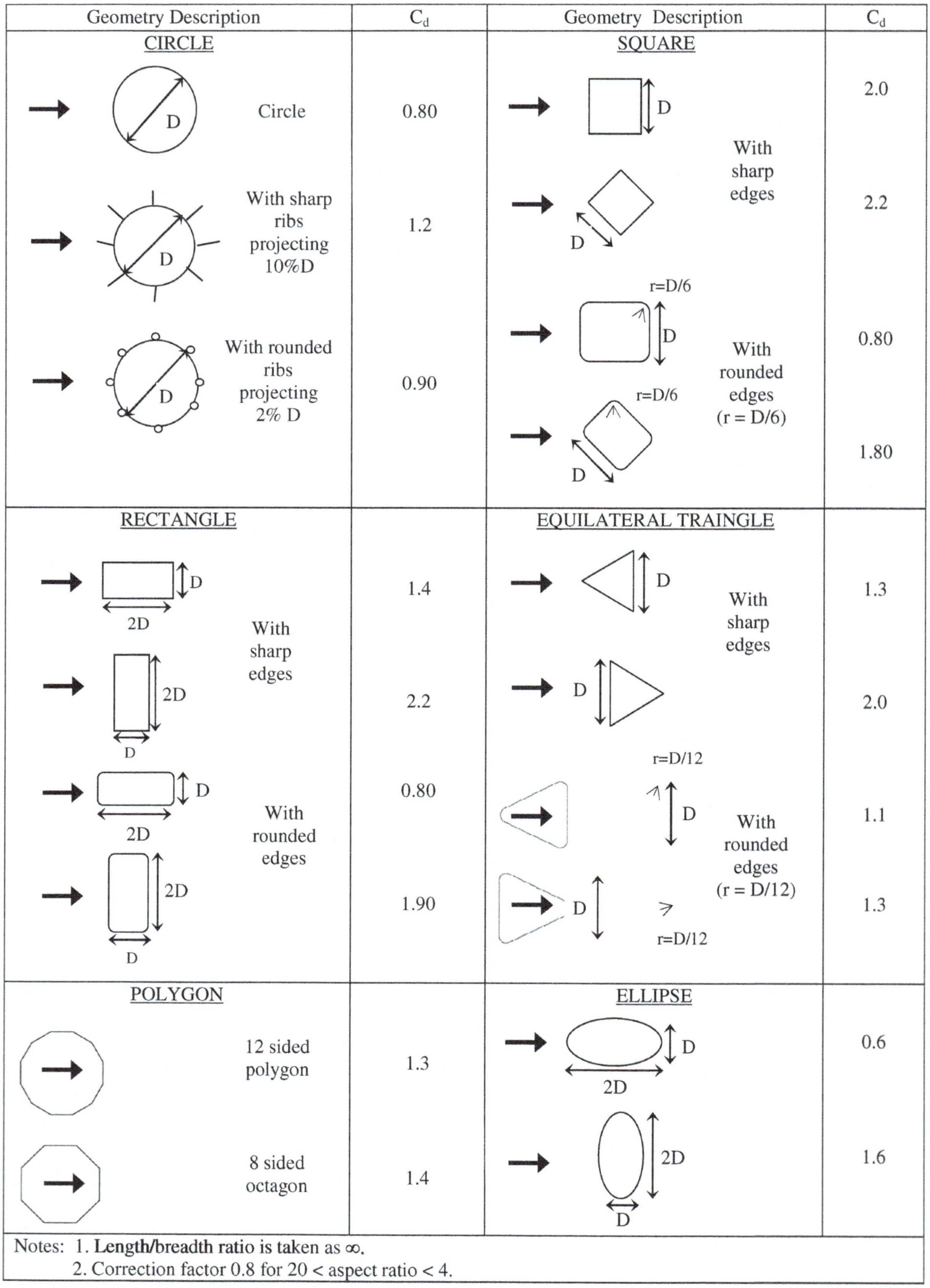

Fig. 7.4 Drag coefficient C_d for various cross sections [2]

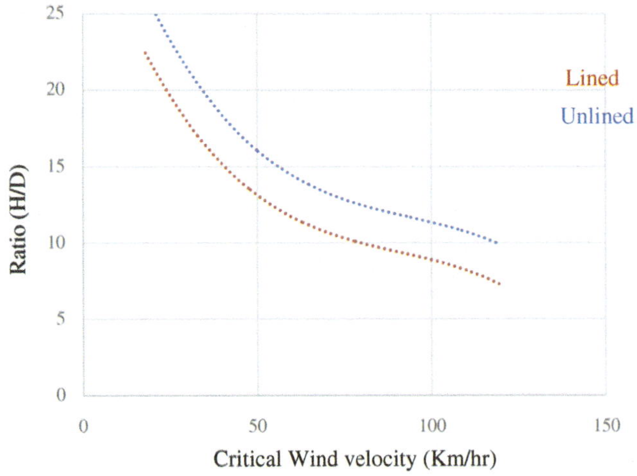

Fig. 7.5 Critical wind velocity in concrete chimneys cylinder [2]

Critical Wind Speed

Due to vortex shedding phenomenon, for the *i*th mode of vibration, critical wind speed (V_{cri}) is calculated by Eq. (7.13):

$$V_{cri} = \frac{f_i d}{S_n} \tag{7.13}$$

where

d effective diameter taken as average diameter over top 1/3 rd of chimney in 'm'

S_n Strouhal number to be taken as 0.20

All the modes which may be excited up to wind speeds 10% above the maximum expected at the height of effective diameter must be considered for the further detailed analysis. If the calculated critical wind speed for any mode of oscillation exceeds the limits specified above, it is assumed that the problem of vortex excited resonance will not be design criteria for that particular mode and subsequent higher modes. In such cases, across-wind load analysis may not be necessary.

Shear Force and Bending Moment

Sectional shear force F_{zoi} and bending moment M_{zoi} at any height z_o, for *i*th mode of vibration are obtained from the following equations:

$$F_{zoi} = 4\pi^2 f_i^2 \eta_{oi} \int_{z_0}^{H} m_z \varphi_{zi} dz \tag{7.14}$$

$$M_{zoi} = 4\pi^2 f_i^2 \eta_{oi} \int_{z_0}^{H} m_z \varphi_{zi} (z - z_0) dz \tag{7.15}$$

7.4.1.4 Combination of Along-Wind and Across-Wind Loads

Across-wind loads shall be computed with coexisting along-wind loads as described above. The combined design moment at any section shall be taken as the root sum square of the moment due to the across-wind loads and the coexisting along-wind loads.

Example 7.2

An example problem is illustrated for the dynamic instability study of a 175 m tall RC chimney. The input parameters considered are given below:

Height of chimney = 175 m
Bottom diameter of chimney = 14.6 m
Top diameter of chimney = 12.6 m
Fundamental frequency = 0.31 Hz
Damping factor (β) = 0.01
Air density = 1.18 kg mass/m^3 and
Kinematic viscosity of air = 180 × 10^{-7} kg mass/m s.

The design wind speed at the site is assumed as 120 km/hr. Find the shear force and bending moments for across-wind effects due to vortex shedding considering the chimney's diameter and magnitude of lumped weights at reasonable elevations including its characteristic shape.

Solution

i. Frequency

$$\begin{aligned} \omega_1 &= 2\pi n \\ &= 2\pi(0.31) \\ &= 1.948 \text{ rad/s} \end{aligned}$$

ii. Diameter

$$\begin{aligned} &\text{Average chimney diameter}(D_{\text{average}}) \\ &= (12.6 + 14.6)/2.0 \\ &= 13.6 \text{ m} \\ &\text{Diameter at one third height from top} \\ &= 12.6 + (14.6 - 12.6) \times 1/3 \\ &= 13.27 \text{ m} \\ &= (12.6 + 13.27)/2.0 \\ &= 12.935 \text{ m} \end{aligned}$$

iii. $$\begin{aligned} H/D_{\text{average}} &= 175/13.6 \\ &= 12.87 \end{aligned}$$

iv. From Fig. 7.5, to make a preliminary investigation, for H/D_{co} = 12.87 the critical wind speed is approximately

52 km/h which is less than that of given design wind speed 120 km/h, and hence, across-wind loads need to be determined.

v. Equivalent Mass

$$m_{ei} = \frac{\int_0^H m_z \varphi_{zi}^2 dz}{\int_0^H \phi_{zi}^2 dz} = \frac{862.0}{46.05} = 18.71 \text{ kN s}^2/\text{m}^2$$

vi. Mass damping parameter

$$K_{si} = 2m_{ei}\delta_s/\sigma d^2 = 2 \times 18714 \times (2\pi \times 0.01)/(1.18 \times 12.935^2) = 11.91$$

vii. Critical wind speed

$$V_{cri} = \frac{f_i d}{S_n} = 0.31 \times 13.27/0.20 = 20.57 \text{ m/s} = 74 \text{ km/h} < 120 \text{ km/h(Design Wind Speed)}.$$

Since the critical wind speed is less than the design wind speed, vortex shedding is possible and hence across-wind loads need to be evaluated. Various parameters required to evaluate the shear force and bending moment at different sections of chimney are presented in Table 7.12.

7.4.2 Wind Load Analysis of RC Chimney as per IS4998: 2015

In this method which is based on random response, HMW speed is used in computing the along-wind response. The pressure due to HMW component will be added to the pressure due to fluctuating wind load component using gust factor as per IS 4998:2015.

7.4.2.1 Along-Wind Response

The along-wind response of a chimney is obtained by Gust factor approach which requires hourly mean wind (HMW) speed at any height 'z' and is obtained as per IS 875 (Part 3): 2015. The along-wind load, F_z per unit height at any height 'z' on a chimney, is calculated from the following equation:

$$F_z = F_{zm} + F_{zf} \tag{7.16}$$

where F_{zm} is the mean along-wind load in N/m height due to HMW at height 'z' and is obtained by the following equation:

$$F_{zm} = \overline{p_z} C_d d_z \tag{7.17}$$

where, $\bar{p}_z$ is the design pressure at height 'z' due to HMW and can be obtained as follows:

$$\overline{p_z} = 0.60\,\overline{V_z}^2 \tag{7.18}$$

C_d is the mean drag coefficient taken as 0.8
d_z is the outer diameter of chimney at height z

Now, F_{zf} is the wind load in N/m height due to fluctuating component of wind at height 'z' and is obtained as below:

$$F_{zf} = \left[\frac{3(G-1)}{H^2}\right]\left(\frac{z}{H}\right)\int_0^H F_{zm} z dz \tag{7.19}$$

where

H is the total height of chimney above ground level, in m
G is the gust response factor which shall be calculated from following equation

$$G = 1 + g_f r_t \sqrt{B + \frac{SE}{\beta}} \tag{7.20}$$

where

β structural damping as a fraction of critical damping to be taken as 0.016 for along-wind loads
r_t twice the turbulence intensity at the top of chimney (= $0.622 - 0.178\log_{10}H$)
g_f peak factor defined as the ratio of the expected peak value to RMS value of the fluctuating load and given by

$$= \sqrt{2 \log_e \nu T} + \frac{0.577}{\sqrt{2 \log_e \nu T}} \tag{7.21}$$

in which

$$\nu T = \frac{3600 f_1}{\sqrt{1 + \frac{B\beta}{SE}}} \tag{7.22}$$

where

ν effective cycling rate
T sample period taken as 3600 s

Table 7.12 Forces due to vortex shedding

Elevation (m)	D_z (m)	Wt. (kN)	Mass = M_r (kN s²/m)	Φ_r (m)	Φ_r^2 (m²)	$\Phi_r^2 dz$ (m³)	$M_r\Phi_r^2$ (kN ms²)	$D_z\Phi_r dz$ (m³)	$F = M_r\Phi_r \omega_1^2\eta_{o1}$ (kN)	Shear Force (kN)	Bending Moment (kN m)
175	12.60	2030	207.0	1.000	1.0000	15.00	207.0	189.0	408.5	408.5	0.000E+00
160	12.77	3002	306.1	0.860	0.7396	11.09	226.4	164.7	519.4	927.9	6.127E+03
145	12.94	3002	306.1	0.728	0.5300	7.95	162.2	141.3	439.7	1367.6	1.392E+04
130	13.11	3002	306.1	0.596	0.3552	5.33	108.7	117.2	360.0	1727.6	2.051E+04
115	13.28	2988	304.7	0.468	0.2190	3.29	66.7	93.2	281.4	2009.0	2.591E + 04
100	13.45	3553	362.3	0.353	0.1246	1.87	45.1	71.2	252.4	2261.4	3.014E+04
85	13.62	4146	422.8	0.250	0.0625	0.94	26.4	51.1	208.6	2469.9	3.392E+04
70	13.79	4867	496.3	0.164	0.0269	0.40	13.3	33.9	160.6	2630.5	3.705E+04
55	13.96	4820	491.5	0.097	0.0094	0.14	4.6	20.3	94.1	2724.6	3.946E+04
40	14.13	5366	547.2	0.046	0.0021	0.03	1.2	9.7	49.7	2774.3	4.087E+04
25	14.30	5599	571.0	0.019	0.0004	0.01	0.2	4.1	21.4	2795.7	4.161E+04
10	14.47	4243	432.7	0.003	0.0000	0.00	0.0	0.4	2.6	2798.3	4.194E+04
0	14.60	2339	238.5	0	0.0000	0.00	0.0	0.0	0.0	2798.3	4.196E+04
					$\sum$=	46.05	862.0	896.25			

B background factor indicating the slowly varying component of wind load fluctuation

$$= \left[1 + \left(\frac{H}{265}\right)^{0.63}\right]^{-0.88} \quad (7.23)$$

S = Size reduction factor

$$= \left[1 + 5.78\left(\frac{f_1}{\overline{V}_{10}}\right)^{1.14} H^{0.98}\right]^{-0.88} \quad (7.24)$$

E a measure of the available energy in the wind at the natural frequency of chimney

$$= \frac{123\left(\frac{f_1}{\overline{V}_{10}}\right)H^{0.21}}{\left[1 + \left(\frac{330 f_1}{\overline{V}_{10}}\right)^2 H^{0.42}\right]^{0.83}} \quad (7.25)$$

in which

$\overline{V}_{10}$ hourly mean wind speed in m/sec at 10 m above ground level, in m/sec$V_b\overline{k}_2$ where k_2 and V_b are as defined in IS 875 (Part 3): 2015

f_1 natural frequency of chimney in the first mode of vibration, in Hz, and may be calculated as per Clause 5.5.8 of IS 4998:2015.

7.4.2.2 Across-Wind Response

Across-wind loads due to vortex shedding for chimney shells in the first and second modes of vibrations shall be considered in the design if the critical wind speed V_{cr} is between 0.5 V(Z_{Ref}) and 1.3 V(Z_{Ref}). Outside this range, it need not to be considered.

Across-wind base bending moment M_{ac} is given by:

$$M_{ac} = \left[g_{ac} S_s C_L \frac{\rho_a V_{cr}^2}{2} d_e H^2\right] \left[\frac{\pi}{4(\beta_s + \beta_a)}\right]^{0.5} S_p \left\{\frac{2L}{\left(\left(\frac{H}{d_e}\right) + C_E\right)}\right\}^{0.5} \quad (7.26)$$

in which

S_s mode shape factor (0.57 for first mode and 0.18 for second mode of vibration)

C_E end effect factor taken as 3

g_{ac} peak factor for across-wind load taken as 4

d_e effective diameter taken as average outer diameter over top one-third height of chimney, in m

C_L RMS lift coefficient and is given by the following expression

$$C_L = C_{LO} F_{1B} \quad (7.27)$$

where

C_{Lo} RMS lift coefficient modified for local turbulence and is given by

$$C_L = -0.243 + 5.648 I_{ref} - 18.182 I_{ref}^2 \quad (7.28)$$

in which

$$I_{\text{Ref}} = \frac{1}{\ln\left(\frac{Z_{\text{ref}}}{Z_o}\right)} \tag{7.29}$$

Z_{ref} = reference height, equals to (5/6)H

$$F_{1B} = -0.089 + 0.337\ln\left(\frac{H}{d_e}\right) \tag{7.30}$$

F_{1B} shall be between 0.2 and 1.0.

$$V_{\text{cr}} = \left(\frac{f_1 d_e}{S_t}\right) \tag{7.31}$$

$$V_{\text{cr}} = 5f_2 d_e \tag{7.32}$$

Equations 7.31 and 7.32 are used for calculating V_{cr} for first and second modes of vibration respectively of unlined chimneys.

Here, f_1 and f_2 are the natural frequencies of unlined chimney for the first and second modes of vibration, respectively, in Hz as per Clause 5.5.8 of IS 4998:2015.

$$S_t = 0.25F_{1A} \tag{7.33}$$

$$F_{1A} = 0.333 + 0.206\ln\left(\frac{H}{d_e}\right) \tag{7.34}$$

F_{1A} shall be between 0.6 and 1.0.

$$\beta_s = 0.01 + \frac{0.10[\overline{V}^* - \overline{V}(Z_{\text{ref}})]}{\overline{V}(Z_{\text{ref}})} \tag{7.35}$$

B_s shall be between 0.01 and 0.04.

Here, $\overline{V}^*$ is to be varied between 0.8 and 1.2 V_{cr} taking at least 10 intervals. The maximum value of $\overline{V}^*$ shall be limited to 1.3 $\overline{V}(Z_{\text{ref}})$.

$$\beta_a = \frac{K_a \rho_a d_e^2}{m_{av}} \tag{7.36}$$

where

m_{av} average mass in top one-third of chimney per unit height, in kg/m

$$K_a = K_{a0}F_{1B} \tag{7.37}$$

in which

$$K_{ao} = \frac{-1}{(1+5I_{\text{Ref}})\left[1 + \frac{|k-1|}{0.1+I_{\text{ref}}}\right]} \tag{7.38}$$

where

$$K = \frac{\overline{V}^*}{V_{\text{cr}}} \tag{7.39}$$

$$S_p = \frac{k^{1.5}}{\sqrt{B_w}(\pi^{0.25})}\exp\left[-\frac{1}{2}\left(\frac{1-k^{-1}}{B_w}\right)^2\right] \tag{7.40}$$

In which

B_w is the bandwidth parameter and given by:

$$B_w = 0.10 + 2I_{\text{ref}} \tag{7.41}$$

L = correlation length coefficient, taken as 1.2

The maximum value of M_{ac} determined in the region of 0.8 and 1.2 V_{cr} shall be considered as design across-wind base bending moment.

When $\overline{V}^* \geq \overline{V}(Z_{\text{Ref}})$; M_{ac} shall be multiplied by following multiplication factor (MF):

$$\text{MF} = \left\{1.0 - 0.95\left[\frac{\overline{V}^* - V(Z_{\text{ref}})}{V(Z_{\text{ref}})}\right]\right\} \tag{7.42}$$

Using the above-calculated M_{ac}, the across-wind load per unit height of chimney at any height, $F_{ac}(Z)$ in N/m shall be calculated based on the corresponding mode shape of chimney as follows:

$$F_{\text{ac}}(Z) = \frac{M_{\text{ac}} m(z)\phi_i(z)}{\int_0^H m(z)\phi_i(z) z\,dz} \tag{7.43}$$

where

$m(z)$ mass per unit height of chimney at level z, in kg/m and

$\phi_i(z)$ mode shape corresponding to ith mode of vibration

Now, using this $F_{ac}(Z)$, the design across-wind bending moments $M_{ac}(Z)$ at any height can be obtained.

7.4.2.3 Combination of Along-Wind and Across-Wind Loads

Across-wind loads shall be computed with coexisting along-wind loads as described above as per IS 4998:2015. The combined design bending moment $M_{comb}(z)$at any section shall be taken as a resultant of across-wind bending moment component ($M_{ac}(Z)$) and coexisting mean along-wind bending moment component $(\overline{M}_{a1}(Z))$. The resultant is obtained using root sum squared technique and can be presented as:

$$M_{\text{comb}}(z) = \left[M_{\text{ac}}^2(z) + \overline{M}_{a1}^2(z)\right]^{0.5} \tag{7.44}$$

where $\overline{M}_{a1}(Z)$ is the bending moment component induced by mean along-wind load $\overline{F}_{a1}(Z)$ which is given by:

$$\overline{F}_{a1}(Z) = \overline{F}(Z)\left[\frac{\overline{V}^*}{\overline{V}(Z_{\text{ref}})}\right]^2 \tag{7.45}$$

Except that $\overline{F}_{a1}(Z)$ shall not exceed $\overline{F}(Z)$.

7.4.2.4 Circumferential Response Due to Wind

The radial distribution of wind pressure on a horizontal section of a chimney depends on R_e whose value is normally in the ultra-critical range. The pressure distribution for a cylinder at high R_e is shown in Fig. 7.6 [11]. It is assumed that the along-wind resultant of such pressures is balanced by the resultant of shear forces induced in the structure. These shear forces, in turn, are assumed to vary sinusoidally along the circumference of the chimney shell. With these assumptions, bending moments in the shell can be obtained using the established analytical methods. The circumferential ring moments due to wind are calculated as follows:

$$M_{oe}(z) \text{ or } M_{oi}(z) = 0.335\, p_z\, r_m^2 \text{ Nm/m height} \tag{7.46}$$

where

$M_{oe}(z)$ or, $M_{oi}(z)$		external and internal ring moments at height z, respectively, in N m/m
p_z		design wind pressure due to 3 s gust wind speed at height z in N/m^2
r_m		mean radius of the shell at the section under consideration, in m

Here, p_z shall be obtained in accordance with IS 875 (Part-3), treating the chimney as Class A structure in Category II terrain. It shall be increased by 50% for a distance of 1.5 d (*H*) but not exceeding 15 m from the top. The hoop force and shear due to ovalling effect need not to be considered.

7.4.2.5 Load Combinations

The various load combinations for design of chimney are presented in Table 7.13.

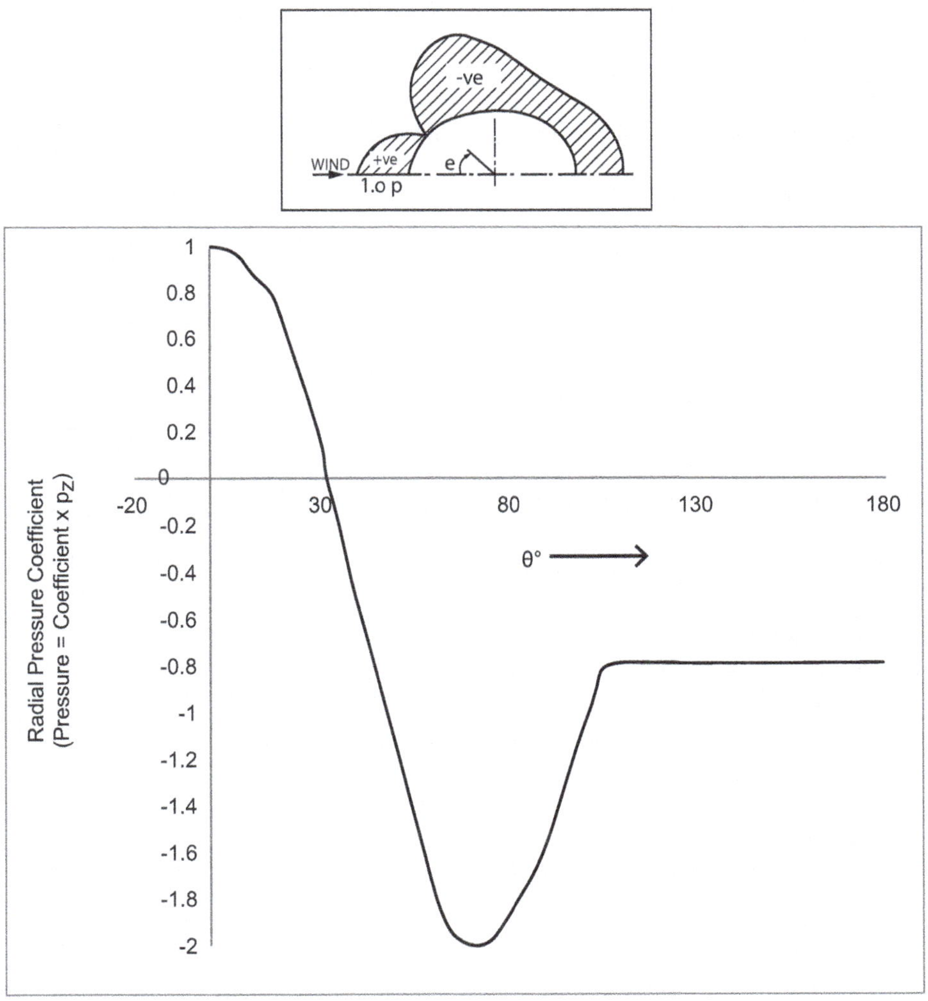

Fig. 7.6 Radial pressure distribution over circular cross section [2]

Table 7.13 Various load combinations for design of chimney shell [IS4998: 2015]

S. N.	Combination of loads
1	Dead load
2	Dead load + Wind load + Temperature effects
3	Dead load + Earthquake load + Temperature effects
4	Circumferential stresses in concrete due to temperature

7.5 Structural Design

Structural design of chimney shall be carried out as per IS 4998:2015 considering limit state design method adopting the appropriate partial safety factors for loads (γ_f) and material strengths (γ_m). The values of γ_f for various load combinations in limit state of collapse are prescribed in Clause 6.1.1 of IS4998: 2015. The values of γ_m shall be 1.5 for concrete and 1.15 for reinforcement steel.

7.5.1 Limit State of Collapse

Sections of chimney structures shall be designed for limit state of collapse considering following cases:

(a) Combined axial load and uniaxial bending,
(b) Combined axial load, uniaxial bending and temperature effects,
(c) Circumferential ring moments due to wind,
(d) Combined circumferential ring moments due to wind and temperature effects.

7.5.2 Limit State of Serviceability

In addition to the limit state of strength requirements, chimney structures shall be designed for limit state of serviceability in order to control the deflection and the crack width, etc., as per IS code of practices.

7.5.2.1 Deflection

The maximum lateral deflection of the top of a chimney structure under all service condition for all load cases and considering γ_m= 1.0 shall be limited to $H/500$, where H is the total height of chimney above ground level.

7.5.2.2 Cracking

The provisions of Clause 6.3.2 of IS 4998:2015 shall apply for crack width criteria.

7.5.3 Minimum Requirements

The minimum thickness of concrete shell and other reinforced concrete components of chimney structures along with minimum vertical and circumferential reinforcement requirements shall be as per Clause 6.4 of IS 4998:2015.

7.5.4 Stability

The stability of chimney shall be ensured as per requirements of Clause 6.4.3.3 of IS 4998:2015.

Exercise Problems

1. Considering the temperature gradient across the shell thickness of a RC chimney with tubular cross section, where reinforcing steel should be adequate to provide for resisting the tensile stress due to effect of temperature alone (caused by emission of hot gases). Justify your answer.
2. A 300-mm-thick concrete shell of a single flue chimney is subjected to a thermal gradient of 38 °C. Due to other loads, the maximum concrete and steel stresses at the mean radius of the chimney are 4.0 and 120.0 N/mm^2 on the leeward and windward sides, respectively. With 0.3% steel reinforcement of which 72% is placed on the outer face of shell, calculate the combined vertical stresses. Concrete cover to the main reinforcement is 60 mm, E_c = 25.5 kN/mm^2, E_s = 200 kN/mm^2, and coefficient of thermal expansion is 11 × 10^{-6} per °C. Evaluate the displacements and accelerations along the height of the chimney considering IS 1893 spectra for Zones III, IX and X and compare.
   Note: Use two modes and 5% damping.
3. Find the design shear force and bending moments for across-wind effects due to vortex shedding of a 150 m tall RC chimney with following input parameters:
   Bottom diameter of chimney = 14 m
   Top diameter of chimney = 12 m
   Fundamental frequency = 0.3 Hz

Damping factor = 0.01
Air density = 1.18 kg mass/m^3
Kinematic viscosity of air = 180 × 10^{-7} kg mass/m s
Design wind speed at the site = 120 km/hr.

4. Find the fundamental frequency of a stack-like structure as per IS1893 (Part 4): 2015, if it is assumed to be fixed at base with following parameter:
   Height of chimney = 100 m
   Area of cross section = 8.2 m^2
   Moment of inertia = 87.9 m^4
   Grade of concrete = M30
   Weight density of concrete = 25.0 kN/m^3.
5. List out the importance of chimney lining.
6. What are the different types of loads generally considered under the circumferential pressure effects of chimney shells.
7. What are the static and dynamic wind effects on tall RC chimneys? What are different forms of dynamic action due to wind on tall stack-like structures? Which is the most important dynamic action due to wind for tall chimney?
8. A square chimney 30 m high has an opening of 1 m 1 m. Find the plan dimension of the chimney and thickness of chimney wall if the maximum stress in construction material of chimney is not to be exceeded by 3.5 MPa. Assume a basic wind speed of 44 m/sec and density of material as 25 kN/m^3. The cross section may be assumed uniform along height of chimney.
9. Adopting the simplified modal analysis technique, evaluate the seismic response of a tapered RC chimney of height 100m above ground level with the input data listed below:
   Total height above Raft (Base) = 103.8 m
   Top inner diameter (ID) = 3.0 m
   Top outer diameter (OD) = 3.4 m
   Bottom ID = 6.2 m
   Bottom OD = 7.4 m
   Thickness at top = 0.2 m
   Thickness at bottom = 0.6 m
   Unit weight of chimney material = 25 kN/m^3
   Young's modulus of elasticity = 3.4×107 kN/m^2
   Seismic zone and zone factor (Z) = 0.16 (Zone III)
   Importance factor (I) = 2.00
   Response reduction factor (R) = 3.0
   Damping factor = 5%
   Founding stratum = Rock

   Assume, the chimney has uniform cross section above Raft (base) up to a height of 3.8 m (up to ground level). The geometric properties at various heights above ground level may be assumed suitably considering uniform variations.

References

1. Pinfold GM (1984) Reinforced concrete chimneys and towers, 2nd edn. View Pant Publication, London
2. Manohar SN (1985) Tall chimneys: design and construction. Tata McGraw –Hill Publishers, New Delhi India
3. Rumman WS (1985) Modal characteristics of linearly tapered reinforced concrete chimneys. ACI J 82(4):531–536
4. Wilson JL (2003) Earthquake response of tall reinforced concrete chimneys. Eng Struct 25(1):11–24
5. ASCE 4-98, Seismic analysis of safety related Nuclear Structures and Commentary, American Society of Civil Engineers Standards
6. ACI307-98, Design and Construction of Reinforced Concrete Chimneys, American Concrete Institute Standards
7. ACI349, Code requirements for nuclear safety related structures, American Concrete Institute Standards
8. ACI318-95, Building code requirements for Reinforced concrete, American Concrete Institute Standards
9. Dryden Hugh H (1984) Problems with in-line stacks: experience with full scale objects. Eng Struct 6(4):340–343
10. ASCE Task committee on wind forces (1961), Wind forces on structures, ASCE, 126(2), 1124–1198, American Society of Civil Engineers Standards
11. Roshko A (1961) Experiments on flow past a circular cylinder at very high Reynolds numbers. J Fluid Mech 10(3):345–356
12. IS875 (Part 1): 1987, Code of practices for design loads (other than earthquakes): Dead loads, Bureau of Indian Standards, New Delhi, India
13. IS875 (Part 2): 1987, Code of practices for design loads (other than earthquakes): Live loads, Bureau of Indian Standards, New Delhi, India
14. IS875 (Part 3): 2015, Code of practices for design loads (other than earthquakes): Wind loads, Bureau of Indian Standards, New Delhi, India
15. IS4998: 2015, Design of reinforced concrete chimneys: Criteria, Bureau of Indian Standards, New Delhi, India
16. IS1893 (Part 1): 2016, Criteria for earthquake resistant design of structures: General provisions & buildings, Bureau of Indian Standards, New Delhi, India
17. IS1893 (Part 4): 2015, Criteria for earthquake resistant design of structures: Industrial structures including stack, Bureau of Indian Standards, New Delhi, India
18. IS2131: 1981 (Reaffirmed 2002), Method for standard penetration tests for soils

Seismic Analysis and Design of Steel Structures

8

R. M. Parmar, Y. M. Parulekar, Praveen Kumar, and G. R. Reddy

Failures of Structures due to earthquakes causes the risk

Symbols

A	Area of cross section
A_c	Area at root of threads
A_e	Effective cross-sectional area
A_f	Total flange area
A_g	Gross cross-sectional area
A_{gf}	Gross cross-sectional area of flange
A_{go}	Gross cross-sectional area of outstanding member
A_n	Net area of total cross section
A_{nb}	Net tensile cross-sectional area of bolt
A_{nc}	Net cross-sectional area of connected leg of a member
A_{nf}	Net cross-sectional area of each flange
d_h	Diameter of bolt/rivet hole
b_w	Width of outstanding leg
E	Modulus of elasticity of steel
α_1	Coefficient of thermal expansion
G	Modulus of rigidity
KL/r	Slenderness ratio
r	Radius of gyration
KL	Effective length
f_{cc}	Euler buckling stress
χ	Stress reduction factor
α	Imperfection factor
K_c, K_b	Effective flexural stiffness of columns and beams meeting at the joint at the ends of columns and rigidly connected at joints
I	Moment of inertia of the member about an axis perpendicular to the frame
L	Length of the member equal to centre-to-centre distance of the intersecting member
C	Correction factor
Z_p and Z_e	Plastic and elastic section of moduli of the cross section
I_t	Torsional constant
I_w	Warping constant
h_f	Centre-to-centre distance between flanges
$M_d y, M_d z$	Design strength under corresponding moment acting alone
C_{my}, C_{mz}	Equivalent moment factor as per table 18 IS800
$P_{dy} P_{dz}$	Design strength under axial compression as governed by buckling about minor axis (y) and major axis (z), respectively
n_y, n_z	Ratio of actual applied axial force to the design axial strength for buckling about minor axis (y) and major axis (z), respectively

R. M. Parmar · Y. M. Parulekar · P. Kumar · G. R. Reddy (✉)
Bhabha Atomic Research Centre, Mumbai, India
e-mail: rssred@barc.gov.in

R. M. Parmar
e-mail: rparmar@barc.gov.in

Y. M. Parulekar
e-mail: yogitap@barc.gov.in

P. Kumar
e-mail: praveen@barc.gov.in

G. R. Reddy et al., *Textbook of Seismic Design*,
https://doi.org/10.1007/978-981-13-3176-3_8

C_{mLT}	Equivalent moment factor for lateral torsional buckling as per table 18 of IS 800:2007
M_k	Modal mass
ϕ_{ik}	Mode shape coefficient at floor i in the mode k
W_i	Seismic weight of floor i
n	Number of modes considered
d	Base dimension of building at the plinth level in m along considered direction of lateral face
M	Torque in N-mm caused by eccentric load
N	Number of rivets in rivet group
F_1	Force in rivet due to direct shear P
F_2	Force in rivet due to torque M
K	Constant
r_1, r_2, r_3, r_4	The distance of rivets from the center of gravity of the rivet group
S	Size of weld

8.1 Introduction

Steel is a material manufactured with an excellent quality control and is shaped in various structural shapes to suit construction of buildings and structures. The strength of steel is substantially higher than other construction material like concrete, masonry material which is most commonly used. In addition to the strength, steel possesses a significantly high ductility and toughness when compared with other construction materials.

Due to its high strength per unit weight, steel requires significantly lesser sizes of structural members when compared with the other construction materials. Hence, usually steel member becomes significantly slender. This possesses a serious issue in the process of structural design. Further, higher is the strength of the steel, the member becomes more slender and more serious is the buckling issue.

8.2 Damages Caused to Steel Structures in the Past Earthquakes

In the past few years, many earthquakes have occurred all over the world. Specifically in India for the last decades, we have witnessed earthquakes. The structures have witnessed beyond design basis earthquakes, i.e., earthquake of larger magnitude for which it is designed. Analyzing the failure helps the engineer to learn and improve the understanding of behavior of the structures subjected to earthquake. Some of the noticeable failures occurred during the earthquakes are shown in Figs. 8.1, 8.2, 8.3, 8.4, 8.5, and 8.6.

Fig. 8.1 Failure of steel frame structure due to ductile deformation [10]

Fig. 8.2 Failure of weak columns [10]

Figure 8.1 shows the failure of a steel structure because of large ductile deformation due to severe ground motion. Some buildings collapsed due to failure of columns at the ground/parking area. This type of damage is shown in Fig. 8.2. The building shown had moved about 60 cm downwards. Several vehicles parked at ground level were damaged. Fortunately, no loss of life occurred because the superstructure was safe. The reasons for this type of failure appear to be due to a heavy, strong floor, strong vertical acceleration and uneven settlement of columns due to liquefaction. Figures 8.3 and 8.4 show a structure which is

Fig. 8.3 Toppled house due to failure of steel column and concrete column at junction [10]

Fig. 8.4 Exaggerated view of column failure [10]

completely toppled due to failure at the junction of steel and concrete column. Figure 8.5 shows a failure of a steel column due to buckling. To avoid failures, this type of construction should be avoided.

8.3 Limit State Design of Steel Structures

In the limit state design method, the structure is designed to withstand all the loads, that will likely to act on it, safely. The structure should not suffer total collapse under

Fig. 8.5 Buckling failure of leg of a container crane [10]

Fig. 8.6 Exaggerated view of leg of a container crane [10]

accidental loads such as from explosions or impact or due to consequences of human error to an extent beyond the local damages. The objective of the limit state design is to achieve a structure that will remain fit for use during its life with acceptable target reliability. The acceptable limit for the safety and serviceability requirements before failure occurs is called a limit state.

In order to achieve the design objectives, the design shall be based on characteristic values of material strength and applied load (action), which take into account the variation in the material strength and in the loads to be supported. If the statistical data is available, that can be used for obtaining the characteristic values, where the statistical data is not available characteristic values shall be based on experience. With the help of partial safety factors, the design values for material strength and for applied loads are derived from the characteristic values. In the absence of special considerations, these factors shall have the values given in this section according to the material, the type of load, and the limit state being considered. The reliability of design is ensured by satisfying the requirement:

$$\text{Design action} \leq \text{Design strength} \tag{8.1}$$

Limit states are the states beyond which the structure no longer satisfies the performance requirements specified. The limit states are classified as:

(a) Limit state of strength and
(b) Limit state of serviceability.

The limit states of strength are those associated with failures (or imminent failure), under the action of probable and most unfavorable combination of loads on the structure using the appropriate partial safety factors, which may endanger the safety of life and property. The limit state of strength includes:

(a) Loss of equilibrium of the structure as a whole or any of its parts or components,
(b) Loss of stability of the structure (including the effect of sway where appropriate and overturning) or any of its parts including supports and foundations,
(c) Failure by excessive deformation, rupture of the structure or any of its parts or components,
(d) Fracture due to fatigue,
(e) Brittle fracture.

The limit state of serviceability includes:

(a) Deformation and deflections, which may adversely affect the appearance or effective use of the structure or may cause improper functioning of equipment or services or may cause damages to finishes and non-structural members.
(b) Vibrations in the structure or any of its components causing discomfort to people, damages to the structure, its contents or which may limit its functional effectiveness. Special consideration shall be given to systems susceptible to vibration, such as large open floor areas free of partitions to ensure that such vibrations are acceptable for the intended use and occupancy.

8.3.1 Mechanical Properties of Structural Steel

The principal mechanical properties of the structural steel like yield stress, ultimate stress, maximum percent elongation on a standard gauge length and notch toughness, Young's Modulus of elasticity which is being used in the design are derived from a tensile test carried out on a specimen. Structural mild steel and high tensile structural steel confirming to IS2062 may be used. In the present chapter, structural mild steel is considered. In the absence of experimental data, the physical properties of structural steel irrespective of its grade may be taken as following as per IS 800:2007.

(a) Unit mass of steel, $\rho = 7850$ kg/m^3
(b) Modulus of elasticity, $E = 2.0 \times 10^5$ MPa
(c) Poisson ratio, $\mu = 0.3$
(d) Modulus of rigidity, $G = 0.769 \times 10^5$ MPa
(e) Coefficient of thermal expansion, $\alpha_t = 12 \times 10^{-6}$/ °C.

Following conventions for member axes are used unless otherwise specified

(a) *x-x* along the member,
(b) *y-y* an axis of the cross section
    (1) perpendicular to the flanges, and
    (2) perpendicular to the smaller leg in an angle section.
(c) *z-z* an axis of the cross section
    (1) axis parallel to flanges, and
    (2) axis parallel to smaller leg in angle section.
(d) *u-u* major axis (when it does not coincide with *z-z* axis),
(e) *v-v* minor axis (when it does not coincide with *y-y* axis).

8.3.2 Loads and Forces

For design of any structural element, member or a structure, the following loads (actions) and their effects need to be considered.

(a) Dead loads,
(b) Imposed loads (live load, crane load, snow load, dust load, wave load, earth pressures, etc.),
(c) Wind loads,
(d) Earthquake loads,
(e) Erection loads,
(f) Temperature effects,
(g) Accidental loads such as those due to blast, impact of vehicles, etc., and
(h) Secondary effects due to contraction or expansion resulting from temperature changes, differential settlements of the structure as a whole or of its components, eccentric connections, rigidity of joints differing from design assumptions.

8.3.3 Load Combinations

8.3.3.1 Actions

The actions (loads) to be considered in design include direct actions (loads) experienced by the structure due to self-weight, external actions, etc., and imposed deformations such as that due to temperature and settlements.

8.3.3.2 Characteristic Actions (Loads)

The characteristic actions, Q_c, are the values of the different actions that are not expected to be exceeded with more than 5% probability, during the life of the structure and they are taken as:

(a) The self-weight, in most cases calculated on the basis of nominal dimensions and unit weights [see 1S 875 (Part 1)],
(b) The variable loads, values of which are specified in relevant standard [see IS 875 (all Parts) and IS 1893 (Part l)],
(c) The upper limit with a specified probability (usually 5%) not exceeding during design life,
(d) Specified by client, or by designer in consultation with client, provided they satisfy the minimum provisions of the relevant loading standard.

For the design purpose, load combinations which produce maximum forces and effects and consequently maximum stresses and deformations are considered. As per IS 800:2007, the following load combinations with appropriate partial safety factors as given in Table 8.1 are considered.

(a) Dead load + imposed load,
(b) Dead load + imposed load + wind or earthquake load,
(c) Dead load + wind or earthquake load, and
(d) Dead load + erection load.

8.3.4 Classification of Cross Sections

Plate elements of a cross section may buckle locally due to compressive stresses. The local buckling can be avoided before the limit state is achieved by limiting the width to thickness ratio of each element of a cross section subjected to compression due to axial force, shear, or moment. When plastic analysis is used, the members shall be capable of forming plastic hinges with sufficient rotation capacity (ductility) without local buckling, to enable the redistribution of bending moment required before formation of the failure mechanism. And when elastic analysis is used, the member shall be capable of developing the yield stress under compression without local buckling. On basis of these, four classes of sections are defined as follows:

(a) **Class 1 (Plastic)**: Cross sections, which can develop plastic hinges and have the rotation capacity required for failure of the structure by formation of plastic mechanism. The width to thickness ratio of plate elements shall be less than that specified under Class 1 (plastic), in Table 8.1.
(b) **Class 2 (Compact)**: Cross sections, which can develop plastic moment of resistance, but have inadequate plastic hinge rotation capacity for formation of plastic mechanism, due to local buckling. The width to thickness ratio of plate elements shall be less than that specified under Class 2 (Compact), but greater than that specified under Class 1 (Plastic), in Table 8.1.

Table 8.1 Partial safety factors for loads, γ_m (Table 5 of IS 800:2007)

Combination	Limit state of strength					Limit state of serviceability			
	DL	LL		WL/EL	AL	DL	LL		WL/EL
		Leading	Accompanying				Leading	Accompanying	
DL+LL+CL	1.5	1.5	1.05	–	–	1.0	1.0	1.0	–
DL+LL+CL+	1.2	1.2	1.05	0.6	–	1.0	0.8	0.8	0.8
WL/EL	1.2	1.2	0.53	1.2	–				
DL+WL/EL	1.5(0.9)[b]	–	–	1.5	–	1.0	–	–	1.0
DL+ER	1.2 (0.9)[b]	1.2	–	–	–	–	–	–	–
DL+LL+AL	1.0	0.35	0.35	–	1.0	–	–	–	–

[a]When action of different live loads is simultaneously considered, the leading live load shall be considered to be the one causing the higher load effects in the member/section

[b]This value is to be considered when the dead load contributes to stability against overturning is critical or the dead load causes reduction in stress due to other loads

Abbreviations: *DL* dead load, *LL* imposed load (live loads), *WL* wind load, *CL* crane load (vertical/horizontal), *AL* accidental load, *ER* erection load, *EL* earthquake load

(c) **Class 3 (Semi-compact)**: Cross sections, in which the extreme fiber in compression can reach yield stress, but cannot develop the plastic moment of resistance, due to local buckling. The width to thickness ratio of plate elements shall be less than that specified under Class 3 (semi-compact), but greater than that specified under Class 2 (Compact), in Table 8.1.

(d) **Class 4 (Slender)**: Cross sections, in which the elements buckle locally even before reaching yield stress. The width to thickness ratio of plate elements shall be greater than that specified under Class 3 (semi-compact), in Table 8.1. In such cases, the effective sections for design shall be calculated either by following the provisions of IS 801 to account for the post-local-buckling strength or by deducting width of the compression plate element in excess of the semi-compact section limit.

When different elements of a cross section fall under different classes, the section shall be classified as governed by the most critical element. The maximum value of limiting width to thickness ratios of elements for different classifications of sections as per Table 2 of IS 800:2007 are given in Table 8.2.

Where B, d, D, t, t_f, t_f for various types of sections are as shown in Fig. 8.7.

8.3.5 Maximum Effective Slenderness Ratio

As per Table 3 of IS 800:2007, effective slenderness ratio (*KL/r*) for beam, strut, or tension member should not exceed the values specified in Table 8.3, where 'KL' *is* the effective length of the member and '*r*' is appropriate radius of gyration based on the effective section (Table 8.4).

Table 8.2 Limiting width to thickness ratio (Table 2 of IS 800:2007)

Compression element			Ratio	Class 1 Plastic	Class 2 Compact	Class 3 Semi-compact
Outstanding element of compression flange	Rolled section		b/t_f	9.4ε	10.5ε	15.7ε
	welded section		b/t_f	8.4ε	9.4ε	13.6ε
Internal element of compression flange	Compression due to bending		b/t_f	29.3ε	33.5ε	42ε
	Axial compression		b/t_f	Not applicable		
Web of an I, H or box section	Neutral axis at mid-depth		d/t_w	84ε	105ε	126ε
	Generally	If r_1 is negative	d/t_w	$\frac{84\varepsilon}{1+r_1}$ But $\leq 42\varepsilon$	$\frac{105\varepsilon}{1+r_1}$	$\frac{126\varepsilon}{1+r_1}$ But $\leq 42\varepsilon$
		If r_1 is positive	d/t_w		$\frac{105\varepsilon}{1+1.5r_1}$ But $\leq 42\varepsilon$	
	Axial compression		d/t_w	Not applicable		But $\leq 42\varepsilon$
Web of a channel						
Angle, compression due to bending (both criteria should be satisfied)			b/t	9.4ε	10.5ε	15.7ε
			d/t	9.4ε	10.5ε	15.7ε
Single angle or double angles with the components separated, axial compression (all three criteria should be satisfied)			d/t	9.4ε	10.5ε	15.7ε
Outstanding leg of an angle in contact back to back in a double angle member			d/t	9.4ε	10.5ε	15.7ε
Stem of a T-section, rolled or cut from a rolled I- or H-section			D/t_f	8.4ε	9.4ε	18.9ε
Circular hollow tube, including welded tube subjected to:						
(a) Moment			D/t	$42\varepsilon^2$	$52\varepsilon^2$	$146\varepsilon^2$
(b) Axial compression			D/t	Not applicable		$88\varepsilon^2$

Notes

1. Elements which exceed semi-compact limits are to be taken as of slender cross section
2. $\varepsilon = \sqrt{250/f_y}$
3. Webs shall be checked for shear buckling in accordance with **8.4.2** when $d/t > 67\varepsilon$, where b is the width of the element (may be taken as clear distance between lateral supports or between lateral support and free edge, as appropriate), t is the thickness of element, d is the depth of the web, D is the outer diameter of the element (refer Fig. 8.7)
4. Different elements of a cross section can be in different classes. In such cases, the section is classified based on the least favorable classification
5. The stress ratio r_1 and r_2 are defined as

$$r_1 = \frac{\text{Actual average axial stress (negative if tensile)}}{\text{Design compressive stress of web alone}}$$

$$r_2 = \frac{\text{Actual average axial stress (negative if tensile)}}{\text{Design compressive stress of overall section}}$$

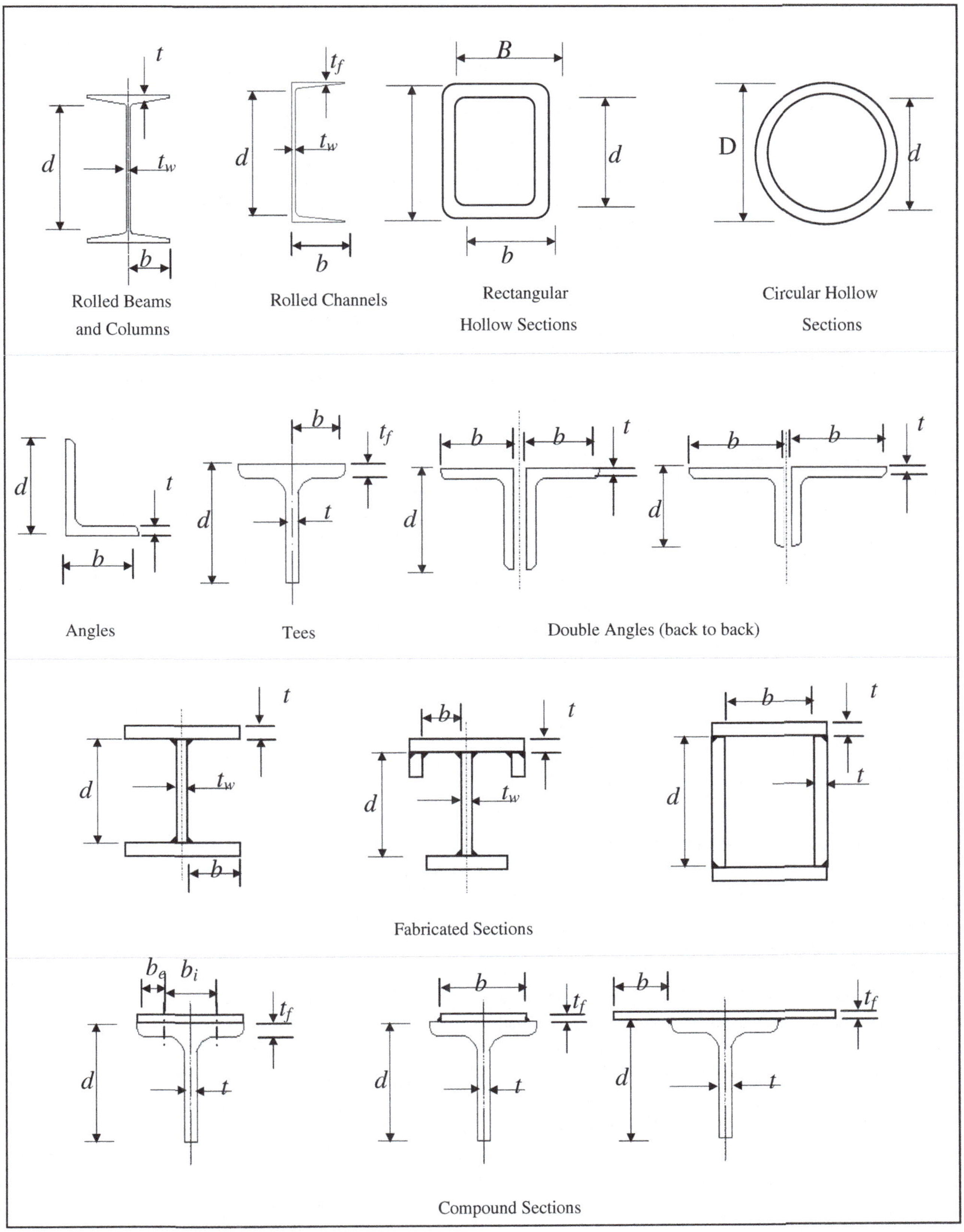

Fig. 8.7 Dimensions of sections

Table 8.3 Maximum values of effective slenderness ratios (Table 3 of IS 800:2007)

Sl. No.	Member	Maximum effective slenderness ratio
(i)	A member carrying compressive loads resulting from dead loads and imposed loads	180
(ii)	A tension member in which a reversal of direct stress occurs due to loads other than wind or seismic forces	180
(iii)	A member subjected to compression forces resulting only from combination with wind/earthquake actions, provided the deformation of such member does not adversely affect the stress in any part of the structure	250
(iv)	Compression flange of a beam against lateral torsional buckling	300
(v)	A member normally acting as a tie in a roof truss or a bracing system not considered effective when subject to possible reversal of stress into compression resulting from the action of wind or earthquake forces	350
(vi)	Members always under tension[a] (other than pre-tensioned members)	400

[a]Tension members, such as bracing's, pre-tensioned to avoid sag, need not satisfy the maximum slenderness ratio limits

Table 8.4 Partial safety factor for materials, γ_m (Table 5 of IS 800:2007)

Sl. No.	Definition	Partial safety factor	
1	Resistance, governed by yielding, γ_{m0}	1.1	
2	Resistance, governed by buckling, γ_{m0}	1.1	
3	Resistance, governed by ultimate stress, γ_{m1}	1.25	
4	Resistance of connection:	Shop fabrications	Field fabrications
	(a) Bolts-friction type, γ_{mf}	1.25	1.25
	(b) Bolts-bearing type, γ_{mb}	1.25	1.25
	(c) Rivets, γ_{mr}	1.25	1.25
	(d) Welds, γ_{mw}	1.25	1.50

Table 8.5 Imperfection factor (α)

Buckling class	*a*	*b*	*c*	*d*
α	0.21	0.34	0.49	0.76

8.3.6 Design of Compression Members

8.3.6.1 Design Strength

The steel members which are used for carrying axial compression such as common hot rolled and built-up steel members generally fail by Flexural buckling. The factors that affect the buckling strength of these members are initial bow, residual stresses, and accidental eccentricities of load. To account for all these factors, the strength of members subjected to axial compression is defined by buckling class *a*, *b*, *c*, or *d*.

The design compressive strength (P_d) of a member is given by:

$$P_d = A_e f_{cd} \tag{8.2}$$

where

A_e effective sectional area, and

f_{cd} design compressive stress

The design compressive stress, f_{cd}, of axially loaded compression members is calculated using Eq. (8.3).

$$f_{cd} = \frac{f_y/\gamma_{m0}}{\phi + \sqrt{\phi^2 + \lambda^2}} = \chi f_y/\gamma_{m0} \le f_y \gamma_{m0} \tag{8.3}$$

where

ϕ $0.5\left[1 + \alpha(\lambda - 0.2) + \lambda^2\right]$

λ non-dimensional effective slenderness ratio = $\sqrt{f_y/f_{cc}} = \sqrt{f_y(\mathrm{KL}/r)^2/\pi^2 E}$

f_{cc} Euler buckling stress = $\frac{\pi^2 E}{(\mathrm{KL}/r)^2}$

KL/*r* effective slenderness ratio or ratio of effective length (KL) to appropriate radius of gyration (*r*)

α imperfection factor as given in Table 8.5

χ stress reduction factor = $\frac{1}{\left(\phi + \sqrt{\phi^2 - \lambda^2}\right)}$

λ_{m0} partial safety factor for material strength

Table 8.6 Design compressive stress, f_{cd}, for various buckling class for f_y = 250 MPa

KL/r	λ	f_{cd} (in MPa)				f_{cd}/f_y			
		Class *a*	Class *b*	Class *c*	Class *d*	Class *a*	Class *b*	Class *c*	Class *d*
10	0.23	227.3	227.3	227.3	227.3	0.909	0.909	0.909	0.909
20	0.338	226.0	225.3	224.4	222.8	0.904	0.901	0.897	0.891
30	0.45	220.1	215.9	211.3	203.6	0.880	0.864	0.845	0.814
40	0.56	213.4	205.8	197.8	185.1	0.854	0.823	0.791	0.740
50	0.68	205.4	194.4	183.5	167.2	0.821	0.778	0.734	0.669
60	0.788	195.2	181.3	168.2	149.9	0.781	0.725	0.673	0.599
70	0.90	182.4	166.4	152.2	133.5	0.730	0.665	0.609	0.534
80	1.01	166.8	150.2	136.3	118.3	0.667	0.601	0.545	0.473
90	1.13	149.2	133.8	121.0	104.7	0.597	0.535	0.484	0.419
100	1.24	131.5	118.2	107.0	92.6	0.526	0.473	0.428	0.371
110	1.35	115.1	104.1	94.6	82.1	0.460	0.416	0.378	0.328
120	1.46	100.7	91.7	83.7	73.0	0.403	0.367	0.335	0.292
130	1.58	88.3	81.0	74.3	65.2	0.353	0.324	0.297	0.261
140	1.69	77.8	71.8	66.2	58.4	0.311	0.287	0.265	0.234
150	1.80	68.9	64.0	59.2	52.6	0.276	0.256	0.237	0.210
160	1.91	61.4	57.3	53.3	47.5	0.246	0.229	0.213	0.190
170	2.03	55.0	51.5	48.1	43.1	0.220	0.206	0.192	0.173
180	2.14	49.5	46.5	43.6	39.3	0.198	0.186	0.174	0.157
190	2.25	44.7	42.2	39.7	35.9	0.179	0.169	0.159	0.144
200	2.36	40.7	38.5	36.3	33.0	0.163	0.154	0.145	0.132
210	2.48	37.1	35.2	33.3	30.4	0.148	0.141	0.133	0.121
220	2.59	34.0	32.3	30.6	28.0	0.136	0.129	0.123	0.112
230	2.70	31.2	29.8	28.3	26.0	0.125	0.119	0.113	0.104
240	2.81	28.8	27.5	26.2	24.1	0.115	0.110	0.105	0.096
250	2.93	26.6	25.5	24.3	22.5	0.106	0.102	0.097	0.090

The calculated values of design compressive stress (f_{cd}) for various buckling class and effective slenderness ratio KL/*r* is presented in Table 8.6.

The classification of different sections under different buckling classes *a*, *b*, *c*, or *d* is given in Table 8.7. The curves corresponding to different buckling classes are presented in non-dimensional form, in Fig. 8.8.

8.3.6.2 Effective Length of Compression Member

In calculating the effective length (l_{eff} = KL) of the compression member, actual length (L) of the member is used by considering the relative translational and rotational boundary conditions at the ends. The actual length of the member is taken as the length from centre to centre of its intersections with the supporting members in the plane of the buckling deformation.

In the case of a member with a free end, the actual length is taken as the freestanding length from the center of the intersecting member at the supported end. As per IS 800:2007, where the boundary conditions in the plane of buckling can be assessed, the effective length, KL, can be calculated on the basis of Table 8.8.

The effective length of compression members, where frame analysis does not consider the equilibrium of a framed structure in the deformed shape (second-order analysis or advanced analysis), can be calculated using the procedure given in IS 800:2007, Appendix D-1.

In the absence of a more exact analysis, the effective length of columns in framed structures may be obtained by multiplying the actual length of the column between the centres of laterally supporting members (beams) as per Annex D of IS 800:2007 with the effective length factor K and calculated by using the equations given below, provided the connection between beam and column is rigid type.

(a) **Non-sway Frames (Braced Frame)**

In the non-sway frames, the relative displacement between the two adjacent floors is restrained by shear

Table 8.7 Buckling class of cross sections (Table 10 of IS 800: 2007)

Cross section	Limits	Buckling about axis	Buckling class
Rolled I section 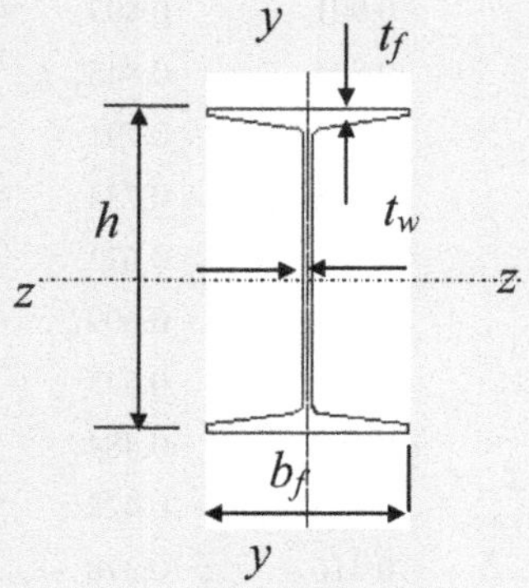	$h/b_f > 1.2$ $t_f \leq 40$ mm	z-z y-y	*a* *b*
	40 mm $\leq t_f \leq$ 100 mm	z-z y-y	*b* *c*
	$h/b_f \leq 1.2$ $t_f \leq 100$ mm	z-z y-y	*b* *c*
	$t_f > 100$ mm	z-z y-y	*d* *d*
Welded I section 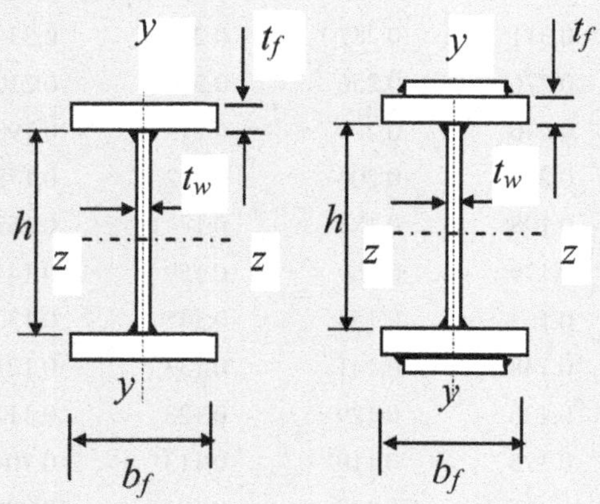	$t_f \leq 40$ mm	z-z y-y	*b* *c*
	$t_f > 40$ mm	z-z y-y	*c* *d*
Hollow section 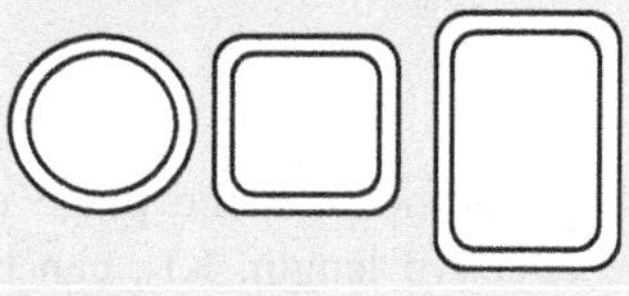	Hot rolled	Any	*a*
	Cold formed	Any	*b*
Welded box section 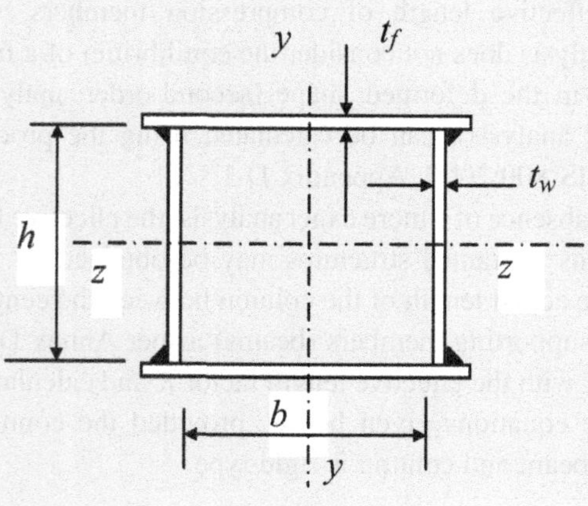	Generally (excepts below)	Any	*b*
	Thick welds and $b/t_f < 30$	z-z	*c*
	$h/t_w \leq 30$	y-y	*c*

(continued)

Table 8.7 (continued)

Cross section	Limits	Buckling about axis	Buckling class
Channel, angle, T and solid sections 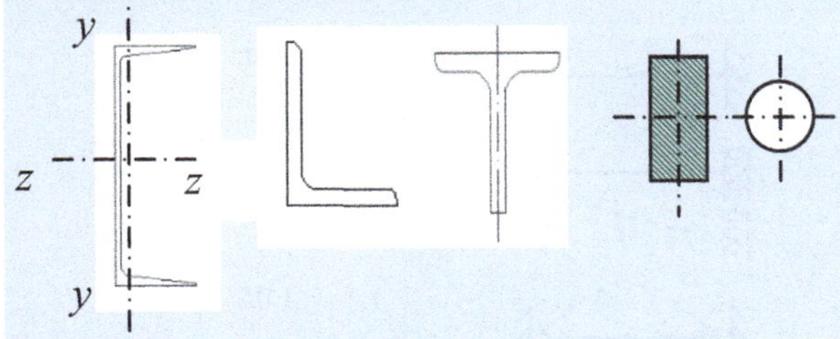	Any	*c*	*c*
Built-up member 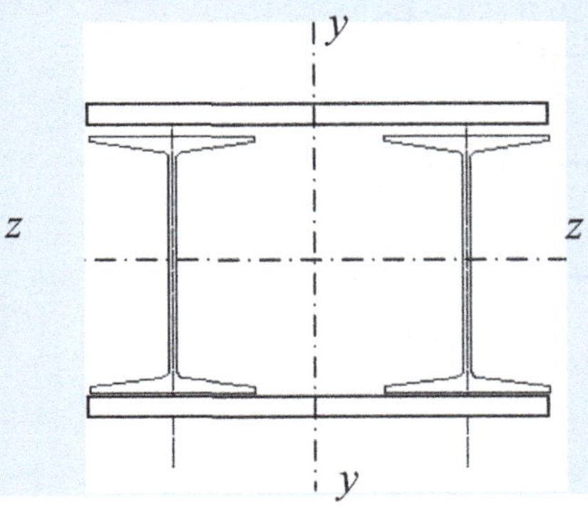	Any	*c*	*c*

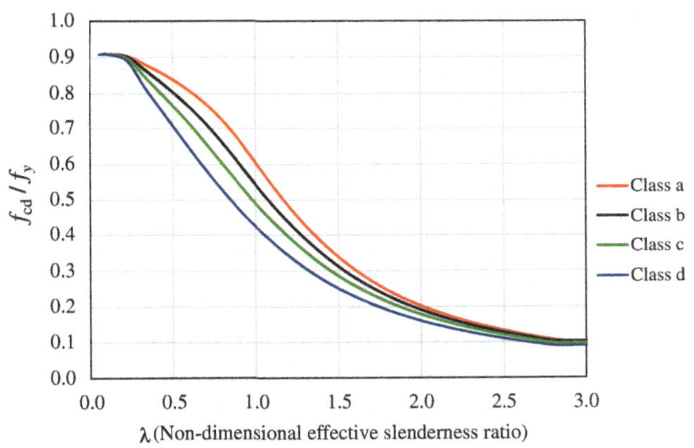

Fig. 8.8 Column buckling curve

walls or bracings. The effective length factor, K, of column in non-sway frames is given by Eq. (8.4).

$$K = \frac{[1 + 0.145(\beta_1 + \beta_2) - 0.265\beta_1\beta_2]}{[2 - 0.364(\beta_1 + \beta_2) - 0.247\beta_1\beta_2]} \quad (8.4)$$

(b) **Sway Frames (Moment-Resisting Frames)**
The effective length factor K of column in sway frame is given by Eq. (8.5)

$$K = \left[\frac{1 + 0.2(\beta_1 + \beta_2) - 0.12\beta_1\beta_2}{1 - 0.8(\beta_1 + \beta_2) - 0.6\beta_1\beta_2}\right]^{0.5} \quad (8.5)$$

where

β_1 and β_2 are given by,

$$\beta = \frac{\sum K}{\sum K_c + \sum K_b} \quad (8.6)$$

K_c, K_b = Effective flexural stiffness of the columns and beams meeting at the joint at the ends of the columns and rigidly connected at the joints, and these are calculated by:

$$K = C(I/L) \quad (8.7)$$

I moment of inertia of the member about an axis perpendicular to the plan of the frame
L length of the member equal to centre-to-centre distance of the intersecting member
C correction factor as shown in Table 8.9

8.3.7 Design of Members Subjected to Bending

8.3.7.1 Design Strength in Bending (Flexure)

A beam which is adequately supported against lateral torsional buckling (laterally supported beam) the design bending strength of that beam is governed by the yield stress. Where the beam is not adequately supported against lateral buckling (laterally unsupported beams), in such cases

Table 8.8 Effective length of prismatic compression members (Table 11 of IS 800: 2007)

Boundary conditions				Schematic representation	Effective length
At one end		At other end			
Translation	Rotation	Translation	Rotation		
Restrained	Restrained	Free	Free	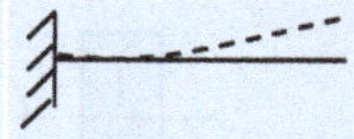	2.0L
Free	Restrained	Free	Restrained	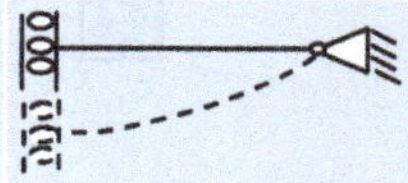	2.0L
Restrained	Free	Restrained	Free	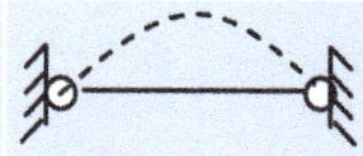	1.0L
Restrained	Restrained	Free	Restrained	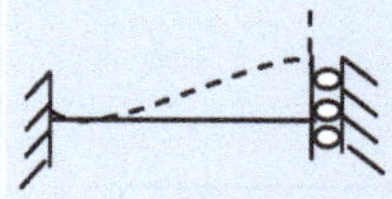	1.2L
Restrained	Restrained	Restrained	Free	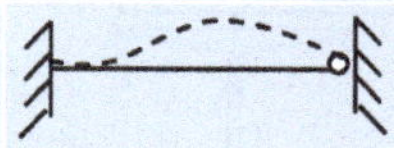	0.8L
Restrained	Restrained	Restrained	Restrained	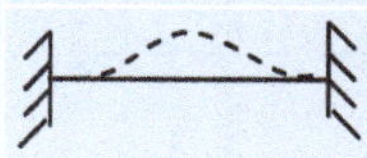	0.65L

Table 8.9 Correction factors for effective flexural stiffness (Table 35 of IS 800:2007)

Sl. No.	Far-end condition	Correction factor, C	
		Braced frame	Unbraced frame
(i)	Pinned	$1.5\,(1-\bar{n})$	$1.5\,(1-\bar{n})$
(ii)	Rigidly connected to column	$1.0\,(1-\bar{n})$	$1.0\,(1-0.2\bar{n})$
(iii)	Fixed	$2\,(1-0.4\bar{n})$	$0.67(1-0.4\bar{n})$
$\bar{n} = P_e/P$ where P_e = elastic buckling load, and P = applied load			

the design bending strength may be governed by lateral torsional buckling strength. The factored design moment, M at any section, in a beam due to external actions, shall satisfy

$$M \leq M_{\mathrm{d}} \tag{8.8}$$

where

M_{d} design bending strength of the section

8.3.7.2 Laterally Supported Beam

If the compression flange has nominal torsional restraint and full lateral restraint at the supports supplied by partial depth end plates, web cleats, or fin plates, it can be assumed that a beam is adequately supported at the supports. Full lateral restraint to compression flange may be assumed to exist if the frictional or other positive restraint of a floor connection to the compression flange of the member is capable of resisting a lateral force not less than 2.5% of the maximum force in the compression flange of the member. The design bending strength of a section which is not susceptible to web buckling under shear before yielding (where $d/t_{\mathrm{w}} \leq 67\varepsilon$) is determined according to Eq. (8.9).

Section with Webs Susceptible to Shear Buckling Before Yielding

The design bending strength shall be calculated using one of the following methods, when the flanges are plastic, compact, or semi-compact, but the web is susceptible to shear buckling before yielding ($d/t_{\mathrm{w}} \leq 67\varepsilon$):

(a) The web is designed only to resist shear and the axial force and bending moment acting on the section maybe assumed to be resisted by flanges only.
(b) The web shall be designed for combined normal and shear stresses by using simple plastic theory in the case of compact and plastic webs and simple elastic theory in

case of semi-compact webs. This design should be used when the axial force and bending moment acting on the section maybe assumed to be resisted by the whole section.

When the factored design shear force does not exceed $0.6V_d$, where V_d is the design shear strength of the cross section, the design bending strength, M_d shall be taken as:

$$M_d = \frac{\beta_b Z_p f_y}{\gamma_{m0}} \quad (8.9)$$

To avoid irreversible deformation under serviceability loads, M_d shall be less than 1.2 $Z_e f_y/\gamma_{m0}$ in case of simply supported and 1.5 $Z_e f_y/\gamma_{m0}$ in cantilever beams;

where

β_b 1.0 for plastic and compact sectionsZ_e/Z_p for semi-compact sections

Z_p, Z_e Plastic and elastic section moduli of the cross section, respectively

f_y yield stress of the material

γ_{m0} partial safety factor

When the design shear force (factored), V exceeds $0.6V_d$, where V_d is the design shear strength of the cross section the design bending strength, M_d shall be taken

$$M_d = M_{dv} \quad (8.10)$$

M_{dv} design bending strength under high shear (as explained in Sect. 8.3.10.1)

8.3.7.3 Laterally Unsupported Beams

In the following cases, it is not needed to check for resistance to lateral torsional buckling (member may be treated as laterally supported)

(a) If the bending is about the minor axis of the section,
(b) If the section is hollow (rectangular/tubular) or solid bars, and
(c) In case of major axis bending, where λ_{LT} (defined as follows) is less than 0.4

The design bending strength of laterally unsupported beam as governed by lateral torsional buckling is given by:

$$M_d = \beta_b Z_p f_{bd} \quad (8.11)$$

where

β_b 1.0 for plastic and compact sections Z_e/Z_p for semi-compact sections

Z_p, Z_e plastic section modulus and elastic section modulus with respect to extreme compression fiber

f_{bd} design bending compressive stress, obtained as given below

$f_{bd} = \chi_{LT} f_y/\gamma_{m0}$

χ_{LT} = bending stress reduction factor to account for lateral torisonal buckling, given by

$$\chi_{LT} = \frac{1}{\left\{\phi_{LT} + \left[\phi_{LT}^2 - \lambda_{LT}^2\right]^{0.5}\right\}} \leq 1.0 \quad (8.12)$$

$$\phi_{LT} = 0.5\left[1 + \alpha_{LT}(\lambda_{LT} - 0.2) + \lambda_{LT}^2\right] \quad (8.13)$$

λ_{LT} is the imperfection parameter is given by

λ_{LT} = 0.21 for rolled steel section, and

λ_{LT} = 0.49 for welded steel section.

The non-dimensional slenderness ratio, λ_{LT}, is given by

$$\lambda_{LT} = \sqrt{\beta_b Z_p f_y/M_{cr}} \leq \sqrt{1.2 Z_e f_y/M_{cr}} = \sqrt{f_y/f_{cr,b}} \quad (8.14)$$

where

M_{cr} elastic critical moment

$f_{cr,b}$ extreme fiber bending compressive stress corresponding to elastic lateral buckling moment

Elastic lateral torsional buckling moment

In case of simply supported, prismatic members with symmetric cross section, the elastic lateral buckling moment, M_{cr}, can be determined from:

$$M_{cr} = \sqrt{\left\{\left(\frac{\pi^2 EI_y}{(L_{LT})^2}\right)\left(GI_t + \frac{\pi^2 EI_w}{(L_{LT})^2}\right)\right\}} = \beta_b Z_p f_{cr,b} \quad (8.15)$$

$f_{cr,b}$ of non-slender rolled steel sections in the above equation may be approximately calculated using the following equation:

$$f_{cr,b} = \frac{1.1\,\pi^2 E}{(L_{LT}/r_y)^2}\sqrt{1 + \frac{1}{20}\left(\frac{L_{LT}/r_y}{h_f/t_f}\right)^2} \quad (8.16)$$

As per IS 800:2007, the following simplified equation can be used in the case of prismatic members made of standard rolled I-sections and welded doubly symmetric I-sections, for calculating the elastic lateral buckling moment, M_{cr},

$$M_{cr} = \frac{\pi^2 EI_y h_f}{2L_{LT}^2}\sqrt{1 + \frac{1}{20}\left(\frac{L_{LT}/r_y}{h_f/t_f}\right)^2} \quad (8.17)$$

where

I_t torsional constant = $\sum \frac{b_i t_i^3}{3}$ for open section

I_w warping constant

I_y, r_y moment of inertia and radius of gyration, respectively, about the weaker axis
L_{LT} effective length for lateral torsional buckling
h_f centre-to-centre distance between flanges
t_f thickness of the flange

8.3.8 Design of Tension Members

8.3.8.1 Tension Members

Tension members are linear members in which axial forces act to cause elongation (stretch). Such members can sustain loads up to the ultimate load, at which stage they may fail by rupture at a critical section. However, if the gross area of the member yields over a major portion of its length before the rupture load is reached, the member may become non-functional due to excessive elongation. Plates and other rolled sections in tension may also fail by block shear of end bolted regions.

The factored design tension, T, in the members shall satisfy the following requirement:

$$T \leq T_d \qquad (8.18)$$

where

T_d design strength of the member = lowest of (T_{dg}, T_{dn}, T_{db})

where

T_{dg} design strength due to yielding of gross section
T_{dn} design strength due to rupture strength of critical section
T_{db} design strength due to block shear

Design Strength Due to Yielding of Gross Section

The design strength of members under axial tension, T_{dg}, as governed by yielding of gross section, is given by

$$T_{dg} = A_g f_y / \gamma_{m0} \qquad (8.18)$$

where

f_y yield stress of the material
A_g gross area of cross section
γ_{m0} partial safety factor for failure in tension by yielding

Design Strength Due to Rupture of Critical section

(a) ***Plates***

The design strength in tension of a plate, T_{dn}, as governed by rupture of net cross-sectional area, An, at the holes is given by

$$T_{dn} = 0.9 A_n f_u / \gamma_{m1} \qquad (8.19)$$

where

γ_{m1} partial safety factor for failure at ultimate stress
f_u ultimate stress of the material
A_n net effective area of the member given by

$$A_n = \left[b - n d_h + \sum_i \frac{p_{si}^2}{4 g_i} \right] t \qquad (8.20)$$

where

b, t width and thickness of the plate, respectively
d_h diameter of the bolt-hole (2 mm in addition to the diameter of the hole, in case the directly punched holes)
g gauge length between the bolt-holes (refer Fig. 8.9)
p_s staggered pitch length between line of bolt-holes (refer Fig. 8.9)
n number of bolt-holes in the critical section
i subscript for summation of all the inclined legs

(b) ***Threaded Rods***

The design strength of threaded rods in tension, T_{dn}, as governed by rupture is given by

$$T_{dn} = 0.9 A_n f_u / \gamma_{m1} \qquad (8.21)$$

where

A_n net root area at the threaded section

(c) ***Single Angles***

The rupture strength of an angle connected through one leg is affected by shear lag. The design strength, T_{dn}, as governed by rupture at net section is given by:

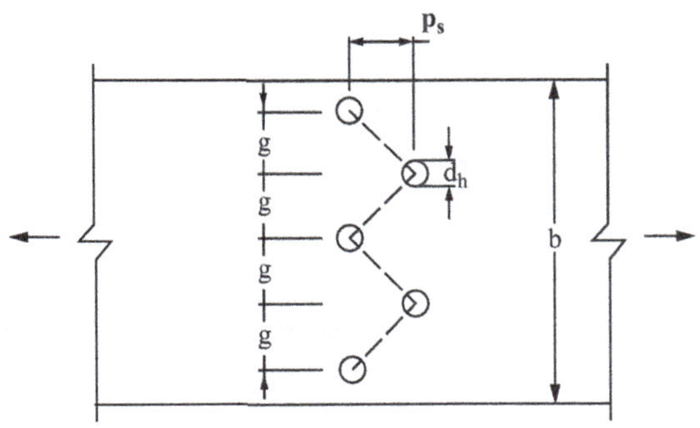

Fig. 8.9 Plates with bolts holes in tension

$$T_{dn} = 0.9 A_{nc} f_u/\gamma_{m1} + \beta A_{go} f_y/\gamma_{m0} \quad (8.22)$$

where

$$\beta = 1.4 - 0.076\left(\tfrac{w}{t}\right)\left(\tfrac{f_y}{f_u}\right)\left(\tfrac{b_s}{L_c}\right) \le \left(\tfrac{f_u \gamma_{m0}}{f_y \gamma_{m1}}\right) \le \tfrac{f_u \gamma_{m0}}{f_y \gamma_{m1}} \ge 0.7 \quad (8.23)$$

where

- w outstand leg width
- b_s shear lag width, as shown in Fig. 8.10
- L_c length of the end connection, that is the distance between the outermost bolts in the end joint measured along the load direction or length of the weld along the load direction

(d) ***Other Section***

The rupture strength, T_{dn}, of the double angles, channels, I-sections, and other rolled steel sections connected by one or more elements to an end gusset is also governed by shear lag effects. The design tensile strength of such sections as governed by tearing of net section may also be calculated using above equation, where β is calculated based on the shear lag distance, b_s, taken from the farthest edge of the outstanding leg to the nearest bolt/weld line in the connected leg of the cross section.

Design Strength Due to Block Shear

The strength as governed by block shear at an end connection of plates and angles is calculated as follows.

(a) ***Bolted Connections***

The block shear strength, T_{db}, of connection shall be taken as the smaller of,

$$T_{db} = \left[A_{vg} f_y/(\sqrt{3}\gamma_{m0}) + 0.9 A_{tn} f_u/\gamma_{m1}\right] \quad (8.24)$$

or

$$T_{db} = (0.9 A_{vn} f_u/(\sqrt{3}\gamma_{m1}) + A_{tg} f_y/\gamma_{m0}) \quad (8.25)$$

- A_{vg}, A_{vn} minimum gross and net area in shear along bolt line parallel to external force, respectively (1–2 and 3–4 as shown in Fig. 8.11a and 1–2 as shown in Fig. 8.11b)
- A_{tg}, A_{tn} minimum gross and net area in tension from the bolt-hole to the toe of the angle, end bolt line, perpendicular to the line of force, respectively (2–3 as shown in Fig. 8.11b
- f_u, f_y ultimate and yield stress of the material, respectively

(b) ***Welded Connection***

The block shear strength, T_{db}, shall be checked for welded end connections by taking an appropriate section in the member around the end weld, which can shear off as a block.

Fig. 8.10 Angles with single leg connections

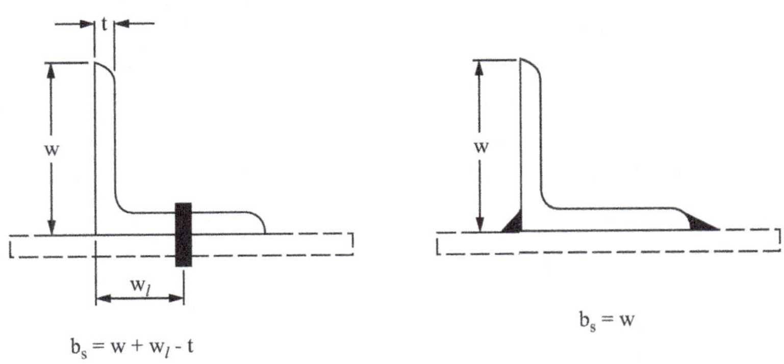

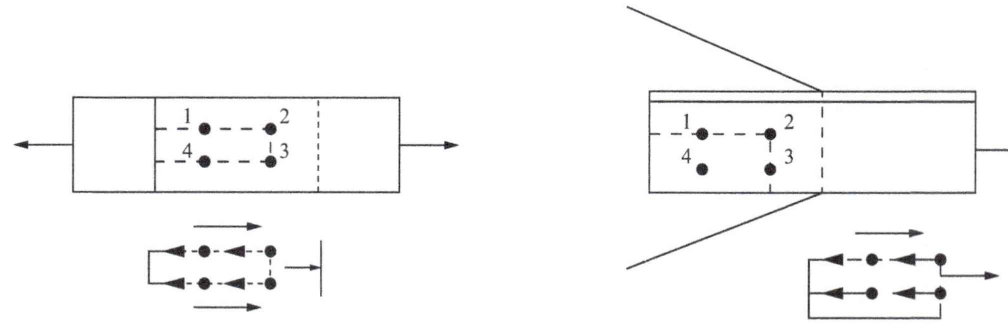

(a) Plate (b) Angle

Fig. 8.11 Block shear failure

8.3.9 Member Subjected to Combined Forces

8.3.9.1 Combined Shear and Bending

No reduction in moment capacity of the section is necessary as long as the cross section is not subjected to high shear force (factored value of applied shear force is less than or equal to 60% of the shear strength of the section). The moment capacity may be taken as, M_d, without any reduction.

When the factored value of the applied shear force is high (0.6 times V_d), the factored moment of the section should be less than the moment capacity of the section under higher shear force, M_{dv}, calculated as given below:

(a) ***Plastic or Compact Section***

$$M_{dv} = M_d - \beta(M_d - M_{fd}) \leq 1.2 Z_e f_y / \gamma_{m0} \quad (8.26)$$

where

- β — $(2V/V_d - 1)^2$
- M_d — plastic design moment of the whole section disregarding high shear force effect considering web buckling effects
- V — factored applied shear force as governed by web yielding or web buckling,
- V_d — design shear strength as governed by web yielding or web buckling
- M_{fd} — plastic design strength of the area of the cross section excluding the shear area, considering partial safety factor γ_{m0}
- Z_e — elastic section modulus of the whole section

(b) ***Semi-compact Section***

$$M_{dv} = Z_e f_y / \gamma_{m0} \quad (8.27)$$

8.3.9.2 Combined Axial Force and Bending Moment

Under combined axial force and bending moment, section strength as governed by material failure and member strength as governed by buckling failure shall be checked.

Section Strength

(a) ***Plastic and compact sections***

Conservatively, the following equation may also be used under combined axial force and bending moment:

$$\frac{N}{N_d} + \frac{M_y}{M_{dy}} + \frac{M_z}{M_{dz}} \leq 1.0 \quad (8.28)$$

where

- M_y, M_z — factored applied moments about the minor and major axis of the cross section, respectively
- N — factored applied axial force (tension, T, or compression, P)
- N_d — design strength in tension or in compression due to yielding given by $N_d = A_g f_y / \gamma_{m0}$
- M_{dy}, M_{dz} — design strength under corresponding moment acting alone
- A_g — gross area of the cross section
- γ_{m0} — partial factor of safety in yielding

(b) ***Semi-compact section***

In the absence of high shear force, semi-compact section design is satisfactory under combined axial force and bending, if the maximum longitudinal stress under combined axial force and bending, f_x, satisfies the following criteria:

$$f_x \leq f_y / \gamma_{m0} \quad (8.29)$$

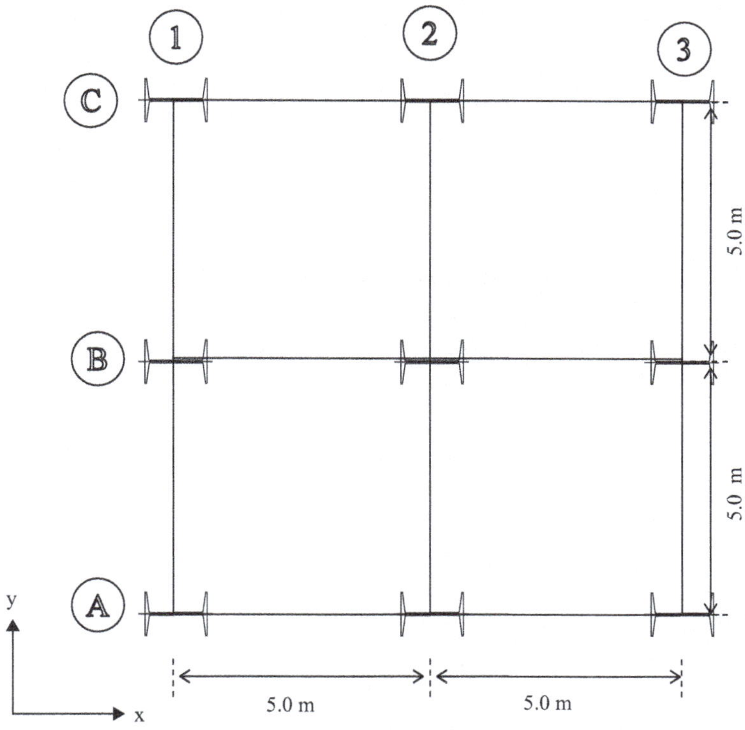

(a) Plan view

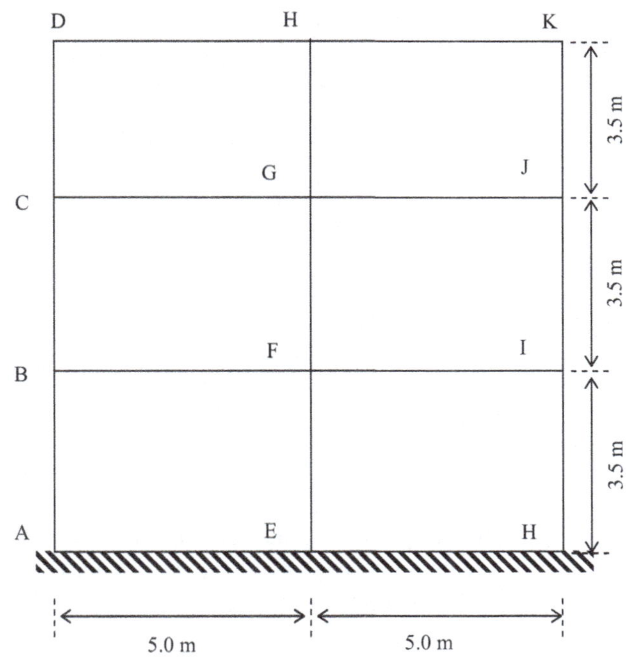

(b) Elevation

Fig. 8.12 Layout of the structure

For cross section without holes, the above criteria reduces to,

$$\frac{N}{N_d} + \frac{M_y}{M_{dy}} + \frac{M_z}{M_{dz}} \leq 1.0 \quad (8.30)$$

where N_d, M_{dy}, M_{dz} are as defined in above section.

8.3.9.3 Bending and Axial Compression

Members subjected to combined axial compression and biaxial bending shall satisfy the following interaction relationships:

$$\frac{P}{P_{dy}} + K_y \frac{C_{my} M_y}{M_{dy}} + K_{LT} \frac{M_z}{M_{dz}} \leq 1.0 \quad (8.31)$$

$$\frac{P}{P_{dz}} + 0.6 K_y \frac{C_{my} M_y}{M_{dy}} + K_z \frac{C_{mz} M_z}{M_{dz}} \leq 1.0 \quad (8.32)$$

where

C_{my}, C_{mz} — equivalent uniform moment factor as per table 18 of IS 800:2007

P — applied axial compression under factored load;

M_y, M_z — maximum factored applied bending moments about y and z axes of the member, respectively

P_{dy}, P_{dz} — design strength under axial compression as governed by buckling about minor (y) and major (z) axes, respectively

M_{dy}, M_{dz} — design bending strength about y (minor) or z (major) axes considering laterally unsupported length of the cross section

K_y — $1 + (\lambda_y - 0.2)n_y \leq 1 + 0.8n_y$

K_z — $1 + (\lambda_z - 0.2)n_z \leq 1 + 0.8n_z$

K_{LT} — $1 - \frac{0.1\lambda_{LT} n_y}{(C_{mLT} - 0.25)} \geq 1 - \frac{0.1 n_y}{(C_{mLT} - 0.25)}$

n_y, n_z — ratio of actual applied axial force to the design axial strength for buckling about the y and z axes, respectively

C_{mLT} — equivalent uniform moment factor for lateral torsional buckling as per Table 18 of IS 800:2007, corresponding to the actual moment gradient between lateral supports against torsional deformation in the critical region under consideration

Example 8.1: Frame Structure

Following steel structure is located in Seismic Zone III and is founded on medium soil. Design the structural members for the self-weight, dead loads, and seismic loads (Fig. 8.12).

Floor height = 3.5 m
Thickness of reinforced concrete floor slab = 200 mm (design of slab is not covered in this chapter)
Floor finish = 1.5 kN/m^3
Live load = 3.5 kN/m^3
Weight density of concrete = 25 kN/m^3.

Since the structure is located in seismic zone III, as per Table 2 of IS 1893:2002, zone factor = 0.16

Assumed section size:

Columns—ISHB 350, beams—ISLB—350. The properties of the section are as follows.

Section	ISHB 350	ISLB 350
Weight in N/m	710.24	486
Sectional area, A (mm^2)	9221	6301
Depth of section, h (mm)	350	350
Width of flange, b_f(mm)	250	165
Thickness of flange, t_f(mm)	11.6	11.4
Thickness of web, t_w(mm)	10.1	7.4
Moment of inertia about z-z, I_{zz}(cm^4)	19,802.8	13,158.3
Moment of inertia about y-y, I_{yy}(cm^4)	2510.5	631.9
Yield stress, f_y(MPa)	250	250
Young's modulus of steel, E (MPa)	200	200
(z-z is major axis and y-y is minor axis)		

Seismic force in the *X*-direction

The stiffness of floor systems in column-framed buildings is considerably higher than the storey stiffness due to columns. Hence, each floor diaphragm composed of slabs and beams is assumed as a rigid and the diaphragm is constrained to move as a single unit in the horizontal plane. The structure is thus idealized as an equivalent spring–mass system. The mass at floor level has been lumped at floor levels and connected with spring with story stiffness.

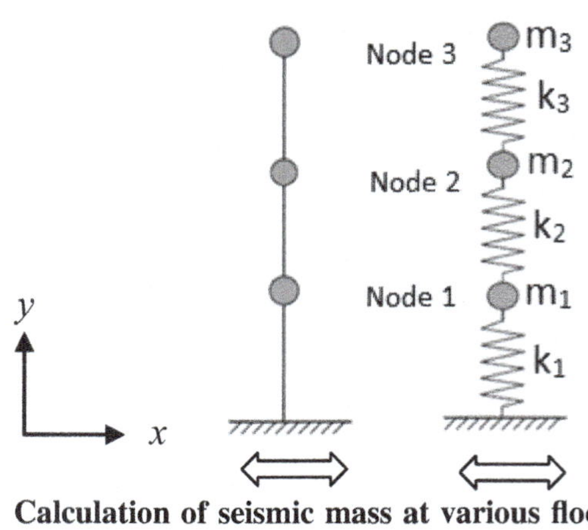

Calculation of seismic mass at various floor levels:

Seismic mass at Level 1

Self-weight of slab = 25 kN/m^3 × 0.2 m × (10 m × 10 m) = 500 kN
Floor finish = 1.5 kN/m^2 × 10 m × 10 m = 150 kN

Lumping 50% of the mass of columns above and below the floor,
Self-weight of nine columns = 9 × (3.5 m×710.24 N/m) = 22,372.56 N = 22.373 kN

Total length of the floor beam = 5 m × 12 = 60 m
Self-weight of the beam = 60 m × 485.60 N/m = 29,136 N = 29.136 kN

Considering 50% live load for seismic weight calculation (as per Table 8 of IS 1893: 2002)
Live load = 0.5 × 3.5 kN/m^2 × 10 m × 10 m = 175 kN

Weight of wall (230-mm-thick brick wall along periphery) = 10 m × 4 × 0.23 × (3.5 − 0.35) m × 20 kN/m^3 = 579.6 kN

Total seismic mass at floor level 1 = 500 + 150 + 22.373 + 29.136 + 175 + 579.6 = 1456.109 kN = 145.611 t

Seismic mass at Level 2

Self-weight of slab = 25 kN/m^3 × 0.2 m × (10 m × 10 m) = 500 kN
Floor finish = 1.5 kN/m^2 × 10 m × 10 m = 150 kN

Lumping 50% of the mass of columns of storey above and below,
Self-weight of nine columns = 9 × (3.5 m × 710.24 N/m) = 22,372.56 N = 22.373 kN

Total length of the floor beam = 5 m × 12 = 60 m
Self-weight of the beam = 60 m × 485.60 N/m = 29,136 N = 29.136 kN

Considering 50% live load for seismic weight calculation (as per Table 8 of IS 1893: 2002)
Live load = 0.5 × 3.5 kN/m^2 × 10 m × 10 m = 175 kN

Weight of wall (230-mm-thick brick wall along periphery) = 10 m × 4 × 0.23 × (3.5 − 0.35) m × 20k N/m^3 = 579.6 kN

Total seismic mass at floor level 1 = 500 + 150 + 22.373 + 29.136 + 175 = 1456.109 kN = 145.611 t

Seismic mass at Level 3 (roof)

Self-weight of slab = 25 kN/m^3 × 0.2 m × (10 m × 10 m) = 500 kN
Floor finish at roof (includes waterproofing) = 3 kN/m^2 × 10 m×10 m = 300 kN

Lumping 50% of the mass of columns,
Self-weight of nine columns = 0.5 × 9 × (3.5 m × 710.24 N/m) = 11186.28 N = 11.186 kN

Total length of the floor beam = 5 m × 12 = 60 m
Self-weight of the beam = 60 m × 485.60 N/m = 29,136 N = 29.136 kN

Weight of wall (230-mm-thick brick wall along periphery) = 0.5 × 10 m × 4 × 0.23 × (3.5 − 0.35) m × 20 kN/m^3 = 289.800 kN

Since the roof is inaccessible, live load is not considered for seismic mass calculation at roof level (Clause 7.3.2 of IS 1893:2002)

Total seismic mass at roof level = 500 + 300 + 11.186 + 29.136 + 289.800 = 1130.122 kN = 113.012 t

Neglecting the rotation, stiffness matrix of each element is given by,

$$[K_e] = \frac{12EI}{L^3}\begin{bmatrix} 1 & -1 \\ -1 & 1 \end{bmatrix}$$

Combined stiffness and mass matrices of the idealized spring–mass system are as follows.

$$[K] = 9 \times \frac{12EI}{L^3}\begin{bmatrix} 1 & -1 & 0 \\ -1 & 2 & -1 \\ 0 & -1 & 2 \end{bmatrix};$$

$$[M] = \begin{bmatrix} m_1 & 0 & 0 \\ 0 & m_2 & 0 \\ 0 & 0 & m_3 \end{bmatrix}$$

Substituting the values, we get

$$[K] = \frac{9 \times 12 \times 2 \times 10^{11} \times 19{,}802.8 \times 10^{-8}}{3.5^3}\begin{bmatrix} 2 & -1 & 0 \\ -1 & 2 & -1 \\ 0 & -1 & 1 \end{bmatrix}$$

$$[M] = \begin{bmatrix} m_1 & 0 & 0 \\ 0 & m_2 & 0 \\ 0 & 0 & m_3 \end{bmatrix} = \begin{bmatrix} 145{,}611 & 0 & 0 \\ 0 & 145{,}611 & 0 \\ 0 & 0 & 113{,}012 \end{bmatrix} \text{kg}$$

Eigenvalues and eigenvectors can be obtained by using any one of the methods explained in Chap. 4.

For the present case, eigenvalues are as follows.

Mode →	Mode 1	Mode 2	Mode 3
Eigenvalues, $\lambda = \omega^2$	154.17	1168.12	2301.07
Natural circular frequencies, ω (rad/s)	12.42	34.18	47.97
Natural frequencies, $f = \omega/2\pi$ (Hz)	1.976	5.440	7.635
Time periods (s)	0.506	0.184	0.131

Corresponding eigenvectors **(Mass-Ortho-normalized)** for each eigenvalues are as follows,

$$\{\phi_1\} = \begin{Bmatrix} -0.00094 \\ -0.00167 \\ -0.00203 \end{Bmatrix}; \quad \{\phi_2\} = \begin{Bmatrix} 0.00199 \\ 0.00059 \\ -0.00182 \end{Bmatrix};$$

$$\{\phi_3\} = \begin{Bmatrix} -0.00142 \\ 0.00193 \\ -0.00120 \end{Bmatrix}$$

(Note: For mass-Ortho-normalized eigenvectors $\{\phi_i\}^T[M]\{\phi_i\} = 1$)

Modal participation factor: Modal participation for each mode can also be obtained as follows.

Mode 1:
Modal Participation factor for Mode 1, Γ_1

$$\Gamma_1 = \{\phi_1\}^T[M]\{1\} = \begin{Bmatrix} -0.00094 \\ -0.00167 \\ -0.00203 \end{Bmatrix}^T \begin{bmatrix} 145{,}611 & 0 & 0 \\ 0 & 145{,}611 & 0 \\ 0 & 0 & 113{,}012 \end{bmatrix} \begin{Bmatrix} 1 \\ 1 \\ 1 \end{Bmatrix} = -609.55$$

Mass participated in first mode $= \frac{[\{\phi_1\}^T[M]\{1\}]^2}{\{\phi_1\}^T[M]\{\phi_1\}} = \frac{(\Gamma_1)^2}{1} = (609.55)^2 = 371{,}547.1$ kg

Mass participated in % $= \frac{371{,}547.1}{404{,}234} \times 100 = 91.9\%$

Mode 2:
Modal Participation factor for Mode 2, Γ_2

$\Gamma_2 = \{\phi_2\}^T[M]\{1\} = 169.97$

Mass participated in second mode $= \frac{[\{\phi_2\}^T[M]\{1\}]^2}{\{\phi_2\}^T[M]\{\phi_2\}} = \frac{(\Gamma_2)^2}{1} = (169.97)^2 = 28{,}890.7$ kg

Mass participated in % $= \frac{28{,}890.7}{404{,}234} \times 100 = 7.1\%$

Mode 3:
Modal Participation factor for Mode 3, Γ_3

$\Gamma_3 = \{\phi_3\}^T[M]\{1\} = -61.61$

Mass participated in second mode $= \frac{[\{\phi_3\}^T[M]\{1\}]^2}{\{\phi_3\}^T[M]\{\phi_3\}} = \frac{(\Gamma_3)^2}{1} = (61.61)^2 = 3795.79$ kg

Mass participated in % $= \frac{3795.79}{404{,}234} \times 100 = 0.94\%$

As per IS 1893:2016, for Soil Type II (medium stiff soil sites), S_a/g is given by,

$$\frac{S_a}{g} = \begin{cases} 1 + 15T & T \le 0.10\text{ s} \\ 2.5 & 0.10\text{ s} \le T \le 0.55\text{ s} \\ 1.36/T & 0.55\text{ s} \le T \le 4.0\text{ s} \\ 0.34 & T > 4.00\text{ s} \end{cases}$$

As per IS 1893:2016, Table 9, for steel building ordinary moment-resisting frame (OMRF), response reduction factor (R) = 3

Calculated value of A_{hk} for different modes is as follows.

Mode 1:	Mode 2:
For $T_1 = 0.506$ s and medium soil $\frac{S_{a1}}{g} = 2.5$ $A_{h1} = \frac{Z}{2}\frac{I}{R}\frac{S_a}{g} = \frac{0.16}{2} \times \frac{1.0}{3} \times 2.5 = 0.067$ Hence $S_{a1} = 0.067\text{ g} = 0.654\text{ m/s}^2$	*For* $T_2 = 0.184$ s and medium stiff soil sites $\frac{S_{a2}}{g} = 2.5$ $A_{h2} = \frac{Z}{2}\frac{I}{R}\frac{S_a}{g} = \frac{0.16}{2} \times \frac{1.0}{3}$ $\times 2.5 = 0.067$ Hence $S_{a2} = 0.067\text{ g} = 0.654\text{ m/s}^2$

Modal displacement for each mode is as follows.

Modal displacements

$$\{x\} = \Gamma\{\phi\}\frac{S_a}{\omega^2}$$

$$\{\phi_1\} = \begin{Bmatrix} -0.00094 \\ -0.00167 \\ -0.00203 \end{Bmatrix}; \{\phi_2\} = \begin{Bmatrix} 0.00199 \\ 0.00059 \\ -0.00182 \end{Bmatrix}; \{\phi_3\} = \begin{Bmatrix} -0.00142 \\ 0.00193 \\ -0.00120 \end{Bmatrix}$$

Substituting the values, we get

$$\begin{Bmatrix} x_{11} \\ x_{21} \\ x_{31} \end{Bmatrix} = \Gamma_1\{\phi_1\}\frac{S_{a1}}{(\omega_1)^2} = -609.55 \times \begin{Bmatrix} -0.00094 \\ -0.00167 \\ -0.00203 \end{Bmatrix} \times \frac{0.654}{154.17} = \begin{Bmatrix} 0.00244 \\ 0.00432 \\ 0.00524 \end{Bmatrix}\text{m} = \begin{Bmatrix} 2.44 \\ 4.32 \\ 5.24 \end{Bmatrix}\text{ mm}$$

$$\begin{Bmatrix} x_{12} \\ x_{22} \\ x_{32} \end{Bmatrix} = \Gamma_2\{\phi_2\}\frac{S_{a2}}{(\omega_2)^2} = 169.38 \times \begin{Bmatrix} 0.00199 \\ 0.00059 \\ -0.00182 \end{Bmatrix} \times \frac{0.654}{1168.12} = \begin{Bmatrix} 0.000189 \\ 0.000056 \\ -0.000173 \end{Bmatrix}\text{m} = \begin{Bmatrix} -0.189 \\ 0.056 \\ 0.173 \end{Bmatrix}\text{ mm}$$

As per IS 1893:2016, Clause 7.7.5.2, the number of modes to be used in analysis for earthquake shaking along a considered direction, should be such that the sum total of modal masses of these modes considered at least 90% of the total seismic mass.

Modal mass for first mode is 91.9% of the total mass. Hence, the number of modes to be considered is satisfied by considering only first mode of vibration. However, for illustration purpose, first two modes of vibrations are considered in this example.

Lateral load for each mode is given by,

$$\{Q_i\} = S_{ai}\Gamma_i\{\phi_i\}^T[M]$$

$$\begin{aligned}\{Q_1\} &= S_{a1}\Gamma_1\{\phi_1\}^T[M]\\ &= 0.654 \times (-609.55)\\ &\quad \times \begin{Bmatrix} -0.00094 \\ -0.00167 \\ -0.00203 \end{Bmatrix}^T \begin{bmatrix} 145{,}611 & 0 & 0 \\ 0 & 145{,}611 & 0 \\ 0 & 0 & 113{,}012 \end{bmatrix}\end{aligned}$$

$$\{Q_1\} = \begin{Bmatrix} 54928.5 \\ 97497.0 \\ 91681.0 \end{Bmatrix} N = \begin{Bmatrix} 54.928 \\ 97.497 \\ 91.681 \end{Bmatrix} \text{kN}$$

$$\{Q_2\} = \begin{Bmatrix} 32361.4 \\ 9549.237 \\ -22929.4 \end{Bmatrix} N = \begin{Bmatrix} 32.361 \\ 9.549 \\ -22.929 \end{Bmatrix} \text{kN}$$

Since all of the modes are well separated (Clause 3.2), the contribution of different modes is combined by the square root of the sum of the square (SRSS) method

$$\{Q\} = \begin{Bmatrix} \sqrt{(54.928)^2 + (32.361)^2} \\ \sqrt{(97.497)^2 + (9.549)^2} \\ \sqrt{(91.681)^2 + (22.929)^2} \end{Bmatrix} = \begin{Bmatrix} 63.752 \\ 97.964 \\ 94.505 \end{Bmatrix} \text{kN}$$

Hence, base shear = V_B = 63.752 + 97.964 + 94.505 = 256.220 kN.

As per Clause 7.7.3 IS 1893(Part 1): 2016, design base shear V_B shall not be less than design base shear, $\bar{V}_B$ calculated using fundamental period.

The approximate fundamental natural period of vibration (T_a), in seconds, of a moment-resisting frame buildings with brick infill panels for steel frame building estimated by the empirical expression is given by,

$$T_a = \frac{0.09h}{\sqrt{d}}$$

where

- h height of the building in m
- d base dimension of the building at the plinth level along the considered direction of earthquake shaking, in m.

$$T_a = \frac{0.09 \times 10.5}{\sqrt{10}} = \frac{0.09}{\sqrt{10}} = 0.0299 \text{ sec}$$

As per IS 1893(Part 1): 2016, for structure located on medium stiff soil site sand T_a = 0.299 sec, S_a/g = 2.5

Design horizontal acceleration coefficient value, A_h is given by,

$$A_h = \frac{Z}{2}\frac{I}{R}\frac{S_a}{g} = \frac{0.16}{2} \times \frac{1.0}{4.0} \times 2.5 = 0.05$$

$$\begin{aligned}\text{Design base shear} = \bar{V}_B = A_h W &= 0.05 \times 4042.34\\ &= 202.117 \text{ kN}\end{aligned}$$

Since $\bar{V}_B$ is less than the base shear obtained by dynamic analysis ($V_B = 256.220$kN); hence, all the response quantities need not to be scaled up.

Solving for lateral load we get,

Distributing the lateral loads with 2/3 at the mid grid and 1/3 at edge grids.
Let us consider middle grid. Lateral loads for middle grid will be as follows.

$$\{Q\} = \begin{Bmatrix} \frac{2}{3} \times 63.752 \\ \frac{2}{3} \times 97.964 \\ \frac{2}{3} \times 94.505 \end{Bmatrix} \text{kN}$$

Let us design column DC of portal frame. The forces in the member are as follows.

<u>Design of members</u>:

For illustration, one column and one beam element design is presented in the chapter. In similar way, all the elements of the portal can be designed and checked against codal provisions.

Design is carried out for the following load combinations (for illustration, only one load combination is considered. In similar way, the capacity has to be checked for all possible loads and their load combination as per the code).

1.5 Dead Load (DL) + 1.5 Seismic Load (EL)

(Note: It is to be noted that self-weight of the members has been ignored while analyzing the structure for dead load and seismic load.)

Design of column CD (upper storey column)

Preliminary sizing of the members:

1. Obtain the initial sizes of columns from the load to be resisted by column and assuming compressive stress of 100–110 MPa.

2. Based on assumed size of column, carry out the analysis and obtain the member end forced for various loads and their load combination as per IS 800: 2007.
3. If the capacity of the member is more than

For the present case,

Height of column, L = 3.5 m.
Based on boundary conditions of column, it is reasonable to consider effective length factor ($K = L_{eff}/L$) as 1.20 [Assumed end condition of column: Column is restrained in direction and position at one end and other end restrained in direction but not in position.]
Grade of steel = Fe 410,
Yield strength, f_y = 250 MPa,
Partial safety factor for material = γ_{m0} = 1.10.

Properties of ISHB 350 @ 710.2 N/m (column AB and CD) are as follows (Indian Standard Handbook 1)

H = 350 mm, b_f = 250 mm, t_f = 11.6 mm, t_w = 10.1 mm, A = 9221 mm2, r_z = 146.5, r_y = 52.2 mm

$$\frac{h}{b_f} = \frac{350}{250} = 1.4 > 1.2 \quad \text{and} \quad t_f = 11.6\ \text{mm} < 40\ \text{mm}$$

Hence as per Table 10 of IS 800: 2007, buckling class for buckling about y-y axis is b and for buckling about z-z axis will be Buckling class a.

Buckling about major axis *y-y* axis

Effective slenderness ratio = $\lambda_y = \frac{KL}{r_y} = \frac{1.2\times3000}{52.2} = 68.97$

As per Table 7 of IS 800:2007, for buckling class b, imperfection factor = 0.34

Considering E = 210 GPa,

Non-dimensional effective slenderness ratio, λ, is

$$\lambda = \sqrt{f_y/f_{cc}} = \sqrt{f_y(KL/r)^2/\pi^2 E}$$
$$= \sqrt{250 \times 68.97^2/\pi^2 \times 2 \times 10^5} = 0.776$$

$$\phi = 0.5\left[1 + \alpha(\lambda + 0.2) + \lambda^2\right]$$
$$= 0.5\ \left[1 + 0.34\,(0.776 - 0.2) + 0.776^2\right] = 0.899$$

$$\chi = \text{stress reduction factor} = \frac{1}{\left(\phi + \sqrt{\phi^2 - \lambda^2}\right)}$$
$$= \frac{1}{\left(0.899 + \sqrt{0.899^2 - 68.96^2}\right)} = 0.7391$$

$$\text{Design compressive stress} = f_{cd} = \frac{\chi f_y}{\gamma_{m0}} = \frac{0.7391 \times 250}{1.1}$$
$$= 167.98\ \text{MPa}$$

$$\left(f_{cd} \le \frac{f_y}{\gamma_{m0}} = \frac{250}{1.1} = 227.27\ \text{MPa}\right)$$

Hence, design compressive stress = f_{cd} = 167.98 MPa
Design compressive strength = $A_e f_{cd} = 9221 \times 167.98 = 1548.94 \times 10^3$ N

Buckling about major axis *z-z* axis

Effective slenderness ratio = $\lambda_z = \frac{\text{KL}}{r_z} = \frac{1.2\times3000}{146.5} = 24.57$

As per Table 7 of IS 800:2007, for buckling class a, imperfection factor = 0.21

Considering E = 210 GPa,

Non-dimensional effective slenderness ratio, λ, is

$$\lambda = \sqrt{f_y/f_{cc}} = \sqrt{f_y(\text{KL}/r)^2/\pi^2 E}$$
$$= \sqrt{250 \times 24.57^2/\pi^2 \times 2 \times 10^5} = 0.277$$

$$\phi = 0.5\left[1 + \alpha(\lambda - 0.2) + \lambda^2\right]$$
$$= 0.5\ \left[1 + 0.34\,(0.277 - 0.2) + 0.277^2\right] = 0.551$$

$$\chi = \text{stress reduction factor} = \frac{1}{\left(\phi + \sqrt{\phi^2 - \lambda^2}\right)}$$
$$= \frac{1}{\left(0.551 + \sqrt{0.551^2 - 0.277^2}\right)} = 0.973$$

$$\text{Design compressive stress} = f_{cd} = \frac{\chi f_y}{\gamma_{m0}} = \frac{0.973 \times 250}{1.1}$$
$$= 221.14\ \text{MPa}$$

$$\left(f_{cd} \le \frac{f_y}{\gamma_{m0}} = \frac{250}{1.1} = 227.27\ \text{MPa}\right)$$

Hence, design compressive stress = f_{cd} = 221.14 MPa
Design compressive strength = $A_e f_{cd} = 9221 \times 221.14 = 2039.1 \times 10^3$ N

Design of beam BF (upper storey beam)

The properties of ISLB 350 @ 486 N/m (Beam BG) are as follows.

The beam is laterally unsupported

h = 350 mm, bf = 165 mm, t_f = 11.4 mm, t_w = 7.4 mm, $I_y = 631.9 \times 10^4$ mm^4, $I_z = 13{,}158.3 \times 10^4$ mm^4

Lateral–torsional buckling moment, $M_{cr} = \frac{\pi^2 EI_y}{(L_{LT})^2}\left(\frac{I_w}{I_y} + \frac{GI_t(L_{LT})^2}{\pi^2 EI_y}\right)^{0.5}$

For simplified form (is generally on the safe side) for I-section is as follows

Elastic lateral buckling moment, $M_{cr} = \frac{\pi^2 EI_y}{(L_{LT})^2}\left(GI_t + \frac{\pi^2 EI_w}{(L_{LT})^2}\right)$

I_y, I_w, I_t = moment of inertia about the minor axis, warping constant and St. Venant's torsion constant of the cross section, respectively
G = modulus of rigidity
L_{LT} = effective length against lateral torsional buckling

Effective span = 3.5 m
Young's modulus of steel = 200 GPa

Using the simplified equation,

$$M_{cr} = \frac{\pi^2 EI_y h_f}{2L_{LT}^2}\sqrt{1 + \frac{1}{20}\left(\frac{L_{LT}/r_y}{h_f/t_f}\right)^2}$$

Substituting the value, we get

$$M_{cr} = \frac{\pi^2 \times 200 \times 10^3 \times 631.9 \times 10^4 \times (350 - 11.4)}{2 \times 3000^2}$$
$$\sqrt{1 + \frac{1}{20}\left(\frac{3000/31.7}{(350 - 11.4)/11.4}\right)^2} = 288.09 \text{ kN m}$$

8.4 Design of Connections

8.4.1 Riveted Shear Connectors

8.4.1.1 Bracket Connections

Bracket type of connections is made whenever two members to be secured together do not intersect. Structurally, this is not a desirable type of connection because a pronounced eccentricity is introduced both in the connection and in one of the members to be joined.

(A) **Bracket Connection Type I**

When the twisting moment is in plane of connection, the bracket connection is termed as Type I bracket connection. This connection may arise when the line of action of load is in the plane of the riveted connection and the center of gravity of the connection is in the center of rotation. The rivet group is subjected to shear and torsion (Fig. 8.13).

Let

M	torque in N-mm, caused by the eccentric load (load x eccentricity)
e	the perpendicular distance measured from the center of gravity of the rivet group to the line of load eccentricity
P	the eccentric load acting over the joint in N
r_1, r_2, r_3, and r_4	the distance of the rivets from the center of gravity of the rivet group
N	number of rivets in the rivet group

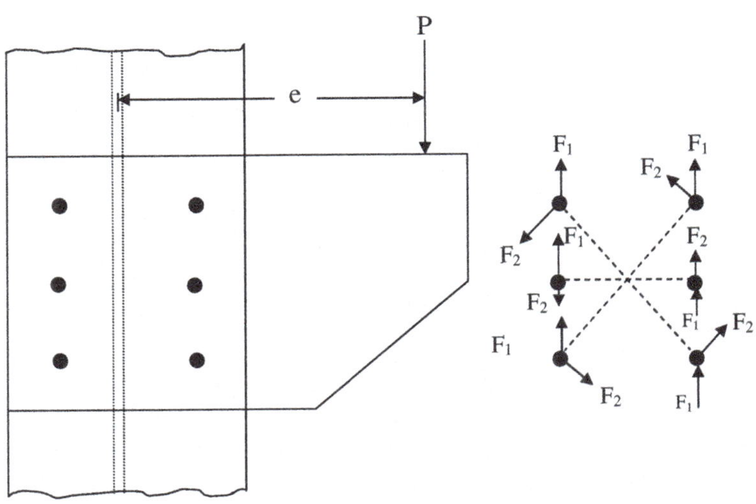

Fig. 8.13 Forces on rivets in bracket connections Type I

F_1 force in a rivet due to direct shear P
F_2 force in a rivet due to torque M
K constant

It is presumed that the rivets will share the shear equally; thereby, force F_1 on any rivet can be calculated as,

$$F_1 = \frac{P}{n} \tag{8.33}$$

Force F_2 in any rivet due to the torque is proportional to it distance from the center of gravity of the connection.

$$F_2 \propto r \quad \text{or} \quad F_2 = Kr \tag{8.34}$$

or

$$K = \frac{F_2}{r} \tag{8.35}$$

Therefore, the torque about the c.g. of the rivet group = $F_2 \times r = K \times r \times r = Kr^2$

$$\text{Total twisting Torque} = \sum Kr^2 = K\sum r^2 = \frac{F_2}{r}\sum r^2 \tag{8.36}$$

This resisting torque should be equal to the torque over the connection.

Hence,

$$M = \frac{F_2}{r}\Sigma r^2$$

or

$$Pe = \frac{F_2}{r}\Sigma r^2$$

or

$$F_2 = \frac{Per}{\Sigma r^2}$$

Force F_2 is maximum, when distance r is maximum. Let the distance of the extreme rivet be r_n, then

$$F_2 = \frac{Per_n}{\Sigma r^2} \tag{8.37}$$

The two forces F_1 and F_2 act at some angle on various rivets in the connection. Let θ be the angle between these forces on the critical rivet. Then, the resultant force F on the critical rivet can be computed by

$$F = \sqrt{F_1^2 + F_2^2 + 2F_1F_2\cos\theta} \tag{8.38}$$

For the connection to be safe, this force F must be less than the rivet value.

From the above equation, it is clear that the resultant force F on a rivet depends on F_1, F_2 and θ. For F to be maximum with F_1 being same for all the rivets and F_2 being maximum for the rivet farthest from the center of gravity of the joint, the angle θ between the two forces has to be minimum.

Design Steps:

1. To design a bracket connection Type I, the number and diameter of rivets for the connection are assumed and these are placed in two or more vertical rows at a suitable pitch and edge distance.
2. The resultant force on the critical rivet is worked by Eq. (8.38)
3. The rivet value is computed. It should be more than the resultant force on the rivet.

(B) Bracket Connection Type II

When the moment is in a plane perpendicular to the plane of connection, the bracket connection is termed as Type II bracket connection. The line of action of load does not lie in the plane of group of rivets and the line of rotation does not pass through the center of gravity of the rivet group.

Let

M moment in N-mm caused by the eccentric load
e the eccentricity of the load P from the rivet plane to the line of action
P the load acting over the joint
M' moment of resistance provided by rivets in tension
F_1 force in a rivet due to direct shear P
F_2 force in a rivet due to bending moment
n number of rivets in the rivet group
Y_1, Y_2, …. Y_n the distance of the rivets in tension from the axis of rotation (Fig. 8.14)

The eccentric load P can be made concentric along with a moment M as shown in Fig. 8.14. The rivets will be

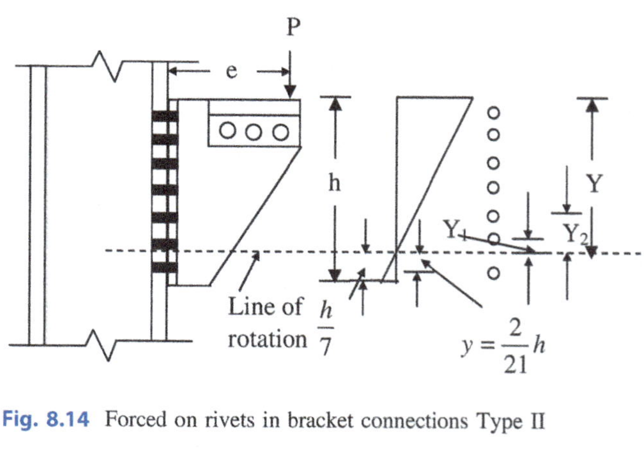

Fig. 8.14 Forced on rivets in bracket connections Type II

subjected to direct shear due to load P and tension due to moment M. The critical rivet is investigated for combined stress due to direct shear and tension.

Design Steps:

1. Assume the diameter, pitch and edge distance of rivets.
2. Compute the rivet value.
3. Compute the number of rivets using the following equation

$$n = \sqrt{6M/n'sR_v} \tag{8.39}$$

where s is the pitch of the rivets and R_v is rivet value; n is the number of columns of rivets, and M is the moment acting.

4. Compute the shear force F_1 using Eq. (8.33) in the extreme critical rivet and calculate the shear value.
5. Calculate the tensile force F_2

$$F_2 = \frac{M'\Sigma Y}{\Sigma Y^2} \tag{8.40}$$

where, M' is the moment resisted by the rivets in tension.

6. Check the connection in combined tension and shear by

$$\frac{\tau_{\text{tf,cal}}}{\tau_{\text{vf}}} + \frac{\sigma_{\text{tf,cal}}}{\sigma_{\text{tf}}} \leq 1.0 \tag{8.41}$$

8.4.1.2 Seat Connections

Two types of seat connections are used, viz. unstiffened seat connections and stiffened seat connections. In the unstiffened seat connection, an angle is provided below the beam flange, and this provides a surface on which the beam flange may rest during erection. The arrangement eliminates the need of erection bolts.

Design of unstiffened seat connections

1. A seat angle is chosen suitably on the following considerations
   (a) The seat angle is assumed to have a length B, equal to the width of the beam flange.
   (b) The length of the outstanding leg of the seat angle may be calculated on the basis of the web crippling of the beam. The seat leg length is kept more than the calculated bearing length.
   The equation may yield a negative value of b for a large beam with a small reaction; therefore, a minimum bearing length is specified:

$$b = \frac{R\,\gamma_{\text{m0}}}{t_{\text{w}} f_{\text{y,w}}} \tag{8.42}$$

where

R end reaction from the beam
t_{w} thickness of the web of the beam in mm
$f_{\text{y,w}}$ yield strength of web
γ_{m0} partial safety factor = 1.10

a dispersion of 45° is assumed from the bearing on the seat to the root line
The length of bearing on the seat is given by,

$$b_1 = b - (t_f + R_1) \tag{8.43}$$

where

t_{f} thickness of flange of beam
R_1 root radius of flange of beam

The distance between the ends of bearing on seat to the root of angle is given by,

$$b_2 = b_1 + c - (t_{\text{a}} + R_{\text{a}}) \tag{8.44}$$

where

t_{a} thickness of seat angle
R_{a} root radius of seat angle
c clearance

   (c) The connected leg is so chosen that at least three horizontal rows of rivets can be accommodated. It can be assumed as 150 mm or more as desirable.
   (d) The thickness of the seat angle is chosen such that the outstanding leg does not fail in bending on a section at the toe of the fillet.

$$M = \frac{R}{b_1} \times b_2 \times \frac{b_2}{2} \tag{8.45}$$

The moment capacity of the angle is given by,

$$M_d = 1.2 Z_e \frac{f_y}{\gamma_{m0}} \tag{8.46}$$

where

Z_{e} section modulus of the angle leg (outstanding leg which provides bearing)

Moment capacity, M_d< Moment M

$$\text{The bending stress} = \frac{M}{Z} = \frac{\text{Re}_1}{Bt^2/6} \tag{8.47}$$

or

$$\sigma_{bc} = \frac{6 \times \text{Re}_1}{Bt^2} \tag{8.48}$$

or

$$t^2 = \frac{6 \times \text{Re}_1}{\sigma_{bc} B} \Rightarrow t = \sqrt{\frac{6 \times \text{Re}_1}{\sigma_{bc} B}} \tag{8.49}$$

As the section of the angle is not known in the beginning, the radius at the root of the fillet is also not known. Hence, eccentricity and thus the thickness cannot be worked out. The angle thickness is, therefore, assumed and the moment of resistance is computed, which should be more than the moment at the critical section.

2. Power-driven rivets are used to connect the seat angle with the column flange. Two or three horizontal rows of rivets are provided. These are subjected to direct shear forces. The rivet diameter is assumed and number of rivets required to connect the seat angle with the flange of column.
   The outstanding leg of the seat angle is connected to the beam flange with two rivets of the same diameter as provided on the connected leg.

$$n = \frac{\text{end reaction}}{\text{rivet value}} \quad (8.50)$$

where n is number of rivets.

3. A cleat angle of nominal size is provided on the top flange of the beam and is connected by two rivets, on each of its legs, of the same diameter as provided on the seat angle.

Design of stiffened seat connections

1. Assume the size of the seat angle on the basis of the bearing length calculate using,

$$b = \frac{R}{\sigma_p \times t} - \sqrt{3}\,h_2 \quad (8.51)$$

2. Provide a suitable stiffener angle. The bearing length is measured from the end of the stiffener leg. The outstanding leg of the stiffener angle must provide the bearing area required. The outstanding leg should not exceed 16 times its thickness to avoid local buckling.

$$\text{Bearing area required} = \frac{R}{\sigma_p} \quad (8.52)$$

where

R end reaction
σ_p permissible bearing in MPa (0.75 f_y)

The thickness of the stiffener angle should not less than the thickness of the web of the beam supported.

3. The seat is not flexible. Hence, the reaction is assumed to have greater eccentricity. The rivets in the connecting legs are subjected to moments in addition to direct shear. The connection behaves as a Type II bracket connection. It is assumed that the reaction from the beam acts at the middle of the bearing length.
4. Compute the bending moment about the face of the column flange.
5. The numbers of rivets are computed and check similar to the bracket connection Type II is done.

$$\frac{\tau_{tf,cal}}{\tau_{vf}} + \frac{\sigma_{tf,cal}}{\sigma_{tf}} \leq 1.0 \quad (8.53)$$

6. A cleat angle of nominal size is provided and is connected with rivets of same diameter as on the seat angle. Two rivets are provided on each leg of the cleat angle.

8.4.2 Welded Shear Connections

8.4.2.1 Bracket Connections

Connections Type I:

Following type of connection is commonly used. In this case, the weld is subjected to a twisting moment in the plane of weld. The type of connection is designed as follows (Fig. 8.15).

Design steps

1. Assuming overlap of the bracket, work out the length of fillet weld.
2. Calculate the distance of the centroid of the weld group ($\bar{x}$) from the edge of the bracket plate.
3. Compute the polar moment of inertia (I_p) of the weld group.
4. Calculate r, the distance of the extreme weld from the center of gravity of the weld group.

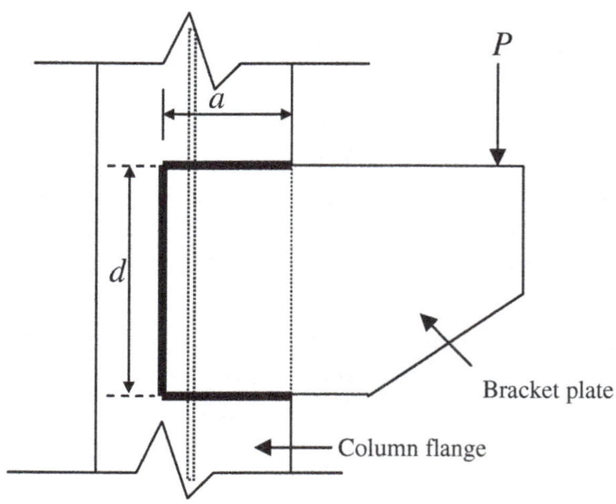

Fig. 8.15 Welded bracket connection Type I

5. Calculate the shear stress τ_{vf1} and τ_{vf2} as follows:

$$\text{Direct Shear Stress}, \tau_{vf1} = \frac{\text{load}}{\text{effective area of weld}} = \frac{P}{(2a + d)t} \quad (8.54)$$

Shear stress due to twisting moment ($M = Pe$) is given by,

$$\tau_{vf2} = \frac{Per}{J} \quad (8.55)$$

J = Polar moment of Inertia of weld group.

6. Calculate the resultant shear stress using the following equation.

$$\tau_{vf} = \sqrt{(\tau_{vf1})^2 + (\tau_{vf2})^2 + 2\tau_{vf1}\tau_{vf2}\cos\theta} \quad (8.56)$$

For a safe design, resultant shear stress $\leq$ design shear strength of weld (f_{wd})
Design shear strength of weld, f_{wd}, is given by,

$$f_{wd} = \text{Design shear stength of weld} = \frac{f_u}{\sqrt{3}\gamma_{mw}} \quad (8.57)$$

7. The size of weld can be obtained by, throat thickness, $=KS$, where $K=$ 0.7 and S is the size of the weld.

Connections Type II

When the moment is in a plane perpendicular to the weld, i.e., the center of gravity of the weld group lies in a plane perpendicular to the plane of line of action of the applied load, the weld is subjected to shear and bending (Fig. 8.16).

Design Steps:

1. Assume the size of the weld and compute the throat thickness. Calculate the depth of the bracket from either of the following appropriate equations.
In case of butt weld,

$$d = \sqrt{6M/t\sigma_b} \quad (8.58)$$

where $f_b = \frac{f_y}{\gamma_{m0}}$; $\gamma_{m0} = 1.1$
In case of fillet weld,

$$d = \sqrt{6M/2tf_{wd1}} \quad (8.59)$$

where f_{wd1} = Design strength of fillet weld per unit length. A reduction factor of 0.8 is considered for the effect of direct shear force.

$$f_{wd1} = 0.8\left(0.7 \times S \times \frac{f_u}{\sqrt{3}\gamma_{m0}}\right) \quad (8.60)$$

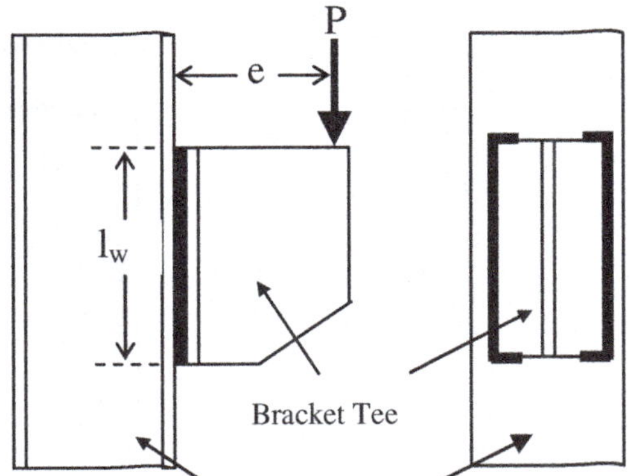

Fig. 8.16 Welded bracket connections Type II

where τ_{vf} = 108 N/mm^2.

2. The direct shear stress is calculated in terms of throat thickness,

$$\text{Direct shear stress in the weld} = \frac{\text{Load}}{\text{Effective area of weld}} \quad (8.61)$$

For fillet weld, $\tau_{vf} = \frac{P}{2l_w t}$
For butt weld, $\tau_{vf} = \frac{P}{l_w t}$

3. The stress due to bending moment computed in terms of throat thickness is as follows:
(Since fillet weld is subjected to flexure, $f_b = My/I$ is used to evaluate the stresses in the weld.)

$$\text{For Fillet Weld}, f_b = \frac{Pe\left(\frac{l_w}{2}\right)}{2\left(\frac{t l_w^3}{12}\right)} = \frac{Pe}{2\left(\frac{t l_w^2}{6}\right)} \quad (8.62)$$

$$\text{For Butt Weld}, f_b = \frac{Pe}{\frac{1}{6} t l_w^2} \quad (8.63)$$

4. The combined stress is calculated using,

$$\text{For fillet weld} f_e = \sqrt{(f_b)^2 + 3(\tau_{vf})^2} \leq \frac{f_u}{\sqrt{3}\gamma_{mw}} \quad (8.64)$$

$$\text{For Butt weld}, f_e = \sqrt{(f_b)^2 + 3(\tau_{vf})^2} \leq \frac{f_y}{\gamma_{m0}} \quad (8.65)$$

8.4.2.2 Seat Connections

Unstiffened Seat Connection

In the welded unstiffened seat connection, two angles are used. In this, vertical welds are provided to connect the seat angle and these are turned at the ends (Figure 8.17).

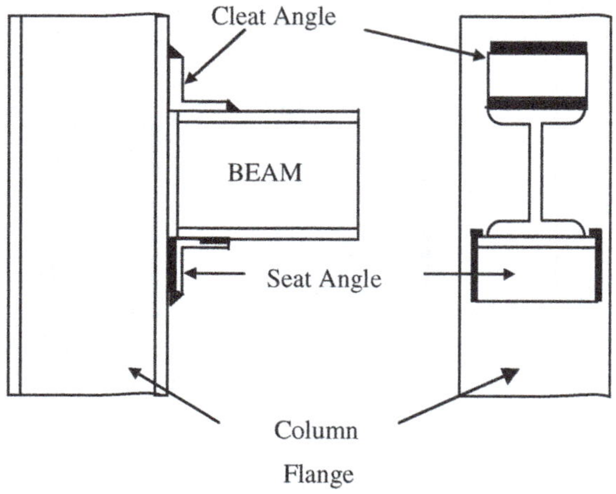

Fig. 8.17 Welded unstiffened seat connection

Design Steps

1. The design of the seat angle is same as that discussed in the design of the unstiffened riveted seat connection.
2. Calculate the vertical shear per mm by

$$\tau_{vf1} = \frac{R}{2d}(\mathrm{t} = \mathrm{unity}) \quad (8.66)$$

3. Calculate the eccentricity of the reaction and compute the bending moment.
4. Calculate the horizontal shear per mm to the bending moment,

$$\tau_{vf2} = \frac{M}{2 \times \frac{1}{6}d^2 \times 1}, (\mathrm{t} = \mathrm{unity}) \quad (8.67)$$

where d = length of the weld.

5. Compute the resultant shear per mm,

$$\tau'_{vf} = \sqrt{(\tau_{vf1})^2 + (\tau_{vf2})^2} \quad (8.68)$$

6. This resultant shear per mm is equated to the strength of the weld per mm to find the size of the weld,

$$\tau'_{vf} = 0.7 \times S \times 1 \times \tau_{vf} \quad (8.69)$$

Stiffened Seat Connection

In the stiffened seat connection, a T-section built-up of two plates is used. The bearing length of the seat plate is calculated as in the case of the unstiffened seat connection. The thickness of seat plate is kept equal to the thickness of the flange of the beam and that of the stiffening plate is kept equal to the thickness of the web of the beam (Fig. 8.18).

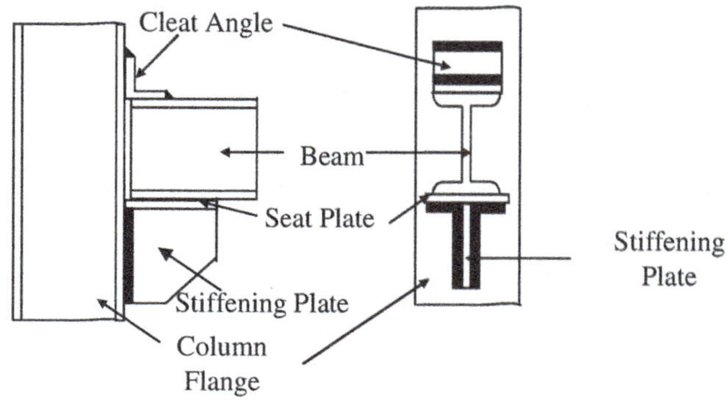

Fig. 8.18 Welded stiffened seat connection

The depth of the stiffening plate is decided depending upon the length of the vertical weld required. The seat plate is kept wider than the flange of the beam by at least twice the size of the weld on each size of the beam flange to facilitate welding.

Design Steps

1. The width of seating plate is calculated in the same way as in the riveted unstiffened seat connection.
2. The thickness of the seat plate is assumed equal to the thickness of beam flange.
3. The thickness of the stiffening plate is assumed equal to the thickness of the web of the beam.
4. The eccentricity of load and moment due to it are calculated.
5. Vertical and horizontal shears per mm length of weld τ_{vf1} and τ_{vf2} are computed.
6. The resultant shear per mm of weld length τ_{vf} is computed by the vector sum if τ_{vf1} and τ_{vf2}.
7. The size of the weld required to make the connection is calculated by equating the strength of the weld per mm length to the resultant shear per mm length.

8.4.2.3 Framed Connections

Beams are connected to the column or to the other beams by means of framing angles or plates. A pair of angles or plates are placed one on each side of the web of the beam to be connected. These angles or plates are used to transmit shear and moment. One of the legs of the angle is welded to the beam in the shop and the other leg is welded to the column in the field (Fig. 8.19).

Design Steps:

1. Calculate the vertical reaction V, on each weld.
2. Compute vertical shear/mm length and horizontal shear/mm length for the weld connecting the framing angle leg to the column.
3. Compute the resultant shear τ_v.

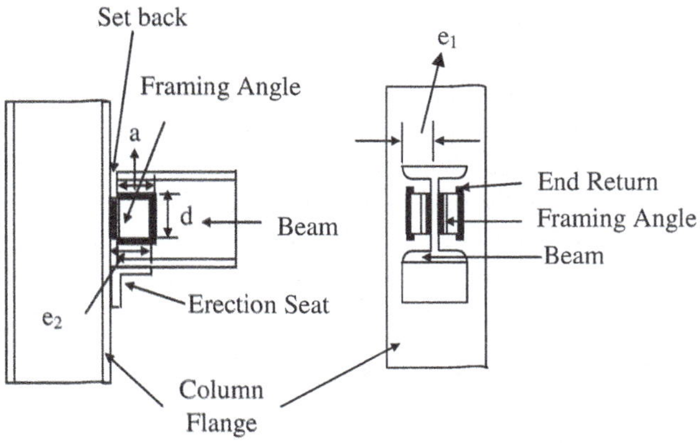

Fig. 8.19 Welded framed connection

4. Determine the size of the weld by equating the strength of weld/unit length to the computed value of τ_v.
5. Find out the center of gravity of the weld group to connect the angle section with the web of the beam.
6. Compute the polar moment of inertia (J).
7. Compute the twisting moment (T).
8. Calculate the horizontal and vertical shear in the weld. Find out the resultant shear and equate it to the allowable shear to obtain the size of the weld.
   Horizontal shear,

$$\tau_{vh} = \frac{Tr}{J} = \frac{Ve_2 r}{J} \tag{8.70}$$

and Vertical shear,

$$\tau_{vs} = \frac{V}{\text{effective area of weld}} = \frac{V}{dt + 2at} \tag{8.71}$$

where r is the extreme distance of the weld from the center of gravity of the weld group and a and d are the weld lengths, and e_1 and e_2 are the eccentric distances.
Resultant shear,

$$\tau_v = \sqrt{(\tau_{vs})^2 + (\tau_{vh})^2} \tag{8.72}$$

8.4.3 Welded Moment Connections

Moment-resistant connections with weld are made using plates only. A plate is provided over the top flange of the beam. The length of the plate is kept about 1.5 times the width of the beam flange. The length should be such as to ensure that the plate may yield before the connection. The plate is connected to the beam flange by a fillet weld and to the column flange by a butt weld as shown in Fig. 8.20. The plate is tapered to facilitate fillet welding. When the beam is connected to the web of the column, the connected top plate is flared to connect the plate with the column flange by side welds.

The connecting plate has little resistance to shear. Therefore, a seat plate is provided to resist it. If the end reactions are large, stiffening plates or angle may also be provided. Under heavy moments, the column flange may deflect away due to the pull and its moment carrying capacity is reduced. To check this, additional plates called stiffener plates are welded between column flanges opposite the beam flanges as shown in Fig. 8.20. Some codes specify the length of stiffener plates to be equal to the depth of the column. Generally, the plates are provided for the full depth (Fig. 8.20).

8.4.4 Semi-rigid Connections

A semi-rigid connection resists end moments but gives a relative rotation between the beam and column as shown in Fig. 8.21. A beam having a flexible connection with a column and subjected to uniformly distributed load, w, per unit length has the greatest moment at the center section (Fig. 8.21).

For the same beam having fully rigid connections (end fixed), the central moment reduces to $wl^2/24$ and the end moments increase to $wl^2/12$. Thus, the beam needs a section modulus just 2/3 of that required for a simple beam.

If a semi-rigid connection is used as shown in Fig. 8.22, the center moment and end moments may be made equal to $wl^2/16$ each, provided, the connections have an end restraint $R = 75\%$. The section modulus requirements for the section reduce to 50% of that needed for a simple beam. Also, this is

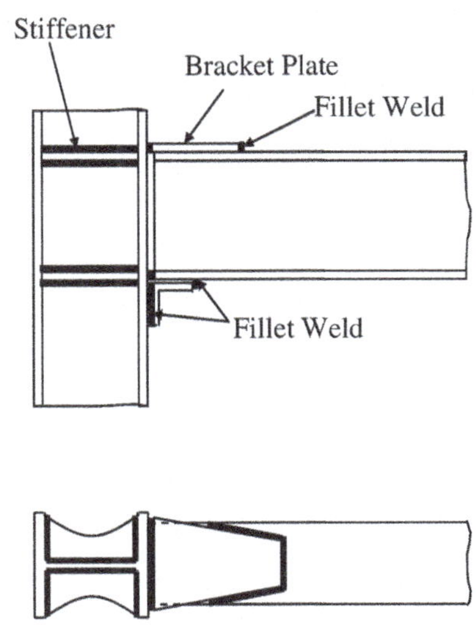

Fig. 8.20 Welded moment connection

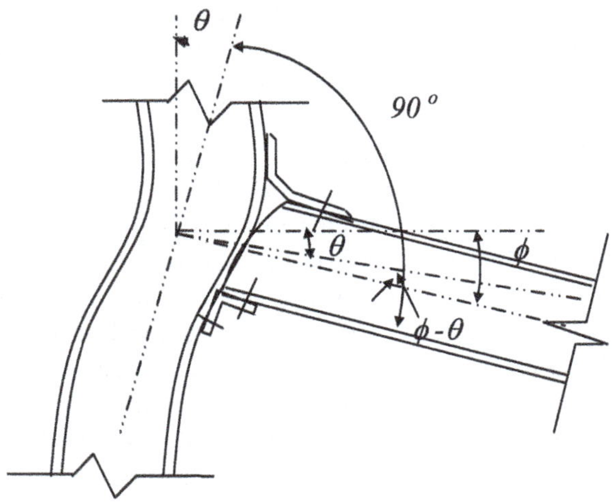

Fig. 8.21 Semi-rigid connection

true provided the supports to which the connections join the beam are unyielding, and the beam is not influenced by adjacent spans from which additional moment might be carried over through the connection. Since the condition of R = 75% does not offer any range of rigidity, a little error may lead to hazard. To take care of this, it is suggested that the beam is designed for an end restraint of R = 50% and the connection for a restraint of R = 75%. This appears to be good until the resulting design moment of $wl^2/12$ is not lower than if fully rigid welded connections were used. So there is no saving in beam requirements by using semi-rigid connections. However, their slight reduction in the amount of connecting welds (Fig. 8.22).

Example 8.2: Truss structure

Following truss structure supports 150-mm-thick slab. Considering live load of 3 kN/m^2, design the structural members of a truss.

Loading calculations:

Loading area to be considered for the truss = 4 m × 2.5 m = 10 m^2

Weight of 150-mm-thick slab = 0.15 × 10 m^2 × 25 = 37.5 kN

Live load = 3.0 kN/m^2 × 10 m^2 = 30 kN.

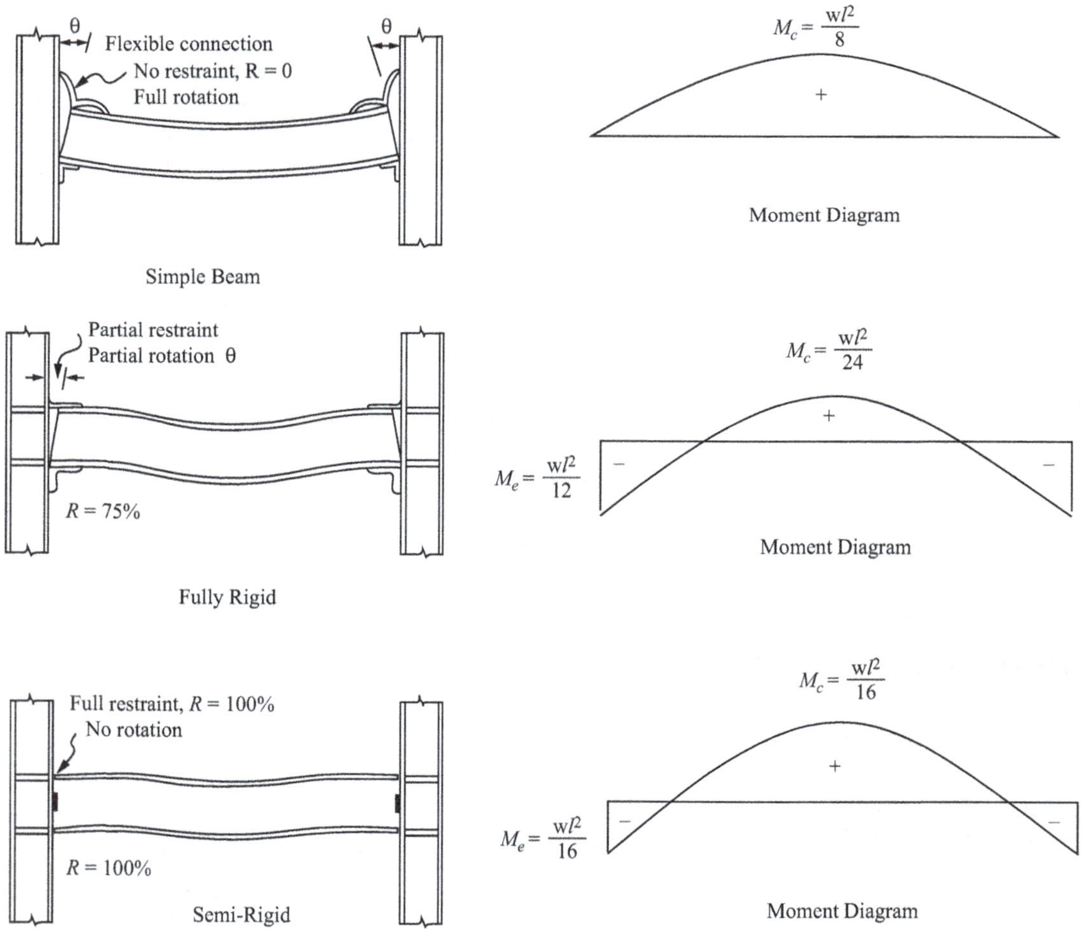

Fig. 8.22 Moment diagrams for beam with different end conditions

Total load = 37.5 + 30 = 67.5 kN say 70 kN

Increase the load structure by 20% to account for self-weight of members.

Total load on the structure = 1.2 × 70 = 84 kN say 90 kN

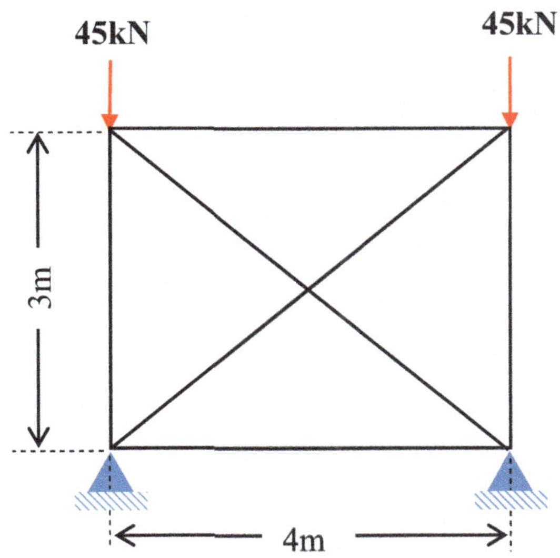

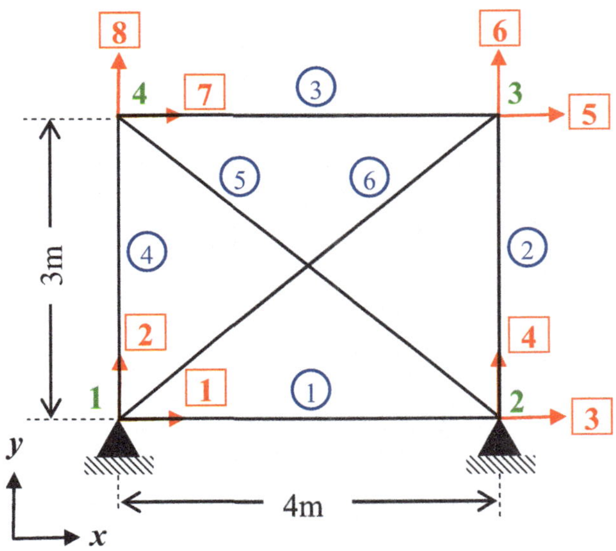

The stiffness matrix $[k_e]$ and mass matrix $[m_e]$ for the linear truss element in plane can be calculated as follows.

$$[k_e] = \frac{AE}{L}\begin{bmatrix} c^2 & cs & -c^2 & -cs \\ cs & s^2 & -cs & -s^2 \\ -c^2 & -cs & c^2 & cs \\ -cs & -s^2 & cs & s^2 \end{bmatrix}$$

where $c = \cos\alpha$ and $s = \sin\alpha$, where α is the angle between local and global coordinate system.

The consistent mass matrix for each element is given by,

$$[m_e] = \frac{\rho AL}{6}\begin{bmatrix} 2 & 0 & 1 & 0 \\ 0 & 2 & 0 & 1 \\ 1 & 0 & 2 & 0 \\ 0 & 1 & 0 & 2 \end{bmatrix}$$

It can be modified into the diagonal mass matrix, which is used in the calculations is as follows

$$[m_e] = \frac{\rho AL}{6}\begin{bmatrix} 3 & 0 & 0 & 0 \\ 0 & 3 & 0 & 0 \\ 0 & 0 & 3 & 0 \\ 0 & 0 & 0 & 3 \end{bmatrix} = \frac{\rho AL}{2}\begin{bmatrix} 1 & 0 & 0 & 0 \\ 0 & 1 & 0 & 0 \\ 0 & 0 & 1 & 0 \\ 0 & 0 & 0 & 1 \end{bmatrix}$$

The mass matrix $[m_e]$ is invariant to the rotation of the element and hence $[m_e] = [m]$.

Considering member size for members 1–4 as ISA (200 × 200 × 12) mm, Cross-sectional area = A = 9322 mm^2 and bracing members 5–6 as ISA (50 × 50 × 5) mm, Cross-sectional Area = 958.0 mm^2.

Young's Modulus of steel, E = 200 GPa.

Element no.	1	2	3	4	5	6
α (degree)	0	90	180	270	323.13	36.87
Length (m)	4	3	4	3	5	5

Stiffness matrix for each element in global coordinate is as follows.

For Element 1
(connected between nodes 1–2)

$$[k_1] = 4.661 \times 10^8 \begin{array}{c} \begin{matrix} 1 & 2 & 3 & 4 \end{matrix} \\ \begin{bmatrix} 1 & 0 & -1 & 0 \\ 0 & 0 & 0 & 0 \\ -1 & 0 & 1 & 0 \\ 0 & 0 & 0 & 0 \end{bmatrix} \end{array} \begin{matrix} 1 \\ 2 \\ 3 \\ 4 \end{matrix} \quad (\text{N/m})$$

For Element 2
(connected between nodes 2–3)

$$[k_2] = 6.215 \times 10^8 \begin{array}{c} \begin{matrix} 3 & 4 & 5 & 6 \end{matrix} \\ \begin{bmatrix} 0 & 0 & 0 & 0 \\ 1 & 0 & 0 & -1 \\ 0 & 0 & 0 & 0 \\ -1 & 0 & 0 & 1 \end{bmatrix} \end{array} \begin{matrix} 3 \\ 4 \\ 5 \\ 6 \end{matrix} \quad (\text{N/m})$$

For Element 3
(connected between nodes 3–4)

$$[k_3] = 4.661 \times 10^8 \begin{array}{c} \begin{matrix} 5 & 6 & 7 & 8 \end{matrix} \\ \begin{bmatrix} 1 & 0 & -1 & 0 \\ 0 & 0 & 0 & 0 \\ -1 & 0 & 1 & 0 \\ 0 & 0 & 0 & 0 \end{bmatrix} \end{array} \begin{matrix} 5 \\ 6 \\ 7 \\ 8 \end{matrix} \quad (\text{N/m})$$

For Element 4
(connected between nodes 4–1)

$$[k_4] = 6.125 \times 10^8 \begin{array}{c} \begin{matrix} 7 & 8 & 1 & 2 \end{matrix} \\ \begin{bmatrix} 0 & 0 & 0 & 0 \\ 0 & 1 & 0 & -1 \\ 0 & 0 & 0 & 0 \\ 0 & -1 & 0 & 1 \end{bmatrix} \end{array} \begin{matrix} 7 \\ 8 \\ 1 \\ 2 \end{matrix} \quad (\text{N/m})$$

For Element 5
(connected between nodes 4–2)

$$[k_5] = 3.832 \times 10^7 \begin{array}{c} \begin{matrix} 7 & 8 & 3 & 4 \end{matrix} \\ \begin{bmatrix} 0.64 & -0.48 & -0.64 & 0.48 \\ -0.48 & 0.36 & 0.48 & -0.36 \\ -0.64 & 0.48 & 0.64 & -0.48 \\ 0.48 & -0.36 & -0.48 & 0.36 \end{bmatrix} \end{array} \begin{matrix} 7 \\ 8 \\ 3 \\ 4 \end{matrix} \quad (\text{N/m})$$

For Element 6
(connected between nodes 1–3)

$$[k_6] = 3.832 \times 10^7 \begin{array}{c} \begin{matrix} 1 & 2 & 5 & 6 \end{matrix} \\ \begin{bmatrix} 0.64 & 0.48 & -0.64 & -0.48 \\ 0.48 & 0.36 & -0.48 & -0.36 \\ -0.64 & -0.48 & 0.64 & 0.48 \\ -0.48 & -0.36 & 0.48 & 0.36 \end{bmatrix} \end{array} \begin{matrix} 1 \\ 2 \\ 5 \\ 6 \end{matrix} \quad (\text{N/m})$$

The degree of freedom of the structure is eight, and hence, the size of the global stiffness matrix will be (8 × 8). Stiffness matrix in global coordinate is as follows.

Assembled stiffness matrix in global coordinate system is as follows.

$$[K] = 10^8 \times \begin{bmatrix} 4.9062 & 0.1893 & -4.661 & 0 & -0.2452 & -0.1839 & 0 & 0 \\ 0.1893 & 6.3526 & 0 & 0 & -0.1839 & -0.138 & 0 & -6.2147 \\ -4.661 & 0 & 4.9062 & -0.1839 & 0 & 0 & -0.2452 & 0.1839 \\ 0 & 0 & -0.1839 & 6.3526 & 0 & -6.2147 & 0.1839 & -0.138 \\ -0.2452 & -0.1839 & 0 & 0 & 4.9062 & 0.1839 & -4.661 & 0 \\ -0.1839 & -0.138 & 0 & -6.2147 & 0.1839 & 6.3526 & 0 & 0 \\ 0 & 0 & -0.2452 & 0.1839 & -4.661 & 0 & 4.9062 & -0.1839 \\ 0 & -6.2147 & 0.1839 & -0.138 & 0 & 0 & -0.1839 & 6.3526 \end{bmatrix} \begin{matrix} 1 \\ 2 \\ 3 \\ 4 \\ 5 \\ 6 \\ 7 \\ 8 \end{matrix}$$

Considering mass density of steel, $\rho = 7850\ \text{kg/m}^3$ and cross-sectional area of member depending upon member sizes, the mass matrix for each element is as follows.

For Element 1

$$[m_1] = \frac{7850 \times 9322 \times 10^{-6} \times 4}{2} \times \begin{array}{c} \begin{matrix} 1 & 2 & 3 & 4 \end{matrix} \\ \begin{bmatrix} 1 & 0 & 0 & 0 \\ 0 & 1 & 0 & 0 \\ 0 & 0 & 1 & 0 \\ 0 & 0 & 0 & 1 \end{bmatrix} \end{array} \begin{matrix} 1 \\ 2\ \text{kg} \\ 3 \\ 4 \end{matrix}$$

For Element 2

$$[m_2] = \frac{7850 \times 9322 \times 10^{-6} \times 3}{2} \times \begin{array}{c} \begin{matrix} 3 & 4 & 5 & 6 \end{matrix} \\ \begin{bmatrix} 1 & 0 & 0 & 0 \\ 0 & 1 & 0 & 0 \\ 0 & 0 & 1 & 0 \\ 0 & 0 & 0 & 1 \end{bmatrix} \end{array} \begin{matrix} 3 \\ 4\ \text{kg} \\ 5 \\ 6 \end{matrix}$$

For Element 3

For Element 4

(continued)

$$[m_3] = \frac{7850 \times 9322 \times 10^{-6} \times 4}{2} \times \begin{bmatrix} 1 & 0 & 0 & 0 \\ 0 & 1 & 0 & 0 \\ 0 & 0 & 1 & 0 \\ 0 & 0 & 0 & 1 \end{bmatrix} \begin{matrix} 5 \\ 6 \\ 7 \\ 8 \end{matrix} \text{ kg} \quad (\text{columns } 5\ 6\ 7\ 8)$$

$$[m_4] = \frac{7850 \times 9322 \times 10^{-6} \times 3}{2} \times \begin{bmatrix} 1 & 0 & 0 & 0 \\ 0 & 1 & 0 & 0 \\ 0 & 0 & 1 & 0 \\ 0 & 0 & 0 & 1 \end{bmatrix} \begin{matrix} 7 \\ 8 \\ 1 \\ 2 \end{matrix} \text{ kg} \quad (\text{columns } 7\ 8\ 1\ 2)$$

For Element 5

$$[m_5] = \frac{7850 \times 958 \times 10^{-6} \times 5}{2} \times \begin{bmatrix} 1 & 0 & 0 & 0 \\ 0 & 1 & 0 & 0 \\ 0 & 0 & 1 & 0 \\ 0 & 0 & 0 & 1 \end{bmatrix} \begin{matrix} 7 \\ 8 \\ 3 \\ 4 \end{matrix} \text{ kg} \quad (\text{columns } 7\ 8\ 3\ 4)$$

For Element 6

$$[m_6] = \frac{7850 \times 958 \times 10^{-6} \times 5}{2} \times \begin{bmatrix} 1 & 0 & 0 & 0 \\ 0 & 1 & 0 & 0 \\ 0 & 0 & 1 & 0 \\ 0 & 0 & 0 & 1 \end{bmatrix} \begin{matrix} 1 \\ 2 \\ 5 \\ 6 \end{matrix} \text{ kg} \quad (\text{columns } 1\ 2\ 5\ 6)$$

Assembled mass matrix in global coordinate is as follows. Mass of 4500 kg is added at degrees of freedom 5, 6, 7, and 8 corresponding to loads at nodes 3 and 4 of the truss.

$$[M] = \begin{bmatrix} 274.923 & 0 & 0 & 0 & 0 & 0 & 0 & 0 \\ 0 & 274.923 & 0 & 0 & 0 & 0 & 0 & 0 \\ 0 & 0 & 274.923 & 0 & 0 & 0 & 0 & 0 \\ 0 & 0 & 0 & 274.923 & 0 & 0 & 0 & 0 \\ 0 & 0 & 0 & 0 & 4774.923 & 0 & 0 & 0 \\ 0 & 0 & 0 & 0 & 0 & 4774.923 & 0 & 0 \\ 0 & 0 & 0 & 0 & 0 & 0 & 4774.923 & 0 \\ 0 & 0 & 0 & 0 & 0 & 0 & 0 & 4774.923 \end{bmatrix}$$

$$[K] = 10^8 \times \begin{bmatrix} 4.9062 & 0.1893 & -4.661 & 0 & -0.2452 & -0.1839 & 0 & 0 \\ 0.1893 & 6.3526 & 0 & 0 & -0.1839 & -0.138 & 0 & -6.2147 \\ -4.661 & 0 & 4.9062 & -0.1839 & 0 & 0 & -0.2452 & 0.1839 \\ 0 & 0 & -0.1839 & 6.3526 & 0 & -6.2147 & 0.1839 & -0.138 \\ -0.2452 & -0.1839 & 0 & 0 & 4.9062 & 0.1839 & -4.661 & 0 \\ -0.1839 & -0.138 & 0 & -6.2147 & 0.1839 & 6.3526 & 0 & 0 \\ 0 & 0 & -0.2452 & 0.1839 & -4.661 & 0 & 4.9062 & -0.1839 \\ 0 & -6.2147 & 0.1839 & -0.138 & 0 & 0 & -0.1839 & 6.3526 \end{bmatrix}$$

Since the structure is hinged at nodes 1 and 2, $u_1 = u_2 = u_3 = u_4 = 0$

Reduced stiffness matrix is as follows.

$$[\bar{K}] = 10^8 \times \begin{bmatrix} 4.9062 & 0.1839 & -4.661 & 0 \\ 0.1839 & 6.3526 & 0 & 0 \\ -4.661 & 0 & 4.9062 & -0.1839 \\ 0 & 0 & -0.1839 & 6.3526 \end{bmatrix}$$

Similarly, the reduced mass matrix is as follows.

$$[\bar{M}] = \begin{bmatrix} 4774.923 & 0 & 0 & 0 \\ 0 & 4774.923 & 0 & 0 \\ 0 & 0 & 4774.923 & 0 \\ 0 & 0 & 0 & 4774.923 \end{bmatrix}$$

Solving for eigenvalues, we get the following

Mode	Mode 1	Mode 2	Mode 3	Mode 4
Eigenvalues, $\lambda = \omega^2$	5019.3	13,2821.3	13,3156.8	20,0583.1
Natural circular frequencies, ω (rad/s)	70.85	364.45	364.91	447.87
Natural frequencies, $f = \omega/2\pi$ (Hz)	11.28	58.00	58.08	71.28
Time periods (s)	0.0887	0.0172	0.0172	0.0140

Corresponding eigenvectors (mode shapes) are as follows.

$$\begin{Bmatrix} \phi_{15} \\ \phi_{16} \\ \phi_{17} \\ \phi_{18} \end{Bmatrix} = \begin{Bmatrix} -0.01023 \\ 0.00031 \\ -0.01023 \\ -0.00031 \end{Bmatrix}; \begin{Bmatrix} \phi_{25} \\ \phi_{26} \\ \phi_{27} \\ \phi_{28} \end{Bmatrix} = \begin{Bmatrix} 0.00058 \\ -0.01022 \\ -0.00058 \\ -0.01022 \end{Bmatrix};$$

$$\begin{Bmatrix} \phi_{35} \\ \phi_{36} \\ \phi_{37} \\ \phi_{38} \end{Bmatrix} = \begin{Bmatrix} 0.00031 \\ 0.01023 \\ 0.00031 \\ -0.01023 \end{Bmatrix}; \begin{Bmatrix} \phi_{45} \\ \phi_{46} \\ \phi_{47} \\ \phi_{48} \end{Bmatrix} = \begin{Bmatrix} 0.01022 \\ 0.00058 \\ -0.01022 \\ 0.00058 \end{Bmatrix}$$

Note: Mode shapes (Eigen vectors for each eigen values) are Mass-Ortho-normalized i.e. $\{\phi_i\}^T[M]\{\phi_i\} = 1$

<u>**Mode 1**</u>:

Modal participation factor:

$$\Gamma_1 = \{\phi_1\}^T[M]\{1\}$$

$$= \begin{Bmatrix} -0.01023 \\ 0.00031 \\ -0.01023 \\ -0.00031 \end{Bmatrix}^T \begin{bmatrix} 4774.923 & 0 & 0 & 0 \\ 0 & 4774.923 & 0 & 0 \\ 0 & 0 & 4774.923 & 0 \\ 0 & 0 & 0 & 4774.923 \end{bmatrix} \times \begin{Bmatrix} 1 \\ 0 \\ 1 \\ 0 \end{Bmatrix} = -97.679$$

$$\textbf{Modal mass} = (\Gamma_1)^2 = \frac{[\{\phi_1\}^T[M]\{1\}]^2}{\{\phi_1\}^T[M]\{\phi_3\}} = (97.679)^2 = 9541.2\ \text{kg}$$

$$\text{Mass participated in}\,\% = \frac{9541.2}{9549.845} \times 100 = 99.909\%$$

Since mass participation in 1st mode is more than 90%, only 1st mode is considered in the calculation.

As per IS 1893:2016, for Soil Type II (medium stiff soil sites), S_a/g is given by,

$$\frac{S_a}{g} = \begin{cases} 1 + 15T & T \le 0.10\ \text{s} \\ 2.5 & 0.10\ \text{s} \le T \le 0.55\ \text{s} \\ 1.36/T & 0.55\ \text{s} \le T \le 4.0\ \text{s} \end{cases}$$

As per IS 1893:2016, in Table 9, for steel building ordinary moment-resisting frame (OMRF), response reduction factor (R) = 3

Calculated value of design horizontal acceleration coefficient, A_h, is as follows.

For T_1 = 0.0887 s and medium stiff soil sites,

$$\frac{S_{a1}}{g} = 1 + 15T = 1 + (15 \times 0.0887) = 2.33$$

$$A_{h1} = \frac{Z}{2}\frac{I}{R}\frac{S_a}{g} = \frac{0.16}{2} \times \frac{1.0}{3} \times 2.33 = 0.062$$

Since $A_{h1} < \frac{Z}{2}$, take $A_{hi} = Z/2 = 0.16/2 = 0.08$ (Z/2 = Design Basis Earthquake which is equal to Peak Ground Acceleration to be considered for design purpose) [IS 1893:2002]

Lateral load can be calculated as follows.

Lateral load for each mode is given by,

$$\{Q_i\} = A_{hi} g \Gamma_i \{\phi_i\}^T [M]$$

$$\{Q_1\} = 0.08 \times 9.81 \times (-97.679) \times \begin{Bmatrix} -0.01023 \\ 0.00031 \\ -0.01023 \\ -0.00031 \end{Bmatrix}^T \begin{bmatrix} 4774.923 & 0 & 0 & 0 \\ 0 & 4774.923 & 0 & 0 \\ 0 & 0 & 4774.923 & 0 \\ 0 & 0 & 0 & 4774.923 \end{bmatrix}$$

$$\{Q_1\} = \begin{Bmatrix} 3742.6 \\ -112.6 \\ 3742.6 \\ 112.6 \end{Bmatrix}$$

$$\text{N} = \begin{Bmatrix} 3.743 \\ -0.113 \\ 3.743 \\ 0.113 \end{Bmatrix} \text{kN}$$

Let's apply the lateral load to the truss and obtain nodal displacement at nodes 3 and 4 as follows.

$$\begin{Bmatrix} u_5 \\ u_6 \\ u_7 \\ u_8 \end{Bmatrix} = [\bar{K}]^{-1} \begin{Bmatrix} P_5 \\ P_6 \\ P_7 \\ P_8 \end{Bmatrix}$$

$$= 10^8 \times \begin{bmatrix} 2.1367 & -0.0619 & 2.0321 & 0.0588 \\ -0.0619 & 0.1592 & -0.0588 & -0.0017 \\ 2.0321 & -0.0588 & 2.1367 & 0.0619 \\ 0.0588 & -0.0017 & 0.0619 & 0.1592 \end{bmatrix} \begin{Bmatrix} 3742.6 \\ -112.6 \\ 3742.6 \\ 112.6 \end{Bmatrix}$$

$$= \begin{Bmatrix} 1.562 \times 10^{-4} \\ -4.7 \times 10^{-6} \\ 1.562 \times 10^{-4} \\ 4.7 \times 10^{-6} \end{Bmatrix} \text{m}$$

Displacements at nodes 1 and 3 in global coordinate is,

$$\begin{Bmatrix} u_1 \\ u_2 \\ u_5 \\ u_6 \end{Bmatrix} = \begin{Bmatrix} 0 \\ 0 \\ 1.562 \times 10^{-4} \\ 4.700 \times 10^{-6} \end{Bmatrix} \text{m}$$

Nodal displacement at nodes 3 and 4 is known. Hence, forces in each member can be calculated as follows.

Let us find out the force in member 6.

$$\text{Force in member } 6(P_6) = P_6 = \frac{AE}{L}\{-c \quad -s \quad c \quad s\}\begin{Bmatrix} u_1 \\ u_2 \\ u_5 \\ u_6 \end{Bmatrix}$$

Substituting A = 958×10^{-6} m^2, E = 2×10^{11} N/m^2 and theta = 36.87 Degrees, L = 5m, we get

$$P_6 = 3.832 \times 10^7\{-0.8 \quad -0.6 \quad 0.8 \quad 0.6\}\begin{Bmatrix} 0 \\ 0 \\ 1.562 \times 10^{-4} \\ -4.7 \times 10^{-6} \end{Bmatrix} = 4680.0 \text{ N (Compression)}$$

Static Analysis

The equilibrium equation (force–displacement equation) is given by,

$$[K]\{u\} = \{P\}$$

$$10^8 \times \begin{bmatrix} 4.906 & 0.189 & -4.661 & 0 & -0.245 & -0.1839 & 0 & 0 \\ 0.189 & 6.353 & 0 & 0 & -0.184 & -0.138 & 0 & -6.215 \\ -4.661 & 0 & 4.906 & -0.184 & 0 & 0 & -0.245 & 0.1839 \\ 0 & 0 & -0.184 & 6.353 & 0 & -6.215 & 0.184 & -0.138 \\ -0.245 & -0.184 & 0 & 0 & 4.906 & 0.184 & -4.661 & 0 \\ -0.184 & -0.138 & 0 & -6.215 & 0.184 & 6.353 & 0 & 0 \\ 0 & 0 & -0.245 & 0.184 & -4.661 & 0 & 4.906 & -0.184 \\ 0 & -6.215 & 0.184 & -0.138 & 0 & 0 & -0.184 & 6.353 \end{bmatrix} \begin{Bmatrix} u_1 \\ u_2 \\ u_3 \\ u_4 \\ u_5 \\ u_6 \\ u_7 \\ u_8 \end{Bmatrix} = \begin{Bmatrix} P_1 \\ P_2 \\ P_3 \\ P_4 \\ P_5 \\ P_6 \\ P_7 \\ P_8 \end{Bmatrix}$$

$$10^8 \times \begin{bmatrix} 4.906 & 0.189 & -4.661 & 0 & -0.245 & -0.1839 & 0 & 0 \\ 0.189 & 6.353 & 0 & 0 & -0.184 & -0.138 & 0 & -6.215 \\ -4.661 & 0 & 4.906 & -0.184 & 0 & 0 & -0.245 & 0.1839 \\ 0 & 0 & -0.184 & 6.353 & 0 & -6.215 & 0.184 & -0.138 \\ -0.245 & -0.184 & 0 & 0 & 4.906 & 0.184 & -4.661 & 0 \\ -0.184 & -0.138 & 0 & -6.215 & 0.184 & 6.353 & 0 & 0 \\ 0 & 0 & -0.245 & 0.184 & -4.661 & 0 & 4.906 & -0.184 \\ 0 & -6.215 & 0.184 & -0.138 & 0 & 0 & -0.184 & 6.353 \end{bmatrix} \begin{Bmatrix} 0 \\ 0 \\ 0 \\ 0 \\ u_5 \\ u_6 \\ u_7 \\ u_8 \end{Bmatrix} = \begin{Bmatrix} P_1 \\ P_2 \\ P_3 \\ P_4 \\ 0 \\ -45{,}000 \\ 0 \\ -45{,}000 \end{Bmatrix}$$

Applying the boundary conditions, $u_1 = u_2 = u_3 = u_4 = 0$ unknown displacements u_5, u_6, u_7, and u_8 can be obtained as follows.

Load on the structure is as follows,

$$P_5 = 0, \quad P_6 = 45\ \text{kN}, \quad P_7 = 0, \quad P_8 = 45\ \text{kN}$$

Substituting the values, we get the following

$$10^8 \times \begin{bmatrix} 4.906 & 0.189 & -4.661 & 0 & -0.245 & -0.1839 & 0 & 0 \\ 0.189 & 6.353 & 0 & 0 & -0.184 & -0.138 & 0 & -6.215 \\ -4.661 & 0 & 4.906 & -0.184 & 0 & 0 & -0.245 & 0.1839 \\ 0 & 0 & -0.184 & 6.353 & 0 & -6.215 & 0.184 & -0.138 \\ -0.245 & -0.184 & 0 & 0 & 4.906 & 0.184 & -4.661 & 0 \\ -0.184 & -0.138 & 0 & -6.215 & 0.184 & 6.353 & 0 & 0 \\ 0 & 0 & -0.245 & 0.184 & -4.661 & 0 & 4.906 & -0.184 \\ 0 & -6.215 & 0.184 & -0.138 & 0 & 0 & -0.184 & 6.353 \end{bmatrix} \begin{Bmatrix} 0 \\ 0 \\ 0 \\ 0 \\ u_5 \\ u_6 \\ u_7 \\ u_8 \end{Bmatrix} = \begin{Bmatrix} P_1 \\ P_2 \\ P_3 \\ P_4 \\ 0 \\ -45{,}000 \\ 0 \\ -45{,}000 \end{Bmatrix}$$

Solving for unknown displacements, we get,

$$10^8 \times \begin{bmatrix} 4.9062 & 0.1839 & -4.661 & 0 \\ 0.1839 & 6.3526 & 0 & 0 \\ -4.661 & 0 & 4.9062 & -0.1839 \\ 0 & 0 & -0.1839 & 6.3526 \end{bmatrix} \begin{Bmatrix} u_5 \\ u_6 \\ u_7 \\ u_8 \end{Bmatrix} = \begin{Bmatrix} 0 \\ -45{,}000 \\ 0 \\ -45{,}000 \end{Bmatrix}$$

$$\begin{Bmatrix} u_5 \\ u_6 \\ u_7 \\ u_8 \end{Bmatrix} = 10^{-9} \begin{bmatrix} 21.3670 & -0.6185 & 20.3212 & 0.5883 \\ -0.6185 & 1.5921 & -0.5883 & -0.0170 \\ 20.3212 & -0.5883 & 21.3670 & 0.6185 \\ 0.5883 & -0.0170 & 0.6185 & 1.5921 \end{bmatrix} \times \begin{Bmatrix} 0 \\ -45{,}000 \\ 0 \\ -45{,}000 \end{Bmatrix}$$

$$\begin{Bmatrix} u_5 \\ u_6 \\ u_7 \\ u_8 \end{Bmatrix} = \begin{Bmatrix} 1.362 \times 10^{-6} \\ -7.087 \times 10^{-5} \\ -1.362 \times 10^{-6} \\ -7.087 \times 10^{-5} \end{Bmatrix} \text{m} = \begin{Bmatrix} 0.0014 \\ -0.0709 \\ -0.0014 \\ -0.0709 \end{Bmatrix} \text{mm}$$

Substituting values of u_5, u_6, u_7, and u_8 in the stiffness matrix, the reactions at the support can be obtained as follows.

$$\{P\} = [K]\{u\}$$

$$\begin{Bmatrix} P_1 \\ P_2 \\ P_3 \\ P_4 \\ P_5 \\ P_6 \\ P_7 \\ P_8 \end{Bmatrix} = \begin{Bmatrix} 1270 \\ 45{,}000 \\ -1270 \\ 45{,}000 \\ 0 \\ -45{,}000 \\ 0 \\ -45{,}000 \end{Bmatrix} \text{N}$$

Element 3

$$P_3 = \frac{AE}{L}\{-c \quad -s \quad c \quad s\}\begin{Bmatrix} u_5 \\ u_6 \\ u_7 \\ u_8 \end{Bmatrix}$$

Substituting A = 9322×10^{-6} m^2, E = 2×10^{11} N/m^2 and theta = 180 Degrees and L = 4m, we get

$$p_3 = 3.832\times10^7\{1 \quad 0 \quad -1 \quad 0\}\begin{Bmatrix} 1.362\times10^{-6} \\ -7.087\times10^{-5} \\ -1.362\times10^{-6} \\ -7.087\times10^{-5} \end{Bmatrix}$$

$$= 1270.0\text{N (Compression)}$$

Element 6

$$p_6 = \frac{AE}{L}\{-c \quad -s \quad c \quad s\}\begin{Bmatrix} u_1 \\ u_2 \\ u_5 \\ u_6 \end{Bmatrix}$$

Substituting A = 958×10^{-6} m^2, E = 2×10^{11} N/m^2 and theta = 36.87 Degrees, L = 5m, we get

$$p_6 = 3.832\times10^7\{-0.8 \quad -0.6 \quad 0.8 \quad 0.6\}$$

$$\times\begin{Bmatrix} 0 \\ 0 \\ 1.362\times10^{-6} \\ -7.087\times10^{-5} \end{Bmatrix} = -1587.7\text{N (Tension)}$$

In similar way, the forces in each element can be calculated. Let us design element 6:

Design of Members

Element 6 is a truss element with length = 5m and section ISA (50 x 50 x 5)

Dead load case

Axial force = 1587.7 N (Tension)

Earthquake load case

Axial Force = 4680.0 N (Compression)

Load Case: 1.5 × (D.L + E.L)

Factored load (Compressive)

= 1.5 × (4680.0)

= 7020.0 N

Properties of ISA (50 x 50 x 5) are as follows:

Cross section area, A = 4.79 cm^2, Radius of gyration about x axis, r_{xx} = 15.2 mm, Radius of gyration about y axis, r_{yy} = 15.2 mm

Effective length about x-x axis = Effective length about y-y axis = KL = 0.7 × 5 = 3.5 m

KL_{xx}/r_{xx} or KL_{xx}/r_{xx} = 3500/15.2 = 230.3,

The section is buckling class $\alpha = c$

Non-dimensional effective slenderness ratio, λ_y

$$\lambda_y = \sqrt{\frac{f_y}{f_{cc}}} = \sqrt{\frac{f_y\left(\frac{KL}{r}\right)^2}{\pi^2 E}} = \sqrt{\frac{250\times230.3^2}{\pi^2\times2\times10^5}} = 2.5918$$

$$\phi = 0.5[1+\alpha(\lambda-0.2)+\lambda^2] = 4.44$$

$$\chi = \text{stress reduction factor} = \frac{1}{(\phi+\sqrt{\phi^2-\lambda^2})} = 0.124$$

$$f_{cd} = \frac{f_y/\gamma_{m0}}{\phi+\sqrt{\phi^2-\lambda^2}} = \chi f_y/\gamma_{m0} \leq f_y\gamma_{m0}$$

Substituting the values, we get

$$f_{cd} = 0.113\times f_y$$

Therefore, f_{cd} = 0.113 × 250 = 28.25 N/mm^2

$$P_d = A\times f_{cd} = 28.25\times4.79\times10^2 = 13531.75\,\text{N} = 13.53\,\text{kN}$$

Capacity, Pd =13.53 kN > 7.02 kN hence safe

Hence, the section is safe.

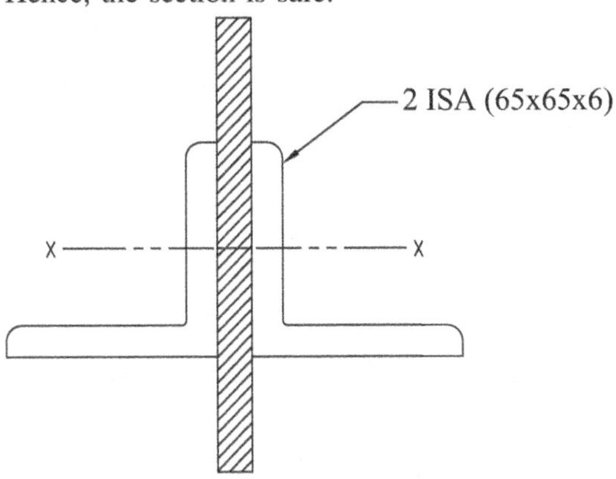

Design of connections:

A tie member consisting of ISA 80mm x 50mm x 8 mm (welding on all three sides as shown in figure) Fe 410 grade steel is welded to 12-mm-thick gusset plate. Design the weld to transmit load equal to design strength of the member (f_u = 410 MPa, f_y = 250 MPa).

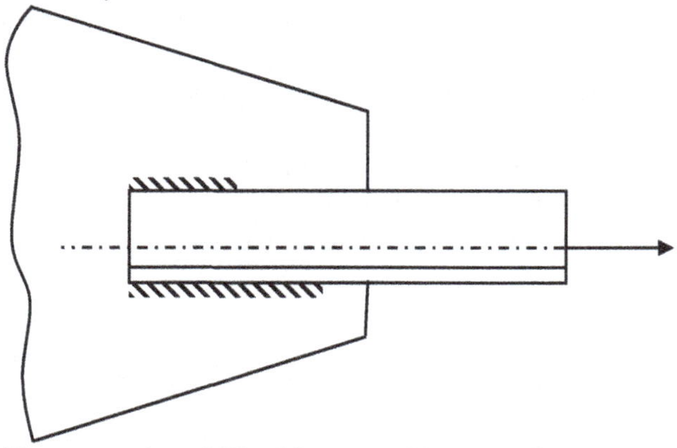

The properties of ISA 80 mm × 50 mm × 8 mm are as follows

Area $= 978\ \text{mm}^2$, $C_{zz} = 27.3\text{mm}$

Partial safety factor for material =

$\gamma_{mw} = 1.10\,(\text{site welding})$

Permissible shear stress in weld $= \frac{f_u}{\sqrt{3}\gamma_{mw}} = 215.19$ MPa

Design strength of the member governed by yielding of the gross section

$T_{dg} = A_g \frac{f_y}{\gamma_{m0}} = 978 \times \frac{250}{1.1} = 222.27 \times 10^3\ \text{N} = 222.27\ \text{kN}$

Hence, the weld will be designed to transmit a force equal to 227.27 kN.

The force to be resisted by weld at bottom side of the angle,

$$P_1 = \frac{222.27 \times (80 - 27.3)}{80} = 146.42\ \text{kN}$$

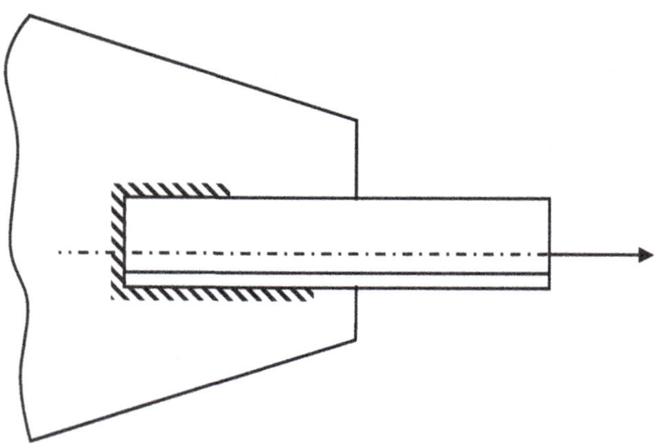

The force to be resisted by weld on upper side of the angle,

$$P_2 = \frac{222.27 \times 27.3}{80} = 75.849\ \text{kN}$$

Assuming size of weld = 6 mm

Effective throat thickness of the weld $= 0.7 \times 6 = 4.2$ mm

The design strength of weld = $P_{dw} = l_w t_t \frac{f_u}{\sqrt{3}\gamma_{mw}}$

Strength of weld per mm = $1.0 \times 4.2 \times \frac{410}{\sqrt{3} \times 1.5} = 662.8$ N

$$l_{w1} = \frac{P_1}{\text{strength of weld per mm}} = \frac{146.42 \times 10^3}{662.8}$$
$$= 220.91\ \text{mm} \approx 221\ \text{mm}$$

$$l_{w2} = \frac{P_2}{\text{strength of weld per mm}} = \frac{75.849 \times 10^3}{662.8}$$
$$= 114.44\ \text{mm} \approx 115\ \text{mm}$$

Design the fillet weld for ISA 80 mm × 50 mm × 8 mm (welding on all three sides as shown in figure).

Equating the strength of weld to the load to be carried by weld, we get the following equation

$$(l_{w1} + l_{w2} + 80\ \text{mm}) \times 662.8\ \text{N/mm} = 222.27 \times 10^3\ \text{N}$$

$$l_{w1} + l_{w2} = 255.35\ \text{mm} \approx 256\ \text{mm}$$

Taking moment about the top edge of the angle section,

$$662.8 \times 80 \times \frac{80}{2} + 662.8 \times l_{w1} \times 80$$
$$= 222.27 \times 10^3 \times (80 - 27.3)$$

$$l_{w1} = 180.91 \approx 181\ \text{mm}\,(\text{Bottom})$$

$$l_{w1} = 256 - l_{w1} = 256 - 181\ \text{mm} = 75\ \text{mm}$$

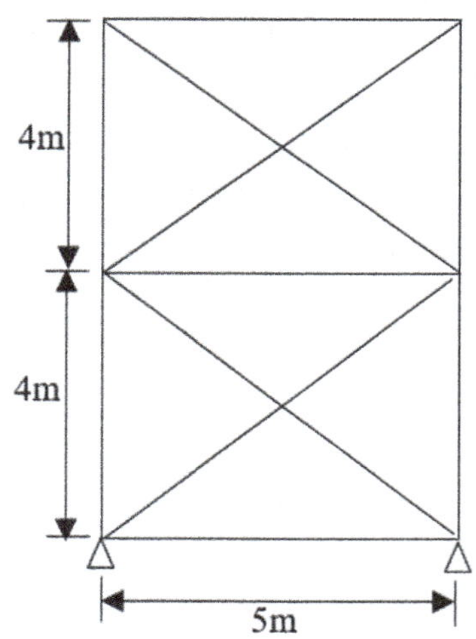

Fig. 8.23 Global stiffness matrix of the 2D truss

Exercise Problems

1. Obtain the global stiffness matrix of the 2D truss as shown in Fig. 8.23 (EA = constant).
2. What are the advantages and disadvantages of steel structures?
3. Neatly define engineering stress–strain curve of steel which is used to derive engineering parameters like Young's modulus, yield strength, ultimate strength, ductility.
4. What is the basic difference between the working stress method and limit state method of design? Define the following terms used in limit state of design (a) limit state of strength (b) limit state of serviceability.
5. What is shape factor of a cross section of a flexural member? Determine the value of shape factor for the following cross-sectional shape having same cross-sectional area and compare the values. Rectangular, triangular, diamond shape, solid circular, hollow circular, wide flanged I beam.
6. How the elastic buckling stress varies with the slenderness ration of column?
7. Define the following terms used in the design of steel structures. plastic section, semi-compact section, compact section, slender section
8. Determine the flexural capacity of a beam (ISMB 300@433.6 N/m) for the following cases (consider effective span of beam as 4 m).
   (a) Laterally supported
   (b) Laterally unsupported
9. A column ISHB 300@618.0 N/m, 5.0 m long has to support a vertical load of 80 kN. Assume effective length factor, K ($K = L_{eff}/L$) as follows. $K_z = 1.20$, $K_y = 2.0$. Carryout the design checks for the section per IS 800: 2007.
10. A bracket is bolted to a column as shown in Fig. 8.24. Design the bolts.
11. A tie member as shown in Fig. 8.25 consisting of angle section ISA 80 mm × 80 mm × 8 mm ($f_y = 250$ MPa) is welded to a 12 mm gusset plate. Assuming 6 mm size of weld, determine the length weld l1 and l2, to transmit a load of full strength of the member. Allowable shear stress in the weld = 108 MPa
12. An 8.0 m long column (ISHB 400@77.4 kg/m) as shown in the Fig. 8.26 is used as a column for an industrial building. It supports a vertical load of 80kN at an eccentricity of 1 m. The ends of the column are restrained in position but not in the direction. Check the

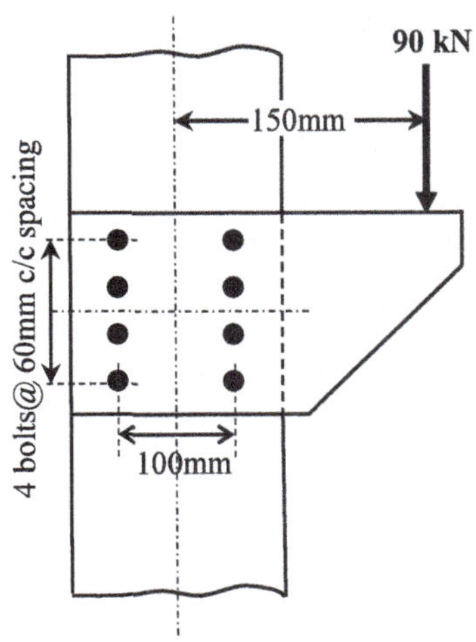

Fig. 8.24 A bracket is bolted to a column

Fig. 8.26 An 8.0 m long column (ISHB 400@77.4 kg/m)

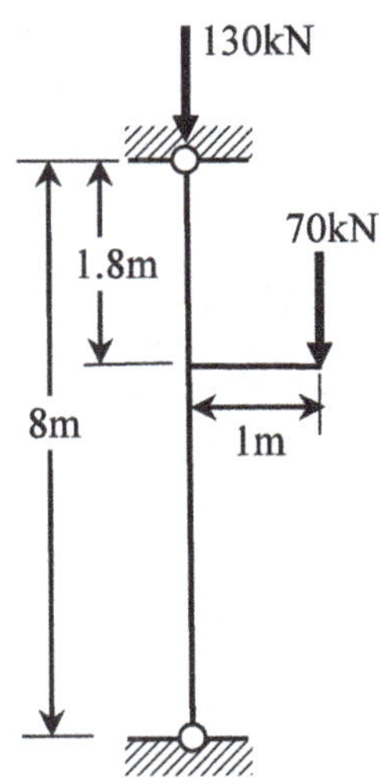

Fig. 8.25 A tie member consisting of angle section ISA 80 mm × 80 mm × 8 mm

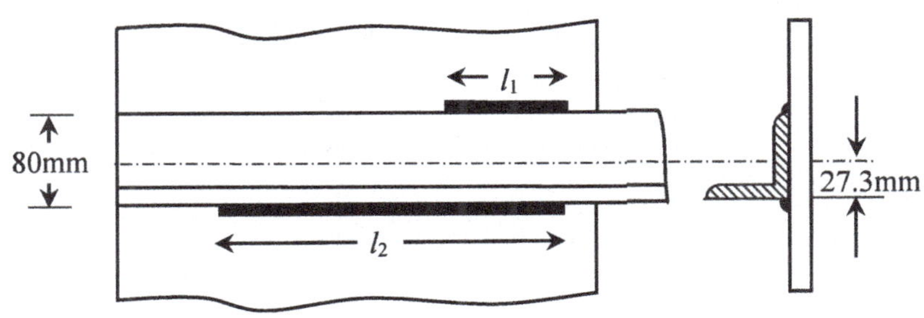

adequacy of the section for combined compression and bending checks as per IS 800: 1984. Use the following equations. (Assume yield stress of steel, $f_y = 250$ MPa.)

13. Web stiffener in a plate girder is used to
    (a) transfer the load from the top flange to the bottom one,
    (b) prevent buckling of web,
    (c) decrease the effective depth of web, and
    (d) prevent excessive deflection

14. For the same cross-sectional area of beam section, which one of the following shape will have the largest value of shape factor?
    (a) Rectangular,
    (b) I section,
    (c) Solid circular,
    (d) Diamond

15. Determine the value of collapse load, wu, per unit length for the fixed beam with uniformly distributed load as shown in the figure with plastic moment capacity Mp
    (a) 0.32Mp
    (b) 0.64 Mp
    (c) 2.0 Mp
    (d) 8.0 Mp

16. The cross section of beam as shown in the figure becomes a plastic hinge. If the yield stress of steel is denoted by σy, obtain the bending compressive force acting on the section for the beam subjected to dead load.
    (a) $bh\sigma y/4$
    (b) $2bh\sigma y/9$
    (c) $bh\sigma y/2$
    (d) $bh\sigma y/3$

17. Match the items of two lists for a good design practice to be followed in design of simply supported plate girder.

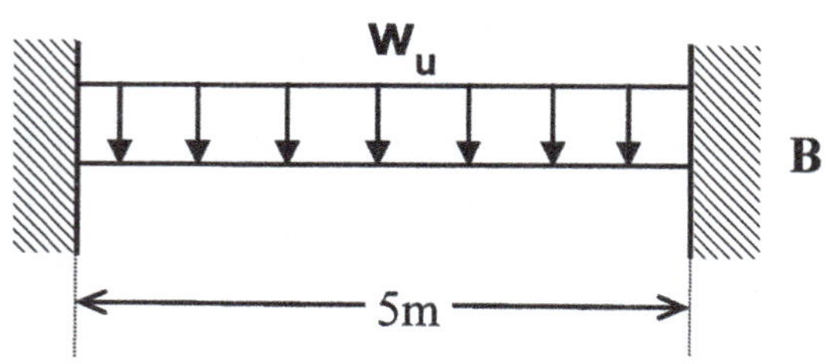

Fig. 8.27 Fixed beam with uniformly distributed load

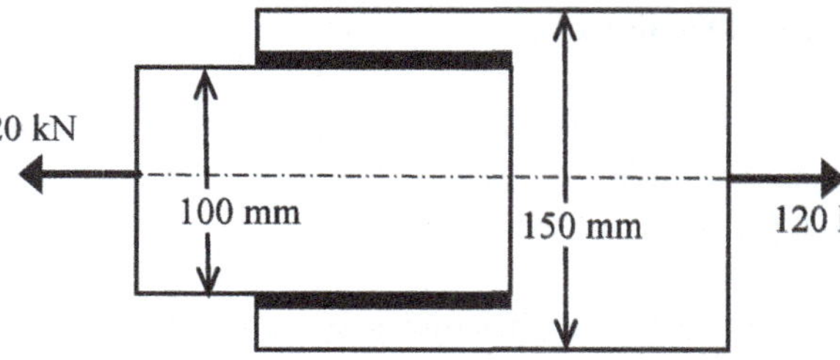

Fig. 8.28 Two 12-mm-thick plates are connected by 6-mm-size fillet welds and are subjected to tensile working load

a.	Flange splice	1	At or near the supports
b.	Web splice	2	Away from centre of span
c.	Transverse stiffeners	3	Away from support
d.	Longitudinal stiffener	4	Near the tension flange
		5	In the compression portion of web

18. Determine the value of collapse load, wu, per unit length for the fixed beam with uniformly distributed load as shown in Fig. 8.27 with plastic moment capacity Mp

19. Which of the following statements are true for bracings in steel structure as per IS800: 2007
    I. Bracings are provided in plan of the structure to resist torsional effects of wind/earthquake forces.
    II. K bracing shall not be provided for structures resisting lateral loads.
    III. The bracing members shall be designed so that net area rupture and not gross area yielding would govern the design tensile strength.
    IV. Bracing shall not be designed as slender section.
        (a) I, III, IV
        (b) I, II, IV
        (c) I, II, III
        (d) I and IV only

20. Two 12-mm-thick plates are connected by 6-mm-size fillet welds and are subjected to tensile working load as shown in Fig. 8.28. Obtain the minimum design length of each weld using limit state method as per IS 800:2007 (consider: $f_y = 250$ MPa, ultimate strength of weld, $f_u = 410$ MPa, partial safety factor for shop weld = 1.25).

21. If depth of two column sections is equal, then the column splice is provided using

(a) Bearing plates,
(b) Filler and packing plates,
(c) Two columns are prepared to butt against each other and load is transferred by bearing with nominal connection,
(d) End plates.

Appendix

Example 8.3 Design a seat connection for a beam end reaction of 130 kN. The beam section I.S.M.B. 300 connected to the flange of the column section I.S.H.B. 250.

Solution
The relevant properties of the section from I.S. Handbook

	I.S.M.B. 300	I.S.H.B. 215
B	140 mm	225 mm
t	7.5 mm	6.5 mm
h_2	29.25 mm	
t	13.1 mm	9.1 mm
G	80 mm	55 mm

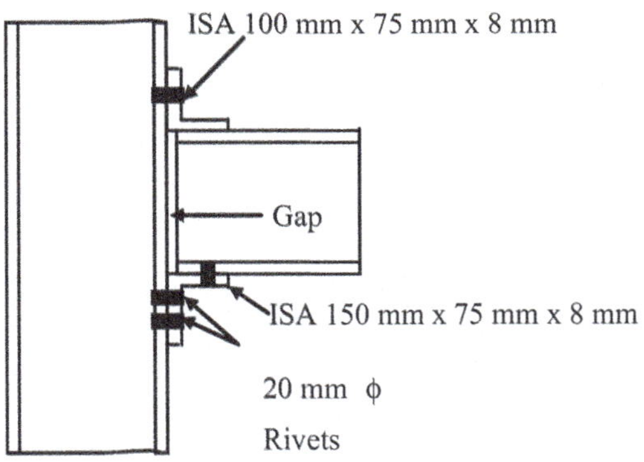

Seat angle:

$$\text{Length of the seat angle} = \text{width of beam flange} = 125\ \text{mm}$$

Bearing length,

$$b = \frac{R}{\sigma_p \times t} - \sqrt{3}h_2 < \frac{1}{2}\frac{R}{\sigma_p t}$$

$\sigma_p = 0.75 f_y = 0.75 \times 250 = 187.5\ \text{N/mm}^2$

$$b = \frac{R}{\sigma_p \times t} - \sqrt{3}h_2 = 41.78\ \text{mm} > 23.11\ \text{mm}$$

Therefore, the bearing length required is 41.78 mm.

Provide a seat leg length of 75 mm. Provide a clearance of 10 mm between the beam and the column flange. Using equation,

$$t = \sqrt{6\frac{\text{Re}_1}{B\sigma_{bs}}}$$

Let us chose an angle of 150 × 115 × 16 mm. The relevant properties of the angle section from I.S. Handbook:

$r_1 = 11$ mm
$A = 2070\ \text{mm}^2$
$e_1 = 10 + (b/2) - t - r_1 = 10 + (41.78/2) - 16 - 11 = 3.89$ mm (Assume a 16-mm-thick seat angle)

$$t = \sqrt{6\frac{\text{Re}_1}{B\sigma_{bs}}} = \sqrt{\frac{6 \times 130 \times 10^3 \times 3.89}{125 \times 185}} = 11.456\ \text{mm} < 16\ \text{mm. Hence safe.}$$

Hence, provide a seat angle of 150 mm × 115 mm × 16 mm.

Connection:

Let us provide power-driven 20 mm ϕ rivets.

$$\text{Gross diameter} = 20 + 1.5 = 21.5\ \text{mm}$$

$$\text{Strength of the rivet in single shear} = (\pi/4) \times (21.5)^2 \times 100 = 36,305.03\ \text{N}$$

$$\text{Strength of the rivet in bearing} = 21.5 \times 6.1 \times 300 = 39,345\ \text{N}$$

$$\text{Value of rivet} = 36,305.03\ \text{N}$$

The number of rivets required to connect the flange of the columns with the seat angle,

$$n = \frac{130 \times 1000}{36{,}305.03} = 3.58 \approx 4\ \text{rivets}$$

The minimum pitch = 2.5ϕ = 2.5 × 20 = 50 mm

Minimum edge distance for 20 mm ϕ rivets is 29 mm. Provide the rivets in two rows. Total number of rivets provided is four.

Provide two rivets, 20 mm ϕ, to connect the seat angle leg with the beam flange. Provide a nominal size cleat angle 100 × 75 × 8 mm. To connect the cleat angle with the beam flange and column flange, provide two 20 mm ϕ rivets on each leg.

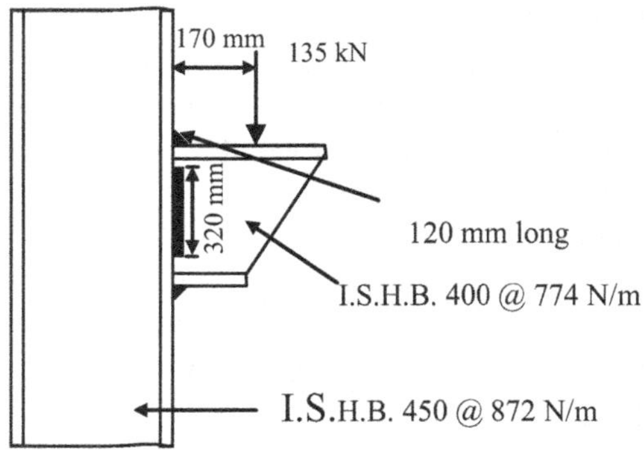

Example 2

A joist cutting is used as a bracket to support a load of 160 kN. It is welded to the column flange as shown in figure. Compute the size of the fillet weld.

Solution

Let

t = effective throat thickness of the weld on the web

$2t$ = effective throat thickness of the weld on the flange

$$I_{xx} \text{ of the weld} = \frac{2 \times 320^3 \times t}{12} + (2 \times t \times 120) \times 2 \times (200)^2$$

$$= 2466.133 \times 10^4 t \text{ mm}^4$$

$$= 546.133 \times 10^4 t + 1920 \times 10^4$$

Direct shear stress in weld:

$$\text{Total area of the welds} = \text{area of the weld on the flanges} + \text{area of weld on the web}$$
$$= 2 \times 120 \times 2t + 2 \times 320 \times t$$
$$= 1120t \text{ mm}^2.$$

Load is shared in the proportion of areas of welds,

$$\text{Load shared by the flange welds} = \frac{2 \times 120 \times 2t}{1120t} \times 160 = 68.571 \text{ kN}$$

$$\text{Load shared by the web welds} = \frac{2 \times 320t}{1120t} \times 160 = 91.43 \text{ kN}$$

$$\text{Shear stress in the flange weld} = \frac{68.571 \times 10^3}{2 \times 120 \times 2t} = \frac{142.86}{t} \text{ N/mm}^2$$

Shear stress in the flange weld

$$= \frac{142.86}{t} \text{ N/mm}^2$$

The moment of inertia shared in proportion of the extreme weld distance from the neutral axis. Horizontal shear stress in the flange weld,

$$\tau_{vf1,cal} = \frac{160 \times 10^3 \times 170}{2466.133 \times 10^4 \times t} \times 200$$

$$= \frac{220.59}{t} \text{ N/mm}^2$$

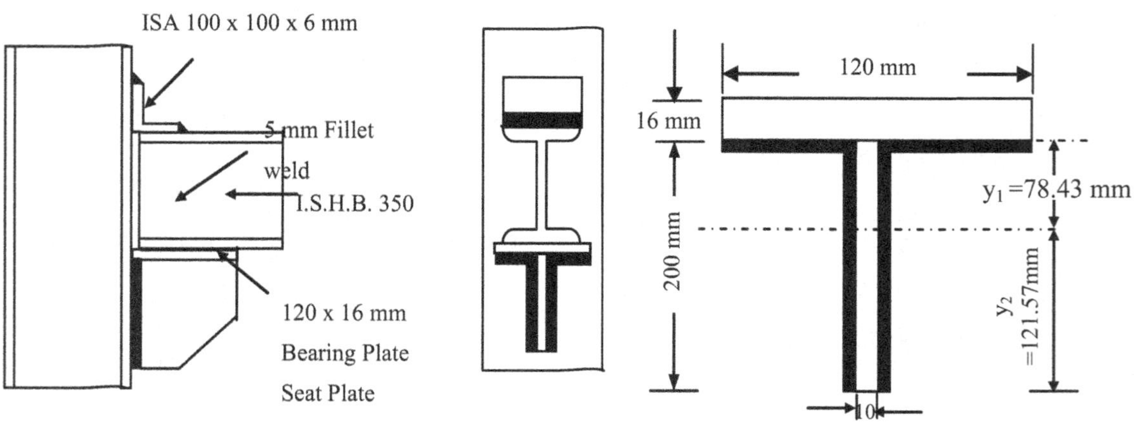

Taking the worst combination,

$$108 = \left[\left(\frac{220.59}{t}\right)^2 + \left(\frac{142.86}{t}\right)^2\right]^{\frac{1}{2}}$$

$$t = 2.433 \text{ mm}$$

$$108 = \frac{262.81}{t}$$

Size of the web weld = (2.433/0.707) = 3.44 mm.
Hence, provide the size of the weld on web as equal to 6 mm.
Size of weld on flange = 2 × 3.44 = 6.88–8 mm.

Example 3

Design a stiffened seat connection to join I.S.M.B. 350@524 N/m with a column section I.S.H.B. 300@588 N/m. The beam has to transmit an end reaction of 230 kN.

Solution

Bearing length b = 97.746 mm
Provide a clearance of 10 mm

$$\text{Width of seat plate} = 97.746 + 10 = 107.746 \text{ mm.}$$

Adopt width of seat as 120 mm.

$$\begin{aligned}\text{Thickness of seat plate} &= \text{thickness of beam flange} \\ &= 14.2 \text{ mm} \\ &= 16 \text{ mm.}\end{aligned}$$

Provide a seat plate 120 × 16 mm

$$\begin{aligned}\text{Thickness of the stiffening plate} &= \text{thickness of beam web} \\ &= 8.1 - 10 \text{ mm}\end{aligned}$$

Provide a stiffener plate 200 mm × 10 mm.

Assume length of weld as shown in figure. Distance of end reaction from outer end of seat plate,

$$\begin{aligned}&= 120 - (1/2) \times 97.746 \\ &= 71.127 \text{ mm}\end{aligned}$$

Bending moment, $M = 230 \times 71.127 =$ 16,359.21 kN mm.

Width of flange of beam

$$\bar{y}_1 = \frac{2 \times 200 \times 100 + 0}{2 \times 200 + (120 - 10)} = 78.43 \text{ mm}$$

C.g. of the weld group,

$$\bar{y}_2 = 200 - 78.43 = 121.57 \text{ mm}$$

$$I_{xx} = \left[1 \times (120 - 10) \times 78.43^2 + 2 \times \frac{200^2}{12} \times 1 + 2 \times 200 \times (121.57 - 100)^2\right]$$

$$= 219.62 \times 10^4 \text{ mm}$$

Vertical shear,

$$\tau_{vf\,1} = \frac{230 \times 10^3}{2 \times 200 + 110} = 450.98 \text{ N/mm}$$

$$\tau_{vf\,2} = \frac{16359.21 \times 10^3}{219.607 \times 10^4} \times 78.43 = 584.249 \text{ N/mm}$$

Horizontal shear,

$$\text{Resultant shear stress/mm} = \text{length of weld/mm}$$

$$\sqrt{450.98^2 + 584.249^2} = 0.7 \times S \times 108 \times 1$$

$$S = 9.76 \text{ mm say } 10 \text{ mm}$$

Provide a 10 mm fillet weld to make connection.

Provide a nominal size flange cleat 10 × 100 × 6 mm and connect it by a 5 mm fillet weld as shown in figure.

Example 8.4: Seismic analysis of Industrial Steel Structure by Working Stress Method

Building Details
Type: Two Storied structure.
Plan Size: 22.0 m × 6.2 m
Height: 3.0 m, First story and 3.5 m, Second story above ground level.

Site Details
Location : Mumbai
Seismic zone: Zone III
Safe bearing capacity: 100 kN/m^2

The steel structure is housed inside the highbay area. The plan and section of steel structure at each level is presented below:

Steps carried out for seismic analysis of steel structure:

Step 1

Finite element modeling of the building based on the drawing and presumed sections. It includes modeling of

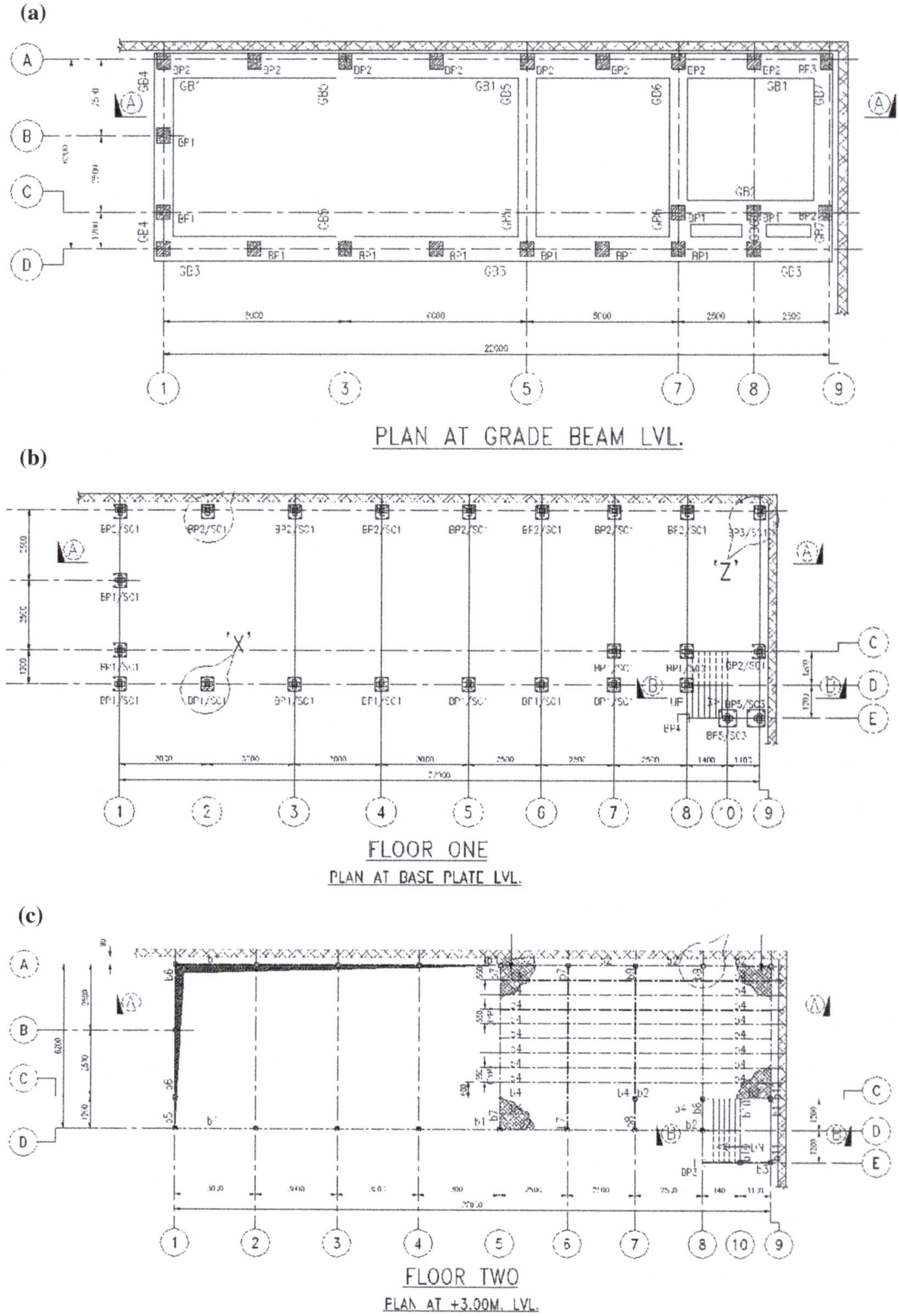

Fig. 8.29 **a** Plan at grade beam LVL. **b** Floor one plan at base plate LVL. **c** Floor two plat at +3.00 M. LVL. **d** Roof plan at +6.50 M. LVL. **e** Section A-A

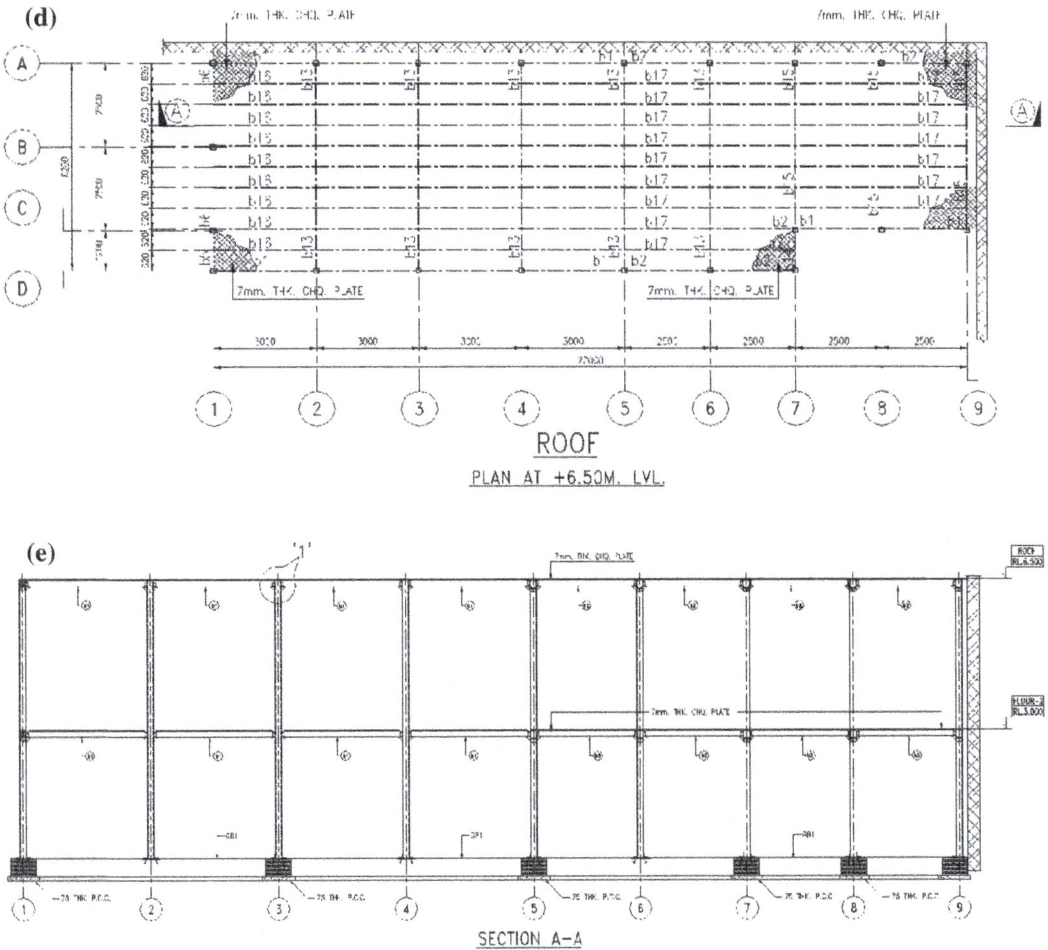

Fig. 8.29 (continued)

geometry, mass, stiffness, and damping. Figure 8.29a–e shows the plan and elevations in the various levels. Finite element model is shown in Fig. 8.30. ISMB and ISMC sections are used for the beam and column, respectively. 3D Beam element is used for modeling the frame.

Step 2

Apply the static loads and thereby perform static analysis (dead loads and live loads).

Live load is taken as 5 kN/m^2 at the floor 1 level and 1.5 kN/m^2 at roof level. There is also a 10 kN capacity of underslung crane for the movement of equipment and material. Floor slab and roof slab is provided with MS chequred plate.

Step 3

Perform free vibration analysis. Figure 8.31a–d shows the fundamental mode in *X*, *Y*, and *Z* directions. The frequencies and mass participated in the associated modes are shown in Table 8.10.

Step 4

Carry out response spectrum analysis. [Use factor (2/3) for vertical direction.]

Response Spectra

As per IS 1893 2002:

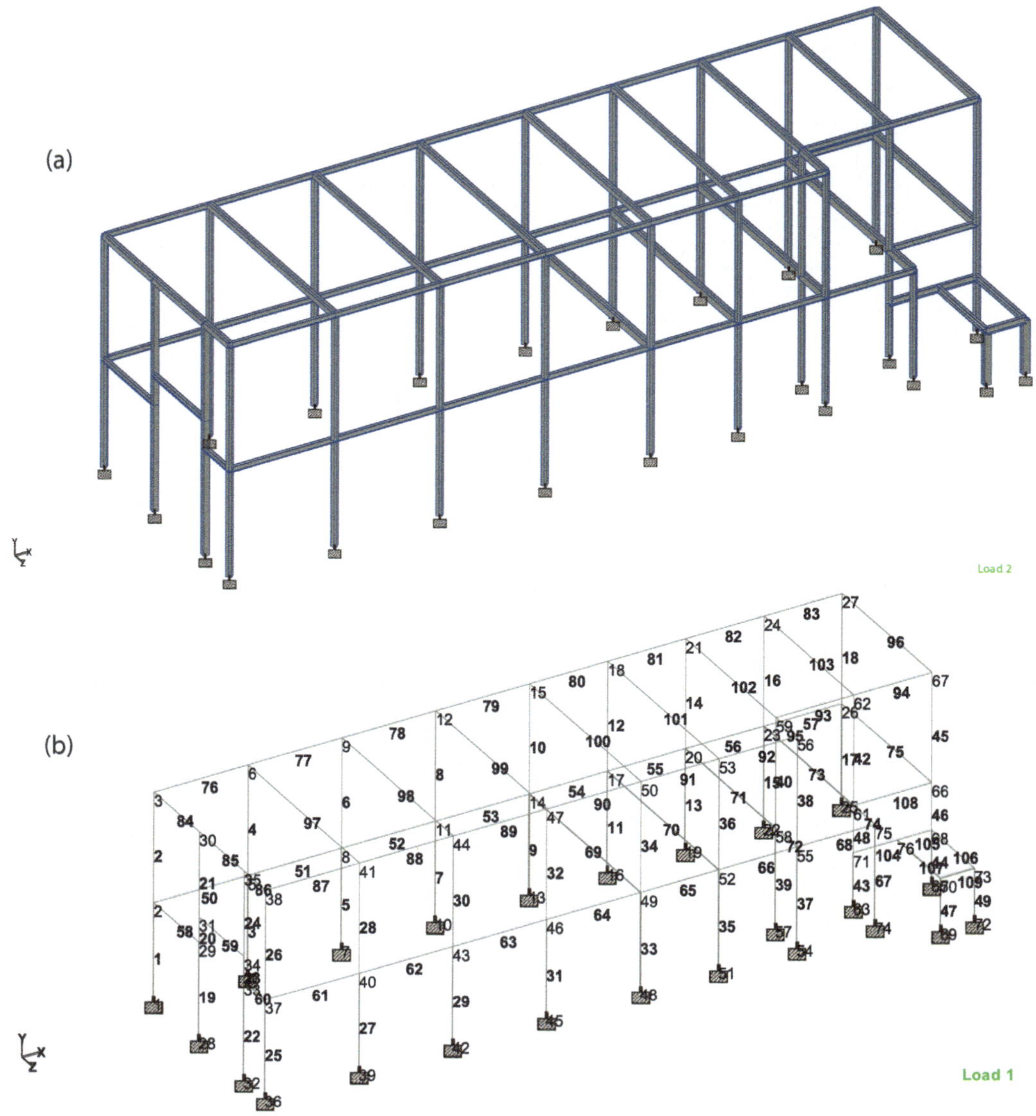

Fig. 8.30 **a** Finite element model of steel structure. **b** Finite element model of steel structure with node and member number

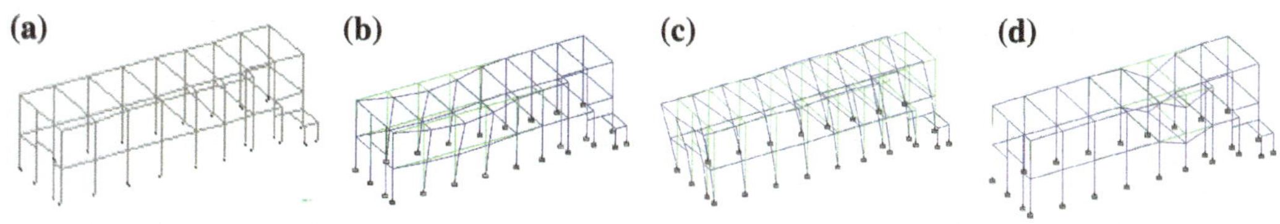

Fig. 8.31 **a** FE model of structure. **b** First mode shape, *Z*-direction. **c** Fourth-mode shape, *X*-direction. **d** 39th mode shape, *Y*-direction

Table 8.10 Frequencies and mass participation of structure

Mode	Frequency (Hz)	Period (sec)	% Mass participation in directions		
			X	*Y*	*Z*
1	0.999	1.001	0.024	0.0	27.75
2	1.242	0.805	0.025	0.0	28.12
4	1.585	0.631	76.65	0.0	34.63
19	5.067	0.197	93.36	0.01	94.36
39	29.695	0.0337	99.75	19.19	99.77
50	33.073	0.0302	99.75	78.93	99.77
80	57.700	0.0173	99.77	91.78	99.92
100	74.330	0.0134	100	96.46	99.98

Reduction Factor: 4 (Moment-Resisting Steel Structure, Table 3: IS 1893 part IV)
Damping: 0.02 (2% for DBE)
Multiplying factor: 1.4 (for 2% damping spectra, see Table 6.5)
Structure Category: 3
Importance Factor: 1.5 (Table 2: IS 1893 part IV)
Soil type 3, Soft soil

Step 5

Combination of Forces:

(1) D.L. +L.L.
(2) D.L + L.L. ± E.Q.
(3) D.L. ± E.Q.

Step 6

Carry out design of sections, connections, base plates MS plate, etc., as per IS 800 2007 and footing as per IS 456:2000.

Design of Footing:

Reaction from the column is considered to calculate the footing. Here, strip footing is taken. Size of footing is calculated as total load divided by SBC of soil and then check for critical load combination in earthquake condition in such way that nowhere in the pressure below the footing is within the permissible SBC.

Design of Base Plate

Critical load combination over the base plate is:

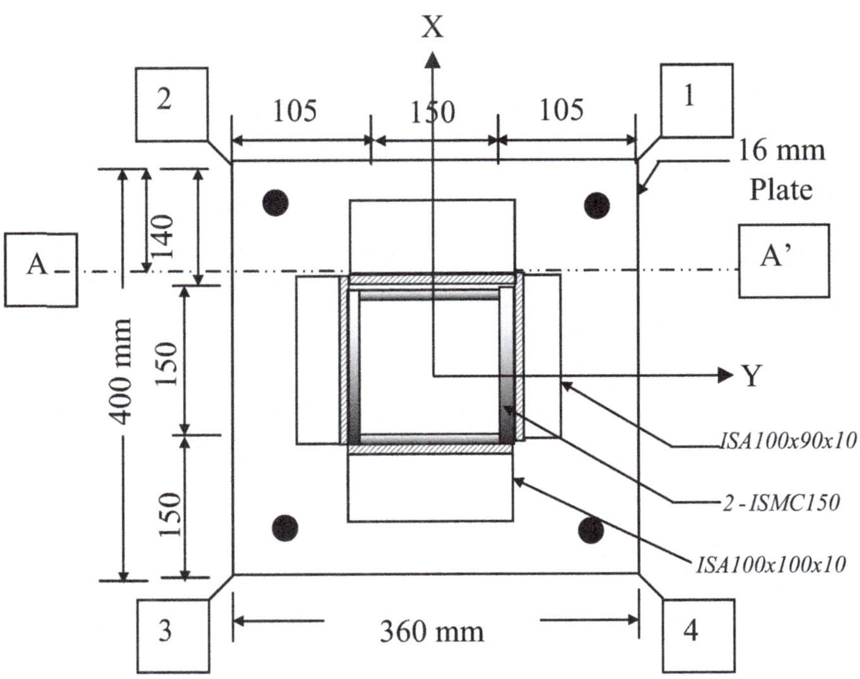

D.L., L.L., and E.Q.

P = 84.44 kN
M_x = 4.77 kN m
M_y = 1.53 kN m

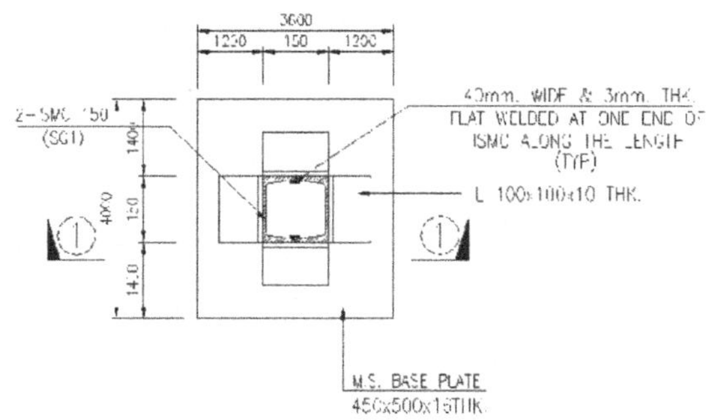

DETAIL' X' (BP1/SC1)

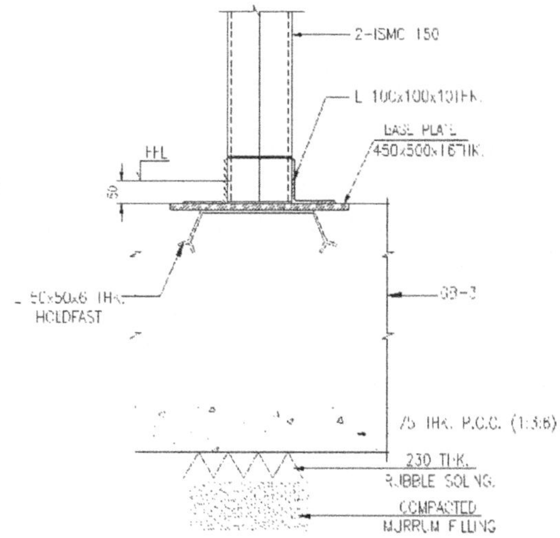

SECTION 1-1

Calculation for Size of base plate,

Size of base plate is calculated based on larger of two criteria: (a) based on the bearing capacity of foundation and (b) based on the minimum dimension required to support the column, gusset plate, gusset angle.

(a) Based on the bearing capacity of foundation

$$\text{Area of the base plate} = \frac{\text{Total axial Load}}{\text{Safe compressive strength of concrete}} = \frac{84{,}440}{4} = 2110\ \text{m}^2$$

$$L = B = \sqrt{21{,}110} = 145.29\ \text{mm say } 146\ \text{mm}$$

(b) Based on the minimum dimension:

$$\text{Length of plate} = 150 + 2 \times 100 + 2 \times 20 = 390\ \text{mm}$$
Say 400 mm.

$$\text{Breadth of plate} = 100 + 2 \times 100 + 2 \times 20 = 340\ \text{mm}$$
Say 360 mm.

$e_x = M_y/P = 1.53 \times 1000/84.44 = 18.155\,\text{mm} < (L/6) = 60$ mm.
$e_y = M_x/P = 4.77 \times 1000/84.44 = 56.49\ \text{mm} < (B/6) = 66.67$ mm.
$P_i\ (i = 1,2,3,4) = (P/A) \times [1 \pm (e_x/L) \pm (e_y/B)]$

$$P_1 = 0.778\ \text{N/mm}^2,\quad P_2 = 0.7275\ \text{N/mm}^2,\quad P_3 = 0.6771\ \text{N/mm}^2,\quad P_4 = 0.4955\ \text{N/mm}^2$$

Hence, no tension.

Provide 4 bolts of size 20 mm nominal diameter.

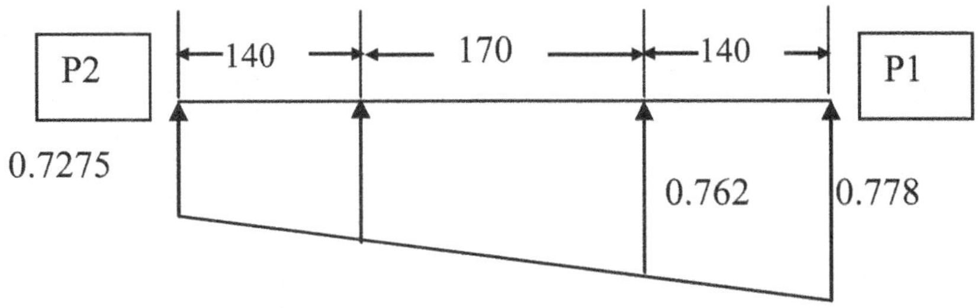

Pressure distribution under base plate

Taking the critical section Along A–A'.

(Permissible bending stress = 165 N/mm^2. In case of seismic load, the permissible stress shall be increased by 33.5% for all steel, and in case of riveted and bolted, the permissible shall be increase by 25%.)

$$\frac{1}{6} \times t^2 \times \sigma_{bs} = 0.762 \times \frac{140^2}{2} + \frac{1}{2} \times (0.778 - 0.762) \times 140 \times \left(\frac{2}{3}\right) \times 140$$

$$\frac{1}{6} \times t^2 \times 165 = 7572.13$$

$$t = \sqrt{\frac{7572.13 \times 6}{165}} = 16.59 \text{ mm}$$

$$\text{Thickness of base plate} = t - \text{thickness of angle} = 16.6 - 10 = 6.6 \text{ mm}$$

Hence provide a 16-mm-thick base plate.

Design of Bolt or Hold fast:

Base plate shall be hold in foundation either by providing Anchor bolts or hold fasts.

In the present case, there is no tension in the base plate; hence, nominal bolt or hold fast will be sufficient to take care of horizontal force.

Hence, provide four anchor bolts at four corners or provide ISA50 × 50 × 6 at the four corner of base plate.

Anchor bolts when there is tension in base plate:

T_b = Allowable tension in bolt
m = modular ratio
n = Depth of neutral axis
a = Distance of anchor bolt from the respective edge
d = Distance of the farther side bolts from the nearer edge.
T = tension in farther side bolt
A_s = Net area of bolt

$$\text{For equilibrium, } P + T = C \qquad \text{(a)}$$

where $T = A_s\, t$ and $C = b\, n\, c/2$

From the stress diagram,

$$\frac{c}{\frac{t_b}{m}} = \frac{n}{d-n} \Rightarrow n = \frac{d}{1+\frac{t_b}{mc}}$$

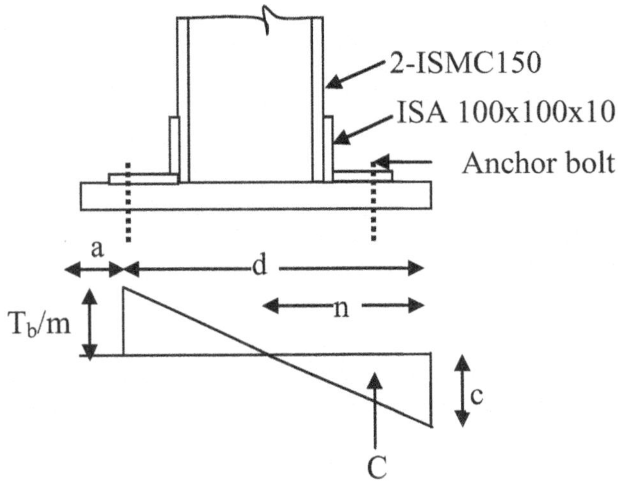

Fig Stress Diagram

This gives the value of n.

Take moments about the tensile force in the farther side anchor bolts

$$C\left(d - \frac{n}{3}\right) = M + P\left(\frac{L}{2} - a\right)$$

This gives C. Now, T can be calculated from Eq. a
Hence, net area of bolt at the root of the thread

$$A_s = T/T_b$$

Length of bolt is calculated based on the tensile force in bolt and it should be greater than the bond strength in the concrete. Normally, length of each bolt is taken as 24 times diameter.

Nominal sizes of the bolts in common use are 20–38 mm.

Design of column

(Element No. 33 along grid D/5)
Critical load from the both node of member no. 33
Axial load = P = 83.60 kN
Moment in M_y = 7.74 kN m
Moment in M_z = 7.26 kN m
Effective length = 3 m

Step 1 Considering buckling class *a*
Stress reduction factor for as per IS800:2007 Table 8(a) for slenderness ratio 52.23 = 0.894
Design compressive stress, f_{cb} = 0.894 × 250 = 223.5 N/mm^2
Permissible compressive stress, $f_{ac} = 0.6 f_{cb}$ = 134.10 N/mm^2
Approximate area of cross section, $A_c = P/f_{ac}$

$$A_c = 83{,}600/134.10 = 623.4\ \text{mm}^2$$

$$A_c\ \text{required} = 1.5 \times 623.4 = 935\ \text{mm}^2$$

Let us try 2-ISMC150

$$A = 4176\ \text{mm}^2$$

Slenderness ratio = Effective length of column/Least radius of gyration = 52.23 < 250 as per IS800:2007 when earthquake load is considered.

Hence safe.

Step 2 Check for the member strength requirement

Actual compressive stress, $f_c = P/A$ = 83,600/4176 = 20.02 N/mm^2

Moment, M_y = 7.74 kN m

Applied compressive stress bending about major Y axis, f_{bcy}

$$= 7{,}740{,}000 \times 75/8{,}770{,}100$$
$$= 66.19\ \text{N/mm}^2$$

Applied compressive stress bending about major Y axis, f_{bcz}

$$= 7{,}260{,}000 \times 75/12{,}288{,}600$$
$$= 44.309\ \text{N/mm}^2$$

Member strength requirement must be satisfied.

$$\frac{f_c}{0.6 f_y} + \frac{f_{bc}y}{f_{abc}y} + \frac{f_{bc}z}{f_{abc}z} \le 1$$

At the support, the values of f_{abcy} and f_{abcz} shall be calculated using laterally supported member and is given as $0.66 f_y$ as per IS800:2007

where f_{abcy} and f_{abcz} allowable bending compressive stress due to bending about minor (y) and major (z) axis of the cross section.

$$\frac{20.02}{150} + \frac{66.19}{165} + \frac{44.309}{165} = 0.133 + 0.401 + 0.268$$
$$= 0.8025 \le 1.\ldots\ldots\textbf{OK}$$

Therefore, adopt 2-ISMC150 as the section of the column.

Design of Beam

Element No. 69 along grid D/5)
Critical load from the both node of member no. 33
Shear force = V = 34.25 kN
Moment in Mz = 29.45 kN m
Effective length = 6.2 m

Step 1 Find the required section modulus of the section, Z_{reqd}

$$Z_{reqd} = \text{M}/\text{f}_{bc}$$
$$= 29450000/165 = 177{,}272\ \text{mm}^3$$

From steel table, choose the Section 2-ISMC200

$I_x = 3.6386 \times 10^7$ mm^4

Step 2 Average shear stress, $f_v = F/t_w d = 29{,}450/2 \times 6.1 \times 200 = 12.07$ N/mm^2

which is less than the permissible shear stress = $0.4f_y$ = 100 N/mm^2. Hence **safe**.

Step 3 Check for deflection

Maximum deflection at the centre,

$$\delta_{\max} = \frac{5}{384} \times \frac{wl^4}{EI} = \frac{5}{384} \times \frac{12 \times 6200^4}{2.1e^5 \times 3.6386e^7} = 12.3 \text{ mm}$$

$$\delta_{\text{perm}} = \frac{l}{325} = \frac{6200}{325} = 19.078 \text{ mm}$$

$\delta_{\max} < \delta_{\text{perm}}$ Hence **safe**

Design of connection between beam and column

Column and beam is connected by the plate from both sides; hence, load on the one side of the connection will be half the total load on the joint. Also, the cross member is connected over the same plate. Hence, the load coming from the cross member shall also be consider in the design of plate connection.

Shear force = V = 34.25 kN

Moment in M_z = 29.45 kN m

Shear force from the cross member = 6.2 kN

Load line falls in the plane of welds, and hence, the weld section undergoes axial and torsional stresses.

Total effective length of weld = 280 + 2 × 130 + 50 + 50 = 650 mm

Axial stress,

$$f_a = \frac{P}{t \times l} = \frac{17{,}120 + 6200}{t \times 650} = \frac{35.8}{t} \text{ N/mm}^2$$

C.g. of weld section

$650X = 2 \times 130 \times 130/2 + 100 \times 130$

$X = 46$ mm

Eccentricity e of the load = 130/2 + 65 + (65 − 46) = 149 mm

$$\begin{aligned}\text{Torque} = T &= P \times e + \text{Moment from the beam} \\ &= 23.32 \times 10^3 * 149 + 29.45 \times 10^6/2 \\ &= 18.19 \times 10^6 \text{ Nmm}\end{aligned}$$

$$\begin{aligned}I_{xx} &= \frac{1}{12}t(280)^3 + 2t130 \times 65^2 + \frac{1}{12}t(100)^3 \\ &= 3{,}011{,}166t \text{ mm}^4\end{aligned}$$

$$\begin{aligned}I_{yy} &= 2\left\{\frac{1}{12}t(130)^3 + 130t(65-46)^2\right\} \\ &\quad + 280t46^2 + 130t(130-46)^2 \\ &= 1{,}969{,}787t \text{ mm}^4\end{aligned}$$

Polar moment of inertia J is given by

$$J = I_{XX} + I_{YY} = 3{,}011{,}166t + 1{,}969{,}787t = 4{,}980{,}953t$$

Maximum torsion stress will occurs at farther end from the c.g. of the weld.

Distance r from the c.g. of weld group

$$r = \sqrt{140^2 + (65-46)^2} = 141.2 \text{ mm}$$

Torsional stress,

$$f_p = \frac{Tr}{J} = \frac{18.19 \times 10^6 \times 141.2}{4{,}980{,}953} = \frac{515}{t}$$

$$\cos\theta = \frac{130-46}{141.2} = 0.595$$

Resultant stress f_R is given as

$$f_R = \sqrt{f_a^2 + f_p^2 + 2f_a f_p \cos\theta}$$

$$f_R = \sqrt{\left(\frac{35.8}{t}\right)^2 + \left(\frac{515}{t}\right)^2 + 2\left(\frac{35.8}{t}\right)\left(\frac{515}{t}\right)0.595}$$

$$f_R = \frac{537}{t}$$

f_R must not exceed the permissible stress 110 N/mm^2 and this value can be increased by 33.3% as in case of earthquake load.

$$\frac{537}{t}\langle 146$$

or

$$t > \frac{537}{146}$$

Size of weld,

$$s > \frac{537}{146 \times 0.7} = 5.25 \text{ mm}$$

Hence, use 6 mm fillet welds of length 650 mm both side in column and continuous weld on three sides to plate and beam. Also, provide seat angle of size 80 × 80x8 thick with 10-mm-thick stiffener plate at the bottom of beam.

Load on MS plate = DL + Finishes + Fixtures + Live load

$$= 0.55 + 0.45 + 0.5 + 5 = 6.5 \text{ kN/m}$$

Considering simply supported over the beam

Max BM = wl^2/2=6.5 * 550^2/8 = 245,781 N mm/m length

$$f_{bc} = \frac{M}{Z} = \frac{6M}{bt^2} \Rightarrow 165 = \frac{6 \times 245{,}781}{1000 \times t^2}$$

t = 2.99 mm > 7 mm. Hence safe.

Check for deflection:

Maximum deflection at the centre,

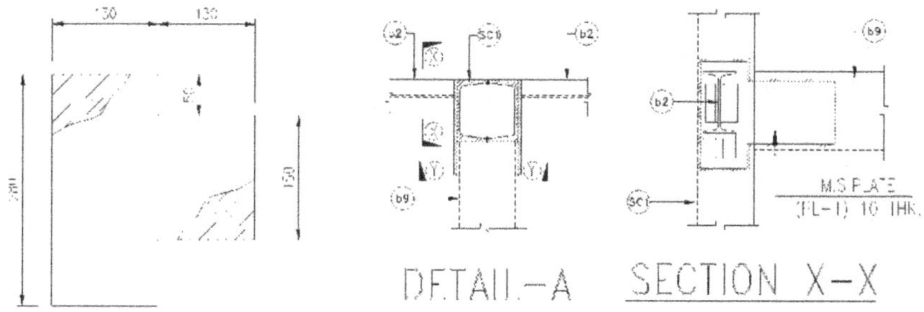

DETAIL-A SECTION X-X

PLATE (PL-1)

Design of MS Plate for flooring and roofing:

MS chequered plate is used as flooring on the floor 2 level, and MS plate is used as roofing at roof level.

Span of MS plate = 550 mm

Assuming 7-mm-thick plate

$$\delta_{max} = \frac{5}{384} \times \frac{wl^4}{EI} = \frac{5}{384} \times \frac{6.5 \times 550^4}{2.1e^5 \times 28{,}583} = 1.29 \text{ mm}$$

$$\delta_{perm} = \frac{l}{325} = \frac{550}{325} = 1.69 \text{ mm}$$

δ_{max} is leass than δ_{perm} of the section. Hence safe.

```
>> K=9*12*2*10^11*19802.8*10^-8*[1 -1 0;-1 2 -1;0 -1 2]

K =

   1.0e+09 *

    4.2774   -4.2774         0
   -4.2774    8.5548   -4.2774
         0   -4.2774    8.5548

>> [V,D]=eig(K,M)

V =

   -0.0020   -0.0018    0.0012
   -0.0017    0.0006   -0.0019
   -0.0009    0.0020    0.0014


D =

   1.0e+04 *

    0.6610         0         0
         0    5.0083         0
         0         0    9.8658

>> w=sqrt(D)

w =

   81.3022         0         0
         0  223.7922         0
         0         0  314.0993

>> K=9*12*2*10^11*19802.8*10^-8/3.5/3.5/3.5*[1 -1 0;-1 2 -1;0 -1 2]

K =

   1.0e+08 *

    0.9976   -0.9976         0
   -0.9976    1.9953   -0.9976
         0   -0.9976    1.9953

>> [V,D]=eig(K,M)

V =

   -0.0020   -0.0018    0.0012
   -0.0017    0.0006   -0.0019
   -0.0009    0.0020    0.0014

D =
```

```
   1.0e+03 *

    0.1542         0         0
         0    1.1681         0
         0         0    2.3011

>> w=sqrt(D)

w =

   12.4165         0         0
         0   34.1777         0
         0         0   47.9695

>> f=w/2/pi

f =

    1.9762         0         0
         0    5.4395         0
         0         0    7.6346

>> T=1/f
Error using  /
Matrix dimensions must agree.

>> T=inv(f)

T =

    0.5060         0         0
         0    0.1838         0
         0         0    0.1310

>> phi1=V(:,1)

phi1 =

   -0.0020
   -0.0017
   -0.0009

>> phi2=V(:,2)

phi2 =

   -0.0018
    0.0006
    0.0020

>> phi3=V(:,3)

phi3 =

    0.0012
```

```
   -0.0019
    0.0014

>> phi1

phi1 =

    0.0020
    0.0017
    0.0009

>> phi1'*M*[1;1;1]

ans =

  609.5466

>> phi2'*M*[1;1;1]

ans =

  169.9726

>> phi3'*M*[1;1;1]

ans =

   61.6138

>> G1=phi1'*M*[1;1;1]

G1 =

609.5466

>> G2=phi2'*M*[1;1;1]

G2 =

  169.9726

>> G3=phi3'*M*[1;1;1]

G3 =

   61.6138

>> M1=G1*G1

M1 =

   3.7155e+05

>> M2=G2*G2
```

```
M2 =

   2.8891e+04

>> M3=G3*G3

M3 =

   3.7963e+03

>> Q1=0.657*G1*phi1'*M

Q1 =

   1.0e+04 *

    9.1681    9.7497    5.4928

>> Q2=0.657*G2*phi2'*M

Q2 =

   1.0e+04 *

   -2.2929    0.9549    3.2361

>> x1=G1*phi1*0.657/154.2

x1 =

    0.0053
    0.0043
    0.0024

>> x2=G2*phi2*0.657/1168.1

x2 =

   1.0e-03 *

   -0.1737
    0.0561
    0.1903

>>
```

National Information Centre of Earthquake Engineering at IIT Kanpur, INDIA

MATLAB script

```
clear all
E=2*10^11; % in Pa
I=19802.8*10^-8; % m4
L=3.5; %in m
K= 9*12*E*I/(L^3)*[2 -1 0;-1 2 -1;0 -1 1];
M=[145611 0 0; 0 145611 0;0 0 113012];
```

```
[Phi,lamda]=eig(K,M);
lamda_1=lamda(1,1);
lamda_2=lamda(2,2);
lamda_3=lamda(3,3);
w1=sqrt(lamda_1);
f1=sqrt(lamda_1)/2/pi;
w2=sqrt(lamda_2);
f2=sqrt(lamda_2)/2/pi;
w3=sqrt(lamda_3);
f3=sqrt(lamda_3)/2/pi;
phi1=Phi(:,1);
phi2=Phi(:,2);
phi3=Phi(:,3);
J=[1;1;1];
G1=phi1'*M*J;
G2=phi2'*M*J;
G3=phi3'*M*J;
Total_mass=sum(M(:));
M1=100*(phi1'*M*J)^2/Total_mass;
M2=100*(phi2'*M*J)^2/Total_mass;
M3=100*(phi3'*M*J)^2/Total_mass;
Q1=0.657*G1*phi1'*M;
Q2=0.657*G2*phi2'*M;
Q3=0.657*G3*phi3'*M;
x1=G1*phi1*0.657/lamda_1 % m
x2=G2*phi2*0.657/lamda_2 % m
x3=G3*phi3*0.657/lamda_3 % m
```

Truss

```
>> K

K =

   1.0e+08 *

    4.9062    0.1839   -4.6610         0
    0.1839    6.3526         0         0
   -4.6610         0    4.9062   -0.1839
         0         0   -0.1839    6.3526

>> M

M =

   1.0e+03 *

    4.7749         0         0         0
         0    4.7749         0         0
         0         0    4.7749         0
         0         0         0    4.7749

>> [V,D]=eig(K,M)

V =
```

```
   -0.0102    0.0006    0.0003    0.0102
    0.0003   -0.0102    0.0102    0.0006
   -0.0102   -0.0006    0.0003   -0.0102
   -0.0003   -0.0102   -0.0102    0.0006


D =

   1.0e+05 *

    0.0502         0         0         0
         0    1.3282         0         0
         0         0    1.3316         0
         0         0         0    2.0058

>> [V,lamda]=eig(K,M)

V =

   -0.0102    0.0006    0.0003    0.0102
    0.0003   -0.0102    0.0102    0.0006
   -0.0102   -0.0006    0.0003   -0.0102
   -0.0003   -0.0102   -0.0102    0.0006


lamda =

   1.0e+05 *

    0.0502         0         0         0
         0    1.3282         0         0
         0         0    1.3316         0
         0         0         0    2.0058

>> f=sqrt(lamda)/2/pi

f =

   11.2757         0         0         0
         0   58.0035         0         0
         0         0   58.0767         0
         0         0         0   71.2799

>> G1=phi1'*M*[1;0;1;0]

G1 =

  -97.6791

>> G2=phi2'*M*[0;1;0;1]

G2 =

  -97.5648
```

```
>> G3=phi3'*M*[0;1;0;1]

G3 =

  -3.9933e-12

>> G4=phi4'*M*[1;0;1;0]

G4 =

     0

>>

Date: 10.08.2017
>> P

P =

           0
      -45000
           0
      -45000

>> u=inv(K)*P

u =

   1.0e-04 *

    0.0136
   -0.7088
   -0.0136
   -0.7088

Undefined function or variable 'unnamed'.

>> Truss_10.082017
 Truss_10.082017
          |
Error: Unexpected MATLAB expression.

>> Truss99

V =

   -0.0102    0.0006    0.0003    0.0102
    0.0003   -0.0102    0.0102    0.0006
   -0.0102   -0.0006    0.0003   -0.0102
   -0.0003   -0.0102   -0.0102    0.0006


lamda =

   1.0e+05 *
```

```
    0.0502         0         0         0
         0    1.3282         0         0
         0         0    1.3316         0
         0         0         0    2.0058

>> Truss99

V =

   -0.0102    0.0006    0.0003    0.0102
    0.0003   -0.0102    0.0102    0.0006
   -0.0102   -0.0006    0.0003   -0.0102
   -0.0003   -0.0102   -0.0102    0.0006


lamda =

   1.0e+05 *
    0.0502         0         0         0
         0    1.3282         0         0
         0         0    1.3316         0
         0         0         0    2.0058


f =

   11.2757         0         0         0
         0   58.0035         0         0
         0         0   58.0767         0
         0         0         0   71.2799

Undefined function or variable 'phi1'.

Error in Truss99 (line 3)
G1=phi1'*M*[1;0;1;0]

>> Truss99

V =

   -0.0102    0.0006    0.0003    0.0102
    0.0003   -0.0102    0.0102    0.0006
   -0.0102   -0.0006    0.0003   -0.0102
   -0.0003   -0.0102   -0.0102    0.0006


lamda =

   1.0e+05 *

    0.0502         0         0         0
         0    1.3282         0         0
         0         0    1.3316         0
         0         0         0    2.0058
```

```
f =

11.2757          0          0          0
      0    58.0035          0          0
      0          0    58.0767          0
      0          0          0    71.2799


phi1 =

   -0.0102
    0.0003
   -0.0102
   -0.0003


phi2 =

    0.0006
   -0.0102
   -0.0006
   -0.0102

phi3 =

    0.0003
    0.0102
    0.0003
   -0.0102


phi4 =

    0.0102
    0.0006
   -0.0102
    0.0006


G1 =

-97.6791


G2 =

  -97.5648


G3 =

  -3.9933e-12


G4 =
```

```
     0

>> Truss99

V =

-0.0102    0.0006    0.0003    0.0102
    0.0003   -0.0102    0.0102    0.0006
   -0.0102   -0.0006    0.0003   -0.0102
   -0.0003   -0.0102   -0.0102    0.0006


lamda =

   1.0e+05 *

    0.0502         0         0         0
         0    1.3282         0         0
         0         0    1.3316         0
         0         0         0    2.0058


f =

   11.2757         0         0         0
         0   58.0035         0         0
         0         0   58.0767         0
         0         0         0   71.2799


phi1 =

   -0.0102
    0.0003
   -0.0102
   -0.0003


phi2 =

0.0006
   -0.0102
   -0.0006
   -0.0102


phi3 =

    0.0003
    0.0102
    0.0003
   -0.0102


phi4 =

    0.0102
    0.0006
   -0.0102
    0.0006


G1 =

  -97.6791


G2 =

  -97.5648


G3 =

  -3.9933e-12


G4 =

     0


M1 =

   9.5412e+03


M2 =

   9.5189e+03


M3 =

   1.5946e-23

>> Truss99

V =

   -0.0102    0.0006    0.0003    0.0102
    0.0003   -0.0102    0.0102    0.0006
   -0.0102   -0.0006    0.0003   -0.0102
   -0.0003   -0.0102   -0.0102    0.0006


lamda =

   1.0e+05 *

    0.0502         0         0         0
```

```
         0    1.3282         0         0
         0         0    1.3316         0
         0         0         0    2.0058


f =

   11.2757         0         0         0
         0   58.0035         0         0
         0         0   58.0767         0
         0         0         0   71.2799


phi1 =

   -0.0102
    0.0003
   -0.0102
   -0.0003


phi2 =

    0.0006
   -0.0102
   -0.0006
   -0.0102


phi3 =

    0.0003
    0.0102
    0.0003
   -0.0102


phi4 =

    0.0102
    0.0006
   -0.0102
    0.0006


G1 =

  -97.6791


G2 =

 -97.5648

G3 =

  -3.9933e-12


G4 =

     0


Total_mass =

   1.9100e+04


M1 =

   49.9548


M2 =

   49.8380


M3 =

   8.3489e-26

>> Truss99

V =

   -0.0102    0.0006    0.0003    0.0102
    0.0003   -0.0102    0.0102    0.0006
   -0.0102   -0.0006    0.0003   -0.0102
   -0.0003   -0.0102   -0.0102    0.0006


lamda =

   1.0e+05 *

    0.0502         0         0         0
         0    1.3282         0         0
         0         0    1.3316         0
         0         0         0    2.0058


f =

   11.2757         0         0         0
         0   58.0035         0         0
         0         0   58.0767         0
         0         0         0   71.2799
```

```
phi1 =

   -0.0102
    0.0003
   -0.0102
   -0.0003


phi2 =

    0.0006
   -0.0102
   -0.0006
   -0.0102


phi3 =

    0.0003
    0.0102
    0.0003
   -0.0102


phi4 =

    0.0102
    0.0006
   -0.0102
    0.0006


G1 =

  -97.6791


G2 =

-97.5648


G3 =

  -3.9933e-12


G4 =

     0


Total_mass =

   9.5498e+03
```

```
M1 =

   99.9096


M2 =

   99.6759


M3 =

   1.6698e-25

>> Truss99

V =

   -0.0102    0.0006    0.0003    0.0102
    0.0003   -0.0102    0.0102    0.0006
   -0.0102   -0.0006    0.0003   -0.0102
   -0.0003   -0.0102   -0.0102    0.0006


lamda =

   1.0e+05 *

    0.0502         0         0         0
         0    1.3282         0         0
         0         0    1.3316         0
         0         0         0    2.0058


f =

   11.2757         0         0         0
         0   58.0035         0         0
         0         0   58.0767         0
         0         0         0   71.2799


phi1 =

   -0.0102
    0.0003
   -0.0102
   -0.0003


phi2 =

    0.0006
   -0.0102
   -0.0006
```

```
   -0.0102


phi3 =

    0.0003
    0.0102
    0.0003
   -0.0102


phi4 =

    0.0102
    0.0006
   -0.0102
    0.0006


G1 =

  -97.6791


G2 =

-97.5648


G3 =

  -3.9933e-12


G4 =

     0


Total_mass =

   9.5498e+03


M1 =

   99.9096


M2 =

   99.6759


M3 =
```

```
   1.6698e-25

>> Truss99

V =

   -0.0102    0.0006    0.0003    0.0102
    0.0003   -0.0102    0.0102    0.0006
   -0.0102   -0.0006    0.0003   -0.0102
   -0.0003   -0.0102   -0.0102    0.0006


lamda =

   1.0e+05 *

    0.0502         0         0         0
         0    1.3282         0         0
         0         0    1.3316         0
         0         0         0    2.0058


f =

   11.2757         0         0         0
         0   58.0035         0         0
         0         0   58.0767         0
         0         0         0   71.2799


phi1 =

   -0.0102
    0.0003
   -0.0102
   -0.0003


phi2 =

    0.0006
   -0.0102
   -0.0006
   -0.0102


phi3 =

    0.0003
    0.0102
    0.0003
   -0.0102


phi4 =
```

```
    0.0102
    0.0006
   -0.0102
    0.0006


G1 =

  -97.6791


G2 =

-97.5648


G3 =

  -3.9933e-12


G4 =

     0


Total_mass =

   9.5498e+03


M1 =

   99.9096


M2 =

   99.6759


M3 =

   1.6698e-25


Q1 =

   1.0e+03 *

    3.1343   -0.0943    3.1343    0.0943

>> Truss99

V =

   -0.0102    0.0006    0.0003    0.0102
    0.0003   -0.0102    0.0102    0.0006
   -0.0102   -0.0006    0.0003   -0.0102
   -0.0003   -0.0102   -0.0102    0.0006


lamda =

   1.0e+05 *

    0.0502         0         0         0
         0    1.3282         0         0
         0         0    1.3316         0
         0         0         0    2.0058


f =

   11.2757         0         0         0
         0   58.0035         0         0
         0         0   58.0767         0
         0         0         0   71.2799


phi1 =

   -0.0102
    0.0003
   -0.0102
   -0.0003


phi2 =

    0.0006
   -0.0102
   -0.0006
   -0.0102


phi3 =

    0.0003
    0.0102
    0.0003
   -0.0102


phi4 =

    0.0102
    0.0006
   -0.0102
    0.0006
```

```
G1 =

  -97.6791


G2 =

-97.5648


G3 =

  -3.9933e-12


G4 =

     0


Total_mass =

   9.5498e+03


M1 =

   99.9096


M2 =

   99.6759


M3 =

   1.6698e-25


Q1 =

   1.0e+03 *

    2.9005   -0.0873    2.9005    0.0873

>> Truss99

V =

   -0.0102    0.0006    0.0003    0.0102
    0.0003   -0.0102    0.0102    0.0006
   -0.0102   -0.0006    0.0003   -0.0102
   -0.0003   -0.0102   -0.0102    0.0006


lamda =

   1.0e+05 *

    0.0502         0         0         0
         0    1.3282         0         0
         0         0    1.3316         0
         0         0         0    2.0058


f =

   11.2757         0         0         0
         0   58.0035         0         0
         0         0   58.0767         0
         0         0         0   71.2799


phi1 =

   -0.0102
    0.0003
   -0.0102
   -0.0003


phi2 =

    0.0006
   -0.0102
   -0.0006
   -0.0102


phi3 =

    0.0003
    0.0102
    0.0003
   -0.0102


phi4 =

    0.0102
    0.0006
   -0.0102
    0.0006


G1 =

  -97.6791


G2 =

-97.5648


G3 =
```

```
  -3.9933e-12

G4 =

     0

Total_mass =

   9.5498e+03

M1 =

   99.9096

M2 =

   99.6759

M3 =

   1.6698e-25

Q1 =

   1.0e+03 *

    2.9005   -0.0873    2.9005    0.0873

>> P1=P

P1 =

           0
      -45000
           0
      -45000

>> P1=Q1

P1 =

   1.0e+03 *

    2.9005   -0.0873    2.9005    0.0873

>> P1=P1'

P1 =

   1.0e+03 *

    2.9005
   -0.0873
    2.9005
    0.0873
```

```
>> u1=inv(K)*P1

u1 =

   1.0e-03 *

    0.1210
   -0.0036
    0.1210
    0.0036

Undefined function or variable 'unnamed'.

>> Truss_forces

theta =

    0.6435
s =

    0.6000

c =

0.8000


T =

    0.8000    0.6000         0         0
   -0.6000    0.8000         0         0
         0         0    0.8000    0.6000
         0         0   -0.6000    0.8000

>> T*u2

ans =

   1.0e-04 *

         0
         0
    0.9463
   -0.7553

>> Truss_forces

A =

   9.5800e-04


E =

   2.0000e+11
```

```
L =

     5


k =

    38320000           0   -38320000           0
           0           0           0           0
   -38320000           0    38320000           0
           0           0           0           0


theta =

    0.6435


s =


    0.6000

c =

    0.8000


T =

    0.8000    0.6000         0         0
   -0.6000    0.8000         0         0
         0         0    0.8000    0.6000
         0         0   -0.6000    0.8000

>> Truss_forces

A =

   9.5800e-04


E =

   2.0000e+11


L =

     5


k =

    38320000           0   -38320000           0
           0           0           0           0
   -38320000           0    38320000           0
           0           0           0           0
```

```
theta =

    0.6435


s =

    0.6000


c =

    0.8000


T =


    0.8000    0.6000         0         0
   -0.6000    0.8000         0         0
         0         0    0.8000    0.6000
         0         0   -0.6000    0.8000
ans =

   1.0e+03 *

   -3.6264
         0
    3.6264
         0

>> Truss_forces

A =

   9.5800e-04


E =

   2.0000e+11


L =

     5


k =

    38320000           0   -38320000           0
           0           0           0           0
   -38320000           0    38320000           0
           0           0           0           0
```

```
theta =

    0.6435


s =

    0.6000


c =

    0.8000


T =

     0.8000    0.6000         0         0
    -0.6000    0.8000         0         0
          0         0    0.8000    0.6000
          0         0   -0.6000    0.8000


f6 =

   1.0e+03 *

   -3.6264
         0
    3.6264
         0

>>
```

References

1. IS 1893 (Part 1): 2002, Criteria for Earthquake Resistant Design of Structures. Bureau of Indian Standards, New Delhi
2. Jain SK (1995) Explanatory examples on Indian Seismic Code IS 1893 (Part I), IIT Kanpur, Document No. IITK-GSDMA-EQ21-V2.0, IITK-GSDMA Project on Building Codes
3. IS: 875 (Part 1)—1987, Code of practice for design loads (other than earthquake) for buildings and structures. Bureau of Indian Standards, New Delhi
4. IS: 875 (Part 2)—1987, Code of practice for design loads (other than earthquake) for buildings and structures. Bureau of Indian Standards, New Delhi
5. IS: 875 (Part 3)—1987, Code of practice for design loads (Other than earthquake) for buildings and structures. Bureau of Indian Standards, New Delhi
6. IS 800: 1984, code of practice for general construction in steel, second revision. Bureau of Indian Standards, New Delhi
7. IS 800: 2007, "General Construction in steel - Code of practice", third revision, Bureau of Indian Standards, New Delhi
8. IS: 226: 1975, Structural steel (standard quality), 5th ed. Bureau of Indian Standards, New Delhi
9. Duggal SK (2016) Limit state design of steel structures, 2nd ed. Tata Mcgraw Hill, Inc. ISBN:13:978-93-5143-349-3
10. Reddy GD (1998) Advanced approaches for the seismic analysis of Nuclear Power Plant Structures, equipment and piping systems. Ph. D. thesis, Tokyo Metropolitan University, Japan
11. Logan DL (2010) A first course in finite element method, 5th ed. Cengage Learning. ISBN:13: 978-0-495-66825-1

9 Generation of Floor Response Spectra and Multi-support Excitations

G. R. Reddy and R. K. Verma

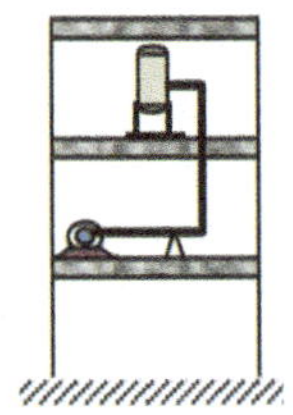

Floor Response spectrum is the filtered motion for design of floor mounted equipment and piping systems

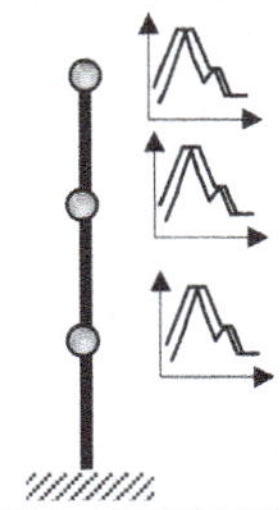

Symbols

PSD	power spectral density
ϕ_i, ϕ_j	Mode shapes
H_i, H_j	Transfer functions
S(ω)	Design power spectral density
M	Mass matrix
K	Stiffness matrix
C	damping coefficient matrix
Γ_{sm}	participation factor

9.1 Introduction

Industrial Systems and Components (SCs) are supported at different elevations of structure. These SCs are designed using floor time history (FTH) or floor response spectra (FRS). From structural analysis, FRS is obtained at various floor levels and at the locations where SCs are supported. FRS is generated for damping of the SCs using time history analysis, stochastic analysis, or direct simplified methods.

9.2 FRS Generation

Generally, time history (time domain) methods are used for generating FRS from FTH. Frequency domain methods, called stochastic methods, are also used sometimes for generating FRS. Designers generally used former method because it is straightforward and simple. For conservative design of SCs, direct method which is simple and less time consuming can be adopted. Details of these methods are discussed in the following sections.

G. R. Reddy (✉) · R. K. Verma
Bhabha Atomic Research Centre, Mumbai, India
e-mail: rssred@barc.gov.in

R. K. Verma
e-mail: rajeev@barc.gov.in

G. R. Reddy et al., *Textbook of Seismic Design*,
https://doi.org/10.1007/978-981-13-3176-3_9

Table 9.1 Frequency steps for FRS generation

Frequency range (Hz)	Increment (Hz)
0.5–3.0	0.10
3.0–3.6	0.15
3.6–5.0	0.20
5.0–8.0	0.25
8.0–15.0	0.50
15.0–18.0	1.0
18.0–22.0	2.0
22.0–34.0	3.0

9.2.1 Time History Analysis

The various steps involved in time history analysis are given below:

1. Generate design basis ground motion called design basis time history.
2. Generate mathematical model of the structure. The model could be beam model or 3D finite element (FE)model.
3. Generate floor time histories from the structural analysis using design basis time history.
4. Generate FRS using floor time histories. While generating FRS, the spectrum ordinates shall be computed at sufficiently small frequency intervals to produce accurate response spectra, including significant peaks normally expected at the natural frequencies of the structure. One acceptable frequency interval to compute FRS is at frequencies listed in Table 9.1 [1]. In addition, it is suggested to include the frequencies of the structure also.

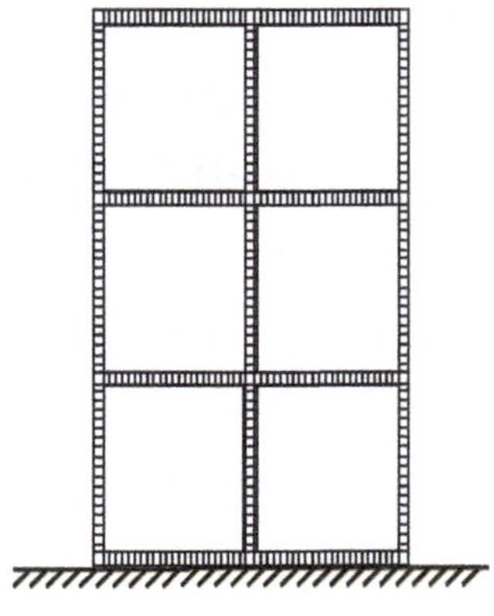

Fig. 9.1 Three-story RCC Structure

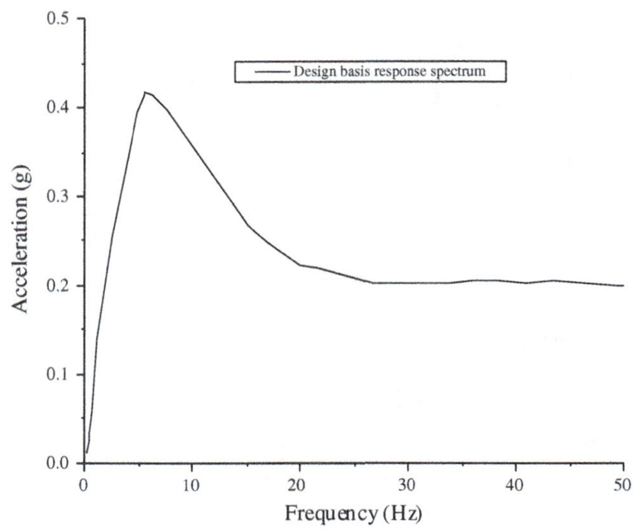

Fig. 9.2 Design basis response spectrum

Example 9.1 Generate FRS for a three-story RCC structure as shown in Fig. 9.1 by time history analysis. The geometrical details and material properties of the structure are given below:

Column sizes—100 × 75 mm.
Modulus of elasticity of concrete (E) = 2.5×10^{10} N/m^2.
Moment of inertia of the column section (I) = 3.52×10^{-6} m^4.

Solution:

Step 1: **Generate design basis time history**

A compatible time history has been generated from design basis response spectrum (DBRS) shown in Fig. 9.2 using the procedure explained in Chap. 12. Data for the graph is given in Appendix 1.

Compatible time history has been shown in Fig. 9.3 (data for the graph is given in Appendix 1). Its compatibility has been checked as per IEEE 344 standard and shown in Fig. 9.4 [2]. It can be seen that in most of the frequency ranges it is enveloping the DBRS.

Step 2: **Generate mathematical model of the structure**

Finite element model of the structure has been generated and is shown in Fig. 9.5a for horizontal excitation and Fig. 9.5b for vertical direction. For further detailed analysis, Fig. 9.5a is considered.

Step 3: **Generate floor time histories by time history analysis**

Floor time histories at different floors of the structure have been generated by time history analysis as explained in

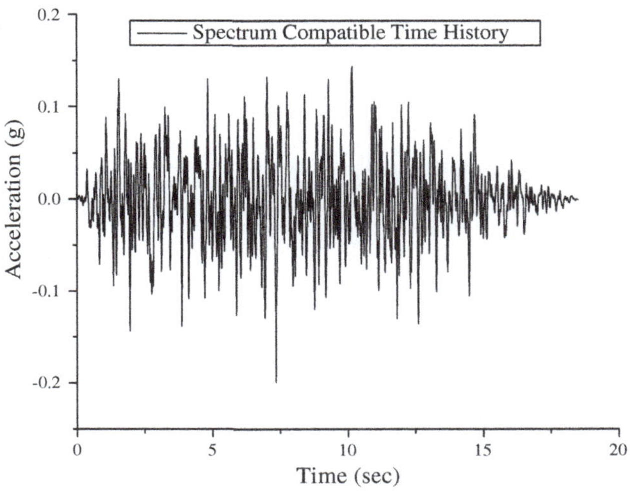

Fig. 9.3 Compatible time history

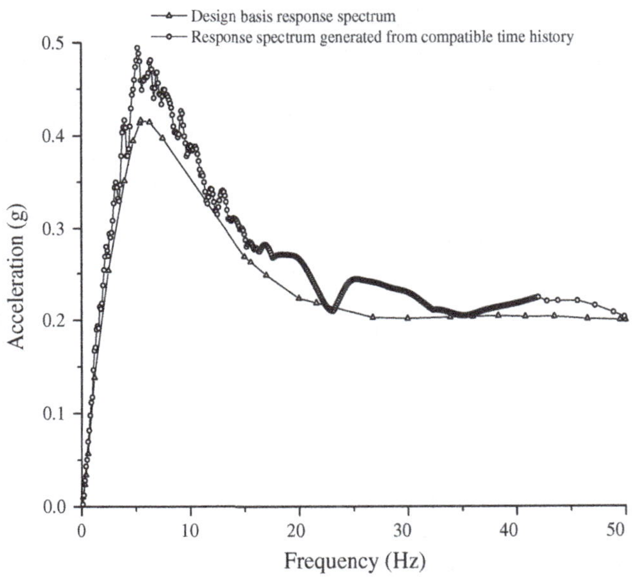

Fig. 9.4 Compatibility of time history as per IEEE Std 344

Fig. 9.5 Finite element model of structure

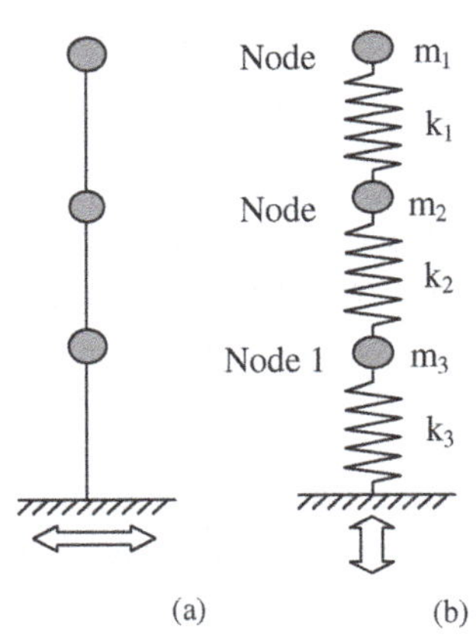

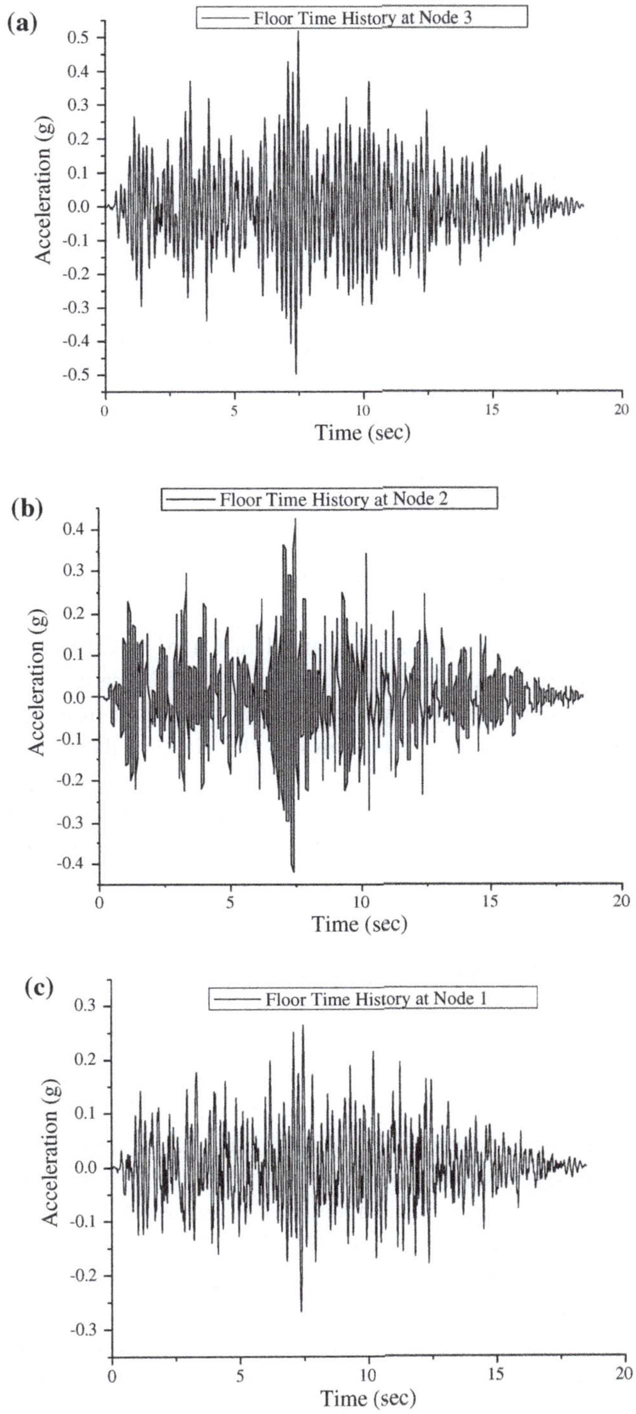

Fig. 9.6 Floor time histories at **a** Node 3 **b** Node 2 **c** Node 1

Chap. 4. Compatible time history has been used as input for time history analysis. Analysis has been performed by considering 7% damping in RCC structure. Floor time histories obtained are shown in Fig. 9.6a–c.

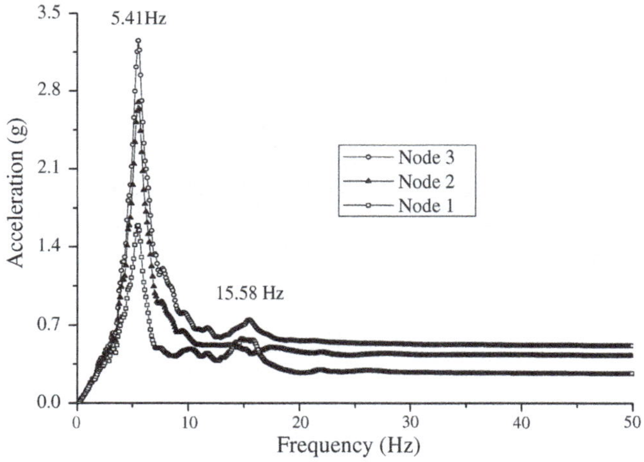

Fig. 9.7 Floor response spectra at 5% damping by time history analysis

Step 4: **Generate FRS using floor time histories**

Using the procedure explained in Chap. 1, FRS at different floor levels of the structure at 5% damping of equipment or piping system has been generated from floor time histories and is shown in Fig. 9.7 (data for the graph is given in Appendix 1). From this figure, it is important to note that the frequencies at which peaks are occurring are around 5.41 and 15.58 Hz. These are natural frequencies of the structure as estimated in Chap. 4.

9.2.2 Stochastic Analysis

The various steps involved in stochastic method are given below:

1. Generate design basis ground motion called design basis power spectral density function (PSDF).
2. Generate mathematical model of the structure. The model could be beam model or 3D FE model.
3. Generate floor power spectral density function from the structural analysis using design basis PSDF.
4. Generate FRS using floor PSDF. The frequency intervals shall be chosen as explained in the previous subsection.

Example 9.2 Generate FRS for a three-story RCC structure shown in Fig. 9.1 by stochastic analysis.

Solution:

Step 1: **Generate power spectral density function (PSDF)**

Power spectral density function compatible to design response spectrum shown in Fig. 9.1 is shown in Fig. 9.8 (data for the graph is given in Appendix 1). Chaps. 1 and 12 explain the method to generate the power spectral density function. Using this procedure, PSDF can be generated for compatible time histories shown in Fig. 9.6. The procedure as per the ASCE 4-98 [1] used to accept the time histories shall be extended to the PSDF also. One procedure is averaging as follows.

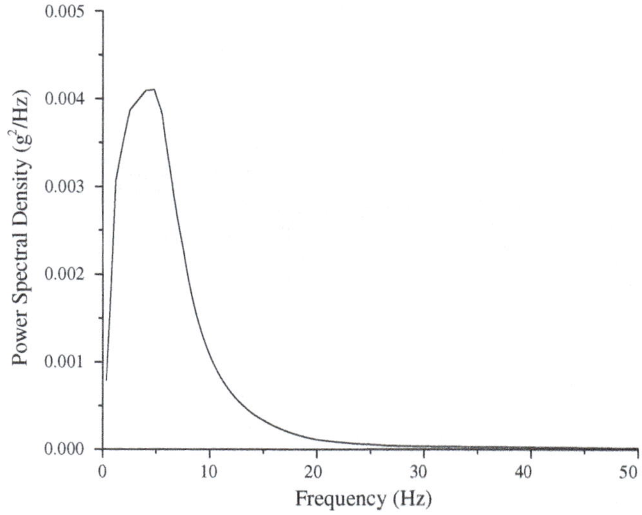

Fig. 9.8 Design response spectrum compatible PSD

$$\text{Compatible PSDF } G(\omega_j) = \frac{\sum_{i=1}^{N} G_i(\omega_j)}{N} \tag{9.1}$$

where

$G_i(\omega_j)$ is the PSDF of ith record at jth frequency.

Also smoothed PSDF of the ground motion proposed by Kanai-Tajimi [6] or modified K-T PSDF presented by Clough and Penzien [8] may also be used. Alternately power spectral density function may also be generated directly from response spectrum [3, 4]. Details of generating PSDF are explained in Appendix 2.

Step 2: **Generate mathematical model of the structure**

Finite element model of the structure has been generated and is shown in Fig. 9.5.

Step 3: **Generate floor power spectral density function**

Power spectral density functions at different floors have been generated using the following relation:

$$S_{oi}(\omega_l) = \sum_{j=1}^{n}\sum_{k=1}^{n} \phi_{ij}\phi_{ik}\Gamma_j\Gamma_k H_j H_k^* S_{in}(\omega_l) \tag{9.2}$$

Note: Since the structure has been idealized as three degree of freedom system, j and k will vary from 1 to 3.

where

ϕ_{ij} and ϕ_{ik}	Mode shapes of *i*th node in *j*th and *k*th modes
Γ_j and Γ_k	Mass participation factors in *j*th and *k*th modes
H_i and H_j	*j*th and *k*th mode transfer functions
$S_{\text{in}}\ (\omega_l)$	Input power spectral density at *l*th frequency
$S_{\text{oi}}\ (\omega_l)$	Output power spectral density at *i*th node and at *l*th frequency

Mode shapes of the structure may be obtained by anyone of the methods explained in Chap. 4. Transfer function H_j and H_k are given as

$$H_j = \frac{1}{\sqrt{\left[\left\{1-\left(\frac{\omega_l}{\omega_j}\right)^2\right\}^2+\left(2\xi\frac{\omega_l}{\omega_j}\right)^2\right]}} \tag{9.3}$$

$$H_k = \frac{1}{\sqrt{\left[\left\{1-\left(\frac{\omega_l}{\omega_k}\right)^2\right\}^2+\left(2\xi\frac{\omega_l}{\omega_k}\right)^2\right]}} \tag{9.4}$$

Using the above relation for multi-degree of freedom system, floor PSDs have been obtained and are shown in Fig. 9.9. Details of the calculation are given in Appendix 2 for $j = 1$ and $k = 1$. Similarly by varying j and k from 1 to 3 and considering all the values at each frequency, floor PSD can be obtained.

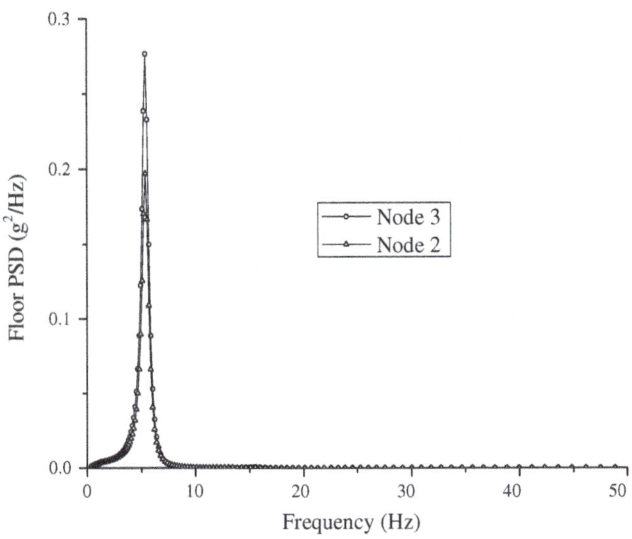

Fig. 9.9 Floor PSD

Step 4: **Generate FRS using floor PSDF**

Once the floor PSDF is obtained, it can be converted to FRS by calculating the maximum response of single-degree of freedom systems using the PSDF as input. Mean square response of a single-degree of freedom system of frequency ω subjected to random vibration expressed in terms of PSDF is given as

$$\ddot{x}^2(\omega_l) = \sum_{j=1}^{n} S_{ol}(\omega_j) \times |H(\omega_j)| \times \Delta f \tag{9.5}$$

Maximum response of single-degree of freedom system = 3 × Mean square response

FRS at different floor levels of structure at 5% damping has been generated from floor PSDs using Eq. (9.5) and is shown in Fig. 9.10.

9.2.3 Simplified Analysis

In simplified procedure, the floor response spectrum may be obtained directly from the two design ground spectra which correspond to the damping value of the structure and the equipment or piping systems, using modal characteristics of the structure obtained in modal analysis. The various steps involved in the simplified analysis are given below:

1. Obtain the design basis ground motion called design basis response spectra corresponding to the damping value of the structure and the equipment or piping systems.
2. Generate mathematical model of the structure. The model could be beam model or 3D FE model.
3. Obtain the eigenvalues and eigenvectors of the structure by modal analysis.
4. Generate FRS based on the procedure outlined in Example 9.3 by using the eigenvalues and eigenvectors of the structure [4].

Spectral acceleration at *i*th building mode and at *j*thnatural frequency of the equipment is given as follows.

$$\text{SA}_{ij} = \frac{1}{\sqrt{\left\{1-\left(\frac{\omega_{\text{E}j}}{\omega_{\text{B}i}}\right)^2\right\}^2+4(\xi_{\text{E}j}+\xi_{\text{B}i})^2\left(\frac{\omega_{\text{E}j}}{\omega_{\text{B}i}}\right)^2}} \times \sqrt{\left\{\left(\frac{\omega_{\text{E}j}}{\omega_{\text{B}i}}\right)^2 \text{SA}(\omega_{\text{B}i},\xi_{\text{B}i})\right\}^2+\text{SA}(\omega_{\text{E}j},\xi_{\text{E}j})^2} \tag{9.6}$$

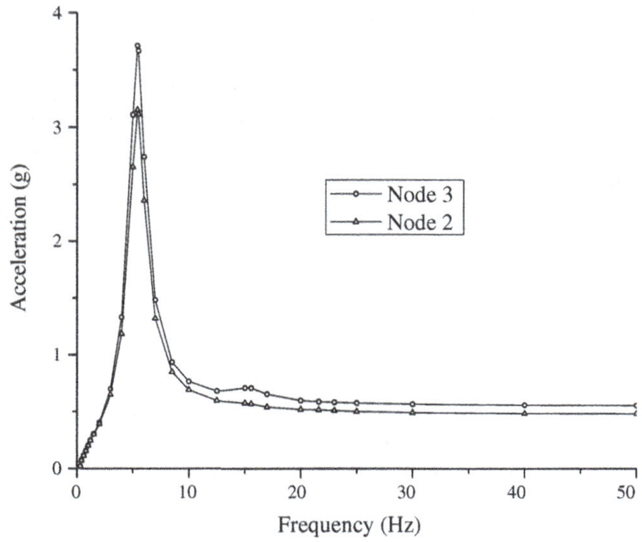

Fig. 9.10 Floor response spectra at 5% damping by stochastic analysis

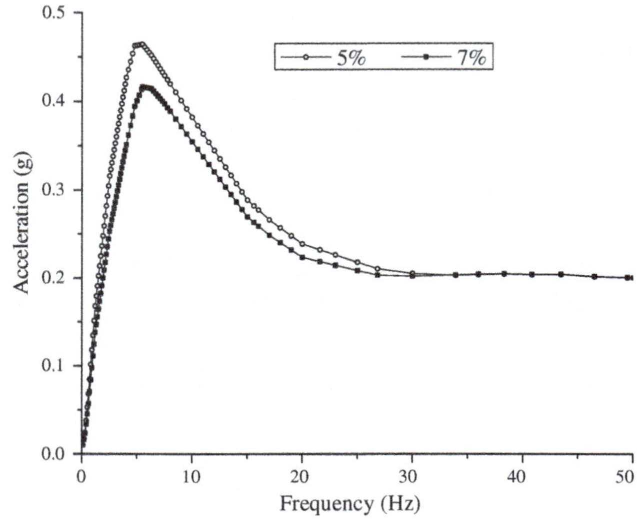

Fig. 9.11 Typical design response spectra at 5 and 7% damping

$$\mathrm{SA}_j = \sqrt{\sum_i^n \left(\Gamma_i \phi_{ik} \times \mathrm{SA}_{ij}\right)} \qquad (9.7)$$

where

SA_j	Floor response spectrum value at *j*th frequency of the equipment or system taking into account all building modes ($i = 1$ to n)
Γ_i	The *i*th modal participation factor
ϕ_{ik}	*k*th floor mode shape in *i*th mode
ζ_{Ej}	Damping factor of equipment or system at *j*th frequency
ω_{Ej}	*j*th frequency of the equipment or system
ζ_{Bi}	Damping factor of the structure in *i*th mode
ω_{Bi}	*i*th modal frequency of the building
$\mathrm{SA}(\omega_{Bi}, \zeta_{Bi})$	The standard design ground spectral value corresponding to ω_{Bi}, ζ_{Bi} of the building.
$\mathrm{SA}(\omega_{Ej}, \zeta_{Ej})$	The standard design ground spectral value corresponding to ω_{Ej}, ζ_{Ej} of the equipment or systems

Notes:

(1) The mass m_A of the systems and components needs to be sufficiently smaller than the mass m_{Bi} of the structure.

(2) The floor response spectra, obtained from the above method, need to be broadened by at least ±15% to account for the uncertainty of frequency analysis of SSCs.

Example 9.3 Generate FRS for a three-story RCC structure as shown in Fig. 9.1 by simplified analysis.

Solution:

Step 1: **Obtain the design basis response spectra**

Design basis response spectra corresponding to damping of structure and components/systems are shown in Fig. 9.11 (data for the graph is given in Appendix 1).

Step 2: **Generate mathematical model of the structure**

Finite element model of structure has been generated and is shown in Fig. 9.5.

Step 3: **Obtain the eigenvalues and eigenvectors by modal analysis**

Eigenvalues (frequencies) and eigenvectors (mode shapes) can be obtained by using anyone of the methods explained in Chap. 4. Natural frequencies, mode shapes, and participation factors of the structure are obtained as

$$f_1 = 5.41\ \text{Hz}, \quad f_2 = 15.58\ \text{Hz}, \quad f_3 = 21.61\ \text{Hz}$$

$$\phi_1 = \begin{Bmatrix} 0.00695 \\ 0.01263 \\ 0.01545 \end{Bmatrix} \quad \phi_2 = \begin{Bmatrix} 0.01492 \\ 0.00724 \\ -0.01407 \end{Bmatrix}$$

$$\phi_3 = \begin{Bmatrix} 0.01465 \\ -0.01337 \\ 0.00699 \end{Bmatrix}$$

$$\Gamma_1 = 78.492, \quad \Gamma_2 = 20.286, \quad \Gamma_3 = 10.363$$

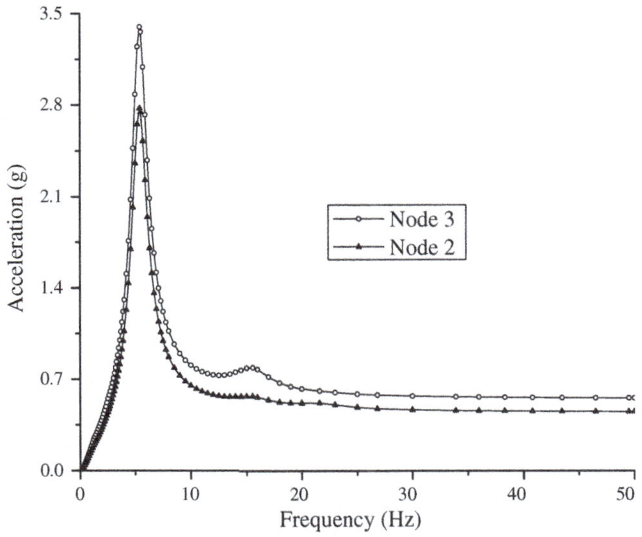

Fig. 9.12 Floor response spectra at 5% damping by simplified analysis

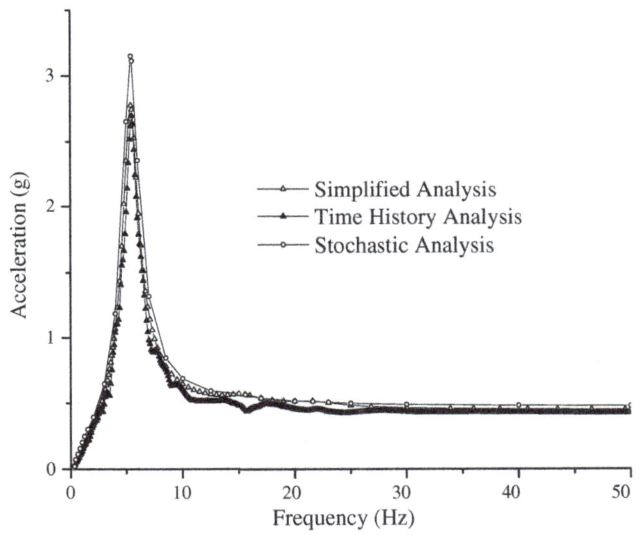

Fig. 9.13 FRS generated by simplified, time history and stochastic analysis at Node 2

Step 4: **Generate FRS**

Using the above procedure, FRS at the two levels of the structure are generated. Spectra have been generated for 7% damping for the structure and 5% damping for the components. FRS generated by simplified analysis are shown in Fig. 9.12.

FRS generated by all the three methods at node 2 and 3 are shown in Figs. 9.13 and 9.14, respectively. It can be seen that the spectra generated using stochastic method are more conservative.

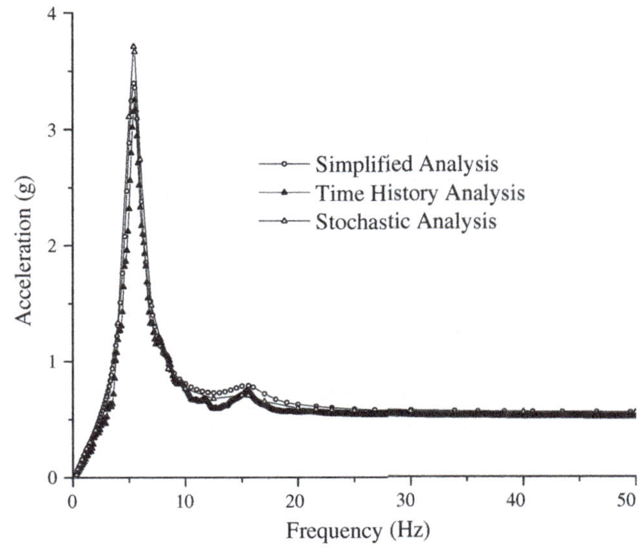

Fig. 9.14 FRS generated by simplified, time history and stochastic analysis at Node 3

9.3 Peak Broadening of Floor Response Spectra

The floor response spectra generated using any one of the methods explained above are smoothed and peaks associated with the structural frequencies are broadened to account for the structural frequency variation due to possible uncertainties in the soil–structure interaction, equipment–structure interaction and the approximations in the modeling technique used in the seismic analysis. Generally, raw spectrum will have peaks at natural frequencies of the structure. In the present case, the fundamental frequency is 5.41 Hz and second frequency is 15.58 Hz, and hence, two distinct peaks at these frequencies are seen. Due to the above uncertainties, these frequencies may shift by 15% and hence locations of the peaks in the raw spectrum may shift within these limits. Hence, as per ASME Section III Division 1 Appendix N, the floor response spectra need to be broadened by at least 15% to account for above uncertainties [5]. The computed and broadened FRS are shown in Fig. 9.15. Steps of spectra broadening are summarized below.

a. Identify the peaks of response spectrum based on the eigenvalues (natural frequencies) of the structure. Conservatively all the peaks of the spectrum may be considered, or precisely, modal frequencies having more than 10% mass participation may be considered.
b. Broaden the peaks by 15% (i.e., draw horizontal line).
c. Extend the end of this line by drawing lines parallel to the raw spectrum.

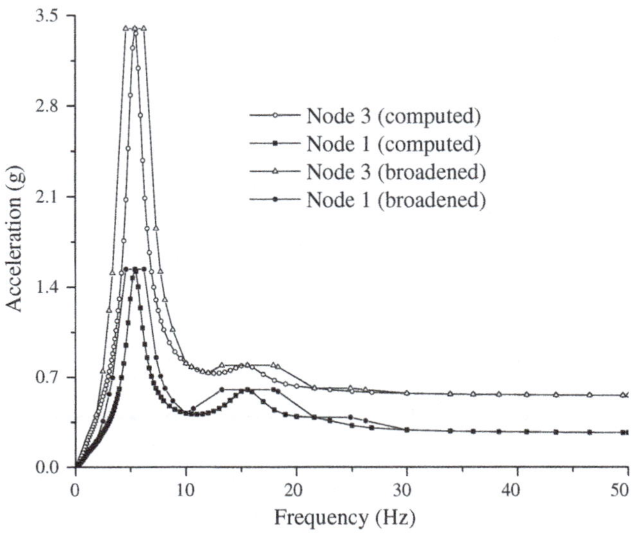

Fig. 9.15 Computed and broadened floor response spectra

d. These lines may intersect between two peaks or intersect raw spectrum or may cut ordinate axis toward zero frequency.

9.4 Analysis of Systems Subjected to Multiple-Support Excitations

Consider a simple spring–mass system as shown in Fig. 9.16. The equation of motion for this system can be written as follows.

$$[M]\{\ddot{x}\} + [C]\{\dot{x}\} + [K]\{x\} = -[M]\ddot{x}_g\{1\} \quad (9.8)$$

where

M is mass matrix (lumped/consistent),
C is damping coefficient matrix, and
K is stiffness matrix

Natural frequencies (ω_n) and mode shapes (ϕ_n) of the system can be obtained by using anyone of the methods explained in Chap. 4.

$$\text{Let } \{x\} = \{\phi\}X \quad (9.9)$$

where

$\{\phi\}$ is the mode shape and
X is the generalized displacement

Substituting Eq. (9.9) into Eq. (9.8), one can get

$$[M]\{\phi\}\ddot{X} + [C]\{\phi\}\dot{X} + [K]\{\phi\}X = -[M]\ddot{x}_g\{1\} \quad (9.10)$$

Multiply Eq. (9.10) with $\{\phi\}^T$ on both sides

Fig. 9.16 Spring–mass system

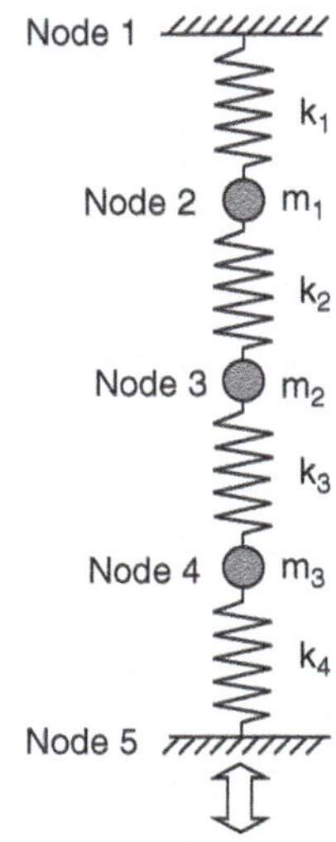

$$\{\phi\}^T[M]\{\phi\}\ddot{X} + \{\phi\}^T[C]\{\phi\}\dot{X} + \{\phi\}^T[K]\{\phi\}X = -\{\phi\}^T[M]\ddot{x}_g\{1\} \quad (9.11)$$

Considering excitations from 's' supports [1, 5], Eq. (9.11) can also be written as

$$\{\phi\}^T[M]\{\phi\}\ddot{X} + \{\phi\}^T[C]\{\phi\}\dot{X} + \{\phi\}^T[K]\{\phi\}X = -\sum_{s=1}^{ns}\{\phi\}^T[M]\ddot{x}_g\{U_s\} \quad (9.12)$$

where U_s is the influence vector. It has unity values along each degree of freedom for uniform support excitation. For the case of multiple-support excitations, it is the displacement vector of the structural system when the support 's' undergoes a unit displacement in the direction of motion of the support, while the other supports remain fixed. Equation 9.12 can be solved by using direct time history method. To solve Eq. 9.12 by using modal superposition technique, Eq. 9.12 can be simplified as follows. Using orthogonal properties of mode shapes, Eq. 9.12 can be written as

$$[1]\ddot{X}_n + [2\varsigma\omega_n]\dot{X}_n + [\omega_n^2]X_n = -\sum_{s=1}^{ns}\Gamma_{sm}\ddot{x}_g \quad (9.13)$$

where Γ_{sm} is the participation factor for support s and mode m. The equation for evaluation of residual rigid response due to the missing mass is changed to Eq. (9.14).

$$[K]\{x_r\} = -[M]\sum_{s=1}^{ns}\left\{\{U_s\} - \sum_{i=1}^{m}\Gamma_{si}\{\phi_i\}\right\}\ddot{x}_g \quad (9.14)$$

Example 9.4 Consider a piping system running from floor levels 1 to 3 of a structure shown in Fig. 9.1. Its simplified version is shown in Fig. 9.17a. This piping will be subjected to the support motion (FRS) at level 1 and at level 3.

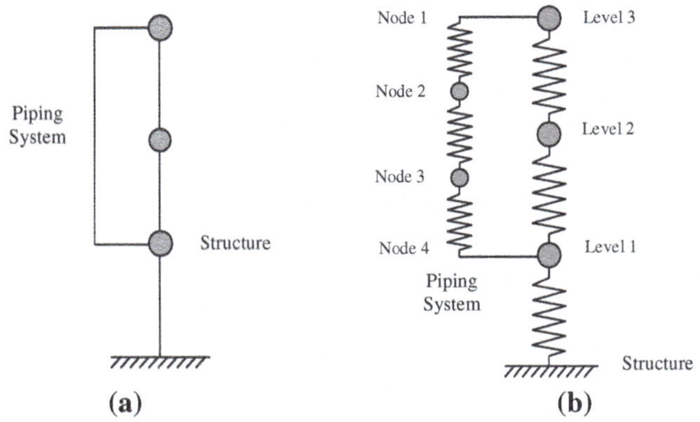

Fig. 9.17 **a** Piping system supported at two locations; **b** finite element model

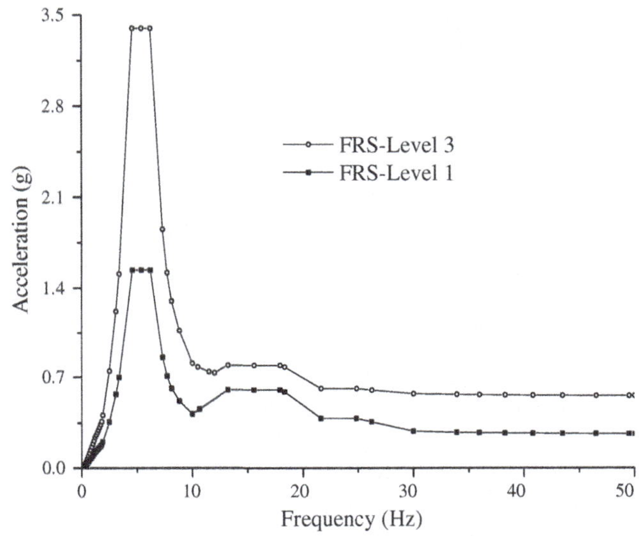

Fig. 9.18 FRS at Level 1 and Level 3 of the Structure

Evaluate the response of piping system considering multi-support excitation.

Finite element model of the structure and piping system is shown in Fig. 9.17b for coupled analysis. Since the mass of the piping system is small as compared to the structure, the piping system can be decoupled as per ASCE 4-98 and response can be obtained using the FRS of the two locations [1]. The spectra at the two locations are shown in Fig. 9.18.

Solution:

The piping system has been idealized as two-degree of freedom system. Idealized model for the piping is shown in Fig. 9.19. The frequencies and mode shapes of the system are evaluated using the similar procedure explained in Chap. 4 and are given below.

1. Frequencies obtained are
   17.6 rad/sec (2.8 Hz) in first mode and
   28.3 rad/sec (4.5 Hz) in second mode.
2. The ortho-normalized mode shapes are given below

$$\begin{Bmatrix} \phi_{12} \\ \phi_{13} \end{Bmatrix} = \begin{Bmatrix} 0.261 \\ 0.465 \end{Bmatrix}$$

$$\begin{Bmatrix} \phi_{22} \\ \phi_{23} \end{Bmatrix} = \begin{Bmatrix} 0.657 \\ -0.185 \end{Bmatrix}$$

In the above vectors, the first suffix indicates mode number and the second suffix indicates the node number.

3. The combined stiffness and mass matrix for the full system can be written as follows

$$[\mathrm{K}] = \begin{bmatrix} 1000 & -1000 & & \\ -1000 & 1500 & -500 & \\ & -500 & 1500 & -1000 \\ & & -1000 & 1000 \end{bmatrix}$$

$$[\mathrm{M}] = \begin{bmatrix} 0 & & & \\ & 2 & & \\ & & 4 & \\ & & & 0 \end{bmatrix}$$

4. Applying unit displacement at support 2 (node 1) and at support 1 (node 4), the influence vectors, respectively, can be obtained as follows.

$$\{U_2\} = \begin{Bmatrix} \frac{3}{4} \\ \frac{1}{4} \end{Bmatrix}$$

$$\{U_1\} = \begin{Bmatrix} \frac{1}{4} \\ \frac{3}{4} \end{Bmatrix}$$

The suffix of the influence vector indicates the support number.

5. Now by using Eqs (9.12) and (9.13), the participation factors can be obtained as follows:
   The participation factor for support 1 excitation in mode 1 is

$$\Gamma_{11} = [\phi]^T [M] \{U_1\}$$

$$\Gamma_{11} = [0.261 \quad 0.465]^T \begin{bmatrix} 2 & \\ & 4 \end{bmatrix} \begin{Bmatrix} \frac{1}{4} \\ \frac{3}{4} \end{Bmatrix} = 1.526\ (2.382)$$

Similarly, the participation factor for support 2 excitation in mode 1 is

$$\Gamma_{21} = [\phi]^T [M] \{U_2\}$$

$$\Gamma_{21} = [0.261 \quad 0.465]^T \begin{bmatrix} 2 & \\ & 4 \end{bmatrix} \begin{Bmatrix} \frac{3}{4} \\ \frac{1}{4} \end{Bmatrix} = 0.8565\ (2.382)$$

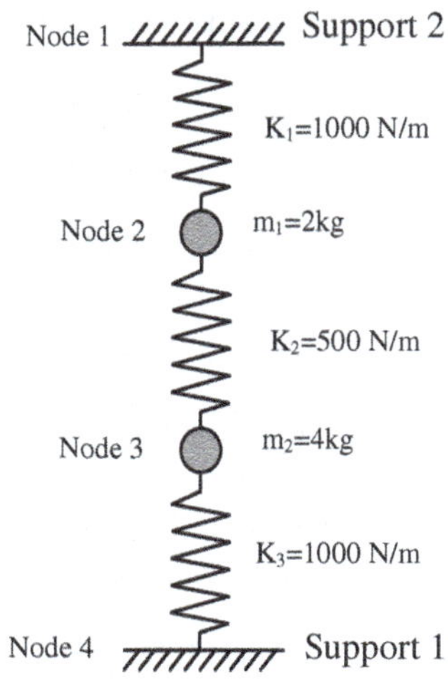

Fig. 9.19 Idealized model for the piping system

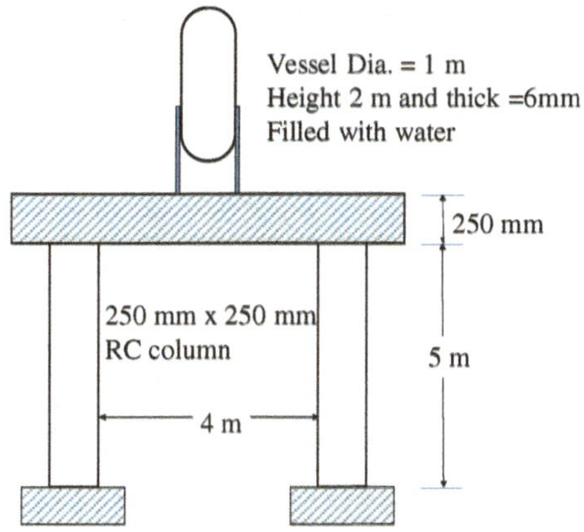

Fig. 9.20 Structure supporting vessel

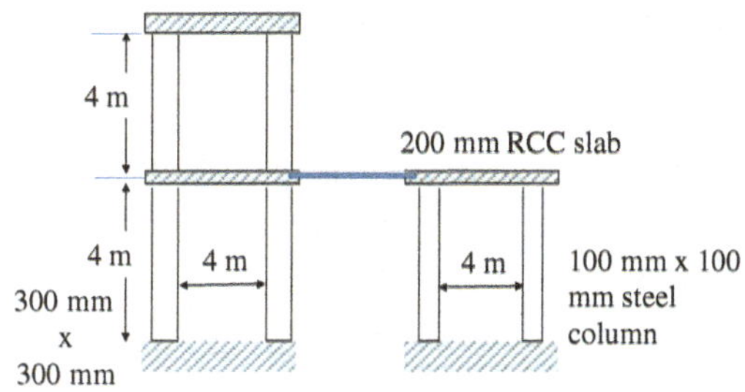

Fig. 9.21 Example on multi-support excitation

The values in the bracket are the participation factors for uniform support excitation.

6. For evaluating the acceleration response in mode 1, spectral acceleration can be obtained from Fig. 9.18 as 0.99 and 0.47 g at level 3(support 2) and level 1 (support 1), respectively. The damping of piping is considered as 5%. Now, the acceleration response at node 2 and 3 can be obtained by using Eq. 9.13 as follows.

$$\ddot{x}_j = -\sum_{i=1}^{m}\sum_{s=1}^{ns}\Gamma_{si}\phi_{ij}X_s$$

Considering one mode, the responses at node 2 and 3 are

$$\begin{aligned}\ddot{x}_2 &= 1.526 \times 0.47g \times 0.261 + 0.8565 \times 0.99g \times 0.261\\ &= 0.409g\ (0.615g)\\ \ddot{x}_3 &= 1.526 \times 0.47g \times 0.465 + 0.8565 \times 0.99g \times 0.465\\ &= 0.728g\ (1.097g)\end{aligned}$$

The values in the bracket are evaluated considering uniform excitation which is basically coming from support 2, and corresponding spectral acceleration is 0.99 g. It is also sometimes called envelope response spectrum analysis. It can be clearly seen that envelope response spectrum analysis gives conservative results compared to the response obtained in multiple-support excitation analysis.

Exercise Problems

1. Explain the significance of peak broadening of floor response spectra.
2. Calculate the acceleration response in mode 2 and total acceleration response for nodes 2 and 3 for the piping system given in Example 9.4.
3. Broaden the floor response spectra generated by time history analysis at node 1 given in Fig. 9.7.
4. Explain the steps to convert the response spectrum to acceleration–displacement spectrum (ADRS).
5. Evaluate the FRS for the design of the vessel supports (Fig. 9.20). Consider the mass of the vessel including water while evaluating spectra. Damping of the structure —5% and Damping of the vessel—3%. Use IS1893 spectra for Zone III.
6. Evaluate the spectra at first floor for Zone III, and using it, evaluate response acceleration of connected structure using multi-support excitation (Fig. 9.21).
7. Evaluate response of a piping system shown in Fig. 9.22 by using multi-support excitation approach, and use the floor response spectra generated in the previous problem. Damping of the piping system is 2%.

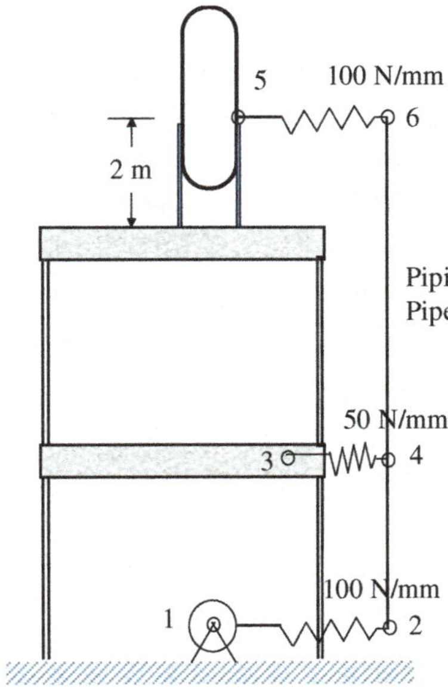

Fig. 9.22 Piping system

Appendix 1: Data for Generation of Floor Response Spectrum

See Tables 9.2, 9.3 and 9.4

Appendix 2: Power Spectral Density Function

A.2.1 Introduction

Power spectral density function (PSDF) may also be generated directly from response spectrum using two methods [3, 4]. The first method is termed as method SIMQKE [3] which is used in the SIMQKE program [3] for artificial motion generation. The second method is termed as method UNRUH_KANA [4], which is described as below.

Table 9.2 Data for design basis response spectrum shown in Fig. 9.2

f (Hz)	*T* (sec)	Acc(*g*)
0.100	10.000	0.010
0.300	3.333	0.023
0.400	2.500	0.034
0.600	1.667	0.057
1.200	0.833	0.138
2.500	0.400	0.253
4.000	0.250	0.351
4.800	0.208	0.395
5.408	0.185	0.414
5.500	0.182	0.417
6.300	0.159	0.415
7.500	0.133	0.398
15.000	0.067	0.269
15.583	0.064	0.263
17.000	0.059	0.248
20.000	0.050	0.223
21.607	0.046	0.218
26.800	0.037	0.203
30.000	0.033	0.202
33.900	0.029	0.203
35.970	0.028	0.204
38.310	0.026	0.205
40.820	0.024	0.204
43.480	0.023	0.204
46.510	0.022	0.201
49.500	0.020	0.200
50.000	0.020	0.200

Power spectral density function (PSDF) $\psi(\omega)$ of single-degree of freedom system subjected to excitation PSDF $\phi(\omega)$ is given as

$$\psi(\omega, \omega_0) = H(\omega, \omega_0) \times H^*(\omega, \omega_0) \times \phi(\omega) \qquad (9.15)$$

where

$$H(\omega, \omega_0) = \left[\frac{\omega_0^2 + 2i\omega_0\,\beta\,\omega}{(\omega_0^2 - \omega^2) + 2i\omega_0\,\beta\,\omega} \right] \qquad (9.16)$$

Table 9.3 Data for compatible time history shown in Fig. 9.3

Time	Acc (g)	Time	Acc (g)	Time	Acc (g)	Time	Acc (g)	Time	Acc (g)	Time	Acc (g)	Time	Acc (g)
0.01	−0.0003	2.66	0.0277	5.31	−0.0166	7.96	−0.0746	10.60	−0.0384	13.24	−0.0102	15.88	−0.0058
0.02	−0.0006	2.67	−0.0095	5.32	−0.0688	7.97	−0.0943	10.61	−0.0014	13.25	−0.0416	15.89	−0.0129
0.03	−0.0003	2.68	−0.0255	5.33	−0.0661	7.98	−0.0916	10.62	0.0094	13.26	−0.0789	15.90	−0.0067
0.04	0.0001	2.69	−0.0653	5.34	−0.0754	7.99	−0.0784	10.63	0.0328	13.27	−0.1006	15.91	−0.0078
0.05	0.0011	2.70	−0.0839	5.35	−0.0384	8.00	−0.0581	10.64	0.0294	13.28	−0.0871	15.92	0.0116
0.06	0.0027	2.71	−0.0909	5.36	−0.0160	8.01	−0.0416	10.65	−0.0105	13.29	−0.0648	15.93	0.0209
0.07	0.0039	2.72	−0.0634	5.37	−0.0065	8.02	−0.0183	10.66	−0.0346	13.30	−0.0054	15.94	0.0314
0.08	0.0036	2.73	−0.0599	5.38	0.0101	8.03	−0.0210	10.67	−0.0603	13.31	0.0106	15.95	0.0259
0.09	0.0029	2.74	−0.0837	5.39	0.0164	8.04	−0.0359	10.68	−0.0758	13.32	−0.0077	15.96	0.0116
0.10	0.0019	2.75	−0.0947	5.40	−0.0128	8.05	−0.0529	10.69	−0.0750	13.33	−0.0153	15.97	0.0008
0.11	0.0019	2.76	−0.1034	5.41	−0.0436	8.06	−0.0835	10.70	−0.0394	13.34	−0.0124	15.98	−0.0116
0.12	0.0005	2.77	−0.0929	5.42	−0.0354	8.07	−0.0517	10.71	−0.0412	13.35	0.0026	15.99	−0.0081
0.13	−0.0024	2.78	−0.1011	5.43	−0.0442	8.08	−0.0160	10.72	−0.0192	13.36	0.0128	16.00	−0.0083
0.14	−0.0014	2.79	−0.0762	5.44	−0.0567	8.09	0.0012	10.73	−0.0066	13.37	0.0333	16.01	−0.0060
0.15	−0.0027	2.80	−0.0669	5.45	−0.0456	8.10	0.0063	10.74	0.0063	13.38	0.0275	16.02	0.0138
0.16	0.0003	2.81	−0.0682	5.46	−0.0418	8.11	−0.0106	10.75	0.0100	13.39	0.0392	16.03	0.0265
0.17	0.0016	2.82	−0.0948	5.47	0.0200	8.12	−0.0067	10.76	0.0256	13.40	0.0232	16.04	0.0427
0.18	0.0031	2.83	−0.0741	5.48	0.0439	8.13	0.0107	10.77	0.0231	13.41	0.0029	16.05	0.0351
0.19	0.0032	2.84	−0.0298	5.49	0.0600	8.14	0.0068	10.78	0.0104	13.42	−0.0122	16.06	0.0282
0.20	−0.0002	2.85	0.0010	5.50	0.0480	8.15	0.0308	10.79	−0.0090	13.43	−0.0154	16.07	0.0230
0.21	−0.0068	2.86	−0.0012	5.51	0.0396	8.16	0.0192	10.80	−0.0240	13.44	−0.0248	16.08	0.0053
0.22	−0.0072	2.87	0.0265	5.52	0.0260	8.17	0.0472	10.81	−0.0584	13.45	−0.0026	16.09	−0.0035
0.23	−0.0043	2.88	0.0559	5.53	0.0148	8.18	0.0290	10.82	−0.0299	13.46	0.0261	16.10	−0.0100
0.24	−0.0022	2.89	0.0544	5.54	0.0023	8.19	−0.0255	10.83	−0.0260	13.47	0.0462	16.11	−0.0097
0.25	−0.0046	2.90	0.0690	5.55	0.0038	8.20	−0.0345	10.84	−0.0256	13.48	0.0529	16.12	−0.0069
0.26	−0.0024	2.91	0.0689	5.56	0.0189	8.21	−0.0226	10.85	0.0198	13.49	0.0160	16.13	0.0018
0.27	0.0000	2.92	0.0595	5.57	0.0357	8.22	−0.0326	10.86	0.0347	13.50	−0.0264	16.14	0.0033
0.28	0.0009	2.93	0.0698	5.58	0.0447	8.23	−0.0275	10.87	0.0731	13.51	−0.0391	16.15	0.0082
0.29	0.0022	2.94	0.0502	5.59	0.0417	8.24	−0.0367	10.88	0.0961	13.52	−0.0421	16.16	0.0026
0.30	−0.0029	2.95	0.0108	5.60	0.0590	8.25	−0.0075	10.89	0.1024	13.53	−0.0383	16.17	0.0003
0.31	−0.0014	2.96	−0.0075	5.61	0.0922	8.26	0.0100	10.90	0.1019	13.54	−0.0246	16.18	0.0006
0.32	−0.0015	2.97	−0.0094	5.62	0.0761	8.27	0.0107	10.91	0.0632	13.55	−0.0203	16.19	0.0065
0.33	0.0029	2.98	−0.0174	5.63	0.0478	8.28	0.0100	10.92	0.0056	13.56	0.0033	16.20	0.0133
0.34	0.0069	2.99	−0.0161	5.64	0.0457	8.29	−0.0006	10.93	−0.0068	13.57	0.0103	16.21	0.0146
0.35	0.0219	3.00	−0.0447	5.65	0.0276	8.30	−0.0324	10.94	−0.0122	13.58	0.0202	16.22	0.0113
0.36	0.0294	3.01	−0.0371	5.66	0.0145	8.31	−0.0525	10.95	0.0198	13.59	0.0075	16.23	0.0044
0.37	0.0328	3.02	−0.0101	5.67	−0.0295	8.32	−0.0603	10.96	0.0523	13.60	0.0073	16.24	−0.0152
0.38	0.0252	3.03	0.0292	5.68	−0.0440	8.33	−0.0292	10.97	0.1055	13.61	−0.0052	16.25	−0.0294
0.39	0.0126	3.04	0.0518	5.69	−0.0215	8.34	−0.0069	10.98	0.1009	13.62	−0.0035	16.26	−0.0253
0.40	0.0054	3.05	0.0763	5.70	0.0187	8.35	−0.0025	10.99	0.0880	13.63	−0.0039	16.27	−0.0099
0.41	−0.0006	3.06	0.0811	5.71	0.0369	8.36	−0.0161	11.00	0.0811	13.64	−0.0123	16.28	0.0046
0.42	−0.0048	3.07	0.0716	5.72	0.0534	8.37	0.0175	11.01	0.0736	13.65	−0.0141	16.29	0.0185
0.43	−0.0124	3.08	0.0424	5.73	0.0231	8.38	0.0351	11.02	0.1014	13.66	−0.0429	16.30	0.0167

(continued)

Table 9.3 (continued)

Time	Acc (*g*)	Time	Acc (*g*)	Time	Acc (*g*)	Time	Acc (*g*)	Time	Acc (*g*)	Time	Acc (*g*)	Time	Acc (*g*)
0.44	−0.0248	3.09	0.0048	5.74	−0.0125	8.39	0.0742	11.03	0.0976	13.67	−0.0731	16.31	0.0278
0.45	−0.0303	3.10	−0.0215	5.75	−0.0303	8.40	0.1085	11.04	0.0857	13.68	−0.0719	16.32	0.0315
0.46	−0.0304	3.11	−0.0503	5.76	−0.0550	8.41	0.1130	11.05	0.0678	13.69	−0.0791	16.33	0.0294
0.47	−0.0306	3.12	−0.0784	5.77	−0.0466	8.42	0.0960	11.06	0.0116	13.70	−0.0514	16.34	0.0274
0.48	−0.0266	3.13	−0.0664	5.78	−0.0208	8.43	0.0391	11.07	−0.0331	13.71	−0.0385	16.35	0.0276
0.49	−0.0241	3.14	−0.0492	5.79	−0.0120	8.44	−0.0162	11.08	−0.0633	13.72	−0.0160	16.36	0.0257
0.50	−0.0225	3.15	−0.0246	5.80	−0.0048	8.45	−0.0563	11.09	−0.0615	13.73	−0.0037	16.37	0.0131
0.51	−0.0314	3.16	−0.0195	5.81	−0.0106	8.46	−0.0789	11.10	−0.0452	13.74	0.0078	16.38	0.0088
0.52	−0.0270	3.17	−0.0139	5.82	−0.0106	8.47	−0.0827	11.11	−0.0652	13.75	0.0048	16.39	0.0061
0.53	−0.0205	3.18	−0.0124	5.83	0.0131	8.48	−0.0729	11.12	−0.0939	13.76	0.0091	16.40	0.0086
0.54	−0.0089	3.19	−0.0219	5.84	−0.0054	8.49	−0.0580	11.13	−0.0996	13.77	0.0045	16.41	0.0061
0.55	−0.0010	3.20	−0.0306	5.85	−0.0326	8.50	−0.0331	11.14	−0.0802	13.78	0.0046	16.42	0.0063
0.56	0.0006	3.21	−0.0101	5.86	−0.0621	8.51	−0.0008	11.15	−0.0239	13.79	−0.0109	16.43	0.0011
0.57	−0.0044	3.22	−0.0048	5.87	−0.1083	8.52	0.0085	11.16	0.0365	13.80	−0.0135	16.44	0.0036
0.58	−0.0172	3.23	0.0351	5.88	−0.1260	8.53	−0.0056	11.17	0.0758	13.81	−0.0087	16.45	0.0030
0.59	−0.0217	3.24	0.0762	5.89	−0.0988	8.54	−0.0084	11.18	0.0667	13.82	−0.0156	16.46	−0.0008
0.60	−0.0244	3.25	0.0912	5.90	−0.0472	8.55	−0.0125	11.19	0.0435	13.83	−0.0165	16.47	−0.0027
0.61	−0.0133	3.26	0.0997	5.91	0.0009	8.56	0.0216	11.20	0.0135	13.84	−0.0260	16.48	−0.0100
0.62	0.0016	3.27	0.0705	5.92	0.0441	8.57	0.0446	11.21	−0.0143	13.85	−0.0157	16.49	−0.0249
0.63	0.0124	3.28	0.0802	5.93	0.0751	8.58	0.0457	11.22	0.0059	13.86	0.0084	16.50	−0.0329
0.64	0.0114	3.29	0.0493	5.94	0.0867	8.59	0.0328	11.23	0.0447	13.87	0.0209	16.51	−0.0374
0.65	−0.0006	3.30	0.0467	5.95	0.0655	8.60	0.0043	11.24	0.0789	13.88	0.0453	16.52	−0.0382
0.66	0.0057	3.31	0.0626	5.96	0.0417	8.61	−0.0160	11.25	0.0502	13.89	0.0290	16.53	−0.0313
0.67	0.0177	3.32	0.0505	5.97	0.0319	8.62	−0.0238	11.26	0.0421	13.90	0.0363	16.54	−0.0267
0.68	0.0206	3.33	0.0694	5.98	0.0552	8.63	−0.0246	11.27	0.0405	13.91	0.0329	16.55	−0.0257
0.69	0.0278	3.34	0.0447	5.99	0.0294	8.64	−0.0129	11.28	0.0262	13.92	0.0426	16.56	−0.0184
0.70	0.0289	3.35	0.0309	6.00	0.0037	8.65	0.0124	11.29	−0.0054	13.93	0.0387	16.57	−0.0099
0.71	0.0204	3.36	0.0308	6.01	−0.0580	8.66	0.0281	11.30	−0.0234	13.94	0.0306	16.58	0.0055
0.72	0.0147	3.37	0.0719	6.02	−0.0592	8.67	0.0305	11.31	−0.0301	13.95	0.0343	16.59	0.0153
0.73	−0.0048	3.38	0.0906	6.03	−0.0797	8.68	−0.0001	11.32	−0.0265	13.96	0.0210	16.60	0.0177
0.74	−0.0254	3.39	0.0693	6.04	−0.0852	8.69	−0.0118	11.33	−0.0358	13.97	0.0096	16.61	0.0119
0.75	−0.0317	3.40	0.0487	6.05	−0.0793	8.70	−0.0002	11.34	−0.0645	13.98	−0.0005	16.62	0.0036
0.76	−0.0314	3.41	0.0229	6.06	−0.0320	8.71	0.0025	11.35	−0.0636	13.99	−0.0124	16.63	0.0010
0.77	−0.0218	3.42	−0.0037	6.07	0.0131	8.72	−0.0060	11.36	−0.0654	14.00	−0.0009	16.64	0.0017
0.78	−0.0161	3.43	−0.0470	6.08	0.0495	8.73	−0.0020	11.37	−0.0638	14.01	−0.0041	16.65	0.0042
0.79	−0.0175	3.44	−0.0772	6.09	0.0496	8.74	−0.0331	11.38	−0.0795	14.02	−0.0007	16.66	0.0003
0.80	−0.0295	3.45	−0.0654	6.10	0.0512	8.75	−0.0787	11.39	−0.0699	14.03	0.0076	16.67	−0.0022
0.81	−0.0442	3.46	−0.0393	6.11	0.0608	8.76	−0.1032	11.40	−0.0438	14.04	0.0040	16.68	−0.0096
0.82	−0.0711	3.47	−0.0124	6.12	0.0464	8.77	−0.1196	11.41	−0.0166	14.05	−0.0061	16.69	−0.0138
0.83	−0.0645	3.48	−0.0306	6.13	0.0134	8.78	−0.1069	11.42	0.0023	14.06	−0.0189	16.70	−0.0095
0.84	−0.0267	3.49	−0.0290	6.14	−0.0120	8.79	−0.0790	11.43	0.0242	14.07	−0.0260	16.71	−0.0010
0.85	0.0033	3.50	−0.0313	6.15	0.0084	8.80	−0.0389	11.44	0.0351	14.08	−0.0322	16.72	0.0083
0.86	0.0257	3.51	−0.0358	6.16	0.0518	8.81	−0.0230	11.45	0.0507	14.09	−0.0379	16.73	0.0133
0.87	0.0273	3.52	−0.0134	6.17	0.0732	8.82	0.0080	11.46	0.0488	14.10	−0.0395	16.74	0.0093

(continued)

Table 9.3 (continued)

Time	Acc (g)	Time	Acc (g)	Time	Acc (g)	Time	Acc (g)	Time	Acc (g)	Time	Acc (g)	Time	Acc (g)
0.88	0.0278	3.53	0.0022	6.18	0.1113	8.83	0.0536	11.47	0.0110	14.11	−0.0372	16.75	0.0017
0.89	0.0361	3.54	−0.0053	6.19	0.0981	8.84	0.0802	11.48	−0.0263	14.12	−0.0289	16.76	−0.0002
0.90	0.0363	3.55	0.0025	6.20	0.1053	8.85	0.0974	11.49	−0.0426	14.13	−0.0214	16.77	−0.0044
0.91	0.0446	3.56	0.0130	6.21	0.0596	8.86	0.0808	11.50	−0.0341	14.14	0.0003	16.78	0.0028
0.92	0.0365	3.57	0.0157	6.22	0.0028	8.87	0.0836	11.51	−0.0055	14.15	0.0034	16.79	0.0101
0.93	0.0352	3.58	0.0093	6.23	0.0225	8.88	0.0709	11.52	0.0316	14.16	0.0261	16.80	0.0135
0.94	−0.0117	3.59	0.0247	6.24	0.0491	8.89	0.0421	11.53	0.0324	14.17	0.0566	16.81	0.0069
0.95	−0.0331	3.60	0.0051	6.25	0.0871	8.90	−0.0113	11.54	0.0292	14.18	0.0760	16.82	0.0015
0.96	−0.0356	3.61	−0.0077	6.26	0.0827	8.91	−0.0460	11.55	−0.0011	14.19	0.0646	16.83	−0.0060
0.97	−0.0086	3.62	−0.0116	6.27	0.0611	8.92	−0.0647	11.56	−0.0469	14.20	0.0478	16.84	−0.0043
0.98	0.0050	3.63	−0.0077	6.28	0.0301	8.93	−0.0926	11.57	−0.0554	14.21	0.0083	16.85	−0.0064
0.99	−0.0060	3.64	0.0168	6.29	0.0356	8.94	−0.0566	11.58	−0.0435	14.22	−0.0152	16.86	−0.0020
1.00	−0.0364	3.65	0.0143	6.30	−0.0220	8.95	−0.0525	11.59	−0.0474	14.23	−0.0249	16.87	−0.0059
1.01	−0.0409	3.66	0.0047	6.31	−0.0618	8.96	−0.0108	11.60	−0.0629	14.24	0.0004	16.88	−0.0045
1.02	−0.0388	3.67	0.0060	6.32	−0.0467	8.97	0.0276	11.61	−0.0518	14.25	0.0282	16.89	−0.0007
1.03	−0.0204	3.68	−0.0111	6.33	−0.0319	8.98	0.0391	11.62	−0.0691	14.26	0.0423	16.90	0.0048
1.04	−0.0077	3.69	0.0007	6.34	0.0122	8.99	0.0375	11.63	−0.0511	14.27	0.0327	16.91	0.0056
1.05	0.0371	3.70	−0.0170	6.35	0.0136	9.00	0.0306	11.64	−0.0162	14.28	0.0170	16.92	−0.0007
1.06	0.0742	3.71	−0.0171	6.36	0.0206	9.01	0.0428	11.65	0.0200	14.29	0.0062	16.93	−0.0057
1.07	0.0883	3.72	0.0105	6.37	−0.0045	9.02	0.0366	11.66	0.0552	14.30	0.0072	16.94	−0.0102
1.08	0.0724	3.73	0.0150	6.38	−0.0056	9.03	0.0470	11.67	0.0828	14.31	−0.0158	16.95	−0.0143
1.09	0.0707	3.74	0.0276	6.39	−0.0446	9.04	0.0009	11.68	0.0278	14.32	−0.0203	16.96	−0.0109
1.10	0.0548	3.75	0.0248	6.40	−0.0635	9.05	0.0213	11.69	0.0035	14.33	−0.0272	16.97	−0.0075
1.11	0.0439	3.76	0.0378	6.41	−0.0770	9.06	0.0039	11.70	0.0033	14.34	−0.0274	16.98	−0.0063
1.12	0.0066	3.77	0.0523	6.42	−0.0326	9.07	0.0290	11.71	0.0016	14.35	−0.0321	16.99	−0.0099
1.13	0.0025	3.78	0.0596	6.43	−0.0268	9.08	0.0488	11.72	0.0239	14.36	−0.0368	17.00	−0.0150
1.14	−0.0039	3.79	0.0519	6.44	−0.0050	9.09	0.0634	11.73	0.0061	14.37	−0.0189	17.01	−0.0174
1.15	0.0110	3.80	0.0540	6.45	−0.0352	9.10	0.0709	11.74	−0.0291	14.38	0.0003	17.02	−0.0122
1.16	0.0047	3.81	0.0515	6.46	−0.0464	9.11	0.0906	11.75	−0.0529	14.39	−0.0002	17.03	−0.0032
1.17	0.0043	3.82	0.0740	6.47	−0.0575	9.12	0.0733	11.76	−0.0548	14.40	−0.0020	17.04	0.0024
1.18	−0.0131	3.83	0.0639	6.48	−0.0614	9.13	0.0906	11.77	−0.0322	14.41	−0.0148	17.05	0.0038
1.19	−0.0004	3.84	0.0208	6.49	−0.0315	9.14	0.0471	11.78	−0.0370	14.42	−0.0202	17.06	0.0054
1.20	−0.0221	3.85	−0.0200	6.50	0.0312	9.15	0.0286	11.79	−0.0308	14.43	−0.0196	17.07	0.0048
1.21	−0.0169	3.86	−0.0849	6.51	0.0783	9.16	−0.0198	11.80	−0.0629	14.44	−0.0345	17.08	0.0047
1.22	−0.0239	3.87	−0.1385	6.52	0.0883	9.17	−0.0822	11.81	−0.0808	14.45	−0.0522	17.09	0.0038
1.23	0.0008	3.88	−0.1240	6.53	0.0564	9.18	−0.1051	11.82	−0.1302	14.46	−0.0669	17.10	0.0055
1.24	−0.0087	3.89	−0.0719	6.54	−0.0083	9.19	−0.1065	11.83	−0.1021	14.47	−0.0982	17.11	0.0072
1.25	0.0143	3.90	−0.0095	6.55	−0.0136	9.20	−0.0896	11.84	−0.0441	14.48	−0.1054	17.12	0.0066
1.26	0.0154	3.91	0.0093	6.56	−0.0370	9.21	−0.0642	11.85	−0.0077	14.49	−0.0866	17.13	0.0055
1.27	0.0307	3.92	0.0099	6.57	−0.0483	9.22	−0.0474	11.86	0.0147	14.50	−0.0424	17.14	0.0095
1.28	0.0217	3.93	0.0130	6.58	−0.0496	9.23	−0.0255	11.87	−0.0120	14.51	0.0086	17.15	0.0136
1.29	0.0075	3.94	0.0228	6.59	−0.0534	9.24	−0.0125	11.88	−0.0387	14.52	0.0315	17.16	0.0155
1.30	−0.0040	3.95	0.0205	6.60	−0.0637	9.25	−0.0101	11.89	−0.0681	14.53	0.0377	17.17	0.0111
1.31	−0.0176	3.96	0.0450	6.61	−0.0608	9.26	−0.0058	11.90	−0.0827	14.54	0.0512	17.18	0.0048

(continued)

Table 9.3 (continued)

Time	Acc (g)	Time	Acc (g)	Time	Acc (g)	Time	Acc (g)	Time	Acc (g)	Time	Acc (g)	Time	Acc (g)
1.32	−0.0283	3.97	0.0331	6.62	−0.0376	9.27	0.0581	11.91	−0.0570	14.55	0.0341	17.19	−0.0060
1.33	−0.0510	3.98	0.0446	6.63	−0.0033	9.28	0.1042	11.92	−0.0314	14.56	0.0281	17.20	−0.0108
1.34	−0.0580	3.99	0.0070	6.64	0.0016	9.29	0.1300	11.93	−0.0256	14.57	0.0058	17.21	−0.0088
1.35	−0.0944	4.00	0.0086	6.65	−0.0070	9.30	0.1031	11.94	−0.0190	14.58	−0.0232	17.22	−0.0029
1.36	−0.0786	4.01	0.0187	6.66	−0.0208	9.31	0.0548	11.95	−0.0068	14.59	−0.0166	17.23	0.0001
1.37	−0.0441	4.02	0.0352	6.67	0.0109	9.32	0.0256	11.96	0.0009	14.60	−0.0171	17.24	0.0019
1.38	−0.0079	4.03	0.0470	6.68	0.0468	9.33	−0.0232	11.97	0.0358	14.61	−0.0146	17.25	0.0020
1.39	0.0207	4.04	0.0436	6.69	0.0480	9.34	−0.0534	11.98	0.0694	14.62	0.0054	17.26	0.0084
1.40	0.0548	4.05	0.0165	6.70	0.0429	9.35	−0.0481	11.99	0.1021	14.63	0.0124	17.27	0.0086
1.41	0.0414	4.06	0.0341	6.71	0.0349	9.36	−0.0350	12.00	0.0898	14.64	0.0284	17.28	0.0092
1.42	0.0312	4.07	0.0437	6.72	0.0302	9.37	−0.0078	12.01	0.0667	14.65	0.0525	17.29	0.0066
1.43	0.0013	4.08	0.0455	6.73	0.0039	9.38	−0.0111	12.02	0.0413	14.66	0.0825	17.30	0.0073
1.44	−0.0290	4.09	0.0380	6.74	−0.0154	9.39	−0.0006	12.03	0.0098	14.67	0.0886	17.31	−0.0031
1.45	−0.0709	4.10	−0.0033	6.75	−0.0109	9.40	0.0418	12.04	−0.0033	14.68	0.0917	17.32	−0.0083
1.46	−0.0794	4.11	−0.0648	6.76	−0.0069	9.41	0.0728	12.05	−0.0219	14.69	0.0869	17.33	−0.0130
1.47	−0.0828	4.12	−0.1064	6.77	−0.0116	9.42	0.0807	12.06	−0.0286	14.70	0.0767	17.34	−0.0153
1.48	−0.0430	4.13	−0.1084	6.78	−0.0356	9.43	0.0695	12.07	−0.0457	14.71	0.0634	17.35	−0.0091
1.49	−0.0275	4.14	−0.0810	6.79	−0.0460	9.44	0.0394	12.08	−0.0626	14.72	0.0350	17.36	−0.0087
1.50	0.0217	4.15	−0.0338	6.80	−0.0645	9.45	−0.0045	12.09	−0.0587	14.73	0.0177	17.37	−0.0026
1.51	0.0749	4.16	−0.0041	6.81	−0.0655	9.46	−0.0243	12.10	−0.0218	14.74	0.0067	17.38	0.0008
1.52	0.0978	4.17	−0.0101	6.82	−0.0954	9.47	−0.0522	12.11	0.0114	14.75	−0.0012	17.39	0.0084
1.53	0.0898	4.18	−0.0200	6.83	−0.0905	9.48	−0.0375	12.12	0.0515	14.76	−0.0181	17.40	0.0138
1.54	0.1018	4.19	−0.0186	6.84	−0.0501	9.49	−0.0082	12.13	0.0225	14.77	−0.0109	17.41	0.0124
1.55	0.1146	4.20	−0.0161	6.85	−0.0179	9.50	0.0160	12.14	0.0035	14.78	−0.0058	17.42	0.0132
1.56	0.1302	4.21	0.0132	6.86	0.0161	9.51	−0.0062	12.15	−0.0187	14.79	0.0188	17.43	0.0067
1.57	0.1156	4.22	0.0156	6.87	0.0271	9.52	−0.0185	12.16	−0.0223	14.80	0.0314	17.44	0.0027
1.58	0.0725	4.23	−0.0117	6.88	0.0163	9.53	−0.0228	12.17	−0.0096	14.81	0.0371	17.45	−0.0060
1.59	0.0441	4.24	−0.0478	6.89	0.0148	9.54	−0.0354	12.18	0.0117	14.82	0.0094	17.46	−0.0088
1.60	0.0017	4.25	−0.0536	6.90	−0.0003	9.55	−0.0388	12.19	0.0396	14.83	0.0073	17.47	−0.0092
1.61	−0.0091	4.26	−0.0304	6.91	−0.0521	9.56	−0.0312	12.20	0.0532	14.84	0.0060	17.48	−0.0078
1.62	−0.0237	4.27	−0.0186	6.92	−0.0767	9.57	−0.0113	12.21	0.0478	14.85	0.0096	17.49	−0.0033
1.63	−0.0237	4.28	0.0065	6.93	−0.1276	9.58	0.0144	12.22	0.0496	14.86	0.0105	17.50	0.0029
1.64	−0.0075	4.29	−0.0145	6.94	−0.1292	9.59	0.0213	12.23	0.0666	14.87	−0.0071	17.51	0.0025
1.65	−0.0251	4.30	−0.0204	6.95	−0.1233	9.60	−0.0019	12.24	0.0964	14.88	−0.0147	17.52	0.0050
1.66	−0.0411	4.31	−0.0296	6.96	−0.1023	9.61	−0.0200	12.25	0.1050	14.89	−0.0213	17.53	0.0057
1.67	−0.0307	4.32	−0.0283	6.97	−0.0734	9.62	−0.0387	12.26	0.0815	14.90	−0.0383	17.54	0.0053
1.68	0.0060	4.33	−0.0342	6.98	−0.0533	9.63	−0.0430	12.27	0.0375	14.91	−0.0364	17.55	0.0062
1.69	0.0013	4.34	−0.0084	6.99	−0.0014	9.64	−0.0434	12.28	0.0092	14.92	−0.0031	17.56	0.0041
1.70	−0.0056	4.35	0.0395	7.00	0.0758	9.65	0.0046	12.29	−0.0424	14.93	0.0162	17.57	0.0051
1.71	−0.0343	4.36	0.0647	7.01	0.1083	9.66	0.0305	12.30	−0.0632	14.94	0.0278	17.58	0.0025
1.72	−0.0417	4.37	0.0454	7.02	0.1245	9.67	0.0361	12.31	−0.0971	14.95	0.0078	17.59	0.0015
1.73	−0.0592	4.38	0.0426	7.03	0.1325	9.68	−0.0379	12.32	−0.0759	14.96	−0.0119	17.60	0.0009
1.74	−0.0605	4.39	0.0297	7.04	0.1127	9.69	−0.0758	12.33	−0.0675	14.97	−0.0297	17.61	0.0052
1.75	−0.0443	4.40	0.0637	7.05	0.0727	9.70	−0.0920	12.34	−0.0793	14.98	−0.0242	17.62	0.0042

(continued)

Table 9.3 (continued)

Time	Acc (*g*)	Time	Acc (*g*)	Time	Acc (*g*)	Time	Acc (*g*)	Time	Acc (*g*)	Time	Acc (*g*)	Time	Acc (*g*)
1.76	0.0004	4.41	0.0685	7.06	0.0440	9.71	−0.0789	12.35	−0.0775	14.99	−0.0395	17.63	0.0033
1.77	0.0475	4.42	0.0675	7.07	0.0385	9.72	−0.0516	12.36	−0.0455	15.00	−0.0428	17.64	0.0015
1.78	0.0487	4.43	0.0363	7.08	0.0621	9.73	−0.0121	12.37	−0.0178	15.01	−0.0304	17.65	0.0019
1.79	0.0685	4.44	0.0285	7.09	0.0744	9.74	0.0538	12.38	0.0198	15.02	−0.0039	17.66	−0.0031
1.80	0.0922	4.45	0.0339	7.10	0.0534	9.75	0.0760	12.39	0.0301	15.03	−0.0091	17.67	−0.0049
1.81	0.0789	4.46	0.0301	7.11	0.0260	9.76	0.0542	12.40	0.0370	15.04	−0.0116	17.68	−0.0099
1.82	0.0752	4.47	0.0557	7.12	−0.0024	9.77	0.0441	12.41	0.0448	15.05	0.0019	17.69	−0.0130
1.83	0.0442	4.48	0.0506	7.13	−0.0362	9.78	0.0280	12.42	0.0288	15.06	0.0150	17.70	−0.0128
1.84	0.0267	4.49	0.0568	7.14	−0.0392	9.79	0.0375	12.43	0.0265	15.07	0.0254	17.71	−0.0094
1.85	0.0023	4.50	0.0254	7.15	−0.0371	9.80	0.0312	12.44	0.0121	15.08	0.0064	17.72	−0.0064
1.86	−0.0041	4.51	0.0225	7.16	−0.0039	9.81	0.0322	12.45	0.0409	15.09	−0.0200	17.73	−0.0039
1.87	−0.0065	4.52	0.0307	7.17	0.0311	9.82	0.0329	12.46	0.0403	15.10	−0.0364	17.74	−0.0030
1.88	−0.0024	4.53	0.0263	7.18	0.0381	9.83	0.0470	12.47	0.0451	15.11	−0.0290	17.75	−0.0023
1.89	−0.0172	4.54	0.0417	7.19	0.0337	9.84	0.0396	12.48	0.0586	15.12	−0.0421	17.76	0.0015
1.90	0.0081	4.55	0.0479	7.20	0.0599	9.85	0.0496	12.49	0.0573	15.13	−0.0431	17.77	0.0035
1.91	0.0430	4.56	0.0360	7.21	0.0681	9.86	0.0506	12.50	0.0627	15.14	−0.0411	17.78	0.0016
1.92	0.0429	4.57	0.0360	7.22	0.0611	9.87	0.0365	12.51	0.0430	15.15	−0.0105	17.79	0.0005
1.93	0.0309	4.58	0.0063	7.23	0.0358	9.88	0.0054	12.52	0.0145	15.16	0.0109	17.80	−0.0004
1.94	−0.0670	4.59	−0.0299	7.24	0.0098	9.89	−0.0429	12.53	0.0275	15.17	0.0033	17.81	0.0003
1.95	−0.1163	4.60	−0.0562	7.25	0.0053	9.90	−0.0717	12.54	0.0376	15.18	0.0144	17.82	0.0013
1.96	−0.1438	4.61	−0.0598	7.26	0.0156	9.91	−0.0730	12.55	0.0538	15.19	0.0302	17.83	0.0075
1.97	−0.0930	4.62	−0.0694	7.27	0.0276	9.92	−0.0678	12.56	0.0270	15.20	0.0284	17.84	0.0093
1.98	−0.0371	4.63	−0.0475	7.28	0.0282	9.93	−0.0501	12.57	−0.0078	15.21	0.0283	17.85	0.0069
1.99	0.0077	4.64	−0.0114	7.29	0.0391	9.94	−0.0198	12.58	−0.0523	15.22	0.0288	17.86	−0.0001
2.00	0.0393	4.65	0.0027	7.30	0.0054	9.95	0.0045	12.59	−0.1052	15.23	0.0295	17.87	−0.0027
2.01	0.0318	4.66	0.0313	7.31	−0.0333	9.96	−0.0035	12.60	−0.1283	15.24	0.0288	17.88	−0.0028
2.02	−0.0166	4.67	−0.0055	7.32	−0.0846	9.97	0.0185	12.61	−0.1358	15.25	0.0188	17.89	−0.0043
2.03	−0.0475	4.68	−0.0388	7.33	−0.1016	9.98	0.0202	12.62	−0.0932	15.26	0.0082	17.90	−0.0065
2.04	−0.0551	4.69	−0.0667	7.34	−0.1290	9.99	0.0123	12.63	−0.0444	15.27	0.0073	17.91	−0.0076
2.05	−0.0469	4.70	−0.0618	7.35	−0.1995	10.00	−0.0057	12.64	−0.0098	15.28	0.0023	17.92	−0.0064
2.06	−0.0420	4.71	−0.1025	7.36	−0.1161	10.01	−0.0189	12.65	0.0006	15.29	0.0134	17.93	−0.0070
2.07	−0.0287	4.72	−0.1077	7.37	−0.1040	10.02	−0.0406	12.66	0.0136	15.30	0.0161	17.94	−0.0100
2.08	−0.0255	4.73	−0.0730	7.38	−0.0896	10.03	−0.0464	12.67	0.0261	15.31	0.0149	17.95	−0.0100
2.09	−0.0116	4.74	−0.0744	7.39	−0.0712	10.04	−0.0486	12.68	0.0507	15.32	0.0078	17.96	−0.0069
2.10	0.0209	4.75	−0.0457	7.40	−0.0384	10.05	−0.0247	12.69	0.0499	15.33	0.0022	17.97	−0.0032
2.11	0.0263	4.76	−0.0481	7.41	−0.0199	10.06	−0.0181	12.70	0.0269	15.34	−0.0010	17.98	−0.0003
2.12	0.0013	4.77	−0.0403	7.42	0.0142	10.07	−0.0037	12.71	−0.0022	15.35	0.0006	17.99	−0.0011
2.13	−0.0320	4.78	−0.0108	7.43	0.0570	10.08	−0.0447	12.72	−0.0325	15.36	0.0000	18.00	−0.0001
2.14	−0.0503	4.79	0.0014	7.44	0.0993	10.09	−0.0307	12.73	−0.0526	15.37	−0.0032	18.01	0.0020
2.15	−0.0578	4.80	0.0267	7.45	0.1014	10.10	0.0066	12.74	−0.0591	15.38	0.0029	18.02	0.0029
2.16	−0.0569	4.81	0.0562	7.46	0.0994	10.11	0.0715	12.75	−0.0487	15.39	0.0024	18.03	0.0027
2.17	−0.0370	4.82	0.1014	7.47	0.0699	10.12	0.0968	12.76	−0.0493	15.40	−0.0086	18.04	0.0029
2.18	−0.0021	4.83	0.1305	7.48	0.0692	10.13	0.1308	12.77	−0.0343	15.41	−0.0121	18.05	0.0025
2.19	0.0346	4.84	0.1195	7.49	0.0477	10.14	0.1355	12.78	−0.0145	15.42	−0.0150	18.06	0.0012

(continued)

Table 9.3 (continued)

Time	Acc (*g*)	Time	Acc (*g*)	Time	Acc (*g*)	Time	Acc (*g*)	Time	Acc (*g*)	Time	Acc (*g*)	Time	Acc (*g*)
2.20	0.0620	4.85	0.0587	7.50	0.0359	10.15	0.1396	12.79	0.0206	15.43	−0.0068	18.07	−0.0006
2.21	0.0398	4.86	0.0070	7.51	0.0500	10.16	0.1443	12.80	0.0545	15.44	−0.0001	18.08	−0.0015
2.22	0.0502	4.87	−0.0233	7.52	0.0542	10.17	0.1248	12.81	0.0641	15.45	0.0043	18.09	−0.0005
2.23	0.0454	4.88	−0.0497	7.53	0.0796	10.18	0.0937	12.82	0.0352	15.46	0.0114	18.10	0.0010
2.24	0.0604	4.89	−0.0495	7.54	0.0713	10.19	0.0655	12.83	0.0150	15.47	0.0169	18.11	0.0019
2.25	0.0438	4.90	−0.0151	7.55	0.0488	10.20	0.0484	12.84	−0.0323	15.48	0.0099	18.12	0.0006
2.26	0.0239	4.91	−0.0069	7.56	0.0396	10.21	0.0085	12.85	−0.0462	15.49	0.0159	18.13	−0.0002
2.27	−0.0010	4.92	−0.0145	7.57	0.0010	10.22	0.0114	12.86	−0.0400	15.50	0.0289	18.14	−0.0011
2.28	−0.0220	4.93	−0.0334	7.58	−0.0251	10.23	0.0094	12.87	−0.0077	15.51	0.0363	18.15	−0.0031
2.29	−0.0243	4.94	−0.0195	7.59	−0.0382	10.24	0.0180	12.88	0.0127	15.52	0.0352	18.16	−0.0039
2.30	−0.0445	4.95	−0.0060	7.60	−0.0188	10.25	0.0298	12.89	0.0216	15.53	0.0236	18.17	−0.0037
2.31	−0.0209	4.96	0.0086	7.61	−0.0122	10.26	−0.0083	12.90	0.0147	15.54	0.0079	18.18	−0.0036
2.32	−0.0084	4.97	−0.0002	7.62	−0.0059	10.27	−0.0100	12.91	0.0194	15.55	−0.0059	18.19	−0.0032
2.33	0.0136	4.98	−0.0440	7.63	0.0023	10.28	−0.0205	12.92	0.0086	15.56	−0.0206	18.20	−0.0024
2.34	0.0214	4.99	−0.0675	7.64	−0.0281	10.29	0.0016	12.93	0.0021	15.57	−0.0286	18.21	−0.0001
2.35	0.0368	5.00	−0.0572	7.65	−0.0414	10.30	−0.0293	12.94	−0.0112	15.58	−0.0267	18.22	0.0016
2.36	0.0469	5.01	−0.0625	7.66	−0.0474	10.31	−0.0189	12.95	−0.0116	15.59	−0.0248	18.23	0.0027
2.37	0.0709	5.02	−0.0141	7.67	−0.0320	10.32	0.0030	12.96	0.0099	15.60	−0.0128	18.24	0.0030
2.38	0.0618	5.03	−0.0154	7.68	−0.0067	10.33	0.0037	12.97	0.0155	15.61	−0.0136	18.25	0.0030
2.39	0.0595	5.04	0.0087	7.69	−0.0053	10.34	−0.0098	12.98	0.0079	15.62	−0.0107	18.26	0.0031
2.40	0.0151	5.05	0.0103	7.70	0.0060	10.35	−0.0239	12.99	−0.0245	15.63	−0.0126	18.27	0.0039
2.41	−0.0037	5.06	0.0290	7.71	0.0183	10.36	−0.0409	13.00	−0.0333	15.64	−0.0138	18.28	0.0032
2.42	−0.0165	5.07	0.0314	7.72	0.0330	10.37	−0.0312	13.01	−0.0590	15.65	−0.0015	18.29	0.0027
2.43	−0.0451	5.08	0.0773	7.73	0.0379	10.38	−0.0224	13.02	−0.0476	15.66	0.0073	18.30	0.0020
2.44	−0.0417	5.09	0.0833	7.74	0.0744	10.39	−0.0263	13.03	−0.0266	15.67	0.0154	18.31	0.0018
2.45	−0.0281	5.10	0.0917	7.75	0.0847	10.40	−0.0223	13.04	0.0226	15.68	0.0154	18.32	0.0007
2.46	−0.0008	5.11	0.0384	7.76	0.1161	10.41	0.0084	13.05	0.0635	15.69	0.0046	18.33	0.0001
2.47	0.0228	5.12	−0.0035	7.77	0.1025	10.42	−0.0169	13.06	0.0827	15.70	0.0008	18.34	0.0004
2.48	0.0521	5.13	−0.0353	7.78	0.0807	10.43	−0.0166	13.07	0.0775	15.71	0.0075	18.35	0.0009
2.49	0.0538	5.14	−0.0477	7.79	0.0779	10.44	−0.0305	13.08	0.0599	15.72	0.0170	18.36	0.0006
2.50	0.0605	5.15	−0.0623	7.80	0.0970	10.45	−0.0397	13.09	0.0734	15.73	0.0181	18.37	0.0000
2.51	0.0502	5.16	−0.0524	7.81	0.0838	10.46	−0.0238	13.10	0.0651	15.74	0.0170	18.38	−0.0007
2.52	0.0124	5.17	−0.0495	7.82	0.0806	10.47	−0.0311	13.11	0.0604	15.75	−0.0043	18.39	−0.0013
2.53	0.0124	5.18	−0.0290	7.83	0.0397	10.48	−0.0604	13.12	0.0700	15.76	−0.0218	18.40	−0.0013
2.54	0.0472	5.19	−0.0360	7.84	−0.0040	10.49	−0.0704	13.13	0.0390	15.77	−0.0394	18.41	−0.0011
2.55	0.0342	5.20	−0.0137	7.85	−0.0219	10.50	−0.0551	13.14	0.0112	15.78	−0.0417	18.42	−0.0008
2.56	0.0123	5.21	−0.0426	7.86	−0.0516	10.51	−0.0351	13.15	−0.0054	15.79	−0.0366	18.43	0.0000
2.57	0.0034	5.22	−0.0487	7.87	−0.0428	10.52	−0.0317	13.16	−0.0002	15.80	−0.0325	18.44	0.0003
2.58	−0.0314	5.23	−0.0996	7.88	−0.0284	10.53	0.0028	13.17	0.0411	15.81	−0.0285	18.45	0.0003
2.59	−0.0342	5.24	−0.0889	7.89	0.0131	10.54	0.0095	13.18	0.0569	15.82	−0.0307	18.46	0.0001
2.60	−0.0459	5.25	−0.0691	7.90	−0.0038	10.55	0.0309	13.19	0.0637	15.83	−0.0395	18.47	−0.0001
2.61	−0.0595	5.26	−0.0253	7.91	−0.0258	10.56	0.0115	13.20	0.0545	15.84	−0.0357	18.48	−0.0002
2.62	−0.0465	5.27	0.0234	7.92	−0.0423	10.57	−0.0181	13.21	0.0372	15.85	−0.0260	18.49	−0.0002
2.63	−0.0281	5.28	0.0701	7.93	−0.0311	10.58	−0.0491	13.22	0.0359	15.86	−0.0145	18.50	−0.0002

(continued)

Table 9.3 (continued)

Time	Acc (*g*)	Time	Acc (*g*)	Time	Acc (*g*)	Time	Acc (*g*)	Time	Acc (*g*)	Time	Acc (*g*)	Time	Acc (*g*)
2.64	−0.0055	5.29	0.0579	7.94	−0.0412	10.59	−0.0612	13.23	0.0147	15.87	−0.0094	18.51	−0.0002
2.65	−0.0024	5.30	0.0333	7.95	−0.0444								

Table 9.4 Data for floor response spectra at 5% damping by time history analysis shown in Fig. 9.7

Freq	Node 3	Node 2	Node 1	Freq	Node 3	Node 2	Node 1	Freq	Node 3	Node 2	Node 1
0.100	0.0025	0.0024	0.0023	16.800	0.6289	0.4804	0.3959	33.400	0.5267	0.4311	0.2750
0.200	0.0114	0.0114	0.0112	16.900	0.6271	0.4851	0.3814	33.500	0.5267	0.4311	0.2748
0.300	0.0293	0.0288	0.0282	17.000	0.6245	0.4895	0.3743	33.600	0.5267	0.4310	0.2747
0.400	0.0437	0.0433	0.0425	17.100	0.6209	0.4934	0.3686	33.700	0.5268	0.4310	0.2745
0.500	0.0560	0.0553	0.0541	17.200	0.6164	0.4964	0.3630	33.800	0.5268	0.4309	0.2744
0.600	0.0732	0.0730	0.0721	17.300	0.6111	0.4984	0.3575	33.900	0.5268	0.4309	0.2742
0.700	0.0868	0.0862	0.0852	17.400	0.6053	0.4994	0.3521	34.000	0.5268	0.4308	0.2741
0.800	0.1080	0.1067	0.1043	17.500	0.5992	0.4995	0.3467	34.100	0.5268	0.4308	0.2739
0.900	0.1199	0.1190	0.1176	17.600	0.5933	0.4988	0.3414	34.200	0.5268	0.4307	0.2738
1.000	0.1365	0.1354	0.1336	17.700	0.5880	0.4976	0.3364	34.300	0.5268	0.4307	0.2737
1.100	0.1587	0.1572	0.1538	17.800	0.5835	0.4963	0.3317	34.400	0.5268	0.4307	0.2736
1.200	0.1846	0.1815	0.1746	17.900	0.5799	0.4951	0.3271	34.500	0.5268	0.4308	0.2735
1.300	0.1918	0.1901	0.1863	18.000	0.5772	0.4941	0.3228	34.600	0.5268	0.4309	0.2734
1.400	0.2191	0.2139	0.2030	18.100	0.5752	0.4933	0.3186	34.700	0.5268	0.4310	0.2733
1.500	0.2224	0.2177	0.2079	18.200	0.5736	0.4925	0.3145	34.800	0.5268	0.4312	0.2732
1.600	0.2239	0.2202	0.2138	18.300	0.5723	0.4917	0.3103	34.900	0.5268	0.4313	0.2731
1.700	0.2514	0.2467	0.2360	18.400	0.5709	0.4906	0.3067	35.000	0.5268	0.4314	0.2730
1.800	0.2879	0.2777	0.2574	18.500	0.5692	0.4891	0.3042	35.100	0.5268	0.4316	0.2729
1.900	0.3419	0.3221	0.2838	18.600	0.5672	0.4870	0.3018	35.200	0.5268	0.4317	0.2729
2.000	0.3681	0.3455	0.2996	18.700	0.5649	0.4845	0.2994	35.300	0.5268	0.4318	0.2728
2.100	0.3900	0.3647	0.3151	18.800	0.5638	0.4815	0.2971	35.400	0.5268	0.4319	0.2727
2.200	0.4054	0.3797	0.3304	18.900	0.5655	0.4782	0.2948	35.500	0.5268	0.4320	0.2727
2.300	0.3847	0.3661	0.3305	19.000	0.5668	0.4747	0.2924	35.600	0.5268	0.4321	0.2726
2.400	0.4107	0.3869	0.3373	19.100	0.5676	0.4742	0.2900	35.700	0.5268	0.4322	0.2726
2.500	0.4480	0.4195	0.3604	19.200	0.5680	0.4733	0.2874	35.800	0.5267	0.4323	0.2725
2.600	0.4803	0.4502	0.3890	19.300	0.5679	0.4719	0.2848	35.900	0.5267	0.4323	0.2724
2.700	0.4862	0.4550	0.3918	19.400	0.5675	0.4703	0.2820	36.000	0.5266	0.4323	0.2724
2.800	0.5107	0.4762	0.4038	19.500	0.5667	0.4684	0.2791	36.100	0.5266	0.4324	0.2723
2.900	0.4649	0.4335	0.3775	19.600	0.5657	0.4663	0.2766	36.200	0.5265	0.4323	0.2723
3.000	0.5247	0.4932	0.4250	19.700	0.5645	0.4643	0.2757	36.300	0.5264	0.4323	0.2722
3.100	0.5884	0.5513	0.4689	19.800	0.5631	0.4622	0.2749	36.400	0.5262	0.4322	0.2721
3.200	0.6241	0.5815	0.4879	19.900	0.5617	0.4603	0.2742	36.500	0.5261	0.4322	0.2721
3.300	0.6334	0.5853	0.4846	20.000	0.5605	0.4585	0.2736	36.600	0.5260	0.4321	0.2720
3.400	0.6094	0.5599	0.4570	20.100	0.5607	0.4568	0.2731	36.700	0.5258	0.4320	0.2719
3.500	0.6344	0.5616	0.4533	20.200	0.5609	0.4553	0.2728	36.800	0.5257	0.4318	0.2719
3.600	0.7189	0.6532	0.5209	20.300	0.5609	0.4539	0.2725	36.900	0.5255	0.4318	0.2718
3.700	0.8552	0.7569	0.5990	20.400	0.5609	0.4527	0.2726	37.000	0.5253	0.4318	0.2717

(continued)

Table 9.4 (continued)

Freq	Node 3	Node 2	Node 1	Freq	Node 3	Node 2	Node 1	Freq	Node 3	Node 2	Node 1
3.800	1.0067	0.8889	0.6449	20.500	0.5608	0.4517	0.2732	37.100	0.5251	0.4318	0.2717
3.900	1.0749	0.9471	0.6968	20.600	0.5609	0.4507	0.2739	37.200	0.5249	0.4318	0.2716
4.000	1.1862	1.0406	0.7407	20.700	0.5609	0.4498	0.2749	37.300	0.5248	0.4318	0.2716
4.100	1.2618	1.1008	0.7749	20.800	0.5611	0.4489	0.2761	37.400	0.5246	0.4318	0.2715
4.200	1.2661	1.1043	0.7710	20.900	0.5613	0.4478	0.2774	37.500	0.5244	0.4318	0.2715
4.300	1.3072	1.1376	0.7971	21.000	0.5615	0.4465	0.2787	37.600	0.5242	0.4318	0.2714
4.400	1.4281	1.2304	0.8489	21.100	0.5618	0.4449	0.2799	37.700	0.5240	0.4318	0.2714
4.500	1.6418	1.4088	0.9388	21.200	0.5619	0.4454	0.2839	37.800	0.5238	0.4318	0.2714
4.600	1.8162	1.5546	1.0163	21.300	0.5619	0.4471	0.2880	37.900	0.5236	0.4318	0.2713
4.700	1.8617	1.5935	1.0386	21.400	0.5617	0.4489	0.2918	38.000	0.5234	0.4317	0.2713
4.800	1.9566	1.6627	1.0567	21.500	0.5612	0.4508	0.2951	38.100	0.5232	0.4317	0.2713
4.900	2.1214	1.7896	1.1221	21.600	0.5605	0.4524	0.2978	38.200	0.5230	0.4317	0.2712
5.000	2.3172	1.9564	1.2140	21.700	0.5596	0.4538	0.2997	38.300	0.5228	0.4317	0.2712
5.100	2.5603	2.1438	1.2907	21.800	0.5584	0.4549	0.3007	38.400	0.5226	0.4317	0.2712
5.200	2.7967	2.3400	1.4164	21.900	0.5571	0.4554	0.3009	38.500	0.5224	0.4316	0.2711
5.300	3.0148	2.5178	1.5096	22.000	0.5558	0.4555	0.3002	38.600	0.5222	0.4316	0.2711
5.400	3.1525	2.6182	1.5846	22.100	0.5544	0.4551	0.2988	38.700	0.5221	0.4316	0.2710
5.500	3.2521	2.6967	1.5924	22.200	0.5530	0.4543	0.2969	38.800	0.5219	0.4315	0.2710
5.600	3.1679	2.6345	1.5493	22.300	0.5518	0.4531	0.2946	38.900	0.5218	0.4315	0.2710
5.700	2.9419	2.4384	1.4479	22.400	0.5506	0.4515	0.2921	39.000	0.5216	0.4315	0.2709
5.800	2.7112	2.2475	1.3066	22.500	0.5496	0.4496	0.2895	39.100	0.5215	0.4315	0.2708
5.900	2.5199	2.0764	1.1820	22.600	0.5488	0.4474	0.2870	39.200	0.5213	0.4314	0.2708
6.000	2.3300	1.9111	1.0825	22.700	0.5480	0.4450	0.2846	39.300	0.5212	0.4314	0.2707
6.100	2.1691	1.7858	1.0143	22.800	0.5475	0.4426	0.2837	39.400	0.5211	0.4314	0.2707
6.200	2.0971	1.7156	0.9514	22.900	0.5470	0.4401	0.2851	39.500	0.5210	0.4313	0.2706
6.300	2.0212	1.6190	0.8801	23.000	0.5467	0.4377	0.2860	39.600	0.5209	0.4313	0.2705
6.400	1.9291	1.5131	0.8103	23.100	0.5464	0.4357	0.2862	39.700	0.5208	0.4313	0.2704
6.500	1.8187	1.4407	0.7346	23.200	0.5462	0.4349	0.2859	39.800	0.5207	0.4313	0.2704
6.600	1.6825	1.3271	0.6606	23.300	0.5460	0.4341	0.2850	39.900	0.5207	0.4312	0.2703
6.700	1.5458	1.2234	0.6061	23.400	0.5457	0.4332	0.2836	40.000	0.5206	0.4312	0.2702
6.800	1.4238	1.1312	0.5662	23.500	0.5455	0.4323	0.2820	40.100	0.5206	0.4312	0.2701
6.900	1.3271	1.0443	0.5338	23.600	0.5453	0.4314	0.2803	40.200	0.5205	0.4311	0.2701
7.000	1.3365	0.9829	0.5036	23.700	0.5450	0.4305	0.2786	40.300	0.5205	0.4311	0.2700
7.100	1.3235	0.9550	0.4964	23.800	0.5448	0.4296	0.2773	40.400	0.5205	0.4310	0.2699
7.200	1.2487	0.9101	0.4936	23.900	0.5445	0.4287	0.2789	40.500	0.5205	0.4310	0.2698
7.300	1.1910	0.8978	0.4943	24.000	0.5442	0.4280	0.2804	40.600	0.5205	0.4309	0.2698
7.400	1.1490	0.8925	0.4961	24.100	0.5439	0.4273	0.2818	40.700	0.5206	0.4308	0.2697
7.500	1.1750	0.9047	0.4958	24.200	0.5435	0.4267	0.2829	40.800	0.5206	0.4308	0.2696
7.600	1.1977	0.9108	0.4914	24.300	0.5432	0.4263	0.2838	40.900	0.5206	0.4307	0.2696
7.700	1.2047	0.9062	0.4824	24.400	0.5428	0.4261	0.2844	41.000	0.5207	0.4306	0.2695
7.800	1.1899	0.8878	0.4705	24.500	0.5424	0.4261	0.2847	41.100	0.5208	0.4305	0.2695
7.900	1.1576	0.8590	0.4583	24.600	0.5420	0.4262	0.2846	41.200	0.5208	0.4304	0.2694
8.000	1.1221	0.8291	0.4516	24.700	0.5417	0.4266	0.2843	41.300	0.5209	0.4303	0.2694
8.100	1.0928	0.8101	0.4453	24.800	0.5413	0.4270	0.2853	41.400	0.5210	0.4302	0.2694

(continued)

Table 9.4 (continued)

Freq	Node 3	Node 2	Node 1	Freq	Node 3	Node 2	Node 1	Freq	Node 3	Node 2	Node 1
8.200	1.0692	0.7978	0.4378	24.900	0.5409	0.4277	0.2872	41.500	0.5211	0.4300	0.2693
8.300	1.0541	0.7874	0.4290	25.000	0.5405	0.4284	0.2888	41.600	0.5212	0.4299	0.2693
8.400	1.0482	0.7765	0.4243	25.100	0.5402	0.4292	0.2902	41.700	0.5213	0.4298	0.2693
8.500	1.0356	0.7618	0.4249	25.200	0.5399	0.4301	0.2915	41.800	0.5214	0.4297	0.2693
8.600	1.0100	0.7402	0.4261	25.300	0.5396	0.4309	0.2925	41.900	0.5214	0.4295	0.2693
8.700	0.9700	0.7117	0.4247	25.400	0.5393	0.4318	0.2934	42.000	0.5215	0.4294	0.2692
8.800	0.9218	0.6801	0.4186	25.500	0.5390	0.4326	0.2941	42.100	0.5216	0.4293	0.2692
8.900	0.8765	0.6515	0.4219	25.600	0.5388	0.4335	0.2947	42.200	0.5217	0.4291	0.2692
9.000	0.8441	0.6377	0.4295	25.700	0.5386	0.4343	0.2951	42.300	0.5218	0.4290	0.2692
9.100	0.8258	0.6383	0.4359	25.800	0.5384	0.4350	0.2954	42.400	0.5219	0.4288	0.2692
9.200	0.8143	0.6404	0.4427	25.900	0.5383	0.4358	0.2956	42.500	0.5219	0.4287	0.2692
9.300	0.8025	0.6442	0.4512	26.000	0.5381	0.4365	0.2957	42.600	0.5220	0.4286	0.2692
9.400	0.8041	0.6472	0.4602	26.100	0.5380	0.4371	0.2957	42.700	0.5220	0.4284	0.2692
9.500	0.8128	0.6476	0.4678	26.200	0.5379	0.4378	0.2956	42.800	0.5221	0.4285	0.2691
9.600	0.8159	0.6444	0.4728	26.300	0.5378	0.4384	0.2955	42.900	0.5221	0.4285	0.2691
9.700	0.8127	0.6386	0.4757	26.400	0.5377	0.4389	0.2953	43.000	0.5221	0.4286	0.2691
9.800	0.8041	0.6307	0.4767	26.500	0.5376	0.4395	0.2950	43.100	0.5221	0.4286	0.2691
9.900	0.7898	0.6208	0.4812	26.600	0.5375	0.4399	0.2947	43.200	0.5221	0.4286	0.2691
10.000	0.7695	0.6091	0.4864	26.700	0.5375	0.4404	0.2943	43.300	0.5221	0.4286	0.2690
10.100	0.7535	0.5963	0.4885	26.800	0.5375	0.4408	0.2939	43.400	0.5221	0.4286	0.2690
10.200	0.7391	0.5834	0.4880	26.900	0.5374	0.4412	0.2933	43.500	0.5220	0.4286	0.2690
10.300	0.7215	0.5712	0.4858	27.000	0.5374	0.4416	0.2928	43.600	0.5220	0.4286	0.2690
10.400	0.7017	0.5600	0.4823	27.100	0.5374	0.4419	0.2921	43.700	0.5219	0.4286	0.2690
10.500	0.6798	0.5496	0.4773	27.200	0.5373	0.4422	0.2914	43.800	0.5219	0.4285	0.2689
10.600	0.6674	0.5402	0.4702	27.300	0.5373	0.4424	0.2906	43.900	0.5218	0.4285	0.2689
10.700	0.6667	0.5323	0.4608	27.400	0.5373	0.4426	0.2898	44.000	0.5217	0.4284	0.2689
10.800	0.6650	0.5261	0.4494	27.500	0.5372	0.4427	0.2890	44.100	0.5216	0.4283	0.2688
10.900	0.6685	0.5241	0.4371	27.600	0.5372	0.4428	0.2881	44.200	0.5215	0.4283	0.2688
11.000	0.6666	0.5242	0.4259	27.700	0.5371	0.4429	0.2871	44.300	0.5213	0.4282	0.2688
11.100	0.6609	0.5227	0.4182	27.800	0.5370	0.4429	0.2862	44.400	0.5212	0.4281	0.2688
11.200	0.6551	0.5206	0.4187	27.900	0.5369	0.4428	0.2852	44.500	0.5211	0.4280	0.2687
11.300	0.6535	0.5187	0.4226	28.000	0.5368	0.4427	0.2843	44.600	0.5209	0.4278	0.2687
11.400	0.6577	0.5178	0.4298	28.100	0.5367	0.4426	0.2833	44.700	0.5208	0.4277	0.2687
11.500	0.6661	0.5178	0.4389	28.200	0.5366	0.4424	0.2824	44.800	0.5206	0.4276	0.2686
11.600	0.6746	0.5181	0.4466	28.300	0.5364	0.4422	0.2815	44.900	0.5205	0.4275	0.2686
11.700	0.6791	0.5179	0.4496	28.400	0.5363	0.4420	0.2810	45.000	0.5203	0.4274	0.2686
11.800	0.6765	0.5166	0.4459	28.500	0.5362	0.4417	0.2809	45.100	0.5201	0.4272	0.2686
11.900	0.6663	0.5159	0.4359	28.600	0.5360	0.4415	0.2808	45.200	0.5200	0.4271	0.2686
12.000	0.6541	0.5176	0.4218	28.700	0.5358	0.4411	0.2806	45.300	0.5198	0.4270	0.2685
12.100	0.6414	0.5187	0.4131	28.800	0.5356	0.4408	0.2805	45.400	0.5197	0.4268	0.2685
12.200	0.6253	0.5190	0.4026	28.900	0.5354	0.4404	0.2804	45.500	0.5195	0.4267	0.2685
12.300	0.6091	0.5187	0.3929	29.000	0.5352	0.4401	0.2802	45.600	0.5193	0.4266	0.2685
12.400	0.5951	0.5181	0.3858	29.100	0.5350	0.4397	0.2801	45.700	0.5192	0.4265	0.2685
12.500	0.5935	0.5174	0.3816	29.200	0.5348	0.4392	0.2801	45.800	0.5191	0.4263	0.2684

(continued)

Table 9.4 (continued)

Freq	Node 3	Node 2	Node 1	Freq	Node 3	Node 2	Node 1	Freq	Node 3	Node 2	Node 1
12.600	0.5979	0.5168	0.3807	29.300	0.5345	0.4388	0.2800	45.900	0.5189	0.4262	0.2684
12.700	0.5981	0.5162	0.3821	29.400	0.5342	0.4383	0.2799	46.000	0.5188	0.4261	0.2684
12.800	0.5926	0.5159	0.3899	29.500	0.5339	0.4379	0.2799	46.100	0.5187	0.4260	0.2684
12.900	0.5923	0.5158	0.4041	29.600	0.5336	0.4374	0.2798	46.200	0.5185	0.4259	0.2684
13.000	0.5965	0.5160	0.4124	29.700	0.5333	0.4369	0.2798	46.300	0.5184	0.4258	0.2684
13.100	0.6015	0.5165	0.4145	29.800	0.5329	0.4364	0.2797	46.400	0.5183	0.4258	0.2684
13.200	0.6066	0.5174	0.4111	29.900	0.5326	0.4359	0.2797	46.500	0.5182	0.4257	0.2683
13.300	0.6111	0.5185	0.4157	30.000	0.5322	0.4354	0.2797	46.600	0.5181	0.4256	0.2683
13.400	0.6142	0.5199	0.4261	30.100	0.5319	0.4349	0.2796	46.700	0.5180	0.4256	0.2683
13.500	0.6154	0.5211	0.4357	30.200	0.5315	0.4344	0.2796	46.800	0.5180	0.4255	0.2683
13.600	0.6142	0.5220	0.4481	30.300	0.5311	0.4339	0.2795	46.900	0.5179	0.4255	0.2683
13.700	0.6219	0.5222	0.4584	30.400	0.5308	0.4334	0.2795	47.000	0.5178	0.4255	0.2683
13.800	0.6346	0.5216	0.4652	30.500	0.5304	0.4330	0.2794	47.100	0.5178	0.4254	0.2683
13.900	0.6454	0.5200	0.4834	30.600	0.5301	0.4325	0.2793	47.200	0.5177	0.4254	0.2683
14.000	0.6539	0.5177	0.5047	30.700	0.5297	0.4321	0.2792	47.300	0.5177	0.4254	0.2683
14.100	0.6606	0.5147	0.5234	30.800	0.5294	0.4317	0.2791	47.400	0.5177	0.4254	0.2682
14.200	0.6663	0.5112	0.5385	30.900	0.5291	0.4314	0.2790	47.500	0.5176	0.4254	0.2682
14.300	0.6715	0.5072	0.5487	31.000	0.5288	0.4312	0.2788	47.600	0.5176	0.4254	0.2682
14.400	0.6766	0.5027	0.5532	31.100	0.5285	0.4312	0.2787	47.700	0.5176	0.4254	0.2682
14.500	0.6817	0.4994	0.5519	31.200	0.5283	0.4312	0.2786	47.800	0.5176	0.4255	0.2682
14.600	0.6870	0.4980	0.5452	31.300	0.5280	0.4312	0.2784	47.900	0.5176	0.4255	0.2682
14.700	0.6928	0.4959	0.5690	31.400	0.5278	0.4312	0.2783	48.000	0.5176	0.4255	0.2682
14.800	0.6990	0.4928	0.5808	31.500	0.5276	0.4313	0.2781	48.100	0.5176	0.4256	0.2682
14.900	0.7050	0.4883	0.5802	31.600	0.5274	0.4313	0.2780	48.200	0.5177	0.4256	0.2682
15.000	0.7093	0.4818	0.5686	31.700	0.5272	0.4313	0.2778	48.300	0.5177	0.4256	0.2682
15.100	0.7109	0.4819	0.5657	31.800	0.5270	0.4313	0.2777	48.400	0.5177	0.4257	0.2682
15.200	0.7277	0.4813	0.5675	31.900	0.5269	0.4313	0.2775	48.500	0.5178	0.4258	0.2681
15.300	0.7409	0.4777	0.5650	32.000	0.5268	0.4313	0.2774	48.600	0.5178	0.4258	0.2681
15.400	0.7488	0.4711	0.5695	32.100	0.5266	0.4313	0.2772	48.700	0.5179	0.4259	0.2681
15.500	0.7503	0.4620	0.5678	32.200	0.5266	0.4313	0.2771	48.800	0.5179	0.4259	0.2681
15.600	0.7457	0.4515	0.5583	32.300	0.5265	0.4313	0.2769	48.900	0.5180	0.4260	0.2681
15.700	0.7363	0.4407	0.5645	32.400	0.5265	0.4313	0.2768	49.000	0.5181	0.4261	0.2681
15.800	0.7238	0.4429	0.5688	32.500	0.5264	0.4313	0.2766	49.100	0.5181	0.4262	0.2681
15.900	0.7100	0.4501	0.5647	32.600	0.5264	0.4313	0.2764	49.200	0.5182	0.4262	0.2681
16.000	0.6963	0.4560	0.5527	32.700	0.5264	0.4313	0.2762	49.300	0.5183	0.4263	0.2681
16.100	0.6837	0.4606	0.5338	32.800	0.5264	0.4313	0.2761	49.400	0.5184	0.4264	0.2681
16.200	0.6726	0.4642	0.5096	32.900	0.5265	0.4313	0.2759	49.500	0.5185	0.4265	0.2680
16.300	0.6631	0.4671	0.4817	33.000	0.5265	0.4312	0.2757	49.600	0.5186	0.4266	0.2680
16.400	0.6551	0.4696	0.4533	33.100	0.5265	0.4312	0.2755	49.700	0.5187	0.4267	0.2680
16.500	0.6480	0.4719	0.4345	33.200	0.5266	0.4312	0.2754	49.800	0.5188	0.4268	0.2680
16.600	0.6414	0.4740	0.4213	33.300	0.5266	0.4311	0.2752	49.900	0.5188	0.4269	0.2680
16.700	0.6347	0.4757	0.4094								

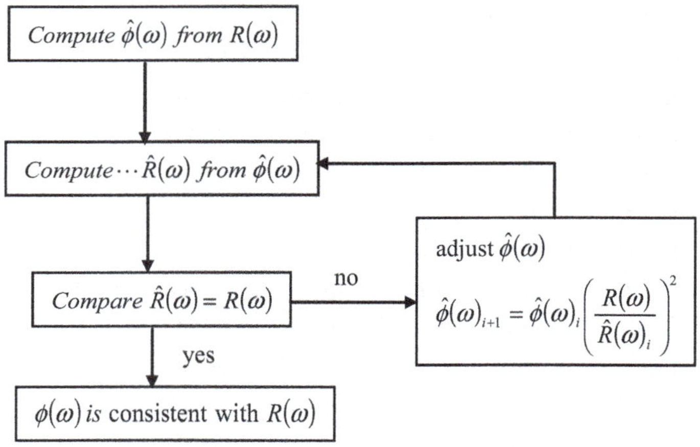

Fig. 9.23 Flowchart for iterative procedure

Table 9.5 Data for design response spectrum compatible PSD as shown in Fig. 9.8

Freq (f)	S(f)	Freq (f)	S(f)	Freq (f)	S(f)	Freq (f)	S(f)	Freq (f)	S(f)	Freq (f)	S(f)	Freq (f)	S(f)
0.3330	0.00081	0.7020	0.00164	1.4380	0.00322	2.9440	0.00394	6.0270	0.00338	12.3400	0.00060	25.2640	0.00005
0.3430	0.00079	0.7230	0.00169	1.4800	0.00325	3.0290	0.00395	6.2020	0.00325	12.6990	0.00055	25.9990	0.00005
0.3530	0.00081	0.7440	0.00175	1.5230	0.00328	3.1180	0.00396	6.3830	0.00310	13.0680	0.00051	26.7550	0.00004
0.3630	0.00083	0.7650	0.00181	1.5670	0.00330	3.2080	0.00398	6.5680	0.00294	13.4480	0.00047	27.5330	0.00004
0.3740	0.00085	0.7880	0.00187	1.6130	0.00333	3.3010	0.00399	6.7590	0.00279	13.8390	0.00043	28.3330	0.00004
0.3850	0.00087	0.8110	0.00193	1.6590	0.00336	3.3970	0.00401	6.9560	0.00265	14.2410	0.00039	29.1570	0.00003
0.3960	0.00089	0.8340	0.00199	1.7080	0.00340	3.4960	0.00402	7.1580	0.00251	14.6550	0.00036	30.0050	0.00003
0.4070	0.00092	0.8580	0.00206	1.7570	0.00343	3.5980	0.00404	7.3660	0.00238	15.0810	0.00033	30.8770	0.00003
0.4190	0.00095	0.8830	0.00213	1.8080	0.00346	3.7030	0.00405	7.5800	0.00224	15.5200	0.00030	31.7750	0.00003
0.4310	0.00097	0.9090	0.00220	1.8610	0.00349	3.8100	0.00407	7.8010	0.00208	15.9710	0.00027	32.6990	0.00003
0.4440	0.00100	0.9350	0.00228	1.9150	0.00353	3.9210	0.00408	8.0280	0.00193	16.4360	0.00024	33.6500	0.00003
0.4570	0.00103	0.9630	0.00235	1.9710	0.00356	4.0350	0.00410	8.2610	0.00179	16.9140	0.00022	34.6280	0.00003
0.4700	0.00106	0.9910	0.00244	2.0280	0.00359	4.1520	0.00410	8.5010	0.00166	17.4050	0.00020	35.6350	0.00003
0.4840	0.00109	1.0190	0.00252	2.0870	0.00363	4.2730	0.00410	8.7480	0.00154	17.9110	0.00018	36.6710	0.00003
0.4980	0.00113	1.0490	0.00261	2.1480	0.00367	4.3970	0.00410	9.0030	0.00143	18.4320	0.00016	37.7380	0.00002
0.5120	0.00116	1.0800	0.00270	2.2100	0.00370	4.5250	0.00410	9.2650	0.00132	18.9680	0.00014	38.8350	0.00002
0.5270	0.00120	1.1110	0.00279	2.2740	0.00374	4.6570	0.00410	9.5340	0.00122	19.5200	0.00012	39.9640	0.00002
0.5430	0.00124	1.1430	0.00289	2.3410	0.00378	4.7920	0.00411	9.8110	0.00113	20.0870	0.00011	41.1260	0.00002
0.5580	0.00128	1.1760	0.00299	2.4090	0.00382	4.9310	0.00405	10.0970	0.00105	20.6710	0.00010	42.3220	0.00002
0.5750	0.00132	1.2110	0.00307	2.4790	0.00386	5.0750	0.00400	10.3900	0.00097	21.2720	0.00009	43.5520	0.00002
0.5910	0.00136	1.2460	0.00309	2.5510	0.00388	5.2220	0.00394	10.6920	0.00090	21.8910	0.00008	44.8190	0.00002
0.6090	0.00140	1.2820	0.00312	2.6250	0.00389	5.3740	0.00389	11.0030	0.00083	22.5270	0.00008	46.1220	0.00001
0.6260	0.00145	1.3190	0.00314	2.7010	0.00390	5.5310	0.00382	11.3230	0.00076	23.1820	0.00007	47.4630	0.00001
0.6440	0.00149	1.3580	0.00317	2.7800	0.00391	5.6910	0.00367	11.6520	0.00070	23.8570	0.00006	48.8430	0.00001
0.6630	0.00154	1.3970	0.00319	2.8610	0.00393	5.8570	0.00352	11.9910	0.00065	24.5500	0.00006	50.2630	0.00001
0.6820	0.00159												

$H^*(\omega, \omega_0)$ is the complex conjugate transfer function
Now,

$$\psi(\omega, \omega_0) = \left[\frac{\omega_0^4 + 4\omega_0^2\beta^2\omega^2}{\left(\omega_0^2 - \omega^2\right)^2 + 4\omega_0^2\beta^2\omega^2}\right]\phi(\omega) \qquad (9.17)$$

RMS value or standard deviation of the response of SDOF having frequency ω_0 is given as

$$\sigma^2(\omega_0) = \int_0^{\omega_{\max}} \psi(\omega, \omega_0)\, d\omega \qquad (9.18)$$

Response of SDOF in terms of standard deviation is given as

$$R(\omega_0) = F_0(\omega_0) \times \sigma(\omega_0) \qquad (9.19)$$

where F_0 is the factor of multiplication with standard deviation to obtain the peak response of SDOF, and it is given as

$$F_0(\omega_0) = \left[-2\ln\left\{-\left(\frac{\pi}{T}\right)\left(\frac{\sigma}{\dot{\sigma}}\right)\ln(1-r)\right\}\right]^{\frac{1}{2}} \qquad (9.20)$$

where

T = Earthquake time duration [1]

$$\dot{\sigma}^2(\omega_0) = \int_0^{\omega_{\max}} \omega^2\psi(\omega, \omega_0)d\omega \qquad (9.21)$$

r = Probability of exceedance (typical value is 0.15)

To have compatible mapping between and earthquake response acceleration and PSDF, an approximate solution is given as

$$\hat{\phi}(\omega_0) = \frac{2\beta}{\pi\omega_0}R^2(\omega_0) \times \left\{-2\ln\left[-\left(\frac{\pi}{\omega_0 T}\right)\ln(1-r)\right]\right\}^{-1} \qquad (9.22)$$

which is used as an initial estimate in an iterative procedure for which flowchart is given as follows (Fig. 9.23 and Table 9.5, 9.6).

A.2.2 MATLAB Program

MATLAB program for evaluation of PSDF from response spectrum using the above-mentioned procedure is given below:

```
%Program starts
clear all; clc; close all;
filename='responsespectrum.xlsx'; % contains response spectrum in the format given in Fig. 9.2
f=xlsread(filename,'A3:A29');
T=xlsread(filename,'B3:B29');
Sa=xlsread(filename,'C3:C29');
f1 =0.1:0.1:3;
f2=3.15:0.15:3.6;
f3=3.8:0.2:5.0;
f4=5.25:0.25:8;
f5=8.5:0.5:15;
f6=16:1:18;
f7=20:2:22;
f8=25:3:34;
f9=35:5:50;
inputfre=[f1 f2 f3 f4 f5 f6 f7 f8 f9]';
Sa_new= interp1(f,Sa,inputfre);
figure (1)
plot(f,Sa,inputfre,Sa_new);
title('Input spectrum')
xlabel('Frequency (Hz)')
ylabel('Spectral Acceleration (g)')
beta=0.05; % damping 5%
r=0.15; % probablity of exceedance
Ttotal=19; % sec
R=Sa_new;
for i=1:1:length(inputfre)
err(i)=1;
omega0(i)=2*pi*inputfre(i);
c1=-2*log(-(pi/(omega0(i)*Ttotal))*log(1-r));
phi_cap(i) =(2*beta*R(i)*R(i)/(pi*omega0(i)/c1));
while (err(i)>=1E-2)
for j=1:1:length(inputfre)
omega(j) =2*pi*inputfre(j);
c2=(omega0(i)^4)+4*(omega0(i)^2)*beta*beta*omega(j)^2;
c3=(omega0(i)^2-omega(j)^2)^2+4*omega0(i)^2*beta*-
beta*omega(j)^2;
chi(i,j)=c2*phi_cap(i)/c3;
chidot(i,j)=chi(i,j)*omega(j)^2;
end
sigma2(i)=trapz(chi(i,:),omega);
sigmadot2(i)=trapz(chidot(i,:),omega);
sigmadot(i)=sqrt(abs(sigmadot2(i)));
sigma(i)=sqrt(abs(sigma2(i)));
F0(i)= sqrt(abs(-2*log(-(pi/Ttotal)*(sigma(i)/sigmadot(i))
*log(1-r))));
R_cap(i)=F0(i)*sigma(i);
err(i)=abs(R_cap(i)-R(i));
phi_cap(i) =(2*beta*R(i)*R(i)/(pi*omega0(i)/c1))*(R(i)/
R_cap(i))^2;
end
end
figure (2)
plot(inputfre,phi_cap)
```

Table 9.6 Data for typical design response spectra at 5 and 7% damping as shown in Fig. 9.11

f (Hz)	Acc(g) (ξ–7%)	Acc(g) (ξ–5%)
0.100	0.010	0.010
0.300	0.023	0.024
0.400	0.034	0.038
0.600	0.057	0.068
1.200	0.138	0.168
2.500	0.253	0.316
4.000	0.351	0.427
4.800	0.395	0.463
5.500	0.417	0.464
6.300	0.415	0.452
7.500	0.398	0.429
15.000	0.269	0.288
17.000	0.248	0.266
20.000	0.223	0.238
26.800	0.203	0.210
30.000	0.202	0.205
33.900	0.203	0.203
35.970	0.204	0.203
38.310	0.205	0.204
40.820	0.204	0.203
43.480	0.204	0.203
46.510	0.201	0.201
49.500	0.200	0.200
50.000	0.200	0.200

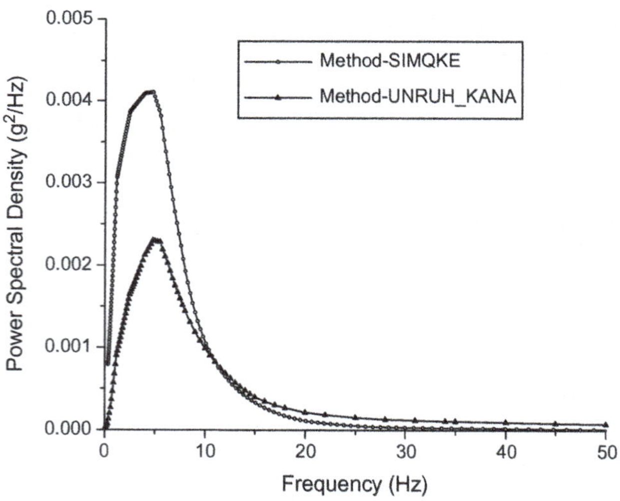

Fig. 9.24 Comparison of PSDF from design response spectrum

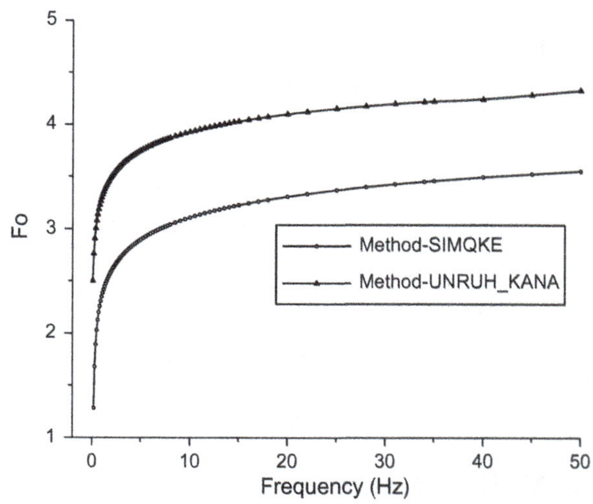

Fig. 9.25 Comparison of F_o using two methods

```
title('PSD');
xlabel('Frequency (Hz)');
ylabel('Power spectrum density');
%Program end
```

The comparison of PSDF from design response spectrum shown in Fig. 9.1 using the two methods is shown in Fig. 9.24. It is observed that the PSDF obtained from method SIMQK is higher than that of method

Table 9.7 Data for calculation of floor PSDs

ω_l	S(f)	ω_1	ω/ω_1	$1-(\omega/\omega_1)^2$	$(1-(\omega/\omega_1)^2)^2$	$(2*\xi*(\omega/\omega_1))^2$	$(1-(\omega/\omega_1)^2)^2+(2*\xi*(\omega/\omega_1))^2$	$H(f)^2$	$H(f)$	$\varphi_{i1}\varphi_{i1}\Gamma_1\Gamma_1 H_1 H_1^* S_{in}(\omega_l)$
0.3330	0.000805	33.978	0.0616	0.99621	0.99243	0.00007432	0.99250491	1.00755	1.003769	0.001193
0.3430	0.000789	33.978	0.0634	0.99598	0.99197	0.00007885	0.99204893	1.00801	1.003999	0.001170
0.3530	0.000808	33.978	0.0653	0.99574	0.99150	0.00008352	0.99157956	1.00849	1.004237	0.001199
0.3630	0.000828	33.978	0.0671	0.99549	0.99101	0.00008832	0.99109683	1.00898	1.004482	0.001229
0.3740	0.000850	33.978	0.0692	0.99522	0.99046	0.00009375	0.99055039	1.00954	1.004759	0.001262
0.3850	0.000872	33.978	0.0712	0.99493	0.98989	0.00009934	0.98998781	1.01011	1.005044	0.001295
0.3960	0.000895	33.978	0.0732	0.99464	0.98930	0.00010510	0.98940908	1.01070	1.005338	0.001330
0.4070	0.000920	33.978	0.0753	0.99434	0.98870	0.00011102	0.98881424	1.01131	1.005640	0.001368
0.4190	0.000946	33.978	0.0775	0.99400	0.98803	0.00011767	0.98814695	1.01200	1.005980	0.001408
0.4310	0.000973	33.978	0.0797	0.99365	0.98734	0.00012450	0.98746051	1.01270	1.006329	0.001450
0.4440	0.001002	33.978	0.0821	0.99326	0.98656	0.00013213	0.98669528	1.01348	1.006719	0.001493
0.4570	0.001031	33.978	0.0845	0.99286	0.98577	0.00013998	0.98590763	1.01429	1.007122	0.001538
0.4700	0.001062	33.978	0.0869	0.99245	0.98495	0.00014805	0.98509758	1.01513	1.007536	0.001586
0.4840	0.001094	33.978	0.0895	0.99199	0.98404	0.00015701	0.98420021	1.01605	1.007995	0.001635
0.4980	0.001128	33.978	0.0921	0.99152	0.98311	0.00016622	0.98327695	1.01701	1.008468	0.001687
0.5120	0.001162	33.978	0.0947	0.99104	0.98215	0.00017570	0.98232781	1.01799	1.008955	0.001740
0.5270	0.001199	33.978	0.0975	0.99050	0.98110	0.00018614	0.98128222	1.01907	1.009492	0.001797
0.5430	0.001237	33.978	0.1004	0.98992	0.97994	0.00019762	0.98013431	1.02027	1.010083	0.001856
0.5580	0.001276	33.978	0.1032	0.98935	0.97882	0.00020869	0.97902761	1.02142	1.010654	0.001917
0.5750	0.001317	33.978	0.1063	0.98869	0.97752	0.00022159	0.97773770	1.02277	1.011321	0.001980
0.5910	0.001359	33.978	0.1093	0.98806	0.97626	0.00023410	0.97648914	1.02408	1.011967	0.002047
0.6090	0.001402	33.978	0.1126	0.98732	0.97480	0.00024858	0.97504457	1.02559	1.012716	0.002115
0.6260	0.001447	33.978	0.1158	0.98660	0.97338	0.00026265	0.97364150	1.02707	1.013446	0.002185
0.6440	0.001493	33.978	0.1191	0.98582	0.97184	0.00027797	0.97211497	1.02868	1.014241	0.002258
0.6630	0.001540	33.978	0.1226	0.98497	0.97016	0.00029461	0.97045807	1.03044	1.015107	0.002334
0.6820	0.001590	33.978	0.1261	0.98409	0.96844	0.00031174	0.96875450	1.03225	1.015999	0.002413
0.7020	0.001641	33.978	0.1298	0.98315	0.96658	0.00033029	0.96691100	1.03422	1.016967	0.002496
0.7230	0.001694	33.978	0.1337	0.98213	0.96457	0.00035035	0.96492000	1.03636	1.018015	0.002583
0.7440	0.001750	33.978	0.1376	0.98107	0.96250	0.00037100	0.96287250	1.03856	1.019097	0.002672
0.7650	0.001807	33.978	0.1415	0.97999	0.96038	0.00039224	0.96076870	1.04083	1.020212	0.002767
0.7880	0.001867	33.978	0.1457	0.97877	0.95798	0.00041618	0.95840017	1.04341	1.021472	0.002865
0.8110	0.001929	33.978	0.1500	0.97751	0.95552	0.00044082	0.95596460	1.04606	1.022773	0.002968
0.8340	0.001994	33.978	0.1542	0.97622	0.95300	0.00046618	0.95346226	1.04881	1.024114	0.003075
0.8580	0.002060	33.978	0.1587	0.97483	0.95029	0.00049340	0.95078025	1.05177	1.025557	0.003187
0.8830	0.002130	33.978	0.1633	0.97334	0.94739	0.00052257	0.94790986	1.05495	1.027109	0.003305
0.9090	0.002202	33.978	0.1681	0.97174	0.94429	0.00055380	0.94484211	1.05838	1.028775	0.003428
0.9350	0.002277	33.978	0.1729	0.97011	0.94110	0.00058593	0.94169065	1.06192	1.030495	0.003556
0.9630	0.002355	33.978	0.1781	0.96829	0.93758	0.00062155	0.93820367	1.06587	1.032408	0.003691
0.9910	0.002436	33.978	0.1833	0.96642	0.93396	0.00065822	0.93462070	1.06995	1.034385	0.003832
1.0190	0.002519	33.978	0.1884	0.96449	0.93025	0.00069594	0.93094235	1.07418	1.036427	0.003980
1.0490	0.002606	33.978	0.1940	0.96237	0.92616	0.00073752	0.92689612	1.07887	1.038686	0.004135
1.0800	0.002696	33.978	0.1997	0.96011	0.92182	0.00078176	0.92260157	1.08389	1.041101	0.004298

(continued)

Table 9.7 (continued)

ω_l	S(f)	ω_1	ω/ω_1	$1-(\omega/\omega_1)^2$	$(1-(\omega/\omega_1)^2)^2$	$(2*\xi*(\omega/\omega_1))^2$	$(1-(\omega/\omega_1)^2)^2+(2*\xi*(\omega/\omega_1))^2$	$H(f)^2$	H(f)	$\varphi_{i1}\varphi_{i1}\Gamma_1\Gamma_1 H_1 H_1^* S_{in}(\omega_l)$
1.1110	0.002790	33.978	0.2054	0.95779	0.91737	0.00082728	0.91819259	1.08910	1.043598	0.004469
1.1430	0.002888	33.978	0.2114	0.95533	0.91265	0.00087562	0.91352233	1.09466	1.046262	0.004649
1.1760	0.002989	33.978	0.2175	0.95271	0.90765	0.00092691	0.90858054	1.10062	1.049103	0.004838
1.2110	0.003068	33.978	0.2239	0.94985	0.90222	0.00098291	0.90320119	1.10717	1.052223	0.004995
1.2460	0.003092	33.978	0.2304	0.94691	0.89664	0.00104054	0.89768114	1.11398	1.055453	0.005065
1.2820	0.003116	33.978	0.2371	0.94380	0.89076	0.00110154	0.89185815	1.12125	1.058893	0.005138
1.3190	0.003141	33.978	0.2439	0.94051	0.88456	0.00116604	0.88572165	1.12902	1.062555	0.005216
1.3580	0.003167	33.978	0.2511	0.93694	0.87785	0.00123601	0.87908895	1.13754	1.066556	0.005298
1.3970	0.003193	33.978	0.2583	0.93326	0.87098	0.00130803	0.87228960	1.14641	1.070705	0.005383
1.4380	0.003220	33.978	0.2659	0.92929	0.86358	0.00138593	0.86496441	1.15612	1.075229	0.005475
1.4800	0.003247	33.978	0.2737	0.92510	0.85581	0.00146807	0.85727509	1.16649	1.080040	0.005571
1.5230	0.003276	33.978	0.2816	0.92068	0.84766	0.00155462	0.84921137	1.17756	1.085156	0.005673
1.5670	0.003305	33.978	0.2898	0.91603	0.83912	0.00164574	0.84076322	1.18940	1.090594	0.005781
1.6130	0.003334	33.978	0.2983	0.91103	0.82998	0.00174378	0.83172209	1.20232	1.096506	0.005896
1.6590	0.003365	33.978	0.3068	0.90588	0.82063	0.00184466	0.82247163	1.21585	1.102655	0.006016
1.7080	0.003396	33.978	0.3158	0.90024	0.81044	0.00195524	0.81239265	1.23093	1.109474	0.006147
1.7570	0.003427	33.978	0.3249	0.89444	0.80002	0.00206903	0.80208677	1.24675	1.116579	0.006284
1.8080	0.003459	33.978	0.3343	0.88822	0.78893	0.00219089	0.79112542	1.26402	1.124287	0.006430
1.8610	0.003492	33.978	0.3441	0.88157	0.77717	0.00232122	0.77948747	1.28289	1.132649	0.006589
1.9150	0.003526	33.978	0.3541	0.87460	0.76492	0.00245788	0.76737911	1.30314	1.141550	0.006757
1.9710	0.003560	33.978	0.3645	0.86716	0.75196	0.00260374	0.75456376	1.32527	1.151203	0.006938
2.0280	0.003595	33.978	0.3750	0.85936	0.73850	0.00275651	0.74125901	1.34906	1.161489	0.007132
2.0870	0.003630	33.978	0.3859	0.85106	0.72430	0.00291923	0.72722159	1.37510	1.172645	0.007342
2.1480	0.003667	33.978	0.3972	0.84223	0.70934	0.00309238	0.71243647	1.40363	1.184751	0.007570
2.2100	0.003704	33.978	0.4087	0.83299	0.69387	0.00327347	0.69713951	1.43443	1.197678	0.007814
2.2740	0.003742	33.978	0.4205	0.82317	0.67761	0.00346581	0.68107952	1.46826	1.211717	0.008080
2.3410	0.003780	33.978	0.4329	0.81260	0.66032	0.00367305	0.66399110	1.50604	1.227210	0.008372
2.4090	0.003819	33.978	0.4455	0.80155	0.64249	0.00388953	0.64637907	1.54708	1.243817	0.008690
2.4790	0.003860	33.978	0.4584	0.78985	0.62387	0.00411886	0.62798843	1.59239	1.261898	0.009039
2.5510	0.003879	33.978	0.4717	0.77747	0.60446	0.00436159	0.60882110	1.64252	1.281608	0.009371
2.6250	0.003890	33.978	0.4854	0.76437	0.58426	0.00461830	0.58888328	1.69813	1.303123	0.009715
2.7010	0.003901	33.978	0.4995	0.75053	0.56330	0.00488960	0.56818608	1.75999	1.326645	0.010098
2.7800	0.003913	33.978	0.5141	0.73572	0.54129	0.00517981	0.54646993	1.82993	1.352748	0.010531
2.8610	0.003926	33.978	0.5291	0.72010	0.51854	0.00548605	0.52402951	1.90829	1.381409	0.011017
2.9440	0.003938	33.978	0.5444	0.70362	0.49509	0.00580897	0.50089537	1.99642	1.412949	0.011562
3.0290	0.003951	33.978	0.5601	0.68626	0.47096	0.00614925	0.47710559	2.09597	1.447747	0.012179
3.1180	0.003964	33.978	0.5766	0.66755	0.44563	0.00651592	0.45214541	2.21168	1.487171	0.012895
3.2080	0.003978	33.978	0.5932	0.64809	0.42002	0.00689751	0.42691304	2.34240	1.530489	0.013705
3.3010	0.003992	33.978	0.6104	0.62739	0.39361	0.00730323	0.40091680	2.49428	1.579330	0.014644
3.3970	0.004007	33.978	0.6282	0.60540	0.36651	0.00773419	0.37424139	2.67207	1.634647	0.015745
3.4960	0.004021	33.978	0.6465	0.58206	0.33880	0.00819156	0.34698913	2.88193	1.697626	0.017043
3.5980	0.004037	33.978	0.6653	0.55732	0.31061	0.00867653	0.31928190	3.13203	1.769754	0.018593

(continued)

Table 9.7 (continued)

ω_l	S(f)	ω_1	ω/ω_1	$1-(\omega/\omega_1)^2$	$(1-(\omega/\omega_1)^2)^2$	$(2*\xi*(\omega/\omega_1))^2$	$(1-(\omega/\omega_1)^2)^2+(2*\xi*(\omega/\omega_1))^2$	H(f)²	H(f)	$\varphi_{i1}\varphi_{i1}\Gamma_1\Gamma_1 H_1 H_1^* S_{in}(\omega_l)$
3.7030	0.004052	33.978	0.6848	0.53111	0.28207	0.00919033	0.29126332	3.43332	1.852922	0.020460
3.8100	0.004068	33.978	0.7045	0.50362	0.25363	0.00972913	0.26335826	3.79711	1.948617	0.022714
3.9210	0.004084	33.978	0.7251	0.47427	0.22493	0.01030428	0.23523781	4.25102	2.061800	0.025530
4.0350	0.004095	33.978	0.7462	0.44326	0.19648	0.01091216	0.20738884	4.82186	2.195873	0.029042
4.1520	0.004097	33.978	0.7678	0.41050	0.16851	0.01155416	0.18006592	5.55352	2.356591	0.033457
4.2730	0.004098	33.978	0.7902	0.37564	0.14111	0.01223741	0.15334446	6.52127	2.553677	0.039297
4.3970	0.004099	33.978	0.8131	0.33888	0.11484	0.01295796	0.12779720	7.82490	2.797302	0.047171
4.5250	0.004101	33.978	0.8368	0.29983	0.08990	0.01372338	0.10362002	9.65065	3.106549	0.058207
4.6570	0.004103	33.978	0.8612	0.25838	0.06676	0.01453571	0.08129698	12.30058	3.507218	0.074227
4.7920	0.004105	33.978	0.8861	0.21476	0.04612	0.01539067	0.06151332	16.25664	4.031953	0.098148
4.9310	0.004052	33.978	0.9118	0.16855	0.02841	0.01629648	0.04470453	22.36910	4.729598	0.133303
5.0750	0.003997	33.978	0.9385	0.11928	0.01423	0.01726219	0.03148894	31.75718	5.635351	0.186670
5.2220	0.003943	33.978	0.9657	0.06752	0.00456	0.01827669	0.02283507	43.79229	6.617575	0.253922
5.3740	0.003889	33.978	0.9938	0.01244	0.00015	0.01935616	0.01951093	51.25331	7.159142	0.293138
5.5310	0.003817	33.978	1.0228	−0.04610	0.00213	0.02050365	0.02262928	44.19054	6.647597	0.248081
5.6910	0.003667	33.978	1.0524	−0.10750	0.01156	0.02170706	0.03326397	30.06256	5.482933	0.162105
5.8570	0.003522	33.978	1.0831	−0.17305	0.02995	0.02299187	0.05293976	18.88940	4.346193	0.097842
6.0270	0.003384	33.978	1.1145	−0.24214	0.05863	0.02434592	0.08297715	12.05151	3.471528	0.059972
6.2020	0.003251	33.978	1.1469	−0.31532	0.09943	0.02578026	0.12520672	7.98679	2.826091	0.038181
6.3830	0.003105	33.978	1.1803	−0.39321	0.15462	0.02730697	0.18192328	5.49682	2.344530	0.025097
6.5680	0.002944	33.978	1.2146	−0.47514	0.22576	0.02891280	0.25467346	3.92660	1.981564	0.017001
6.7590	0.002791	33.978	1.2499	−0.56219	0.31605	0.03061884	0.34667162	2.88457	1.698403	0.011841
6.9560	0.002647	33.978	1.2863	−0.65458	0.42847	0.03242970	0.46090042	2.16967	1.472979	0.008445
7.1580	0.002509	33.978	1.3237	−0.75207	0.56561	0.03434055	0.59994804	1.66681	1.291050	0.006151
7.3660	0.002379	33.978	1.3621	−0.85537	0.73166	0.03636531	0.76802803	1.30204	1.141068	0.004556
7.5800	0.002238	33.978	1.4017	−0.96474	0.93073	0.03850900	0.96924182	1.03173	1.015743	0.003396
7.8010	0.002078	33.978	1.4426	−1.08098	1.16852	0.04078725	1.20930935	0.82692	0.909350	0.002527
8.0280	0.001928	33.978	1.4845	−1.20385	1.44926	0.04319551	1.49245601	0.67004	0.818558	0.001900
8.2610	0.001789	33.978	1.5276	−1.33364	1.77858	0.04573925	1.82432277	0.54815	0.740371	0.001442
8.5010	0.001659	33.978	1.5720	−1.47120	2.16443	0.04843551	2.21286339	0.45190	0.672237	0.001103
8.7480	0.001538	33.978	1.6177	−1.61689	2.61433	0.05129103	2.66562139	0.37515	0.612492	0.000849
9.0030	0.001426	33.978	1.6648	−1.77167	3.13883	0.05432483	3.19315635	0.31317	0.559616	0.000657
9.2650	0.001321	33.978	1.7133	−1.93534	3.74555	0.05753269	3.80307895	0.26294	0.512781	0.000511
9.5340	0.001223	33.978	1.7630	−2.10827	4.44478	0.06092200	4.50570411	0.22194	0.471106	0.000399
9.8110	0.001132	33.978	1.8143	−2.29150	5.25099	0.06451347	5.31550175	0.18813	0.433738	0.000313
10.0970	0.001047	33.978	1.8671	−2.48620	6.18120	0.06832955	6.24952731	0.16001	0.400015	0.000246
10.3900	0.000969	33.978	1.9213	−2.69147	7.24399	0.07235273	7.31634159	0.13668	0.369703	0.000195
10.6920	0.000895	33.978	1.9772	−2.90918	8.46333	0.07661993	8.53994828	0.11710	0.342194	0.000154
11.0030	0.000827	33.978	2.0347	−3.13990	9.85898	0.08114207	9.94012274	0.10060	0.317179	0.000122
11.3230	0.000763	33.978	2.0938	−3.38420	11.45284	0.08593040	11.53876918	0.08666	0.294388	0.000097
11.6520	0.000705	33.978	2.1547	−3.64268	13.26912	0.09099652	13.36011183	0.07485	0.273587	0.000078
11.9910	0.000649	33.978	2.2174	−3.91676	15.34097	0.09636840	15.43733878	0.06478	0.254515	0.000062

(continued)

Table 9.7 (continued)

ω_l	S(f)	ω_1	ω/ω_1	$1-(\omega/\omega_1)^2$	$(1-(\omega/\omega_1)^2)^2$	$(2*\xi*(\omega/\omega_1))^2$	$(1-(\omega/\omega_1)^2)^2+(2*\xi*(\omega/\omega_1))^2$	$H(f)^2$	$H(f)$	$\varphi_{i1}\varphi_{i1}\Gamma_1\Gamma_1 H_1 H_1^* S_{in}(\omega_l)$
12.3400	0.000598	33.978	2.2819	−4.20713	17.69991	0.10205967	17.80196908	0.05617	0.237010	0.000049
12.6990	0.000551	33.978	2.3483	−4.51451	20.38079	0.10808437	20.48887199	0.04881	0.220923	0.000040
13.0680	0.000507	33.978	2.4165	−4.83964	23.42211	0.11445693	23.53656529	0.04249	0.206124	0.000032
13.4480	0.000466	33.978	2.4868	−5.18419	26.87588	0.12121022	26.99708729	0.03704	0.192460	0.000025
13.8390	0.000427	33.978	2.5591	−5.54903	30.79177	0.12836105	30.92012985	0.03234	0.179837	0.000020
14.2410	0.000392	33.978	2.6334	−5.93504	35.22466	0.13592671	35.36058331	0.02828	0.168167	0.000016
14.6550	0.000359	33.978	2.7100	−6.34411	40.24778	0.14394464	40.39172825	0.02476	0.157345	0.000013
15.0810	0.000328	33.978	2.7888	−6.77729	45.93160	0.15243480	46.08403536	0.02170	0.147307	0.000010
15.5200	0.000298	33.978	2.8700	−7.23666	52.36927	0.16143856	52.53070414	0.01904	0.137973	0.000008
15.9710	0.000269	33.978	2.9534	−7.72232	59.63422	0.17095746	59.80517530	0.01672	0.129310	0.000007
16.4360	0.000243	33.978	3.0393	−8.23762	67.85836	0.18105733	68.03942014	0.01470	0.121233	0.000005
16.9140	0.000219	33.978	3.1277	−8.78274	77.13649	0.19174166	77.32822827	0.01293	0.113718	0.000004
17.4050	0.000197	33.978	3.2185	−9.35895	87.58998	0.20303546	87.79301982	0.01139	0.106726	0.000003
17.9110	0.000177	33.978	3.3121	−9.97002	99.40131	0.21501240	99.61632024	0.01004	0.100192	0.000003
18.4320	0.000157	33.978	3.4084	−10.61750	112.73131	0.22770301	112.95901600	0.00885	0.094089	0.000002
18.9680	0.000140	33.978	3.5076	−11.30300	127.75770	0.24113870	127.99883721	0.00781	0.088389	0.000002
19.5200	0.000124	33.978	3.6096	−12.02949	144.70861	0.25537799	144.96399271	0.00690	0.083056	0.000001
20.0870	0.000110	33.978	3.7145	−12.79742	163.77399	0.27042946	164.04442393	0.00610	0.078076	0.000001
20.6710	0.000101	33.978	3.8225	−13.61136	185.26921	0.28638272	185.55559747	0.00539	0.073411	0.000001
21.2720	0.000092	33.978	3.9336	−14.47335	209.47793	0.30327771	209.78120908	0.00477	0.069043	0.000001
21.8910	0.000084	33.978	4.0481	−15.38698	236.75921	0.32118484	237.08039388	0.00422	0.064946	0.000001
22.5270	0.000077	33.978	4.1657	−16.35300	267.42051	0.34011874	267.76062430	0.00373	0.061112	0.000000
23.1820	0.000070	33.978	4.2868	−17.37679	301.95271	0.36018502	302.31289814	0.00331	0.057514	0.000000
23.8570	0.000063	33.978	4.4116	−18.46254	340.86524	0.38146571	341.24670417	0.00293	0.054133	0.000000
24.5500	0.000057	33.978	4.5398	−19.60966	384.53864	0.40394928	384.94259402	0.00260	0.050969	0.000000
25.2640	0.000052	33.978	4.6718	−20.82589	433.71776	0.42778748	434.14555039	0.00230	0.047993	0.000000
25.9990	0.000046	33.978	4.8077	−22.11432	489.04300	0.45304061	489.49603927	0.00204	0.045199	0.000000
26.7550	0.000041	33.978	4.9475	−23.47810	551.22111	0.47977073	551.70088546	0.00181	0.042574	0.000000
27.5330	0.000039	33.978	5.0914	−24.92238	621.12492	0.50807861	621.63300336	0.00161	0.040108	0.000000
28.3330	0.000037	33.978	5.2393	−26.45067	699.63776	0.53803306	700.17578874	0.00143	0.037792	0.000000
29.1570	0.000035	33.978	5.3917	−28.07056	787.95651	0.56978304	788.52629217	0.00127	0.035612	0.000000
30.0050	0.000033	33.978	5.5485	−29.78613	887.21326	0.60340805	887.81666468	0.00113	0.033561	0.000000
30.8770	0.000031	33.978	5.7098	−31.60153	998.65662	0.63898996	999.29560762	0.00100	0.031634	0.000000
31.7750	0.000030	33.978	5.8758	−33.52541	1123.95335	0.67669810	1124.63004469	0.00089	0.029819	0.000000
32.6990	0.000029	33.978	6.0467	−35.56257	1264.69637	0.71662637	1265.41299459	0.00079	0.028111	0.000000
33.6500	0.000028	33.978	6.2226	−37.72023	1422.81561	0.75891647	1423.57452162	0.00070	0.026504	0.000000
34.6280	0.000027	33.978	6.4034	−40.00366	1600.29254	0.80367167	1601.09620940	0.00062	0.024991	0.000000
35.6350	0.000026	33.978	6.5896	−42.42315	1799.72341	0.85109368	1800.57450332	0.00056	0.023566	0.000000
36.6710	0.000025	33.978	6.7812	−44.98469	2023.62246	0.90129995	2024.52375512	0.00049	0.022225	0.000000
37.7380	0.000024	33.978	6.9785	−47.69962	2275.25329	0.95451246	2276.20780448	0.00044	0.020960	0.000000
38.8350	0.000023	33.978	7.1814	−50.57205	2557.53216	1.01081216	2558.54297488	0.00039	0.019770	0.000000
39.9640	0.000022	33.978	7.3901	−53.61421	2874.48371	1.07043855	2875.55414407	0.00035	0.018648	0.000000

(continued)

Table 9.7 (continued)

ω_l	S(f)	ω_1	ω/ω_1	$1-(\omega/\omega_1)^2$	$(1-(\omega/\omega_1)^2)^2$	$(2*\xi*(\omega/\omega_1))^2$	$(1-(\omega/\omega_1)^2)^2+(2*\xi*(\omega/\omega_1))^2$	$H(f)^2$	$H(f)$	$\varphi_{i1}\varphi_{i1}\Gamma_1\Gamma_1 H_1 H_1^* S_{in}(\omega_l)$
41.1260	0.000020	33.978	7.6050	−56.83633	3230.36818	1.13359203	3231.50177069	0.00031	0.017591	0.000000
42.3220	0.000020	33.978	7.8262	−60.24916	3629.96125	1.20048353	3631.16173513	0.00028	0.016595	0.000000
43.5520	0.000019	33.978	8.0536	−63.86105	4078.23374	1.27127658	4079.50501477	0.00025	0.015657	0.000000
44.8190	0.000017	33.978	8.2879	−67.68978	4581.90565	1.34631959	4583.25196800	0.00022	0.014771	0.000000
46.1220	0.000015	33.978	8.5289	−71.74180	5146.88560	1.42573924	5148.31134079	0.00019	0.013937	0.000000
47.4630	0.000014	33.978	8.7769	−76.03324	5781.05292	1.50985142	5782.56277483	0.00017	0.013150	0.000000
48.8430	0.000012	33.978	9.0320	−80.57788	6492.79525	1.59892651	6494.39417559	0.00015	0.012409	0.000000
50.2630	0.000011	33.978	9.2946	−85.39022	7291.48975	1.69324832	7293.18299689	0.00014	0.011710	0.000000

UNRUH_KANA. This is due to higher F_o values used in method UNRUH_KANA. The comparison of F_o values for both methods is shown in Fig. 9.25, which indicates higher F_o values for method UNRUH_KANA compared to that of method SIMQKE (Tables 9.2 and 9.3).

A.2.3 Calculation of Floor PSDs

$$S_{oi}(\omega_l) = \sum_{j=1}^{n}\sum_{k=1}^{n} \phi_{ij}\phi_{ik}\Gamma_j\Gamma_k H_j H_k^* S_{in}(\omega_l) \quad (9.23)$$

where

ω_1 2 × π × 5.408 (first natural frequency of the structure)
ξ 0.07 (damping of structure)

See Table 9.7

References

1. ASCE 4-98 (1998) Seismic Analysis of Safety-Related Nuclear Structures and Commentary, American Society of Civil Engineers
2. IEEE Std 344 (1987) IEEE Recommended Practice for Seismic Qualification of Class 1E Equipment for Nuclear Power Generating Stations
3. SIMQKE (1976) A Program for Artificial Motion Generation-User's Manual and Documentation's, NISEE
4. Unruh JF, Kana DD (1981) An iterative procedure for the generation of Consistent Power/Response Spectrum. Nucl Eng Des 66:427–435
5. IAEA-TECDOC-1347 (2003) Consideration of external events in the design of nuclear facilities other than nuclear power plants, with emphasis on earthquakes
6. ASME Boiler & Pressure Vessel Code (2004) Rules for Construction of Nuclear Facility Components Section III Division 1 Appendix N
7. Clough RW, Penzien J (1993) Dynamics of structures. McGraw-Hill, NewYork

Further Reading

9. Lin CW, Loceff E (1980) A new Approach to compute System Response with Multiple Support Response Spectra Input. Nucl Eng Des 60:345–352

10. J. K. Biswas, "Seismic Analysis of Equipment Supported at Multiple Levels", Proceedings of ASME Pressure Vessel and Piping Conference, PVP-Vol.65, Oriando, Florida, (1982)

10 Seismic Analysis and Design of Equipment

G. R. Reddy, A. R. Kiran, M. K. Agrawal, and M. Eswaran

Equipment stores water, gases, chemicals etc. Failure may affect the life line services and may spread hazardous gases

Symbols

N	Number of nodes
M	Number of modes
α and β	Proportional damping coefficients
ξ	Damping ratio
$\omega_{\min}$ and $\omega_{\max}$	Undamped circular frequencies (minimum and maximum frequencies)
$M\sum\phi^T\Gamma$	Mass participation (nodewise)
dof	Degree of freedom
R_f	Ratio of frequencies of secondary and primary systems
R_m	The ratio of mass of secondary and primary systems
F_Y	Maximum vertical force response due to empty tank shell, per unit length
C	Longitudinal compressive stress per unit length in tank shell
M_B	Combined overturning moment at the base
ρ_m and t	Mass density of tank wall and wall thickness, respectively
$d_{\max}$	Maximum slosh height
T	Fundamental period of the structure, seconds
T_L	Long-period transition period, seconds
l	Length of the beam
μ	Mass per unit length of the beam
I	Moment of inertia of the structure about its axis of bending
E	Young's modulus of elasticity
Z	Zone factor
I	Importance factor
R	Response reduction factor
V_b	Wind speed
V_Z	Design wind speed at any height z in m/s
k_2	Terrain, height and structure size factor
k_3	Topography factor
p_z	Design wind pressure in N/m^2 at height z
V_Z	Design wind velocity in m/s at height z
P_Z	Modified design wind pressure
C_s	Shape factor, for cylindrical geometry = 0.7 (Table 23 of IS 875 Part 3)
G	Gust response Factor

10.1 Introduction

The industrial equipment consists of vessels, heat exchangers, valves, pumps, motors, fans, and blowers. The equipment is very important component of an industry which holds and circulates fluids and gases. Safety of the industry hinges on the performance of the equipment under normal and accidental loads such as earthquakes. A systematic design, quality construction, and quality maintenance ensure

G. R. Reddy (✉) · A. R. Kiran · M. K. Agrawal · M. Eswaran
Bhabha Atomic Research Centre, Mumbai, India
e-mail: rssred@barc.gov.in

A. R. Kiran
e-mail: arkiran@barc.gov.in

M. K. Agrawal
e-mail: mkagra@barc.gov.in

M. Eswaran
e-mail: eswaran@barc.gov.in

G. R. Reddy et al., *Textbook of Seismic Design*,
https://doi.org/10.1007/978-981-13-3176-3_10

high safety of the plant and protect the industrial personnel and public around the plant. First step of the design is to categorize the equipment based on the hazard potential and consequences in the case of failures under normal and accidental loads.

10.2 Industrial Equipment–Safety and Seismic Categorization

Depending on the seismic importance, the equipment can be appropriately classified into various safety classes. For each specific class, it is necessary to make sure that it is safe with respect to the designed seismic force. Safety classification and seismic categorization for industrial equipment are described in this section.

10.2.1 Safety Classification

The objective of safety classification is to identify the equipment that is important for the safety considerations. These safety demands are defined below:

a. To design equipment with adequate margins,
b. To prevent accidents,
c. To minimize the accident consequences.

Typically, the equipment, whose failure leads to release of toxic gases, release of contents that cause fire, shall be put into safety Class I. Those supporting safety Class I equipment shall be classified as safety Class II, and others shall be put into industrial safety class.

10.2.2 Seismic Categorization

Seismic categorization is done in accordance with safety importance under earthquake. The purpose of this categorization is to facilitate the protection of public and environment against release of toxic gases and fire.

Seismic Category I
Equipment under this category is designed as well as demonstrated to withstand the consequences of ground motion under earthquakes of maximum considered earthquake (MCE) and design basis earthquake (DBE). Category I shall include:

a. Failure of these items can result into accident conditions,
b. Items that are required to prevent releases of toxic gases or to extinguish the fire or to maintain release below regulatory limits.

Seismic Category II
The items categorized under this category shall be designed to withstand ground motion associated with DBE.Equipment under this category shall include:

a. Items not in Category I but to be used for preventing toxic gases release beyond normal operation limits.
b. Items, not in Category I, required to minimize those accident conditions which may extend for long periods.

Seismic Category III
Seismic Category III consists of items which are not in seismic Category I or seismic Category II.

Safety and seismic categorization and service levels for equipment of nuclear facilities are given in Appendix 1.

10.3 Development of Analytical Model

Analyzing the equipment using continuum mechanics approach may be an ideal approach. However, this approach becomes difficult to represent complex geometry and discontinuities to formulate and solve the problem using differential equations. Therefore, practicing engineers prefer numerical techniques such as finite element approach for the analysis. The important parameters which are to be represented mathematically in analytical model (finite element model) to evaluate proper seismic loads can be explained considering the equation of motion which represents the balancing of inertial, damping, and elastic forces with applied earthquake loading.

Mass, stiffness, and damping of the structure, system, and components (SSCs) have to be represented accurately to evaluate the seismic loads for a given design input ground motion.

Development of analytical model is the key to the efficiency of analysis process and the success of seismic design. An analytical model is developed by appropriately ascertaining the degrees of freedom, evaluating lumped/consistency masses and stiffness properties of the connecting structural elements, etc. Different types of models that can be developed for the SSCs are discussed below.

10.3.1 Single Mass Model

In this model, the full mass is lumped at a single point on a single element of equivalent stiffness restraining the mass. Typically, pumps, valves, motors, fans, and heat exchangers can be put into the category of single mass model. For example, a schematic of blower on isolating pad is shown in Fig. 10.1a. Its idealization as SDOF system for horizontal

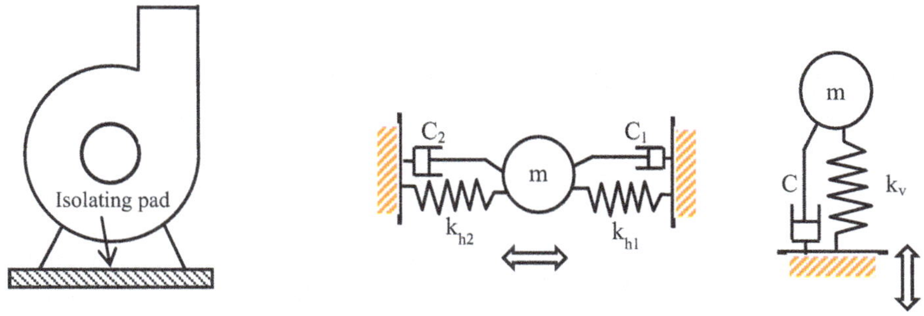

a. Blower on isolating pad

b. Idealization for horizontal and vertical excitations

Fig. 10.1 Idealization of blower on isolating pad as a SDOF system

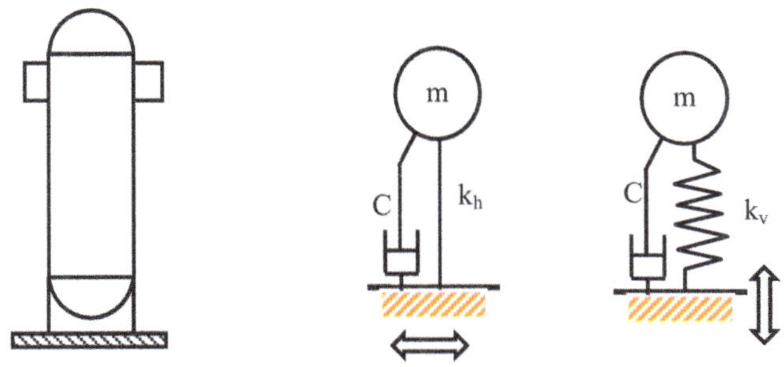

a. Skirt supported vessel with heavy mass at the top

b. Idealization for horizontal and vertical excitations

Fig. 10.2 Idealization of skirt-supported vessel with heavy mass at the top as a SDOF system

and vertical excitations is shown in Fig. 10.1b. The mass of blower m is considered as lumped mass connected with horizontal and vertical spring–damper systems as shown in the figure. The spring–damper in horizontal direction is split into tension–compression systems with stiffness and damping as k_{h1}, k_{h2} and C_1, C_2, respectively. In vertical direction, the stiffness and damping are k_v and C, respectively. Consider a skirt-supported vessel with heavy mass at the top as shown in Fig. 10.2a. This can be idealized as SDOF system as shown in Fig. 10.2b for excitations in horizontal and vertical directions. The mass of vessel m is considered as lumped mass connected with horizontal and vertical spring–damper systems with stiffness and damping as k_h (horizontal stiffness), k_v (vertical stiffness), and C, respectively.

10.3.2 One-Dimensional Finite Element or Beam Model

This model is best suited to structures with beams, columns, cable trays, frames, ducts, conduits, tanks, cabinets, and storage racks. This model can be defined as one-dimensional finite element in a two- or three-dimensional space. Masses are defined by lumped parameters, which develop a diagonal elemental mass matrix. Pressure vessels and heat exchangers can also be modeled this way, especially when using the equivalent static method. This model has less approximation than point model. Distribution of mass and element stiffness are more idealized.

10.3.3 Two-/Three-Dimensional Finite Element Model

This model represents the complete geometry of the structure. This type of modeling is well suited for items whose dominant mode of failure is biaxial bending stress, plane stress, or plane strain. Under this category, cabinets, tanks, pressure vessels, and heat exchangers can be included. The eccentric loads supported on equipment which would tend to excite shell or local modes of vibration can be captured. Beam and shell models for a typical pressure vessel, water storage tank, and piping system are shown in Figs. 10.3, 10.4 and 10.5, respectively.

Plate and shell model and beam model of pressure vessel are shown in Fig. 10.3. Base of the pressure vessel is assumed to be fixed in the model. In Fig. 10.4, plate and shell model and beam model of the vertical water storage tank are shown. Effects of water are included by changing the density vessel in the analysis. In Fig. 10.5, 3D piping system is modeled using shell element and line element for dynamic analysis under earthquake loading. It consists of straight pipes and elbows. Ends of the pipe are anchored. The other supports are u-clamp to restrict the lateral movement of the piping.

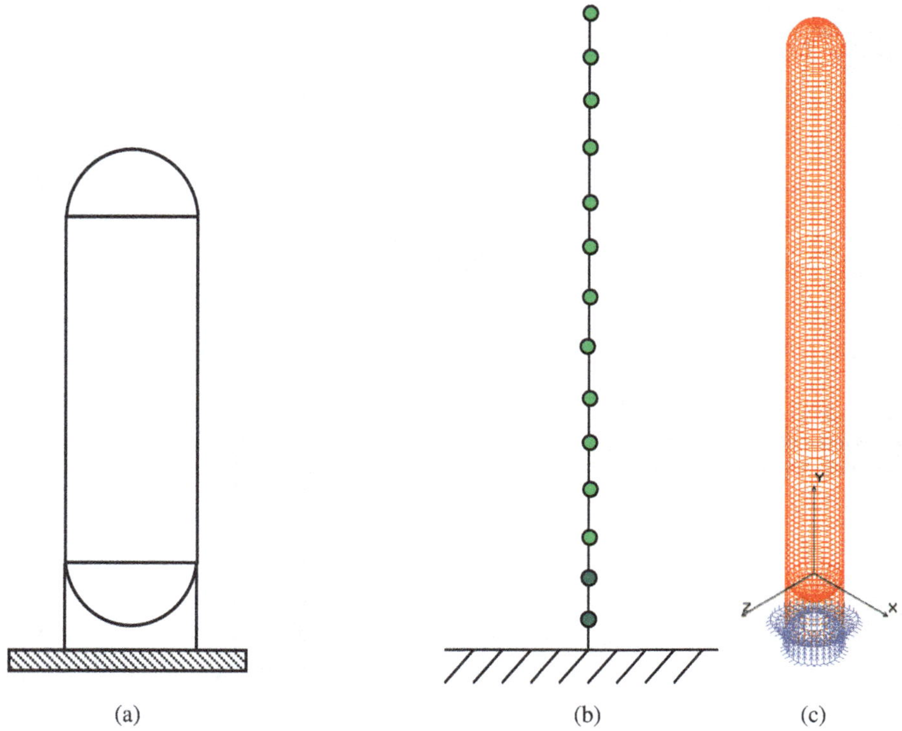

Fig. 10.3 **a** Pressure vessel and its, **b** beam model and **c** plate and shell model

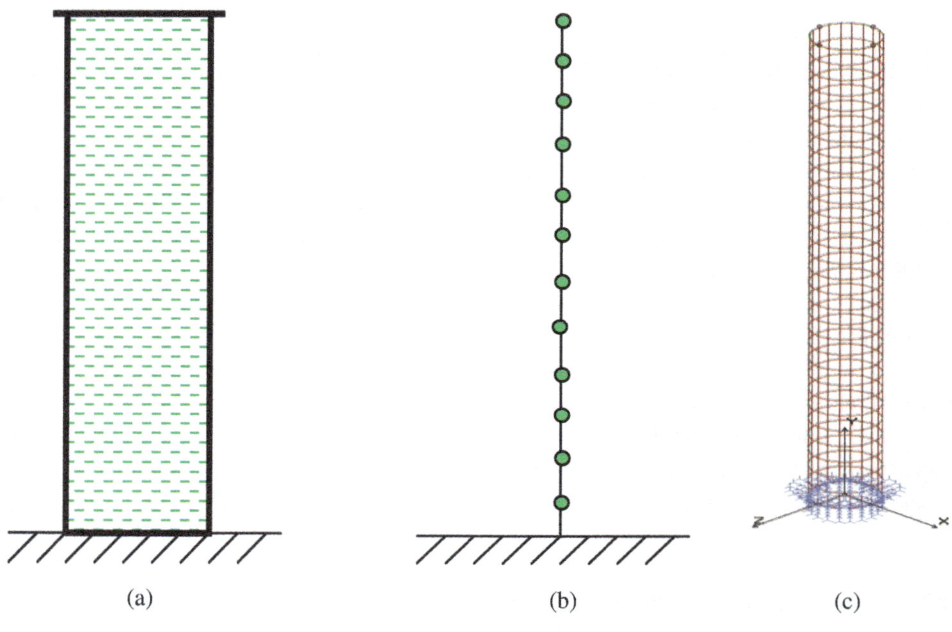

Fig. 10.4 **a** Water storage tank and its, **b** beam model and **c** plate and shell model

10.3.4 Selection of Number of Nodes

In finite element model, one can discretize the structure into different number of nodes based on the response output desired and accuracy of the results. As the number of nodes increases, response will be approaching close to idealized solution. To limit the computation time and size of problem, without compromising the response at desired location, optimization is required in selection of number of nodes. For example, in the stick models a node is required where the piping or equipment is supported, at floor level and along height. The number of nodes is selected based on the number of modes to be evaluated within the rigid body frequencies. The rigid body frequency is generally considered as ≥ 33 Hz. The number of modes required within the 33 Hz can be evaluated approximately using the classical formulae. For example, in the piping system, span between the two supports can be isolated, number of modes within 33 Hz can be evaluated, and number of nodes can be fixed. Similarly in the cantilever type of structures and equipment or

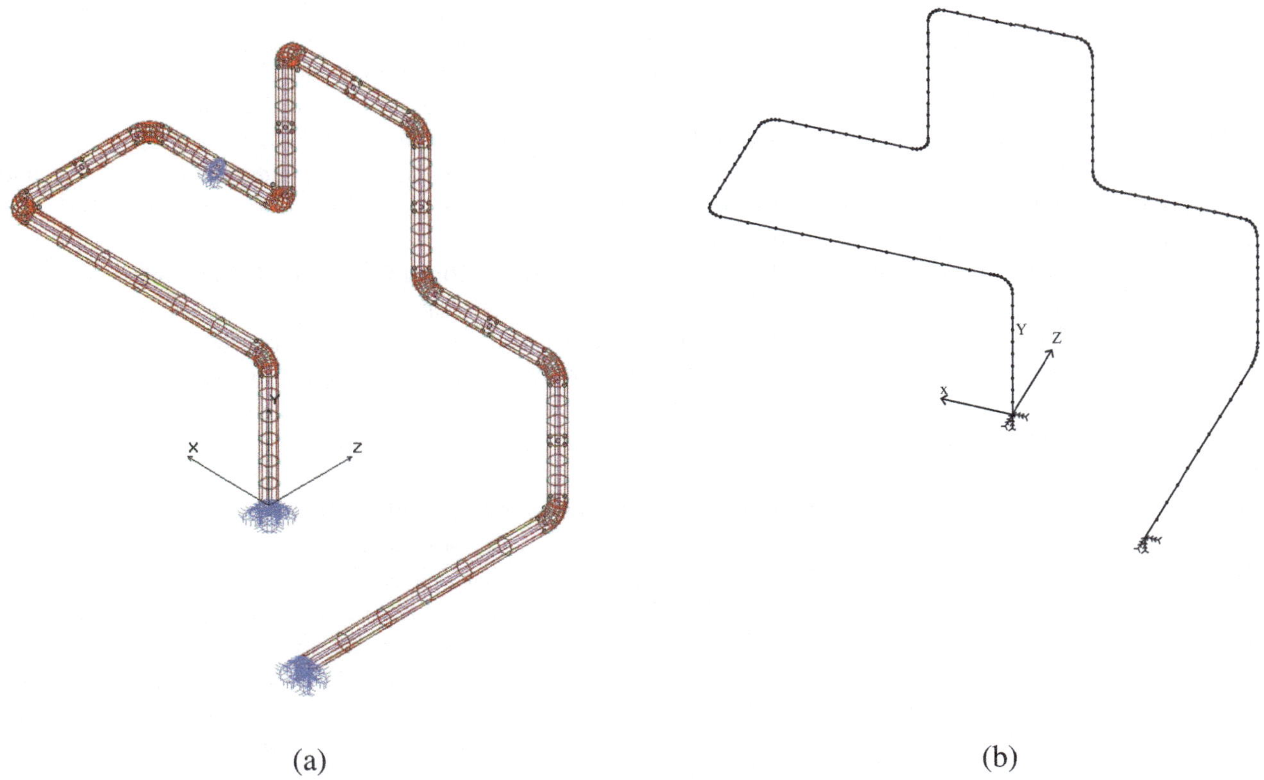

Fig. 10.5 **a** Shell and **b** beam model of piping system

components, number of nodes can be fixed for evaluating appropriate frequencies within 33 Hz. The relation between number of nodes and number of modes is given in Eq. (10.1).

$$N = 2m + 1 \tag{10.1}$$

where,

N number of nodes,
m number of modes.

10.3.5 Modeling of Mass

Mass can be represented as a lumped mass or as a consistent mass in the model. In lumped mass approach, the inertial mass of a structure may be modeled by lumping at the node points of the model. The second approach is the consistent mass formulation. Normally, three translational and rotational degrees of freedom shall be used at each node point. However, rotation degree of freedom may be neglected provided it does not affect response significantly. In lumped mass approach, conditions to be checked are as below:

a. Total structural mass to be conserved, and center of gravity is preserved,
b. The number of degrees of freedom shall be selected which is also number of lumped mass so that all significant vibration modes of the structure are captured.

10.3.6 Modeling of Damping

Every real structure dissipates energy under vibratory/cyclic motion. The dissipation of energy is defined as damping.

10.3.6.1 Damping Properties of Equipment

Damping is a common designation for all kinds of energy absorption of vibratory SSC. In SSC, energy dissipation takes place due to internal material friction, sliding at the joints and slipping of structural member. Damping can be classified into three categories, namely

a. Under damped,
b. Over damped, and
c. Critical damped.

The critical damping of a material is defined as a value for which the oscillatory motion gets seized.Damping ratios for structural materials are generally less than 20% and vary with materials.

Damping values are defined in various codes based on level of earthquake used for the design purpose and types of structure. Damping ratio given in Table 10.1 is for operating basis earthquake(OBE) and safe shutdown level of earthquake (SSE) for SSCs of nuclear facilities as defined in regulatory guide RG 1.61 [34]. While the IS-1893-2005 Part 4 [IS 1895, 2005] Indian code gives damping ratio for two types of materials under DBE(design basis earthquake)and MCE (maximum credible earthquake) level of earthquake as given Table 10.2. However, the values under OBE can be used for design condition, i.e., for DBE, values under SSE can be used for MCE condition if the values are not specified in IS 1893 for other structures. This may be justified due to fact that return period of OBE level of earthquake and DBE is of same order.

In the dynamic analysis, response is evaluated in different modes in mode superposition method. When modal analysis is to be performed for structures with only one type of material (hence a single value of damping), the damping values given in Table 10.1 can directly be used for modal damping. For structural systems that consist of substructures with different damping properties, the equivalent modal damping values may be obtained using the method given in Sect. 10.3.6.3, which is based on strain energy equivalence.

10.3.6.2 Proportional Damping (Rayleigh Damping)

In time history analysis, the damping matrix [C] formed by a linear combination of the mass and stiffness matrices may be used as given in Eq. (10.2) [11].

$$[C] = \alpha[M] + \beta[K] \tag{10.2}$$

where α and β are proportional damping coefficients and are given by Eqs. (10.3) and (10.4).

$$\alpha = \frac{2\,\xi\omega_{\max}\omega_{\min}}{\omega_{\max} + \omega_{\min}} \tag{10.3}$$

$$\beta = \frac{2\xi}{\omega_{\max} + \omega_{\min}} \tag{10.4}$$

where ξ is the damping ratio which can be obtained from Tables 10.1 and 10.2. The $\omega_{\max}$ and $\omega_{\min}$ are the undamped circular frequencies to define the range of frequencies which contribute to the response of the structure. The $\omega_{\min}$ is the first modal frequency of the SSCs, and the $\omega_{\max}$ is the maximum frequency corresponding to final significant mode which is usually considered to be 207 rad/s (33 Hz).

10.3.6.3 Composite Modal Damping

Structural systems that consist of major substructures with different damping values, composite modal damping values obtained from Eq. (10.6) may be used without further justification, as long as the resultant damping values are less than 20%.

Evaluation of Modal Damping

In case the structural systems contain substructures with different materials/damping properties, the equivalent modal damping values may be obtained using the procedure given below. The different damping values may be because of variation in material or due to soil damping. The example of composite damping is illustrated in Fig. 10.6. Schematic of a vessel supported on steel frame is shown in Fig. 10.6a, and its idealization is shown in Fig. 10.6b.

The evaluation of percentage of critical damping in each mode using the weighted strain energy principle is described below [11]. Strain energy of the equivalent system in mode *j* is summation of the strain energy of the individual element as given in Eq. (10.5).

$$\xi_J = \frac{\{\theta_J\}^T\left[\sum_{i=1}^{N}[\xi K]_i\right]\{\theta_J\}}{\omega_j^2} \tag{10.5}$$

The following Eq. (10.6) can be used for the mass-weighted damping based on kinetic energy principle:

$$\xi_j = \{\Phi\}^T \sum_{i=1}^{N} [\xi M]_i \{\Phi\} \tag{10.6}$$

where

ξ_i damping ratio of the element (subsystem)
$[K]_i$ stiffness matrix of the *i*th element (subsystem)
$[M]_i$ mass matrix of the *i*th element (subsystem).

Table 10.1 Damping values for various structures in (% of critical damping) [34]

Structure type	OBE (%)	SSE (%)
Pipe system	3	4
Welded steel str.	2	4
Bolted steel str.	5	7

Table 10.2 Damping ratio for various structures as given in IS-1893-2002–Part 4

Material	Damping ratio (under DBE)	Damping ratio (under MCE)
Steel	2%	4%

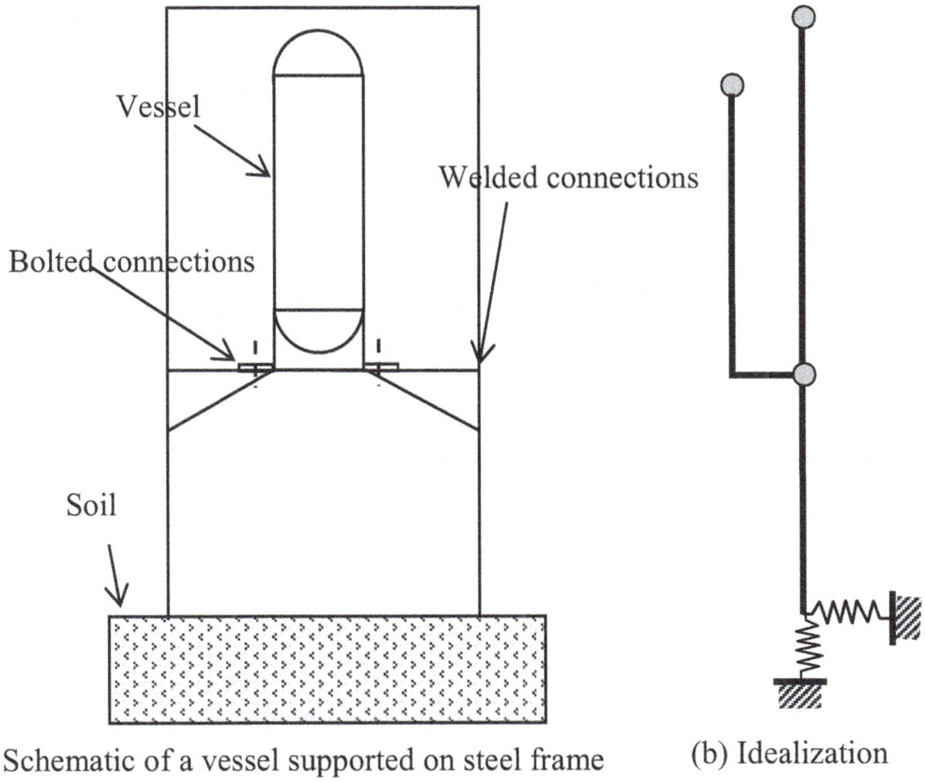

Fig. 10.6 Different cases of composite damping

10.4 Analysis of Structures

In general, following methods are used for seismic response analysis:

i. Time history method,
ii. Response spectrum method (RSM),
iii. Equivalent static method (ESM).

The details of time history method and response spectrum method are discussed in Chap. 3. However, consideration for number of modes in the analysis, missing mass correction, and modal combination are discussed in following sections. Equivalent static method is also discussed in this chapter.

10.4.1 The Number of Modes Considered in the Mode Superposition Method

In response spectrum method, mode superposition method is used for solving the equation of motion. The number of total modes represents the number of degree of freedom in the model. Ideally, all the modes to be included in the analysis for complete solution. However, due to practical difficulty in handling the size of the problem and computation time, codal provisions are provided for selection of number of modes in the solution. Number of modes to be included in the analysis shall be sufficient to ensure that all remaining modes do not contribute more than 10% in total responses. Alternatively, ASCE standard (4-98) permits to include all the modes in the analysis having frequencies less than the zero period acceleration (ZPA) frequency or cutoff frequency provided that the residual rigid response due to the missing mass is included. The following criteria to be adopted while selecting the minimum number of modes in the analysis [1].

1. Number of modes evaluated is such that highest mode frequency is greater than or equal to 33 Hz,
2. The number of modes evaluated is such that the sum of all the modal mass is more than 90% in each of the three directions.

Any one of the two criteria can be used to determine the number of modes to be considered in mode superposition analysis.

10.4.2 Combination of Modal Response

In response spectrum method, response obtained in each mode is the maximum. However, the maximum response will not occur simultaneously. Absolute summation of response is highly conservative. To reduce the conservatism, following methods are used for the combination of modal response [1, 15].

i. Square root of sum of squares (SRSS) method
ii. 10% method
iii. Complete quadratic combination (CQC) method.

In SRSS method, the modes are assumed to be statistically uncorrelated. However, in reality the modes will always not be uncorrelated when the frequencies are close to each other. The modes whose frequencies are less than 10% apart are called closely spaced modes and more than 10% apart are defined as non-closely spaced modes. Based on the mode spacing, the response is combined by using SRSS, 10% method or CQC methods.

10.4.2.1 With Non-closely Spaced Modes (SRSS)

In a response spectrum modal dynamic analysis, if the modes are spaced more than 10% apart, the maximum response for design should be calculated by taking the square root of the sum of the squares (SRSS). Mathematically, this can be defined as [1]:

$$R = \left[\sum_{k=1}^{N} R_k^2\right]^{1/2} \qquad (10.7)$$

where R is the maximum response of a given element to a given excitation of an earthquake, R_k is the peak response of the element due to the kth mode, and N is the number of significant modes considered in the modal response combination.

10.4.2.2 With Closely Spaced Modes (10% Methods)

The modes are called closely spaced if the difference between two adjacent modal frequencies is less than 10%. Mathematically, this is written as:

$$\frac{\omega_j - \omega_i}{\omega_i} \leq 0.1 \quad \text{also } 1 \leq i \leq j \leq N \qquad (10.8)$$

The maximum response is obtained by taking the square root of the sum of the squares (SRSS). Mathematically, this can be defined as follows:

$$R = \left[\sum R_k^2 + 2\sum |R_i R_j|\right]^{1/2} i \neq j \qquad (10.9)$$

where R is the response of a given element to a given excitation of an earthquake, R_k is the peak response of the element due to the kth mode, and N is the number of significant modes considered in the modal response combination.

10.4.2.3 Complete Quadratic Combination (CQC) Method

The complete quadratic combination method takes care of coupling effects of modes [15]. The modal coupling in CQC is function of both the modal damping and ratios of frequency between modes. Therefore, there is no need to define separately closely spaced mode criteria in this method. In this method, if modes are non-closely spaced the combination is done as per the SRSS method and if modes are closely spaced combination will be done by 10% methods.

$$R = \left[\sum_{k=1}^{N}\sum_{s=1}^{N} R_k R_s \varepsilon_{ks}\right]^{1/2} \qquad (10.10)$$

For constant damping ratio (uniform modal damping)

$$\varepsilon_{ks} = \left[\frac{(\omega_k + \omega_s)^2 \xi^2}{(\omega_k - \omega_s)^2 + (\omega_k + \omega_s)^2 \xi^2}\right] \qquad (10.11)$$

For different damping ratio (non-uniform modal damping)

$$\varepsilon_{ks} = \left[\frac{2\sqrt{\xi_k \xi_s}\left[(\omega_k+\omega_s)^2(\xi_k+\xi_j)+(\omega_k-\omega_s)^2(\xi_k-\xi_j)\right]}{4(\omega_k^2-\omega_s^2)+(\omega_k+\omega_s)^2(\xi_k+\xi_j)^2}\right] \qquad (10.12)$$

10.4.3 Missing Mass Correction

In response spectrum analysis method, analysis is done for finite number of modes with cutoff frequency as 33 Hz. The truncation of the mode results in ignoring some mass of the system and is called as missing mass. Forces associated with these inertial masses are significant for system where mass participation at cutoff frequency is not enough. This missing mass is evaluated, and forces are calculated by static method based on the following Eqs. (10.13), (10.14) and (10.15) [1].

Mass participated in the analysis for m modes

$$Mp = M\sum_{i=1}^{m} \Gamma_i\{\Phi_i\} \qquad (10.13)$$

Mass not participated in the analysis (Missing Mass)

$$Ms = M\left\{\{1\} - \sum_{i=1}^{m} \Gamma_i\{\Phi_i\}\right\} \qquad (10.14)$$

Force due to missing mass

$$F = M\left\{\{1\} - \sum_{i=1}^{m} \Gamma_i\{\Phi_i\}\right\} S_{A\max} \quad (10.15)$$

where

$S_{A\max}$ highest spectral acceleration at the cutoff frequency
M Total mass

A case study of 3DOF system is shown in Example 10.1 to describe the mass participation and concept of missing mass.

Example 10.1 A vertical hollow cylinder of 1 m outer diameter, 10 mm thickness, and height of 4 m is supported at bottom. It has a total mass of 399.6 kg. Idealize this cylinder using three elements, and evaluate the free vibration characteristics and missing mass.

Solution

i. Total mass = 399.6 kg. The total mass is divided into three elements equally as shown in Fig. 10.7,
ii. From Eq. (10.1), minimum three nodes are required to obtain first mode correctly. Then mass is discretized nodewise between the elements. Mass at the base is assigned as 66.6 kg. However, this mass will not participate due to fixed boundary condition.
m1 = 133.2 kg
m2 = 133.2 kg
m3 = 66.6 kg
iii. Mass matrix can be given in this form:

$$M = \begin{bmatrix} 133.2 & 0 & 0 \\ 0 & 133.2 & 0 \\ 0 & 0 & 66.6 \end{bmatrix}$$

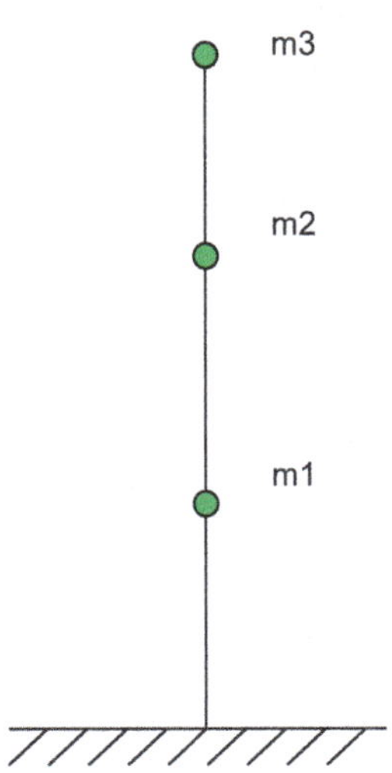

Fig. 10.7 FE model of 3DOF system

iv. First mode shape coefficients obtained by using code provided in Chap. 4 are

$$\phi = \begin{bmatrix} 1.550 \times 10^{-2} \\ 5.175 \times 10^{-2} \\ 9.581 \times 10^{-2} \end{bmatrix}$$

v. The following check suggests that the mode shape coefficients are correct and also meeting the orthogonality condition of shape function with mass matrix.

$$\phi^T M \phi = 1$$

vi. First natural frequency of vibration and mass participation has been calculated using equations provided in Chap. 4. These are provided in Table 10.3. The first natural frequency is 29 Hz, and corresponding mass participation is 58.85%,

Analytical checks for mass participation factor and missing mass concepts are provided below.

vii. Mass Participation in First mode
Modal participation factor in first mode

$$\begin{aligned}\Gamma &= \sum \phi_1^T M \\ &= 1.55 \times 10^{-2} \times 133.3 + 5.175 \times 10^{-2} \\ &\quad \times 133.3 + 9.581 \times 10^{-2} \times 66.6 \\ &= 15.338\end{aligned}$$

Mass participation

$$\Gamma^2 = 235.22$$

Percentage participation = 235.22/399.6 = 58.85%

Concept of missing mass nodewise (up to one mode). Numerical procedure for calculation of modal mass in different nodes is described below:

viii. As the analysis has been performed only for first mode, calculation has been done for mass participation in nodewise for first mode. The difference of mass is called missing mass, as second and third modes are excluded in the calculation.

Table 10.3 Frequency of vibration and cumulative participation factor

Mode no	Frequency (Hz)	% Cumulative mass participation
1	0.2907×10^{02}	58.85

ix. Mass participation nodewise (up to first mode)
Mass participation (nodewise) = $M \sum \phi^T \Gamma$
For Node 3
Mass participated in first mode = 66.6 ($9.58 \times 10^{-2} \times 66.6 \times 9.58 \times 10^{-2}$) = 40.7
Missing mass at node 3 = 66.6 − 40.7 = 25.9 kg
For Node 2
Mass participated in first mode = 90.9 kg
Missing mass at node 2 = 133.3 − 90.9 = 42.4 kg
For Node 1
Mass participated up to 2 modes = 50.6
Missing mass at node 1 = 133.3 − 50.6 = 82.7 kg

These missing masses will be considered in the analysis by equivalent static analysis. Force on different node can be applied based on missing mass, and acceleration corresponds to cutoff frequency from the response spectrum.

10.4.4 Combination of Response Due to Missing Mass with Dynamic Response

Residual response due to missing mass is combined with response due to dynamic analysis. The above response will be combined as an additional mode having frequency equal to the ZPA or cutoff frequency and will be combined using the SRSS rule.

10.4.5 Spatial Combination

In the previous section, responses are combined for modes with respect to one particular direction of excitation. The dynamic response which is triaxial in nature is combined as SRSS or by 100:30:30 methods as described in the code [1]. Spatial combination of response due to three components of earthquake is carried out by SRSS method as given in Eq. (10.16).

$$R = \left(\mathrm{R}_x^2 + \mathrm{R}_y^2 + \mathrm{R}_z^2\right)^{1/2} \qquad (10.16)$$

As per IS-1893, spatial combination due to three earthquake components is carried out by 100:30:30 method. Maximum response due to earthquake will be highest of three as given in Eq. (10.17).

$$\begin{aligned} R &= \pm \mathrm{R}x \pm 0.3\mathrm{R}y \pm 0.3\mathrm{R}z \\ R &= \pm \mathrm{R}z \pm 0.3\mathrm{R}x \pm 0.3\mathrm{R}x \\ R &= \pm \mathrm{R}z \pm 0.3\mathrm{R}x \pm 0.3\mathrm{R}y \end{aligned} \qquad (10.17)$$

10.4.6 Equivalent Static Method (ESM)

The equivalent static method is a simplified method as compared to any other dynamic analysis methods. This method is best suited for cantilever models with uniform mass distribution. The equivalent static load is determined by multiplying the equipment mass by acceleration equal to 1.5 times the peak spectral acceleration of the site or floor. Smaller value may be used if justified, or the floor ZPA value may be used if it is shown that fundamental frequency is so high, typically 33 Hz. No dynamic amplification will occur at this frequency [4]. The significance of utilizing factor 1.5 attributed to account for higher modes and add conservatism in the equivalent static method.

$$F_\mathrm{h} = K\, Sa_{\max} W \qquad (10.18)$$

$$F_\mathrm{v} = K\, Sa_{\max} W \qquad (10.19)$$

where

- F_h Equivalent static inertia force applied to the component in the horizontal direction
- F_v Equivalent static inertia force applied to the component in the vertical direction
- K Load coefficient applied to the model. The value of K is taken as 1.5
- $Sa_{\max}$ peak acceleration of applicable amplified or floor response spectra in the ith direction (in g's)
- W Total dead load (weight) which exist during the postulated seismic event. This includes piping weight, water weight, and insulation. The units of the term W must be consistent with the terms F_h and F_v above.

It has also been explained in detail in Numerical Example 10.2.

Example 10.2 Consider the 3DOF system as defined in Example 10.1. Calculate the forces for the system using equivalent static method. Figure 10.8a describes how the forces are applied in static manner. Use the site-specific spectra as shown in Fig. 10.8b.

Solution

i. Peak spectral acceleration from the IS spectra can be calculated from the following Eq.

$$A_\mathrm{h} = \frac{\left(\frac{Z}{2}\right)\left(\frac{S_a}{g}\right)}{\left(\frac{R}{I}\right)}$$

Fig. 10.8 **a** Application of force in ESM with, **b** site-specific spectrum

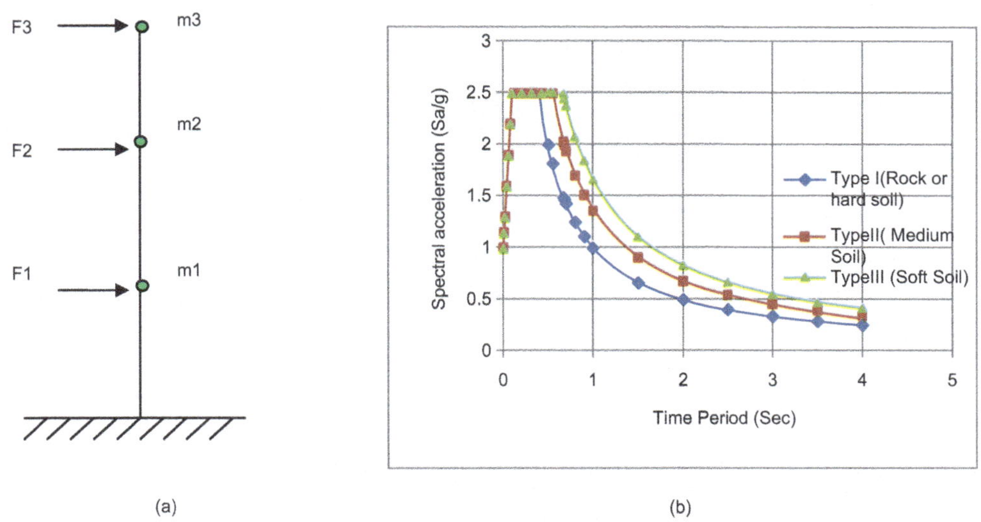

where

Z	zone factor = 0.16 (Assuming zone 3)
R	Response reduction factor (1)
I	Importance factor (1)
S_a/g	Spectral acceleration coefficient for rock and soil sites (2.5)

$$A_h = \frac{\left(\frac{0.16}{2}\right)(2.5)}{\left(\frac{1}{1}\right)} = 2.0\,\mathrm{m/s^2}$$

$$Sa_{max} = A_h = 2.0\,\mathrm{m/s^2}$$

ii. Forces applied at various locations in two horizontal directions are:
F1 = 1.5 × 33.3 × 2.0 = 99.9 N
F2 = 1.5 × 133.3 × 2.0 N = 399.9 N
F3 = 1.5 × 66.6 × 2.0 N = 199.8 N

iii. Force in vertical direction will be applied 2/3 of the horizontal force. Now this problem can be solved as static problem for dynamic load.

iv. **Significance of factor 1.5 in ESM:** As it can be seen from the above analysis, participation in first mode in cantilever type of structure is about 60% (2/3) of total mass. To compensate for the higher modes in the analysis, 1.5 factor is multiplied to excite full mass in the same mode.

10.5 Analysis of Equipment Located at Other Than Ground Floor

Equipment can be located either at ground level or at some floor in the building structure. If equipment is located on the ground, the dynamic method as described in the previous section can be utilized with the use of response spectra or time history defined at ground level. However, for the equipment which is located on other than ground level, floor response spectrum (FRS) is used for seismic response analysis.

FRS is used for the decoupled analysis of the systems and components. FRS is determined at the floor level of the structure where the systems, components, and equipment of interest are located. At first step, floor time histories are obtained from the time history analysis of the structure. These floor time histories are used to obtain FRS. The FRS so obtained is called the raw spectrum, which has number of peaks and valleys. The major peaks occur at natural frequency of the structure.

Peak broadening of the FRS is done to account for the structural frequency variation due to possible uncertainties in the soil structure interaction and the approximations in the modeling technique used in the seismic analysis. The minimum broadening shall be ±15% at each frequency in the amplified response region for the best estimate soil shear modulus case. The details of the peak broadening are given in Chap. 3. The flowchart shown in Fig. 10.9 defines the analysis for the floor-mounted equipment to be analyzed.

10.6 Structure–Equipment Interaction Due to Earthquake

Structural systems are said to be in interaction when response of one system is affected by other system. Structure and equipment interaction cannot be overlooked if the two systems are tuned or the equipment to the structure mass ratio is large. Schematic of primary and secondary systems is shown in Fig. 10.10. Primary and secondary systems can be either SDOF systems as shown in Fig. 10.10a or MDOF systems as shown in Fig. 10.10b. The interaction effects can be avoided by coupling all structure and equipment together and analyzing for a given ground response spectrum.

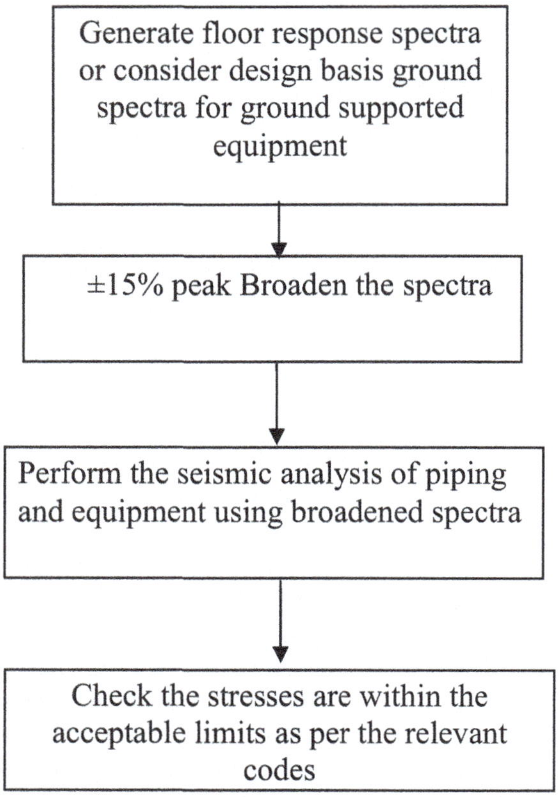

Fig. 10.9 Flow chart for seismic analysis and design of equipment

However, coupling all the equipment to the structure is not possible for three main reasons. These are: Firstly, at the initial design stage of the plant all the details of equipment may not be available, secondly, the coupling all the equipment may lead to numerical problems in computation because of large difference in stiffness, and thirdly, the cost of the computation may become high and time consuming due to coupling because of large problem size. Some codes have defined conservatively criteria to account for the structure interaction effects (USNRC 1992). The decoupling criteria are important if coupled model frequency differs more than 15% than uncoupled model frequency. In such cases, the equipment design may be unconservative.

10.6.1 Decoupling Criteria

Decoupling criteria are based on the frequency ratio (R_f) and mass ratio (R_m) of the secondary system to the primary system. For SDOF systems, R_f is defined as the ratio of frequencies of secondary and primary systems and R_m is defined as the ratio of mass of secondary and primary systems. For MDOF systems, these are defined as in Eqs. (10.20) and (10.21) as follows:

Fig. 10.10 Primary and secondary system

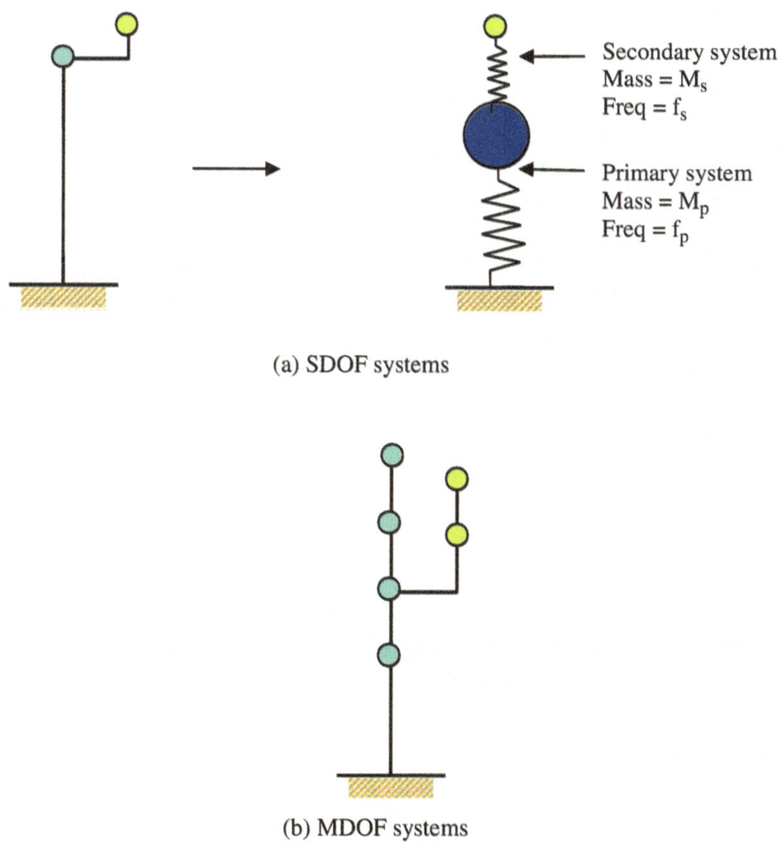

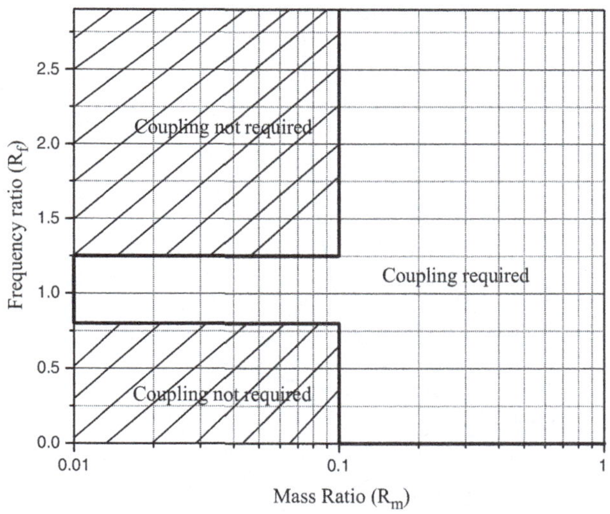

Fig. 10.11 Decoupling criteria for primary and secondary systems based on USNRC

$$R_{\mathrm{f}} = \frac{\text{Dominant frequency of secondary system}}{\text{Dominant frequency of primary system}} \quad (10.20)$$

$$R_{\mathrm{m}} = \frac{\text{Modal mass of secondary system}}{\text{modal mass of Primary system}} \quad (10.21)$$

where the dominant frequency is defined as the lowest natural frequency with more than 20% mass participation. Figures 10.11 and 10.12 describe the graphical representation of decoupling by USNRC and ASCE 4-98, respectively. The decoupling procedure based on USNRC is explained as follows:

(i) Decoupling can be done for any R_{f}, if $R_{\mathrm{m}} < 0.01$
(ii) If $0.01 \leq R_{\mathrm{m}} \leq 0.1$ decoupling can be done provided $0.8 \geq R_{\mathrm{f}} \geq 1.25$
(iii) If $R_{\mathrm{m}} \geq 0.1$ and $R_{\mathrm{f}} \geq 3$ (i.e., rigid secondary structure), it is sufficient to include only the mass of the system in the primary structure
(iv) If $R_{\mathrm{m}} \geq 0.1$ and $R_{\mathrm{f}} < 0.33$ (flexible secondary system), decoupling can be done.
(v) If $R_{\mathrm{m}} \geq 0.1$ and $0.33 < R_{\mathrm{f}} < 3$, coupled system analysis is required.

10.7 Procedure for Hydrodynamic load of Tank Filled with Water/Liquid

The power plants, manufacturing industries, and many oil and gas industries require the liquid storage tanks for storing liquids like water, oil, chemical fluids, and wastes of different forms. Liquid storage tanks are also vital components of industrial processes and safety facilities. These tanks are widely used in municipal water supply, firefighting systems and storage of liquefied natural gas, chemical fluids, and hazardous fluids and wastes of different forms in industries.

10.7.1 Classification of Various Tanks for Hydrodynamic Load

In general, the tanks are classified based on following parameters:

a. Based on orientation
   i. Horizontal
   ii. Vertical
b. Based on geometry of tank
   i. Cylindrical shape
   ii. Rectangular shape
   iii. Spherical shape
   iv. Asymmetric shape
c. Based on location
   i. Ground mounted
   ii. Elevated tank.

10.7.2 Common Tank Failures

The common failure modes of the tanks are explained in this section.

10.7.2.1 Buckling of the Shell

This bucking is caused by excessive axial compression due to overall bending or beam-like action of the structure. Shell buckling typically takes the form of “elephant’s foot” (or elastic–plastic buckling of the tank wall) or diamond-shaped buckling that appear a short distance above the base and usually elongate around most or all of the circumference (as shown in Figs. 10.13, 10.14 and 10.15). Dorninger et al. [12] is reported an elephant foot bucking in a tank wall during the Chile 1985 earthquake. Elephant’s foot-type buckle formation results from the high circumferential tensile stresses due to the internal pressure, in combination with the axial membrane stresses due to the over tuning moment instigated by the horizontal earthquake excitation. The local bending stress due to the tank restraints on the tank bottom has also been reinvigorated this failure. Diamond shaped is most serious with respect to the elastic buckling due to axial compression forces. A diamond-shaped bucking is shown in a photograph by Niwa and Clough [9], Rammerstrfer et al. [32].

Fig. 10.12 **a** Decoupling criteria for primary and secondary systems based on ASCE 4-98, **b** decoupling criteria for primary and secondary systems based on ASCE 4-98

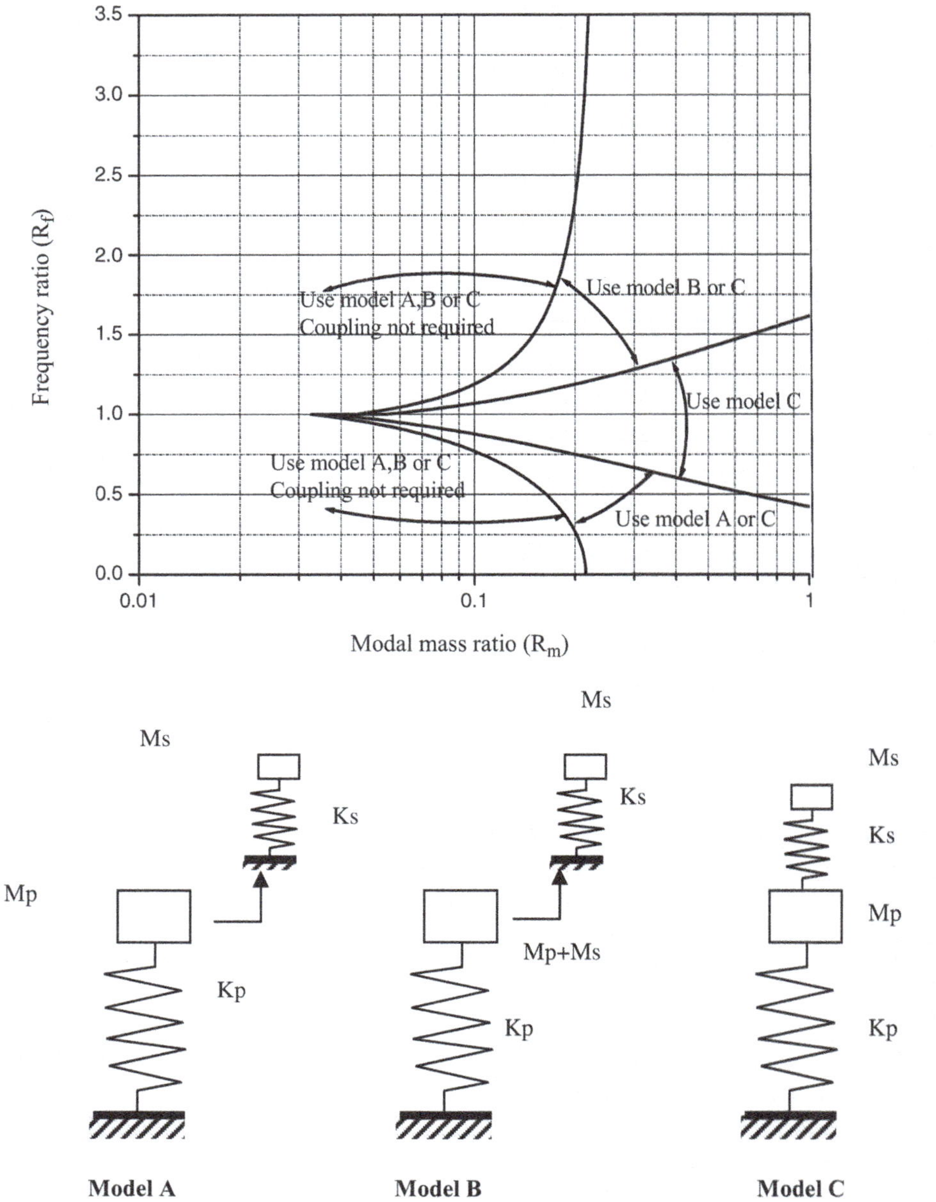

Fig. 10.13 Tank wall buckling [16]

(a) Elephant boot type buckling (b) Buckled water tank

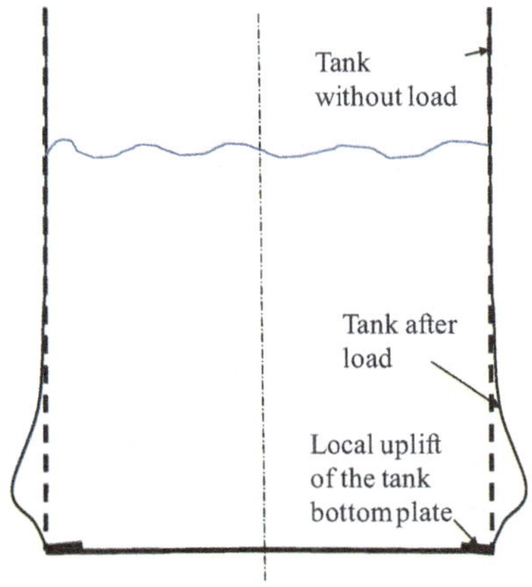

Fig. 10.14 Elephant's foot buckling

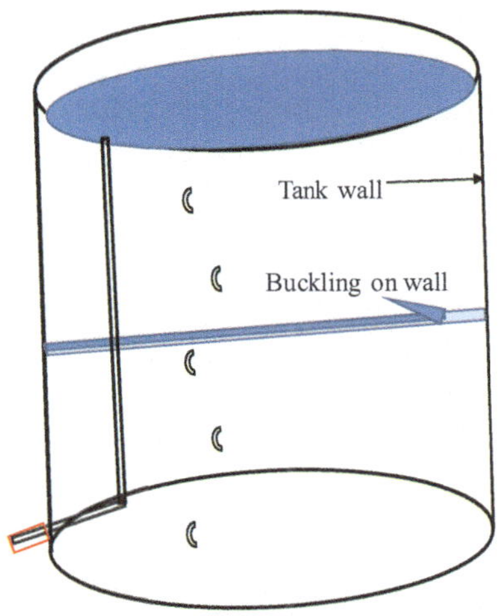

Fig. 10.15 Diamond-shaped buckling

10.7.2.2 Damage to the Roof and Bottom Plate

The inadequate freeboard between the liquid surface and the roof may lead to roof damage [13]. Insufficient freeboard causes upward load on the roof due to impacts from the sloshing wave and increase in impulsive mass due to constraining action of the roof. The upward force on the roof can damage the roof, break the roof–shell connection, or tear the shell [30].

10.7.3 Impulsive and Convective Masses

When a tank containing liquid with a free surface is subjected to horizontal earthquake ground motion, the tank wall and liquid are subjected to horizontal acceleration. Total mass of liquid gets divided into two parts, i.e., impulsive mass and convective mass. During the ground acceleration, a limited fraction of liquid is able to move into the phase of tank motion, which is known as impulsive mass. The liquid in the lower region of tank behaves like an impulsive mass that is rigidly connected to the tank and oscillates with it. The remaining fraction of liquid is oscillating with different frequency, i.e., it is not oscillating in the phase of excitation, but oscillating with a slightly different frequency which may be lower or higher than the excitation frequency. This fraction of liquid is termed as convective mass. The liquid mass in the upper region is generally considered as convective mass. The insufficient freeboard affects the free movement of convective mass.

10.7.4 Seismic Design Parameters for Liquid Tanks

The code provides the individual provision for designing the water storage tank of circular and rectangular geometry which may be ground mounted or at some elevation. These parameters are depending on the seismic zone and the corresponding response spectra and the amount of water to be stored. On the basis of these inputs, the design parameters are calculated. The designed structure should be able to resist them for safe design.

It is convenient to replace the liquid conceptually by an equivalent linear mechanical system to incorporate the dynamic effects of sloshing in the pools. The equations of motion of oscillating masses and rigid masses are simpler than the equations of fluid dynamics. The equations of motion can be derived from static and dynamic properties of spring–mass model. As per static properties, the sum of all the masses must be equal to the liquid mass and center of mass of the model must be at the same elevation as the liquid. The dynamic equations can be derived by inserting the acceleration terms and applying static properties into the force equation. The common codes for seismic design of liquid tanks are ACI 350.3 [8], TID-7024, Eurocode 8 [14], IS 1893-Part 1 [20], BIS [10] and NZS 3106 [28]. Most of these codes are based on equivalent spring–mass system. In the following section, the procedure given by TID 7024 and BIS [10] has been described. Comparison of various parameters evaluated by these codes is given in Table 10.4.

10.7.5 Procedure for Hydrodynamic Load by TID 7024

In industry, tanks are invariably used for the storage of water or liquid for different purpose. When these tanks are subjected to ground excitation, the water inside the tank exerts pressure on the walls of the tank. The pressure gives rise to shear force at the bottom of the tank and also generates bending moment just above the tank. Dynamic fluid also acts on the bottom of the tank which generates a vertical couple and is additive to the bending moment at the base. This

Table 10.4 Comparison of hydrodynamic response along *x*- and *y*-directions

Sl. No	Term	Guidelines from BIS [10]		Guidelines from TID-7024		Unit
		Longitudinal (x) direction	Lateral (y) direction	*Longitudinal* (x) direction	Lateral (y) direction	
1	Impulsive mass (m_i)	510,551	573,496	510,541	573,486	kg
2	Convective mass (m_c)	361,576	309,295	360,891	308,710	kg
3	Height of the impulsive mass above the bottom of the tank wall (h_i) (without considering base pressure)	2.63	2.63	2.63	2.63	m
4	Height of the convective mass above the bottom of the tank wall (h_c) (without considering base pressure)	4.24	4.46	4.24	4.46	m
5	Height of the impulsive mass above the bottom of the tank wall (h_i*) (with considering base pressure)	4.88	4.25	4.88	4.25	m
6	Height of the convective mass above the bottom of the tank wall (h_c*) (with considering base pressure)	5.49	5.17	5.47	5.16	m
7	Wall deflection (d) due to load	0.0329	0.0088	0.16	0.19	m
8	Impulsive frequency (f_i)	2.751	5.307			hz
9	Convective frequency (f_c)	0.238	0.261	0.25	0.28	hz
10	Seismic coefficient (A_h)					
	Impulsive $(A_h)_i$	0.36	0.36	0.36	0.36	
	Convective $(A_h)_c$	0.04	0.07	0.04	0.06	
11	Total shear force (V) at bottom of the wall	2751.78	3620.81	20.21	22.93	KN
	Impulsive V_i	2748.12	3614.58			KN
	Convective V_c	141.88	212.39			KN
12	Total bending moment at bottom of the wall (M)	8.30	10.92	0.04957	0.06367	MN-m
	Impulsive M_i	8.28	10.88			MN-m
	Convective M_c	0.60	0.95			MN-m
13	Over turning moment at bottom of base slab. (M*)	13.75	17.84	0.09185	0.09922	MN-m
14	Impulsive time period (T_i)	0.36	0.19			sec
15	Convective time period (T_c)	4.20	3.84	4.01	3.61	sec
16	Slosh height (h_s)	0.240	0.350	0.21	0.28	m
17	Impulsive pressure on wall (y = 0) (P_{iw})	110.32	18.08	110.32	18.08	KN/m^2
	Impulsive pressure on top of base (y = 0) (P_{ib})	8.66	8.72			KN/m^2
18	Convective pressure on wall (y = 0) (P_{cw})	0.61	0.62	1.96	2.45	KN/m^2
	Convective pressure on wall ($y = h$) (P_{cwt})	1.96	2.86	1.86	2.39	KN/m^2
	Convective pressure on top of base (y = 0) (P_{cb})	0.61	0.62	2.46	4.77	KN/m^2
19	Pressure due to wall inertia (P_{ww})	2.65	4.41			KN/m^2
	Pressure due to vertical excitation (P_v)	15.45	15.45			KN/m^2
20	Maximum hydrodynamic pressure (P)	26.86	27.30	21.52	21.91	KN/m^2

moment causes overturning moment acting on the supporting structures. The tank wall and supports are required to be designed for these forces and bending moment. The supporting structures are also required to be designed for shear forces and overturning moment.

A simplified design approach based on the TID 7024[TID 7024, 1963] and ASCE-4-98 [1] is explained in the section for ground-supported tank. The method described in TID 7024 is applicable for flat-bottom tank, vertically oriented tank of uniform rectangular or circular geometry. The tank containing liquid under dynamic excitation is shown in Fig. 10.16a, b which describes the motion of fluid in tank and its mathematical model. The tank base dimension for circular tank is 2R, or rectangular tank base dimension is

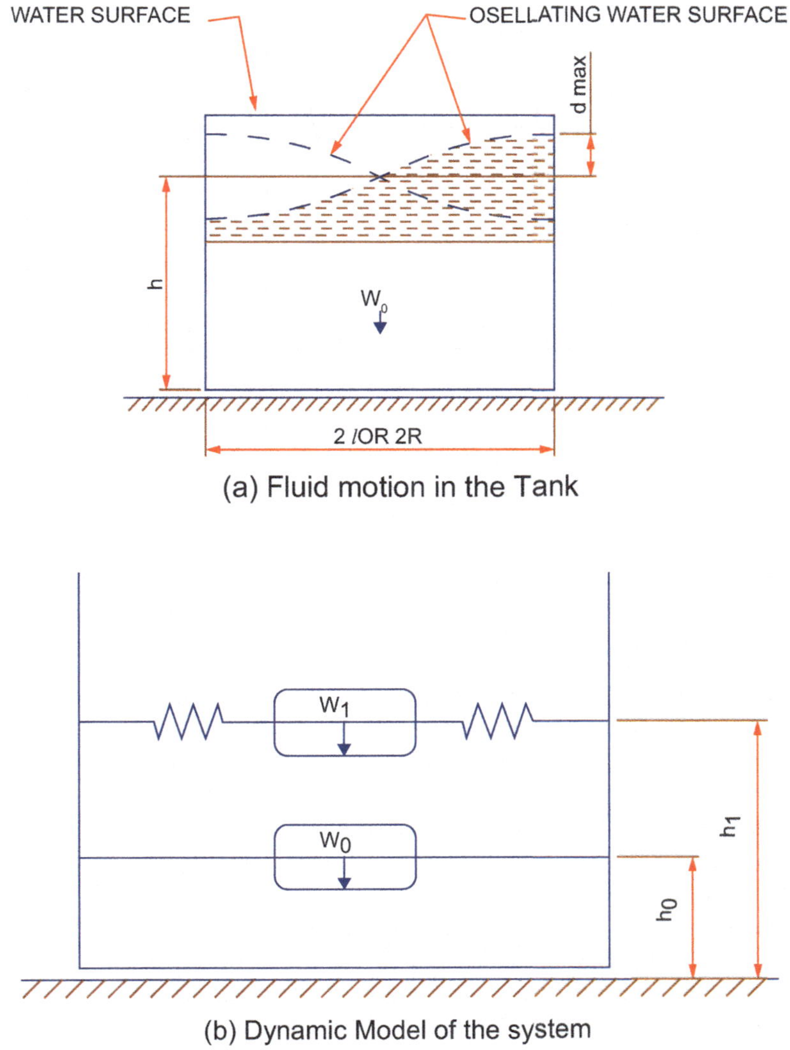

Fig. 10.16 Mathematical idealization of seismic response of fluid in tank

2 l. The height of the water is h. Total weight of the liquid is taken as W. The water mass located at the bottom portion W_0 moves rigidly with tank wall. However, the top portion of water W_1 oscillates with different frequency and behaves as flexible mass attached to tank. These masses along with height are shown in Fig. 10.16b. The rigid mass is known as impulsive mass, and flexible mass is known as convective or sloshing mass. The tank itself behaves as rigid tank. The horizontal excitation of mass relative to walls determines the water slosh height and forces exerted on walls. In Fig. 10.17a, the schematic of the impulsive mass W_0, convective mass W_1, and their respective height from the bottom of the tank are shown. The formula developed for hydrodynamic forces is valid for short height tank with h/l (cylindrical tank h/r) ratio less than 1.5. The slender tank in which h/l ratio is greater than 1.5 is dealt differently. The top portion height up to 1.5 l is used for the calculation of hydrodynamic forces, and remaining mass below this height is considered constrained mass. Figure 10.17b describes the schematic of distribution of mass and height.

10.7.5.1 Hydrodynamic Force Calculation for Vertical Tank

The equations for calculation of horizontal impulsive and convective forces and their related quantities for both rectangular and cylindrical tanks were developed by Housner [17]. It is assumed that liquid contained in the tank is incompressible and frictionless and its displacements are small compared to the tank dimension. The tank wall is assumed to be rigid for calculation of impulsive and convective pressures on the tank wall. The impulsive pressure is proportional to the acceleration of the tank wall, whereas the convective pressure is caused by sloshing of fluid and depends on acceleration corresponding to the sloshing frequency. The formulation provided for hydrodynamic forces by TID is valid for the tank height to base ratio less than 0.75($h/2R < 0.75$ or $h/R < 1.5$). Calculation method and evaluation of different parameters for rectangular tank are provided in this section. Similarly, calculation procedure for circular tank is described in Example 10.3.

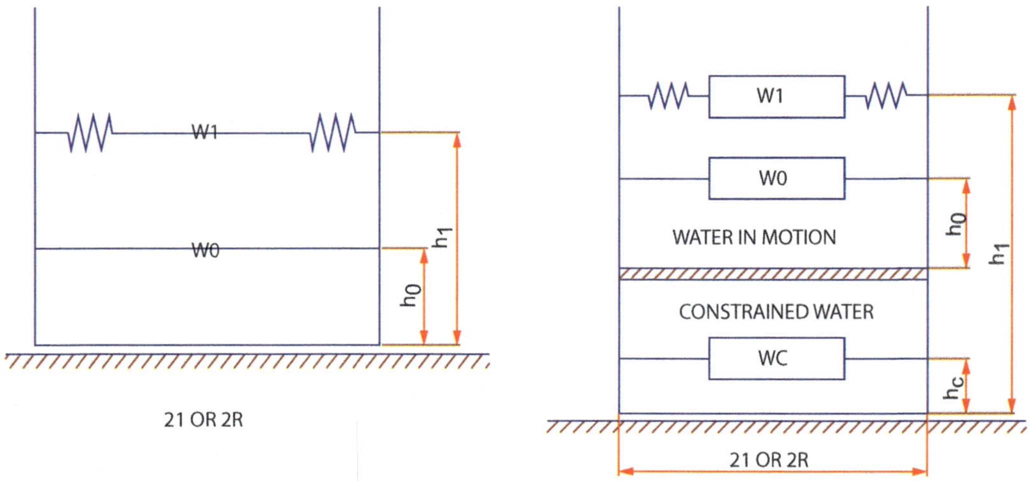

(b) SHORT HEIGHT TANK (h/l OR h?R < 1.5)

(c) SLENDER TANK (h/l OR h/R > 1.5)

Fig. 10.17 Mathematical idealization of fluid for short, height, and slender tanks

The effect of overturning moment due to the dynamic pressure on the tank is evaluated by increasing the vertical arms of the impulsive and convective forces P_0 and P_1, respectively. The arm length h_0 and h_1 is used to calculate the bending moment on a plane just above the bottom, and large value of these arms is used to calculate overturning moment on a plane just below the bottom. Since the smaller arms of these forces are not used to calculate the dynamic fluid pressure on the bottom, these are designated as excluding the bottom pressure (EBP). The larger value is designated as including the bottom pressure (IBP). The tank whose height-to-base ratio is higher than 0.75 ($h/2R > 0.75$ or $h/R > 1.5$) is dealt as slender tank. Separate provisions are explained for the slender tank in Sect. 10.7.5.1 (B) given below.

A. **Horizontal impulsive mode**

i. **For rectangular tank**

(a) **Impulsive mass**

$$\frac{W_o}{W} = \frac{\tanh\left(\sqrt{3}\frac{l}{h}\right)}{\sqrt{3}\frac{l}{h}} \tag{10.22}$$

(b) **Height of impulsive mass**

$$h_0 = \frac{3}{8}h \quad \text{(EBP)} \tag{10.23}$$

Excluding effect of dynamic pressure on tank bottom

$$\frac{h_0}{h} = \frac{1}{8}\left[\frac{4}{\frac{\tanh\left(\sqrt{3}\frac{l}{h}\right)}{\sqrt{3}\frac{l}{h}}} - 1\right] \quad \text{(IBP)} \tag{10.24}$$

Including effect of dynamic pressure on tank bottom

(c) **Impulsive force**

$$P_0 = \ddot{u}_0 \frac{W_0}{g} \tag{10.25}$$

(d) **Impulsive pressure on wall**

$$p_w = \rho \ddot{u}_0 h \sqrt{3}\left[\frac{y}{h} - \frac{1}{2}\left(\frac{y}{h}\right)^2\right] \tanh\left(\sqrt{3}\frac{l}{h}\right) \tag{10.26}$$

(e) **Impulsive pressure on tank bottom**

ii. **For cylindrical tank**

$$p_b = \rho \ddot{u}_0 h \frac{\sqrt{3}}{2} \frac{\sinh\left(\sqrt{3}\frac{x}{h}\right)}{\cosh\left(\sqrt{3}\frac{l}{h}\right)} \tag{10.27}$$

(a) **Impulsive mass**

$$\frac{W_o}{W} = \frac{\tanh\left(\sqrt{3}\frac{R}{h}\right)}{\sqrt{3}\frac{R}{h}} \tag{10.28}$$

(b) **Height of impulsive mass**

$$h_0 = \frac{3}{8}h \quad \text{(EBP)} \tag{10.29}$$

$$\frac{h_0}{h} = \frac{1}{8}\left[\frac{4}{\frac{\tanh\left(\sqrt{3}\frac{R}{h}\right)}{\sqrt{3}\frac{R}{h}}} - 1\right] \quad \text{(IBP)} \tag{10.30}$$

(c) **Impulsive force**

$$P_0 = \ddot{u}_0 \frac{W_0}{g} \tag{10.31}$$

(d) **Impulsive pressure on wall**

$$p_w = -\rho \ddot{u}_0 h\sqrt{3}\left[\frac{y}{h} - \frac{1}{2}\left(\frac{y}{h}\right)^2\right]\tanh\left(\sqrt{3}\frac{R}{h}\cos\phi\right) \quad (10.32)$$

(e) **Impulsive pressure on tank bottom**

$$p_b = -\rho \ddot{u}_0 h \frac{\sqrt{3}}{2}\frac{\sinh\left(\sqrt{3}\frac{x}{h}\right)}{\cosh\left(\sqrt{3}\frac{R}{h}\right)} \quad (10.33)$$

B. **Horizontal convective mode (sloshing)**

i. **For rectangular tank**

(a) **Convective mass**

$$\frac{W_1}{W} = 0.527\frac{l}{h}\tanh\left(1.58\frac{h}{l}\right) \quad (10.34)$$

(b) **Height of convective mass**

$$\frac{h_1}{h} = 1 - \frac{\cosh\left(1.58\frac{h}{l}\right) - 1}{1.58\frac{h}{l}\sinh\left(1.58\frac{h}{l}\right)} \text{ (EBP)} \quad (10.35)$$

$$\frac{h_1}{h} = 1 - \frac{\cosh\left(1.58\frac{h}{l}\right) - 2}{1.58\frac{h}{l}\sinh\left(1.58\frac{h}{l}\right)} \text{ (IBP)} \quad (10.36)$$

(c) **Sloshing frequency**

$$\omega^2 = \frac{1.58g}{l}\tanh\left(1.58\frac{h}{l}\right) \quad (10.37)$$

(d) **Amplitude of angular oscillation**

$$\theta_h = 1.58\frac{A_1}{l}\tanh\left(1.58\frac{h}{l}\right) \quad (10.38)$$

(e) **Sloshing force**

$$P_1 = W_1\theta_h\sin(\omega t) \quad (10.39)$$

(f) **Maximum vertical displacement of water surface (slosh height)**

$$d_{max} = \frac{0.527l\coth\left(1.58\frac{h}{l}\right)}{\frac{g}{\omega^2\theta_h l} - 1} \quad (10.40)$$

For second and higher modes, d_{max} shall be calculated using the formula below:

$$d_{max} = \theta_h \cdot l \quad (10.41)$$

(g) **Hydrodynamic pressure on wall and bottom of tank**

$$p_w = 0.527\rho l^2\omega^2\theta_h\frac{\cosh\left(1.58\frac{h-y}{l}\right)}{\sinh\left(1.58\frac{h}{l}\right)} \quad (10.42)$$

$$p_b = 0.791\rho l^2\omega^2\theta_h\left[\frac{x}{l} - \frac{1}{3}\left(\frac{x}{l}\right)^3\right]\frac{1}{\sinh\left(1.58\frac{h}{l}\right)} \quad (10.43)$$

ii. **For cylindrical tank**

(a) **Convective mass**

$$\frac{W_1}{W} = 0.318\frac{R}{h}\tanh\left(1.84\frac{h}{R}\right) \quad (10.44)$$

(b) **Height of convective mass**

$$\frac{h_1}{h} = 1 - \frac{\cosh\left(1.84\frac{h}{R}\right) - 1}{1.84\frac{h}{R}\sinh\left(1.84\frac{h}{R}\right)} \text{ (EBP)} \quad (10.45)$$

$$\frac{h_1}{h} = 1 - \frac{\cosh\left(1.84\frac{h}{R}\right) - 2.01}{1.84\frac{h}{R}\sinh\left(1.84\frac{h}{R}\right)} \text{ (IBP)} \quad (10.46)$$

(c) **Sloshing frequency**

$$\omega^2 = \frac{1.84g}{R}\tanh\left(1.84\frac{h}{R}\right) \quad (10.47)$$

(d) **Amplitude of angular oscillation**

$$\theta_h = 1.534\frac{A_1}{R}\tanh\left(1.84\frac{h}{R}\right) \quad (10.48)$$

(e) **Sloshing force**

$$P_1 = 1.2W_1\theta_h\sin(\omega t) \quad (10.49)$$

(f) **Maximum vertical displacement of water surface (slosh height)**

$$d_{max} = \frac{0.408R\coth\left(1.84\frac{h}{R}\right)}{\frac{g}{\omega^2\theta_h R} - 1} \quad (10.50)$$

For second and higher modes, d_{max} shall be calculated using Eq. (10.51) as given below:

$$d_{max} = \theta_h \cdot R \quad (10.51)$$

(g) **Hydrodynamic pressure on wall and bottom**

$$p_w = 0.408\rho R^2\omega^2\theta_h\left(1 - \frac{1}{3}\cos^2\phi - \frac{1}{2}\sin^2\phi\right)\cos\phi\frac{\cosh\left(1.84\frac{h-y}{R}\right)}{\sinh\left(1.84\frac{h}{R}\right)} \quad (10.52)$$

$$p_b = 0.408\rho l^2\omega^2\theta_h\left(1 - \frac{1}{3}\cos^2\phi - \frac{1}{2}\sin^2\phi\right)\cos\phi\frac{1}{\sinh\left(1.84\frac{h}{R}\right)} \quad (10.53)$$

C. **Vertical fluid response mode**

The hydrodynamic pressure, p_v, on the tank shell at depth 'y' from the top of the fluid surface can be calculated from Eq. (10.54).

$$p_v = (S_{av})\rho y \tag{10.54}$$

where

ρ fluid mass density

The spectral acceleration S_{av} is determined for the damping of the tank shell corresponding to the frequency of the vertical response mode. The vertical fluid response mode natural frequency is generally influenced by the breathing flexibility of the tank wall and can be computed from Eq. (10.55).

$$\omega_v = \frac{\pi}{2h}\left[\rho\left(\frac{D}{tE}+\frac{1}{K}\right)\right]^{-1/2} \tag{10.55}$$

where

- K fluid bulk modulus
- E elastic modulus of tank wall
- t thickness of tank wall.

D. **Overturning moment**

The maximum overturning moment at the base of the tank shall be calculated considering the effects of impulsive and sloshing horizontal overturning moment by using SRSS method of combination.

E. **Longitudinal compressive stress in tank shell**

The seismic-induced longitudinal compressive force for tank shell buckling evaluation shall be calculated for both horizontal and vertical response modes using SRSS method. When the tank is anchored sufficiently to prevent uplift, the seismic-induced compressive stress in the tank shell can be calculated by using the formula,

$$c = \sqrt{(F_v)^2 + \left(\frac{1.273M_B}{D^2}\right)^2} \tag{10.56}$$

where

- F_v Maximum vertical force response due to empty tank shell, per unit length.
- M_B Combined overturning moment at the base.
- C Longitudinal compressive stress per unit length in tank shell.

F. **Hoop tension in tank shell**

Seismic-induced hydrodynamic pressure at any level is calculated from the pressure resulting from horizontal impulsive, sloshing, and vertical hydrodynamic pressures. SRSS (Square Root of Sum of Squares) method is used to combine the pressures. The resultant pressure is sum of hydrodynamic pressure and hydrostatic pressure at same level to determine the hoop tension in the tank shell.

G. **Freeboard requirement**

The freeboard is the height of the tank above the water in the tank. If the tank is open, the freeboard requirement is to ensure the spilling of liquid due to slosh. The freeboard height must be kept higher than the slosh height. If the tank is closed at the top, the tank top is required to be analyzed for the pressure and impact force resulting from the sloshing. The tank is considered to be full, and 100% mass is considered as impulsive mass if the freeboard is less than half the slosh height and sloshing mode calculation is not required.

H. **Damping values**

The damping associated with sloshing is considered as 0.5% of critical damping. The first mode involves sloshing almost entirely, whereas the second mode is primarily the motion of the supporting structure. Therefore, it is reasonable to use 0.5% damping in the first mode, and a value in the second mode appropriate for the supporting structure shall be used.

10.7.5.2 Steps for Calculating the Hydrodynamic Loads

A. **Vertical tanks supported on the ground (h/l or $h/R \leq 1.5$)**

When a tank is supported on the ground, with $h/R \leq 1.5$ or $h \leq 1.5\,R$, the procedure for calculating the forces and water-surface displacement caused by ground motion is as follows:

i. Calculate the impulsive mass and two heights, one excluding the liquid pressure on the bottom and the other including the liquid pressure on the bottom.
ii. Add the impulsive mass and the tank mass, and obtain the gross equivalent weight $\mathbf{W_0''}$ and the corresponding $\mathbf{h_0''}$ values.

iii. Calculate the impulsive force $\mathbf{P''_0}$ from $\mathbf{W''_0}$, determining the maximum seismic horizontal acceleration, $\ddot{u}_0$, from a spectrum curve for $T = 0$ of the site.
iv. Calculate the impulsive bending moment at the base of the tank and the impulsive overturning moment using the $\mathbf{h''_0}$ values.
v. Calculate convective mass $\mathbf{W_1}$ and the two $\mathbf{h_1}$ values, one excluding the dynamic fluid pressure on the bottom and the other including the dynamic fluid pressure on the bottom.
vi. Calculate the natural frequency, ω for slosh mass, and corresponding spectral acceleration (S_a) may be obtained for the given damping from the response spectrum curve of the site.
vii. Using S, compute the maximum amplitude, $\mathbf{A_1}$, of the displacement; the angle of free oscillation, $\emptyset_h$, at the water surface; and the convective force, $\mathbf{P_1}$.
viii. Compute the maximum water-surface displacement, $\mathbf{d_{max}}$ (slosh height above its original level) from the values of ω and $\emptyset_h$; and the maximum bending moment, overturning moment and shear at the base of the tank.

The use of formulae and the calculation steps for a short height tank are illustrated in Example 10.3 (Sect. 10.7.7).

B. **Slender vertical tanks supported on ground (h/l or h/R $\geq$ 1.5)**

In case of a slender tank, the depth of the liquid exceeds 1.5 times of l, or in a circular tank where the depth exceeds 1.5 times of the radius, and the entire mass of the liquid below this depth is considered as a rigid body for the calculation of impulsive pressures. The schematic of representative mass is shown in Fig. 10.17b.

The impulsive mass W_0 and its level h_0 are calculated using the equation given in Sect. 10.7.5.1. The constraint mass W'_0 acts at its center of gravity. The constraint masses W'_0 and W_0 are combined into single mass W''_0 with height h''_0 above the actual bottom of the tank. However, the concept of dividing the tank into upper and lower zones does not apply in the case of convective mass and related quantities. They are calculated as a function of the full depth of water.

C. **Elevated vertical tanks**

When the tank is mounted on an elevated supporting structure, the coupling to the ground is no more rigid, as was the case with ground-supported tanks, and the flexibility of the supporting structure must be considered. The resulting model is a two degrees of freedom system. The fictitious springs joining W_1 to the tank walls in Fig. 10.9a are replaced with a hypothetical column of same stiffness k_1, forming a direct coupling between W_1 and W_0. The weight W_0 is connected to the ground through a similar hypothetical column representing the support structure having stiffness k_0.

The properties W_0, W_1, h_0, h_1, and ω are independent of support conditions and are calculated from equations given in Sect. 10.7.5.1 for ground-supported tanks. The value of spring constant k_1 is calculated from the following relation,

$$k_1 = \omega_2 \times \frac{W_1}{g} \qquad (10.57)$$

It should be noted that, as in the case of ground-supported tanks, the weight W''_0 replaces W_0 in the calculations when the tank weight or that of the structure is taken into account. The resulting 2DOF system model can be solved either by hand calculation or using FEM technique. The results are obtained by summing the contributions of the two modes. For each mode, the amplitude A1 (max. deflection of W_1 relative to W_0) used to calculate the maximum angle of free oscillation, θ_h, of water surface is the algebraic difference of the maximum horizontal deflections of the two masses. The vertical displacement of water surface contributed by second mode is calculated by Eqs. (10.58) and (10.59).

$$d_{\max} = \theta_h \cdot l \quad \text{(For rectangular tank)} \qquad (10.58)$$

$$d_{\max} = \theta_h \cdot R \quad \text{(For circular tank)} \qquad (10.59)$$

D. **Horizontal Cylindrical Tank**

The horizontal cylindrical tank is assumed to be equivalent to a rectangular tank with width equal to the width of free surface of water, and equivalent height is calculated by equating the cross-sectional area of the tank up to the depth of water. For a cylindrical tank with depth of water H and diameter D, the dimensions of equivalent rectangular tank (depth of water (h) and width (2 l)) are calculated as follows Kobayashi et al. [23]:

$$2l = D\cos\theta \qquad (10.60)$$

where

$$\sin\theta = \frac{(H - D/2)}{(D/2)} \qquad (10.61)$$

$$h = \frac{1}{2}\left(H - \frac{D}{2}\right) + \frac{D}{2}\frac{\left[\sin^{-1}\left\{\frac{2H-D}{D}\right\} + \frac{\pi}{2}\right]}{\left[8\sqrt{H(D-H)}\right]} \qquad (10.62)$$

10.7.6 The Procedure Given as per the BIS [10], ACI 350.3 (2001), and IS 1893 (Part 2): 2014

The procedure for calculation of hydrodynamic loads as per BIS [10], ACI 350.3 (2001), and IS 1893 (Part 2): 2014 codes is given below:

10.7.6.1 Time Period

The time period is known as the time required to complete an oscillation. The time period is also divided in the similar way of impulsive and convective masses, since the time taken to complete the oscillation is difficult to compute for each layer of liquid. Thus, time period of impulsive and convective masses is called as impulsive and convective time periods. The expression for impulsive and convective mode time periods of tank is taken from the ACI 305.3 (2001).

$$T_{\mathrm{I}} = 2\pi\sqrt{\frac{d}{g}} \tag{10.63}$$

$$T_{\mathrm{C}} = \sqrt{\frac{m_{\mathrm{C}}}{K_{\mathrm{C}}}} = C_{\mathrm{C}}\sqrt{\frac{D}{g}} \quad (D = L \text{ for rectangular tank}) \tag{10.64}$$

where

T_{i} impulsive time period,
T_{c} convective time period,
K_{s} tank stiffness,
m_{c} convective mass,
K_{c} convective spring stiffness,
D tank diameter,
d deflection of tank wall,
g gravity constant, and
C_{c} convective coefficient of time period.

Here, d is the deflection of the tank wall on the vertical centerline at a height of $\bar{h}$,

$$\bar{h} = \frac{\left(\frac{m_{\mathrm{I}}h_{\mathrm{I}}}{2} + \frac{\bar{m}_{\mathrm{w}}h}{2}\right)}{(m_{\mathrm{I}}/2) + \bar{m}_{\mathrm{w}}} \tag{10.65}$$

when loaded by uniformly distributed pressure of intensity q,

$$q = \frac{\left(\frac{m_{\mathrm{I}}}{2} + \bar{m}_{\mathrm{w}}\right)g}{Wh} \tag{10.66}$$

where $\bar{m}_{\mathrm{w}}$ = Mass of one tank wall perpendicular to the direction of seismic force.

Assuming that wall takes the pressure 'q' by cantilever action, one can find the deflection (d), by considering wall strip of unit width and height $\bar{h}$ which is subjected to concentrated load,

$$P = q\,h \tag{10.67}$$

One can obtain d as,

$$d = \frac{P\bar{h}^3}{3EI_W}$$

$$I_W = \frac{1}{12}t^3 \tag{10.68}$$

The convective coefficient of time period is depending on the ratio of liquid height and tank diameter.

$$C_{\mathrm{C}} = \frac{2\pi}{\sqrt{3.68\tanh(3.16h/D)}} \quad \text{for cylindrical tank}$$

$$C_{\mathrm{C}} = \frac{2\pi}{\sqrt{3.16\tanh(3.16\,h/L)}} \quad \text{for rectangular tank} \tag{10.69}$$

10.7.6.2 Base Shear

During base excitation, the induced lateral force by the system inertia produces shear near the fixed support. However, in actual practice, the lateral acceleration originates from the ground motion under seismic conditions. For numerical study, the ground acceleration is used as an input and calculated by using the response spectrum of that location. This base shear should be less than allowable shear stress for the structure. This condition should be ensured for safe design process. Further, they are categorized by the impulsive and convective base shear.

Base shear in impulsive mode (V_{i}) just above the base of staging is given by

$$V_{\mathrm{i}} = (SA_{\mathrm{h}})_{\mathrm{I}}(m_{\mathrm{I}} + m_{\mathrm{S}})g \tag{10.70}$$

Convective mode base shear (V_{C}) is given by

$$V_{\mathrm{C}} = (SA_{\mathrm{h}})_{\mathrm{C}}(m_{\mathrm{C}})\,g \tag{10.71}$$

where,

$(SA_{\mathrm{h}})_{\mathrm{I}}$ Design horizontal seismic coefficient for impulsive mode,
$(SA_{\mathrm{h}})_{\mathrm{C}}$ Design horizontal seismic coefficient for convective mode

Total base shear V can be obtained from square root of sum of squares of base shear in impulsive and convective modes as follows:

$$V = \sqrt{V_{\mathrm{I}}^2 + V_{\mathrm{C}}^2} \tag{10.72}$$

10.7.6.3 Overturning Moment

The overturning moment is generated due to the combined dynamic pressure of impulsive and convective modes on the tank wall. The inertial and reaction forces are developed between the liquid and wall, when the liquid displaces by boundary wall due to excitation. The average values of reaction forces (such as impulsive force, convective force, and wall inertia) at their center of gravity (CG) and multiply with their corresponding heights of CG from the foundation yields the overturning moment.

Overturning moment in impulsive mode (M_{I}^{*}) is given by

$$M_{\mathrm{i}}^{*} = (SA_{\mathrm{h}})_{\mathrm{I}}\left(m_{\mathrm{I}}\left(h_{\mathrm{I}}^{*}+h_{\mathrm{S}}\right)+m_{\mathrm{S}}h_{\mathrm{CG}}\right)g \tag{10.73}$$

Overturning moment in convective mode (M_{C}^{*}) is given by

$$M_{\mathrm{c}}^{*} = (SA_{\mathrm{h}})_{\mathrm{c}}\left(m_{\mathrm{c}}\left(h_{\mathrm{c}}^{*}+h_{\mathrm{s}}\right)\right)g \tag{10.74}$$

where h_{I}^{*}, h_{C}^{*},h_{CG}, and h_{S} are

- h_{I}^{*} Height of impulsive mass above bottom of tank wall
- h_{C}^{*} Height of convective mass above bottom of tank wall
- h_{CG} Height of CG of empty tank from staging base
- h_{S} Structural height of staging

Total moment shall be obtained by combining the moment in impulsive and convective modes through SRSS which is given as follows:

$$M^{*} = \sqrt{M_{\mathrm{I}}^{*2}+M_{\mathrm{C}}^{*2}} \tag{10.75}$$

10.7.6.4 Hydrodynamic Pressure

The pressure at a location has both hydrostatic and hydrodynamic components. The difference in kinetic energy arises due to hydrodynamic components only. The fluid itself gets excited due to the ground excitation, and the hydrodynamic pressure is generated inside the fluid due to non-uniform kinetic energy around the wall.

Impulsive hydrodynamic pressure: The lateral impulsive hydrodynamic pressure exerted by the liquid on the tank wall and base is given by,

$$p_{\mathrm{IW}} = Q_{\mathrm{IW}}(y)(SA_{\mathrm{h}})_{\mathrm{I}}\rho\, gh\cos\phi \tag{10.76}$$

Coefficient of impulsive hydrodynamic pressure on the wall is given by

$$Q_{\mathrm{IW}}(y) = 0.866\left[1-\left(\frac{y}{h}\right)^{2}\right]\tanh\left(0.866\frac{D}{h}\right) \tag{10.77}$$

where

- ρ Mass density of liquid
- ϕ Circumferential angle
- y Vertical distance of a point on tank wall from the bottom of the tank wall

Convective hydrodynamic pressure: The lateral convective hydrodynamic pressure exerted by oscillating liquid on the tank wall and tank base is given by

$$p_{\mathrm{CW}} = Q_{\mathrm{CW}}(y)(SA_{\mathrm{h}})_{\mathrm{C}}\rho\, gD\left[1-\frac{1}{3}\cos^{2}\phi\right]\cos\phi \tag{10.78}$$

$$Q_{\mathrm{CW}}(y) = 0.5625\frac{\cosh(3.674y/D)}{\cosh(3.674h/D)} \tag{10.79}$$

The impulsive and convective pressure distributions on side wall of tank are shown in Fig. 10.18a, b, respectively.

10.7.6.5 Pressure Due to Wall Inertia

Pressure due to wall inertia will act in the same direction as that of seismic force. The wall inertia may be substantial for concrete tanks than steel tanks. Pressure due to wall inertia, which is constant along the wall height for walls of uniform

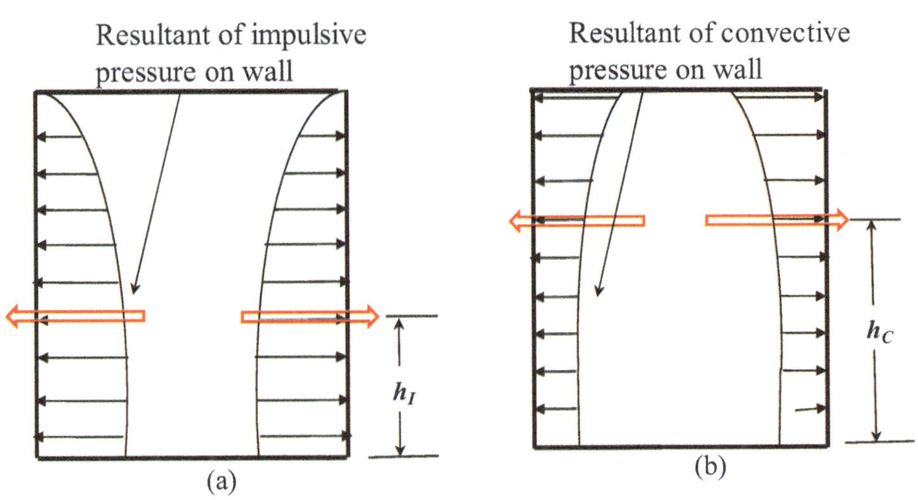

Fig. 10.18 a Impulsive and b convective pressure distribution on side wall of tank

thickness, should be added to the impulsive pressure (IS 1893, Part 2). Pressure on tank wall due to its inertia is given by

$$p_{WW} = (SA_{\mathrm{h}})_{\mathrm{I}} t \rho_{\mathrm{m}} g \quad (10.80)$$

where

ρ_{m} Mass density of tank wall
t Wall thickness.

10.7.6.6 Sloshing Wave Height

The liquid oscillation inside the tank about the mean liquid level is termed as sloshing or the natural oscillation of liquid surface. These standing waves show different behavior with excitation frequencies. The maximum possible vertical height achieved by the free surface of liquid during oscillation is given by

$$d_{\max} = (SA_{\mathrm{h}})_{\mathrm{C}} R \frac{D}{2} \quad (10.81)$$

where

$d_{\max}$ Maximum slosh height

The ratio of maximum seismic force on a structure during specified ground motion to the design seismic force is known as response reduction factor (R). Based on strength, redundancy, and ductility of structure, the actual seismic forces are reduced by a factor R to get the design load. This reduction depends on strength, redundancy, and ductility of the structure. The values of R given in the present guideline are based on studies of Jaiswal et al. [24]. The highest value of R is 2.5, and lowest value is 1.3.

10.7.6.7 Design of Horizontal Cylindrical Tank

The sloshing frequencies of a horizontally resting cylindrical tank (Fig. 10.19) are difficult to obtain, even for an ideal frictionless liquid. The analytical solution is not available for sloshing in arbitrary height liquid-filled horizontal cylindrical vessels. However, an analytical solution for the half-full horizontal cylinder is available [31]. The equivalent rectangle method could be used for approximating the half-full horizontal cylinder sloshing response for tank design (i.e., the response of the horizontal liquid cylinder can be compared with the response of an equivalent rectangular container, which has the same free surface dimensions and contains the same amount of liquid) (Fig. 10.20).

For sloshing in arbitrary height, liquid-filled horizontal cylindrical vessels, an approximate procedure is generally used as suggested by Eurocode 8. According to Eurocode 8, when liquid height-to-radius ratio (H/r) is between 0.5 and 1.6, the horizontal tanks can be analyzed as the rectangular

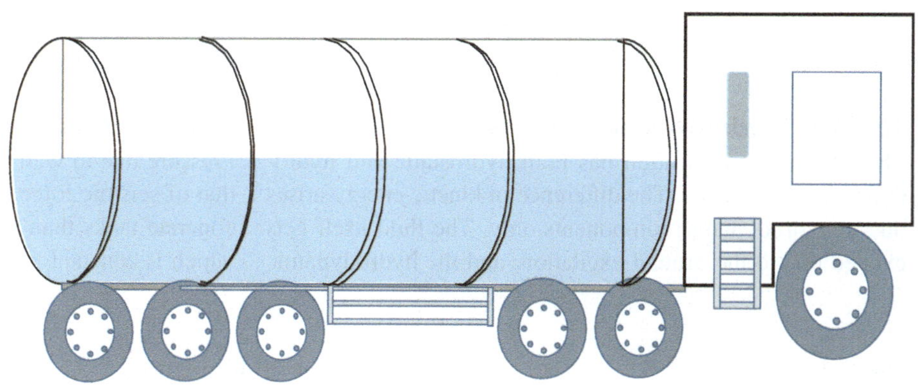

Fig. 10.19 A typical road vehicle carrying liquid container

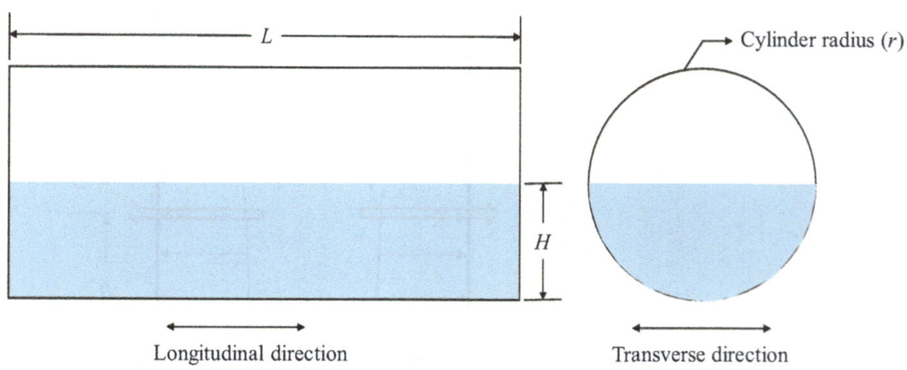

Fig. 10.20 Impulsive and convective pressure distribution on side wall of tank

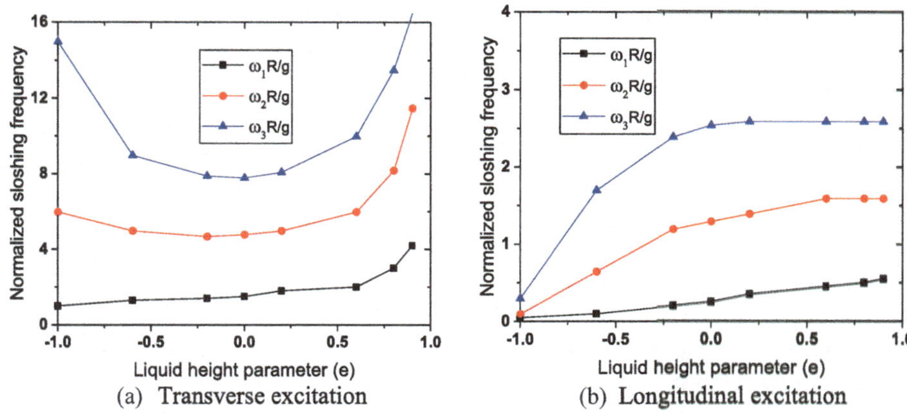

Fig. 10.21 Variation of the first three sloshing frequencies of a horizontal cylinder with respect to the liquid height (Karamanosx et al. 2006)

tank of equal dimension at the liquid level and in the direction of motion, and of depth required to give equal liquid volume. When *H/r* ratio *is* greater than 1.6, the tank should be presumed to act as if it were full with the total fluid mass rigidly connected to the tank [7]. The liquid height parameter (*e*) is difference between *H* and *r*. Variation of the first three sloshing frequencies under excitation in transverse and longitudinal directions of a horizontal cylinder is shown in Fig. 10.21a, b, respectively [27].

10.7.7 Comparison of Response with Two Codal Provisions

One example has been considered for comparing the results for the design provision given in different codes.

Example 10.3 Consider a rectangular RC water tank of 840 m^3 capacity has plan dimensions of 12 m × 10 m, height of 8 m, and liquid still level 7 m as shown in Fig. 10.22. The wall and base slab have thickness of 300 and

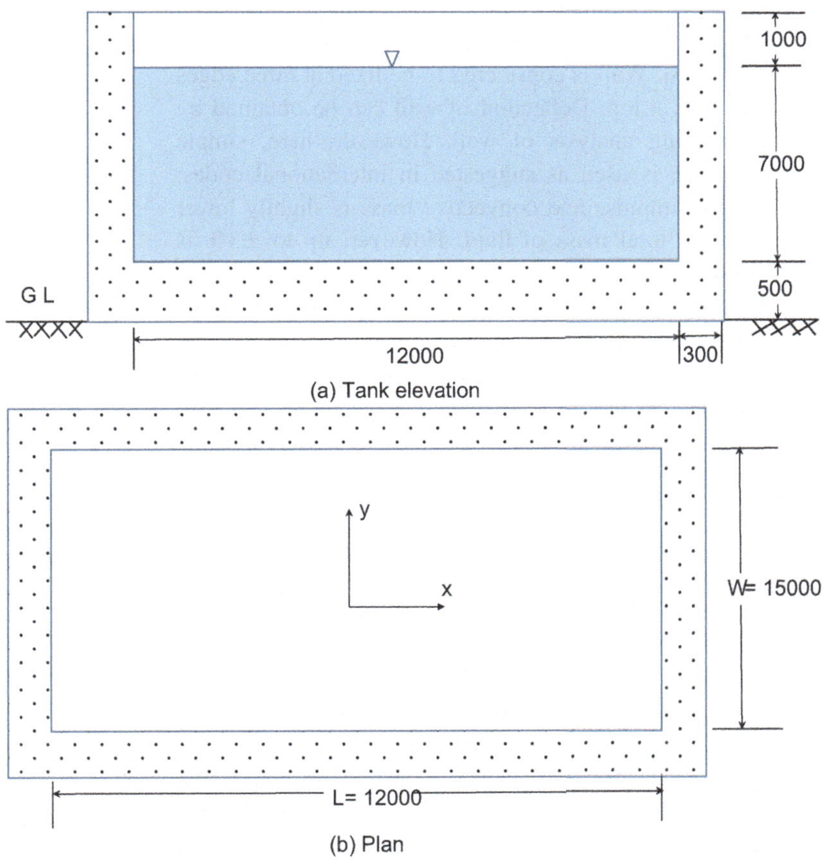

Fig. 10.22 Tank geometry details (all dimensions are in mm)

500 mm, respectively. There is no roof slab on the tank. This ground-mounted tank is placed in Zone V, hard soil. Grade of concrete is M45. Analyze the tank for seismic loads.

Solution

Length of the tank (L) = 12,000 mm
Width of the tank (W) = 10,000 mm
Height of the tank (H) = 8000 mm
Height of the liquid (H_s) = 7000 mm
Liquid density (ρ_l) = 1000 kg/m^3
Thick wall (t) = 0.3 m
Concrete density (ρ_s) = 2500 kg/m^3
Height of the base (h_b) = 0.5 m

Weight calculations:

Mass of water (m_l) = 840,000 kg
Mass of the tank wall (m_{wall}) = 267,600 kg
Mass of base (m_{base}) = 54,000 kg
Total seismic mass ($m_{seismic}$) = 1,161,600 kg

i. For rectangular tank, seismic analysis is to be performed for loading in X- and Y-directions. Formulas used here are taken from BIS [10] (Fifth revision) in which the formulations were based on IS 1893 (Part 1): [20], ACI 350.3, Housner [17] Eurocode 8, and NZS 3106 [28]. Wall is considered to be fixed at three edges and free at top. Deflection of wall can be obtained by performing analysis of wall. However, here, simple approach is used as suggested in international codes. Sum of impulse and convective mass is slightly lower than the total mass of fluid. However, up to ±3% is allowed.
ii. Damping in the convective mode for all types of liquids can be taken as 0.5% of the critical. Damping in the impulsive mode shall be taken as 2 and 5% of the critical for concrete and masonry tanks, respectively [25, 26]. The tank is located on hard soil in Zone V. Here, seismic coefficient value for convective and impulsive masses is taken as 0.04 and 0.36 g, respectively, in X-direction while 0.07 and 0.36 g in Y-direction, respectively. Comparison of hydrodynamic response along x- and y-directions is given in Table 10.4.

Observations: The hydrodynamic parameters are evaluated by two different codes. The BIS [10] values are little higher than the TID-7024 values especially in convective time period and hydrodynamic pressure. In BIS [10], code parameters of mechanical model are evaluated using rigid tank model. The flexibility of tank is also considered in the evaluation of impulsive time period while TID-7024 uses only rigid tank model formulation.

Example 10.4 One detail calculation for the cylindrical tank has been demonstrated using TID-7024 procedure. Consider a cylindrical tank of 8 m in diameter and containing 4.5 m of water is supported on the ground. Determine the impulsive and convective forces and moments and displacement of the water surface. Response spectrum with 0.5% critical damping is to be used. Given density of water as 1000 kg/m^3 and the weight of empty tank is 9 tons (90 kN).

Impulsive force

$$\frac{W_0}{W} = \frac{\tanh\left(\sqrt{3}\frac{R}{h}\right)}{\sqrt{3}\frac{R}{h}} = \frac{\tanh\left(\sqrt{3}\frac{4}{4.5}\right)}{\sqrt{3}\frac{4}{4.5}} = 0.59$$

$$W = \pi(4^2)(4.5)(1000) = 226\,T(2260\,\text{kN}).$$

$$W_0 = 0.59(2260) = 1333.4\,\text{kN}.$$

For simplicity, assume the tank weight of 90 kN acting at the center of gravity of W_0. The gross value of W_0 is

$$W_0'' = 1333.4 + 90 = 1423.4\,\text{kN}$$

$$h_0 = \frac{3}{8}h$$

$$h_0 = \frac{3}{8}(4.5) = 1.6875\,\text{m(EBP)}$$

$$h_0 = \frac{h}{8}\left[\frac{4}{\frac{\tanh\left(\sqrt{3}\frac{R}{h}\right)}{\sqrt{3}\frac{R}{h}}} - 1\right]$$

$$h_0 = \frac{4.5}{8}\left[\frac{4}{\tanh(1.539)/1.539} - 1\right] = 3.25\,\text{m(IBP)}$$

$$\ddot{u}_0 = 0.12\,\text{g}\ (\text{for T} = 0\ \text{for}\ 2\%\ \text{damping})$$

Use W_0'' instead of W_0

$$P_0'' = \frac{\ddot{u}_0}{g}W_0''$$

P_0'' 0.12 (1423.4) = 170.8 kN
M_0'' 170.8 × 1.6875 = 288.2 kNm (EBP)
M_0'' 170.8 × 3.25 = 555.1 kNm (IBP)

Convective force

$$W_1 = (0.318)\frac{R}{h}\tanh\left(1.84\frac{h}{R}\right)W$$

$$W_1 = (0.318)\frac{4}{4.5}\tanh\left(1.84\frac{4.5}{4}\right) \times 2260 = 618\,\text{kN}.$$

$$\frac{h_1}{h} = 1 - \frac{\cosh\left(1.84\frac{h}{R}\right) - 1}{1.84\frac{h}{R}\sinh\left(1.84\frac{h}{R}\right)}$$

$$\frac{h_1}{h} = 1 - \frac{\cosh\left(1.84\frac{4.5}{4}\right) - 1}{1.84\frac{4.5}{4}\sinh\left(1.84\frac{4.5}{4}\right)} = 0.625$$

$$h_1 = 0.625 \times 4.5 = 2.8125\,\text{m} \quad (EBP)$$

For IBP calculation

$$\frac{h_1}{h} = 1 - \frac{\cosh\left(1.84\frac{h}{R}\right) - 2.01}{1.84\frac{h}{R}\sinh\left(1.84\frac{h}{R}\right)}$$

$$\frac{h_1}{h} = 1 - \frac{\cosh\left(1.84\frac{45}{4}\right) - 2.01}{1.84\frac{4.5}{4}\sinh\left(1.84\frac{4.5}{4}\right)} = 0.75$$

$$h_1 = 0.75 \times 4.5 = 3.376\,\text{m} \quad (IBP)$$

Calculation for frequency of slosh

$$\omega^2 = \frac{1.84\,g}{R}\tanh\left(1.84\frac{h}{R}\right)$$

$$\omega^2 = \frac{(1.84)(9.8)}{4}\tanh\left(1.84\frac{4.5}{4}\right) = 4.5$$

$$\varpi = 2.12\,\text{rad/s}(0.337\,\text{Hz})$$

$$T = \frac{2\pi}{\omega} = \frac{2\pi}{2.12} = 2.96\,\text{s}$$

$$S_a = 0.072\,\text{g} = 0.7056\,\text{m/s}^2$$

(using 0.5% critical damping, spectral acceleration)
Horizontal amplitude of sloshing motion,

$$y_{max} = A_1 = \frac{S_a}{\omega^2} = 0.7056/4.5 = 0.1568\,\text{m}$$

$$\theta_h = 1.534\frac{A_1}{R}\tanh\left(1.84\frac{h}{R}\right)$$

$$\theta_h = 1.534\frac{0.1568}{4}\tanh\left(1.84\frac{4.5}{4}\right) = 0.058$$

Calculation for pressure load on wall

$P_1 = 1.2\ W_1 Y_h \sin \varpi t$
$P_1 = 1.2\ (516)\ (0.058) \sin \varpi t$
Maximum $P_1 = 43$ kN (for $\sin \omega t = 1$)
$M_1 = 43 \times 2.8125 = 120.96$ kNm (EBP)
$M_1 = 43 \times 3.376 = 145.2$ kNm (IBP)

Displacement of water surface (slosh height)

$$d_{max.} = \frac{0.408R\coth\left(1.84\frac{h}{R}\right)}{\frac{g}{\omega^2\theta_h R} - 1}$$

$$d_{max.} = \frac{0.408(4)\coth\left(1.84\frac{4.5}{4}\right)}{\frac{9.8}{4.5(0.058)4} - 1} = 0.2\,\text{m}$$

Maximum bending moment on the tank at a section just above the bottom:

$$M_0'' + M_1 = 288.2 + 120.96 = 409\,\text{kNm (EBP)}$$

Maximum overturning moment:

$$M_0'' + M_1 = 555.1 + 145.2 = 700.3\,\text{kNm (IBP)}$$

Maximum shear at base:

$$P_0'' + P_1 = 170.8 + 43.01 = 213.81\,\text{kN}$$

Longitudinal compressive stress in tank shell:
Considering the vertical spectral acceleration as two-thirds of the horizontal PGA,

$$S_{av} = 2/3 \times 0.12 = 0.08\,\text{g}$$

Vertical force response due to empty tank (seismic)

$$F = W \times S_{av} = 90 \times 0.08 = 7.2\,\text{kN}$$

Force per unit length

$$F_v = 1.68/(3.14 \times \text{D})$$
$$F_v = 1.68/(3.14 \times 8) = 0.286\,\text{kN/m}$$

$$M_B = 409\,\text{kNm (EBP)}$$

$$c = \sqrt{F_v^2 + 2\left(\frac{1.273 M_B}{D^2}\right)^2}$$

$$c = \sqrt{(0.286)^2 + 2 \times \left(\frac{1.273 \times 409}{8^2}\right)^2} = 11.5\,\text{kN/m}$$

Force per unit length due to self-weight = 90/(3.14 × 8) = 3.58 kN/m

Compressive force per unit length due to (DW + EQ) = 11.5 + 3.58 = 15.08 kN/m

If thickness of the tank shell is considered as 6 mm, then Longitudinal comp. Stress in tank shell

$$(\sigma_c) = 15.08 \times 1000/0.006 = 2.513\,\text{MPa}$$

Hoop tension in tank shell

(i) Hydrostatic pressure, p_s = wh = 1000 × 4.5 × 10 = 45,000 Pa

(ii) Impulsive pressure on wall near the bottom (p_i)
For $y = h$ and $\Phi = 0$

$$p_i = \rho \ddot{u}_0 h\sqrt{3}\left[1 - \frac{1}{2}\right]\tanh\left(\sqrt{3}\frac{R}{h}\right)$$

$$p_i = 1000 * 0.12 * 9.8 * 4.5 * \sqrt{3} * \frac{1}{2}\tanh\left(\sqrt{3}\frac{4}{4.5}\right)$$

$$p_i = 4265\,\text{Pa}$$

(iii) Convective pressure on wall near the bottom (p_c)
for $y = h$ and $\Phi = 0$

$$p_c = 0.408\rho R^2\omega^2\theta_h\left(1 - \frac{1}{3}\right)\frac{\cosh\left(1.84\frac{h-h}{R}\right)}{\sinh\left(1.84\frac{h}{R}\right)}$$

$$p_c = 0.408 \times 1000 \times 4^2 \times 4.5 \times 0.058 \times \frac{2}{3} \times \frac{1}{\sinh\left(1.84\frac{4.5}{4}\right)}$$

$$p_c = 291\,\text{Pa}$$

(iv) Hydrodynamic pressure due to vertical response mode

$$\omega_v = \pi/2h[\rho(D/tE + 1/K)]^{-1/2}$$

$$= \frac{3.14}{2 * 4.5}\left[1000\left(\frac{8 * 1000}{6 * 2.0 \times 10^{11}} + \frac{1}{1.63 \times 10^9}\right)\right]^{-1/2}$$

$$= 42.77\,\text{rad/s}\,(6.8\,\text{Hz})$$

Considering vertical acceleration as two-thirds of horizontal,

$$S_{av} = 2/3 * 0.26\,\text{g}(2\%\text{damping spectra at } 6.8\,\text{Hz}) = 0.1716\,\text{g}$$

For $y = h$ = 4.5 m,

$$p_v = (Sav)\rho y = (0.1716 * 9.8) * 1000 * 4.5 = 7567.5\,\text{Pa}$$

Total pressure (hydrostatic + hydrodynamic),

$$p = p_s + \text{sqrt}\left(p_i^2 + p_c^2 + p_v^2\right) = 45,000 + 8691 = 53,691.5\,\text{Pa}$$

$$= 45,000 + 8691 = 53,691.5\,\text{Pa}$$

Therefore,
Hoop tension stress in shell $= pD/2t$

$$= 53,691.5 * 8/(2 * 6/1000)$$

Hoop tension stress in shell $= 35.79 \times 10^3$ kPa(35.79 MPa)

Observations:

i. If whole of the fluid mass is considered to respond as impulsive mass (sloshing is ignored), then

Max. shear force at the base = (2260 + 90) * 0.12 = 282 kN.

Max. bending and overturning moment at the base = (2260 + 90) * 0.12 * 4.5/2 = 634.5 kNm

Max. longitudinal compressive stress in the tank shell = 2.513 Mpa

Max. hoop tension in the tank shell = 35.79 MPa

ii. In this particular case, the sloshing mass is about 27% of the total fluid mass and impulsive mass constitutes about 60.2%.

iii. The bending moment and shear force at the base of the tank is higher if the whole mass of fluid is considered to move as rigid mass with the tank. However, the overturning moment just below the base of the tank is under estimated compared to the case when sloshing of fluid is considered.

10.8 Simplified Codal Procedures for Design/Analysis of Industrial Equipment

Industrial facilities contain several equipment such as vertical vessels, horizontal vessels and exchangers, stacks and towers. This equipment may be self-supported or supported on a braced or moment-resisting frame. Simplified codal procedures for design of such equipment are given in this section.

10.8.1 Industrial Equipment Categorization Based on Boundary Conditions

Based on boundary conditions, the industrial equipment can be categorized as follows:

a. **Rigid equipment**: Equipment having first natural frequency higher than 17 Hz, such as a horizontal pressure vessel or heat exchanger on relatively rigid supports.
b. **Unanchored equipment**: Example of such equipment is unanchored flat-bottom tanks supported at or below grade. Effects of sloshing and vessel uplift are required to be considered during an earthquake.
c. **Flexible equipment**: Vertical pressure vessels with skirt supports, spherical vessels with leg supports, horizontal pressure vessels or heat exchangers on long piers, and cooling towers are examples of flexible equipment. Typically, an equipment is treated as flexible if it first natural frequency of about 17 Hz or less.
d. **Combination of structure and equipment:** An example of such combination of structure and equipment is a tall vertical pressure vessel with skirt supports on a moment-resisting frame. The type of analysis depends on whether the equipment is rigid or flexible, and whether its mass exceeds or is less than about one-fourth of the mass of the supporting structure.

In an industrial facility, various non-structural systems and components are supported within equipment. The weight of such components should only be a smaller than the total weight of the support structure and non-structural systems and components. Examples of non-structural components in an industrial facility are:

a. Horizontal pressure vessels and heat exchangers supported on a frame, with a mass less than one-fourth of the total weight of the support frame and non-structural component,
b. Equipment supported within supporting frame,
c. Cable trays, ducting, and piping systems within equipment,
d. False ceilings, electrical cabinets, and lighting fixtures.

10.8.2 Selection of Analysis Type for Industrial Equipment

The description of analysis methods is given in Sect. 10.3. However, a brief review of the procedures is provided below. Two types of seismic analysis procedures are commonly used and are given below:

a. Equivalent static method (ESM)
b. Dynamic analysis.

The selection of the type of analysis depends on the equipment's seismic category, dynamic properties, and cost of analysis. For most number of cases, the ESM is suitable to determine the distribution of lateral forces. However, few equipment, which have shape irregularities, mass, or stiffness or are affected by interaction with neighboring equipment or structures require dynamic analysis. If a detailed dynamic analysis is carried out, design codes typically require that seismic base shear force should also be calculated by the ESM and compared with the shear force that obtained from a dynamic analysis.

10.8.2.1 Equivalent Static Method (ESM)

In this analysis, seismic loads are determined using static methods. Design codes typically restrict the use of the ESM for building-like equipment, if the shape of the equipment is irregular and has different stiffness along its length. In such situation, dynamic analysis is required to evaluate the lateral force distribution. In an ideal situation, the mass is uniformly distributed over the height of equipment. However, this ideal situation is rare for industrial equipment. But for all practical purposes, the simplified force distribution is justified.

10.8.2.2 Dynamic Analysis

Dynamic analysis is suitable for all industrial equipment of regular or irregular shapes for all seismic categories. More details of equivalent lateral force procedure are provided in the next section.

10.8.3 Equivalent Static Method (ESM) for Industrial Equipment

The concept of ESM for equipment is to evaluate the seismic load in terms of base shear, which depends on equipment mass distribution, peak ground acceleration, free vibration characteristics of equipment, importance factor of equipment, and the ductility. The base shear is then distributed vertically along the height of the equipment. Then static analysis of the equipment will be carried out to determine the response of equipment.

10.8.3.1 Determination of Base Shear "*V*" as Per Indian (IS-1893) and International (ASCE-7) Codes

The determination of total horizontal base shear (V) in any of the two orthogonal horizontal directions as per Indian code [IS 1893, 2005] and international code [3] is given in Table 10.5.

Table 10.5 Comparison of base shear evaluation as per Indian and international codes

As per Indian code	As per International code
$V = A_h\ W$ (10.83)	$V = C_S\ W$ (10.85)
where	$C_s = S_{DS}/(R/I)$ (10.86)
$A_h = \frac{(\frac{Z}{2})(\frac{S_a}{g})}{(\frac{R}{I})}$ (10.84)	

The procedure to obtain S_{DS} is explained below:

The design parameters for short period and 1 s period are calculated from Maximum Considered Earthquake (MCE) parameters using the following equation.

$$S_{DS} = 2/3S_{MS}; \quad (10.87)$$

$$S_{D1} = 2/3S_{M1} \quad (10.88)$$

where,

S_{MS} MCE, 5% damped, spectral acceleration for short period adjusted for site class

S_{M1} MCE, 5% damped, spectral acceleration at 1 s period adjusted for site class

S_{MS}, S_{M1} can be calculated as follows:

$$S_{MS} = F_a S_S \quad (10.89)$$

$$S_{M1} = F_v S_1 \quad (10.90)$$

S_1 and S_S are the mapped MCE spectral acceleration parameters at periods of 1 and 0.2 s, respectively, for Class B site with 5% damping.

S_1 and S_S are multiplied by F_V and F_A, respectively, to obtain S_{M1} and S_{MS}. Local soil conditions are taken into account by these factors, where F_V is velocity-related soil factor and F_A, the acceleration-related soil factor. The soil conditions are classified into six categories—A, B, C, D, E, and F as per NEHRP site classification which is given in Table 10.6.

Table 10.6 NEHRP site classification

Site class	Description	Shear wave velocity (V_s) in m/s
A	Hard rock	>1500
B	Rock	$760 < V_s \leq 1500$
C	Very dense soil and soft rock	$360 < V_s \leq 760$
D	Stiff soil	$180 < V_s \leq 360$
E	Soft soil	<180

10.8.3.2 Total Seismic Weight for Industrial Equipment

Total seismic weight, W of equipment includes the total self-weight of the equipment, self-weight of attachments such as equipment, piping, electrical cable trays and valves and live load for storage equipment. Usually, live load is not considered for seismic weight except for storage equipment. This is due to non-attachment of live load to the equipment, which results in non-contribution to its inertia. For storage equipment, usually one-fourth of live load is considered in the total seismic weight.

10.8.3.3 Response Modification/Reduction Factor R

The response modification/reduction factor, R, includes the combination of ductility and capacity of the equipment, which reduces the seismic design forces. Ductility accounts for the energy dissipation in the inelastic range and depends on material and type of equipment. Equipment capacity comes from several factors such as material strength, factors of safety in design. It is to be observed that usage of higher R-values in design tend to result in flexible equipment which will result in higher displacements.

10.8.3.4 Plotting of Response Spectrum

The design response spectrum as per IS-1893 for rock and soil sites is given in Sect. 10.3.2. The IS-1893 code also permits the use of a site-dependent design response spectrum if available. As per international codes [18], a typical design response spectrum is shown in Fig. 10.23. This can be plotted based on time period and spectral coefficients, S_{D1} and S_{DS} using the following equations:

$$T_S = S_{D1}/S_{DS} \quad (10.91)$$

$$T_0 = 0.2T_S \quad (10.92)$$

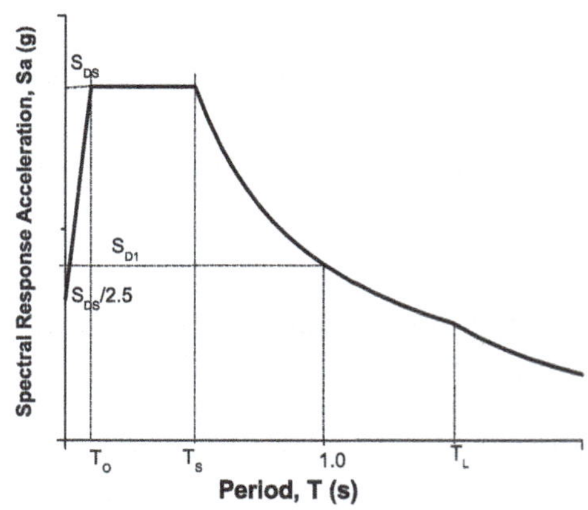

Fig. 10.23 Typical design response spectrum as per international code

In Fig. 10.23, T_L is the long-period transition period in seconds and T is the fundamental period of the structure in seconds.

10.8.4 Guidelines for Seismic Design of Non-structural Components

Various non-structural components and systems are supported within a support frame or equipment. Usually, the weight of non-structural components is less than one-fourth of the total seismic weight of the equipment or structure. If any component or equipment is directly supported on ground or shared the foundation with other equipment is not treated as non-structural component. Non-structural components should be designed, with the following considerations:

1. Same seismic category should be assigned to the non-structural components as that of equipment or structure to which they are attached.
2. Usually, the importance factor, *Ip*, of the components is equal to 1.0. The importance factor, *Ip*, for the components should be considered as 1.5 if any one of the following conditions are met.
   i. The function of the component is to meet safety purpose such as fire protection.
   ii. The component handles hazardous chemicals, or
   iii. The failure of the component hampers the continued operation of the facility.
3. To evaluate the seismic capacity of equipment, the test data or experience-based data may be used as an alternative to the analytical method.
4. Proper anchoring of non-structural components such as heavy bookshelves and individual computers may be ensured to prevent their impact to nearby critical components (i.e., to prevent them to become flying objects during an earthquake).

10.8.4.1 Design Provision for Non-structural Components in International Code (UBC-1997)

The seismic design provisions for non-structural components as per 1997 UBC code [33] are given below:

$$F_p = 4.0 C_a I_p W_p \qquad (10.93)$$

$$F_p(\text{minimum}) = 0.7 C_a I_p W_p \qquad (10.94)$$

In the absence of floor response spectrum (FRS) for the design of a non-structural component, the following equation can be used to include the effect of floor response:

$$F_p(\text{alternative}) = \frac{a_p c_a I_p}{R_p}\left(1 + 3\frac{h_x}{h_r}\right) W_p \qquad (10.95)$$

where the height-above-grade factor $\left(1 + 3\frac{h_x}{h_r}\right)$ is a simplification to account for floor amplification.

Example 10.5 The example for design calculation using the 1997 UBC is given below. The component is a vessel of mass 5 tons mounted on the roof (Fig. 10.24). Calculate the anchorage force if the vessel is (a) horizontal vessel with fundamental period of 0.02 s and (b) vertical vessel with fundamental period of 0.2 s.

Case (a): Rigid equipment

For this case, the equipment is considered as a rigid component as the first natural period of vibration is less than 0.06 s. The parameters are:

$h_x/h_r = 1.0$;
$I_p = 1.0$;
$a_p = 1.0$;
$R_p = 3.0$;
$C_a = 0.4$

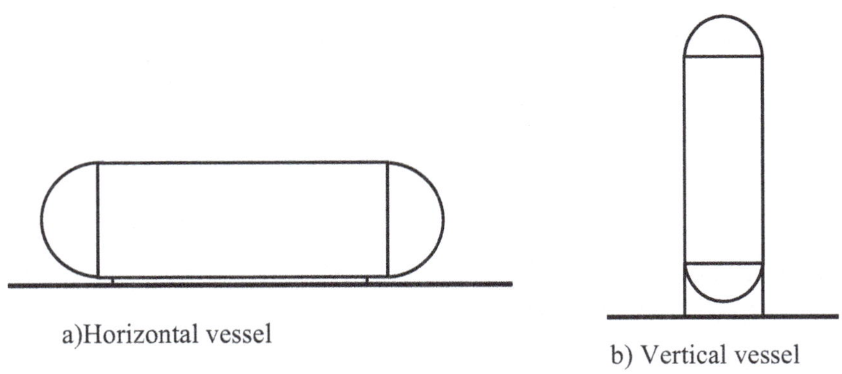

Fig. 10.24 Schematic of a vessel mounted on the roof

$F_p = (1.0 \times 0.4 \times 1.0/3.0) \times (1 + 3)\ W_p$
F_p (considering floor response) = 0.533 Wp = 26.16 kN

Case (b): Flexible equipment

For this case, the equipment is considered as a flexible component as the first natural period of vibration is more than 0.06 s. In case of flexible equipment (fundamental period is greater than 0.06 s), component amplification factor is high (2.5) and hence the anchorage force increases as follows:

$h_x/h_r = 1.0$;
$I_p = 1.0$;
$a_p = 2.5$;
$R_p = 3.0$;
$C_a = 0.4$

$F_p = (2.5 \times 0.4 \times 1.0/3.0) \times (1 + 3)\ W_p$
$F_p = 1.336\ W_p = 65.55$ kN.

10.8.5 Qualification Procedure of a Process Vessel for Earthquake and Wind Loads

This is explained with example problems. Example 10.6 illustrates the qualification procedure of a process vessel for earthquake loads.

Example 10.6: Seismic Analysis of Vessel Using IS-1893

Analysis of vertical vessel as given in Example 10.3 has been analyzed for dynamic load using IS 1893 spectrum for particular site. A typical tank process plant is supported on skirt-type support, which is of same thickness as that of tank hence it will behave like a cantilever beam of cylindrical cross section. The schematic of process vessel is shown in Fig. 10.25.

Solution

i. The first fundamental frequency for bending of the cantilever beam is found out from the formulae given below:

$$f_b = \left(\lambda_i^2/2\,\pi\, l^2\right) \times (EI/\mu)^{0.5} \tag{10.96}$$

where

λ_I	1.875 for first mode
E	Young's modulus of elasticity
I	Moment of inertia of the structure about its axis of bending
μ	Mass per unit length of the beam
l	Length of the beam

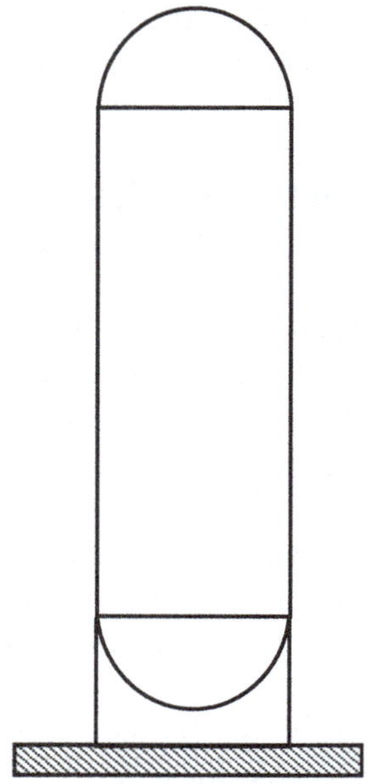

Fig. 10.25 Schematic of process vessel

Data required for the calculation is given below:

E	2.1×10^{11} N/m^2
l	22,000 mm
Length	22,000 mm
Inner diameter	4250 mm
Outer diameter	4334 mm

$$\begin{aligned} I &= (1/64) \times \pi \times \left(4334^4 - 4250^4\right) \\ &= 1.30415 \times 10^{12}\,\text{mm}^4 \\ &= 1.30415\,\text{m}^4 \end{aligned}$$

$$\begin{aligned} \text{Total Mass of the Tank} &= \text{Volume} \times \text{density} \\ &= (1/4) \times \pi \times \left(4.334^2 - 4.250^2\right) \\ &\quad \times 22 \times 7850 \\ &= 97{,}802.78\,\text{kg} \end{aligned}$$

$$\begin{aligned} \text{Total Mass of water in the Tank} &= \text{Volume} \times \text{density} \\ &= (1/4) \times p \times \left(4.250^2\right) \\ &\quad \times 22 \times 1000 \\ &= 312{,}097.6\,\text{kg} \end{aligned}$$

$$\begin{aligned} \text{Total Mass per unit length} &= \mu \\ &= (97{,}802.78 + 312{,}097.6)/22 \\ &= 18{,}631.83\,\text{kg/m} \end{aligned}$$

Therefore, frequency $f_b = (1.875^2/2\pi \times 22^2) \times (2.1 \times 10^{11} \times 1.30415/18{,}631.83)^{0.5}$
= 4.43 Hz.

Therefore, time period of the structure $T = 1/\ 4.43 = 0.23$ s

ii. **According to IS1893 (2002)**

Design horizontal seismic coefficient is given by

$$A_h = \frac{ZIS_a}{2Rg}$$

where

- Z Zone factor for maximum considered earthquake (MCE). The factor 2 in denominator is used to reduce MCE to DBE.
- Z 0.16 (For Zone III, i.e., Mumbai)
- I Importance factor = 2 (for chemical tanks)
- R Response reduction factor = 4 (for steel structure)

$$\frac{S_a}{g} = \begin{cases} 1 + 15T & (0 \le T \le 0.1) \\ 2.5 & (0.1 \le T \le 0.4) \\ 1.00/T & (0.4 \le T \le 4.00) \end{cases}$$

Substituting the values,

$$A_h = 0.16 \times 2 \times 2.5/(2 \times 4) = 0.1$$

$$V_b = A_h \times W$$

where

W = Seismic weight of the building = 4,099,000 N

$$V_b = 0.1 \times 4,099,000 = 4,099,000\,\text{N}$$

Considering the base shear uniformly distributed along the height:

$$w = \text{weightper unit length} = 18,631.8\,\text{N/m} = 18,631.8\,\text{N/m}$$

$$\text{Moment} = \text{wh}^2/2 = 18,631.8 \times 22^2/2 = 4,508,895.6\,\text{N m}$$

iii. **Stress due to seismic load**

$$\text{Bending stresss}_b = \text{My/I} = 7.49\,\text{MPa}$$

$$\text{Shear stress}\ \tau = F/\text{As} = 1.45\,\text{MPa}$$

Axial compressive stress due to seismic load, $\sigma_a = P_v/A$.

Where

$$P_v = (2/3)V_b$$

$$\sigma_a = 0.482\,\text{MPa}$$

iv. **Stress due to dead load**

$$\text{Axial stress} = 4,099,000/[(\pi/4) \times (4.334^2 - 4.25^2)] = 7.24\,\text{MPa}$$

v. **Total stress**

Total compressive stress due to Dead load and earthquake= 7.49 + 7.24 + 0.48 =15.21 MPa

Total stress is found to be much lower than allowable stress limits. Hence, it is safe.

The following example illustrates the qualification procedure of a process vessel for earthquake and wind loads.

Example 10.7 The process column has a height of 53.59 m and diameter of 4.25 m and is provided with cross-flow sieve trays. The tower is made of carbon steel, SA516 Gr 70 and is supported on skirt with base plate. The tower is made of carbon steel, SA516 Gr 70 and is supported on skirt with base plate. The thickness of cylindrical shell is 40 mm, and the top and bottom hemispherical portions have a thickness of 25 and 28 mm, respectively. The thickness of the skirt support is 40 mm. The details of process column are given in Table 10.7.

The schematic of process vessel is shown in Fig. 10.25.

i. **Natural frequency of process column**

The process vessel is idealized as a cantilever beam with cylindrical cross section. It has inner and outer diameters 4.25 and 4.33 m, respectively, with a height of 53.59 m. The cross-sectional area, moment of inertia, and section modulus are 0.539 m^2, 1.24 m^4, and 0.573 m^3, respectively. The total mass per unit length is 5.22 T/m.

The first natural frequency of a cantilever beam is given by

$$f = 1.875^2\sqrt{\frac{EI}{\rho AL^4}} \tag{10.97}$$

By substituting corresponding values, the first natural frequency of the process vessel is obtained as 1.3 Hz.

ii. **Seismic load approximation**

Approximation of seismic load for the process vessel is shown in Fig. 10.26. The base shear is resolved into linear lateral force distribution (inverted triangular) as shown.

Table 10.7 Details of process column

Design code	ASME Section VIII Div. 2
Design data	
Operating pressure	110.35 kg/cm^2 g
Design pressure	22.4 kg/cm^2 g
Design temperature	132 °C
Corrosion allowance	6 mm
Radiography and joint efficiency	Full & 100%
Major material of constructions (MOCs)	
Shell/heads/RF pads	SA 516 Gr 70
Nozzle necks and flanges	SA 106 Gr B & SA 105
Skirt support	SA516 Gr 70 / IS 2062
Anchor bolts	A 193 -B7
Major dimensions	
Total height	53,590 mm
Inside diameter	4250 mm
Equipment Weight	
Fabricated weight (w/o internals)	250 tons
Weight of trays/internals	30 tons

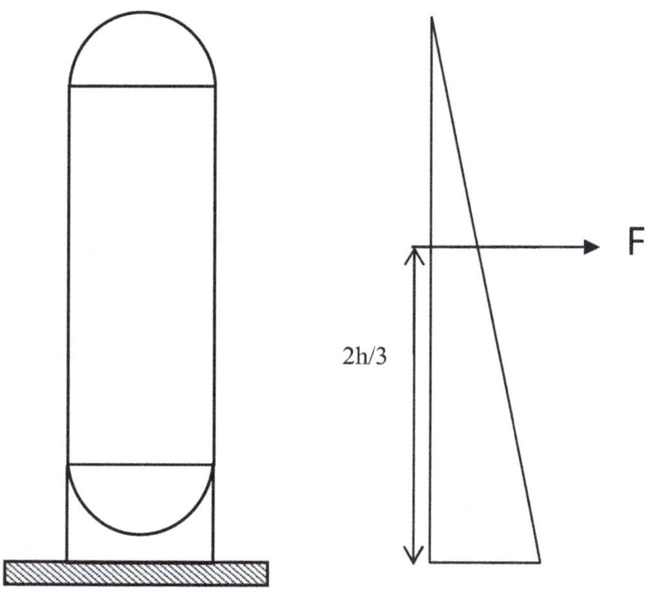

Fig. 10.26 Approximation of seismic load for the process vessel

The concept of lateral load distribution is valid for uniform mass distribution, i.e., it is assumed that the mass distribution is uniform from the support to the top of the vessel.

From the design response spectrum of IS-1893, the spectral acceleration corresponding to first natural frequency = 0.13 g

Total base shear is given by

$$\begin{aligned} F &= m \times a \\ &= 280 \times 10^3 \times 0.13 \times 9.81 \\ &= 357.1\,\text{kN} \end{aligned}$$

where

'a' is the spectral acceleration corresponding to first natural frequency

Total moment is given by

$$\begin{aligned} M &= F \times 2h/3 \\ &= 12.76\,\text{MN m} \end{aligned}$$

Maximum bending stress at the bottom of shell due to earthquake load = M/Z

=22.26 MPa

Note: In this calculation, the height of skirt is neglected.

Longitudinal stress in the shell due to operating weight

$$\begin{aligned} &= 280{,}000 \times 9.81/0.539 \\ &= 5.1\,\text{MPa} \end{aligned}$$

Total stress due to combination of earthquake and operating weight

$$\begin{aligned} \sigma &= \frac{-W}{A} \pm \frac{M}{Z} \\ &= 27.36\,\text{MPa} \end{aligned}$$

iii. **Wind load calculation for process vessel**

Wind load profile is generated using procedure described in Chap. 7 in accordance with IS 875 Part 3 [21]. A basic wind speed (V_b) of 47 m/s is used [21]. The design wind velocity (V_z) is obtained using Eq. (10.98):

$$V_z = V_b \times k_1 \times k_2 \times k_3 \tag{10.98}$$

where

- V_z Design wind speed at any height z in m/s;
- V_b Basic wind speed in m/s;
- k_1 Probability factor (risk coefficient);1.07 for 100 years return period and1.35 for 1000 years return period
- k_2 Terrain, height and structure size factor;
- k_3 Topography factor; a value of 1.0 is considered.

Process column with 54 m height comes under the Class C equipment, and its natural frequency is 1.34 Hz. Therefore gust response factor is considered as 1. K_2 factor used to calculate design wind velocity is given in Table 10.8 below: As per the definition, the equipment falls under Category I terrain.

The design wind pressure at any height above mean ground level is obtained by the following relationship:

$$p_z = 0.6V_z^2 \tag{10.99}$$

where

P_z design wind pressure in N/m^2 at height z, and

The design wind pressure calculated using Eq. (10.99) is further modified to account for dynamic effects and geometry of the structure:

$$P_z = C_s \times G \times p_z \tag{10.100}$$

where

- P_z Modified design wind pressure
- C_s Shape factor0.7 (for cylindrical geometry, Table 23 of IS 875 Part 3)
- G Gust response factor1 (When gust response factor is not used, for equipment frequency more than 1 Hz)

Process column with 54 m height comes under the Class C equipment, and its natural frequency is 1.34 Hz. Therefore gust response factor is considered as 1. Accordingly, wind load profile is generated up to 100 m equipment height for 100 years and 1000 years of wind return period and same are given in Tables 10.9 and 10.10, respectively. However, these coefficients can be interpolated to the actual tower height to determine respective wind load.

iv. **Design wind pressure calculation (for 100-year return period)**

The distribution of wind load for 100-year return period is shown in Fig. 10.27. It is assumed that a constant wind pressure of 1228 N/m^2 acts on the process vessel up to 25 m elevation and 1400 N/m^2 for the remaining portion. The force, F_1 is calculated by multiplying the constant wind pressure of 1228 N/m^2 with projected area of 25 m elevation (106.25 m^2). The force, F_2 is calculated by multiplying the constant wind pressure of 1400 N/m^2 with projected area of 28.6 m elevation (121.55 m^2).

Table 10.8 K_2 factors for Category I terrain

Height, m	Class C
10	0.99
15	1.03
20	1.06
30	1.09
50	1.14
100	1.20

Table 10.9 Design wind pressure for 1000 years of wind return period without considering gust response factor

For 1000-year return period and Class C Equipment (without Gust Response Factor)								
Height Z (m)	Factor			Basic wind speed, V_b (m/s)	Design Wind speed (V_z m/s)	Shape Factor C_s	Design wind pressure (P_Z N/m^2)	Design wind pressure ($P_Z$kgf/m^2)
	K_1	K_2	K_3					
10	1.35	0.99	1	47	62.82	0.70	1657.23	168.99
15	1.35	1.03	1	47	65.35	0.70	1793.85	182.92
20	1.35	1.06	1	47	67.26	0.70	18,910.87	193.73
30	1.35	1.09	1	47	610.16	0.70	2008.93	204.85
50	1.35	1.14	1	47	72.33	0.70	2197.47	224.08
100	1.35	1.2	1	47	76.14	0.70	2434.87	248.29

Table 10.10 Design wind pressure for 100 years of wind return period without considering gust response factor

For 100-year return period and Class C Equipment (without Gust Response Factor)								
Height	Factor	Factor	Factor	Basic wind	Design wind	Shape factor	Design wind	Design wind
Z meter	K_1	K_2	K_3	Speed, V_b (m/s)	Speed, V_z (m/s)	C_s	Pressure (P_Z N/m^2)	Pressure ($P_Z$kgf/m^2)
10	1.07	0.99	1	47	410.79	0.70	1041.08	106.16
15	1.07	1.03	1	47	51.80	0.70	1126.90	114.91
20	1.07	1.06	1	47	53.31	0.70	1193.51	121.70
30	1.07	1.09	1	47	54.82	0.70	1262.02	128.69
50	1.07	1.14	1	47	57.33	0.70	1380.46	140.77
100	1.07	1.2	1	47	60.35	0.70	15,210.59	155.98

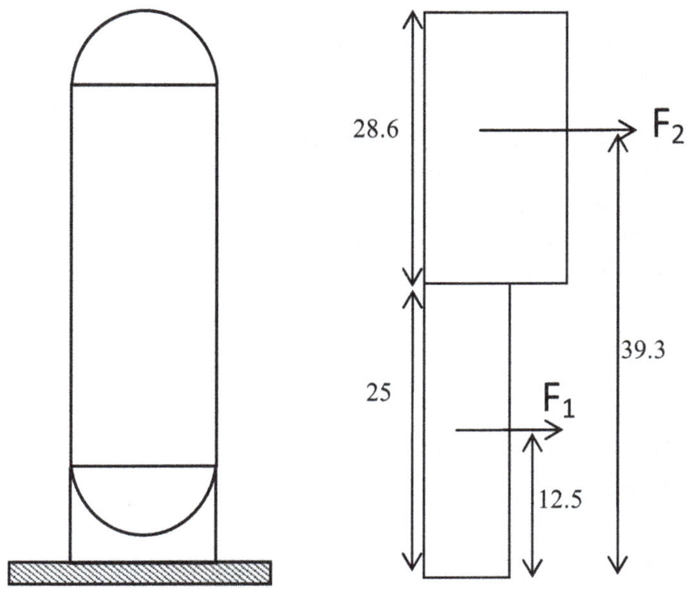

Fig. 10.27 **Distribution of wind load**

Hence,

$$F_1 = 1228 \times 106.25$$
$$= 130.475\,\text{kN}$$

$$F_2 = 1400 \times 121.55$$
$$= 170.17\,\text{kN}$$

$$\text{Moment at the base} = 12.5\,F_1 + 26.8\,F_2$$
$$= 8.32\,\text{MN m}$$

Note: In this calculation, the height of skirt is neglected.

Maximum bending stress at the bottom of shell due towind load
$= 14.52\,\text{MPa}$

Longitudinal stress in the shell due to operating weight $= W/A$
$= 280{,}000 \times 10.81/0.539$
$= 5.1\,\text{MPa}$

Total stress due to combination of wind load and operating weight

$$\sigma = \frac{-W}{A} \pm \frac{M}{Z}$$
$$= 19.62\,\text{MPa}$$

Table 10.11 Allowable stress values for vessel and supports

Parts	Load Combinations	Allowable stress	Allowable stress (MPa)
Vessel	Internal Pressure + Operating Weight	1S	156
	Internal Pressure + Operating Weight + Wind Load	1.5S	234
	Internal Pressure + Operating Weight + Seismic Load		
Skirt	Operating Weight + Wind Load	1.5S	234
	Operating Weight + Seismic Load		
Anchor bolts	Operating Weight + Wind Load	1S	420
	Operating Weight + Seismic Load		

v. **Allowable stresses for vessels and supports**

The vessels are subjected to various loading conditions such as operating weight, internal pressure, seismic load, and wind loads. A typical load combination and allowable stress limits for vessels and supports as per ASME Section VIII Div. 2 [4] are given in Table 10.11. It is assumed that the allowable tensile stresses at design temperature 'S' for SA 516, Gr-70 material and A 193-B7 are 156 and 420 MPa, respectively. The load combination and allowable stresses for vessel, skirt, and anchor bolts are given in Table 10.11.

Hence, the total stresses in the vessel due to various load combinations are less than corresponding allowable stress.

vi. **Design check for vessel support**

The vessel support is intended to support the pressure vessel on the base. The vessel support has to be designed to withstand the operating weight and seismic load from the pressure vessel.

Skirt support vessel

Most vertical vessels are supported by skirts as shown in Fig. 10.28. These supports transfer the loads from the vessel to foundation through anchor bolts and bearing plates by shear action.

To design a skirt support, the vessel self-weight, W and bending moment, M due to seismic, wind, and other mechanical loads are considered. The stress in the skirt is then a combination of axial and bending stresses and is given by

$$\sigma = \frac{-W}{A} \pm \frac{M}{Z} \tag{10.101}$$

where A is the cross-sectional area of the support and approximately equals πDt, with D the diameter and t the thickness. Z is the section modulus and is approximately equals to $\pi D^2 t/4$ for circular cross section. Equation (10.101) then becomes

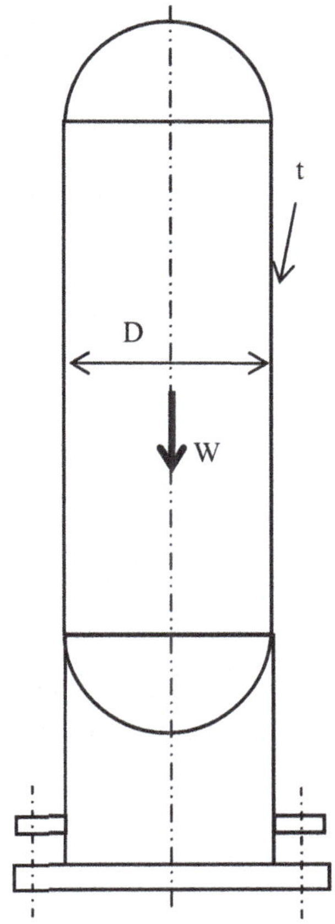

Fig. 10.28 Schematic of a skirt-supported vessel

$$\sigma = \frac{-W}{\pi Dt} \pm \frac{4M}{\pi D^2 t} \tag{10.102}$$

After designing the skirt, anchor bolts are required to be designed. If the total number of anchor bolts is N, then the load on a single bolt, P, is given by

$$P = \frac{-W}{N} \pm \frac{4M}{ND} \tag{10.103}$$

As the skirt also has a thickness of 40 mm and inner diameter is 4.25 m, the cross-sectional area and sectional modulus of the skirt are same as those for the vessel.

Hence, the combined stresses in the skirt support are same as those for the vessel. It is observed that the total stresses in the skirt due to various load combinations are less than corresponding allowable stress.

vii. **Design check for anchor bolts**

If the total number of anchor bolts is N, area of one bolt is A, then the load on a single bolt, P is obtained by using the following relations:

The combined stress in bolts is given by

$$\sigma = \frac{-W}{A} \pm \frac{M}{Z} \quad (10.104)$$

The moment of inertia for all bolts is given by

$$I = \frac{NAR^2}{2} \quad (10.105)$$

The load on a single bolt is given by

$$P = \frac{-W}{N} \pm \frac{2M}{NR} \quad (10.106)$$

or

$$P = \frac{-W}{N} \pm \frac{4M}{ND} \quad (10.107)$$

where

$$\begin{aligned} \text{M} &= \text{Maximum moment for earthquake and wind load} \\ &= 10.86\,\text{MN m} \end{aligned}$$

The process vessel of example has 20 number of anchor bolts of size M50 with base ring diameter of 5 m.

$$\begin{aligned} \text{Load on each bolt} &= -(2,80,000 \times 10.81/20) \\ &\quad \pm (4 \times 10.86 \times 1\text{E}6/(20 \times 5)) \\ &= -137.34 \pm 434.4\,\text{kN} \\ &= -571.74\,\text{kN} \end{aligned}$$

$$\begin{aligned} \text{Area of each bolt} &= 17.1\,\text{cm}^2 \\ &= 17.1 \times 10^{-4}\,\text{m}^2 \\ \text{Axial stress in bolt} &= P/A \\ &= 334.35\,\text{MPa} \end{aligned}$$

This stress is less than allowable stress for various load combinations.

Example 10.8 Skirt support and anchor bolt design

Design a skirt support for process column with a total self-weight of 720 kN and an overturning moment of 2050 kN-m due to earthquake. The bolt circle diameter of the skirt support is 4.5 m. Consider a thickness of 10 mm for the skirt and the mean diameter of the support as 4.25 m. Also assume the permissible compressive stress in skirt is 200 MPa.

Solution

i. The skirt support is to be designed to prevent buckling under the compressive load. With the given dimensions

D 4.25 m,
t 10 mm,
W 720 kN, and
M 2050 kNm.

The stress in the skirt is

$$\sigma = \frac{-720 \times 10^3}{\pi(4250)(10)} \pm \frac{4 \times 2050 \times 10^3 \times 10^3}{\pi(4250)^2(10)} = -19.84\,\text{MPa}$$

The skirt does not buckle for the compressive stress of 110.84 MPa

ii. The anchor bolts are designed as follows.

Considering 12 bolts, the load per bolt is calculated as

$$P = \frac{720 \times 10^3}{12} + \frac{4 \times 2050 \times 10^6}{12 \times 4500} = 211.85\,\text{kN}$$

For steel bolts, for proof strength of 225 MPa, bolt area of 941 mm^2 is obtained by dividing the load per bolt with proof strength. Therefore, 12 bolts with 36 mm diameter can be used as the anchor bolts.

10.8.6 Dynamic Analysis Methods

The description of dynamic analysis methods are given in Sect. 10.3. However, a brief review of the methodology is provided below.

10.8.6.1 Description of Dynamic Analysis Methods

Ground motion

Design basis ground motion parameters for a dynamic analysis are usually design response spectrum (DRS) or spectrum (DRS) compatible time history.

Mathematical model

The mathematical model of the equipment should be adequate to evaluate its dynamic characteristics properly by considering reasonable distribution of mass and stiffness.

Response spectrum Analysis

A response spectrum analysis is an elastic dynamic analysis of equipment in which the peak responses correspond to different modes that are combined to obtain the resultant equipment response. Peak modal responses are obtained from the ordinates of the design response spectrum corresponding to different frequencies. These modal responses are then combined to obtain the peak response of the equipment. The following aspects should be considered while carrying out a response spectrum analysis:

a. Total number of modes for analysis: The total mass participation of considered modes should be at least 90% of total mass in each direction. Otherwise, missing mass correction has to be applied. This correction is applied by carrying out a static analysis of equipment for a force equals to the missing mass multiplied with peak ground acceleration.
b. Modal combination: Initially, maximum response of equipment during each mode is obtained in a response spectrum analysis. These modal responses are to be combined to obtain the peak response of equipment. Modal responses are combined using either SRSS method or CQC method for closely spaced modes.

Exercise Problems

1. A steel tank with a radius of 10 m and height of 9.6 m. The tank is filled with water to a height of 8 m. Assume the response spectrum value for impulsive (for 2% damping) and convective (for 0.5% damping) is 0.87 and 0.07 g, respectively. Calculate slosh height and impulsive and convective frequencies. Take the material properties appropriately. (Ans: Slosh height 0.7 m; impulsive frequency = 8.13 Hz; convective frequency = 0.2 Hz).
2. A ground-supported rectangular RC water tank has plan dimensions of 18 m × 10 m and height of 6.6 m (including a freeboard of 0.3 m). Estimate the first few mode convective frequencies in x- and y-directions. (Ans: Given in following Table 10.12).
3. A slender cylindrical tank 7.92 m in diameter and containing 4.57 m of water is supported on the ground. Determine impulsive and convective forces and slosh height. (Ans: Impulsive force = 468.8 KN; convective force = 100 KN; slosh height = 0.56 m).
4. A vertical vessel of mass 10 tons and is mounted on roof with importance factor and component response modification factors as 1.0. Assume the compressor is relatively rigid and seismic coefficient as 0.2. Evaluate total strength design lateral seismic force on the compressor. (Ans. 196 kN).
5. A pump of mass 3 tons is mounted without vibration isolators on roof with importance factor and component response modification factors as 1.0. Assume the first natural frequency of pump is 10 Hz and seismic coefficient as 0.3. Evaluate total strength design lateral seismic force on the compressor. (Ans. 147 kN).
6. Design a skirt support for a pressure vessel with a total vertical load of 1500 kN, and an overturning moment of 3000 kNm. The bolt circle diameter of the support may be assumed to be 5 m. Assume a thickness of 20 mm for the support skirt and the mean diameter of the support as 4.5 m. Also assume the permissible compressive stress in skirt is 200 MPa. (Ans. 14.7 MPa, 12 number of M20 bolts).
7. Using frequency response curve, explain how damping is evaluated using half-power bandwidth method.
8. What are operating basis earthquake (OBE) and safe shutdown earthquake (SSE)? Describe in brief?
9. What is response spectrum? Explain the need and method for spectral peak broadening as per ASME Sec-III, Appendix N?
10. A tall vessel of height 53.59 m has a cylindrical shell of outer diameter of 4.33 m and thickness of 40 mm and hemispherical heads of same outer diameter and thickness. The vessel has a total mass of 280 tons and supported on skirt support of same outer diameter and thickness. The design response spectrum for 4% damping is shown in Fig. 10.29. Assuming the tower as a cantilever beam, calculate the first natural frequency of the tower. Also calculate the spectral acceleration to which the tower is subjected to corresponding to this frequency. (Ans. 1.3 Hz, 0.11 g).

Table 10.12 Convective frequencies in x- and y-directions

Mode no	Convective frequency	
	x	y
1	0.186	0.274
2	0.291	0.395
3	0.360	0.484
4	0.416	0.559

Appendix 1: Safety and Seismic Categorization and Service Levels for Equipment of Nuclear Facilities

Integrity and functionality for equipment of nuclear facilities are to be ensured under earthquake condition. Safety classification and seismic categorization for equipment of nuclear facilities are described below.

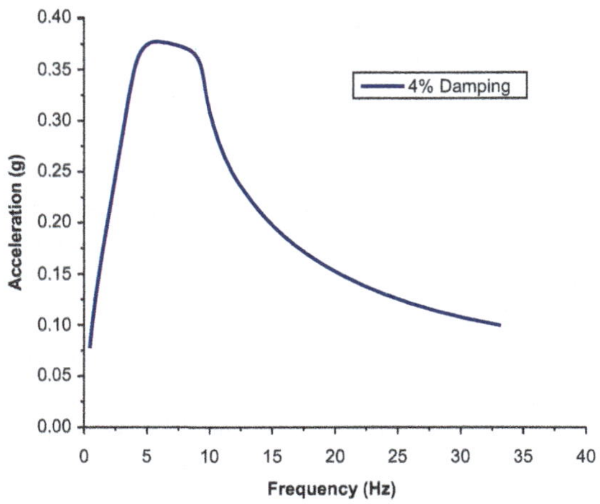

Fig. 10.29 Response spectrum for 4% damping

A.10.1.1 Safety Classification of Equipment of Nuclear Facilities

The objective of safety classification of nuclear facilities is to ensure pubic and plant safety by identifying those systems and components that are important from the point of view of safety and grade them in importance [5].

Safety requirement: The defense in depth philosophy in ensuring nuclear safety consists of three important steps:

a. Design of components and systems with adequate margins,
b. Prevention of accidents,
c. Mitigation of consequences of accidents.

Safety Class I: It includes those functions necessary to prevent release of substantial fraction of radioactive material to the containment.

Safety Class II: It includes those functions necessary to mitigate the consequence of accidents which would otherwise lead to substantial radioactive release.

Safety Class III: It includes those safety functions performing support roles to safety functions in safety classes 1, 2, or 3. Failure of this would not lead to a direct increase in radiation exposure.

Safety Class IV: It includes all safety functions which do not fall within safety classes 1, 2, or 3.

A.10.1.2 Service Levels

Service levels are defined to categorize all loads into various levels according to their occurrence and severity of load. The service limits are defined in ASME Section III [4] are as follows. Table A.10.13 defines the graded approach for different service level and frequency of occurrences [22].

Level A service limits: (normal)
The piping components or supports must satisfy these sets of limits in the performance of their specified service function.

Level B service limits: (abnormal)
The piping components or support system must withstand these loadings without damage requiring repair.

Level C service limits: (emergency condition)
The occurrence of the stress up to these limits may necessitate the removal of the piping component from service for inspection or repair of damage.

Level D service limits: (faulted condition)
These sets of limits permit gross general deformations with some consequent loss of dimensional stability and damage

Table A.10.2 Load combinations for design condition and service levels for equipment, piping and supports

Plant classification	Design/service level	Load combination as per ASME
Design	Design condition	Pressure dead weight, sustained loads, temperature[a]
Normal	Service Level A	Normal operating transients (pressure, temperature, mechanical)
Upset	Service Level B	Pressure, dead weight, sustained loads, upset condition transients (pressure, temperature[b], mechanical), including operating basis earthquake (OBE)[c]
Emergency	Service Level C	Pressure, dead weight, sustained loads, temperature[a], emergency condition transients (pressure, mechanical)
Faulted	Service Level D	Pressure, dead weight, sustained loads, temperature[a], faulted condition transients (pressure, mechanical), pipe rupture loads, safe shutdown earthquake (SSE)[c]

Note
[a]Temperature is used to determine allowable stress only
[b]Thermal transients under Level C shall also be considered in Level B
[c]Operating basis earthquake (OBE) and safe shutdown earthquake (SSE) include both inertial and seismic anchor movements (SAM)

Table A.10.13 Service level and frequency of occurrences

Service Level A (normal operation)	Frequency of occurrence > 1/reactor year
Service Level B (upset condition)	Frequency of occurrence < 1 & > 10-2 /reactor year
Service Level C (emergency condition)	<10-2 & > 10-4 /reactor year
Service Level D (faulted condition)	<10-4 & > 10-6/reactor year

requiring repair which may require removal of the piping components from service.

Table A.10.2 defines that the various loads can be categorized in different service levels. The relation between plant classifications and service conditions is also correlated in following table.

A.10.1.3 Seismic Categorization for Equipment of Nuclear Facilities

The purpose of seismic categorization of equipment of nuclear facilities is to facilitate the protection of public and environment against radioactive releases [19]. It is categorized in terms of their importance to safety in the event of an earthquake.

Seismic Category I

Item under this category shall be designed and demonstrated to withstand the consequences of ground motion associated with earthquakes of levels S-2 (SSE) and S-1 (OBE). Category I shall include:

(a) Items whose failure could directly or indirectly cause accident conditions,
(b) Item required for shutting down the reactor, maintaining the reactor in a shutdown condition, and removing residual heat over a long period,
(c) Item that are required to prevent radioactive releases are to maintain release below limits stabilized by the regulatory body for accidental conditions (e.g., containment systems).

Seismic Category II

The items categorized under this category shall be designed to withstand with ground motion associated with earthquake of level S-I (OBE). Category II should include:

1. That an earthquake of defined severity will occur during this period. Items not in Category I which are required to prevent the escape of radioactivity beyond normal operation limits,
2. Items, not in Category I, required to mitigate those accident conditions which last for such long periods that there is a reasonable likelihood.

Seismic Category III

Seismic Category 3 should include all items that could pose a radiological hazard but that are not related to the nuclear reactor (e.g., the spent fuel building and the radioactive waste building). In some states, these items are required to have safety margins consistent with their potential for radiological consequences, which are expected to be different from the potentials associated with the reactor, as they would be in general related to different release mechanisms (e.g., leakage from waste, failure of spent fuel casks).

Seismic Category IV

Seismic Category 4 should include all items that are not in seismic Category 1 or 2 or 3. Nuclear power plant items in seismic Category IV should be designed as a minimum in accordance with national practice for non-nuclear applications, and therefore for facilities at conventional risk. For some items of this seismic category important to the operation of the plant, it may be reasonable to choose more stringent acceptance criteria based only on operational targets.

A.10.4 Qualification of Equipment of Nuclear Facilities for Seismic Load

The first step in qualification of equipment of nuclear facilities is to obtain the design basis ground motion (DBGM) parameters of the site and floor response spectra (FRS) of location where the equipment is mounted. The generation of DBGM parameters for a site is explained in Chap. 2. Then the equipment is analyzed for corresponding seismic load. Qualification of equipment is carried out by comparing the resulting response with allowable limits provided by design codes.

References

1. ASCE Standard, ASCE 4-98, "Seismic Analysis of Safety Related Nuclear Structures"
2. ASME SectionIII,Div. 1, Appendices, 2004
3. ASCE 7, 2005,"Minimum Design Loads for Buildings and Other Structures, ASCE/SEI 7-05 including Supplement No. 2", American Society of Civil Engineers, Reston, VA

4. American Society of Mechanical Engineers (2007) ASME SECTION VIII Division 1 and 2: Rules for Construction of Pressure Vessels. ASME, New York
5. AERB Safety Guide, "Seismic Qualification of Structures, Systems and Components of Pressurized Heavy Water Reactors", AERB/NPP-PHWR/SG/D-23 (2009)
6. ASME Boiler & Pressure Vessel Code, "Rules for Construction of Nuclear Facility Components", Section III Division 1-Subsection NC (2004)
7. Di Carluccio, G. Fabbrocino, E. Salzano, G. Manfredi, 2008, Analysis of pressurized horizontal vessels under seismic excitation, The 14th World Conference on Earthquake Engineering, October 12–17, 2008, Beijing, China
8. ACI 350.3, 2001, "Seismic design of liquid containing concrete structures", American Concrete Institute, Farmington Hill, MI, USA
9. Akira Niwa, Ray W. Clough, 1982, Buckling of cylindrical liquid-storage tanks under earthquake loading, Earthquake Engineering & Structural Dynamics 10(1):107–122. Association of New Zealand, Wellington
10. BIS (2014) Indian Standard Criteria for Earthquake Resistant Design of Structures: General Provisions and Buildings. Bureau of Indian Standards, New Delhi
11. Chopra A. K.,"Dynamics of structures", Prentice Hall of India, 1998
12. Dorninger K, Fischer FD, Rammerstorfer F G and Seeber R, 1986, Progress in the analysis of earthquake loaded tanks, Proceedings of the 8theuropian conference on earthquake engineering 8ECEE, Lisabon, Portugal, pp- 73–80
13. Eswaran M, 2011, Numerical and experimental investigations of capturing liquid free surface characteristics in externally induced sloshing tanks, PhD thesis, IIT Guwahati
14. Eurocode 8, 1998, "Design provisions for earthquake resistance of structures, Part 1- General rules and Part 4 – Silos, tanks and pipelines", European Committee for Standardization, Brussels
15. E.L Wilson, A Der Kiureghian and E. P Bayo, " A Replacement for the SRSS Method in Seismic Analysis", Earthquake Engineering and Structural Dynamics, Vol. 9, 1981, PP 187-192.IS 1893 Part-4, 2005, Criteria for Earthquake resistant Design of Structure
16. G.R. Reddy, 1998, Advanced Approaches for the Seismic Analysis of Nuclear Power Plant Structures, Equipnlent and Piping Systems, PhD thesis, Tokyo Metropolitan University, Tokyo, Japan
17. Housner, G. W., 1963a, "Dynamic analysis of fluids in containers subjected to acceleration", Nuclear Reactors and Earthquakes, Report No. TID 7024, U. S. Atomic Energy Commission, Washington D.C
18. IBC (2006) International Building Code. International Code Council, Country Club Hills, IL
19. INTERNATIONAL ATOMIC ENERGY AGENCY, Seismic Design and Qualification for Nuclear Power Plants, Safety Guides Series No.NS-G-1.6, IAEA, Vienna (2003)
20. IS 1893 (Part 1):2002, "Indian Standard Criteria for Earthquake Resistant Design of Structures: General Provisions and Buildings", Bureau of Indian Standards, New Delhi
21. IS 875: Part 3: "Code of Practice for Design Loads (Other than Earthquake) for Buildings and Structures: Wind Loads", 1987
22. Arlaud JC (1975) Application of Reliability Analysis Methods to Rapsodie Reactor. IAEA, Proc. Reliability of Nuclear power Plants
23. Kobayashi, N. et al., "A Study of the Liquid Slosh Response in Horizontal Cylindrical Tanks", Transactions of ASME, Vol. 111, February, (1989)
24. Jaiswal, O. R. Rai, D. C. and Jain, S.K., 2004, "Codal provisions on design seismic forces for liquid storage tanks: a review", Report No. IITK-GSDMA-EQ-01-V1.0, Indian Institute of Technology Kanpur, Kanpur
25. Jaiswal OR, 2004, Review of code provisions on seismic analysis of liquid storage tanks, Document No: IITK-GSDMA-EQ04-V1.0, Final Report: A - Earthquake Codes IITK-GSDMA Project on Building Codes
26. JSME international journal, Vol.33(2), pp 111–124
27. Karamanos A.,Lazaros A. Patkas, Manolis A. Platyrrachos, 2006, "Sloshing effects on the seismic design of horizontal-cylindrical and spherical industrial vessels", Journal of Pressure Vessel Technology, Vol. 128, 329–340
28. NZS 3106, 1986, "Code of practice for concrete structures for the storage of liquids", Standards
29. P K. Malhotra and M Eeri,2005, Earthquake Induced Sloshing in Tanks with Insufficient Freeboard, Earthquake Spectra 21(4), November 2005
30. Papaspyrou S, Valougeorgis D, Karamanos A (2004) Sloshing effects in half-full horizontal cylindrical vessels under longitudinal excitation. J Appl Mech 71:255–265
31. Rammerstorfer FG, Scharf K, Fischer FD (1990) Storage tanks under earthquake loading. Appl. Mech. Reviews 43:261–281
32. UBC, 1997, "Uniform Building Code", International Conference of Building Officials, Whittier, CA
33. USNRC, Regulating Guide 1.61, Damping Values for Seismic Design of Nuclear Power Plant, 2007

Design and Analysis of Piping and Support 11

P. N. Dubey, R. K. Verma, Gaurav Verma, and G. R. Reddy

Piping are the energy carrying systems and its failures may lead to economy loss, spread of hazards and can affect the lives

Symbols

S	Schedule number
Q	Flow rate in m^3/hr
V	Average flow velocity
D_i	Internal diameter of pipe in mm
P_a	Pressure
t_m	Thickness
L	Developed length of piping (ft)
U	Anchor to anchor distance (ft)
Y	Total thermal movement to be absorbed(in)
W	Weight per unit length of pipe
ΔP	Pressure drop in N/m^2
V	Flow velocity in m/s
g	Acceleration due to gravity
h	Geodetic height in meter
λ	Friction factor
γ	Specific weight in kg/m^3
ζ	Coefficient of resistance
l_0	It is the length of straight pipe which causes the same pressure drop as fitting
θ	Angle of deviation of bend
v	Mean velocity in m/s
$m^{\cdot}$	Mass flow rate in kg/s
h_{max}	Pressure rise in m
V_{sw}	Velocity of sound in water under existing condition or velocity of pressure wave propagation
V_0	Normal velocity of flow before closure of valve
K	Bulk modulus of fluid
t	Thickness of pipe
d	Diameter of pipe
P	Design pressure (gauge)
D_0	Outside diameter of pipe
t	Nominal thickness of product
I	Moment of inertia
M_i	Resultant moment due to design mechanical loads
S_m	Allowable design stress intensity value (N/m^2)
P_0	Range of service pressure
E_{ab}	Average modulus of elasticity of two sides of gross structural discontinuity or material discontinuity at room temperature
T_a, T_b	Range of average temperature on side a or b of gross structural discontinuity
S_c	Expansion stress
T_i	Internal surface temperature
T_0	External surface temperature
t	Wall thickness
υ	Poisons ratio of material

P. N. Dubey · R. K. Verma · G. Verma · G. R. Reddy (✉)
Bhabha Atomic Research Centre, Mumbai, India
e-mail: rssred@barc.gov.in

P. N. Dubey
e-mail: pndubey@barc.gov.in

R. K. Verma
e-mail: rajeev@barc.gov.in

G. Verma
e-mail: gverma@barc.gov.in

G. R. Reddy et al., *Textbook of Seismic Design*,
https://doi.org/10.1007/978-981-13-3176-3_11

ΔT_1	Absolute value of the range of the temperature difference between outside surface and inside surface of pipe wall
ΔT_2	Absolute value of the range for that portion of the nonlinear thermal gradient through the wall thickness not included in ΔT_1
R	Bend radius
r_m	Mean radius of pipe
S	Maximum allowable stress for the material at design temperature
A	Additional thickness to compensate for threading or grooving
t_n	Nominal thickness
Z	Section modulus
M_A	Moment due to sustained loads
B_1, B_2	Stress indices
i	Stress intensification factor
P_{max}	Design max pressure
W	Weld joint strength reduction factor
Y	Material coefficient

11.1 Introduction

Piping systems are used to transport liquids, gases, slurries, or fine solid particles in process industries. The use of pipe was very common even in the prehistoric era, particularly bamboo pipes used for irrigation purposes.

A piping system is complete interconnection of straight pipes, elbows, tees, and flanges. Pumps, heat exchanges, valves, and storage tanks are also considered as a part of piping system. Piping system is a very important part of an industry as it accounts for a significant part of total plant budget. A nuclear power plant is more critical because it has numbers of piping systems housed in a much smaller confinement as compared to chemical or other process plants. Piping systems arranged within a very confined area are a major challenge to piping and support designer.

Piping systems need to be designed considering normal loads and accidental loads such as extreme wind and earthquake. Improper design may lead to failure and affects the safety, non-availability of the plant, and loss of economy. Figure 11.1a shows the failure of elbow, and Fig. 11.1b shows the support failure of the piping system.

In piping design [1–3], major loads considered are pressure, dead weight, seismic, and reaction due to resistance to thermal expansion. Design requirements for thermal and seismic loads are contrary in nature because to minimize the thermal stresses in the piping it should have sufficient flexibility and to reduce the seismic loads piping should be stiff. It is tedious and tricky process to include these contrary characteristics in same piping system.

The preliminary design of piping system is mainly governed by functional requirements, i.e., transport of fluid from one point to another, and the detailed design is governed by the type of fluid being transported, allowable pressure drop/losses (pumping power), desired velocity, and availability of construction material. Since piping systems constitute the major part of plant budget, therefore economy is also a prime concern in selection of material and drawing the layout. Nuclear power plants are an exception to this because minimization of environmental hazards and safety of the personnel working at the site are major concern [4].

11.2 Piping Design Codes and Standards

The need of safe, economical and efficient operation of plants increased the requirement of regulation for piping systems, especially in nuclear power plants. Standardization is required to reduce the cost, inconvenience, and confusion that result from unnecessary and undesirable differences in equipment, systems, materials, and procedures. It also facilitates the adoption of accepted industrial practices in the area of safety, testing, and installation. The most commonly used codes for piping are:

Fig. 11.1 a Failure of elbows

Fig. 11.1 b failure of support

a. ASME B 31.1: power piping code,
b. ASME B 31.3: chemical plant piping code,
c. ASME B 31.5: refrigeration plant piping code,
d. ASME Sec-III: boiler and pressure vessels code for nuclear system,
e. ASME Sec-III, Div-1, Subsection-NB: safety class 1 piping,
f. ASME Sec-III, Div-1, Subsection-NC: safety class 2 piping,
g. ASME Sec-III, Div-1, Subsection-ND: safety class 3 piping,
h. ASME Sec-III, Div-1, Subsection-NF: piping support system,
i. ASME Sec-III, Div-1, Appendix N: dynamic analysis.

Different parts of ASME code cover other design requirements like material, fabrication welding, and testing for a particular safety class component. In ASME Sec-III, Div-1, Subsection-NB/NC/ND-3600 [5–8] covers the design and qualification of nuclear piping.

11.3 Sizing of Pipe

All steel pipes are designated by their outer diameter and wall thickness or by their nominal inside diameter. Due to manufacturing conditions, the outside diameters are normally constant and variation in the wall thickness changes the internal diameter. This is the reason why a pipe in general is not referred to by its inside diameter.

In common engineering practice, the size of pipe is represented by nominal diameter and schedule number. Nominal size and schedule numbers are representative numbers of a pipe with fixed outer diameter and thickness, respectively. For a particular nominal size, outer diameter of pipe is constant, whereas inner diameter reduces with increase in schedule number. Approximately, schedule number (S) is related to design pressure (P) and allowable stresses ($\sigma_{\rm all}$) by formula given by (11.1):

$$S = \frac{1000 \times P}{\sigma_{\rm all}} \tag{11.1}$$

For example, 8″ pipe will have 8.625″ outer diameter, and its thickness and internal diameter have been tabulated for different schedule numbers in Table 11.1.

Outer diameter of pipe up to 12″ size is more than nominal diameter. The pipe size corresponding to schedule 40 is called standard pipe, because the nominal size is nearly equal to the inner diameter of pipe. This is valid up to 10″ nominal size, for more than 10″ size, standard pipes have 3/8″ constant thickness. Traditionally, three pipe sizes are referred:

Standard wall thickness pipe	≤ 10″ same as schedule 40
Extra strong pipe (XS)	≤ 8″ same as schedule 80 <8″ thickness = 0.5″
Double extra strong pipe (XXS)	≤ 8″ thickness varies <8″ < 12″ thickness = 1.0″ >12″ thickness not specified

Selection of pipe diameter is mainly governed by the expected rate of flow and available pressure head. The pressure head is necessary to maintain the flow by overcoming the losses in the flow rate due to internal friction in the moving flow and pipe wall. The pressure drop may further increase due to turbulence and separation of flow at elbows and valves, etc. The diameter of pipe can be determined from following Eq. (11.2):

$$V = \frac{354\,Q}{D_{\rm i}^2} \tag{11.2}$$

where

Q flow rate m^3/hr,
V average flow velocity (m/s) and
$D_{\rm i}$ internal diameter of pipe in mm

11.4 Size of Elbow and Tee

There are two sizes of elbows, long radius, and short radius elbows. These elbows are called standard elbows, and elbows other than these radii are termed as non-standard

Table 11.1 Thickness of 8″ pipe for different schedules

Schedule no.	Outer diameter	Thickness	Inner diameter
20	8.625″	0.25″	8.125″
30		0.277″	8.071″
40		0.322″	7.981″
60		0.406″	7.813″
80		0.50″	7.625″

Table 11.2 Thickness of straight pipe prior to bending for making elbows

Bend radius (D = nominal dia.)	Minimum thickness prior to bending
$6 \times D$	$1.06 \times t_m$
$5 \times D$	$1.08 \times t_m$
$4 \times D$	$1.16 \times t_m$
$3 \times D$	$1.25 \times t_m$

elbows. The mean radius of long radius elbow is equal to 1.5 times the nominal diameter of pipe. Long radius elbows are stiffer and cause lesser pressure drop as compared to short radius elbows. The mean radius of short radius elbow is same as nominal diameter of pipe.

Generally, elbows are manufactured by bending a straight pipe to required bend radius. Pipe wall thickness at any point in the bend should not be less than the minimum allowable thickness (t_m) given by Eq. (11.3) for maximum allowable pressure (P_a). Therefore, higher thickness pipe is selected prior to bending. The thicknesses of pipe prior to bending have been tabulated in Table 11.2, for different bend radius of elbows.

$$t_m = \frac{P_a D_o}{2(S_m + P_a y)} \tag{11.3}$$

where

S_m allowable design stress intensity value, N/m^2
y 0.4

For short radius elbows, the pipe thickness prior to bending should be 20% more than the minimum required thickness.

11.5 General Design Guidelines for Piping Design

In order to enhance the resistance of piping system against the vibrations due to earthquake, the following considerations should be kept into mind:

i. While designing the supports for piping system, all the joints and connections should be designed to behave in similar fashion (e.g. hinge, fixed, guided) as considered in the analysis of piping system. They should cater support for the full range of loads expected in the connected members. In case of integral supports, design, fabrication, and installation of these supports should be such that any unexpected failure or initiation of crack in the body of supporting system should not propagate in the supported systems like pressure vessel and piping system.
ii. In the design of anchors for equipment and piping, all the potential forces and moments need to be properly evaluated and also suitability of anchoring materials should be thoroughly evaluated. To avoid the prying of base plates, they should be stiff enough. Anchor bolts need to be adequately tightened to avoid rocking effects. In case of over-designed or redundant anchors, preloaded bolts close to their yield point are recommended.
iii. Cross-braced legs should be used; otherwise, use of different types of supports should be warranted by suitable design calculation. In most cases, stiffness of supports can be increased to avoid resonance. In case of stiff supporting system the effect of thermal stresses, resonance with external dynamic loads and effect of differential supporting point movement should be addressed properly.
iv. As far as practically possible, resonance of equipment, piping, instrumentation, and core internals should be avoided with the dominant frequency range of external excitation. Dampers are recommended at the places where the response of equipment, although significant, cannot be reduced by other means.
v. As a consequence of dynamic displacement, collision between adjacent components, structures, and building should be avoided by following the suitable design and spacing practices.

11.6 Practical Considerations for the Design of Piping System

While designing the piping system, the following requirements need to be considered:

i. The weight of the piping, pipe fittings and valves (with or without insulation), weight of the operating medium or test medium, and weight of the hanger parts to be carried by the supports,
ii. The support should provide adequate stiffness to the piping against external loads such as wind load, ice, snow, seismic loads,
iii. The resistance to free thermal expansion/contraction of the piping due to the supports should be minimal,
iv. Overstressing of pipes at supporting location should be avoided,
v. Supporting should prevent sagging of piping system, to avoid draining problem,

vi. Supports should withstand and dampen vibration forces applied to the piping by connected equipment,
vii. The frequency of the piping system should be controlled, if possible, to avoid resonance with structural and equipment vibrations,
viii. Supports should cater the resistance against the occasional loads, viz. earthquake, relief valve discharge, and water hammer,
ix. Interaction with adjacent piping and equipment,
x. Limiting deflections in critical piping systems.

11.7 Various Pipe Supporting Systems

Pipe supporting systems can be broadly divided into three categories based on their required functions:

i. Hangers or supports	used to sustain the dead weight of the piping system
ii. Restraints	used to restrict the movement of pipelines due to thermal and dynamic loading
iii. Vibration Absorbers	used to restrict the movement due to vibration caused by earthquake, flow-induced vibration, wind effects, etc.

Selection of most suitable support/hanger for any given application is mainly governed by the individual piping configuration and the functional requirements. Piping stress analyst decides the location and effect of any restraint, guide, or anchor on a high-temperature and high-pressure system. Indiscriminate placing of a restraint on a piping system could alter the piping stresses and reaction forces significantly which may change a conservatively designed piping system into a system where stresses exceed the limits of good design practices.

11.8 Hangers Supports

Hangers and supports are used basically for carrying the dead weight of the piping system along with its contents. The primary difference between a hanger and a support is that the hanger sustains the dead weight in tension, while the support sustains it in compression. Consequently, hanger supports the piping system usually from the top, while supports are placed below the piping. The types of support or hanger assemblies from where appropriate kind of support is selected are classified as follows:

i. Rigid hangers and supports, such as rod hangers and stanchions,
ii. Flexible hangers (constant hangers, variable spring hangers),
iii. Rollers.

A brief description of various types of hangers/supports is given below.

11.8.1 Rigid Hanger Supports

Rigid hangers are normally placed at locations where vertical movement of piping is negligible. For designing, a rigid hanger material selection for clamps is governed by pipe temperature and dimension of rigid hanger is governed by the expected loads to be supported.

The hanger assembly uses, besides the eye nuts/eye rods or simple rods, a variety of small items like brackets, clamps, bolts, and turnbuckles. The turnbuckles provide the length adjustment for the hanger based on the actual location of the piping. Though the hangers prevent vertical movement of the piping, the horizontal movement is allowed to a limited degree, due to the assortment of hinges provided within the hanger assembly in the form of eye nuts or eye rods. The inclination from vertical for a hanger assembly is limited to four degrees. The compressive force imposed on the hanger due to thermal expansion of the piping also needs to be considered, while accepting a hanger length, to avoid buckling of the hanger rod.

The advantages of rigid hangers over other types of hangers are mainly their lower requirement of space and the higher stiffness and the resulting higher frequency of the piping system causing less vibration problems. Also, their maintenance and inspection requirements are less compared to flexible hanger. In some cases, not only the rigid hanger is used for supporting the vertical loads, but also it acts as a restraint against vertical piping movements. In these cases, the piping engineer should take care in selecting the location of the rigid hanger.

Supports have the advantage of occupying less space than the rigid hangers. They provide friction against the horizontal movement, which has to be negated by the use of rollers or low friction surfaces, etc. Pipe shoes, pipe chairs, etc., are examples for pipe supports. While using such supports, the lifting of pipe at the support location due to thermal expansion should be checked to ensure that this does not nullify the support function.

While discussing supports, pipe racks demands a special mention, in that they are the most convenient method of supporting a multiple numbers of piping running in parallel, laid out in such a manner that their bottom surfaces are at the elevation of racks. The rack can be prepared in precast concrete or by steel members. The intermediate members used to strengthen the rack are usually prepared in steel and covered with concrete for fireproofing. Piping is either directly supported on racks or they transfer the load through a shoe welded to the rack. While supporting insulated pipes, special care should be taken to avoid crushing of the insulation. Pipe shoes and insulation protection saddles are used to achieve the same.

11.8.2 Flexible Hanger Support

To allow the thermal expansion up to a certain extent in modern high-temperature piping installations, use of flexible support is a necessity. In this case, spring hangers are employed. Equipment manufacturers such as the turbine and boiler manufacturers are especially concerned about the transfer of pipe weight on their equipment and sometimes specify that the loads at pipe connections shall be limited to a particular value (Figs. 11.2 and 11.3).

In that case, the hanger support designer must ensure that the loads on the equipment–pipe juncture do not exceed the limits specified by the equipment manufacturer.

Considering above, the majority of supports for a high-temperature system are spring type. Support designer must accurately work out the supporting force required at each hanger location to assure balanced support, in order to select the appropriate size and type of spring support. The flexible hangers are of two types:

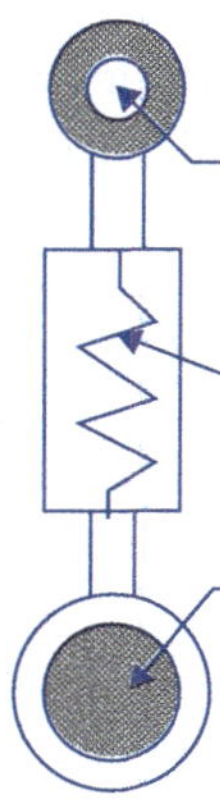

Fig. 11.2 Flexible hanger

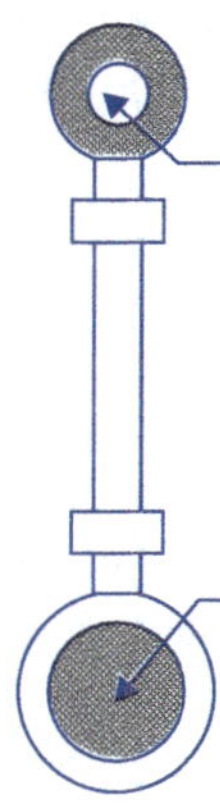

Fig. 11.3 Rod hanger

i. Variable Spring hangers,
ii. Constant support hangers.

These types of hangers are used when the requirement is to sustain the primary loads without adversely affecting the flexibility of the system to ensure lower stresses during thermal loading. This is achieved generally by the use of spring, which sustains the primary load by its resistance against compression while allowing for free thermal expansion.

11.8.2.1 Variable Spring Hanger

Variable spring hangers are used to support the piping subjected to vertical movement where constant supports are not required. Variable spring hanger is so called because its supporting force varies with the spring deflection.

The pipe weight is balanced by reaction that is equal to the spring stiffness multiplied by the compression which spring undergoes due to the supported weight. During thermal loading, the vertical movement of the pipe tends to reduce the compression on spring due to this supporting force is also reduced. The supporting force exerted by the spring is reduced by the amount of the spring travel multiplied by spring rate. Since the pipe weight is same during all conditions envisaged during pipe life, the variation in supporting force results in transfer of pipe weight to the adjacent supports and/or equipment nozzles. Consequently, additional stresses in the system are developed. In view of this, the effect of load variation must be considered while using variable spring hangers. To quantify this variation, a factor called variability factor is introduced which is defined as given in Eq. (11.4),

$$\text{Variability factor}(\%) = \frac{|\text{Hot load} - \text{Cold load}\,|}{\text{HotLoad}} \times 100 \tag{11.4}$$

where hot load is the load supported by the hanger during hot conditions and cold load is the load supported by the hanger during cold conditions

It can be seen that the difference between the hot and cold loads is the product of spring rate and spring travel. Variable spring hangers are preferred for general purpose, non-critical pipelines. These hangers can also be employed in critical systems where expected vertical movement is small. In acceptable practices, variability in supporting force should be limited to maximum 25% of supporting force. The hangers are usually preset at factory, for the specified load and travel, and comes with a preset bar. This bar is meant to prevent excessive loading of spring during installation and hydro testing, by providing an alternate load path. Variable spring hanger may be used for transfer of load to adjacent hangers or non-critical equipment where the vertical movement of the piping is less than 3/4 inch, provided the variation in supporting effect is limited to 25% of the calculated piping load throughout the total vertical travel.

11.8.2.2 Constant Support Hanger

These supports provide constant supporting force throughout the full operating range of temperature, i.e., both in cold and hot conditions. Due to vertical expansion and contraction, they transfer very less/negligible pipe reactions at equipment connections and anchors. The magnitude of total supporting force variation should be checked so that it may not impose stresses and forces on the system beyond a certain limit that may cause failure. Constant support hangers are most preferable and economical type of flexible support in the cases piping stresses and reactions are very close to allowable; i.e., safety margin is very low.

For non-critical piping systems, variable spring hangers can perform satisfactorily at the places where piping stresses and end reactions are low. Variable flexible supports can be employed for supporting the critical piping systems where piping movement in vertical direction is small.

Constant support should be used for critical, high-temperature lines where vertical movement in piping at hanger locations is 3/4 inch or more or at the necessary instances to avoid the transfer of load to adjacent hangers or connected equipment. These types of supports are used when percentage variation in support force is to be controlled within ±6% in critical piping and within ±25% in the case of non-critical piping. There are two types of constant support hanger designs available:

i. constant weight hanger,
ii. constant spring hanger.

In case of constant weight hangers, the bearing load is kept constant by the use of a counterweight. To reduce the size of counterweight, it is often the practice to apply the principles of pulley or of lever and fulcrum. Besides the undesired aesthetic appearance of the plant, these hangers unavoidably require the entire support to be larger since either the counterweight should be heavier, or the pulley or lever should be larger requiring more space as the bearing load increases. On the other hand, the constant spring hangers use a helical spring in conjunction with a bell crank lever so proportioned that the spring force times its distance from the lever point is always equal to the pipe force times its distance from the lever point. In other words, the increasing load required to extend (or compress) the spring is provided by an increase in the effective lever arm.

The load supported by the hanger will be almost constant if the position and proportions of the bell crank levers are properly engineered. These hangers are widely accepted by various plants because of their lower weight as compared to the constant weight hangers, compactness, lower cost, and their availability as a completely engineered product for all the load ranges. These hangers also are supplied with presets to lock the spring during installation and hydrotest.

11.9 Roller Supports

Basically, pipe roller supports are of two types, one attached to overhead structure and other are placed beneath the pipe as base supports. Rollers supports are required where operating temperature is sufficiently high and thermal insulation is used to reduce heat loss and for personnel protection as well. In these cases, a pipe covering protection saddle is used along with the rollers supports to protect the insulation from crushing. Pipe will directly rest on the rollers in case of bare pipes.

In common practice, roller supports are used to support very long gas or fluid transmission lines where fluid being transported is not at elevated temperatures, but change in the length is significant due to its long span due to the seasonal temperature variation, viz. difference of temperature in summer to winter.

11.10 Restraints

Supports can become fairly complicated when expansion has to be catered for. It should be remembered that not only will steam and hot liquids cause a pipe to expand; the climate also gives considerable changes in temperature. There have been references of expansion as high as 10 inches due to climatic conditions only.

There are three ways of dealing with expansion:

i. By installing special fittings, which absorb movement such as the expansion joints etc.,
ii. By designing the pipe route so that the pipeline is sufficiently flexible in itself to adopt a new route under the influence of expansion, this method being used with high temperature all welded steam and feed piping in modern power stations,
iii. By incorporating restraints in the piping systems in such a way that the thermal expansion can be suitably diverted/divided or controlled.

Restraints are used to restrict thermal movement of the pipe, and so they are not basically meant to support the dead weight. Restraints can be categorized into three types based on the nature of restriction imposed on the piping:

i. Anchor fully fixes the point; neither translation nor rotation is permitted
ii. Stop restricts translational movement
iii. Guide restricts rotations

Anchors are used to decouple two systems so that their behavior can be analyzed independently and also to increase the frequency of the piping system. Stops prevent excessive deflection in the piping system, which may induce higher stresses in the system. Stops are used to protect the equipment nozzles from excessive stresses due to thermal movement of the connected piping. Guides are used where large deformations occur due to rotation of large pipe lengths. In the next few paragraphs, we will look at the various types of restraints and their uses.

11.11 Anchors

Anchors do not allow the movement or rotation of support points and completely fix the line in its place. From the point of view of bearing the dead weight, they are nothing but rigid supports. Using anchors, it is possible to divide the pipeline into two independent pipelines. It is recommended, therefore, to use them at the location of branch or in the neighborhood of it on the main pipeline when it refers to, for example, a piping system in which the thermal expansion is produced independently from that of its main pipeline, such as an emergency pipeline.

In addition to providing the piping engineers a simple and unobstructed view with respect to thermal expansion problem, they have a further advantage of restraining vibration and oscillation when they are used at the intermediate point in a piping system as they increase the rigidity of the whole system.

11.12 Limit Stops

Stops restrict the displacement of the pipelines at least along a directional axis, and they are named after the axis in which they are acting; for example, vertical stops are called Y-stops. When two-way displacement is restrained, they are sometimes called double stops. They prevent excessive reaction or stress arising to the piping system, thereby contributing to keep the force balance in the system. They are generally used for protecting equipment nozzles from the excessive thermal expansion caused by its pipelines, from the reaction force of blowing safety valve and also from the thrust force when expansion joints are used.

11.13 Guide Supports

A guide (Fig. 11.4) is fundamentally defined as a device to restrict the rotation of a pipe. Because of its lower lateral rigidity, a long pipeline tends to be unstable against the wind effects and earthquakes, and further, it is difficult to maintain its external appearance due to a large lateral deformation caused by thermal expansion. This large deformation may cause some problems of interfering with adjoining pipelines or equipment or causing a necessity of more sophisticated pipe support design. Application of guides is very effective to control such large lateral deformations and consequently to resolve the said problems at the same time.

11.14 Vibration Absorbers

Besides the thermal expansion, pipelines in a plant may suffer from vibrations caused by various sources such as:

i. Mechanical vibrations transmitted through pumps, compressors, turbines, and other rotating and reciprocating machinery,
ii. Impact and vibrations due to the sudden opening/closing of valves,
iii. Surging of compressed gas or two-phase fluid,
iv. Periodic wind forces against the outdoor pipelines,
v. Earthquakes,

Protection of the system against these can be accomplished by:

i. Minimising the effect at source by smoothing the flow, eliminating unbalance, etc.,
ii. Controlling the deflections of the pipelines by bracing,
iii. Controlling the movement of the pipeline by damping, i.e., energy dissipation,
iv. Limiting deflection or rotation by limit stops.

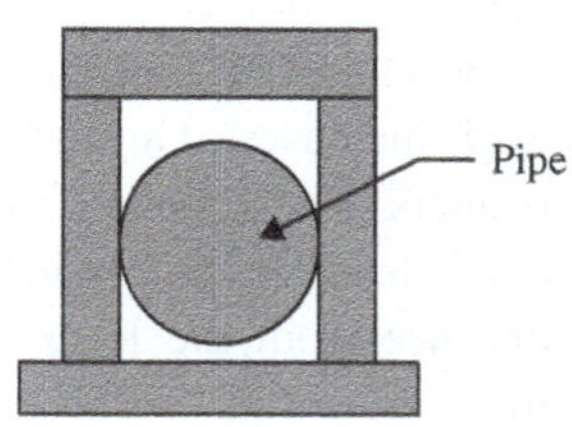

Fig. 11.4 Guide support

11.15 Modifying the Natural Frequency of the Pipeline or Supporting Structure

The best and fundamental way to resolve the problems associated with the vibrations is to find out the sources of vibration in the particular pipeline and then to remove them away or reduce them to acceptable levels, nevertheless, in many cases, they are practically not avoidable. Adequate control may be provided by the thermal expansion restraints and hanger or supports which modify the natural frequency of the pipeline, or by sliding supports, which have a damping effect due to friction. There are various kinds of supports in use in the industry for mitigating the effects of vibrations, which range from the use of swing struts, sway braces and, in the requirement of heavy damping, hydraulic shock absorbers (like the one used in vehicle suspension) and the snubbers.

One of the major vibration absorbers used for supporting the piping systems is the snubber. The requirement of a snubber is to provide little resistance to the movement associated with the thermal expansion of the piping, while providing stiff supporting reaction against the vibratory motion associated with dynamic loading such as an earthquake load. Snubber goes to restraint mode based on a particular velocity or acceleration of motion depending on its design concept and is known as its activation level. Snubbers are basically of two types:

i. Mechanical snubbers,
ii. Hydraulic snubbers.

Typical snubber locations involve:

a. Points of large concentrated loads such as valves, large components, flanges, and other components that are not anchored,
b. Changes in piping direction,
c. Within long vertical risers, and
d. On long straight pipe runs, as determined by piping analysis.

Snubbers should not be used at more than one point of restraint on straight pipe runs to absorb an axial dynamic force unless simultaneous activation is provided. Multiple snubbers may be used at one point of restraint if simultaneous activation is provided.

11.16 Mechanical Snubbers

Mechanical snubbers (Fig. 11.5) generally consist of a motion sensing device, an activating rod, and a fixed housing with a connector. A braking device is employed to limit the motion of the movable rod end. The snubber is activated either by the development of progressively higher resisting forces as increased rod motion is passed through a geared mechanical system, or a breaking action is performed on a rotating mass attached to the active rod.

The snubber with the brake and rotating mass system has a motion sensor, which responds to changes in rate of angular rotation of the mass. The braking torque is produced when axial movement of the ball nut is converted to the rotation of the ball screw and the brake attached to its head is put to work. It scarcely has any resistance to the slow motion like the thermal expansion, as the brake does not work in such conditions. On the contrary, as the brake acts against the vibrations, a great resistance force is developed under vibrating conditions.

These snubbers have a distinct advantage over the hydraulic snubbers wherein the oil change is required periodically and also the makeup for oil leakage is required. However, they are at a disadvantageous position, in case the snubber inadvertently locks in its position, giving rise to high thermal stresses.

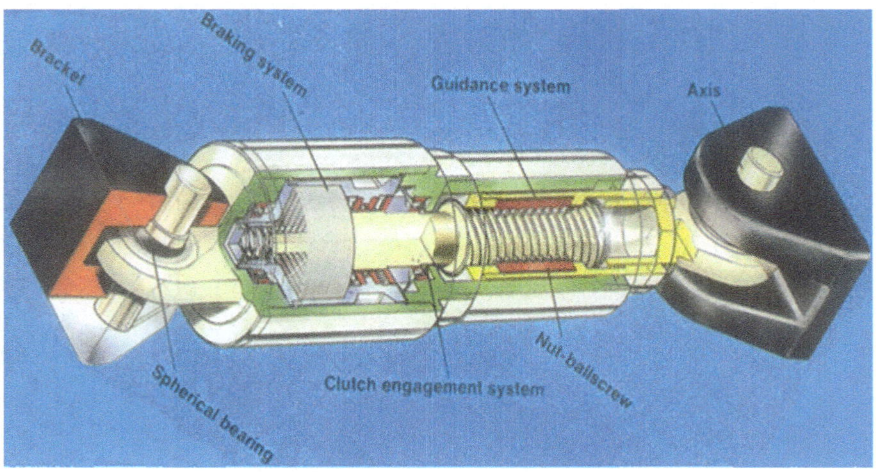

Fig. 11.5 Mechanical snubber

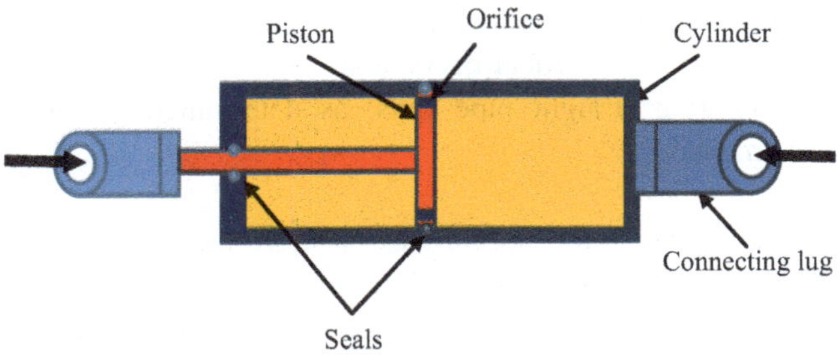

Fig. 11.6 Hydraulic snubber

11.16.1 Hydraulic Snubbers

Hydraulic snubber (Fig. 11.6) consists of a double acting cylinder, a flow control device, and, in some cases, a flow control reservoir. Simpler construction of a hydraulic snubber involves use of an orifice in piston, without very distinct activation levels. In hydraulic snubbers using external flow controls, the flow control device permits the fluid to flow at a predetermined rate. This snubber works based on the principle of pressure difference between the two chambers separated by the piston when the oil flows through the orifice at high speed. The resistance force in this case is generally proportional to the square of piston speed, but may vary depending upon the shape and size of the orifice and also upon the combination of valves. Thus, when the piston velocity is low, the resistance force is also less. In case of vibratory loading, the piston velocity is high enough to cause a sufficient resisting force to this motion. It is possible to obtain a great capacity of vibration restricting force with a relatively small size hydraulic snubber by simply narrowing its orifice opening. The release of snubber at particular dynamic loads is brought about by a bypass circuit or inherent leakage in flow control device. A reservoir caters for thermal expansion of fluid, volumetric change during snubber motion, and the reserve fluid. Piston design can be single ended or double ended. Double-ended piston is longer in length, but it gives even response to tension and compressive load compared to the single-ended type.

11.17 Support Spacing from Dead Weight Considerations

Support locations are decided, depending upon the size of pipe, layout, the location of concentrated masses, viz. massive valves and fittings, and the availability/nearness of supporting structure. There are well-defined rules or limits available that can accurately fix the location of a piping system. Decision to place the supports at appropriate locations is completely based on the experience and engineering judgment of the designer. A pipeline is a continuous structure partially restrained by multiple locations depending upon the magnitude and type of loading at the piping. Due to partial restraint, actual tresses and deflections in the piping lie between deflection of a free end beam and a fully restrained one. Table 11.3 gives the suggested support spacing for sustaining the dead weight of piping as per ASME B31.1 code.

The basic assumptions based on which the support spacing in the above table are suggested are as follows:

i. The support spacing mentioned in Table 11.3 is based on a fixed beam support where bending stress does not exceeding 15.86 MPa and the sag does not exceed 2.5 mm between the supports of insulated pipe filled with water or the equivalent weight of steel pipe for steam, gas, or air service.
ii. Maximum support spacing given in Table 11.3 is applicable in case of horizontal straight pipes of standard and heavier pipe, operating at 750 °F (400 °C) temperature.

Table 11.3 Suggested piping support spacing as ASME SEC-III-NF/ASME B31.1

Nominal pipe size NPS	Suggested maximum span			
	Water service		Steam service	
	Feet	Meter	Feet	Meter
1	7	2.1	9	2.7
2	10	3.0	13	4.0
3	12	3.7	15	4.6
4	14	4.3	17	5.2
6	17	5.2	21	6.4
8	19	5.8	24	7.3
12	23	7.0	30	9.1
16	27	8.2	35	10.7
20	30	9.1	39	11.9
24	32	9.8	42	12.8

iii. The spacing values are not applicable in case of concentrated loads such as valves and fittings, etc., between the supports. It also does not suit well in case of change of direction between the supports.
iv. To keep the stresses low, the supports should be put preferably beneath the concentrated mass; otherwise, it be very close to it.
v. At the instance of change in the direction of any critical piping system, the spacing between supports/hangers should be kept less than three-fourths of the full spans given in Table 11.3. It is good practice to keep a hanger/support immediate after any change in direction of the piping.

In long span steam lines, expected sag may be large enough to cause accumulation of condensate at the lower points of the pipeline. Partially drained water lines can get damaged due to freezing. Such occurrences can be avoided by providing a downward slope in the direction of flow. This can obviously be achieved by placing the supports in such a way that succeeding supports are lower than the points of maximum deflection in preceding spans.

11.18 Special Considerations

Pipelines above ground level can either be supported from below or hung from brackets, and where there is no expansion to contend with, the design of supports and hangers presents few problems. Where cleanliness is a definite requirement, as in a food processing plant, the use of tubing for support brackets is recommended because there are no flat surfaces to accumulate dust and dirt.

The pipeline needs to be supported and guided between anchors. If an expansion bend is used, it could well be that the pipe would find it easier to lift away from the supports and become bowed rather than compress the bend; this could also apply when a steel bellows-piece is used. Not only must the supports hold and guide the pipe, but they must also allow for the progressive growth of the pipe between the anchors in the desired manner. If the hangers are erected when the pipe is cold, they may be out of plumb when the pipe expands, or vice versa, and when out of plumb, they may get wrongly stressed or may not be doing the desired job at all. As much ingenuity as possible should be brought to bear on the work to ensure that the supports are doing their job properly.

Wherever possible, pipe clamps shall be used for horizontal piping. Generally, for temperatures up to 300 °C, pipe clamp material is carbon steel, and beyond 300 °C, alloy steel clamps are preferred. Up to 232 °C, malleable iron piping clamps may be used. Hanger rods should be employed to support the tensile loads only because it may buckle under compressive load due to its slender dimension. A suitable linkage shall be provided to allow swing at hanger locations where lateral or axial movement is expected. The instances where horizontal movement of piping is beyond 0.5 inch, or the angularity from the vertical direction is higher than 4° from the cold to hot condition of the pipe, offset between hanger rod and structural attachments should be kept such that the rod comes in vertical position under hot operating condition. Supports, guides, and anchors shall be designed such that excessive heat should be transmitted to the supporting structure/building. For the design of various elements of supports, the temperature of supporting parts shall be based on a temperature gradient of 100 °F per inch from the outside surface of the pipe.

11.19 Criteria for Support Design

Design of pipe supports should cater for the various loads transferred onto them by the piping system.

ASME B31.3 Section-321 provides a discussion of the types of loads and conditions that must be considered. Design of piping layout, its support, and surrounding elements should aim to avoid the following conditions:

i. Over shooting of stresses in the piping beyond the allowable value of the code,
ii. Leakage of flange and other pipe joints,
iii. Excessive shock loads and moments at the nozzles of connected equipment, viz. pumps, vessels, and turbines,
iv. Overstressing of the supporting or restraining) elements,
v. Resonance with external excitation or vortex shedding frequency induced by flow,
vi. Excessive hindrance with thermal expansion and contraction of the piping which is otherwise adequately flexible,
vii. Unwanted disengagement of piping from its supports,
viii. Excessive piping sag resulting fluid accumulation and need for slope for drainage,
ix. Excessive distortion or sag of piping (e.g., thermoplastics) subject to creep under conditions of repeated thermal cycling, and
x. Excessive heat flow from pipe to surrounding/supporting elements, resulting in exposure to a temperature beyond outside their design limits.

These are generally requirements for sustained and thermal loads, stated elsewhere in the code as well as in the section on supports. The pipe support elements must be designed for all the loads that they can be subjected to, including surge, thermal expansion, and weight.

As far as the design of piping supports is concerned, all the loads such as those transferred due to the dead weight, thermal expansion, pressure or due to the dynamic loadings by the piping system are considered as primary loads for the design of supports. In addition to this, the designer should also account for the primary loads within the support, e.g., due to the weight of the elements of support, and the secondary loads such as those generated by the expansion of the support itself, in the design of piping supports. The basic design of piping supports is mostly carried out by following the fundamental principles of mechanics and strength of materials. However, in order to arrive at a safe supporting system design for the piping systems, the designer has to look for the various support design criteria.

In this paragraph, we would discuss the support design criteria as provided in ASME B31.1 code for power piping. Similar criteria should be looked at for the other piping systems also in the relevant codes. ASMEB31.1 requires that the support elements should be designed in accordance with MSS specification 58 except as superseded by the code itself. Allowable stress values as tabulated in MSS SP-58 shall be used except when it is different from the allowable stress values given in the code itself. When the allowable stress values of a material used is not tabulated in both MSS SP-58 and the code, the values can be taken from ASME Section II, Part D. If it is not available in ASME Section II, Part D also, then 25% of the minimum tensile strength as given in material specification can be used as the allowable stress value for temperature not exceeding 650 °F (345 °C).

For steel material of unknown specification or of a specification not listed in the code or in MSS SP-58, the allowable stress value can be taken as 30% of the yield strength at room temperature, corresponding to 0.2% offset, as found out by a tensile test on a specimen of the material up to a temperature of 650 °F (345 °C). The allowable stress value, however, shall not exceed 9500 psi (65.5 MPa) in such cases.

Having established the allowable stresses as per the preceding paragraphs, the following limits shall be met for the stresses of various kinds:

i. The allowable shear stress shall not exceed 80% of the allowable stress,
ii. The allowable compressive stress shall not exceed the allowable stress,
iii. The allowable bearing stress shall not exceed 160% of the allowable stress,
iv. The allowable stress established shall be reduced by 25% for threaded hanger rods,
v. For welds, the allowable stresses shall be arrived at by reducing the established allowable stresses as per above paragraphs or the allowable shear stress in a) above by 25%, for the weaker of the metals joined,
vi. Increase in the allowable stress values shall be permitted as follows:
    a. An increase of 20% for short time overloading during operation,
    b. An increase up to 80% of the minimum yield strength at room temperature during hydrotest.

For threaded hanger rods, calculation of safe loads shall be based on the root diameter of the threads members and allowable stress of the rod material. In no condition, diameter of hanger rod used for supporting the pipe NPS 2 shall be less than 3/8th inch (9.5 mm), or less than 0.5 inch (12.5 mm) for supporting pipe NPS 2.5 and larger. Pipe, straps, or bars of strength and effective area equal to the equivalent hanger rod may be used instead of hanger rods.

11.20 Optimization of Supports

Design of nuclear piping system such as the PHT piping is usually performed keeping in view the established conventional design practices with additional requirements for safety. Selection and location of supports play a vital role in this exercise. Adopting the methodology outlined below the optimization of various supports can be performed:

i. Start with an isometric of restrained piping system, add the mandatory anchors such as equipment nozzles, etc., and mark the areas where piping restraints cannot be provided due to overcrowding or due to its being too away from the supporting structure.
ii. Perform a thermal analysis, and optimize the layout for thermal stresses. Provide additional flexibility wherever required. Mark the areas of low thermal movement on the piping. Keep the thermal stresses as low as possible and not more than $2S_m$.
iii. Place the gravity supports preferably at the areas of low thermal movements. Perform a dead weight analysis and ensure that the gravity supports should not significantly increase the thermal stresses. If the simple supports increase the thermal stresses, use the spring hangers. Keep the gravity stresses limited to 0.5–0.6 S_m so as to leave a good margin for seismic stresses.
iv. Perform a first seismic analysis based on a few modes only. Avoid conservatism by stretching piping to its allowable limits. If seismic stresses are within allowable limit, perform the complete stress analysis including ASME code evaluation. If the code evolution fails, continue with (v) below.

v. If the seismic stresses or displacement are excessive, identify the areas of high stress or displacement. Check whether the frequencies associated with these areas attract considerable acceleration from the applicable floor response spectra (FRS). Restrain to increase the areas at locations of small thermal movements to increase the frequency of these segments. Optimize the restraint arrangement to obtain low gravity and thermal stresses and seismic stresses within allowable limits. If successful, complete the stress analysis including code evaluation.
vi. If optimization process does not finally lead to compliance with codal requirements, i.e., rigid restraints cannot be used in all the locations, continue with the following step.
vii. Consider changes in the pipe routing in certain portions of the layout to change the frequency as well as flexibility of the system. Relocate the supports and analyze the system for various loadings. If successful, complete the stress analysis including code qualification. If not, continue with step (vii) below.
viii. Replace the restraints near the areas of high stress with snubbers or engineered gap-type restraints. Optimize the number of such restraints and qualify the piping.

This approach of optimization of supports along with the sound design practices can be implemented during the early stages of design that will result in a flexible piping system layout with very few simple and sturdy supports and with bare minimum number of snubbers.

11.21 Basic Steps in Piping Design

The steps in pipeline design are as follows:

i. Gathering information about the system that may include following points:
   a. The characteristics of the working fluid, flow rate, and permissible pressure drop,
   b. Routing path of pipeline from source to destination and marking probable support locations,
   c. Type of application and applicable codes and standards,
   d. Selection of material accounting availability, compatibility with working fluid, and design temperature.
ii. Deciding the pipe diameter based on flow rate and allowable head loss,
iii. Structural analysis:
   a. Designing for pressure: Calculation of pipe wall thickness accounting corrosion and machining allowances,
   b. Stress analysis accounting the static, dynamic, and thermal loads,
iv. Load combination as per applicable code provisions and comparing the stresses with code allowable values. The stress analysis is performed until compliance with the code is achieved,
v. Support and anchor design considering reactions at supporting locations,
vi. Preparation of drawings, specification, and the design report.

11.22 Piping Stress Analysis

Stress analysis of a piping system generally covers the flexibility analysis, determination wall thickness of pipe to encounter the internal/external pressure, quantification of reinforcement required, allowable sag to avoid the fluid pockets in a drained piping system prior to repair, etc. Piping stress analysis more or less is limited to mainly linear, static, and dynamic analysis in international and national piping design codes. Therefore, all the modern piping design codes are based on the Traditional Allowable Stress Design (ASD) methodology.

Most of the pipe stress analysis is carried by using finite element analysis software that is based on the beam element theory in addition to use of stress indices, stress intensity, and stress concentration factors. Local design checks may include analysis of non-standard branch connections, pressure vessel nozzle to shell analysis, additional pipe wall membrane stresses caused by local interaction from pipe supports, special flanges, high frequency (acoustic) fatigue calculations. At present for pipe stress analysis using FEA techniques static analysis, equivalent static analysis and dynamic analysis are performed.

The provisions in the code for those piping system which do not require detailed analysis are as follows:

i. Those pipe which are duplicate copy of successfully operating installed pipes,
ii. Those pipelines which can be judged properly by comparing with the systems, previously analyzed, and
iii. Systems of uniform size that have no more than two anchor points, no intermediate restraints, and fall within the limitation of Eq. (11.5) given in ASME B31.1:

$$\frac{DY}{(L-U)^2}\left(\frac{\text{in}\times\text{in}}{(\text{ft}-\text{ft})^2}\right)\leq 0.03 \qquad (11.5)$$

where

Y total thermal movement to be absorbed (in)
D diameter of piping (in)
L developed length of piping (ft)
U anchor to anchor distance (ft)

11.23 Static Analysis

Static analysis is carried out to evaluate the stresses due to sustained (primary) loading, pipe support loads, internal pressure, deadweight of the piping including fluid contents, pipe insulation, snow and ice accumulation, weight of valves, etc. and other sustained and displacement loads. Static analysis is considered mandatory for all piping systems requiring a detailed analysis.

11.24 Dynamic Analysis

Dynamic analysis is carried for calculating the natural characteristics of the system, viz. natural frequency, mode shape and stresses due to reversing and non-reversing dynamic events. The dynamic analysis of piping systems consists of the following:

11.24.1 Modal Analysis

Modal analysis of piping system is performed in order to determine the natural frequency of different spans of piping between the supports. Modal analysis helps in determining the correct pipe support spacing in order to avoid the resonance with vibrations caused by internal flow, pressure transients, earthquake, and vortex shedding oscillations from wind or sea currents interacting with the piping. In modal analysis, selection of element size and a number of elements plays a very important role while making the finite element model. The method of FE modelling of piping is explained in Sect. 11.25.

11.24.2 Response Spectrum Analysis

Response spectra method is used in the seismic analysis in which peak values of particular responses like displacement and moment are determined for each significant mode. Then, the total response is obtained by combining the peak modal responses. This is the most commonly used because of the following reasons:

(a) Design ground motion is mostly depicted in the form of response spectra.
(b) This method yields conservative estimation of responses such as force, moments, and displacements.
(c) Method is simple.
(d) Uncertainties concerning to FE modeling, soil–structure interaction, and structure–equipment interaction, etc., can easily be overcome due to inherent conservatism.
(e) Modal responses derived from responses spectra are combined to give the total response under an earthquake scenario.

11.24.3 Time History Analysis

The response of a system is time-variant phenomenon. The force or acceleration as a function of time describes the event most accurately. The points in the piping are selected, where forcing function is applicable. The response of the piping is studied, at discrete time intervals.

11.25 Finite Element Modeling of Piping Systems

The FE model of piping system is mathematical representative of static and dynamic characteristics of the system. For finite element analysis, commonly lumped mass model is used. Piping system can be modeled as pipe or beam element. A pipe element is same as beam element with additional internal/external pressure-applying characteristics. The element length and node positions are the important factors of finite element modeling. In lumped mass model, a finite number of discrete concentrated masses are used to represent a continuous system of piping. In order to evaluate the seismic stresses more accurately, each lumped mass has to theoretically respond in the same characteristics as the portion of the continuous pipe to be represented, due to seismic disturbance. In order to get an accurate finite element model, the following FE modeling guidelines should be followed:

i. The spacing of the lumps can be determined from Eq. (11.6) of the fundamental frequency of transverse vibration of uniform pipe for simply supported case.

$$f = \frac{0.5}{L^2}\sqrt{\frac{EIg}{W}} \tag{11.6}$$

where

- E Young's modulus of the material
- f natural frequency
- I moment of inertia
- W weight per unit length of the pipe (Including insulation weight and fluid weight)
- L distance between consecutive lump masses

The spacing of lumped masses can be obtained by using $f = 33$ Hz.

ii. There must be at least one lump mass between two supports. Similarly, there must be at least one lump mass between two bends also.
iii. Lump mass is required wherever there is concentrated or heavyweight such as valve or flange is located in the system.
iv. The distance between the first lump mass and a fixed end shall be L/2. Similarly, the masses next to simple support shall be about a distance of L/2 away from support.

11.26 Boundary Points of Piping System

Piping model is generally started from anchor point and also ended at anchor points. Generally, stiff equipment nozzles or fully constrained wall penetrations are treated as anchors. In most of the piping system, there are no well-defined anchor points as above, and this sometimes may result in very large FE model and difficult to analyze. In that scenario, piping can be separately modeled by decoupling or overlapping the model.

i. Branch lines can be separately modeled by decoupling from the model of the main line if the diameter of branch (d) pipe is much smaller as compared to the diameter of run pipe (D) and Eq. (11.7) condition is satisfied:

$$D^4 > 25d^4 \tag{11.7}$$

After decoupling as above, the following points should be also accounted in the separated models:

(a) In the analysis of branch pipe, displacement at junction with the run pipe should be applied at decoupled point.
(b) In the analysis of run pipe and the branch pipe analysis, it is necessary to use the branch stress intensification factor at decoupled node.
(c) Response spectrum at the elevation of support attachments on the run pipe close to the branch should be used.

ii. If the model is terminated at support point 'A' and the next model starts at support point 'B', the pipe section 'AB' should be common to both models. It is an overlap region that should contain at least two bilateral supports.

11.27 Seismic Anchor Movement (SAM)

This differential movement of piping system supported from two seismically independent structures that can move out of phase during a seismic event is termed as Seismic Anchor Movement (SAM). Due to SAM, piping will experience stresses, which is secondary in nature.

The contribution of SAM is more significant where:

i. The piping is supported from two seismically independent structures.
ii. The piping is connected to a large equipment which has its own modes shapes, e.g. steam generator, pressuriser, reactor vessel, or reactor coolant pump.

Conventional static analysis is performed to evaluate the component seismic stresses due to Seismic Anchor Movement (SAM). The resulting stresses due to SAM can be very large in case of large differential/relative movement of the supports. SAM may not exist if all the supports of a piping system supported at two or more points have uniform excitation. For multiple connected/supported systems with differential seismic excitation, support displacement can be obtained either from the response calculations of the supporting structure or from the spectral displacements determined from floor response spectra. The effect of relative SAM shall be obtained by using the most conservative combination of peak displacement. Stress due to SAM is cyclic in nature; it should be combined with other cyclic-type secondary stresses, viz. thermal expansion stresses by SRSS combination. However, ASME code gives freedom to combine stresses due to SAM with primary or secondary stress. The total secondary stress range should include the thermal and SAM stress range.

11.28 Required Inputs for Piping Stress Analysis

Piping stress analysis includes the calculation of stresses due to static and dynamic loading resulting from the effects of gravity, temperature changes, internal and external pressures, changes in fluid flow rate, and seismic event. For performing the stress analysis of a piping system, the following basic information is required:

i. Piping isometric drawing with following information:
   (a) Size (diameter, wall thickness and mass per unit length) of elbows, valve, flanges, reducers/expanders, and other non-standard fittings,
   (b) Material and material properties (density, Young's modulus, yield strength, ultimate tensile strength, allowable stress value, and coefficient of thermal expansion) of piping components at design and operating temperatures.
ii. Design, operating and transient pressures, and temperatures,
iii. Damping of piping system,
iv. Insulation thickness and insulation density or mass per unit length,
v. Applicable code,
vi. Specific gravity of working fluid,
vii. Seismic input (time history or response spectrum),
viii. Location of anchor points,
ix. Seismic/thermal anchor movement data if any.

With above information, Finite Element (FE) modelling of piping can be started, which is explained under following heading.

11.29 Hydraulic Consideration in Piping Design

11.29.1 Pressure Drop in Straight Pipe

Pressure drop in horizontally laid straight pipe with inside incompressible fluid, viz. water, can be calculated as given in Eq. (11.8):

$$\Delta P = \lambda \frac{lV^2\gamma}{2d} \tag{11.8}$$

Pipelines changing their level or elevation, pressure drop can be calculated by Eq. (11.9):

$$\Delta P = \lambda \frac{lV^2\gamma}{2d} \pm \gamma hg \tag{11.9}$$

where

ΔP	pressure drop in N/m^2
l	length of pipe in m
V	flow velocity in m/s
γ	specific weight in kg/m^3
g	acceleration due to gravity
h	geodetic height in m
λ	friction factor, to derived from Reynolds number

11.29.2 Pressure Drop in Valves and Fittings

The friction factor 'λ' exclusively applies to straight pipe but where the direction of flow is changed or flow is affected owing to the resistance of the bend, valves and other fittings losses are accounted by substituting the coefficient of resistance 'ζ' and pressure drop can be calculated by Eq. (11.10):

$$\Delta P = \zeta \frac{V^2\gamma}{2} \tag{11.10}$$

where

Coefficient of resistance $(\zeta) = \frac{l_0}{d}$

l_0 = length of straight pipe which causes same pressure drop as the fitting.

The typical values of coefficient of resistance for valves have been listed in Table 11.4; these values vary from design to design of valves.

11.30 Thrusts in Pipeline

Thrust is generated due to change in flow direction (bends), flow divisions (branches), and flow obstruction (valves, blind flanges). Thrust force (Fig. 11.7) due to internal pressure can be calculated by multiplying the pressure with area on which it acts.

$$\text{Pressure thrust for closed valves and blind flanges} = P \times A$$
$$\text{Pressure thrust for change of flow direction and pipe bends} = P \times A \times 2\sin(\theta/2) \tag{11.11}$$

where

θ angle of deviation of bend

Table 11.4 Coefficient of resistance ζ for valves

NB in mm	25	32	40	50	65	80	100	125	150	200
Straight way valves	4.0	4.2	4.4	4.5	4.7	4.8	4.8	4.5	4.1	3.6
Angle valves	2.8	3.0	3.3	3.5	3.7	3.9	3.8	3.3	2.7	2.0
NRV/butterfly valves	1.9	1.6	1.5	1.4	1.4	1.3	1.2	1.0	0.9	0.8

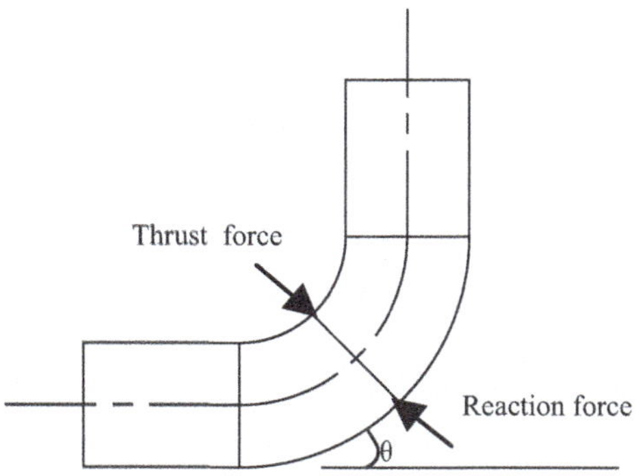

Fig. 11.7 Thrust force on elbow

$$\text{Dynamic thrust} = (2\sin(\theta/2) \times \dot{m} \times v\,(\text{kgf})$$

where

$\dot{m}$ mass flow rate (kg/s)
v mean velocity of fluid (m/s)

Since both the above thrusts act in same direction, the total thrust can be calculated by adding. Normally, dynamic thrust does not create any problem in welded pipes, but with dense fluids at high velocities, it becomes appreciable at change of direction and may require rigid supports or anchors.

11.30.1 Water Hammer in Pipelines

Due to sudden obstruction in flow passage caused by sudden closure of valves may lead to pressure rise in the pipe beyond normal working pressure is termed as water hammer. It propagates a series of shocks due to pressure waves moving back and forth, sounding like hammer blows, which may have sufficient energy which can rupture the pipe or damage the connected fittings. The water hammer causes rise in the pressure beyond normal working pressure, which can be calculated as:

$$h_{\max} = \frac{V_{\text{sw}} V_0}{g} \quad (11.12)$$

where

$h_{\max}$ pressure rise in m
V_{sw} velocity of sound in water under existing condition or velocity of pressure wave propagation and
V_0 normal velocity of flow before closure of valve

The velocity of pressure wave propagation can be calculated by Eq. (11.13) as:

$$V_{\text{sw}} = \frac{1}{\sqrt{\frac{\gamma}{g}\left(\frac{1}{k} + \frac{d}{tE}\right)}} \quad (11.13)$$

where

γ specific weight of fluid
K bulk modulus of fluid. K increases with increase in water pressure and temperature.2100 kg/cm^2 (for cold water up to 35 kg/cm^2 gauge pressure)
t wall thickness of pipe
d diameter of pipe
E modulus of elasticity of pipe material (for steel $E = 2.1 \times 10^6$ kg/cm^2)

For cold water,

$$V_{\text{sw}} = \frac{1440}{\sqrt{\left(1 + \frac{d}{100t}\right)}}\ \text{m/s} \quad (11.14)$$

Water hammer pressure is maximum, when the valve closing time 't_c' is critical. If actual time is $t_a > t_c$, the actual water hammer is reduced in proportion to t_c/t_a. Therefore, actual water hammer in m is given by

$$h_{\max} = \frac{V_{\text{sw}} V_0}{g} \times \frac{t_c}{t_a} \quad (11.15)$$

where

t_c 2 l/V_{sw} and
l length of pipe in m

Now, water hammer in principle can be reduced by

i. Reducing flow velocity,
ii. Selection of pipe diameter,
iii. Increasing t_a, i.e., closing valves slowly,
iv. Air chambers or accumulators are used for absorbing the shock.

11.31 Safety Classification of Piping of Nuclear Reactors

Nuclear power plant components and system that are important to safety are grouped into four safety classes depending upon the nature of nuclear safety function performed by them. The four safety classes of piping and equipments based on above criteria are safety class 1, safety class 2, safety class 3, and safety class 4 systems. The ASME Sec-III and associated subsections provide the design rules for safety-related systems and components.

11.32 Seismic Categorization

As per design safety guide AERB/SG/D-1, depending on the importance to safety during seismic event, the various piping and equipments of nuclear power plant (NPP) are classified into three seismic categories, viz. seismic category I, II, and III. In nuclear industry, dual earthquake philosophy is followed for qualifying the nuclear piping for seismic loading; i.e., qualification of piping will be carried out for both OBE as well as SSE.

11.33 Service Levels and Load Combinations for NPP Piping

In NPPs, the safety systems are subjected to a combination of loadings, which introduce different levels of stresses arising out of different service levels. For each service levels, system should be qualified to meet the stress requirement imposed by ASME code. According to the probability of their occurrences, there are four service levels described in ASME Sec-III, Subsection NCA, viz. Level A (normal), Level B (upset), Level C (emergency), and Level D (faulted). Required load combinations are tabulated in Table 11.5.

The combined stresses shall meet the relevant codal requirements of ASME Section III, Division-1, Subsection-NB, NC, or ND, whichever is applicable. The combined stresses in supports shall meet the requirements of ASME Section III, Division-1, and Subsection-NF. The combined stresses in the nozzle as calculated using WRC-297/WRC-107 or any other justifiable method shall meet the requirements of ASME Section-III, Division-1, Subsection NB, NC, or ND.

11.34 Damping Values Prescribed by Codes

Damping of a piping system includes combined effect of piping and its support system. In case of linear non-deforming type of supports, the major contribution to the damping comes from supports because of friction between different links and connection of supports. Damping value prescribed by codes for piping system has been listed in Table 11.6.

11.35 Piping Design

Aim of piping design is to ensure the safe operation of piping system under loading conditions postulated to occur during the lifetime of the plant, which is ensured by stress analysis. Stress analysis involves the calculation and comparison of permissible values of stress in the pipe wall, piping expansion, equipment nozzle loads, and system natural frequencies. In addition to above, the stress analysis encounters the determination of design loads for supports, so that system may be restrained safely. Design of every system is governed by the corresponding applicable codes and standards, which sets the minimum requirements for safe

Table 11.5 Load combinations for different service levels

Load classification	Design/service level	Load combination
Design	Design condition	Pressure, dead weight, sustained loads, temperature[a]
Normal	Service level A	Normal operating transients (pressure temperature, mechanical)
Upset	Service level B	Pressure, dead weight, sustained loads upset condition transients (pressure, temperature mechanical), along with OBE inertia and SAM
Emergency	Service level C	Pressure, dead weight, sustained loads, temperature[a], emergency condition transients (pressure, mechanical)
Faulted	Service level D	Pressure, dead weight, sustained loads, temperature[a], faulted condition transients (pressure, mechanical), pipe rupture loads, SSE

[a]Temperature effect is accounted for selecting the material properties at corresponding temperature

Table 11.6 Damping values (percentage of critical damping) for piping

System	USNRC RG 1.61		ASME Sec-III, Appendix N	
	OBE (%)	SSE (%)	OBE (%)	SSE (%)
Piping system	3	4	5	5
Bolted steel structures	3	4	4	7
Welded steel structure	3	4	4	4

Table 11.7 Common loads on piping system

Pressure	Internal pressure, external pressure
Weight	Material, fitting, insulation, operating, and test fluid and snow.
Thermal	Resistance to thermal expansion, thermal gradient load (time gradient, gradient at pipe cross section).
Dynamic	Vibration, impact, earthquake, relief valve discharge, and wind load.

design, construction, and operation of the concerned system. In piping industry, there are series of codes, administered under the authentication of ASME.

In the design and analysis of piping system, calculation of support loads may vary from very simple hand calculation to very sophisticated finite element analysis depending upon the complexity of the problem and criticalness of the piping system. The major loads considered for a piping system design have been tabulated in Table 11.7.

In addition to above, selection of material is also a very important factor of piping system design, which is mainly governed by nature of fluid working and environment (corrosive/non-corrosive, acidic/basic) and working fluid. Sometimes, selection of material is also governed by economy and availability of material.

11.36 Piping Qualification as per ASME Section-III, Div. 1, Subsection-NB

ASME Boiler and Pressure Vessel Code, Section- III, Div-1, Subsection NB details the requirements pertaining to those sections of nuclear piping designated as safety class 1. The loading requiring consideration in the design of piping under subsection NB are pressure, weights (live and dead loads), impact, earthquake, vibration, and loading induce by thermal expansion and contraction. The code design ensures pressure boundary integrity, not functionality of the pipe or other components. As per ASME code, design stress intensity for piping material under consideration should be the minimum of:

i. 1/3rd of the specified minimum ultimate tensile strength (UTS) at room temperature,
ii. 1/3rd of the UTS at temperature,
iii. 2/3rd of the specified minimum yield strength (YS) at room temperature,
iv. 2/3rd of the yield strength (YS) at temperature.

The stress due to different load combinations must be satisfied to meet the safety class 1 code criterion, which is as follows.

11.36.1 Primary Stress Intensity Check

The primary stress intensity limit is satisfied when requirement of Eq. (11.16) is met.

$$B_1 \frac{PD_0}{2t} + B_2 \frac{M_i D_o}{2I} \leq kS_m \tag{11.16}$$

where

B_1 and B_2	primary stress indices for the specific product under investigation.
P	internal design pressure (gauge), N/m^2
D_o	outside diameter of pipe, m (NB-3683)
t	nominal wall thickness, of product, m (NB-3683)
M_i	resultant moment due to a combination of design mechanical loads, N-m
I	moment of inertia, m^4.
K	1.5 for Level A,1.8 for Level B, and 2.25 for Level C
S_m	allowable design stress intensity value (N/m^2)

For branch connection or Tee, the moment term of Eq. (11.16) will be replaced by B_{2b} (M_b/Z_b) + B_{2r} (M_r/Z_r).

11.36.2 Primary Plus Secondary Stress Intensity Range Check

This is to account a stress range as the system goes from one load set (pressure, temperature, moment, and force loading) to any other set of which follows it in time. It is the range of pressure, temperature, and moments between two load sets, which is to be used in the calculations. For each specified pair of load sets, S_n is calculated by Eq. (11.17):

$$S_n = C_1 \frac{P_o D_0}{2t} + C_2 \frac{M_i D_o}{2I} + C_3 E_{ab} \times |\alpha_a T_a - \alpha_b T_b| \leq 3S_m \tag{11.17}$$

where

M_i resultant range of moment which occurs when system goes from one service load set to another

T_a, T_b range of average temperature on side ‘a’ or ‘b’ of gross structural discontinuity or material discontinuity
$\alpha_{a,b}$ coefficient of thermal expansion on side a or b of gross structural discontinuity or material discontinuity at room temperature
E_{ab} average modulus of elasticity of two sides of gross structural discontinuity or material discontinuity at room temperature
P_o range of service pressure

If Eq. (11.17) is not satisfied for all pairs of load sets, then the component may still be qualified by using the simplified elastic-plastic discontinuity analysis, explained below; otherwise, peak stress intensity range should be calculated by Eq. (11.23).

If S_n exceeds its limit for some pairs of load sets, simplified elastic plastic analysis may be performed if thermal stress ratchet is not present. This analysis is required only for the specific load sets that exceed the primary plus secondary stress intensity range check. The following two sets of Eq. (11.18) and (11.19) must be satisfied:

$$S_e = C_2 \frac{M_i D_o}{2I} \leq 3S_m \tag{11.18}$$

where

S_e expansion stress, and
M_i resultant range of moments due to thermal expansion and thermal anchor movements

$$C_1 \frac{P_o D_0}{2t} + C_2 \frac{M_i D_o}{2I} + C_3' E_{ab} \times |\alpha_a T_a - \alpha_b T_b| \leq 3S_m \tag{11.19}$$

where

M_i resultant range of moment which occurs when system goes from one service load set to another, excluding moments due to thermal expansion and thermal anchor movements
C_3' stress index for component under investigation

If $S_n > 3\ S_m$, the thermal stress ratchet must be evaluated and demonstrated to be satisfactory before a simplified elastic-plastic discontinuity analysis can be performed. This ratchet is function of $|\Delta T_1|$ range only. The following Eq. (11.20) must be checked for thermal ratchet.

$$|\Delta T_1|\text{range} \leq \frac{y' S_y C_4}{0.7E\alpha} \tag{11.20}$$

where

y' 3.33, 2.00, 1.20, and 0.80 for $x = 0.3, 0.5, 0.7$, and 0.8, respectively

$$x = \frac{PD_0}{2t} \times \frac{1}{S_y}, \tag{11.21}$$

where

P maximum pressure for condition under consideration
C_4 1.1 for ferrite and 1.3 for austenitic material
E modulus of elasticity at room temperature

11.36.3 Peak Stress Intensity Range and Fatigue Analysis

Average temperature of pipe wall can be calculated as follows:

$$\begin{aligned} T_{ave} &= \left(-\frac{1}{t}\right) \int_0^t T(x)\mathrm{d}x \\ \Delta T_1 &= T_i - T_o \\ \Delta T_2 &= (T_o + T_i)/2 \end{aligned} \tag{11.22}$$

where

t wall thickness, and
$T(x)$ temperature as function of distance through wall
T_i internal surface temperature
T_o external surface temperature

For each loading condition, specified peak stress value can be calculated by Eq. (11.23),

$$\begin{aligned} S_p = {} & K_1 C_1 \frac{P_o D_0}{2t} + K_2 C_2 \frac{M_i D_o}{2I} + \frac{1}{2(1-v)} K_3 E\alpha |\Delta T_1| \\ & + K_3 C_3 E_{ab} \times |\alpha_a T_a - \alpha_b T_b| + \frac{1}{1-v} E\alpha |\Delta T_2| \end{aligned} \tag{11.23}$$

where

v Poisson’s ratio of the material
$|\Delta T_1|$ absolute value of range of temperature difference between temperature of outside surface and inside surface of pipe wall, assuming moment generating equivalent linear temperature distribution
$|\Delta T_2|$ absolute value of range that portion of nonlinear thermal gradient through the wall thickness not

included in ΔT_1, assuming moment generating equivalent linear temperature distribution

For each S_p, alternating stress intensity can be calculated by Eq. (11.24):

$$S_{\text{alt}} = K_e \frac{S_p}{2} \tag{11.24}$$

where

$$K_e = \begin{cases} 1.0 & \text{For } S_n \leq 3S_m \\ 1.0 + \frac{1-n}{n(m-1)} \times \left(\frac{S_n}{S_m} - 1\right) & \text{For } 3S_m < S_n < 3_m S_m \\ \frac{1}{n} & For\ S_n \geq 3mS_m \end{cases}$$

Stress analysis needs to be performed using appropriate FE models for various loads, viz. pressure, thermal, mechanical, impact loads, weights, wind, vibrations, earthquake, reactions from supports. The load combinations to be adopted for ASME Class 1, 2, and 3 piping systems for design and service levels are given in Table 11.5.

11.36.4 Consideration of Faulted (Level D) Condition

The system loadings associated with the faulted condition refers SSE loading and to those dynamic loads which result from the occurrence of a postulated rupture of any piping which directly or indirectly endangers the integrity of reactor coolant pressure boundary. The individual loads that are to be considered for faulted condition are

(i) Accidental pressure loading: The permissible pressure shall not exceed 2.0 Pa.
(ii) Resultant moment M_i due to
   (a) Dead weight of piping, fluid weight,
   (b) Inertial loads due to SSE (E-W, N-S, vertical),
   (c) Pipe rupture loads.

The conditions of following Eq. (11.25) shall be met using Service Level D coincident pressure P and moment M_i, which results in the maximum calculated stress. The allowable stress to be used for this condition is 3.0 S_m, but not greater than 2.0 S_y.

$$B_1 \frac{PD_0}{2t} + B_2' \frac{M_i D_o}{2I} \leq \min(3S_m, 2S_y) \tag{11.25}$$

In year 1995, code permissible limit for Service Level D due to reversible dynamic loading typically experienced during SSE was revised upward from 3 to $4.5S_m$ (for above Eq. (11.25), and it is given by Eq. (11.26).

$$B_1 \frac{PD_0}{2t} + B_2 \frac{M}{Z} \leq 4.5S_m \tag{11.26}$$

where
For straight pipe, B_1 = 0.5 and B_2 = 1
For bends, B_1 = −0.1 +0.4 h and $B_2 = \frac{1.3}{h^{2/3}}$
where

h $= \frac{tR}{r_m^2}$
R bend radius
r_m mean radius of the pipe

However, year 2001 version of code has further brought it down to $3S_m$ limit with revised stress indices in code and is given by Eq. (11.27).

$$B_1 \frac{PD_0}{2t} + B_2' \frac{M}{Z} \leq 3S_m \tag{11.27}$$

where

B_2' $\frac{0.87}{h^{2/3}}$ for butt welding elbows

The sustained stress due to weight loading shall not exceed the following condition:

$$B_2 \frac{M_w}{Z} \leq 0.5S_m \tag{11.28}$$

where

M_W resultant moment due to weight effects

The range of the resultant moment M_{AM} and the amplitude of the longitudinal force F_{AM} resulting from the anchor motions due to earthquake and other reversing-type dynamic loading shall not exceed the following:

$$C_2 \frac{M_{AM}}{Z} \leq S_1 \tag{11.29}$$

where

S_1 $6S_m$

$$\frac{F_{AM}}{A_M} \leq S_2 \tag{11.30}$$

where

S_2 $1.0S_m$ and
A_M cross-sectional area of metal in the piping component wall

11.36.5 Test Condition

i. Test pressure is considered 25% more than design pressure.
ii. If actual test pressure < $1.25 \times P_{\mathrm{Des}} \times \frac{S_m\,(\text{at test Temp})}{S_m\,(\text{at design Temp})}$, then there is no need to perform test analysis separately.
iii. If test pressure exceeds by 6% of the value given by the code, then following checks need to be performed.

$$\text{(a)}\quad t_{\mathrm{m}} = \frac{P_{\mathrm{test}} D_{\mathrm{o}}}{2(S_m + yt)} + \mathrm{A},$$

where S_m = 0.9 S_y at test temperature

$$\text{(b)}\quad B_1 \frac{PD_0}{2t} + B_2 \frac{M_i D_{\mathrm{o}}}{2I} \le 1.35 S_y$$

Example 11.1 Typical industrial SSCs are shown in Fig. 11.8a. A straight SS 316 LN pipe (O.D. = 0.15 m, nominal wall thickness = 0.0065 m, length = 5 m) is running from a pump to vessel as shown in Fig. 11.8b. The following loads are acting on the pipe:

Design pressure	8.336 MPa
Design temperature	80 °C
Operating pressure	6.865 MPa
Operating temperature	75 °C

This pipe has been categorized as class 1. Design the pipe as per applicable code. Internal fluid density is 1000 kg/m^3.

Solution

For safety class 1 piping system, applicable design code is ASME Sec-III, Div 1, Subsection NB and system has to be designed for OBE and SSE both. Table 11.8 gives the loads and allowable stress as per ASME Sec-III, Div 1, Subsection NB.

Since the pipe is welded to the nozzles of the pump and vessel, pipe has been assumed as anchored at both ends.

For the analysis, the following material properties have been used at design temperature as per ASME Sec-II, Part D.

Tensile strength	:	515 MPa
Yield strength	:	205 MPa
Young's modulus	:	191.2 GPa
Allowable stress intensity	:	138 MPa
Coefficient of thermal expansion	:	15.93 × 10^{-6} mm/mm/°C

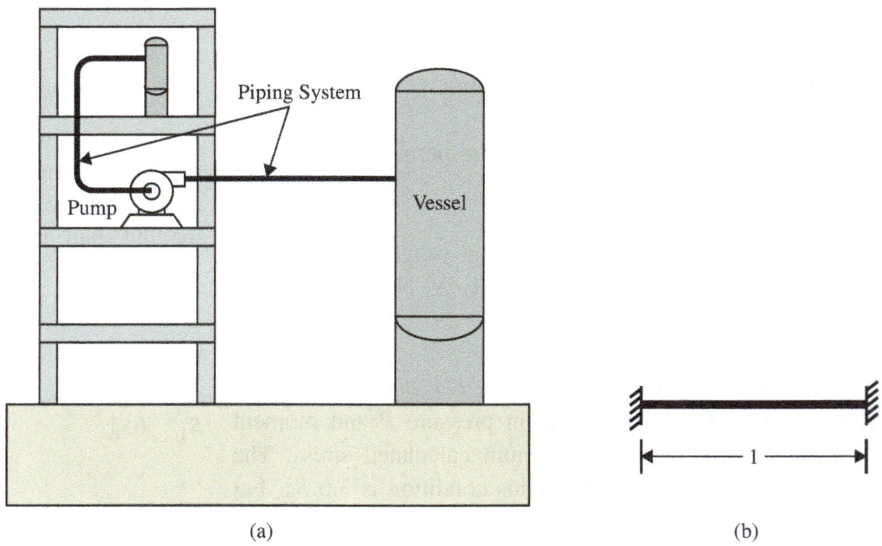

Fig. 11.8 **a** Typical industrial SSCs, **b** piping system

Table 11.8 Load combinations as per ASME Sec-III, Div 1, Subsection NB for different service levels

Load Case	Loads	Allowable	Equation
Design condition	Design pressure + dead weight	1.5 S_m	(11.16)
Normal operation/Level A	Operating pressure + thermal	3.0 S_m	(11.17)
Upset condition/Level B	Operating pressure + dead weight + OBE Operating Pressure + thermal + OBE	1.8 S_m 3.0 S_m	(11.16) (11.17)
Emergency condition/level C	Design pressure + dead weight + SSE	2.25 S_m	(11.16)

Table 11.9 Stress indices for pipe fittings

Piping component	h	B_1	B_2	C_1	C_2	C_3
Straight pipe	1.0	0.5	1.0	1.0	1.0	1.0

The allowable stress intensity at operating temperature is 138 MPa.

1. **Check for minimum thickness**:

$$t_m = \frac{PD_o}{2(S_m + Py)} + A$$

$$= \frac{8.336 \times 10^6 \times 0.15}{2(138 \times 10^6 + 8.336 \times 10^6 \times 0.4)} + 0 = 0.004424\,\text{m}$$

Actual thickness of pipe is more than the minimum required thickness.

2. **Design condition**:

For design condition, Eq. (11.16) need to be satisfied.

$$B_1 \frac{PD_0}{2t} + B_2 \frac{M_i D_o}{2I} \leq 1.5 S_m$$

Table 11.9 gives the stress indices for pipe fittings.

Total weight of the pipe including internal fluid is 190.92 kg. Maximum bending moment due to weight on pipe is

$$\frac{wl^2}{12} = \frac{374.59 \times 25}{12} = 780.39\,\text{Nm}$$

where

w weight per unit length and
l length

$$B_1 \frac{PD_0}{2t} + B_2 \frac{M_i D_o}{2I} \leq 1.5 S_m$$

$$0.5 \times \frac{8.336 \times 10^6 \times 0.15}{2 \times 0.0065} + 1 \times \frac{780.39 \times 0.15}{2 \times 7.56 \times 10^{-6}}$$

$$= 48.09 + 7.74 = 55.83\,\text{MPa}$$

Allowable stress for the design condition is 207 MPa.

3. **Service Level A**:

i. Since the pipe is straight, there will be no moment due to thermal expansion. However, thermal expansion will cause direct stress in pipe.

Direct stress due to thermal expansion

$$= E \times \alpha \times \Delta T$$

$$= 191.2 \times 10^9 \times 15.93 \times 10^{-6} \times 50$$

$$= 152.29\,\text{MPa}$$

ii. For Service Level A, Eq. 11.17 need to be satisfied.

$$C_1 \frac{P_o D_0}{2t} + C_2 \frac{M_i D_o}{2I} + C_3 E_{ab} \times |\alpha_a T_a - \alpha_b T_b| \leq 3 S_m$$

$$1 \times \frac{6.865 \times 10^6 \times 0.15}{2 \times 0.0065} + 0 + 1 \times 191.2 \times 10^9 \times 15.93 \times 10^{-6} \times 50$$

$$= 79.21 + 152.29 = 231.50\,\text{MPa}$$

Allowable stress for Service Level A is 414 MPa.

4. **Service Level B**:

i. Natural frequencies of pipe are (assumed as fixed-fixed beam)

$$f_1 = \frac{1}{2\pi} \times \frac{22.373}{l^2} \times \sqrt{\frac{EI}{\rho}} = 27.71\,\text{Hz}$$

$$f_2 = 2.757 \times f_1 = 76.40\,\text{Hz}$$

$$f_3 = 5.404 \times f_1 = 149.76\,\text{Hz}$$

Response spectra for SSE is shown in Fig. 11.9. OBE is taken half of SSE.

The acceleration attracted by the pipe obtained from Fig. 11.9 is 0.37/2 g in first mode and 0.36/2 g in second and other higher modes.

Maximum bending moment on pipe due to OBE excitation in lateral direction, assuming all the mass is excited in first mode, is

$$\frac{wl^2}{12} = \frac{38.18 \times 0.185 \times 9.81 \times 25}{12} = 144.36\,\text{Nm}$$

Bending moment on pipe due to OBE excitation in axial direction will be zero. Bending moment on pipe due to OBE excitation in all three direction is $\sqrt{M_1^2 + M_2^2 + M_3^2}$ (SRSS for spatial combination).

For Service Level B, Eq. (11.16) and Eq. (11.17) need to be satisfied.

$$B_1\frac{PD_0}{2t}+B_2\frac{M_iD_o}{2I}\leq 1.8S_m$$
$$0.5\times\frac{6.865\times10^6\times0.15}{2\times0.0065}+1\times\frac{(780.39+204.16)\times0.15}{2\times7.56\times10^{-6}}$$
$$=39.61+9.77=49.38\,\text{MPa}$$

Allowable stress for Service Level B is 248.4 MPa.

$$C_1\frac{P_oD_0}{2t}+C_2\frac{M_iD_o}{2I}+C_3E_{ab}\times|\alpha_aT_a-\alpha_bT_b|\leq 3S_m$$

$$1\times\frac{6.865\times10^6\times0.15}{2\times0.0065}+1\times\frac{204.16\times0.15}{2\times7.56\times10^{-6}}$$
$$+1\times191.2\times10^9\times15.93\times10^{-6}\times50$$
$$=79.21+2.03+152.29=233.53\,\text{MPa}$$

Allowable stress for Service Level B is 414 MPa.

5. **Service Level C**:

For Service Level C, Eq. (11.16) needs to be satisfied.

$$B_1\frac{PD_0}{2t}+B_2\frac{M_iD_o}{2I}\leq 2.25S_m$$
$$0.5\times\frac{8.336\times10^6\times0.15}{2\times0.0065}+1\times\frac{(780.39+408.31)\times0.15}{2\times7.56\times10^{-6}}$$
$$=48.09+11.79=59.88\,\text{MPa}$$

Allowable stress for Service Level C is 310.5 MPa.

Example 11.2 An SS 316 LN piping system (O. D. = 0.15 m, nominal wall thickness = 0.0065 m) shown in Fig. 11.10 is running from a pump to vessel. The following loads are acting on the piping system:

Design pressure	8.336 MPa
Design temperature	80 °C
Operating pressure	6.865 MPa
Operating temperature	75 °C

This pipe has been categorized as class 1. Design the pipe as per applicable code. Internal fluid density is 1000 kg/m^3.

Solution

For safety class 1 piping system, applicable design code is ASME Sec-III, Div 1, Subsection NB and system has to be designed for OBE and SSE both. Table 11.9 gives the loads and allowable stress as per ASME Sec-III, Div 1, Subsection NB.

For the analysis, the following material properties have been used at design temperature as per ASME Sec-II, Part D.

Tensile Strength	:	515 MPa
Yield strength	:	205 MPa
Young's modulus	:	191.2 GPa
Allowable stress intensity	:	138 MPa
Coefficient of thermal expansion	:	15.93×10^{-6} mm/mm/°C

The allowable stress intensity at operating temperature is 138 MPa.

1. **Check for Minimum Thickness**:

$$t_m=\frac{PD_o}{2(S_m+Py)}+A$$
$$=\frac{8.336\times10^6\times0.15}{2(138\times10^6+8.336\times10^6\times0.4)}$$
$$+0=0.004424\,\text{m}$$

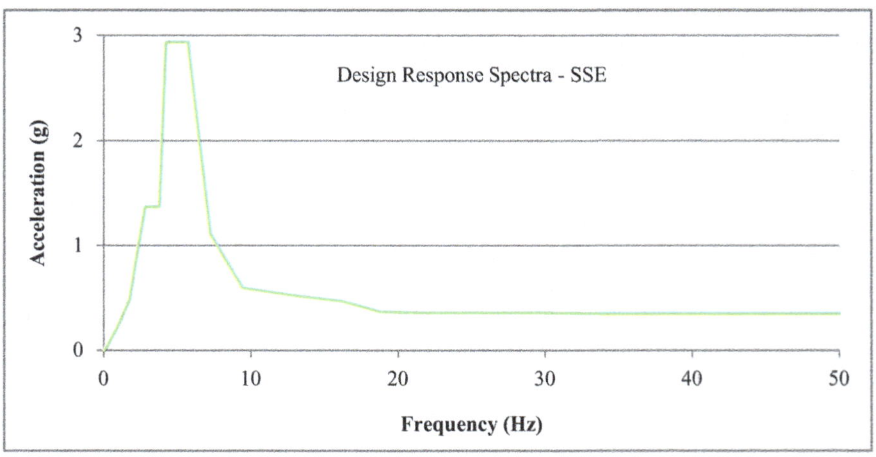

Fig. 11.9 Response spectra for SSE

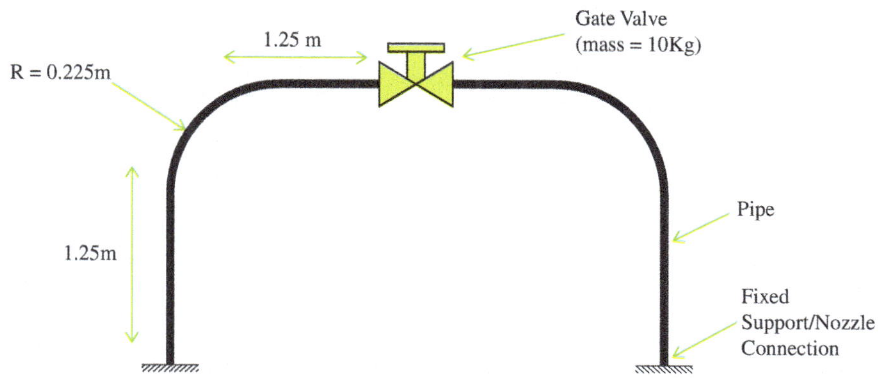

Fig. 11.10 Piping system

Table 11.10 Stress indices for pipe fittings

Piping component	h	B_1	B_2	C_1	C_2
Straight pipe	1.0	0.5	1.0	1.0	1.0
Elbow	0.284	0.014	3.008	1.234	4.512

Actual thickness of pipe is more than the minimum required thickness.

2. **Design Condition**:

For design condition, Eq. (11.16) needs to be satisfied

$$B_1 \frac{PD_0}{2t} + B_2 \frac{M_i D_o}{2I} \leq 1.5 S_m$$

Table 11.10 gives the stress indices for pipe fittings.

Bending moment due to weight on piping system is estimated by finite element analysis. Finite element model of the piping system is shown in Fig. 11.11. Bending moments at different elements of the piping system are listed in Table 11.11.

Straight Pipe:

$$0.5 \times \frac{8.336 \times 10^6 \times 0.15}{2 \times 0.0065} + 1 \times \frac{1094.5 \times 0.15}{2 \times 7.56 \times 10^{-6}}$$
$$= 48.09 + 10.86 = 58.95\,\text{MPa}$$

Elbow:

$$0.014 \times \frac{8.336 \times 10^6 \times 0.15}{2 \times 0.0065} + 3.008 \times \frac{657.80 \times 0.15}{2 \times 7.56 \times 10^{-6}}$$
$$= 1.35 + 19.63 = 20.98\,\text{MPa}$$

Allowable stress for the design condition is 207 MPa.

3. **Service Level A**:

For Service Level A, Eq. (11.17) needs to be satisfied.

$$C_1 \frac{P_o D_0}{2t} + C_2 \frac{M_i D_o}{2I} + C_3 E_{ab} \times |\alpha_a T_a - \alpha_b T_b| \leq 3 S_m$$

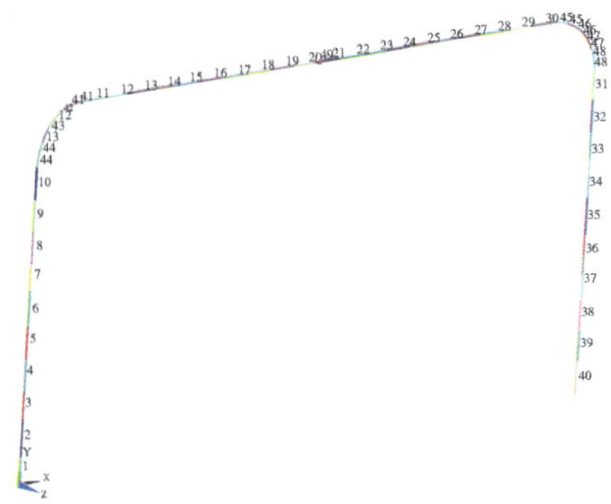

Fig. 11.11 Finite element model of piping system

Straight Pipe:

$$1 \times \frac{6.865 \times 10^6 \times 0.15}{2 \times 0.0065} + 1 \times \frac{2227.8 \times 0.15}{2 \times 7.56 \times 10^{-6}} + 0$$
$$= 79.21 + 22.10 = 101.31\,\text{MPa}$$

Elbow:

$$1.234 \times \frac{6.865 \times 10^6 \times 0.15}{2 \times 0.0065} + 4.512 \times \frac{2256.0 \times 0.15}{2 \times 7.56 \times 10^{-6}} + 0$$
$$= 97.75 + 100.98 = 198.73\,\text{MPa}$$

Allowable stress for Service Level A is 414 MPa.

4. **Service Level B**:

Modal analysis has been performed to find out natural frequencies, mode shapes, and mass participation in each mode. Some of the dominant modes have been listed in

Table 11.11 Bending moments at different elements of the piping system

Element	*Mx*	*My*	*Mz*	Element	*Mx*	*My*	*Mz*
1	−1094.50	0.00	91.84	25	−189.70	0.00	216.52
2	−1047.70	0.00	68.13	26	−189.70	0.00	181.13
3	−1000.80	0.00	44.41	27	−189.70	0.00	139.88
4	−954.01	0.00	20.70	28	−189.70	0.00	92.78
5	−907.19	0.00	−3.01	29	−189.70	0.00	39.82
6	−860.37	0.00	−26.72	30	−189.70	0.00	−18.99
7	−813.54	0.00	−50.43	31	−673.08	0.00	−121.57
8	−766.72	0.00	−74.15	32	−719.90	0.00	−97.86
9	−719.90	0.00	−97.86	33	−766.72	0.00	−74.15
10	−673.08	0.00	−121.57	34	−813.54	0.00	−50.43
11	−189.70	0.00	−18.99	35	−860.37	0.00	−26.72
12	−189.70	0.00	39.82	36	−907.19	0.00	−3.01
13	−189.70	0.00	92.78	37	−954.01	0.00	20.70
14	−189.70	0.00	139.88	38	−1000.80	0.00	44.41
15	−189.70	0.00	181.13	39	−1047.70	0.00	68.13
16	−189.70	0.00	216.52	40	−1094.50	0.00	91.84
17	−189.70	0.00	246.06	41	−190.96	0.00	−50.43
18	−189.70	0.00	269.75	42	−389.82	0.00	−93.30
19	−189.70	0.00	287.59	43	−553.01	0.00	−125.56
20	−189.70	0.00	299.57	44	−657.80	0.00	−140.60
21	−189.70	0.00	299.57	45	−190.96	0.00	−50.43
22	−189.70	0.00	287.59	46	−389.82	0.00	−93.30
23	−189.70	0.00	269.75	47	−553.01	0.00	−125.56
24	−189.70	0.00	246.06	48	−657.80	0.00	−140.60

Table 11.12 Frequencies and mass participation

S. no.	Frequency	*X*-direction		*Y*-direction		*Z*-direction	
		MP[a]	%MP[b]	MP	%MP	MP	%MP
1.	19.49	–	–	–	–	12.873	72.72
2.	27.76	13.394	78.72	–	–	–	–
3.	44.24	–	–	9.7837	42.00	–	–
4.	75.82	–	–	–	–	3.8188	06.40
5.	205.33	–	–	–	–	−4.2335	07.86
6.	255.44	−4.5305	09.01	–	–	–	–
7.	314.92	–	–	4.5708	09.17	–	–

[a]*MP* Mass participation in kg
[b]*%MP* percentage mass participation

Table 11.12. Some of the dominant modes of piping system are shown in Figs. 11.12, 11.13, 11.14 and 11.15.

For Service Level B, Eqs. (11.16) and (11.17) need to be satisfied.

$$B_1 \frac{PD_0}{2t} + B_2 \frac{M_i D_o}{2I} \leq 1.8 S_m$$

Maximum bending moment due to OBE on piping system is estimated by finite element analysis. Response spectra for SSE is shown in Fig. 11.9. OBE is taken half of SSE.

Straight Pipe:

$$0.5 \times \frac{6.865 \times 10^6 \times 0.15}{2 \times 0.0065} + 1 \times \frac{(1094.5 + 204.34) \times 0.15}{2 \times 7.56 \times 10^{-6}}$$
$$= 39.61 + 12.89 = 52.50\,\text{MPa}$$

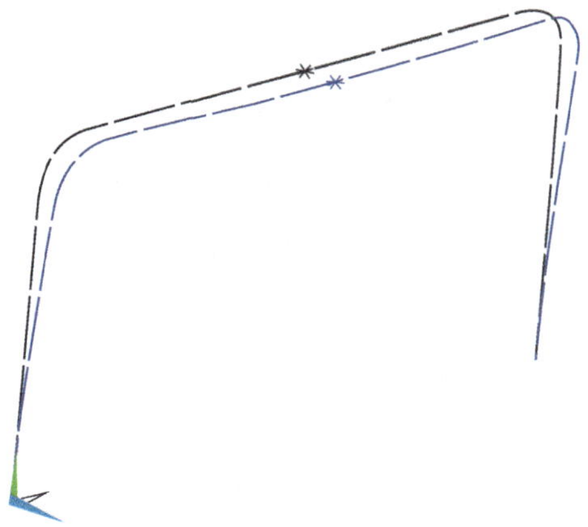

Fig. 11.12 Mode shape (frequency 19.49 Hz)

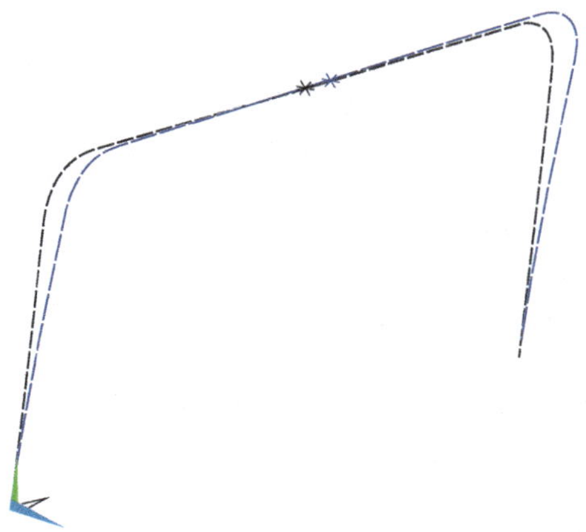

Fig. 11.13 Mode shape (frequency 27.76 Hz)

Elbow:

$$0.014 \times \frac{6.865 \times 10^6 \times 0.15}{2 \times 0.0065} + 1 \times \frac{(657.80 + 124.93) \times 0.15}{2 \times 7.56 \times 10^{-6}}$$
$$= 1.11 + 7.77 = 8.88\,\text{MPa}$$

Allowable stress for Service Level B is 248.4 MPa.

$$C_1 \frac{P_o D_0}{2t} + C_2 \frac{M_i D_o}{2I} + C_3 E_{ab} \times |\alpha_a T_a - \alpha_b T_b| \le 3S_m$$

Straight Pipe:

$$1 \times \frac{6.865 \times 10^6 \times 0.15}{2 \times 0.0065} + 1 \times \frac{(2227.8 + 204.34) \times 0.15}{2 \times 7.56 \times 10^{-6}} + 0$$
$$= 79.21 + 24.13 = 103.34\,\text{MPa}$$

Elbow:

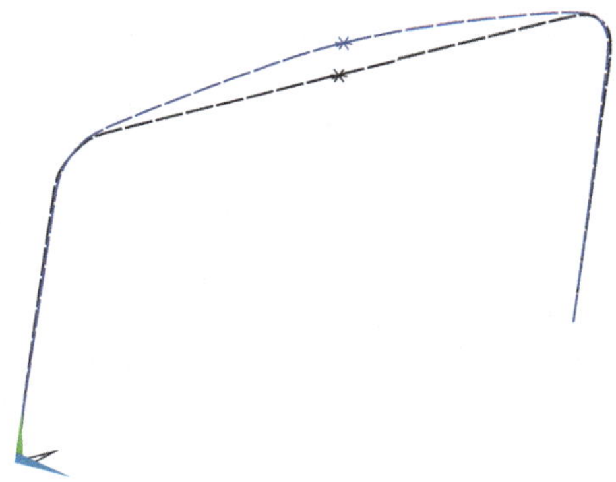

Fig. 11.14 Mode shape (frequency 44.24 Hz)

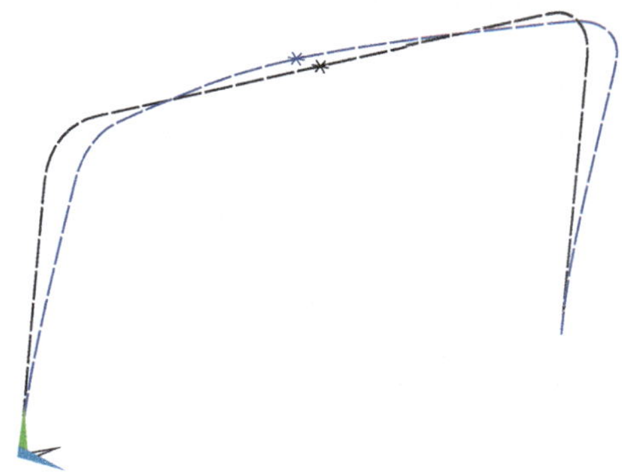

Fig. 11.15 Mode shape (frequency 75.82 Hz)

$$1.234 \times \frac{6.865 \times 10^6 \times 0.15}{2 \times 0.0065} + 4.512 \times \frac{(2256.0 + 124.93) \times 0.15}{2 \times 7.56 \times 10^{-6}} + 0$$
$$= 97.75 + 106.57 = 204.32\,\text{MPa}$$

Allowable stress for Service Level B is 414 MPa.

5. ***Service Level C***.

For Service Level C, Eq. (11.16) needs to be satisfied.

$$B_1 \frac{PD_0}{2t} + B_2 \frac{M_i D_o}{2I} \le 2.25 S_m$$

Straight Pipe:

$$0.5 \times \frac{6.865 \times 10^6 \times 0.15}{2 \times 0.0065} + 1 \times \frac{(1094.5 + 408.68) \times 0.15}{2 \times 7.56 \times 10^{-6}}$$
$$= 39.61 + 14.91 = 54.52\,\text{MPa}$$

Elbow:

$$0.014 \times \frac{6.865 \times 10^6 \times 0.15}{2 \times 0.0065} + 1 \times \frac{(657.80 + 249.86) \times 0.15}{2 \times 7.56 \times 10^{-6}}$$
$$= 1.11 + 9.00 = 10.11\,\text{MPa}$$

Allowable stress for Service Level C is 310.5 MPa.

Example 11.3 The following are the loads acting on a straight portion of a piping system. Estimate the stresses and check the design as per ASME Sec III, Div. 1, Subsection NB.

Data

Seamless pipe SS 304, OD = 24″, nominal wall thickness = 2.5″

Design pressure = 2485 psi

S_m = 20,000 psi at 70 °F, and $E\alpha$ = 258 at 70 °F

S_m = 15,600 psi at 600 °F

$y = 0.4$ and $A = 0$

Hydrotest pressure = 3590 psi

Stress indices for straight pipe

B_1	C_1	K_1	B_2	C_2	K_2	C_3	K_3
0.5	1.1	1.2	1.0	1.0	1.8	1.0	1.7

Moment	Dead weight (lb-in)	OBE (lb-in)	Max (DW + OBE) lb-m
M_x	+59.7 × 10^3	±29 × 10^3	88.7 × 10^3
M_y	−281.2 × 10^3	±58.9 × 10^3	340.1 × 10^3
M_z	527.8 × 10^3	±458.9 × 10^3	986.4 × 10^3

Note These are the moments acting on the straight pipe, obtained by finite element analysis

Solution

Design Check as per ASME Sec III, Div. 1, Subsection NB:

$$\text{Thickness of pipe excluding tolerances} = 0.875 \times 2.5''$$
$$= 2.1875''(\text{allowances } 12.5\%)$$

Moment of inertia = 9890in^4

Check 1: min wall thickness = $t_m = \frac{PD_0}{2(S_m + Py)} + A = \frac{2485 \times 24}{2(15600 + 2485 \times 0.4)} + 0 = 1.797''$

⇒ Here, 2.188″ > 1.797″: **Design for pressure is ok**

Check 2: Test Condition: P_{test}(theoretical) = $1.25 \times P_{Des} \times \frac{S_m(\text{at test Temp})}{S_m(\text{at design Temp})}$

$$1.25 \times 2485 \times \frac{20000}{15600} = 3968\,\text{psi}$$

- ⇒ P_{test} (actual) < P_{test} (theoretical): No special analysis needs to be performed for test condition separately.

Check Service Level A (Eq. (11.16)):

$$(M_i) = 10^3 \times \sqrt{88.7^2 + 340.1^2 + 986.4^2}$$
$$= 1047.15 \times 10^3\,\text{lb} - \text{in}$$

$$B_1 \frac{PD_0}{2t} + B_2 \frac{M_i D_o}{2I} \leq 1.5 S_m$$

$$0.5 \frac{2485 \times 24}{2 \times 2.5} + 1.0 \frac{1047.15 \times 10^3 \times 24}{2 \times 9890} = 5964 + 1270.55$$
$$= 7234.55\,\text{psi}$$

$$1.5 S_m = 1.5 \times 15,600 = 23,400\,\text{psi}$$

$$1.5 S_m = 1.5 \times 15,600 = 23,400\,\text{psi}$$

$1.5S_m$ = 1.5×15,600 = 23,400 psi

Here, 7234.55 psi < $1.5S_m$ hence Ok

Check Service Level A (Eq. (11.17)):

Let us take one case where range of moment from one service level to other is given as follows:

M_x = 1014.2 × 10^3 lb-in,
M_y = 4104.2 × 10^3 lb-in, and
M_z = 5025 × 10^3 lb-in.

Pressure range = 3,590 psi.
ΔT_1 = 56 and ΔT_2 = 10

$$(M_i) = 10^3 \times \sqrt{1014.2 + 4104.5^2 + 5025^2}$$
$$= 65670 \times 10^3\,\text{lb} - \text{in}$$

$$S_n = C_1 \frac{P_o D_o}{2t} + C_2 \frac{M_i D_o}{2I} + C_3 E_{ab} \times |\alpha_a T_a - \alpha_b T_b| \leq 3S_m$$

$$S_n = 1.1 \frac{3590 \times 24}{2 \times 2.5} + 1.0 \frac{3567 \times 10^3 \times 24}{2 \times 9890}$$
$$+ 1.0 \times 258 \times 0 \leq 18950 + 7968 + 0$$
$$= 26918\,\text{psi}, < 3 \times S_m = 3 \times 15600 = 46800\,\text{psi}$$

Hence, requirement of Eq (11.17) is met.

Check- Service Level A (Peak stresses-fatigue check)

$$S_p = K_1 C_1 \frac{P_o D_0}{2t} + K_2 C_2 \frac{M_i D_o}{2I} + \frac{1}{2(1-\nu)} K_3 E\alpha |\Delta T_1|$$
$$+ K_3 C_3 E_{ab} \times |\alpha_a T_a - \alpha_b T_b| + \frac{1}{1-\nu} E\alpha |\Delta T|_2$$

$$S_p = 1.2 \times 1.1 \frac{3590 \times 24}{2 \times 2.5} + 1.8 \times 1.0 \frac{6567 \times 10^3 \times 24}{2 \times 9890}$$
$$+ \frac{1}{2(1 - .3)} \times 1.7 \times 258 \times 56 + 0$$
$$+ \frac{1}{1 - .3} \times 258 \times 10 = 58308\,\text{psi}$$

Calculate S_{alt} stress from Eq. (11.24)

$$S_{alt} = K_e \frac{S_p}{2} = 1.0 \frac{58308}{2} = 29154\,\text{psi}$$

Read from S–N curve number of allowable cycles (N) corresponding to S_{alt} = 34,000 and say actual cycles (n) are 18,495; then, cumulative usage factor = n/N = 0.05

$n/N \ll 1$ hence component is safe under fatigue loading.

11.37 Piping Design and Qualification as per ASME, Sec-III, Div-1, Subsection-NC/ND

ASME Section-III, Div-1, Subsection-NC, and ND cover the design and qualification of NPP systems components and piping which falls under safety class 2. Under the following subheadings, design and qualification of piping have been explained as per ASME Sec-III, Div-1, Subsection-NC/ND.

11.37.1 Design Pressure

The design, service pressure shall be determined by considering anticipated or postulated conditions to occur during the intended service life of the component. The specified internal and external design pressure shall not be less than the maximum difference in pressure between the inside and outside of the item, or between any two chambers of a combination unit, which exists under the most severe loadings for which the Level A service limits are applicable. The design pressure shall include allowances for pressure surges, control system error, and system configuration effects such as static pressure heads. Minimum required thickness of pipe can be calculated using following code formula:

$$t_m = \frac{PD_o}{2(S + Py)} + A \qquad (11.31)$$

where

- P allowable pressure
- D_o outer diameter of pipe
- S maximum allowable stress for the material at the design temperature
- A additional thickness to compensate for threading or grooving, to provide for mechanical strength of the pipe

Since small diameter, thin wall pipe or tubing is susceptible to mechanical damage during erection, operation, and maintenance; appropriate means must be employed to protect such piping against these types of loads if they are not considered as design loads. Increased wall thickness is one way of contributing to resistance against mechanical damage. Since corrosion and erosion vary widely from installation to installation, it is the responsibility of designers to determine the proper amounts which must be added for either or both of these conditions.

Example 11.4 Design pressure (P) = 105 bar,

Design temperature = 171.1 °C

Outside diameter (D_o) = 168.275 mm for 150 NB pipe, y = 0.4,

Material: SS 304L

Allowable stress at design temperature (S) = 112.38 MPa at 171.1 °C

From above data, calculated t_m = 7.58 mm. After adding 12.5% recommended tolerance, required thickness (t_{req}) = 8.53 mm.

- For above-required thickness, available pipe schedule is 80S whose thickness is 10.97 mm.

In addition to above-selected pipe schedule must meet following requirements:

i. For Level $B < 1.1\ P_a$,
ii. For Level $D < 2.0\ P_a$.

P_a = maximum allowable pressure calculated

t = actual wall thickness—allowances (tolerance and corrosion/erosion/threading). Here, t = 10.97 × 0.875 = 9.6 mm

$$P_a = \frac{2St}{(D_0 - 2yt)}$$
$$= \frac{2 \times 112.38 \times 10^6 \times 9.6 \times 10^{-3}}{(168.27 \times 10^{-3} - 2 \times 0.4 \times 9.6 \times 10^{-3})}$$
$$= 134.36\,\text{bar}$$

<210 bar (i.e., 105 bar × 2 = 210 bar)

- Selecting next schedule 160 for which thickness = 18.26 mm. Now, Pa = 267.1 bar > 210 bar; hence, 150 NB schedule 160 pipe will be suitable for this application.

11.37.2 Design Temperature

The specified design temperature shall not be less than the expected maximum mean metal temperature through the thickness of the part considered for which Level A service limits are specified. Where a component is heated by trace heating, such as induction coils, jacketing, or by internal heat generation, the effect of such heat input shall be considered in establishing the design temperature. The design temperature shall consider control system error and system configuration effects.

11.37.3 Design Mechanical Loads

The specified design mechanical loads shall be selected such that, when combined with the effects of design pressure, they produce the highest primary stresses of any coincident combination of loadings for which Level A service limits are designated in the design specification. The design specification for flanges and fittings shall meet ASME B16.5 and B16.9 requirements, respectively.

11.38 ASME Sec-III, Div-1, Subsection-NC/ND Code Compliance Checks

ASME Sec-III, Div-1, Subsection-NC and ND, respectively, govern the qualification procedure for nuclear safety class 2 and safety class 3 piping and equipments:

11.38.1 Design Condition

The effects of pressure, weight, and other sustained mechanical loads must meet the requirements of following Eq. (11.32):

$$B_1 \frac{PD_0}{2t_n} + B_2 \frac{M_A}{Z} \leq 1.5S_h \tag{11.32}$$

where

D_o	nominal OD
t_n	nominal thickness
Z	section modulus
B_1, B_2	stress indices
M_A	moment due to weight, other sustained loads
S_h	allowable stress at temperature
P	design pressure

11.38.2 Occasional Loads

The effects of pressure, weight, other sustained loads, and occasional loads, including non-reversing dynamic loads, for which Level B service limits are designated, must meet the following requirements given by Eq. (11.33).

$$B_1 \frac{P_{max} D_0}{2t_n} + B_2 \frac{M_A + M_B}{Z} \leq \min(1.8S_h, 1.5S_y) \tag{11.33}$$

where

P_{max}	design or maximum pressure
M_B	moment due to Service Level B occasional loads
S_y	yield stress at temperature

11.38.3 Thermal Expansion

For service loadings for which Level A and B service limits are designated, the requirements of either of Eqs. (11.34), or (11.36) and Eq. (11.35) are given as follows.

(a) The effects of thermal expansion must meet the requirements of Eq. (11.34):

$$i \frac{M_c}{Z} \leq S_A \tag{11.34}$$

where

i	stress intensification factor
S_A	allowable stress range ($S_A = f\,(1.25S_h + 0.25S_c)$)

(b) The effects of any single non-repeated anchor movement shall meet the requirements of Eq. (11.35):

$$i \frac{M_D}{Z} \leq 3S_C \tag{11.35}$$

where

M_D	resultant moment due to any single non-repeated anchor movement (e.g., predicted building settlement)

(c) The effects of pressure, weight, other sustained loads, and thermal expansion shall meet the requirements of Eq. (11.36):

$$\frac{P_{max} D_0}{4t_n} + 0.75i \frac{M_A}{Z} + i \frac{M_c}{Z} \leq S_h + S_A \tag{11.36}$$

$0.75i$ shall not be less than 1.0.

(d) The effects of reversing dynamic loads must meet the requirements of Eq. (11.37):

$$i\frac{M_R}{Z} \leq 2.0S_A \qquad (11.37)$$

where

M_R range of resultant moments due to inertia and anchor motion effects of reversing dynamic loads

11.38.4 Consideration of Level D Service Limits

(a) The permissible pressure shall not exceed 2.0 times the pressure P_a calculated in accordance with Eq. (11.3).

(b) The conditions of Eq. (11.38) shall be met using service Level D coincident pressure 'P' and moment (M_A + M_B), which result in the maximum calculated stress. The allowable stress to be used for this condition is $3.0S_h$, but not greater than $2.0S_y$.

$$B_1\frac{P_{max}D_0}{2t_n} + B_2\frac{M_A + M_B}{Z} \leq \min(3.0S_h, 2.0S_y) \qquad (11.38)$$

(c) The sustained stress due to weight loading shall not exceed the limit of the following Eq. (11.39):

$$B_2\frac{M_W}{Z} \leq 0.5S_m, \qquad (11.39)$$

where

M_W resultant moment due to weight effects.

(d) The stress due to weight and inertial loading due to reversing dynamic loads in combination with the Level D coincident pressure shall not exceed the limit of Eq. (11.40):

$$B_1\frac{P_D D_0}{t} + B_2'\frac{M_E}{Z} \leq 3S_m \qquad (11.40)$$

where

B_2' $0.87/h^{2/3}$ for curved pipe or butt welding elbows (≥ 1.0)

B_2' $0.27(R_m/T_r)^{2/3}$ and $B_{2r}' = 0.33(R_m/T_r)^{2/3}$ for butt welding tees (≥ 1.0)

P_D the pressure occurring coincident with the reversing dynamic load

M_E the amplitude of the resultant moment due to the inertial loading from the earthquake, other reversing-type dynamic events, and weight

(e) The range of the resultant moment M_{AM} and the amplitude of the longitudinal force F_{AM} resulting from the anchor motions due to earthquake and other reversing-type dynamic loading shall not exceed the following:

$$C_2\frac{M_{AM}}{Z} \leq S_1 \qquad (11.41)$$

where

S_1 $6S_m$

$$\frac{F_{AM}}{A_M} \leq S_2 \qquad (11.42)$$

where

S_2 $1.0S_m$ and

AM cross-sectional area of metal in the piping component wall

Table 11.13 provides the comparison of ASME Sec III, Div-1, Subsection-NB/NC/ND and ASME Sec-VIII, Div-1.

11.39 Stress Indices

They are defined as ratio of significant stress in the product and nominal stress in the component under consideration.

$$\text{Stress Index} = \frac{\text{Significant Stress}}{\text{Nominal Stress}} \qquad (11.43)$$

For example, in case of thermal stress, significant stress will be maximum stress in the component and nominal stress equals $E\alpha\Delta T$.

Stress indices are categorized into three. Type one indices are called the primary stress indices (B_1 and B_2) used for primary loads such as pressure, dead weight. They are obtained by limit load analysis. Type two indices are called secondary stress indices (C_1 and C_2) and used for secondary for secondary loading such as thermal. Type three indices are called peak stress indices (K_1, K_2, and K_3) and used for fatigue analysis. Table 11.14 provides the stress indices for piping components.

Table 11.13 Comparison of ASME Sec-III, Div-1, Subsection-NB/NC/ND and ASME Sec-VIII, Div-1

Code	ASME, Sec-III, NB	ASME, Sec-III, NC	ASME, Sec-III, ND	ASME Sec-VIII, Div-1
Applicable for	For nuclear class 1 components	For nuclear class 2 components	For nuclear class 3 components	For non-nuclear components (fired/unfired)
Maximum allowable pressure	Not given	Not given	Not given	3000 psi
Maximum permissible temp	≤ 800 °F	≤ 800 °F	≤ 800 °F	≤ 1500 °F
Discontinuity stress	Stress indices used	Stress indices used	Stress indices used	Not evaluated
Fatigue analysis	Required	Not performed	Not performed	Not performed
Failure theory	Maximum shear stress	Maximum principal stress	Maximum principal stress	Maximum principal stress
Design by	Analysis	Analysis	Analysis	Rules
Failure mode	Fatigue, creep, tearing, crushing	Fatigue, creep, tearing, crushing	Fatigue, creep, tearing, crushing	Yielding and buckling
Exp joints	Not permitted	Can be used	Can be used	Can be used
Allowable stress	Min ($2/3\sigma_y$, $1/3\ \sigma_u$)	Min ($2/3\sigma_y$, ¼ σ_u)	Min ($2/3\sigma_y$, ¼ σ_u)	Min ($2/3\sigma_y$, $1/3.5\ \sigma_u$)
Weld joint inspection	100% radiography	Not mandatory	Not mandatory	Not mandatory

Table 11.14 Stress indices for piping components

Index type	Formula		Remarks
	Straight pipe	Elbow	
Primary	$B_1 = 0.5$ $B_2 = 1.0$	$B_1 = -0.1 + 0.4$ h $B_2 = 1.3/\mathrm{h}^{(2/3)}$	$0 \leq B_1 \leq 0.5$, $B_2 \geq 1.0$ and valid for $D_o/t \leq 50$
Secondary	$C_1 = 1.0$ $C_2 = 1.0$	$C_1 = (2R - r_m)/(2(R - r_m)$ $C_2 = 1.95/\mathrm{h}^{2/3}$	$C_1 \geq 1.0$, $C_2 > 1.5$, for elbow and valid for $D_o/t \leq 100$
Intensification factor (i)	1.0	$i = 0.9/\mathrm{h}^{(2/3)}$	

11.40 Support Qualification

Supports are qualified as per ASME Sec–III, Division-1; Subsection–NF–3320. Forces and moments obtained from finite element analysis of piping have been used together with supports sectional properties to check the stresses in the support for different loading conditions.

11.40.1 Stress in Shear

Allowable stress in shear can be calculated as follows:

$$F_V = 0.4 S_y \tag{11.44}$$

11.40.2 Stresses in Axial Compression

The allowable stress in axial compression for other than austenitic stainless steel is computed as follows:

$$F_a = \frac{\left\{\left[1 - (kl/r)^2/2C_c^2\right]\right\}S_y}{5/3 + \left[3(kl/r)/8C_c\right] - \left[(kl/r)^3/8C_c^3\right]} \tag{11.45}$$

where (For $kl/r < C_c$) $C_c = \sqrt{2\pi^2 E/S_y}$

$$F_a = \frac{12\pi^2 E}{23(kl/r)^2} \tag{11.46}$$

For $kl/r > 120$.

$$F_{as} = \frac{F_a[(\mathrm{Eq.}(4)\mathrm{or}(5))]}{1.6 - \left(\frac{1}{200r}\right)} \tag{11.47}$$

For $kl/r > 120$.
where

kl/r (buckling factor)	(effective length/min. radius of gyration)
K	effective length factor
l	length of support
r	radius of gyration $(I_{min}/A)^{1/2}$

S_y basic material yield strength
E modulus of elasticity

11.40.3 Stress in Axial Tension

The allowable stress in axial tension is computed as follows:

$$F_t = 0.6\,S_y \tag{11.48}$$

11.40.4 Combined Compression and Bending Stresses

For the supports subjected to both axial compression and bending, including bending moment resulting in secondary stress shall be proportioned to satisfy the requirement of Eqs. (11.49), (11.50), and (11.51):

$$F_b = 0.66\,S_y \tag{11.49}$$

$$\frac{f_a}{F_a} + \frac{C_{mx} f_{bx}}{(1 - f_a/F'_{ex})F_{bx}} + \frac{C_{my} f_{by}}{(1 - f_a/F_{ey})F_{by}} \leq 1 \tag{11.50}$$

$$\frac{f_a}{0.60 S_y} + \frac{f_{bx}}{F_{bx}} + \frac{f_{by}}{F_{by}} \leq 1 \tag{11.51}$$

After evaluating primary stresses, the right-hand side of Eq. (11.51) may be replaced by 1.5 when both primary and secondary bending stresses are considered. When $f_a/F_a \leq 0.15$, Eq. (11.51) may be used in lieu of Eqs. (11.49) and (11.50):

$$\frac{f_a}{F_a} + \frac{f_{bx}}{F_{bx}} + \frac{f_{by}}{F_{by}} \leq 1 \tag{11.52}$$

In Eqs. (11.49), (11.50), and (11.51), the subscript x and y indicate the axis of bending about which a particular stress or design property applies, and:

$$F'_e = \frac{12\pi^2 E}{23(kl_b/r_b)^2} \tag{11.53}$$

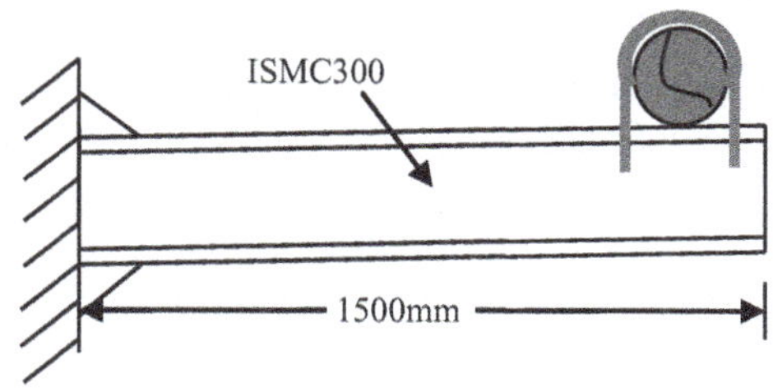

Fig. 11.16 Pipe support

where

l_b actual length in plane of bending
r_b radius of gyration
C_m is a coefficient whose value is 0.85 for compression in frames subjected to joint translation (sideways)

The effect of combined axial tension and bending is checked by Eq. (11.52) where f_b is the computed axial tensile stress. The allowable stress values are multiplied by appropriate stress limit factors for different loading categories as per Table NF- 3623 (b) –1 (Table 11.15).

Example 11.5 Qualification of support has been done by taking maximum force on the support (Fig. 11.16).

Data:
Maximum force is acting
F_X = 1159 N, F_Y = 3138 N, F_Z = 2007 N.
Resultant force = 3901 N.
Length of support = 1.50 m
Cross section = ISMC300
Area of cross section = 45.64 cm^2
Thickness of web = 7.6 mm.
I_{XX} = 6362.6 cm^4
I_{YY} = 310.8 cm^4
Yield strength of material: 250 × 10^6 N/m^2
Young's modulus: 200 × 10^9 N/m^2

Solution

Bending Stress

Actual bending stress = $\frac{MY}{I} = \frac{(3901\times1.5)\times0.150}{6362.6\times10^{-8}}$
f_a = 9.20 × 10^6 N/m^2

Table 11.15 Stress limit factors (NF –3623 (b) –1)

Stress category	Design condition	Service level A	Service level B	Service level C
Primary Stresses	K_s = 1.0	K_s = 1.0	K_s = 1.33	K_s = 1.5
	K_v = 1.0	K_v = 1.0	K_v = 1.33	K_v = 1.5
	K_{bk} = 1.0	K_{bk} = 1.0	K_{bk} = 1.33	K_{bk} = 1.5

where
K_s = stress limit factor applicable to the design allowable tensile and bending stresses
K_v = stress limit factor applicable to the design allowable shear stresses
K_{bk} = stress limit factor applicable to the design allowable compressive axial and bending stresses to determine buckling limit

Actual bending stress is much below the allowable stress; hence, support is safe under bending load.

Axial tension check

Allowable stress $= 0.6S_y$

$= 150 \times 10^6\,\mathrm{N/m^2}$

Actual tensile stress $= \frac{F}{A} = \frac{3901}{45.64 \times 10^{-4}}$

$= 0.855 \times 10^6\,\mathrm{N/m^2}$

Actual tensile stress is much below the allowable stress; hence, support is safe under tensile load.

Combined axial compression and bending

Axial compression

Allowable stress in axial compression: $C_c = \sqrt{\frac{2\pi^2 E}{S_y}} = 125.66$

Slenderness ratio $kl/r = 0.25 \times 1.5/0.02605$

$= 14.40$

$C_c \gg kl/r$, so applicable equation is (15) NF-3322.1

$$F_a = \frac{\left\{\left[1 - (kl/r)^2/2C_c^2\right]\right\}S_y}{5/3 + \left[3(kl/r)/8C_c\right] - \left[(kl/r)^3/8C_c^3\right]}$$

$$= \frac{\left\{\left[1 - (14.40)^2/125.66^2\right]\right\} \times 250 \times 10^6}{5/3 + [3 \times 14.40/8 \times 125.66] - [14.40^3/8 \times 125.66^3]}$$

$$= 144.33X10^6\,\mathrm{N/m^2}$$

Axial Bending

$$F_b = 0.66S_y$$

$$= 165 \times 10^6\,\mathrm{N/m^2}$$

$$f_a/F_a = 9.20 \times 10^6/144.33 \times 10^6 = 0.064$$

$f_a/F_a = 0.064 \ll 0.22$, hence Eq. (21) NF-3322.1 will be used, i.e.,

$$\frac{f_a}{F_a} + \frac{f_{bx}}{F_{bx}} + \frac{f_{by}}{F_{by}} \le 1$$

$$\frac{9.20 \times 10^6}{144.33 \times 10^6} + 0.00 + 0.00 = 0.064$$

$$\lll 1$$

Hence, support 7141-PS-2 is safe under combined compression and bending load.

11.41 Design and Analysis of Non-nuclear Piping

A piping system is essentially a network of piping of various sizes along with different types of pipe fittings, e.g., valves, flanges, reducers and supports. The piping layout need to be prepared considering equipment layout, system requirements, operating conditions, etc., while taking other factors like space constraints and maintenance and access requirements also into account. Since these requirements vary from plant to plant, the piping systems also differ from one another in terms of their layout, pipe sizes, pipe fittings and valves and in terms of space constraints and approach for maintenance. Thus, standardization of neither a piping layout nor its supporting arrangement is practically possible, which is in contrast to the design of pressure vessels.

The designer of the industrial piping system needs to fulfill the requirements imposed by the applicable design code. In addition, contradictory requirements are imposed on the piping systems, needing it to have, on one hand, sufficient stiffness to withstand the loads (due to dead weights and seismic loads, etc.), and, on the other hand, to have sufficient flexibility to absorb the thermal expansions of the piping and connected equipment without overstressing the equipment or piping. Since the piping layout is essentially dictated by plant requirements, judicious supporting arrangement appears as the major factor in achieving the requirements imposed on piping. Thus, the task for piping designer is quite challenging, in the sense that he has to manage the location and nature of supporting to cater to the various loads imposed on the piping, within the available space and constraints.

For a designer, the design of supports mainly is limited to selection process. The finalisation of supports and their location is an iterative process. As a first step, supporting arrangement is planned to cater to the major primary load, taking clue from the behavior of unsupported system under the load. The adequacy of these supports to cater to other loads such as secondary loads, seismic load and occasional loads are verified subsequently. If the supporting is found inadequate to cater to these loads, the designer needs to redefine the supporting by providing additional supports, readjustment of existing support locations, modifying the nature of support at some locations or by a combination of the above. Satisfactory behavior of the system under the various loads is the essence for the designer to arrive at an optimum support arrangement, and hence, his knowledge of the behavior of piping system needs to be good enough to enable him to arrive at the best supporting arrangement.

11.42 Piping Qualification as per ASME, B 31.1 Code

Industrial piping systems are subjected to internal pressure, temperature, and weight loadings under normal operating conditions. The loadings have been categorized into two groups:

(a) Primary stresses due to dead weight and internal pressure, and
(b) Thermal stresses due to temperature loadings.

The stresses due to these loadings are to be limited to the allowable limits as per the power piping code ASME B31.1. Equations used for the stress checking are as given below:

11.42.1 Stress Due to Sustained Loads

The effects of pressure, weight, and other sustained mechanical loads shall meet the requirement of Eq. (11.54)

$$SL = \frac{PD_o}{4t_n} + 0.75i\frac{M_A}{Z} \leq 1.0S_h \quad (11.54)$$

where

P design pressure
D_o outside pipe dia
M_A resultant moment loading due to weight and other sustained loads
Z section modulus
i stress intensification factor (the product $0.75i$ shall never be taken <1.0)
S_L sum of the longitudinal stresses due to pressure, wt, and other sustained loads
S_h allowable stress of the piping material at operating temperature

11.42.2 Thermal Expansion Stress Range

The effects of thermal expansion range shall meet the following requirement Eq. (11.55)

$$S_E = i\frac{M_C}{Z} \leq [S_A + f(S_h - S_L)] \quad (11.55)$$

where

S_A is allowable stress for thermal stresses and defined as $S_A = f\,(1.25S_C + 0.25\ S_h)$
M_C range of resultant moments due to thermal Expansion
Z section modulus
i stress intensification factor
f stress range reduction factor for cycling conditions for total number N of full temperature cycle
S_L sum of the longitudinal stresses due to pressure, weight, and other sustained mechanical loads
S_h allowable stress of the piping material at operating temperature
S_c allowable stress of the piping material at room temperature

If Eqs. (11.54) and (11.55) are satisfied, then the layout will be flexible enough to take care of stresses due to sustained loads and thermal expansion loads.

11.42.3 Seismic Analysis

This code specifies an increase of 15% on allowable stress for sustained loadings to take care of occasional loads for the loads acting for no more than 8 h at any one time and no more than 800 h/year. Further, it specifies an increase of 20% on allowable stress for sustained loadings to take care of occasional loads for the loads acting for no more than 1 h at any one time and no more than 80 h/year. Illustrative example on piping qualification as per SME B31.1 is given in Annexure I.

11.43 Piping Qualification as per ASME, B 31.3

Design and qualification of process piping are covered by ASME B31.3. Design requirement of petroleum refineries, chemical, pharmaceutical, textile, paper, semiconductor, and cryogenic plants is mainly covered by this code. This code has vast scope of application as compared to any other codes of the ASME B31 for pressure piping. It covers the design of pipelines carrying nonhazardous fluids such as water and as hazardous as mustard gas. It also covers for vast range of design temperatures, i.e., subzero (cryogenic conditions) to 815 °C and beyond. For pressure design, the pressure ranges vacuum and atmospheric to 3400 bar and higher. Some very low-pressure piping system is excluded from the scope of this code if they meet certain pressure, temperature, and inflammability requirements.

11.43.1 Materials

Applicable materials in this code are those which are listed in the allowable stress tables of the code and those permitted in component standards listed in the code in Table 326.1 (or A326.1 or K326.1, of the code as appropriate).

11.43.2 Design Conditions and Criteria

Design conditions in ASME B31.3 are specifically intended for pressure design. The design pressure and temperature are the most severe coincident conditions. Design conditions are not intended to be a combination of the highest potential pressure and the highest potential temperature, unless such conditions occur at the same time.

11.43.3 Allowable Stress

i. 1/3rd of specified minimum tensile strength at room temperature,
ii. 1/3rd of tensile strength at temperature,
iii. 2/3rd of specified minimum yield strength at room temperature,
iv. Average stress for a minimum creep rate of 0.01%/ 1000 h,
v. 2/3rd of average stress for creep rupture in 100,000 h,
vi. 80% of minimum stress for creep rupture in 100,000 h,
vii. For austenitic stainless steel and nickel alloys, allowable value may be as high as 90% of the yield strength at temperature. (to account the strain hardening).

11.43.4 Allowances for Pressure and Temperature Variations

Increases in pressure and temperature above the design conditions are permitted for short-term events as long as several conditions are satisfied, one of which is that this maximum allowable working pressure is not exceeded by more than some percentage. Thus, the variation can be much higher than the design conditions, yet remain permissible. The following conditions are requirements for use of the variations:

i. For pressure-containing piping system and components, cast iron or other non-ductile materials should not be used.
ii. The nominal hoop stress for straight pipe must be less than the yield strength of the material.
iii. The longitudinal stresses must be within the normally permitted limits.
iv. Over the life of the system, the total number of pressure–temperature variations above the design conditions must be less than 1000.
v. The maximum pressure must be less than the test pressure; this can be a limitation if pneumatic or alternative leak testing was used.

Meeting above conditions, the pressure rating may be exceeded by 33% for events that are not more than 10 h at any one time not more than 100 h per year and by 20% for events that are not more than 50 h at any one time nor more than 500 h per year.

11.43.5 Pressure Design (Internal Pressure)

The required minimum thickness to be taken for a particular piping system, considering manufacturer's tolerance, must be at least equal to t_m, ($t_m = t + c$)
where

$$t = \frac{PD}{2(SEW + PY)}, \quad \text{for pipe with } t < D/6 \qquad (11.56)$$

where

c sum of the mechanical allowances plus corrosion and erosion allowances
t pressure design thickness
D outer diameter of pipe
E quality factor (for electric resistance weld pipe $E = 0.85$)
P internal design gauge pressure
S allowable stress value
W weld joint strength reduction factor
Y material coefficient (for austenitic steel, $Y = 0.4$, up to 566 °C)

11.44 Design for Sustained and Occasional Loads

11.44.1 Longitudinal Load

i. Longitudinal stresses in the piping due to pressure, weight, and other sustained loading are denoted by 'S_L.' The sum of the longitudinal stress should not exceed S_h.
ii. The thickness used for calculating the longitudinal stresses should be obtained by subtracting the mechanical, corrosion, and erosion allowances from the nominal thickness of pipe. Unlike pressure design, it is not required to account for mill tolerance of the piping.

11.44.2 Occasional Loads

The combination of the longitudinal stresses caused by pressure, weight, and other sustained loads and of the stresses due to occasional loads such as earthquake or wind shall not exceed $1.33S_h$. Since earthquake and wind loads are low probable events, therefore they need not be combined.

11.44.3 Design Criteria for Thermal Expansion

Allowable stress for stresses due to resistance to thermal expansion and other secondary stresses is substantially higher than that of sustained loads. This is due to the differences in the nature of stresses produced by these two loading cases. The former is load-controlled, non-self-limiting, whereas later is deformation-controlled and self-limiting. The displacement stress range (S_E) shall not exceed S_A, i.e.,

$$S_E < S_A \quad (11.57)$$

where

- S_E $(S_b^2 + 4S_t^2)^{1/2}$
- S_b resultant bending stress, $[(i_iM_i)^2 + (i_oM_o)^2]^{1/2}/Z$
- M_i in-plane bending moment
- M_o out-of-plane bending moment
- i_i in-plane stress intensification factor
- i_o out-of-plane stress intensification factor obtained
- S_t torsional stress $(M_t/(2Z)$,
- M_t torsional moment,

The allowable thermal expansion stress is designed to result in shakedown to elastic behavior after a few numbers of operating cycles. The allowable stress value for stresses induces due to resistance to thermal expansion can be calculated as follows:

$$S_A = f(1.25S_c + 0.25S_h) \quad (11.58)$$

where

- S_A allowable displacement stress range
- S_c basic allowable stress at the minimum metal temperature expected during the displacement cycle under analysis.
- S_h basic allowable stress at the maximum metal temperature expected during the displacement cycle under analysis
- f Stress range reduction factor stress range reduction factor

The values of S_c and S_h do not include weld joint quality factors; however, casting quality factors, E_c, must be included.

The following equation is detailed equation that considers the magnitude of the sustained longitudinal stress:

$$S_A = f[1.25(S_c + S_h) - S_L)] \quad (11.59)$$

where

- S_L longitudinal stress due to sustained loadings

Here, allowable thermal expansion stress range can exceed the yield strength for the material, because both S_c and S_h may be as high as 2/3rd of the yield strength. However, it is anticipated that the piping system will shake down to elastic behavior if the stress range is within this limit. Even though the stress range is limited to maximum yield strength of the material due to elastic shakedown, there remains the potential for fatigue failure if there are sufficient numbers of a stress cycles. Therefore, the stress range reduction factor 'f' is used to reduce the allowable stress range when numbers of stress cycles exceed 7000. This number is based on one stress cycle per day for twenty years.

11.44.4 Pressure Testing

Hydrotest is generally the preferred alternative because it is conducted at a higher pressure, which has beneficial effects such as crack blunting and warm pressurising, and entails substantially less risk than a pneumatic test. These reduce the risk of crack growth and brittle fracture after the hydrotest when the pipe is placed in service. The test is generally conducted at a pressure of P_T and P_T is given by following equation:

$$P_T = 1.5P\frac{S_T}{S} \quad (11.60)$$

where

- S_T allowable stress at test temperature
- S allowable stress at design temperature
- P design pressure

Here, ratio of S_T/S cannot be less than 1.0 and cannot exceed 6.5.

Exercise Problems

1. What is difference between tube and pipe?
2. What standard method to denote the size of pipe?
3. From which size onwards NB of pipe is equal to OD of pipe?
4. Define standards and non-standard elbows.
5. Explain in the major differences between code ASME B31.1 and code ASME B31.3

6. Explain the significance of "Stress Intensification Factor (SIF)"? How many types of these SIFs exist?
7. What is the effect of pressure on flexibility of a piping?
8. How water hammer is detrimental to piping? Explain various ways to minimize the pressure shock due to water hammer.
9. Explain pipe friction and different reasons of pressure drop in piping due to flow.
10. Which ASME code is followed for design of piping systems in process piping (refineries and chemical industries)?
11. Normally where do we use (a) eccentric reducers (b) concentric reducers
12. What are the various standard requirements to be considered during layout?
13. What do you mean by jacketed piping and steam traced piping?
14. What is the minimum distance to be maintained between two welds in a pipe?
15. What is advantage of pressure hydrotest? How much should be the time and pressure for hydrotest?
16. How do you calculate the pipe spacing?
17. What is maximum sagging allowed in piping due to self-weight?
18. What are the steps involved in stress analysis (or any stress package carries out)?
19. What are the different types of stresses that may get generated within pipe during normal operation?
20. What are the sources of sustained loads generated in piping system?
21. Mention some of primary loads
22. Mention some of secondary loads
23. What is the failure theory subscribed under ASME B31.3?
24. How do you calculate the stress developed due to thermal expansion?
25. What is cumulative damage fraction and where it is used in piping design?
26. What is seismic anchor movement (SAM) in piping? Explain the method of calculating the stresses due to SAM and combining the same with other loads as per ASME Sec-III, NB code?
27. What is the basic span of supports for 4″,5″,8″,12″ water pipe?
28. What is the difference between variable spring hanger and constant spring hanger in their design and application?
29. Differentiate between static load and dynamic load.
30. Enumerate different types of dynamic loads expected on piping with examples.
31. Define primary and secondary loads and enumerate expected primary and secondary loads in piping.
32. How to calculate the basic allowable stress for a material in ASME B31.3 code?
33. A pipe with 250 mm diameter delivers a liquid with specific gravity and viscosity 0.95 and 0.23, respectively, at a vessel located 70 m away from the pump. Find the pressure difference from one end to the other if the velocity of flow is 1.4 m/s.
34. Define the working principle of snubbers and associated drawbacks of using snubber support in piping
35. Explain the method of checking the fatigue strength of safety class 1 piping.
36. What is simplified elasto-plastic analysis of Class 1 piping?
37. Explain the method considering the occasional loads in petrochemical piping?
38. What is difference between consideration of occasional loads for pressure piping and petrochemical piping?
39. Define the method for calculating the allowable stress on a piping support subjected to only axial load?
40. What is vortex shedding in flow-induced vibration? Why flow-induced vibration is more dangerous for tall structures as compared to short structures? With the help of sketches, explain any four methods of controlling flow-induced vibrations in structures?
41. Explain briefly on flow-induced vibrations with examples.
42. What are the most commonly used performance levels? Briefly explain the performance-based design for a structure or equipment or piping system.

Appendix 1: Piping Qualification as per ASME B31.1

Example 11.6 A typical piping is shown in Fig. 11.17, and the design and operation conditions are given below:

Pipe size	= 2″ Sch 40
Outer diameter (D_o)	= 60.3 mm and wall thickness(t) = 3.91 mm
Sectionmodulus	$= 9.1757 \times 10^{-6}\,\mathrm{m}^3$
Mass per unit length	= 5.44 kg/m
Radius of elbow (R)	= 76.2 mm
Material	SS304L(ASME Class 2, SA − 312, seamless & welded pipe)

(Tables 11.16 and 11.17)

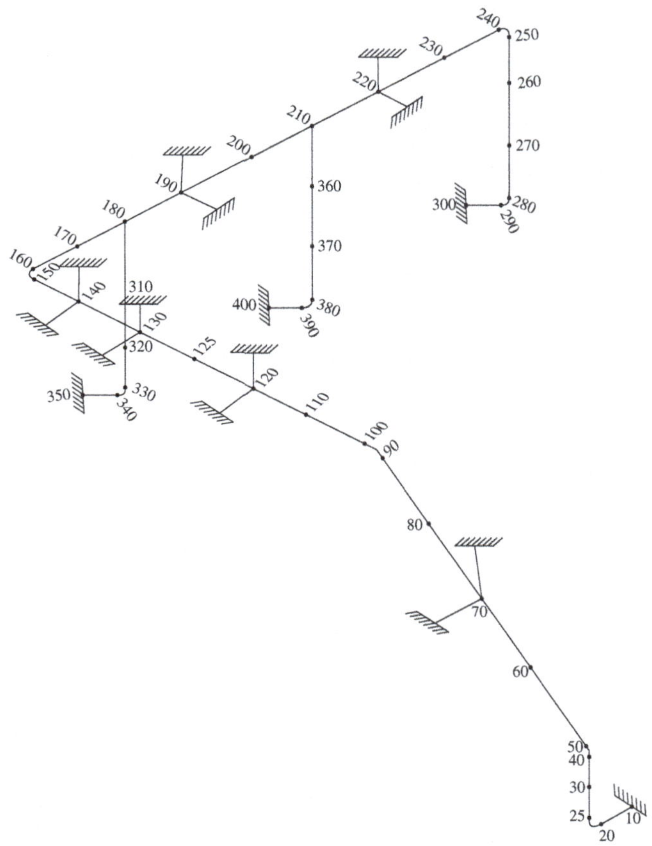

Fig. 11.17 FE model of a typical piping

Solution Elbow parameter $(h) = tR/(r_m)^2$

r_m = mean pipe radius = $(D_o - t)/2 = (60.3 - 3.91)/2 = 28.195$ mm

$h = 0.375$

Stress intensification factor $(i) = 0.9/h^{2/3} = 1.73$

Qualification Check for sustained load:

$$SL = \frac{PD_o}{4t_n} + 0.75i\frac{M_A}{Z} \leq 1.0S_h$$

$$\frac{1.18 \times 10^5 \times 0.0603}{4 \times 0.00391} + 0.75 \times 1.73 \times \frac{6}{9.1757 \times 10^{-6}}$$

$$= 454948.85 + 848436.63$$

$$= 1303385.48\,\text{N/m}^2 = 1.30\,\text{MPa}\,(S_L)$$

Actual stress is much lower than the allowable stress ($S_h = 115 \times 10^6$ N/m^2); hence, piping is safe under sustained loads.

Qualification Check for Thermal Load:

$$S_E = i\frac{M_C}{Z} \leq [S_A + f(S_h - S_L)]$$

$$\text{Allowable stress for thermal loads } (S_A) = f\,(1.25S_C + 0.25\,S_h)$$
$$= 1 \times (1.25 \times 115 + 0.25 \times 115)$$
$$= 172.5\,\text{MPa}$$

$S_A + f(S_h - S_L) = 172.5 + 1 \times (115 - 1.30) = 286.2$ MPa

Stresses in piping due to thermal expansion

$$i\frac{M_C}{Z} = 1.73 \times \frac{\left\{(-114)^2 + (18)^2 + (-190)^2\right\}^{\frac{1}{2}}}{9.1757 \times 10^{-6}}$$

$$= 1.73 \times \frac{222.31}{9.1757 \times 10^{-6}} = 41.91 \times 10^6\,\text{N/m}^2$$

Actual stress is lower than the allowable stress; hence, piping is safe under thermal loads.

Table 11.16 Material properties

Temperature (°C)	Young's modulus (GPa)	Coefficient of thermal expansion (mm/mm/°C)	Allowable stress intensity (S_h)(MPa)
30	195.1	15.23×10^{-6}	115
93	190.3	15.82×10^{-6}	115
149	186.2	16.20×10^{-6}	115
204	182.7	16.54×10^{-6}	110
260	177.9	16.87×10^{-6}	103
316	174.4	17.15×10^{-6}	97

Table 11.17 Forces and moment at critical section of pipe

Node no.	El. type	M_x	M_y	M_z	Load case
390	LR elbow	0	0	6	Dead weight
		−114	18	−190	Thermal

References

1. Thomas JVL, Smith PR (1987) Piping and pipe support systems: design and engineering. McGraw-Hill
2. Mohinder L, Nayyar PE (2000) Piping hand book. McGraw-Hill
3. Sahu GK (1998) Hand book of piping design. New Age International
4. Atomic Energy Regulatory Board Safety Guide AERB/SG/D-1 (2002)
5. ASME Section III, Division-1, Subsections-NB-3000, NC-3000 or ND-3000 (1995)
6. ASME Section III, Division-1, Subsection-NF (2001)
7. ASME Section III, Division-1, Subsection-NB-3000, NC-3000 or ND-3000 (2001)
8. ASME Section-III-Appendix-N-Article N 1230 (2001)

Seismic Qualification of Structures, Systems, and Components by Test

12

G. R. Reddy and R. K. Verma

In general testing is the qualification method for active equipment, to validate numerical methods of analysis and margin assessment

Symbols

ω_a, ω_b	Circular frequency
x	Displacement
c	Damping coefficient
k	Stiffness
τ_d	Time period
f	Frequency
δ	Logarithmic decrement
ϕ	Phase angle

12.1 Introduction

The general methods of seismic qualification of structures, systems, and components (SSCs) are grouped into the following categories:

a. Analysis,
b. Testing,
c. Combination of analysis and testing,
d. Use of experienced data.

The choice should be based on the practicality of the method for the type, size, shape, and complexity of the SSCs. Structural integrity and certain serviceability requirements such as leak tightness can be fulfilled by analysis, but the functional requirements of equipment especially rotating mechanical components, electrical components may be difficult in most cases. In addition to this, understanding the performance levels of the structures and equipment in plastic conditions becomes sometimes difficult. In these situations, testing or combination of analysis and testing of structures, systems, and components have become one of the important activities in the complete design process of the industrial facilities.

The standard IEEE 344 discusses the details of testing for seismic qualification of Class 1E equipment for nuclear power generating stations [1]. The procedures described in the standard can be followed in general for all the SSCs, and the important aspects of testing are discussed here.

Structures are generally very large in size and weight, and it is not possible to test full-scale structure on existing shake tables in the world. The largest shake table existing in the world is of 1200 ton capacity at Kobe, Japan. Even on this table, full-scale small residential buildings may be possible to test but not the industrial buildings. For example, the important safety-related structure in the nuclear power plant is the containment, which will have weight to the level of 20,000 tons. Frame-type safety-related structure will be to the tune of 2000 tons and above. In addition, simulation of boundary conditions on the test table is difficult which depend on the foundations and the soil/rock supporting it. Due to these reasons, structures are generally qualified using well-established and proven analytical procedures.

However, to find out the ultimate capacity of the structures, a procedure established by testing scaled models is used. While scaling, the dynamic characteristics like frequencies, mode shapes have to be maintained. The natural frequencies

G. R. Reddy (✉) · R. K. Verma
Bhabha Atomic Research Centre, Mumbai, India
e-mail: rssred@barc.gov.in

R. K. Verma
e-mail: rajeev@barc.gov.in

G. R. Reddy et al., *Textbook of Seismic Design*,
https://doi.org/10.1007/978-981-13-3176-3_12

of structures depend on the stiffness and mass distribution which has been discussed in detail in earlier chapters.

The natural frequencies of the scaled model are generally higher than the actual. In these situations, if the frequencies of the structure need to be matched with the actual structure for having the same response acceleration, sufficient mass may be added to the model. Otherwise, the response spectra shape is modified to see the same response acceleration as that of full structure at the model frequencies.

The disadvantage of the first case is that it may end up with the requirement of large mass which may exceed the capacity of the table. In the second case, it may not be possible to excite higher modes. In any case, one needs to take proper care while scaling the model. Figure 12.1 shows a train of glove boxes tested on a 3 × 3 m shake table. In case of active equipment, systems like electrical and control systems, it may be possible to test full-scale model. Otherwise, same methodology as explained above can be adopted.

12.2 Vibration Measurement and Signal Analysis

The topic measurement and signal processing play very important role in the subject of vibration and seismic engineering. Vibrations in structures, systems, and components occur due to either or combined of the following loads:

1. Impact loads due to missiles, shipping, and handling or due to accidental drops,
2. Sinusoidal loads due to unbalanced masses in rotating equipment,
3. Random loads on vehicles traveling over rough roads or aerodynamic loads in aircraft/spacecraft or rail loads in rail transportation,
4. Seismic loads during earthquakes.

As explained in the earlier chapters, signals such as displacement, velocity, and accelerations are very important to be measured to understand the vibration caused by above loadings in structures, systems, and components. The various sensors to record these signals are called displacement, velocity, and acceleration sensors. Figure 12.2 explains various steps involved in signal acquiring and processing.

12.2.1 Accelerometers

An accelerometer is a device that measures the vibration, or acceleration of motion of a structure, system, or component. If the natural frequency of the instrument is very high compared to the disturbance frequency, it can be used to measure the acceleration. Figure 12.3 shows the variation of dynamic response factor with frequency ratio. Variation of dynamic response factor with frequency ratio for a damping ratio of 0.7 is insignificant up to 0.2; hence, the useful frequency ratio is about 0.2. Thus, an instrument with a natural frequency of 100 Hz has a useful frequency of about 0–20 Hz.

Fig. 12.1 Train of glove boxes tested on 3 × 3 m shake table

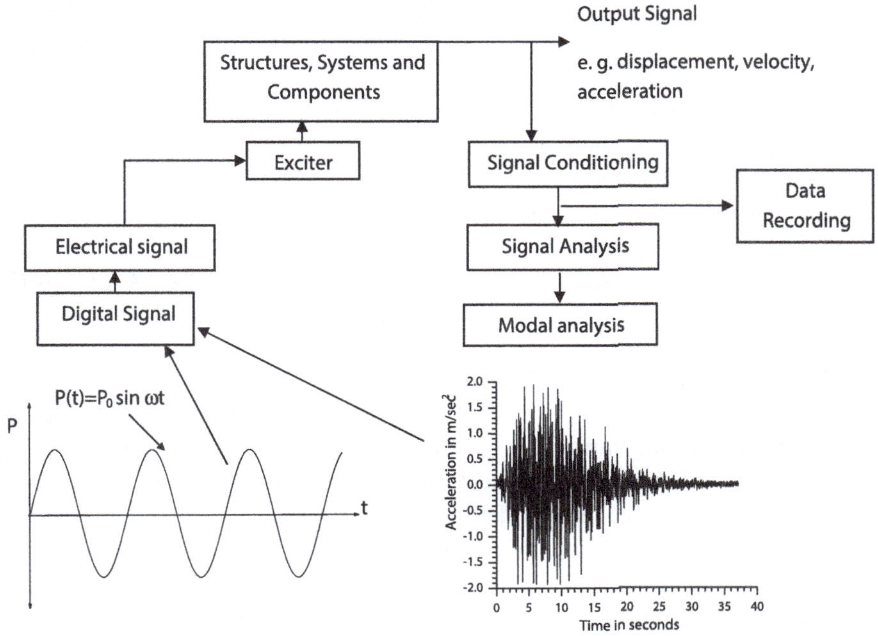

Fig. 12.2 Steps involved in signal acquiring and processing

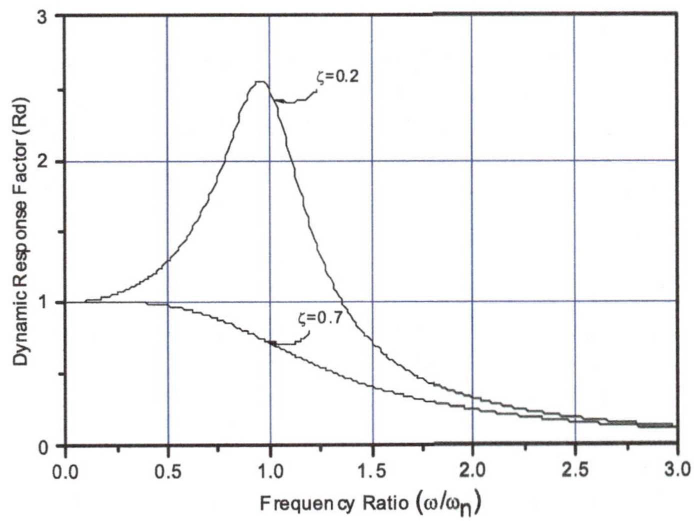

Fig. 12.3 Variation of dynamic response factor with frequency ratio

Electromagnetic-type accelerometers generally utilize damping ratio around 0.7. Very high natural frequency instruments such as piezoelectric crystal accelerometers operate without distortion up to frequency of 0.06 times natural frequency. The natural frequency of such instruments is about 50,000 Hz.

12.2.2 Seismometer

It is an instrument with low natural frequency. When the natural frequency ω_n of the instrument is low in comparison with the vibration frequency ω to be measured, the ratio approaches large number and the relative displacement approaches to the excitation regardless of the damping. The mass supported on spring becomes stationary, while the supporting case moves with the vibrating body. Such instruments are called seismometers. It is of large size. The relative motion is usually converted to an electric voltage by making the seismic mass a magnet moving relative to coils fixed at the case as shown in Fig. 12.4.

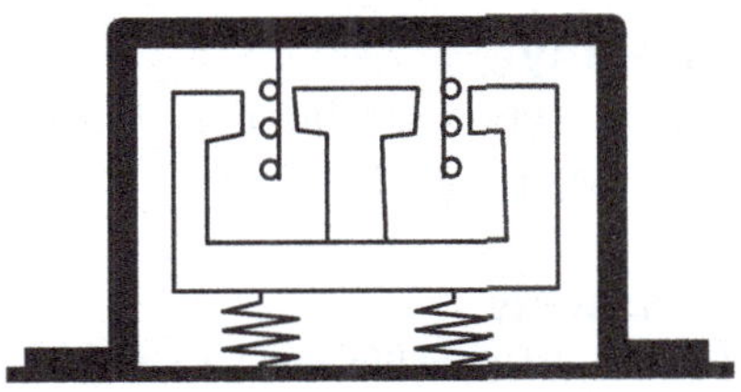

Fig. 12.4 Seismometer

12.2.3 Linear Variable Differential Transformer (LVDT)

An alternating current is driven through the primary circuit, causing a voltage to be induced in each secondary proportional to its mutual inductance with the primary. The frequency is usually in the range 1–10 kHz, range: 0.01–24 in non-contact type, nonlinearity: 0.10–0.25%, resolution: 1 micro-inch, cost: medium, life: high.

12.2.4 Specifications of Sensor

12.2.4.1 Dynamic Range

It is the ratio of maximum recordable input amplitude to minimum input amplitude, i.e., DR = 20 log (maximum input amplitude/minimum input amplitude) dB.

12.2.4.2 Linearity

It is the deviation of the output from a best-fit straight line for a given range of the sensor.

12.2.4.3 Transfer Function (Frequency Response Function)

The ratio of the frequency domain representation of the output to the frequency domain representation of the input is called the complex frequency response function, or transfer function.

12.2.4.4 Bandwidth

It is the frequency range between the lower and upper cut-off frequencies, within which the sensor transfer function is constant gain or linear.

12.2.4.5 Noise

It is the random fluctuation in the value of input that causes random fluctuation in the output value.

12.2.4.6 Repeatability

Sensor's repeatability indicates the range of output that the user can expect when the sensor measures the same input values several times. On the other hand, it means the ability to give the same output for repeated application of the same input value.

12.2.4.7 Linearity

Linearity often quoted on specification sheets as a ± value for the sensor's output signal.

12.2.4.8 Sensitivity

It is defined as the relationship indicating how much output to get per unit input.

12.2.4.9 Stability

It is the ability to give the same output when used to measure a constant input over a period of time.

12.2.4.10 Accuracy

It is defined as the extent which the value indicated by a measurement system might be wrong.

12.2.4.11 Periodic Motion

It is the motion which is repeated in equal intervals of time. Figure 12.5 shows periodic motions.

12.2.4.12 Random Signal

Non-deterministic data is referred as random time functions. Good example is ground motion during an earthquake, road surface unevenness, etc. Figure 12.6 shows random signal.

12.2.4.13 Correlation

It is a measure of similarity between two quantities. Considering the following two time-dependent quantities shown in Fig. 12.7, the term is explained clearly.

The correlation is computed by multiplying the ordinates of the two records at each time t and determining the average value. The value so obtained will be largest if the two records are similar or identical.

12.2.4.14 Auto- and Cross-Correlation

Considering two identical signals shown in Fig. 12.8, but shifted to the left by a time τ as shown above, then at time t when x_1 is $x(t)$, the value of x_2 is $x(t + \tau)$ and the correlation is given by $< x(t)x(t + \tau)$. It can be noticed that as $\tau = 0$, complete correlation can be seen and as τ increases the correlation decreases. The average value of these computations is called autocorrelation and is given as follows

$$\begin{aligned} C_{ij}(\tau) &= \sum_{k=1}^{N} x(t_k)y(t_k+\tau) \\ C_{ii}(\tau) &= \sum_{k=1}^{N} x(t_k)x(t_k+\tau) \end{aligned} \tag{12.1}$$

where τ is the time increment and N is the number of samples in the record. The above equations refer to "cross-covariance" which is obtained by analysis of the covariance of the signal with another signal and "auto-covariance" which is obtained by analysis of the covariance of the signal with itself. This shows the similarity of the motions in time domain only.

12.2.4.15 Fourier Spectrum

The periodic function shown in Fig. 12.9 can be represented as follows.

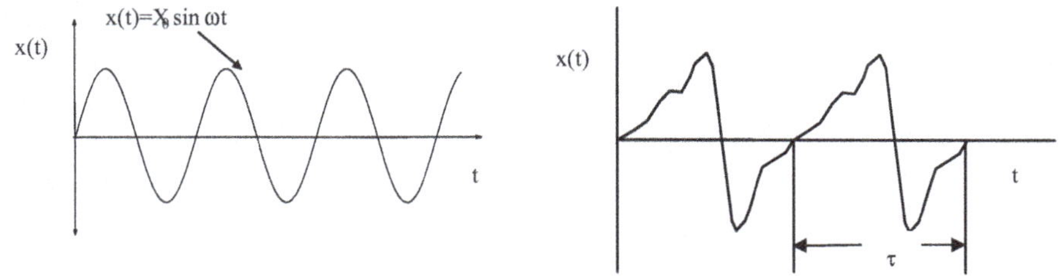

Fig. 12.5 Periodic motions

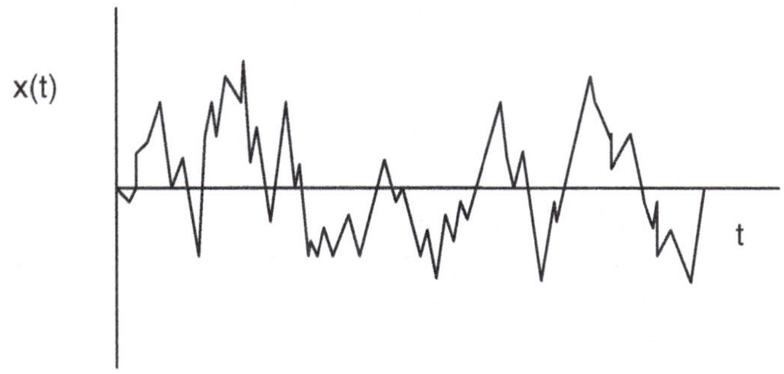

Fig. 12.6 Random signal

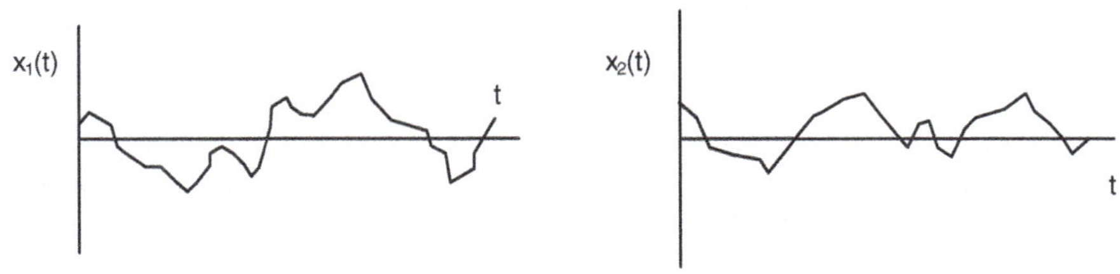

Fig. 12.7 Correlation between two signals

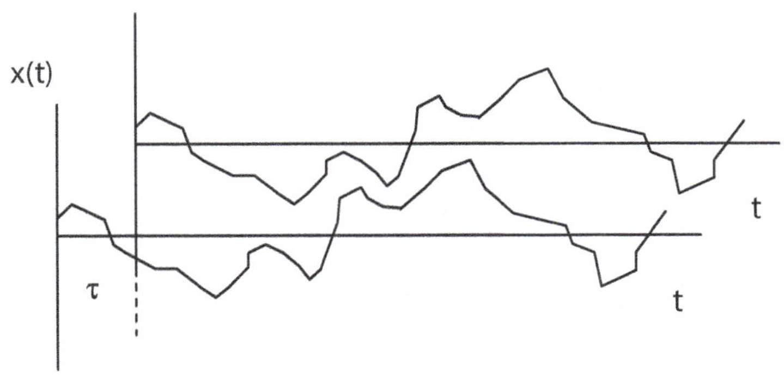

Fig. 12.8 Auto- and cross-correlation

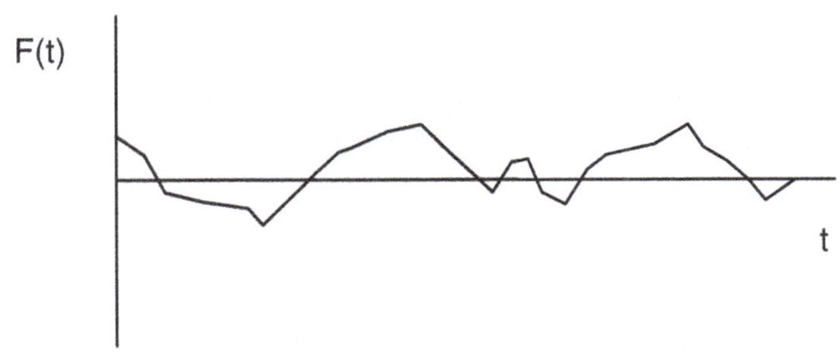

Fig. 12.9 Periodic function

$$F(t) = a_0 + \sum_{n=1}^{\infty} a_n \cos \omega_n t + b_n \sin \omega_n t$$
$$\text{where } \omega_1 = \frac{2\pi}{T} \tag{12.2}$$

In finite series

$$a_0 = \frac{2}{T}\int F(t)dt$$
$$a_n = \frac{2}{T}\int F(t) \cos \omega_n t \, dt \tag{12.3}$$
$$b_n = \frac{2}{T}\int F(t) \sin \omega_n t \, dt..$$

where n = 1,2,3,4, …

Now, the Fourier spectrum is given as shown in Fig. 12.10.

The exponential form of the Fourier series can be written as follows.

$$F(t) \text{ or } x(t) = c_0 + \sum_{n=1}^{\infty} (c_n e^{in\omega_n t} + c_n^* e^{-in\omega_n t}) \tag{12.4}$$

where

$$c_n = a_n - ib_n$$
$$c_n^* = a_n + ib_n$$

Then, the power spectrum is defined as follows.

$$G(f_n) = \frac{1}{2} c_n c_n^* \tag{12.5}$$

Now, the power spectral density is defined as follows.

$$S(f_n) = \frac{G(f_n)}{\Delta f} = \frac{c_n c_n^*}{2\Delta f} \tag{12.6}$$

Then, the mean square value is given as follows.

$$\overline{x^2} = \int_0^f s(f)df \tag{12.7}$$

Note: Numerical computation of Fourier coefficients for a periodic signal is given in Appendix 1.

12.2.4.16 Coherence Function

The coherence function is a measurement of the noise in the signal. If it is zero, then the measurement is of pure noise. If the value of the coherence is 1, the input and output are not contaminated. Mathematically, it is defined as follows.

$$\gamma^2 = \frac{|S_{xy}(\omega)|^2}{S_{xx}(\omega)S_{yy}(\omega)} \tag{12.8}$$

where $S_{yy}(\omega)$ is the PSDF of the input and $S_{xx}(\omega)$ is the PSDF of the response.

Another way of defining coherency is by taking the Fourier transform of the cross- and autocorrelation at different frequencies, and using the expression given below, we obtain the similarity of motions in frequency domain, as given below:

$$\gamma_{ij}(\omega) = \frac{C_{ij}(\omega)}{\sqrt{C_{ii}(\omega)C_{jj}(\omega)}} \tag{12.9}$$

in which $C_{ij}(w)$ is the Fourier transform of the cross-covariance and $C_{ii}(w)$ is the Fourier transform of the auto-covariance. An example illustrating the above phenomena is given in Annexure III for shake table testing on piping system.

12.3 Inputs to Be Collected for Seismic Qualification of SSCs

Before planning the tests on SSCs, the following inputs of SSCs in detail need to be collected.

1. Description of the SSCs,
2. Mounting and connection details,
3. Procedure for accelerated aging,
4. The operational and environmental conditions to be simulated,
5. Performances and acceptance criteria to be specified, and environmental variables to be measured,
6. Maintenance and repairs allowed during testing,
7. Performance limits or failure definitions,
8. Input motion,

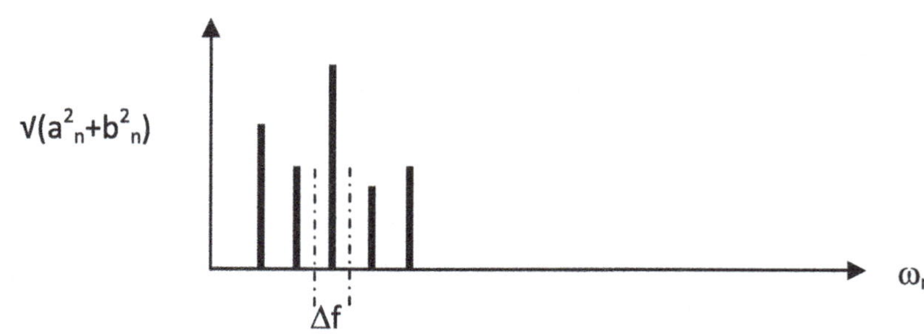

Fig. 12.10 Fourier spectrum

9. Operational loads and operating conditions,
10. Exploratory vibration tests,
11. Monitoring of output response and functional performance of the equipment,
12. Demonstration of operability (acceptance criteria, malfunction, qualification criteria, etc.),
13. Test sequences.

In the testing of SSCs, there are mainly two steps. First one is exploratory tests where the frequencies and damping values are evaluated for the SSCs to be tested. In second step, SSCs are subjected to required response spectra/desired signal.

12.4 Exploratory Tests

These are generally performed at input levels well below the seismic qualification levels. These are performed in several ways. The most common method is resonance search. A second method is by impacting the SSCs at critical points and capturing the impact and response data and computing the transfer function between the impact and response. Associated phase information is also obtained. A third method involves the use of the broadband input and simultaneous measurement of response at locations of interest. Resonance search method is explained below in detail.

12.4.1 Resonance Search Method

Generally, a slowly swept low-level sinusoidal vibration is employed in this method. The sweep rate should be two octaves per minute or less to ensure the resonance build up. A 0.2 g peak input is the conventional input, but it may be adjusted to avoid equipment damage. It is recommended that the resonance search be carried out beyond 33 Hz, for example 50 Hz, or to the cutoff frequency. Figure 12.11 shows the sine sweep signal used for evaluating natural characteristics.

Structural resonance is normally detected by observing amplifications of the input motion in the test item. Phase relation between the sinusoidal input signal and the structural response at the point of measurement may also be helpful in defining the resonance. A higher degree of confidence in establishing resonance is obtained by combining amplification and phase angle.

The details are explained considering testing of glove boxes as shown in Fig. 12.1 using 10-ton shake table existing at CPRI, Bangalore. The size of the table is 3 × 3 meters. At first, resonance search tests were carried out using the sine sweep signal as shown in Fig. 12.11. The sweep rate used is 1 octave per minute. Starting initial frequency is 1 Hz; the sweep was carried out till 50 Hz which is the range of interest contained in earthquake motion.

The response accelerations were recorded at the top of the glove boxes, and its transfer function and phase angle with respect to the base excitation are given, respectively, along *X*, *Y*, and *Z* directions in Fig. 12.12. It can be seen that the fundamental frequency in *X*-direction is about 4.5 Hz, along *Y*-direction is 4.8 Hz, and along *Z*-direction is 15 Hz.

Similar tests were carried out for the double module glove box as shown in Fig. 12.13, and its transfer functions and phase angles at the top of it along *X* and *Y* directions are shown in Fig. 12.14. Based on the resonance peaks and phase angle shift, the frequencies along *X* and *Y* directions are obtained and they are 10.5 and 12 Hz, respectively. Details of resonance peak and phase angle change are discussed further in detail.

12.5 Evaluation of Frequency and Damping of the SSCs from the Sine Sweep Tests

The transfer function and phase angle of a single-degree-of-freedom system when it is subjected to sine/sine sweep tests will have characteristics as shown in Fig. 12.15. It can be seen that at resonance the phase angle changes which is the indicator of the natural frequencies of the SSCs. Based on this explanation, the frequency of the glove boxes is obtained from Figs. 12.12 and 12.14. Next important parameter of the SSCs to be evaluated is damping.

Damping can be evaluated using the following approaches. First is logarithmic decrement method as shown in Fig. 12.16, and second is half-power method as shown in Fig. 12.17. The first method is simple and applicable for such SSCs, which behave like single-degree-of-freedom system. Second method can be applied for transfer functions recorded for glove boxes as shown in Figs. 12.12 and 12.14, and damping can be evaluated.

12.5.1 Logarithmic Decrement Method

Logarithmic decrement is a method to determine the amount of damping present in a system. In this, rate of decay is measured to evaluate damping. The rate of decay will be higher for larger damping. Logarithmic decrement is defined as the natural logarithm of the ratio of any two successive amplitudes of a real system under free vibration. Displacement of a freely vibrating system is shown in Fig. 12.16.

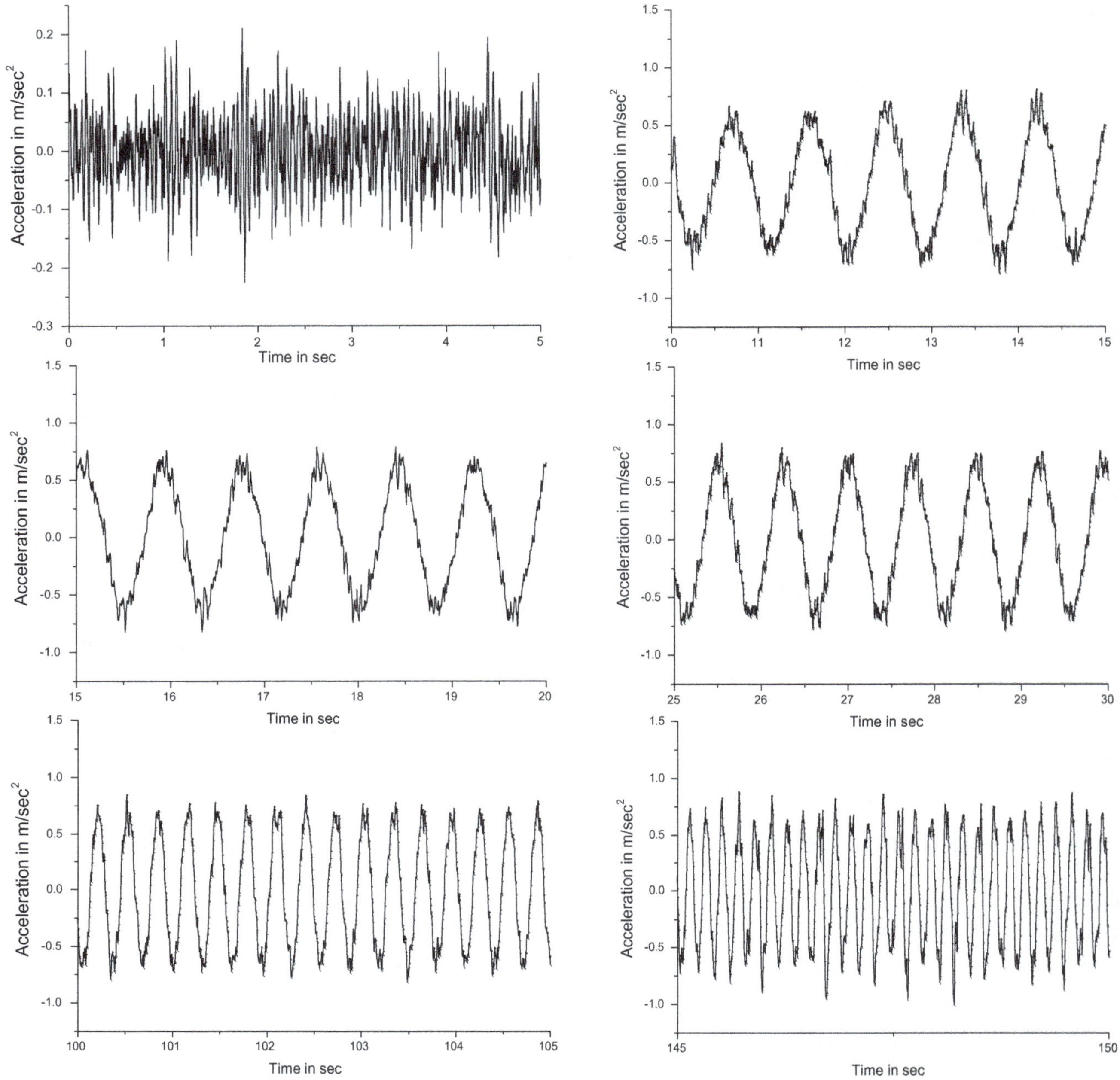

Fig. 12.11 Sine sweep signal used for evaluating natural characteristics

Logarithmic decrement

$$\delta = \ln\frac{x_1}{x_2} = \ln\frac{e^{-\xi\omega_n t_1}\sin(\sqrt{1-\xi^2}\omega_n t_1 + \phi)}{e^{-\xi\omega_n (t_1+\tau_d)}\sin[\sqrt{1-\xi^2}\omega_n (t_1+\tau_d) + \phi]} \tag{12.10}$$

Since values of sines are equal when the time is increased by the time period τ_d,

$$\delta = \ln\frac{x_1}{x_2} = \ln\frac{e^{-\xi\omega_n t_1}}{e^{-\xi\omega_n (t_1+\tau_d)}} = \ln e^{\xi\omega_n t_d} = \xi\omega_n t_d \tag{12.11}$$

by substituting

$$\tau_d = \frac{2\pi}{\omega_n\sqrt{1-\xi^2}} \tag{12.12}$$

One can have

$$\delta = \frac{2\pi\xi}{\sqrt{1-\xi^2}} \tag{12.13}$$

If the value of δ is known damping present in the system can be calculated.

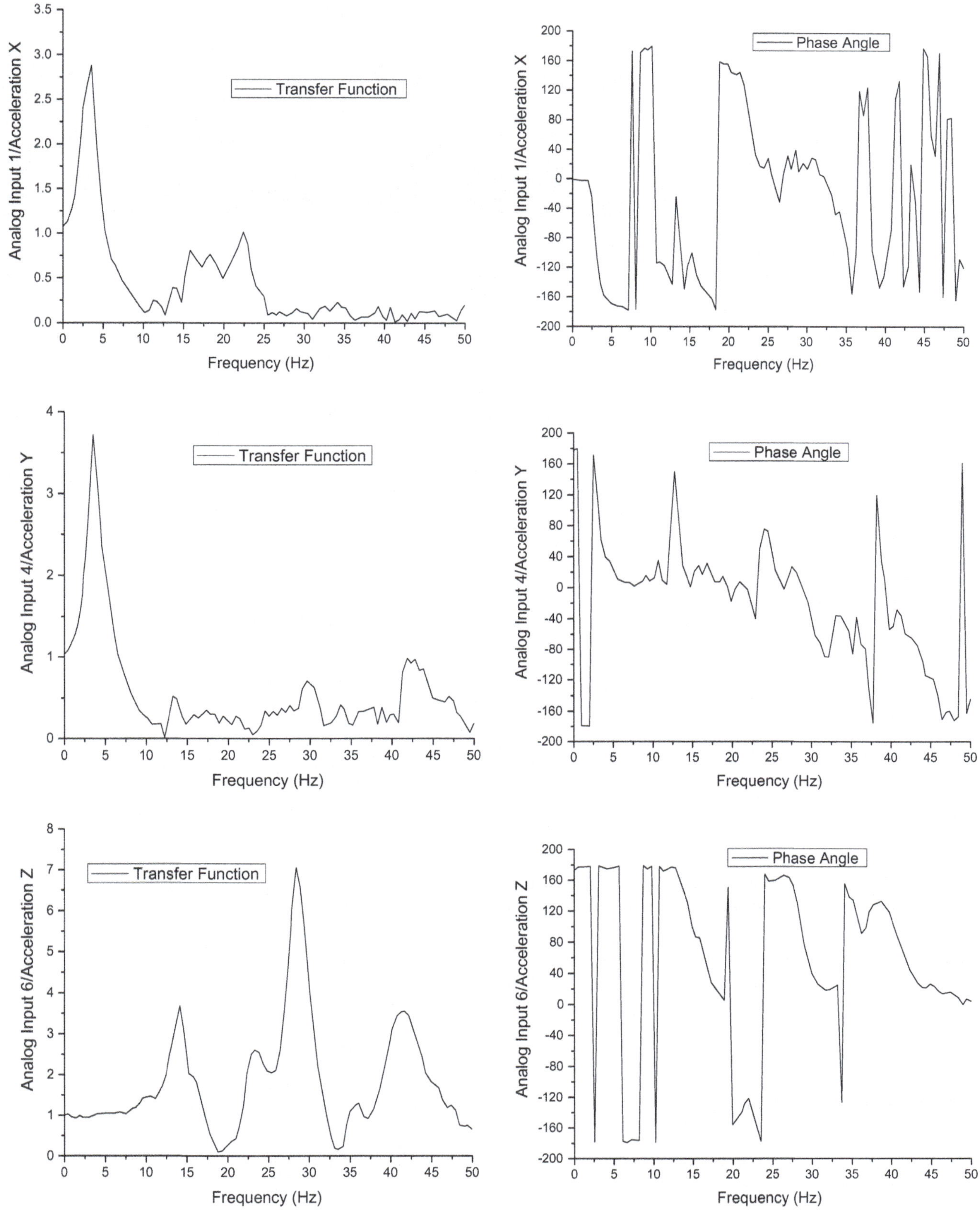

Fig. 12.12 Transfer functions and phase angle for train of glove boxes

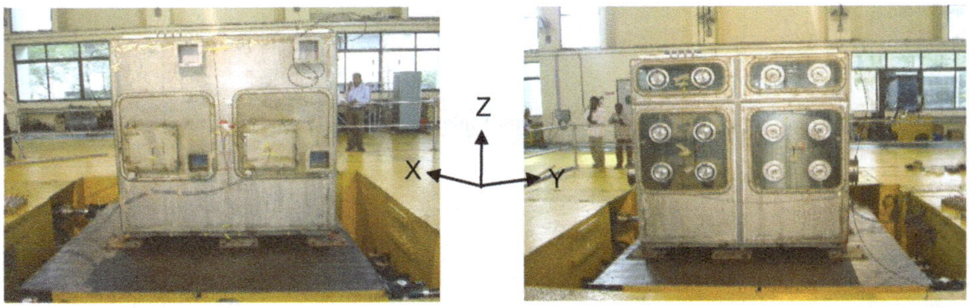

Fig. 12.13 Double module glove box (back and front views)

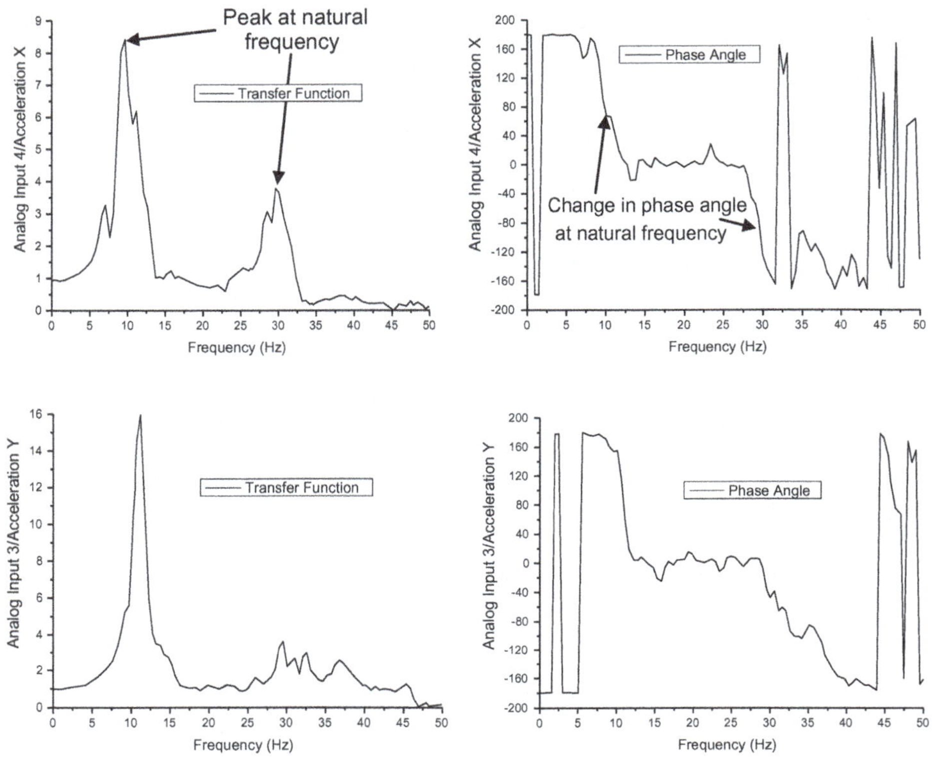

Fig. 12.14 Transfer functions and phase angles along three directions

Example 12.1: Measurement of Damping by Logarithmic Decrement Method Consider two cantilever structures fixed on a plate as shown in Fig. 12.17. Free vibration tests are performed to measure the damping present in both the systems. Free vibration responses of cantilever structures are shown in Fig. 12.18. Table 12.1 shows the damping calculations.

12.5.2 Half-Power Method

The points R_1 and R_2, on frequency response curve (R_d) where the amplification factor falls to $Q/\sqrt{2}$, are called half-power points because the power absorbed (ΔW) by the damper responding harmonically at a given frequency is proportional to the square of the amplitude. Q is value of the amplitude ratio at resonance. This property of frequency response curve (R_d) is used for evaluating the damping present in a system.

If ω_a and ω_b are the frequencies corresponding to half-power points on either side of resonant frequency, then for small ξ (Fig. 12.19)

$$\frac{\omega_b - \omega_a}{\omega_n} = 2\xi \Rightarrow \xi = \frac{\omega_b - \omega_a}{2\omega_n} \tag{12.14}$$

It can be seen from Fig. 12.15 that at resonance the phase angle changes from 0 to 180 degrees. This concept has been adopted to identify the frequencies for the glove boxes above. Coming to the damping evaluation, the logarithmic decrement is generally adopted for SSCs which are close to the single-degree-of-freedom system whereas the concept of half-power and magnification factor method can be adapted to complex SSCs such as glove boxes. Using half-power method, the damping values of the glove boxes were evaluated.

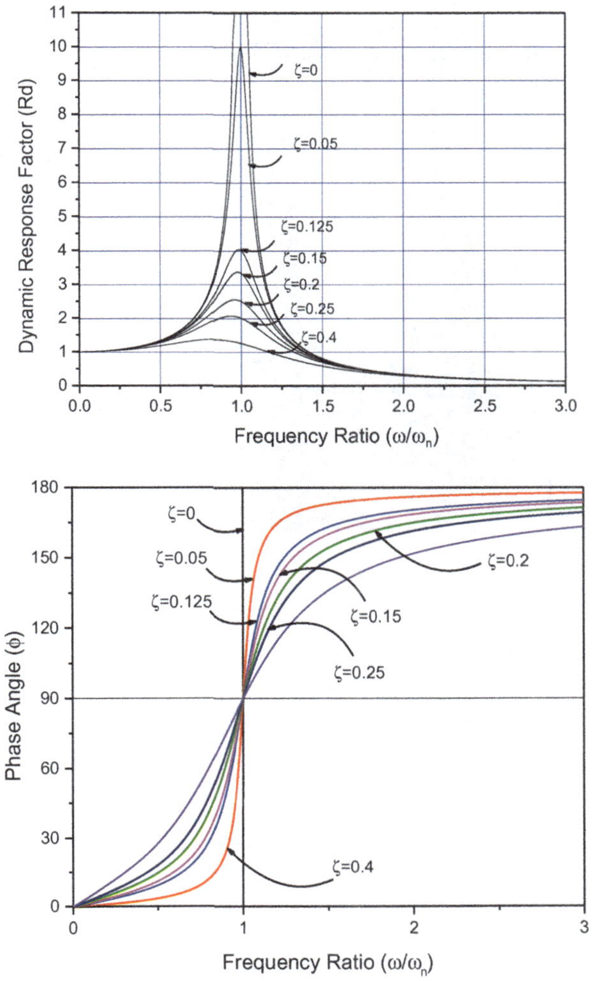

Fig. 12.15 Variation of dynamic response factor and phase angle with frequency ratio

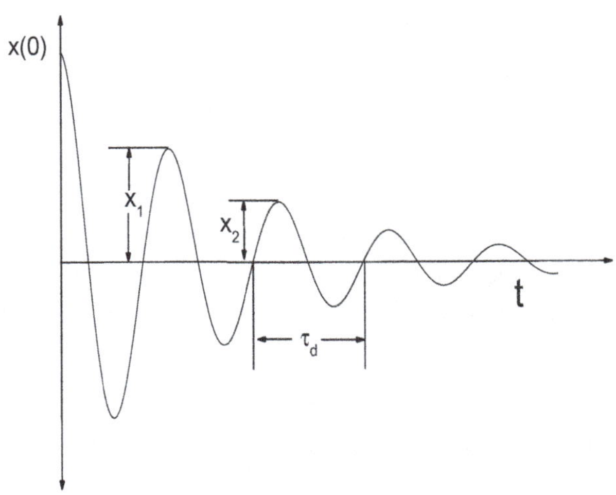

Fig. 12.16 Displacement of a freely vibrating system

The damping values for the train of glove boxes are 17.8, 14, and 13%, respectively, along *X*, *Z*, and *Y* directions, whereas for the double module glove box it is 5, 3, and 6.2% along *X*, *Z*, and *Y* directions, respectively. In the first case, the damping is high because the glove box is simply resting on the floor, whereas in the second case, it is fixed to the table. After evaluating the damping and frequencies, qualification tests were performed for the given RRS.

Example 12.2: Measurement of Damping by Half-Power Method Figure 12.20 shows the transfer function of the response acceleration with respect to the base excitation, recorded at the top of the glove boxes in *Z*-direction. Half-power method has been used to evaluate the damping present in the system.

$$\xi = \frac{\omega_b - \omega_a}{2\omega_n} = \frac{2\pi \times (11.74 - 10.35)}{2\pi \times (2 \times 11.21)} = 0.062$$

Similarly, damping in other directions can be estimated.

12.6 Generation of Time Histories Compatible with the Required Response Spectrum

For testing of structures, systems, and components mounted at the ground and at different elevations of the structure, artificial time history compatible with required response spectrum (RRS) is to be generated. The RRS at different floors of the structure, which supports equipment and systems, is generally called floor response spectrum. The characteristics of the RRS are different from that of design ground response spectrum (DGRS) which are explained in earlier chapters. Considering the DGRS of a typical site as RRS, the procedure to generate time histories is explained here after.

12.6.1 Requirements of Artificial Time Histories

i. One or more recorded, modified recorded or synthetic earthquake ground motion time histories may be used as explained in Chap. 2.
ii. Time histories shall be selected or developed, so that they reasonably represent the ground motion expected for the site. (e.g., amplitude and duration appropriate for the magnitude and distance). Ground motion parameters as defined in Fig. 12.21 and Table 12.2 should be used unless specified [2]. Figure 12.22 shows the RRS and TRS for typical site at the ground level.

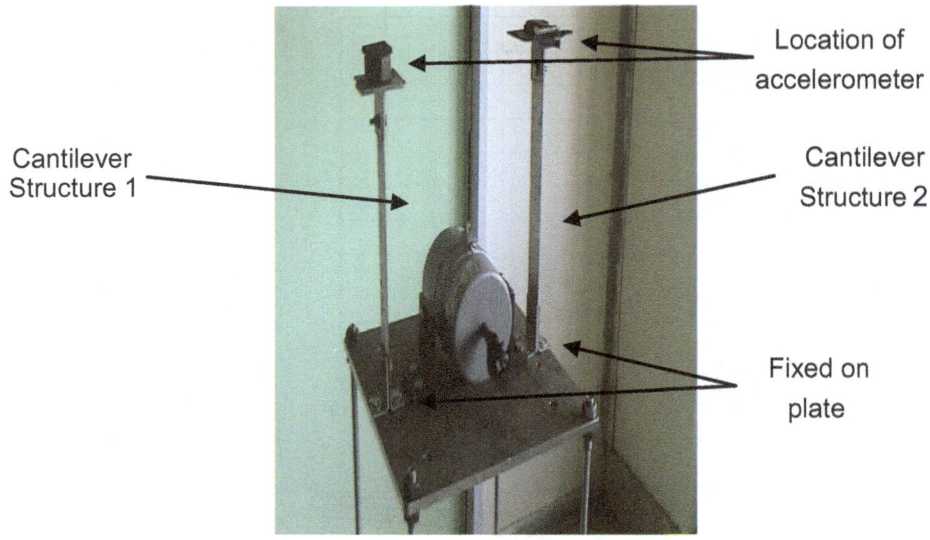

Fig. 12.17 Cantilever structures for damping measurement

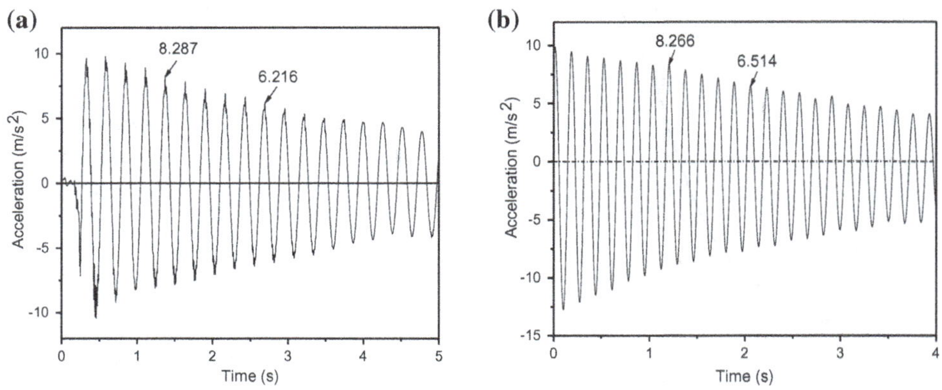

Fig. 12.18 Free vibration response of **a** cantilever structure 1 and **b** cantilever structure 2

Table 12.1 Measurement of damping

	Cantilever structure 1	Cantilever structure 2
x_1	8.287	8.266
x_2	6.216	6.514
δ	0.058	0.048
ξ	0.00923	0.00764
ξ (%)	0.923	0.764

Note Since these are aluminum structures, they have lower damping value

After fixing the above parameters, power spectral density function (PSDF) is evaluated for the given RRS using the procedure given below.

12.6.2 Power Spectral Density Functions for the Given Time History

The PSD computed from an accelerogram $x(t)$ is defined in terms of the Fourier amplitudes of the time history as

$$G(\omega) = \frac{|F(\omega)|^2}{\pi t_m} \tag{12.15}$$

where $F(\omega)$ is the Fourier amplitude which is given as

$$F(\omega) = \int_0^t x(t)\mathrm{e}^{-\mathrm{i}\omega t} dt \tag{12.16}$$

where $G(\omega)$ is the one-sided PSD, ω is the circular frequency, $F(\omega)$ is the Fourier amplitude of the time history computed over the duration tm, and tm is the equivalent strong motion duration.

12.6.3 Time History Generation Procedure

Periodic function can be expanded into a series of sinusoidal waves

$$X(t) = \sum_n A_n \sin(\omega_n t + \phi_n) \tag{12.17}$$

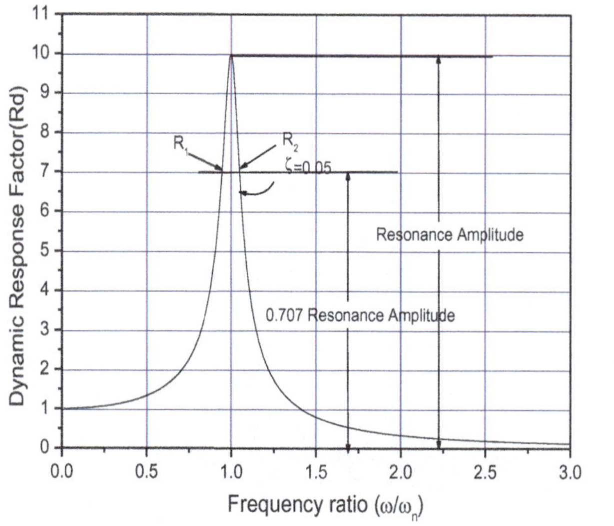

Fig. 12.19 Variation of dynamic response factor with frequency ratio

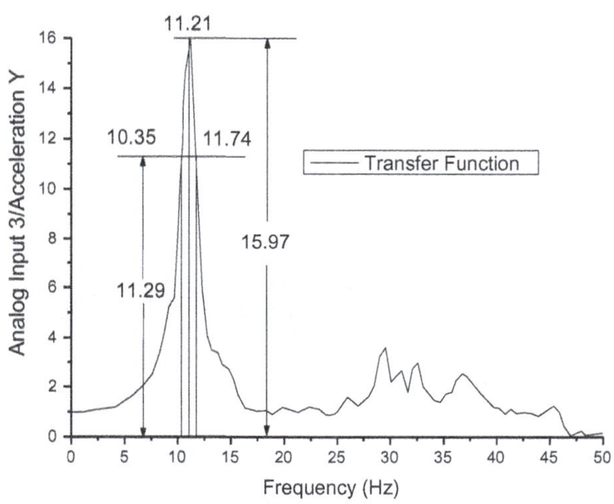

Fig. 12.20 Transfer function of the response accelerations

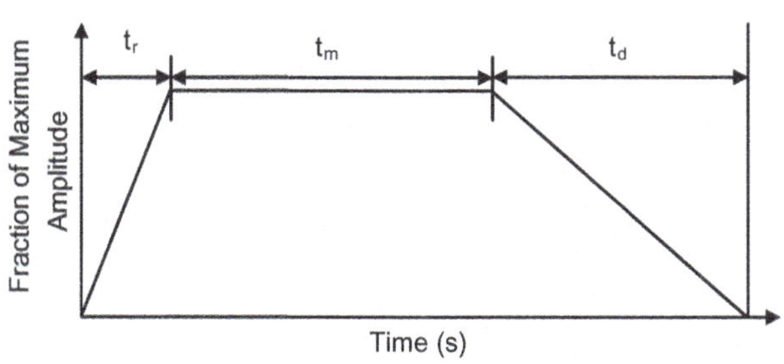

Fig. 12.21 Duration envelope function

where A_n is the amplitude, and ϕ_n is the phase angle of the nth contributing sinusoid. Random number generator is used to produce strings of phase angles in the range between 0 and 2π. The amplitudes An are related to the (one-sided) spectral density function $G(\omega)$ as:

$$G(\omega_n)\Delta\omega = \frac{A_n^2}{2} \tag{12.18}$$

To simulate the transient character of real earthquakes, the steady-state motions are multiplied by deterministic envelop function $I(t)$. The artificial motion becomes:

$$Z(t) = I(t)\sum_n A_n \sin(\omega_n t + \phi_n) \tag{12.19}$$

To smoothen the calculated spectrum and to improve the matching, an iterative procedure is generally used. To do this, the PSD is modified as follows:

$$G(\omega)_{i+1} = G(\omega)\left[\frac{S_{v(\omega)v}}{S^i_{v(\omega)}}\right]^2 \tag{12.20}$$

where S_v is the target spectral value. Using the similar procedure, time histories for the given RRS are generated and shown in Figs. 12.23, 12.24, and 12.25 for *X*, *Y*, and *Z* directions, respectively.

12.6.4 Test Response Spectra (TRS) Acceptability Criteria as per ASCE 4-98

(a) Test response spectra (TRS) Acceptability Criteria
   (i) The TRS envelops the RRS over the frequency range for which the particular test is designed or the applicable portion of the RRS taking into account the dynamic characteristics of the equipment under test, i.e., the natural resonance frequencies.
   (ii) For comparison of the TRS and the RRS, the TRS is computed with a damping value equal to or greater than that of the RRS.
   (iii) It is recommended that the TRS be computed with 1/6 octave or narrower bandwidth resolution.
   (iv) Total test duration is given in Table 12.2.
   (v) The time history should indicate frequency contents at least as broad as that defined by the amplified region of the RRS.

(b) The RRS occasionally requires high acceleration levels at the lowest frequencies that require very high test-table displacement capability. The general requirement for enveloping the RRS by the TRS can be modified under the following criteria.
   (i) In those cases, where it can be shown by a resonance search that no resonance response phenomena exist below 5 Hz, it is required to envelop the RRS only down to 3.5 Hz.

Table 12.2 Duration enveloping function parameters

Magnitude	Rise time (t_r)	Duration of strong motion (t_m)	Decay time (t_d)
7.0–7.5	2	13	9
6.5–7.0	1.5	10	7
6.0–6.6	1	7	5
5.5–6	1	6	4
5.0–5.5	1	5	4

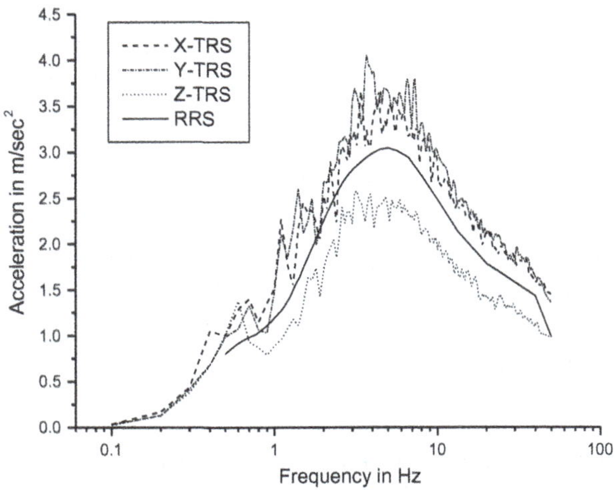

Fig. 12.22 Required and test response spectra

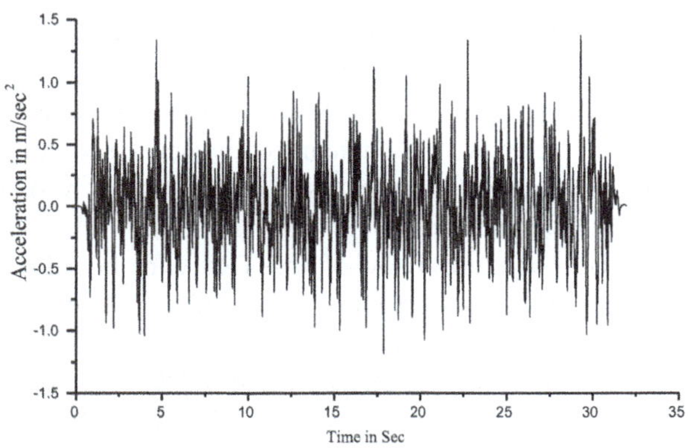

Fig. 12.23 Time history (*X*-direction)

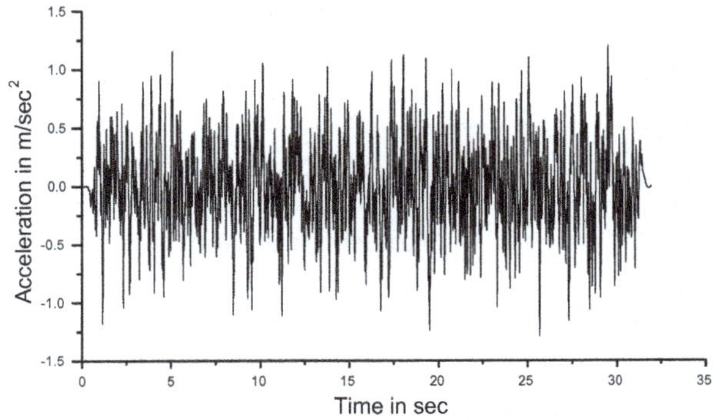

Fig. 12.24 Time history (*Y*-direction)

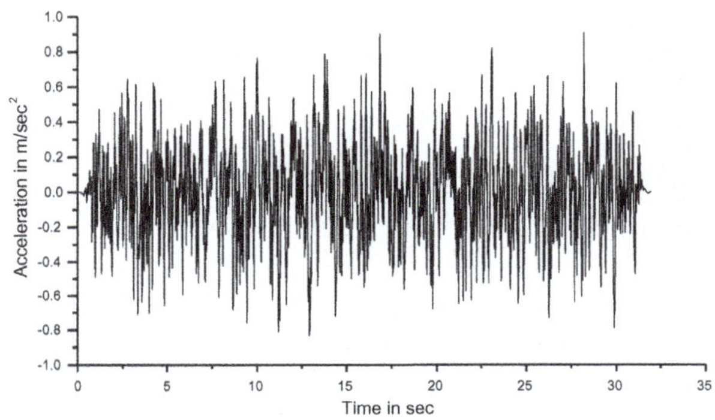

Fig. 12.25 Time history (*Z*-direction)

Excitation must continue to be maintained in the 1–3.5 Hz range to the capability of the test facility.

(ii) When resonance phenomena exist below 5 Hz, it is required to envelop the RRS only down to 70% of the lowest frequency of resonance.

(iii) When the absence of resonance response phenomena or malfunction below 5 Hz cannot be justified, the general requirement applies and the low-frequency enveloping should be satisfied down to 1 Hz.

(iv) Under any circumstances, failure to envelop the RRS at, or above, 3.5 Hz must be justified.

(c) In the performance of a test program, the TRS may, on occasion, not fully envelop the RRS. The general requirement for a retest may be exempted if the following criteria are met:

(i) A point of the TRS may fall below the RRS by 10% or less, provided the adjacent 1/6 octave points are at least equal to the RRS and the adjacent 1/3 octave points are at least 10% above.

(ii) A maximum of 5 of the 1/6 octave analysis points may be below the RRS, as above, provided they are at least one octave apart.

(iii) It is acceptable that the TRS stays below the RRS according to the limit of the vibration table provided that additional tests are performed using equivalent excitation methods (such as discrete frequency tests), and their equivalence is demonstrated.

12.7 Test Documentation

The test report shall be prepared as per the guidelines given in Clause 10 of IEEE-344 and shall consist the following as a minimum:

(i) Test procedure,
(ii) Description of test facility,
(iii) Input motion documentation in terms of time history plots and TRS/RRS,
(iv) Statement that TRS envelopes RRS,
(v) Photographs of test arrangement and mounting positions,
(vi) Monitored and performance data,
(vii) Equipment identification, date, time, venue of the test, and signatures of the witnessing personnel,
(viii) Maintenance and repairs carried out during test.

The response accelerations recorded at the top of double module glove boxes are shown in Figs. 12.26, 12.27 and 12.28 along *X*, *Y*, and *Z* directions, respectively.

Example 12.3: Mode Shape Measurement Linear behavior of the structures, systems, and components (SSCs) subjected to the earthquake loading is well understood. However, the behavior of SSCs designed for earthquake loading is nonlinear and needs to be understood in detail and develop more rational design and analysis methods. How to get mode shapes from the experiment is being discussed here. Shake table testing of the structure (as shown in Fig. 12.29) has been performed at CPRI, Bangalore. At first, resonance search tests were carried out using the sine sweep signal as shown in Fig. 12.11. The sweep rate used is 1 octave per minute. Starting initial frequency is 1 Hz; the sweep was carried out till 50 Hz which is the range of interest contained in earthquake motion. Transfer functions and phase angles have been measured at three locations A9, A10, and A11 of the structure as shown in Fig. 12.29. Transfer functions and phase angles are given in Fig. 12.30.

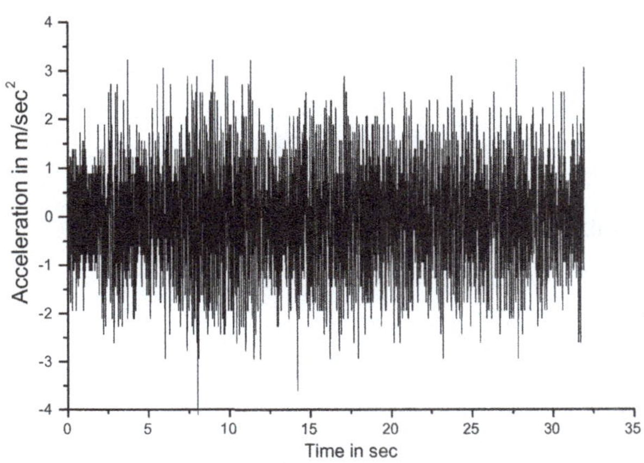

Fig. 12.26 Response at the top of double module glove box (*X*-direction)

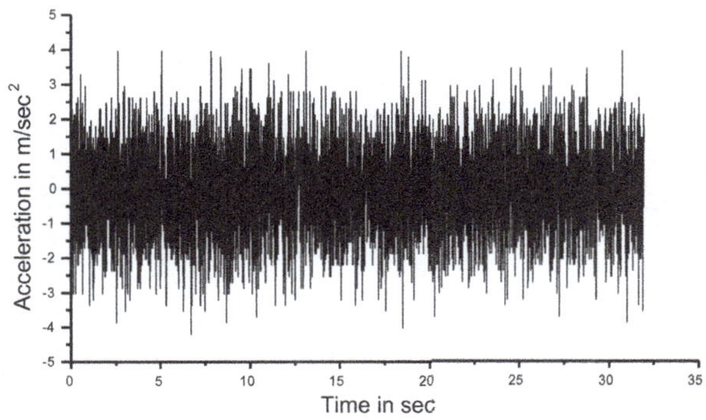

Fig. 12.27 Response at the top of double module glove box (*Y*-direction)

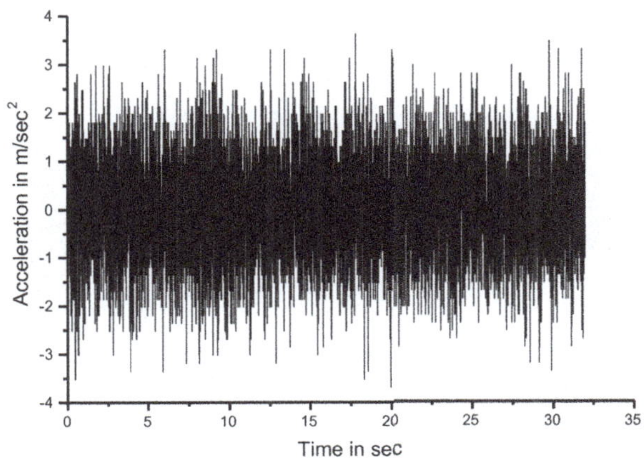

Fig. 12.28 Response at the top of double module glove box (*Z*-direction)

From the transfer function, the values of modal damping are obtained using half-power method as explained above. Natural frequencies are obtained by observing the peak in the transfer function together with the change in phase angles.

$$\omega_1 = 25.132 \text{ rad/s} \quad \xi_2 = 0.065$$
$$\omega_2 = 106.81 \text{ rad/s} \quad \xi_3 = 0.064$$
$$\omega_3 = 179.07 \text{ rad/s} \quad \xi_1 = 0.150$$

Transfer function matrix $|H_{sr}(\omega_i)|$ can be represented as [3]

Fig. 12.29 Three-story RCC structure

$$|H_{sr}(\omega_i)| = \frac{|u_i u_i^T|_{sr}}{2\xi\omega_i^2} \quad (12.21)$$

$$|u_i u_i^T|_{sr} = |2\xi_i\omega_i^2|_{sr}|H_{sr}(\omega_i)| \quad (12.22)$$

where u_i are mode shape vectors, $|H_{sr}(\omega_i)|$ is magnitude of frequency response function measured between points s and r and evaluated at the ith natural frequency. $|u_i u_i^T|$ results in a n × n matrix. The values of $|H_{sr}(\omega_i)|$ can be obtained from the transfer function as shown in Fig. 12.27. The values of $|H_{sr}(\omega_i)|$ are given below:

$$\begin{array}{lll} |H_{11}(\omega_1)| = 4 & |H_{11}(\omega_2)| = 2.62 & |H_{11}(\omega_3)| = 0.3 \\ |H_{21}(\omega_1)| = 3.35 & |H_{21}(\omega_2)| = 0.9 & |H_{21}(\omega_3)| = 0.6 \\ |H_{31}(\omega_1)| = 2.35 & |H_{31}(\omega_2)| = 3 & |H_{31}(\omega_3)| = 0.76 \end{array}$$

From Eq. (12.22) one can have the matrix

$$|u_i u_i^T| = \begin{bmatrix} 757.45 & 634.64 & 445.28 \\ 3880.78 & 1334.11 & 4441.85 \\ 1231.25 & 2462.50 & 3117.09 \end{bmatrix} \quad (12.23)$$

If we represent the mode shape vector as $u_i = \begin{bmatrix} a_1 \\ a_2 \\ a_3 \end{bmatrix}$, then $|u_i u_i^T|$ can be given as follows

$$|u_i u_i^T| = \begin{bmatrix} a_1^2 & a_1 a_2 & a_1 a_3 \\ a_2 a_1 & a_2^2 & a_2 a_3 \\ a_3 a_1 & a_3 a_2 & a_3^2 \end{bmatrix} \quad (12.24)$$

Comparing the matrices (12.23) and (12.24), one can get

$$u_1 = \begin{bmatrix} a_1 \\ a_2 \\ a_3 \end{bmatrix} = \begin{bmatrix} 27.52 \\ 23.06 \\ 16.15 \end{bmatrix}$$

Similarly,

$$u_2 = \begin{bmatrix} 106.25 \\ 36.53 \\ 121.59 \end{bmatrix} \quad u_3 = \begin{bmatrix} 22.05 \\ 44.12 \\ 55.83 \end{bmatrix}$$

Using the phase as either + or – sign, the vectors u_1, u_2, and u_3 become

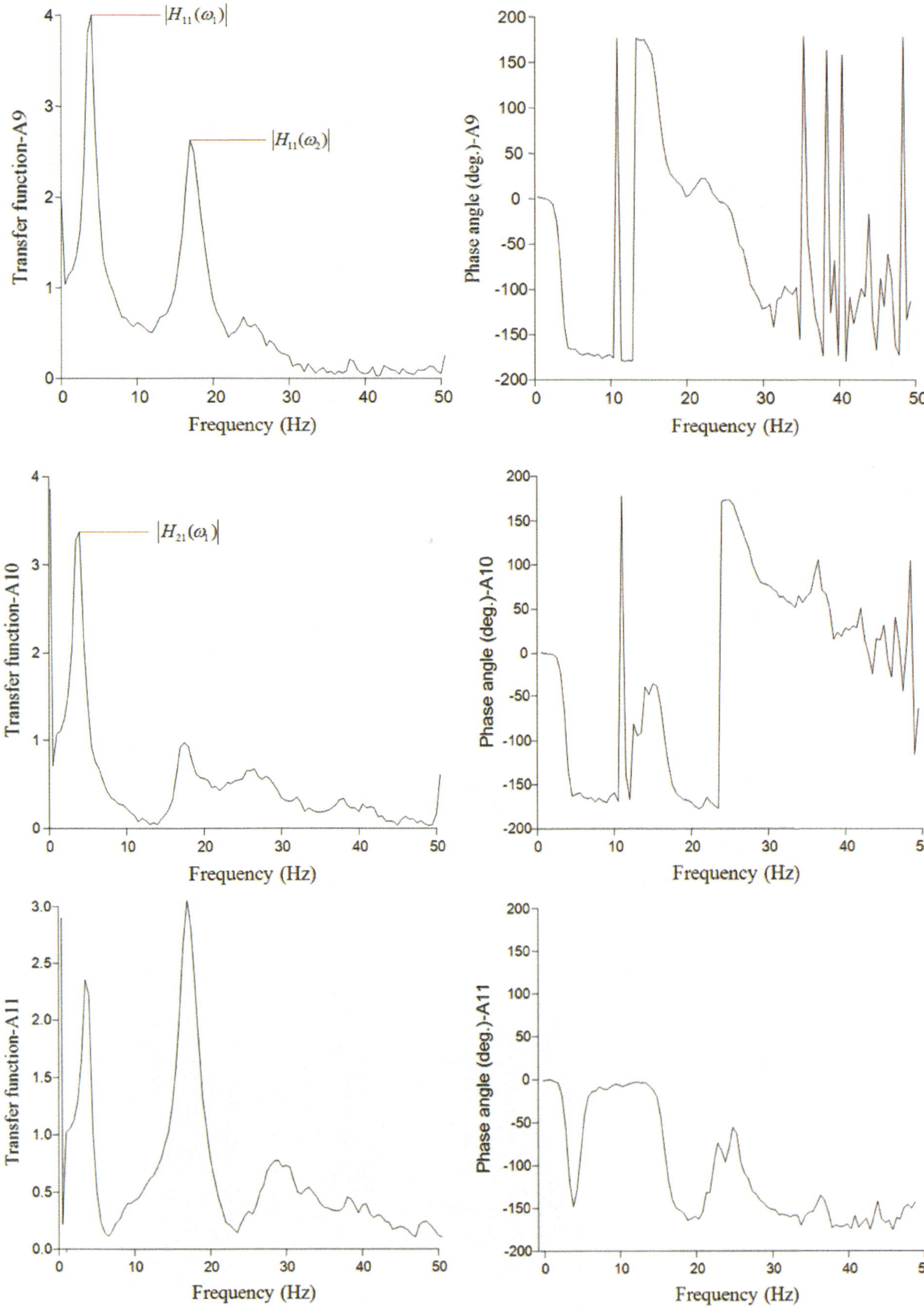

Fig. 12.30 Transfer function and phase angle plots at locations A9, A10, and A11

$$u_1 = \begin{bmatrix} -27.52 \\ -23.06 \\ -16.15 \end{bmatrix} \quad u_2 = \begin{bmatrix} +106.25 \\ -36.53 \\ -121.59 \end{bmatrix} \quad u_3 = \begin{bmatrix} -22.05 \\ +44.12 \\ -55.83 \end{bmatrix}$$

u_1, u_2, and u_3 are the mode shapes of the structure with natural frequencies 4, 17, and 28.5 Hz, respectively.

Example 12.4: Shake Table Testing of Piping Loop With/Without Damper To study the effectiveness of commercially available viscoelastic dampers in controlling the seismic response of piping systems, shake table testing has been performed at CPRI, Bangalore. A 6" NB Schedule 40 piping loop has been designed to accommodate it on the shake table of 3 × 3 m in size. Figure 12.31 shows the piping layout for shake table testing. Masses of 250 kg each have been attached at two locations to bring the frequency down to spectral peak in order to attract maximum seismic excitation. Testing of piping loop with/without viscoelastic dampers has been carried out.

Piping loop was subjected to a sinusoidal loading in the form of sine sweep with a sweep rate of 1 octave per minute to get the response of the piping. Sweep was 1–20 Hz, which was the range of interest as frequencies of the most of the piping systems lie in this range. Acceleration time histories (responses) have been measured at different locations. Figure 12.32 shows the response of the piping loop with/without viscoelastic dampers for a sinusoidal loading of 0.2 g amplitude at location 6.

An artificial time history corresponding to a typical site-specific ground response spectrum has been generated as shown in Fig. 12.32. This time history has been used as an input to shake table for testing of the piping loop with/without viscoelastic dampers. Testing has been done for two levels of earthquake, viz. 0.15 g PGA and 0.45 g PGA to get the response of the piping due to earthquake loading.

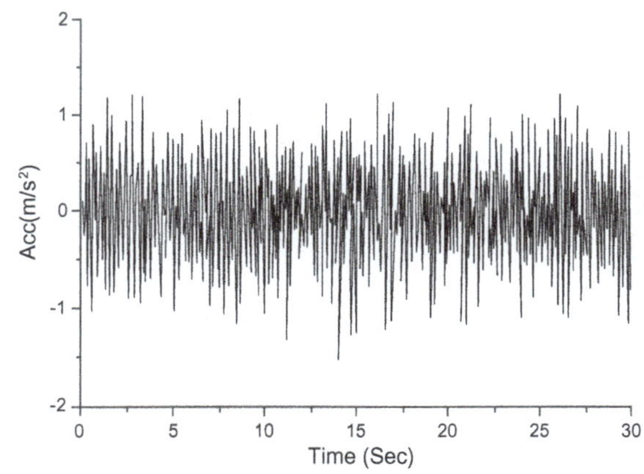

Fig. 12.32 Input time history (0.15 g PGA)

Figures 12.33 and 12.34 show the response of the piping loop with/without viscoelastic dampers for an earthquake load corresponding to 0.15 g PGA and 0.45 g PGA at location 6, respectively.

Based on the above experimental study, it can be concluded that the commercially available viscoelastic dampers are effective in controlling the response of the piping systems subjected to sinusoidal and earthquake loadings.

Example 12.5: Resonance Search Test of a Piping Loop Resonance search test was carried out using the sine sweep signal as shown in Fig. 12.11 to estimate the natural frequencies of a piping loop shown in Fig. 12.31. The sweep rate used is 1 octave per minute. Starting initial frequency is 1 Hz; the sweep was carried out till 20 Hz which is the range of interest. Acceleration time histories (responses) have been measured at different locations. Finite element analysis has also been performed to find out natural frequencies of the piping loop. Finite element model of the piping loop is shown in Fig. 12.35.

Fig. 12.31 Test setup for shake table testing of piping system

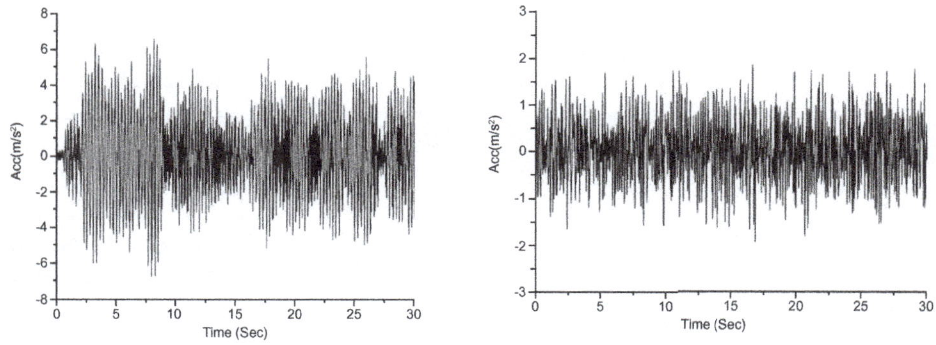

Fig. 12.33 Response of piping loop for an earthquake load corresponding to 0.15 g PGA at location 6 **a** without dampers; **b** with dampers

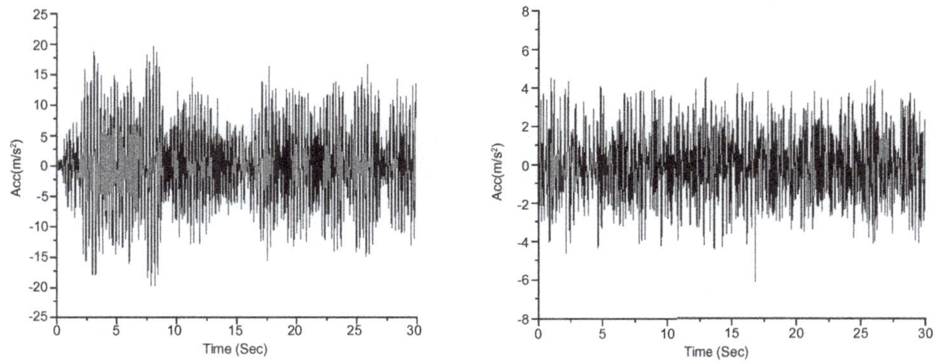

Fig. 12.34 Response of piping loop for an earthquake load corresponding to 0.45 g PGA at location 6 **a** without dampers; **b** with dampers

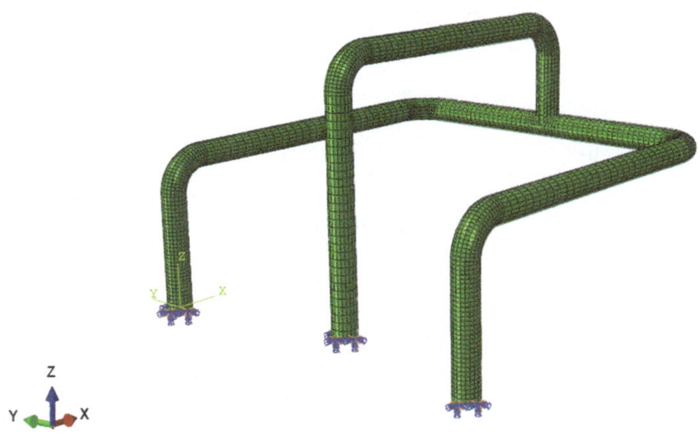

Fig. 12.35 Finite element model of the piping loop

Transfer functions at three locations A4, A5, and A6 are shown in Fig. 12.36. Table 12.3 shows the natural frequencies obtained from tests and finite element analysis.

Natural frequencies obtained from tests and finite element analysis are closely matching as observed from Table 12.3.

Exercise Problems

1. Numerate the methods of seismic qualification of structures, systems, and components (SSCs).
2. What are the inputs required to be collected for seismic qualification of SSCs?
3. Explain the significance of exploratory tests.
4. What is logarithmic decrement method?
5. Explain the acceptability criteria of test response spectra (TRS).
6. Explain the procedure of testing the equipment shown in the following figure for its natural frequencies, damping, and mode shapes (Fig. 12.37).
7. Explain the procedure of testing the pump for its performance for a given design basis response spectrum (e.g., Zone II ground spectra) (Fig. 12.38).

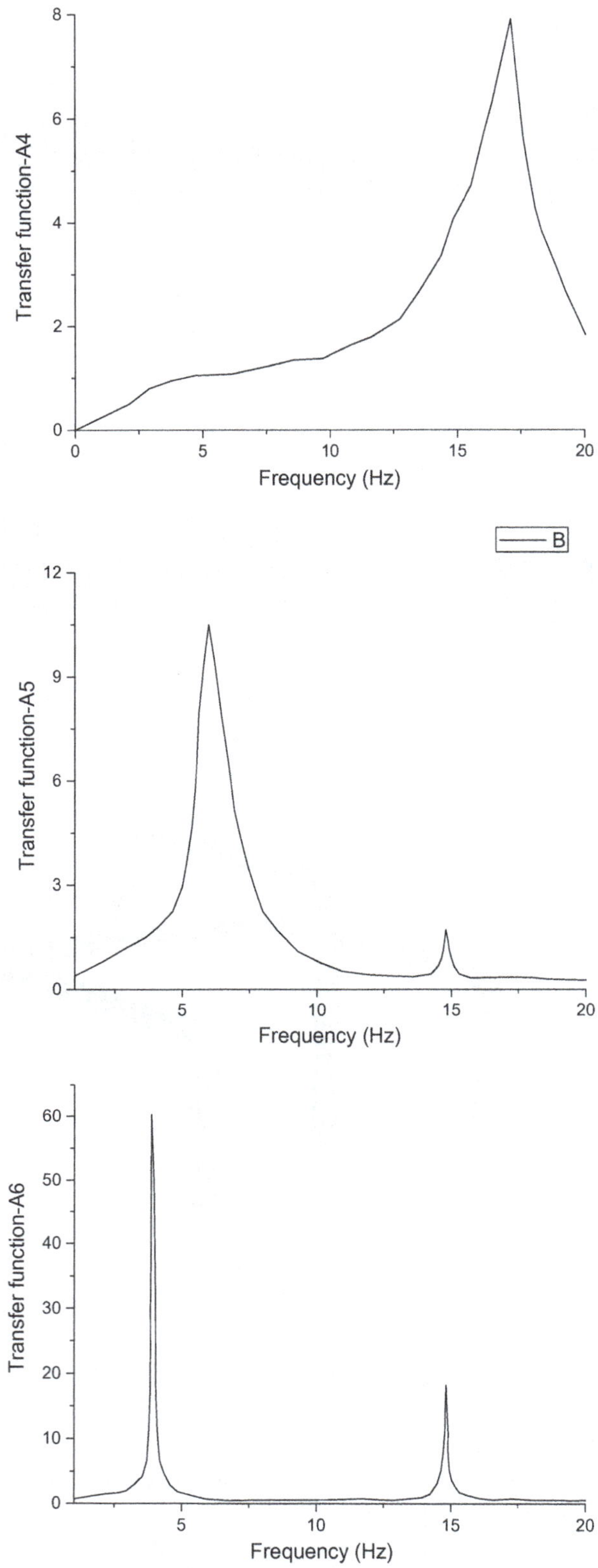

Fig. 12.36 Transfer function plots at locations A4, A5, and A6

8. The structure shown below is not to be mounted on shake table. Explain the procedure to evaluate frequencies, mode shapes, and damping using in situ tests (Fig. 12.39).

Appendix 1: Numerical Computation of Fourier Coefficients

For every simple forms of function $x(t)$, the integrals can be evaluated easily.

$$a_0 = \frac{2}{T}\int_0^T x(t)dt \quad (12.25)$$

$$a_n = \frac{2}{T}\int_0^T x(t)\cos n\omega t\, dt \quad (12.26)$$

$$b_n = \frac{2}{T}\int_0^T x(t)\sin n\omega t\, dt \quad (12.27)$$

However, the integration becomes involved if $x(t)$ does not have a simple form. In some practical applications as with the case of experimental determination of the amplitudes of vibration transducers, the function $x(t)$ is not available in the form of a mathematical expression; only the values of $x(t)$ at a number of points t_1, t_2, … t_n are available as shown in Fig. 12.40.

In these cases, the coefficients a_0, a_n, and b_n of Eqs. (12.25)–(12.27) can be evaluated using a numerical integration procedure like trapezoidal or Simpson's rule.

If t_1, t_2, … t_n are assumed to be an even number of equidistant point over period τ (N is also even) with the corresponding values $x_1 = x(t_1)$, $x_2 = x(t_2), \ldots x_n = x(t_n)$, respectively. The application of trapezoidal rule gives the coefficients a_0, a_n, and b_n (by setting $T = N\,\Delta t$)

$$a_0 = \frac{2}{N}\sum_{i=1}^{N} x_i \quad (12.28)$$

$$a_n = \frac{2}{N}\sum_{i=1}^{N} x_i \cos\frac{2n\pi t_i}{\tau} \quad (12.29)$$

$$b_n = \frac{2}{N}\sum_{i=1}^{N} x_i \sin\frac{2n\pi t_i}{\tau} \quad (12.30)$$

Example 12.6 The pressure fluctuations of water in a pipe measured at 0.01 s intervals are given in Table 12.4. These

Table 12.3 Natural frequencies obtained from tests and finite element analysis

S. no.	Natural frequencies	
	Tests	Finite element analysis
1.	3.86	3.75
2.	5.96	5.84
3.	14.81	15.07
4.	17.12	17.53

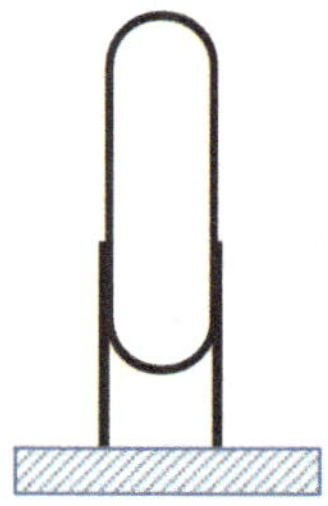

Fig. 12.37 The procedure of testing the equipment

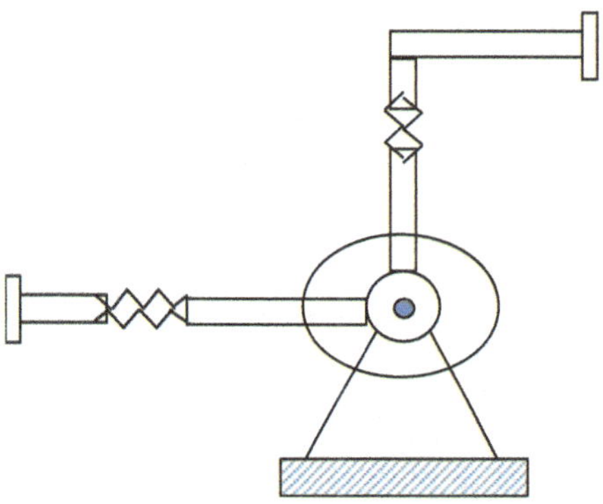

Fig. 12.38 Explain the procedure of testing the pump for its performance for a given design basis response spectrum

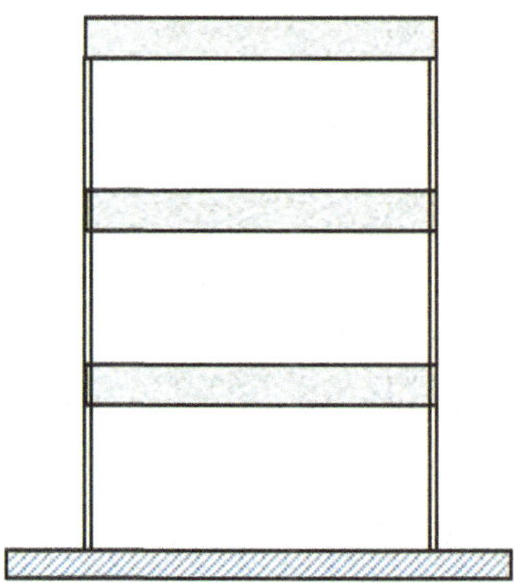

Fig. 12.39 Explain the procedure to evaluate frequencies, mode shapes and damping

fluctuations are repetitive in nature. Make a harmonic analysis of the pressure fluctuations, and determine the first three harmonics of the Fourier series expansion.

Solution:

Since the given fluctuations repeat every 0.12 s, the period is τ = 0.12 s and circular frequency of the first harmonics is 2π radians per 0.12 s or $\omega = 2\pi/0.12$. As the observed values in each wave is 12, a_0 can be obtained from Eq. (12.3)

$$a_0 = \frac{2}{N}\sum_{i=1}^{N} p_i = \frac{1}{6}\sum_{i=1}^{12} p_i = 79.50$$

The coefficients a_n and b_n can be determined from Eqs. (12.4) and (12.5), respectively.

$$a_n = \frac{2}{N}\sum_{i=1}^{N} p_i \cos\frac{2n\pi t_i}{\tau} = \frac{1}{6}\sum_{i=1}^{12} p_i \cos\frac{2n\pi t_i}{0.12}$$

$$b_n = \frac{2}{N}\sum_{i=1}^{N} p_i \sin\frac{2n\pi t_i}{\tau} = \frac{1}{6}\sum_{i=1}^{12} p_i \sin\frac{2n\pi t_i}{0.12}$$

With these coefficients and using Table 12.5, the Fourier series expansion of the pressure fluctuations $p(t)$ can be obtained as

$$\begin{aligned} p(t) &= 79.50/2 - 9.80\cos 52.36t - 0.82\,\sin\,52.36t \\ &\quad - 27.67\cos 104.72t \\ &\quad + 0.29\sin 104.72t - 0.50\cos 157.08t + 0.0\sin 157.08t + \ldots \end{aligned}$$

Considering three harmonics, approximate $p(t)$ can be obtained as follows (Table 12.6).

Comparison of actual pressure and approximate pressure with three Fourier harmonics considered is shown in Fig. 12.41.

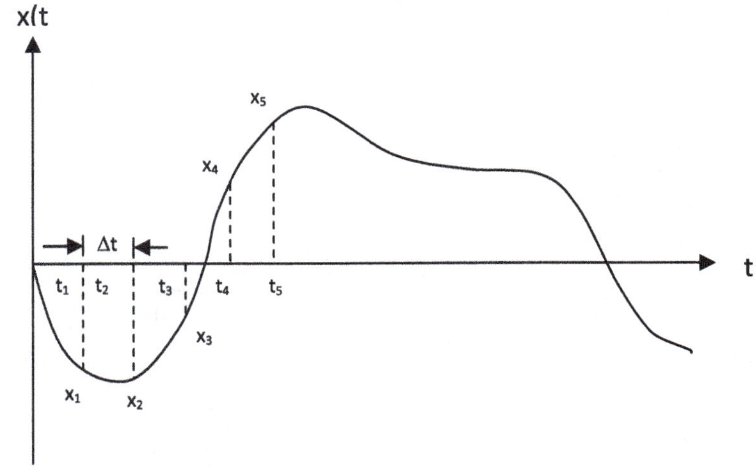

Fig. 12.40 Random signal

Table 12.4 Pressure fluctuations

Time station, i	Time (s), t_i	Pressure (kN/m^2), p_i
0	0.00	0.00
1	0.01	20.00
2	0.02	45.00
3	0.03	70.00
4	0.04	55.00
5	0.05	35.00
6	0.06	20.00
7	0.07	40.00
8	0.08	52.00
9	0.09	75.00
10	0.10	45.00
11	0.11	20.00
12	0.12	0.00

Table 12.5 Evaluation of various terms

i	t_i	p_i	$p_i \cos\frac{2\pi t_i}{0.12}$	$p_i \sin\frac{2\pi t_i}{0.12}$	$p_i \cos\frac{4\pi t_i}{0.12}$	$p_i \sin\frac{4\pi t_i}{0.12}$	$p_i \cos\frac{6\pi t_i}{0.12}$	$p_i \sin\frac{6\pi t_i}{0.12}$
1	0.01	20.00	17.32	10.00	10.00	17.32	0.00	20.00
2	0.02	45.00	22.50	38.97	−22.50	38.97	−45.00	0.00
3	0.03	70.00	0.00	70.00	−70.00	0.00	0.00	−70.00
4	0.04	55.00	−27.50	47.63	−27.50	−47.63	55.00	0.00
5	0.05	35.00	−30.31	17.50	17.50	−30.31	0.00	35.00
6	0.06	20.00	−20.00	0.00	20.00	0.00	−20.00	0.00
7	0.07	40.00	−34.64	−20.00	20.00	34.64	0.00	−40.00
8	0.08	52.00	−26.00	−45.03	−26.00	45.03	52.00	0.00
9	0.09	75.00	0.00	−75.00	−75.00	0.00	0.00	75.00
10	0.10	45.00	22.50	−38.97	−22.50	−38.97	−45.00	0.00
11	0.11	20.00	17.32	−10.00	10.00	−17.32	0.00	−20.00
12	0.12	0.00	0.00	0.00	0.00	0.00	0.00	0.00
$\sum_{i=1}^{12}()$		477.00	−58.81	−4.90	−166.00	1.73	−3.00	0.00
$\frac{1}{6}\sum_{i=1}^{12}()$		79.50	−9.80	−0.82	−27.67	0.29	−0.50	0.00

Table 12.6 Approximate pressure variation with time

Time (s), t_i	Pressure (kN/m^2), $p(t_i)$
0.00	0.00
0.01	17.27
0.02	48.72
0.03	66.60
0.04	57.03
0.05	33.75
0.06	22.39
0.07	35.06
0.08	58.94
0.09	68.23
0.10	49.64
0.11	17.59
0.12	1.78

Appendix 2: Experimental Modal Analysis

A12.2.1 Introduction

Modal testing is the form of vibration testing of an object whereby the natural (modal) frequencies, modal masses, modal damping ratios, and mode shapes of the object under test are determined. A modal test consists of an acquisition phase and an analysis phase. The complete process is often referred to as a modal analysis or experimental modal analysis. The modal analysis provides a set of modal parameters that characterize the dynamic behavior of a structure.

There are several ways to do modal testing, but impact hammer testing and shaker testing are most common. The fundamental problem of modal parameter estimation consists of adjusting (estimating) the parameters in the model, so that the data predicted by the model approximates the measured data as closely as possible.

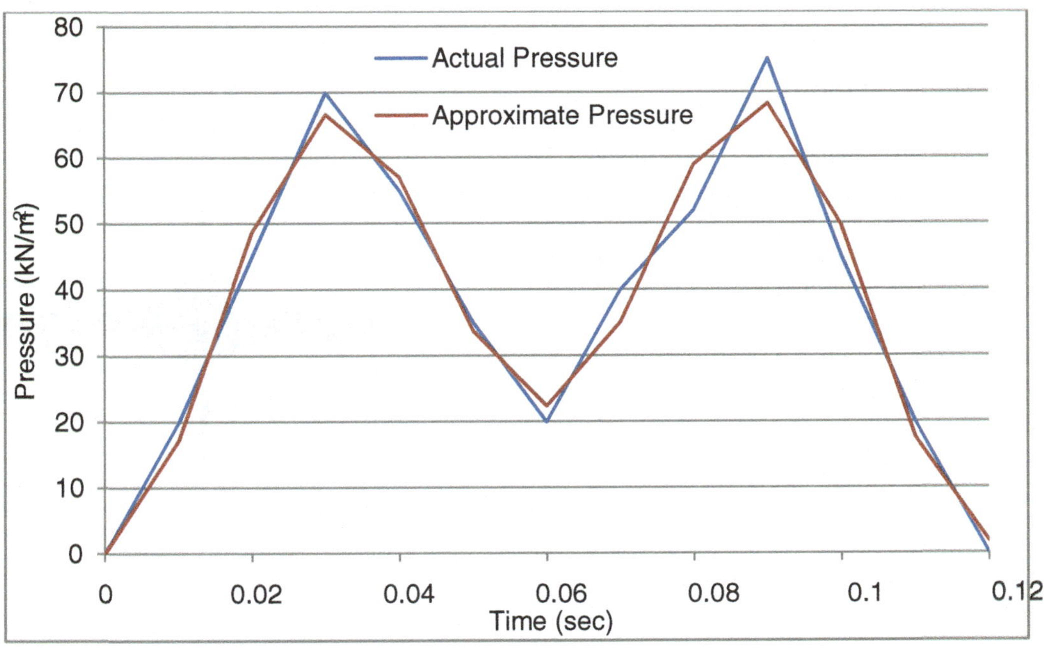

Fig. 12.41 Actual pressure and approximate pressure with three Fourier harmonics

Fig. 12.42 Test setup for shake table testing of piping loop

A12.2.2 Curve Fitting Methods for Estimation of Modal Parameters

Curve fitting is a numerical process that is typically used to represent a set of experimentally measured data points by some assumed analytical function. The results of this curve fitting process are the coefficients, or parameters, that are used in defining the analytical function. With regard to the frequency response function, the parameters that are calculated are its modal parameters (i.e., modal frequency, damping, and residue).

Although there are several ways in which curve fitting methods can be categorized, the most straightforward is single-mode versus multiple-mode classification.

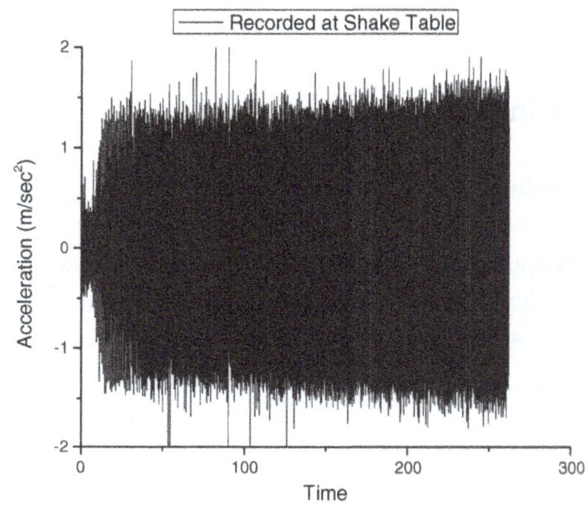

Fig. 12.43 Sine sweep signal

A12.2.3 Single-Mode Methods

The basic assumption for single-mode approximations is that in the vicinity of a resonance, the response is due primarily to that single mode. The resonant frequency can be estimated from the frequency response data by observing the frequency at which any of the following trends occur:

- The magnitude of the frequency response is a maximum.
- The imaginary part of the frequency response is a maximum or minimum.
- The real part of the frequency response is zero.
- The response lags the input by 90° phase.

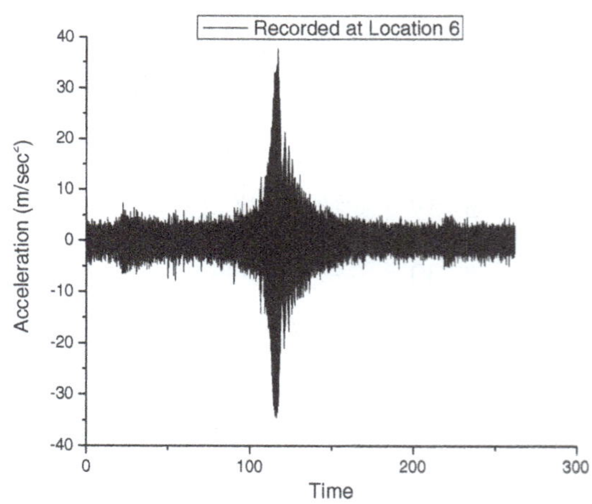

Fig. 12.44 Response of the piping loop at location 6

Example 12.7 Estimate the natural frequency of piping loop shown in Fig. 12.42 by testing.

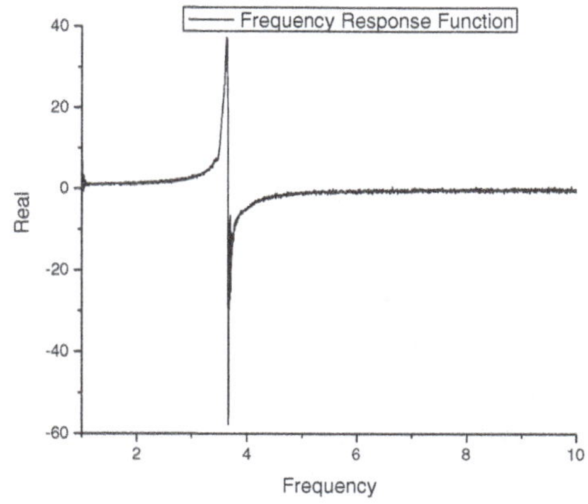

Fig. 12.45 Real part of frequency response function (FRF)

Solution:

Resonance search tests were carried out using the sine sweep signal as shown in Fig. 12.43 to estimate the natural frequency. The sweep rate used is 1 octave per minute. Starting initial frequency is 1 Hz; the sweep was carried out till 20 Hz which is the range of interest. Acceleration time histories (responses) have been measured at different locations. Response of the piping loop at location 6 is shown in Fig. 12.44.

Single-mode methods have been used to estimate the natural frequency of the piping loop. The real part of frequency response function (FRF) is shown in Fig. 12.45.

Fig. 12.46 Location of accelerometers on piping system and view of piping system at shake table

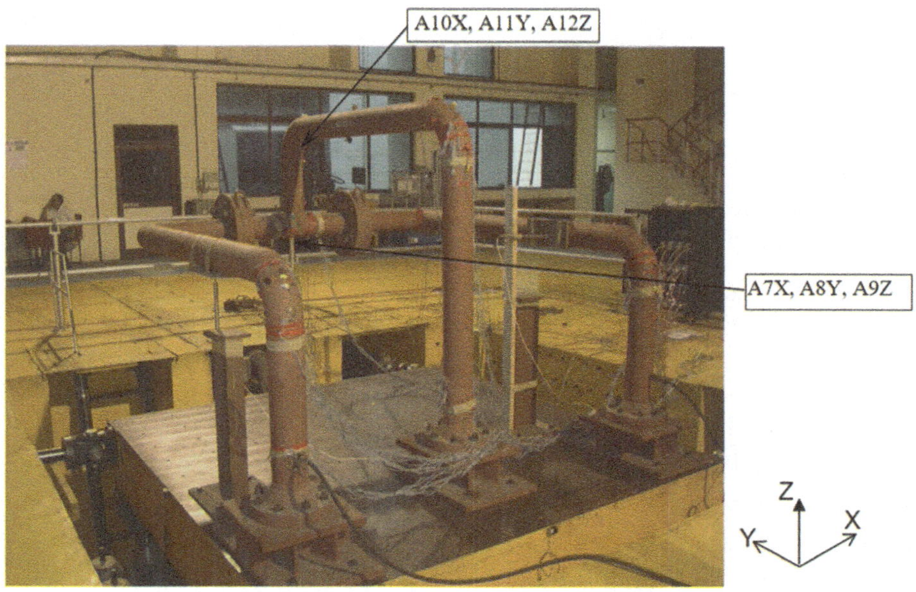

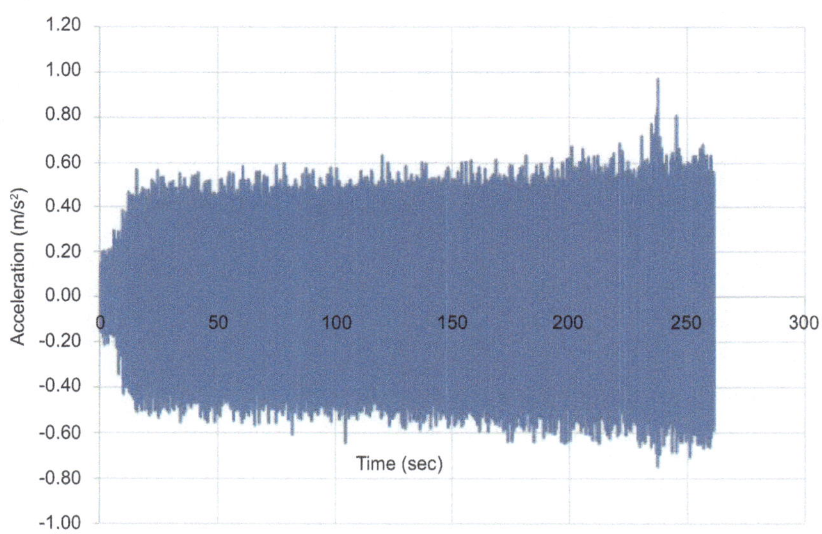

Fig. 12.47 **a** Input acceleration along *X*-direction **b** Response acceleration along *X*-direction **c** Power spectrum of input acceleration along *X*-direction **d** Power spectrum of response acceleration along *X*-direction **e** Coherency between input and response acceleration along *X*-direction

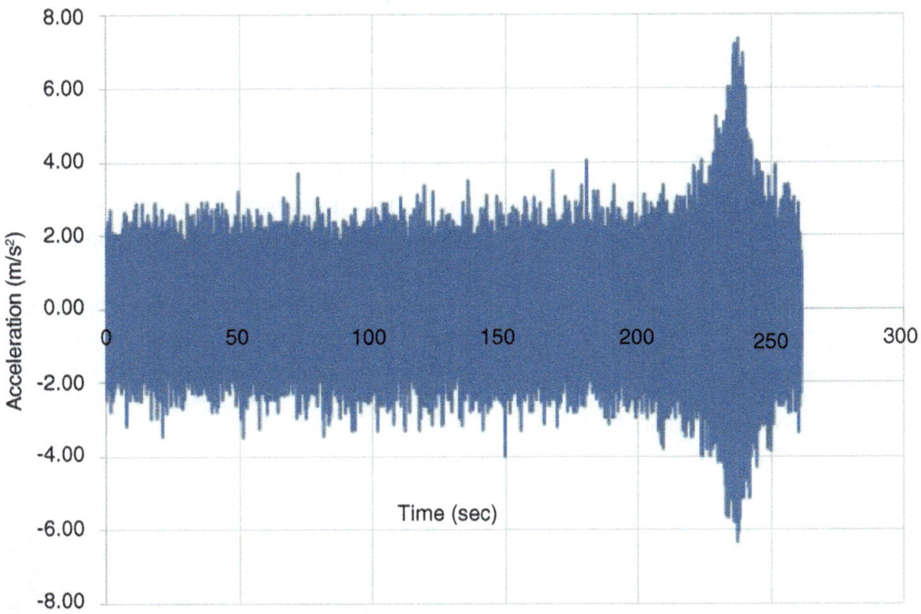

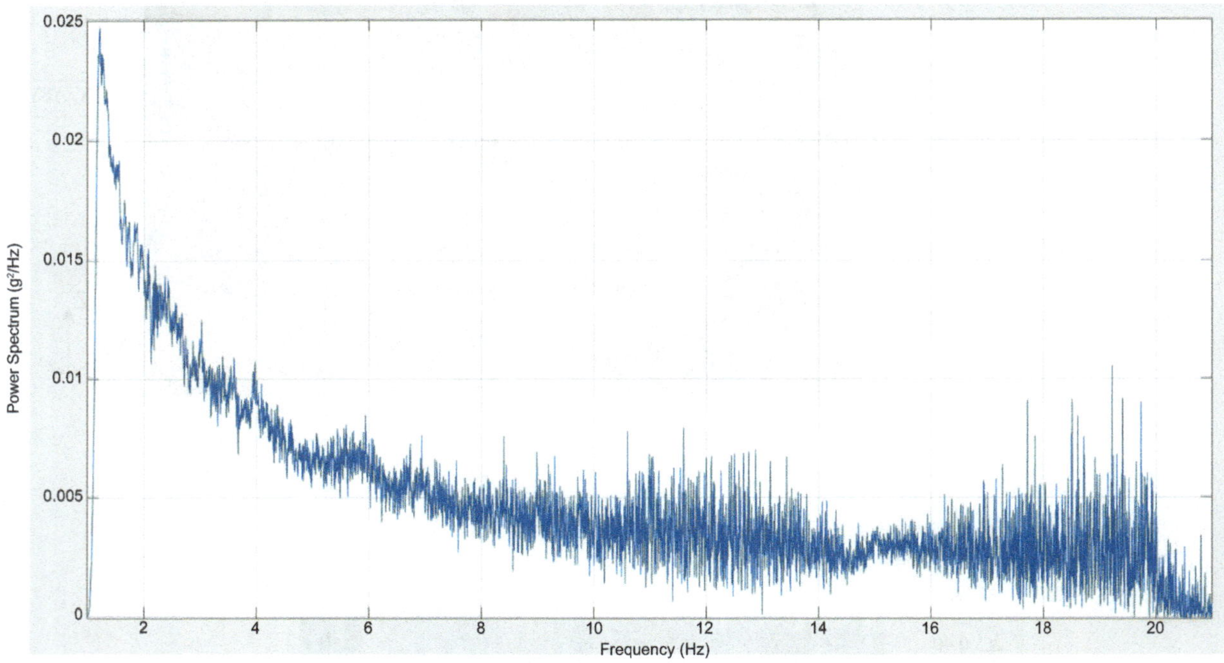

Fig. 12.47 (continued)

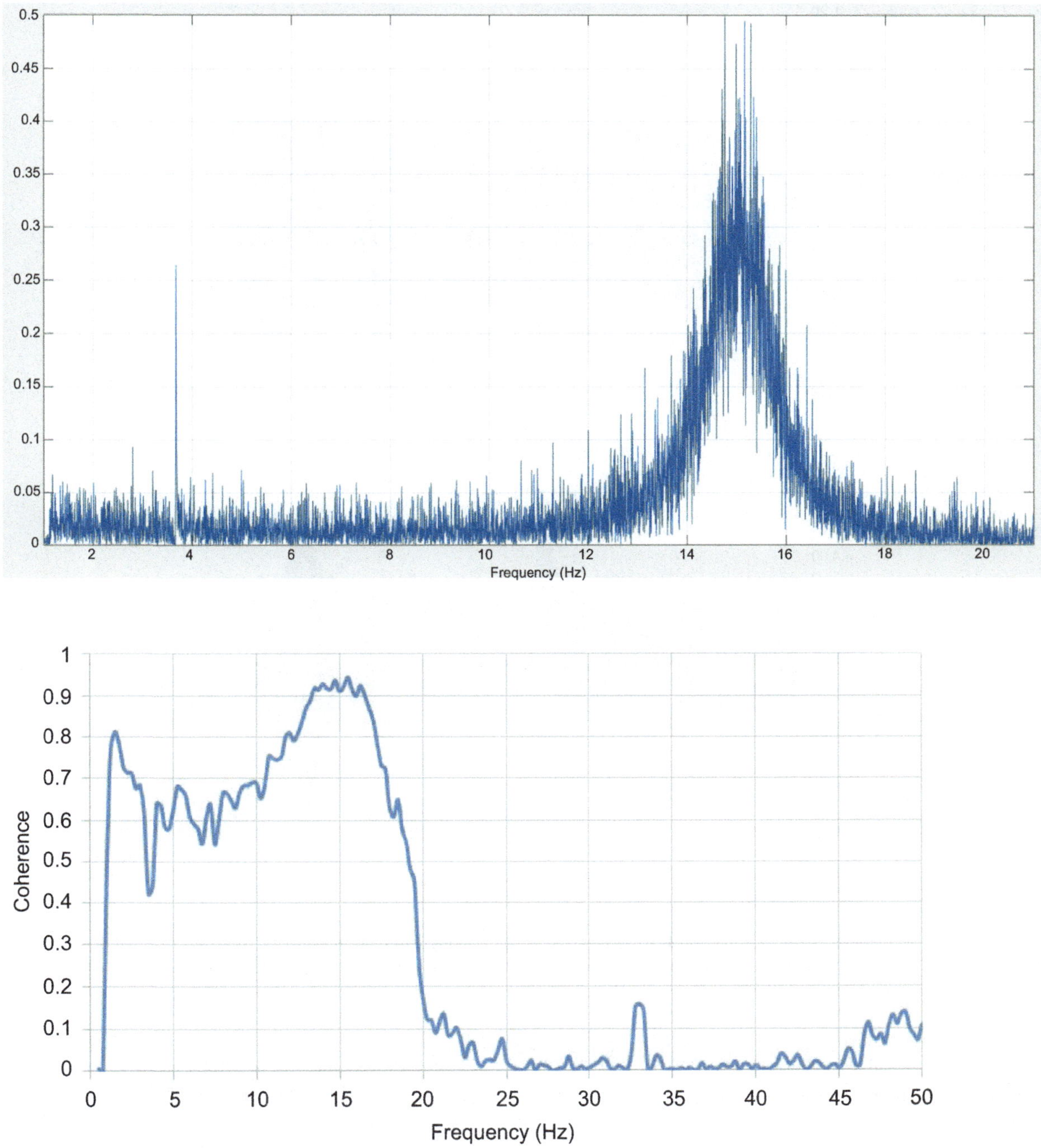

Fig. 12.47 (continued)

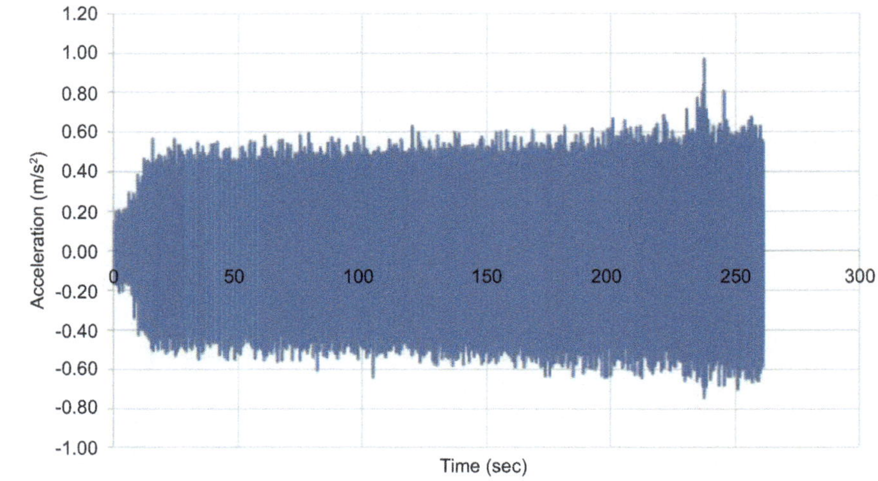

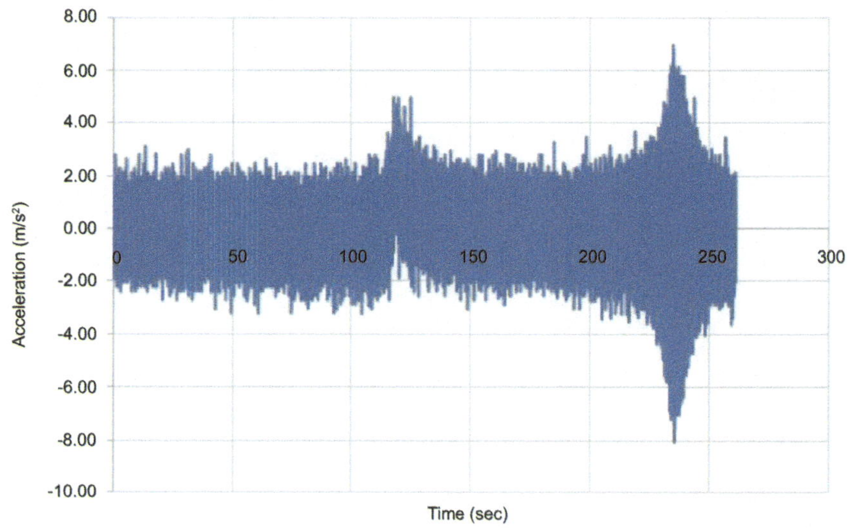

Fig. 12.48 **a** Input acceleration along *X*-direction **b** Response acceleration along *X*-direction **c** Power spectrum of input acceleration along *X*-direction **d** Power spectrum of response acceleration along *X*-direction **e** Coherency between input and response acceleration along *X*-direction

From Fig. 12.45, it can be seen that real part of FRF is zero at 3.65 Hz, so the natural frequency is 3.65 Hz.

Appendix 3: Experimental Study for Evaluating Coherency of Response and Input Signal

An example of piping system shown in Fig. 12.46 is considered for evaluating coherence of response signal and excitation signal. Response acceleration was measured at various points as shown in Fig. 12.46.

The input signal has a sine sweep nature with starting frequency of 1 Hz and ending at 20 Hz with sweep rate of 1 octave per minute. Correlations and coherence between input signals along *X* and *Y* directions for locations A7X, A8Y, A10X, and A11Y have been found out, and the results are plotted in Figs. 12.47, 12.48, 12.49 and 12.50.

A.3.1 Coherence Between Input and A7X

Time history of input is shown in Fig. 12.47a.

Time history of A7X is shown in Fig. 12.47b:

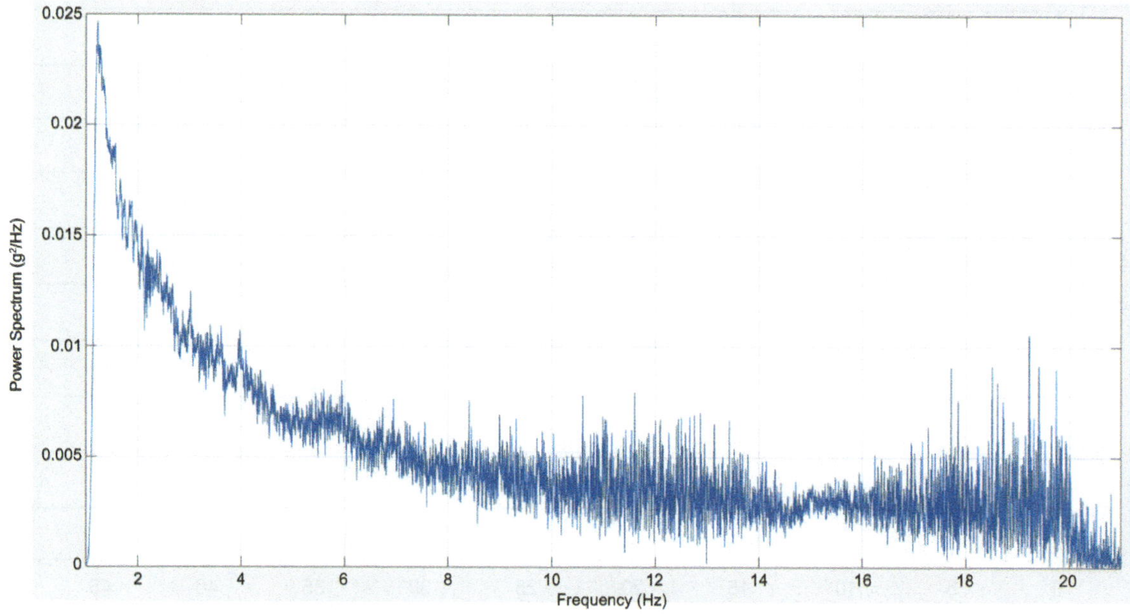

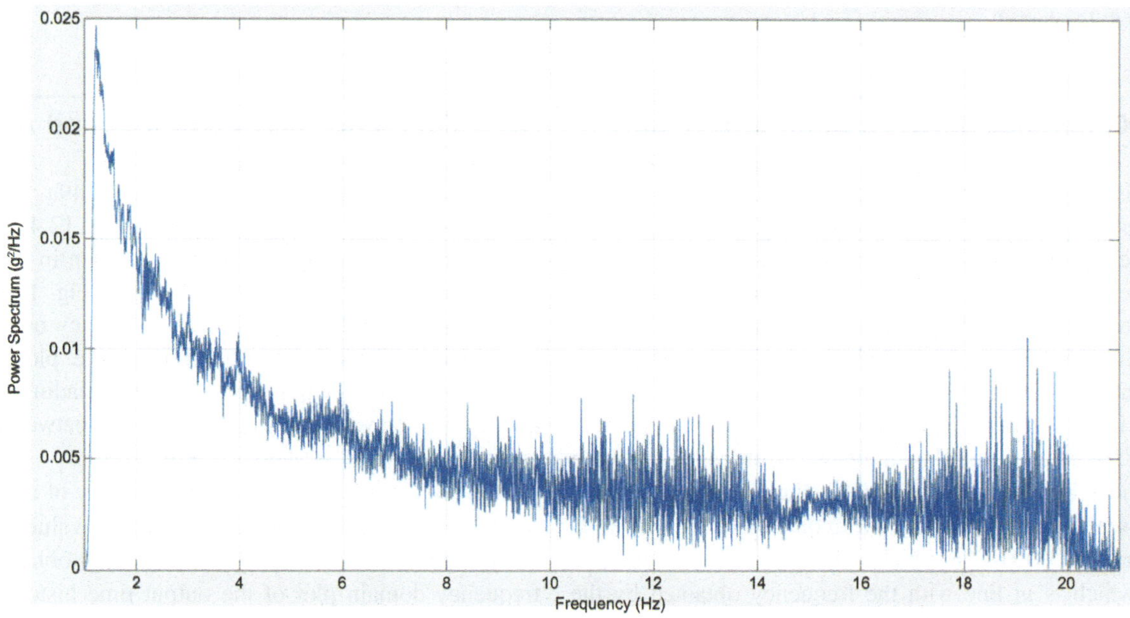

Fig. 12.48 (continued)

Power spectrum of input signal is shown in Fig. 12.47c.
Power spectrum of A7X is shown in Fig. 12.47d.

Observation: The predominant frequency of the system is around 15.0 Hz as identified from the plot of **Fourier** spectrum for the output accelerometer reading. This frequency is to be verified using "coherency" between the input and output signals as shown in Fig. 12.47e.

It is seen that the predominant frequency of the system is found to be 15.0 Hz (in which coherency value is close to unity) which is in line with the frequency obtained by the frequency domain plot of the output time history. Hence, the fundamental frequency of the system is validated by both the approaches, which confirms the correctness of the approach.

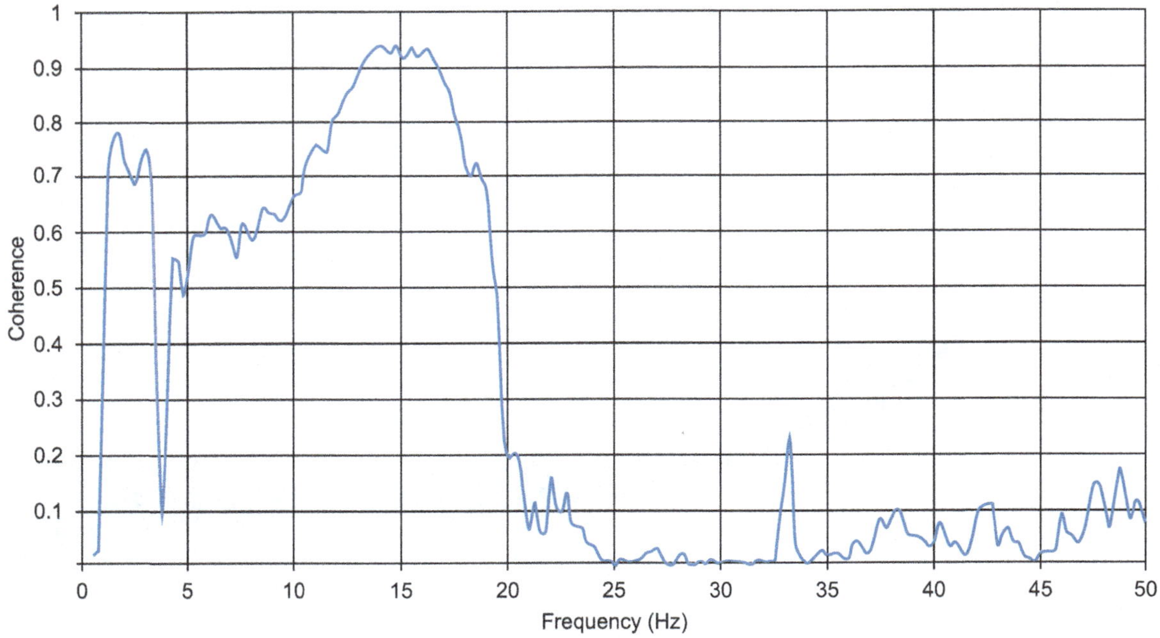

Fig. 12.48 (continued)

A.3.2 Coherence Between Input and A10X

Time history of input is shown in Fig. 12.48a.

Time history of A10X is shown in Fig. 12.48b.

Power spectrum of input signal is shown in Fig. 12.48c.

Power spectrum of A10X is shown in Fig. 12.48d.

Observation: The predominant frequency of the system is around 15.0 Hz as identified from the plot of **power** spectrum for the output accelerometer reading. This frequency is to be verified using "coherency" between the input and output signals as shown below. Coherence between input and A10X signals is shown in Fig. 12.48e.

It is seen that the predominant frequency of the system is found to be 15.0 Hz (in which coherency value is close to unity) which is in line with the frequency obtained by the frequency domain plot of the output time history. Hence, the fundamental frequency of the system is validated by both the approaches, which confirms the correctness of the approach.

A.3.3 Coherence Between Input and A11Y

Time history of input is shown in Fig. 12.49a.

Time history of A11Y is shown in Fig. 12.49b.

Power spectrum of input signal is shown in Fig. 12.49c.

Power spectrum of A11Y is shown in Fig. 12.49d.

Observation: The predominant frequency of the system is around 5.78 Hz as identified from the plot of power spectrum for the output accelerometer reading. This frequency is to be verified using "coherency" between the input and output signals as shown in Fig. 12.49e.

It is seen that the predominant frequency of the system is found to be 5.0 Hz (in which coherency value is close to unity) which is in line with the frequency obtained by the frequency domain plot of the output time history. Another peak is observed at 13.25 Hz which represents the second mode of the piping system in *Y*-direction which is also seen in the power spectrum plot.

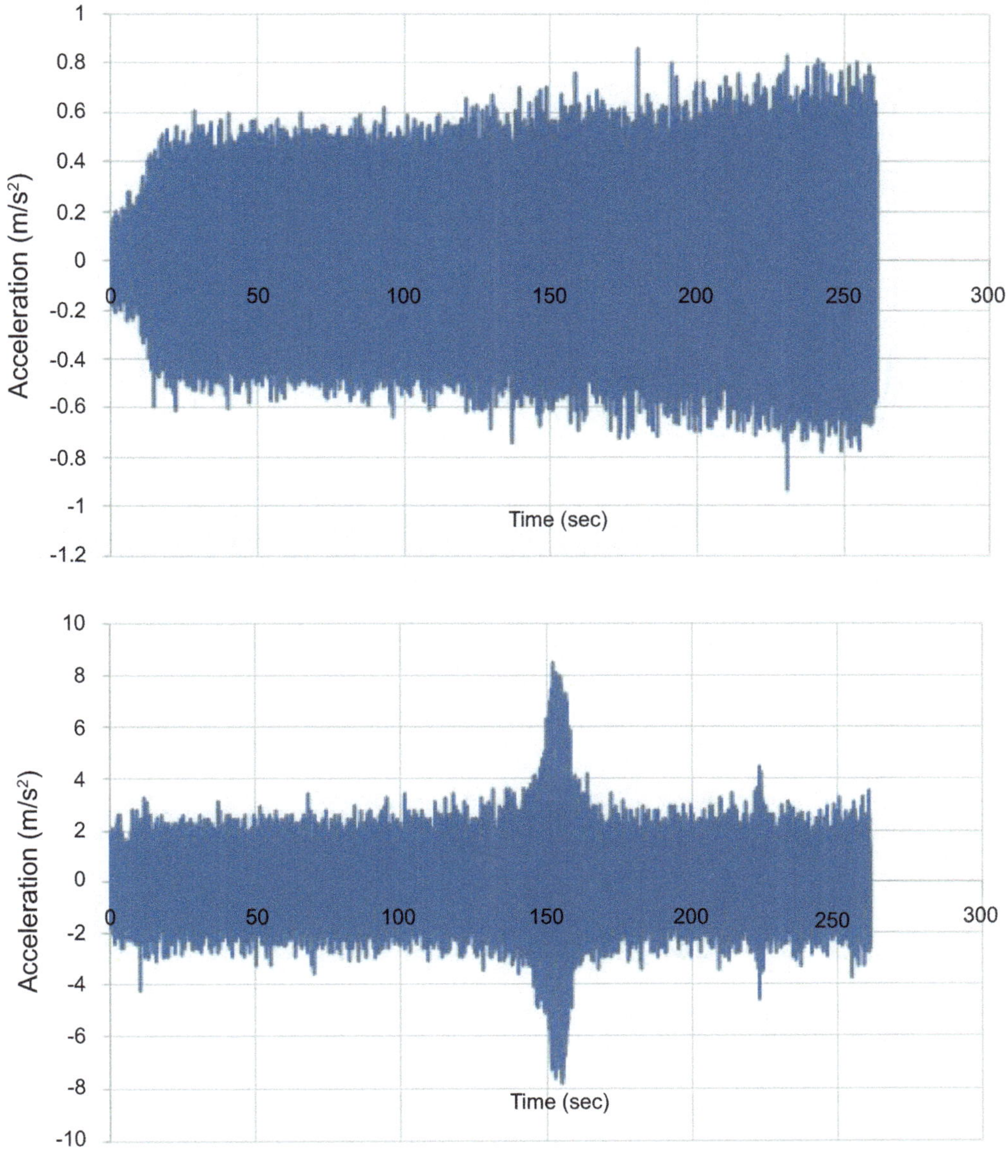

Fig. 12.49 **a** Input acceleration along *Y*-direction **b** Response acceleration along *Y*-direction **c** Power spectrum of input acceleration along *Y*-direction **d** Power spectrum of response acceleration along *Y*-direction **e** Coherency between input and response acceleration along *Y*-direction

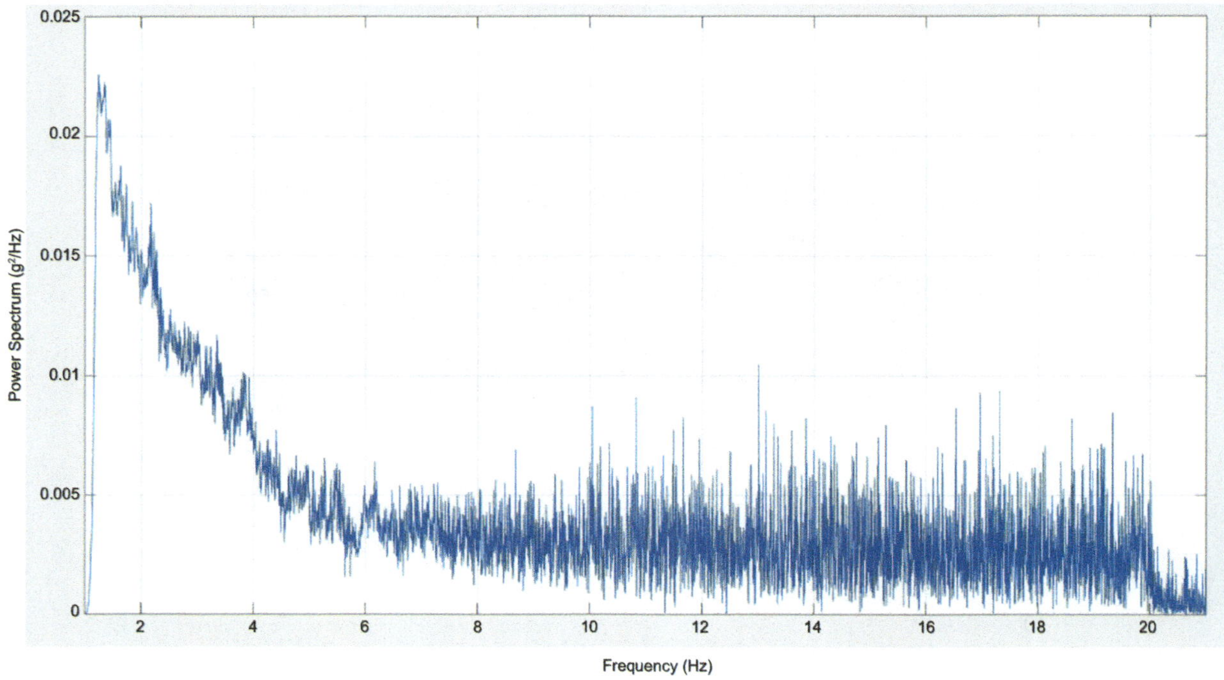

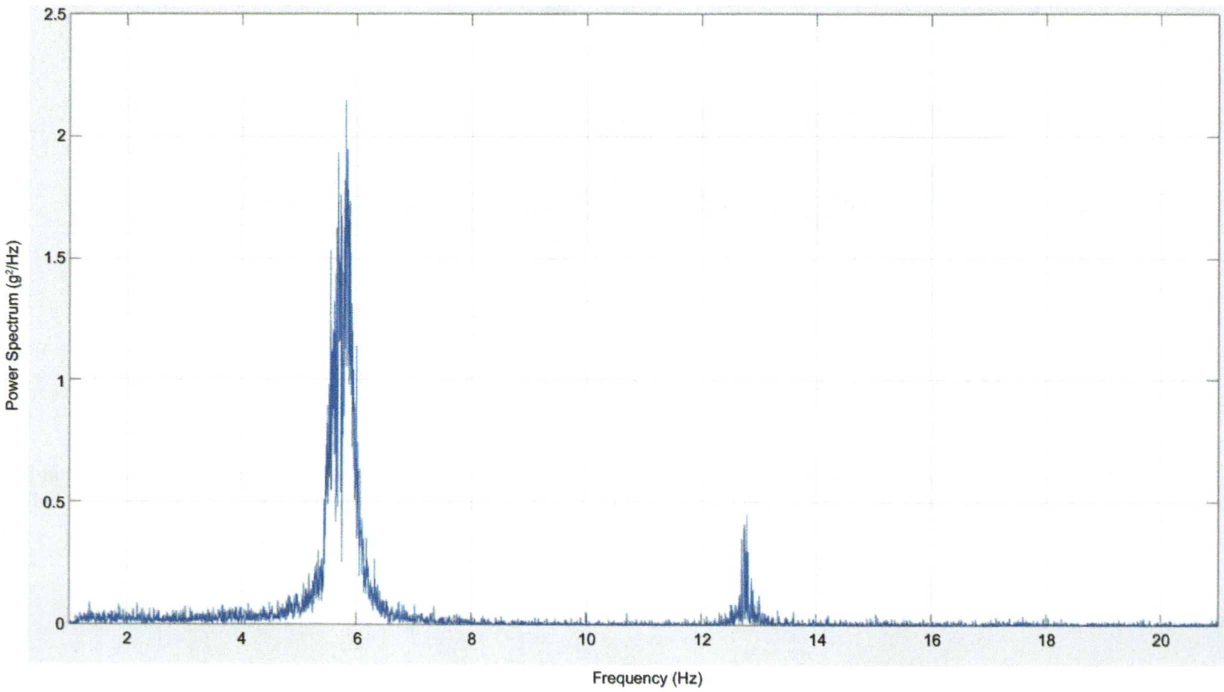

Fig. 12.49 (continued)

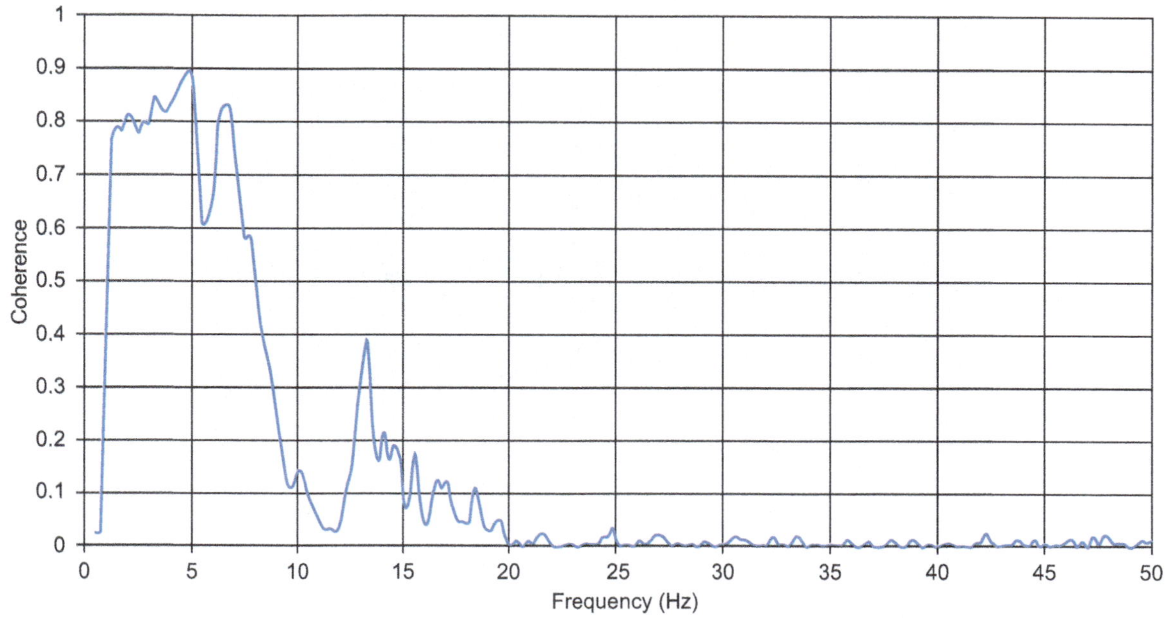

Fig. 12.49 (continued)

A.3.4 Coherence Between Input and A8Y

Time history of input is shown in Fig. 12.50a.

Time history of A8Y is shown in Fig. 12.50b.

Power spectrum of input signal is shown in Fig. 12.50c.

Power spectrum of A8Y is shown in Fig. 12.50d.

Observation: The predominant frequency of the system is around 5.0 Hz as identified from the plot of power spectrum for the output accelerometer reading. This frequency is to be verified using "coherency" between the input and output signals as shown in Fig. 12.50e.

It is seen that the predominant frequency of the system is found to be 5.0 Hz (in which coherency value is close to unity) which is in line with the frequency obtained by the frequency domain plot of the output time history. Another peak is observed at 13 Hz which represents the second mode of the piping system in *Y*-direction which is also seen in the **power** spectrum plot. Cross-correlation (or coherency) between signals A8Y and A10X in the piping system is shown in Fig. 12.51.

As it can be seen that there is no single frequency at which there is coherence greater than 0.9. It is because these modes are mutually perpendicular in direction; hence, there is little correlation between them as can be seen in the plot above as the modes are significantly uncoupled. The need of this study is to identify and validate the natural frequencies of the piping system which is given in Table 12.7.

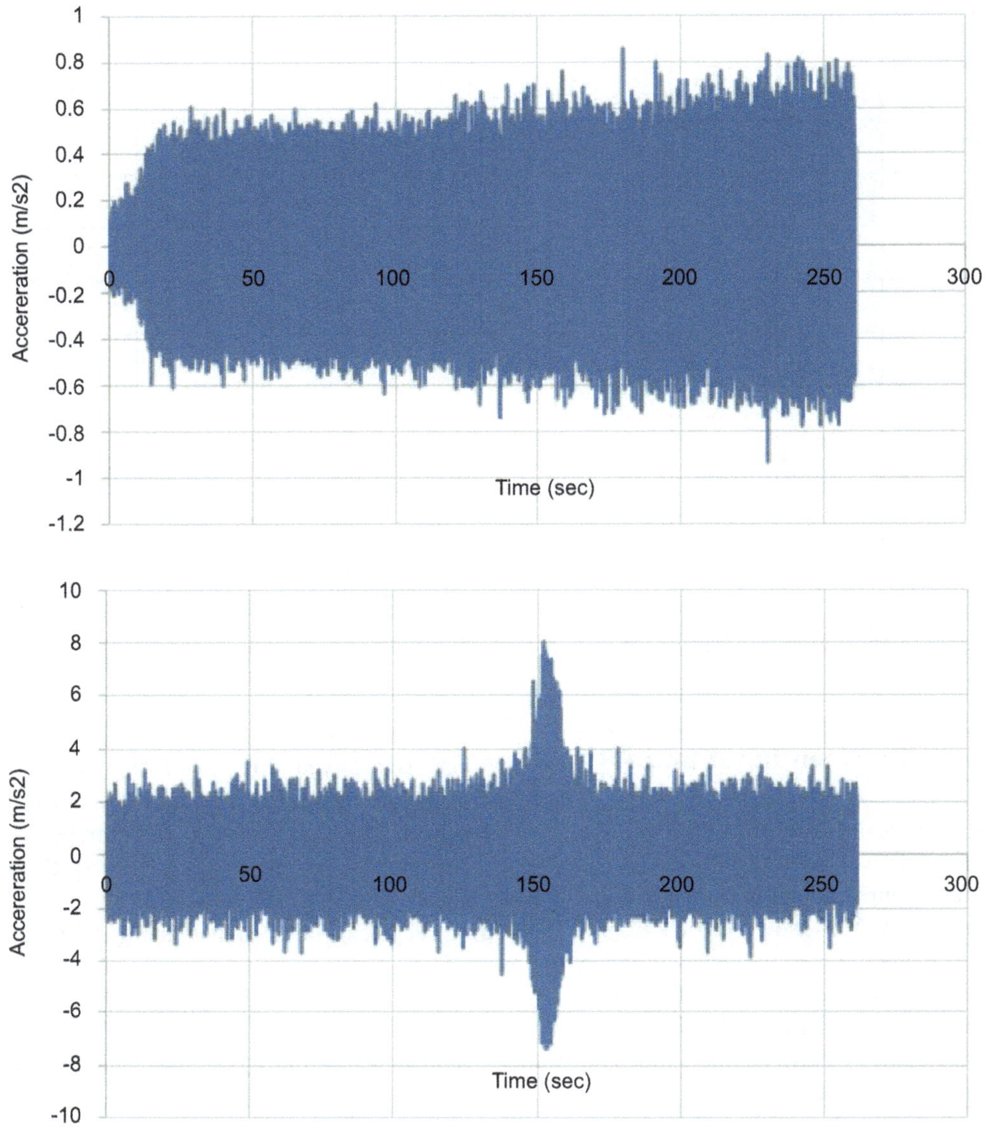

Fig. 12.50 **a** Input acceleration along *Y*-direction **b** Response acceleration along *Y*-direction **c** Power spectrum of input acceleration along *Y*-direction **d** Power spectrum of response acceleration along *Y*-direction **e** Coherency between input and response acceleration along *Y*-direction

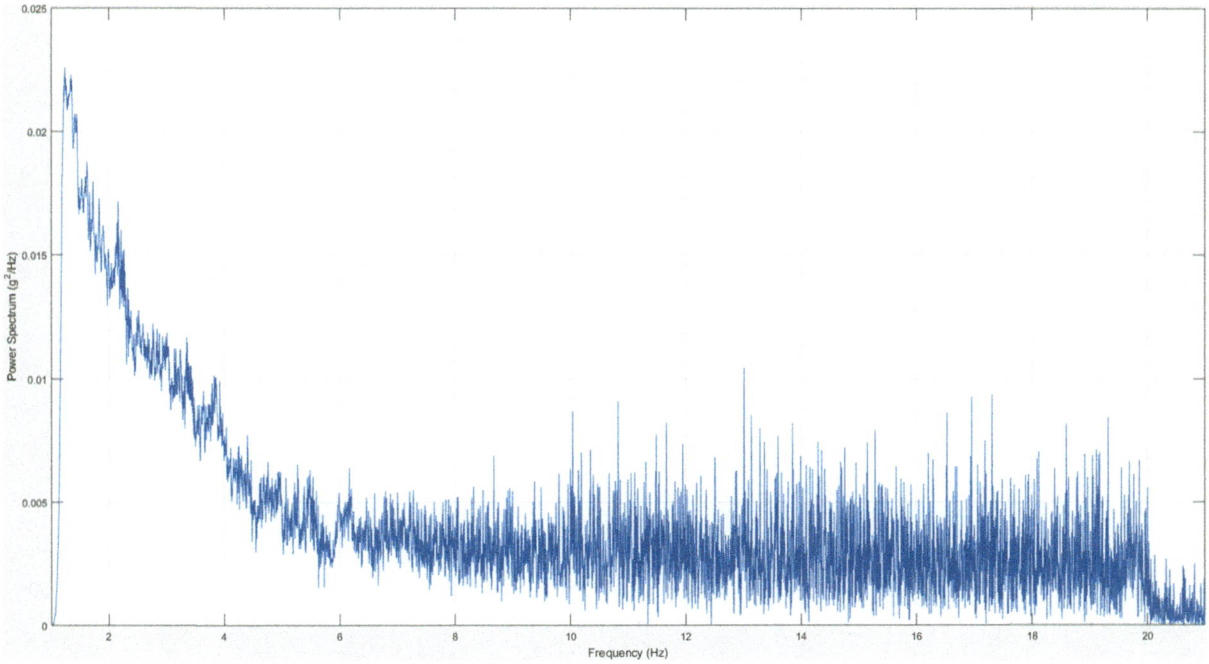

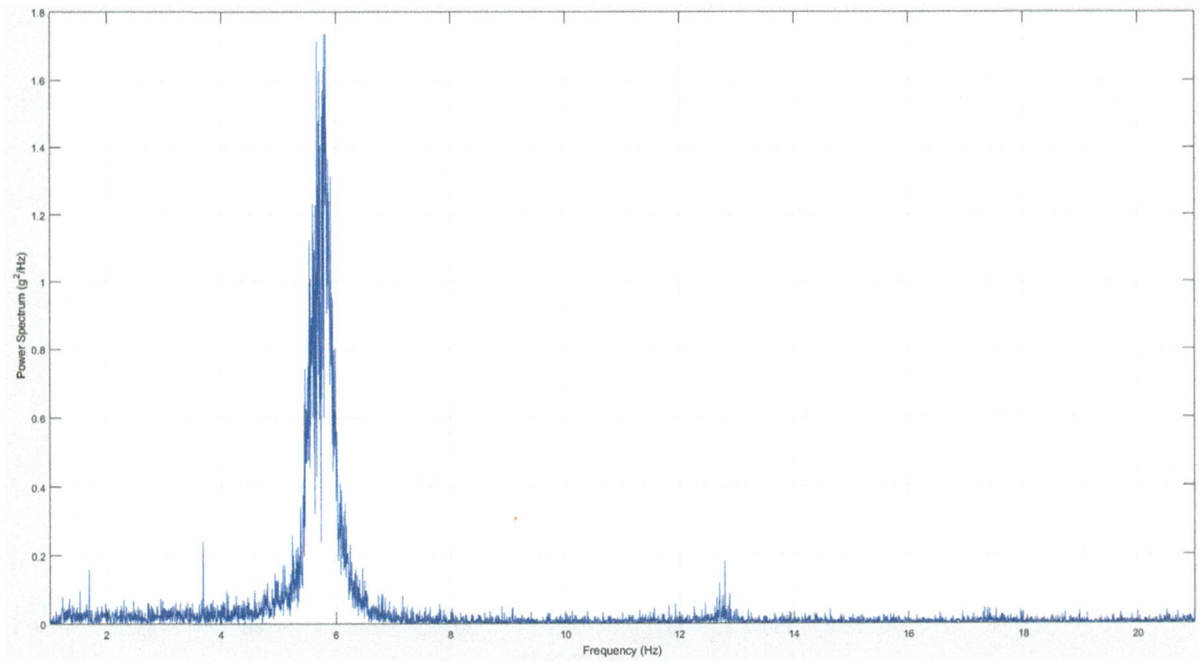

Fig. 12.50 (continued)

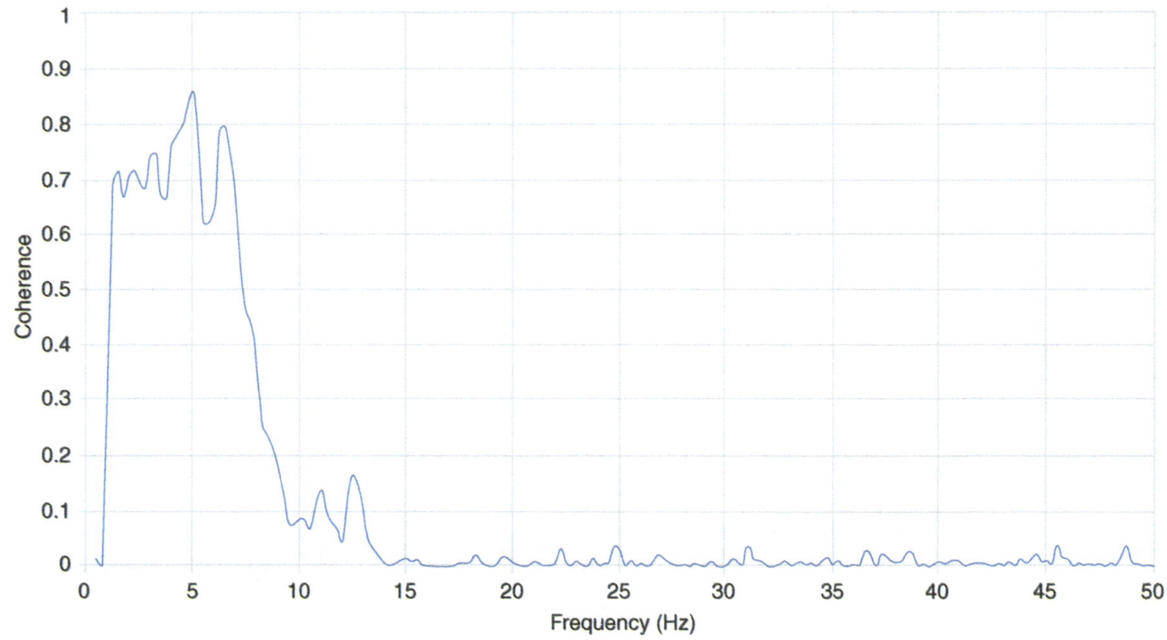

Fig. 12.50 (continued)

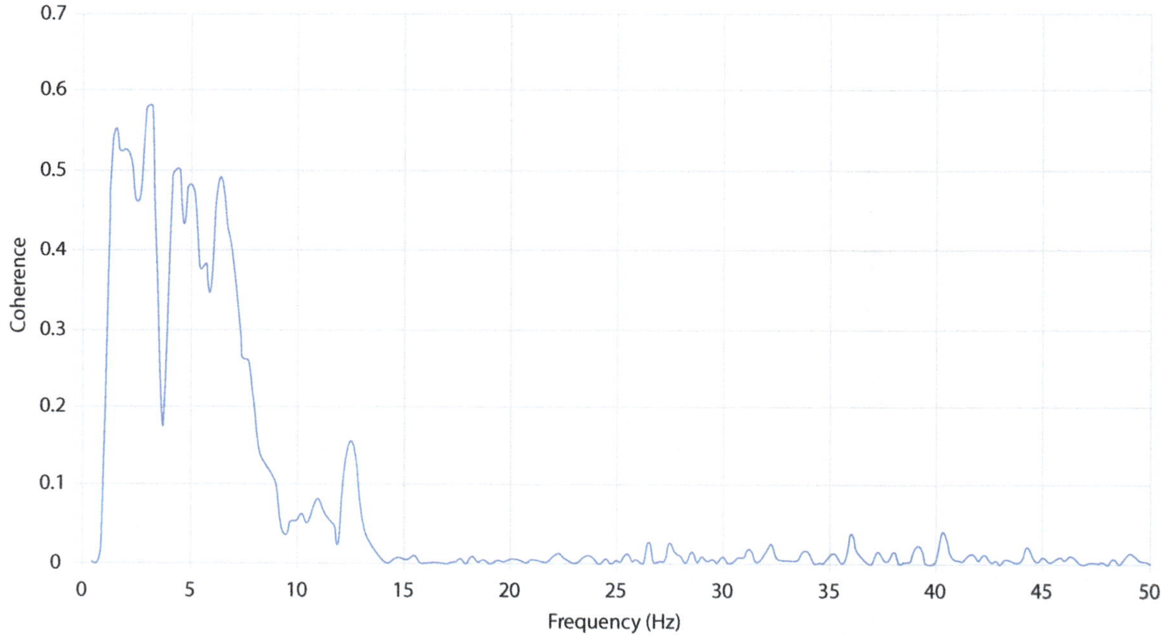

Fig. 12.51 Coherency between response acceleration along X and Y directions

Table 12.7 Natural frequencies of the piping system

Direction	Frequency (Hz) by analysis
X-direction	15.00
Y-direction	5.87

References

1. IEEE Std 344 (1987) IEEE recommended practice for seismic qualification of class 1E equipment for nuclear power generating stations
2. ASCE 4-98 (1998) Seismic analysis of safety-related nuclear structures and commentary
3. Inman DJ (1996) Engineering vibration

Further Reading

4. Thomson WT et al (1998) Theory of vibrations with applications, 5th edn
5. Kostarev VV, Berkovski AM, Schukin AJ (1999) Upgrading of dynamic reliability and life extension of piping by means of high viscous damper technology. Transactions of PVP ASME Conference, Boston
6. Gemant A (1936) A method of analyzing experimental results obtained from elasto viscous bodies. Physics 7(8):311–317

13 Retrofitting of Structures and Equipments

G. R. Reddy, D. K. Jha, and Gaurav Verma

Health of the structure need to be ensured during it service life for its intended seismic performance. Requalification and retrofitting of structures is a continuous process to ensure good seismic performance

Symbols

$\sum$	Summation
$\int$	Integration
$\dot{x}_s$	Shaft or rod velocity
γ	Proportionality coefficient for viscous dampers
M	Mass
k	Stiffness
f	Natural frequency
A_1/A_2	Amplitude ratio
Z	Damping ratio
ζ_s	Damping ratio
ζ_d	Damping ratio after first iteration
f_d	Damped frequency
c	Damping coefficient
c_{cr}	Critical damping coefficient
ω	Circular natural frequency
X	Displacement
$\dot{x}$	Velocity
$\ddot{x}_g$	Ground acceleration
$\ddot{x}$	Acceleration
$\ddot{x}_t$	Total acceleration
R_d	Response reduction factor/dynamic acceleration factor
$\bar{\omega}$	Excitation frequency
T	Torque
μ_b	Friction factor (bolt and steel plate)
F_b	Normal force on the bolt
D_b	Diameter of the bolt
F	Slip load
μ_p	Friction factor (liner and steel plate)
ΔE	Energy dissipated by a damper in one hysteretic cycle
Σ	Strain energy of the system
E_{KE}	Kinetic energy
E_{PE}	Potential energy
E_D	Energy dissipated due to structural damping
E_d	Additional energy dissipation due to passive damper
∞	Infinity
a_1	Original cross section of lead
a_2	Extruded area of lead
μ	Friction factor on the bulge
α	Angle on the bulge
L	Length of the rod
g	Proportionality constant for lead extrusion damper
a	Length
b	Breadth
t	Thickness of the X-plate
I	Second area moment of inertia.
σ_y	Bending stress at yield
M_y	Bending moment at yield
F_y	Force at yield
d_y	Displacement at yield
ε	Strain
H	Strain hardening rate

G. R. Reddy (✉) · G. Verma
Bhabha Atomic Research Centre, Mumbai, India
e-mail: rssred@barc.gov.in

G. Verma
e-mail: gverma@barc.gov.in

D. K. Jha
Bhabha Atomic Research Centre, Mumbai, India
e-mail: dkjhabarc@gmail.com

G. R. Reddy et al., *Textbook of Seismic Design*,
https://doi.org/10.1007/978-981-13-3176-3_13

Y	Distance from neutral axis
δ	Displacement (Ramberg–Osgood relation)
δ_y	Displacement at yield (Ramberg–Osgood relation)
k_1	Stiffness of the plate
K	Stiffness of the base structure
X	Displacement of the base structure
$\dot{X}$	Velocity of the base structure
$\ddot{X}$	Acceleration of the base structure
η	Mass ratio
ψ, u	Independent variables
H	Height of water level
σ, ϕ, T_h	Dependent variables
λ	Damping parameter
v	Kinematic viscosity
τ	Wave Number
S	Surface contamination factor

13.1 Introduction

As time passes, similar to human beings, the health of a structure deteriorates and capacity of the structure decreases considerably. A number of structures existing today were designed and constructed well before the advent of seismic design philosophy, and the failures of these structures due to recent earthquakes have exposed their vulnerability to the seismic forces. As per the current design codes and standards, after about every ten years, the health of the structure needs to be assessed and necessary repairs and retrofitting needs to be undertaken. Many of these structures such as bridges, hospitals, schools, pipelines, electric supply units tend to incur large direct and indirect losses to the society. Thus, in order to retain these structures, seismic re-qualification and retrofitting become necessary which further increases its chances of surviving future earthquakes [42, 55].

Seismic retrofitting of a structure refers to the technology of making an existing structure earthquake safe. It is a science that involves seismic assessment of a structure, comparing the capacity of the structure against the seismic demand laid on the structure, establishing the performance, identification of weak links of a structure, designing a suitable remedy for the weakness, and implementation of the same on site. Today, there are many possible alternative methods available for retrofitting, and therefore, it becomes important for a designer to be aware of the strength and limitations of each method in order to take a right decision that is technically and economically suitable and viable.

13.2 Need for Re-qualification and Retrofitting

The retrofitting of a structure may be needed in any one or more of the following cases:

(a) The structure was not originally designed for seismic loads. In most cases, old structures were not originally built against seismic demands. For those, it is important to check the existing strength toward the current seismic demands and if needed to retrofit the same.
(b) The design basis ground motion for the site gets revised. As the knowledge of seismic activity of a particular region grows, the authorities may allow a higher seismic demand associated with a particular region. Change of seismic zones from IS 1893:1984 to IS 1893:2002 where Zone I was merged with Zone II is a typical example. In such cases, the structures whose seismic zone gets upgraded might need retrofitting as the design and maximum considered earthquake has changed.
(c) Structure of building to be extended (e.g., extra floor to be constructed). In such cases, the dynamic characteristics of the structure change and it may attract higher forces. In such cases, also retrofitting may be needed.
(d) Functionality (importance) of the building changes. Consider a case where a building designed as a residential building is required to be used for the purpose of running school or an office building converted to hospital. In such cases, the importance of the building becomes higher and it might need retrofitting (Table 13.1 and Fig. 13.1).

Table 13.1 Levels of performance

Level	Collapse prevention and life safety (L4)	Enhanced life safety (L3)	Enhanced damage control (L2)	Immediate occupancy (L1)
Level of damage expected due to EQ	No collapse but serious damages requiring major repairs/irrepairable, e.g., residential and office structures	Some structural damage. Building would be unusable temporarily, e.g., theaters, cinemas, etc.	Minor repairable damage would be expected after a major earthquake, e.g., historical structures	Structure within elastic range, e.g., hospitals, schools, and lifeline structures

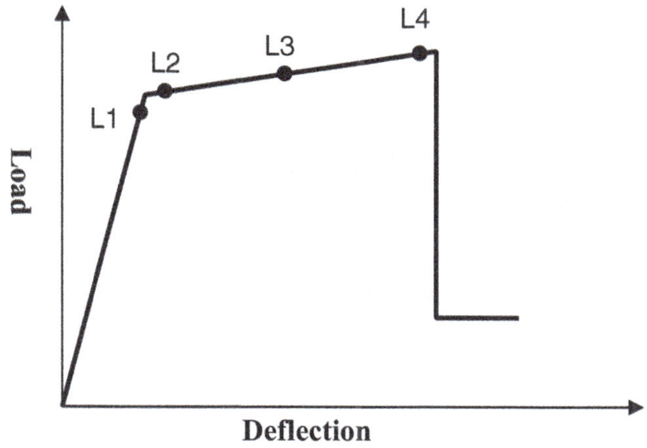

Fig. 13.1 Levels of seismic performance

(e) Changes in design methodology. As the knowledge of seismic behavior of systems grows, the prevalent design methodologies may be replaced by a better and more suitable methodology. For example, in structures: Soft story design which was accepted earlier is now unacceptable after the Northridge earthquake in 1994, Kobe earthquake 1995, etc., in the present design because soft story buildings failed severely and caused large damages to the vehicles parked in the ground floor; in piping and components, method of design for high stiffness is becoming obsolete and new designs are aiming for high flexibility and energy absorbing capacity and performance-based design in the place of conventional design. In such cases, the new design philosophy may render the existing structure unsuitable against seismic loads and may recommend retrofit.

(f) Re-qualification and Retrofitting become essential if the hazard classification of a facility changes.

13.3 Steps Involved in Seismic Re-qualification and Retrofitting

Generally, the various steps involved in the seismic re-qualification of a structure are as follows:

(a) Collection of complete information of the plant, structure, and system. The information includes design reports, drawings, and loadings including equipment, systems and checking of the hazard classification, if adequate or not.

(b) Estimation of the review basis earthquake and compare with the design basis.

(c) Performing the seismic walk down and generating a report which is a very important step of re-qualification. It helps in conducting a quick visual inspection to ascertain the deterioration taking place in the structure due to creep, shrinkage, corrosion, etc. It also helps in locating any abuses by cutting small portions of structural members and removing members for extra space settlement, foundation settlement and to accommodate for changes in equipment loading, leakages, seepages, etc.

Based on the findings of the walk down report, a detailed health assessment is then carried out. The following tests are important from the point of view of carrying out the health assessment of a structure:

(a) Visual inspection,
(b) Rebar location,
(c) Rebound hammer test,
(d) Ultrasonic pulse velocity test,
(e) Corrosion and carbonation test,
(f) Core cutting and strength evaluation, and
(g) Measure of tilt or large deflection, if any.

13.4 Methods of Retrofitting

There are numerous methods that can be employed and utilized for retrofitting of existing structures. The most suitable method depends on certain factors such as:

(a) Level of retrofitting aimed for,
(b) Technical constraints,
(c) Financial constraints,
(d) Time constraints.

The level of retrofitting refers to the performance level that sought to be achieved by the structure post-retrofitting. This is mainly decided by the importance of the structure. Technical constraints refer to the basic configuration of the structure, space availability, etc. Financial constraints refer to the amount of money that can be spent on retrofitting a structure. Time constraints refer to how much time can be allowed to spend over retrofitting. Mostly, the above-mentioned factors clash with each other, e.g., in general the time-saving techniques increase the financial burden. In such cases, an optimum solution has to be arrived at.

Almost all of the available retrofitting techniques available today can be divided into two categories, namely

(a) Strengthening methods/techniques,
(b) Response control methods/techniques.

Methods of Retrofitting

Increase the capacity/strength of the system

(Seismic Resistance Based Design)

e.g., (i) Concrete Jacketing, (ii) Steel Jacketing, (iii) FRP Wrapping.

Reduce the demand/forces on the system

(Seismic Response Control Based Design)

e.g., (iv) Elasto-Plastic Dampers, (v) Base Isolators, (vi) Lead Extrusion Dampers, (vii) Tuned Liquid Dampers, (viii) Friction Dampers.

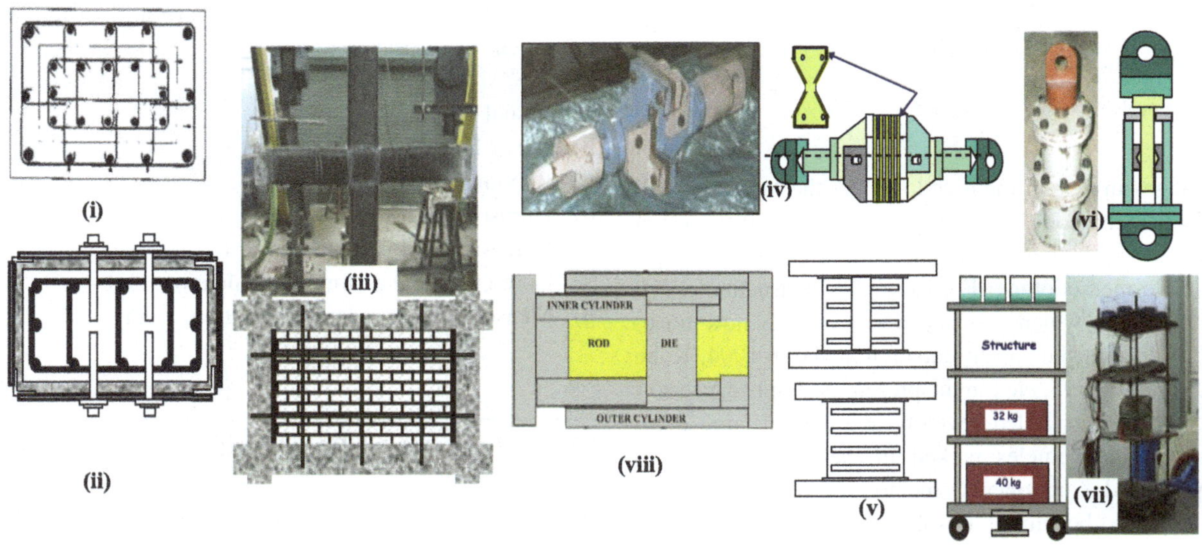

Fig. 13.2 Different methods of retrofitting

13.5 Retrofitting of Structures and Equipment Using Strengthening Techniques

The structures and equipment can be retrofitted using various strengthening techniques to be made suitable for the prevalent seismic demand. Broadly, the techniques may be categorized into two groups, namely global and local techniques.

13.5.1 Global Strengthening Methods

Global strengthening methods modify the dynamic characteristics of the whole structure or/and equipment by stiffening rather than strengthening. These methods include bracing, addition of shear walls, addition of masonry shear walls. Bracing (Fig. 13.3) is popular in removing the deficiencies in a structure due to soft story mechanism, where the bottom most story stand bare framed to act as parking space, whereas upper stories are stiffer due to masonry infill panels. Similarly, shear walls may be added in a framed or frame–shear wall structure to enhance the stiffness of the structure.

13.5.2 Local Strengthening Methods

Local strengthening methods involve in enhancing the strength of the members of the structural system so that they can withstand higher forces. Some of the techniques used for local strengthening include concrete jacketing, steel jacketing, external prestressing, strengthening with fiber composites. These techniques are presented in detail in this section.

13.5.2.1 Concrete Jacketing

Concrete jacketing is a popular and conventional method for strengthening of reinforced concrete (RC) members such as beams, columns, shear walls, footings. IS 13935:1993 [47] provides guidelines for retrofitting using this method and refers to it as casing that is providing additional cage of longitudinal and lateral tie reinforcement around the member and casting a concrete ring (Fig. 13.4). Both the strength and ductility of the section can be improved using this method.

Fig. 13.3 Bracing as global strengthening method

The procedure of concrete jacketing involves the addition of an adequate layer of concrete section as a jacket over existing RC original sections using longitudinal reinforcement and transverse ties. This additional concrete and reinforcement contribute to increase in strength and stiffness. It may require adequate doweling to the existing column. The longitudinal bars added need to be anchored to the foundation and made continuous through the slab if existing in the original structure. The process requires drilling of holes in existing column, slab, beams, and footings. For faster construction, precast concrete jackets are also available nowadays with customized design and installation technique. New longitudinal reinforcement is set around the existing column, and precast concrete segments are set around the new reinforcement. All segments are tied together by strands. After injecting non-shrinkage mortar between the existing concrete and precast concrete segment, prestressed force is introduced in the strands to assure the contact of the segments.

Example 13.1 An elevated RCC cylindrical tank of 15 m internal diameter, 200 mm wall thickness, and 5 m height is supported on a cylindrical (hollow) shaft of 4 m internal diameter, 200 mm wall thickness, and 20 m high above foundation level as shown in Fig. 13.5. The tank has a base slab of 250 mm and top slab of 150 mm thickness. The structure is rested on rock-type founding strata. The tank has a freeboard of 470 mm and the effective storage capacity of 800,000 L (800 m^3) of domestic/process water. The tank was initially designed for static load cases only. Now this particular tank structure was to be qualified for prevailing seismic demand as per IS 1893:2002 standard. The SDF representation of the tank is shown in Fig. 13.6. [Grade of Concrete: M20 ($E = 2.24 \times 10^{10}$ N/m^2, Unit Weight = 25 kN/m^3).]

Analysis for Dead Load:

Weight of stored water in the tank

$$= [\text{Volume of tank}] \times [\text{Unit weight of water}]$$

$$= \left[\left\{\pi(7.5)^2 \times 4.53\right\}\text{m}^3\right] \times \left[10\,\text{kN/m}^3\right] = 8005\,\text{kN}$$

Calculation of self-weight of the various components of RCC tank structure

Fig. 13.4 Concrete jacketing

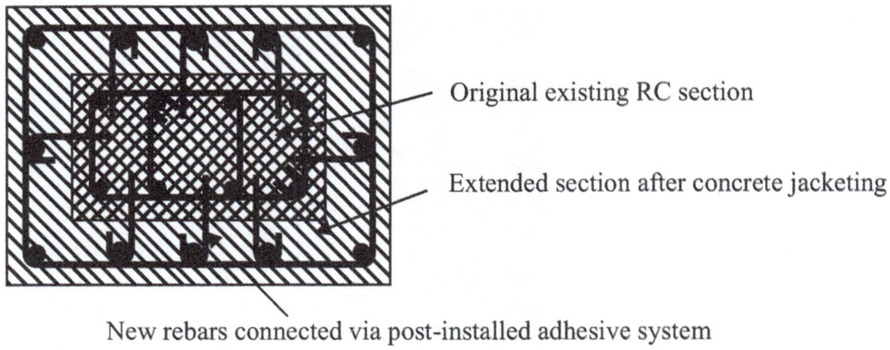

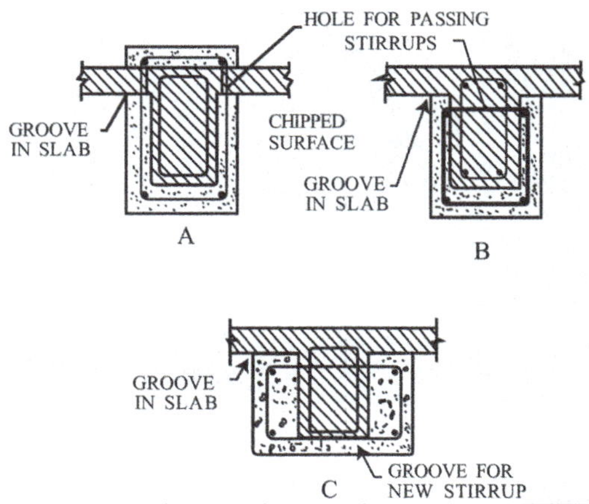

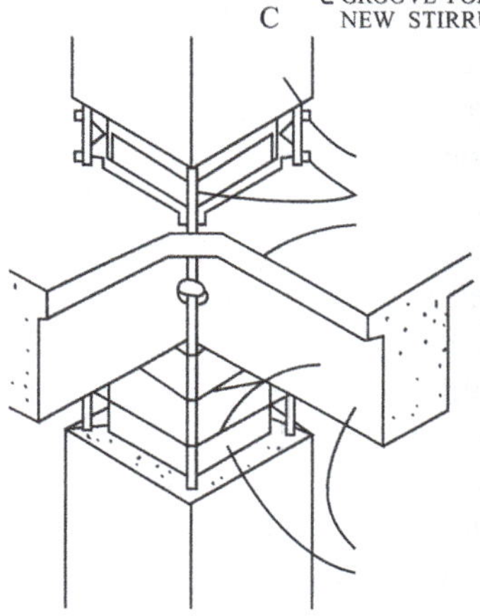

Self-weight of components of structure	Calculation	Weight (kN)
200-mm-thick RCC tank wall	[π(7.7^2 − 7.5^2) × 5 m^3] × [25 kN/m^3]	1194
150-mm-thick RCC top slab	[π(7.5)^2 × 0.15 m^3] × [25 kN/m^3]	663
250-mm-thick RCC base slab	[π(7.5)^2 × 0.25 m^3] × [25 kN/m^3]	1104
200-mm-thick, 20-m-high, and 4-m-dia shaft	[π(2.2^2 − 2.0^2) × 20 m^3] × [25 kN/m^3]	1319
Total		4280

Hence, total weight of RCC tank structure including the stored water; W = 12,285 kN

Maximum stress of shaft wall (at bottom)

$$= \left[\left(12{,}285 \times 10^3\right)/\left\{\pi\left(2200^2 - 2000^2\right)\right\}\right]\ \mathrm{N/mm^2}$$
$$= 4.66\,\mathrm{MPa}$$
$$< 5\,\mathrm{MPa}\ (\text{Permissible stress in direct compression of} \times \text{M20 Concrete})$$
$$< 20\,\mathrm{MPa}\ (\text{Characteristic compressive strength of} \times \text{M20 Concrete})$$

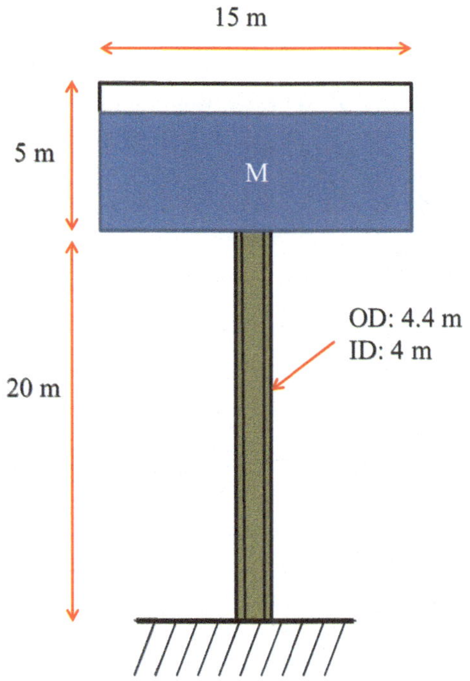

Fig. 13.5 Elevated RC cylindrical tank on a cylindrical hollow shaft

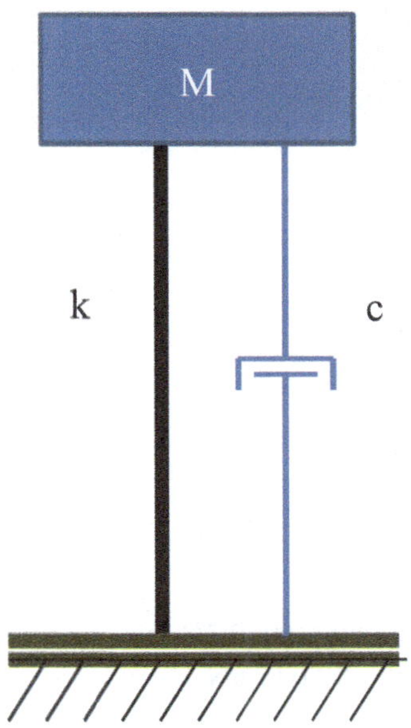

Fig. 13.6 SDF representation of elevated RC cylindrical tank

So, the tank structure is found to be safe for dead load case.

Analysis for Seismic Load:

Now the same tank structure has been re-evaluated for seismic load case with following seismic input:

Zone factor (z) = 0.36 (Zone V, Guwahati),
Importance factor (I): 1.5,
Response reduction factor (R): 3 (Considering ordinary moment-resisting frame (OMRF) detailing of RCC Structure),
Damping = 5%,
Response spectra.

Seismic mass,

$$M = (1/g) \times \left[8005 \times 10^3 + 1194 \times 10^3 + 663 \times 10^3 + 1104 \times 10^3 + 0.5{*}1319 \times 10^3\right] = 1185 \times 10^3\,\text{Kg}$$

Moment of inertia,

$$I = \frac{\pi}{64}\left(4.4^4 - 4.0^4\right) = 5.83\,\text{m}^4 = 5.83 \times 10^{12}\,\text{mm}^4$$

Stiffness

$$K = 3\,\text{EI}/L^3 = \left(3{*}2.24 \times 10^{10}{*}5.83\right)/20^3\,\text{N/m} = 48.97 \times 10^6\,\text{N/m}$$

Fundamental natural frequency (Angular);

$$\omega = \sqrt{\frac{K}{M}} = 6.42\,\text{rad/s}$$

Hence, fundamental natural frequency (Linear) $f = \frac{\omega}{2\pi} = 1.02\,\text{Hz}$ andFundamental time period = $T = 1/f = 0.98$ s.

Hence, spectral acceleration = S_a/g = 1.0 (refer response spectra given in Fig. 13.7)

$$A_h = z/2(I/R)S_a/g = (0.36/2)^*(1.5/3)^*1.0 = 0.09$$

Lateral seismic force on structure, $F = A_h{}^*W = 0.09^*[(1185 \times 10^3)^*9.81]\,\text{N} = 1.046 \times 10^6\,\text{N} = 1046\,\text{kN}$.

Hence base shear, V = 1046 kN and base moment, $M = F^*L = 1046^*20\,\text{kN m} = 20.92 \times 10^3\,\text{kN m} = 20.92 \times 10^9\,\text{N mm}$.

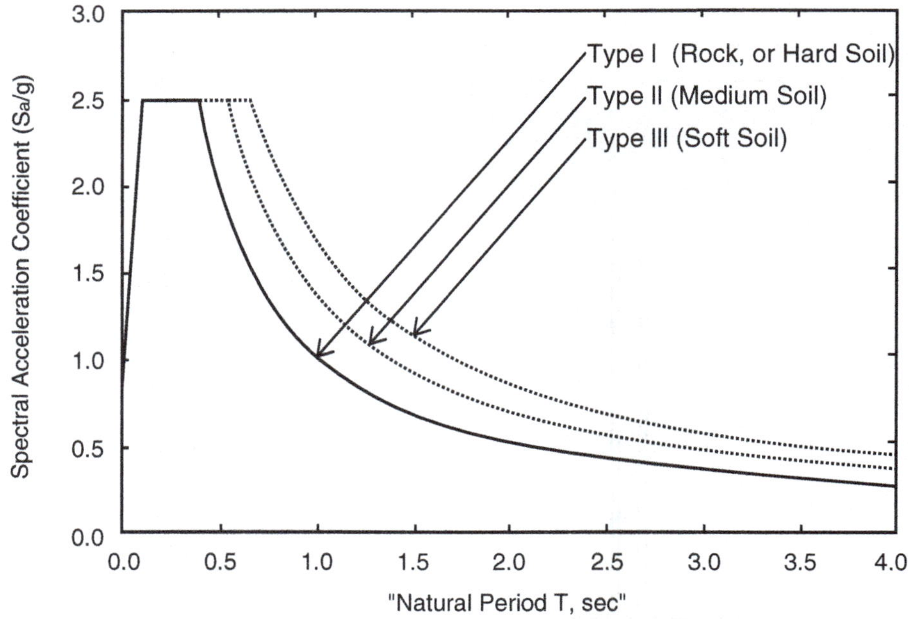

Fig. 13.7 Response spectra as per IS 1983: 2002

Section modulus of shaft, $Z = \frac{\pi}{32D}\left(D^4 - d^4\right) = \frac{\pi}{32^*4400}\left(4400^4 - 4000^4\right) = 2.65 \times 10^9\,\text{mm}^3$

Bending Stress (Maximum)

$$= M/z = \left(20.92 \times 10^9\right)/2.65 \times 10^9 = 7.9\,\text{MPa}$$

$$> \text{Allowable bending stress in compression for M20}$$

$$\times \text{ Concrete as per IS456 : 2000 is 7 MPa}$$

The bending stress in the concrete section (shaft wall) is found to be exceeding the allowable limit as per IS 456:2000. Hence, the section needs to be retrofitted by suitable technique. The concrete jacketing which increases the thickness of shaft wall sections by 50 mm is first tried.

Reanalysis for Dead Load:

Weight of stored water in the tank = 8005 kN (same as earlier)

Calculation of self-weight of the various components of RCC tank structure

Self-weight of components of structure	Calculation	Weight (kN)
200-mm-thick RCC tank wall	$[\pi(7.7^2 - 7.5^2) \times 5\ \text{m}^3] \times [25\ \text{kN/m}^3]$	1194
150-mm-thick RCC top slab	$[\pi(7.5)^2 \times 0.15\ \text{m}^3] \times [25\ \text{kN/m}^3]$	663
250-mm-thick RCC base slab	$[\pi(7.5)^2 \times 0.25\ \text{m}^3] \times [25\ \text{kN/m}^3]$	1104
250-mm-thick, 20-m-high, and 4-m-dia shaft	$[\pi(2.25^2 - 2.0^2) \times 20\ \text{m}^3] \times [25\ \text{kN/m}^3]$	1669
Total		4629

Hence, total weight of RCC tank structure including the stored water; $W = 12{,}635$ kN

Maximum stress of shaft wall (at bottom)

$$= [(12{,}635 \times 10^3)/\{\pi(2250^2 - 2000^2)\}] \text{N/mm}^2$$
$$= 3.79 \text{ MPa}$$
$$< 5 \text{ MPa (Permissible stress in direct compression of} \times \text{M20 Concrete)}$$
$$< 20 \text{ MPa (Characteristic compressive strength of} \times \text{M20 Concrete)}$$

So, the tank structure is found to be safe for dead load case.

Reanalysis for Seismic Load:

Seismic mass,

$$M = (1/g) \times [8005 \times 10^3 + 1194 \times 10^3 + 663 \times 10^3 + 1104 \times 10^3 + 0.5^* 1669 \times 10^3] = 1203 \times 10^3 \text{ Kg}$$

Moment of inertia,

$$I = \frac{\pi}{64}(4.5^4 - 4.0^4) = 7.56 \text{ m}^4 = 7.56 \times 10^{12} \text{ mm}^4$$

Stiffness,

$$K = 3 \text{EI}/L^3 = (3^*2.24 \times 10^{10*}7.56)/20^3 \text{ N/m} = 63.50 \times 10^6 \text{ N/m}$$

Fundamental natural frequency (angular);

$$\omega = \sqrt{\frac{K}{M}} = 7.26 \text{ rad/s}$$

Hence, fundamental natural frequency (Linear) $f = \frac{\omega}{2\pi} = 1.16$ Hz and,

Fundamental time period = T = $1/f$ = 0.86 s.

Hence, spectral acceleration = S_a/g = 1.1 (refer response spectra given above)

$$A_h = z/2(I/R)S_a/g = (0.36/2)^*(1.5/3)^*1.1 = 0.099$$

Lateral Seismic Force on structure, $F = A_h{}^*W = 0.099^*[(1203 \times 10^3)^*9.81]$ N = 1.168×10^6 N = 1168 kN.

Hence, base shear, V = 1168 kN and base moment, $M = F^*L$ = 1168*20 kN m = 23.36 × 103 kN m = 23.36 × 109 N mm.

Modified section modulus of shaft,

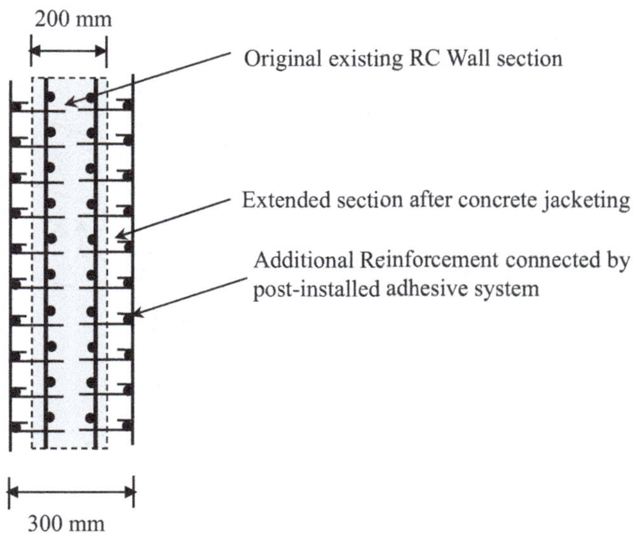

Fig. 13.8 Typical retrofitting scheme of a RC wall section adopting concrete jacketing

$$Z^* = \frac{\pi}{32D}(D^4 - d^4) = \frac{\pi}{32^*4500}(4500^4 - 4000^4) = 3.36 \times 10^9 \text{ mm}^3$$

Bending stress (Maximum)

$$= M/z = (23.36 \times 10^9)/3.36 \times 10^9 = 6.95 \text{ MPa}$$
$$< \text{Allowable bending stress in compression for} \times \text{M20 Concrete (9.33 MPa)}$$

(As per IS 456:2000 Allowable bending stress in compression is 1.333*7 MPa, i.e., 9.33 MPa, considering 33.33% increase in allowable limit in the case of earthquake load.)

Hence Safe

Following retrofitting scheme as shown in Fig. 13.8 involving concrete jacketing of tank wall has been designed and may be implemented.

After successful implementation of the Concrete Jacketing procedure in reducing the developed stress, steel jacketing is tried as shown in Fig. 13.9.

Reanalysis for Dead Load:

Weight of stored water in the tank = 8005 kN (same as earlier)

Calculation of self-weight of the various components of RCC tank structure

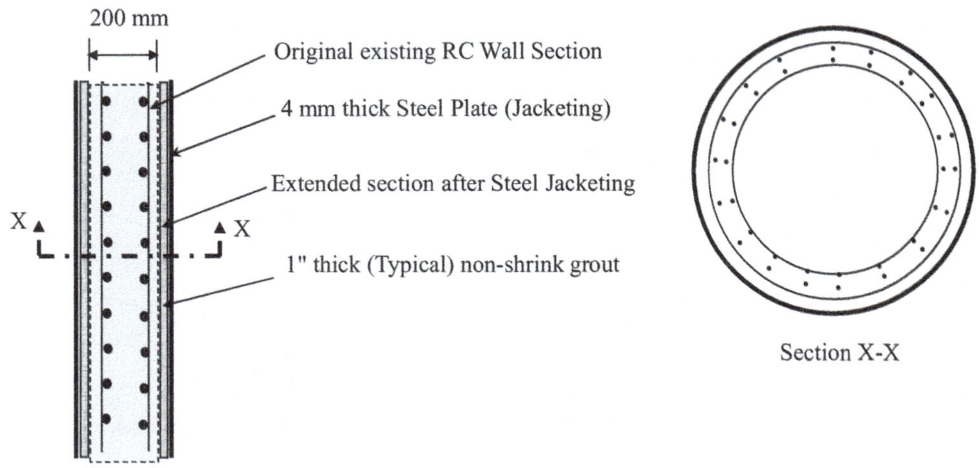

Fig. 13.9 Typical retrofitting scheme of a RC wall section adopting steel jacketing

Self-weight of components of structure	Calculation	Weight (kN)
200-mm-thick RCC tank wall	$[\pi(7.7^2 - 7.5^2) \times 5\ m^3] \times [25\ kN/m^3]$	1194
150-mm-thick RCC top slab	$[\pi(7.5)^2 \times 0.15\ m^3] \times [25\ kN/m^3]$	663
250-mm-thick RCC base slab	$[\pi(7.5)^2 \times 0.25\ m^3] \times [25\ kN/m^3]$	1104
200-mm-thick, 20-m-high, and 4-m-dia shaft	$[\pi(2.2^2 - 2.0^2) \times 20\ m^3] \times [25\ kN/m^3]$	1319
4-mm-thick plate all around shaft	$[\pi(2.204^2 - 2.200^2) \times 20\ m^3] \times [78.5\ kN/m^3]$	87
Total		4367

Hence, total weight of RCC tank structure including the stored water; $W = 12{,}372$ kN

$$\text{Maximum stress in concrete section on shaft wall (at bottom)} = \left[(12{,}372 \times 10^3)/\{\pi(2200^2 - 2000^2)\}\right] \text{N/mm}^2$$
$$= 4.69\,\text{MPa}$$
$$< 5\,\text{MPa (Permissible stress in direct compression of} \times \text{M20 Concrete)}$$
$$< 20\,\text{MPa (Characteristic compressive strength of} \times \text{M20 Concrete)}$$

So, the tank structure is found to be safe for dead load case.

Reanalysis for Seismic Load:

Seismic mass,

$$M = (1/g) \times \left[8005 \times 10^3 + 1194 \times 10^3 + 663 \times 10^3 + 1104 \times 10^3 + 0.5^*(1319 + 87) \times 10^3\right]$$
$$= 1190 \times 10^3\,\text{Kg}$$

Moment of inertia,

$$I_{\text{Concrete}} = 5.83\,\text{m}^4$$
$$I_{\text{Steel}} = 0.13\,\text{m}^4$$
$$I_{\text{Composite}} = 5.98\,\text{m}^4$$

Stiffness,

$$K = K_{\text{Concrete}} + K_{\text{Steel}} = \left(3^* E_{\text{Concrete}}{}^* I_{\text{Concrete}}/L^3\right) + \left(3^* E_{\text{Steel}}{}^* I_{\text{Steel}}/L^3\right)$$
$$= \left[\left(3^* 2.24 \times 10^{10*} 5.83\right)/20^3 + \left(3^* 2.00 \times 10^{11*} 0.13\right)/20^3\right] \text{N/m}$$
$$= \left[48.97 \times 10^6 + 9.75 \times 10^6\right] \text{N/m}$$
$$= 58.72 \times 10^6\,\text{N/m}$$

Fundamental natural frequency (Angular) = 7.91 rad/s.

Hence, fundamental natural frequency (Linear) = 1.26 Hz and

Fundamental time period = $T = 1/f = 0.8$ s,
Hence, spectral acceleration = $S_a/g = 1.2$ (refer response spectra given),
$A_h = z/2\,(I/R) S_a/g = (0.36/2)^*(1.5/3)^* 1.2 = 0.108$,
Lateral seismic force on structure,
$F = A_h{}^* W = 0.108^*[(1190 \times 10^3) \times 9.81]\,\text{N}$
$= 1.26 \times 10^6\,\text{N} = 1260\,\text{kN}$,
Hence, base shear, $V = 1260$ kN,
and base moment, $M = F^* L = 1260^* 20\,\text{kN m} = 25.20 \times 10^3\,\text{kN m} = 25.20 \times 10^9\,\text{N mm}$.

Modified section modulus of shaft,

$$Z^* = I_{\text{Composite}}/y = 5.98/2.204 = 2.71\,\text{m}^3$$
$$= 2.71 \times 10^9\,\text{mm}^3$$

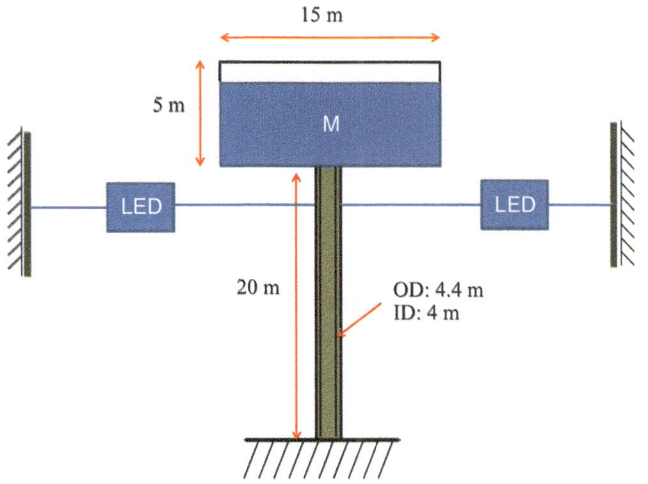

Fig. 13.10 Elevated RC cylindrical tank on a cylindrical hollow shaft retrofitted with lead extrusion dampers

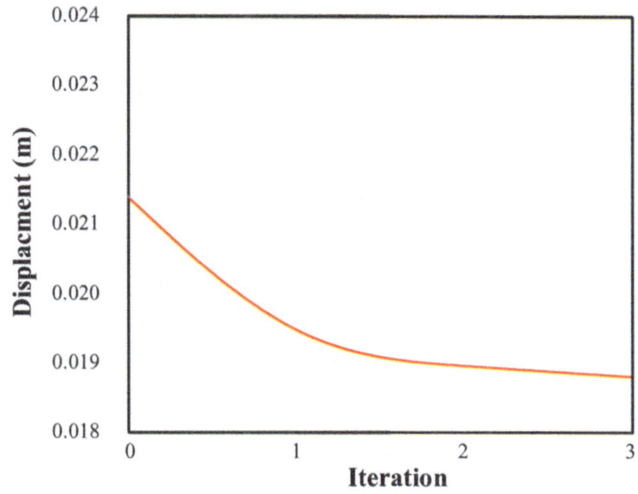

Fig. 13.11 Variation of displacement with iteration

$$
\begin{aligned}
&\text{Bending Stress (Maximum)} \\
&\quad = M/Z^* = \left(25.20 \times 10^9\right)/\left(2.71 \times 10^9\right) = 9.298\,\text{MPa} \\
&\quad < \text{Allowable bending stress in compression for} \\
&\qquad \times \text{M20 Concrete (9.33 MPa)}
\end{aligned}
$$

(As per IS 456:2000 Allowable bending stress in compression is 1.333*7 MPa, i.e., 9.33 MPa, considering 33.33% increase in allowable limit in the case of earthquake load.)

Hence Safe

Retrofitting using Passive Devices for Seismic Load:

The maximum bending stress is observed to be higher than the allowable stress if no jacketing is done; therefore, retrofitting can also be performed using passive energy dissipation devices. Lead Extrusion Damper is employed for the present situation as shown in Fig. 13.10. Now, the maximum bending stress exceeding the allowable stress is 0.9 MPa. The corresponding base moment is calculated using the same section modulus of the shaft as 2.385 MN-m. Subsequently, the lateral force is estimated to be 119.2 kN. Thus, a force of 119.2 kN is to be dissipated through the dissipating devices. The energy of dissipation (E_d) is calculated from the damping ratio required to reduce the additional lateral force. Two LED dampers are designed for total of 36 kN (18 kN each). Figures 13.11 and 13.12 plot the variation of displacement and damping ratio with increase in iteration number.

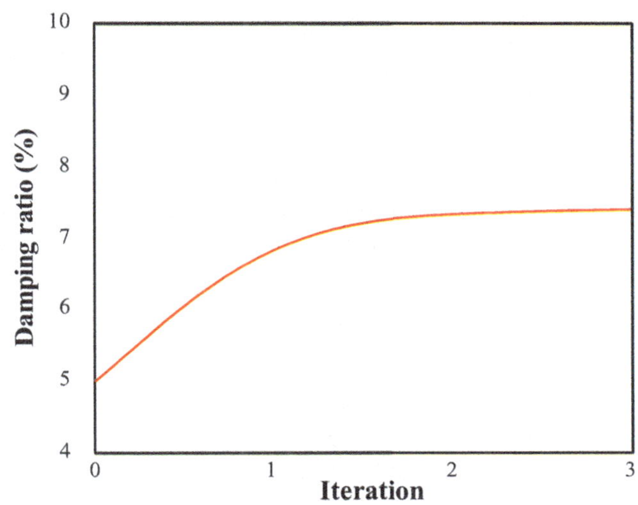

Fig. 13.12 Variation of bending stress with iteration

MATLAB code for solving the above problem:

```
clear all
clc

m=1185000;                  % Mass of the Structure
I=5.83;                     % Section Moment of Inertia
E=2.24*10^10;               % Modulus of Elasticity
L=20;

k=(3*E*I)/L^3; % Stiffness
w=sqrt(k/m);
f=w/(2*pi);                 %Natural Frequency
T=1/f;
Z=0.36;                     %Zone Factor
IF=1.5;                %Importance Factor
R=3;                   %Response Reduction Factor
i=1;
zeta(1)=0;
zeta(2)=0.05;
Sag(1)=0;
Ah(1)=0;
F(1)=0;
Ed(1)=0;
M(1)=0;
BS(1)=0;
zetad(1)=0;

if T>=0 && T<0.1

  Sag(2)=1+15*T;

elseif T>=0.1 && T<0.4
  Sag(2)=2.5;

else
  Sag(2)=1/T;
end
```

```
SM=2.65                          %Section Modulus

for i=2:1:10

  if zeta(i)>=0.05 && zeta(i)<=0.07
      q=1-(((zeta(i)-0.05)/(0.07-0.05))*0.1);

  elseif zeta(i)>0.07 && zeta(i)<=0.1
    q=0.9-(((zeta(i)-0.07)/(0.1-0.07))*0.1);

  elseif zeta(i)>0.1 && zeta(i)<=0.15
    q=0.8-(((zeta(i)-0.1)/(0.15-0.1))*0.1);

  elseif zeta(i)>0.15 && zeta(i)<=0.2
    q=0.7-(((zeta(i)-0.15)/(0.2-0.15))*0.1);

  elseif zeta(i)>0.2 && zeta(i)<=0.25
    q=0.6-(((zeta(i)-0.2)/(0.25-0.2))*0.05);

  else zeta(i)>0.25 && zeta(i)<=0.30
    q=0.55-(((zeta(i)-0.25)/(0.30-0.25))*0.05);

  end

  Ah(i)=(Z/2)*(IF/R)*Sag(2)*q;
  F(i)= Ah(i)*m*9.81;            % Lateral Force
  u(i)=(Ah(i)*9.81)/w^2;                % Displacement
  M(i)=F(i)*L;                   % Base Moment
  BS(i)=M(i)/SM;                 % Bending Stress
  Ed(i)=4*36000*u(i);                   % Energy Dissipation
  zetad(i)=Ed(i)/(2*pi*k*u(i)^2);
  zeta(i+1)=zeta(i)+zetad(i);

end
```

After retrofitting with the Lead Extrusion Damper, the base moment and displacement reduced from 20.92×10^3 kN m and 21.4 mm to 18.52×10^3 kN m and 19.15 mm, respectively, whereas jacketing resulted in the base moment and displacement values of 23.36×10^3 kN m and 18.43 mm. The bending stress evaluated after jacketing is 6.95 MPa, whereas it is estimated to be 6.98 MPa for the damper. A procedure exhibiting retrofitting by steel jacketing is also presented.

Example 13.2 A RCC framed structure of following configuration and dimension was to be qualified for prevailing seismic load case as per IS 1893:2002. The geometry of the frame is shown in Fig. 13.13. The structure is rested on rock-type founding media.

Dimension of RC columns: 230 mm × 230 mm,
Dimension of RC beams (plinth and floor beams): 200 mm × 230 mm,

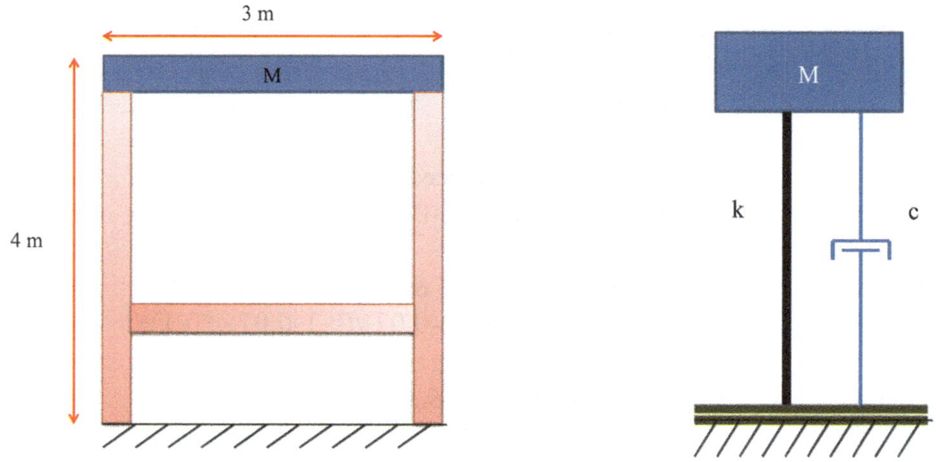

Fig. 13.13 RCC framed structure and its mathematical representation with equivalent mass and stiffness

Dimension of RC slab: 150 mm,
Floor to floor height: 3 m,
Depth of foundation = 1 m from plinth level.

Analysis for Dead Load:

Floor live load on RCC slab (say, 3 kN/m^2) = [3 × 3 × 3] = 27 kN

Calculation of self-weight of the various components of RCC framed structure

Self-weight of components of structure	Calculation	Weight (kN)
150-mm-thick RCC slab	1 × [(3 × 3) × 0.15 m^3] × [25 kN/m^3]	34
200 mm × 230 mm floor beam, 3 m span	4 × [(0.2 × 0.23 × 3) m^3] × [25 kN/m^3]	14
230 mm × 230 mm column section, 3 m high	4 × [(0.23 × 0.23 × 3) m^3] × [25 kN/m^3]	16
200 mm × 230 mm plinth beam, 3 m span	4 × [(0.2 × 0.23 × 3) m^3] × [25 kN/m^3]	14
230 mm × 230 mm column section, 1 m high	4 × [(0.23 × 0.23 × 1) m^3] × [25 kN/m^3]	5
Total		83

Hence, total load on one column at the base; W = (83 + 27)/4 kN = 27.5 kN

Maximum stress of column (at bottom)

$$= \left[(27.5 \times 10^3)/(230 \times 230)\right] \text{N/mm}^2 = 0.52\,\text{MPa}$$
$$< 5\,\text{MPa (Permissible stress in direct compression of} \times \text{M20 Concrete)}$$
$$< 20\,\text{MPa (Characteristic compressive strength of} \times \text{M20 Concrete)}$$

So, the framed structure is found to be safe for dead load case.

Analysis for Seismic Load:

Now the same tank structure has been evaluated for seismic load case with following seismic input:

Zone factor (z) = 0.36 (Zone V, Guwahati),
Importance factor (I): 1.5,
Response reduction factor (R): 3 (considering ordinary moment-resisting frame (OMRF) detailing of RCC Structure),
Damping = 5%,
Response spectra: As in Example 13.7.

Seismic mass,

$$\begin{aligned} M &= (1/g) \times \left[27 \times 10^3 + 34 \times 10^3 + 14 \times 10^3 \right. \\ &\quad \left. + 0.5^{*}16 \times 10^3 + 0.5^{*}14 \times 10^3 + 0.5^{*}5 \times 10^3\right] \\ &= 9.43 \times 10^3\,\text{kg} \end{aligned}$$

Moment of inertia,

$$\begin{aligned} I &= 4^{*}\left[(0.23^{*}0.23^3)/12\right] = 9.32 \times 10^{-4}\,\text{m}^4 \\ &= 9.32 \times 10^8\,\text{mm}^4 \end{aligned}$$

Stiffness

$$\begin{aligned} K &= 12\text{EI}/L^3 = [(12^{*}2.24 \times 10^{10*}9.32 \times 10^{-4})]/4^3\,\text{N/m} \\ &= 3.915 \times 10^6\,\text{N/m} \end{aligned}$$

Fundamental natural frequency (Angular);

$$\omega = \sqrt{\frac{K}{M}} = 20.375\,\text{rad/s}$$

Hence, fundamental natural frequency (Linear) $f = \frac{\omega}{2\pi} =$ 3.24 Hz and,Fundamental time period = T = $1/f$ = 0.31 s.

Hence, spectral acceleration = S_a/g = 2.5 (refer response spectra given above)

$$A_h = z/2(I/R)S_a/g = (0.36/2)/(1.5/3)^*2.5 = 0.225$$

Lateral seismic force on structure,

$$F = A_h{}^*W = 0.225^*\left[(9.43 \times 10^3)^*9.81\right]\text{N} = 20{,}814\,\text{N} = 20.81\,\text{kN}$$

Hence, lateral seismic force on one single column, $F_{\text{column}} = 1/4\,(20.81)\,\text{kN} = 5.20\,\text{kN}$.

Hence, base shear on single column, V_{column} = 5.20 kN and base moment, $M = F_{\text{column}} * L_{\text{column}}$ = 5.20*4 kN m = 20.80 kN m = 20.80 × 10^6 N mm.

Section modulus of column, Z = (230*230^2)/6 = 2.03 × 10^6 mm^3

$$\text{Bending Stress (maximum)} = M/z = \left(20.80 \times 10^6\right)/2.03 \times 10^6 = 10.25\,\text{MPa}$$
$$> \text{Allowable bending stress in compression for} \times \text{M20 Concrete (9.33 MPa)}$$

(As per IS 456:2000, allowable bending stress in compression is 1.333*7 MPa, i.e., 9.33 MPa, considering 33.33% increase in allowable limit in the case of earthquake load).

The bending stress in the column section is found to be exceeding the allowable limit as per IS 456:2000. Hence, the section needs to be retrofitted by suitable technique. The concrete jacketing which increases the thickness of shaft wall sections by 50 mm is first tried.

Reanalysis for Dead Load:

Floor Live Load on RCC slab (say, 3 kN/m^2) = [3 × 3 3] = 27 kN (Same as earlier).

Calculation of self-weight of the various components of RCC framed structure

Self-weight of components of structure	Calculation	Weight (kN)
150-mm-thick RCC slab	1 × [(3 × 3) × 0.15 m^3] × [25 kN/m^3]	34
200 mm × 230 mm floor beam, 3 m span	4 × [(0.2 × 0.23 × 3) m^3] × [25 kN/m^3]	14
330 mm × 330 mm column section, 3 m high	4 × [(0.33 × 0.33 × 3) m^3] × [25 kN/m^3]	33
200 mm × 230 mm plinth beam, 3 m span	4 × [(0.2 × 0.23 × 3) m^3] × [25 kN/m^3]	14
330 mm × 330 mm column section, 1 m high	4 × [(0.33 × 0.33 × 1) m^3] × [25 kN/m^3]	11
Total		106

Hence, total load on one column at the base; W = (106 + 27)/4 kN = 33.25 kN

$$\text{Maximum stress of column(at bottom)} = \left[(33.25 \times 10^3)/(330 \times 330)\right]\text{N/mm}^2 = 0.31\,\text{MPa}$$
$$< 5\,\text{MPa (Permissible stress in direct compression of} \times \text{M20 Concrete)}$$
$$< 20\,\text{MPa (Characteristic compressive strength of} \times \text{M20 Concrete)}$$

So, the framed structure is found to be safe for dead load case.

Reanalysis for Seismic Load:

Seismic mass,

$$M = (1/g) \times \left[27 \times 10^3 + 34 \times 10^3 + 14 \times 10^3 + 0.5^*33 \times 10^3 + 0.5^*14 \times 10^3 + 0.5^*11 \times 10^3\right] = 10.6 \times 10^3\,\text{kg}$$

Moment of inertia,

$$I = \left(4^*0.33^*0.33^3\right)/12 = 39.52 \times 10^{-4}\,\text{m}^4 = 39.52 \times 10^8\,\text{mm}^4$$

Stiffness

$$K = 12\text{EI}/L^3 = [4^*(12^*2.24 \times 10^{10*}39.52 \times 10^{-4})]/4^3\,\text{N/m} = 16.6 \times 10^6\,\text{N/m}$$

Fundamental natural frequency (Angular);

$$\omega = \sqrt{\frac{K}{M}} = 39.57\,\text{rad/s}$$

Hence, fundamental natural frequency (Linear) $f = \frac{\omega}{2\pi} =$ 6.3 Hz and,

Fundamental time period = T = $1/f$ = 0.16 s.

Hence, spectral acceleration = S_a/g = 2.5 (refer response spectra given above)

$$A_h = z/2(I/R)S_a/g = (0.36/2)/(1.5/3)^*2.3 = 0.225$$

Lateral Seismic Force on structure,

$$F = A_h{}^*W = 0.225^*\left[(10.6 \times 10^3)^*9.81\right]\text{N} = 21,525\,\text{N} = 23.4\,\text{kN}$$

Hence, lateral seismic force on one single column, $F_{\text{column}} = 1/4\,(23.4)\,\text{kN} = 5.85\,\text{kN}$.

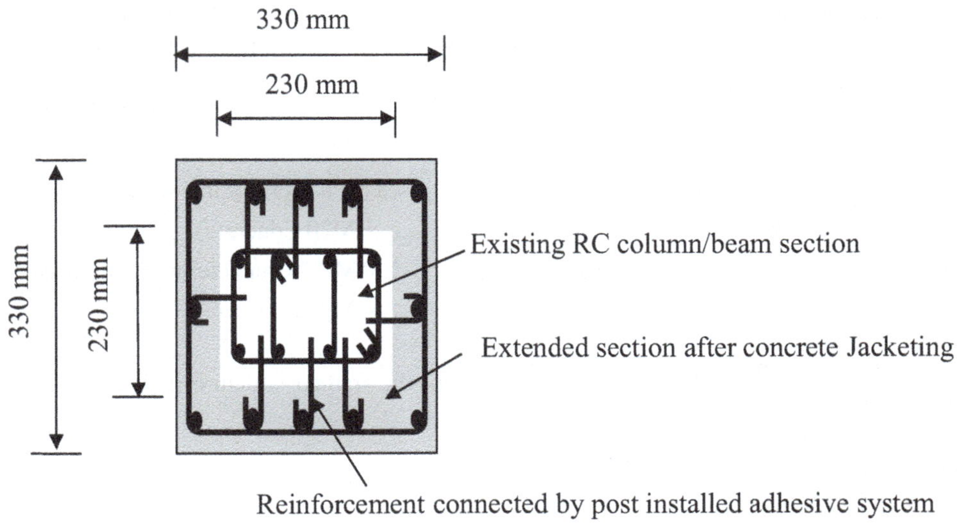

Fig. 13.14 Typical retrofitting scheme of a RC column section adopting concrete jacketing

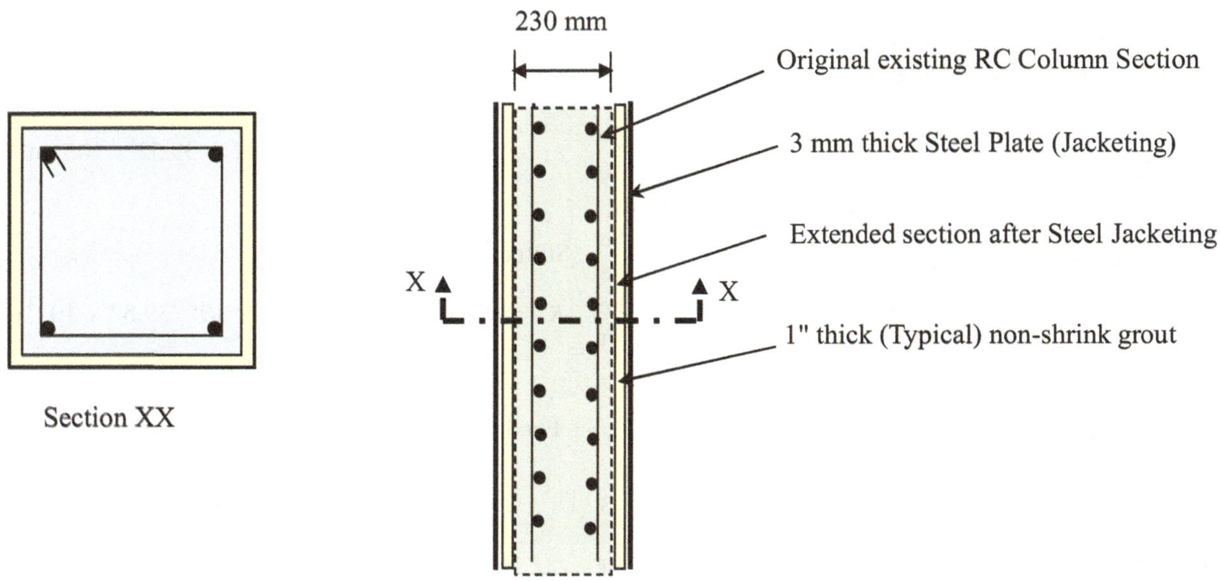

Fig. 13.15 Typical retrofitting scheme of a RC column section adopting steel jacketing

Hence, base shear on single column, V_{column} = 5.85 kN and base moment, $M = F_{\text{column}} * L_{\text{column}}$ = 5.38*4 kN m = 23.4 kN m = 23.4 × 10^6 N mm.

Modified section modulus of column, $Z = (330*330^2)/6 = 5.99 \times 10^6$ mm^3

Bending Stress (Maximum)

$$= M/z = (23.4 \times 10^6)/5.99 \times 10^6 = 3.91\,\text{MPa}$$

< Allowable bending stress in compression for × M20 Concrete (9.33 MPa)

(As per IS 456:2000, allowable bending stress in compression is 1.333*7 MPa, i.e., 9.33 MPa, considering 33.33% increase in allowable limit in the case of earthquake load.)

Hence Safe

Following retrofitting scheme as shown in Fig. 13.14 involving concrete jacketing of tank wall has been designed and may be implemented.

After successful implementation of the Concrete Jacketing procedure in reducing the developed stress, steel jacketing is tried as shown in Fig. 13.15.

Reanalysis for Dead Load:

Floor live load on RCC slab (say, 3 kN/m^2) = [3 × 3 × 3] = 27 kN (same as earlier)

Calculation of self-weight of the various components of RCC framed structure

Self-weight of components of structure	Calculation	Weight (kN)
150-mm-thick RCC slab	1 × [(3 × 3) × 0.15 m^3] × [25 kN/m^3]	34
200 mm × 230 mm floor beam, 3 m span	4 × [(0.2 × 0.23 × 3) m^3] × [25 kN/m^3]	14
230 mm × 230 mm column section, 3 m high	4 × [(0.23 × 0.23 × 3) m^3] × [25 kN/m^3]	16
200 mm × 230 mm plinth beam, 3 m span	4 × [(0.2 × 0.23 × 3)m^3] × [25 kN/m^3]	14
230 mm × 230 mm column section, 1 m high	4 × [(0.23 × 0.23 × 1)m^3] × [25 kN/m^3]	5
3 mm plate all around column section, 3 m high	4 × [4 × (0.003 × 0.236 × 3)m^3] × [78.5 kN/m^3]	3
3 mm plate all around column section, 1 m high	4 × [4 × (0.003 × 0.236 × 1) m^3] × [78.5 kN/m^3]	1
Total		87

Hence, total load on one column at the base; W = (87 + 27)/4 kN = 28.5 kN

$$\begin{aligned}&\text{Maximum stress of column (at bottom)}\\&= \left[(28.5\times 10^3)/(230\times 230)\right]\,\text{N/mm}^2 = 0.54\,\text{MPa}\\&<5\,\text{MPa (Permissible stress in direct compression of}\\&\quad\times\text{ M20 Concrete)}\\&<20\,\text{MPa (Characteristic compressive strength of}\\&\quad\times\text{ M20 Concrete)}\end{aligned}$$

So, the framed structure is found to be safe for dead load Case.

Reanalysis for Seismic Load:

Seismic mass,

$$\begin{aligned}M &= (1/g)\times\left[27\times 10^3 + 34\times 10^3 + 14\times 10^3\right.\\&\quad\left.+\,0.5^*14\times 10^3 + 0.5^*3\times 10^3 + 0.5^*1\times 10^3\right]\\&= 8.6\times 10^3\,\text{kg}\end{aligned}$$

Moment of inertia,

$$\begin{aligned}I_{\text{Concrete}} &= 9.32\times 10^{-4}\,\text{m}^4\\I_{\text{Steel}} &= 1.01\times 10^{-4}\,\text{m}^4\\I_{\text{Composite}} &= 11.28\times 10^{-4}\,\text{m}^4\end{aligned}$$

Stiffness,

$$\begin{aligned}K &= K_{\text{Concrete}} + K_{\text{Steel}}\\&= \left[\left(12^*E_{\text{Concrete}}{}^*I_{\text{Concrete}}/L^3\right) + \left(12^*E_{\text{Steel}}{}^*I_{\text{Steel}}/L^3\right)\right]\\&= [(12^*2.24\times 10^{10}{}^*9.32\times 10^{-4})/4^3\\&\quad + (12^*2.00\times 10^{11}{}^*1.01\times 10^{-4})/4^3]\ \text{N/m}\\&= (12/64)^*\left[(2.24^*9.32^*\times 10^6) + (2.00^*10.1^*\times 10^6)\right]\\&= (12/64)^*[20.88 + 20.2]\times 10^6\,\text{N/m} = 7.70\times 10^6\,\text{N/m}\end{aligned}$$

Fundamental natural frequency (Angular) = 29.9 rad/s.

Hence, fundamental natural frequency (Linear) = 4.76 Hz and

Fundamental time period = T = 1/f = 0.21 s.

Hence, spectral acceleration = S_a/g = 2.5 (refer response spectra given in Fig. 13.7)

$$A_h = z/2\,(I/R)\,S_a/g = (0.36/2)^*(1.5/3)^*2.5 = 0.225$$

Lateral seismic force on structure,

$$\begin{aligned}F &= A_h{}^*W = 0.225^*\left[(8.6\times 10^3)^*9.81\right]\,\text{N} = 18{,}982\,\text{N}\\&= 18.98\,\text{kN}\end{aligned}$$

Hence, lateral seismic force on one single column, $F_{\text{column}} = 1/4\,(18.98)\,\text{kN} = 4.75\,\text{kN}$.

Hence, base shear on single column, V_{column} = 4.75 kN and base moment, M = F_{column} * L_{column} = 4.75*4 kN m = 18.98 kN m = 18.98 × 10^6 N mm.

Modified section modulus of retrofitted column,

$$\begin{aligned}Z^* &= I_{\text{Composite}}/y = (11.28\times 10^{-4})/(0.233/2)\,\text{m}^3\\&= 9.68\times 10^{-3}\,\text{m}^3 = 9.68\times 10^6\,\text{mm}^3\end{aligned}$$

$$\begin{aligned}&\text{Bending stress (Maximum)}\\&= M/z = \left(18.98\times 10^6\right)/\,9.68\times 10^6 = 1.96\,\text{MPa}\\&<\text{Allowable bending stress in compression for}\\&\quad\times\text{ M20 concrete (9.33 MPa)}\end{aligned}$$

(As per IS 456:2000, allowable bending stress in compression is 1.333*7 MPa, i.e., 9.33 MPa, considering 33.33% increase in allowable limit in the case of earthquake load.)

Hence Safe

Retrofitting using Passive Devices for Seismic Load:

The maximum bending stress on each column is observed to be higher than the allowable stress if no jacketing is done; therefore, retrofitting can also be performed using passive energy dissipation devices. Double-sliding friction damper is employed for the present situation as shown in Fig. 13.16.

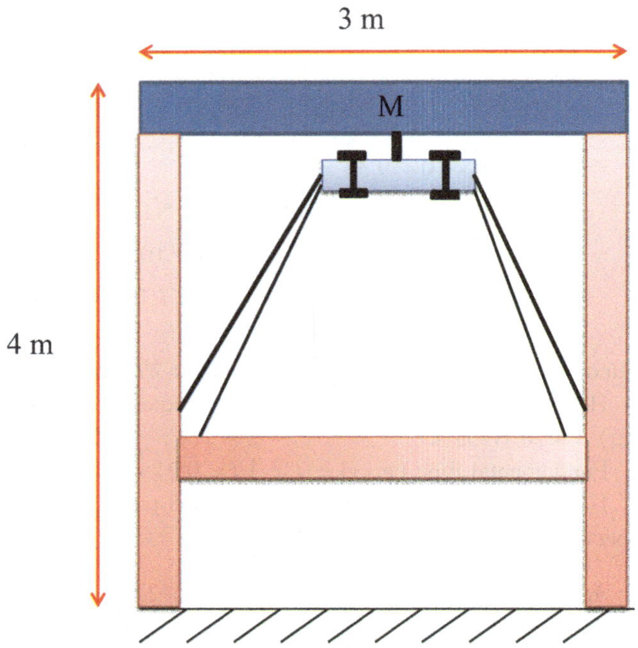

Fig. 13.16 RCC framed structure retrofitted with friction damper

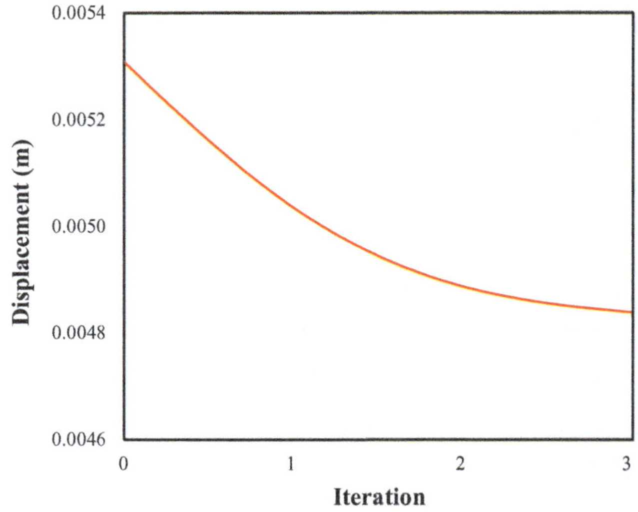

Fig. 13.17 Variation of displacement with iteration

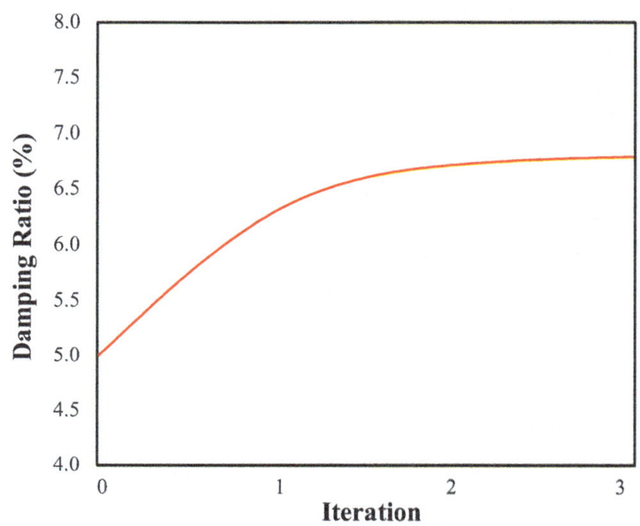

Fig. 13.18 Variation of damping ratio with iteration

Now, the maximum bending stress exceeding the allowable stress is 0.92 MPa. The corresponding base moment is calculated using the same section modulus as 1.867 kN m. Subsequently, the lateral force is estimated to be 466.9 N per column. Thus, a force of 466.9 N is to be dissipated through the dissipating devices. The energy of dissipation (E_d) is calculated from the damping ratio required to reduce the additional lateral force. The damper is designed for 500 N. Figures 13.17 and 13.18 plot the variation of displacement and damping ratio with increase in iteration number.

MATLAB code for solving the above problem:

```
clear all
clc

m=9430;                 % Mass of the Structure
I=9.32E-4;                       %Moment of Inertia
E=2.24*10^10;                    % Modulus of Elasticity
L=4;

k=(12*E*I)/L^3;                  % Stiffness
w=sqrt(k/m);
f=w/(2*pi);                      % Natural Frequency
T=1/f;

Z=0.36;                           %Zone Factor
IF=1.5;                          %Importance Factor
R=3;                             %Response Reduction Factor
i=1;
zeta(1)=0;
zeta(2)=0.05;
Sag(1)=0;
Ah(1)=0;
F(1)=0;
Ed(1)=0;
M(1)=0;
BS(1)=0;
zetad(1)=0;
if T>=0 && T<0.1
   Sag(2)=1+15*T;

elseif T>=0.1 && T<0.4
   Sag(2)=2.5;

else
   Sag(2)=1/T;
end
```

After retrofitting with the friction damper, the combined base moment and displacement reduced from 20.8 kN m and 5.32 mm to 18.89 kN m and 4.84 mm, respectively, whereas jacketing resulted in the base moment and

```
SM=2.03E-3                    %Section Modulus

for i=2:1:10

  if zeta(i)>=0.05 && zeta(i)<=0.07
        q=1-(((zeta(i)-0.05)/(0.07-0.05))*0.1);

  elseif zeta(i)>0.07 && zeta(i)<=0.1
     q=0.9-(((zeta(i)-0.07)/(0.1-0.07))*0.1);

  elseif zeta(i)>0.1 && zeta(i)<=0.15
     q=0.8-(((zeta(i)-0.1)/(0.15-0.1))*0.1);

  elseif zeta(i)>0.15 && zeta(i)<=0.2
     q=0.7-(((zeta(i)-0.15)/(0.2-0.15))*0.1);

  elseif zeta(i)>0.2 && zeta(i)<=0.25
     q=0.6-(((zeta(i)-0.2)/(0.25-0.2))*0.05);

  else zeta(i)>0.25 && zeta(i)<=0.30
     q=0.55-(((zeta(i)-0.25)/(0.30-0.25))*0.05);

 end

  Ah(i)=(Z/2)*(IF/R)*Sag(2)*q;
  F(i)= Ah(i)*m*9.81;
  u(i)=(Ah(i)*9.81)/w^2;
  M(i)=F(i)*L;
  BS(i)=M(i)/(4*SM);
  Ed(i)=4*500*u(i);
  zetad(i)=Ed(i)/(2*pi*k*u(i)^2);
  zeta(i+1)=zeta(i)+zetad(i);

end
```

displacement values of 23.4 kN m and 1.41 mm. The bending stress evaluated after jacketing is 3.91 MPa, whereas it is estimated to be 9.31 MPa for the friction damper. Retrofitting by steel jacketing also resulted in increase in the natural frequency of the structure (however less than the concrete jacketing), whereas the damper does not affect the frequency.

13.5.2.2 Steel Jacketing

Steel jacketing (Fig. 13.19) is also one of the popular methods of strengthening [48, 49]. The lap splices of column longitudinal reinforcement and the shear resistance of the sections can be improved by using steel jackets (through confinement). The steel jackets provide passive confinement to the member and can be considered as continuous hoop reinforcement.

Steel jacketing is basically a process of encasing the column with steel plates and filling the gap with a non-shrink grout and hence providing passive confinement to core concrete in this process. Rectangular steel jackets on rectangular columns are not generally recommended and a use of an elliptical jacket is more appropriate. However, steel jackets over concrete sections are vulnerable to corrosion and impact with floating materials, hence not recommended to be used for columns in river, lake, and seas.

A Case Study on Steel Jacketing

A spherical prestressed concrete water storage tank structure is shown in Fig. 13.20. It was constructed in 1959 in India. It has a storage capacity of 850,000 gallons of demineralized water. Prior to the seismic re-evaluation of subject structure, the following nondestructive tests were carried out at selected locations to find out the health of the structure.

- Visual examination: to determine the cracks if any,
- Impact hammer test: for checking the quality of cover concrete cover,

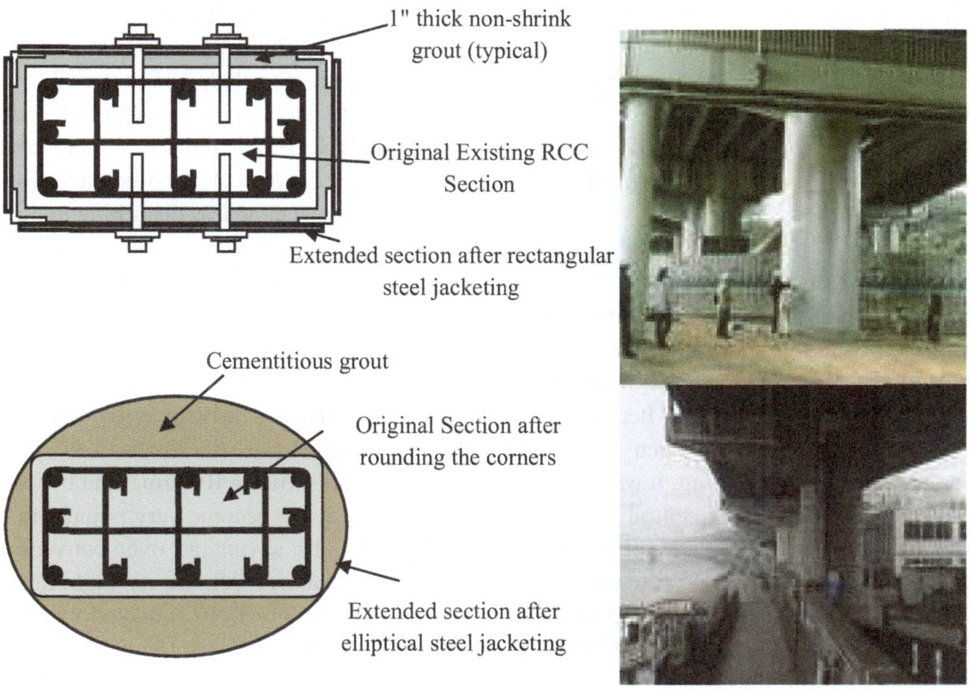

Fig. 13.19 Steel jacketing

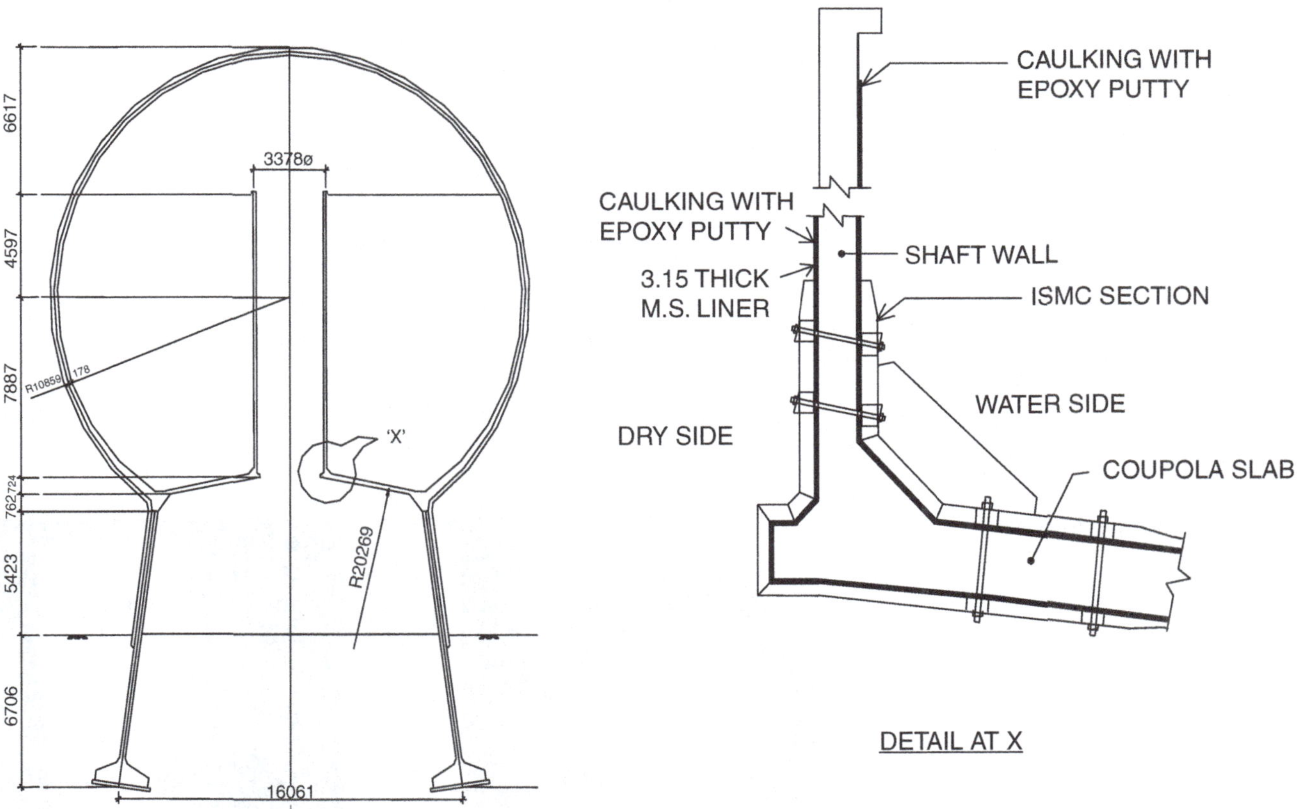

Fig. 13.20 Typical section of water tank structure with retrofitting measure

- Ultrasonic pulse velocity test: to test the quality of concrete,
- Half-cell potential test to detect corrosion in reinforcement,
- Corrosion test: to determine the corrosion rate.
- Concrete core sampling: to measure aging effects on compressive strength.

After confirming the strength of the concrete, seismic analysis was performed for the ground motion as shown in Fig. 13.21. It was found that the bottom of central shaft is unsafe for the postulated earthquake load. After, reviewing many different retrofitting suggestions, such as internal prestressing, concrete repairs, and jacketing, it was proposed to strengthen the bottom portion of the central shaft using 3.15-mm-thick-stiffened carbon steel plates. The plate is provided on both faces of the cupola and central shaft covering up to 1.5 m on cupola (radially), over full height of shaft on wet side and 1.75 m height on dry side. With this arrangement, the tank is found to be safe.

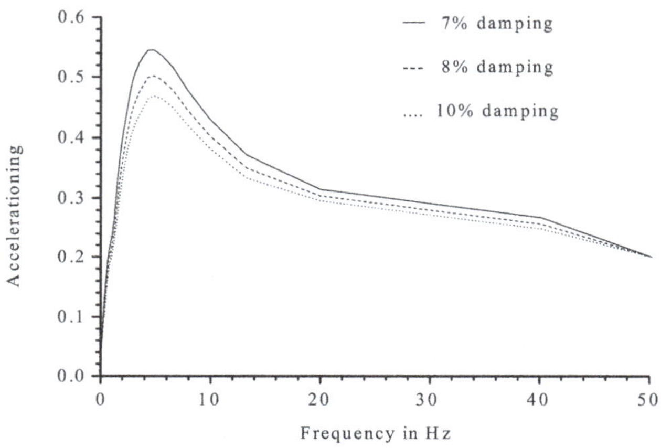

Fig. 13.21 Design basis ground motion

13.5.2.3 External Prestressing

It is the process of applying prestress force to the existing columns with the help of external strands (group of high-strength steel wires) to provide active confinement (Fig. 13.22). It is quite efficient strengthening technique and can be more economical than steel jacketing. The installation of such system can be less disturbing to the existing building occupants. In this technique, the shear strength of members increases only due to increase in concrete strength due to prestressing against the jacketing where the jacket contributes significantly toward shear strength.

13.5.2.4 Strengthening with Fiber Composites

The application of fiber-reinforced polymer (FRP) to strengthen the existing RC structural members as an external reinforcement has become very popular in recent years [44–46, 50, 51]. Its advantage over conventional construction materials is many such as its high tensile strength over weight ratio, ability to be modeled into various shapes and potential resistance to environmental exposure conditions. These advantages of FRP-based strengthening methodology result into a lower maintenance cost technique. These special properties of FRP composites make it a suitable alternative for innovative construction technology. The main advantages of fiber-reinforced composite laminates can be listed as follows:

- No corrosion,
- No transportation problem as it is available in rolls,
- High ultimate strength,
- High Young's modulus,
- Very good fatigue properties,
- Low weight,
- Endless tapes available therefore no joints.

Some of the major disadvantages of FRP composite laminates are following:

Fig. 13.22 External prestressing

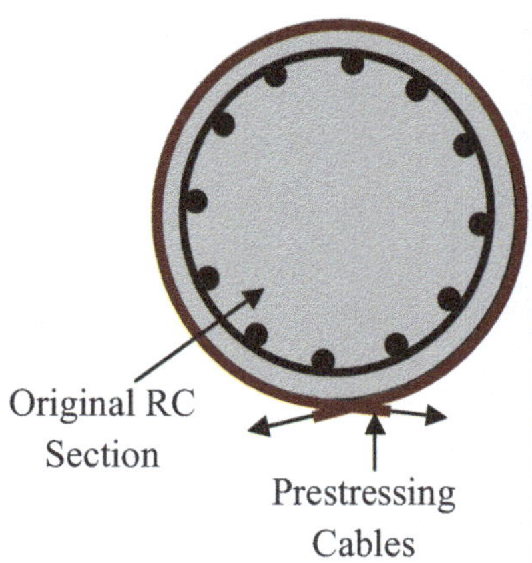

- Erratic plastic behavior and less ductility,
- Susceptibility to local unevenness,
- High cost,
- Use in small quantity is difficult.

FRP Material

FRP composites consist of high-strength fibers embedded in a matrix of polymer resin as shown in Fig. 13.23. In FRP, the fibers provide both load-carrying capacity and stiffness to the composite while the matrix is to ensure sharing of the load among fibers and to protect the fibers themselves from the environment.

Most of the FRP materials are made of fibers with high strength and stiffness, while their strain at failure is lower than that of the matrix. Fibers typically used in FRP are glass, carbon, and aramid. Typical values for properties of the fibers used in FRP composites are listed in Table 13.2. Carbon fibers are the stiffest, most durable, and most expensive fibers. Carbon is quite resistant to most environmental impact. Glass fibers have lower strength and significantly lower stiffness but also a lower cost. Unprotected glass fibers degrade in most environments. Finally, aramid fibers have mechanical characteristics between those of glass and carbon. The engineering behavior of fiber is linear elastic up to failure, with no significant yielding compared to steel. Typical stress–strain relationships for fiber, matrix and the resulting FRP material are shown in Fig. 13.24. Before the yielding of the matrix, the strain in fiber and matrix is the same. After the yielding of the matrix, a knee will appear in the stress–strain curve due to the fact that the matrix no longer contributes to the stiffness. The mechanical properties of composites depend upon the properties of fibers, matrix, and fiber–matrix bond, amount of fiber and orientation distribution of fibers. A composite with all fibers in one direction is designated as unidirectional. If the fibers are woven or oriented in many directions; the composite is called multi-directional. However, it is mainly the fibers that provide stiffness and strength composites are often anisotropic with high stiffness in the fiber direction. In strengthening applications, unidirectional FRP composites as shown in Fig. 13.23 are predominantly used. The approximate stiffness and strength of an unidirectional FRP with a 65% volume fraction of carbon fiber is listed in Table 13.2 along with the corresponding properties of steel for the comparison sake.

The FRP composites mainly applied for strengthening of RC structures can be further categorized as follows:

- Laminates (used mainly for flexural strengthening),
  Laminates generally consist of carbon fibers blended in an epoxy matrix resin. It acts as external tension reinforcements to increase the flexural strength of the RC members.
- Fiber wraps (used mainly for shear and axial strengthening),
  Fiber wraps are made up of three different materials, namely carbon, aramid, and glass. Carbon fiber is the

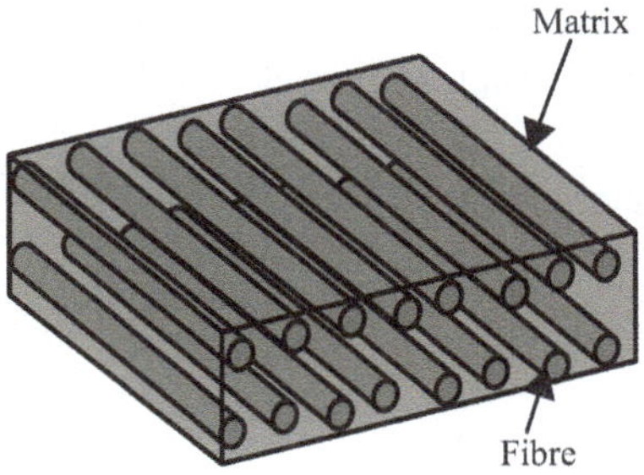

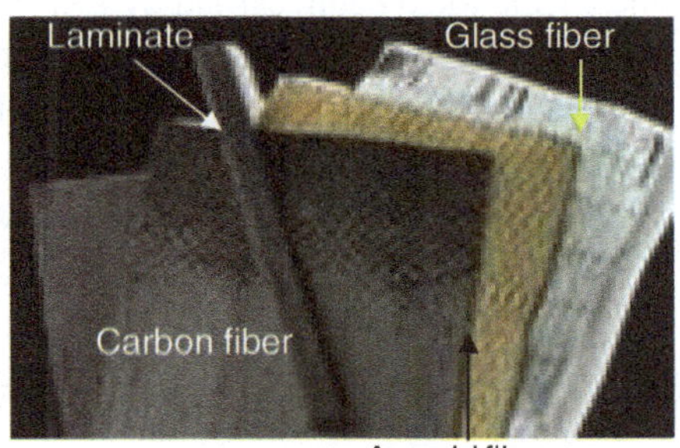

Fig. 13.23 Unidirectional FRP composites used for strengthening applications

Table 13.2 Typical strength and stiffness values for FRP composite materials

Typical property	Glass		Carbon		Aramid		Steel
	E-glass	*S*-glass	Kevlar 29	Kevlar 49	High strength	High modulus	Mild steel
Density (Kg/m^3)	260	250	144	144	180	190	7850
Young's modulus (GPa)	72	87	100	124	230	370	200
Tensile strength (MPa)	172	253	227	227	248	179	250

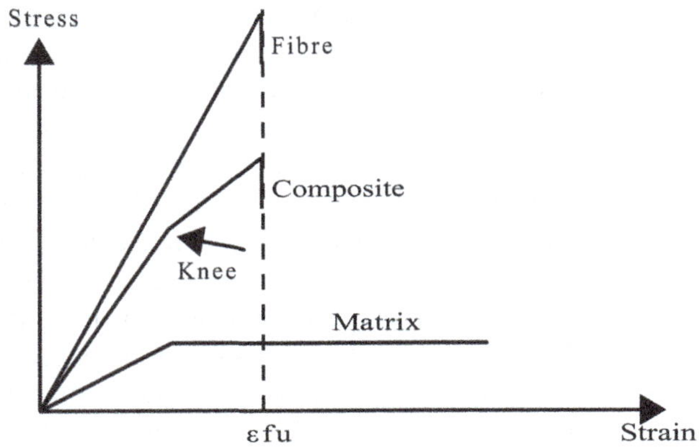

Fig. 13.24 Stress–strain curves for typical fiber, resin, and FRP composite

strongest, most inert but expensive one; glass is the cheapest and has low elastic modulus and strength. Aramid fiber is used mainly for improving the impact resistance.

Flexural Strengthening

Flexural strengthening with FRP composites involves in applying the laminates using the epoxy at the soffit of the beam/slab to act as external reinforcement. The stresses and effective strains for a strengthened section are shown in Fig. 13.25.

Since the section being strengthened is generally stressed and hence deflected before strengthening, while calculating the tensile force being carried by FRP, it is important to first estimate the existing strains at the soffit due to sustained loads since the laminates will become effective only beyond this particular value of strain, e_o.

The modified neutral axis depth, c, can then be obtained by solving the following equation:

$$C_c + C_s = T_s + T_f \tag{13.1}$$

The moment carrying capacity of the strengthened section, M_{str}, may be obtained by taking the moment about midsection depth as

$$\begin{aligned} M_{str} = {} & C_c(D/2 - bc) + C_s(D/2 - d') \\ & + T_s(d - D/2) + T_f(D - D/2) \end{aligned} \tag{13.2}$$

where

- C_c compressive force in concrete
- C_s force in compression steel
- T_s force in tension steel
- d distance from extreme compression fiber to the centroid of tension reinforcement
- d' distance from extreme compression fiber to the centroid of comp. reinforcement
- c depth of neutral axis
- β ratio of the depth of rectangular compression block to the depth of the neutral axis
- T_f tension force in wrap
- D height of the externally strengthened beam section.

Figure 13.26 shows the application of FRP for flexural strengthening of RC beams and slabs. Also, due consideration must be given to prevent the de-bonding of the laminates; otherwise, a premature failure of the laminate may occur (Fig. 13.26). To prevent this, either the stress in the laminate may be limited, recommended by ACI 440-02 or mechanical end anchorage may suitably be provided (Fig. 13.27). In order to utilize the large strength of the laminates, sometimes the laminates are prestressed before being applied to the soffit (Fig. 13.28).

Shear Strengthening

Existing RC members are strengthened for shear using FRP sheets is carried out by bonding the external FRP sheets with

Fig. 13.25 Stress and effective strains in strengthened RC member

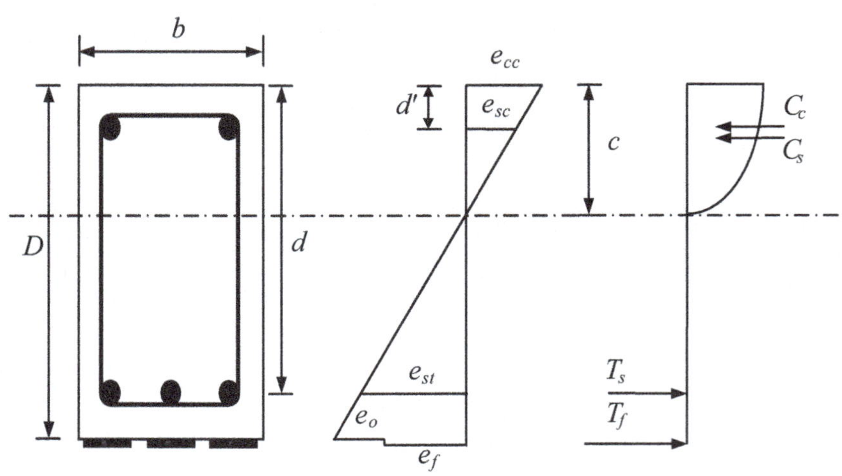

Fig. 13.26 Flexural strengthening of slabs and beams

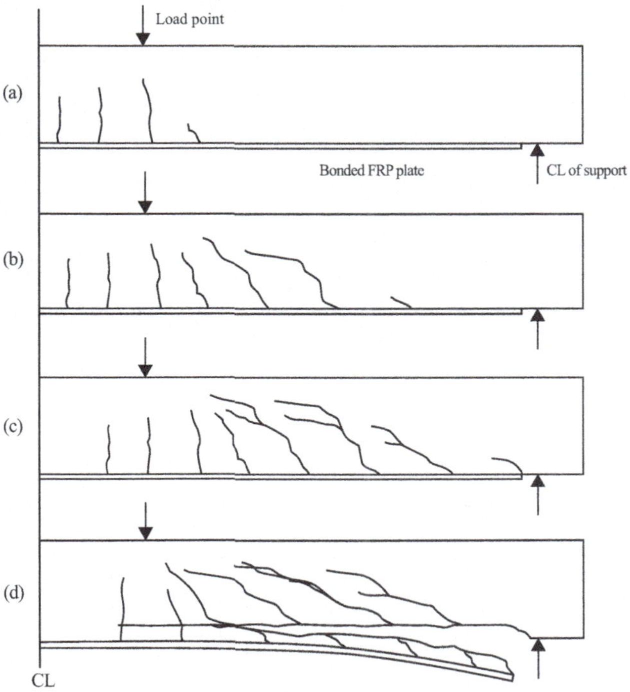

Fig. 13.27 De-bonding of laminate

its principal fiber direction as parallel as practically possible to that of the direction of maximum principal tensile stresses. In this way, the effectiveness of FRP can be maximized. However, it is normally more practical to attach the external FRP with the principal fiber direction perpendicular to the member axis either continuous or intermittently (Fig. 13.29). The FRP can be wrapped around the member and applied as U-shaped jacket or only on sides of the member as shown in Fig. 13.29.

The shear contribution of FRP shear reinforcement, V_f, is given by

$$V_f = A_{fv} f_{fe} d_f / s_f \tag{13.3}$$

where

Area of fiber wrap as shear reinforcement is given by,

$$A_{fv} = 2 n t_f b_f \tag{13.4}$$

n number of layers of fiber
t_f thickness of one fiber layer
b_f width of fiber sheet.

Effective fiber stress is given by

$$f_{fe} = \varepsilon_{fe} E_f \tag{13.5}$$

ε_{fe} is the effective strain in fiber
E_f is modulus of elasticity of fiber
d_f depth of fiber wrap
s_f spacing of fiber wraps.

For continuous fiber wrap, $b_f = s_f$.

Axial Strengthening

Axial strengthening/confinement of concrete using FRP composites (Fig. 13.30) is the process of wrapping of existing RC columns by fiber wraps (high strength and low weight) to achieve passive confinement in the existing structural members. This process basically increases strength and ductility both. The process of FRP wrapping around the columns is carried out using epoxy resin in which fibers of FRP are placed in perpendicular direction to the longitudinal axis of existing column. By this process, significant strength against shear can also be achieved in addition to effective passive confinement. The original shape and size of existing structural members are unchanged (unlike in any other jacketing process discussed above), and this method is more suitable for strengthening historic structures. As concrete is compressed uniaxially in this process, transverse strains are induced in the members due to Poisson's effect causing some radial expansion of the concrete (volumetric expansion). Further, by confining the concrete using a continuous FRP wrap, the fibers resist the transverse expansion of the concrete. This confining pressure developed by continuous FRP wraps induces a triaxial state of stress in the original concrete members. That is why FRP wrap concrete sections show superior behavior in both strength and ductility than concrete under uniaxial compression (Fig. 13.31). However,

Fig. 13.28 Mechanical anchorage to prevent de-bonding of laminate

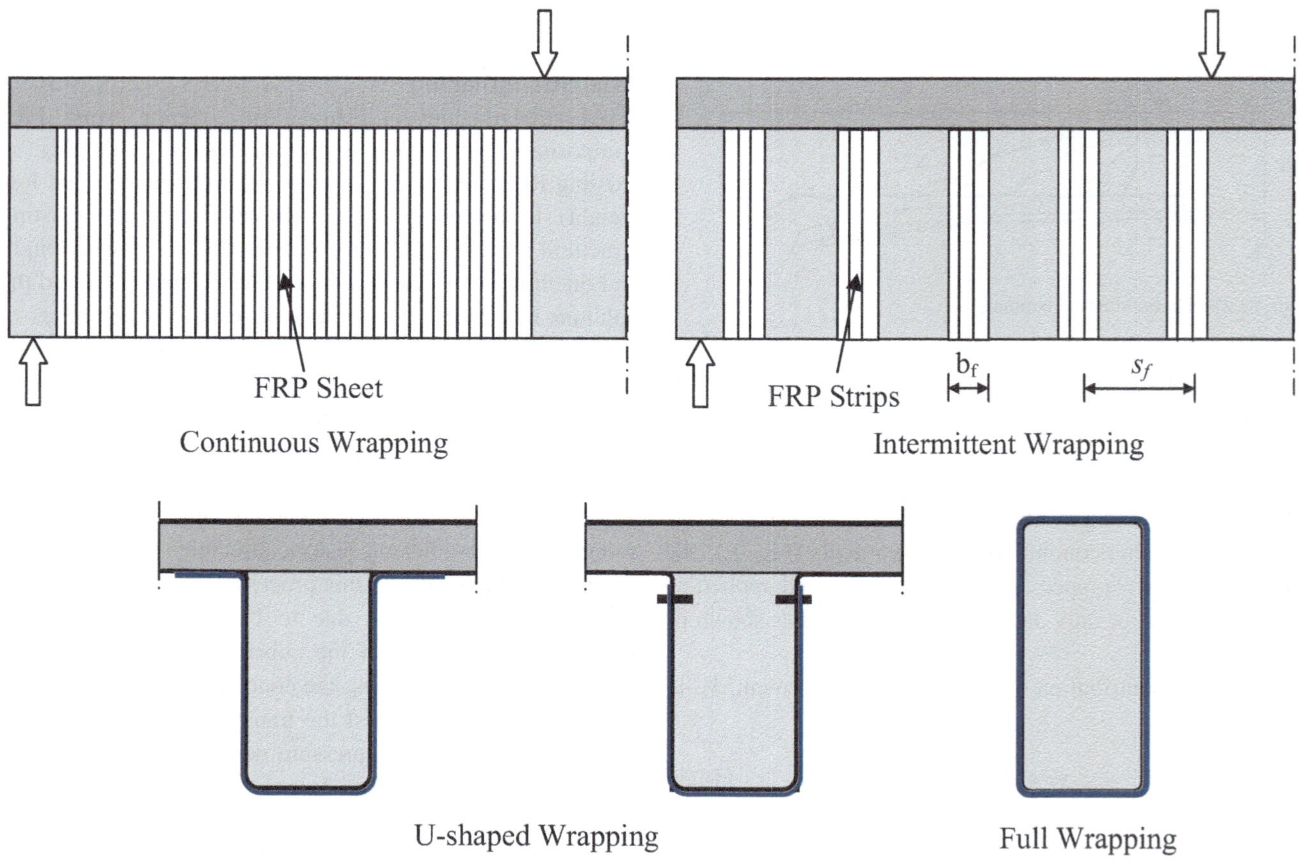

Fig. 13.29 Shear strengthening of sections using FRP composite

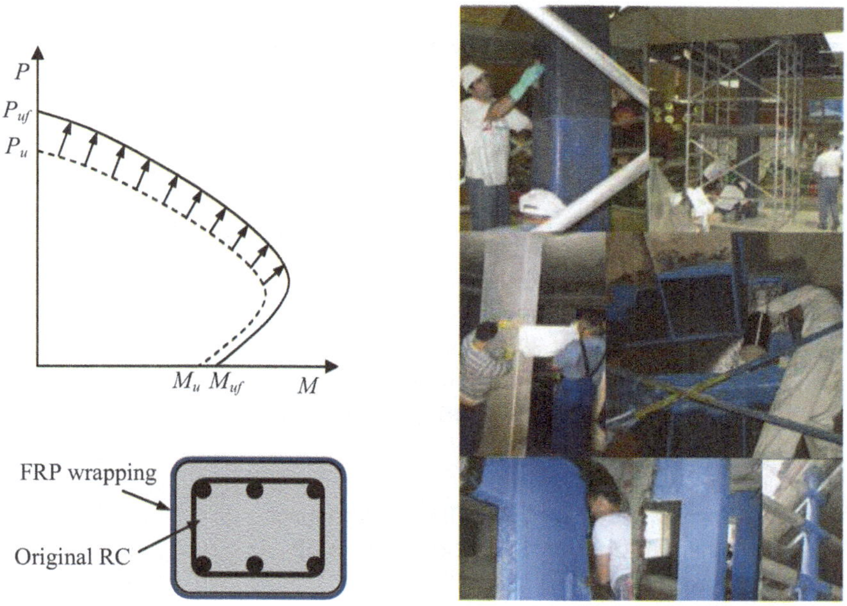

Fig. 13.30 Axial strengthening using FRP composites

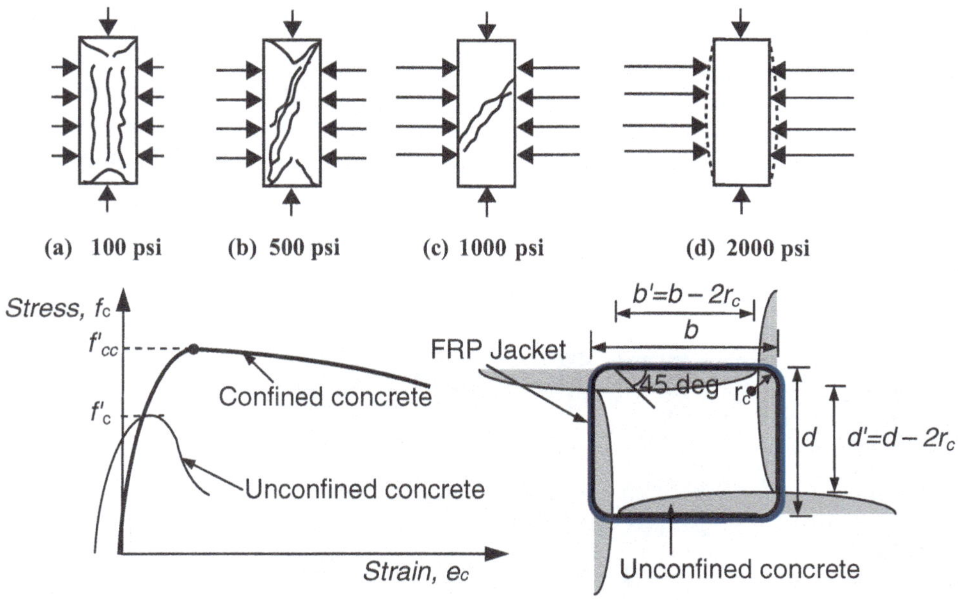

Fig. 13.31 Concept of confinement

to increase the effectiveness of FRP wrap, sharp edges of original rectangular sections should be practically rounded.

Design of FRP for Axial Strengthening

The design of FRP strengthening is performed on the well-established principles of mechanics. Most major codes like ACI, CEB-FIP, Eurocode, Japanese code, Swedish bridge code, Chinese Standard, Turkish code give guidelines for the design of FRP system for wrapping of concrete columns to increase their capacity. Almost all codes of practices recommend Mander equation to evaluate the effective compressive strength of confined concrete given by:

$$f'_{cc} = f'_c\left(2.254\sqrt{\left(1 + 7.94 f_l/f'_c\right)} - 2f_l/f'_c - 1.254\right) \quad (13.6)$$

where

f'_{cc} effective cylinder compressive strength of confined concrete

f'_c actual cylinder compressive strength of unconfined concrete = $0.8 f_{ck}$

f_l confining stress provided by FRP jacket.

By strain compatibility, the strain in the jacket is equal to the transverse strain in the concrete. The confining pressure may then be found by analyzing the statics of a thin-walled cylindrical cylinder as

$$f_l = \kappa_a \rho_f f_{fe}/2 = \kappa_a \rho_f \varepsilon_{fe} E_f/2 \quad (13.7)$$

where

- κ_a efficiency factor, based on section geometry
- ρ_f FRP reinforcement ratio = $4t_j/h$
- ε_{fe} effective strain level in FRP reinforcement
- f_f stress level in FRP reinforcement
- E_f tensile elastic modulus for FRP reinforcement
- t_j total thickness of jacket = nt_f
- h overall thickness of member
- n number of plies (layers) of FRP reinforcement
- t_f nominal thickness of one ply (layer) of FRP reinforcement.

The apparent increase in the compressive strength of concrete under the confining pressure supplied by the jacket is again quantified by Mander equation given above. For rectangular sections, in which the corners have been rounded to a radius, r_c, to allow wrapping around them, CEB-FIP recommends following formula for effective confining pressure, f'_{lx} and f'_{ly} in x- and y-directions, respectively:

$$f'_{lx} = k_e f_{lx} \quad \text{and} \quad f'_{ly} = k_e f_{ly} \quad (13.8)$$

where

- k_e efficiency factor = $1 - (b'^2 + d'^2)/3A_c$
- b' $b - 2r_c$,
- d' $d - 2r_c$
- f_{lx} $2E_f\varepsilon_{fe}t_j/d$,
- f_{ly} $2E_f\varepsilon_{fe}t_j/b$.

Strengthening of Beam–Column Joints

Experiments were performed on RC beam–column joints before and after repair with FRP. The joints were first tested till failure under cyclic loads and then repaired and strengthened with carbon FRP and retested. The spalled concrete was replaced by epoxy mortar. Flexural and shear strengthening and confinement were provided with FRP. Figure 13.32 shows the failure mode and typical hysteretic loops obtained for control and repaired specimen. The enhancement in strength, ductility, and improvement in failure mode is clearly demonstrated.

Retrofitting of Masonry Walls with FRP

Composite fiber wraps are widely used world over to strengthen and add ductility to masonry and reinforced concrete components against both in-plane and out-of-plane loads. It has been proved by several researchers and is recommended that "When FRP is bonded to the surface of masonry infill, the strength of the composite section increases." In terms of design, masonry strengthened with

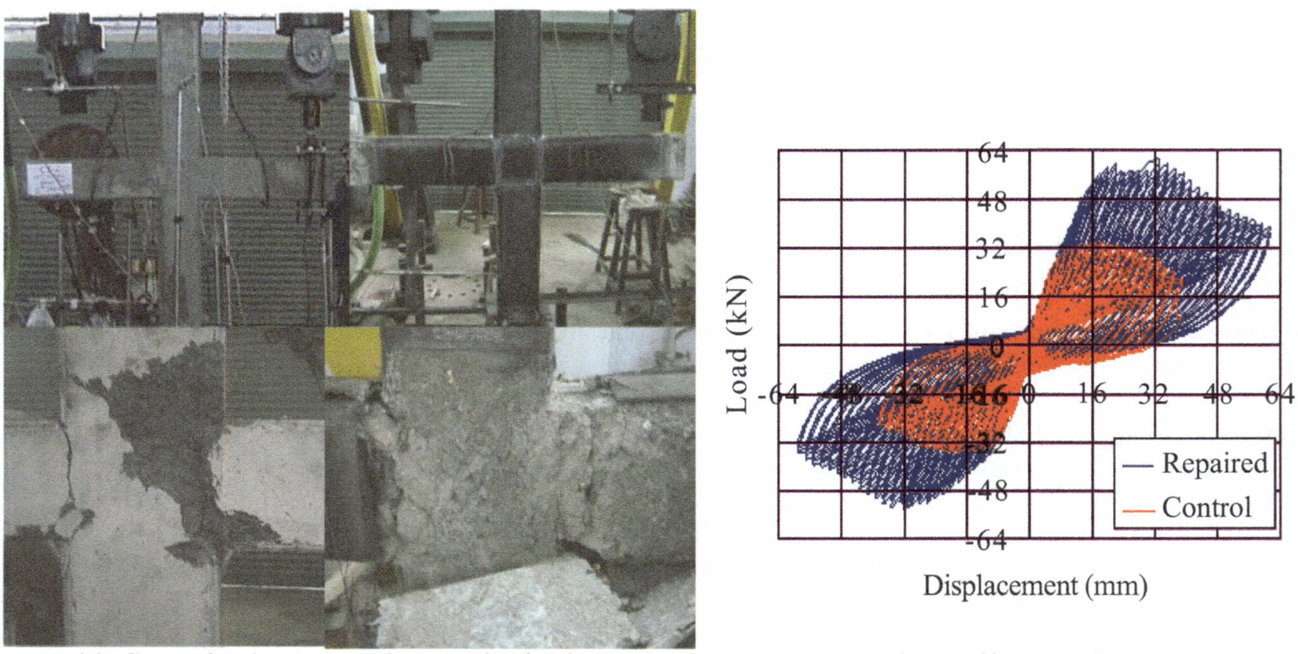

(a) Control Joint (b) Repaired Joint (c) Comparison of hysteresis

Fig. 13.32 Strengthening of beam–column joints using fiber composites

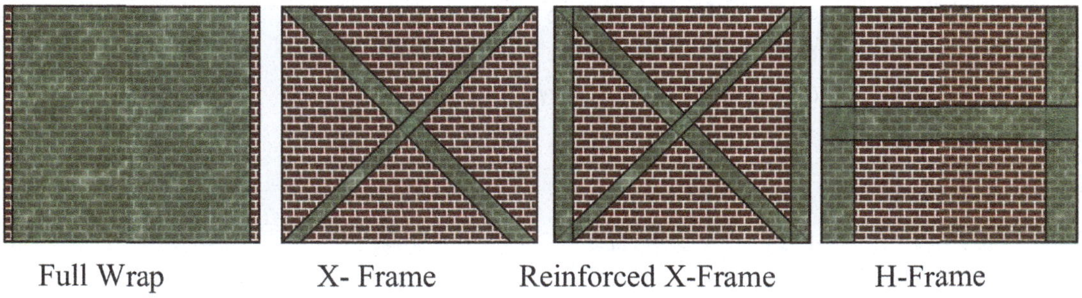

Fig. 13.33 Various FRP layouts studied for strength increase against in-plane loads

FRP strips or sheets may be treated in the same manner as RC members, following the procedures of prevailing design codes.

Strengthening Against In-Plane Loads

Figure 13.33 shows several different bonding schemes used to strengthen masonry walls against in-plane loads. Depending on the FRP strip orientation and the number of layers used, a different strength increase factor for FRP must be utilized. Based on research performed, a table of FRP strength increase factors, ξ_1, was developed and provided (Table 13.3). These factors were developed from testing of URM panels, with fairly uniform aspect ratios. The strength of a masonry wall strengthened with a particular FRP layout can be obtained by multiplying the factor, ξ_1, with the strength of original wall.

Similarly, Table 13.4 as per TR-02-33 gives the stiffness increase factor, ξ_2, for the walls with different FRP layouts.

Table 13.3 Strength increase factor, ξ_1, recommended by TR-02-33

FRP sheet layout	ξ_1
2 layer full wrap	1.51
1 layer reinforced X-frame	1.48
1 layer full wrap	1.41
2 layer X-frame	1.33
1 layer X-frame	1.29
1 layer H-frame	1.20

Table 13.4 Stiffness increase factor, ξ_2, recommended by TR-02-33

FRP sheet layout	ξ_2
Full wrap	1.53
Reinforced X-frame	1.43
H-frame	1.27
X-frame	1.23

Strengthening Against Out-of-Plane Loads

FRP is highly effective in strengthening the masonry walls against out-of-plane loads since it acts as external reinforcement to otherwise unreinforced wall. The procedure of seismic evaluation and design of strengthening for walls against out-of-plane loads is explained below with the help of a case study.

In this study, a masonry wall has been evaluated to obtain the maximum flexural stresses using simplified elastic method and the retrofitting is suggested using FRP wraps. The simplified elastic method involves following steps [Paks, Hungary].

- Determine boundary conditions (simply supported, fixed, free),
- Predict natural frequency,
- Determine horizontal seismic acceleration,
- Estimate maximum out-of-plane bending stress and check with allowable stress.

The frequency of the wall can be estimated as

$$f_w = F\,\alpha_{B,f}\alpha_E\alpha_D \tag{13.9}$$

where

F	is frequency factor, given by,
F	$(EI/m)^{1/2}/H^2$
H	wall height
E	elastic modulus typically taken as 1000 $f_{m,c}$
$f_{m,c}$	compressive strength of masonry = 7.5 MPa in this case
I	moment of inertia
m	distributed mass per unit surface area
$\alpha_{B,f}$	boundary condition coefficient for frequency calculation (Table 13.5)
α_E	elastic modulus coefficient (Table 13.6)
α_D	mass density coefficient (Table 13.7).

Corresponding to the given frequency, spectral acceleration value is obtained. Maximum flexural stress can be obtained as

Table 13.5 Boundary condition coefficient for fundamental frequency calculation $\alpha_{B,f}$

H/L	Free-free	SS-free	Fixed-free	SS-SS	SS-fixed	Fixed-fixed
Simple support top and bottom with specified combination of side supports						
≤ 0.20	1.571	1.571	1.571	1.571	1.571	1.571
0.40	1.571	1.612	1.622	1.822	1.870	1.931
0.67	1.571	1.698	1.748	2.270	2.480	2.765
1.00	1.571	1.859	2.020	3.142	3.764	4.608
1.50	1.571	2.182	2.677	5.106	6.769	8.968
2.50	1.571	2.992	4.875	11.390	16.540	23.160
H/L	Free-free	SS-free	Fixed-free	SS-SS	SS-fixed	Fixed-fixed
Fixed top and bottom with specified combination of side supports						
≤ 0.20	3.561	3.561	3.561	3.561	3.561	3.561
0.40	3.561	3.587	3.594	3.706	3.731	3.764
0.67	3.561	3.638	3.664	3.986	4.116	4.299
1.00	3.561	3.734	3.823	4.608	5.066	5.730
1.50	3.561	3.944	4.254	6.221	7.666	9.672
2.50	3.561	4.545	5.994	12.070	17.050	23.520
H/L	Free-free	SS-free	Fixed-free	SS-SS	SS-fixed	Fixed-fixed
Simple support top/fixed bottom (or vice versa) with specified combination of side supports						
≤ 0.20	2.454	2.454	2.454	2.454	2.454	2.454
0.40	2.454	2.491	2.499	2.646	2.682	2.727
0.67	2.454	2.558	2.593	3.008	3.175	3.407
1.00	2.454	2.685	2.804	3.764	4.307	5.066
1.50	2.454	2.951	3.349	5.579	7.144	9.260
2.50	2.454	3.672	5.344	11.690	16.760	23.320
H/L	Free-free	SS-free	Fixed-free	SS-SS	SS-fixed	Fixed-fixed
Free top/fixed bottom with specified combination of side supports						
≤ 0.20	0.560	0.560	0.560	0.560	0.560	0.560
0.40	0.560	0.613	0.634	0.780	0.855	0.959
0.67	0.560	0.704	0.793	1.190	1.488	1.891
1.00	0.560	0.897	1.105	2.020	2.804	3.823
1.50	0.560	1.103	1.786	3.932	5.833	8.243
2.50	0.560	1.607	3.965	10.140	15.620	22.460
H/L	Free-free[a]	SS-free	Fixed-free	SS-SS	SS-fixed	Fixed-fixed
Free top/simple support bottom with specified combination of side supports						
≤ 0.20	0.000	0.107	0.159	0.224	0.258	0.285
0.40	0.000	0.210	0.257	0.479	0.587	0.727
0.67	0.000	0.356	0.491	0.971	1.313	1.755
1.00	0.000	0.536	0.854	1.859	2.685	3.734
1.50	0.000	0.800	1.585	3.821	5.75	8.186
2.50	0.000	1.313	3.834	10.080	15.570	22.420

[a]Mode representing rigid body motion

Table 13.6 Elastic modulus coefficient α_E

E (GPa)	α_E
3.45	0.71
4.80	0.84
6.20	0.95
6.90	1.00
8.60	1.12
10.30	1.22
12.00	1.32
13.70	1.41
15.40	1.50
17.20	1.58
18.80	1.66
20.60	1.73

Table 13.7 Mass density factor α_D

ρ (kg/m^3)	α_D
3200	0.87
2800	0.93
2400	1.00
2000	1.10
1600	1.22
1200	1.41

$$\sigma_{Max} = S\,\alpha_{Bs}(S_{af}/g)\alpha_D \tag{13.10}$$

where

S is stress factor, given by:

$$S = H^2 m\, g\, t_{act}/(2I)\ [\text{in Pa}] \tag{13.11}$$

$\alpha_{B,S}$ boundary condition coefficient for maximum bending stress calculation (Table 13.8)

$S_{a,f}$ spectral acceleration (average of floors above and below wall)

g gravity acceleration.

If $\sigma_{max} \leq \sigma_{all}$ (allowable flexural tensile stress), then the wall is safe otherwise retrofit is required. Based on the above-given formulations, a masonry wall of an existing old structure was analyzed and was found to be requiring retrofit. The calculations are given below

Length of the wall, L = 6780 mm,
Height of the wall, H = 3300 mm,
Thickness of the wall, t = 230 mm.
Boundary conditions: simply supported on top and bottom, free on sides.
ρ = 2000 kg/m^3,
E = 7.5 Gpa [Paks, Hungary],
σ_{all} = 0.16 MPa [Paks, Hungary].

Based on the values of H/L ratio, ρ and E, the various coefficients are found out

$\alpha_{B,f}$ = 1.571,
α_E = 1.0424,
α_D = 1.10.

The frequency factor F is found as

$$F = 10.338$$

The fundamental natural frequency of the wall is calculated as

$$f_w = 10.338^*1.571^*1.0424^*1.10 = 18.622\,\text{Hz}$$

The corresponding S_{af} for the specified ground motion = 0.31 g.

The stress factor, S, is calculated as

$$S = 6.69\,\text{MPa}$$

For the given boundary conditions and H/L ratio, $\alpha_{B,s}$ = 0.125.

Maximum flexural tensile stress for the wall is calculated as

$$s_{max} = 6.69^*0.125^*0.31/1.1^2 = 0.214\,\text{MPa}$$

Since σ_{max} (0.214 MPa) > σ_{all} (0.16 MPa), the wall is unsafe and requires retrofit.

The retrofitting involves the design of the spacing of the carbon FRP laminates. This is done in a similar way as we design the reinforcement for a slab for a given moment value, by considering the stress–strain curve for masonry instead of concrete and that of FRP instead of steel. The proposed retrofit using carbon FRPs is shown in Fig. 13.34.

13.6 Retrofitting of Structures and Equipment Using Response Control Devices

Often tall buildings are subjected to vibrations due to wind loads, earthquake excitations, machine vibrations, etc. Despite having huge progresses in seismic design methods, extensive structural damage has been observed during most of the recent earthquakes. In addition, strong wind storms have also affected the serviceability of tall and flexural structures by causing excessive deflection leading to occupant's discomfort. Generally, structures possessing high

Table 13.8 Boundary condition coefficient for maximum bending stress calculation $\alpha_{B,S}$

H/L	Free-free	SS-free	Fixed-free	SS-SS	SS-fixed	Fixed-fixed
Simple support top/bottom with specified combination of side supports						
≤0.20	0.125	0.125	0.125	0.125	0.125	0.125
0.40	0.125	0.125	0.125	0.110	0.110	0.105
0.67	0.125	0.122	0.118	0.076	0.101	0.095
1.00	0.125	0.106	0.110	0.042	0.078	0.064
1.50	0.125	0.078	0.093	0.034	0.046	0.034
2.50	0.125	0.040	0.057	0.018	0.019	0.013
H/L	Free-free	SS-free	Fixed-free	SS-SS	SS-fixed	Fixed-fixed
Fixed top/fixed bottom with specified combination of side supports						
≤0.20	0.083	0.083	0.083	0.083	0.083	0.083
0.40	0.083	0.083	0.083	0.083	0.083	0.083
0.67	0.083	0.083	0.083	0.082	0.079	0.075
1.00	0.083	0.083	0.083	0.070	0.061	0.051
1.50	0.083	0.083	0.083	0.047	0.042	0.034
2.50	0.083	0.083	0.059	0.020	0.020	0.014
H/L	Free-free	SS-free	Fixed-free	SS-SS	SS-fixed	Fixed-fixed
Simple support top/fixed bottom (or vice versa) with specified combination of side supports						
≤0.20	0.125	0.125	0.125	0.125	0.125	0.125
0.40	0.125	0.125	0.125	0.125	0.125	0.119
0.67	0.125	0.125	0.125	0.110	0.102	0.095
1.00	0.125	0.125	0.125	0.084	0.069	0.061
1.50	0.125	0.125	0.109	0.050	0.046	0.034
2.50	0.125	0.105	0.063	0.020	0.020	0.013
H/L	Free-free	SS-free	Fixed-free	SS-SS	SS-fixed	Fixed-fixed
Free top/fixed bottom with specified combination of side supports						
≤0.20	0.500	0.500	0.500	0.500	0.500	0.500
0.40	0.500	0.480	0.468	0.356	0.356	0.328
0.67	0.500	0.451	0.404	0.220	0.236	0.184
1.00	0.500	0.395	0.300	0.118	0.126	0.088
1.50	0.500	0.311	0.175	0.054	0.056	0.037
2.50	0.500	0.224	0.076	0.021	0.020	0.013
H/L	Free-free[a]	SS-free[b]	Fixed-free	SS-SS	SS-fixed	Fixed-fixed
Free top/simple support bottom with specified combination of side supports						
≤0.20	1	2	2	0.780	0.780	0.780
0.40	1	2	2	0.340	0.722	0.582
0.67	1	2	0.672	0.187	0.172	0.123
1.00	1	2	0.373	0.112	0.120	0.083
1.50	1	2	0.194	0.057	0.056	0.036
2.50	1	2	0.072	0.021	0.022	0.014

[a]Unstable conditions, outlier

[b]Almost unstable conditions, outlier

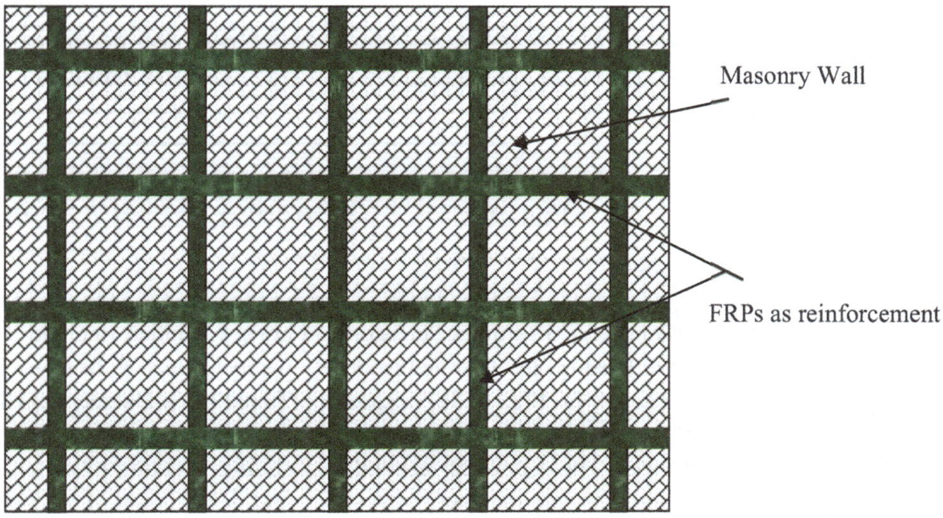

Fig. 13.34 Retrofitting of masonry walls using FRP strips as reinforcement

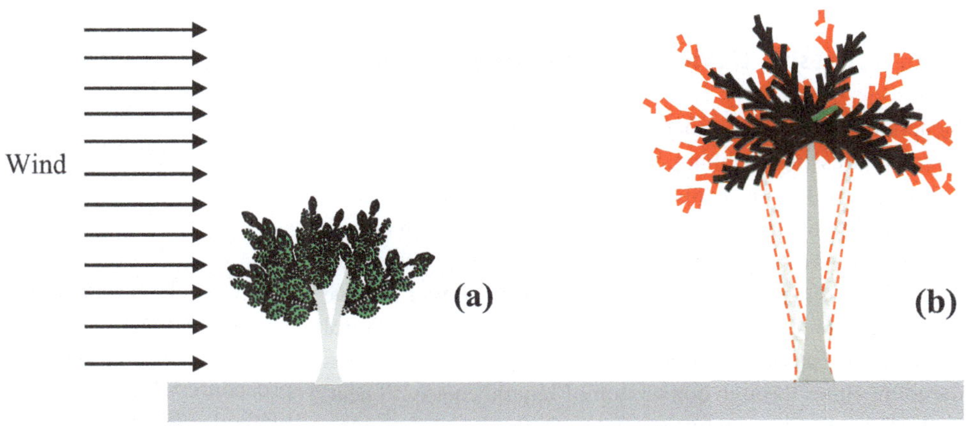

Fig. 13.35 Energy dissipation of **a** short tree; **b** tall tree

inherent natural damping are less prone to damage. However, structures which are subjected to strong vibrations, response mitigation just by the aid of the natural damping of the structure may not be viable. In such situations, additional damping devices are frequently used to control the response.

Let us understand this through an example. One of the most beautiful gifts from nature to mankind is trees. Figure 13.35 illustrates a short tree and a tall tree subjected to wind loads. At high wind velocities, in order to dissipate the energy, the branches of short tree move or vibrate rigorously, whereas in the case of the tall tree the whole tree along with the stem oscillates. In the former case, all the energy is shared by the branches and less is transferred to the stem, whereas in the tall tree due to the flexibility of the stem, it deforms more along with the branches to dissipate the energy. It is important to understand that if the tall tree had branches along the height, the same behavior that of the former would have been observed. With this one can infer that in any structural system subjected to vibratory loads, if flexible members are introduced that takes up large energy, resulting in low load transfer to the main static load-bearing members.

Recent earthquakes have revealed that with conventional construction procedures, structural damage is unavoidable, especially of tall buildings, due to high displacement of top stories. However, it has been realized that by increasing the energy dissipation characteristics of these structures, the damage risk can be significantly reduced and can further improve the serviceability in case of natural calamities like earthquakes. In other words, if the amount of energy received by the structure is controlled and if a significant portion of the energy is dissipated mechanically independent of primary structure, the seismic response of the structure and damage can be controlled considerably. An efficient means is to alter its damping characteristics. This objective can be achieved by adopting new techniques such as base isolation [43] and use of energy dissipation devices [39, 41]. These external or auxiliary damping devices can be attached to the structure either during its construction or subsequently retrofitted after construction, thus increasing its damping.

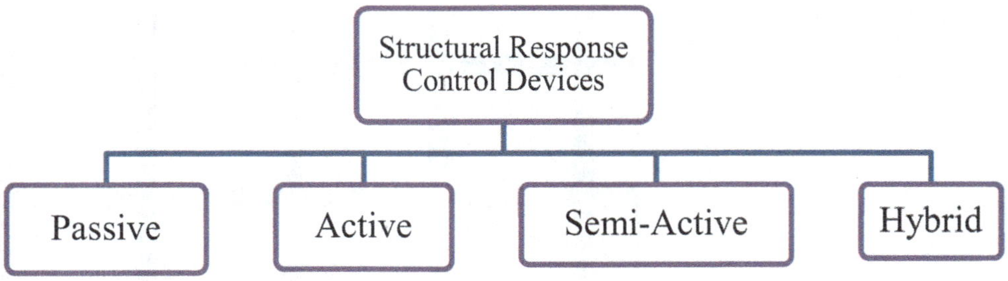

Fig. 13.36 Classification of structural response control devices

Figure 13.36 shows classification of different categories of dampening devices often utilized for the seismic response control of structures. As shown in Fig. 13.36, response control devices can be widely classified into passive, active, semi-active, and hybrid devices. These devices respond as per the structural response and dissipate the energy eventually reducing the response.

13.6.1 Passive Devices for Response Control

Generally, vibrating structures dissipate energy through internal stressing, rubbing, cracking, plastic deformation, and various other mechanisms. With introduction of external devices, the rate of energy dissipation is enhanced which is an effective way of reducing the structural response under dynamic loadings. This is achieved by incorporating passive devices in the structure. Passive Devices or Passive Control Devices or passive energy dissipation devices imparts damping either through conversion of kinetic energy to heat, or by transferring energy among other vibrating modes. Block diagram of Passive Control Devices is shown in Fig. 13.37. The importance of Passive Control Devices in seismic design is paramount as it not just aids in mitigating seismic hazards but also helps in restoration and rehabilitation of aging and deficient structures.

In general, such devices are characterized by their principle and mechanism by which energy is dissipated. The energy dissipation devices involving direct dissipation of the energy applied include Fluid Viscous Damper, friction damper, Lead Extrusion Damper, Elasto-Plastic Damper, Tuned Mass Damper, and Tuned Liquid Dampers to name a few. These devices work on the principles of frictional sliding, yielding of metals, phase transformation in metals, and deformation of viscous–elastic (VE) solids or fluids.

The equation of motion of a vibrating system under ground motion is given as Eq. (13.12)

$$m\ddot{x} + c\dot{x} + kx = -m\ddot{x}_g \tag{13.12}$$

where m, k, c have their usual meanings. However, when a passive device is installed, the equation of motion governing the system is described through the Eq. 13.13 as

$$m\ddot{x} + c\dot{x} + kx + k_n x = -m\ddot{x}_g \tag{13.13}$$

where k_n is the nonlinear stiffness of the damper, e.g., yielding damper. Using the characteristics of these dampers, equations of motion can be solved using the numerical procedures like Newmark-beta method of Wilson-Theta method. However, the design inputs are generally in the form of a spectrum.

Let us consider a vibrating structure with natural frequency f_1 and is marked in Fig. 13.38. Generally, it is assumed that the stiffness of the damper is much lower than

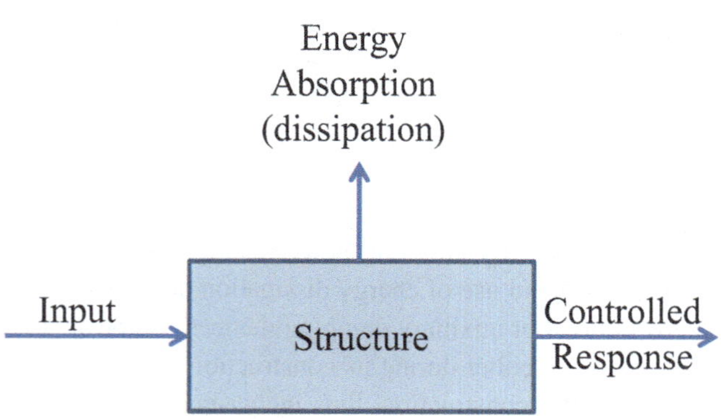

Fig. 13.37 Block diagram for passive control devices

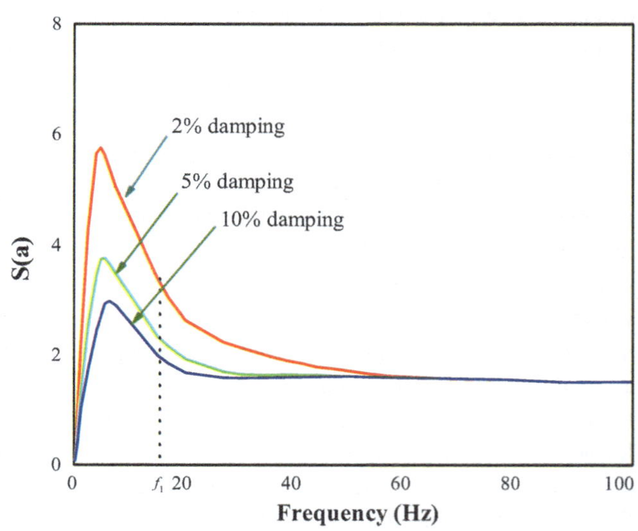

Fig. 13.38 Spectrum

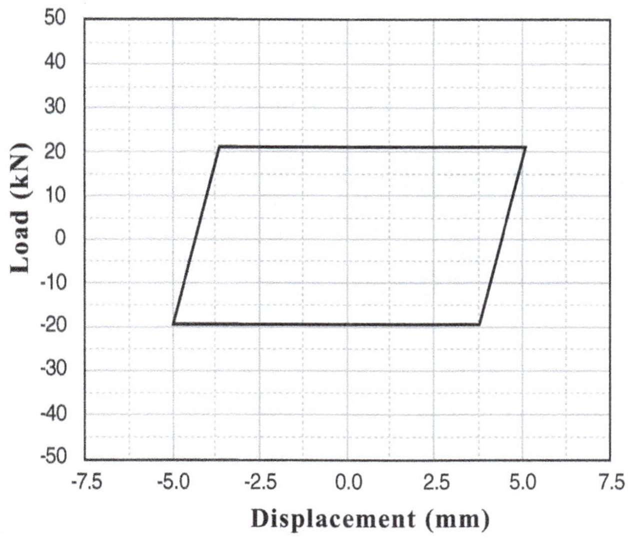

Fig. 13.39 Typical hysteresis curve of a damper

the stiffness of the structure itself; hence, it is assumed that it does not affect the natural frequency of the system except dissipating energy. The energy balance for an external undamped system can be written as

$$E_{KE} + E_D + E_{PE} = E_I \tag{13.14}$$

where E_{KE} is the kinetic energy, E_D is the structural damping energy, E_{PE} is the potential energy, and E_I is the total energy. By introducing a damper, the equation becomes

$$E_{KE} + E_D + E_d + E_{PE} = E_I \tag{13.15}$$

where E_d is the dissipation energy due to damper. It is worth noticing that by adding an additional term in Eq. 13.14, the value of the other terms would be reduced keeping the total energy term constant, and hence, the response will reduce. With this assumption, the response can be obtained initially considering damping due to damper. After evaluating the response, energy dissipated in the damper is evaluated as area of the hysteresis curve (Fig. 13.39). From this, the damping ratio can be estimated using Eq. 13.16.

$$\zeta_d = \frac{\Delta E}{4\pi\Sigma} \tag{13.16}$$

Now with the increased total damping, reduced seismic response is obtained from the spectrum shown in Fig. 13.38 and this process is repeated for successive iterations until the response is differed by 1% and the final reduced response is used for the design. A flowchart illustrating the methodology is shown in Fig. 13.40. The methodology presented in Fig. 13.40 is applicable for all types of passive energy dissipating devices under influence of seismic excitation.

13.6.1.1 Fluid Viscous Dampers

The working principle of Fluid Viscous Damper is based on the movement of high-velocity fluid through orifices. These dampers are frequently utilized for shock absorption and vibration isolation in aerospace and defense industries. Figure 13.41 shows the schematic of a typical Fluid Viscous Damper. The damper shown here consists of a steel piston with orifice head (made of bronze) and an accumulator and is filled with silicon oil as the fluid. The piston head uses special shaped passages which alters the flow characteristics

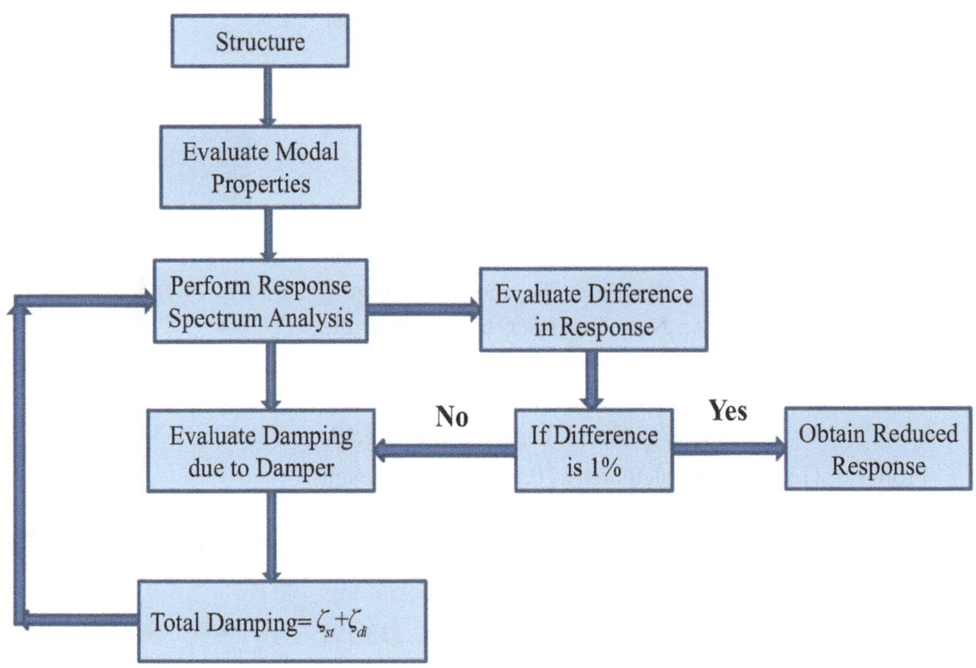

Fig. 13.40 Flowchart of the methodology for seismic analyses using passive device

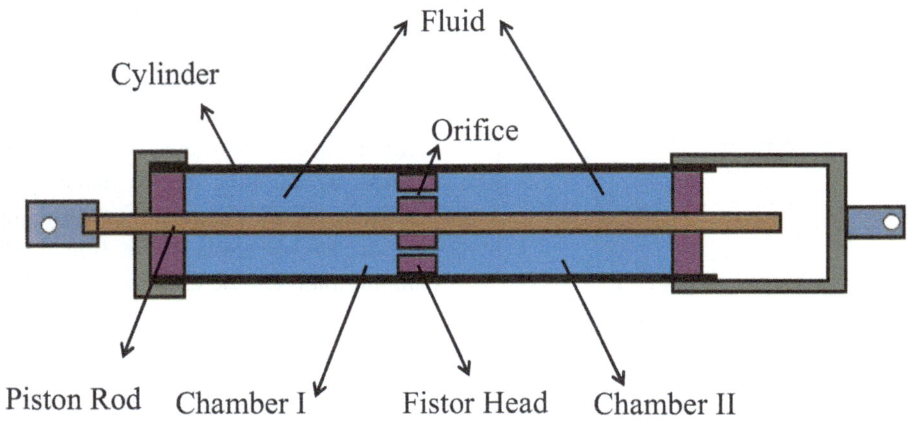

Fig. 13.41 Schematic of a fluid viscous damper

with fluid speed such that the force output is proportional to the piston rod velocity ($\propto \dot{x}_s^{\gamma}$) where γ is a predetermined coefficient within the range of 0.5–2. The proportionality constant is the damping constant which has a small dependence on temperature. The damper behavior is linear in case $\gamma = 1$ and nonlinear otherwise [1].

Example 13.3 Consider a cylindrical water tank with mass 45.45 ton and lateral stiffness of 1.4677 MN/m having a natural frequency of 0.9 Hz. Now, calculate the damping ratio, damped natural frequency, and the damping coefficient when viscous dampers are attached to the structure on the top as shown in Fig. 13.42 such that the successive free vibration amplitude ratio is 1.4. Also, estimate the maximum response of the structure in terms of the displacement and acceleration of the structure when a sinusoidal ground motion of 0.1 g of 0.3 Hz is applied on the structure.

Solution Given: Mass, m = 45,450 kg; stiffness, K = 1.4677 MN/m; natural frequency, f = 0.9 Hz;

Amplitude ratio, A_1/A_2 = 1.4

Now, damping ratio can be calculated from the following logarithmic Eq. (13.12),

$$\ln(A_1/A_2) = 2\pi\zeta/\sqrt{(1-\zeta^2)} \tag{13.17}$$

where ζ is the damping ratio. From the above equation, ζ comes out to be 5.34%. Now, in order to calculate the damped natural frequency, f_d,

$$f = f_d/\sqrt{(1-\zeta^2)} \tag{13.18}$$

From Eq. (13.18), f_d is calculated as 0.899 Hz. This result explains that the effect of damping on the natural frequency is negligible for the lower value of damping ratios. Thus, the damping coefficient is estimated as,

$$c = c_{cr}\zeta \tag{13.19}$$

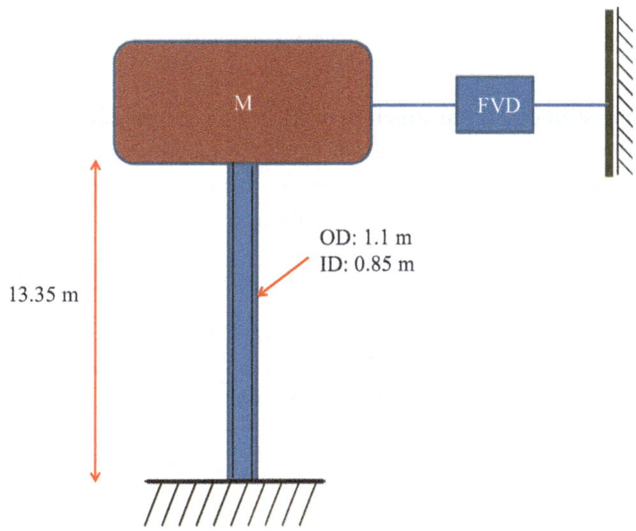

Fig. 13.42 Schematic of a water tank with fluid viscous damper

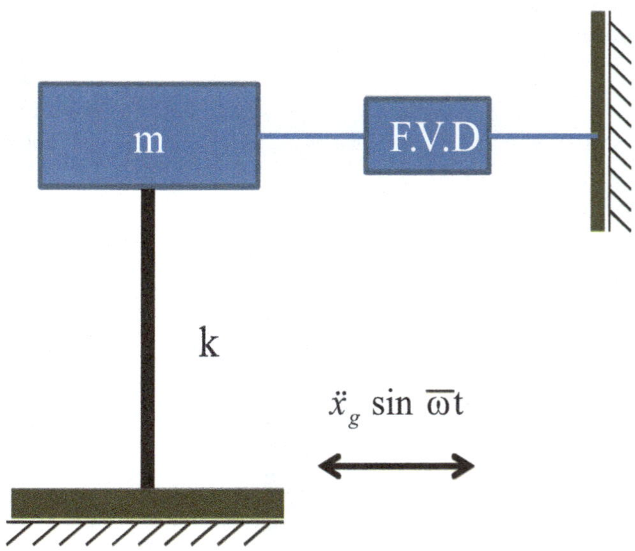

Fig. 13.43 Schematic SDOF representation of the water tank system with fluid viscous damper under sinusoidal ground motion

where ζ is 5.34% and c_{cr} is given as

$$c_{cr} = 2m\omega \quad (13.20)$$

Therefore, c_{cr} is 514,027.4 N-s/m and c is 27,449.1 N-s/m. For the second part of the problem, a sinusoidal ground motion is given with acceleration 0.1 g and excitation frequency 0.3 Hz. Now the maximum lateral displacement of the damped structure is given as

$$|x| = \ddot{x}_g/\omega^2 \quad (13.21)$$

By solving the above equations, maximum displacement is found to be 3.06 cm.

MATLAB code for solving the above problem:

```
% A function named "funcfvd" is formed

title ('prob13.6.1.1');
function f= funcfvd (t,x)
f= zeros (2, 1);
f(1)=x(2);
f(2)= -(1457700/45450)*x(1)-(27449.1/45450)*x(2)-0.1*9.81*sin(1.89*t);
end
% The differential equation is solved by calling the "funcfvd"
tspan = [0:0.01:20];
x0= [0; 0.1];

[t, x] = ode23 ('funcfvd', tspan, x0);
disp ('t      x(t)    xd(t)');
disp([t x]);
plot (t,x (:,1));
xlabel ('t');
gtext ('x(t)');
```

13.6.1.2 Friction Damper

The friction damper is a passive energy dissipation device which absorbs vibration energy by friction forces between contacting plates. Friction dampers are extensively used in civil engineering structures, and one such example is shown in Fig. 13.44. In the present text, double-sliding friction dampers (DSFD) are explained [2]. The characterization of these types of dampers is done by means of cyclic tests using hydraulic actuator. A photograph of DSFD damper is shown in Fig. 13.45, and the schematic is shown in Fig. 13.46.

The DSFD friction damper comprises of an inner (plate I) and two outer steel plates (plate II). Teflon liners are provided between the contacting plates and frictional forces which are produced between them. The normal force required to induce desired friction force between the contacting plates is provided by means of pretensioning of the bolts between plate I and plate II. Oblong holes are provided in the inner plate to facilitate sliding. The pretensioning of bolts is done by tightening bolts to a specified value of torque using wrench. The picture of the test setup used for the characterization of damper is shown in Fig. 13.47a and its schematic in Fig. 13.47b.

As shown in Fig. 13.47, one end of the damper is fixed rigidly through bracket and other end is connected to a hydraulic actuator. For the given type of damper, initially all bolts between plate II and plate III are tightened rigidly and a torque of 40 lb-ft is applied to the bolts between plate I and plate II. Subsequently, a quasi-static cyclic displacement of 5 mm amplitude is applied to the damper and is repeated for ten cycles. Further, the test is carried out for various torque values ranging from 40 lb-ft to 100 lb-ft in the intervals of 10 lb-ft and the resulting cyclic load–displacement curves for various torque values in lb-ft (T40 to T100) is shown in Fig. 13.48. From this figure, it is observed that a stable hysteretic behavior is obtained over the repeated cycles of loading. Finally, the slip load for each applied torque value is obtained and is plotted in Fig. 13.49. It is observed from this figure that the variation of slip load is linear with respect to applied torque.

The coefficient of friction between the liner and steel plate is calculated as follows:

Torque applied for each bolt,

$$T = \mu_b F_b D_b \quad (13.22)$$

Fig. 13.44 Friction dampers for seismic control are used in La Gardenia Towers, New Delhi

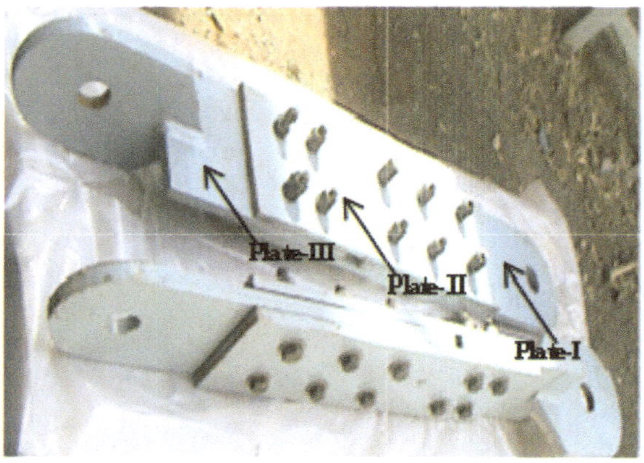

Fig. 13.45 Photograph of friction dampers

Normal force in each bolt,

$$F_b = T/(\mu_b D_b) \tag{13.23}$$

Coefficient of friction between the liner and plate,

$$\mu_p = \frac{F}{nF_b} \tag{13.24}$$

where n represents the number of contact points, and F is the slip force or slip load.

Case Study: Seismic Retrofitting of process tanks of Heavy Water Plant with the application of friction damper [2, 3]

During seismic re-evaluation of one of the process tank of Heavy Water Plant, it was found that existing foundation bolts were not sufficient to meet the present seismic load requirements of maximum considered earthquake (MCE) condition as per the IS code. Based on the background studies, it was evident that friction dampers could be effectively utilized in reducing the seismic response of structures, equipment and piping systems. Hence, it was proposed to use friction dampers of double-sliding friction damper (DSFD) of 30 kN capacity for the retrofitting of the process tank.

The scheme of retrofitting for the process tank using friction dampers is shown in Fig. 13.50. The process tank retrofitted has a height of 22.55 m and inner diameter of 4.25 m and is provided with cross-flow sieve trays. The tank is made of carbon steel, SA 516 Gr 70 and is supported on skirt with base plate. The upper cylindrical portion has 40 mm, and the lower portion has 42 mm thicknesses. The top and bottom hemispherical portions have thickness of 25 and 28 mm, respectively. The tower has eight number of M24 foundation bolts, and it was found that these bolts are not sufficient to meet present seismic load requirements of maximum considered earthquake (MCE) condition. Thus, two friction dampers were provided in orthogonal directions at 19.5 m/19.2 m elevations, and the slip load for each damper was set to 30 kN. For both the dampers, one end of the damper is attached to process tank, whereas the other end is connected to the adjacent supporting structure.

In order to qualify the process tank for the seismic excitation, an input seismic load was conservatively taken by considering the envelope of excitation at base and damper location. The analysis was carried out by using the envelope of ground spectrum (MCE) for 4% damping and floor response spectrum (FRS) of supporting structure at 19.5 m elevation. The base moment and maximum displacement at 19.5 m elevation of the process tank before retrofitting were estimated to be 4.886 MN-m and 3.108 mm, respectively. Iterative response spectrum method was used to evaluate the seismic response of the process tank with friction dampers. It was also observed that an additional damping of 33% was

Fig. 13.46 Schematic of a double-sliding friction damperfriction Damper

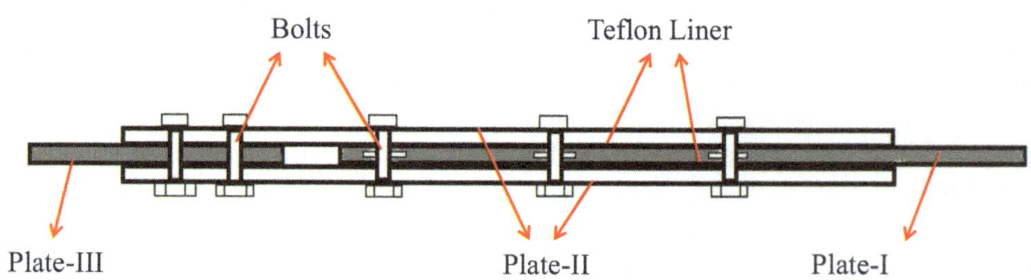

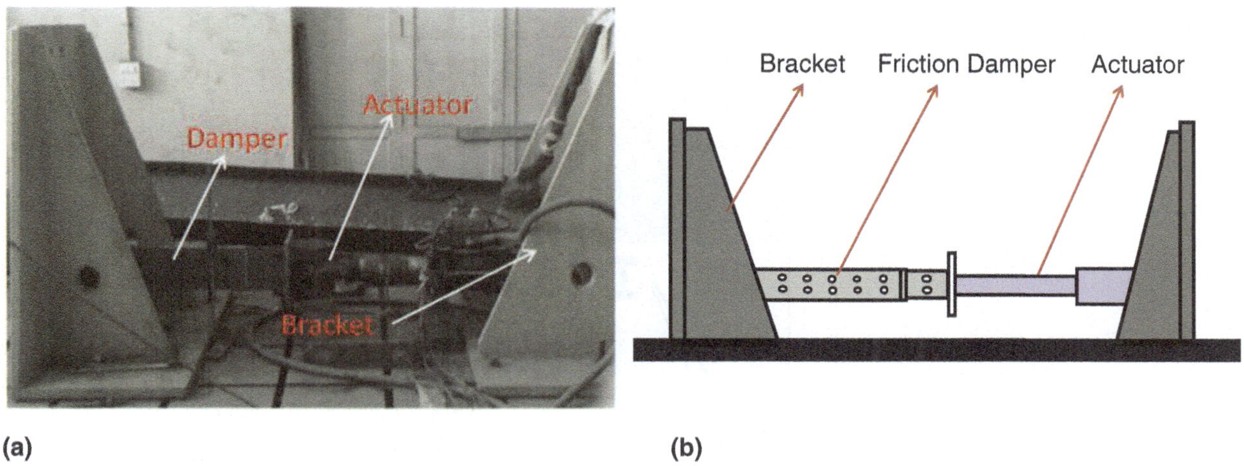

Fig. 13.47 **a** Photograph of test setup used for characterization of DSFD dampers. **b** schematic of the test setup used for characterization of DSFD dampers

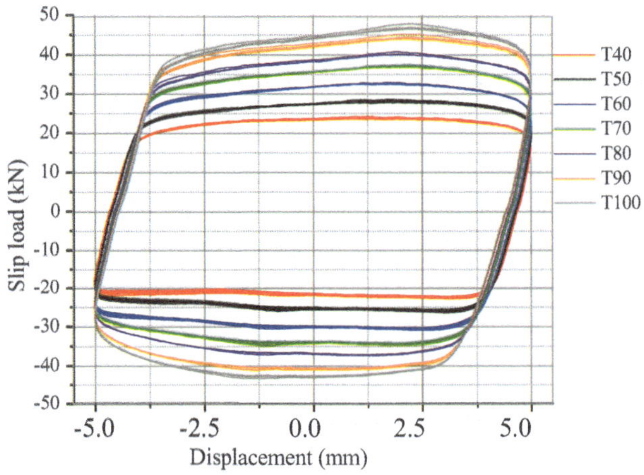

Fig. 13.48 Cyclic load-displacement curves

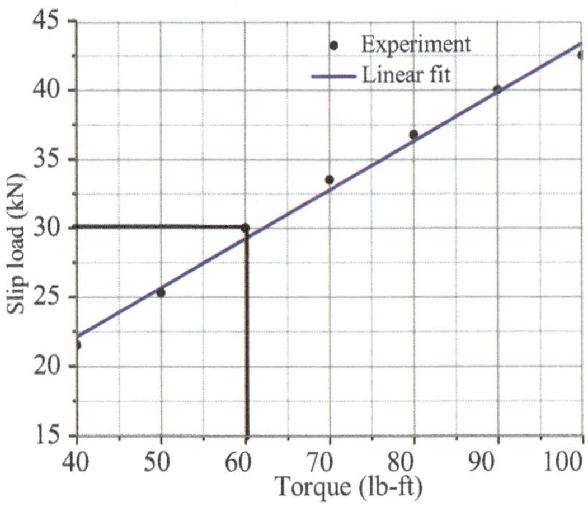

Fig. 13.49 Variation of slip load with applied torque

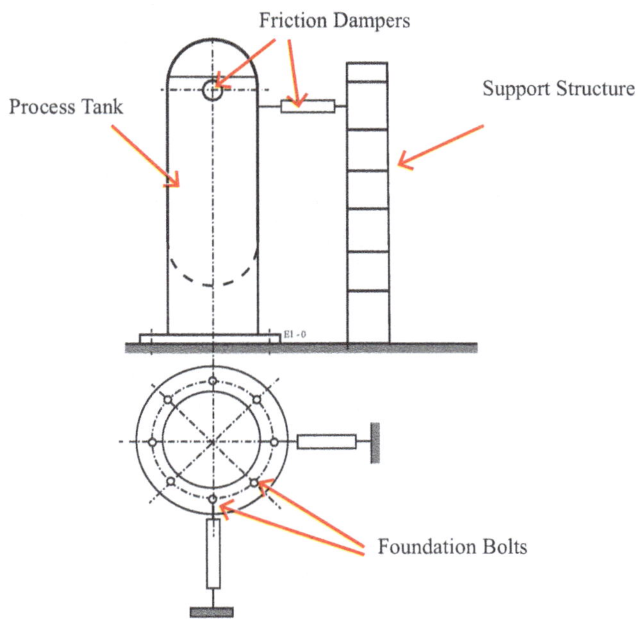

Fig. 13.50 Scheme of retrofitting for process tank using friction dampers

obtained from friction damper and the displacement at 19.5 m elevation reduced to 0.7 mm. The base moment was reduced to 1.168 MN-m for one-directional spectral loading, and for triaxial excitation, the resultant base moment is estimated to be 1.65 MN-m.

Example 13.4 Consider the problem of a water tank system having a mass of 25.2 tons and stiffness of 1.79 MN/m with structural damping of 4% (Fig. 13.51). The tank is supported on four vertical rectangular columns with outer

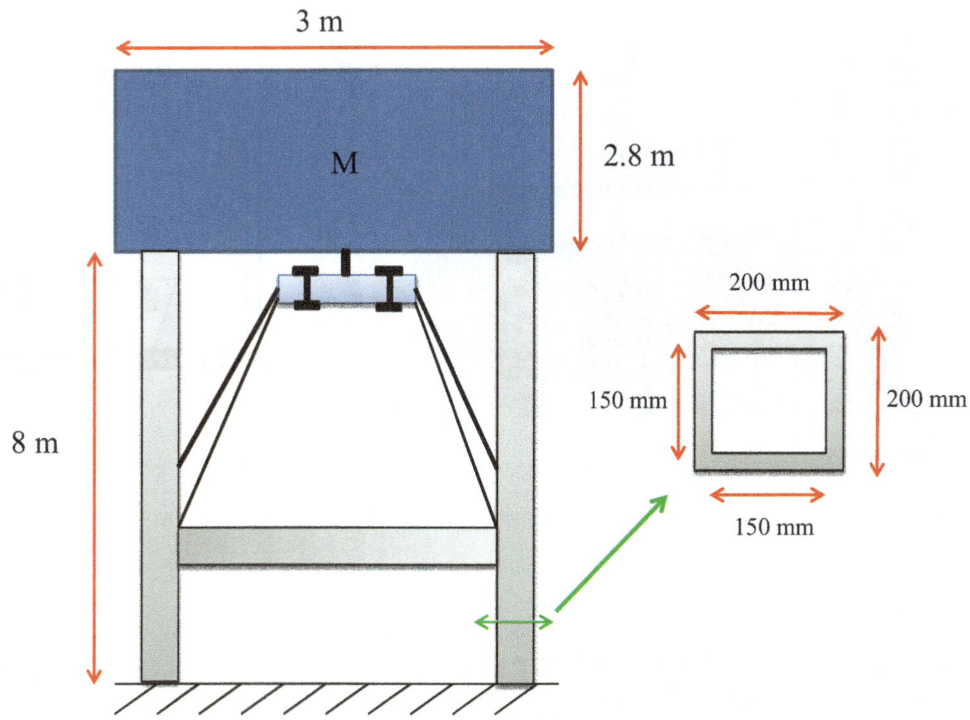

Fig. 13.51 Schematic of a SDOF system with friction damper

edge of 0.2 m and inner edge of 0.15 m. The length of the system is 8 m. It is excited with a ground motion shown in Fig. 13.52. In order to reduce the response of the structure, a friction damper is attached to it (Fig. 13.51). Estimate the response and damping of the structure.

Solution Given: mass of the structure, m = 25.2 tons; stiffness of the structure, k = 1.79 MN/m; natural frequency, f = 1.34 Hz; damping in the structure, $\zeta_s = 4\%$.

For the given ground motion, the maximum lateral force in the structure assuming no friction damper

$$F = m\ddot{x}_g$$

The ground acceleration is obtained from the response spectrum for 4% damping as 3.32 m/s^2. The maximum deflection of the structure is estimated as 46.8 mm. From the above equation, the maximum force on each column is calculated as 20.89 kN. The base moment on each column is calculated from $M = FL$ as 167,115.31 N-m. The section modulus of the shaft is 9.114 × 10^{-4} m^3. Therefore, the maximum bending stress is 183.36 MPa. The allowable bending stress is 115 MPa. Thus, an additional stress of 68.36 MPa is induced on the columns.

Now, in order to reduce the load on the structure due to ground acceleration, a double-sliding friction damper is attached to it as shown in Fig. 13.46. The size of the friction damper (f_s) is 10 kN.

Iteration 1:

Now, for the first iteration, damping due to friction damper (ζ_d) is calculated as Eq. 13.16,

$$\zeta_d = \frac{\Delta E}{4\pi\Sigma}$$

where ΔE the area under the load–displacement is curve given by 4 $f_s\delta$ and Σ is the strain energy of the vibrating structure given by $k\delta^2/2$. For the present case, $\Delta E =$ 1.872 kJ and Σ is 1.964 kJ. Thus, the ζ_d is calculated as 0.0758. In the next iterations, response for both structural damping added with friction damping ($\zeta_s + \zeta_d$), i.e., 11.58% is estimated.

Iteration 2:

Now, with this increased damping, the response of the structure is estimated. For the damping ratio of 11.58%, the

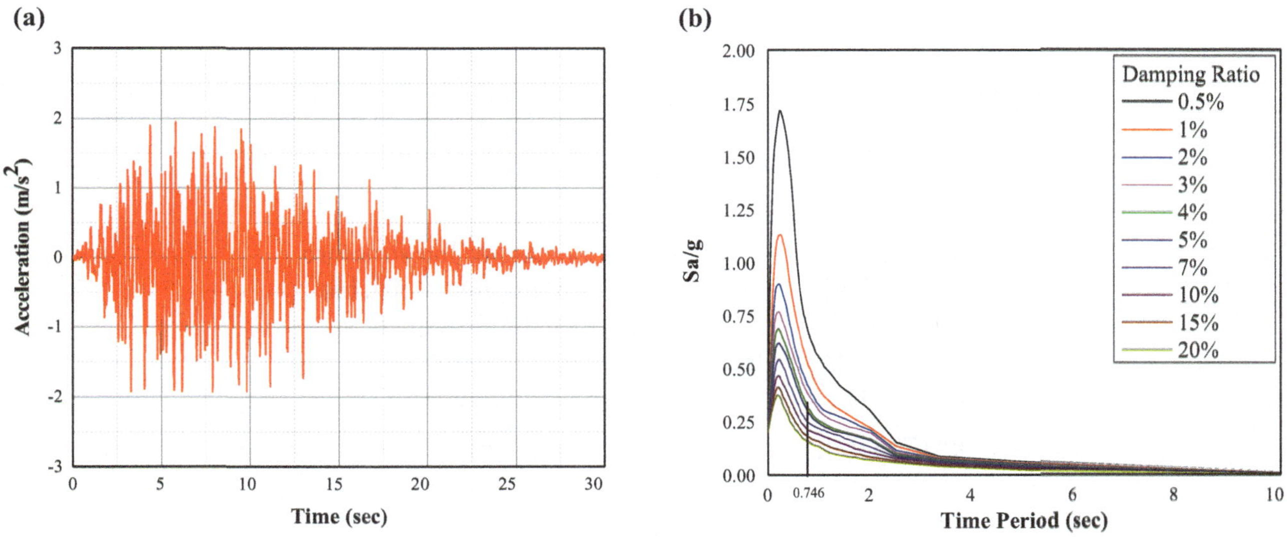

Fig. 13.52 **a** Applied ground motion. **b** response spectrum

ground acceleration is 2.046 m/s^2. The peak displacement is observed as 28.86 mm. For the present iteration, $\Delta E =$ 84.37J and Σ is 746 J. Thus, the ζ_d is calculated as 0.009. The value of $(\zeta_s + \zeta_d)$ is 12.48%.

Iteration 3:

With the increased damping of 12.48%, the ground acceleration is calculated as 1.988 m/s^2. The peak displacement is observed as 28.04 mm. Since the difference in displacement is below 5%, subsequent iterations are not needed. Therefore, with this damping ratio of 12.48%, the response comes under the allowable limits. The lateral force is 50.1 kN, i.e., at each column, the lateral force is evaluated as 12.525 kN. The base moment at each column is estimated 100, 195.2 N-m. For the given section modulus of 9.114×10^{-4} m^3, the bending stress comes within the allowable limits.

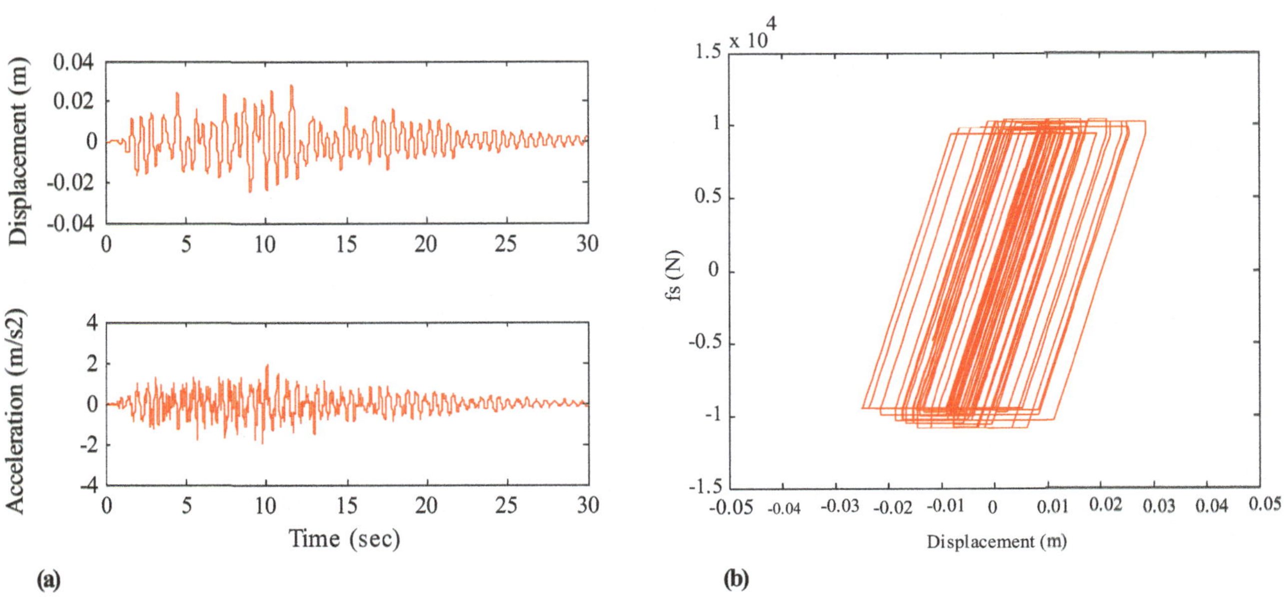

Fig. 13.53 **a** Response of the tank with time. **b** variation of damper force with displacement

MATLAB code for solving the above problem:

```
clear all;
close all;
clc;

m=25200;                  % mass of the structure
c=16973.65;               % damping coefficient
nfreq=1.34;               % natural frequency
omegan=2*pi*nfreq;
zeta=0.04;                % damping ratio
k=1793500;                % stiffness

gamma = 0.5;              % Newmark Constant
beta = 1/4;               % Newmark Constant
dt = 0.01;
tfinal=30;
noofiter=tfinal/dt;

d = gamma/(beta*dt);
e = 1/(beta*dt^2);
t = 0:dt:tfinal;
mu = 0.3;
g = 9.81;
fs = 10000;               % Damper Load
kd = 10000;
acel=xlsread('Input2');
for i=1:noofiter+1
  p(i)=m*acel(i,2)
end
```

```
u = 0;
v = 0;
a = 0;
aa = (1/(beta*dt))*m + (gamma/beta)*c;
bb = (1/(2*beta))*m + dt*(gamma/(2*beta)-1)*c;

for i = 1:1:noofiter
  dp(:,i) = p(i+1)-p(i);


  if (i<=1)
       fd(i) = 0;
       df(i) = 0;
       kc(i) = k + kd + d*c + e*m;
       dpc(i) = dp(i)+aa*v(i)+bb*a(i);

  end

  if i>1
       if (sign(v(i-1))/sign(v(i)))<0

         kd = fs/0.008;
         kc(i) = k + kd + d*c + e*m;
         dpc(:,i) = dp(:,i)+aa*v(:,i)+bb*a(:,i);
           if abs(fd(i))>fs
              fd(i) = -fs*sign(v(i));

              else
                fd(i) = fd(i);
              end
         else
           if abs(fd(i))<fs;
              kd = fs/0.008;
              kc(i) = k + kd + d*c + e*m;
              dpc(:,i) = dp(:,i)+aa*v(:,i)+bb*a(:,i);
           else
              kd = 0;
              kc(i) = k + d*c + e*m;
              dpc(:,i) = dp(:,i)+aa*v(:,i)+bb*a(:,i);
           end
         end
    end
    kcinv(i) = inv(kc(i));
    du(:,i) = kcinv(i)*dpc(:,i);
    dv(:,i) = (gamma/(beta*dt))*du(:,i)-(gamma/beta)*v(:,i)+dt*(1-(gamma/(2*beta)))*a(:,i);
    da(:,i) = (1/(beta*dt^2))*du(:,i)-(1/(beta*dt))*v(:,i)-(1/(2*beta))*a(:,i);
    u(:,i+1) = u(:,i)+du(:,i);
```

```
    v(:,i+1) = v(:,i)+dv(:,i);
    a(:,i+1) = a(:,i)+da(:,i);
    dfd(i) = kd*du(i);
    fd(i+1) = fd(i)+dfd(i);
    fdnormal(i+1)=fd(i+1);

end
```

Approach	Displacement (mm)	Base moment (N-m)	Bending stress (MPa)
Response spectrum method	28.04	100,195.2	109.9
Time history method	27.53	98,357.2	107.9

13.6.1.3 Lead Extrusion Damper

The Lead Extrusion Damper works on the principle of extrusion and utilizes hysteretic energy dissipation properties of lead. The Lead Extrusion Damper absorbs vibration energy by plastic deformation of lead and thereby converting mechanical energy into heat. This process makes the Lead Extrusion Damper suitable for absorbing excitation energy of a piping or equipment during a seismic event [53, 54]. Following extrusion, lead possesses tendency to recrystallize immediately, and thus, in this way, it recovers its original mechanical properties before successive extrusion. However, it is important to note that the amount of energy absorbed is not restricted by work hardening or fatigue of lead [4]. Commonly, two types of Lead Extrusion Dampers are employed: rotary type and cylinder type. Figure 13.54a shows the schematic of a cylindrical type of Lead Extrusion Damper.

The rotary-type dampers were first developed and reported by Fujita et al. in 1989 by the Mitsubishi Industries in Japan [5, 6]. A rotary-type Lead Extrusion Damper consists of housing, a rotational shaft with bulges and the lead filling the space between them. With the rotation of the shaft relative to the housing, the extrusion of lead occurs through the orifice path formed because of the bulge and the housing. A key present prevents the relative movement between the housing and the lead. The effective operation of the Lead Extrusion Dampers depends on the quality of lead, which recovers and recrystallizes rapidly. Cylinder-type dampers were first developed as a passive energy device and reported

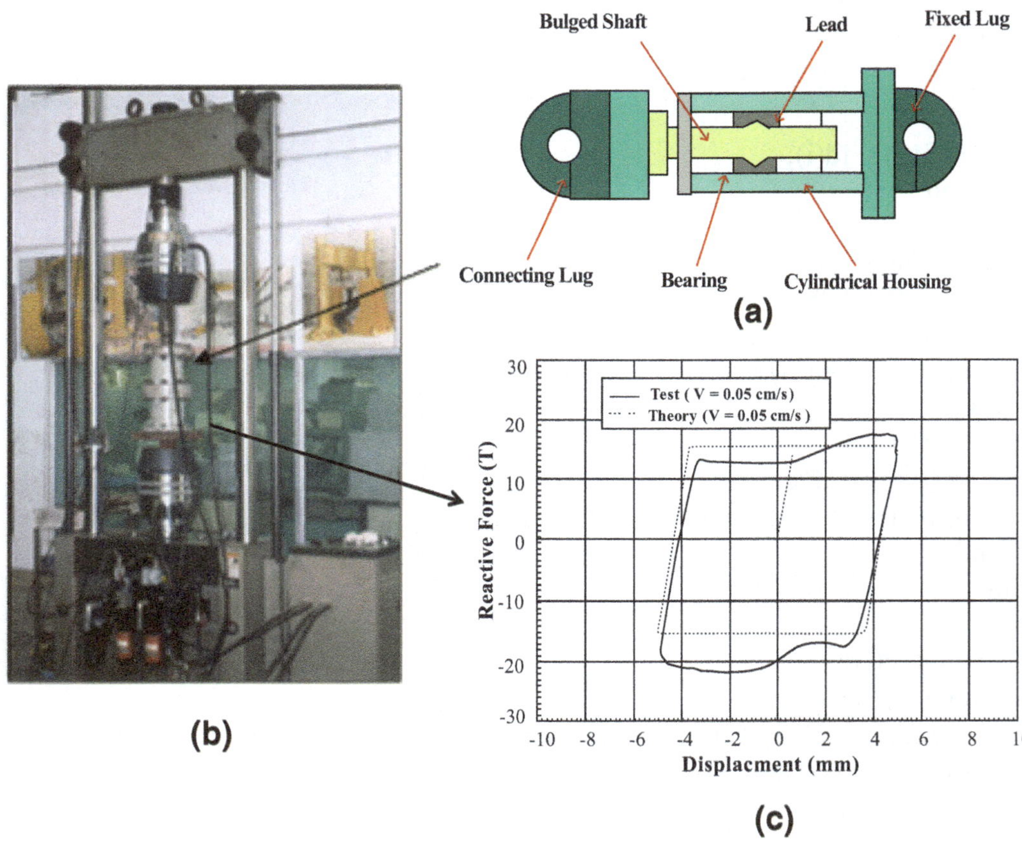

Fig. 13.54 **a** Schematic of a cylinder-type lead extrusion damper; **b** testing of a cylinder-type lead extrusion damper; **c** hysteresis characteristic of a cylinder-type lead extrusion damper

by Robinson and Greenbank [4, 7]. In case of cylinder-type dampers, the rotational shaft is replaced by a translational rod with bulges. Equation 13.25 shows the analytical formulation of the Lead Extrusion Damper. The force in the Lead Extrusion Damper can be calculated as by taking into account the rod displacement or vice versa.

$$F = E(1 + \mu \cot \alpha) \ln \frac{a_1}{a_2} + E\mu \left(\frac{L}{d}\right)^2 \tag{13.25}$$

where P is the damper force, E is the yield strength of lead, μ is the friction factor on the bulge, d is the diameter of the rod, L is the stroke length of the rod, α is the angle on the bulge, a_1 is the original cross section are of lead, and a_2 is the extruded area of the lead. Figure 13.54b shows the testing of a Lead Extrusion Damper, and its hysteretic curve is illustrated in Fig. 13.54c [8, 9, 52].

It is important to note that the reaction force of the Lead Extrusion Dampers is velocity dependent. It was observed through various dynamic tests by Kokubo et al. [10] that reaction force developed was comparatively small at low velocities (thermal expansion) than at high velocities (earthquakes). For low-velocity regions, the relation between the reaction force and the shaft velocity of the Lead Extrusion Damper is described by Kokubo et al. [10] as

$$F_r = F_{ro}(\dot{x}_s/\dot{x}_{so})^g \tag{13.26}$$

where F is the reaction force (tons) corresponding to the velocity V (cm/s); the value $\dot{x}_{so}$ is 0.01 cm/s; F_{ro} is 90 ton-force (i.e., 882 kN), and g is 0.085 when $\dot{x}_s > 0.01$ cm/s or 0.15 if otherwise. The relationship between the reaction force and the shaft velocity of a cylinder-type Lead Extrusion Damper is shown in Fig. 13.55.

Example 13.5 Calculate the exerted pressure of extrusion for a Lead Extrusion Damper for which the casing diameter is 100 mm, rod diameter is 55 mm, and the stroke length is 40 mm. The diameter at the bulge is 60 mm, and the bulge angle is 12° with the friction factor on the bulge being 0.14. The yield strength of lead to be taken is 90 MPa.

Solution Given diameter of casing, D = 100 mm; diameter of rod, d = 55 mm; stroke length, L = 40 mm; diameter at the bulge, d_o = 60 mm; angle of the bulge, $\alpha = 12°$; friction factor on the bulge, $\mu = 0.14$; yield Strength of lead, E = 90 MPa.

The exerted pressure of extrusion is calculated from Eq. 13.26 for which the a_1 and a_2 are calculated as

$$a_1 = \frac{\pi}{4}(D^2 - d^2) = 0.785\left(0.1^2 - 0.055^2\right) = 0.005467\ \text{m}^2 \tag{13.27}$$

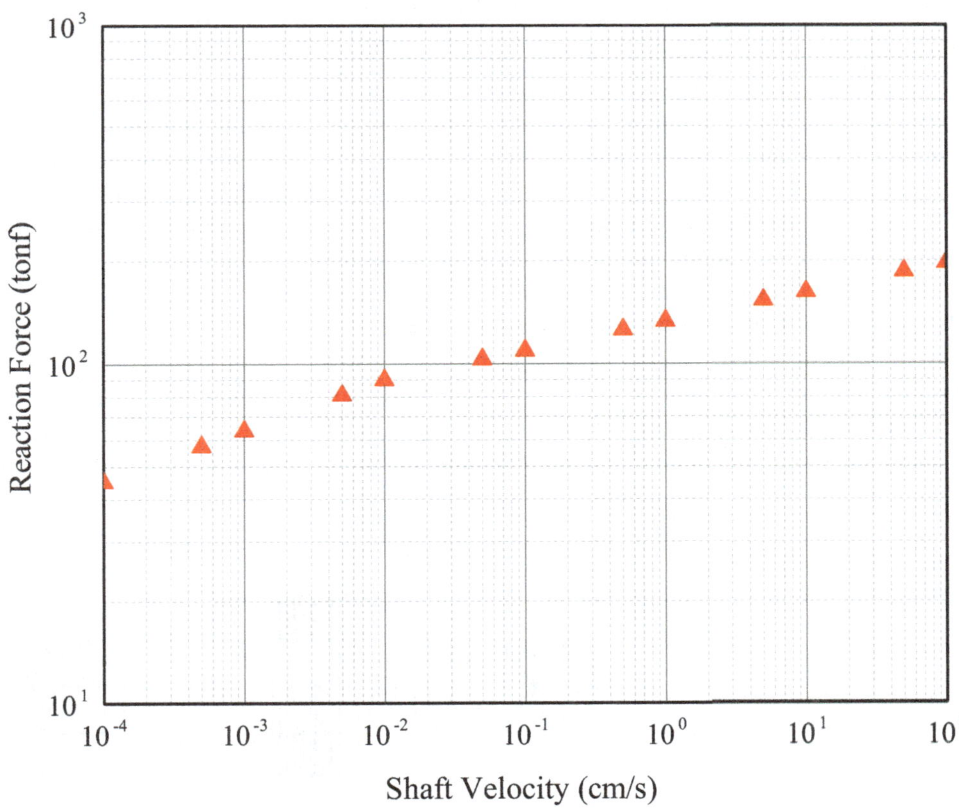

Fig. 13.55 Relation between reaction force and shaft velocity of lead extrusion damper (Kokubo et al.)

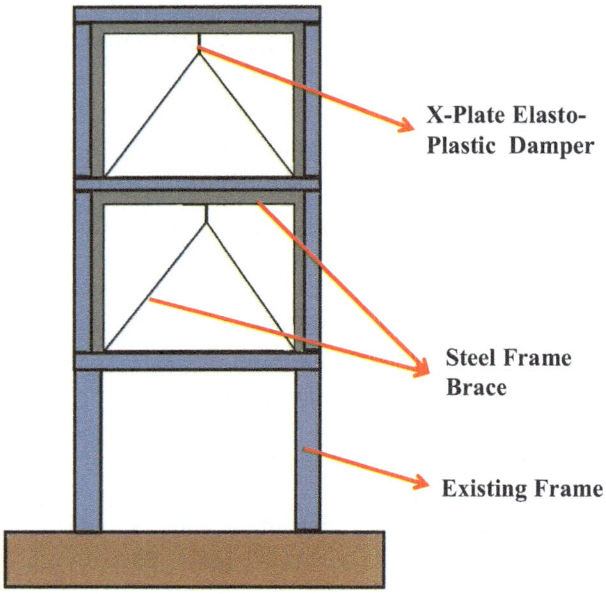

Fig. 13.56 Retrofitting of existing structure with elasto-plastic dampers

$$a_2 = \frac{\pi}{4}(D^2 - d_o^2) = 0.785\left(0.1^2 - 0.06^2\right) = 0.005027\,\text{m}^2 \tag{13.28}$$

Now, according to Eq. 13.26,

$$P = 90^* 10^6 (1 + 0.14 \cot 12^\circ) \ln \frac{0.005467}{0.005027} + 90^* 10^{6*} 0.14 \left(\frac{0.04}{0.055}\right)^2$$

Thus, the extrusion pressure comes out to be 18.98 MPa. By multiplying the area at the bulged location with the pressure, the load is estimated to be 9.7 tons.

13.6.1.4 Elasto-Plastic Damper

The numerous limitations posed by snubbers were frequently utilized in piping systems paved the way for the development of Elasto-Plastic Dampers as energy absorbing device under seismic excitation. Although snubber allowed for free thermal expansion in case of thermal loads and provided rigid support during dynamic loadings (earthquakes, etc.), the structure of snubber remained complex, provided less damping, was expensive and required frequent maintenance. Additionally, mechanical snubber posed locking problems and hydraulic snubber, if used, leaked and not worked when required. Hence, Elasto-Plastic Damper was found to be a better substitute in reducing the response of piping systems [11]. Figure 13.56 shows the schematic of frame structure retrofitted using Elasto-Plastic Damper.

Supports with Elasto-Plastic Dampers reduced the seismic response of piping elements subjected to earthquake load by absorbing the energy in the form of hysteretic deformation. During its development, multiple tests were performed [12, 13] which revealed that these devices could sustain many cycle of stable point yielding deformation, resulting in high level of energy dissipation or damping. To understand its static and dynamic characteristics, numerous tests were conducted by BARC and SERC (Madras).

The mechanism of Elasto-Plastic Damper is based on plastic deforming of steel components or layered laminated plates in bending, shear, torsion or in a combination. As mentioned earlier, these dampers sustain multiple cycles of stable yielding, hence leading to high energy dissipation and damping. Generally, an X-shaped plate is chosen as an Elasto-Plastic energy absorber, such that the strain is constant over the height of the device, ensuring simultaneous and uniform yielding over the full height of the damper. A typical schematic of an Elasto-Plastic Damper is shown in Fig. 13.57 [13].

Force–Displacement Characteristics of an X-plate used in Elasto-Plastic Dampers

The force–displacement characteristics of an X-plate are obtained through analytical formulations and also by carrying out static and dynamic tests on plates of different thickness. In the present text, the analytical formulation and the nature of the static and dynamic tests are discussed.

(a) *Analytical Formulations*

The force–displacement curve for an X-shaped plate is obtained using the beam theory. The schematic of the X-plate is shown in Fig. 13.58. The following expressions are derived for three different cases:

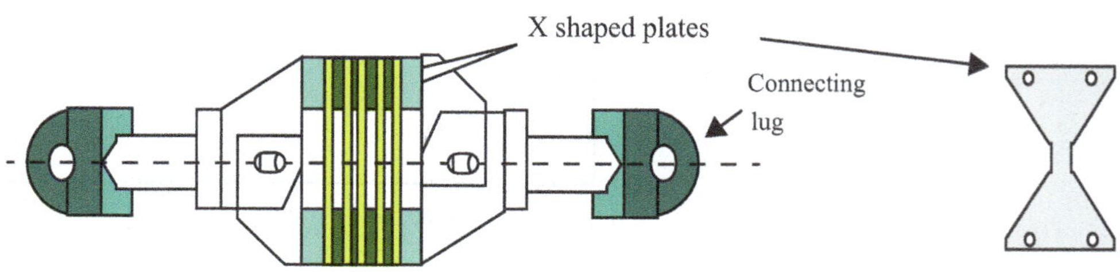

Fig. 13.57 Schematic of an elasto-plastic damper

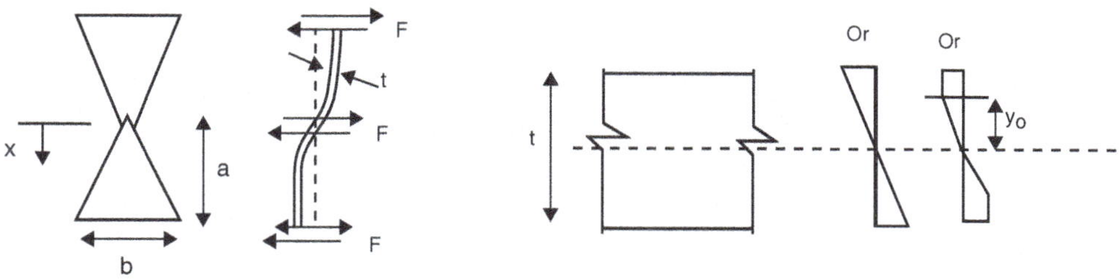

Fig. 13.58 Characteristics of an X-shaped plate

Case 1: Elastic bending stress in the X-plate

The X-shaped plate connects the top of two triangular plates. Thus, considering only triangular plate

$$EI\frac{d^2y}{dx^2} = -Fx \tag{13.29}$$

$$F = \frac{Ebt^3d}{6a^3} \tag{13.30}$$

For the X-plate, $k = \frac{F}{2d} = \frac{Ebt^3}{12a^3}$ where d is the displacement, b is the width, t is the thickness, a is the height of the triangular plate, and E is Young's modulus of elasticity.

Case 2: Bending stress just reaching yield stress in the X-plate,

For triangular plate, $M_y = \frac{\sigma_y bt^2}{6}$

$$F_y = \frac{\sigma_y bt^2}{6a} \tag{13.31}$$

Substituting Eqs. 13.31 in Eq. 13.30, we get

$$d_y = \frac{\sigma_y a^2}{Et} \tag{13.32}$$

For X-plate, $d_y = 2\frac{\sigma_y a^2}{Et}$

Case 3: Bending stress for which elastic depth of the X-plate reaches $2y_o$

The stress of the triangular plate before yield for $0 < y < y_0$ is given as follows

$$\sigma_1 = \frac{y\sigma_y}{y_0} \tag{13.33}$$

The stress of the triangular plate after yield using the strain hardening rate H is given as follows

$$\sigma_2 = H\varepsilon + \frac{\sigma_y(E-H)}{E} \tag{13.34}$$

Using the balance of the moment at the fix point of the triangular plate,

$$Fa = 2b\left(\int_0^{y_0} \sigma_1 y dy + \int_{y_0}^{t/2} \sigma_2 y dy\right) \tag{13.35}$$

Substituting values of σ_1 and σ_2 in Eq. 13.35, we get

$$F = \frac{b\sigma_y}{12Ea}\left\{(4y_0^2 - 3t^2)(H-E) + \frac{Ht^3}{y_0}\right\} \tag{13.36}$$

The force in the X-plate is the same as that in the triangular plate,

$$d = \frac{\sigma_y a^2}{2Ey_0} \tag{13.37}$$

For X-plate,

$$d = \frac{\sigma_y a^2}{Ey_0} \tag{13.38}$$

Using the above equations, the force–displacement curve is obtained for a 3-mm-thick X-plate of carbon steel having the following properties $E = 2.1 \times 10^5$ N/mm^2, $H = 2.58\%$ of $E = 5418$ N/mm^2, and $\sigma_y = 220$ N/mm^2.

(b) *Static Test*

Static test was conducted to check the mechanical property of the X-shaped plate material [11]. The test consisted of applying a known displacement at one end and measuring the reaction force at the other end of X-plate. The test was carried out for 3-mm-, 4-mm-, and 6-mm-thick X-plates. The force–displacement characteristics of a 3 mm X-plate obtained from the tests and beam theory are shown in Fig. 13.59. It is observed that the test results are found to be in good agreement with that of the theoretical results. Dynamic test also included fatigue test details [12].

(c) *Hysteretic Modeling of X-plate*

The dynamic characteristics of the X-plates are evaluated analytically by using the Ramberg–Osgood model. Ramberg–Osgood model [11], given in Eq. 13.39, approximates the shape of force–displacement relation for cyclic loading of a material by appropriate selection of the parameters n and

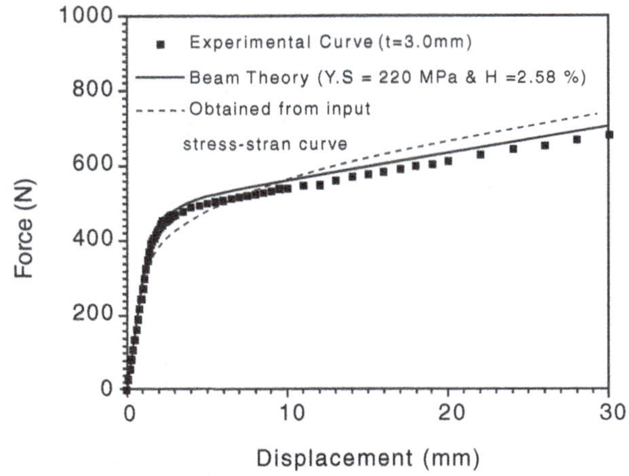

Fig. 13.59 Comparison between experimental and theoretical force-displacement curve for 3-mm-thick plate under static load

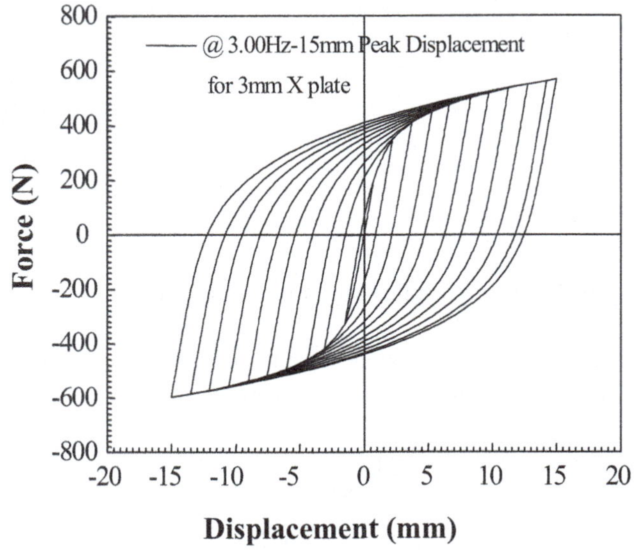

Fig. 13.61 Theoretical force–displacement curve for 3-mm-thick plate using R-O model

α. A schematic of a typical hysteresis curve is shown Fig. 13.60. As can be seen from Fig. 13.60, during initial loading, the force–displacement curve is shown by O–A–B. The material is first loaded in tension and it yields at A. At B, the loading direction is reversed. Unloading occurs along the elastic line B–C till the load becomes compressive. While unloading, yielding starts to initiate at lower compressive force values (marked as C) due to Bauschinger effect. In Bauschinger effect, the material is seen to soften in case of inverted loading. It is important to note that had there been no directional effect on the loading, and the material would have started to deform plastically. Also, if the material did not show any dependence on loading direction, the compressive curve would have been similar to the tensile curve (not shown in Fig. 13.60). Subsequently, in re-loading under tensile forces, a hysteresis loop is seen following the path D–E–B.

$$\frac{\delta}{\delta_y} = \frac{P}{P_y}\left[1 + \alpha\left(\frac{P}{P_y}\right)^{n-1}\right] \tag{13.39}$$

where δ_y is the yield displacement and P_y is the yield force. P and δ are the force and the displacement, respectively, at any point on the curve. The values of n and α are evaluated from the experimental results of the static tests conducted on the 3-mm-thick X-plates. For the 3 mm plate considering $\alpha = 0.39$ and $n = 5.66$, a typical hysteresis curve when the

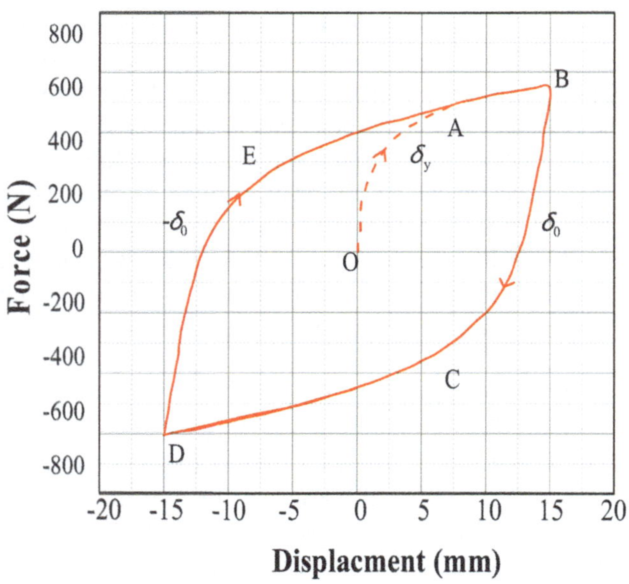

Fig. 13.60 Typical force–displacement hysteresis curve

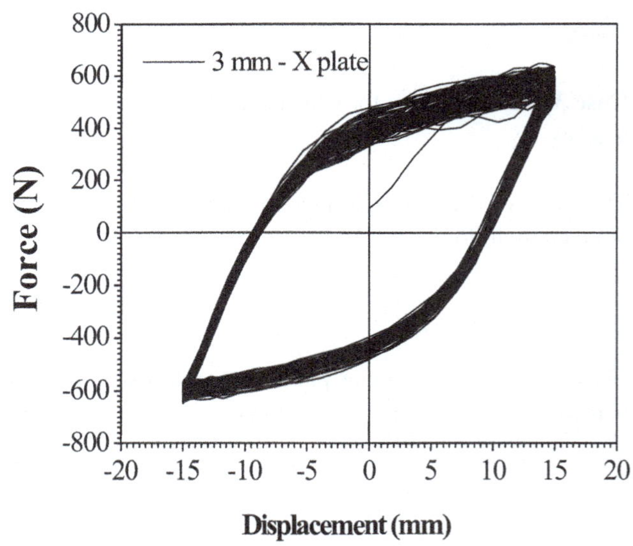

Fig. 13.62 Force-displacement curve for 3-mm-thick plate at 3 Hz with 15 mm peak displacement for 100 cycles

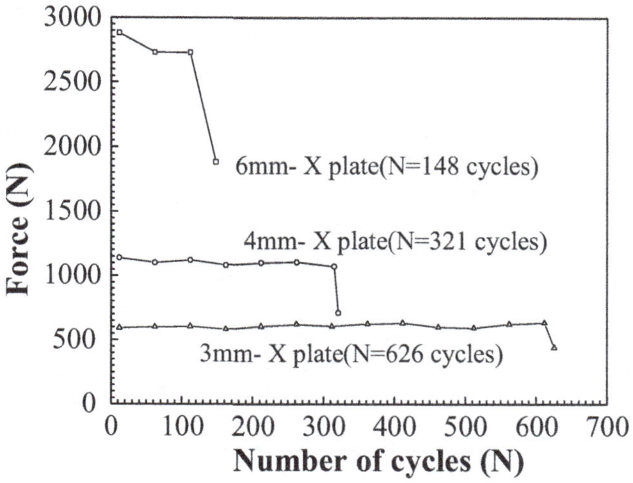

Fig. 13.63 Fatigue test results on different X-plates at 3.0 Hz with 15 mm peak displacement

applied force is increased step by step as shown in Fig. 13.61.

For path OB, the Ramberg–Osgood model is described Eq. 13.40

$$\frac{\delta}{\delta_y} = \frac{P}{P_y}\left[1 + \alpha\left(\frac{P}{P_y}\right)^{n-1}\right], \quad \alpha > 0, n > 1; \tag{13.40}$$

For path BD, Eq. 13.41 describes the model as

$$\frac{\delta - \delta_0}{-2\delta_y} = \frac{P_0 - P}{2P_y}\left[1 + \alpha\left(\frac{P_0 - P}{2P_y}\right)^{n-1}\right] \tag{13.41}$$

and for path DB,

$$\frac{\delta + \delta_0}{2\delta_y} = \frac{P + P_0}{2P_y}\left[1 + \alpha\left(\frac{P + P_0}{2P_y}\right)^{n-1}\right] \tag{13.42}$$

(d) *Cyclic Behavior of X-plates*

In cyclic testing of X-plates, a sinusoidal displacement of 15 mm peak at a frequency of 3.0 Hz was applied at one end and the reactive force was measured at the other end. These tests were conducted for 3-mm-, 4-mm-, and 6-mm-thick X-plates. The force–displacement characteristics obtained from the test for a typical 3-mm-thick X-plate are shown in Fig. 13.62.

(e) *Fatigue Tests of X-plates*

Large number of cyclic load tests was carried out on X-plates of different thickness, i.e., 3, 4, and 6 mm in order the study their fatigue behavior. Figure 13.63 shows maximum force v/s number of cycles for these plates. It can be observed from the figure that 3 mm plates are capable of taking about 626 cycles.

Example 13.6 To understand the effect of Elasto-Plastic Dampers under dynamic loading, an example showing a cantilever piping model is considered which is excited using a shake table [13]. The piping model consists of a 3 in. schedule 40 cantilever pipe of length 1.6 m as shown in Fig. 13.64a, b. One end of the pipe element is rigidly connected to the shake table using an *L* fixture, whereas the other end is connected to an *L* platform, attached with three friction-free linear motion bearings to facilitate the free

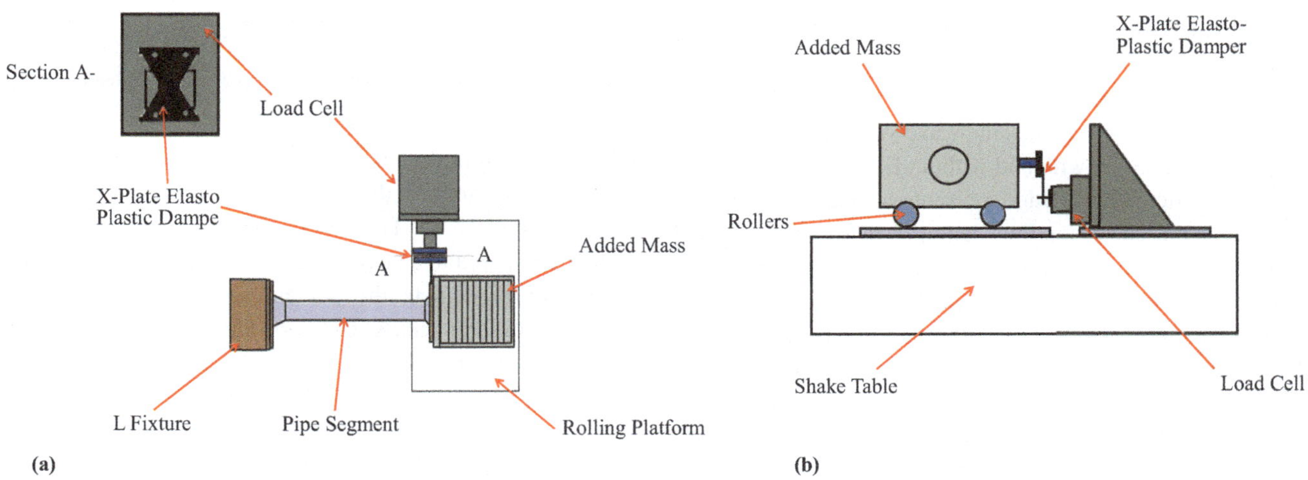

Fig. 13.64 **a** Schematic view of the test arrangement for the shake table test on the SDOF pipe system (plan), **b** schematic view of the test arrangement for the shake table test on the SDOF pipe system (elevation)

movement. Different masses varying from 115 to 495 kg are added at the L platform using steel plates with the purpose of achieving different frequencies of the piping system. Subsequently, an X-plate Elasto-Plastic Damper is added to the system in such a way that it rigidly connects its one end to the freely moving L platform, while the other end of the X-plate Elasto-Plastic Damper is rigidly connected to the L fixture, which in turn is fixed to the shake table.

Solution

Outer diameter, d_o= 0.0889 m,
Thickness, t = 0.00,549 m,
Modulus of rigidity, E = 210 GPa,
Moment of Inertia, $I = \frac{\pi}{64}(d_o^4 - d_i^4) = 1.26 \times 10^{-6}\,\text{m}^4$
Weight of the pipe per unit length, w = 11.29 kg/m;
Length, L =1.6 m,
Thus, the stiffness of the pipe is calculated as, $K_p = \frac{3EI}{L^3} = 192{,}806\,\text{N/m}$,
Now, the natural frequency of the pipe is calculated using Eq. 13.42

$$f = \frac{1}{2\pi}\sqrt{\frac{K_p}{\frac{33}{140}(m_p)}} = 33.87\,\text{Hz}, \qquad (13.42)$$

And, when an additional mass is added at the tip of the pipe, Eq. 13.42 changes to Eq. 13.43 given below

$$f = \frac{1}{2\pi}\sqrt{\frac{K_p}{\frac{33}{140}(m_p) + m_e}} \qquad (13.43)$$

For the first case, where mass at the tip is, m_e= 115.33 kg, the natural frequency is calculated as 6.39 Hz. For other values of the tip masses, the natural frequency of the pipe is plotted in Fig. 13.65.

Harmonic analysis using sine wave sweep was carried out for the piping model with and without Elasto-Plastic Damper using analytical formulation present in chapter 2, and the results for the corresponding damping ratio were obtained. Sine wave sweep excitation test was performed with constant base acceleration of 0.1 g peak with frequency sweep from 0.1 to 30 Hz with constant sweep rate of 0.01 Hz/sec. The results obtained by testing for the cantilever piping model without damper, with 2 mm and 3 mm X-plate for five different mass levels are shown in Figs. 13.65 and 13.66 and compared with the analytical values. It can be seen from Fig. 13.65 that the frequency values when no damper was attached are lower in comparison with when X-plate damper is attached. The resonant frequency results obtained by testing the cantilever piping model for three different cases (and each case for five different mass levels respectively) are shown in Fig. 13.65. These obtained experimental values is further compared to analytical results as shown in Fig. 13.65. Figure 13.66 depicts the percentage damping ratio with respect to different mass levels for the three different cases. As can be seen from Fig. 13.66, the damping increases when 3 mm X-plate Elasto-Plastic Damper is attached in comparison with its 2 mm counterpart. Also, as the mass level increases, the damping effect increases.

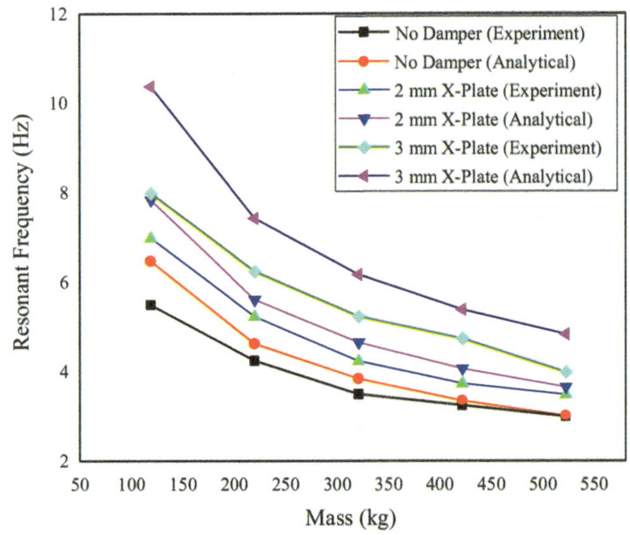

Fig. 13.65 Comparison of resonant frequency-mass levels for different cases

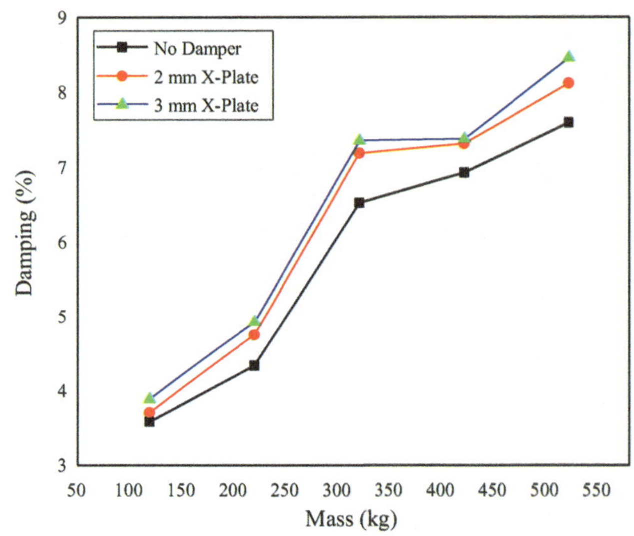

Fig. 13.66 Comparison of damping-mass levels for different cases

The stiffness of the X-plate Elasto-Plastic Damper is calculated using Eq. 13.42. The properties taken for the 2 mm X-plate is given as,

X-plate thickness, t_l = 2 mm = 0.002 m,
Width of the X-plate, b = 0.04 m,
Length of the X-plate, a = 0.04 m.
The stiffness of the plate is,

$$K_l = \frac{Ebt^3}{12a^3} = 85034\,\mathrm{N/m}, \tag{13.44}$$

And similarly for 3 mm X-plate, the stiffness is estimated to be 286,697 N/m.

Subsequently, the natural frequency expression shown in Eq. 13.43 changes to Eq. 13.45,

$$f = \frac{1}{2\pi}\sqrt{\frac{K_p + K_l}{\frac{33}{140}(m_p) + m_e}} \tag{13.45}$$

In order to estimate the geometrical properties of the X-plate such that natural frequency of the cantilever pipe is not affected, the thickness of the X-plate is kept 2 mm

Width of the new X-plate, b_n= 0.08 m,
Length of the X-plate, a_n = 0.08 m.

And using Eq. 13.44, the stiffness of the plate is 21,875 N/m. Putting this value in Eq. 13.45, we get the natural frequency for the tip mass of 115.33 kg as 6.74 Hz, which is very close to the undamped natural frequency of 6.39 Hz. Thus, using an X-plate Elasto-Plastic Damper of length and width 80 mm and thickness 2 mm does not alter the natural frequency of the pipe much.

In order to calculate the effective damping developed due to the X-plate damper of length and width 80 mm and thickness 2 mm, Ramberg–Osgood model is used. Using the equations of the model, a characteristic hysteresis curve is generated as shown in Fig. 13.67a. By calculating the area under the characteristic curve, the energy dissipated per unit cycle by the Elasto-Plastic Damper is estimated and the damping (damping ratio) is numerically computed through Eq. 13.22. The plot of the damping ratio (percentage) with respect to the tip mass is shown in Fig. 13.67b. Here, it is worth reminding that these damping ratios are estimated for such an X-plate damper which does not alter the natural frequency of the cantilever pipe.

13.6.1.5 Tuned Mass Damper

Tuned Mass Damper is a classical device consisting of a mass, spring, and dashpot which when added to a vibrating system reduces its dynamic response. The frequency of the damper is tuned to a particular structural frequency (usually the fundamental frequency) such that when the frequency is excited, the damper resonates out of phase, dissipating energy through its own inertia. Earlier, it was used to suppress the undesirable vibrations taking place in the mechanical systems, but now the concept is often utilized to reduce the wind-induced vibration in tall structures. Tuned Mass Dampers are generally utilized to dampen the response of a structure in a particular mode.

Figure 13.68 shows the schematic of an undamped base structure of mass (M) and stiffness (K) together with a damper oscillator of mass (m), stiffness (k), and damping (c), on top of it. Let us assume X is the absolute position of base

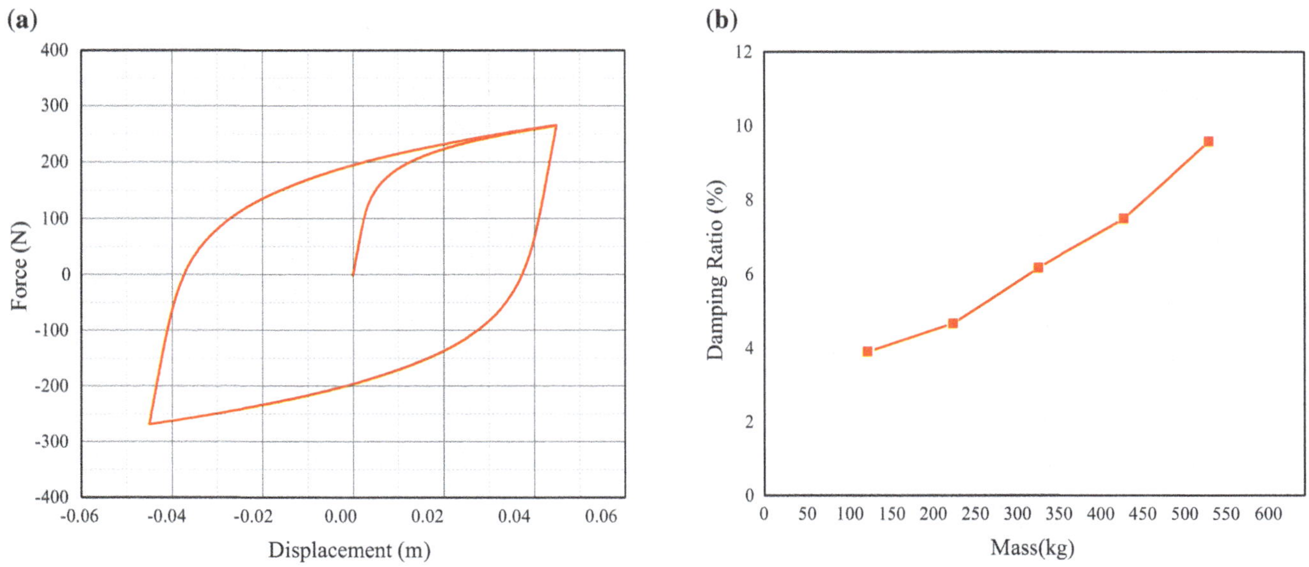

Fig. 13.67 **a** Theoretical force-displacement curve for X-plate elasto-plastic damper of length and width 80 mm and thickness 2 mm using R-O model. **b** theoretical damping values with different mass levels for 3D cantilever pipe with X-plate elasto-plastic damper (length and width 80 mm and thickness 2 mm)

structure and x is the position of the damper oscillator. The relative position of m is given by $X - x$. Let us assume that there is a periodic force $F\cos\omega t$ on mass M. The equations of motion of the complete system can be described as [14]

$$M\ddot{X} + KX + k(X - x) + c(\dot{X} - \dot{x}) = F\cos\omega t \quad (13.46)$$

$$m\ddot{x} + k(x - X) + c(\dot{x} - \dot{X}) = 0 \quad (13.47)$$

By replacing Eq. 13.46 by the sum of Eqs. 13.46 and 13.47, we get

$$M\ddot{X} + KX + m\ddot{x} = F\cos\omega t \quad (13.48)$$

$$m\ddot{x} + k(x - X) + c(\dot{x} - \dot{X}) = 0 \quad (13.49)$$

The periodic solutions of displacements X and x could be written in the form

$$X = e\cos\omega t \quad (13.50)$$

$$x = n\cos(\omega t + \phi) \quad (13.51)$$

Now, by substituting Eqs. 13.50 and 13.51 in Eqs. 13.48 and 13.49, we get

$$\begin{bmatrix} (K - m\omega^2 - M\omega^2) & 0 \\ -(k + c\omega) & (k - m\omega^2 - c\omega) \end{bmatrix} \begin{bmatrix} e \\ n \end{bmatrix} = \begin{bmatrix} F \\ 0 \end{bmatrix} \quad (13.52)$$

Combining together, we get,

$$e = \frac{F}{(K - m\omega^2 - M\omega^2)} \quad (13.53)$$

$$n = \frac{F(k + c\omega)}{(K - m\omega^2 - M\omega^2)(k - m\omega^2 - c\omega)} \quad (13.54)$$

The amplitude (U) of X is given as,

$$U = \sqrt{e^2} = \frac{F}{(K - m\omega^2 - M\omega^2)} \quad (13.55)$$

and amplitude (V) of x is given as,

$$V = \sqrt{n^2} = \frac{F(k + c\omega)}{(K - m\omega^2 - M\omega^2)(k - m\omega^2 - c\omega)} \quad (13.56)$$

The natural frequency of the base structure (ω_b) and the damper oscillator (ω_0) is given as

$$\omega_b = \sqrt{\frac{K}{M}}, \quad \omega_o = \sqrt{\frac{k}{m}} \quad (13.57)$$

The mass ratio is given as $\eta = m/M$; damping ratio, $\zeta = \frac{c}{2m\omega_o}$ and static deformation, $u_{st} = \frac{F}{K}$.

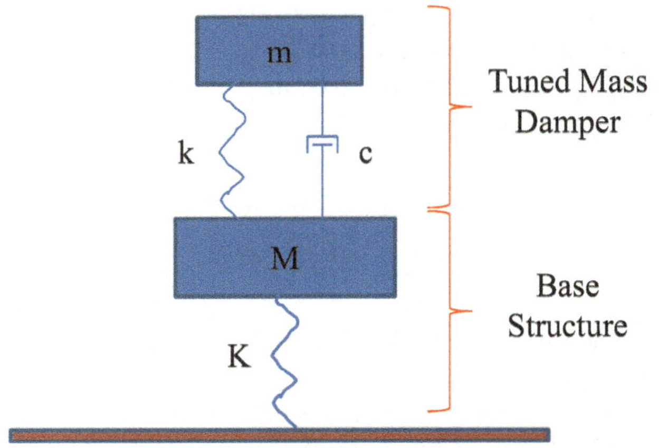

Fig. 13.68 Schematic of a tuned mass damper with undamped base structure

So now, the optimal tuning of the system is obtained when the two responses, i.e., $U = V$. In that case,

$$\left(\frac{\omega_0}{\omega_b}\right)_{opt} = \frac{1}{1 + \eta} \quad (13.58)$$

and the optimal absorber damping is estimated from averaging the damping values that maximize the amplitude at ω_b and ω_o,

$$\zeta_{opt} = \sqrt{\frac{3\eta^3}{8(1 + \eta)^3}} \quad (13.59)$$

In the similar lines, equations for damped base structure with damper oscillator can also be obtained by including damping coefficient (c) term in the equation of motion of the base structure.

To increase the efficacy of the mass dampers for a vibrating structure and increase its effectiveness to overcome the various uncertainties, concept of Multiple Tuned Mass Dampers or Multiple Mass Dampers was introduced as

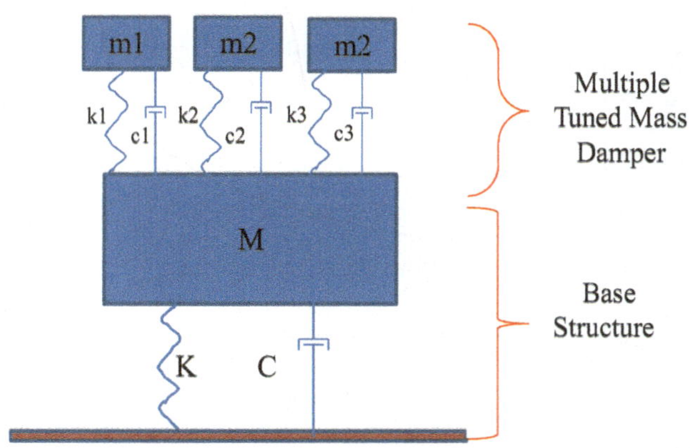

Fig. 13.69 Schematic of a Multiple tuned mass damper with damped base structure

shown in Fig. 13.69. The early pioneers Iwanami and Seto [15] observed that two Tuned Mass Dampers are more effective in energy dissipation compared to that a structure having a single Tuned Mass Damper. Further, Xu and Igusa [16], Yamaguchi and Harpornchai [17], Abe and Fujino [18], Jangid [19], Zuo and Nayfeh [20, 21] and Bandivadekar and Jangid [22] advocated Multiple Tuned Mass Dampers each of which is tuned to different natural frequencies and distributed throughout. This kind of strategic spatial distribution of tuned multiple dampers is allowed for enhanced mitigation of complex buildings and structures motions. For the same reason, multiple damper configurations are found to be more effective with slightly detuned systems. Also for a system having same total mass, the equivalent damping introduced by multiple dampers is more effective compared to a single damper.

13.6.1.6 Tuned Liquid Damper

Wind-induced and earthquake-induced excitations on tall structures can be effectively controlled by using Tuned Liquid Damper (TLD). Tuned Liquid Dampers (TLDs) control the response of the structure by absorbing energy using the sloshing characteristics of the liquid in a low elevation tank. A TLD is essentially a rigid tank with lower water level and is connected rigidly to the vibrating structure. By tuning the fundamental sloshing frequency of the TLD to the structures natural frequency, large energy is absorbed by the sloshing liquid. TLD absorbs the major portion of the energy of external excitation and results in damping of the structural vibrations. TLDs are widely used to control the vibrations of high rise structures as these are highly flexible. The TLDs are incorporated in various structures around the world. In Japan, TLD was installed in the 42 m tall Nagasaki Airport Tower, in 1987 and also in the 149 m Shin Yokohama Prince Hotel, Yokohoma, in 1991. A total of 80 TLD units were installed in the 105 m Hobart Tower in Tasmania, Australia [23, 24].

In general, TLDs can be categorized into Single Tuned Liquid Dampers (STLDs) and Multiple Tuned Liquid Dampers (MTLDs). STLDs are quite effective in reducing the response of a structure when input excitation is close to the fundamental frequency of the TLD as in case of wind loads. However, for seismic excitations, MTLDs prove to be more effective. This can be simply explained as MTLDs have TLDs tuned for distributed frequencies over a certain range, whereas this is not the case for STLDs. The effectiveness of MTLDs under wind vibration and earthquake excitation was studied and its ability to suppress the response under earthquake excitation was subsequently recognized [25].

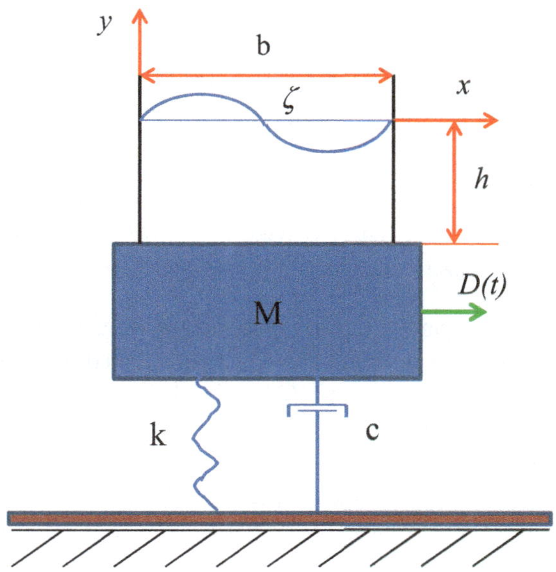

Fig. 13.70 Schematic of single tuned sloshing damper system

Analytical Formulation

Figure 13.70 shows the schematic of single tuned sloshing damper system (a SDOF) structure with a TLD. Here, m represents the mass of the structure, k represents its stiffness, and c represents its damping. The TLD has tank of length b and is filled with water up to a height h. When Fig. 13.70 is subjected to a ground motion, denoted by $\ddot{x}_g$, the governing equation of motion for this SDOF structure is given by [23, 24]

$$m\ddot{x} + c\dot{x} + kx = -m\ddot{x}_g + F \qquad (13.58)$$

where $\ddot{x}$ is the response acceleration of the structure, $\dot{x}$ is the response velocity of the structure, and x is the response displacement of the structure with respect to the ground. The force F denotes the shear force developed at the base of the TLD due to water sloshing. Equation 13.48 after normalization with respect to the structure mass is given as

$$\ddot{x} + 2\zeta\omega\dot{x} + \omega^2 x = -\ddot{x}_g + \frac{F}{m} \qquad (13.59)$$

where $\boldsymbol{\omega}$ is the circular natural frequency and $\boldsymbol{\zeta}$ is the damping ratio of the structure. The value of F is obtained by solving the equations of motion of water in the TLD.

Subsequently, the base excitation of the TLD is equal to the response of the top of the structure. Since the depth of water is assumed low, the equations of motion of the water inside the tank are defined in terms of the free surface motion. The governing equation of water motion for strong motion earthquake resulting in large amplitude TLD excitation is given as

$$\frac{\partial \psi}{\partial t} + h\sigma \frac{\partial (\phi u)}{\partial x} = 0 \quad (13.60)$$

$$\frac{\partial u}{\partial t} + (1 - T_{\mathrm{h}}^2) u \frac{\partial u}{\partial x} + C_{\mathrm{f}}^2 g \frac{\partial \psi}{\partial x} + gh\sigma\phi \frac{\partial^2 \psi}{\partial x^2} \frac{\partial \psi}{\partial x} = -C_{\mathrm{d}} \lambda u - \ddot{x} \quad (13.61)$$

where $\psi(x, t)$ and $u(x, t)$ are independent variables denoting the free surface elevation above the undisturbed water level. The parameters σ, ϕ and T_{h} are in Eqs. 13.60 and 13.61 which are given as

$$\sigma = \frac{\tanh(\tau h)}{(\tau h)} \quad (13.62)$$

$$\phi = \frac{\tanh[\tau(h + \psi)]}{\tanh(\tau h)} \quad (13.63)$$

$$T_{\mathrm{h}} = \tanh[\tau(h + \psi)] \quad (13.64)$$

where τ is the wave number. The λ in Eq. 13.61 is a damping parameter which accounts for the effects of the boundary layer on the bottom and the side wall and can be expressed as

$$\lambda = \frac{1}{(\psi + h)} \frac{1}{\sqrt{2}} \sqrt{\omega_1 v} \left[1 + \frac{2h}{d} + s\right] \quad (13.65)$$

Here, ω_1 is the fundamental sloshing frequency, v denotes kinematic viscosity, d denotes the width of the tank, and s denotes the surface contamination factor whose general value could be taken as unity. The fundamental sloshing frequency is calculated as shown in Eq. 13.66.

$$\omega_1 = \sqrt{\frac{\pi g}{b} \tanh(\pi \Delta)} \quad (13.66)$$

where $\Delta = \frac{h}{b}$ is the water depth ratio. The coefficients C_f and C_d are used to modify the water wave phase velocity and damping when waves are unstable ($\psi > h$) and break. In case when ($\psi \leq h$), the coefficients take the value of unity. Now, by solving Eqs. 13.60 and 13.61 simultaneously for the free surface, shear force F is estimated as shown in Eq. 13.67.

$$F = \frac{\rho g b}{2} [(\psi_{\mathrm{h}} + h)^2 - (\psi_{\mathrm{o}} + h)^2] \quad (13.67)$$

where η_{h} and η_{o} denote the free surface elevation at right and left walls, respectively, of the tank.

However, linear analysis of sloshing in the Tuned Liquid Damper can also be carried out effectively by considering an equivalent Tuned Mass Damper (TMD). The linear model

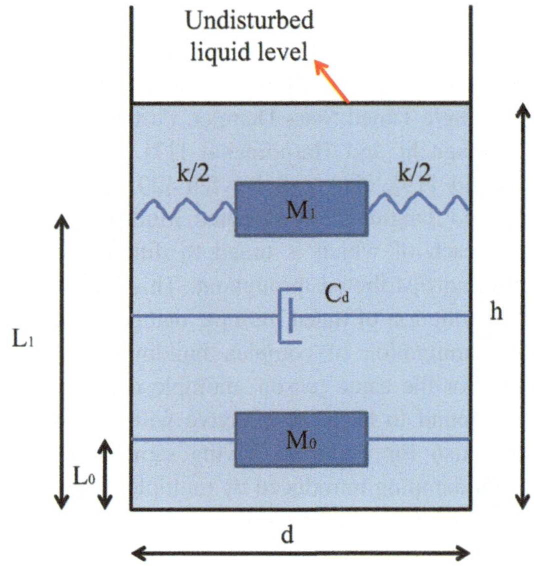

Fig. 13.71 Equivalent TMD model for linear sloshing TLD

proposed by Housner [26] is essentially for cylindrical and rectangular tanks. Water response on tank walls is modeled using two lumped masses: a fixed one moving with the acceleration of the structure, rigidly connected to the tank "impulsive mass" and the other "convective mass" connected to the tank by a spring and damper and determining the sloshing effect which is responsible for the energy dissipation. Figure 13.71 shows the equivalent TMD model for linear sloshing in which the various parameters are evaluated based on conditions such as

1. The equivalent masses and moments of inertia is conserved.
2. The center of gravity is same for small oscillations.
3. The system possesses the same modes of oscillations, and the damping forces generated is same.
4. The force and moment components under certain excitations of the model are equivalent to that are produced by the actual system [27–29].

The model properties such as convective mass, impulsive mass, and the spring stiffness along with their locations are evaluated as follows:

Sloshing frequency,

$$f_{\mathrm{sl}} = \frac{1}{2\pi} \sqrt{\frac{3.8317g}{D} \tanh(3.8317\Delta)} \quad (13.68)$$

Convective mass,

$$m_1 = m_{\mathrm{T}} \left(0.318 \frac{R}{h} \tanh(3.8317\Delta)\right) \quad (13.69)$$

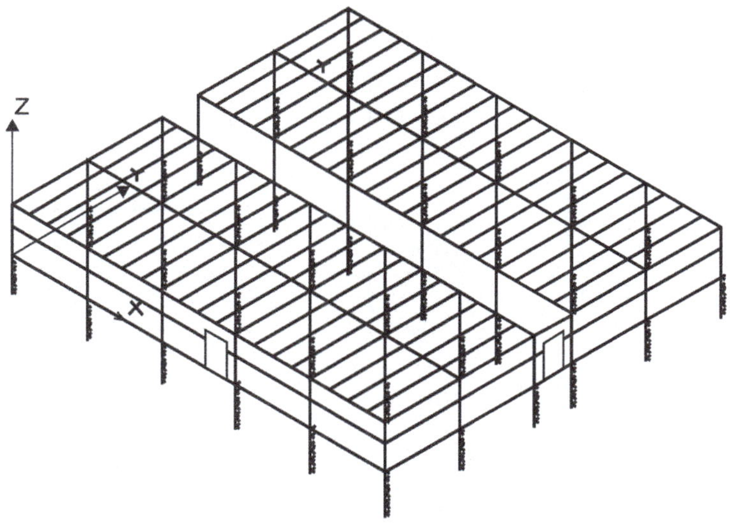

Fig. 13.72 Finite element model of RC framed structure

where $R = D/2$.

Impulsive mass,

$$m_0 = m_T - m_1$$

Total liquid mass,

$$m_T = dbh\rho \text{ (for a rectangular tank)}$$

Location of convective mass,

$$L_1 = h\left(1 - \frac{\cosh(3.8317\Delta) - 1}{1.916\frac{h}{R}\sinh(3.8317\Delta)}\right) \quad (13.70)$$

Location of impulsive mass,

$$L_0 = h - L_1$$

Damping ratio,

$$\varsigma = \frac{2.89}{\pi}\sqrt{\frac{v}{R^{3/2}g^{1/2}}\left[1 + \frac{0.318}{\sinh(1.84\Delta)}\left(\frac{1-\Delta}{\cosh(1.84\Delta)} + 1\right)\right]} \quad (13.71)$$

Stiffness,

$$K_1 = m_T\left(\frac{g}{1.19h}\right)[\tanh(3.68\Delta)]^2 \quad (13.71)$$

Example 13.7 Consider a single-degree-of-freedom system structure attached with a Tuned Liquid Damper as shown in Fig. 13.70. The mass of the structure is 91 kg, and its natural frequency is 1.1 Hz. The estimated value of structural damping is 1.6%. The Tuned Liquid Damper is in the form of a rectangular tank with length 153 mm and width 228 mm and has a depth ratio of 0.083 and mass ratio of 0.49. For an equivalent Tuned Mass Damper, obtain the fundamental sloshing frequency, height and value of the mass and stiffness of the convective component.

Solution

Given is the mass of the structure, M_s= 91 kg; natural frequency, f_s= 1.1 Hz; structural damping, $\zeta_s = 1.6\%$; tank length, d = 153 mm; tank width, w = 228 mm; depth ratio, $\Delta = 0.083$; and mass ratio, $\mu = 0.49\%$ (ratio of modal mass of water to the structural mass).

Now using the expression of the sloshing frequency (Eq. 13.67),

$$f_{sl} = \frac{1}{2\pi}\sqrt{\frac{3.8317g}{0.153}\tanh(3.8317{*}0.083)}$$

f_{sl} comes out to be 1.38 Hz. Now, multiplying the mass ratio to the structure mass gives the total mass of the water in the Tuned Liquid Damper, m_t= 0.446 kg. The water depth, h, is calculated by multiplying tank length with the depth ratio giving, h = 12.7 mm Using the total mass, convective mass is calculated using following expression:

$$m_1 = 0.446\left(0.318\frac{0.0765}{0.0127}\tanh(3.8317{*}0.083)\right)$$

The value of the convective mass is calculated as 0.262 kg. Subsequently, the location of the convective mass can be calculated as in Eq. 13.69.

$$L_1 = 0.0127\left(1 - \frac{\cosh(3.8317{*}0.083) - 1}{1.916\frac{0.0127}{0.0765}\sinh(3.8317{*}0.083)}\right)$$

Thus, the location of the convective mass is 6.4 mm above the tank bottom. The stiffness is calculated as from Eq. 13.71:

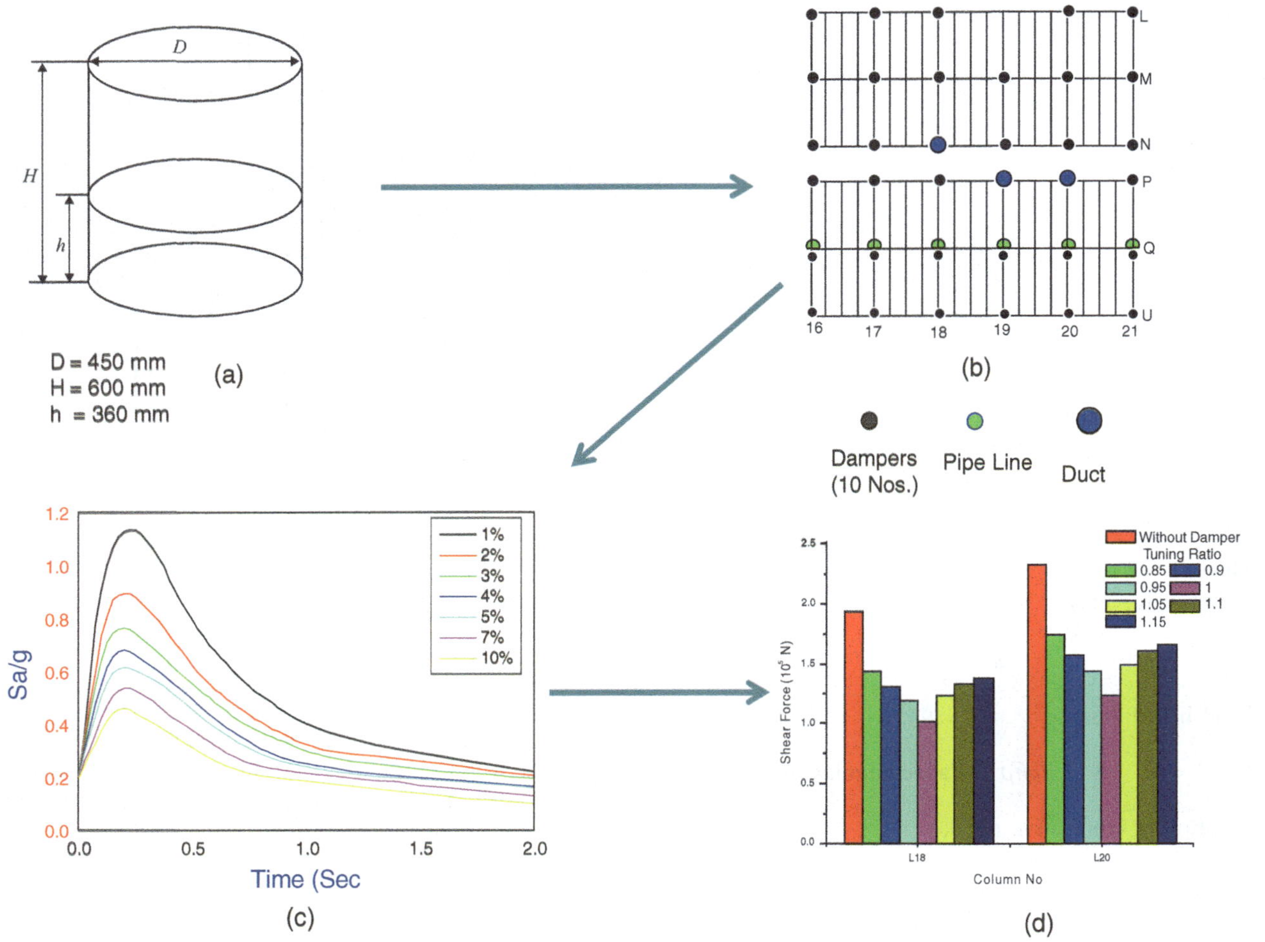

Fig. 13.73 **a** Geometrical details of the TLD, **b** Location of the dampers attached, **c** ground response spectrum of Tarapur site and, **d** comparison of shear force distribution of the columns with and without dampers for various tuning ratios

Fig. 13.74 Experimental setup of a multiple tuned liquid damper

$$K_1 = 0.446\left(\frac{g}{1.19*0.0127}\right)[\tanh(3.68*0.083)]^2$$

Therefore, the stiffness is calculated as 27.415 N/m.

Case study: Retrofitting of RCC Frame Structure of Tarapur Atomic Power Plant using Tuned Liquid Dampers

The idealized model of the RCC frame structure to be retrofitted using TLD is shown in Fig. 13.72. This building was found not meeting the present seismic requirement. The columns were found week and were not meeting the seismic demand. Due to the space limitations, it was found difficult to do retrofitting using strengthening methods and Elasto-Plastic dampers, and hence, Tuned Liquid Mass Dampers are selected for qualification.

The dampers of dimension as shown in Fig. 13.73a were placed on the columns of the building as shown in Fig. 13.73b. But a duct is running on the roof which obstructs the placing of dampers on columns N18, P19, and P20. Hence, there are total of 32 columns available to place dampers, and it was decided to place 10 dampers on each column giving a mass ratio of 1.32%.

Using the response spectrum method, seismic response of the building was obtained with seismic input motion for 7% damping of TAPP spectra as shown in Fig. 13.73c. From the response spectra analysis, it was found that the columns L18 and L20 are the weakest. Figure 13.73d shows the comparison of shear force distribution for these columns with and without dampers for various tuning ratios.

The concept of Multiple Tuned Mass Damper has been extended to Tuned Liquid Dampers in the form of Multiple

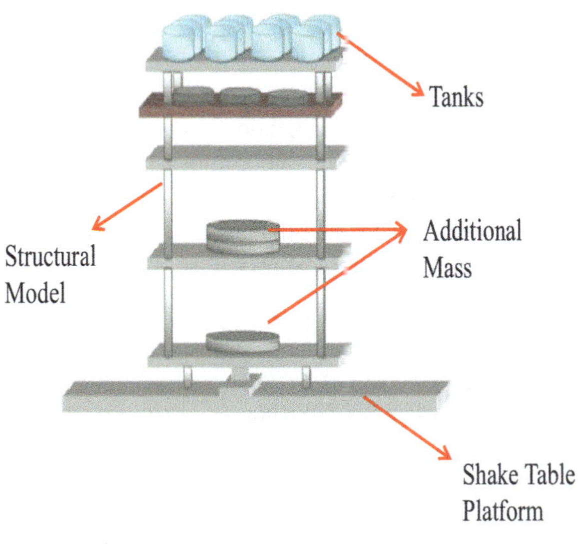

Fig. 13.75 Schematic of the experimental setup of multiple tuned liquid damper shown in Fig. 13.74

Tuned Liquid Dampers. It has been observed by Fujino and Sun [27] that Multiple Tuned Liquid Dampers have similar characteristics to that of Multiple Tuned Mass Dampers but has more efficiency and is less frequency sensitive. However, it is important to note that the effectiveness of the Multiple Tuned Liquid Dampers decreases and is comparable to a Single Tuned Liquid Damper in case of strongly excited motions where large liquid displacements are possible (in case of wave breaking) [30]. Figure 13.74 shows the experimental setup of a Multiple Tuned Liquid Damper conducted at BARC facility, and its corresponding schematic is shown in Fig. 13.75. Kareem and Kline [31, 32] extended the concept to Tuned Liquid Column Dampers which paved the way for their improved construction, installation, and maintenance offering a variety in spatial distributions in the structure.

Case Study: Use of Ferro-Fluids in Tuned Liquid Damper (Vaithianathan [33])

It has been observed that the response reduction capabilities of Tuned Liquid Damper with increases with increase in viscosity of liquid used. Figure 13.76 displays this viscosity effect of the Tuned Liquid Damper for various liquids. From Fig. 13.76, it can be seen that response reduction capabilities are higher for the ferro-fluid (without the magnetic field) than compared to PVA gel [40] and water. Ferro-fluids (Fe_3O_4) are colloidal liquids made of nanoscale ferromagnetic or ferrimagnetic particles (particle sizes in the nanometer range) suspended in a carrier fluid usually an organic solvent or water. This happens because of the presence of magnetic nanoparticles in the ferro-fluids which at the time of sloshing and wave breaking dissipate energy by collision and friction between particles and carrier liquid. This, in a way suggests that the viscosity of the ferro-fluid is higher than the PVA gel and water even without the magnetic field. The presence of the magnetic field would further enhance the viscous properties bringing down the response. It can also be observed that the response reduction due to water and PVA gel at lower excitation amplitude (up to 0.5 mm) is not much due to linear sloshing of liquid inside tank. However, at higher amplitudes, the response reduction increases for water, PVA gel, and ferro-fluid.

It was observed in the experiments that in cases where there was no TLD attached to the structure the response was found to be maximum. When Tuned Liquid Damper using water is attached with structure, the response reduces. This reduction in response is mainly attributed to the sloshing and wave breaking of the water inside the TLD. However, in case of ferro-fluid as liquid in Tuned Liquid Damper, response reduction obtained is more. It is observed that change in viscosity from water to ferro-fluid as measured by ford

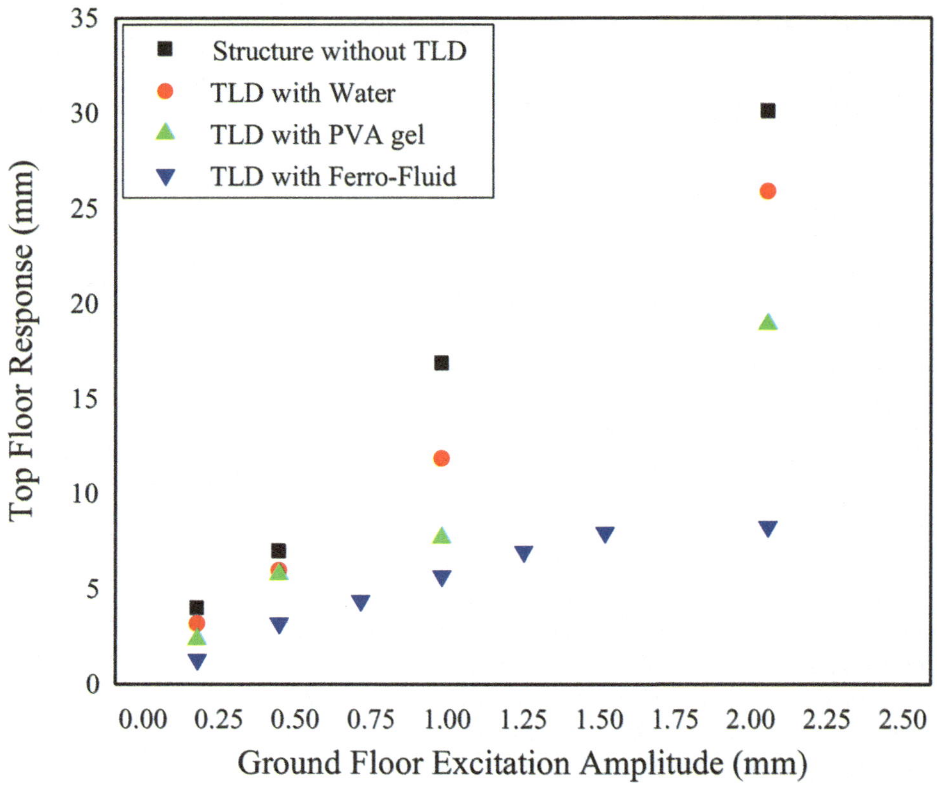

Fig. 13.76 Effect of viscosity of the liquid used in tuned liquid damper on the response of the structure (Vaithianathan)

cup viscometer method is less. This difference in viscosity, in fact, does not explain the response reduction of structure when ferro-fluid is used. The viscosity of ferro-fluid is measured in the static conditions where liquid is flowing in steady state flow. In case of linear sloshing of liquid, due to the presence of nanoparticles in ferro-fluid, there is more energy dissipation due to collision between particles, friction between particles, and friction between particles and liquid; hence, the presence of nanoparticles in ferro-fluid causes more response reduction.

To account for this unusual viscous property of the ferro-fluid in the stiffness and the damping characteristics, a new mathematical TLD model was realized by Yu et al. [34]. Jin-Kyu Yu et al. modeled TLD as an equivalent solid mass damper with nonlinear stiffness and damping known as the nonlinear stiffness damping (NSD) model. This model was a significant expansion of those typically used for mass dampers and the nonlinear characteristics which are calibrated based upon shaking table tests [35]. This mechanical model adequately characterized the performance of the TLD for a wide range of amplitudes of excitation and the stiffness and damping parameters for a SDOF were determined keeping the energy dissipation provided by NSD and TLD same.

The variable E'_{w} is the non-dimensional energy dissipation for the TLD determined from the shake table experiments and is given by [34–37]

$$E'_{\mathrm{w}} = \frac{E_{\mathrm{w}}}{(1/2)m_{\mathrm{w}}(\omega A)^2} \tag{13.72}$$

the denominator of Eq. 13.72 represents the maximum kinetic energy of the water mass when treated as a solid mass. Here, m_{w} is the mass of the liquid, ω is the angular excitation frequency, and A is the sinusoidal amplitude of excitation. The variable E_{w} is the energy dissipation per cycle and is defined as

$$E_{\mathrm{w}} = \int_{T_{\mathrm{s}}} F_{\mathrm{w}} \mathrm{d}x \tag{13.73}$$

where F_{w} is the force generated due to liquid sloshing in the tank and is integrated over the complete shake table displacement for each cycle.

The variable E'_{d} is the non-dimensional dissipation energy for the NSD model and is expressed as

$$E'_{\mathrm{d}} = 2\pi \left|F'_{\mathrm{d}}\right| \sin\phi \tag{13.74}$$

Under harmonic excitation having frequency ratio β, amplitude F_{d}, and phase φ of the control force of the NSD are defined as (in non-dimensional form):

$$\left|F'_{\mathrm{d}}\right| = \frac{\sqrt{(1+(4\varsigma_{\mathrm{d}}^2-1)\beta^2)^2 + 4\varsigma_{\mathrm{d}}^2\beta^6}}{1+(4\varsigma_{\mathrm{d}}^2-2)\beta^2+\beta^4} \tag{13.75}$$

$$\phi = \tan^{-1}\left[\frac{2\varsigma_d\beta^3}{-1+(1-4\varsigma_d^2)\beta^2}\right] \quad (13.76)$$

Excitation frequency,

$$\beta = \frac{f_e}{f_d} \quad (13.77)$$

Natural frequency (NSD model),

$$f_d = (1/2\pi)\sqrt{k_d/m_d} \quad (13.78)$$

Damping ratio of NSD model,

$$\varsigma_d = c_d/c_{cr} \quad (13.79)$$

Critical damping coefficient,

$$c_{cr} = 2m_d\omega_d \quad (13.80)$$

Mass of NSD model,

$$m_d = m_w \quad (13.81)$$

Fundamental natural angular frequency,

$$\omega_d = 2\pi f_d \quad (13.82)$$

Frequency shift ratio,

$$\xi = f_d/f_w \quad (13.83)$$

Linear fundamental frequency for liquid,

$$f_w = \frac{1}{2\pi}\sqrt{\frac{\pi g}{L}\tanh\left(\frac{\pi h_0}{L}\right)} \quad (13.84)$$

Stiffness hardening ratio,

$$\kappa = \xi^2 \quad (13.85)$$

Stiffness hardening ratio,

$$\kappa = k_d/k_w \quad (13.86)$$

Stiffness of water,

$$k_w = m_w(2\pi f_w)^2 \quad (13.87)$$

Sun et al. [36] experienced similar liquid behavior under small amplitude excitation where initially 80% of the total mass was effectively involved in wave breaking, however, that percentage increased with the amplitude of excitation subsequently leading to the observation that mass participation depended on the geometry of the tank and liquid viscosity. Further, it is observed that results obtained by Yu et al. for damping and frequency ratios lie close to those obtained by Sun et al. for water and the higher viscosity liquid results. It was inferred that with increase in the excitation amplitude, the effective mass of 80% for Sun's model increased to 100% participation. It was also realized from Sun et al. study that weak wave breaking results cannot be simply extended for the strong wave breaking region stressing on laboratory tests for large amplitude excitation to understand the governing phenomenon.

In the design of TLDs, the primary objective is to effectively dampen peak displacement for which the required control force is based on the selection of tank size and water depth. The NSD model helps estimate the nonlinearly tuned water depth at a given peak displacement for a particular structural frequency for maximum effectiveness. In addition, it also helps the designer to fully characterize the tank parameters for subsequent numerical analyses using the damping relationships previously derived. The linearly tuned water depth is calculated from equation as

$$h_0 = \frac{L}{\pi}\tanh^{-1}\left(\frac{4\pi L f_s^2}{g}\right) \quad (13.88)$$

The nonlinear tuning frequency,

$$f_d = \frac{\xi}{2\pi}\sqrt{\frac{\pi}{L}\tanh\left(\frac{\pi h_0}{L}\right)} \quad (13.89)$$

This includes the stiffness hardening parameter. Solving equation for the water depth yields the following expression:

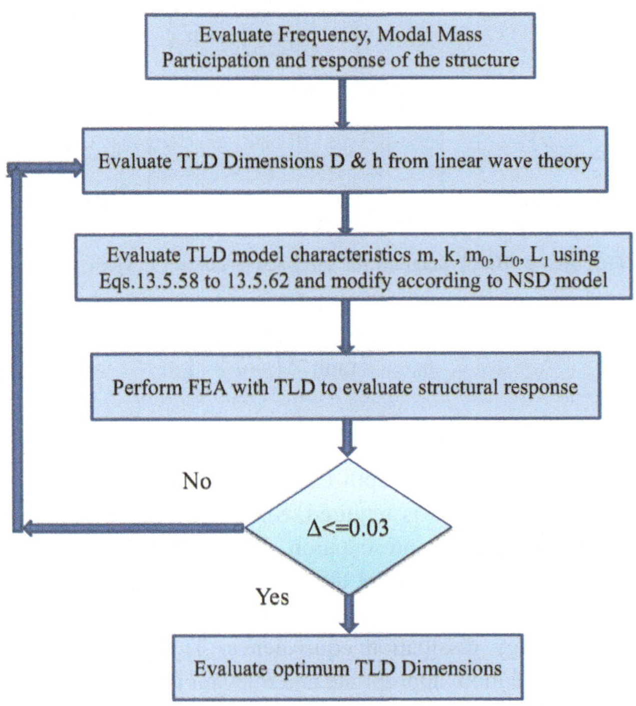

Fig. 13.77 Flowchart for determining optimum TLD dimensions

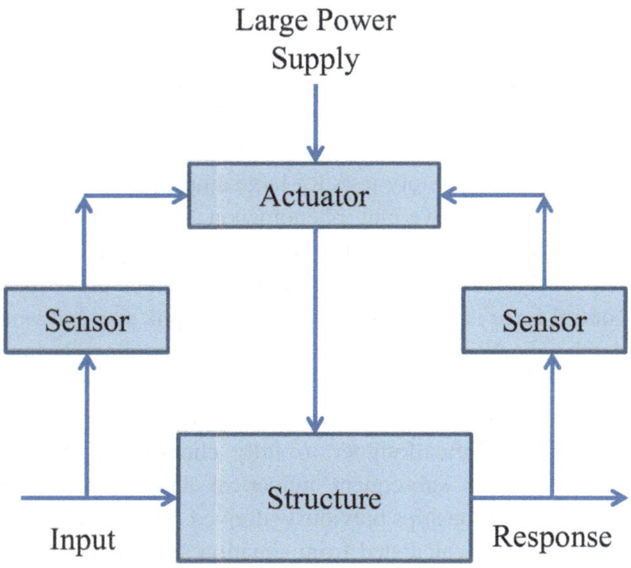

Fig. 13.78 Block diagram for active control devices

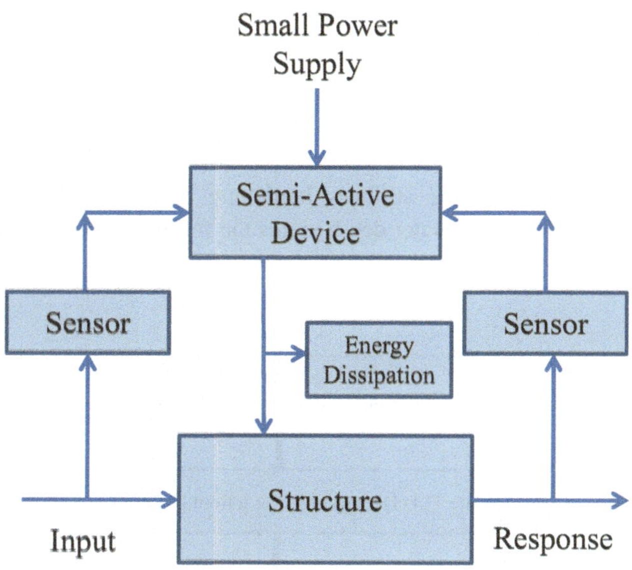

Fig. 13.79 Block diagram for semi-active control devices

$$h_0 = \frac{L}{\pi}\tanh^{-1}\left(\frac{4\pi L f_s^2}{g\xi^2}\right) \quad (13.90)$$

It is to be noted that if the peak structural amplitude is estimated, the water depth can be evaluated; otherwise, an iterative procedure is required. Since the NSD model is an equivalent TMD representation of the TLD, the damping parameter is a function of the peak excitation amplitude and tank length as the model is obtained from a scheme based upon energy dissipation equivalence. The NSD model is found to be more appropriate and relevant in representing the behavior for large seismic excitations than those previously obtained as its parameters are obtained from various tests of tanks subjected to large amplitudes of excitation. In addition, these large amplitude results have been prominent in the development of various design guidelines. For optimum TLD design, considering hardening behavior, an iterative procedure as shown in Fig. 13.77 is followed to evaluate the optimum TLD dimensions by Savyanavar et al. [37].

13.6.2 Active Devices [38]

Apart from using passive devices to check the response of the structure under seismic excitation, active devices (or Active Control Systems/Devices) are also utilized; however, their principle of operation is different. These devices resist the response of the excited structure through external energy sources. These systems consist of real-time processors which under any dynamic loading help generate a balance force (using actuators—hydraulic or electrical) in real time subsequently improving the dynamic performance of the structure. Another advantage of this system is the space requirement. Figure 13.78 describes the block diagram for the Active Control System. Active Mass Damper (AMD) is one such system where the active mass is configured to be either in sliding or pendulum mode.

13.6.3 Semi-active Devices [38]

The demand of large space posed by passive systems and high energy demands of active system led to the advent of semi-active systems/devices. The energy required by these systems is much less as compared to active systems. Further, their ease of installation and workability is advantageous over passive systems. Semi-active system consists of a passive device installed in the structure and is tuned to the optimum level as per the real-time excitation signals generated by structure. This principle of response control is successfully employed with Tuned Liquid Column Damper (TLCD) and Tuned Mass Damper (TMD) and is further extended to other energy dissipaters like (MR), electrorheological (ER). Figure 13.79 shows the block diagram of the semi-active devices.

13.6.4 Hybrid Devices [38]

Often, it has been observed that a single type of control system alone is not effective enough to govern the response of the structure under different types of dynamic loading conditions. In such cases, it is more rational to use more than one type of control methodologies to achieve the required effectiveness. Hence, combinations of such systems acting

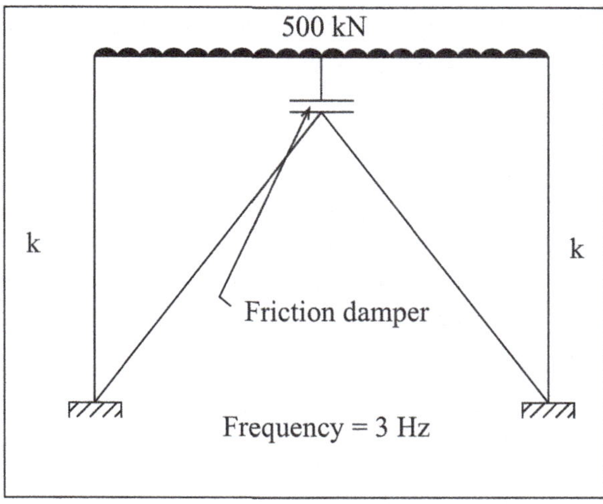

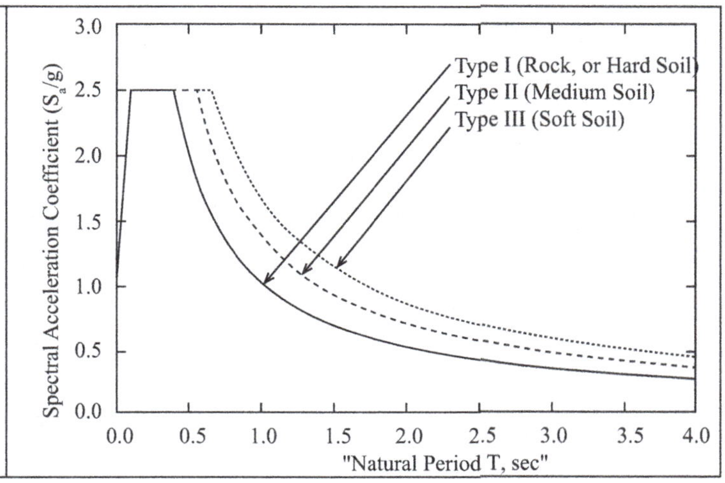

Fig. 13.80 Portal frame and response spectra

simultaneously are employed to resist the structural response. It is well known that passive systems have limited ability when it comes to high frequency constituents of seismic loads. This shortcoming of the passive systems is overcome using hybrid systems/devices. In this kind of systems, passive energy dissipation (PED) devices work along with Active Control Systems where the loads are shared between the two systems in a way that a part of the control is accomplished by passive system, leading to lower the burden on the Active Control Systems. For example, for a Tuned Mass Damper, the building may be equipped with a passive secondary mass damper system and a tertiary small mass connected to the secondary mass with a spring, damper, and actuator. The secondary (passive) system is set in motion by the active tertiary mass; thus, more effective and efficient structural control is achieved.

Exercise Problems

1. Consider the portal frame shown in Fig. 13.80 with friction damper having limiting friction force of 5 kN. Damping of the structure is 5%.

$$\text{Additional damping due to damper} = \frac{\Delta E}{4\pi(\text{StrainEnergy})}$$

Consider IS 1893 response spectra for 5% damping shown in Fig. 13.80 for hard soil. Evaluate reduction in response acceleration and increase in damping after first iteration.

2. A three-story-reinforced concrete frame building as shown in Fig. 13.81 is situated at Mumbai. The height between the floors is 3 m, and total height of building is 9 m. The total load of 3000 kN is lumped at each floor. The soil below the foundation is assumed to be hard rock. Determine the total base shear and shear force distribution along the height of building as per IS 1893–2002. Zone factor for Mumbai = 0.16, important factor for building = 1, response reduction factor = 3.0.)
   (Hint: Evaluate fundamental natural period as per IS 1893–2002, use equivalent static lateral force method and use response spectra shown in figure of prob. 1.)

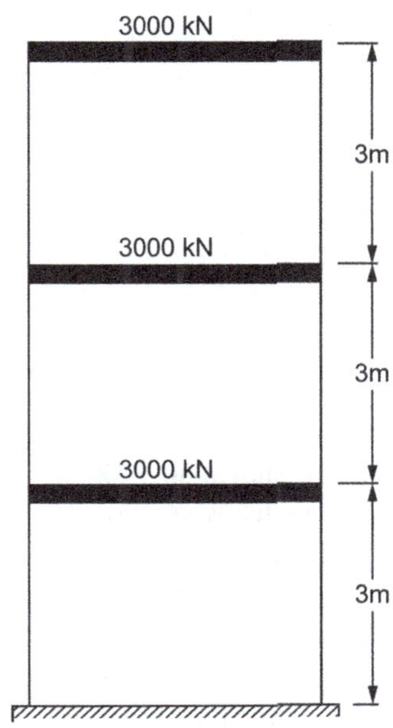

Fig. 13.81 Three-story-reinforced concrete frame building

3. Consider the symmetric frame structure shown in Fig. 13.82 located in Seismic Zone III. Its natural frequency $f = 4$ Hz, mass of the floor = 10,000 kg and damping is 5%. Designer would like to investigate the response acceleration considering passive devices and isolators. Compare the response considering Elasto-Plastic dampers of 6 kN (yield force) and

Table 13.9 Multiplication factors for obtaining values for other damping

Damping (%)	0	2	5	7	10	15	20	25	30
Factors	3.20	1.40	1.00	0.90	0.80	0.70	0.60	0.55	0.50

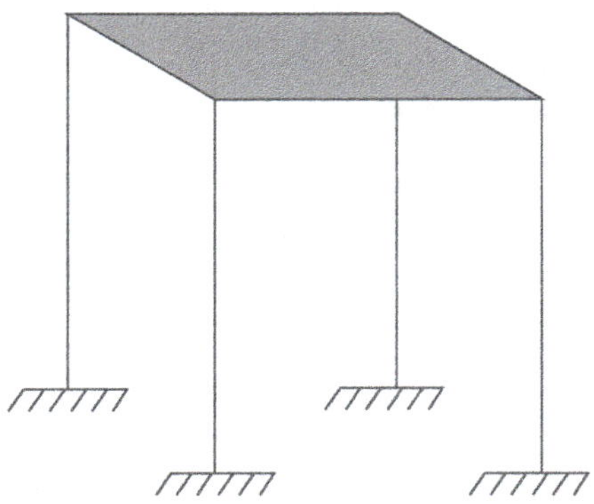

Fig. 13.82 Symmetric frame structure

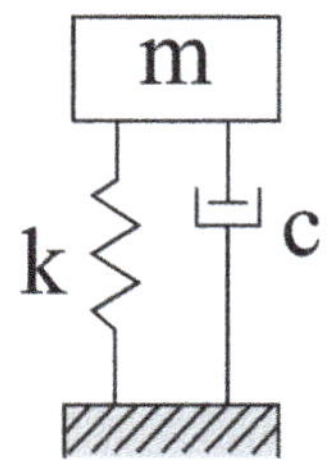

Fig. 13.83 Mean square response acceleration and force of a single DOF system

laminated rubber bearings separately for IS 1893 (Part I): 2002 rock spectra shown in Prob. 1 (Note I. Assume damper is elastic perfectly plastic and consider only two iterations, ii. Consider suitable locations for isolators and dampers, iii. Neglect mass of columns and assume structure is in elastic state).

4. What are the reasons for re-qualification and retrofitting of structures of nuclear facilities?
5. Write brief notes on seismic re-qualification and retrofitting of existing facilities.
6. Explain various aspects need to be considered for seismic design optimization.
7. Explain the following: a. strength-based design, b. deformation-based design.
8. Evaluate mean square response acceleration and force of a single DOF system equivalent to a structure located in Mumbai using following data (Fig. 13.83).
   Frequency = 2 Hz, m = 100,000 kg, $\zeta = 0.05$

$$\mathrm{PSDF}\,(g^2/\mathrm{Hz}) = S(\omega) = S_0 \frac{1 + \left[2\zeta_g(\omega/\omega_g)\right]^2}{\left[1 - (\omega/\omega_g)\right]^2 + \left[2\xi_g(\omega/\omega_g)\right]^2}$$

Frequency range = 1–5 Hz, $S_0 = 0.07$, $\omega_g = 27$, $\zeta_g = 0.34$

Note: Consider frequency interval of 1 Hz and design peak ground acceleration of 0.16 g.

9. Explain briefly methodologies for controlling the response of a structure.

References

1. Verma RK (2010) Iterative response spectrum method for complex piping systems. M Tech thesis. HBNI, Mumbai, India
2. RaviKiran A, Agrawal MK, Sakhrodia DK, Sahu J, Tale KV, Reddy GR, Biswas B, Singh RK, Vaze KK, Bhowmick A, Gupta RV (2014) Re-evaluation of equipment of heavy water plant-kota for earthquake and wind loads. RSD Report, BARC, Mumbai, India
3. Reddy GR (2016) Retrofitting of equipment subjected to dynamic loads. In: WSRR 2016 conference on structural rehabilitation and retrofitting. Mumbai, India
4. Robinson WH, Greenbank LR (1975) Properties of an extrusion energy absorber. Bull NZ Soc Earthq Eng 8(3):187–191
5. Fujita et al. (1989) Study on lead extrusion damper as a seismic support. In: Transactions of the 10th international conference on structural mechanics in reactor technology. Anaheim, K2, pp 733–737
6. Fujita K, Ito T, Kajii S, Kokubo E, Nakatogawa T, Kato T (1991) Study on a rotary type lead extrusion damper as a high damping support for nuclear piping systems. In: Transactions of the 11th SMIRT, Tokyo, K33, pp 475–480
7. Robinson WH, Greenbank LR (1976) Extrusion energy absorber suitable for the protection of structures during an earthquake. Earthq Eng Struct Dynam 4(3):251–259
8. Kokubo E, Fujita K, Fuyama H, Ichihasi I (1995) Development of lead extrusion damper for PWR reactor coolant loop system. In: Seismic engineering conference proceedings. ASME PVP, vol 312, pp 425–431
9. Namita Y, Yoshinaga T, Shibata H, Kunieda M, Hara F, Suzuki K, Ichihashi I, Iiyama K, Matsuda T, Murota M (1991) Development of energy absorber and its application to piping system in nuclear power plants. ASME PVP 211:51–57
10. Shibata H, Kunieda M, Hara F, Suzuki K, Ichihashi I, Fukuda T, Satoh A, Takata K, Furukawa S, Kobayashi H (1991) Development of elasto-plastic damper as a seismic support for piping system in nuclear power plants. ASME PVP 211:63–69
11. Namita Y, Ono S, Suzuki K, SasakiY, Abe H, Kuroda K (1997) Proving test on the seismic reliability of the main steam piping system: part 3—analytical evaluation of BWR feed water piping supported by energy absorbing support. In: Seismic engineering conference proceedings, ASME PVP, vol 345, pp 125–131
12. Parulekar YM, Verma V, Reddy GR, Vaze KK (2002) Retrofitting of existing structures for seismic requalification. J Struct Eng 29 (1):77–82
13. Satishkumar K, Muthumani K, Sivarama Sarma B, Gopalakrishnan N (2000) Evaluation of X type metal passive energy absorbers. In: An report on experiments prepared by SERC for BARC
14. http://web.mit.edu/jorloff/www/jmoapplets/secondorder/tunedmassdamper.pdf

15. Iwanami K, Seto K (1984) Optimum design of dual tuned mass dampers and their effectiveness. Proc JSME 50:44–52
16. Xu K, Igusa T (1992) Dynamic characteristics of multiple substructures with closely spaced frequencies. Earthq Eng Struct Dyn 21:1059–1070
17. Yamaguchi H, Harnpornchai N (1993) Fundamental characteristics of multiple tuned mass dampers for suppressing harmonic forced oscillations. Earthq Eng Struct Dyn 22:51–62
18. Abe M, Fujino Y (1994) Dynamic characteristics of multiple tuned mass dampers and some design formulas. Earthq Eng Struct Dyn 23:813–835
19. Jangid RS (1995) Dynamic characteristics of structure with multiple tuned mass dampers. Struct Eng Mech 28:1041–1049
20. Zuo L, Nayfeh SA (2004) Minimax optimization of mutiple-degree-of-freedom tuned mass dampers. J Sound Vib 27:893–908
21. Zuo L, Nayfeh SA (2006) Two-degree-of-freedom tuned mass damper for suppression of single-mode vibration under random and harmonic excitation. J Vib Acous 28(1):56–65
22. Bandivadekar TP, Jangid RS (2012) Mass distribution of multiple tuned mass dampers for vibration control of structures. IJCSE 3:2012
23. Samanta A, Banerji P (2008) Structural control using modified tuned liquid dampers. In: 14th WCEE, Beijing, China
24. Banerji P (2004) Tuned liquid dampers for the control of earthquake response. Paper no. 1666, 13th WCEE, Vancouver, Canada
25. Fujino Y, Sun LM (1993) Vibration control by multiple tuned liquid dampers (MTLDs). J Struct Eng ASCE 119(12):3482–3502
26. Housner GW (1963) Dynamic analysis of fluids in containers subjected to acceleration. In: Nuclear reactors and earthquakes, Report no. TID 7024, U.S. Atomic Energy Commission, Washington DC
27. Fujino Y, Sun LM, Pacheco BM, Chaiseri P (1992) Tuned liquid dampers (TLD) for suppressing horizontal motion of structures. J Eng Mech ASCE 118(10):2017–2030
28. Fujita T (ed) (2005) Seismic isolation, energy dissipation and active vibration control of structures. In: Proceedings of the 9th world seminar, JAVIT, Tokyo, Japan
29. Rahman MM (2007) The use of tuned liquid damper to enhance seismic performance of concrete rigid frame buildings. Ph.D. thesis. University of western Ontario, Canada
30. Gao H, Kwok KSC, Samali B (1999) Characteristics of multiple tuned liquid column dampers in suppressing structural vibration. Eng Struct 21:316–331
31. Kareem A (1994) The next generation of tuned liquid dampers. In: Proceedings of the 1st world conference on structural control. Los Angeles
32. Kline S, Kareem A (1995) Performance of multiple mass dampers under random loading. J Struct Eng ASCE 121(2):348–361
33. Vaithianathan JRK (2008) Seismic response control of structures using semi-active tuned liquid dampers with ferro-fluids. M.Tech. thesis. HBNI, Mumbai, India
34. Yu JK, Wakahara T, Reed DA (1999) A non-linear numerical model of the tuned liquid damper. Earthq Eng Struct Dyn 28: 671–686
35. Yu JK (1997) Non-linear characteristics of tuned liquid dampers. Ph.D. thesis. Department of Civil Engineering, University of Washington, Seattle, WA
36. Sun LM, Fujino Y, Chaiseri P, Pacheco BM (1995) The properties of tuned liquid damper using TMD analogy. Earthq Eng Struct Dyn 24:967–976
37. Savyanavar RC, Reddy GR, Parulekar YM, Kumar P, Rama Rao A, Chavan SP, Vaze KK, Ghosh AK, Kushwaha HS (2006) Experimental evaluation and retrofitting of structure subjected to earthquake loading with circular Tuned Liquid Damper (TLD). J Struct Eng SERC
38. Rai NK, Reddy GR, Ramanujam S, Venkatraj V, Agarwal P (2009) Seismic response control systems for structures. Def Sci J 59(3):239–251
39. Reddy GR, Sharma A, Parulekar YM, Vaze KK, Ghosh AK, Kushwaha HS (2005) Seismic retrofitting of existing concrete structures using energy absorbers. In: 2nd national conference on advances in materials and mechanics of concrete structures (NCAMMCS). Indian Institute of Technology Madras, Chennai, 12–13 Aug 2005
40. Sharma A, Singh R, Reddy GR, Vaze KK, Ghosh AK, Kushwaha HS (2008) Tuned water and gel dampers for response control of structures. In: Proceedings of the sixth structural engineering convention (SEC-2008), 18–20 Dec 2008, pp 885–892
41. Soong TT, Dargush GF (1997) Passive energy dissipation systems in structural engineering. Wiley, New York
42. ATC-40 (1996) Seismic evaluation and retrofit of concrete buildings. Report no. SSC 96-01. Applied Technology Council, California Seismic Safety Commission, California
43. Kelly JM, Naeim F (1999) Design of seismic isolated structures, from theory to practice. Wiley
44. Hollaway LC, Leeming MB (2001) Strengthening of reinforced concrete structures Using externally-bonded FRP composites in structural and civil engineering. CRC Press, Boca Raton, Boston, New York, Washington, DC
45. CEB-FIP (2001) Bonded FRP reinforcement for RC structures. Technical report on the design and use of externally bonded fibre reinforced polymer reinforcement (FRP EBR) for reinforced concrete structures. Technical report. Bulletin, vol 14
46. American Concrete Institute (2002) Guide for the design and construction of externally bonded FRP systems for strengthening concrete structures. ACI 440.2R.02, ACI Michigan
47. IS 13935 (1993): Seismic evaluation, repair and strengthening of masonry buildings-guidelines; Bureau of Indian Standards
48. Aboutaha R, Engelhardt M, Jirsa J, Kreger M (1999) Rehabilitation of shear critical concrete columns by use of rectangular steel jackets. ACI Struct J 96:68–78
49. Aboutaha R, Engelhardt M, Jirsa J, Kreger M (1996) Retrofit of concrete columns with inadequate lap splices by the use of rectangular steel jackets. Earthq Spectra 12(4):693–714
50. Chakraborty S, Jha DK, Reddy GR, Srinivas K (2016) Nonlinear analysis of fiber wrapped reinforced concrete beams with cohesive model. In: Proceedings of the international congress on computational mechanics and simulation (ICCMS 2016). IIT Bombay, Mumbai, India
51. Bencardino F, Spadea G, Swamy RN (2002) Strength and ductility of reinforced concrete beams externally reinforced with carbon fiber fabric. ACI Struct J 99:163–171
52. Muthumani K, Gopalakrishnan N, Kumar KS, Avinash S, Reddy GR, Parulekar YM (2003) Lead extrusion damper device for passive energy absorption. In: Proceedings of the national seminar on seismic design of nuclear power plants. SERC, Chennai
53. Ichihashi I (1995) Development of a lead extrusion damper for PWR reactor coolant loop system. In: Seismic engineering conference proceedings, ASME PVP, vol 312, pp 425–430
54. Parulekar YM, Reddy GR, Vaze KK, Kushwaha HS (2004) Lead extrusion dampers for reducing seismic response of coolant channel assembly. Nucl Eng Des 227:175–183
55. US Department of Energy (USDOE) (1997) Seismic evaluation procedures for equipment in US department of energy facilities. DOE/EH-0545
56. IS1893 (Part 1): 2002, Criteria for earthquake resistant design of structures: General provisions & buildings, Bureau of Indian Standards, New Delhi, India
57. IS456: 2000, Plain and reinforced concrete: Code of practice, Bureau of Indian Standards, New Delhi, India

Seismic Base Isolation of Structures

14

G. R. Reddy, T. Nagender, and P. N. Dubey

	Base isolation intends to isolate the ground motion reaching to the structure and floor mounted equipment and piping systems	

Symbols

A_r	Cross-sectional area of rubber
A_{rd}	Overlap area of top and bottom face of bearing
A_L	Cross-sectional area of lead
A_s	Effective shear area
A_ϕ	Bearing cross-sectional area
$G_{0.5}$	Shear modulus at 50% strain
D_s	Displacement
D	Diameter or side length of cross section
D_{cr}	Critical displacement
D_M	Target displacement
D_y	Yield displacement
E_c	Compression modulus of rubber
f_v	Vertical frequency
f_h	Horizontal frequency
F_y	Yield force
G_L	Shear modulus of lead
H	Total thickness of isolator including rubber layers and steel shims
I_S	Moment of inertia of bearing cross section
K	Bulk modulus of rubber
K_1	Initial stiffness
K_2	Post-yield stiffness
K_H	Lateral stiffness
$K_{effective}$	Effective stiffness
K_V	Vertical stiffness of isolator
P_E	Euler buckling load for a column with no shear deformation
P_s	Shear stiffness per unit length
Q	Characteristic strength
t_r	Thickness of individual rubber layer
T_M	Target time period
W	Seismic weight
X_c	Maximum vertical displacement in compression
W_{max}	Maximum vertical load
X_c	Maximum vertical displacement in compression
δ_v	Vertical deflection
ξ_M	Target damping ratio
σ_{yL}	Shear yield strength of lead
ε_v	Vertical strain
$\sum t_r$	Cumulative thickness of rubber layers

14.1 Introduction

Seismic design of structures has been one of the most important and interesting issues in the last century. Eventually, the fundamental frequency of conventional medium-rise buildings lies in the frequency range of earthquake that has maximum energy content. Consequently, these buildings attract maximum earthquake response due to resonance. The response of the building further amplifies with floor rise. This increases the stresses in the structural members and results in large interstorey drifts. This leads either to damage or loss of functionality of the various

G. R. Reddy (✉) · T. Nagender · P. N. Dubey
Bhabha Atomic Research Centre, Mumbai, India
e-mail: rssred@barc.gov.in

T. Nagender
e-mail: nagender@barc.gov.in

P. N. Dubey
e-mail: pndubey@barc.gov.in

G. R. Reddy et al., *Textbook of Seismic Design*,
https://doi.org/10.1007/978-981-13-3176-3_14

service components mounted on different floor levels. The ground acceleration expected during an earthquake has a great deal of uncertainty in its assessment. Moreover, the magnitudes of the forces during an earthquake are very large, and the use of more popular design in elastic range of material behavior results in very large sections of structural members which is uneconomical and sometimes impractical. Further, the low probability of occurrence of large earthquake events does not justify the use of elastic design and the practice of inelastic design has evolved for earthquake-resistant design. In conventional earthquake-resistant design, an attempt is made to arrest all the brittle mode of failure by introducing special detailing schemes that have been well established from experiments world over. These detailing schemes have been prescribed as ductile detailing procedures in various codes (IS-13920 [25] in case of India).

Although the structural members do not collapse in the event of the maximum considered earthquakes, the resulting inelastic deformations are large enough to disrupt the functionality of the structure and making it unavailable after a major earthquake. Thus, in structures expected to be functional even after the severest of earthquakes, certain non-conventional methods are there, which modifies the dynamic characteristics of the structure.

14.2 Non-conventional Earthquake Design

The demand force acting on the structure in the event of an earthquake is directly proportional to the spectral acceleration of the structure which is related to the time period and damping of the structure. This relationship is represented by a response spectrum. The response spectra obtained for hard soil according to IS-1893-2002 [26] for different damping ratios are presented in Fig. 14.1. Most medium-rise structures have frequency in the range of 1–10 Hz. In this range, the spectral accelerations are higher than the ground acceleration. However for structures having fundamental frequency less than 0.5 Hz are subjected to significantly low values of spectral accelerations. It can also be observed in Fig. 14.1 that with increase in damping, the magnitudes of the spectral accelerations and spectral displacements are reduced significantly. In a non-conventional earthquake design, the above characteristics of the earthquake demand force are ingeniously used for earthquake-resistant design.

Different kinds of base isolation methodologies used for reducing the forces due to earthquake in structural members have been discussed in brief in the following section. In the present chapter, the base isolation using elastomeric bearing has been discussed in detail.

14.3 Seismic Base Isolation

In this method, a flexible member is introduced at the base of the structure which results in shift of horizontal frequency below 0.5 Hz. In this process, the superstructure and the foundation are decoupled from the strong motion horizontal components of ground motion because of which the devastating strong motion signals are filtered out. The isolators also possess inherent damping characteristics which will be added to the overall damping of the structure. A considerable reduction in demand forces as well as in the interstorey drifts is achieved. The base isolation system concept mainly focuses on the shift of the structure's fundamental period out of the dominant earthquake energy frequencies range and also increases the energy absorbing capability. The concept is illustrated in Figs. 14.2 and 14.3. Following are some of the commonly used base isolation methods.

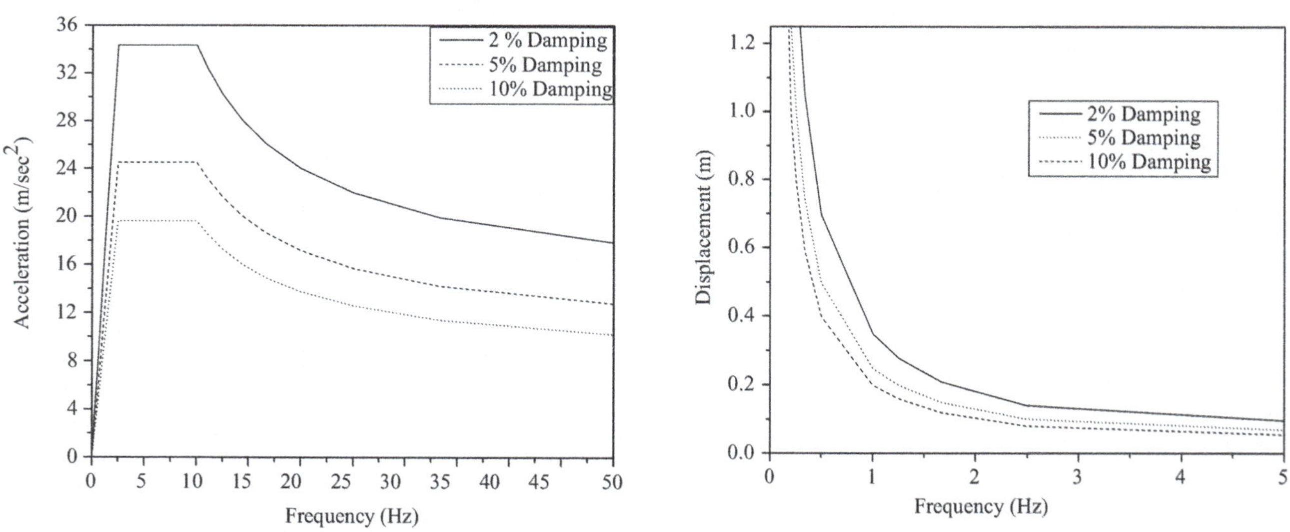

Fig. 14.1 IS-1893 spectra for hard soil

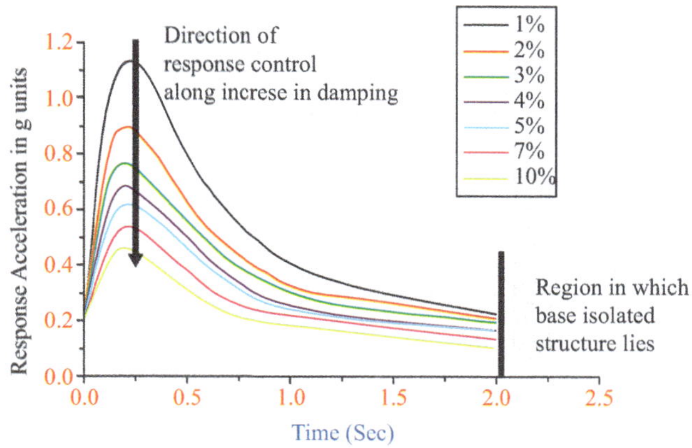

Fig. 14.2 Concept of base isolation and dampers

14.3.1 Passive Base Isolation Systems

Mud Layer underneath the Structure

Idea of base isolation technique was first implemented by Frank Lloyd Wright. He constructed short-length piles which were spaced closely in a thick soil layer underlain by a thick mud layer above hard strata to isolate Imperial Hotel, Tokyo. This hotel could be able to survive an earthquake in 1923 [1].

Flexible First Storey

Martel in 1929 [2] was the first to propose the concept of flexible first storey and it was studied further by Green in 1935 [3] and Jacobson in 1938 [4] to reduce the load on upper storey members. Chopra et al. [5] in their further studies with the help of computers proved that the concept is not practical. Even the recent earthquakes occurred at Bhuj (India) and Kobe (Japan) had revealed that the extensive damage was seen in most of the buildings having soft storey.

Roller Bearings in Foundations

The structures provided with the roller bearing systems for isolation had a serious drawback as the rollers were subjected to and fro motion in particular direction, whereas earthquake has three-direction motion resulting in the ineffective isolation against the earthquake forces. And the major problem was that for keeping the device in good operation throughout its service life, it needed consistent maintenance.

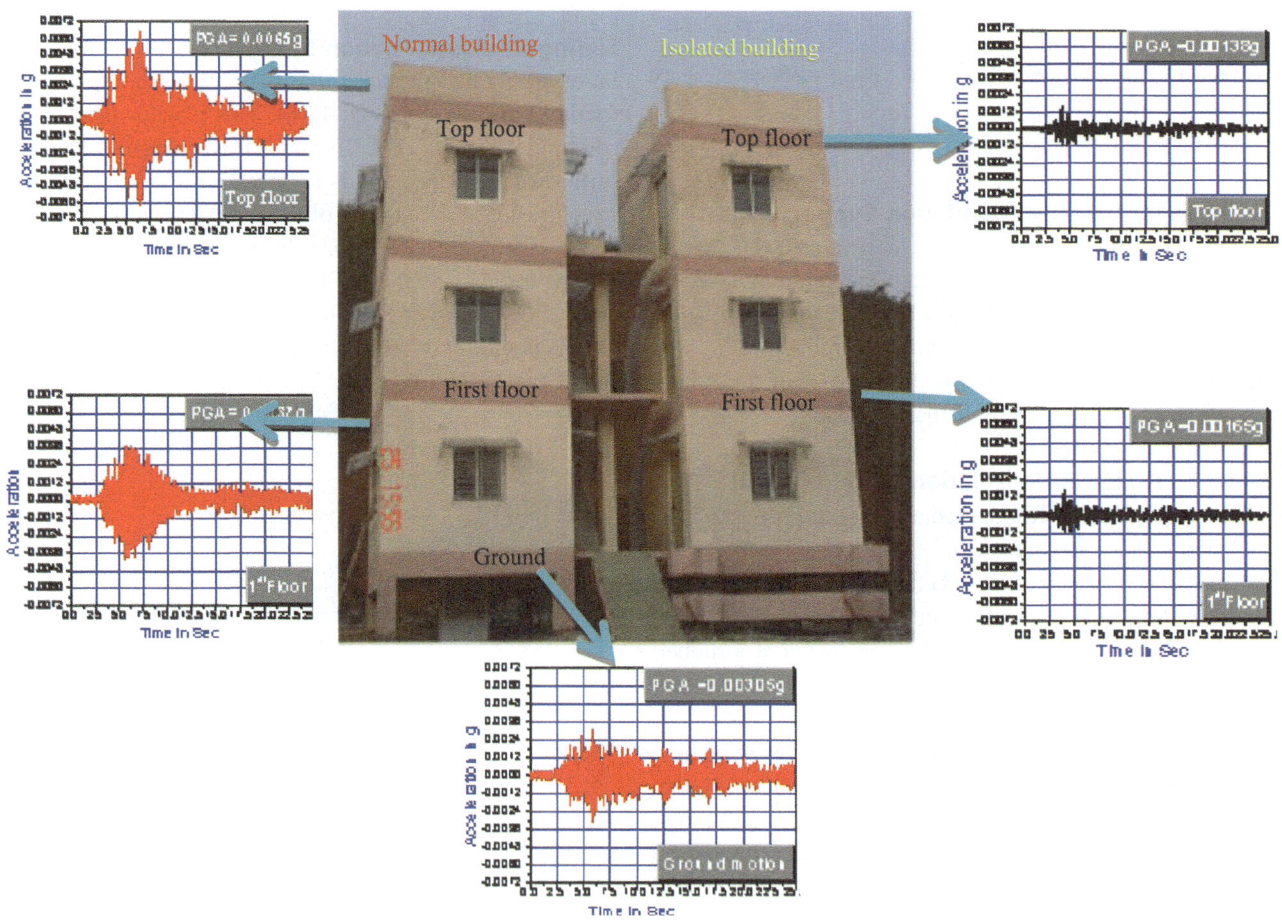

Fig. 14.3 Identical conventional and base-isolated buildings at IIT Guwahati

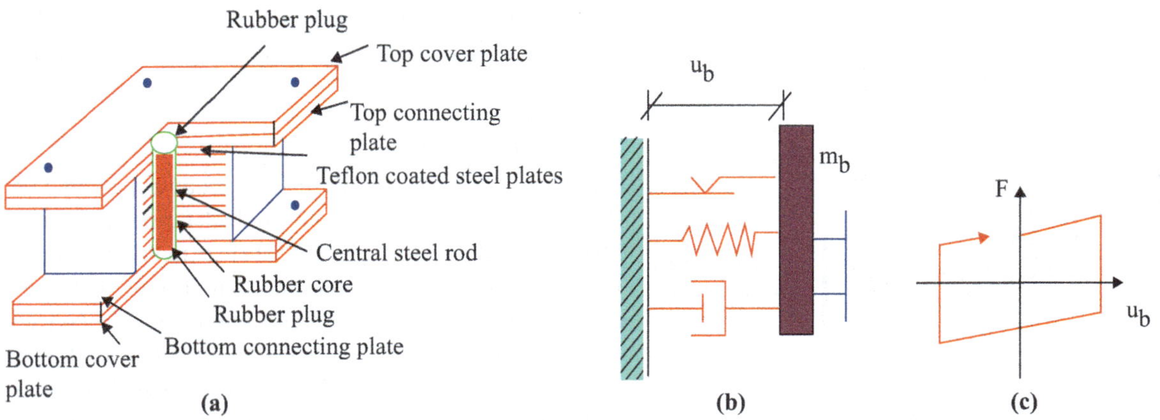

Fig. 14.4 Resilient–friction base isolation system: **a** cross-sectional details and **b** system representation) force–deformation relation

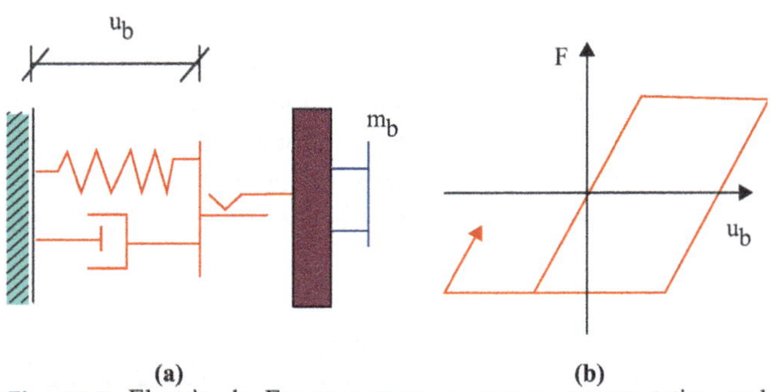

Fig. 14.5 Electric de France system: **a** system representation and **b** force–deformation relation

Rubber Layer used as Foundation Support

In 1969, a school building was constructed on rubber foundations in Skopje, Yugoslavia. The building used to rock forward and backward and also bounce during earthquake because of uniform stiffness of rubber in all the directions. Another problem was the bulging of the rubber foundation under the weight of the building.

Resilient–Friction Base Isolation System (R-FBI)

N. Mostaghel and M. Khodaverdian [6] had proposed resilient–friction base isolation (Fig. 14.4) system which consists of concentric layers of Teflon-coated plates which provide sliding resistance and a central core of rubber will provide additional beneficial effect of resilience of a rubber.

Electric de France system (EDF)

In 1970, Electric de France had developed friction-type base isolation concept called Electric de France system (Fig. 14.5). For the nuclear power plants located in the high seismicity regions, this system is standardized. The system comprises of a laminated neoprene pad provided with lead bronze plate at the top, which will be in frictional contact with the steel plate anchored to the structure's base raft. Hence, its cross section is more or less similar to that of the laminated rubber bearings (LRB) system. Due to very low displacement capacity of the neoprene pad (approximately 5 cm), the sliding element provides the needed movement, when this capacity is exceeded. This system does not provide with any restoring device, and hence, permanent displacement could occur. The system had been installed in the nuclear power plant at Koeberg in South Africa.

Sliding Resilient–Friction System

Su et al. [7] had proposed the concept of sliding resilient–friction base isolator (Fig. 14.6). This isolator has the good combined features of EDF and R-FBI systems. The upper surface of R-FBI system is replaced with the friction plates which result in the sliding of the structure on its foundation in a similar manner of the EDF base isolator system. And the system behaves as the R-FBI system for low level of seismicity. Only during high level of ground acceleration, the top plate slides which provides the additional safety against the severe ground motion.

Pure Friction System

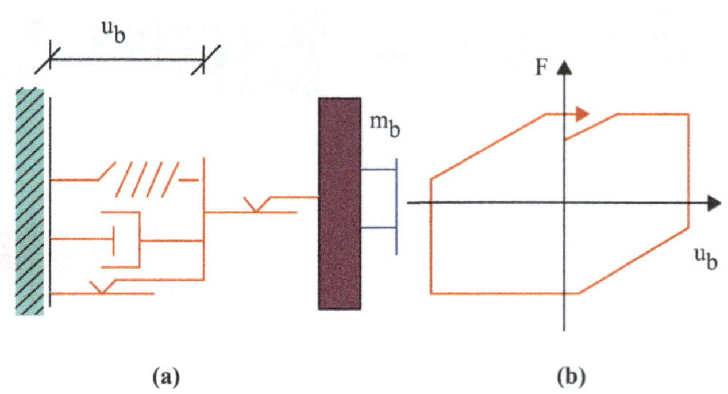

Fig. 14.6 Sliding resilient–friction system: **a** system representation and **b** force–deformation relation

The concept of this base isolator is to develop frictional force using a sand layer or rollers at the base, which can dissipate the energy of seismic force. In China, for low-rise structures the system is developed. For a wide range of frequency input, this system is useful. The main advantage of this concept is that it is economical. And the major problem with this system is that it is unable to recover the displacement after a seismic event and sand layer is too sensitive for settlement of foundation.

Friction Pendulum System

Working principle of friction pendulum bearings is similar to a simple pendulum. During the seismic event, the slider moves along the profile of mating concave surface which causes the supported structure to move in simple harmonic motions with predefined amplitude. The working principle of friction pendulum is shown in Fig. 14.7. Due to free sliding of the structure along the concave profile of the bearing, the frequency is lowered, and thus, the high frequency earthquake signals having comparatively higher damaging potential are filtered out. Due to sliding movement between the concave frictional surface and slider imparts additional damping to the structure. Added frictional damping reduces lateral displacement of the structure. The frictional pendulum has inherent auto-re-centring capability that allows structure to restore its original position after the earthquake.

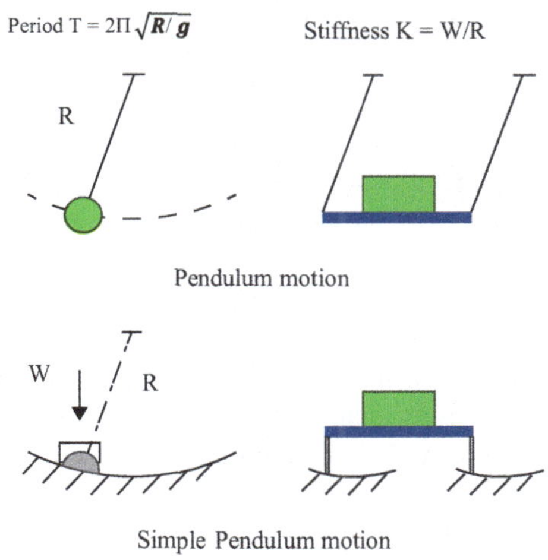

Fig. 14.7 **a** Friction pendulum concept, **b** schematic of friction pendulum bearing

The period (T) of the friction pendulum bearing is selected simply by choosing the radius of curvature (R) of the concave surface. It is independent of the mass (W) of the supported structure. The damping is selected by choosing the friction coefficient. Torsion motions of the structure are minimized because the center of stiffness of the bearings automatically coincides with the center of mass of the supported structure. Bearings can be designed to accommodate different magnitudes of displacement simply by adjusting the curvature and diameter of the bearing surface. To allow the lateral displacement to slider surface during earthquake, the mating surface of the bearing is coated with special material to reduce the friction between the slider and the concave surface.

Parameters considered for the design of friction pendulum bearings are weight of the structure (W) to be isolated, allowed movement during earthquake, soil type, and the dimensions of the structure to be isolated.

Spring-Type Systems

For isolating the structure from lateral seismic forces, elastomeric and sliding isolation systems are very effective. But whenever three-dimensional isolation is required, spring-type systems would be used. This type of system of GERB brand was developed with large helical steel springs which have flexibility in both the horizontal and vertical directions. The vertical frequency of the system would be around 3–5 times that of the horizontal frequency. The steel springs had been provided with GERB visco-damper.

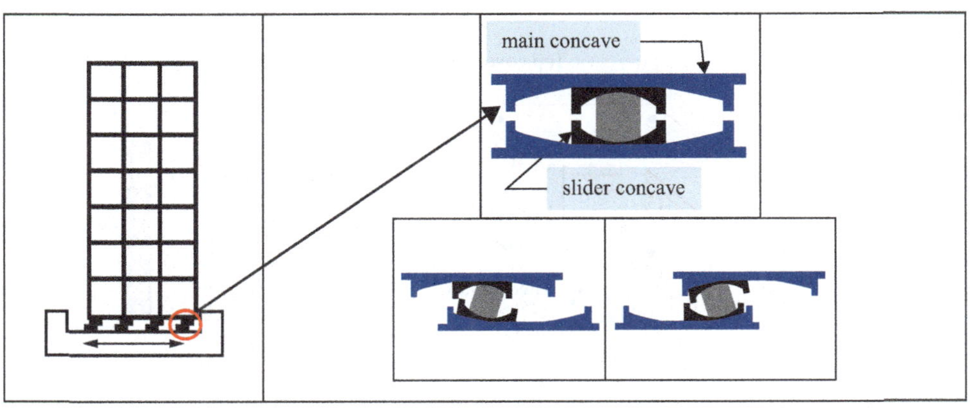

Fig. 14.7 (continued)

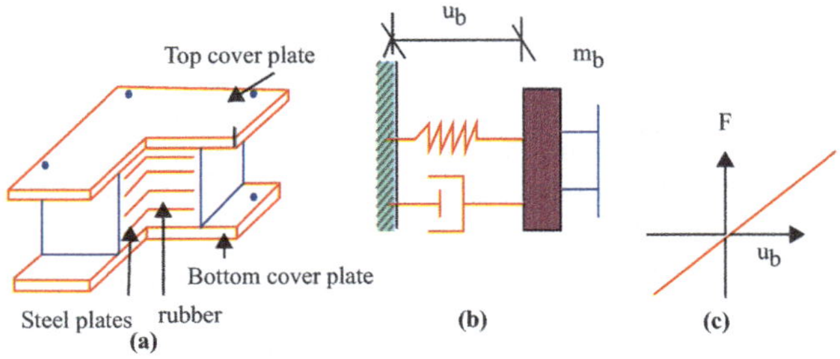

Fig. 14.8 Laminated rubber bearing system: **a** sectional details, **b** system representation, **c** force–deformation behavior

This system was used in two steel-framed houses in Santa Monica, California. But those houses were severely damaged by the 1994 Northridge earthquake. The responses of those buildings were monitored and it was found that the reduction in the accelerations in these buildings was ineffective due to rocking motion.

Base Isolation Using Geo-Synthetic Materials

The concept of using geo-synthetic material for isolating the base of the structures has been developed by Yegian and Kadakal [8]. A high strength, non-woven geo-textile placed over an ultra-high-molecular-weight poly-ethylene (UHMWPE) liner was used. These two materials had static and dynamic friction coefficients of 0.1 and 0.07, respectively. Therefore when a geo-synthetic material is placed underneath, the foundation of a structure and over a liner will allow the dissipation of seismic energy through sliding friction. It was recommended by them that the range of the sliding friction should be 0.05–0.15 between the two materials.

BS Cushion

In Hangzhou, China, new type of base isolation system called BS cushion was invented in 1999 (Chinese Patent Number ZL99202381.5). This base isolation cushion is made from treated asphalt–fiber. The main advantage of this kind of isolator is most economical and safety while its isolation effect is moderate.

The BS cushion resembles laminated steel plate rubber bearing. Fiber and treated asphalt in BS cushion provide similar effect as of steel plate and elastomer in laminated rubber bearing. In Hangzhou, China, two 7-storey masonry–concrete residential buildings isolated with BS cushion were built before 2001. One of the buildings is isolated by replacing some of the depth of base soil below mattress foundation with alternative piling of four layers of BS cushion and four layers of sand. The fundamental period of this building is increased from 0.3 to 1 s (0.3 s is obtained from a similar building and 1 s is obtained from this building).

Elastomeric Bearings

Elastomeric bearings are most commonly used for base isolation. They consist of alternate lamination of thin elastomer (rubber) layers and steel plates (shims) which are bonded to each other to provide the horizontal flexibility and vertical stiffness. Vertical rigidity provides the support for the weight of structure, while horizontal flexibility converts destructive lateral movement into gentle movement. The seismic response of the isolated structures in the horizontal direction is not influenced by strong motion signals above 0.5 Hz frequency. The elastomeric base may be of either circular or square cross section. Schematic of elastomeric bearings are shown in Fig. 14.8. Natural rubber has a low material damping about 3–5% of critical, thereby resulting in a linear force–deformation relationship. The laminated bearing made using natural rubber having low damping is called laminated rubber bearing (LRB). As a consequence of low frequency and damping of isolated system with LRB, the horizontal displacement increases. To bring the undesired large horizontal displacement within the acceptable limit, lead core is introduced in the LRB. The elastomeric bearing with lead core is called lead plug bearing (LPB) as

Fig. 14.9 New Zealand bearing system (LPB) **a** sectional details, **b** system representation, **c** force–deformation

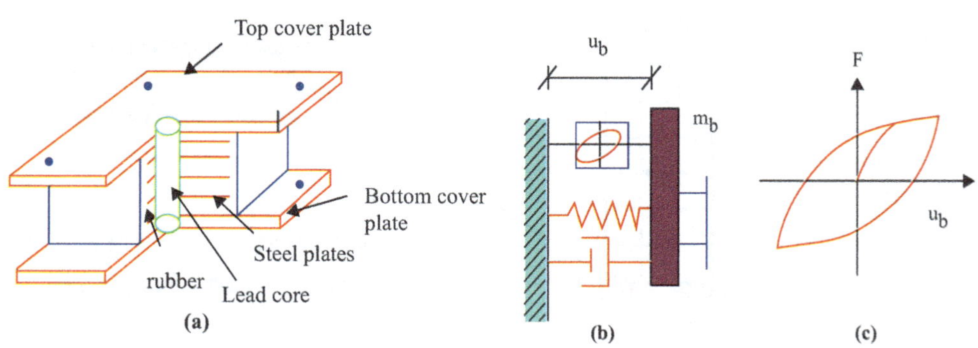

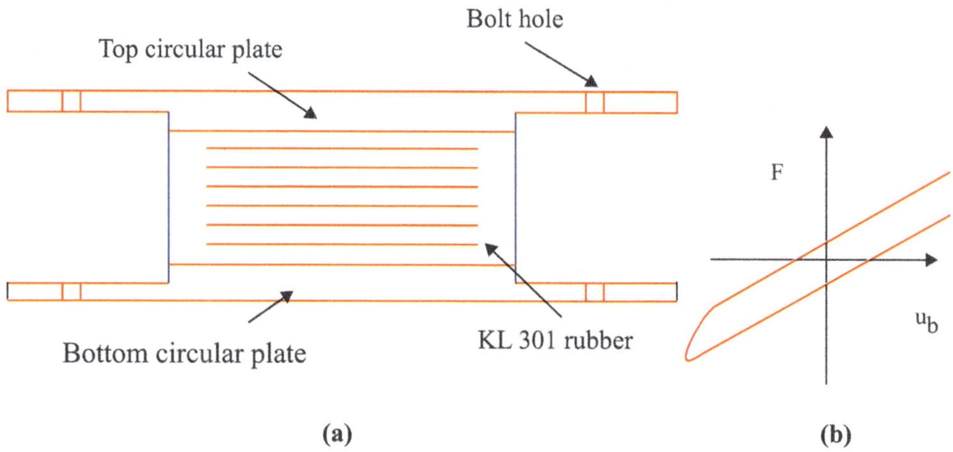

Fig. 14.10 High damping rubber bearing: **a** sectional details and **b** force deformation behavior

shown in Fig. 14.9. The lead core yields during strong earthquake and provides additional damping.

High Damping Rubber Bearing (HDRB)

In these bearings, a blend of high damping rubber is used. A high damping elastomer, called KL301 is manufactured by the Bridgestone Corporation Limited (BCL), Japan. KL301 [9] has a shear modulus of 4300 kPa at small strains, which reduces to 650 kPa for 50% strain, 430 kPa for 100% strain, and 340 kPa for 150% strain. The typical bearing (as shown in Fig. 14.10) of this kind consists of 20 layers of 2.2-mm-thick rubber of 176 mm diameter, nineteen no. of 1-mm steel shims, and 12-mm-thick top and bottom plates. Its design axial pressure is about 3.23 MPa. The bearings are provided with flange-type end plates to provide bolted connection and foundation connection. The construction features of high damping rubber bearings (HDRB) are same as of LRB, but HRDBs are generally fabricated using special high-density rubber. These bearings provide additional damping without any use of lead core.

14.3.2 Hybrid Isolation System with Semi-Active Devices

Both passive isolation systems and semi-active or active controlling devices are used in hybrid isolation system. The Medical Centre of the Italian Navy located at Ancona, had analyzed the behavior of a hybrid system comprises of low damping rubber bearing (LDRB) which is passive seismic isolator, and magnetorheological (MR) damper, which is semi-active controlling device. The analytical results showed that there is a significant reduction up to 50% in the building accelerations.

14.3.3 Hybrid Base Isolation with Passive Energy Dissipaters

The energy dissipating device (Fig. 14.11) generally dissipates the seismic energy and reduces the seismic effect on the structure. These devices can be provided at the base of

Fig. 14.11 Use of energy dissipating devices at base level

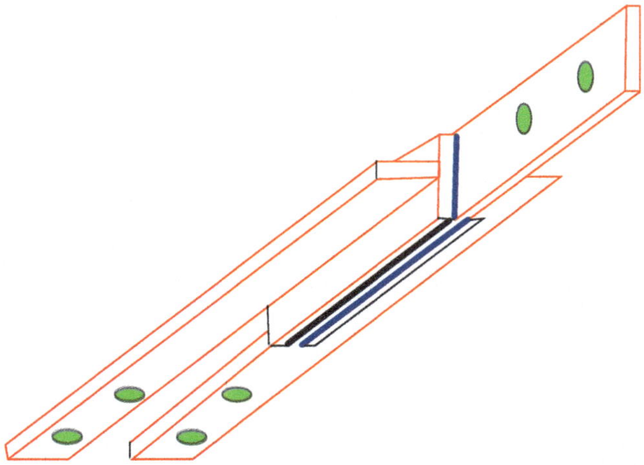

Fig. 14.12 Visco-elastic damper

the structure or also in superstructure at appropriate locations. These devices can be used in combination with passive base isolation systems. Visco-elastic dampers (VEDs) (Fig. 14.12) can control the structure responses. They are made of linear springs and dashpots arranged in parallel and are commonly used in bracings of building frame or at ground level. Similarly, suitable dampers such as Elasto-Plastic Dampers, lead extrusion dampers, tuned liquid dampers, shape memory alloy dampers, Tuned Mass Damper may be used.

14.4 Components of a Base Isolation System

Efficiency and design of base isolation system are governed by different structural and non-structural components. Considering these parameters, design goal of isolation system is set. The main components of base isolation system are shown in Fig. 14.13 and are described as follows:

Structural System

Structural system includes structural members of the superstructure including the foundation. Due to the implementation of flexible member, i.e., isolator at the base, the interstorey drifts are very low and the superstructure behaves as a linear elastic system.

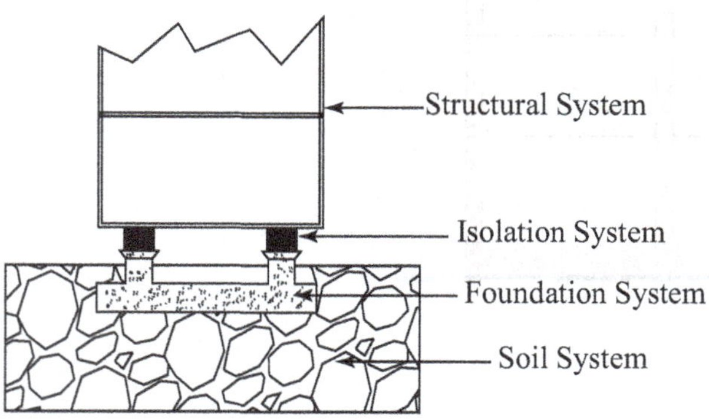

Fig. 14.13 Components of Base isolation system

Isolation System

Isolation system consists of isolating device designed for target isolation efficiency and its connecting members.

Foundation System

For effective base-isolated system, the foundation of the structure should be strong.

Soil System

The subsoil stratum exhibits their own stiffness and damping properties which may or may not affect the response of the structures they bear. This influence of the interaction between the soil and the structural system becomes significant in case of loose subsoil strata. The hard soil strata give maximum response reduction on base isolation because loose soil possesses inherently low frequency system and benefits of isolation of such structures will not be significant.

14.5 Design Philosophy of LRB

In the present section, design calculations of laminated rubber bearings (LRB) have been presented. A mathematical model for the base isolation system can be obtained by implementing the springs with equivalent stiffness at the base of the structure in the direction of isolation. The isolator devices employed in a base isolation system may be of the same type or of different types. Either way, the behavior of the base isolation system may be represented by a bilinear force–deformation relationship as illustrated in Fig. 14.14. Mathematical model for the base isolation system is obtained by superimposition of the mathematical models representing individual isolator devices used in the isolation plane. The lateral stiffness K_H of LRB can be calculated by using Eq. (14.1)

$$K_H = \frac{G_{0.5}\, A_r}{\sum t_r} \tag{14.1}$$

Here,

$G_{0.5}$ Shear modulus of rubber at 50% strain
A_r Cross-sectional area of rubber and
$\sum t_r$ Cumulative thickness of rubber layers.

14.5.1 Stability Check Criteria of Isolator

For stability of base isolation system using LRB, following criteria must be satisfied:

Fig. 14.14 Bilinear characteristics of an isolator

- Q: Characteristic Strength
- F_y: Yield Force
- D_y: Yield Displacement
- K_1: Initial Stiffness
- K_2: Post-yield Stiffness
- K_{eff}: Effective Stiffness
- D: Displacement

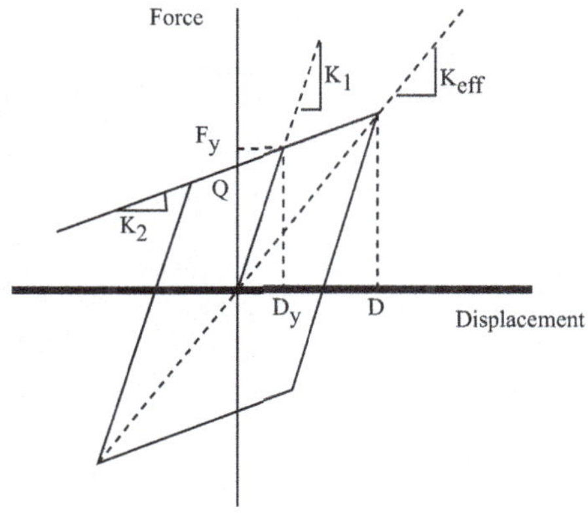

1. **Vertical strain**

The total vertical strain under compressive loading should be less than 20%, which can be calculated using following procedures.

Each individual rubber layer sandwiched between the steel shims acts as vertical springs. For vertical load transfer, these springs are in series. Geometrically, all the rubber layers have same dimensions; therefore, vertical as well as horizontal stiffness of each are same. The vertical stiffness of a single rubber layer is calculated by Eq. (14.2).

$$K_V = \frac{E_C A_r}{\sum t_t} \tag{14.2}$$

where E_C = Compression modulus of rubber (shape-dependent)

$$E_C = \frac{6G_{0.2} S^2 K}{K + 8G_{0.2} S^2} \tag{14.3}$$

S = Primary shape factor (ratio of loaded area and unloaded area of rubber layer as shown in Fig. 14.15b)

$$S = \frac{\frac{\pi}{4} D^2}{\pi D t_r} = \frac{D}{4t_r} \quad \text{for circular section} \tag{14.4}$$

$$S = \frac{D^2}{4Dt_r} = \frac{D}{4t_r} \quad \text{for square section} \tag{14.5}$$

K_V Vertical stiffness of isolator
D Diameter or side length of cross section
t_r Thickness of individual rubber layer
K Bulk modulus of rubber.

Further the vertical deflection (δ_v) under vertical load is calculated using Eq. (14.6), and finally, the vertical strain (ε_v) is calculated by using Eq. (14.7).

$$\delta_V = \frac{W_{max}}{K_V} \tag{14.6}$$

$$\varepsilon_V = \frac{\delta_V}{\sum t_r} \tag{14.7}$$

where W_{max} = Maximum vertical load.

2. **Critical Buckling Load**

For proper stability of isolator, critical buckling load should be five times or more than the maximum vertical load capacity. In order to assess the stability of the bearing under vertical loads, a method much similar to linear elastic analysis of Euler buckling of column as described by Naeim and Kelly [10] has been adopted. An increase in the effective shear area A_s is considered because steel will not deform in the composite system which is given by Eq. (14.8).

$$A_s = A_\phi \frac{H}{\sum t} \tag{14.8}$$

where

A_ϕ Bearing cross-sectional area and
H Total thickness of isolator including rubber layers and steel shims

In a similar manner, the bending stiffness is also modified as suggested in Eq. (14.9).

$$EI_s = \frac{E_C I H}{3 \sum t} \tag{14.9}$$

where

E_C Compression modulus for the bearing as obtained from Eq. (14.3)
I_s Moment of inertia of bearing cross section

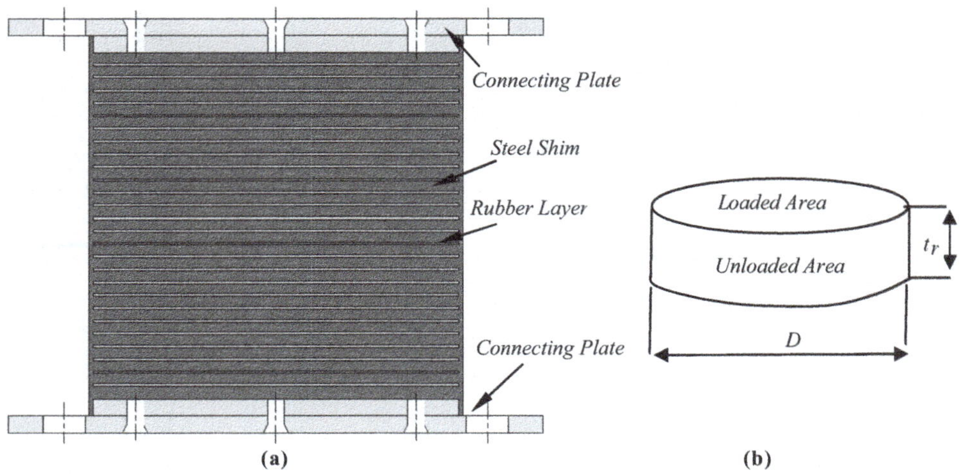

Fig. 14.15 **a** Details of laminate rubber bearing and **b** single layer of rubber

The shear stiffness per unit length P_S and the Euler buckling load for a column with no shear deformation 'P_E' are given by Eqs. (14.10) and (14.11), respectively. The critical bulging load is then calculated by Eq. (14.12).

$$P_S = G_{0.2} A_s \tag{14.10}$$

$$P_E = \frac{\pi^2 E I_s}{H^2} \tag{14.11}$$

$$P_{cr} = \frac{\sqrt{P_S^2 + 4 P_S P_E} - P_S}{2} \tag{14.12}$$

The calculated critical load should be about five or more times greater than the maximum vertical load.

3. **Critical Displacement**

Calculated displacement for stability at large strains should be higher than expected displacements. At large displacement, the effective area available for carrying the vertical load reduces. This can affect the stability of the isolation system. To assess the stability, a critical displacement is evaluated, beyond which horizontal stiffness vanishes. Physically, this is the displacement beyond which the horizontal displacement goes on increasing without any increase in the applied force. The critical displacement may be calculated by Eq. (14.13) as suggested by Naeim and Kelly [10].

$$\frac{A_{rd}}{A} = \left(\frac{P}{P_{cr}}\right)^2 \tag{14.13}$$

$$A_{rd} = \underbrace{\frac{1}{2}(\phi - D_{cr})\sqrt{\phi^2 - D_{cr}^2}}_{\text{circular}} = \underbrace{\phi(\phi - D_{cr})}_{\text{square}} \tag{14.14}$$

where

A_{rd} Overlap area of the top and bottom face of the bearing
A Area of the bearing face
P = Vertical load on the bearing
D_{cr} Critical displacement
ϕ Diameter for circular bearing or side length for the square bearing

The overlap area for square and circular bearings in terms of the critical displacement (D_{cr}) is given by Eq. (14.14). The expected displacement in case of the worst earthquake should never exceed the critical displacement. The stability criteria for laminated rubber bearings as explained above should be also satisfied.

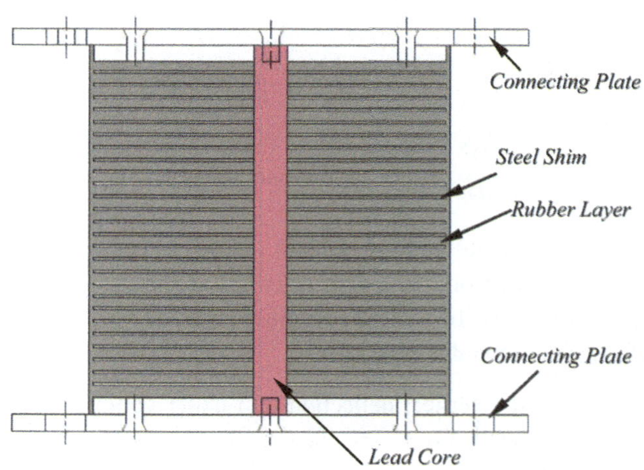

Fig. 14.16 Details of lead plug bearing

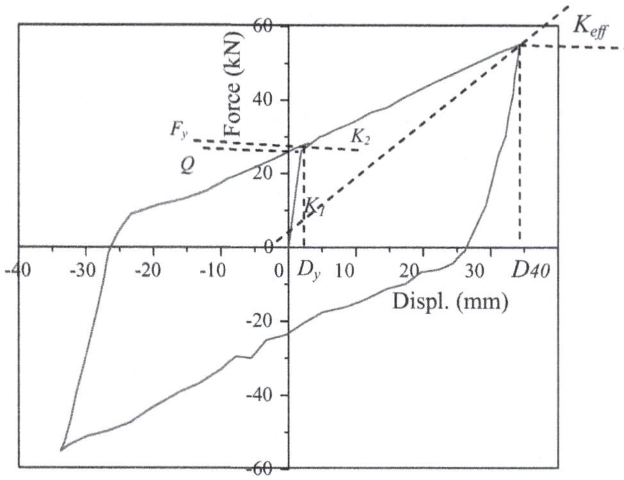

Fig. 14.17 Typical force–displacement characteristic of LPB

14.6 Design Philosophy of Lead Plug Bearings (LPB)

LPBs are similar in design to natural rubber bearings, but they contain a lead plug at the center to gain hysteretic energy dissipation characteristics. Design features of LPB are shown in Fig. 14.16. The central lead plug provides an initial higher stiffness and hysteretic energy dissipation. A characteristics of load–deformation curve for a typical LPB are shown in Fig. 14.17. The various bilinear characteristics as illustrated in Fig. 14.17 are evaluated using the following equations for any given LPB.

The characteristic strength Q is evaluated on the basis of the lead cross-sectional area A_L by using Eq. (14.15) where σ_{YL} is the shear yield strength for lead ($\approx$10 MPa).

$$Q = \sigma_{YL} \times A_L \tag{14.15}$$

The initial stiffness K_1 is governed mainly by the lead plug characteristics and is given by Eq. (14.16). The post-yield stiffness K_2 is contributed by the laminated rubber layers and is given by Eq. (14.17).

$$K_1 = \frac{G_L A_L}{H} + \frac{G_{0.5} A_R}{\sum t} \tag{14.16}$$

$$K_2 = \frac{G_{0.5} A_R}{\sum t} \tag{14.17}$$

where

G_L Shear modulus of lead
H Height of the lead plug
Σt Total rubber thickness
A_r Cross-sectional area of rubber
A_L Cross-sectional area of lead

The yield strength 'F_y' and the yield displacement 'D_y' are evaluated using the relations Eqs. (14.18) and (14.19).

$$F_y = Q + K_2 D_y \tag{14.18}$$

$$D_y = \frac{Q}{K_1 - K_2} \tag{14.19}$$

The stability criteria applicable for LRB are also applicable in the case of LPB.

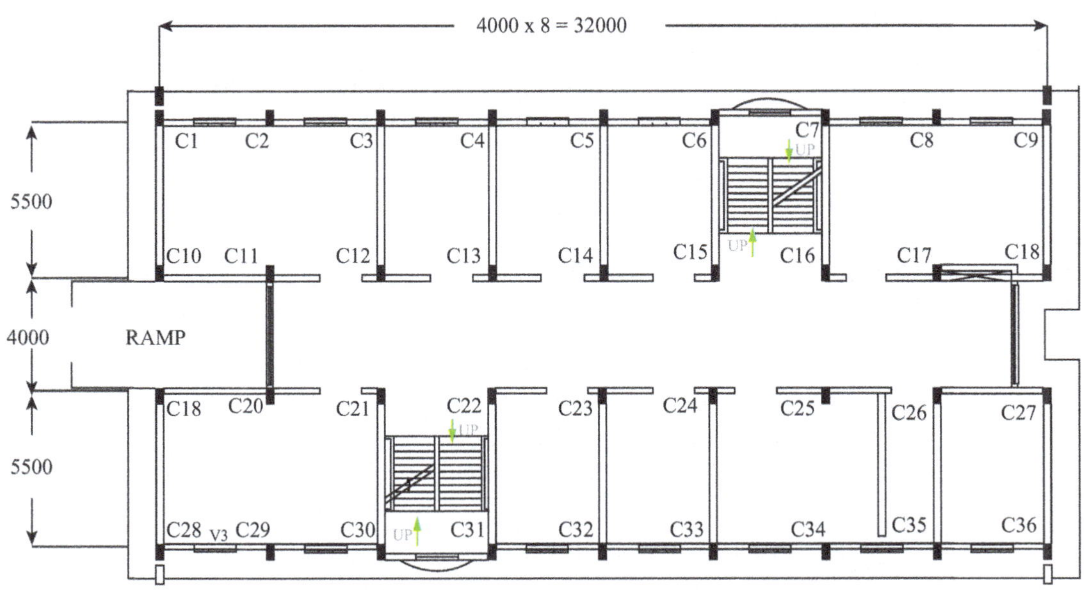

Fig. 14.18 Plan for the example Building

14.7 Design Requirements for Seismically Isolated Structures

The latest recommendations of design requirements are given in ASCE-7-10. The salient features are discussed in the following. The recommendations have been adopted in the design example as well. ASCE-7-10, Chap. 17 provides in detail all the design requirements related to seismic base isolation. A few of these are mentioned here for discussion. For details, the code may be referred.

i. The importance factor shall be taken as 1.0 for a seismically isolated structure, regardless of its risk category assignment.
ii. In addition to the requirements for vertical and lateral loads induced by wind and earthquake, the isolation system shall provide for other environmental conditions including aging effects, creep, fatigue, operating temperature, and exposure to moisture or damaging substances.
iii. Isolated structures shall resist design wind loads at all levels above the isolation interface. At the isolation interface, a wind-restraint system shall be provided to limit lateral displacement in the isolation system to a value equal to that required between floors of the structure above the isolation interface.
iv. Fire resistance for the isolation system shall meet that required for the columns, walls, or other such gravity-bearing elements in the same region of the structure.
v. The isolation system shall be configured to produce a restoring force such that the lateral force at the total design displacement is at least 0.025 W (W—vertical reaction of the column) greater than the lateral force at 50 percent of the total design displacement.

14.8 Step-by-Step Design Procedure

In order to explain a step-by-step design procedure (as illustrated in UBC-97 [11]/ASCE-07-10 [12]), an example of 2-storey building (as shown in Fig. 14.18) has been used. The total weight of the superstructure is 30,000 kN, which is transferred to the foundations by 36 columns. It is desired to design a suitable base isolation system for this building located in Seismic Zone V in India on hard stratum and compare the base shears for fixed base and isolated design.

In the design of seismic base isolation system, the behavior of the isolators is considered by an equivalent bilinear model as shown in Fig. 14.13. For a given site, the zero period acceleration (ZPA) may be determined by a suitable analysis or from codal provisions. Using this value, the variation of characteristic strength may be plotted against the effective damping ratio for a given isolator type (governed by the value of $n = K_1/K_2$ the post-yield stiffness ratio), and a given soil type and seismic weight. Figure 14.19 shows such variation for various time period values, where characteristic strength is plotted as a fraction of seismic weight on the Y-axis against damping ratio on X-axis, for lead rubber bearings ($n = 10$) at a site (Zone factor $Z = 0.36$) with hard soil.

In the first step of the design process, the service horizontal loads (wind loads) are evaluated. In seismic isolation, it is desired that the isolation level should not have excess deformations for service loads. This can be assured by selecting a characteristic strength Q equal to the horizontal service loads expected. For the value of Q, the appropriate degree of isolation can be decided by assuming a suitable time period T and damping ratio ξ using Fig. 14.19. A step-by-step design procedure has been outlaid here. The aforementioned design example is presented and the values evaluated at every step are cited.

Step 1: Assumption of Appropriate Target Dynamic Characteristics

Table 14.1 Design calculations—MCE parameters

Quantity evaluated	Value obtained	Units
Seismic weight 'W'	30,000	kN
Service horizontal loads 'Q'	900	kN
MCE time period 'T_M' (assumed)	2.25	s
MCE damping ratio 'ξ_M'	0.15	–
Effective stiffness at MCE 'K_{eff} Eq. (14.20)	23,848	kN/m
MCE displacement 'D_M' Eq. (14.22)	0.140	m
DBE displacement 'D_D' Eq. (14.23)	0.070	m

The values for the target time period 'T_M' and the damping ratio'ξ_M' for the maximum considered earthquake (MCE) level and the ratio of elastic stiffness to post-yield stiffness 'n' for the type of isolator in consideration are assumed. For a rational assumption of the time period and damping, the curves developed in Fig. 14.19 may be used. The wind load for the structure is evaluated (typically = 0.03 W), and the same is considered as Q for the assumed time period and damping. The value of effective stiffness 'K_{eff}' for an individual isolator is evaluated on the basis of the column load 'W' using Eq. (14.20).

$$K_{eff} = \frac{4\pi^2 W}{T_M^2 g} \tag{14.20}$$

With the site parameters like the zone factor, the soil type, and the relations for response spectrum parameters, the spectral acceleration for MCE level can be evaluated. For Indian condition, this is given by Eq. (14.21). S_a (spectral acceleration) and β = I/2Rg(I—importance factor, R—response reduction factor, g—acceleration due to gravity) can be obtained from IS-1893.

$$\text{Accel}_{MCE} = ZS_a\beta \tag{14.21}$$

From the spectral acceleration, the spectral displacement at MCE level is evaluated by using Eq. (14.22). As per Indian code, the displacement for the design basis earthquake (DBE) level is half of that for MCE level (as given by Eq. (14.23)).

$$D_M = \frac{T_M^2}{4\pi^2}\text{Accel}_{MCE} \tag{14.22}$$

$$D_D = \frac{D_M}{2} \tag{14.23}$$

The calculations of Step 1 are tabulated in Table 14.1.

Step 2: Evaluating Characteristic Constants for the Base Isolation System

The bilinear characteristics of the individual isolators are evaluated by using the iterative procedure illustrated as below. In the iterative procedure, the characteristic strength 'Q', the initial or elastic stiffness 'K_1', and the displacement at yield 'D_y' are evaluated by using Eqs. (14.24), (14.25), (14.26), and (14.27).

$$Q = \frac{\pi K_{eff} D_M^2 \xi_M}{2(D_M - D_y)} \tag{14.24}$$

An initial estimate of 'Q' is obtained by neglecting 'D_y' in Eq. (14.24).

$$K_2 = K_{eff} - \frac{Q}{D_M} \tag{14.25}$$

$$D_y = \frac{Q}{K_1 - K_2} \tag{14.26}$$

$$K_1 = nK_2 \tag{14.27}$$

In this step, the converged (1% convergence) values for 'Q', 'K_1', 'K_2', and 'D_y' are obtained. The calculations of Step 2 are tabulated in Table 14.2.

Step 3: Evaluating DBE Stiffness and Damping

The constants that determine the isolator characteristics are obtained in Step 2. The effective stiffness and the effective damping ratio of the isolator are displacement dependent. The values of effective isolator stiffness 'K_D' and the effective damping ratio 'ξ_D' for any displacement 'D_s' are obtained by using Eqs. (14.28) and (14.29).

$$K_D = K_2 + \frac{Q}{D_s} \tag{14.28}$$

Table 14.2 Design calculations—characteristic constants for base isolation system

Quantity evaluated	Value obtained	Units
Ratio initial to post-yield stiffness 'n'	10	–
Characteristic strength 'Q' (three iterations)	815.80	kN
Elastic stiffness 'K_1' (three iterations)	180,208.6	kN/m
Post-yield stiffness 'K_2' (three iterations)	18,020.86	kN/m
Displacement at yield 'D_y' (three iterations)	0.005	m

Table 14.3 Design calculations—DBE stiffness and damping

Quantity evaluated	Value obtained	Units
DBE stiffness 'K_D' Eq. (14.28)	29,675	kN/m
Damping ratio at DBE 'ξ_D' Eq. (14.29)	0.2322	–

Table 14.4 Values of *RI* as obtained from Table-A-16-E of UBC-97 [11]

Basic structural system	Lateral force resisting system	RI
Bearing wall system	1: Shear walls A: Masonry B: Concrete	2.0 2.0
Moment-resisting frame system	1: Special moment-resisting frame (SMRF) A: Steel B: Concrete 2: Ordinary moment-resistant frame (OMRF) A: Steel B: Concrete	2.0 2.0 2.0 2.0
Dual systems	Shear wall with OMRF Shear wall with SMRF	2.0 2.0
Cantilevered column building systems	Cantilevered column elements	1.4
Shear wall-frame interaction system	Concrete	2.0

Table 14.5 Design calculations—base shear

Quantity evaluated	Value obtained	Units
DBE base shear–fixed base	2700	kN
DBE foundation base shear Eq. (14.30)	2077.25	kN
DBE base shear-isolated Eq. (14.31)	1038.6	kN
MCE base shear–fixed base	5400	kN
MCE base shear–isolated	1670	kN
MCE foundation base shear	3340	kN

$$\xi_D = \frac{2Q(D_s - D_y)}{\pi K_D D^2} \tag{14.29}$$

The values of 'K_D' and 'ξ_D' are calculated for the design displacement obtained using Eq. (14.23). In case of computer-aided analysis using commercial software packages, the above design values are specified to model the isolators. The calculations of Step 3 are tabulated in Table 14.3.

Step 4: Determination of Base Shear

The shear forces acting at the base are the base isolation system stiffness times the displacement. This full horizontal force has to be transferred to the subsoil through the foundation. Thus, the design forces for the foundation can be given by Eq. (14.30) [12].

$$V_f = K_D D_D \tag{14.30}$$

For the superstructure, a reduction in the base shear is permitted. This reduction is allowed because of over strength of the member elements. The values of this reduction factor 'RI' as given in UBC-97 are tabulated in Table 14.4. The base shear for the superstructure is given by Eq. (14.31)

$$V_s = \frac{K_D D_D}{\mathrm{RI}} \tag{14.31}$$

The values of horizontal base shear obtained using Eqs. (14.30) and (14.31) are for the DBE conditions. The values for the MCE conditions may be obtained by using the MCE values 'K_M' and 'D_M' in place of 'K_D' and 'D_D', respectively. These values should be evaluated to assess the performance of the structure at the DBE and the MCE level of displacements (Table 14.5). In the design example, the base shear is evaluated using the seismic coefficient method elaborated in IS-1893.

14.9 Selection Criteria and Design of Base Isolators

Selection of a particular type or a combination of different types of isolators is based on the design requirement (i.e., the bilinear characteristics of isolation system as evaluated in Step 2). The various isolators should be selected so that their load–deformation curves superpose to give effective bilinear characteristics of the isolation system matching with what is demanded by the design. Different combinations of isolators units are also possible.

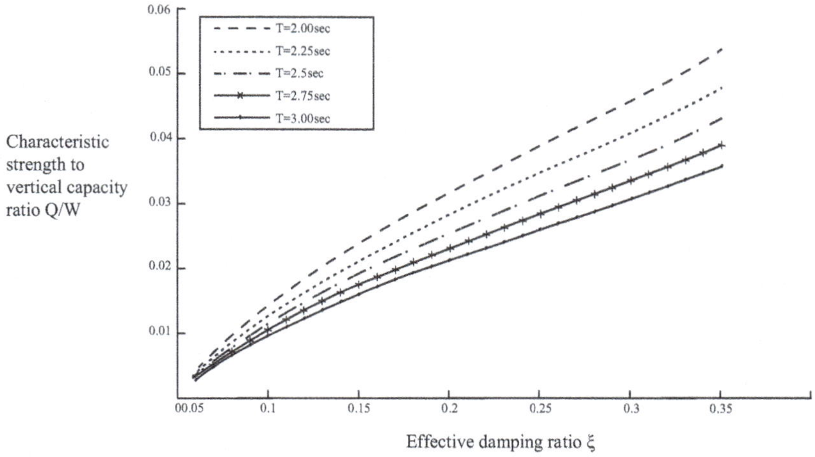

Fig. 14.19 Variation of characteristic strength against effective damping ratio for hard soil (Z = 0.36, n = 10)

Step 5: Designing the Isolators Comprising the Base Isolation System

In the present design example, a single lead rubber bearing (LRB) is designed under each of the 36 column locations. The natural rubber with G = 0.5 MPa has been employed for the LRB. The post-yield horizontal stiffness required is 18,020.86 kN/m. This stiffness corresponds to the isolator stiffness contributed by the rubber layers [i.e., 'K_2' for lead rubber bearings given by Eq. (14.17)]. Since identical isolator units are used for all the columns, equal stiffness is contributed by each of the 36 isolators. Hence, the rubber stiffness for individual isolator works out as 503.35 kN. The average vertical load on the isolator units is 833 kN. The individual isolators shall be designed to sustain a minimum of 1000kN vertical load. The allowable compressive strain is calculated by using Eq. (14.32), and the shape factor S is evaluated on the basis of required vertical frequency using Eq. (14.33).

$$\sigma_{\text{all}} = \frac{\min([6.895\text{ MPa}], [0.6895G_{0.5}S])}{\text{SF}} \tag{14.32}$$

$$S = \sqrt{\frac{(f_V/f_H)^2 - 3}{6}} \tag{14.33}$$

In the present case, horizontal frequency 'f_H' is 0.44 Hz. The vertical frequency is assumed to be 15 Hz to ensure high vertical stiffness and thereby low compressive strains. The shape factor is calculated as 13.9. The allowable compressive stress is then evaluated as 3.15 MPa with a safety factor as 1.5. Thus for square-shaped bearings, the side length required to satisfy the allowable stress criteria for 1000 kN vertical load is evaluated as 563.4 mm by using Eq. (14.34). Hence, square bearings of side length 570 mm are selected.

$$\phi = \sqrt{\frac{W_{\max}}{\sigma_{\text{all}}}} \tag{14.34}$$

The characteristic strength is required to be 815.8kN. Since this will be contributed by the lead plug, the total lead area needed is worked out by using Eq. (14.15) as 81,580 mm^2. Since 36 LRBs are provided, the required diameter of lead plug in each of the LRB is 53.7 mm (provide lead plug of diameter 55 mm). The total rubber thickness required is evaluated as 325 mm by using Eq. (14.1). For a shape factor of 13.9, the thickness of individual rubber layer shall be 10.5 mm. Thus, 31 rubber layers and 30 steel shims shall be needed. Total height of the bearing is 435 mm including the top and the bottom plates. A cross section of the isolators is shown in Fig. 14.20.

Step 6: Evaluating the Final Properties of the Base-Isolated Building

In the design values, the geometrical dimensions of the isolators have been rounded off. In this process, the value of the constants 'Q', 'K_1', and 'K_2' changes slightly. This has some effect on the effective stiffness and damping. The revised values of the bilinear characteristic and the other related values accordingly calculated have been tabulated in Table 14.6. From the results, it is clear that there is a reduction in base shear of about 60.9 and 68.8% in superstructure when compared with the fixed base design for DBE and MCE, respectively. This considerable reduction allows for slender sections in superstructure saving a lot of material. Moreover, the cumbersome ductile detailing need not be

Table 14.6 Design calculations—final isolator properties

Quantity evaluated	Value obtained	Units	
Actual lead plug diameter	0.055	m	
Actual 'Σt'	0.325	m	
Actual lead plug area 'A_p'	0.0024	m^2	
Actual rubber area 'A_r'	0.3249	m^2	
Actual value of 'Q' (single Pb isolator)	24.0	kN	
Actual value for 'K_R' (single isolator)	499.8	kN/m	
Actual bilinear characteristics and properties for the base isolation system			
Property	Early	Actual	Units
Characteristic strength 'Q'	815.8	864.0	kN
Post-yield stiffness 'K_2'	18,020.86	17,964	kN/m
Post-yield stiffness ratio 'n'	10	10	–
Yield displacement 'D_y'	0.0050	0.0053	m
DBE stiffness 'K_D' Eq. (14.28)	29,675	30,306.8	kN/m
Damping ratio at DBE 'ξ_D' Eq. (14.29)	0.2322	0.2382	–
MCE Stiffness 'K_M' Eq. (14.28)	23,848	24,135.4	kN/m
Damping ratio at MCE 'ξ_M' Eq. (14.29)	0.150	0.155	–
Time period at MCE	2.25	2.24	s
DBE base shear–fixed base	2700	2700	kN
DBE base shear–isolated	1038.6	1056.4	kN
DBE foundation base shear Eq. (14.30)	2077.25	2112.8	kN
MCE base shear–fixed base	5400	5400	kN
MCE base shear–isolated	1670	1685.1	kN
MCE foundation base shear	3340	3370.3	kN

followed since the superstructure will be subjected to very low interstorey drifts keeping it in elastic range.

14.10 Base Isolation Using Elastomeric Bearings

To study the behavior of isolated building under actual earthquake scenario, two 3-storey RCC buildings with similar construction were built at IIT Guwahati, India [37]. One of the buildings is provided with the conventional foundation and other with base isolation system. The normal building and base-isolated buildings are shown in Fig. 14.3.

14.10.1 Case Study on High Damping Laminated Rubber Bearing

The purpose of lead in case of lead plug bearings is to provide hysteretic damping characteristics to the isolation system, along with the high initial stiffness. Another alternative to LRB is the use of high damping laminated rubber bearings (HDRB). Building supported over HDRB of vertical load capacity 120kN is presented in this section. The design process involved has been illustrated. It was desired to design HDRB isolator units of 120 kN capacity having a target time period of 2 s and a target damping ratio of 10% for an earthquake of peak ground acceleration 0.00305 g (Z = 0.00305). Accordingly, the relevant design steps for HDRB are provided.

Step 1: Assumption of Appropriate Target Dynamic Characteristics

For the present problem, the target time period and damping are considered same as discussed earlier. The calculations of effective stiffness needed and the displacement expected for the target earthquake are tabulated in Table 14.7.

Step 2: Evaluating Characteristic Constants for the HDRB

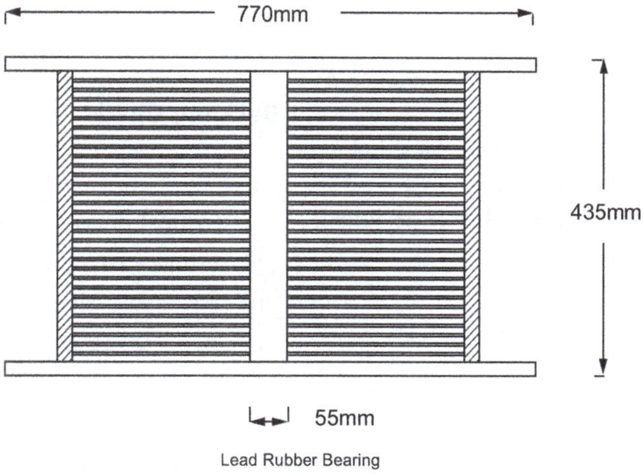

Fig. 14.20 Cross section for the isolators

The horizontal stiffness 'K_{eff}' in case of HDRB shall be contributed by the rubber layers. Hence, the required horizontal stiffness of the bearings is 120 kN/m.

Step 3: Designing the Isolators

i. The desired vertical frequency is 10 Hz, from which the shape factor is evaluated by using Eq. (14.4) as 8.13 (assume $S = 9$).
ii. Rubber material with G = 0.57 MPa has been selected for the design.
iii. The allowable compressive stress is 3.53 MPa with a safety factor of 1.
iv. A circular cross section is selected for the bearing.
v. The diameter for safety in compression for 120 kN vertical load capacity works out to be 208 mm.
vi. A bearing of diameter 208 mm is selected.
vii. Based on the shape factor, the thickness of individual rubber layer using Eq. (14.4) is found to be 5.8 mm.

Table 14.7 Design calculations—target earthquake

Quantity evaluated	Value obtained	Units
Seismic weight 'W'	120	kN
Target time period 'T_M' (assumed)	2.0	s
Target damping ratio 'ξ_M'	0.10	–
Effective stiffness 'K_{eff}' Eq. (14.20)	120	kN/m
Target displacement 'D_M' Eq. (14.22)	0.231	m

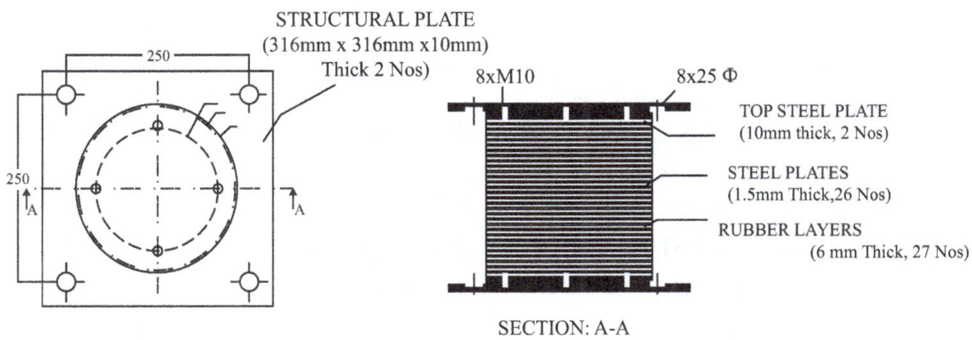

Fig. 14.21 Details of the high damping rubber bearing

The thickness of individual rubber layer is selected as 6 mm.

viii. The total thickness of rubber required is evaluated by using Eq. (14.1) as 164 mm.

ix. 27 rubber layers of 6 mm thickness are selected along with 26 steel shims of 1.5 mm thickness.

x. Fig. 14.21 shows the details of the HDRB designed.

Check on the Vertical Frequency

i. Young's modulus of rubber $E_C = \frac{6G_{0.2}S^2K}{K+8G_{0.2}S^2} = 2.33 \times 10^8\ \mathrm{N/m^2}$

ii. Vertical stiffness of isolator, $K_v = \frac{AE_C}{t_r} = 4.89 \times 10^7\ \mathrm{N/m}$

iii. Vertical frequency $f_v = \sqrt{\frac{K_v}{m}} = 10.15\ Hz$

iv. Vertical Frequency = 10.15 Hz, which is >10. 0 Hz. Hence, the requirement is met.

Check on the Horizontal Frequency

i. New horizontal stiffness $(K_h) = \frac{GA}{t_r} = \frac{5.6898 \times 10^5 \times 0.034}{0.162} = 1.194 \times 10^5\mathrm{N/m}$

ii. Now horizontal frequency $(f_h) = \frac{1}{2\pi}\sqrt{\frac{K_h}{m}} = \frac{1}{2\pi}\sqrt{\frac{1.194 \times 10^5}{12 \times 10^3}} = 0.502\ \mathrm{Hz} \approx \sim 0.50\ \mathrm{Hz}$

iii. Horizontal frequency (f_h) is = 0.5 Hz, and the requirement is met.

Check on the Compressive Strain

i. Maximum vertical displacement in compression: $X_c = \frac{P_C}{K_v} = 0.0023\ \mathrm{m} \approx 2.3\ \mathrm{mm}$

ii.
Percentage compression $= \frac{X_c}{t_r} \times 100 = \frac{2.3}{162} \times 100 = 1.42\%$

iii. Allowable percentage compression is 20% (requirement is met)

iv. Compressive stress $= \frac{P_C}{A} = \frac{12 \times 10^3 \times 9.81}{0.034} = 3.46\,\mathrm{MPa}$ (which is within allowable limit)

v. If no spacers are available, then shape factor $(S) = \frac{\Phi}{4 \times t_r} = \frac{208}{4 \times 162} = 0.321$

vi. Compression modulus $E_C = \frac{6G_{0.2}S^2K}{K+8G_{0.2}S^2} = 3.515 \times 10^6\ \mathrm{N/m^2}$

vii. Compressive stiffness $K_1C = \frac{AE_C}{t_r} = 7.38 \times 10^5\ \mathrm{N/m}$

viii. Ratio $= \frac{K_v}{K_1V} = 66.26$

ix. Hence, sandwiched spring is 66.26 times stiffer than the whole spring made of rubber.

Buckling Check

i. Critical pressure $(P_{crit}) = \frac{\pi GS\lambda}{\sqrt{6}}$

ii. Primary shape factor (S) = 9

iii. Secondary shape factor, $\lambda = \frac{\Phi}{t_r} = \frac{208}{162} = 1.284$

iv. $P_{crit} = \frac{\pi \times 5.6898 \times 10^5 \times 9 \times 1.284}{9.81 \times \sqrt{6}} = 8.596 \times 10^5\mathrm{kg/m^2}$

v. Critical load $(P_{crit} \times A) = 8.596 \times 10^5 \times 0.034 = 29.226 \times 10^3\ \mathrm{kg} = 29.226\ \mathrm{T}$

vi. Allowable bucking load is very high to the actual axial load applied on the bearing, so bearing is safe for buckling load.

14.10.2 Observed Behavior of the HDRB

A 12-ton capacity laminated rubber bearings (LRBs) with above design have been indigenously manufactured. In order to study the effect of base isolation and performance of indigenously developed base isolators, a three-storey framed building as shown in Fig. 14.22 has been tested on shake table at with and without base isolation. The details of the building tested are as follows:

i. Both beam and column sections are 150 mm × 200 mm.

ii. The slab is 50 mm thick.

iii. Base slab is 4000 mm × 4000 mm and 300 mm thick.

iv. Mass of the building without base slab = 11.07 T.

v. Mass of the base slab = 12 T.

vi. Total mass of the building with base slab = 23.07 T.

vii. Experimental building shown in Fig. 14.22 has been analyzed and frequencies in first modes in two lateral directions were found to be 5.09 and 6.00 Hz with 85% mass participation.

viii. The calculated frequency of isolated building using four numbers of 12-ton capacity base isolators is 0.724 Hz. The frequency of isolated building is higher than 0.5 Hz because the mass of the building is only 23.07 ton, whereas the combined capacity of all four base isolators is 12 ton × 4 ton = 48 tons.

From Fig. 14.22, it can be observed that the recorded acceleration on base-isolated building is only 0.085 and 0.086 g at first floor and roof of the building. Whereas, for same seismic input excitation, the recorded accelerations are 0.107 and 0.157 g at first and roof of the building without

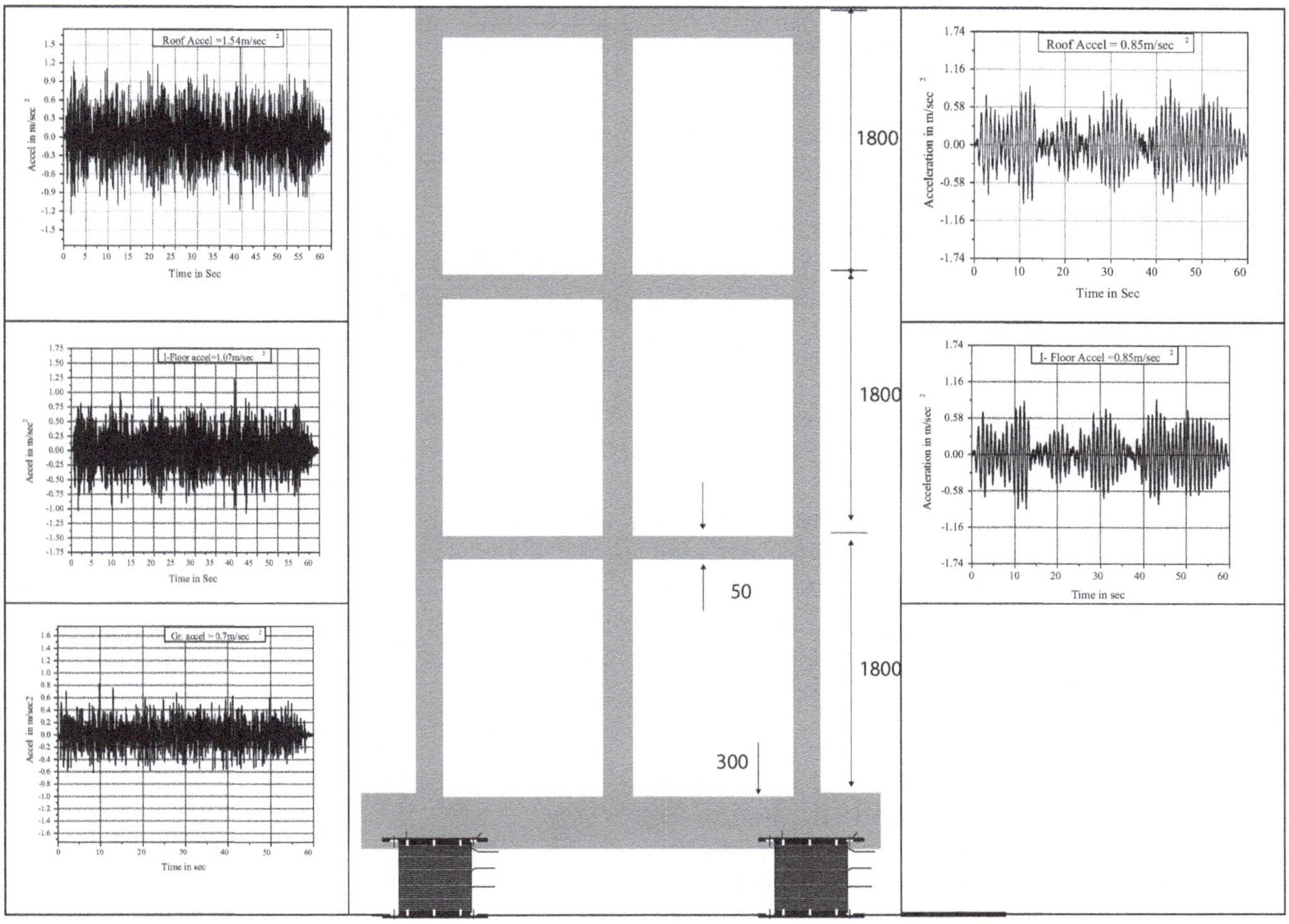

Fig. 14.22 Shake table test data of building without and with isolators

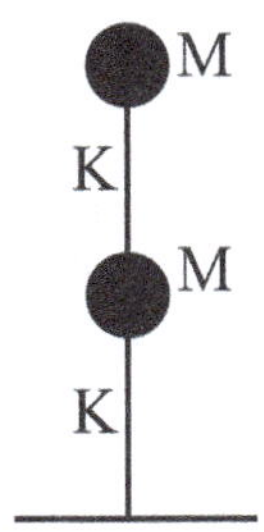

Fig. 14.23 Multi degree freedom system

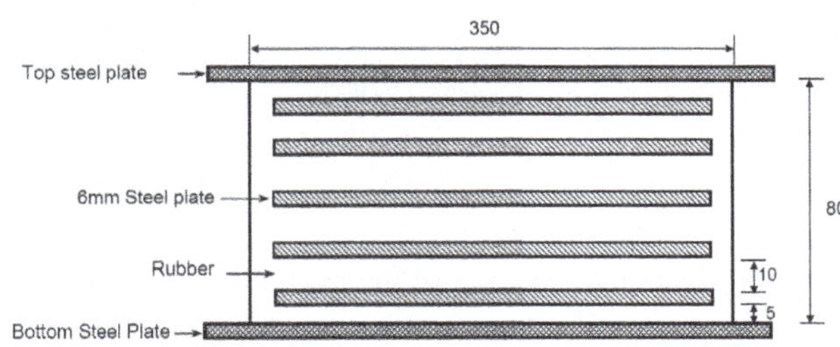

Fig. 14.24 Elastomeric bearing

base isolation. The recorded acceleration is constant along the height of the isolated building, whereas there is amplification in seismic response along the height of the building without isolation. The mass of the building (23.07 T) is lesser than the capacity of the isolators (48 T); therefore, frequency of the isolated system is higher than 0.5 Hz. More effective base isolation effects can be achieved out of same base isolation system by increasing the mass of the building to 48 T, which eventually shifts the natural frequency of isolated building to 0.5 Hz. Even at higher frequency of base-isolated system, the response of base-isolated building is lesser than the unisolated building, but the effectiveness is less than the desired due to high frequency.

Exercise Problems

1. Briefly explain the principle of base isolation system and its advantages in the design of seismic resistant structures.
2. Briefly explain the differences in the behavior of the laminated rubber bearing and lead-laminated rubber bearing.

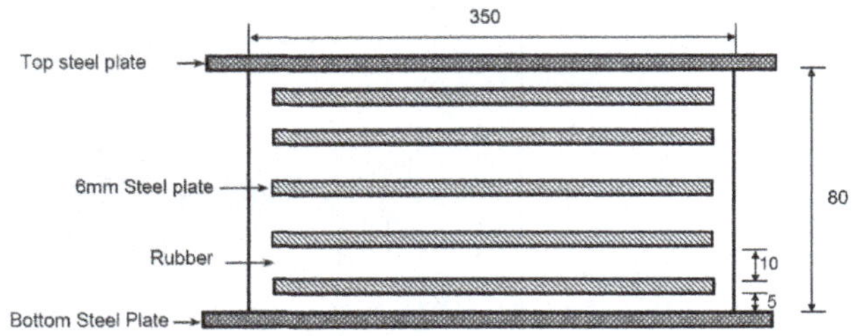

Fig. 14.25 Circular laminated rubber bearing

3. Calculate the base shear and bending moment for the given multi-degree freedom system as shown in Fig. 14.23 with stiffness (K) and lumped mass (M) of 6 × 105 kN/m and 150 T, respectively, for both fixed support and isolated support conditions.
4. Determine the preliminary bilinear characteristics of lead-laminated rubber like initial stiffness, post-yield stiffness, yield displacement, and ultimate displacement for the given data:
seismic weight = 1500 kN; desired time period = 2 s; spectral acceleration coefficient S_a/g = 0.105; damping ratio 'ξ' = 0.1.
5. Determine the allowable compressive stress in the laminated rubber bearing for the desired horizontal and vertical frequencies of 0.5 Hz and 15 Hz, respectively, for the given shear modulus of rubber G0.5 = 0.5 MPa.
6. Determine the horizontal and vertical stiffness of square (350 mm × 350 mm) and circular (diameter = 350 mm) bearing as shown in the figure. The height of the bearing between top and bottom steel plates are 80 mm. The five number of 6-mm-thick steel plates are provided in the bearings. Take the shear modulus of the rubber as 1.06 N/mm^2 (Fig. 14.24). Also determine the horizontal and vertical frequencies for both the circular and square bearings, for the given vertical load of 2500 kN.
7. Determine the diameter of the lead plug required for the lead-laminated rubber bearing for the given data:
seismic weight = 2000 kN; desired time period = 2 s; spectral acceleration coefficient S_a/g = 0.105; damping ratio 'ξ' = 0.125.
8. Briefly explain the design criteria and design checks for the laminated rubber bearings.
9. Calculate the critical buckling load of the circular laminated rubber bearing as shown in Fig. 14.25. Take the shear modulus of the rubber as 0.6 N/mm^2. Also determine the allowable vertical load.
10. How is isolation achieved? Define isolation efficiency?
11. Briefly explain how to control the vibrations (Hint: damping, isolation)

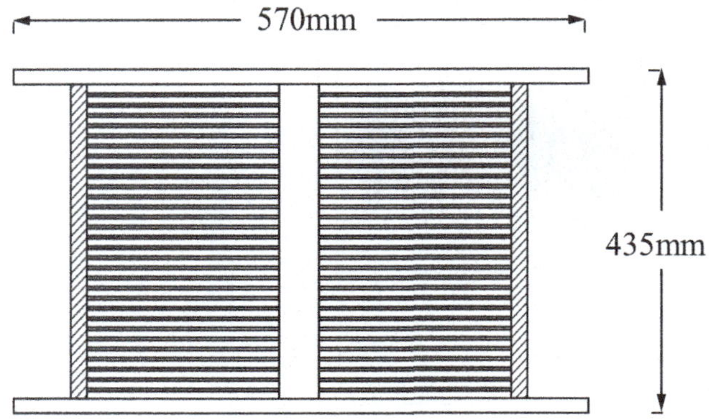

Fig. 14.26 Lead plug laminated rubber bearing

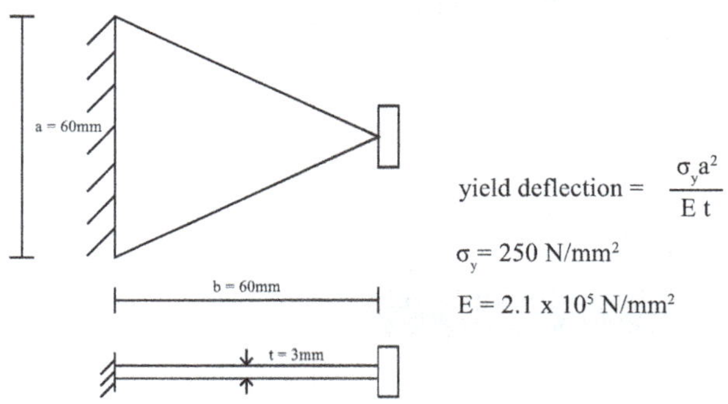

Fig. 14.27 Damper

12. What are transmissibility and isolation? With the help of graph, explain these terms
13. What is the vibration absorber and how is different from vibration base isolation?
14. What are active and passive response control devices? Why it is usually recommended to go for passive devices in nuclear industry? Name the passive device used for response control of machinery and structures
15. Briefly explain the principle of base isolation system and the types of isolators.
16. Design the lead plug for the circular laminated rubber bearing (with square cross section of 570 mm × 570 mm) shown in the below figure for the given data.
Seismic weight = 1000 kN; wind load = 20 kN; shear modulus of the rubber = 0.5 N/mm^2; shear modulus of lead = 4000 N/mm^2; yield strength of lead = 10 N/mm^2; number of rubber layers with 10.5 mm thickness = 31; number of shim layers with 3 mm thickness = 30 (Fig. 14.26).

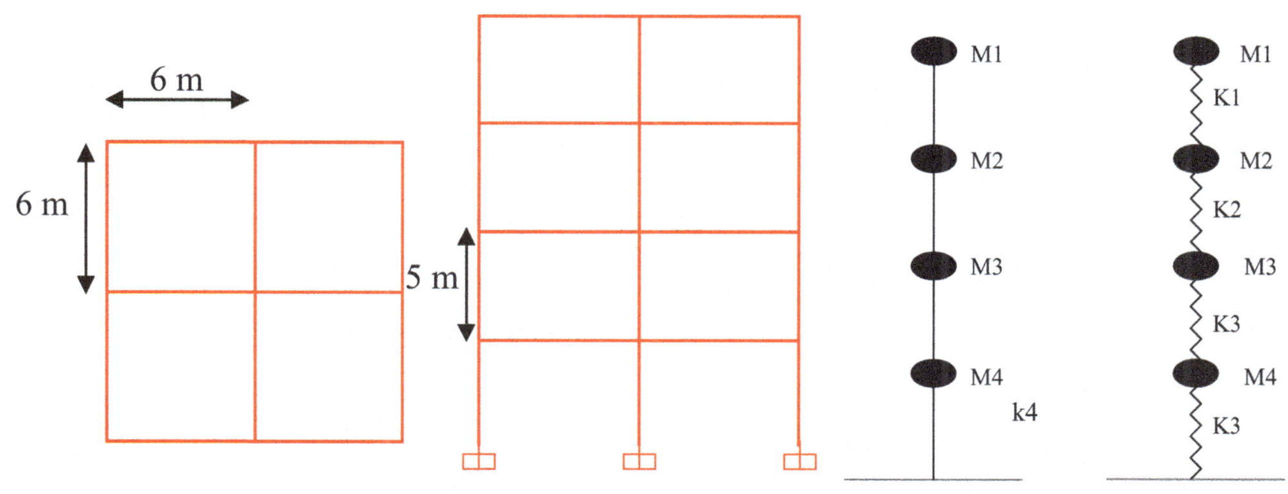

Fig. 14.28 Four-storey frame structure

17. Considering yielding damper element shown in Fig. 14.27, estimate the force required for complete yielding of the section. Assume elastic-perfectly plastic condition and draw the hysteresis for 10 mm deflection.

18. What are the methods of controlling the dynamic response of structures, equipment, and piping systems? Explain briefly about passive control.

19. A single-degree-of-freedom system with stiffness of 100 kN/m and mass of 100 kg has an initial damping 5%. Spectral acceleration (S_a) for 5% damping is 0.05 g. It is given that Sa is inversely proportional to damping. A friction damper with force of 250 N capacity is introduced. Calculate the additional damping due to damper using iterative procedure.

Appendix 1: A Case Study on Comparison of 3D RCC Frame of Fixed Base System with Isolated Base System

A 3D RCC frame of two bays in both the directions and four storeys is considered for the comparison as shown in Fig. 14.28. The frame is idealized into multi-degree-of-freedom system by considering mass and spring system as shown in Fig. 2.

Specifications of RCC Frame
Column dimensions: 0.8 m × 0.8 m
Beam dimension: 0.35 m × 0.6 m
Slab thickness: 0.25 m
Brick wall thickness: 0.23 m
Plan dimensions: 12 m × 12 m
Height of the column: 5 m
Grade of concrete: M35
Grade of steel: Fe500

Evaluation of mass lumped at various levels

i. **Load on floors 1, 2, and 3**
Dead load of RCC slab and beam (D_f) = 42850 kg
Live load (L_f) = 36222.22 kg
Equipment Load (E_f) = 72444.44 kg
Brick wall load (B_f) = 99360 kg
ii. **Load on Roof**
Dead load (D_r) = 24850 kg
Live load (L_r) = 7200 kg
Brick wall load (B_r) = 49680 kg
iii. **Columns' load**
For floors (C_f) = 72,000 kg
For roof (C_r) = 36,000 kg
iv. **Masses on floors and roof**
Mass on floors 1, 2, and 3 (M_2, M_3, M_4) = (D_f + L_f + E_f + B_f + C_f) = 3, 22,876 kg
Mass on roof (M_1) = (D_r + L_r + B_r + C_r) = 1, 17,750 kg

Response Spectrum Method

Case 1: Fixed base model

i. Young's modulus of concrete $E = 5000 \times \sqrt{35} = 2.958 \times 10^{10}$ N/m^2
ii. Moment of inertia of column $I = 0.0341$ m^4
iii. Column stiffness = $12\, EI/L^3 = 7.28 \times 10^7$ N/m

iv. Global Stiffness matrix =

$$[K] = 9 \times C \times \begin{bmatrix} 1 & -1 & 0 & 0 \\ -1 & 2 & -1 & 0 \\ 0 & -1 & 2 & -1 \\ 0 & 0 & -1 & 2 \end{bmatrix}$$

v. Global Mass matrix = $[M]$ =

$$\begin{bmatrix} m1 & & & \\ & m2 & & \\ & & m3 & \\ & & & m4 \end{bmatrix} = \begin{bmatrix} 117750 & & & \\ & 322876 & & \\ & & 322876 & \\ & & & 322876 \end{bmatrix}$$

vi. The natural frequencies and mode shapes are estimated using one of the methods as explained in Chap. 4 and are given below:

$f_1 = 4.521$ Hz, $f2 = 12.875$ Hz, $f3 = 19.27$ Hz and $f_4 = 22.73$ Hz

$$\varphi_1 = \begin{Bmatrix} -0.00273 \\ -0.00252 \\ -0.00193 \\ -0.00105 \end{Bmatrix} \quad \varphi_2 = \begin{Bmatrix} 0.002729 \\ 0.001072 \\ -0.00192 \\ -0.00253 \end{Bmatrix}$$

$$\varphi_3 = \begin{Bmatrix} 0.002734 \\ -0.00098 \\ -0.00197 \\ 0.002519 \end{Bmatrix} \quad \varphi_4 = \begin{Bmatrix} 0.002823 \\ -0.00252 \\ 0.001888 \\ -0.00101 \end{Bmatrix}$$

$\Gamma_1 = -1073.64$, $\Gamma2 = -319.443$, $\Gamma3 = 142.27$ and $\Gamma4 = -41$

First-Mode Acceleration Response

Node	Mass participation factor	Mode shape	Spectral value	Modal response (acceleration in 'g')
1	−1073.64	−0.00273	0.618	0.777
2	−1073.64	−0.00252	0.618	0.719
3	−1073.64	−0.00193	0.618	0.551
4	−1073.64	−0.00105	0.618	0.298

Second-Mode Acceleration Response

Node	Mass participation factor	Mode shape	Spectral value	Modal response (acceleration in 'g')
1	−319.443	0.002729	0.444	−0.166
2	−319.443	0.001072	0.444	−0.065
3	−319.443	−0.00192	0.444	0.116
4	−319.443	−0.00253	0.444	0.154

Third-Mode Acceleration Response

Node	Mass participation factor	Mode shape	Spectral value	Modal response (acceleration in 'g')
1	142.27	0.002734	0.371	0.062
2	142.27	−0.00098	0.371785	−0.022
3	142.27	−0.00197	0.371785	−0.044
4	142.27	0.002519	0.371785	0.057

Fourth-Mode Acceleration Response

Node	Mass participation factor	Mode shape	Spectral value	Modal response (acceleration in 'g')
1	−41	0.002823	0.354	−0.017
2	−41	−0.00252	0.354	0.015
3	−41	0.001888	0.354	−0.012
4	−41	−0.00101	0.354	0.006

vii. Total response using SRSS combination

Node	Total acceleration (g)	Lateral force (kN)
1	0.797	600.93
2	0.722	1112.11
3	0.564	870.23
4	0.340	524.63

viii. Base shear = 31.07.9 kN

ix. Bending moment = 400,40 kN m

Case 2: Isolated base model

Column stiffness (a) = 6.347×10^7 N/m

Isolator stiffness (b) = 1.0717×10^7 N/m

$$\text{Global Stiffness matrix} = [\mathrm{K}] = \begin{bmatrix} a & -a & 0 & 0 \\ -a & 2a & -a & 0 \\ 0 & -a & 2a & -a \\ 0 & 0 & -a & a+b \end{bmatrix}$$

$$\text{Global Mass matrix} = [\mathrm{M}] = \begin{bmatrix} m1 & & & \\ & m2 & & \\ & & m3 & \\ & & & m4 \end{bmatrix} = \begin{bmatrix} 95812.5 & & & \\ & 279{,}001 & & \\ & & 279{,}001 & \\ & & & 279{,}001 \end{bmatrix}$$

The natural frequencies and mode shapes are estimated using one of the methods as explained in Chap. 4 and are given below:

$f_1 = 0.498\,\mathrm{Hz}$, $f2 = 4.908\,\mathrm{Hz}$, $f3 = 8.779\,\mathrm{Hz}$ and $f_4 = 10.936\,\mathrm{Hz}$

$$\varphi_1 = \begin{Bmatrix} 0.002397 \\ 0.002388 \\ 0.00236 \\ 0.002313 \end{Bmatrix} \varphi_2 = \begin{Bmatrix} -0.00332 \\ -0.00207 \\ 0.000774 \\ 0.003025 \end{Bmatrix} \varphi_3 = \begin{Bmatrix} 0.003341 \\ -0.00067 \\ -0.00302 \\ 0.002096 \end{Bmatrix} \varphi_4 = \begin{Bmatrix} 0.003467 \\ -0.00299 \\ 0.002024 \\ -0.00072 \end{Bmatrix}$$

$\Gamma_1 = 1038.114$, $\Gamma_2 = 13.98$, $\Gamma_3 = 3.0282$ and $\Gamma_4 = -0.6713$

First-Mode Acceleration Response

Node	Mass participation factor	Mode shape	Spectral value	Modal response (acceleration in 'g')
1	1038.114	0.0023973	0.164	0.166
2	1038.114	0.002388	0.164	0.166
3	1038.114	0.0023597	0.164	0.164
4	1038.114	0.0023127	0.164	0.161

-Second-Mode Acceleration Response

Node	Mass participation factor	Mode shape	Spectral value	Modal response (acceleration in 'g')
1	13.98	−0.0033209	0.62050324	−0.012
2	13.98	−0.0020748	0.62050324	−0.007
3	13.98	0.00077398	0.62050324	0.003
4	13.98	0.0030249	0.62050324	0.011

(continued)

Third-Mode Acceleration Response

Node	Mass participation factor	Mode shape	Spectral value	Modal response (acceleration in 'g')
1	3.0282	0.0033405	0.5393565	0.002
2	3.0282	−0.00067014	0.5393565	−0.0004
3	3.0282	−0.0030246	0.5393565	−0.002
4	3.0282	0.0020959	0.5393565	0.0014

Fourth-Mode Acceleration Response

Node	Mass participation factor	Mode shape	Spectral value	Modal response (acceleration in 'g')
1	−0.6713	0.0034674	0.486780378	−0.0005
2	−0.6713	−0.0029925	0.486780378	0.0004
3	−0.6713	0.0020238	0.486780378	−0.00026
4	−0.6713	−0.00072105	0.486780378	9.61E−05

Total response using SRSS combination

Node	Total acceleration (g)	Lateral force (kN)
1	0.167	112.4
2	0.166	229.54
3	0.164	226.66
4	0.161	222.58

Base shear = 791.95 kN
Bending moment = 9070.9 kN m

Column design

Response spectrum analysis and design of columns were carried out in an iterative process by optimizing the column dimensions, starting with 0.6 × 0.6 m (width × depth) for the fixed base system and 0.4 m × 0.4 m for isolated base system. The columns' dimensions got converged at 0.8 m 0.8 m and 0.55 m × 0.55 m for fixed base system and isolated base system, respectively.

Fixed base:

Column size: 0.8 m × 0.8 m
Reinforcement percentage: 2.903%
Interaction ratio: 0.96

Isolated base:

Column size: 0.55 m × 0.55 m
Reinforcement percentage: 2.573%
Interaction ratio: 0.91

Comparison between fixed base frame and isolated base frame

S. No	Design parameter	Percentage reduction (%)
1	Base shear	75
2.	Bending moment	78.4
3.	Concrete quantity	53.8
4.	Steel quantity	59.5

Conclusion

The system with base isolators is more flexible when compared to fixed base system. As the system has become more flexible with base isolators, it is subjected to low spectral acceleration, which results in the overall reduction of forces experienced by the system. There is reduction of about 75 and 78.4% in the base shear and bending moment, respectively. As the forces are coming down, the design sections of the columns also got significantly reduced.

Isolator Design Procedure

For the isolated base frame, the columns are mounted on base isolators with an axial capacity of 1200 kN. (The average vertical reaction of all the nine columns individually is approximately 1200 kN.)

1. **Design calculations of Laminated Rubber Bearings (LRB)**

Step 1: Assumption of Appropriate Target Dynamic Characteristics

Seismic weight $W = 1100$ kN
Service horizontal load $Q = 0.03 \times W = 36$ kN
Time period 'T_M' (assumed) = 2 s
From site-specific spectra $S_a/g = 0.105$
Damping ratio 'ξ_M' = 0.1

Step 2: Evaluating Characteristic Constants for the Base Isolation System

Effective stiffness $K_{eff} = \frac{4\pi^2 W}{T_M^2 g} = 1108\,\text{kN/m}$

Displacement at SSE $D_{sse} = \frac{T_M^2}{4\pi^2} S_a = 104.366$ mm

Displacement at OBE $D_{obe} = \frac{D_{sse}}{2} = 52.1829$ mm

Evaluation of bilinear characteristics using the following equations (using iterative procedure till 1% of convergence obtained). In the iterative procedure, the characteristic strength 'Q', the initial or elastic stiffness 'K_1', post-yield stiffness 'K_2', and the displacement at yield 'D_y' are evaluated

$$D_y = \frac{Q}{K_1 - K_2} Q = \frac{\pi K_{eff} D_M^2 \xi_M}{2(D_M - D_y)}$$

$$K_2 = K_{eff} - \frac{Q}{D_M} K_1 = n K_2$$

Sl	D_y	Q (kN)	K_2 (kN/m)	K_1 (kN/m)	D_y (m)
1	0.00000	36.000	863.060	8630.597	0.00463
2	0.00463	20.724	1009.430	10094.297	0.00228
3	0.00228	20.246	1014.008	10140.077	0.00222
4	0.00222	20.234	1014.127	10141.266	0.00222
		20.234	1014.127	10141.266	0.00222

Design calculations—characteristic constants for base isolation system

Quantity evaluated	Value obtained	Units
Ratio initial to post-yield stiffness'n'	10	–
Characteristic strength'Q'(4 iterations)	20.234	kN
Elastic stiffness 'K_1' (4 iterations)	1141.266	kN/m
Post-yield stiffness 'K_2' (4 iterations)	1014.127	kN/m
Displacement at yield 'D_y' (4 iterations)	0.00222	m

Step 3: Evaluating OBE Stiffness and Damping

OBE stiffness $K_{sse} = K_2 + \frac{Q}{D_{sse}} = 1401.87\,\text{KN/m}$

OBE damping ratio $\xi_D = \frac{2Q(D_{obe} - D_y)}{\pi K_{obe} D_{obe}^2} = 16.86\%$

OBE base shear $V_f = K_{obe}\, D_{obe}$

OBE design base shear $V_s = V_f/\text{RI}$

For site-specific spectra $RI = 1$ as per Table-A-16-E of UBC-97

So $V_s = 80$ KN

Step 4: Evaluating SSE Stiffness and Damping

SSE stiffness $K_{sse} = K_2 + \frac{Q}{D_{sse}} = 1108\,\text{kN/m}$

SSE damping ratio $\xi_D = \frac{2Q(D_{sse}-D_y)}{\pi K_{sse} D_{sse}^2} = 10.06\%$

SSE base shear $V_f = K_{sse}\ D_{sse}$

SSE design base shear $V_s = V_f/RI$

For site-specific spectra $RI = 1$ as per Table-A-16-E of UBC-97

So $V_s = 130$ kN.

A.1.2 Selection Criteria and Design of Base Isolators

Type of isolator used = Lead rubber bearing
Shear modulus of rubber at higher strain = 0.5 MPa
Required post-yield horizontal stiffness $K_2 = 1014.12$ KN/m
Minimum vertical load on isolator = 1100 kN
Required vertical frequency f_V = 15 Hz
Required horizontal frequency f_H = 0.5 Hz

Shape factor $= S = \sqrt{\frac{(f_V/f_H)^2 - 3}{6}} = 12.227$

Safety factor SF = 1.5

Allowable compressive stress $= \sigma_{all} = \frac{\min([6.895\ \text{MPa}],[0.6895 G_{0.5} S])}{\text{SF}} = 2.81\,\text{MPa}$

Assume shape of the bearing—circular

Bearing area required $A = \frac{W_{max}}{\sigma_{all}} = 427{,}019.4275\,\text{mm}^2$

Diameter of circular bearing provide = ϕ = 750 mm

Required characteristic strength Q = 20.234 KN (This will be contributed by the lead plug)

Shear yield strength of lead σ_{yL} = 10 MPa

Lead area required = $A_L = Q/\sigma_{yL}$ = 2023.38 mm^2

Diameter of lead plug ϕ_L = 50.757 mm (provide lead diameter of 55 mm)

Total Rubber thickness required $= \sum t_r = \frac{G_{0.5} A_r}{K_2} = 217\,\text{mm}$

A_r—Cross-sectional area of the rubber

Thickness of individual rubber layer $t_r = \phi/4S$ = 15.4 mm

Provided thickness = t_r = 16 mm

Total rubber thickness $\sum t_r$ = 224 mm

Bulk modulus K = 2000 Mpa

Result summary

Number of rubber layers n_R = 13.56 Nos

Provided number of layers n_R = 14 Nos

Number of steel shims n_s = 13 Nos

Max. thickness of steel shim $t_s = t_s * n_s + t_r * n_R \leq \phi$

Maximum thickness of steel shims t_s = 41 mm

Thickness of steel shims provided = 3 mm

Total thickness (rubber + shims) h = 263 mm

Top/bottom plate's thickness t_p = 51 mm

Total height of bearing $H = (h + 2t_p)$ = 365 mm

A.1.2.1 Design Checks

Primary shape factor $S = \phi/4t = 11.72$

Strain due to compressive loading (should be less than 20%)

VERTICAL STIFFNESS OF BEARING

$$K_{vi} = A_r E_c / \sum t_r$$

Compression modulus $E_c = 6GS^\wedge 2K/(6GS^\wedge 2 + K)$

$$E_c = 3.42E+08\ \ \text{N/m}^2$$

Therefore, $K_{vi} = 6.74E+08\,\text{N/m}$

Vertical stiffness of single rubber layer $K_v = K_{vi}/n_R = 4.81E+07\,\text{N/m}$

Vertical displacement $D_v = W_{max}/K_{vi} = 1.78106\,mm$

Compressive strain percentage $\varepsilon_v = D_v/\sum t = 0.7951 < 20\%$

SAFE

Check for Critical Buckling Load—Option 1

(It should be about five or more times greater than the maximum vertical load.)

Effective shear area $A_s = A_\phi^*(H/\sum t) = 5.19E+05$

Bearing cross-sectional area $A_\phi = 4.42E+05\,\text{mm}^2$

Total thickness of rubber and steel shim H = 263 mm

Moment of Inertia of bearing $I = 1.55E+10\,\text{mm}^4$

Bending stiffness $EI_s = E_C^* I^*(H/3\sum t) = 2.08E+12$

Shear stiffness per unit length $P_S = G_{0.2}A_s = 2.59E+05$ N/mm

Euler buckling load $P_E = \pi^2 EI_s/H^2 = 2.96E+08$ N

Critical Load $P_{cr} = (\text{Sqrt}(P_s{}^2 + 4P_sP_E - P_s))/2 = 8767.12\,\text{kN}$

Ratio of critical to actual vertical load = 7.31

SAFE

Check for critical load calculation (Option 2)

Secondary shape factor $\lambda = \phi/\sum t_r = 3.35$

$P_{cr} = \pi GS\lambda/\text{sqrt}(6) = 25.16\,\text{N/mm}^2$

Load $= P_{cr}^* A_r = 11116.07$ kN

Ratio of critical to actual vertical load = 9.26

SAFE

Note:

Allowable buckling load is very high to the actual vertical load applied on the bearing.

Hence, bearing is safe for buckling load

Design Check for frequency

Average mass on isolator m = 112.36 T
$K_h = G^* A_s / \sum t_r = 9.86E+05\,N/m$
Horizontal Frequency ($f_h \leq$ 0.5 Hz) $f_h = 1/2\pi^*(\sqrt{K_h}/m)$
OBE = 0.45 SAFE
SSE = 0.45 SAFE
Vertical frequency ($f_v \leq$ 10 Hz) OBE = $1/2\pi^*(\sqrt{K_v}/m)$
OBE = 11.81 SAFE
SSE = 11.81 SAFE

Quantity evaluated	Value obtained	Units
Actual lead plug diameter "Φ_L"	0.060	m
Actual total rubber thickness "$\sum t$"	0.224	m
Actual lead plug area "A_p"	0.00283	m^2
Actual rubber area "A_r"	0.4418	m^2
Actual value of "Q"	28.274	KN
Actual value for "K_R"	1108.042	KN/m

1. **Design calculations of Laminated Rubber Bearings (LRB)**

Step 1: Assumption of appropriate target dynamic characteristics

Seismic weight W = 1100 KN
Time period 'T_M' (assumed) = 2 s
From site-specific spectra S_a/g = 0.105
Damping ratio 'ξ_M' = 0.1

Step 2: Evaluating characteristic constants for the base isolation system

Effective stiffness $K_{eff} = \frac{4\pi^2 W}{T_M^2 g} = 1108\,kN/m$

Displacement at SSE $D_{sse} = \frac{T_M^2}{4\pi^2} S_a = 104.366$ mm

Displacement at OBE $D_{obe} = \frac{D_{sse}}{2} = 52.1829$ mm

A.1.3 Selection Criteria and Design of Base Isolators

Type of isolator used = Laminated rubber bearing
Shear modulus of rubber at higher strain = 0.5 MPa
Required post-yield horizontal stiffness K_2 = 1108 KN/m
Minimum vertical load on isolator = 1100 kN
Required vertical frequency f_V= 15 Hz
Required horizontal frequency f_H = 0.5 Hz

Shape factor = $S = \sqrt{\frac{(f_V/f_H)^2}{6}} = 12.47$

Safety factor SF = 1.5
Allowable compressive stress = $\sigma_{all} = \frac{\min([6.895\,\text{MPa}],[0.6895 G_{0.5} S])}{\text{SF}} = 2.81\,\text{MPa}$
Assume shape of the bearing—circular
Bearing area required $A = \frac{W_{max}}{\sigma_{all}} = 427019.4275\,mm^2$
Diameter of circular bearing provide = ϕ = 750 mm
Total rubber thickness required = $\sum t_r = \frac{G_{0.5} A_r}{K_2}$ = 217 mm
A_r—cross-sectional area of the rubber
Thickness of individual rubber layer t_r = ϕ/4S = 15.4 mm
Provided thickness = t_r = 16 mm
Total rubber thickness $\sum t_r$ = 224 mm
Bulk modulus K = 2000 Mpa
Result summary
Number of rubber layers n_R = 13.56 Nos
Provided number of layers n_R = 14 Nos
Number of steel shims n_s = 13 Nos
Max. thickness of steel shim $t_s = t_s{*}n_s + t_r{*}n_R \leq \phi$
Maximum thickness of steel shims t_s = 41 mm
Thickness of steel shims provided = 3 mm
Total thickness (rubber + shims) h = 263 mm
Top/bottom plates' thickness t_p = 51 mm
Total height of bearing $H = (h + 2t_p)$ = 365 mm

A.1.3.1 Design Checks

Primary shape factor $S = \phi/4t = 11.72$

Strain due to compressive loading (should be less than 20%)

VERTICAL STIFFNESS OF BEARING

$$K_{vi} = A_r E_c / \sum t_r$$

Compression modulus $E_c = 6GS^2K/(6GS^2 + K)$

$$E_c = 3.42E+08\,N/m^2$$

Therefore, $K_{vi} = 6.74E+08\,N/m$

Vertical stiffness of single rubber layer $K_v = K_{vi}/n_R = 4.81E+07\,N/m$

Vertical displacement $D_v = W_{max}/K_{vi} = 1.78106$ mm

Compressive strain percentage $\varepsilon_v = D_v/\sum t = 0.7951 < 20\%$

SAFE

Check for critical buckling load (Option 1)

(It should be about five or more times greater than the maximum vertical load.)

Effective shear area $A_S = A\phi^*(H/\sum t) = 5.19E+05$

Bearing cross-sectional area $A_\phi = 4.42E+05\,mm^2$

Total thickness of rubber and steel shim H = 263 mm

Moment of inertia of bearing $I = 1.55E+10\ mm^4$

Bending stiffness $EI_S = E_C{}^*I^*(H/3\sum t) = 2.08E+12$

Shear stiffness per unit length $P_S = G_{0.2}A_S = 2.59E+05\,N/mm$

Euler buckling load $P_E = \pi^2 EI_s/H^2 = 2.96E+08\,N$

Critical load $P_{cr} = \left(\mathrm{Sqrt}\left({P_S}^2 + 4P_sP_E - P_s\right)\right)/2 = 8767.12\,kN$

Ratio of critical to actual vertical load = 7.31

SAFE

Check for critical load calculation (Option 2)

Secondary shape factor $\lambda = \phi/\sum t_r = 3.35$

$P_{cr} = \pi GS\lambda/\mathrm{sqrt}(6) = 25.16\,N/mm^2$

Load $= P^*_{cr}A_r = 11116.07\,kN$

Ratio of critical to actual vertical load = 9.26

SAFE

Note:

Allowable buckling load is very high to the actual vertical load applied on the bearing.

Hence, bearing is safe for buckling load

Design check for frequency

Average mass on isolator m = 112.36 T

$$K_h = G^*A_s/\sum t_r = 9.86E+05\,N/m$$

Horizontal frequency ($f_h \le 0.5$ Hz) $f_h = 1/2\pi^*(\sqrt{K_h}/m)$

OBE = 0.45 SAFE

SSE = 0.45 SAFE

Vertical frequency ($f_v \le$ 10 Hz) OBE = $1/2\pi^*(\sqrt{K_v}/m)$

OBE = 11.81 SAFE

SSE = 11.81 SAFE

Quantity evaluated	Value obtained	Units
Actual total rubber thickness "$\sum t$"	0.224	m
Actual rubber area "A_r"	0.4418	m^2
Actual value for "K_2"	1108	kN/m

References

1. Wright FL (1955) An American architecture, p 150
2. Martel RR (1929) The effects on earthquakes on buildings with a flexible first story. Bull Seismol Soc Am 19(3)
3. Green NB (1935) Flexible first story construction for earthquake resistance. Trans Am Soc Civ Eng 100:645–674
4. Jacobsen LS (1938) Effects of a flexible first story in a building located on vibrating ground. In: S. Timoshenko 60th anniversary volume. The Macmillan Co., New York, pp 93–103
5. Chopra AK, Clough DP, Clough RW (1973) Earthquake resistance of buildings with a soft first story. Earthquake Eng Struct Dyn 1:347–355
6. Mostaghel N, Khodaverdian M (1987) Dynamics of resilient-friction base isolator (R-FBI). Earthquake Eng Struct Dyn 15:379–390
7. Su L, Ahmadi G, Tadjbakhsh IG (1991) Performance of sliding resilient-friction base-isolation system. J Struct Eng 117(1)
8. Yegian MK, Kadakal U (2004) Foundation isolation for seismic protection using a smooth synthetic liner. J Geotech Geoenviron Eng 1121–1130
9. Patill SJ, Reddy GR (2012) State of art review—base isolation systems for structures. Int J Emerg Technol Adv Eng 2(7). ISSN 2250-2459
10. Naeim F, Kelly JM (1999) Design of seismic isolated structures from theory to practice. Wiley, New York
11. UBC-1997 Uniform building code, Vol 2, pp 9–38, 405–416
12. ASCE-7-10 (2010) Minimum design loads for buildings and other structures. American Society for Civil Engineers
13. Agarwal P, Shrikhande MS (2007) Earthquake resistant design of structures. Prentice-Hall, India
14. ATC-17-1 (1993) Proceedings of seminar on seismic isolation, passive energy dissipation, and active control. Applied Technology Council, California
15. Brown RP, Butler T (2000) Natural ageing of rubber changes in physical properties over 40 years. Rapra Technology Limited Shawbury, Shropshire SY4 4NR, UK
16. Clough RW, Penzien J (2003) Dynamics of structures. Computers and Structures Inc., Berkeley
17. Deb SK, Dutta A (2007) Verification test of a prototype two storeyed RCC frame building subjected to actual earthquakes. In: Project Report for 8th Technical Pragrame Discussion Meeting
18. Dutta TK (2007) Structural control of seismic disaster mitigation. International workshop on earthquake hazards & mitigations, Paper No. 4. IIT, Guwahati, India, 7–8 Dec 2007
19. FEMA-356 (2000) Prestandard and commentary for the seismic rehabilitation of buildings. Federal Emergency Management Agency, Washington, DC
20. Franco B, Michelangelo L (2004) Field testing of low rise base isolated building. Eng Struct 26:1599–1610
21. Gazetas George (1991) Formulas and Charts for impedances of surface and embedded foundations. J Geotech Eng ASCE 117 (9):1363–1381
22. Gobel EF (1974) Rubber springs design. Newnes Butterworths, London, pp 22–101
23. Hwang JS, Chiou JM (1996) An equivalent linear model of lead-rubber seismic isolation bearing. Eng Struct 18(5):528–536
24. ISO 11346: 2004(E), International Standard Second edition (2004) Rubber, vulcanized or thermoplastic—estimation of life-time and maximum temperature of use. ISO copyright office, Case postale Geneva, Switzerland
25. IS-13920 (1993) Ductile detailing of reinforced concrete structures subjected to seismic forces—code of practice. Bureau of Indian Standards, New Delhi

26. IS-1893 (Part-1)-2002. Criteria for earthquake resistant design of structures-general provisions and buildings. Bureau of Indian Standards, New Delhi
27. Jangid RS, Datta TK (1995) Seismic behavior of base isolated buildings: a-state of the art review. Proc Inst Civil Eng Struct Build 110(2):186–203
28. Kelly JM (1986) Aseismic base isolation: review and bibliography. Soil Dyn Earthquake Eng 5(3):202–216
29. Kelly JM (1997) Earthquake-resistant design with rubber. Springer, London
30. Kelly JM, Aiken ID (1991) Experimental studies of the seismic response of structures incorporating base isolation systems. Nucl Eng Des 127:329–338
31. Kelly E, Trevor SE (2001) Design guidelines for base isolation of structures. Holmes Consulting Group Ltd.
32. Komodromos P (2000) Seismic isolation for earthquake-resistant structures. WIT Press, Southampton
33. Kato M, Watanabe Y et. al (1996) Investigation of ageing effects for laminated rubber bearings of Pelham Bridge. In: 11 world conference on earthquake engineering (WCEE), Japan, Paper No. 1450
34. Masahiko H, Shin O (2006) Response control and seismic isolation of buildings. Taylor & Francis, New York
35. Mo YL, Chang YF (1995) Application of base isolation concept to soft first storey buildings. Comput Struct 55(5):883–896
36. Paulay T, Priestley MJN (1992) Seismic design of reinforced concrete and masonry buildings. Wiley, New York
37. Dubey PN, Reddy GR, Vaze KK, Ghosh AK, Kushwaha HS, Deb SK (2008) Performance of base isolated RCC framed building under actual earthquake. J Struct Eng 35(3):147–153
38. Pradeep Kumar TV, Paul DK (2007) Force-deformation behaviour of isolation bearings. J Bridge Eng ASCE, pp 527–529
39. Brown RP, Butler T (2000) Natural ageing of rubber, changes in physical properties over 40 years. Rapra Technology Limited, Shawbury
40. Brown Roger (1999) Hand book of polymer testing. Marcel Dekker Inc., New York
41. Skinner RI, Robinson WH, McVerry GH (1993) An introduction to seismic isolation. Wiley, England
42. Spyrakos CC, Koutromanos IA, Maniatakis CA (2009) Seismic response of base isolated building including soil structure interaction. Soil Dyn Earthquake Eng 29:658–668

Index

G. R Reddy et al., *Textbook of Seismic Design*,
https://doi.org/10.1007/978-981-13-3176-3

T

U

V

W

The manufacturer's authorised representative in the EU is Springer Nature Customer Service Centre GmbH, Europaplatz 3, 69115 Heidelberg, Germany. If you have any concerns regarding our products, please contact ProductSafety@springernature.com

Printed and bound by CPI Group (UK) Ltd, Croydon, CR0 4YY
02/10/2026
02227131-0001